José Rolim et al. (Eds.)

Parallel and Distributed Processing

15 IPDPS 2000 Workshops
Cancun, Mexico, May 1-5, 2000
Proceedings

Springer

Series Editors

Gerhard Goos, Karlsruhe University, Germany
Juris Hartmanis, Cornell University, NY, USA
Jan van Leeuwen, Utrecht University, The Netherlands

Managing Volume Editor

José Rolim
Université de Genève, Centre Universitaire d'Informatique
24, rue Général Dufour, CH-1211 Genève 4, Switzerland
E-mail: Jose.Rolim@cui.unige.ch

Cataloging-in-Publication Data applied for

Die Deutsche Bibliothek - CIP-Einheitsaufnahme

Parallel and distributed processing : 15 IPDPS 2000 workshops, Cancun,
Mexico, May 1 - 5, 2000, proceedings / José Rolim et al. (ed.). -
Berlin ; Heidelberg ; New York ; Barcelona ; Hong Kong ; London ;
Milan ; Paris ; Singapore ; Tokyo : Springer, 2000
  (Lecture notes in computer science ; 1800)
  ISBN 3-540-67442-X

CR Subject Classification (1998): C.1-4, B.1-7, D.1-4, F.1-2, G.1-2, E.1, H.2

ISSN 0302-9743
ISBN 3-540-67442-X Springer-Verlag Berlin Heidelberg New York

Springer-Verlag is a company in the BertelsmannSpringer publishing group.
© Springer-Verlag Berlin Heidelberg 2000
Printed in Germany

Typesetting: Camera-ready by author, data conversion by Boller Mediendesign
Printed on acid-free paper   SPIN: 10720149   06/3142   5 4 3 2 1 0

Volume Editors

Foreword

This volume contains the proceedings from the workshops held in conjunction with the IEEE International Parallel and Distributed Processing Symposium, *IPDPS 2000*, on 1-5 May 2000 in Cancun, Mexico.

The workshops provide a forum for bringing together researchers, practitioners, and designers from various backgrounds to discuss the state of the art in parallelism. They focus on different aspects of parallelism, from run time systems to formal methods, from optics to irregular problems, from biology to networks of personal computers, from embedded systems to programming environments; the following workshops are represented in this volume:

- Workshop on Personal Computer Based Networks of Workstations
- Workshop on Advances in Parallel and Distributed Computational Models
- Workshop on Par. and Dist. Comp. in Image, Video, and Multimedia
- Workshop on High-Level Parallel Prog. Models and Supportive Env.
- Workshop on High Performance Data Mining
- Workshop on Solving Irregularly Structured Problems in Parallel
- Workshop on Java for Parallel and Distributed Computing
- Workshop on Biologically Inspired Solutions to Parallel Processing Problems
- Workshop on Parallel and Distributed Real-Time Systems
- Workshop on Embedded HPC Systems and Applications
- Reconfigurable Architectures Workshop
- Workshop on Formal Methods for Parallel Programming
- Workshop on Optics and Computer Science
- Workshop on Run-Time Systems for Parallel Programming
- Workshop on Fault-Tolerant Parallel and Distributed Systems

All papers published in the workshops proceedings were selected by the program committee on the basis of referee reports. Each paper was reviewed by independent referees who judged the papers for originality, quality, and consistency with the themes of the workshops.

We would like to thank the general co-chairs Joseph JaJa and Charles Weems for their support and encouragement, the steering committee chairs, George Westrom and Victor Prasanna, for their guidance and vision, and the finance chair, Bill Pitts, for making this publication possible. Special thanks are due to Sally Jelinek, for her assistance with meeting publicity, to Susamma Barua for making local arrangements, and to Danuta Sosnowska for her tireless efforts in interfacing with the organizers.

We gratefully acknowledge sponsorship from the IEEE Computer Society and its Technical Committee of Parallel Processing and the cooperation of the ACM SIGARCH. Finally, we would like to thank Danuta Sosnowska and Germaine Gusthiot for their help in the preparation of this volume.

February 2000   José D. P. Rolim

Contents

3rd Workshop on Personal Computer based Networks Of Workstations (PC-NOW 2000)

Clusters composed of fast personal computers are now well established as cheap and efficient platforms for distributed and parallel applications. The main drawback of a standard NOWs is the poor performance of the standard inter-process communication mechanisms based on RPC, sockets, TCP/IP, Ethernet. Such standard communication mechanisms perform poorly both in terms of throughput as well as message latency.

Several prototypes developed around the world have proved that re-visiting the implementation of the communication layer of a standard Operating System kernel, a low cost hardware platform composed of only commodity components can scale up to several tens of processing nodes and deliver communication and computation performance exceeding the one delivered by the conventional high-cost parallel platforms.

This workshop provides a forum to discuss issues related to the design of efficient NOW/Clusters based on commodity hardware and public domain operating systems as compared to custom hardware devices and/or proprietary operating systems.

Workshop Organizers

G. Chiola (DISI, U. Genoa, I)
G. Conte (CE, U. Parma, I)
L.V. Mancini (DSI, U. Rome, I)

Sponsors

IEEE TFCC
(Task Force on Cluster Computing)

J. Rolim et al. (Eds.): IPDPS 2000 Workshops, LNCS 1800, pp. 1-3, 2000.
© Springer-Verlag Berlin Heidelberg 2000

Program Commitee

Referees

Accepted Papers

Session 1: Cluster Interconnect Design and Implementation

- M. Trams, W. Rehm, D. Balkanski, and S. Simeonov "Memory Management in a combined VIA/SCI Hardware"
- M. Fischer, et al. "ATOLL, a new switched, high speed Interconnect in comparison to Myrinet and SCI"
- R.R. Hoare "ClusterNet: An Object-Oriented Cluster Network"

Session 2: Off-the-shelf Clusters Communication

- M. Baker, S. Scott, A. Geist, and L. Browne "GigaBit Performance under NT"
- H.A. Chen, Y.O. Carrasco, and A.W. Apon "MPI Collective Operations over IP Multicast"

Session 3: Multiple Clusters and Grid Computing

- S. Lalis, and A. Karipidis 'An Open Market-Based Architecture for Distributed Computing"
- M. Barreto, R. Avila, and Ph. Navaux "The MultiCluster Model to the Integrated Use of Multiple Workstation Clusters"

Session 4: Data Intensive Applications

- S.H. Chung, et al. "Parallel Information Retrieval on an SCI-Based PC-NOW"
- M. Exbrayat, and L. Brunie 'A PC-MOW Based Parallel Extension for a Sequential DBMS"

Other Activities

In addition to the presentation of contributed papers an invited talk will be scheduled at the workshop.

Memory Management in a combined VIA/SCI Hardware

Mario Trams, Wolfgang Rehm, Daniel Balkanski and Stanislav Simeonov *
{mtr,rehm}@informatik.tu-chemnitz.de
DaniBalkanski@yahoo.com, stan@bfu.bg

Technische Universität Chemnitz
Fakultät für Informatik**
Straße der Nationen 62, 09111 Chemnitz, Germany

Abstract In this document we make a brief review of memory management and DMA considerations in case of common SCI hardware and the Virtual Interface Architecture. On this basis we expose our ideas for an improved memory management of a hardware combining the positive characteristics of both basic technologies in order to get one completely new design rather than simply adding one to the other. The described memory management concept provides the opportunity of a real zero–copy transfer for Send–Receive operations by keeping full flexibility and efficiency of a nodes' local memory management system. From the resulting hardware we expect a very good system throughput for message passing applications even if they are using a wide range of message sizes.

1 Motivation and Introduction

PCI–SCI bridges (Scalable Coherent Interface [12]) become a more and more preferable technological choice in the growing market of Cluster Computing based on non–proprietary hardware. Although absolute performance characteristics of this communication hardware increases more and more, it still has some disadvantages. Dolphin Interconnect Solutions AS (Norway) is the leading manufacturer of commercial SCI link chips as well as the only manufacturer of commercially available PCI–SCI bridges. These bridges offer very low latencies in range of some microseconds for their distributed shared memory and reach also relatively high bandwidths (more than 80MBytes/s). In our clusters we use Dolphins PCI–SCI bridges in junction with standard PC components [11]. MPI applications that we are running on our cluster can get a great acceleration from low latencies of the underlying SCI shared memory if it is used as communication medium for transferring messages. MPI implementations e.g. such as [7] show a

---

* Daniel Balkanski and Stanislav Simeonov are from the Burgas Free University, Bulgaria.
** The work presented in this paper is sponsored by the SMWK/SMWA Saxony ministries (AZ:7531.50-03-0380-98/6). It is also carried out in strong interaction with the project GRANT SFB393/B6 of the DFG (German National Science Foundation).

J. Rolim et al. (Eds.): IPDPS 2000 Workshops, LNCS 1800, pp. 4-15, 2000.

bandwidth of about 35MByte/s for a message size of 1kByte which is quite a lot (refer also to figure 1 later).

The major problem of MPI implementations over shared memory is big CPU utilization on long message sizes due to copy operations. So the just referred good MPI performance [7] is more an academic peak performance which is achieved with more or less total CPU consumption. A standard solution for this problem is to use a block–moving DMA engine for data transfers in background. Dolphins PCI–SCI bridges implement such a DMA engine. Unfortunately, this one can't be controlled directly from a user process without violating general protection issues. Therefore kernel calls are required here which in end effect increase the minimum achievable latency and require a lot of additional CPU cycles.

The Virtual Interface Architecture (VIA) Specification [16] defines mechanisms for moving the communication hardware closer to the application by migrating protection mechanisms into the hardware. In fact, VIA specifies nothing completely new since it can be seen as an evolution of U–Net [15]. But it is a first try to define a common industry–standard of a principle communication architecture for message passing — from hardware to software layers. Due to its DMA transfers and its reduced latency because of user–level hardware access, a VIA system will increase the general system throughput of a cluster computer compared to a cluster equipped with a conventional communication system with similar raw performance characteristics. But for very short transmission sizes a programmed IO over global distributed shared memory won't be reached by far in terms of latency and bandwidth. This is a natural fact because we can't compare a simple memory reference with DMA descriptor preparation and execution.

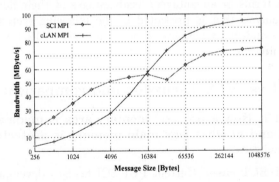

Figure1. Comparison of MPI Implementations for Dolphins PCI–SCI Bridges and GigaNets cLAN VIA Hardware

Figure 1 shows bandwidth curves of MPI implementations for both an SCI and a native VIA implementation (GigaNet cLAN). The hardware is in both cases based on the PCI bus and the machines where the measurements were taken are comparable. The concrete values are based on ping–pong measurements and where taken from [7] in case of SCI, and from [10] (Linux case) for the cLAN hardware.

As expected, the bandwidth in case of SCI is looking better in the range of smaller message sizes. For larger message sizes the cLAN implementation demonstrates higher bandwidth because of its advanced DMA engine. But not less important is the fact that a DMA engine gives the CPU more time for computations. Details of such CPU utilization considerations are outside the scope of this paper and are already discussed in [14] and [8].

As summarization of these motivating facts we can state that besides a powerful DMA engine controllable from user–level a distributed shared memory for programmed IO is an important feature which shouldn't be missed in a communication system.

2   What are the Memory Management Considerations?

First of all we want to make a short definition what belongs to memory management regarding this document.
This can be stated by the following aspects expressed in the form of questions:

1. How a process' memory area is made available to the Network Interface Controller (NIC) and in what way main memory is protected against wrong accesses?
2. At which point in the system a DMA engine is working and how are the transactions of this DMA engine validated?
3. In which way memory of a process on a remote node is made accessible for a local process?

Based on these questions we can classify the different communication system architectures in terms of advantages/disadvantages of their memory management. In the analysis that is presented in the following sections we'll reveal these advantages and disadvantages arisen from common PCI–SCI architecture and the VI Architecture.

3   PCI–SCI vs. VIA discussion and comparison

3.1   Question 1: How a process' memory area is made available to the NIC and in what way main memory is protected against wrong accesses?

Common PCI–SCI case: Current PCI–SCI bridges developed by Dolphin realize a quiet static memory management [4] to get access to main memory or rather PCI address space. To avoid unwanted accesses to sensitive locations, the PCI–SCI bridge is set up to allow accesses only to a dedicated memory window. Memory access requests caused by remote machines are only allowed if they fall within the specified window. This causes two big disadvantages:

– Continuous exported regions must also be continuous available inside the physical address space. Additionally, these regions must be aligned to the minimum exportable block size which is typically quite large (512kB for Dolphin's bridges).

– Exported Memory must reside within this window.

To handle these problems it is required to reserve main memory only for SCI purposes. This, in practice, 'wastes' a part of memory if it is not really exported later.

In consequence these disadvantages of common PCI–SCI bridge architecture make their use with MPI applications very difficult. Especially in view of zero–copy transfer operations. Because data transfers can be processed using the reserved memory region only, it would require that MPI applications use special `malloc()` functions for allocating data structures used for send/receive purposes later. But this violates a major goal of the MPI standard: Architecture Independence.

VIA case: The VI Architecture specifies a much better view the NIC has on main memory. Instead of a flat one–to–one representation of the physical memory space it implements a more flexible lookup–table address translation. Comparing this mechanism with the PCI–SCI pendant the following advantages become visible.

– Continuous regions seen by the VIA hardware are not required to be also continuous inside the host physical address space.
– Accesses to sensitive address ranges are prevented by just not including them into the translation table.
– The NIC can get access to **every** physical memory page, even if this may not be possible for all physical pages at once (when the translation table has less entries than the number of physical pages).

The translation table is not only for address translation purposes, but also for protection of memory. To achieve this a so–called *Protection Tag* is included for each translation and protection table entry. This tag is checked prior to each access to main memory to qualify the access. For more information about this see later in section 3.2.

Conclusions regarding question 1: It is clear, that the VIA approach offers much more flexibility. Using this local memory access strategy in a PCI–SCI bridge design will eliminate all of the problems seen in current designs.
Of course, the drawback is the more complicated hardware and the additional cycles to translate the address.

3.2 Question 2: At which point in the system a DMA engine is working and how are the transactions of this DMA engine validated?

Common PCI–SCI case: The DMA engine accesses local memory in the same way as already discussed in section 3.1. Therefore it inherits also all disadvantages when dealing with physical addresses on the PCI–SCI bridge.

For accesses to global SCI memory a more flexible translation table is used. This *Downstream Translation Table* realizes a virtual view onto global SCI memory — similar as the view of a VIA NIC onto local memory. Every page of the virtual SCI memory can be mapped to a page of the global SCI memory.

Regarding validation, the DMA engine can't distinguish between regions owned by different processes (neither local nor remote). Therefore the hardware can't make a check of access rights on–the–flow. Rather it is required that the DMA descriptor containing the information about the block to copy is assured to be right. In other words the operating system kernel has to prepare or at least to check any DMA descriptor to be posted to the NIC. This requires OS calls that we want to remove at all cost.

VIA case: A VIA NIC implements mechanisms to execute a DMA descriptor from user–level while assuring protection among multiple processes using the same VIA hardware. An user process can own one or more interfaces of the VIA hardware (so–called *Virtual Interfaces*). In other words, a virtual interface is a virtual representation of a virtual unique communication hardware. The connection between the virtual interfaces and the VIA hardware is made by *Doorbells* that represent a virtual interface with its specific control registers. An user–level process can insert a new DMA descriptor into a job queue of the VIA hardware by writing an appropriate value into a doorbell assigned to this process. The size of a doorbell is equal to the page size of the host computer and so the handling which process may access which doorbell (or virtual interface) can be simply realized by the hosts' virtual memory management system. Protection during DMA transfers is achieved by usage of *Protection Tags*. These tags are used by the DMA engine to check if the access of the current processed virtual interface to a memory page is right. The protection tag of the accessed memory page is compared with the protection tag assigned to the virtual interface of the process that provided this DMA descriptor. Only if both tags are equal, the access is legal and can be performed. A more detailed description of this mechanism is outside the scope of this document (refer to [13] and [16]).

Conclusions regarding question 2: The location of the DMA engine is in both cases principally the same. The difference is that in case of VIA a real lookup–table based address translation is performed between the DMA engine and PCI memory. That is, the VIA DMA operates on a virtual local address space, while the PCI–SCI DMA operates directly with local physical addresses. The answer for the access protection is simple: The common PCI–SCI DMA engine supports no protection in hardware and must trust on right DMA descriptors. The VIA hardware supports full protection in hardware where the DMA engine is only one part of the whole protection mechanism.

3.3   Question 3: In which way memory of a process on a remote node is made accessible for a local process?

Common PCI–SCI case: Making remote memory accessible is a key function in a SCI system, of course. Each PCI–SCI bridge offers a special PCI memory window which is practically the virtual SCI memory seen by the card. So the same SCI memory the DMA engine may access can be also accessed via memory references (also called programmed IO here). The procedure of making globally available SCI memory accessible for the local host is also referred as *importing global memory into local address space.*

On the other side, every PCI–SCI bridge can open a window to local address space and make it accessible for remote SCI nodes. The mechanism of this window is already described in section 3.1 regarding question 1. The procedure of making local memory globally accessible is also called *exporting local memory into global SCI space.*

Protection is totally guaranteed when dealing with imported and exported memory in point of view of memory references. Only if a process has got a valid mapping of a remote process' memory page it is able to access this memory.

VIA case: The VI Architecture offers principally no mechanism to access remote memory as it is realized in a distributed shared memory communication system such as SCI. But there is an indirect way by using a so–called Remote DMA (or RDMA) mechanism. This method is very similar to DMA transfers as they are used in common PCI–SCI bridges. A process that wants to transfer data between its local memory and memory of a remote process specifies a RDMA descriptor. This contains an address for the local VIA virtual address space and an address for the remote nodes' local VIA virtual address space.

Conclusions regarding question 3: While a PCI–SCI architecture allows processes to really share their memory globally across a system, this is not possible with a VIA hardware. Of course, VIA was never designed for realizing distributed shared memory.

4   A new PCI–SCI Architecture with VIA Approaches

In our design we want to combine the advances of an ultra–low latency SCI Shared Memory with a VIA–like advanced memory management and protected user–level DMA. This combination will make our SCI hardware more suitable for our message passing oriented parallel applications requiring short as well as long transmission sizes.

4.1   Advanced Memory Management

In order to eliminate the discussed above restrictions with continuous and aligned exported memory regions that must reside in a special window, our PCI–SCI architecture will implement two address translation tables — for both local and remote memory accesses. In contrast, common PCI–SCI bridges use only one translation table for accesses to remote memory. This new and more flexible memory management combined with reduced minimal page size of distributed shared memory leads to a much better usage of the main memory of the host system.

In fact, our targeted amount of imported SCI memory is 1GB with a page granularity of 16kB. With a larger downstream address translation table this page size may be reduced further to match exactly the page size used in the host systems (such as 4kB for x86 CPUs).

In case of the granularity of memory to be exported in SCI terminology or to be made available for VIA operations there's no question: It must be equal to the host system page size. In other words, 4kB since the primary target system is a x86 one. 128MB is the planned maximum window size here.

4.2   Operation of Distributed Shared Memory from a memory–related point of view

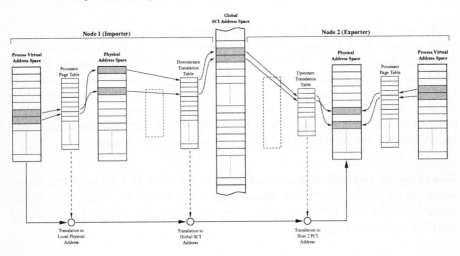

Figure2. Address Translations between exporting and importing Processes for programmed IO

Figure 2 gives an overall example of exporting/importing memory regions. The example illustrates the address translations performed when the importing process accesses memory exported by a process on the remote node.

The exporting process exports some of its previously allocated memory by registering it within its local PCI–SCI hardware. Registering memory is done on a by–page basis. Remember that in case of a common PCI–SCI system it would be required that this exported memory is physically located inside this special memory area reserved for SCI purposes. But here we can take the advantage of the virtual view onto local memory similar to this in VI Architecture.

Once the upstream address translation table entries are adjusted, the exported memory can be accessed from remote machines since it became part of the global SCI memory. To access this memory, the remote machine must import it first. The major step to do here is to set up entries inside its downstream address translation table so that they point to the region inside the global SCI memory that belongs to the exporter. From now, the only remaining task is to map the physical PCI pages that correspond to the prepared downstream translation entries into the virtual address space of the importing process.

When the importing process accesses the imported area, the transaction is forwarded through the PCI–SCI system and addresses are translated three times. At first the host MMU translates the address from the process' virtual address space into physical address space (or rather PCI space). Then the PCI–SCI bridge takes up the transaction and translates the address into the global SCI address space by usage of the downstream translation table. The downstream address translation includes generation of the remote node id and address offset inside the remote nodes' virtual local PCI address space. When the remote node receives the transaction, it translates the address to the correct local physical (or rather PCI) address by using the upstream address translation table.

4.3   Operation of Protected User–Level Remote DMA from a memory–related point of view

Figure 3 shows the principle work of the DMA engine of our PCI–SCI bridge design. This figure shows principally the same address spaces and translation tables as shown by figure 2. Only the process' virtual address spaces and the corresponding translation into physical address spaces are skipped to not overload the figure.

The DMA engine inside the bridge is surrounded by two address translation tables, or more correct said by two address **translation and protection** tables. On the active node (that is, where the DMA engine is executing DMA descriptors — node 1 here) both translation tables are involved. However, on the remote node there has practically nothing changed compared to the programmed IO case. Hence the remote node doesn't make any difference between transactions whether they were generated by the DMA engine or not.

Both translation tables of one PCI–SCI bridge incorporate protection tags as described in section 3.2. But while this is used in VIA for accesses to local memory, here it is also used for accesses to remote SCI memory. Together with VIA mechanisms for descriptor notification and execution the DMA engine is unable to access wrong memory pages — whether local (exported) nor remote (imported) ones. Note that a check for right protection tags is really made only for the DMA engine and only on the active node (node 1 in figure 3). In all other cases the same translation and protection tables are used, but the protection tags inside are ignored.

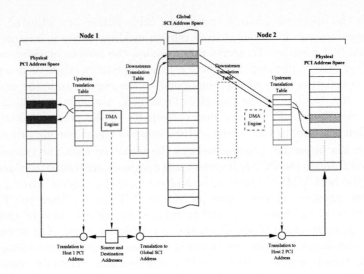

Figure3. Address Translations performed during RDMA Transfers

4.4   A free choice of using either Programmed I/O or User–Level Remote DMA

This kind of a global memory management allows applications or more exactly communication libraries to decide on–the–fly depending on data size in which way it should be transferred. In case of a short message a PIO transfer may be used, and in case of a longer message a RDMA transfer may be suitable. The corresponding remote node is not concerned in this decision since it doesn't see any differences. This keeps the protocol overhead very low.

And finally we want to remember the VIA case. Although we already have the opportunity of a relatively low–latency protected user–level remote DMA mechanism without the memory handling problems as in case of common PCI–SCI, there's nothing like a PIO mechanism for realizing a distributed shared memory. Hence the advantages of an ultra–low latency PIO transfer are not available here.

5   Influence on MPI Libraries

To show the advantages of the presented advanced memory management we want to take a look at the so–called *Rendezvous Protocol* that is commonly used for Send–Receive operations.

Figure 4 illustrates the principle of the Rendezvous protocol used in common MPI implementations [7] based on Dolphins PCI–SCI bridges. One big problem in this model is the copy operation that takes place on the receivers' side to take data out of the SCI buffer. Although the principally increasing latency can be hidden due to the overlapping mechanism a lot of CPU cycles are burned there.

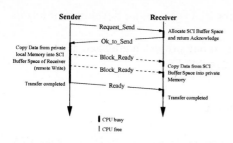
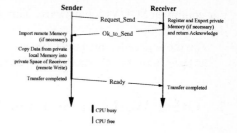

Figure4. Typical Rendezvous–Protocol in common PCI–SCI Implementations

Figure5. Improved Rendezvous–Protocol based on advanced PCI–SCI Memory Management

With our proposed memory management there's a chance to remove this copy operation on the receivers' side. The basic operation of the Rendezvous protocol can be implemented as described in figure 5. Here the sender informs the receiver as usual. Before the receiver sends back an acknowledge it checks if the data structure the data is to be written to is already exported to the sender. If not, the memory region that includes the data structure is registered within the receivers' PCI–SCI bridge and exported to the sender. The sender itself must also import this memory region if this was not already done before. After this the sender copies data from private memory of the sending process directly into private memory of the receiving process. As further optimization the sender may decide to use the DMA engine to copy data without further CPU intervention. This decision will be typically based on the message size.

6   State of the project (November 1999)

We developed our own FPGA–based PCI–SCI card and have prototypes of this card already running. At the moment they only offer a so–called *Manual Packet Mode* for now that is intended for sideband communication besides the regular programmed IO and DMA transfers.

The card itself is a 64Bit/33MHz PCI Rev.2.1 one [8]. As SCI link controller we are using Dolphins LC–2 for now, and we are looking to migrate to the LC–3 as soon as it is available. The reprogrammable FPGA design leads to a flexible reconfigurable hardware and offers also the opportunity for experiments.

Linux low–level drivers for Alpha and x86 platforms and several configuration/test programs were developed. In addition our research group is working on an appropriate higher–level Linux driver for our card [5, 6]. This offers a software–interface (advanced Virtual Interface Provider Library) that combines SCI and VIA features such as importing/exporting memory regions, VI connection management etc. Also it emulates parts of the hardware so that it is possible to run other software on top of it although the real hardware is not available. As an example, a parallelized MPI–version of the popular raytracer *POVRAY* is already running over this emulation. This program uses an MPI–2 library for our combined SCI/VIA hardware. This library is also under development at our department [3].

For more details and latest news refer to our project homepage at
http://www.tu-chemnitz.de/~mtr/VIA_SCI/

7   Other Works on SCI and VIA

Dolphin already presented some performance measurements in [1] for their VIA implementation which is a emulation over SCI shared memory. Although the presented VIA performance is looking very good, it's achieved by the cost of too big CPU utilization again.

The number of vendors of native VIA hardware is growing more and more. One of these companies is GigaNet [17] where performance values are already available. GigaNet gives on their web pages latencies of $8\mu s$ for short transmission sizes. Dolphin gives a latency for PIO operations (remote memory access) of $2.3\mu s$. This demonstrates the relatively big performance advantage a distributed shared memory offers here.

University of California, Berkeley [2] and the Berkeley Lab [9] are doing more open research also in direction of improving the VIA specification. The work at the University of California, Berkeley is concentrated more on VIA hardware implementations based on Myrinet. In contrast, the work at the Berkeley Lab is targeted mainly to software development for Linux.

8   Conclusions and Outlook

The combined PCI–SCI/VIA system is not just a simple result of adding two different things. Rather it is a real integration of both in one design. More concrete it is an integration of concepts defined by the VIA specification into a common PCI–SCI architecture since major PCI–SCI characteristics are kept. The result is a hardware design with completely new qualitative characteristics. It combines the most powerful features of SCI and VIA in order to get highly efficient messaging mechanisms and high throughput over a broad range of message lengths.

The advantage that MPI libraries can take from a more flexible memory management was illustrated for the case of a Rendezvous Send–Receive for MPI. The final proof in practice is still pending due to lack of a hardware with all implemented features.

References

1. Torsten Amundsen and John Robinson: *High–performance cluster–computing with Dolphin's CluStar PCI adapter card.* In: Proceedings of SCI Europe '98, Pages 149–152, Bordeaux, 1998

2. Philip Buonadonna, Andrew Geweke: *An Implementation and Analysis of the Virtual Interface Architecture.* University of California at Berkeley, Dept.of Computer Science, Berkeley, 1998. www.cs.berkeley.edu/~philipb/via/

3. *A new MPI–2–Standard MPI Implementation with support for the VIA.* www.tu-chemnitz.de/informatik/RA/projects/chempi-html/

4. Dolphin Interconnect Solutions AS: *PCI–SCI Bridge Spec. Rev. 4.01.* 1997.

5. Friedrich Seifert: *Design and Implementation of System Software for Transparent Mode Communication over SCI.*, Student Work, Dept. of Computer Science, University of Technology Chemnitz, 1999. See also: www.tu-chemnitz.de/~sfri/publications.html

6. Friedrich Seifert: *Development of System Software to integrate the Virtual Interface Architecture (VIA) into Linux Operating System Kernel for optimized Message Passing.* Diploma Thesis, TU–Chemnitz, Sept. 1999. See also: www.tu-chemnitz.de/informatik/RA/themes/works.html

7. Joachim Worringen and Thomas Bemmerl: *MPICH for SCI–connected Clusters.* In: Proceedings of SCI–Europe'99, Toulouse, Sept. 1999, Pages 3–11. See also: wwwbode.in.tum.de/events/sci-europe99/

8. Mario Trams and Wolfgang Rehm: *A new generic and reconfigurable PCI–SCI bridge.* In: Proceedings of SCI–Europe'99, Toulouse, Sept. 1999, Pages 113–120. See also: wwwbode.in.tum.de/events/sci-europe99/

9. *M–VIA: A High Performance Modular VIA for Linux.* Project Homepage: http://www.nersc.gov/research/FTG/via/

10. MPI Software Technology, Inc. *Performance of MPI/Pro for cLAN on Linux and Windows.* www.mpi-softtech.com/performance/perf-win-lin.html

11. *The Open Scalable Cluster ARchitecture (OSCAR) Project.* TU Chemnitz. www.tu-chemnitz.de/informatik/RA/projects/oscar_html/

12. *IEEE Standard for Scalable Coherent Interface (SCI).* IEEE Std. 1596-1992. SCI Homepage: www.SCIzzL.com

13. Mario Trams: *Design of a system–friendly PCI–SCI Bridge with an optimized User–Interface.* Diploma Thesis, TU-Chemnitz, 1998. See also: www.tu-chemnitz.de/informatik/RA/themes/works.html

14. Mario Trams, Wolfgang Rehm, and Friedrich Seifert: *An advanced PCI–SCI bridge with VIA support.* In: Proceedings of 2nd Cluster–Computing Workshop, Karlsruhe, 1999, Pages 35–44. See also: www.tu-chemnitz.de/informatik/RA/CC99/

15. The U-Net Project: *A User–Level Network Interface Architecture.* www2.cs.cornell.edu/U-Net

16. Intel, Compaq and Microsoft. *Virtual Interface Architecture Specification V1.0.*, VIA Homepage: www.viarch.org

17. GigaNet Homepage: www.giganet.com

ATOLL, a new switched, high speed Interconnect in Comparison to Myrinet and SCI

Markus Fischer, Ulrich Brüning, Jörg Kluge, Lars Rzymianowicz, Patrick Schulz, Mathias Waack

University of Mannheim, Germany,
markus@atoll-net.de

Abstract. While standard processors achieve supercomputer performance, a performance gap exists between the interconnect of MPP's and COTS. Standard solutions like Ethernet can not keep up with the demand for high speed communication of todays powerful CPU's. Hence, high speed interconnects have an important impact on a cluster's performance. While standard solutions for processing nodes exist, communication hardware is currently only available as a special, expensive non portable solution. ATOLL presents a switched, high speed interconnect, which fulfills the current needs for user level communication and concurrency in computation and communication. ATOLL is a single chip solution, additional switching hardware is not required.

1 Introduction

Using commodity off the shelf components (COTS) is a viable option to build up powerful clusters not only for number crunching but also for highly parallel, commercial applications. First clusters already show up in the Top500 [6] list and it is expected to see the number of entries continuously rising. Powerful CPU's such as the Intel PIII Xeon with SMP functionality, achieve processing performance known from supercomputers. Currently a high percentage of existing clusters is equipped with standard solutions such as Fast Ethernet. This is mainly for compatibility reasons since applications based on standardized TCP/IP are easily portable. This protocol however is known to cause too much overhead [7]. Especially lowering latency is an important key to achieve good communication performance. A survey on message sizes shows that protocols and hardware have to be designed to handle short messages extremely well [14]:

- in seven parallel scientific applications 30% of the messages were between 16 bytes and a kilobyte
- the median message sizes for TCP and UDP traffic in a departmental network were 32 and 128 bytes respectively
- 99% of TCP and 86% of the UDP traffic was less than 200 bytes
- on a commercial database all messages were less than 200 bytes
- the average message size ranges between 19 - 230 bytes

J. Rolim et al. (Eds.): IPDPS 2000 Workshops, LNCS 1800, pp. 16-27, 2000.

Recent research with Gigabit/s interconnects, such as Myrinet and SCI, has shown that one key to achieve low latency and high bandwidth is to bypass the operating system, avoiding a trap into the system: User Level Communication (ULC) gives the user application full control over the interconnect device (BIP, HPVM, UNET, AM). While ULC shortens the critical path when sending a message, a global instance such as the kernel, is no longer involved in scheduling outgoing data. This has the disadvantage, that security issues have to be discussed, if different users are running their application. But also trashing and context switching through multiple processes can lower performance. Current research examines how to multiplex a network device efficiently [8], if this is not supported by the NI hardware itself. Therefore, a unique solution would be to support multiple NI's directly in hardware. Designing interconnects for the standard PCI interface cuts down production costs, due to higher volume. Nevertheless, necessary additional switching hardware increases the total cost per node significantly. While PCI is a standard interface designed for IO, current PCI bridges are limited by a bandwidth of 132 MB/s running at 32bit/33Mhz. Upcoming mainboards will run at 64bit/66Mhz and achieve a maximum bandwidth of 528MB/s. The paper is organized as follows. The design space for network interfaces is evaluated and an overview on key functionality to achieve good communication performance is described in the next section. Section 3 will describe the design issues of ATOLL in comparison to Myrinet and SCI. In section 4 software layers, such as low level API and message passing interfaces for ATOLL and other NIC's, are discussed. Finally, section 5 concludes our paper.

2   Design Space for Network Interfaces

In this section we would like to evaluate current NICs and characterize the design space of IO features in general, differentiating between hardware and software issues. From the hardware's point of view, features like special purpose processor on board, additional (staging) memory, support of concurrency by allowing both, PIO and DMA operations, or support for shared memory at lowest level are of interest. The requirement for additional switching hardware to build up large scaling clusters is another concern. From the software's point of view it is interesting to examine which protocols are offered and how they are implemented, whether MMU functionality is implemented allowing RDMA, or how message delivery and arrival are detected. The latter will have a major impact on performance. We would like to break down the design space into the following items:

- **Concurrency with PIO and DMA Transactions, MMU Functionality to support RDMA**

Basically, when sending a message, the NIC's API chooses PIO or DMA for transfer, depending on the message size. PIO has the advantage of low start-up costs to initiate the transfer. However since the processor is transferring data

directly to the network, it is busy during the entire transaction. To allow concurrency, the DMA mode must be chosen in which the processor only prepares a
message by creating a descriptor pointing to the actual message. This descriptor
is handed to the DMA engine which picks up the information and injects the
message into the network. It is important to know that the DMA engine relies
on pinned down memory since otherwise pages can be swapped out of memory
and the engine usually can not page on demand by itself. The advantage of using
DMA is to hide latency (allowing for multiple sends and receives). However it has
a higher start-up time than PIO. Typically, a threshold values determines which
protocol is chosen for the transaction. Both mechanisms also play an important
role when trying to avoid memory copies.

– Intelligent Network Adapter, Hardware and Software Protocols

The most important feature having an intelligent network adapter (processor
and SRAM on board) is to be flexible in programming message handling functionality. Protocols for error detection and correction can be programmed in
software, but also new techniques can be applied (VIA). Support for concurrency is improved as well. Additional memory on board lowers congestion and
the possibility of deadlocks on the network decreases. It has the advantage to
buffer incoming data, thus emptying the network links on which the message has
been transferred. However, the memory size is usually limited and expensive, also
the number of data copies rises. Another disadvantage of this combination is that
the speed of an processor on board can not cope with the main processing unit.
Finally, programming the network adapter is a versatile task.

– Switches, Scalability and Routing

A benchmark of a point to point routine typically only shows the best performance for non-standard situations. Since a parallel application usually consists
of dozens of processes communicating in a more or less fixed pattern, measuring
the bisection bandwidth generates better information of the underlying communication hardware. A cost-effective SAN has bidirectional links and allows
sending and receiving concurrently. A key factor for performance is scalability,
when switches are added for a multistage connection network to allow larger
clusters. Here blocking behavior becomes the major concern. Another point of
interest is the connection from NIC to NIC: Data link cables must provide a
good compromise between data path width and transfer speed.

– Hardware support for Shared Memory (Coherency) and NI loca-
tions

Currently a trend can be seen in clustering bigger SMP nodes. Within an SMP
node, a cache coherent protocol like MESI synchronizes to achieve data consistency. To add this functionality to IO devices (such as the NIC), they would
have to participate on the cache coherent protocol, being able to snoop on the
system bus. However, this would involve a special solution for every processor
type and system and can not be propagated as a commodity solution. With the

growing distance between the NI and the processor, the latency of the communication operations raises and, at the same time, the bandwidth declines. The only position that results in a wide distribution and, thus, necessary higher production volumes, is the standardized PCI bus. This leads to the loss of a number of functions, like e.g., the cache coherent accesses to the main memory of the processor. As the NI on the PCI card is independent from the used processor (and has to be), functions like the MMU in the NI cannot be recreated, as they differ according to which processor is being used. For this purpose an adaptable hardware realization of the basic mechanisms or an additional programmable processor on the PCI card can be used.

– Performance Issues: Copy Routines and Notification Mechanisms

Once a message is ready for sending, the data has to be placed at a location where the NIC can fetch the data. Using the standard memcpy routines however may show poor performance. The reason is that the cache of the CPU is ruined when larger messages have been injected into the network. Modern CPU's like the Pentium III or Ultrasparc offer special MMX or VIS instructions which copy the data without polluting the cache. Another critical point is the software overhead caused by diverse protocols to guarantee data transfer. Nowadays cables are almost error free. Thus heavy protocols like TCP/IP are no longer necessary. Since an error may occur, an automatic error detection and correction implemented directly in hardware would improve efficiency. Performance is also sensitive to message arrival detection. A polling method typically wastes a lot of CPU cycles and an interrupt causes too much overhead, since contexts have to be switched. Avoiding the interrupt mechanism is very important as each new interrupt handling leads to a latency of approximately 60 μs [8].

3   NIC Hardware Layout and Design

In the ATOLL project, all design space features have been carefully evaluated and the result is an implementation of a very advanced technology.

3.1   ATOLL

Overview The ATOLL cluster interface network, is a future communication technology for building cost-effective and very efficient SAN's using standard processing nodes. Due to an extremely low communication start-up time and very broad hardware support for processing messages, a much higher performance standard in the communication of parallel programs is achieved. Unique is the availability of four links of the interface network, an integrated 8 x 8 crossbar and four independent host ports. They allow for creating diverse network topologies without additional external switches and the ATOLL network is one of the first network on a chip implementations. This design feature especially supports SMP nodes by assigning multiple processes their dedicated device. Figure 1 depicts an overview on hardware layout and data flow of ATOLL.

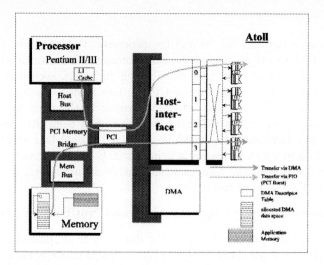

Fig. 1. ATOLL Hardware Layout and Data Flow

Design Features ATOLL's special and new feature in comparison to other NIC's is the availability of multiple and independent devices. ATOLL integrates four host and network interfaces, an 8x8 crossbar and 4 link interfaces into one single ASIC. The chip is mounted on a standard PCI board and has a 64Bit/66Mhz PCI interface with a theoretical bandwidth of 528MBytes/s at the PCI bridge. Choosing this interface, ATOLL addresses commodity solutions with a high volume production. The crossbar has a fall through latency of 24ns and a capacity of 2GBytes/s bisection bandwidth. A message is broken down by hardware into 64Byte link packets, protected by CRC and retransmitted automatically upon transmission errors. Therefore, protocol overhead for data transfer is eliminated and it has been achieved to implement error detection and correction directly in hardware. The chip itself, with crossbar, host- and network interfaces, runs at 250 Mhz. Standard techniques for the PCI bus such as write-combining and read-prefetching to increase performance are supported. Sending and receiving of messages can be done simultaneously without involving any additional controlling instances. The ATOLL API is responsible for direct communication with each of the network interfaces, providing ULC and giving the user complete control of "his" device.

In contrast to other SAN's, most of data flow control is directly implemented in hardware. This results in an extremely low communication latency of less than 2 μs. ATOLL offers Programmed IO (PIO mode) and Direct Memory Access (DMA mode), respectively. A threshold value determines which method to choose. The latter requires one pinned down DMA data space for each device. This data space is separated into send and receive regions. For starting a transmission in DMA mode, a descriptor is generated and entered into the job queue of the host interface. Injecting the message into the network is initiated by raising the descriptor write pointer, which triggers the ATOLL card to fetch the message. Basically, the descriptor contains the following information: The message length, the destination id, a pointer to the message in DMA memory space and a message tag. The completion of a DMA task is signaled through writing a

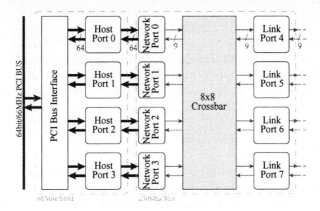

Fig. 2. ATOLL Chip

data word into main memory, which makes the time consuming interrupt handling by the processor unnecessary. Figure 3 depicts the three operations of a DMA send process.

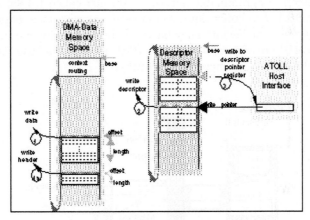

Fig. 3. Process of a DMA send job

DMA data and descriptor memory space are implemented as ring buffers. When receiving a message, the descriptor for the received message is assembled by the NI and copied into main memory. There it can be seen cache coherently by the processor. In this mode the DMA engine can also be seen as a message handler in hardware. If PIO mode is used for very short messages, it is kept in the receive FIFO of the host interface and the processor is informed of the received message through an update of the FIFO's entries in main memory. Just like in DMA mode an expensive interrupt is avoided. To overcome deadlocks, a time barrier throws an interrupt to empty the receive buffer. In this mode busy waiting of the processor on the FIFO entries leads to the extremely short receive latency. As this value is also mirrored cache-coherently into main memory the processor does not waste valuable memory or IO bandwidth. Routing is done via source path routing, identifying sender and receiver by a system wide unique identifier, the Atoll ID. Routing information is stored in a status page resiging in pinned DMA memory space. For each communication partner, a point-to-point connection is created. If two communication partners are within one SMP node, the ATOLL-API transparently maps the communication to shared memory. Finally, the ATOLL NIC supports multithreaded applications. A special register accessable in user mode can be used as a semaphore 'test-and-set'. Typical standard processor's like the PIII restrict locking mechanism to superuser level.

3.2   Myrinet

Overview The Myrinet is a high-speed interconnection technology for cluster computing. Figure 4 depicts the layout of the Myrinet NIC. A complete network consists of three basic components: a switch, the Myrinet card per host and cables which connect each card to the switch. The switch transfers variable-length packets concurrently at 1.28 Gbit/s using wormhole routing through the network. Hardware flow control via back-pressure and in-order delivery is guaranteed. The NI card connects to the PCI bus of the host and holds three DMA engines, a custom programmable network controller called LANai and up to 2 Mbyte of fast SRAM to buffer data. Newer cards improve some parameters, but do not change the basic layout. They have a 64 bit addressing mechanism allowing to address 1Gbyte of memory, a faster RISC processor at 100 Mhz accessing the SRAM which has been increased to 4 Mbytes.

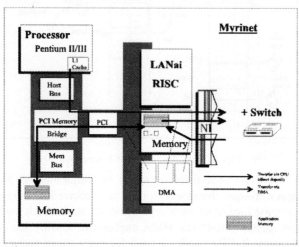

Fig. 4. Myrinet Hardware Layout and Data Flow

Design Features Under the supervision of the RISC, the DMA engines are responsible for handling data for the following interfaces: host memory/NIC's SRAM and SRAM/network, respectively. In detail, one DMA engine moves data from host memory to SRAM and vice-versa, the second stores incoming messages from the network link to the SRAM, and the third injects data from SRAM into the network. The LANai processor runs at 100 MHz, controls the DMA operations, and can be programmed individually by an Myrinet Control Program

(MCP). The SRAM serves primarily for staging communication data, but also stores the code of the MCP. To simplify the software, the NI memory can be mapped into the host's virtual address space. As research shows [1], the limited amount of memory on the NIC is not a bottleneck, but the interaction of DMA engines and LANai. The Myrinet card retrieves five prioritized data streams into the SRAM. However, at a cycle of 10ns only 2 of them can be addressed whereas 3 are stalling. This leads to a stalling LANai, which does not get access to the staging memory. When sending a message with Myrinet, first the user copies data to a buffer in host memory, which is accessible by the NI device. The next step is to provide the MCP with the (physical) address of the buffer position. The LANai starts the PCI bridge engine to copy the data from host memory to NIC memory. Finally the LANai starts up the network DMA engine to inject the data from NIC memory into the network. On the receiving side, the procedure is vice versa. First, the LANai starts the receive DMA engine to copy the data to NIC memory and starts the PCI bridge engine to copy the data to an address in host memory (which was previously specified via a rendez-vous protocol). Finally, after both copies are performed, the receiver LANai notifies the polling processor of the message arrival by setting a flag in host memory.

3.3   Scalable Coherent Interface (SCI)

Overview Compared to Myrinet, SCI is not just another network interface card for message passing, but offers shared memory programming in a cluster environment as well. SCI intends to enable a large cache coherent system with many nodes. Besides its own private cache / memory, each node has an additional SCI cache for caching remote memory. Unfortunately, the caching of remote memory is not possible for PCI bus based systems. This is because transactions on the system bus are not visible on the PCI bus. Therefore an important feature defined in the standard is not available on standard clusters and SCI is no longer coherent when relying solely on its hardware.

Design Features One of the key features of SCI is that by exporting and importing memory chunks, a shared memory programming style is adopted. Remote memory access (RMA) is directly supported at hardware level (Figure 2 depicts an overview of SCI address translations). By providing a unique handle to the exported memory (SCI Node ID, Chunk ID and Module ID) a remote host can import this 'window' and create a mapping. To exchange messages, data has to be copied into this region and will be transferred by the SCI card, which detects data changes automatically.

Packet sizes of 64 Bytes are send immediately, otherwise a store barrier has to be called to force a transaction. In order to notify other nodes when messages have been sent they either can implement their own flow control and poll on data or create an interrupter which will trigger the remote host. However, the latter has a bad performance with a latency of 36 μs on a Pentium II450. One major drawback of SCI is that a shared memory programming style can not easily

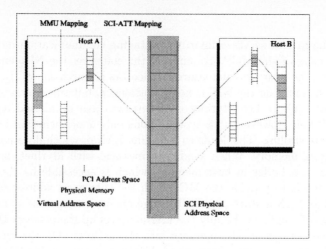

Fig. 5. SCI Address Translation

be achieved because of lacking functionality to cache regions of remote memory in the local processor cache. Furthermore, SCI uses read and write buffers to speed up communication which brings along a certain amount of inconsistency. Finally SCI is not attractive to developers who have to keep in mind the big gap between read and write bandwidth in order to achieve highest performance (74MB/s remote write vs. 7.5 MB/s remote read using a Pentium II450). When looking at concurrency then the preferred method is to use the processor to copy data. In this case however, the processor is busy and can not be used to overlap computation and communication as when DMA would be used. Using the processor, a remote communication in SCI takes place as just a part of a simple load or store opcode execution in a processor. Typically the remote address results in a cache miss, which causes the cache controller to address remote memory via SCI to get the data, and within the order of a microsecond the remote data is fetched to cache and the processor continues execution.

4   Software

4.1   Low Level API

The old approach, moving data through I/O-channel or network-style paths, requires assembling an appropriate communication packet in software, pointing the interface hardware at it, and initiating the I/O operation, usually by calling a kernel subroutine. When the data arrives at the destination, hardware stores them in a memory buffer and alerts the processor with an interrupt. Software then moves the data to a temporary user buffer before it is finally copied to its destination. Typically this process results in latencies that are tens to thousands of times higher than user level communication. These latencies are the main limitation on the performance of Clusters or Networks of Workstations.

ATOLL The ATOLL API is providing access to the device at user level. It offers function calls to establish direct point to point communication between two ATOLL id's. An ATOLL id and one corresponding hostport is assigned to a process when opening the device. A connection between two ATOLL ids is needed in order to call non blocking send and receive operations. Send, receive and multicast have a message passing style in the form of tuples (destination, src, length). Threshold values for PIO and DMA can be adjusted during runtime. In PIO mode, the ATOLL API offers zero level communication. Besides the message passing functionality the ATOLL API offers lock and unlock primitives to the semaphore, which is available for each hostport. The ATOLL API is open source.

Myrinet and SCI Well known API's for the Myrinet NIC are the PM and GM libraries. Both API's are open source and offer send and receive functions in a message passing style. For SCI, Dolphin and Scali offer low level API's to create, map and export memory to remote nodes. The implementation of sending and receiving of data is left to the user. Typically ring buffers in the mapped memory regions are implemented. This allows for simple data flow control. Writing data to this region is detected by the SCI card which transfers the updated data to the remote node.

4.2 Upper Software Layer for Communication

Open Source projects can be seen as a key to the success of a project. Myrinet and ATOLL are open source projects in which ports to standard message passing environments such as MPI and PVM are available to application developers. Drivers for SCI are not in an open format. This makes it difficult to fix bugs, but also the widespread usage of the software is limited. For all API's, devices for MPICH [11] have been written. Especially the Score 2.4 implementation [9] based on PM achieves a good performance supporting intra-node communication at 1150 Mbit/s and inter-node communication at 850 Mbit/s using Pentium II 400's. It is to mention, that Score also allows to have multiple processes from different users using the Myrinet interface. Another device written for MPICH is BIP-MPI [10]. This software also achieves good performance, however is restricted to a single job per node. The Scampi MPI implementation from Scali also achieves high bandwidths, however using the BEFF [12] benchmark from Pallas, the latency on clusters with more than 32 nodes, increased up to 60 μs. ATOLL which will be available 1Q/00 is an Open Source project and first message passing environments will be based on MPICH and PVM. With a hardware latency of 1.4 μs and a link utilization of 80% at 128 Bytes, the achievable performance with MPI looks promising. A loop back device shows a one way roundtrip time of 2.4 μs.

4.3 Communication Overhead

In this section we would like to discuss current techniques to avoid memcpy's when injecting data into the network. Figure 4 depicts necessary steps involved during a transaction for ATOLL.

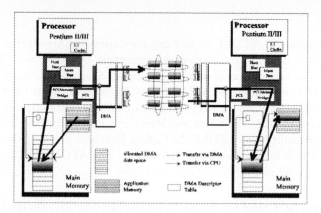

Fig. 6. DMA Copy

Recent research tries to avoid unnecessary data copies which results in a so called zero copy mechanism. In this case the data is directly fetched from its position in application memory and directly deposited in remote application memory. It is expected to decrease latency and increase bandwidth for data transfer using this method. Basically, if PIO is available, this communication mode can be used for zero copy. When sending, data is directly injected by the CPU into the network. On the receiving side, the message can again be delivered directly with PIO. The disadvantage is that the processor will be involved during the entire transaction and can not be used for computation during that time. To enable the DMA engine to perform this task, a virtual-to-physical address translation must be implemented, which increases hardware complexity significantly. Sharing the page tables between the OS and the device is complex and time consuming too. The TLB handling is usually performed by the OS. Pages for translation have to be pinned down, and virtual addresses now represent physical ones. The TLB can be placed and managed at NI memory, the host memory, or both. Using this method, zero-copy can be achieved via remote memory writes using the information provided with the TLB. Send and receive operations carry the physical address of the destination buffer and the DMA engine copies the (contiguous) data directly to the destination address. Typically, a rendezvous model is needed before such operation can take place, since the location at the receiver side is not know a priori. A requirement for the NIC is touch dynamically pinned down data. This DMA method also only makes sense, if the data to be transferred is locked down once and the segment can be re-used. Otherwise expensive lock and unlock sequences will lower performance making a trap into the system. Another problem coming along with zero copies is that of message flow control. It is not obvious when a message has been transferred and the space can be used again. On the other hand support for RDMA eases the implementations of one sided communication. Myrinet features PIO and DMA transactions, however data to be send is first stored in the staging memory of the NIC. In a second step, the LANai then injects the message which will be again stored in the SRAM on the remote card. This may be the reason that latest research shows only a performance increase of 2% when using zero copy instead of DMA copy [8]. Porting issues for using zero copy mechanisms are another point of concern.

5   Conclusion

We have given a description of ATOLL, a new high speed interconnect combining attractive, efficient design issues with new features which fulfill the needs of today's high speed interconnects. ATOLL is a cost effective, affordable SAN with fixed expenses per node, even for large clusters. A more expensive solution is Myrinet, which has the highest number of installed cards, or SCI. Both need additional switches for building larger clusters. In terms of performance and robustness, currently Myrinet seems to be the best choice, however this may change with the availability of ATOLL and 64Bit/66Mhz PCI bridges. Here, the integration of the most important functions for a SAN into one chip shows a high level of performance and extremely low latency for cluster communication. Next development steps of the ATOLL network project will include optical link interconnects for increasing distance. Under investigation is also MMU functionality implemented in hardware. It is also planned to adapt the link technology to the concepts of System IO, since major parts are easily adaptable. This will provide the user with a high speed low latency unified communication infrastructure for the next generation of clusters.

References

[1]   Warschko, Blum and Tichy. *On the Design and Semantics of User-Space Communication Subsystems*, PDPTA 99, Las Vegas, Nevada.

[2]   Santos, Bianchini and Amorim. *A Survey of Messaging Software Issues and System on Myrinet Based Clusters*

[3]   IEEE 345, 47th Street New York. *IEEE Standard for Scalable Coherent Interface (SCI)*, 1993

[4]   O' Caroll, Tezuka, Hori and Ishikawa. *The Design and Implementation of Zero Copy MPI using ...* , In International Conference on Supercomputing '98, pages 243-250, July 1998.

[5]   Rzymianowicz, Bruening, Kluge, Schulz and Waack. *Atoll, A Network on a Chip*, PDPTA 99, Las Vegas, Nevada

[6]   http://www.top500.org

[7]   Kay and Pasquale. *Profiling and Tuning Processing overheads in TCP/IP*. IEEE/ACM Transactions on Networking, Dec. 1996

[8]   Warschko. *Efficient Communication in Parallel Computers*. PhD thesis, University of Karlsruhe, 1998

[9]   Tezuka, O'Caroll, Hori, and Ishikawa. *Pin-down Cache: A Virtual Memory Management Technique for Zero-copy Communication*, IPPS98, pages 308-314, 1998

[10]   Prylli and Tourancheau. *BIP: A new protocol designed for high performance networking on Myrinet*, PCNOW Workshop, IPPS/SPDP98, 1998

[11]   O'Carroll, Tezukua, Hori, and Ishikawa. *MPICH-PM: Design and Implementation of Zero Copy MPI for PM*, Technical Report TR-97011, RWC, March 1998.

[12]   http://www.pallas.com

[13]   Scholtyssik and Dormanns. *Simplifying the use of SCI shared memory by using software SVM techniques*, 2nd Workshop Cluster Computing, Karlsruhe, 1999

[14]   Mukherjee and Hill. *The Impact of Data Transfer and Buffering Alternatives on Network Interface Design*, HPCA98, Feb. 1998

ClusterNet: An Object-Oriented Cluster Network

Raymond R. Hoare

Department of Electrical Engineering, University of Pittsburgh
Pittsburgh, PA 15261
hoare@pitt.edu

Abstract. Parallel processing is based on utilizing a group of processors to efficiently solve large problems faster than is possible on a single processor. To
accomplish this, the processors must communicate and coordinate with each
other through some type of network. However, the only function that most networks support is message routing. Consequently, functions that involve data
from a group of processors must be implemented on top of message routing.
We propose treating the network switch as a function unit that can receive data
from a group of processors, execute operations, and return the result(s) to the
appropriate processors. This paper describes how each of the architectural resources that are typically found in a network switch can be better utilized as a
centralized function unit. A proof-of-concept prototype called ClusterNet$_{4EPP}$ has
been implemented to demonstrate feasibility of this concept.

1   Introduction

In the not-so-distant past, it was common for groups of people to pool their resources
to invest in a single, high-performance processor. The processors used in desktop
machines were inferior to the mainframes and supercomputers of that time. However,
the market for desktop computers has since superceded that of mainframes and
supercomputers combined. Now the fastest processors are first designed for the
desktop and are then incorporated into supercomputers. Consequently, the fastest
processors, memory systems, and disk controllers are packaged as a single circuit
board. Thus, the highest performance "processing element" is a personal computer
(PC).

  Almost every individual and company uses computers to help them be more efficient. Networks enable seemingly random connections of computers to communicate
with each other and share resources. Computational and data intensive applications
can utilize the resources of a cluster of computers if the network is efficient enough.
If the network is inefficient, the added communication and coordination cost reduces,
or even removes, the benefit of using multiple computers. As more computers are
used to execute an application in parallel, the extra overhead eventually removes the
performance benefit of the additional resources. Thus, for a cluster of computers to be

J. Rolim et al. (Eds.): IPDPS 2000 Workshops, LNCS 1800, pp. 28-38, 2000.

used as a single parallel processing machine, they must be able to efficiently communicate and coordinate with each other.

Stone, in his popular book on high-performance computer architecture [1], states that peak performance of a parallel machine is rarely achieved. The five issues he cited are:

1. Delays introduced by interprocessor communications
2. Overhead in synchronizing the work of one processor with another
3. Lost efficiency when one or more processors run out of tasks
4. Lost efficiency due to wasted effort by one or more processors
5. Processing costs for controlling the system and scheduling operations

These issues are relevant for all computer architectures but are particularly troublesome for clusters. Clusters typically use commodity network switches that have been designed for random configurations of computers and routers. Thus, packets are used to encapsulate data and provide routing information. This software overhead accounts for 40-50% of the total communication time[2].

Network switches are designed for typical network traffic. Rarely will every incoming packet be destined for the same output port for a sustained period of time. The outgoing network link would become saturated, its buffers will become full, the switch will have to drop incoming packets, and the packets will have to be resent. While this is extremely rare for a typical network, the *gather* and *all-to-all* communication operations require this type of communication pattern[3].

Processor coordination and synchronization are group operations that require information from each processor. While such operations can be executed using message passing communications, a total of N messages must be sent through the network to gather the data and broadcast the result(s). These communication operations can ideally be overlapped to require only $(\log_2 N + 1)$ communication time steps.

Processor scheduling and control are also operations that require data from each of the processors. However, this information must be maintained to ensure even load distribution. The algorithms used for scheduling and controlling clusters are not computationally intensive but require efficient access to every computer's status.

Operations that involve data from a collection of computers are defined as *aggregate operations* [4]. Ironically, the architectural characteristics of a typical switch are well suited for executing aggregate operations. A typical modern switch interconnects 8 to 80 computers, contains a processor (or routing logic), and stores routing information in a lookup table. For a 32 or 64-processor cluster, a single switch is capable directly interconnecting all of the computers.

Rather than assuming that a cluster is a purely distributed memory architecture that communicates through a point-to-point network, this paper examines the entire cluster architecture, including the network switch to demonstrate how a better cluster architecture can be created. Specifically, the architectural resources contained in a typical switch will be examined, reallocated and/or changed, to facilitate efficient communication and control of the entire cluster. As shown in the following table, the architectural features of a network switch and a cluster are almost exact opposites. In fact, the architectural characteristics are almost exact opposites. The proposed ClusterNet architecture utilizes these differences to form a new architecture that is a complement of both distributed and shared memory, as well as a complement of parallel and serial program execution.

Table 1. Architectural features of network switches and a cluster of computers.

Architectural Feature	Network Switch	Cluster
Number Of Processors	1	16, 32, 64
I/O Ports Per Processor	8-80	1-3
Memory Architecture	Shared	Distributed
Storage	Lookup Table	RAM & Disk
Functionality	Fixed	Programmable
Execution Model	Serial	Parallel
Topology	Unknown	Star Topology
Performance Criteria	Packets per Second	Seconds per Communication
Communication Pattern	Point-To-Point	Point-To-Point & Collective

2 ClusterNet

While network switches can be used to facilitate cluster communications, there are a number of architectural differences between the network switch and the rest of the cluster. By combining these two architectures, a more efficient cluster architecture called ClusterNet, can be built.

Rather than limiting the network switch to routing packets, we propose expanding the role of the switch to execute functions on data gathered from a group of processors. Furthermore, because of the switch's memory architecture, it should also be able to store data. Thus, by combining data storage with computation, an *object-oriented* cluster network can be created. To simplify our discussion, our new object-oriented switch will be labeled an *aggregate function unit (AFU)* and the unmodified network switch will just be called a *switch*.

The goal of this paper is to demonstrate how the resources of a switch can be more efficiently utilized when placed within the context of a cluster architecture. Table 2 shows how ClusterNet's usage of architectural resources differs from a switch.

Table 2. ClusterNet's usage of architectural resources.

Architectural Resource	Switch Usage	AFU Usage
Routing Logic	Route Messages	Execute Functions
Switch Memory	Address Lookup Table	Data Structures
Switch Port	Input/Output Packet Queue	Register Interface
Physical Link	Send/Receive Packets	Send/Receive Data
Software Interface	Send/Receive Messages	Access To AFU Port
Application Interface	MPI	Aggregate Functions

The remainder of this section will discuss each of the resources listed above and how they can be used to provide a more robust cluster architecture called ClusterNet. Section 3 describes a proof-of-concept four-processor prototype that was built. Section 4 describes related work and section 5 offers conclusions and future directions.

2.1 Functionality: Router vs. Aggregate Function Execution

The routing logic (or processor) can collect and distributed information from every processor because most network switches interconnect between 4 and 80 computers. However, cluster implementations have maintained a distributed-memory architecture in which the processors communicate through message passing.

Ironically, group operations such as *Global Sum* are implemented by sending N messages through the same switch in $\log_2 N$ time steps. Each time step requires a minimum of a few microseconds, over 1000 processor cycles. Rather than performing the computation, the network switch is busy routing packets.

Instead of using the network switch's processor to route messages, the processor can be used to execute functions within the network. Because the switch is directly connected to each of the processors, data from every processor can be simultaneously sent into the network switch. Upon arrival, the specified function is computed and the result is returned to each processor.

To quantify this proposition, we define the following variables:

N - The number of processors in the cluster (2 - 64).
α - The communication time between a processor and the switch (1μs).
$k * (N\text{-}1)$ - The number of instructions to be executed.
ε - The amount of time required to compute a single instruction (5 ns).

If an associative computation is executed using N processors and a point-to-point network, the amount of time required is approximately $(2\alpha + k\varepsilon) * \log_2 N$ because computation can be overlapped. If the switch's processor is used to execute the same function, the amount of time required is $(2\alpha + (N\text{-}1) k\varepsilon)$. From an asymptotic perspective, it is better to use all N processors rather than the AFU's processor. However, when typical values are used ($\alpha = 1\mu s$, $\varepsilon = 5ns$) the resulting graphs show the performance tradeoffs as we change k and N, shown in Fig. 1.

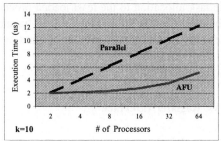
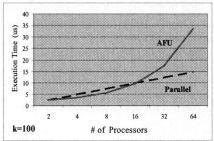

Fig. 1 Collective computation using the AFU verses using all N processors for k=10 and 100, $\alpha = 1\mu s$, $\varepsilon = 5ns$.

2.2 Network Storage: Routing Tables vs. Network-Embedded Data Structures

To enable a switch to be used for any network topology, it must be able to change how it routes different packets. This is typically implemented through a lookup table. When a packet is received, its destination address is used as an index into the lookup table to determines which port the packet should be routed to. This information can also be changed because network configurations change.

In a cluster architecture, the routing lookup table is of minimal use because each processor is directly connected to the switch. If we require that processor i be attached to port i then there is no need for a routing table. The network-embedded memory can then be used as a cluster resource.

For example, the lookup table could be used to track cluster-wide resources. If a resource is needed, the lookup table could be used to determine where the resource is located. This concept can be used to implement a dynamically-distributed shared-memory cluster architecture. In a distributed shared-memory architecture (i.e. Cray T3D) there is a single memory address range that is distributed across all of the processors. Each processor can access any portion in memory by simply specifying a memory address. However, this results in non-uniform memory access times. Direct memory access was not built into the Cray T3E. A *dynamically*-distributed shared-memory still uses a single address range but allows blocks of memory to migrate to the processor that needs them. When a memory request is made, the entire block of memory is relocated and placed in the local memory of the requesting processor. For regular access patterns, this drastically improves performance.

However, there is an inherent contradiction within the dynamically-distributed shared-memory architecture. A shared resource table is needed to determine where each block is located. To share this location table, it too must be placed in shared memory. The location table can be distributed across the processors but requires two requests for every memory access. If the switch's lookup table is used for the location table, memory requests could be sent to the network and the network could forward them to the processor that currently owns the block.

In addition to a lookup table, the network-embedded memory can be used to represent any number of useful data structures. Synchronization data structures can be used to implement static, dynamic and directed synchronization. A processor load table can be kept in the network to facilitate dynamic task allocation to the least loaded processor. Queues and priority queues can also be used for task allocation and load balancing. Even shared linked-lists can be implemented with a small amount of additional control logic.

2.3 Network Port Interface: I/O Queues vs. Register Interface

Because all networks use packets, they also contain I/O queues to store the packets until the router logic is able to handle them. The drawback to this is that the queues become full and overflow. Our design does not require queues because it does not route packets.

The AFU does, however, execute functions and does transmit data. As shown in Fig. 4, the interface to the AFU appears as four registers. The OpCode register is used to specify which function is to be executed. The Data registers are used to move data between the PC and the AFU. Function parameters and function results are passed through these registers using the

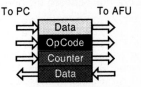

Fig. 3. The AFU Interface Port

full/empty bits to indicate valid data. The Counter register can be used as a function parameter and is useful when accessing the network-embedded data structures described earlier. The Counter is particularly useful when accessing adjacent locations in memory. When a word has been read from memory the counter automatically increments. In this way, streaming read and write operations can be implemented easily by setting the appropriate OpCode and sending/receiving an entire block of data.

Table 3. Latencies (in us) for point-to-point messages for several architectures.

Platform	Latency (in μs)	Send Overhead	Receive Overhead	Ref.
IBM SP2	39	-	-	[5]
Intel Paragon	6.5	21.5%	33.8%	[2]
Meiko 7CS-2	7.5	22.7%	21.3%	[2]
Cray T3D	2.2	-	-	[6]
Memory Channel	5-20	-	-	[7]
Myrinet	11.2	17.9%	23.2%	[2]
SHRIMP	10+	-	-	[8]
ParaStation	5+	-	-	[9]
PAPERS	3-5	-	-	[10]
ClusterNet_{4EPP}	1.7-5.2	-	-	[11]

2.4 Software Interface: Packet vs. Direct Read and Write

As was shown in the Table 3, the software overhead for sending and receiving a message consumes 40-50% of the overall message latency. This is due to the time spent encoding and decoding packets. Rather than accepting this overhead, we propose expanding the functionality of the network and simplifying the network interface. Most architectures layer their communication libraries on top of point-to-point primitives that encode, send and decode packets. ClusterNet executes functions within the network and can be used to execute collective communications within the network.

As a result of executing functions within the network hardware, the software interface is very simple and only requires seven assembly-level instructions listed below. Lines 1 and 2 are used to set the OpCode and Counter registers. The OpCode is used to specify which function should be executed. If the OpCode has not changed, these registers do not need to be set. After data is placed into the network, the function is executed and the results are returned. This architecture relies on the fact that the network link between the processor and the AFU perform error detection and correction.

```
1.  I/O Write (OpCode) /* Optional */
2.  I/O Write (Count)  /* Optional */
3.  I/O Write (Data)
4.  I/O Read (Result)
5.  if ( Result == NOT_A_NUMBER) goto line 4
6.  if ( Result != PREFIX_TOKEN ) goto line 8
7.  I/O "Data" Read (Result)
8.  /* The Aggregate Function has completed. */
```

3 The ClusterNet$_{4EPP}$ Proof-of-Concept Prototype

The four-processor Object-Oriented Aggregate Network[11], called ClusterNet$_{4EPP}$, demonstrates that the simplified network interface is feasible and performs very well using a small FPGA (Altera 10K20). The PCs' parallel ports were used as the network interface and require approximately 1μs to access. Experimental results were performed and a PCI device was accessible in approximately 450 ns. For ClusterNet$_{4PP}$, read and write access time to each of the four registers (Data In, OpCode, Counter, Data Out) was found to be 1.7 μs. IEEE 1284 in EPP mode was used for cable signaling.

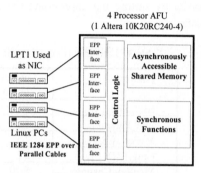

Fig. 4. ClusterNet$_{4EPP}$

To demonstrate that network-embedded data structures are feasible and beneficial, an embedded RAM block was placed inside the FPGA. The control logic for the RAM block was modified and the synchronization/ arithmetic operations shown below were implemented. Each operation is executed on a single memory location. While a processor was not placed in the network, these operations can be used to perform simple global operations.

Experimentation was performed to determine the effect of memory contention but due to the small number of processors and a 120 ns memory-access time, no effect could be detected. All memory accesses required approximately 1.7 μs. If the Op-Code and the Counter need to be set, the total execution time is 5.2 μs. All of the memory operations can be executed on any word in the embedded memory. In addition to memory operations, barrier synchronization and a number of reduction operations were implemented. These operations are described in Table 4.

Table 4. Memory operations for the RAM embedded within ClusterNet$_{4EPP}$

Memory Operations
Non-blocking Memory Read / Exchange
Wait for Lock=1 (or 0) then Read
Wait for Lock=0 (or 1), Exchange and set Lock=1
Non-blocking Write and Unlock/Lock
Wait for Lock=0, OR with RAM
Wait for Lock=0, XOR with RAM
Wait for Lock=0, Decrement/ Decrement RAM
Wait for Lock=0, RAM = RAM -/+ Data

4. Related Research

The NYU Ultracomputer [12, 13] and the IBM RP3 [14] are both dance-hall shared memory architectures. The Ultracomputer was the first architecture to propose that combining be performed within the processor-to-memory interconnection network. Messages that reference identical memory locations are combined if both messages are buffered within the same switch at the same time. The computations that can be performed within the interconnection network are Fetch-and-Add, Fetch-and-Increment, and other Fetch-and-Op functions, where Op is associative.

Active Messages from Berkeley [2] allow functions to be executed at the network interface on the local or remote node. Active Networks perform operations on data values that are passed through the network [15, 16]. Fast Messages[17] modify the network interface drivers to reduce the overhead for sending and receiving messages. Sorting networks were introduced in [18] and have continued to remain a topic of interest [19, 20]. Multistage data manipulation networks are discussed in [21].

A number of commercial architectures have included direct support for various associative aggregate computations. The Cray T3D directly supports, through Cray-designed logic circuits, barrier synchronization, swap, and Fetch-and-Increment [22]. The TMC Connection Machine CM-5 has a control network that supports reduction operations, prefix operations, maximum, logical OR and XOR [22]. These architectures can be considered aggregate networks but they are very specific in the functions that they are designed to execute.

PAPERS, Purdue's Adapter for Parallel Execution and Rapid Synchronization, is a network that allows a number of aggregate computations to be performed within a custom network hub that is attached to a cluster of Linux PCs [23-25]. This design uses a combination of barrier synchronization with a four-bit wide global NAND to construct a robust library of aggregate computations and communications.

A number of cluster projects have employed different approaches to reduce the communication cost of point-to-point and broadcast messages. SHRIMP [8, 26] uses memory bus snooping to implement a virtual memory interface. Point-to-point messages in SHRIMP have a 10+ μs latency and remote procedure calls have a 3+ μs latency. Myrinet [27] provides gigabit bandwidth with a 0.55 μs worst-case latency through its pipelined crossbar switch.

5. Conclusions and Future Directions

This paper has proposed the concept of combining the architectural characteristics of a network switch with that of a cluster of desktop computers. Rather than using the resources of the switch for message routing, this paper has proposed using them to create a function unit that is capable of performing computations on data that is aggregated from a group of processors. Specifically, the following switch resources can better serve the architectural needs of a cluster in the following way:

- The switch lookup table should be used as network-embedded shared memory.
- The functionality of the switch should be expanded to include aggregate functions. This reduces the total amount of time required for group computations.
- The functionality of the switch should be configurable. This will enable greater utilization of the architectural resources of the entire cluster rather than just the processors.
- Packets are not needed if each processor has direct access to a set of registers within the "switch". This remove the need to encode and decode packets and reduces the software overhead to less than ten assembly-level instructions.

ClusterNet$_{4EPP}$ was described and implements numerous instructions that access the shared memory in as little as 1.7μs. A number of functions were implemented that involved data from all of the processors. These functions included OR, XOR, AND and ADD. While ClusterNet$_{4EPP}$ has demonstrated that it is possible to implement functions within the network, there are still a number of issue that have not been addressed. Scalability to large systems has not been demonstrated and the performance of complex functions is still unknown.

Scalability and function performance are currently being examine using an Altera 10K100 that is five times larger that the 10K20 and is currently able to interconnect 8 processors. The figure to the left shows the 10K100 prototype in the left portion of the picture and four of the connectors in the right portion of the picture. The cable in the middle of the picture with the Altera label is the FPGA configuration cable. The EPP in currently working and the remainder of the design is expected to be completed by SuperComputing '99 in the middle of November.

Future directions include using a higher bandwidth physical layer and a PCI network interface card. Each of these areas are under development but experimental results have not been obtained yet. Additionally, embedding a DSP or RISC processor into the network would enable rapid experimentation with system-level resource management. After that is achieved, user-level programmability of the AFU will be approached.

References

1. H. S. Stone, High-Performance Computer Architecture, Third ed. Reading, MA: Addison-Wesley Publishing Company, 1993.
2. D. Culler, L. Liu, R. Martin, and C. Yoshikawa, "Assessing Fast Network Interfaces," IEEE Micro, vol. 16, pp. 35-43, 1996.
3. M. Snir, S. Otto, S. Huss-Lederman, D. Walker, and J. Dongarra, MPI, The Complete Reference. Cambridge, Massachusetts: The MIT Press, 1996.
4. R. Hoare and H. Dietz, "A Case for Aggregate Networks," Proceedings of the 12th International Parallel Processing Symposium and 9th Symposium on Parallel and Distributed Processing, Orlando, FL, 1998.
5. C. Stunkel and e. al., "The SP2 High-Performance Switch," IBM Systems Journeal, vol. 34, pp. 185-204, 1994.
6. R. Kessler and J. Schwarzmeier, "Cray T3D: a New Dimension for Cray Research," Proceedings of the In Digest of Papers. COMPCON Spring '93, San Francisco, CA, 1993.
7. M. Fillo and R. Gillett, "Architecture and Implementation of Memory Channel 2," Digital Equipment Corporation High Performance Technical Computing, pp. 34-48, 1997.
8. M. Blumrich and e. al., "Virtual Memory Mapped Network Interfaces for the SHRIMP Multicomputer," Proceedings of the The 21st Annual International Symposium on Computer Architecture, 1994.
9. T. Warshko, W. Tichy, and C. Herter, "Efficient Parallel Computing on Workstation Clusters," University of Karlsruhe, Dept. of Informatics, Karlsruhe, Germany Technical Report 21/95, 1995.
10. R. Hoare, T. Mattox, and H. Dietz, "TTL-PAPERS 960801, The Modularly Scalable, Field Upgradable, Implementation of Purdue's Adapter for Parallel Execution and Rapid Synchronization," Purdue University, W. Lafayette, Internet On-line Tech Report:, 1996.
11. R. Hoare, "Object-Oriented Aggregate Networks," in School of Electrical Engineering. W. Lafayette: Purdue University, 1999.
12. A. Gottlieb and e. al., "The NYU Ultracomputer, Designing a MIMD Shared Memory Prallel Computer," IEEE Transactions on Computers, pp. 175-189, 1983.
13. R. Bianchini, S. Dickey, J. Edler, G. Goodman, A. Gottlieb, R. Kenner, and J. Wang, "The Ultra III Prototype," Proceedings of the Parallel Systems Fair, 1993.
14. G. Pfister and V. Norton, "'Hot Spot' Contention and Combining in Multistage Interconnection Networks," Proceedings of the 1985 International Conference on Parallel Processing, 1985.
15. D. Tennenhouse and D. Wetherall, "Towards an Active Network Architecture," Computer Communications Review, vol. 26, 1996.
16. D. Tennenhouse and e. al., "A Survey of Active Network Research," IEEE Communications Magazine, vol. 35, pp. 80-86, 1997.
17. H. Bal, R. Hofman, and K. Verstoep, "A Comparison of Three High Speed Networks for Parallel Cluster Computing," Proceedings of the First International Workstion on Communication and Architectural Support for Network-Based Parallel Computing, San Antonio, TX, 1997.
18. K. Batcher, "Sorting Networks and Their Applicaitons," Proceedings of the Spring Joint Computer Conference, 1968.
19. J. Lee and K. Batcher, "Minimizing Communication of a Recirculating Bitonic Sorting Network," Proceedings of the the 1996 International Conference on Parallel Processing, 1996.

20. Z. Wen, "Multiway Merging in Parallel," IEEE Transactions on Parallel and Distributed Systems, vol. 7, pp. 11-17, 1996.
21. H. J. Siegel, Interconnection Networks for Large-Scale Parallel Processing: Theory and Case Studies, Second Edition ed. New York, NY: McGraw-Hill, 1990.
22. G. Almasi and A. Gottlieb, Highly Parallel Computing, Second Edition. Redwood City, CA: The Benjamin/Cummings Publishing Company, Inc., 1994.
23. H. Dietz, R. Hoare, and T. Mattox, "A Fine-Grain Parallel Architecture Based on Barrier Synchronization," Proceedings of the International Conference on Parallel Processing, Bloomington, IL, 1996.
24. R. Hoare, H. Dietz, T. Mattox, and S. Kim, "Bitwise Aggregate Networks," Proceedings of the Eighth IEEE Symposium on Parallel and Distributed Processing, New Orleans, LA, 1996.
25.T. Mattox, "Synchronous Aggregate Communication Architecture for MIMD Parallel Processing," in School of Electrical and Computer Engineering. W. Lafayette, IN: Purdue University, 1997.
26. E. Felten and e. al., "Early Experience with Message-Passing on the SHRIMP Multicomputer," Proceedings of the The 23rd Annual International Symposium on Comuter Architecture, Philadelphia, PA, 1996.
27. N. Boden and e. al., "Myrinet: A Gigabit per Second Local Area Network," in IEEE-Micro, vol. 15, 1995, pp. 29-36.
28. R. Brouwer, "Parallel algorithms for placement and routing in VLSI design", Ph. D. Thesis, University of Illinois, Urbana-Champaign, 1991.
29. J. Chandy, et. al. "Parallel Simulated Annealing Strategies for VLSI Cell Placement", in Proceedings of the 1996 International Conference on VLSI Design, Bangalore, India, January 1996.
30. T. Stornetta, et. al., "Implementation of an Efficient Parallel BDD Package", Proc. 33rd ACM/IEEE Design Automation Conference, 1996.
31. R. Ranjan, et. al., "Binary Decision Diagrams on Network of Workstations", In Proceedings of the International Conference on Computer-Aided Design, pp. 358-364, 1996.

GigaBit Performance under NT

Mark Baker
University of Portsmouth
Hants, UK, PO4 8JF, UK
`Mark.Baker@computer.org`

Stephen Scott and Al Geist
Oak Ridge National Laboratory
Oak Ridge, TN 37831-6367, USA
`{scottsl,gst}@ornl.gov`

Logan Browne
Hiram College
Hiram, OH44234, USA
BrownLC@hiram.edu

January 13, 2000

Abstract

The recent interest and growing popularity of commodity-based cluster computing has created a demand for low-latency, high-bandwidth interconnect technologies. Early cluster systems have used expensive but fast interconnects such as Myrinet or SCI. Even though these technologies provide low-latency, high-bandwidth communications, the cost of an interface card almost matches that of individual computers in the cluster. Even though these specialist technologies are popular, there is a growing demand for Ethernet which can provide a low-risk and upgradeable path with which to link clusters together. In this paper we compare and contrast the low-level performance of a range of Giganet network cards under Windows NT using MPI and PVM. In the first part of the paper we discuss our motivation and rationale for undertaking this work. We then move on to discuss the systems that we are using and our methods for assessing these technologies. In the second half of the paper we present our results and discuss our findings. In the final section of the paper we summarize our experiences and then briefly mention further work we intend to undertake.

Keywords: cluster interconnect, communication network, Gigabit Ethernet, PVM, MPI, performance evaluation.

J. Rolim et al. (Eds.): IPDPS 2000 Workshops, LNCS 1800, pp. 39-50, 2000.

1. Introduction

The concept of a cluster of computers as a distinguished type of computing platform evolved during the early 1990's[1]. Prior to that time, the development of computing platforms composed of multiple processors was typically accomplished with custom-designed systems consisting of proprietary hardware and software. Supercomputers, or high-performance multiprocessor computers, were designed, developed, and marketed to customers for specialized grand challenge applications. Typically, the applications that ran on these supercomputers were written in Fortran or C, but used proprietary numerical or messaging libraries, that were generally not portable.

However, rapid advances in commercial off-the-shelf (COTS) hardware and the shortening of the design cycle for COTS components made the design of custom hardware cost-ineffective. By the time a company designed and developed a supercomputer, the processor speed and capability was out-paced by commercial processing components. In addition to the rapid increase in COTS hardware capability that led to increased cluster performance, software capability and portability increased rapidly during the 1990's. A number of software systems that were originally built as academic projects led to the development of standard portable languages and new standard communication protocols for cluster computing.

The programming paradigm for cluster computing falls primarily into two categories: message passing and distributed shared memory (DSM). Although DSM is claimed to be an easier programming paradigm as the programmer has a global view of all the memory, early efforts instead focused on message passing systems. Parallel Virtual Machine[2] (PVM) started as a message passing research tool in 1989 at Oak Ridge National Laboratory (ORNL). Version 2, written at the University of Tennessee, was publicly released in early 1991. As a result of this effort and other message passing schemes, there became a push for a standardized message passing interface. Thus, in 1994 MPI Version 1 was approved as a *de jure* standard for message-passing parallel applications[3]. Many implementations of MPI-1 have been developed. Some implementations, such as MPICH, are freely available. Others are commercial products optimized for a particular system, such as SUN HPC MPI. Generally, each MPI implementation is built over faster and less functional low-level interfaces, such as BSD Sockets, or the SGI SHMEM interface.

2. Message Passing

2.1 MPI Overview

The MPI standard[4] is the amalgamation of what were considered the best aspects of the most popular message-passing systems at the time of its conception. The standard only defines a message passing library and leaves, amongst other things, process initialisation and control to individual developers to define. MPI is available on a wide range of platforms and is fast becoming the *de facto* standard for message passing.

The design goals of the MPI were portability, efficiency and functionality. Commercial and public domain implementations of MPI exist. These run on a range of systems from tightly coupled, massively-parallel machines, through to networks of workstations. MPI has a range of features including: point-to-point, with synchronous and asynchronous communication modes; and collective communication (barrier, broadcast, reduce).

MPICH[5,6] developed by Argonne National Laboratory and Mississippi State University, is probably the most popular of the current, free, implementations of MPI. MPICH is a version of MPI built on top of Chameleon[7]. MPICH and its variants are available for most commonly used distributed and parallel platforms.

2.2 PVM Overview

The Parallel Virtual Machine[8] (PVM) system provides an environment within which parallel programs can be developed and run. PVM is a continuing research and development project between ORNL, Emory University and the University of Tennessee.

PVM transparently handles all message routing, data conversion and task scheduling across a network of heterogeneous computer architectures. PVM is available for most computer architectures, including Linux and NT. The PVM system consists of:

o   A PVM daemon (or NT service) which is installed on each PVM host computer – this daemon is used to initiate and manipulate the PVM environment.

o   A set of libraries to perform parallel communication between PVM tasks, an initiation method for the parallel environment.

o   A console that allows users to manipulate their PVM environment by, for example, adding, deleting hosts as well as starting and monitoring, and stopping PVM programs.

o   A set of functions for debugging both the PVM environment and a PVM program.

3. Gigabit Ethernet

Gigabit Ethernet offers an upgrade path for current Ethernet installations and allows existing installed stations, management tools and training to be reused. It is anticipated that the initial applications for Gigabit Ethernet are for campuses or buildings requiring greater bandwidth between routers, switches, hubs, repeaters and servers[9]. At some time in the near future Gigabit Ethernet will be used by high-end desktop computers requiring a higher bandwidth than Fast Ethernet can offer.

Gigabit Ethernet is an extension of the standard (10 MBps) Ethernet and Fast Ethernet (100 MBps) for network connectivity. The Gigabit Ethernet standard, IEEE 802.3z, was officially approved by the IEEE standards board in June 1998. Gigabit Ethernet employs the same Carrier Sense Multiple Access with Collision Detection (CSMA/CD) protocol, frame format and size as its predecessors.

Much of the IEEE 802.3z standard is devoted to the definition of physical layer of the network architecture. For Gigabit Ethernet communications, several physical layer standards are emerging from the IEEE 802.3z effort – these standards are for different link technologies as well as short and long distant interconnects. The differences between the technologies are shown in Table 1[10].

	Ethernet 10 BaseT	Fast Ethernet 100 BaseT	Gigabit Ethernet 1000 Base X
Data Rate	10 Mbps	100 Mbps	1000 Mbps
Cat 5 UTP	100 m (min)	100 m	100 m
STP/Coax	500 m	100 m	25 m
Multimode Fiber	2 km	412 m (half duplex) 2 km (full duplex)	550 m
Single-mode Fiber	25 km	20 km	5 km

Table 1: Ethernet segment limitations

4. MPI NT Environments

There are now six MPI environments for NT[11]. These range from commercial products, such a MPI/Pro and PaTENT, to the standard release of MPICH with a WinSock devise. The MPI environments used to evaluate Gigabit network performance are described briefly in sections 4.1 – 4.3.

4.1 MPI/PRO for Windows NT

MPI/Pro[12] is a commercial environment released in April 1998 by MPI Software Technology, Inc. The current version of MPI/Pro is based on WinMPIch[13] but has been fairly radically redesigned to remove the bottlenecks and other problems that were present. MPI/Pro supports both Intel and Alpha processors and is released to be used with Microsoft Visual C++ and Digital Visual Fortran. The MPI/Pro developers are currently working on a new source base for MPI that does not include any MPICH code and supports the VI Architecture[14].

4.2 PaTENT WMPI 4.0

PaTENT[15] is the commercial version of WMPI funded by the European project WINPAR[16]. PaTENT differs from WMPI in a number of small ways which includes: sanitized release, easier installation, better documentation and full user support. PaTENT is available for Microsoft Visual C++ and Digital Visual Fortran and consists of libraries, header files, examples and daemons for remote starting. PaTENT includes ROMIO, ANL's implementation of MPI-IO, configured for UFS. PaTENT uses the Installshield software mechanisms for installation and configuration.

4.3 WMPI

WMPI[17] from the Department of Informatics Engineering of the University of Coimbra, Portugal is a full implementation of MPI for Microsoft Win32 platforms. WMPI is based on MPICH and includes a P4[18] device. P4 provides the communication internals and a startup mechanism (that are not specified in the MPI standard). For this reason WMPI also supports the P4 API. The WMPI package is a set of libraries (for Borland C++, Microsoft Visual C++ and Microsoft Visual FORTRAN). The release of WMPI provides libraries, header files, examples and daemons for remote starting.

5. Performance Tests

5.1 Test Equipment

The aim of these tests is restricted to gathering data that helps indicate the expected communications performance (peak bandwidth and message latency) of MPI on NT. The benchmark environment consisted of two dual-processor Pentium's (450 MHz PIII) with 512 MBytes of DRAM running NT 4 (SP5), Windows 2000β3[1] with individual links between each pair of network cards. The technical details of the network cards assessed is given in Table 2.

Card Make	Technical Details	Cost
NetGear[19] FA310TX 100Mbps	IEEE 802.3u 100BASE-TX Fast Ethernet and 802.3i 1	MSRP $24.95 ($17.50 in qty 50)
GigaNet[20] Clan GNN1000	32/64-bit 33MHz, PCI 2.1 compliant, 1.25Gbps full duplex[2].	MSRP $795
Packet Engine[21] GNIC II	32/64-bit 33MHz, PCI 2.1 compliant, 2 Gbps full duplex	$995 No longer available - out of NIC business
SysKonnect[22] SK-9841	32/64-bit 33/66MHz PCI 2.2 complaint, 2 Gbps full duplex	MSRP $729
NetGear GA620	32/64-bit 33/66MHz PCI 2.1 complaint, 2 Gbps full duplex	MSRP $299.99

Table 2: Network Card Specification

5.2 Multi-processor Benchmark - PingPong

In this program, increasing sized messages are sent back and forth between processes. PingPong is an SPMD program written in C using the PVM, MPI and WinSock message passing APIs. These codes have been carefully developed so that all three versions as closely as possible match each others behaviour. PingPong provides information about the latency of send/receive operations and

---

[1] Our references to NT 5 and Windows 2000 are synonymous.

[2] GigaNet uses a proprietary protocol for communications, rather than Ethernet

the uni-directional bandwidth that can be attained on a link. To ensure that anomalies in message timings do not occur the PingPong is repeated for all message lengths.

5.2.1 MPI Version

The MPI version of the code uses the blocking send/receive on both processes.

```
MPI_Send(A,nbyte,MPI_BYTE,0,10,MPI_COMM_WORLD);
MPI_Recv(A,nbyte,MPI_BYTE,0,20,MPI_COMM_WORLD, &status);
```

5.2.2 PVM Version

The PVM version of the code is slightly more complicated as data needs to be packed into buffers before being sent and unpacked at the receiving end.

Master:

```
pvm_initsend(ENCODING);
for (length = 0, length < maximum; increment message length)
{
    pvm_pkbyte(send buffer, length, 1);
    pvm_send(slave ID, 1)
    pvm_recv(-1, -1)
}
```

Slave:

```
pvm_initsend(ENCODING);
while (true) {
    bufid = pvm_recv(-1, -1);
    pvm_bufinfo(bufid, (int*)0, (int*)0, &dtid);
    pvm_send(parent ID, 2);
}}
```

5.3 Differences of the MPI and PVM versions of PingPong

A comparison of the MPI and PVM codes shows that there are some potential differences in how user data is handled and this may cause some performance differences. The one obvious difference is the way user data is handled. In particular the PVM Master leaves the received user data in a temporary buffer space. This and other effects will be investigated and reported upon in the final workshop presentation.

6. Results

6.1 Introduction

In this section we present and discuss the results that were obtained from running the various performance tests under MPI and PVM. It should be noted that not all the PVM results were available at the time of submission of this paper – but will be available for the actual workshop. It should also be noted that due to design restrictions, PaTENT or WMPI are unable to use alternative network interfaces, other than that pointed at by the local host name. This problem was pointed out to both sets of developers (Genias and Coimbra), but unfortunately a "fix" was provided in time to incorporate the results in this paper.

	System	Latency (μs)
1.	MPI/Pro 1.2.3, SMP NT4	106.3
2.	WSOCK 32, SMP NT4	74.0
3.	WMPI 1.2, SMP NT4	44.2
4.	PaTENT 4.014, SMP NT4	32.8
5.	MPI/Pro 1.2.3, SMP NT5	98.2
6.	WSOCK 32, SMP NT5	76.4
7.	PaTENT SMP NT5	35.5
8.	MPI/Pro 1.2.3, TCP 100 Mbps	207.6
9.	WSOCK 32, TCP 100 Mbps	97.5
10.	WMPI 1.2, TCP 100 Mbps	283.4
11.	MPI/Pro 1.2.3, TCP NT5 100 Mbps	244.1
12.	WSOCK 32 TCP NT5 100 Mbps	112.7
13.	MPI/Pro 1.2.3, TCP GigaNet	207.8
14.	WSOCK 32, GigaNet	96.9
15.	MPI/Pro 1.2.3, TCP Packet Engine	335.6
16.	WSOCK 32, TCP Packet Engine	298.4
17.	MPI/Pro 1.2.3, TCP SysKonnect	178.8
18.	WSOCK 32, TCP SysKonnect	90.6
19.	MPI/Pro 1.2.3, TCP NetGear	585.5
20.	WSOCK 32, NetGear	666.2

Table 3: Measured 1 Byte Message Latency

6.2 Latency Results (Table 3)

SM[3] – PaTENT and WMPI clearly have the lowest latencies under NT4 – approximately half the time taken by WinSock and MPI/Pro. Under NT5 WinSock and PaTENT latencies are slightly slower than under NT4 (~8%). However, MPI/Pro under NT5 is slightly faster (~8%) than under NT4.

TCP[4] **(100 Mbps)** – WinSock has more than half the latency of the MPI environments – both under NT4 and NT5. MPI/Pro is about 25% faster than WMPI. Under NT5 all systems exhibit a 10 – 15% increase in latency.

TCP (GigaBit) – The WinSock results for GigaNet (53%), Packet Engine (11%) and SysKonnect (50%) network cards are all faster than the MPI/Pro results. However, for NetGear performance WinSock (14%) is slower than MPI/Pro. This particular result is unexpected as MPI/Pro is built on top of the WinSock API. Overall, the SysKonnect card exhibits the lowest latencies, closely followed by GigaNet and Packet Engine. The latencies for NetGear are more than double of those for the other network cards.

---

[3] SM is where two processes are running on one computer and potentially communicating via Shared-Memory.

[4] TCP is where two processes are running on separate computers and communicating via TCP/IP.

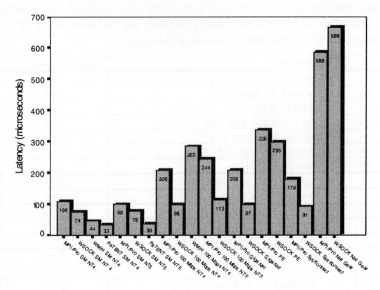

Figure 1 - One Byte Network Latencies

6.3 Network Bandwidths

6.3.1 Shared Memory Results (Figure 2)

PaTENT and WMPI exhibit the best overall performance under NT4 and NT5. Under NT4, PaTENT and WMPI have a peak bandwidth of just over 100 MBytes/s and under NT5 PaTENT peaks at 122 Mbytes/s. MPI/Pro under NT 4 and NT5 has a similar bandwidth to WinSock up until message lengths of 8K. MPI/Pros bandwidth then continues to increase, peaking at 107 Mbytes/s under NT4 and at 122 Mbytes/s under NT5. Winsock peaks at 31 Mbytes/s under NT 4 and 39 Mbytes/s under NT 5 – here it also exhibits a huge performance dip between 16K and 64K message lengths. It should be noted that higher peak bandwidths were achieved under NT5 compared to NT 4.

6.3.2 Distributed Memory

MPI/Pro Results (Figure 3)
The bandwidth results from the 100 Mbps and GigaNet network cards between 1 and 512 Bytes are very similar. Thereafter the GigaNet results continue to increase up to 256K length messages where a peak of 37 Mbytes/s is reached. The 100 Mbps network card outperforms the Packet Engine, SysKonnect and NetGear network cards up until message lengths of about 1K. The 100 Mbps technology peaks at 8.8 Mbytes/s. The bandwidth of NetGear is much poorer than all the other technologies up until 2K message lengths. The peak bandwidths for Packet Engine, SysKonnect and NetGear are 12 Mbytes/s, 17 Mbytes/s and 19 Mbytes/s respectively.

Bandwidth (Log) versus Message Length

(In Shared Memory)

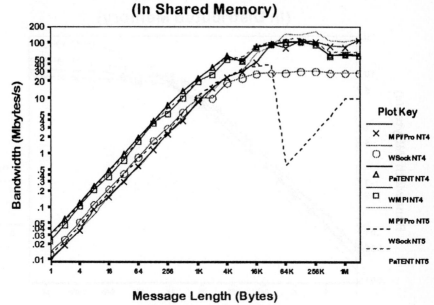

Figure 2 - **PingPong Shared Memory Results**

WinSock Results (Figure 4)

The bandwidth results from the 100 Mbps and GigaNet network cards between 1 and 128 Bytes are very similar. Thereafter the GigaNet results continue to increase up to 8K length messages where a peak of 38 Mbytes/s is reached. The 100 Mbps network card outperforms the SysKonnect network card up until 256 Bytes message length. The 100 Mbps network card outperforms the Packet Engine and NetGear network cards up until 4K message lengths. The 100 Mbps technology peaks at 10 Mbytes/s. The bandwidth of NetGear is much poorer than all the other technologies up until 8K message lengths. The peak bandwidths for Packet Engine, SysKonnect and NetGear are 10.6 Mbytes/s, 17.4 Mbytes/s and 17 Mbytes/s respectively.

7. Summary and Conclusions

7.1 Summary

In this paper we have presented and discussed the results from our simple network performance tests on NT using the MPI, PVM and WinSock message passing APIs on six different network interface technologies. At the date of submission, we have been unable to complete the PVM tests, so our discussion on the performance differences is limited at this moment.

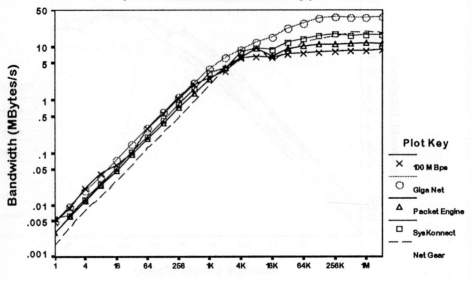

Figure 3 - MPI/Pro Bandwidth Results

Our experiences with the performance of MPI under NT 4 and Windows 2000 are inconclusive. Currently, it appears that in shared-memory mode that the latencies under Windows 2000 may be marginally lower than NT 4. The measured peak bandwidths of Windows 2000 were greater than NT4. In distributed-memory mode the measured latencies under Windows 2000 were approximately 20% higher than the equivalent under NT 4. The measured bandwidths for Windows 2000 and NT 4 were very similar however.

It is interesting to note that the measured network latencies for 100 Mbps Ethernet cards and Giga Net under WinSock and MPI/Pro are almost equivalent. The performance of the Packet Engine Gigbit card is between 7% and 13% faster respectively. However, the performance of the SysKonnect and Net Gear cards are significantly slower that standard 100 Mbps Ethernet.

7.2 Price/Performance Considerations

Table 4 shows the price/performance ratios calculated using the network card costs in September 1999 versus the peak measured bandwidth and minimum latency. It should be noted that the calculated ratios shown are only an approximate indicator as the price of the network cards varies significantly based on the quantity bought and the discounts given. The smaller the price/performance ratio the better value for money that can be expected from a network card. The choice of what is the most appropriate card is often not based solely on the price/performance, but also other factor such as desired performance, compatibility or availability.

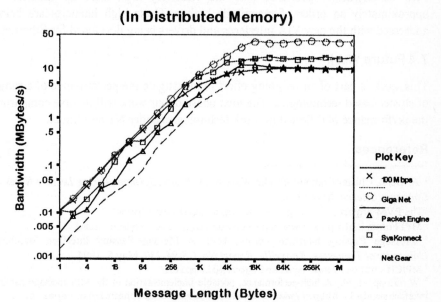

Bandwidth (Log) versus Message Length
(In Distributed Memory)

Figure 4 - WinSock Bandwidth Results

The ratios shown in Table 4 indicate that the 100 Mbps Fast Ethernet cards provide significantly better price/performance than the other network cards. However, the ratios for the NetGear Gigabit card are significantly better than the other price/performance ratios available.

Card Make and speed	Price/Performance ($/Mbytes/s)	Price/Performance ($/μs)
NetGear FA310TX 100Mbps	$24.95/8.8 = 2.835	$24.95/208 = 0.12
GigaNet - Clan GNN1000	$795/37 = 2149	$$795/208 = 3.82
Packet Engine – GNIC II	$995/12 = 82.92	$995/336 = 2.96
SysKonnect – SK-9841	$729/17 = 42.88	$729/179 = 4.07
NetGear - GA620	$299.99/19 = 15.79	$299.99/585 = 0.51

Table 4: Network Card Cost versus Performance (MPI/Pro)

7.3 Summary of Conclusions

Our work has shown that release 1.2.3 of MPI/Pro imposes an approximate additional 1 Byte latency of 25% and 50% over WinSock under shared and distributed-memory modes respectively. We have shown that the Giga Net Gigabit Ethernet provides the highest bandwidth of those tested. We suspect, as currently we do not have a concrete price for this card, that the price/performance of this card will be poorer that that of Net Gear but better than Packet Engine and NetGear. Our price/performance figures do, however, strongly suggest that the current performance and costs of the Gigabits cards makes standard 100 Mbps a much sounder technology investment at the moment. Obviously, other factors, like required peak bandwidth, may make the decision of what technology to choose not one purely based on price/performance. Another factor that puts the Gigabit Ethernet at a disadvantage compared to other network technologies, such as Myrinet[23] and SCI[24], is the relatively high start up latencies – approximately an order of magnitude higher. These high latencies are being addressed with the new VIA interfaces and drivers being developed for Ethernet.

7.4 Future Work

This work is part of an on going effort to investigate the performance of a range of cluster-based technologies. The next phase of our work will involve comparing the performance of different network technologies under NT and Linux.

References

[1] A. Geist, Cluster Computing: The Wave of the future, *Springer Verlag Lecture Notes in Computer Science*, May 1994.

[2] The PVM project – http://www.epm.ornl.gov/pvm/

[3] MPI Forum - http://www.mpi-forum.org/docs/docs.html

[4] Message Passing Interface Forum, MPI: A Message-Passing Interface Standard, *University of Tennessee, Knoxville, Report No. CS-94-230,* May 5, 1994

[5] MPICH - http://www.mcs.anl.gov/mpi/mpich/

[6] W. Gropp, et. al., A high-performance, portable implementation of the MPI message passing interface standard - http://www-c.mcs.anl.gov/mpi/mpicharticle/paper.html

[7] W. Gropp and B. Smith, Chameleon parallel programming tools users manual. *Technical Report ANL-93/23*, Argonne National Laboratory, March 1993.

[8] PVM: A Users' Guide and Tutorial For Networked Parallel Computing – http://www.netlib.org/pvm3/book/pvm-book.html

[9] Gigabit Ethernet Alliance - Gigabit Ethernet: Accelerating the standard for speed, http://www.gigabit-ethernet.org/technology/whitepapers, September 1999.

[10] Ethernet Segment Limits. - http://www.gigabit-ethernet.org/technology/

[11] TOPIC – http://www.dcs.port.ac.uk/~mab/TOPIC/

[12] MPI Software Technology, Inc. – http://www.mpi-softtech.com/

[13] WinMPICh - http://www.erc.msstate.edu/mpi/mpiNT.html

[14] VIA – http://www.viaarch.com

[15] PaTENT - http://www.genias.de/products/patent/

[16] WINdows based PARallel computing - http://www.genias.de/

[17] WMPI - http://dsg.dei.uc.pt/w32mpi/

[18] R. Buttler and E. Lusk, User's Guide to the p4 Parallel Programming System, *ANL-92/17*, Argonne National Laboratory, October 1992.

[19] NetGear - http://netgear.baynetworks.com/

[20] GigaNet - http://www.giga-net.com/

[21] Packet Engine - http://www.packetengines.com/index4.html

[22] SysKonnect - http://www.syskonnect.de/

[23] N. Boden, et. al. Myrinet - A Gbps LAN. *IEEE Micro*, Vol. 15, No.1, February 1995. http://www.myri.com/

[24] Dolphin Interconnect Solutions - http://www.dolphinics.no/

MPI Collective Operations over IP Multicast *

Hsiang Ann Chen, Yvette O. Carrasco, and Amy W. Apon

Computer Science and Computer Engineering
University of Arkansas
Fayetteville, Arkansas, U.S.A
{hachen,yochoa,aapon}@comp.uark.edu

Abstract. Many common implementations of Message Passing Interface (MPI) implement collective operations over point-to-point operations. This work examines IP multicast as a framework for collective operations. IP multicast is not reliable. If a receiver is not ready when a message is sent via IP multicast, the message is lost. Two techniques for ensuring that a message is not lost due to a slow receiving process are examined. The techniques are implemented and compared experimentally over both a shared and a switched Fast Ethernet. The average performance of collective operations is improved as a function of the number of participating processes and message size for both networks.

1 Introduction

Message passing in a cluster of computers has become one of the most popular paradigms for parallel computing. Message Passing Interface (MPI) has emerged to be the *de facto* standard for message passing. In many common implementations of MPI for clusters, MPI collective operations are implemented over MPI point-to-point operations. Opportunities for optimization remain.

Multicast is a mode of communication where one sender can send to multiple receivers by sending only one copy of the message. With multicast, the message is not duplicated unless it has to travel to different parts of the network through switches. Many networks support broadcast or multicast. For example, shared Ethernet, token bus, token ring, FDDI, and reflective memory all support broadcast at the data link layer.

The Internet Protocol (IP) supports multicast over networks that have IP multicast routing capability at the network layer. The goal of this paper is to investigate the design issues and performance of implementing MPI collective operations using multicast. IP multicast is used to optimize the performance of MPI collective operations, namely the MPI broadcast and MPI barrier synchronization, for this preliminary work. The results are promising and give insight to work that is planned on a low-latency network. The remainder of this paper describes IP multicast, design issues in the implementations, experimental results, conclusions, and future planned work.

---

* This work was supported by Grant #ESS-9996143 from the National Science Foundation

2   IP Multicast

Multicast in IP is a receiver-directed mode of communication. In IP multicast, all the receivers form a group, called an IP multicast group. In order to receive a message a receiving node must explicitly join the group. Radio transmission is an analogy to this receiver-directed mode of communication. A radio station broadcasts the message to one frequency channel. Listeners tune to the specific channel to hear that specific radio station. In contrast, a sender-directed mode of communication is like newspaper delivery. Multiple copies of the paper are delivered door-to-door and the newspaper company must know every individual address of its subscriber. IP multicast works like radio. The sender only needs to send one copy of the message to the multicast group, and it is the receiver who must be aware of its membership in the group.

Membership in an IP multicast group is dynamic. A node can join and leave an IP multicast group freely. A node can send to a multicast group without having to join the multicast group. There is a multicast address associated with each multicast group. IP address ranges from 224.0.0.0 through 239.255.255.255 (class D addresses) are IP multicast addresses. Multicast messages to an IP multicast group will be forwarded by multicast-aware routers or switches to branches with nodes that belong to the IP multicast group. IP multicast saves network bandwidth because it reduces the need for the sender to send extra copies of its message and therefore lowers the latency of the network.

In theory, IP multicast should be widely applicable to reduce latency. However, one drawback of IP multicast is that it is unreliable. The reliable Transmission Control Protocol(TCP) does not provide multicast communication services. The User Datagram Protocol (UDP) is used instead to implement IP multicast applications. UDP is a "best effort" protocol that does not guarantee datagram delivery. This unreliability limits the application of IP multicast as a protocol for parallel computing.

There are three kinds of unreliability problems with implementing parallel collective operations over IP multicast. One comes with unreliability at the hardware or data link layer. An unreliable network may drop packets, or deliver corrupted data. In this work, we assume that the hardware is reliable and that packets are delivered reliably at the data link layer. It is also possible that a set of fast senders may overrun a single receiver. In our experimental environment we have not observed these kind of errors. However, a third problem is related to the software design mismatch between IP multicast and parallel computing libraries such as MPI. In WAN's, where IP multicast is generally applied, receivers of a multicast group come and go dynamically, so there is no guarantee of delivery to all receivers. The sender simply does not know who the receivers are. However, in parallel computing all receivers must receive.

With IP multicast, only receivers that are ready at the time the message arrives will receive it. However, the asynchronous nature of cluster computing makes it impossible for the sender know the receive status of the receiver without some synchronizing mechanism, regardless of how reliable the underlying hardware is. This is a paradigm mismatch between IP multicast and MPI. This paper explores two synchronizing techniques to ensure that messages are not lost because a receiving process is slower than the sender.

This work is related to other efforts to combine parallel programming and broadcast or multicast messaging. In work done on the Orca project [8], a technique was developed for ensuring the reliability of a broadcast message that uses a special sequencer node. In research done at Oak Ridge National Laboratory, parallel collective operations in Parallel Virtual Machine (PVM) were implemented over IP multicast[2]. In that work, reliability was ensured by the sender repeatedly sending the same message until ack's were received from all receivers. This approach did not produce improvement in performance. One reason for the lack of performance gain is that the multiple sends of the data cause extra delay.

The goal of this work is to improve the performance of MPI collective calls. This work focuses on the use of IP multicast in a cluster environment. We evaluate the effectiveness of constructing MPI collective operations, specifically broadcast and barrier, over IP multicast in a commodity-off-the-shelf cluster.

3   MPI Collective Operations

The Message Passing Interface (MPI) standard specifies a set of collective operations that allows one-to-many, many-to-one, or many-to-many communication modes. MPI implementations, including LAM[6] and MPICH[7], generally implement MPI collective operations on top of MPI point-to-point operations. We use MPICH as our reference MPI implementation.

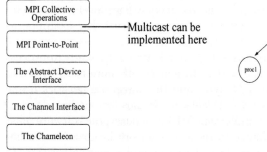

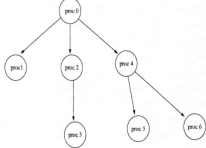

Fig. 1. MPICH Layers

Fig. 2. MPICH Broadcast mechanism with 4 nodes

MPICH[3] uses a layered approach to implement MPI. The MPICH layers include the Abstract Device Interface (ADI) layer, the Channel Interface Layer, and the Chameleon layer. Portability is achieved from the design of the ADI layer, which is hardware dependent. The ADI provides an interface to higher layers that are hardware independent. The MPICH point-to-point operations are built on top of the ADI layer. To avoid implementing collective operations over MPICH point-to-point functions, the new implementation has to bypass all the MPICH layers, as shown in Fig. 1.

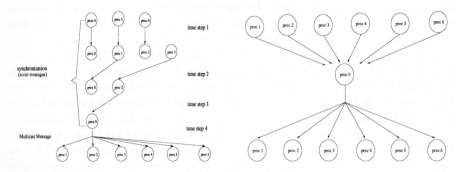

Fig. 3. MPI broadcast using IP multi-cast (Binary Algorithm)

Fig. 4. MPI broadcast using IP multi-cast (Linear Algorithm)

3.1   MPI Broadcast

Since the new layer for MPI collective operations using multicast is compared experimentally with the original MPICH implementation, it is helpful to under-stand how these functions are implemented in MPICH. MPICH uses a tree struc-tured algorithm in its implementation of MPI broadcast operation (MPI_Bcast). In the broadcast algorithm, the sender sends separate copies of the message to some of the receivers. After they receive, the receivers at this level in turn send separate copies of the message to receivers at the next level. For example, as illustrated in Fig. 2, in an environment with 7 participating processes, process 0 (the root) sends the message to processes 4, 2, and 1. Process 2 sends to process 3 and process 4 sends to processes 5 and 6. In general, if there are N partici-pating processes, the message size is M bytes and the maximum network frame size is T bytes, it takes $(\frac{M}{T} + 1) \times (N - 1)$ network frames for one broadcast.

When IP multicast is used to re-implement MPI broadcast, the software must ensure that all receivers have a chance to receive. Two synchronization mecha-nisms have been implemented, a binary tree algorithm and a linear algorithm. In the binary tree algorithm, the sender gathers small scout messages with no data from all receivers in a binary tree fashion before it sends. With K processes each executing on a separate computer, the height of the binary tree is $log_2 K + 1$. In the synchronization stage at time step 1, all processes at the leaves of binary tree send. Scout messages propagate up the binary tree until all the messages are finally received at the root of the broadcast. After that, the root broadcasts the data to all processes via a single send using IP multicast. For example, as illustrated in Fig. 3 in an environment with 7 participating processes, processes 4, 5, and 6 send to processes 0, 1, and 2, respectively. Next, process 1 and pro-cess 3 send to processes 0 and 2, respectively. Then process 2 sends to process

0. Finally, process 0 sends the message to all processes using IP multicast. In general, with N processes, a total of $N-1$ scout messages are sent. With a message size of M, and a maximum network frame size of T, $\frac{M}{T}+1$ network frames need to be sent to complete one message transmission. Adding the $N-1$ scout messages, it takes a total of $(N-1)+\frac{M}{T}+1$ frames to send one broadcast message.

The linear algorithm makes the sender wait for scout messages from all receivers, as illustrated in Fig. 4. Then the message with data is sent via multicast. With K processes in the environment, it takes $K-1$ steps for the root to receive all the scout messages since the root can only receive one message at a time. As illustrated in Fig. 4 with N processes, the root receives $N-1$ point-to-point scout messages before it sends the data. With 7 nodes, the multicast implementation only requires one-third of actual data frames compared to current MPICH implementation. Since the binary tree algorithm takes less time steps to complete, we anticipate it to perform better than the linear algorithm.

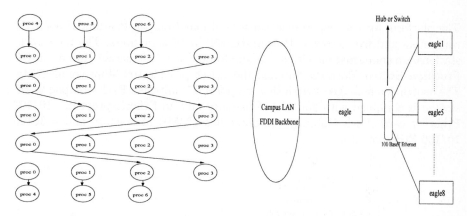

Fig. 5. MPICH barrier synchronization with 7 processes

Fig. 6. The Eagle Cluster

3.2   MPI Barrier Synchronization

Another MPI collective operation re-implemented was MPI_Barrier. MPI_Barrier is an operation that synchronizes processes. All processes come to a common stopping point before proceeding. The MPICH algorithm for barrier synchronization can be divided into three phases. In the first phase, processes that cannot be included in sending pair-wise point-to-point operations send messages to processes who can. In the second phase, point-to-point sends and receives are performed in pairs. In the third phase, messages are sent from the processes in the second phase to processes from the third phase to release them. Figure 5 illustrates MPICH send and receive messages for synchronization between 7 processes. In this example, processes 4, 5, and 6 send messages to processes 0, 1 and 2. In the second phase, point-to-point message are sent between processes 0, 1, 2, and 3. In the third phase, process 0, 1, and 2, send messages to 4, 5, and 6 to release them. If there are N participating processes, and K is the biggest power of 2 less than N, a total of $2 \times (N - K) + log_2 K \times K$ messages need to be sent.

By incorporating IP multicast into the barrier algorithm, we were able to reduce the number of phases by two. The binary algorithm described above is used to implement MPI_Barrier. First, point-to-point messages are reduced to process 0 in a binary tree fashion. After that, a message with no data is sent using multicast to release all processes from the barrier. In general, with N processes in the system, a total of $N - 1$ point-to-point messages are sent. One multicast message with no data is sent.

4   Experimental Results

The platform for this experiment consists of four Compaq PentiumIII 500MHZ computers and five Gateway PentiumIII 450 MHZ computers. The nine workstations are connected via either a 3Com SuperStack II Ethernet Hub or an HP ProCurve Switch. Both the hub and the switch provide 100 Mbps connectivity. The switch is a managed switch that supports IP multicast. Each Compaq workstation is equipped with 256 MB of memory and an EtherExpress Pro 10/100 Ethernet card. Each Gateway computer has 128MB of memory and a 3Com 10/100 Ethernet card.

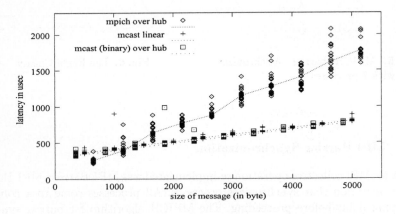

Fig. 7. MPI_Bcast with 4 processes over Fast Ethernet Hub

The performance of the MPI collective operations is measured as the longest completion time of the collective operation. among all processes. For each message size, 20 to 30 different experiments were run. The graphs show the measured

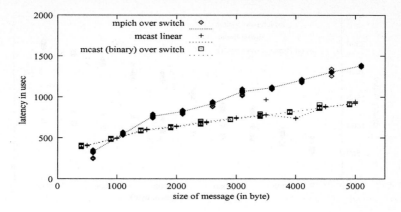

Fig. 8. MPI_Bcast with 4 processes over Fast Ethernet Switch

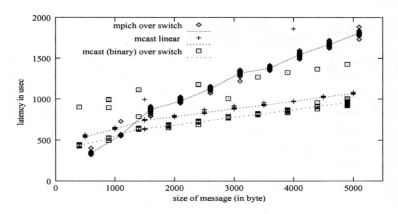

Fig. 9. MPI_Bcast with 6 processes over Fast Ethernet Switch

time for all experiments with a line through the median of the times. The graphs illustrate the sample distribution of measured times.

Figure 7 shows the performance of MPI_Bcast of both implementations over the hub with 4 processes. The figure shows that the average performance for both the linear and the binary multicast implementation is better for message sizes greater than 1000 bytes. With small messages, the cost of the scout messages causes the multicast performance to be worse than MPICH performance. The figure also shows variations in performance for all implementations due to collisions on the Fast Ethernet network. The variation in performance for MPICH is generally higher than the variation in performance for either multicast implementation.

Figures 8, 9, and 10 describe the performance with the switch for 4, 6, and 9 processes respectively. Both the linear and the binary algorithm using multicast show better average performance for a large enough message size. The crossover point of average MPICH performance and the average performance of using

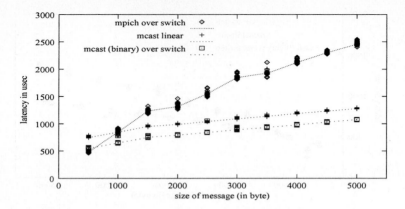

Fig. 10. MPI_Bcast with 9 processes over Fast Ethernet Switch

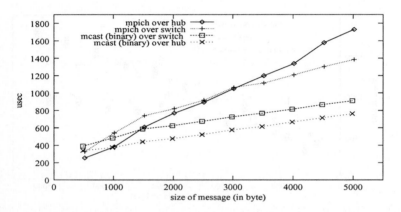

Fig. 11. Performance Comparison with MPI_Bcast over hub and switch for 4 processes

multicast is where the extra latency of sending scout messages becomes less than the latency from sending extra packets of data when the data is large. For some numbers of nodes, collisions also caused larger variance in performance with the multicast implementations. For example, this is observed for 6 nodes as shown in Fig. 9. With 6 nodes using the binary algorithm, both node 2 and node 1 attempt to send to node 0 at the same time, which causes extra delay.

Figure 11 compares the average performance of the switch and the hub for 4 processes. When using IP multicast, the average performance of the hub is better than the switch for all measured message sizes. As for the original MPICH implementation, the average performance of hub becomes worse than the switch when the size of the message is bigger than 3000. The MPICH implementation puts more messages into the network. As the load of the network gets larger, the extra latency of the switch become less significant than the improvement gained with more bandwidth. The multicast implementation is better than MPICH for message sizes greater than one Ethernet frame.

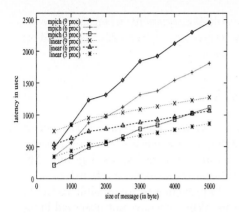

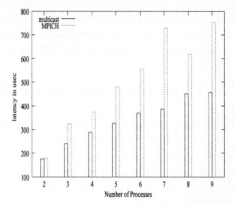

Fig. 12. Performance Comparison with MPI_Bcast over 3, 6, and 9 processes over Fast Ethernet switch

Fig. 13. Comparison of MPI_Barrier over Fast Ethernet hub

Figure 12 compares MPICH and the linear multicast implementation for 3, 6, and 9 processes over the switch. The results show that the linear multicast algorithm scales well up to 9 processes and better than MPICH. With the linear implementation, the extra cost for additional processes is nearly constant with respect to message size. This is not true for MPICH.

Figure 13 describes the results of MPI_Barrier operation over the hub. The results for MPI_Barrier show that IP multicast performs better on the average than the original MPICH implementation. The performance improvement increases as the size of the message gets bigger.

In a Single Program Multiple Data (SPMD) environment, message passing using either the linear algorithm or the binary algorithm is correct even when there are multiple multicast groups. However, since the IP multicast implementation requires the receive call to be posted before the message is sent, it is required that each process execute the multicast calls in the same order. This restriction is equivalent to requiring that the MPI code be safe[5]. If several processes broadcast to the same multicast group (in MPI terms, this is the same process group of same context), the order of broadcast will be correctly preserved. For example, suppose in an environment including the 4 processes with ids 4, 6, 7 and 8, processes 6, 7, and 8 all belong to the same multicast group and the broadcast is called in the following order.

MPI_Bcast(&buffer, count, MPI_INT, 6, MPI_COMM_WORLD);
MPI_Bcast(&buffer, count, MPI_INT, 7, MPI_COMM_WORLD);
MPI_Bcast(&buffer, count, MPI_INT, 8, MPI_COMM_WORLD);

Using either the binary algorithm or the linear algorithm, process 7 cannot proceed to send the the second broadcast until it has received the broadcast message from process 6, and process 8 cannot send in the third broadcast until it has received the broadcast message from process 7. The order of the three broadcasts is carried out correctly. Using a similar argument, when there are two or more multicast groups that a process receives from, the order of broadcast will be correct as long as the MPI code is safe.

5   Conclusions and Future Work

Multicast reduces the number of messages required and improves the performance of MPI collective operations by doing so. Its receiver-directed message passing mode allows the sender to address all the receivers as a group. This experiment focused on a particular implementation using IP multicast.

Future work is planned in several areas. Improvements are possible to the binary tree and linear communication patterns. While we have not observed buffer overflow due to a set of fast senders overrunning a single receiver, it is possible this may occur in many-to-many communications and needs to be examined further. Additional experimentation using parallel applications is planned. Also, low latency protocols such as the Virtual Interface Architecture[9] standard typically require a receive descriptor to be posted before a mesage arrives. This is similar to the requirement in IP multicast that the receiver be ready. Future work is planned to examine how multicast may be applied to MPI collective operations in combination with low latency protocols.

References

[1] D. E. Comer. *Internetworking with TCP/IP Vol. I: Principles, Protocols, and Architecture* . Prentice Hall, 1995.

[2] T. H. Dunigan and K. A. Hall. PVM and IP Multicast. Technical Report ORNL/TM-13030, Oak Ridge National Laboratory, 1996.

[3] W. Gropp, E. Lusk, N. Doss, and A. Skjellum. A High-Performance, Portable Implementation of the MPI Message Passing Interface Standard. Technical Report Preprint MCS-P567-0296, Argonne National Laboratory, March 1996.

[4] N. Nupairoj and L. M. Ni. Performance Evaluation of Some MPI Implementations on Workstation Clusters. In *Proceedings of the 1994 Scalable Parallel Libraties Conference*, pages 98–105. IEEE Computer Society Press, October 1994.

[5] P. Pacheo. *Parallel Programming with MPI* . Morgan Kaufmann, 1997.

[6] The LAM source code. http://www.mpi.nd.edu/lam.

[7] The MPICH source code. www-unix.mcs.anl.gov/mpi/index.html.

[8] A. S. Tannenbaum, M. F. Kaashoek, and H. E. Bal. Parallel Programming Using Shared Objects and Broadcasting. *Computer*, 25(8), 1992.

[9] The Virtual Interface Architecture Standard. http://www.viarch.org.

[10] D. Towsley, J. Kurose, and S. Pingali. A Comparison of Sender-Initiated and Receiver-Initiated Reliable Multicast Protocols. *IEEE JSAC*, 15(3), April 1997.

An Open Market-Based Architecture for Distributed Computing

Spyros Lalis and Alexandros Karipidis

Computer Science Dept.,
University of Crete, Hellas
{lalis,karipid}@csd.uoc.gr
Institute of Computer Science,
Foundation for Research and Technology, Hellas
{lalis,karipid}@ics.forth.gr

Abstract. One of the challenges in large scale distributed computing is to utilize the thousands of idle personal computers. In this paper, we present a system that enables users to effortlessly and safely export their machines in a global market of processing capacity. Efficient resource allocation is performed based on statistical machine profiles and leases are used to promote dynamic task placement. The basic programming primitives of the system can be extended to develop class hierarchies which support different distributed computing paradigms. Due to the object-oriented structuring of code, developing a distributed computation can be as simple as implementing a few methods.

1   Introduction

The growth of the Internet has provided us with the largest network of interconnected computers in history. As off-the-shelf hardware becomes faster and gains Internet access, the network's processing capacity will continue increasing. Many of these systems are often under-utilized, a fact accentuated by the globe's geography since "busy" hours in one time-zone tend to be "idle" hours in another. Distributing computations over the Internet is thus very appealing.

However, several issues must be resolved for this to be feasible. The obstacle of platform heterogeneity must be overcome and security problems arising from the execution of code from untrusted parties must be confronted. Further inconveniences arise when installing and maintaining the corresponding programming environments. And then, distributed computations must be designed and implemented on top of them, a challenging task even for experienced programmers.

In this paper we present a system that addresses these problems, simplifying distributed computing over the Internet considerably. Through a maintenance-free, web-based user interface any machine can be safely connected to the system to act as a host for remote computations. A framework that promotes code reuse and incremental development through object-oriented extensions is offered to the application programmer. Writing computations for the system can be as trivial as implementing a few routines. We feel that the ease of deploying the system

J. Rolim et al. (Eds.): IPDPS 2000 Workshops, LNCS 1800, pp. 61-70, 2000.

and developing applications for it is of importance to the scientific community since most of the programming is done by scientists themselves with little or no support from computer experts.

The rest of the paper is organized as follows. Section 2 summarizes the general properties of the system. Details about the resource allocation mechanism are given in Sect. 3. In Sect. 4 we look into the system architecture, giving a description of components and communication mechanisms. In Sect. 5 we show how our system can be used to develop distributed computations in a straightforward way. A comparison with related work is given in Sect. 6. Section 7 discusses the advantages of our approach. Finally, future directions of this work are mentioned in the last section.

2   System Properties

When designing the system, the most important goal was to achieve a level of simplicity that would make it popular both to programmers and owners of lightweight host machines, most notably PCs. Ease of host registration was thus considered a key issue. Safety barriers to shield hosts from malicious behavior of foreign code were also required. Portability and inter-operability was needed to maximize the number of host platforms that can be utilized. A simple yet powerful programming environment was called for to facilitate the distribution of computations over the Internet. All these features had to be accompanied by a dynamic and efficient mechanism for allocating resources to applications without requiring significant effort from the programmer.

In order to guarantee maximal cross-platform operability the system was implemented in Java. Due to Java's large scale deployment, the system can span across many architectures and operating systems. Host participation is encouraged via a web based interface, which installs a Java applet on the host machine. This accommodates the need for a user friendly interface, as users are accustomed to using web browsers. Furthermore, the security manager installed in Java enabled browsers is a widely trusted firewall, protecting hosts from downloaded programs. Finally, due to the applet mechanism, no administration nor maintenance is required at the host – the majority of users already has a recent version of a web browser installed on their machines.

On the client side we provide an open, extensible architecture for developing distributed applications. Basic primitives are provided which can in turn be used to implement diverse, specialized processing models. Through such models it is possible to hide the internals of the system and/or provide advanced programming support in order to simplify application development.

3   Resource Allocation

Host allocation is based on profiles, which are created by periodically benchmarking each host. A credit based [1] mechanism is used for charging. Credit can be translated into anything that makes sense in the context where the system is deployed. Within a non-profit institution, it may represent time units to facilitate quotas. Service-oriented organizations could charge clients for using hosts by converting credit to actual currency.

Both hosts (sellers) and clients (buyers) submit orders to a market, specifying their actual and desired machine profile respectively. The parameters of an order are listed in table 1. The *performance vectors* include the host's mean score and variance for a set of benchmarks over key performance characteristics such as integer and floating point arithmetic, network connection speed to the market server etc. The host *abort ratio* is the ratio of computations killed by the host versus computations initiated on that host (a "kill" happens when a host abruptly leaves the market). The host performance vectors and abort ratio are automatically produced by the system. Host profiles can easily be extended to include additional information that could be of importance for host selection.

Table 1. Parameters specified in orders

Parameter	Description	
	Sell Orders	**Buy Orders**
price/sec	The minimum amount of credit required per second of use of the host.	The maximum amount of credit offered per second of use of the host.
lease duration	The maximum amount of usage time without renegotiation.	The minimum amount of usage time without renegotiation.
granted/demanded compensation	Credit granted/demanded for not honoring the lease duration.	
performance statistics vectors	The host's average score and variance for each of the benchmarks (measured).	The average performance score and variance a buyer is willing to accept.
abort ratio	The host's measured abort ratio.	The abort ratio a buyer is willing to accept.

An economy-based mechanism is employed to match the orders that are put in the market. For each match, the market produces a lease, which is a contract between a host and a client containing their respective orders and the price of use agreed upon. Leases are produced periodically using continuous double auction [8]. A lease entitles the client to utilize the host for a specific amount of time. If the client's task completes within the lease duration, then the buyer transfers an amount of credit to the seller as a reward, calculated by multiplying actual duration with the lease's price per second. If the lease duration is not honored, an amount of credit is transfered from the dishonoring party to the other.

4   System Architecture

4.1   Overview of System Components

An overview of the system's architecture is depicted in Fig. 1. The basic components of our system are the market server, hosts, the host agent, schedulers, tasks and client applications.

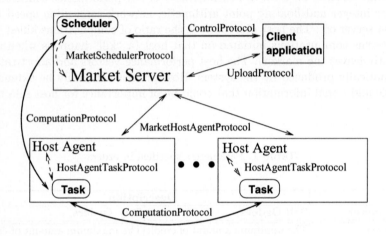

Fig. 1. Overview of architecture

The *Client Application* is a program which needs to perform computations that require considerable processing power. Through the system, it may either distribute a computation across a number of machines or just delegate the execution of an entire computation to a faster machine to speed up execution.

The *Market Server* is the meeting place for buyers and sellers of processing power. It collects orders from clients and hosts. Using the host profiles, it then matches buy with sell orders and thus allocates resources.

A *Host* is a machine made available to be used by clients. A host participates in the market through the *Host Agent*, a Java applet. The user visits a URL with a Java enabled web browser and the agent is downloaded to his system. The agent communicates with the market server, takes care of placing orders on behalf of the user and executes tasks assigned to the host. It also provides the market server with the benchmark scores needed for the host's profile.

A computation in our system consists of a *Scheduler* and one or more *Tasks*. The application installs the scheduler on the market server. The scheduler then places orders in the market for acquiring machines to complete the computation. New orders can be issued at any time in order to adapt to fluid market conditions. When a lease is accepted by the scheduler, a task is launched in the host machine to assist in completing the computation.

4.2   Basic System Services and Communication

There are six protocols used for communication by the system.

The *UploadProtocol* is a fixed, published Remote Method Invocation (RMI) interface used by the client application to upload a computation to the market server and to instantiate it's scheduler. A client application may instantiate multiple schedulers to simultaneously launch the same code with multiple data.

The *ControlProtocol* is a published RMI interface for the client application to control a scheduler. Through this interface the application performs tasks such as starting a computation with new parameters, altering the computation's budget for acquiring hosts, instructing the scheduler to kill all tasks and exit, etc. The basic functions are implemented in the system classes. The programmer can introduce computation specific control functions by extending this interface.

The *ComputationProtocol* is used within the bounds of a single computation for communication among tasks and their scheduler. It is application dependent and thus unknown to the system. We do, however, provide message passing support (not further discussed in this paper) that can be used by application developers to implement flexible, safe and efficient data exchange.

The *MarketSchedulerProtocol* is used for local communication between the market server and schedulers. The market server implements a standard published interface for servicing requests from schedulers such as placing orders and retrieving host and market status information. Respectively, schedulers provide methods for being notified by the market of events such as the opportunity to acquire a new lease, a change in the client's account balance, the completion of a task's work and the failure of a host that was leased to them. Similarly, the *HostAgentTaskProtocol* provides local communication among a host agent and the task it is hosting. The agent implements a published interface for servicing requests from tasks, such as retrieving information about a host's performance.

The *MarketHostAgentProtocol* is a proprietary protocol used by the market server and the host agent. It allows orders to be placed in the market by the host. It is also used to retrieve tasks from the market, ask for "payment" when tasks complete and to post benchmarking data to the market server.

5   Supporting Distributed Computing Paradigms

Through the set of primitives offered by the system, it is possible to develop a wide range of applications. More importantly generic support can be provided for entire classes of distributed computations. Applications can then be developed by extending these classes to introduce specific functionality. This incremental development can greatly simplify programming. As an example, in the following we describe this process for embarrassingly parallel computations requiring no communication between tasks. Other distributed computation paradigms can be supported in similar fashion.

5.1   The Generic Master – Slave Model

In this model work is distributed among many processors by a distinguished processor referred to as the "master". The other processors, referred to as "slaves", complete the work assigned to them and return the results to the master. In order to process its workload a slave does not need to communicate with any other slave. This model is used in image processing, genetics algorithms, brute force search and game tree evaluation. One possible implementation of this model is sketched below. For brevity, only the methods a programmer has to be aware of are shown.

```
public interface MS_Control extends Control {
  void start(Object pars); // inherited by superclass
  void stop();             // inherited by superclass
  Object[] getResults(boolean all, boolean keep);
}
public abstract class MS_Scheduler
extends Scheduler implements MS_Control {
  public abstract Object[] doPartitions(Object pars);
  public void receiveResult(Object result);
}
public abstract class MS_Task extends Task {
  public abstract Object processPartition(Object partition);
}
```

The *MS_Control.start* method starts a new computation. *MS_Control.start* triggers *MS_Scheduler.doPartitions* to produce the various partitions of the computation. These are forwarded to instances of *MS_Task* residing on hosts allocated to the computation and *MS_Task.processPartition* is invoked to process them. The results are returned to the scheduler where post-processing is performed via calls to the *MS_Scheduler.receiveResult* method.

It is important to notice that programmers need to implement just three methods in order to complete a computation following this model. All other implementation issues, including the resource allocation strategy of the scheduler, remain hidden. The *MS_Control* interface, which defines the primitives for controlling and retrieving the results of the computation, is implemented by the base *MS_Scheduler* class and thus does not concern the programmer. This master/slave model could be further extended to introduce additional functionality such as check-pointing and restarting of tasks for fault tolerance. Programmers would exploit this functionality without effort.

5.2   A Sample Client Application

Based on this model, we show how a specific application, e.g. for computing the Mandelbrot set, can be implemented. We assume that the area to be calculated is partitioned in bands, processed in parallel to speed up execution. The user selects an area and the computation is started to zoom into the selected area.

The parameters, partitions and results of the fractal application must be extensions of the *Object* class. The classes must implement the *Serializable* interface in order to be successfully transported across machine boundaries.

```
class FractalParameters extends Object implements Serializable {
  // ... fractal computation parameters
}
class FractalPartition extends Object implements Serializable {
  // ... parameters for calculating a slice
}
class FractalResult extends Object implements Serializable {
  // ... results of a slice calculation
}
```

Assuming the parameter and result objects have been appropriately defined, a *FractalScheduler* class must be programmed as a subclass of *MS_Scheduler* to produce partitions via the *doPartitions* method. The *MS_Scheduler.receiveResult* method is not overridden because individual results are not merged by the scheduler. Also, the basic *MS_Control* interface needs no extension since it already offers the necessary routines for controlling and monitoring the computation. Analogously, a *FractalTask* class must be provided that implements the *MS_Task.processPartition* method to perform the calculation of slices.

```
class FractalScheduler extends MS_Scheduler {
  Object[] doPartitions(Object comp_pars) {
    FractalPartition partitions[];
    FractalParameters pars=(FractalParameters)comp_pars;
    // ... split calculation and produce partitions
    return (partitions);
  }
}
class FractalTask extends MS_Task {
  Object processPartition(Object partition) {
    FractalResult result;
    FractalPartition pars=(FractalPartition)partition;
    // ... perform the computation
    return(result);
  }
}
```

Finally, to run the application, the computation's classes must be uploaded to the market server using the *UploadProtocol* and a scheduler instance must be created. The *MS_Control* interface is used to control the scheduler and periodically retrieve the computation's results.

6   Related Work

Popular distributed programming environments such as PVM [9] and MPI [9] lack advanced resource allocation support. PVM allows applications to be notified when machines join/leave the system, but the programmer must provide code that investigates hosts' properties and decides on proper allocation. MPI, using a static node setup, prohibits dynamic host allocation: the programmer must make a priori such decisions. Both systems require explicit installation of their runtime system on participating hosts. A user must therefore have access to all participating machines, as she must be able to login to them in order to spawn tasks. This is impractical and may result in only a few number of hosts being utilized, even within a single organization. Finally, the choice of C as the main programming language, compared to Java, is an advantage when speed is concerned. But to be able to exploit different architectures, the user must provide and compile code for each one of them, adding to the complexity and increasing development time due to porting considerations. The maturation of Java technology ("just in time" compilation, Java processors, etc.) could soon bridge the performance gap with C. Notably, a Java PVM implementation is underway [6], which will positively impact the portability of the PVM platform.

Condor is a system that has been around for several years. It provides a comparative "matchmaking" process for resource allocation through its "classified advertisment" matchmaking framework [11]. A credit-based mechanism could be implemented using this framework, but is currently unavailable. Condor too requires extensive administration and lacks support for easy development.

Newer systems such as Legion [10] and Globus [7] address the issues of resource allocation and security. They provide mechanisms for locating hosts and signing code. However, both require administration such as compiling and installing the system as well as access to the host computer. They do not support the widely popular Windows platform (though Legion supports NT) and do little to facilitate application development for non-experts. Globus merely offers an MPI implementation whereas Legion provides the "Mentat" language extensions. Legion's solution is more complete but also complicated for inexperienced programmers. It requires using a preprocessor, an "XDR" style serialization process and introduces error-prone situations since virtual method calls will not work as expected in all cases. Stateful and stateless objects are also handled differently. Finally, adding hosts to a running computation is done from the command line and additional hosts are assigned to the computation at random – no matching of criteria is performed.

Several other systems using Java as the "native" programming language have been designed for supporting globally distributed computations, such as Charlotte [3], Javelin [4] and Challenger [5]. These systems automatically distribute computations over machines. However, they do not employ market-based principles to allocate hosts and do not maintain information about hosts' performance.

The market paradigm has received considerable attention in distributed systems aiming for flexible and efficient resource allocation. A system operating on the same principles as ours is Popcorn [12]. Popcorn also uses auction mechanisms to allocate hosts to client computations and exploits Java applet technology to achieve portability, inter-operability and safety. However it does not provide "host profiling", nor promotes incremental development.

7   Discussion

Besides the fact that the allocation strategies used in most systems don't take into account "behavioral patterns" of hosts, there is also virtually no support for leasing. We argue that both are invaluable for efficient resource allocation in open computational environments.

Providing information about the statistical behavior of participating hosts can assist schedulers in taking task placement decisions, avoiding hosts that will degrade performance (and waste credit). For example, assume a scheduler has two tasks to allocate. Blind allocation on two hosts is not a good idea; unless two machines exhibit comparable performance, the faster machine will be wasted since the computation will be delayed by the slower one. Similarly, using the abort ratio, schedulers can avoid unstable hosts for placing critical parts of a computation. Those can be assigned to perhaps more "expensive" but stable hosts. Computations implementing check-pointing and crash-recovery could utilize less credible hosts.

The lack of leasing is also a drawback in open environments: a client could obtain many processors when there is no contention and continue to hold them when demand rises. This is unacceptable in a real world scenario where credit reflects priorities or money. This would imply that prioritized or wealthy computations can be blocked by "lesser" ones. To guarantee quality of service, some form of leasing or preemption must be adopted. Leases are also practical in non-competitive environments. The lease duration allows users to indicate the time during which hosts are under-utilized. Based on this knowledge, tasks can be placed on hosts that will be idle for enough time, and checkpoints can be accurately scheduled, right before a host is about to become unavailable.

Finally, it is generally acknowledged that incremental development increases productivity by separation of concerns and modular design. Distributed computing can benefit from such an approach. Modern object-oriented programming environments are a step towards this direction, but significant programming experience and discipline are still required. We feel that with our system's design, it is possible even for inexperienced programmers to write computations rapidly.

8   Future Directions

New versions of the Java platform will offer more fine grained control in the security system. Using the new mechanisms we expect to be able to provide more efficient services, such as access to local storage for task checkpoints, invocation of native calls to exploit local, tuned libraries such as [2] [13]. Logging mechanisms along with the signing of classes, will further increase the security of the system.

We also wish to experiment with schedulers capable of recording the performance of previous allocations. Accumulated information can perhaps be converted into "experience", leading towards more efficient allocation strategies.

Lastly the issue of scalability needs to be addressed. The current architecture is limited by the market server. A single server could not handle the millions or billions of hosts connecting to a truly world-wide version of this service. It would also be impossible to have all schedulers running on the machine. We intend to overcome this problem by introducing multiple market servers that will allow traffic to be shared among several geographically distributed servers.

References

[1] Y. Amir, B. Awerbuch, and R. S. Borgstrom. A cost-benefit framework for online management of a metacomputing system. In *Proceedings of the First International Conference on Information and Computation Economies*, pages 140–147, October 1998.

[2] M. Baker, B. Carpenter, G. Fox, S. H. Ko, and S. Lim. mpiJava: An Object-Oriented Java Interface to MPI. Presented at International Workshop on Java for Parallel and Distributed Computing, IPPS/SPDP 1999, April 1999.

[3] A. Baratloo, M. Karaul, Z. M. Kedem, and P. Wyckoff. Charlotte: Metacomputing on the web. In *Ninth International Conference on Parallel and Distributed Computing Systems*, September 1996.

[4] P. Cappello, B. Christiansen, M. F. Ionescu, M. O. Neary, K. E. Schauser, and D. Wu. Javelin: Internet-based parallel computing using java. In *Proceedings of the ACM Workshop on Java for Science and Engineering Computation*, June 1997.

[5] A. Chavez, A. Moukas, and P. Maes. Challenger: A multiagent system for distributed resource allocation. In *Proceedings of the First International Conference on Autonomous Agents '97*, 1997.

[6] A. Ferrari. JPVM – The Java Parallel Virtual Machine. *Journal of Concurrency: Practice and Experience*, 10(11), November 1998.

[7] I. Foster and C. Kesselman. Globus: A metacomputing infrastructure toolkit. *Intl J. Supercomputer Applications*, 11(2), 1997.

[8] D. Friedman. The double auction market institution: A survey. In D. Friedman and J. Rust, editors, *Proceedings of the Workshop in Double Auction Markets, Theories and Evidence*, June 1991.

[9] G. A. Geist, J. A. Kohl, and P. M. Papadopoulos. PVM and MPI: a Comparison of Features. *Calculateurs Paralleles*, 8(2):137–150, June 1996.

[10] A. S. Grimshaw and W. A. Wulf. The legion vision of a worldwide computer. *CACM*, 40(1):39–45, 1997.

[11] R. Raman, M. Livny, and M. Solomon. Matchmaking: Distributed resource management for high throughput computing. In *Proceedings of the Seventh IEEE International Symposium on High Performance Distributed Computing*, July 1998.

[12] O. Regev and N. Nisan. The POPCORN Market – an Online Market for Computational Resources. In *Proceedings of the First International Conference on Information and Computation Economies*, pages 148–157, October 1998.

[13] The Java Grande Working Group. Recent Progress of the Java Grande Numerics Working Group. http://math.nist.gov/javanumerics/ reports/jgfnwg-02.html.

The MultiCluster Model to the Integrated Use of Multiple Workstation Clusters

Marcos Barreto*, Rafael Ávila**, and Philippe Navaux***

Institute of Informatics — UFRGS
Av. Bento Gonçalves, 9500 Bl. IV
PO Box 15064 — 90501-910 Porto Alegre, Brazil
E-mail: {barreto,bohrer,navaux}@inf.ufrgs.br

Abstract. One of the new research tendencies within the well-established cluster computing area is the growing interest in the use of multiple workstation clusters as a single virtual parallel machine, in much the same way as individual workstations are nowadays connected to build a single parallel cluster. In this paper we present an analysis on several aspects concerning the integration of different workstation clusters, such as Myrinet and SCI, and propose our MultiCluster model as an alternative to achieve such integrated architecture.

1 Introduction

Cluster computing is nowadays a common practice to many research groups around the world that search for high performance to a great variety of parallel and distributed applications, like aerospacial and molecular simulations, Web servers, data mining, and so forth. To achieve high performance, many efforts have been devoted to the design and implementation of low overhead communication libraries, specially dedicated to fast communication networks used to interconnect nodes within a cluster, which is the case of Fast Ethernet [14], Myrinet [3] and SCI [12]. The design of such software is a widely explored area, resulting in proposals like BIP [21], GM [9], VIA [24] and Fast Messages [19].

Currently, there are other research areas being explored, such as administrative tools for cluster management and what is being called *Grid Computing*, with the objective of joining geographically distributed clusters to form a Metacomputer and taking benefit of the resulting overall computational power [4].

The work presented here is not focused on these areas directly, because our goal is to discuss a practical situation in which a Myrinet cluster must be interconnected with a SCI cluster to form a single parallel machine, which can be used to verify the application's behaviour when it runs on a shared memory cluster or on a message passing cluster, efficiently distribute tasks from an application according to their communication needs, offer a complete environment destined to teach parallel and distributed

---

* M.Sc. student at PPGC/UFRGS (CAPES fellow)
** M.Sc. (PPGC/UFRGS, 1999); RHAE/CNPq researcher at PPGC/UFRGS
*** Ph.D. (INPG, Grenoble — France, 1979); Professor at PPGC/UFRGS

J. Rolim et al. (Eds.): IPDPS 2000 Workshops, LNCS 1800, pp. 71–80, 2000.

programming, allowing the user to express, through the same API, message passing and shared memory interactions.

This paper is organised as follows: Section 2 exposes an analysis on the problems that arise from integrating multiple workstation clusters; in Section 3 we present the MultiCluster model and the DECK environment as our contribution towards this objective; Section 4 brings some comments on related research efforts and finally Section 5 presents our conclusions and current research activities.

2   Integrating Multiple Clusters

When computer networks were an emergent platform to parallel and distributed programming, many efforts were dispended to solve problems related to joining individual PCs in a single virtual parallel machine. From these efforts, communication libraries such as PVM [8] and MPI [17] arose to allow individual network nodes to be identified within the parallel environment.

The integration of multiple workstation clusters presents a similar problem. Individual clusters of workstations are nowadays fairly well managed by communication libraries and parallel execution environments. When we start to think on clusters of clusters, again we have the same problems regarding the connection of elements that run independently from each other and still meet the compromise of offering to the user an appropriate environment for parallel and distributed programming. What we mean by appropriate is to provide an intuitive programming interface and offer enough resources to meet the programmer's needs.

As the purpose of this paper is to identify these problems and propose possible solutions to them, we have divided our study in hardware and software analysis.

2.1   Hardware Aspects

There are no major problems in the hardware point of view to achieve such integration, since the networks considered (Myrinet and SCI) could co-exist within the same node and use different techniques to communicate. Figure 1 presents the most simple cluster interconnection that could be realised.

Each individual cluster could have any number of physical nodes connected through a switch (in the Myrinet case) or directly as a ring (in the SCI case). To allow the integration, each cluster must have a "gateway" node configured with two network interfaces (two Myrinet NIs or a Myrinet + SCI NIs), where the additional Myrinet NI is used to link clusters. For the moment we do not consider SCI a suitable technology as a linking media, since a message-passing paradigm seems more adequate for this purpose.

2.2   Software Aspects

Several points have been discussed by the community in order to identify problems and solutions related to the design and implementation of communication libraries for cluster-based applications, with a main objective: provide high bandwith at small latencies. Besides this, the development of cluster middleware tools to furnish high availability and single system image support is an ongoing task [4, 11].

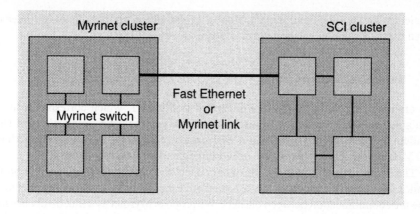

Fig. 1. The simplest way to interconnect two workstation clusters.

In the case of clusters of clusters, performance is not a key point due to the drawbacks implicitly imposed by the loosely coupled integration. There are other problems regarding such integration that must be attended first and performance will then be the consequence of the techniques used to solve them.

The first point to consider is how to combine message passing with distributed shared memory. A desirable solution would be to offer a single communication abstraction that could be efficiently implemented over message passing and shared memory architectures. In practice, however, it is easier to have an individual mechanism to each one and allow the user to choose between them, depending on his application needs.

Another point to treat is the routing problem, which arises when a task needs to exchange data with another task running in a remote cluster. It is necessary that the communication layer identifies what is the location of a communication endpoint and knows how to map physical nodes from separate clusters to be capable of routing messages between them.

Finally, heterogeneity could be a problem. Although most individual workstation clusters are internally homogeneous, there may be cases where multiple clusters could be heterogeneous in relation to each other. In these cases, problems regarding "endianisms" and floating-point data representation have to be addressed.

If the previous problems can be efficiently treated, it is also possible to provide the user with the capacity of deciding where to place a specific set of tasks, according to their communication needs. If the application granularity can be modelled considering the underlying platform, it is still possible to achieve good performance.

3   The MultiCluster Model

The MultiCluster model is an approach to join independent clusters and provide a simple programming interface which allows the user to configure and utilize such an integrated platform. With this model we intend to address and provide solution to the problems mentioned in the previous Section, while still keeping a well structured and efficient programming environment. To best explain the proposed model, we have divided the discussion in hardware and software aspects.

3.1   Hardware Platform

We are assuming the configuration illustrated in Figure 1, which corresponds to our available hardware platform. We currently have a Myrinet cluster, composed by 4 Dual Pentium Pro 200 MHz nodes, and a SCI cluster, composed by 4 Pentium Celeron 300 MHz nodes. These clusters are linked through a Fast Ethernet network.

The choice of the media used to interconnect the clusters depends mostly on the application needs. It is possible to use a standard Ethernet link instead of Myrinet to realise the communication between clusters. We propose Myrinet as a link media because it could minimize the loss in performance originated by the integration of different platforms; for our model, however, it is enough that some node in each cluster plays the role of a gateway.

It is important to say that questions related to cost and scalability are out of the scope of this paper. In a near future, many companies and universities are likely to own a small number of cluster platforms, and so these questions are particular to each of them. We are assuming the situation where at least two clusters are available and have to be used together.

3.2   Software Structure

We have studied each problem mentioned in Section 2.2, trying to find the best solution to each one and structuring our software layer to carry out such solutions. As a result, the MultiCluster model follow some conceptual definitions which rule the way such integration must be handled.

Figure 2 shows the user-defined descriptor file to a MultiCluster application. In this file, the user must specify a list of machines within the clusters he wants to use, the communication subnets identifiers (used to inter-cluster communication), a set of logical nodes with its correspondents machines and the gateway nodes.

Physical and Logical Nodes. A physical node corresponds to each available machine plugged in any individual cluster and only matters to physical questions. Logical nodes are the set of available nodes from the application's point of view. In the case of message-passing clusters, each physical node corresponds to one logical node (this is mandatory). In shared-memory clusters, a logical node can be composed of more than one physical node. The distinction between logical nodes for Myrinet and SCI is made by the node id field. For example, "node 1:0" means the second node within the subnet 0 (which is Myrinet in our example), while "node 4:1" means the first node within the subnet 1 (which is SCI). It is important to notice that this numbering scheme, although complex, is entirely processed by the environment in a transparent manner; the user only knows how many logical nodes he has and what are the physical machines within each logical node.

```
// DECK user-defined descriptor file
// virtual machine
verissimo, quintana, euclides, dionelio,
scliar, ostermann, meyer, luft
// communication subnets
myrinet: 0
sci: 1
// logical nodes
node 0:0 machines: verissimo
node 1:0 machines: quintana
node 2:0 machines: euclides
node 3:0 machines: dionelio
node 4:1 machines: scliar, luft
node 5:1 machines: ostermann, meyer
// gateway nodes
gateways: quintana, scliar
```

Fig. 2. Descriptor file for a MultiCluster application.

Intra- and Inter-node Communication. As the application only sees logical nodes, it is relatively easy to adapt the different communication paradigms: inside a logical node, communication is made by shared memory; between logical nodes, communication is made by message passing. From the user's point of view, there is only one programming interface furnishing both mechanisms to specify communication over Myrinet or SCI clusters; the underlying communication layer is in charge of implementing one or another paradigm.

Heterogeneity. Although a less frequent problem, heterogeneity may arise depending on the availability of clusters that have to be interconnected. Here, we are considering different data representations and the need to indicate to the message receiver what is the architecture type of the message sender. This problem is implicitly treated by the communication software.

Even occuring some performance loss due to such integration, it is possible to the user to define the best location for his application tasks, creating communication resources according to each task location (i.e. communication subnets). Through this facility, the granularity of communication could be balanced among clusters, avoiding as long as possible the traffic across the link network.

3.3   The Programming Environment—DECK

The interface between the programmer and the MultiCluster architecture is the DECK environment. DECK (*Distributed Executive Communication Kernel*) is composed of a runtime system and a user API which provides a set of services and abstractions for the development of parallel and distributed applications. A DECK application runs in an SPMD style, split in terms of logical nodes.

DECK is divided in two layers, one called μDECK, which directly interacts with the underlying OS and a service layer, where more elaborate resources (including the support for multiple clusters) are made available. Figure 3 shows the layered structure of DECK.

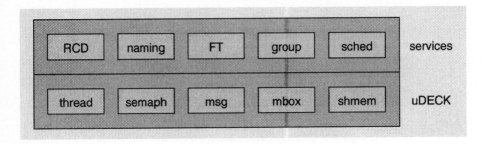

Fig. 3. Internal structure of DECK.

μDECK is the platform-dependent part of DECK. This layer implements the five basic abstractions provided within the environment: *threads, semaphores, messages, mailboxes* and *shared segments*. Each of these abstractions is treated by the application as an object, and has associated primitives for proper manipulation.

Messages present pack/unpack primitives, which do not necessarily perform marshalling/unmarshalling actions. When a message object is created, one of its attributes holds the identification of the host architecture. At the time of a pack no marshalling is performed; at the time of an unpack, if the receiving host is of a different architecture, the proper data conversion is made[1]. Messages can be posted to or retrieved from mailboxes. Only the creator of a mailbox is allowed to retrieve messages from it, but any other thread knowing the mailbox can post to it. To use a mailbox, the creator must register it in a naming server. There are two ways to obtain a mailbox address: fetching it in the name server or receiving it in a message.

The service layer is built on top of μDECK and aims to furnish additional, more sophisticated mechanisms that might be useful to the development of parallel applications, such as naming, group communication and fault tolerance support. In the scope of this paper, two elements of this layer must be analysed: the naming service and the Remote Communication Daemon (RCD).

The name server is a dedicated thread which runs in the first node within each cluster. For example, in the configuration illustrated in Figure 2, there will be a naming server running on "verissimo" and another running on "scliar". Each naming server is responsible to register mailboxes created within its cluster. The name server is automatically executed when the application starts and has a well-known mailbox to allow other threads to communicate.

---

[1] It is important to observe that we only expect this to happen for messages crossing cluster boundaries, since clusters are assumed to be internally homogeneous.

The DECK/Myrinet Implementation. In the implementation of DECK on top of Myrinet, we are currently using BIP (*Basic Interface for Parallelism*) [21] as a communication protocol to efficiently use the underlying hardware and deliver high performance to applications. As BIP utilizes reception queues labeled with tags within each node, our mailbox implementation assigns a specific tag to each mailbox. To create a mailbox, the programmer uses a deck_mbox_create() primitive, passing as arguments the mailbox name and the communication subnet (defined in the descriptor file) in which this mailbox will be used.

The communication is made by post and retrieve operations, passing as arguments the corresponding mailbox and the message object, which contains the DECK supported datatypes. Posting a message is an asynchronous operation, while retrieving a message is a synchronous operation. To achieve this behaviour, we use the bip_tisend() and bip_trecv() primitives, respectively.

The implementation of μDECK mailboxes and messages on top of BIP is straightforward, since both are based on message passing. Shared segments, however, need an additional software DSM support to be implemented with the same library. For the moment we are studying the introduction of a DSM library, such as TreadMarks [25], to allow the usage of shared segments over Myrinet. The primitives for threads and semaphores are trivial and follow the Pthreads standard [13].

The DECK/SCI Implementation. We base our DECK/SCI implementation on two SCI programming libraries: Yasmin [23], which provides basic primitives for creation, mapping and synchronisation of shared segments, and Sthreads [22], which offers a Pthread-like environment on top of Yasmin.

A μDECK shared segment object offers primitives for creation, naming, mapping and locking. To the difference of Myrinet, SCI allows an easier implementation of both communication paradigms, so DECK/SCI offers mailboxes and messages as well as shared segments.

The creation of threads in DECK/SCI follows a simple round-robin placement strategy, according to the number of physical nodes that compose a logical node, which means that placement is still transparent to the end user. Notice that local memory can still be used for communication by local threads (i.e. threads in the same physical node), but it is up to the programmer to keep this kind of control. This means that, within SCI clusters, memory is only guaranteed to be correctly shared between remote threads if it is mapped into a μDECK shared segment.

RCD–Remote Communication Daemon. In order to support the MultiCluster model, the *Remote Communication Daemon* has been designed as a DECK service responsible for communicating to remote clusters. As each cluster must have a "gateway" node, the RCD is automatically executed inside this node when the application starts and follows the same semantic of the name server, i.e., it also has a well-known mailbox.

The RCD acts upon demand on two special cases: when fetching names defined remotely (i.e. on another cluster) and when posting messages to remote mailboxes. When a DECK primitive fails to fetch a mailbox address in a local name server, it contacts the RCD, which then broadcasts the request to other RCDs in the system and wait for an answer, returning it to the caller. In the second case, when a DECK primitive sees a remote mailbox address when posting a message, it contacts the RCD, which then forwards the message to the RCD responsible for the communication subnet in which the mailbox is valid.

It is important to emphasize that communication between threads in different logical nodes, as well as different clusters, must always be made by message passing. Even in the case of a SCI cluster, there must be at least one mailbox to allow the communication with the RCD and, eventually, retrieve messages. For the moment we are disconsidering the utilisation of a global shared memory space to establish communication among clusters due to the lack of this support in the DECK/Myrinet implementation.

Our intention in designing DECK in three parts is to make it usable without changes in both single- and multi-clustered environments. In the first case, the RCD will simply not be brought into action by the application, since all the objects will be local to a specific cluster.

4   Related Work

Since the purpose of this paper is to discuss practical questions involved in the integration of multiple clusters and propose our model to achieve such integration, we tried to identify similar proposals regarding this subject.

There is a great number of research projects concerning the integration of multiple workstation clusters, such as NOW [1], Beowulf [2], Globus [7] and Legion [10]. The goal of these projects is to allow parallel and distributed programming over geographically distributed, heterogeneous clusters that corresponds to a "global computational grid". The differential characteristic of our MultiCluster model is that we are assuming the simultaneous use of different network technologies, while these projects plans to use a common network technology to connect clusters, providing high scalability.

In terms of programming environments, there are also some efforts concentrated in joining message passing and distributed shared memory facilities, such as Stardust [5] and Active Messages II [16]. The main goal is to provide support for both message passing and distributed shared memory paradigms and, at same time, offer mechanisms to fault tolerance and load balancing support, as well as, portability. There are also some important contributions based on Java, such as JavaNOW [15], JavaParty [20] and Javelin [6]. All these contributions aims to provide distributed programming across networks of workstations or Web-based networks, differing in the communication model they used.

The idea behind MultiCluster is similar in some aspects with the objectives found in the projects/environments mentioned here, though in a smaller scale. Our research goal is to identify and propose solutions to problems related to specific integration of Myrinet and SCI clusters, while the goals of such projects comprise a larger universe, including fast communication protocols, cluster tools, job scheduling and so on.

Nevertheless, it is possible to state brief comparisons: our RCD is a simplest implementation when compared with Nexus, the communication system used inside Globus; it is just a way to give remote access to mailboxes defined in another clusters and allow us to separate the functionality of DECK when it runs in a single cluster platform.

The combination of message passing and distributed shared memory we offer is not so different than the usual mechanisms provided by the others environments. We want to efficiently implement these mechanisms in both clusters, without changing the programming interface. To accomplish this, our choice is to provide a mailbox object and a shared segment object to express message passing and memory sharing, respectively.

5   Conclusions and Current Work

In this paper we exposed some problems related to the integration of two different cluster platforms and proposed our MultiCluster model to achieve such desirable integration. We are developing our software environment aiming to accomplish a number of objectives, such as joining two specific cluster platforms (Myrinet and SCI) and providing a uniform API for parallel and distributed programming on both platforms, as well as opening research activities concerning such integration.

The integration is easier in terms of hardware because many solutions are already implemented within the OS kernel (e.g. co-existence of network device drivers). In terms of software, we have to decide what is the abstraction degree we want to offer to the programmer. It is important that the user be aware of the characteristics of each individual cluster to best adapt his application to take benefit of them. On the other hand, the DECK layer must abstract as much as possible implementation details, offering to the users a complete and simple API able to express the application needs. Currently, the descriptor file is the key point to configure the MultiCluster platform, because it represents the communication contexts and the logical nodes the user wants to use. Although this configuration is not so transparent, it is the most suitable way to adapt the execution environment according to the user needs. We consider that there are no problems in this task, since the execution environment guarantees the expected functionality.

Our work has been guided towards the design of a complete set of programming resources, enclosed in a software layer. Through the modularisation of DECK, we have divided our work in such way that we can parallelize our efforts to cover all problems exposed and to make available, as soon as possible, the MultiCluster model. At the moment we already have an implementation of DECK based on Pthreads and UNIX sockets, available at our Web page [18]. This implementation has played an important role to define the DECK structure and behaviour. At the time of this writing, we are concluding the implementation on top of BIP and collecting some performance results and, at same time, starting the implementation of DECK objects on top of SCI. The next step is to join both clusters and develop the RCD communication protocol.

References

1. T. Anderson, D. Culler, and D. Patterson. A case for NOW - Network of Workstations. Available by WWW at http://now.cs.berkeley.edu, Out. 1999.
2. Beowulf. The Beowulf project. Available by WWW at http://www.beowulf.org, Jun. 1999.
3. N. Boden et al. Myrinet: A gigabit-per-second local-area network. *IEEE Micro*, 15(1):29–36, Feb. 1995.

4. Rajkumar Buyya. *High Performance Cluster Computing*. Prentice Hall PTR, Upper Saddle River, NJ, 1999.
5. Gilbert Cabillic and Isabelle Puaut. Stardust: an environment for parallel programming on networks of heterogeneous workstations. *Journal of Parallel and Distributed Computing*, 40:65–80, 1997.
6. B. Christiansen et al. Javelin: Internet-based parallel computing using Java. Available by WWW at http://www.cs.ucsb.edu/research/javelin/, Nov. 1999.
7. Ian Foster and Carl Kesselman.   The Globus project.   Available by WWW at http://www.globus.org, Jul. 1999.
8. Al Geist et al. *PVM: Parallel Virtual Machine*. MIT Press, Cambridge, MA, 1994.
9. GM message passing system. Available by WWW at http://www.myri.com, Nov. 1999.
10. A. Grimshaw et al. The Legion vision of a worldwide virtual computer. *Communications of the ACM*, 40(1), Jan. 1997.
11. Kai Hwang and Zhiwei Xu. *Scalable Parallel Computing: Technology, Architecture, Programming*. McGraw-Hill, New York, NY, 1997.
12. IEEE. IEEE standard for Scalable Coherent Interface (SCI). IEEE 1596-1992, 1992.
13. IEEE. Information technology—portable operating system interface (POSIX), threads extension [C language]. IEEE 1003.1c-1995, 1995.
14. IEEE. Local and metropolitan area networks-supplement—media access control (MAC) parameters, physical layer, medium attachment units and repeater for 100Mb/s operation, type 100BASE-T (clauses 21–30). IEEE 802.3u-1995, 1995.
15. Java and High Performance Computing Group. The JavaNOW project. Available by WWW at http://www.jhpc.org/projects.html, Nov. 1999.
16. Steven S. Lumetta, Alan M. Mainwaring, and David E. Culler. Multi-protocol Active Messages on a cluster of SMP's. In *Proc. of SuperComputing 97*, 1997.
17. MPI FORUM. Document for a standard message passing interface. *International Journal of Supercomputer Applications and High Performance Computing Technology*, 8(3/4), 1994.
18. The MultiCluster project. Available by WWW at http://www-gppd.inf.ufrgs.br/projects/mcluster, Nov. 1999.
19. S. Pakin, M. Lauria, and A. Chien. High performance messaging on workstations: Illinois Fast Messages for Myrinet. In *SuperCOmputing '95*. IEEE Computer Society Press, 1996.
20. Michael Philippsen and Matthias Zenger. JavaParty: A distributed companion to Java. Available by WWW at http://wwwipd.ira.uka.de/JavaParty, Nov. 1999.
21. Loic Prylli and Bernard Tourancheau. BIP: A new protocol designed for high performance networking on Myrinet. In José Rolim, editor, *Parallel and Distributed Processing*, number 1388 in Lecture Notes in Computer Science, pages 472–485. Springer, 1998.
22. Enno Rehling. Sthreads: Multithreading for SCI clusters. In *Proc. of Eleventh Symposium on Computer Architecture and High Performance Computing*, Natal - RN, Brazil, 1999. Brazilian Computer Society.
23. H. Taskin. Synchronizationsoperationen für gemeinsamen Speicher in SCI-Clustern. Available by WWW at http://www.uni-paderborn.de/cs/ag-heiss/en/veroeffentlichungen.html, Aug. 1999.
24. VIA – Virtual Interface Architecture. Available by WWW at http://www.via.org, Nov. 1999.
25. Willy Zwaenepoel et al. TreadMarks distributed shared memory (DSM) system. Available by WWW at http://www.cs.rice.edu/~willy/TreadMarks/overview.html, Dez. 1998.

Parallel Information Retrieval on an SCI-Based PC-NOW

Sang-Hwa Chung, Hyuk-Chul Kwon, Kwang Ryel Ryu, Han-Kook Jang, Jin-Hyuk Kim, and Cham-Ah Choi

Division of Computer Science and Engineering, Pusan National University,
Pusan, 609-735, Korea
{shchung, hckwon, krryu, hkjang, variant, cca}@hyowon.pusan.ac.kr

Abstract. This paper presents an efficient parallel information retrieval (IR) system which provides fast information service for the Internet users on low-cost high-performance PC-NOW environment. The IR system is implemented on a PC cluster based on the Scalable Coherent Interface (SCI), a powerful interconnecting mechanism for both shared memory models and message passing models. In the IR system, the inverted-index file (IIF) is partitioned into pieces using a greedy declustering algorithm and distributed to the cluster nodes to be stored on each node's hard disk. For each incoming user's query with multiple terms, terms are sent to the corresponding nodes which contain the relevant pieces of the IIF to be evaluated in parallel. According to the experiments, the IR system outperforms an MPI-based IR system using Fast Ethernet as an interconnect. Speed- up of up to 4.0 was obtained with an 8-node cluster in processing each query on a 500,000-document IIF.

1. Introduction

As more and more people are accessing the Internet and acquiring a vast amount of information easily, more people consider that the problem of information retrieval (IR) resides no longer in the lack of information, but in how we can choose from a vast amount the right information with speed. Many of us have already experienced that some IR systems provide information service much faster than others. How fast an IR system can respond to users' queries mostly depends on the performance of the underlying hardware platform. Therefore, most of the major IR service providers have been urged to spend several hundred thousand dollars to purchase their hardware systems. However, for many small businesses on the Internet, that cost is too high.

In this paper, as a cost-effective solution for this problem, a PC cluster interconnected by a high-speed network card is suggested as a platform for fast IR service. With the PC cluster, a massive digital library can be efficiently distributed to PC nodes by utilizing local hard disks. Besides, every PC node can act as an entry to process multiple users' queries simultaneously.

It is extremely important to select a network adapter to construct a high-speed system area network (SAN). For a message passing system, the Fast Ethernet card or the Myrinet card can be used. For a distributed shared memory (DSM) system, the SCI card can be considered. Fast Ethernet developed for LAN is based on complicated protocol software such as TCP/IP, and its bandwidth is not high. The Myrinet[1] card is a high-speed message passing card with a maximum bandwidth of 160Mbyte/sec. However, the network cost is relatively high because Myrinet

J. Rolim et al. (Eds.): IPDPS 2000 Workshops, LNCS 1800, pp. 81-90, 2000.

requires crossbar switches for the network connection.   Besides, its message-passing mechanism is based on time consuming operating system calls.   For applications with frequent message-passing, this can lead to performance degradation.   To overcome the system call overhead, systems based on user-level interface for message-passing without intervention of operating system have been developed.   Representative systems include AM[2], FM[3], and U-Net[4].   Recently, Myrinet is also provided with a new message-passing system called GM[5], which supports user-level OS-bypass network interface access.

The SCI (Scalable Coherent Interface: ANSI/IEEE standard 1596-1992) is designed to provide a low-latency (less than 1μs) and high bandwidth (up to 1Gbyte/sec) point-to-point interconnect. The SCI interconnect can assume any topology including ring and crossbar.   Once fully developed, the SCI can connect up to 64K nodes.   Since the SCI supports DSM models that can feature both of NUMA and CC-NUMA variants, it is possible to make transparent remote memory access with memory read/write transactions without using explicit message-passing.   The performance of the SCI-based systems has been proven by the commercial CC-NUMA servers such as Sequent NUMAQ 2000[6] and Data General's Aviion[7].

In this research, the SCI is chosen as an underlying interconnecting mechanism for clustering.   The Parallel IR system is implemented on an SCI-based PC cluster using a DSM programming technique.   In the IR system, the inverted-index file(IIF) is partitioned into pieces using a greedy declustering algorithm and distributed to the cluster nodes to be stored on each node's hard disk.   An IIF is the sorted list of terms (or keywords), with each term having links to the documents containing that term. For each incoming user's query with multiple terms, terms are sent to the corresponding nodes which contain the relevant pieces of IIF to be evaluated in parallel.   An MPI-based IR system using Fast Ethernet as an interconnect is also constructed for comparison purpose.

2. PC Cluster-based IR System

2.1   Typical IR System on Uniprocessor

Figure 1 shows the structure of a typical IR system implemented on a uniprocessor. As shown in the figure, once a user's query with multiple terms is presented to the system, for each query term in turn the IR engine retrieves relevant information from the IIF in the hard disk.   When all the information is collected, the IR engine performs necessary IR operations, scores the retrieved documents, ranks them, and sends the IR result back to the user.   For the efficient parallelization of the system, it is important to find out the most time consuming part in executing the IR system. Using the sequential IR system developed previously[8], the system's execution time is analyzed as shown in Figure 2.   In the sequential system, the most time consuming part is disk access.   Thus, it is necessary to parallelize disk access.   This can be done by partitioning the IIF into pieces and distributing the pieces to the processing nodes in a PC cluster.

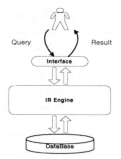
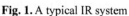

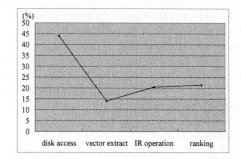

Fig. 1. A typical IR system   **Fig. 2.** Execution time analysis in the sequential IR system

2.2 Declustering IIF

Most current IR systems use a very large lookup table called an inverted index file (IIF) to index relevant documents for given query terms. Each entry of the IIF consists of a term and a list of ids of documents containing the term. Each of the document ids is tagged with a weight of the term for that document. Given a query, all the query terms are looked up from the IIF to retrieve relevant document ids and the corresponding term weights. Next, the documents are scored based on the term weight values and then ranked before they are reported back to the user.

Since our IR system processes user's query in parallel on a PC cluster, it is desirable to have the IIF appropriately declustered to the local hard disks of the processing nodes. We can achieve maximum parallelism if the declustering is done in such a way that the disk I/O and the subsequent scoring job are distributed as evenly as possible to all the processing nodes. An easy random declustering method would be just to assign each of the terms (together with its list of documents) in the IIF lexicographically to each of the processing nodes in turn, repeatedly until all the terms are assigned. In this paper, we present a simple greedy declustering method which performs better than the random method.

Our greedy declustering method tries to put together in the same node those terms which have low probability of simultaneous occurrence in the same query. If the terms in a query all happen to be stored in the same node, the disk I/O cannot be done in parallel and also the scoring job cannot readily be processed in parallel. For an arbitrary pair of terms in the IIF, how can we predict the probability of their co-occurring in the same query? We conjecture that this probability has a strong correlation with the probability of their co-occurrence in the same documents. Given a pair of terms, the probability of their co-occurrence in the same documents can be obtained by the number of documents in which the two terms co-occur divided by the number of all the documents in a given document collection. We calculate this probability for each of all the pairs of terms by preprocessing the whole document collection.

When the size of the document collection is very large, we can limit the calculation of the co-occurrence probabilities only to those terms which are significant. The reason is that about 80% of the terms in a document collection usually exhibits only a single or double occurrences in the whole document collection and they are unlikely to appear in the user queries. Also, since the number of terms in a document collection is known to increase in log scale as the number of documents increases, our method will not have much difficulty in scaling up.   As more documents are added to the collection, however, re-calculation of the co-occurrence probabilities would be needed for maintenance.   But, this would not happen frequently because the statistical characteristics of a document collection does not change abruptly.

In the first step of our greedy declustering algorithm, all the terms in the IIF are sorted in the decreasing order of the number of documents each term appears.   The higher this number the more important the term is in the sense that it is quite likely to be included in many queries.   This is especially true when the queries are modified by relevance feedback[9].   This type of terms also have a longer list of documents in the IIF and thus causes heavier disk I/O.   Therefore, it is advantageous to store these terms in different nodes whenever possible for the enhancement of I/O parallelism. Suppose there are n processing nodes.   We assign the first n of the sorted terms to each of the n nodes in turn.   For the next n terms, each term is assigned to the node which contains a term with the lowest probability of co-occurrence.   From the third pass of the term assignment, a term is assigned to such a node that the summation of the probabilities of co-occurrence of the term with the terms already assigned to the node is the lowest.   This process repeats until all the terms in the IIF are assigned.

2.3 Parallel IR System Model

The PC cluster-based parallel IR system model is shown in Figure 3.   The IR system consists of an entry node and multiple processing nodes.   The participating nodes are PCs with local hard disks and connected by an SCI-based high-speed network.   The working mechanism of the parallel IR system model can be explained as follows. The entry node accepts a user' query and distributes query terms to processing nodes (including itself) based on the declustering information described in the previous subsection.   Each processing node consults the partitioned IIF using the list of query terms delivered from the entry node, and collects the necessary document list for each term from the local hard disk.   Once all the necessary document lists are collected, they are transmitted to the entry node.   The entry node collects the document lists from the participating processing nodes (including itself), performs required IR operations such as AND/OR and ranks the selected documents according to their scores.   Finally the sorted document list is sent back to the user as an IR result.

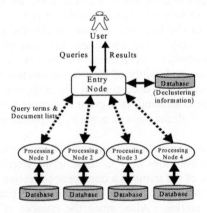

Fig. 3. Parallel IR system model

2.4   Experimental PC Cluster System

In this research, an 8-node SCI-based PC cluster system is constructed as shown in Figure 4.   Each node is a 350MHz Pentium II PC with 128Mbyte main memory and 4.3Gbyte SCSI hard disk, and operated by Linux kernel 2.0.36.   In the cluster, any PC node can be configured as an entry node.   As shown in the figure, each PC node is connected to the SCI network through the Dolphin Interconnect Solution (DIS)'s PCI-SCI bridge card.   There are 4 rings in the network, and 2 nodes in each ring. The rings are interconnected by the DIS's 4×4 SCI switch.   For DSM programming, the DIS's SISCI (Software Infrastructure for SCI) API[10] is used.   With this configuration, the maximum point-to-point bulk transfer rate obtained is 80 Mbyte/sec approximately.

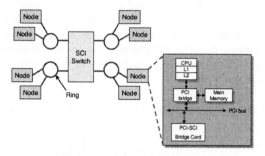

Fig. 4. SCI-based 8 node PC cluster system

For comparison purpose, an 8-node Fast Ethernet-based PC cluster system is also constructed.   Each PC node has the same configuration as the SCI network's node except that a PCI Fast Ethernet Adapter is used for networking.   A switching hub is used to interconnect PC nodes in the cluster.   For message-passing programming, MPICH 1.1.1[11] is used.   In this case, the maximum point-to-point bulk transfer rate obtained is 10 Mbyte/sec approximately.

2.5   SCI-based DSM Programming

The SCI interconnect mechanism supports DSM programming.   By using SISCI, a node in the SCI-based PC cluster can establish a mapping between it's local memory address space and a remote node's memory address space.   Once the mapping is established, the local node can access the remote node's memory directly.   In DSM programming, the communication between PC nodes in the cluster is done using remote read and remote write transactions instead of message-passing.   These remote read/write transactions are actually carried out using the remote read/write functions provided by SISCI.   When the IR program is actually coded, most of the remote memory transactions are implemented using the remote write function.   This is because the remote write function performs about 10 times faster than the remote read function in the DIS's PSI-SCI bridge card.

3. Performance of PC Cluster-based IR System

3.1 Performance Comparison between SCI-based System and MPI-based System

In this experiment, average query processing times are measured for the 8-node SCI-based system, the 8-node MPI-based system and a single node system.   The IIF is constructed from 100,000 documents collected from articles in a newspaper.   A user's query consists of 24 terms.   Each query is made to contain a rather large number of terms because the queries modified by relevance feedback usually have that many terms.   The IIF is randomly declustered to be stored on each processing node's local disk.

As shown in Table 1, the disk access time is reduced for both the SCI-based system and the MPI-based system when compared with the single node system.   However, the MPI-based system is worse than the single node system in total query processing time because of the communication overhead.   The SCI-based system has much less communication overhead than the MPI-based system, and performs better than the single node system.   The speed-up improves with further optimizations presented in the following subsections.

Table 1. Query processing times of 8-node SCI-based system and 8-node MPI-based system (unit : sec)

	SCI-based system	MPI-based system	Single-node System
Send query term	0.0100	0.0251	0
Receive document list	0.0839	0.2097	0
Disk access	0.0683	0.0683	0.2730
IR operation	0.0468	0.0468	0.0468
Total	0.2091	0.3500	0.3198

3.2 Effect of Declustering IIF

The greedy declustering method is compared with the random method on a test set consisting of 500 queries each containing 24 terms.   To generate the test queries we randomly sampled 500 documents from a document collection containing 500,000 newspaper articles.   From each document, the most important 24 terms are selected to make a query.   The importance of a term in a document is judged by the value $tf \times idf$, where tf is the term's frequency in that document and idf is the so called inverse document frequency.   The inverse document frequency is given by $\log_2(N/n) + 1$, where N is the total number of documents in the collection and n is the number of documents containing the term.   Therefore, a term in a document is considered important if its frequency in that document is high enough but at the same time it does not appear in too many other documents.   Table 2 shows the experimental results comparing the random clustering and the greedy declustering methods using those 500 queries on our 500,000 document collection.

Table 2. Comparison of random declustering and greedy declustering (unit: sec)

	Random declustering	Greedy declustering
Average query processing time	0.5725	0.5384
Accumulated query processing time for 500 queries	286.2534	269.1919

3.3 Performance with Various-sized IIF

In this subsection, the performance of the SCI-based parallel IR system is analyzed with the number of documents increased up to 500,000. These documents are collected from a daily newspaper, and 500,000 documents amount to the collection of the daily newspaper articles for 7 years. The size of IIF proportionally increases as the number of documents increases. For example, the size of IIF is 300 Mbytes for 100,000 documents, and 1.5 Gbytes for 500,000 documents. The 8-node PC cluster and the greedy declustering method are used for the experiment.

The experimental result is presented in Figure 5. It takes 0.1805 seconds to process a single query with the 100,000 document IIF, while it takes 0.2536 seconds with the 200,000 document IIF and 0.5398 seconds with 500,000 document IIF. As the IIF size increases, the document list for each query term becomes longer, and the time spent for IR operations (AND/OR operations) increases considerably. As a result, the IR operation eventually takes more time than the disk access, and becomes the major source of bottleneck.

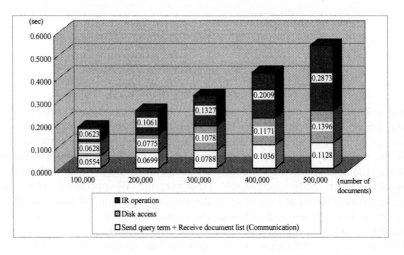

Fig. 5. IIF size vs. query processing time

3.4 Reducing IR Operation Time

As presented in the previous subsection, the IR operation time turns out to be a new overhead as the IIF size increases. In the IR system, AND/OR operations are performed by the entry node after all the necessary document lists are collected from the processing nodes. However, it is possible to perform AND/OR operations partially to the document lists collected in each processing node. So, each processing node can transmit only the result to the entry node. This helps in reducing not only the IR operation time but also the communication time.

The performance of the improved system in comparison with the original system is shown in Figure 6. In the experiment, the 8-node PC cluster, the greedy declustering method and 500,000 document IIF are used. In the original system, the IR operation takes 0.2873 seconds which is more than 53% of the total query processing time. However in the improved system, the IR operation takes only 0.1035 seconds which is about 35% of the total time. Thus, the IR operation takes less time than the disk access again. The communication time is also reduced from 0.1128 seconds to 0.0500 seconds, and the total time is reduced to almost half when compared with the original system.

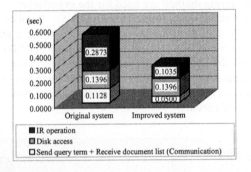

Fig. 6. Query processing time with reduced IR operation time

Figure 7 shows the speed-up of the parallel IR system. The maximum speed-up obtained from the 8-node system when compared with the single node system is 4.0. As shown in the figure, the speed-up of the parallel IR system is saturated rapidly from the 4-node system. As the number of the processing nodes in the system increases, the disk access time[1] is reduced because the average number of query terms assigned to each node decreases. However, the IR operation time and the communication time rather increase as the number of document lists transmitted to the entry node increases, and attenuate the overall speed-up. The problem may be alleviated by applying the following idea. Instead of sending all the document lists to the entry nodes, intermediate nodes can be utilized to merge the document lists by performing AND/OR operations in advance as shown in Figure 8. Thus the entry node finally handles only two document lists. This will help in reducing both the IR

---

[1] The disk access time includes the time spent for partial AND/OR operations in the processing nodes.

operation time and the communication time.   Experiments need to be performed to verify the above idea .

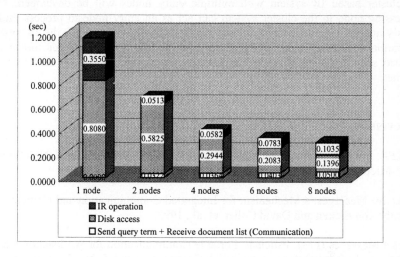

Fig. 7. Number of processing nodes vs. query processing time

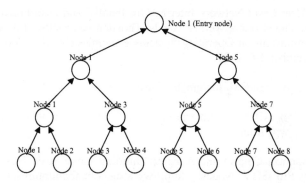

Fig. 8. Merging document lists in intermediate nodes

4. Conclusions

In this paper, as a cost-effective solution for fast IR service, an SCI-based PC cluster system is proposed.   In the parallel IR system developed on the PC cluster, the IIF is partitioned into pieces using a greedy declustering algorithm and distributed to the cluster nodes to be stored on each node's hard disk.   For each incoming user's query with multiple terms, terms are sent to the corresponding nodes which contain the relevant pieces of IIF to be evaluated in parallel.   The IR system is developed using a DSM programming technique based on SCI.   According to the experiments, the IR system outperforms an MPI-based IR system using Fast Ethernet as an interconnect. Speed-up of 4.0 was obtained with the 8-node cluster in processing each query on a

500,000-document IIF.

Currently, the parallel IR system has a single entry node. In the future research, a PC cluster based IR system with multiple entry nodes will be developed. Each processing node in the cluster system can act as an entry node to process multiple users's queries simultaneously. This will help in improving both the IR system's utilization and throughput. With more research effort, we hope this model to be evolved as a practical solution for low-cost high-performance IR service on the Internet.

References

1. IEEE, "MYRINET: A GIGABIT PER SECOND LOCAL AREA NETWORK", IEEE-Micro, Vol.15, No.1, February 1995, pp.29-36.

2. "Active Messages: a Mechanism for Integrated Communication and Computation", Thorsten von Eicken and David Culler, et. al., 1992.

3. "Fast Messages (FM): Efficient, Portable Communication for Workstation Clusters and Massively-Parallel Processors", IEEE Concurrency, vol. 5, No. 2, April-June 1997, pp. 60-73. (Pakin, Karamcheti & Chien)

4. "U-Net: A User-Level Network Interface for Parallel and Distributed Computing", Anindya Basu, Vineet Buch, Werner Vogels, Thorsten von Eicken, Proceedings of the 15th ACM Symposium on Operating Systems Principles (SOSP), Copper Mountain, Colorado, December 3-6, 1995.

5. http://www.myri.com/GM/doc/gm_toc.html

6. "NUMA-Q: An SCI based Enterprise Server", http://www.sequent.com/products/highend_srv/sci_wp1.html

7. "SCI Interconnect Chipset and Adapter: Building Large Scale Enterprise Servers with Pentium Pro SHV Nodes", http://www.dg.com/about/html/sci_interconnect_chipset_and_a.html

8. S.H.Park, H.C.Kwon, "An Improved Relevance Feedback for Korean Information Retrieval System", Proc. of the 16th IASTED International Conf. Applied Informatics, IASTED/ACTA Press, pp.65-68, Garmisch-Partenkirchen, Germany, February 23-25, 1998

9. Salton, G. and Buckley, C., "Improving retrieval performance by relevance feedback", American Society for Information Science, 41, 4, pp. 288-297, 1990.

10. http://www.dolphinics.no/customer/software/linux/index.html

11. "A High-Performance, Portable Implementation of the MPI Message Passing Interface Standard", http://www-unix.mcs.anl.gov/mpi/ mpich/docs.html

A PC-NOW Based Parallel Extension for a Sequential DBMS

Matthieu Exbrayat and Lionel Brunie

Laboratoire d'Ingénierie des Systèmes d'Information
Institut National des Sciences Appliquées, Lyon, France
Matthieu.Exbrayat@lisi.insa-lyon.fr, Lionel.Brunie@insa-lyon.fr

Abstract. In this paper we study the use of networks of PCs to handle the parallel execution of relational database queries. This approach is based on a parallel extension, called *parallel relational query evaluator*, working in a coupled mode with a sequential DBMS. We present a detailed architecture of the parallel query evaluator and introduce Enkidu, the efficient Java-based prototype that has been build according to our concepts. We expose a set of measurements, conducted over Enkidu, and highlighting its performances. We finally discuss the interest and viability of the concept of parallel extension in the context of relational databases and in the wider context of high performance computing.
Keywords: Networks of workstations, Parallel DBMS, Java

1 Introduction

Parallelizing Database Management Systems (DBMS) has been a flourishing field of research for the last fifteen years. Research, experiment and development have been conducted according to three main goals. The first one is to accelerate heavy operations, such as queries involving the confrontation of huge amounts of data (by parallelizing elementary operations over the nodes and distributing data among the disks – I/O parallelism). The second one is to support a growing number of concurrent users (by dispatching connections and queries among the processors). The third goal is to offer a high level of fault tolerance, and therefore to guarantee the availability of data, for instance in the context of intensive commercial transactions (e.g. by using RAID techniques).

The very first parallel DBMSs (PDBMSs) were based on specific machines, such as Gamma [1] and the Teradata Parallel Database Machine [2]. The next logical step appeared in the middle of the 90's, with such PDBMSs as Informix On Line XPS [3], IBM DB2 Parallel Edition [4] and Oracle 7 Parallel Server [5], which were designed to work on standard (parallel) machines. Some of these systems (e.g. Informix), were defined as running on "Networks of Workstation". Nevertheless, this definition was quite erroneous, as they were mainly designed to work on high-end architectures, such as the IBM SP2 machine. The very last developments, like Oracle 8 Parallel Server [6] take advantage of recent cluster architectures, and partially hide the management of parallelism (the administrator only has to define the list of nodes and disks to be used). It is in fact

J. Rolim et al. (Eds.): IPDPS 2000 Workshops, LNCS 1800, pp. 91-100, 2000.

noticeable, that the use of a network of PCs to support a PDBMS has been poorly studied. We can cite Midas [7] (parallel port of a sequential DBMS to a LAN of PCs), and the 100 Node PC Cluster [8] database (developed from scratch).

Nevertheless, while the very large majority of studies and products consist in fully porting sequential DBMSs to parallel architectures, we estimate that networks of PCs could lead to a new approach of DBMS parallelization, considering the network of PCs as a parallel extension for an existing sequential DBMS. This extension, named *coupled query evaluator*, consists of a parallel execution component (on the network of PCs), which works together with a sequential DBMS, in order to offer both high performance for query evaluation (on the parallel component) and coherency for data creation and modification (on the sequential DBMS).

In section 2, we will detail the architecture of our proposal. Its implementation will then be introduced in section 3. In section 4 we will present some measurements conducted over our prototype. In section 5 we will discuss the relevance and impact of the concept of parallel extension. Finally, in section 6 we will present some application domains of our extension.

2   Architecture

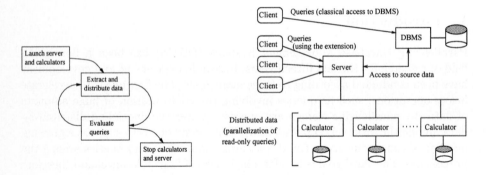

Fig. 1. Extension's basic phases   **Fig. 2.** General overview

2.1   General Overview

The coupled query evaluator works through two successive phases (see fig. 1). First, data is extracted from the pre-existing relational DBMS and distributed among a network of workstations. Second, this distribution is used for the parallel processing of relational queries.

The overall architecture consists of two main components (see fig. 2): the *server* and the *calculators*. The server is the access point. All tasks are submitted to and treated by it. This *server* is connected to several *calculators* which are in charge of storing and processing redistributed data. In our architecture we assume that only one component, i.e. one calculator or the server, is running on each station (we must underline that such a choice does not bring any limitation, for instance on a SMP station, as far as a single calculator can handle several computing threads – see section 3.3).

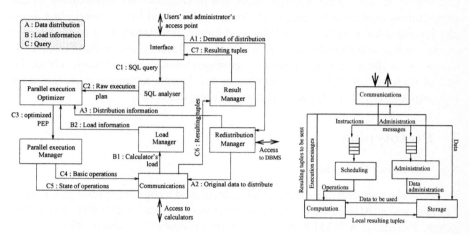

Fig. 3. Server module   **Fig. 4.** Calculator module

2.2   The Server Module

The server module (see fig. 3) consists of eight components in charge of data distribution (circuit A), collection of load information (circuit B) and parallel query execution (circuit C).

Data distribution is done through the *redistribution manager* (A1), which extracts the requested data from the DBMS. Extracted data is sent to the calculators through the *communication* module (A2). Redistribution parameters are then stored by the *parallel execution optimizer* (A3).

Processor load information is regularly sent by each calculator (B1 and B2). Distribution and load information is used by the *parallel execution optimizer* in order to determine the best suited location for each operation.

Query execution is triggered by submitting a SQL query through the *interface*. This query is then translated into an internal format by the *SQL analyzer* (C1). This raw parallel execution plan (PEP) is then improved by the *parallel execution optimizer* (C2). This optimized PEP (C3) consists of basic (elementary) operators connected by flows of data and pre- and post-conditions, e.g. scheduling decisions [9]. The *parallel execution manager* analyses the PEP so that each calculator only receives the operators which take place on it (C4). The *parallel execution manager* receives (C5) processing information (e.g. end of an operator). Resulting tuples are grouped and stored by the *result manager* (C6), and then returned to the user (C7).

2.3   The Calculator Module

The calculator module consists of five components (see fig. 4). The *communication* module is similar to the one of the server module. It allows networking with the server and with all other calculators. Incoming data is transmitted to and stored by the *storage* module. Incoming instructions are transmitted to the *computation* module according to the order determined by the *scheduling* module. Intermediate results that will be used locally are transmitted to the *storage* module, while other results are sent to other calculators (intermediate results) or to the server (final results). Execution messages are also sent to the server at the end of each operator. Finally, calculators can handle administration messages (e.g. suppression of relations, shutdown of the calculator).

3   Prototyping

3.1   General Overview

Based on the architecture above, we have developed a complete prototype, named Enkidu, and written in Java, owing to the robustness and portability of this language. Enkidu is a RAM-based parallel query evaluator, which offers various distribution and execution strategies. It can be used with real data (downloaded from an existing database, or from a storage file) or with self-generated data (according to given skew parameters). Thanks to its Java implementation, Enkidu has already been used under Solaris, Linux and Windows 95.

3.2   Implementation of the Server Module

The server module mainly consists of Java code. Nevertheless, the MPO parallel execution plan optimizer [10], an external component developed in C, is currently being adapted through the Java Native Interface [11]. The server module can simulate concurrent users. This is rather important, as far as the large majority of existing academic PDBMS prototypes do not really care about concurrent queries (though DBMSs are generally supposed to support and optimize the combined load of concurrent users). Data extraction is done by the server, through the jdbc interface. Enkidu first loads the data dictionnary. Then the administrator can distribute data. Extraction is done with a SQL "Select" method, due to the portability and ease of use of this method (see also section 5.1).

3.3   Implementation of the Calculator Module

The calculator module is a pure Java software. The computation module is multithreaded: several computation threads are working on different operators. Their priority is determined according to the precedence of queries and operators. The thread with highest priority runs as long as input data remains available. If no more data is temporarily available (in a pipelined mode), secondary priority threads can start working (i.e. no time is lost waiting). Thread switching is limited by using a coarse grain of treatment: tuples are grouped in packets, and a computation thread can not be interrupted until it finishes its current packet. Thread switching is based on a gentlemen's agreement (i.e. when a packet has been computed, the current thread lets another one start –in its priority level, or on an upper level if exists). This multi-threaded approach offers direct gains (optimized workload), and could also be useful in the context of a SMP machine, as threads could be distributed amongst nodes. With such a hardware architecture, a single calculator module could handle the whole SMP. Storage and I/O management would be managed on a single node, while other nodes would only run one (on some) computation thread(s).

We must also highlight the fact that our calculators have been designed to store data within RAM. Disks remain unused in order to avoid I/O overcosts. While this choice limits the volume of data that can be extracted and distributed, we must notice that the parallel extension is supposed to be an intermediate solution between sequential and PDBMSs. Thus, we can argue that the volume of data should remain reasonable (some GBytes at most).

3.4   Communication Issues

We chose to work at the Java sockets level, owing to their ease of use, and also because existing Java ports of MPI did not offer satisfying performance. The main problem we met did concern serialization. Serialization is a major concept of Java, which consists in automatically transforming objects into byte vectors and vice-versa. Thus, objects can be directly put on a flow (between two processes or between a process and a file). The generic serialization mecanism is powerful, as it also stores the structure of objects within the byte vector, and thus guarantees the file readability across applications. Nevertheless, this structural data is quite heavy, and introduces tremendous overcosts in the context of NOW computing. For this reason we choose to develop a specific light serialization, which only serializes data. This approach is quite similar to the one of [12], and both methods should be compared in a forthcoming paper.

4   Current Performance of the Extension Prototype

4.1   Underlying Hardware

Enkidu is currently tested over the Popc machine. Popc is an integrated network of PCs, which has been developed by *Matra Systems & Information*, as a prototype of their *Peakserver* cluster [13]. It consists of 12 Pentium Pro processors running under Linux, with 64 MByte memory each, connected by both an Ethernet and a Myrinet [14] network. The PopC machine is a computing server, which is classically used as a testbed for low- and high-level components (Myrinet optimization, thread migration, parallel programs...). In the following tests we use the Ethernet network, as it corresponds to the basic standard LAN of an average mid-size company. We are currently studying a Myrinet optimized interface. Simultaneous users are simulated by threads running concurrently on the server. To obtain reliable values, each test has been run at least ten times.

4.2 Speed-up

We realized several speed-up tests over our prototype. The one presented in this paper consists of a single hash join involving two relations (100 and 100 000 tuples). We ran these tests with 1, 5 and 10 simultaneous users. We can see on figure 5 that Enkidu offers the linear speed-ups expected with a hash join. On this figure, speed-ups seem to be "super-linear". This comes both from the structure of the hash-join algorithm and from the fact that networking delays between the server and the calculators are included within our measurements. For this second reason, multi-user tests offer better speed-ups, as networking delays are overlapped by computation.

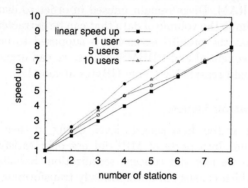

Fig. 5. Speed-up measures

4.3 Real Database Tests

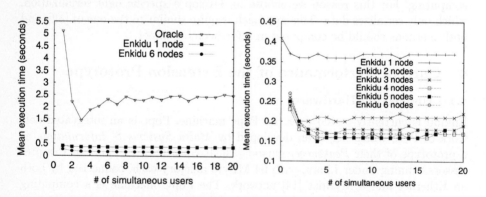

Fig. 6. Enkidu vs. oracle   **Fig. 7.** Details of Enkidu execution time

Basic Database As a good speed-up could hide poor absolute performance, we also compared Enkidu to a real DBMS. The following tests are based on a

Table 1. Comparison of computation time between Oracle 7 and Enkidu

System	# of users	global exec. time (s)	mean exec. time (s)	mean exec. time * # of nodes
Oracle (1 node)	1	350	350	350
Enkidu 6 nodes	1	7.4	7.4	44.1
Enkidu 6 nodes	5	36.3	7.3	44.6
Enkidu 6 nodes	10	72.7	7.3	43.6
Enkidu 6 nodes	15	120.1	8.0	48.0
Enkidu 6 nodes	20	143.1	7.2	42.9

medicine database: the **Claude Bernard Data bank** [15]. The relatively small size of this latter (some MBytes) is counterbalanced by the lack of index (in order to simulate non pre-optimized queries). We ran our tests both on Enkidu and on our source DBMS (Oracle 7 on a Bull Estrella – PowerPC, 64 MByte RAM, AIX). The set of queries consisted in retrieving the name of medicines containing a given chemistry, for 1 to 20 concurrent users. Figures 6 and 7 highlight the good performance of Enkidu. The indicated time consists of the global response time divided by the number of users. In the context of this test, we can notice that, due to the limited size of the database, the observed speed-up is not linear (around 1.7 for 2 machines and 2.3 for 5 machines), as communication and initialization are not negligible compared to computation time.

Extended Database As the first database was quite small, we conducted a similar test with an extended database (10 times bigger for chemistries and medicines, and then 100 times bigger for the links between chemistries and medicines). As our prototype is only using RAM, we could not run this test on a single-node configuration, due to the amount of hashed data and intermediate results (performance would have suffered from the resulting swap). Thus we used a 6 nodes configuration. Concerning the Oracle platform, we only ran one user, due to both the need for very big temporary files, and the resulting swap overcosts. We can see in table 1, that using only RAM allows Enkidu (6 nodes) to compute nearly 50 times faster than Oracle. Considering the ratio computation time / number of nodes, Enkidu remains 8 times faster.

5   Discussion

5.1   Parallel Extension vs. Porting

Providing a parallel extension constitutes a specialized alternative to parallel porting. We can especially notice the differences according to the following axes:

- data storage and access: within the extension, data is loaded from a remote system (the DBMS) and stored in main memory. Within a PDBMS, data is stored on the local disks, from which it is accessed as needed;
- transaction management: the extension does not directly offer transaction management, and updates are limited.

Porting does effectively offer a complete solution, with no data load delays. Updates are automatically and relatively simply managed. Nevertheless, we see several drawbacks inherently linked to parallelization:

- development time: a complete port to parallel architectures is a heavy task, while the extension can be developed in a much faster way, due to its intentionally limited functions;
- persistence of the parallel components: a PDBMS, once initialized, uses a set of disks in a permanent manner. In the context of a network of PCs, this means that a given set of machines is dedicated to the DBMS.

As updates are managed by the DBMS, the extension must regularly update its data. As far as we mostly work with off-line applications, updates can be delayed, as long as their frequency offers a sufficient "freshness" of data (e.g. once a day). We propose to re-extract data rather than using an incremental update (which would need extra development, especially if triggers are used by the sequential DBMS – as tracking updates is then much more difficult). As an example, extracting and distributing the extended database of section 4.3 is done in less than 15 minutes: about 10 minutes for extracting data (from Oracle to the server) and 1.5 minutes for distributing it (from the server to the calculators). Technically speaking, we use a temporary file in order to handle these two phases independantly. Thus, calculators are only locked during distribution.

5.2   Toward a Generalization of the Parallel Extension Concept

The parallel extension concept could fit in a wider perspective of high performance processing. In effect, many similarities exist between our extension and recent developments, for instance in the field of scientific computing, to *extend* sequential applications by parallelizing *some* of their algorithms. Various applications, such as numerical simulation and imaging, could use sequential components during data input and light operations, and could benefit from parallel components during heavy computations. Concerning numerical simulation, parallel computing can be used for heavy computation, such as crash simulation, or fluid mechanics, while designing the structure is usually done in a sequential way. Considering image computing, the input and annotation of pictures should be done sequentially, while image fusion or filtering can benefit from parallel algorithms. In a more general way, a parallel extension could be used whenever a software alternatively executes light and heavy treatments.

6   Context

As our extension appears to be mainly interesting in the context of "read-mostly" applications, we will now give three examples of such applications: decision support systems, data mining and multimedia databases.

Within decision support systems (DSS), data are frequently manipulated off-line(i.e. data generation and manipulation are two distinct and independant tasks). Thus, an extension can be used. As an example we will cite the well known TPC-R and TPC-H benchmarks. TPC-R [16] is a business reporting benchmark, adapted to repetitive queries, concerning a large amount (1 TByte) of data, and thus oriented toward the (static) optimization of data structures and queries. TPC-H [17] works on ad-hoc queries, i.e. various and theoretically non-predictable queries. It can thus be used with various dataset size (from 1 GByte to 1 TByte). Our system can *a priori* be used in both cases, and at least with TPC-H, as small amounts of data can be manipulated. Although our current implementation is not adapted to large datasets (data is stored in RAM), it could anyway work on databases ranging from 1 to 10 GByte, by using a cluster of PCs handling enough memory (e.g. 6 to 10 PCs, each having 256 MByte memory, could easily handle a 1 GByte database). Of course, we could also implement some existing algorithms using disks to store temporary data, such as the well known hybrid hash-join algorithm [18].

Concerning data mining, our extension could at least be used during pre-processing phases, in order to provide a fast and repetitive access to source data. It could even be used during processing, as far as [19] showed that knowledge extraction could also be done through a classical DBMS using SQL.

Concerning multimedia databases, both academic researchers [20] and industrial software developers [21, 22] are deeply implicated in delivering multimedia DBMSs. The read-mostly nature of such databases is trivial. For instance, the Medical Knowledge Bank project [23], and especially its initial and continuing medical education section, mainly involves read-only accesses.

7 Summary

In this paper we proposed and discussed the use of a parallel extension in the context of Database Management Systems and we presented the prototype we built according to our proposal. Through our tests in appeared, that this extension is a valuable alternative to the classical parallel porting of DBMSs, especially in the context of read-mostly applications.

Future work should follow two mains goals: getting even better performance and developing specifically adapted algorithms. From the performance point of view, we plan to develop high performance (Myrinet-based) Java components. We also wish to upgrade our packet-based techniques toward a real macro-pipelining approach. Finally, we are trying to get faster extraction and distribution algorithms. From the applications point of view, we are currently studying some multimedia and information retrieval algorithms working over our architecture. Another important direction of research, from our point of view, consists in testing the concept of parallel extension in various fields, and to propose a global and generic definition for it.

References

1. D. Dewit, S. Ghandeharizadeh, D. Schneider, *et al.*, "The Gamma Database Machine Project," *IEEE TKDE*, vol. 2, pp. 44–62, Mar. 1990.

2. J. Page, "A Study of a Parallel Database Machine and its Performance the NCR/Teradata DBC/1012," in *Proceedings of the 10th BNCOD Conference*, (Aberdeen, Scotland), pp. 115–137, July 1992.

3. B. Gerber, "Informix On Line XPS," in *Proceedings of ACM SIGMOD '95*, vol. 24 of *SIGMOD Records*, (San Jose, Ca, USA), p. 463, May 1995.

4. C. Baru, G. Fecteau, A. Goya, *et al.*, "DB2 Parallel Edition," *IBM Systems Journal*, vol. 34, no. 2, pp. 292–322, 1995.

5. R. Bamford, D. Butler, B. Klots, *et al.*, "Architecture of Oracle Parallel Server," in *Proceedings of VLDB '98*, (New York City, NY, USA), pp. 669–670, Aug. 1998.

6. Oracle, "Oracle Parallel Server: Solutions for Mission Critical Computing," tech. rep., Oracle Corp., Redwood Shores, CA, Feb. 1999.

7. G. Bozas, M. Jaedicke, A. Listl, *et al.*, "On transforming a sequential sql-dbms into a parallel one : First results and experiences of the MIDAS project," in *EuroPar'96*, (Lyon), pp. 881–886, Aug. 1996.

8. T. Tamura, M. Oguchi, and M. Kitsuregawa, "Parallel Database Processing on a 100 Node PC Cluster: Cases for Decision Support Query Processing and Data Mining," in *SC'97*, 1997.

9. L. Brunie and H. Kosch, "Optimizing complex decision support queries for parallel execution," in *PDPTA '97*, (Las Vegas, AZ, USA), July 1997.

10. L. Brunie and H. Kosch, "ModParOpt : a modular query optimizer for multi-query parallel databases," in *ADBIS'97*, (St Petersbourg, RU), 1997.

11. S. Liang, *The Java Native Interface: Programmer's Guide and Specification*. Java Series, Addison Wesley, June 1999.

12. M. Philippsen and B. Haumacher, "More Efficient Object Serialization," in *International Workshop on Java for Parallel and Distributed Computing*, (San Juan, Porto Rico, USA), Apr. 1999.

13. MatraSI, "Peakserver, the Information Server." [On-Line], Available on Internet :<http://www.matra-msi.com/ang/savoir_serv_d.htm>, 1999.

14. N. Boden, D. Cohen, R. Felderman, *et al.*, "Myrinet - a gigabit-per-second local-area network," *IEEE-Micro*, vol. 15, pp. 29–36, 1995.

15. A. Flory, C. Paultre, and C. Veilleraud, "A relational databank to aid in the dispensing of medicines," in *MEDINFO '83*, (Amsterdam), pp. 152–155, 1983.

16. TPC, *TPC Benchmark R (Decision Support) Standard Specification*. San Jose, CA: Transaction Processing Performance Council, Feb. 1999.

17. TPC, *TPC Benchmark H (Decision Support) Standard Specification*. San Jose, CA: Transaction Processing Performance Council, June 1999.

18. D. Schneider and D. DeWitt, "A Performance Evaluation of Four Parallel Join Algorithms in a Shared-Nothing Multiprocessor Environment," in *Proceedings of ACM SIGMOD '89*, (Portland, Oregon, USA), pp. 110–121, June 1989.

19. I. Pramudiono, T. Shintani, T. Tamura, *et al.*, "Mining Generalized Association Rule Using Parallel RDB Engine on PC Cluster," in *DaWak'99*, (Florence, Italy), pp. 281–292, Sept. 1999.

20. H. Ishikawa, K. Kubota, Y. Noguchi, *et al.*, "Document Warehousing Based on a Multimedia Database System," in *ICDE'99*, (Sydney, Australia), pp. 168–173, Mar. 1999.

21. Oracle, "Oracle Intermedia: Managing Multimedia Content," tech. rep., Oracle Corp., Redwood Shores, CA, Feb. 1999.

22. Informix, "Informix Media 360," tech. rep., Informix, Menlo Park, CA, Aug. 1999.

23. W. Sterling, "The Medical Knowledge Bank: A Multimedia Database Application," *NCR Technical Journal*, Aug. 1993.

Workshop on Advances
in Parallel and Distributed Computational Models

In recent years, new parallel and distributed computational models have been proposed in the literature, reflecting advances in new computational devices and environments such as optical interconnects, FPGA devices, networks of workstations, radio communications, DNA computing, quantum computing, etc. New algorithmic techniques and paradigms have been recently developed for these new models.

The main goal of this workshop is to provide a timely forum for the dissemination and exchange of new ideas, techniques and research in the field of the new parallel and distributed computational models. The workshop will bring together researchers and practitioners interested in all aspects of parallel and distributed computing taken in an inclusive, rather than exclusive, sense.

Workshop Chair:
Oscar H. Ibarra (University of California Santa Barbara)
Program Co-Chairs:
Koji Nakano (Nagoya Institute of Technology), Stephan Olariu (Old Dominion University)
Steering Committee
Narsingh Deo (University of Central Florida, USA), Joseph JáJá (University of Maryland, USA), Ernst W. Mayr (Technical University Munich, Germany), Lionel Ni (Michigan State University, USA), Sartaj Sahni (University of Florida, USA), Behrooz Shirazi (University of Texas, USA), Peter Widmayer (ETH, Zurich, Switzerland)
Program Committee
Jik Hyun Chang (Sogang University, Korea), Chuzo Iwamoto (Hiroshima University, Japan), Omer Egecioglu (University of California, USA), Hossam ElGindy (University of New South Wales, Australia), Akihiro Fujiwara (Kyushu Institute of Technology, Japan), Ju-wook Jang (Sogang University, Korea), Rong Lin (SUNY Geneseo, USA), Toshimitsu Masuzawa (Nara Institute of Science and Technology, Japan), Rami Melhem (University of Pittsburgh, USA), Eiji Miyano (Kyushu Institute of Design, Japan), Michael Palis (Rutgers University, USA), Sanguthevar Rajasekaran (University of Florida, USA), Nicola Santoro (Carleton University, Canada), James Schwing (Central Washington University, USA), Hong Shen (Griffith University, Australia), Ivan Stojmenovic (University of Ottawa, Canada), Jerry L. Trahan (Louisiana State University, USA), Ramachandran Vaidyanathan (Louisiana State University, USA), Biing-Feng Wang (National Tsinhua University, Taiwan), Jie Wu (Florida Atlantic University, USA), Masafumi Yamashita (Kyushu University, Japan), Tao Yang (University of California, USA), Si Qing Zheng (University of Texas at Dallas, USA), Albert Y. Zomaya (University of Western Australia, Australia)

J. Rolim et al. (Eds.): IPDPS 2000 Workshops, LNCS 1800, pp. 101-101, 2000.
© Springer-Verlag Berlin Heidelberg 2000

The Heterogeneous Bulk Synchronous Parallel Model

Tiffani L. Williams and Rebecca J. Parsons

School of Computer Science
University of Central Florida
Orlando, FL 32816-2362
{williams,rebecca}@cs.ucf.edu

Abstract. Trends in parallel computing indicate that heterogeneous parallel computing will be one of the most widespread platforms for computation-intensive applications. A heterogeneous computing environment offers considerably more computational power at a lower cost than a parallel computer. We propose the Heterogeneous Bulk Synchronous Parallel (HBSP) model, which is based on the BSP model of parallel computation, as a framework for developing applications for heterogeneous parallel environments. HBSP enhances the applicability of the BSP model by incorporating parameters that reflect the relative speeds of the heterogeneous computing components. Moreover, we demonstrate the utility of the model by developing parallel algorithms for heterogeneous systems.

1   Introduction

Parallel computers have made an impact on the performance of large-scale scientific and engineering applications such as weather forecasting, earthquake prediction, and seismic data analysis. However, special-purpose massively parallel machines have proven to be expensive to build, difficult to use, and have lagged in performance by taking insufficient advantage of improving technologies. Heterogeneous computing [8, 14] is a cost-effective approach that avoids these disadvantages. A heterogeneous computing environment can represent a diverse suite of architecture types such as Pentium PCs, shared-memory multiprocessors, and high-performance workstations. Unlike parallel computing, such an approach will leverage technologies that have demonstrated sustained success, including: computer networks; microprocessor technology; and shared-memory platforms.

We propose a framework for the development of parallel applications for heterogeneous platforms. Our model is called Heterogeneous Bulk Synchronous Parallel (HBSP), which is an extension to the BSP model of parallel computation [17]. BSP provides guidance on designing applications for good performance on homogeneous parallel machines. Experiments [5] indicate that the model also accurately predicts parallel program performance on a wide range of parallel

J. Rolim et al. (Eds.): IPDPS 2000 Workshops, LNCS 1800, pp. 102-108, 2000.

machines. HBSP enhances the applicability of the BSP model by incorporating parameters that reflect the relative speeds of the heterogeneous computing components.

Our starting point for the development of algorithms for HBSP are efficient BSP or HCGM [10, 11] applications. Specifically, we develop three HBSP algorithms—prefix sums, matrix multiplication, and randomized sample sort—that distribute the computational load according to processor speed without sacrificing performance. In fact, the cost model indicates that wall clock performance is increased in many cases. Furthermore, these algorithms can execute unchanged on both heterogeneous and homogeneous platforms.

The rest of the paper proceeds as follows. Section 2 reviews related work. Section 3 describes the HBSP model. Section 4 presents a sampling of algorithms for HBSP. Concluding remarks and future directions are given in Section 5.

2   Related Work

The theoretical foundations of the BSP model were presented in a series of papers by Valiant [15, 16, 17, 18, 19], which describe the model, how BSP computers can be programmed either in direct mode or in automatic mode (PRAM simulations), and how to construct efficient BSP computers. Other work presents theoretical results, empirical results, or experimental parameterization of BSP programs [1, 2, 3, 4, 5, 21]. Many alternative models of parallel computation have been proposed in the literature—a good survey on this topic are papers by Maggs, Matheson, and Tarjan [9] and Skillicorn and Talia [13].

Several models exist to support heterogeneous parallel computation. However, they are either primarily of theoretical interest or are basically languages/runtime systems without a solid theoretical foundation. For an overview of these approaches, we refer the reader to the surveys by Siegel *et al.* [12] and Weems *et al.* [20]. One notable exception is the the Heterogeneous Coarse-Grained Multicomputer (HCGM) model, developed by Morin [10, 11]. HBSP and HCGM are similar in structure and philosophy. The main difference is that HGCM is not intended to be an accurate predictor of execution times whereas HBSP attempts to provide the developer with predictable algorithmic performance.

3   Heterogeneous BSP

The Heterogeneous Bulk Synchronous Parallel (HBSP) model is a generalization of the BSP model [17] of parallel computation. The BSP model is a useful guide for parallel system development. However, it is inappropriate for heterogeneous parallel systems since it assumes all components have equal computation and communication abilities. The goal of HBSP is to provide a framework that makes parallel computing a viable option for heterogeneous systems. HBSP enhances the applicability of BSP by incorporating parameters that reflect the relative speeds of the heterogeneous computing components.

An HBSP computer is characterized by the following parameters:

- the number of processor-memory components p labeled $P_0, ..., P_{p-1}$;
- the *gap* g_j for $j \in [0..p-1]$, a bandwidth indicator that reflects the speed with which processor j can inject packets into the network;
- the *latency* L, which is the minimum duration of a superstep, and which reflects the latency to send a packet through the network as well as the overhead to perform a barrier synchronization;
- processor parameters c_j for $j \in [0..p-1]$, which indicates the speed of processor j relative to the *slowest* processor, and
- the total speed of the heterogeneous configuration $c = \sum_{i=0}^{p-1} c_i$.

For notational convenience, $P_f(P_s)$ represents the fastest (slowest) processor. The communication time and the computation speed of the fastest (slowest) processor are $g_f(g_s)$ and $c_f(c_s)$, respectively. We assume that c_s is normalized to 1. If $c_i = j$, then P_i is j times faster than P_s.

Computation consists of a sequence of supersteps. During a superstep, each processor performs asynchronously some combination of local computation, message transmissions, and message arrivals. A message sent in one superstep is guaranteed to be available to the destination processor at the beginning of the next superstep. Each superstep is followed by a global synchronization of all the processors.

Execution time of an HBSP computer is as follows. Each processor, P_j, can perform $w_{i,j}$ units of work in $\frac{w_{i,j}}{c_j}$ time units during superstep i. Let $w_i = \max(\frac{w_{i,j}}{c_j})$ represent the largest amount of local computation performed by any processor during superstep i. Let $h_{i,j}$ be the largest number of packets sent or received by processor j in superstep i. Thus, the execution time of superstep i is:

$$w_i + \max_{j \in [0..p-1]} \{g_j \cdot h_{i,j}\} + L \tag{1}$$

The overall execution time is the sum of the superstep execution times.

The HBSP model leverages existing BSP research. The more complex cost model does not change the basic programming methodology, which relies on the superstep concept. Furthermore, when $c_j = 1$ and $g_j = g_k$, where $0 \leq j, k < p$, HBSP is equivalent to BSP.

4   HBSP Algorithms

This section provides a sampling of applications for the HBSP model based on those proposed by Morin for the HCGM model [10, 11]. Our algorithms, which include prefix sums, matrix multiplication, and randomized sample sort, illustrate the power and elegance of the HBSP model. In each of the applications, the input size is partitioned according to a processor's speed. If c_i is the speed of processor P_i, then P_i holds $\frac{c_i}{c}n$ input elements. When discussing the performance of the algorithms, we will often make use of a coarse-grained assumption, $p \ll n$, i.e., the size of the problem is significantly larger than the number of processors. Our interpretation of "significantly larger" is $p \leq \frac{n}{p}$.

4.1   Prefix Sums

Given a sequence of n numbers $\{x_0, x_1, ..., x_{n-1}\}$, it is required to compute their prefix sums $s_j = x_0 + x_1 + ... + x_j$, for all j, $0 \leq j \leq n-1$. Under HBSP, each processor locally computes its prefix sums and sends the total sum to P_f. Next, P_f computes the prefix sums of this sequence and sends the $(i-1)$st element of the prefix to P_i. Lastly, P_i adds this value to each element of the prefix sums computed in the first step to obtain the prefix sums of the overall result. The prefix sums algorithm is shown below.

1. Each processor locally computes the prefix sums of its $\frac{c_i}{c}n$ input elements.
2. Each processor, P_i, sends the total sum of its input elements to P_f.
3. P_f computes the prefix sums of the p elements received in Step 2.
4. For $1 \leq i \leq p-1$, P_f sends the $(i-1)$st element computed in Step 3 to P_i.
5. Each processor computes its final portion of the prefix sums by adding the value received in Step 4 to each of the values computed in Step 1.

Analysis. In Step 1 and Step 5, each processor P_i does $O(\frac{c_i}{c}n)$ work and this can be done in $O(\frac{n}{c})$ time. Steps 2 and 4 require a communication time of $\max\{g_s \cdot 1, g_f \cdot p\}$. Step 3 takes $O(\frac{p}{c_f})$ computation time. Since $c_f \geq \frac{c}{p}$ and $p \leq \frac{n}{p}$, $O(\frac{p}{c_f}) \leq O(\frac{n}{c})$. Thus, the algorithm takes time

$$O(\frac{n}{c}) + 2 \cdot \max\{g_s \cdot 1, g_f \cdot p\} + 3L. \qquad (2)$$

If $g_s \leq pg_f$, the communication time is $2pg_f$, otherwise it's $2g_s$.

4.2   Matrix Multiplication

Matrix multiplication is perhaps one of the most common operations used in large-scale scientific computing. Given two $n \times n$ matrices A and B, we define the matrix $C = A \times B$ as $C_{i,j} = \sum_{k=0}^{n-1} A_{i,k} \times B_{k,j}$. We assume that matrix A is partitioned among the processors so that each processor, P_i, holds $\frac{c_i}{c}n$ rows of A and $\frac{n}{p}$ columns of B. At the completion of the computation, P_i will hold $\frac{c_i}{c}n$ rows of C. We denote the parts of A, B, and C held by P_i as A_i, B_i, and C_i, respectively. The matrix multiplication algorithm consists of circulating the columns of B among the processors. When P_i receives column j of B, it can compute column j of C_i. Once P_i has seen all columns of B, it will have computed all of C_i. The matrix multiplication algorithm is given below.

1. repeat p times.
2. P_i computes $C_i = A_i \times B_i$.
3. P_i sends B_i to $P_{(i+1) \bmod p}$.
4. end repeat

Analysis Step 3 requires P_i to perform $O(\frac{c_i}{c}n \cdot \frac{n}{p} \cdot n) = O(\frac{n^3 c_i}{cp})$ amount of work. Over p rounds, the total computation time is $O(\frac{n^3}{c})$. During Step 4, each processor sends and receives $\frac{n}{p}$ columns of matrix B. Therefore, the total time of HBSP matrix multiplication is

$$O(\frac{n^3}{c}) + g_s n^2 + pL. \tag{3}$$

4.3 Randomized Sample Sort

One approach for parallel sorting that is suitable for heterogeneous computing is randomized sample sort. It is based on the selection of a set of $p-1$ "splitters" from a set of input keys. In particular, we seek splitters that will divide the input keys into approximately equal-sized buckets. The standard approach is to randomly select pr sample keys from the input set, where r is called the oversampling ratio. The keys are sorted and the keys with ranks $r, 2r, ..., (p-1)r$ are selected as splitters. By choosing a large enough oversampling ratio, it can be shown with high probability that no bucket will contain many more keys than the average [7]. Once processors gain knowledge of the splitters, their keys are partitioned into the appropriate bucket. Afterwards, processor i locally sorts all the keys in bucket i.

When adapting this algorithm to the HBSP model, we change the way in which the splitters are chosen. To balance the work according to the processor speeds $c_0, ..., c_{p-1}$, it is necessary that $O(\frac{c_i}{c}n)$ keys fall between s_i and s_{i+1}. This leads to the following algorithm.

1. Each processor randomly selects a set of r sample keys from its $\frac{c_i}{c}n$ input keys.
2. Each processor, P_i, sends its sample keys to P_f.
3. P_f sorts the pr sample keys. Denote these keys by $sample_0, ..., sample_{pr-1}$ where $sample_i$ is the sample key with rank i in the sorted order. P_f defines $p-1$ splitters, $s_0, ..., s_{p-2}$, where $s_i = sample_{\lceil (\sum_{j=0}^{i} \frac{c_j}{c})pr \rceil}$.
4. P_f broadcasts the $p-1$ splitters to each of the processors.
5. All keys assigned to the ith bucket are sent to the ith processor.
6. Each processor sorts its bucket.

Analysis In Step 1, each processor performs $O(r) \leq O(n)$ amount of work. This requires $O(\frac{n}{c_s})$ time. Step 2 requires a communication time of $\max\{g_s \cdot r, g_f \cdot pr\}$. To sort the pr sample keys, P_f does $O(pr \lg pr) \leq O(n \lg n)$ amount of work. This can be done in $O(\frac{n}{c_f} \lg n)$ time. Broadcasting the $p-1$ splitters requires $\max\{g_s \cdot (p-1), g_f \cdot p(p-1)\}$ communication time. Since each processor is expected to receive approximately $\frac{c_i}{c}n$ keys [11], Step 5 uses $O(\frac{n}{c})$ computation time and $\max\{g_i \cdot \frac{c_i}{c}n\}$ communication time, where $i \in [0..p-1]$. Once each processor receives their keys, sorting them requires $O(\frac{n}{c} \lg n)$ time. Thus, the total time is

$$O\left(\frac{n}{c_f} \lg n\right) + X(r + (p-1)) + \max_{i \in [0..p-1]} \left\{g_i \frac{c_i}{c} n\right\} + 4L, \text{where} \qquad (4)$$

$$X = \begin{cases} pg_f \text{ if } g_s \leq pg_f \\ \\ g_s \quad \text{otherwise.} \end{cases}$$

5 Conclusions and Future Directions

The HBSP model provides a framework for the development of parallel applications for heterogeneous platforms. HBSP enhances the applicability of BSP by incorporating parameters that reflect the relative speeds of the heterogeneous computing components. Although the HBSP model is somewhat more complex than BSP, it captures the most important aspects of heterogeneous systems. Existing BSP and HCGM algorithms provide the foundation for the HBSP algorithms presented here. These algorithms suggest that improved performance under HBSP results from utilizing the processor speeds of the underlying system. However, experimental evidence is needed to corroborate this claim.

We plan to extend this work in several directions. First, a library based on BSP*lib* (a small, standardized library of BSP functions) [6] will provide the foundation for HBSP programming. Experiments will be conducted to test the effectiveness of the model on a network of heterogeneous workstations. These experiments will test the predictability, scalability, and efficiency of applications written under HBSP. Currently, the HBSP model only addresses a heterogeneous collection of uniprocessor machines. We are investigating variants to the model to address multiprocessor systems.

In conclusion, the goal of HBSP is to offer a framework that makes parallel computing a viable option for a wide range of tasks. We seek to demonstrate that it can provide a simple programming approach, portable and efficient application code, predictable execution, and scalable performance.

References

[1] R. H. Bisseling. Sparse matrix computations on bulk synchronous parallel computers. In *Proceedings of the International Conference on Industrial and Applied Mathematics*, Hamburg, July 1995.

[2] R. H. Bisseling and W. F. McColl. Scientific computing on bulk synchronous parallel architectures. In B. Pehrson and I. Simon, editors, *Proceedings of the 13th IFIP World Computer Congress*, volume 1, pages 509–514. Elsevier, 1994.

[3] A. V. Gerbessiotis and C. J. Siniolakis. Deterministic sorting and randomized mean finding on the BSP model. In *Eighth Annual ACM Symposium on Parallel Algorithms and Architectures*, pages 223–232, June 1996.

[4] A. V. Gerbessiotis and L. G. Valiant. Direct bulk-synchronous parallel algorithms. *Journal of Parallel and Distributed Computing*, 22(2):251–267, August 1994.

[5] M. W. Goudreau, K. Lang, S. Rao, T. Suel, and T. Tsantilas. Towards efficiency and portability: Programming with the BSP model. In *Eighth Annual ACM Symposium on Parallel Algorithms and Architectures*, pages 1–12, June 1996.

[6] J. M. D. Hill, B. McColl, D. C. Stefanescu, M. W. Goudreau, K. Lang, S. B. Rao, T. Suel, T. Tsantilas, and R. Bisseling. BSPlib: The BSP programming library. *Parallel Computing*, 24(14):1947–1980, 1998.

[7] J. Huang and Y. Chow. Parallel sorting and data partitioning by sampling. In *IEEE Computer Society's Seventh International Computer Software & Applications Conference (COMPSAC'83)*, pages 627–631, November 1983.

[8] A. Khokhar, V. Prasanna, M. Shaaban, and C. Wang. Heterogeneous computing: Challenges and opportunities. *Computer*, 26(6):18–27, June 1993.

[9] B. M. Maggs, L. R. Matheson, and R. E. Tarjan. Models of parallel computation: A survey and synthesis. In *Proceedings of the 28th Hawaii International Conference on System Sciences*, volume 2, pages 61–70. IEEE Press, January 1995.

[10] P. Morin. Coarse-grained parallel computing on heterogeneous systems. In *Proceedings of the 1998 ACM Symposium on Applied Computing*, pages 629–634, 1998.

[11] P. Morin. Two topics in applied algorithmics. Master's thesis, Carleton University, 1998.

[12] H. J. Siegel, H. G. Dietz, and J. K. Antonio. Software support for heterogeneous computing. In A. B. Tucker, editor, *The Computer Science and Engineering Handbook*, pages 1886—1909. CRC Press, 1997.

[13] D. B. Skillicorn and D. Talia. Models and languages for parallel computation. *ACM Computing Surveys*, 30(2):123–169, June 1998.

[14] L. Smarr and C. E. Catlett. Metacomputing. *Communications of the ACM*, 35(6):45–52, June 1992.

[15] L. G. Valiant. Optimally universal parallel computers. *Philosophical Transactions of the Royal Society of London*, A 326:373–376, 1988.

[16] L. G. Valiant. Bulk-synchronous parallel computers. In M. Reeve and S. E. Zenith, editors, *Parallel Processing and Artificial Intelligence*, pages 15–22. John Wiley & Sons, Chichester, 1989.

[17] L. G. Valiant. A bridging model for parallel computation. *Communications of the ACM*, 33(8):103–111, 1990.

[18] L. G. Valiant. General purpose parallel architectures. In J. van Leeuwen, editor, *Handbook of Theoretical Computer Science*, volume A: Algorithms and Complexity, chapter 18, pages 943–971. MIT Press, Cambridge, MA, 1990.

[19] L. G. Valiant. Why BSP computers? In *Proceedings of the 7th International Parallel Processing Symposium*, pages 2–5. IEEE Press, April 1993.

[20] C. C. Weems, G. E. Weaver, and S. G. Dropsho. Linguistic support for heterogeneous parallel processing: A survey and an approach. In *Proceedings of the Heterogeneous Computing Workshop*, pages 81–88, 1994.

[21] T. L. Williams and M. W. Goudreau. An experimental evaluation of BSP sorting algorithms. In *Proceedings of the 10th IASTED International Conference on Parallel and Distributed Computing Systems*, pages 115–118, October 1998.

On stalling in LogP[⋆]

(Extended Abstract)

Gianfranco Bilardi[1,2], Kieran T. Herley[3], Andrea Pietracaprina[1], and Geppino Pucci[1]

[1] Dipartimento di Elettronica e Informatica, Università di Padova, Padova, Italy.
{bilardi,andrea,geppo}@artemide.dei.unipd.it
[2] T.J. Watson Research Center, IBM, Yorktown Heights, NY 10598, USA.
[3] Department of Computer Science, University College Cork, Cork, Ireland.
k.herley@cs.ucc.ie

Abstract. We investigate the issue of stalling in the LogP model. In particular, we introduce a novel quantitative characterization of stalling, referred to as *δ-stalling*, which intuitively captures the realistic assumption that once the network's capacity constraint is violated, it takes some time (at most δ) for this information to propagate to the processors involved. We prove a lower bound that shows that LogP under δ-stalling is strictly more powerful than the stall-free version of the model where only strictly stall-free computations are permitted. On the other hand, we show that δ-stalling LogP with $\delta = L$ can be simulated with at most logarithmic slowdown by a BSP machine with similar bandwidth and latency values, thus extending the equivalence (up to logarithmic factors) between stall-free LogP and BSP argued in [1] to the more powerful L-stalling LogP.

1 Introduction

Over the last decade considerable attention has been devoted to the formulation of a suitable computational model that supports the development of efficient and portable parallel software. The widely-studied BSP [6] and LogP [2] models were conceived to provide a convenient framework for the design of algorithms, coupled with a simple yet accurate cost model, to allow algorithms to be ported across a wide range of machine architectures with good performance. Both models view a parallel computer as a set of p processors with local memory that exchange messages through a communication medium whose performance is essentially characterized by two key parameters: *bandwidth* (g for BSP and G for LogP) and *latency* (ℓ for BSP and L for LogP).

A distinctive feature of LogP is that it embodies a *network capacity constraint* stipulating that at any time the total number of messages in transit towards any specific destination should not exceed the threshold $\lceil L/G \rceil$. If this constraint is respected, then every message is guaranteed to arrive within L steps of its submission time. If, however, a processor attempts to submit a message with destination d whose injection into the network would violate the constraint, then the processor is forced to stall until the delivery of some outstanding messages brings the traffic for d below the $\lceil L/G \rceil$ threshold. It seems clear that the intention of the original LogP proposal [2] was strongly to

---

[⋆] This research was supported, in part, by the Italian CNR, and by MURST under Project *Algorithms for Large Data Sets: Science and Engineering*.

J. Rolim et al. (Eds.): IPDPS 2000 Workshops, LNCS 1800, pp. 109-115, 2000.
© Springer-Verlag Berlin Heidelberg 2000

encourage the development of stall-free programs. Indeed, the delays incurred in the presence of stalling were not formally quantified within the model, making the performance of stalling programs an issue difficult to assess with any precision. At the same time, adhering strictly to the stall-free mode might make algorithm design artificially complex, e.g., in situations involving randomization where stalling is unlikely but not impossible. Hence, ruling out stalling altogether might not be desirable.

The relation between BSP and LogP has been investigated in [1], where it is shown that the two models can simulate one another efficiently, under the reasonable assumption that both exhibit comparable values for their respective bandwidth and latency parameters. These results were obtained under a precise specification of stalling behaviour, that attempted to be faithful to the original formulation of the model. Interestingly, however, while the simulation of stall-free LogP programs on the BSP machine can be accomplished with constant slowdown, the simulation of stalling programs incurs a higher slowdown. This difference appears also in subsequent results of [5], where work-preserving simulations are considered. Should stalling programs turn out inherently to require a larger slowdown, it would be an indication that stalling adds power to the LogP model, in contrast with the objective of discouraging its use.

The definition of stalling proposed in [1] states that at each step the network accepts submitted messages up to the capacity threshold for each destination, forcing a processor to stall *immediately* upon submitting a message that exceeds the network capacity, and subsequently awakening the processor *immediately* when its message can be injected without violating the capacity constraint. Although consistent with the informal descriptions given in [2], the above definition of stalling implies the somewhat unrealistic assumption that the network is able to detect and react to the occurrence of a capacity constraint violation *instantaneously*. More realistically, some time lag is necessary between the submission of a message and the onset of stalling, to allow information to propagate through the network.

In this paper we delve further into the issue of stalling in LogP along the following directions:

- We generalize the definition of stalling, by introducing the notion of δ-*stalling*. Intuitively, δ captures the time lag between the submission of a message by a processor which violates the capacity constraint, and the time that the processor "realizes" that it must stall. (A similar time lag affects the "unstalling" process.) The extreme case of $\delta = 1$ essentially corresponds to the stalling interpretation given in [1]. While remaining close to the spirit of the original LogP , δ-stalling LogP has the potential of reflecting more closely the behaviour of actual platforms, without introducing further complications in the design and analysis of algorithms.

- We prove that allowing for stalling in a LogP program enhances the computational power of the model. In particular, we prove a lower bound which separates δ-stalling LogP from stall-free LogP computations by a non-constant factor.

- We devise an algorithm to simulate δ-stalling LogP programs in BSP, which achieves at most logarithmic slowdown under the realistic assumption $\delta = L$. This result, combined with those in [1], extends the equivalence (up to logarithmic factors) between LogP and BSP to L-stalling computations.

The rest of the paper is organized as follows. In Section 2 the definitions of BSP and LogP are reviewed and the new δ-stalling rule is introduced. In Section 3 a lower bound is shown that separates δ-stalling LogP from stall-free LogP computations. In Section 4 the simulation of δ-stalling LogP in BSP is presented.

2   The models

Both the BSP [6] and the LogP [2] models can be defined in terms of a virtual machine consisting of p serial processors with unique identifiers. Each processor i, $0 \leq i < p$, has direct and exclusive access to a private memory and has a local clock. All clocks run at the same speed. The processors interact through a communication medium, typically a network, which supports the routing of messages. In the case of BSP, the communication medium also supports global barrier synchronization. The distinctive features of the two models are discussed below. In the rest of this section we will use P_i^{B} and P_i^{L} to denote, respectively, the i-th BSP processor and the i-th LogP processor, with $0 \leq i < p$.

BSP   A BSP machine operates by performing a sequence of *supersteps*, where in a superstep each processor may perform local operations, send messages to other processors and read messages previously delivered by the network. The superstep is concluded by a barrier synchronization which informs the processors that all local computations are completed and that every message sent during the superstep has reached its intended destination. The model prescribes that the next superstep may commence only after completion of the previous barrier synchronization, and that the messages generated and transmitted during a superstep are available at the destinations only at the start of the next superstep. The performance of the network is captured by a *bandwidth* parameter g and a *latency* parameter ℓ. The running time of a superstep is expressed in terms g and ℓ as $T_{superstep} = w + gh + \ell$, where w is the maximum number of local operations performed by any processor and h the maximum number of messages sent or received by any processor during the superstep. The overall time of a BSP computation is simply the sum of the times of its constituent supersteps.

LogP   In a LogP machine, at each time step, a processor can be either *operational* or *stalling*. If it is operational, then it can perform one of the following types of operations: execute an operation on locally held data (*compute*); submit a message to the network destined to another processor (*submit*); receive a message previously delivered by the network (*receive*). A LogP program specifies the sequence of operations to be performed by each processor.

As in BSP, the behaviour of the network is modeled by a bandwidth parameter G (called *gap* in [2]) and a *latency* parameter L with the following meaning. At least G time steps must elapse between consecutive submit or receive operations performed by the same processor. If, at the time that a message is submitted, the total number of messages in transit (i.e., submitted to the network but not yet delivered) for that destination is at most $\lceil L/G \rceil$, then the message is guaranteed to be delivered within L steps. If, however, the number of messages in transit exceeds $\lceil L/G \rceil$, then, due to congestion, the message may take longer to reach its destination, and the submitting processor may *stall* for some time before continuing its operations. The quantity $\lceil L/G \rceil$ is referred to as the network's *capacity constraint*. Note that message delays are unpredictable, hence different executions of a LogP program are possible. If no stalling occurs, then every message arrives in at most L time steps after its submission.

Upon arrival, a message is promptly removed from the network and buffered in some input buffer associated with the receiving processor. However, the actual acquisition of the incoming message by the processor, through a receive operation, may occur at a later time. LogP also introduces an *overhead* parameter o to represent both the time required to prepare a message for submission and the time required to unpack the message after it has been received. Throughout the paper we will assume that $\max\{2, o\} \leq G \leq L \leq p$. The reader is referred to [1] for a justification of this assumption.

2.1 LogP's stalling behaviour

The original definition of the LogP model in [2] provides only a qualitative description of the stalling behaviour and does not specify precisely how the performance of a program is affected by stalling. In [1], the following rigorous characterization of stalling was proposed. At each step the network accepts messages up to saturation, for each destination, of the capacity limit, possibly blocking the messages exceeding such a limit at the senders. From a processor's perspective, the attempt to submit a message violating the capacity constraint results in immediate stalling, and the stalling lasts until the message can be accepted by the network without capacity violation.

The above characterization of stalling, although consistent with the intentions of the model's proposers, relies on the somewhat unrealistic assumption that the network is able to monitor at each step the number of messages in transit for each destination, blocking (resp., unblocking) a processor *instantaneously* in case a capacity constraint violation is detected (resp., ends). In reality, the stall/unstall information would require some time to propagate through the network and reach the intended processors. Below we propose an alternative, yet rigorous, definition of stalling, which respects the spirit of LogP while modelling the behaviour of real machines more accurately.

Let $1 \leq \delta \leq L$ be an integral parameter. Suppose that at time step t processor P_i^L submits a message m destined to P_j^L, and let $c_j(t)$ denote the total number of messages destined to P_j^L which have been submitted up to (and including) step t and are still in transit at the beginning of this step. If $c_j(t) \leq \lceil L/G \rceil$, then m reaches its destination at some step t_m, with $t < t_m \leq t + L$. If, instead, $c_j(t) > \lceil L/G \rceil$ (i.e., the capacity constraint is violated), the following happens:

1. Message m reaches its destination at some step t_m, with $t < t_m \leq t + Gc_j(t) + L$.
2. P_i^L may be signalled to stall at some time step t', with $t < t' \leq t + \delta$. Until step t' the processor continues its normal operations.
3. Let $\bar{t}$ denote the latest time step when a message that caused P_i^L to stall during steps $[t, t')$ arrives at its destination. Then, the processor is signalled to revert to operational state at some time t'', with $\bar{t} < t'' \leq \bar{t} + \delta$. (Note that if $t' > \bar{t} + \delta$ no stalling takes place.)

Intuitively, parameter δ represents an upper bound to the time the network takes to inform a processor that one of the messages it submitted violated the capacity constraint, or that it may revert to operational state as the result of a decreased load in the network.

We refer to the LogP model under the above stalling rule as δ-*stalling LogP*, or δ-*LogP* for short. A *legal execution* of a δ-LogP program is one where message delivery times and stalling periods are consistent with the model's specifications and with the above rule.[1]

In [1] a restricted version of LogP has also been considered, which regards as correct only those programs whose executions never violate the capacity constraint, that is, programs where processors never stall. We refer to such a restricted version of the model as *stall-free LogP*, or *SF-LogP* for short.

3   Separation between δ-stalling LogP and stall-free LogP

In this section, we demonstrate that allowing for δ-stalling in LogP makes the model strictly more powerful than SF-LogP. We prove our claim by exhibiting a simple problem Π such that *any* SF-LogP algorithm for Π requires time which is asymptotically higher than the time attained by a simple δ-LogP algorithm for Π.

Let Π be the problem of *2-compaction* [4]. On a shared memory machine, the problem is defined as follows: given a vector $x = (x_0, x_1, \ldots x_{p-1})$ of p integer components with at most two nonzero values x_{i_0} and x_{i_1}, $i_0 < i_1$, compact the nonzero values at the front of the array. On LogP, we recast the problem as follows. Vector x is initially distributed among the processors so that P_i^L holds x_i, for $0 \leq i < p$. The problem simply requires to make (i_0, x_{i_0}) and (i_1, x_{i_1}) known, respectively, to P_0^L and P_1^L.

On δ-LogP the 2-compaction problem can be solved by the following simple deterministic algorithm in $O(L)$ time, for any $\delta \geq 1$: each processor that holds a 1 transmits its identity and its input value first to P_0^L and then to P_1^L. Observe that if $G = L$ such a strategy is illegal for SF-LogP, since it generates a violation of the capacity constraint (since, in this case, $\lceil L/G \rceil = 1$). The following theorem shows that, indeed, for $G = L$, 2-compaction cannot be solved on SF-LogP in $O(L)$ time, thus providing a separation between SF-LogP and δ-LogP.

Theorem 1. *For any constant ϵ, $0 < \epsilon < 1$, solving 2-compaction with probability greater than $(1 + \epsilon)/2$ on SF-LogP with $G = L$ requires $\Omega\left(L\sqrt{\log n}\right)$ steps.*

Proof (Sketch). In [4] it is proved that solving 2-compaction with probability greater than $(1 + \epsilon)/2$ on the EREW-PRAM requires $\Omega\left(\sqrt{\log n}\right)$ steps, even if each processor is allowed to perform an unbounded amount of local computation per step. The theorem follows by showing that when $G = L$, any T-step computation of a p-processor SF-LogP can be simulated in $O\left(\lceil T/L \rceil\right)$ steps on a p-processor EREW-PRAM with unbounded local computation. (Details of the simulation will be provided in the full version of the paper.)

---

[1] Note that the characterization of stalling proposed in [1] corresponds to the one given above with $\delta = 1$, except that in [1] a processor reverts to the operational state as soon as the capacity constraint violation ends, which may happen before the message causing the violation reaches its destination.

It must be remarked that the above theorem relies on the assumption $G = L$. We leave the extension of the lower bound to arbitrary values of G and L as an interesting open problem.

4   Simulation of LogP on BSP

This section shows how to simulate δ-LogP programs efficiently on BSP. The strategy is similar in spirit to the one devised in [1] for the simulation of SF-LogP programs, however it features a more careful scheduling of interprocessor communication in order to correctly implement the stalling rule.

The algorithm is organized in *cycles*, where in a cycle P_i^B simulates $C = \max\{G, \delta\} \le L$ consecutive steps (including possible stalling steps) of processor P_i^L, using its own local memory to store the contents of P_i^L's local memory, for $0 \le i < p$. In order to simplify bookkeeping operations, the algorithm simulates a particular legal execution of the LogP program where all messages reach their destinations at cycle boundaries. (From what follows it will be clear that such a legal execution exists.)

Each processor P_i^B has a program counter ρ that at any time indicates the next instruction to be simulated in the P_i^L's program. It also maintains in its local memory a pool for outgoing messages $Q_{out}(i)$, a FIFO queue for incoming messages $Q_{in}(i)$ (both initially empty), and two integer variables t_i and w_i. Variable t_i represents the clock and always indicates the next time step to be simulated, while w_i is employed in case of stalling to indicate when P_i^L reverts to the operational state. Specifically, P_i^L is stalling in the time interval $[t_i, w_i - 1]$, hence it is operational at step t_i, if $w_i \le t_i$. Initially both t_i and w_i are set to 0. The undelivered messages causing processors to stall are retained in a global pool S, which is evenly distributed among the processors.

We now outline the simulation of the k-th cycle, $k \ge 0$, which comprises time steps $C \cdot k, C \cdot k + 1, \ldots C \cdot (k+1) - 1$. At the beginning of the cycle's simulation we have that $t_i = C \cdot k$ and $Q_{in}(i)$ contains all messages delivered by the network to P_i^L at the beginning of step $C \cdot k$, for $0 \le i < p$. Also, S contains messages that have been submitted in previous cycles and that will reach their destination at later cycles, that is, at time steps $C \cdot k'$ with $k' > k$. The simulation of the k-th cycle proceeds as follows.

1. For $0 \le i < p$, if $w_i < C \cdot (k+1)$ then P_i^B simulates the next $x = C \cdot (k+1) - \max\{t_i, w_i\}$ instructions in the P_i^L's program. A submit is simulated by inserting the message into $Q_{out}(i)$, and a receive is simulated by extracting a message from $Q_{in}(i)$. The processor also increments ρ by x and sets $t_i = C \cdot (k+1)$.
2. All messages in $\bigcup_i Q_{out}(i)$ together with those in S are sorted by destination and, within each destination group, by time of submission.
3. Within each destination group, messages are ranked and a message with rank r is assigned delivery time $C \cdot (k + \lceil r/\lceil L/G \rceil \rceil)$ (i.e., the message will be delivered at the beginning of the $(\lceil r/\lceil L/G \rceil \rceil)$-th next cycle).
4. Each message to be delivered at cycle $k + 1$ is placed in the appropriate $Q_{in}(i)$ queue (that of its destination), while all other messages are placed in S.
   Comment: Note that S contains only those messages for which a violation of the capacity constraint occurred.
5. For $0 \le i < p$, if one of the messages submitted by P_i^L is currently in S then
   (a) w_i is set to the maximum delivery time of P_i^L's messages in S;

(b) If $\delta < G$, then all operations performed by P_i^L in the simulated cycle subsequent to the submission of the first message that ended up in S are "undone" and ρ is adjusted accordingly.

Comment: Note that when $\delta < G$ processor P_i^L submits only one message in the cycle, hence the operations to be undone do not involve submits and their undoing is straightforward.

6. Messages in S are evenly redistributed among the processors.

Theorem 2. *For any δ, $1 \leq \delta \leq L$, the above algorithm correctly simulates a cycle of $C = \max\{G, \delta\}$ arbitrary LogP steps in time*

$$O\left(C\left(1 + \log p \left(\frac{1}{G} + \frac{g/G}{1 + \log(C/G)} + \frac{\ell/C}{1 + \log\min\{C/G, \ell/g\}}\right)\right)\right).$$

Proof (Sketch). Consider of the simulation of an arbitrary cycle. The proof of correctness, which will be provided in the full version of the paper, entails showing that the operations performed by the BSP processors in the above simulation algorithm do indeed mimic the computation of their LogP counterparts in a legal execution of the cycle. As for the running time, Steps 1 and 5.(b) involve $O(C)$ local computation. Step 2 involves the sorting of $O((C/G)p)$ messages, since $|Q_{out}(i)| = O(C/G)$, for $0 \leq i < p$, and there can be no more than $\lceil \delta/G \rceil = O(C/G)$ messages in S sent by the same (stalling) processor. Finally, the remaining steps are dominated by the cost of prefix operations performed on evenly distributed input sets of size $O((C/G)p)$ and by the routing of $O(C/G)$-relations. The stated running time then follows by employing results in [3, 6].

The following corollary is immediately established.

Corollary 1. *When $\ell = \Theta(L)$, $g = \Theta(G)$ an arbitrary LogP program can be simulated in BSP with slowdown $O((L/G)\log p)$, if $\delta = 1$, and with slowdown $O(\log p / \min\{G, 1 + \log(L/G)\})$, if $\delta = \Theta(L)$.*

The corollary, combined with the results in [1], shows that LogP, under the reasonable L-stalling rule, and BSP can simulate each other with at most logarithmic slowdown when featuring similar bandwidth and latency parameters.

References

1. G. Bilardi, K.T. Herley, A. Pietracaprina, G. Pucci and P. Spirakis. BSP vs. LogP. *Algorithmica*, 24:405–422, 1999.
2. D.E. Culler, R. Karp, D. Patterson, A. Sahay, K.E. Schauser, E. Santos, R. Subramonian, and T.V. Eicken. LogP: A practical model of parallel computation. *Communications of the ACM*, 39(11):78–85, November 1996.
3. M.T. Goodrich. Communication-Efficient Parallel Sorting. In *Proc. of the 28th ACM Symp. on Theory of Computing*, pages 247–256, Philadelphia PA, 1996.
4. P.D. MacKenzie. Lower bounds for randomized exclusive write PRAMs. *Theorey of Computing Systems*, 30(6):599–626, 1997.
5. V. Ramachandran, B. Grayson, and M. Dahlin. Emulations between QSM, BSP and LogP: a framework for general-purpose parallel algorithm design. TR98-22, Dept. of CS, Univ. of Texas at Austin, November 1998. (Summary in *Proc. of ACM-SIAM SODA*, 1999.)
6. L.G. Valiant. A bridging model for parallel computation. *Communications of the ACM*, 33(8):103–111, August 1990.

Parallelizability of some *P*-complete problems*

Akihiro Fujiwara[1], Michiko Inoue[2], and Toshimitsu Masuzawa[2]

[1] Kyushu Institute of Technology, JAPAN
fujiwara@cse.kyutech.ac.jp
[2] Nara Institute of Science and Technology, JAPAN
{kounoe, masuzawa}@is.aist-nara.ac.jp

Abstract. In this paper, we consider parallelizability of some *P*-complete problems. First we propose a parameter which indicates parallelizability for a convex layers problem. We prove *P*-completeness of the problem and propose a cost optimal parallel algorithm, according to the parameter. Second we consider a lexicographically first maximal 3 sums problem. We prove *P*-completeness of the problem by reducing a lexicographically first maximal independent set problem, and propose two cost optimal parallel algorithms for related problems. The above results show that some *P*-complete problems have efficient cost optimal parallel algorithms.

1   Introduction

In parallel computation theory, one of primary complexity classes is the class *NC*. Let n be the input size of a problem. The problem is in the class *NC* if there exists an algorithm which solves the problem in $T(n)$ time using $P(n)$ processors where $T(n)$ and $P(n)$ are polylogarithmic and polynomial functions for n, respectively. Many problems in the class *P*, which is the class of problems solvable in polynomial time sequentially, are also in the class *NC*. On the other hand, some problems in *P* seem to have no parallel algorithm which runs in polylogarithmic time using a polynomial number of processors. Such problems are called *P-complete*. A problem in the class *P* is *P*-complete if we can reduce any problem in *P* to the problem using *NC*-reduction. (For details of the *P*-completeness, see [9].) Although there are some efficient *probabilistic* parallel algorithms for some *P*-complete problems, it is believed that the *P*-complete problems are inherently sequential and hard to be parallelized.

Among many *P*-complete problems, only some graph problems are known to be asympotically parallelizable. Vitter and Simons[12] showed that the unification, path system accessibility, monotone circuit value and ordered depth-first search problems have cost optimal parallel algorithms if their input graphs are dense graphs, that is, the number of edges is $m = \Omega(n^{1+\epsilon})$ for a constant ϵ where the number of vertices is n.

* Research supported in part by the Scientific Research Grant-in-Aid from Ministry of Education, Science, Sports and Culture of Japan (Scientific research of Priority Areas(B)10205218)

J. Rolim et al. (Eds.): IPDPS 2000 Workshops, LNCS 1800, pp. 116–122, 2000.

In this paper, we consider parallelizability of two P-complete problems. First we consider *a convex layers problem*. For the problem, we propose a parameter d which indicates parallelizability of the problem. Using the parameter, we prove that the problem is still P-complete if $d = n^\epsilon$ with $0 < \epsilon < 1$. Next we propose a parallel algorithm which runs in $O(\frac{n \log n}{p} + \frac{d^2}{p} + d \log d)$ time using p processors $(1 \leq p \leq d)$ on the EREW PRAM. From the complexity, the problem is in NC if $d = (\log n)^k$ where k is a positive constant, and has a cost optimal parallel algorithm if $d = n^\epsilon$ with $0 < \epsilon \leq \frac{1}{2}$.

Second P-complete problem is *a lexicographically first maximal 3 sums problem*. We prove the P-completeness of the problem, and propose a parallel algorithm, which runs in $O(\frac{n^2}{p} + n \log n)$ using p processors $(1 \leq p \leq n)$ on the CREW PRAM, for the problem. The above algorithm is cost optimal for $1 \leq p \leq \frac{n}{\log n}$. In addition, we propose a cost optimal parallel algorithm for a related P-complete problem. These results show that some P-complete problems have efficient cost optimal parallel algorithms.

2 Parameterized convex layers

First we give some definitions for convex layers.

Definition 1 (Convex layers). *Let S be a set of n points in the Euclidean plane. The convex layers is a problem to compute a set of convex hulls, $\{CH_0, CH_1, \ldots, CH_{m-1}\}$, which satisfies the following two conditions.*

(1) $CH_0 \cup CH_1 \cup \ldots \cup CH_{m-1} = S$.
(2) *Each CH_i $(0 \leq i \leq m-1)$ is a convex hull of a set of points $CH_i \cup CH_{i+1} \cup \ldots \cup CH_{m-1}$.*   □

Dessmark et al.[5] proved P-completeness of the convex layers problem, and Chazelle[1] proposed an optimal sequential algorithm which runs in $O(n \log n)$ time. The sequential algorithm is time optimal because computation of a convex hull, which is the first hull of convex layers, requires $\Omega(n \log n)$ time[13].

In this paper, we consider an additional parameter d for the problem, and restrict its input points on d horizontal lines.

Definition 2 (Convex layers for d lines). *The convex layers for d lines is a convex layers problem whose input points are on d horizontal lines.*   □

The parameter d is at most n if there is no restrictions for positions of input points. In the following, $CL(d)$ denotes the convex layers for d lines problem. We can solve the problem sequentially in $O(n \log n)$ time using the algorithm[1], and prove the lower bound $\Omega(n \log n)$ by reduction from the sorting.

We can prove the following theorem for the problem $CL(d)$. (We omit the proof because of space limitation. The proof is described in [7].)

Theorem 1. *The problem $CL(n^\epsilon)$ with $0 < \epsilon \leq 1$ is P-complete.*   □

Next we propose a cost optimal parallel algorithm for $CL(d)$.

Algorithm for computing $CL(d)$

Input: A set of points $\{u_0, u_1, \ldots, u_{n-1}\}$ on lines $\{l_0, l_1, \ldots, l_{d-1}\}$.

Step 1: Set variables $TOP = 0$ and $BOT = d - 1$. (l_{TOP} and l_{BOT} denote top and bottom lines respectively.) Compute a set of points on each line l_i ($0 \leq i \leq d - 1$), and store them in a double-ended queue Q_i in order of x coordinates.

Step 2: For each line l_i ($TOP \leq i \leq BOT$), compute the leftmost point u^i_{left} and the rightmost point u^i_{right}.

Step 3: Let U_{left} and U_{right} denote sets of points $\{u^{TOP}_{left}, u^{TOP+1}_{left}, \ldots, u^{BOT}_{left}\}$ and $\{u^{TOP}_{right}, u^{TOP+1}_{right}, \ldots, u^{BOT}_{right}\}$ respectively. Compute a left hull of U_{left} and a right hull of U_{right}, and store the obtained points on each hull in CH_{left} and CH_{right}, respectively. (The left hull of U_{left} consists of points on a convex hull of U_{left}, which are from u^{BOT}_{left} to u^{TOP}_{left} in clockwise order. The right hull of U_{right} is defined similarly.)

Step 4: Remove points in Q_{TOP}, Q_{BOT}, CH_{left} and CH_{right} as the outmost convex hull.

Step 5 Compute top and bottom lines on which there is at least one point. Set TOP and BOT to obtained top and bottom lines respectively.

Step 6: Repeat Step 2, 3, 4 and 5 until no point remains.

We discuss complexities of the above parallel algorithm on the EREW PRAM. We use at most p processor ($1 \leq p \leq d$) in the algorithm except for Step 1. Step 1 takes $O(\frac{n \log n}{p} + \log n)$ using Cole's merge sort[4] and primitive operations, and Step 2 takes $O(\frac{d}{p})$ time obviously. We can compute the left hull and the right hull in Step 3 using a known parallel algorithm[2, 3] for computing a convex hull of sorted points. The algorithm runs in $O(\frac{d}{p} + \log d)$ time for each hull. Step 4 takes $O(\frac{d}{p})$ time to remove the points. (Points in Q_{TOP}, Q_{BOT} are automatically removed by changing TOP and BOT in Step 5.) We can compute top and bottom lines in Step 5 in $O(\frac{d}{p} + \log d)$ time using a basic parallel algorithm computing the maximum and the minimum. Since the number of the repetition of Step 6 is $\lceil \frac{d}{2} \rceil$, we can compute $CL(d)$ in $O(\frac{n \log n}{p} + \log n + (\frac{d}{p} + \log d) \times \lceil \frac{d}{2} \rceil) = O(\frac{n \log n}{p} + \frac{d^2}{p} + d \log d)$, and obtain the following theorem.

Theorem 2. *We can solve $CL(d)$ in $O(\frac{n \log n}{p} + \frac{d^2}{p} + d \log d)$ time using p processors ($1 \leq p \leq d$) on the EREW PRAM.* □

We can show that the class of the problem changes according to the number of lines d from the above complexity. (Details are omitted.)

Corollary 1. *We can solve $CL((\log n)^k)$, where k is a positive constant, in $O(\log n \log \log n)$ time using n processors on the EREW PRAM, that is, $CL((\log n)^k)$ is in NC.* □

Corollary 2. *We can solve $CL(n^\epsilon)$ with $0 < \epsilon \leq \frac{1}{2}$ in $O(\frac{n \log n}{p})$ time using p processors ($1 \leq p \leq n^\epsilon$) on the EREW PRAM.* □

3   Lexicographically first maximal 3 sums

We first define the lexicographically first maximal 3 sums problem as follows.

Definition 3 (Lexicographically first maximal 3 sums). *Let I be a set of n distinct integers. The lexicographically first maximal 3 sums is a problem to compute the set of 3 integers $LFM3S = \{(a_0, b_0, c_0), (a_1, b_1, c_1), \ldots, (a_{m-1}, b_{m-1}, c_{m-1})\}$, which satisfies the following three conditions.*

1. *The set $S = \{a_0, b_0, c_0, a_1, b_1, c_1, \ldots, a_{m-1}, b_{m-1}, c_{m-1}\}$ is a subset of I.*
2. *Let $s_i = \{a_i, b_i, c_i\}$ $(0 \leq i \leq m - 1)$. Then, (a_i, b_i, c_i) is the lexicographically first set of 3 integers which satisfies $a_i + b_i + c_i = 0$ for $I - (s_0 \cup s_1 \cup \ldots \cup s_{i-1})$.*
3. *There is no set of three integers (a', b', c') which satisfies $a', b', c' \in I - S$ and $a' + b' + c' = 0$.* □

Next we prove P-completeness of LFM3S. We show reduction from *the lexicographically first maximal independent set (LFMIS) problem* to LFM3S. Let $G = (V, E)$ be an input graph for LFMIS. We assume that all vertices in $V = \{v_0, v_1, \ldots, v_{n-1}\}$ are ordered, that is, v_i is less than v_j if $i < j$.

In [11], Miyano proved the following lemma for LFMIS.

Lemma 1. *The LFMIS restricted to graphs with degree at most 3 is P-complete.*

□

Using the above lemma, we can prove the P-completeness of LFM3S. (Details are described in [7].)

Theorem 3. *The problem LFM3S is P-complete.*

(Outline of proof)

It is obvious that LFM3S is in P.

Let $G = (V, E)$ with $V = \{v_0, v_1, \ldots, v_{n-1}\}$ be an input graph with degree at most 3. First we define *a vertex value $VV(i)$* for each vertex v_i. The vertex value is a negative integer and defined as $VV(i) = i - n$. Thus vertices $v_0, v_1, \ldots, v_{n-1}$ have vertex values $-n, -(n - 1), \ldots, -1$ respectively. We also difine a key set of integers $Q = \{q_0, q_1, \ldots, q_{12}\} = \{-64, -61, -32, -31, -29, -15, -14, -13, -10, -8, 23, 46, 93\}$. Using the vertex value and the key set, we define the following 4-tuples for each vertex v_i in V $(0 \leq i \leq n - 1)$ as inputs for LFM3S.

1. **Vertex tuple for v_i:** $VT(i) = [VV(i), q_0, VV(i), 0]$
2. **Auxiliary tuples for v_i:**
   (a) $AT_1(i) = [VV(i), q_1, 0, VV(i)]$ (b) $AT_2(i) = [VV(i), q_2, VV(i), 0]$
   (c) $AT_3(i) = [VV(i), q_3, 0, VV(i)]$ (d) $AT_4(i) = [VV(i), q_4, 0, VV(i)]$
   (e) $AT_5(i) = [VV(i), q_5, VV(i), 0]$ (f) $AT_6(i) = [VV(i), q_6, 0, VV(i)]$
   (g) $AT_7(i) = [VV(i), q_7, 0, VV(i)]$ (h) $AT_8(i) = [VV(i), q_8, VV(i), 0]$
   (i) $AT_9(i) = [VV(i), q_9, 0, VV(i)]$
   (j) $AT_{10}(i) = [2 * |VV(i)|, q_{10}, |VV(i)|, |VV(i)|]$
   (k) $AT_{11}(i) = [2 * |VV(i)|, q_{11}, |VV(i)|, |VV(i)|]$
   (l) $AT_{12}(i) = [2 * |VV(i)|, q_{12}, |VV(i)|, |VV(i)|]$

3. **Link tuples for** v_i: For each adjacent vertex v_j of v_i, which satisfies $i < j$, add one of the following tuples.
   (a) $LT_1(i,j) = [|VV(i)| + |VV(j)|, |q_0| + |q_1|, |VV(j)|, |VV(i)|]$
   (b) $LT_2(i,j) = [|VV(i)| + |VV(j)|, |q_0| + |q_3|, |VV(j)|, |VV(i)|]$
   (c) $LT_3(i,j) = [|VV(i)| + |VV(j)|, |q_0| + |q_7|, |VV(j)|, |VV(i)|]$
   (In case that v_i has only one adjacent vertex v_j which satisfies $i < j$, add $LT_1(i,j)$ for v_j. In case that v_i has the two adjacent vertices v_{j_1}, v_{j_2}, add $LT_1(i,j_1)$ and $LT_2(i,j_2)$ for each vertex. In case that v_i has the three adjacent vertices, add all three tuples similarly.)

The above 4-tuples have the following special feature. Let $\{VT(i), AT_1(i), AT_2(i), \ldots, AT_{12}(i), LT_1(i,s), LT_2(i,t), LT_3(i,u), VT(s), VT(t), VT(u)\}$ be the input for LFM3S[1]. (We assume v_s, v_t and v_u are adjacent vertices which satisfy $i < s < t < u$.) Then the solution of LFM3S is as follows. (We call the solution *TYPE A sums.*)

$$\{(VT(i), AT_4(i), AT_{12}(i)), (AT_2(i), AT_6(i), AT_{11}(i)), (AT_5(i), AT_9(i), AT_{10}(i)),$$
$$(AT_1(i), VT(s), LT_1(i,s)), (AT_3(i), VT(t), LT_2(i,t)), (AT_7(i), VT(u), LT_3(i,u))\}$$

Note that vertex tuples, $VT(s)$, $VT(t)$ and $VT(u)$, are in the sums. In other words, the above vertex tuples are not in the remaining inputs after the computation.

Next, we consider the solution without $VT(i)$ in the input. (We call the solution *TYPE B sums.*)

$$\{(AT_1(i), AT_2(i), AT_{12}(i)), (AT_3(i), AT_5(i), AT_{11}(i)), (AT_7(i), AT_8(i), AT_{10}(i))\}$$

In this case, the vertex tuples, $VT(s)$, $VT(t)$ and $VT(u)$, remain in the inputs.

We give the above 4-tuples for all vertices in V of LFMIS, and compute LFM3S. Then the vertex $v_i \in V$ is in the solution of LFMIS if and only if there exists a sum of three 4-tuples (T_1, T_2, T_3) which satisfies $T_1 = VT(i)$ in the solution of LFM3S. (Proof of correctness is omitted.)

It is easy to see that the above reduction is in NC. Although we define that inputs of $LFM3S$ are distinct integers, inputs of the above reduction are 4-tuples. We can easily reduce each 4-tuple to an integer without loss of the features. Let $2^g \le n < 2^{g+1}$ and $h = \max\{g, 6\}$. Then we can reduce each 4-tuple $[\alpha_0, \alpha_1, \alpha_2, \alpha_3]$ to $\alpha_0 * 2^{3(h+1)} + (\alpha_1 - 65) * 2^{2(h+1)} + \alpha_2 * 2^{h+1} + \alpha_3$. □

Finally, we consider a parallel algorithm for LFM3S on the CREW PRAM. We can propose a sequential algorithm which solves LFM3S in $O(n^2)$ by modifying an algorithm computing the 3 sum problem[8]. The algorithm is the known fastest sequential algorithm for LFM3S. Note that strict lower bound of LFM3S is not known. However the 3 sum has no $o(n^2)$ algorithm and has an $\Omega(n^2)$ lower bound on a weak model of computation[6].

Algorithm for computing $LFM3S$
Input: A set of n integers I.

---

[1] The sum of tuples $A = [\alpha_0, \alpha_1, \alpha_2, \alpha_3]$ and $B = [\beta_0, \beta_1, \beta_2, \beta_3]$ is defined as $A + B = [\alpha_0 + \beta_0, \alpha_1 + \beta_1, \alpha_2 + \beta_2, \alpha_3 + \beta_3]$, and $A < B$ if A is lexicographically less than B. We assume that the sum is zero if the sum of tuples is $[0, 0, 0, 0]$.

Step 1: Sort all elements in I. (Let $S = (s_0, s_1, \ldots, s_{n-1})$ be the sorted sequence.)

Step 2: Repeat the following substeps from $i = 0$ to $i = n - 3$.

(2-1) Create the following two sorted sequences S' and S'_R from S.

$$S' = (s_{i+1}, s_{i+2}, \ldots, s_{n-1}), S'_R = (-s_{n-1} - s_i, -s_{n-2} - s_i, \ldots, -s_{i+1} - s_i)$$

(For $b \in S'$ and $c \in S'_R$ which satisfy $b = s_g$ and $c = -s_h - s_i$ respectively, $b = c$ if and only if $s_i + s_g + s_h = 0$.)

(2-2) Merge S' and S'_R into a sorted sequence $SS = (ss_0, ss_1, \ldots, ss_{2(n-i-1)-1})$.

(2-3) Compute the smallest element ss_j in SS which satisfies $ss_j = ss_{j+1}$.

(2-4) If the above ss_j is obtained, compute s_g and s_h in S such that $s_g = ss_j$ and $s_h = -s_g - s_i$, respectively. (It is obvious that $s_g \in S'$ and $-s_g - s_i \in S'_R$ since all elements in S are distinct.) Delete s_i, s_g, s_h from I, and output (s_i, s_g, s_h), whenever they exist.

We assume the number of processors p is restricted to $1 \leq p \leq n$. We can sort n elements in $O(\frac{n \log n}{p} + \log n)$ time using Cole's merge sort[4] in Step 1. In Step 2, we can compute a substep (2-1) in $O(\frac{n}{p})$ time easily. We can compute substeps (2-3) and (2-4) in $O(\frac{n^2}{p} + \log n)$ time using simple known algorithms and basic operations. In a substep (2-2), we can merge two sorted sequence in $O(\frac{n}{p} + \log \log n)$ time using a fast merging algorithm[10]. Since repetition of Step 2 is $O(n)$, we obtain the following theorem.

Theorem 4. *We can solve LFM3S in $O(\frac{n^2}{p} + n \log n)$ time using p processors $(1 \leq p \leq n)$ on the CREW PRAM.* □

In the case of $1 \leq p \leq \frac{n}{\log n}$, the time complexity becomes $O(\frac{n^2}{p})$. Therefore the above algorithm is cost optimal for $1 \leq p \leq \frac{n}{\log n}$.

As generalization of LFM3S, we can also obtain the similar results for the following problem.

Definition 4 (Lexicographically first maximal set of 3 arguments (LFMS3A)). *Let E be a totally ordered set of n elements. The lexicographically first maximal set of 3 arguments is a problem to compute the set of 3 elements $LFMS3A = \{(a_0, b_0, c_0), (a_1, b_1, c_1), \ldots, (a_m, b_m, c_m)\}$, which satisfies the following three conditions for a given function $f(x, y, z)$ whose value is $TRUE$ or $FALSE$.*

1. *The set $S = \{a_0, b_0, c_0, a_1, b_1, c_1, \ldots, a_m, b_m, c_m\}$ is a subset of E.*
2. *Let $e_i = \{a_i, b_i, c_i\}$ $(0 \leq i \leq m)$. Then, (a_i, b_i, c_i) is the lexicographically first set of 3 elements which satisfies $f(a_i, b_i, c_i) = TRUE$ for $I - (e_0 \cup e_1 \cup \ldots \cup e_{i-1})$.*
3. *There is no set of three elements (a', b', c') which satisfies $a', b', c' \in I - S$ and $f(a', b', c') = TRUE$.* □

Corollary 3. *The problem LFMS3A is P-complete.* □

Theorem 5. *We can solve LFMS3A with an unresolvable function f in $O(\frac{n^3}{p} + n \log n)$ time using p processors $(1 \leq p \leq n^2)$ on the CREW PRAM.* □

4   Conclusions

In this paper, we proved that two problems are P-complete, and proposed cost optimal algorithms for the problems. The results imply that some P-complete problems are parallelizable within the reasonable number of processors.

In the future research, we investigate other parallelizable P-complete problems. The result may imply new classification of problems in P. Another future topic is proposition of fast parallel algorithms which run in $O(n^\epsilon)$ time where $0 < \epsilon < k$ for P-complete problems. Only a few P-complete problems are known to have such algorithms[12].

References

1. B. Chazelle. On the convex layers of a planar set. *IEEE Transactions on Information Theory*, IT-31(4):509–517, 1985.
2. D. Z. Chen. Efficient geometric algorithms on the EREW PRAM. *IEEE transactions on parallel and distributed systems*, 6(1):41–47, 1995.
3. W. Chen. *Parallel Algorithm and Data Structures for Geometric Problems*. PhD thesis, Osaka University, 1993.
4. R. Cole. Parallel merge sort. *SIAM Journal of Computing*, 17(4):770–785, 1988.
5. A. Dessmark, A. Lingas, and A. Maheshwari. Multi-list ranking: complexity and applications. In *10th Annual Symposium on Theoretical Aspects of Computer Science (LNCS665)*, pages 306–316, 1993.
6. J. Erickson and R. Seidel. Better lower bounds on detecting affine and spherical degeneracies. In *34th Annual IEEE Symposium on Foundations of Computer Science (FOCS '93)*, pages 528–536, 1993.
7. A. Fujiwara, M. Inoue, and M. Toshimitsu. Practical parallelizability of some P-complete problems. Technical Report of IPSF, Vol. 99, No. 72 (AL-69-2), September 1999.
8. A. Gajentaan and M. H. Overmars. On a class of $O(n^2)$ problems in computational geometry. *Computational geometry*, 5:165–185, 1995.
9. R. Greenlaw, H.J. Hoover, and W.L. Ruzzo. *Limits to Parallel Computation: P-Completeness Theory*. Oxford university press, 1995.
10. C. Kruskal. Searching, merging and sorting in parallel computation. *IEEE Transactions on Computers*, C-32(10):942–946, 1983.
11. S. Miyano. The lexicographically first maximal subgraph problems: P-completeness and NC algorithms. *Mathematical Systems Theory*, 22:47–73, 1989.
12. J.S. Vitter and R.A. Simons. New classes for parallel complexity: A study of unification and other complete problems for P. *IEEE Transactions of Computers*, C-35(5):403–418, 1986.
13. A. C. Yao. A lower bound to finding convex hulls. *Journal of the ACM*, 28(4):780–787, 1981.

A New Computation of Shape Moments via Quadtree Decomposition [*]

Chin-Hsiung Wu[1], Shi-Jinn Horng[1,2], Pei-Zong Lee[2], Shung-Shing Lee[3], and Shih-Ying Lin[3]

[1] National Taiwan University of Science and Technology, Taipei, Taiwan, R. O. C.
`horng@mouse.ee.ntust.edu.tw`
[2] Institute of Information Science, Academia Sinica, Taipei, Taiwan, R. O. C.
[3] Fushin Institute of Technology and Commerce, I-Lain, Taiwan, R. O. C.

Abstract. The main contribution of this paper is in designing an optimal and/or optimal speed-up algorithm for computing shape moments. We introduce a new technique for computing shape moments. The new technique is based on the quadtree representation of images. We decompose the image into squares, since the moment computation of squares is easier than that of the whole image. The proposed sequential algorithm reduces the computational complexity significantly. By integrating the advantages of both optical transmission and electronic computation, the proposed parallel algorithm can be run in $O(1)$ time. In the sense of the product of time and the number of processors used, the proposed parallel algorithm is time and cost optimal and achieves optimal speed-up.

1 Introduction

Moments are widely used in image analysis, pattern recognition and low-level computer vision [6]. The computation of moments of a two-dimensional (2-D) image involves a significant amount of multiplications and additions in a direct method. Previously, some fast algorithms for computing moments had been proposed using various computation methods [2, 3, 5, 8, 14, 15]. For an $N \times N$ binary image, Chung [2] presented a constant time algorithm for computing the horizontal/vertical convex shape's moments of order up to 3 on an $N \times N$ reconfigurable mesh. Chung's algorithm is unsuitable for complicated objects. In this paper, we will develop a more efficient algorithm to overcome the disadvantage of Chung's algorithm.

The array with a reconfigurable optical bus system is defined as an array of processors connected to a reconfigurable optical bus system whose configuration can be dynamically changed by setting up the local switches of each processor, and messages can be transmitted concurrently on a bus in a pipelined fashion.

[*] This work was partially supported by the National Science Council under the contract no. NSC-89-2213-E011-007. Part of this work was carried out when the second author was visiting the Institute of Information Science, Academia Sinica, Taipei, Taiwan, July - December 1999.

J. Rolim et al. (Eds.): IPDPS 2000 Workshops, LNCS 1800, pp. 123–129, 2000.

More recently, two related models have been proposed, namely the array with reconfigurable optical buses (AROB) [10] and linear array with a reconfigurable pipelined bus system (LARPBS) [9]. The AROB model is essentially a mesh using the basic structure of a classical reconfigurable network (LRN) [1] and optical technology. A 2-D AROB of size $M \times N$, denoted as 2-D $M \times N$ AROB, contains $M \times N$ processors arranged in a 2-D grid. The processor with index (i_1, i_0) is denoted by P_{i_1, i_0}. For more details on the AROB, see [10].

The main contribution of this paper is in designing an optimal speed-up algorithm for computing the 2-D shape moments. The idea of our algorithm is based on the summation of the contribution of each quadtree node where each quadtree node represents a square region. We first represent the image by quadtree decomposition. After that, the image is divided into squares. Then we derive the relationship between the quadtree and the computation of shape moments. Finally, using this representation, an efficient sequential algorithm (SM) and an optimal parallel algorithm (PSM) for shape moment computations are developed. For a constant c, $c \geq 1$, the proposed algorithm PSM can be run in $O(1)$ time using $N \times N^{1+\frac{1}{c}}$ processors when the input image is complicated. If the image is simple (i.e., the image can be represented by a few quadtree nodes), the proposed algorithm PSM can be run in $O(1)$ time using $N \times N$ processors. In the sense of the product of time and the number of processors used, the proposed algorithm PSM is time and cost optimal and achieves optimal speed-up.

2 Basic Data Manipulation Operations

Given N integers a_i with $0 \leq a_i < N$, $0 \leq i < N$, let *sum* stand for

$$\sum_{i=0}^{N-1} a_i. \tag{1}$$

For computing Eq. (1), Pavel and Akl [11] proposed an $O(1)$ time algorithm on a 2-D $N \times \log N$ AROB. In the following, we will use another approach to design a more flexible algorithm for this problem on a 1-D AROB using $N^{1+\frac{1}{c}}$ processors, where c is a constant and $c \geq 1$. Since $a_i < N$ and $0 \leq i < N$, each digit has a value ranging from 0 to $\omega - 1$ for the radix-ω system and a ω-ary representation $\cdots m_3 m_2 m_1 m_0$ is equal to $m_0 \omega^0 + m_1 \omega^1 + m_2 \omega^2 + m_3 \omega^3 \cdots$. The maximum of *sum* is at most $N(N-1)$. With this approach, a_i and *sum* are equivalent to

$$a_i = \sum_{k=0}^{T-1} m_{i,\,k}\,\omega^k, \tag{2}$$

$$sum = \sum_{l=0}^{U-1} S_l\,\omega^l, \tag{3}$$

where $T = \lfloor \log_\omega N \rfloor + 1$, $0 \leq i < N$, $U = \lfloor \log_\omega N(N-1) \rfloor + 1$, and $0 \leq m_{i,\,k}$, $S_l < \omega$.

As $sum = \sum_{i=0}^{N-1} \sum_{k=0}^{T-1} m_{i,\,k}\, \omega^k = \sum_{k=0}^{T-1} \sum_{i=0}^{N-1} m_{i,\,k}\, \omega^k$, let d_k be the sum of N coefficients $m_{i,\,k}$, $0 \leq i < N$, which is defined as

$$d_k = \sum_{i=0}^{N-1} m_{i,\,k}, \tag{4}$$

where $0 \leq k < T$. Then sum can be also formulated as

$$sum = \sum_{k=0}^{T-1} d_k\, \omega^k, \tag{5}$$

where $0 \leq d_k < \omega N$.

Let $C_0 = 0$ and $d_u = 0$, $T \leq u < U$. The relationship between Eqs. (3) and (5) is described by Eqs. (6)-(8).

$$e_t = C_t + d_t,\ 0 \leq t < U, \tag{6}$$

$$C_{t+1} = e_t\ \mathbf{div}\ \omega,\ 0 \leq t < U, \tag{7}$$

$$S_t = e_t\ \mathbf{mod}\ \omega,\ 0 \leq t < U, \tag{8}$$

where e_t is the sum at the tth digit position and C_t is the carry to the tth digit position. Hence, S_t of Eq. (8) corresponds to the coefficient of sum of Eq. (3) under the radix-ω system. Since the carry to the tth digit position of e_t is not greater than N, we have $C_t \leq N$, $0 \leq t < U$. Since $sum \leq N(N-1)$, the number of digits representing sum under radix-ω is not greater than U, where $U = \lfloor \log_\omega N(N-1) \rfloor + 1$. Therefore, instead of computing Eq. (1), we first compute the coefficient $m_{i,\,k}$ for each a_i. Then each S_t can be computed by Eqs. (4), (6)-(8). Finally, sum can be computed by Eq. (3). For more details, see [13].

Lemma 1. *The N integers each of size $O(\log N)$-bit, can be added in $O(1)$ time on a 1-D $N^{1+1/c}$ AROB for a constant c and $c \geq 1$.*

Consequently, given an $N \times N$ integer matrix each of size $O(\log N)$-bit, the sum of these N^2 integers can be computed by the following three steps. First, apply Lemma 1 to compute the partial sum of each row in parallel. Then, route the partial sums located on the first column to the first row. Finally, apply Lemma 1 to accumulate these N partial sums.

Lemma 2. *The N^2 integers each of size $O(\log N)$-bit, can be added in $O(1)$ time on a 2-D $N \times N^{1+\frac{1}{c}}$ AROB for a constant c and $c \geq 1$.*

3   The Quadtree Decomposition

The quadtree is constructed by recursively decomposing the image into four equal-sized quadrants in top-down fashion. Given an $N \times N$ image ($N = 2^d$ for some d), the quadtree representation of it is a tree of degree four which can be defined as follows. The root node of the tree represents the whole image. If the whole image has only one color, we label that root node with that color and stop; otherwise, we add four children to the root node, representing the four quadrants of the image. Recursively we apply this process for each of the four nodes, respectively. If a block has a constant color, then its corresponding node is a leaf node; otherwise, its node has four children.

Recently, Lee and Horng *et al.* [7] addressed a constant time quadtree building algorithm for a given image based on a specified space-filling order.

Lemma 3. *[7] The quadtree of an $N \times N$ image can be constructed in constant time on an $N \times N$ AROB.*

Let the data structure of a quadtree node consist of four fields r, c, I and sz, respectively. The row and column coordinates of the top-left corner of a quadtree node are represented by r and c, the image color of it is represented by I and sz represents the index of the block size of a quadtree node; if the block size is 4^s then sz is s. For a binary image, the third field I can be omitted. In this paper, only the leaves of the quadtree which represent black blocks are useful for computing shape moments; the non-terminal nodes are omitted.

4   Computing Shape Moments

For a 2-D digital image $A = a(x, y)$, $1 \leq x, y \leq N$, the moment of order (p, q) is defined as:

$$m_{pq} = \sum_{x=1}^{N} \sum_{y=1}^{N} x^p y^q a(x, y), \tag{9}$$

where $a(x, y)$ is an integer representing the intensity function (gray level or binary value) at pixel (x, y).

Delta algorithm [15] and Chung's algorithm [2] were based on the summation of the contribution of each row. Ours is based on the summation of the contribution of each quadtree node where each quadtree node represents a square region. For an object represented by a quadtree with α leaves, exactly α non-overlapped squares, $Q_1, Q_2, \cdots, Q_\alpha$, are defined. From the definition of moments, computing the double summations in Eq. (9) of a square is easier than that of an arbitrary shape. Thus, compared to a direct method, the computational complexity can be reduced significantly.

Since the double summations in Eq. (9) are linear operations, the moments of the whole object can be derived from the summations of the moments of these squares. The (p, q)th order moments of theses squares can be computed as follows. From the data structure of quadtree nodes, we can easily find the location of the four corners of the corresponding square. For a square Q_i, assume the coordinates of its top-left corner are (r, c) and its size is 4^s. Let $u = 2^s$, denote the length of each side of the square. Then the coordinates of the other three corners of Q_i are $(r+u-1, c)$, $(r, c+u-1)$ and $(r+u-1, c+u-1)$, respectively.

For a binary digital image, the moment computation of a quadtree node Q_i reduces to the separable computation

$$m_{pq,i} = \sum_{x=r}^{r+u-1} x^p \sum_{y=c}^{c+u-1} y^q = \sum_{x=r}^{r+u-1} x^p h_{q,i} = h_{q,i} \sum_{k=0}^{u-1} (r+k)^p = g_{p,i} \cdot h_{q,i}, \quad (10)$$

where $g_{p,i}$ and $h_{q,i}$ are the p-order and q-order moments for dimension x and dimension y, respectively and they are defined as:

$$g_{p,i} = \sum_{x=r}^{r+u-1} x^p = \sum_{k=0}^{u-1} (r+k)^p,$$

$$h_{q,i} = \sum_{y=c}^{c+u-1} y^q = \sum_{k=0}^{u-1} (c+k)^q. \quad (11)$$

Similarly, the corresponding moments of other quadtree nodes can be obtained from Eqs. (10)-(11) by replacing r, c and u with their corresponding values since they are also represented as squares. Thus, the 2-D shape moments of order (p,q) can be obtained by summing up the corresponding moments of all α square regions:

$$m_{pq} = \sum_{i=1}^{\alpha} m_{pq,i}. \quad (12)$$

Let us conclude this section by stating a sequential algorithm for computing shape moments from the above derivations.

Algorithm SM;

1: For each quadtree node Q_i, compute the 2-D shape moments $m_{pq,i}$, $1 \le i \le \alpha$, according to Eqs. (10)-(11).

2: Compute the 2-D shape moments m_{pq} by summing up $m_{pq,i}$, $1 \le i \le \alpha$, according to Eq. (12).

Theorem 1. *Given an $N \times N$ binary image A, the 2-D shape moments up to order 3 can be computed in $O(\alpha)$ time on a uniprocessor, where α is the number of quadtree nodes.*

Proof : The correctness of this algorithm directly follows from Eqs. (9)-(12). The time complexity is analyzed as follows. Step 1 and 2 each take $O(\alpha)$ time, where α is the number of quadtree nodes. Hence, the time complexity is $O(\alpha)$.

If we consider an $N \times N$ binary image whose entire image has only 1-valued, the comparison of the computational complexity in computing all the moments of order up to $p + q \le 3$ is shown in Table 1. From Table 1, we see that the proposed method reduces the computational computation significantly.

In addition to the computing operations shown in Table 1, contour following, which needs a few comparison operations per pixel, is required for all the non-direct methods to identify the shape of all objects and it takes $O(N^2)$ time. Our algorithm also needs a preprocessing time to create the quadtree nodes for the given image and this can be done in $O(N^2)$ time by the optimal quadtree construction algorithm proposed by Shaffer and Samet [12].

Table 1: Comparison of computational complexity for shape moment methods.

Method	Direct [6]	Delta [15]	Green's [8]	Integral [3]	This paper
Multiplication	$20N^2$	$25N$	0	$8N$	8
Addition	$10N^2$	$N^2 + 6N$	$128N$	$22N$	22

5   Parallel Moment Computation Algorithm

From Eqs. (9)-(12), the algorithm for computing 2-D shape moments m_{pq} includes the following three steps. First build the quadtree for the given image. Then for each quadtree node, compute its corresponding 2-D shape moments by multiplying the two dimensional moments derived from Eqs. (10)-(11). Finally the 2-D shape moments can be obtained by summing up the corresponding moments which were computed by Step 2.

Initially, assume that the given image A is stored in the local variable $a(i, j)$ of processor $P_{i, j}$, $1 \leq i, j \leq N$. Finally, the results are stored in the local variable $m_{pq}(1, 1)$ of processor $P_{1,1}$. Following the definitions of moments, quadtree, and the relationship between them, the detailed moments algorithm (PSM) is listed as follows.

Algorithm PSM;

1: Apply Lemma 3 to build the quadtree for the given image. After that, the results Q_i, $1 \leq i \leq \alpha$, are stored in local variable $Q(x, y)$ in processor $P_{x, y}$, where $i = xN + y$.

2: //For each quadtree node computes its 2-D shape moments. //

    2.1: For each quadtree node Q_i, $1 \leq i \leq \alpha$, computes its 1-D shape moments $g_p(x, y)$ and $h_q(x, y)$ of dimension x and dimension y respectively according to Eq. (11).

    2.2: For each quadtree node Q_i, $1 \leq i \leq \alpha$, compute its 2-D shape moments by computing Eq. (10) (i.e., $m_{pq,i}(x, y) = g_p(x, y) \times h_q(x, y)$).

3: Compute the 2-D shape moments m_{pq} by summing up $m_{pq,i}$, $1 \leq i \leq \alpha$, using Lemmas 1 or 2 according to the value of α. After that, the 2-D moments m_{pq} are stored in the local variable $m_{pq}(1, 1)$ of processor $P_{1, 1}$.

Theorem 2. *Given an $N \times N$ binary image A, the 2-D shape moments up to order 3 can be computed in $O(1)$ time either on an $N \times N$ AROB if A is simple (i.e., α is bounded by $O(N)$), or on an $N \times N^{1+\frac{1}{c}}$ AROB for a constant c and $c \geq 1$ if A is complicated.*

Proof : The time complexity is analyzed as follows. Step 1 takes $O(1)$ time using $N \times N$ processors by Lemma 3. Step 2 takes $O(1)$ time. Step 3 takes $O(1)$ time using $N \times N$ or $N \times N^{1+\frac{1}{c}}$ processors by Lemmas 1 and 2. Hence, the time complexity is $O(1)$.

For computing high order shape moments, Steps 2 and 3 will take $\max\{p, q\}$ times. If both p and q are constant, then the expression for $g_{p,i}$ (or $h_{q,i}$) defined in Eq. (11) will have a constant number of terms with a constant number of powers. Therefore, the results of Theorem 2 can be extended.

6 Concluding Remarks

In this paper, we introduce a new technique based on the quadtree decomposition for computing shape moments. The quadtree decomposition divides the image into squares, where the number of squares is dependant on the image complexity. In the most application, the $N \times N$ image can be decomposed into $O(N)$ squares by quadtree decomposition. As a result, the shape moments can be parallelized and computed in $O(1)$ time on an $N \times N$ AROB.

References

1. Ben-Asher, Y., Peleg, D., Ramaswami, R., Schuster, A.: The Power of Reconfiguration. Journal of Parallel and Distributed Computing **13** (1991) 139–153
2. Chung, K.-L.: Computing Horizontal/vertical Convex Shape's Moments on Reconfigurable Meshes. Pattern Recognition **29** (1996) 1713-1717
3. Dai, M., Batlou, P., Najim, M.: An Efficient Algorithm for Computation of Shape Moments from Run-length Codes or Chain Codes. Pattern Recognition **25** (1992) 1119-1128
4. Guo, Z., Melhem, R. G., Hall, R. W., Chiarulli, D. M., Levitan, S. P.: Pipelined Communications in Optically Interconnected Arrays. Journal of Parallel and Distributed Computing **12** (1991) 269–282
5. Hatamian, M.: A Real Time Two-dimensional Moment Generation Algorithm and Its Single Chip Implementation. IEEE Trans. ASPP **34** (1986) 546-553
6. Hu, M.-K.: Visual Pattern Recognition by Moment Invariants. IRE Trans. Inform. Theory **IT-8** (1962) 179-187
7. Lee, S.-S., Horng, S.-J., Tsai, H.-R., Tsai, S.-S.: Building a Quadtree and Its Applications on a Reconfigurable Mesh. Pattern Recognition **29** (1996) 1571-1579
8. Li, B.-C., Shen, J.: Fast Computation of Moment Invariants. Pattern Recognition **24** (1991) 8071-813
9. Pan, Y., Li, K.: Linear Array with a Reconfigurable Pipelined Bus System— Concepts and Applications. Information Sciences – An Int. Journal **106** (1998) 237-258
10. Pavel, S., Akl, S. G.: On the Power of Arrays with Reconfigurable Optical Bus. Proc. Int. Conf. Parallel and Distributed Processing Techniques and Applications (1996) 1443-1454
11. Pavel, S., Akl, S. G.: Matrix Operations Using Arrays with Reconfigurable Optical Buses. Parallel Algorithms and Applications **8** (1996) 223-242
12. Shaffer, C. A., Samet, H.: Optimal Quadtree Construction Algorithms. Computer Vision, Graphics, Image processing **37** (1987) 402-419
13. Wu, C.-H., Horng, S.-J., Tsai, H.-R.: Template Matching on Arrays with Reconfigurable Optical Buses. Proc. Int. Symp. Operations Research and its Applications (1998), 127-141
14. Yang, L., Albregtsen, F.: Fast and Exact Computation of Cartesian Geometric Moments Using Discrete Green's Theorem. Pattern Recognition **29** (1996) 1061-1073
15. Zakaria, M. F., Zsombor-Murray, P. J. A., Kessel, J. M. H. H.: Fast Algorithm for the Computation of Moment Invariants. Pattern Recognition **20** (1987) 639-643

The Fuzzy Philosophers

Shing-Tsaan Huang

Department of Computer Science and Information Engineering
National Central University
Chung-Li, Taiwan 32054, R.O.C.
E-mail: sthuang@csie.ncu.edu.tw

Consider a network of nodes; each node represents a philosopher; links represent the neighboring relationship among the philosophers. Every philosopher enjoys singing so much that once getting the chance, he always sings a song within a finite delay. This paper proposes a protocol for the philosophers to follow. The protocol guarantees the following requirements: (1) No two neighboring philosophers sing songs simultaneously. (2) Along any infinite time period, each philosopher gets his chances to sing infinitely often. Following the protocol, each philosopher uses only one bit to memorize his state.

Sometimes the philosophers may be fuzzy enough to forget the state. So, a self-stabilizing version of the protocol is also proposed to cope with this problem. However, the philosophers may need additional bits to memorize their states.

1. Introduction

Consider a network of nodes; each node represents a philosopher; links represent the neighboring relationship among the philosophers. This paper proposes a protocol for the philosophers to follow. The protocol guarantees the following two requirements: (1) No two neighboring philosophers sing songs simultaneously. (2) Along any infinite time period, each philosopher gets his chances to sing infinitely often. Following the protocol, each philosopher only uses a boolean variable to memorize his state.

Sometimes the philosophers may be fuzzy enough to forget the state. The fuzzy behavior of the philosophers is modeled as *transient faults*. A transient fault may perturb the values of the variables of a program but not the constants and the program code. To cope with all kinds of possible transient faults, Dijkstra [3] introduced the *self-stabilizing* (SS in short) concept into computer systems. Provided that no more transient faults may occur afterwards, an SS system must be able to stabilize eventually to states which fulfil the desired requirements no matter what current state it is.

Singing a song by the philosophers can be modeled as executing the *critical section* (CS in short). Then, the formulated problem is closely related to the dinning philosophers by Dijkstra [4] and the drinking philosophers by Chandy and Misra [2] although the dinning philosophers and the drinking philosophers do not handle

J. Rolim et al. (Eds.): IPDPS 2000 Workshops, LNCS 1800, pp. 130-136, 2000.

transient faults. That no two neighboring philosophers are allowed to execute the CS simultaneously is the common requirement. The major issue faced in the philosophers in fulfilling the requirement is the symmetry problem. It would be impossible to have a deterministic solution if the system is in a state of which no node is distinguishable from the others. Here in this paper, a simple and elegant approach is proposed which allows a node use only one bit to resolve the conflicts. The result should be interesting to those who might design distributed protocols to resolve the conflicts among the requests from neighboring processes.

There are two versions of the proposed protocol: *A-protocol* and *B-protocol*. A-protocol has the SS property if the network is acyclic, but not otherwise. B-protocol can cope with the transient faults; i.e., it is an SS protocol. Provided that the philosophers are not fuzzy any more, B-protocol eventually guarantees the two requirements. However, the philosophers may need more boolean variables to memorized their states.

An SS protocol is usually presented in rules. Each rule has two parts: the *guard* and the *action*. The guard is a boolean function of the states of the node and its neighbors. If the guard is true, its action is said to be *enabled* and can then be executed. In proving the correctness of an SS protocol, the following three assumptions may be considered:

(1). *Serial execution*: Enabled actions are executed one at a time.

(2). *Concurrent execution*: Any nonempty subset of enabled actions are executed all at a time.

(3) *Distributed execution*: A node may read the states of its neighbors at some different times and evaluate its guards and execute the enabled actions at a later moment.

A distributed-correct protocol is also concurrent-correct, in turn, is also serial-correct; but, not vice versa. Because it is easier to design and prove serial protocols, most of the SS protocols[3] are design in such a way.

The result reported in this paper is inspired by *the alternator* studied by Gouda and Haddix [5]. One major difference between their result and the current one is that their protocol supports correct concurrent execution of serial-correct SS protocols, whereas the proposed B-protocol supports not only correct concurrent execution but also correct distributed execution. Correct-distributed execution is commonly believed more difficult.

A rule is said to be *non-interfering* if once it is enabled, it remains so until the action part is executed. It has been shown that a serial-correct protocol is also distributed-correct provided that its rules are non-interfering [1]. The non-interfering property of the rules makes the proposed B-protocol can support correct distributed execution for the serial-correct SS protocols. Other attempts made to support correct distributed execution for serial-correct SS protocols can also be found in [6],[7].

The rest of the paper is organized as follows. Section 2 presents A-protocol. Next, Section 3 gives its correctness proof. B-protocol and its correctness discussion are then presented in Section 4. The efficiency of A-protocol is discussed in Section 5.

2. A-protocol

The first issue we face is the symmetry problem. To solve the problem, in A-protocol, each link is assigned a *static direction* such that the directed network is acyclic. The directed link is then called the *base edge* and is denoted by (B→). The directed networked induced by the base edges is called the *Bnetwork*. Note that the

Bnetwork is static in the sense that all the directions of its edges are fixed. Hence, the Bnetwork is always acyclic.

Associated with each link, there is another edge called the *control edge*, denoted by $(C\rightarrow)$. The direction of the control edge is dynamically controlled by two control bits maintained by the two nodes incident to the edge, respectively, via the following four rules:

$$0(B\rightarrow)0 \text{ then } (C\rightarrow)$$
$$0(B\rightarrow)1 \text{ then } (C\leftarrow)$$
$$1(B\rightarrow)0 \text{ then } (C\leftarrow)$$
$$1(B\rightarrow)1 \text{ then } (C\rightarrow)$$

Let the control bit maintained by node i be denoted as $C.i$. For two neighboring nodes i and j, the rules imply that if $C.i \oplus C.j = 1$, then the control edge has the reversed direction of the base edge, otherwise they have the same direction. Where $\oplus$ is the exclusive OR operator. The directed network induced by the control edges is called the *Cnetwork*.

According to the four rules, a node can reverse all the directions of its adjacent control edges simply by reversing its control bit. Figure 1 gives an example for the Bnetwork and the Cnetwork. The following A-protocol is a direct consequence of this surprisingly simple result.

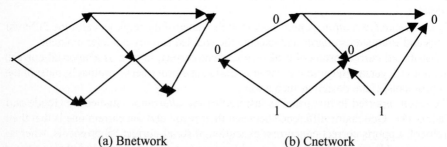

(a) Bnetwork   (b) Cnetwork
Figure 1. Example for the base network and the control network.

Let Csink.i (or Csource,i, respectively) denote that all the control edges of node i are incoming to (or outgoing from, respectively) i. A-protocol consists of one guarded rule only.

[R0.A] Csink.i $\rightarrow$ Execute CS, $C.i := \neg C.i$.

The idea behind A-protocol is very simple. The control edge is used as an arbitrator to decide which one of the two nodes incident to the edge has the priority to execute the CS: the one pointed to has the priority. Csink.i implies that all the neighbors of i agree with that node i has the priority. After executing the CS, node i yields the priority to all its neighbors by reversing its control bit. With common knowledge in the mutual exclusion field [2], A-protocol obviously has the *safety* property, i.e., no two neighboring nodes execute the CS simultaneously.

[R0.A] is *non-interfering*: once the guard is true, it remains true until the action part is executed. The non-interfering property of the rule makes A-protocol distributed-correct, provided that it can be proved serial-correct [1]. Therefore, the correctness proof in the next section only considers serial execution.

3. Correctness of A-protocol

We prove A-protocol correct by showing that it has the following two properties.

[P0]: (Safety Property) No two neighboring nodes execute the CS simultaneously.

[P1]: (Fairness Property) Along any infinite computation, each node executes the CS infinitely often.

As discussed in the previous section, the following Theorem 1 is true.

Theorem 1: A-protocol has the property [P0].

In order to show that A-protocol also has the property [P1], we show that A-protocol is deadlock-free, first. Let all the control bits maintained by the nodes be initialized as zero and assume that the system faces no transient faults. Later, in B-protocol, we will discuss how to handle the transient faults. Under this assumption, we have the following invariant.

[I0] The Cnetwork is acyclic.

Lemma 1: [I0] is an invariant.

Proof: First, [I0] is true at the time when the network is initialized. This is because the Cnetwork is exactly the Bnetwork at the beginning.

Secondly, if [I0] is true before a system state transition, it is also true after the transition. Note that a node changes from a Csink node to a Csource node when it executes the action part of the rule. This is because all the control edges of the node reverse their direction by the action part of the rule. Also, it's not hard to see that an acyclic network remains acyclic if some sink node is replaced with a source node.

All those together implies that [I0] is an invariant. This ends the proof.

Lemma 2: A-protocol is deadlock-free.

Proof: By Lemma 1, the Cnetwork is always acyclic; hence, at any state there exists at least one Csink node, which is enabled. This ends the proof.

Theorem 2: A-protocol has the [P1] property.

Proof: [R0.A] is non-interfering; hence, an enabled node executes the rule eventually. Then, by Lemma 2, along any infinite computation, some node, say node i, must execute the CS infinitely often. By the rule, between two successive CS steps of node i, all its neighbors must execute the CS once. Hence, all the neighbors of node i must execute the CS infinitely often along the computation. Then, because the network is finite, this theorem is proved.

We have proved the correctness of A-protocol under the assumption that no transient faults may occur. However, when transient faults are taken into consideration, [I0] is no longer an invariant. To see this, consider a three-node ring with the following Bnetwork configuration: $i(B{\rightarrow})j(B{\rightarrow})k(B{\leftarrow})i$. At some moment, the Cnetwork configuration may be as $0(C{\rightarrow})0(C{\leftarrow})1(C{\rightarrow})0$. Then, a transient fault may perturb C.i changing its value from 0 to 1 and make the configuration as $1(C\leftarrow)0(C\leftarrow)1(C\leftarrow)1$. A cycle exists. In the next section, B-protocol is modified from A-protocol to cope with the transient faults. Note that A-protocol has the SS property if the original network is acyclic; this is because in such a case the invariant [I0] is valid even after the transient faults.

4. B-protocol

The idea behind B-protocol is to color the links of the network into different colors: color-1, color-2, ..., color-m. The subnetwork induced by links with color-x is called the Cx-subnetwork. The coloring must be carried out in such a way that each of the C1-subnetwork, C2-subnetwork, ..., and Cm-subnetwork is acyclic but may be disconnected. Here we assume the colors are initially given.

According to the colors of the links, nodes are classified into non-mutually-exclusive, different color sets. A node is said to belong to Cx-set (or said to be a Cx node) if the node is incident to at least one link with color-x. Note that a node may belong to several different color sets. As an example, one may color a mesh with two colors: the vertical links with color-1, and the horizontal links with color-2. In such a coloring, each node belongs to two color sets.

In B-protocol, if node i is a Cx node then i maintains a control bit for those links with color-x, denoted as Cx.i, to control the control edges over those links. Hence, for a node belongs to k different color sets, k control bits are needed. Similar to A-protocol, associated with each link, there is a base edge. However, the direction of the base edges can be arbitrary. The requirement that Bnetwork induced by the base edges is acyclic is no longer necessary. The requirement is needed in A-protocol because the invariant [I0] must be initially true. The direction of the control edge over a link with color-x is decided by the direction of the associated base edge and the four rules in the same way as in A-protocol.

B-protocol consists of two rules. The notation Cx-sink in the rules is corresponding to Cx-subnetwork.

[R0.B]: $\forall$Cx: node i $\in$ Cx-set: Cx-sink.i $\rightarrow$ Execute CS, Cx.i := $\neg$Cx.i.
[R1.B]: $\exists$Cx, Cy: x < y, node i $\in$ Cx-set, node i $\in$ Cy-set: $\neg$Cx-sink.i $\land$ Cy-sink.i
          $\rightarrow$ Cy.i := $\neg$Cy.i.

Rule [R0.B] guarantees that B-protocol has the safety property [P0] because its guard guarantees that all the control edges incident to node i point to i. The rules imply that each node waits for executing CS by waiting to hold needed sink status of the control subnetworks one by one from lower color to higher color. By [R1.B], when a node does not hold the needed sink status of a lower color control subnetwork, it does not keep the sink status of a higher color control subnetwork to avoid deadlock.

A-protocol is proved serial-correct. It is also distributed-correct because the only rule [R0.A] is non-interfering. This is not the case for B-protocol because [R1.B] is not non-interfering. The guard of [R1.B] may change from true to false if the action part of it does not execute in time. However, what we really care is the execution of the CS by the nodes, that is, the action part of rule [R0.B]. Rule [R0.B] is obviously non-interfering. Therefore, we conclude that B-protocol is also distributed-correct as long as the two properties [P0] and [P1] are the only concerns.

B-protocol can support correct distributed execution of the serial-correct application protocol in a very simple way. The rules of the application protocol are simply attached into the CS part of B-protocol. The node holding the CS privilege according to B-protocol then executes the rules of the application protocol.

5. Efficiency of A-protocol

This section discusses the efficiency of A-protocol. We are unable to derive good results regarding the efficiency of B-protocol. In the discussion, a *maximal concurrent execution* of A-protocol is assumed as in [5]. That is, the nodes are executed in locked steps; in each step(y,z), which bring the system from state y to state z, all enabled actions at state y are executed in the step.

A-protocol assumes no transient faults. Hence, the Cnetwork is acyclic. A *Cpath* is defined as a directed path from a non-sink node (the *head* node of the Cpath) to a sink node (the *tail* node of the Cpath) in the Cnetwork. Note that the head node is not necessary a source node, also from a non-sink node, many Cpaths may exist. Hence, a Cpath may include many shorter Cpaths. For example, Cpath (h, i, j, …,w) includes Cpath (j, …,w).

The *length* of a Cpath is defined as the number of edges in it, which can only become shorter because the head is fixed and the tail can only shrink. The maximum length of all the Cpaths from a node is the lowest possible number of steps that the node needs to wait for its turn to execute the CS. Therefore, the maximum length of all directed paths in the network, denoted as Xlength of the network, is used as the metric in discussing the efficiency.

Lemma 3. In each step(y,z), an existing Cpath at state y becomes one edge shorter or disappears at state z.

Proof: At state y, except the tail node, which is a sink, all other nodes, including the head node and the middle nodes, of the path are not enabled, and so they remain in the path at state z. Whereas, the tail node is enabled at state y, and hence its action part is executed in the step and reverses the direction of all the control edges incident to it. In other words, the tail node of the path at state y is no longer part of it at state z. This ends the proof.

Note that two or more Cpaths with the same head node may merge into one when they are getting shorter and shorter. For example, Cpath(h, …,u,v,w) and Cpath(h, …u,s) merge into one as Cpath(h, …u,v) at the next state. Also, a new Cpath may be created because a sink node becomes a non-sink node at the next state.

Lemma 4. In each step(y,z), a newly created Cpath has length one, or can only be as long as some Cpath existing in state y.

Proof: Let Csink.i at state y. Node i becomes a non-sink node at state z. Then, (i,j) may be a new Cpath of length one. Or, (i,j, …h) may be a new Cpath because Cpath(j, …h,k) exists at state y; both have the same length. This ends the proof.

Lemma 5. The Xlength of the Cnetwork is non-increasing.

Lemma 5 is a direct consequence of Lemmas 3 and 4. By Lemma 5, the Xlength of the Cnetwork can only become smaller during the computation. Recall that at the beginning, the Cnetwork is exactly the Bnetwork; therefore, A-protocol can be made very efficient by assigning direction of the links of the network in such a way that the Xlength is made as small as possible. For example, in a ring network, the Xlength can be at most two.

A-protocol is an SS protocol when it applies on an acyclic network (viz. tree network) as mentioned before.   When the network is acyclic, A-protocol is very efficient according to the following Theorem 3.

Theorem 3. The Xlength of the Cnetwork stabilizes to one when A-protocol applies on a finite tree network.

Proof: Eventually, any newly created Cpath in each step(y,z) can only be of length one at state z.   This is because the Cnetwork on a finite tree network is finite and acyclic; the creation of a new Cpath of the form (i,j, ...h) eventually becomes impossible.

When this happens, each step afterwards, existing Cpaths become one edge shorter and newly created Cpaths are of length one. Hence, the Xlength of the Cnetwork stabilizes to one. This ends the proof.

Theorem 3 implies that when A-protocol applies on a tree network, the system stabilizes to states in which a node can execute the CS once every two steps.

Acknowledgement:
This research was supported in part by the National Science Council of the Republic of China under the Contract NSC 89-2213-E-007-043.

References:
1.   Brown, G. M., Gouda, M. G., and Wu, C. L.: Token systems that stabilize. IEEE Transaction on Computers, Vol. 38, No. 6 (1989) 845-852.
2.   Chany, K. M. and Misra, J.: The drinking philosophers problem. ACM Transaction on Programming Languages and Systems, Vol. 6, No. 4, Oct. (1984) 632-646.
3.   Dijkstra, E. W.: Self stabilizing systems in spite of distributed control. Communications of the ACM, Vol. 17, No. 6 (1974) 643-644.
4.   Dijkstra, E. W.: Hierarchical ordering of sequential processes. In Operating Systems Techniques, Hoare, C.A. R. and Perrott, R.H., Eds., Academic Press, New York (1972).
5.   Gouda, M. and Haddix, F.: The alternator. Proceedings of the 1999 Workshop on Self-Stabilizing Systems (WSS-99) 48-53.
6.   Huang, S.T., Wuu, L.C., and Tsai, M. S.: Distributed execution model for self-stabilizing systems. Proceedings of the 14th International Conference of Distributed Computing Systems. (ICDCS-94) (1994) 432-439.
7.   Mizuno, M., Nesterenko, M., and Kakugawa, H.: Lock based self-stabilizing distributed mutual exclusion algorithms. Proceedings of the 16th International Conference of Distributed Computing Systems. (ICDCS-96) (1996) 708-716.

A Java Applet to Visualize Algorithms on Reconfigurable Mesh

Kensuke Miyashita and Reiji Hashimoto

Okayama University of Science, Ridai-cho, Okayama 700-0005, Japan
{miyasita, reiji}@ee.ous.ac.jp

Abstract. Recently, many efficient parallel algorithms on the reconfigurable mesh have been developed. However, it is not easy to understand the behavior of a reconfigurable mesh. This is mainly because the bus topology can change dynamically during the execution of algorithm. In this work, we have developed JRM, a Java applet for visualizing parallel algorithm on the reconfigurable mesh to help on understanding and evaluating it. This applet accepts an algorithm written in C-like language and displays the behavior of it graphically. It also displays much information to evaluate the exact performance of an algorithm. Because this software has been developed as a Java applet, it can run on any operating system with standard web browser.

1   Introduction

A reconfigurable mesh (RM) is one of models of parallel computation based on reconfigurable bus system and mesh of processors. Many results and efficient algorithms on the RM are known, such as sorting, graph related problem, computational geometry and arithmetic computation[5].

We make use of the channel of data flow (bus topology) to solve problems on an RM, but it is not easy to follow data since the bus topology changes dynamically during the execution of algorithm on RM. For same reason, it is also hard to find collisions of data on bus. Therefore it is very hard to design, analysis or understand the algorithms on an RM by hand.

Algorithm visualization is very useful to understand the behavior of computational model. It also assists the user to design, analysis and debug the algorithms. Normally, the user can specify the data and control the execution of algorithm via some user interfaces.

As far as we know, there are three tools to visualize algorithms on RM, the Simulator presented by Steckel et al.[6], the Visualization System presented by us[3] and the VMesh by Bordim et al.[1].

The first one displays the behavior of algorithm graphically and the results of simulation are stored in a file. However the algorithm to be visualize in this simulator must be written in assembler language. Hence a sophisticated algorithm is implemented as a large file and there is more probability of existence of miss-codings or bugs. In addition, it cannot display the data that each processor holds neither statistic information.

J. Rolim et al. (Eds.): IPDPS 2000 Workshops, LNCS 1800, pp. 137–142, 2000.

The second one is written in Java and accepts an algorithm written in C-like language. It displays the behavior of algorithm graphically with the source code of algorithm in another window. However it can visualize only a constant-time algorithm (the algorithm without loops), and it runs very slowly.

The third one is written in C language and use X Window System as graphical interface so that it can run on the operating system with X Window System. It accepts an algorithm written in C-like language and displays the behavior of algorithm graphically with the source code of it in another window. An algorithm written in C-like language is translated into a real C source file and is compiled and linked with the VMesh objects. Thus it provides flexible programming interface. However the user has to maintain all source files of the VMesh, libraries used by it, C compiler and linker to recompile it.

In this paper, we present the JRM, a Java applet to visualize the behavior of algorithm on an RM. An applet can run in a standard web browser with Java Virtual Machine (JVM) without recompiling. Thus the JRM can run on Windows, Mac OS and many other UNIX-like operating systems. The JRM can visualize the behavior of algorithm and the data that each processor holds. It can also report bus length, the number of inner-bus configurations and some statistic information. Another important feature is that the JRM accepts an algorithm file specified by the URL (Uniform Resource Locator). Thus it can read an algorithm from anywhere in the Internet.

2   Reconfigurable Mesh

A reconfigurable mesh (RM) is formalized as follows. An RM of size $m \times n$ has m rows and n columns, and there are mn SIMD processors located in a 2-dimensional grid. A processor located at the intersection between ith row and jth column is denoted by $PE(i,j)$ ($0 \leq i < m$, $0 \leq j < n$). Each processor has its own local memory, while no shared memory is available.

A static (outer) bus connects 2 adjacent processors. An outer bus is attached to a processor via a port. Each processor has 4 ports denoted by N (North), S (South), E (East) and W (West). Each port can be connected by reconfigurable (inner) buses with any possible configurations. A connected component made of outer buses and inner buses is called a sub-bus.

All processors work synchronously and a single step of an RM consists of the following 4 phases: **Phase 1:** change the configuration of inner buses, **Phase 2:** send data to a port (the data which is sent at this phase is transferred through a sub-bus), **Phase 3:** receive data from a port (all processors connected to same sub-bus can receive same data sent at the previous phase), **Phase 4:** execute constant-time local computations. The configuration of inner buses is fixed from phase 2 till phase 4.

In this paper, we employ the CREW (Concurrent Read, Exclusive Write) model of bus. That is, at most one processor connected to a sub-bus is allowed to send data at any given time. When more than one processor send data along same sub-bus, a collision occurs and the data is discarded.

3   Specification of Software

In this section, we present the JRM. The JRM is an applet and it is implemented in Java. An applet can be included in an HTML (Hyper Text Markup Language) page, and it can run inside a standard web browser with JVM.

3.1   User Interface

The JRM has some text fields, 2 buttons and an area to draw graphics as shown in Figure 1.

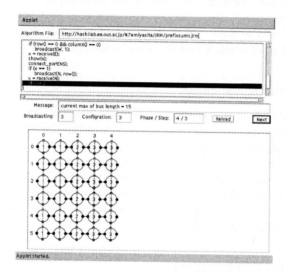

Fig. 1. The appearance of JRM

The algorithm file is specified in the editable text field at the top of window. The JRM accepts an algorithm file specified by the URL. Hence it can read an algorithm file from anywhere in the Internet just like reading a local file and browsing a web page. For this reason, the user can easily publish algorithms and get the other user's algorithms with the JRM.

The JRM accepts an algorithm written in the programming language which has C-like grammar. There are some pre-defined functions to assist the user to write algorithm. The contents of algorithm file are visualized in the main text field and the JRM highlights the line(s) being executed.

The user can execute the algorithm block by block with clicking mouse button on the "Next" button. A block means either a phase or a statement of constant-time computations. Hence the number of blocks is independent of the number of lines and the theoretical computation time of the algorithm.

When each block has been executed, the statistical information is updated and displayed. That information includes the number of broadcasts, the number of configurations used, current number of phase, the number of steps being executed, the length of bus (the number of inner and outer buses) used by current broadcast and the maximum length of bus used by the former broadcasts.

Whenever the user clicks on the "Reload" button, the algorithm file specified in the top text field is re-loaded into the JRM. Thus the user can easily re-execute the algorithm.

The RM is visualized in the main graphic area. The RM of specified size is displayed with row id and column id. The buses and ports on which data flow are painted by some colors to help the user to follow them. Using **show** function in the algorithm file, the user can visualize the specified data being held by each processor at any step.

When a collision of data occurs, a warning message is displayed in the text field at the middle of window and the data is discarded.

3.2 Programming Language

An algorithm file which is accepted by the JRM consists of two parts.

In the first (setup) part, the user must declare the size of the RM on which the algorithm runs and the size of local memory within each processor. And the input of algorithm is stored by each local memory in this part. As well as the user specified value, random value can be used to set up each local memory.

A theoretical algorithm is translated into C-like language and written in the second (main) part of algorithm file. These codes are interpreted and processed by each processor synchronously.

4   Execution of the JRM

In this section, we explain how the JRM works through an implementation of Prefix Sums Algorithm to the JRM for example. Then we show the statistical information of some algorithms by the JRM.

4.1   Visualization of the Prefix Sums Algorithm

The Prefix Sums problem is to compute prefix sums of a binary sequence and defined as follows:

[Prefix Sums Problem]

Input: $a = \langle a_0, a_1, \ldots, a_{n-1} \rangle$ $(a_k \in \{0, 1\}, 0 \le k < n)$.
Output: $s = \langle s_0, s_1, \ldots, s_{n-1} \rangle$ $(s_k = \sum_{i=0}^{k} a_i)$.

This problem can be solved on an RM of size $(n + 1) \times n$ in $O(1)$ time[7].

The algorithm to compute the prefix sums of a binary sequence of length 6 is implemented as follows:

```
 1   # Prefix Sums Algorithm          16      if (x == 0)
 2   BEGIN {                          17         connect_port(EW);
 3      COLUMNS = 5;                  18      else
 4      ROWS = COLUMNS + 1;           19         connect_port(NE, SW);
 5      MEMORYSIZE = 1;               20      if (row() == 0 &&
 6      make_input(0, 0, 0,                           column() == 0)
           COLUMNS - 1, 0, 0, 1);     21         broadcast(W, 1);
 7      alias(x, 0);                  22      x = receive(E);
 8   }                                23      show(x);
 9   {                                24      connect_port(NS);
10      show(x);                      25      if (x == 1)
11      connect_port(NS);             26         broadcast(N, row());
12      if (row() == 0)               27      x = receive(N);
13         broadcast(N, x);           28      show(x);
14      x = receive(N);               29   }
15      show(x);
```

When the user specifies this algorithm file by URL, the JRM loads it and displays the contents, some information and an RM of specified size.

When the user clicks on the "Next" button, the function show in line 10 is executed and each value of 0th local memory is visualized. Then the user clicks on the "Next" button four times, the commands in line 11 through 15 have been done.

Next two clicks make the processor PE(0, 0) broadcast "1" (line 16 through 21), and the length of bus is informed in the message field. And finally the user clicks eight times and the algorithm has been done. The results and some statistical information of this algorithm are shown as Figure 1 (The binary sequence $\langle 1, 0, 1, 1, 0 \rangle$ is given as input).

Now the user can re-execute the algorithm by clicking on the "Reload" button, or execute another algorithm by specifying it by URL.

4.2   Some Algorithms Implemented on the JRM

We have implemented some algorithms on the JRM and the statistical information of them are shown in Table 1. The brief description of them are as follows:

Prefix Sums: It is shown in Section 4.1.

Prefix Remainders: A binary sequence $a = \langle a_0, a_1, \ldots, a_{n-1} \rangle$ and an integer w are given, it computes $r_k = (a_0 + a_1 + \cdots + a_k) \bmod w$ $(0 \leq k < n)$ on an RM of size $(w + 1) \times 2n$ in $O(1)$ time [4].

Sum of Integers: Simulating a look-ahead carry generator, it computes the sum of n d-bit binary values on an RM of size $2n \times 2dn$ in $O(1)$ time [2].

5   Conclusion

We have presented the JRM, a Java applet to visualize the algorithms on Reconfigurable Mesh, in this paper.

Table 1. Statistical information of some algorithms by the JRM

Algorithm	RM Size	Steps	Broadcasts	Configurations	Max. bus length
Prefix Sums	6×5	3	3	3	15
Prefix Remainders	4×12	3	3	8	37
Sum of Integers	8×24	4	6	5	63

The JRM equips the methods to visualize graphically the behavior of algorithm and data. The way of control the execution of algorithm is also provided by the JRM. Furthermore, the JRM can find a collision of data and report statistical information. Thus the JRM helps the user to understand, develop, improve and also study algorithms on the RM.

Because the JRM is implemented as an applet, it can run on any operating system using a standard web browser without re-compiling. The algorithm files can be specified in the URL form, so that the JRM can read them from anywhere in the Internet. And the user can easily publish and share algorithms with the JRM.

We are improving the JRM continuously and planning now to do as follows:

Add pop-up menu: when the user clicks on a processor, a pop-up menu which includes values of its local memory and some statistical information appears.

Make the JRM a multi-threaded applet: each processor parses and executes an algorithm on each thread in parallel.

References

1. J. L. Bordim, T. Watanabe, K. Nakano, and T. Hayashi, A Tool for Algorithm Visualization on the Reconfigurable Mesh, *Proceedings of ISPAN'99*, 1999.
2. J. -W. Jang, and V. K. Prasanna, An Optimal Sorting Algorithm on Reconfigurable Mesh, *Proceedings of 6th International Parallel Processing Symposium*, pp. 130–137, 1992.
3. K. Miyashita, Y. Shimizu, and R. Hashimoto, A Visualization System for Algorithms on PARBS, *Proceedings of International Conference on Computers and Information Technology (ICCIT)*, pp. 215–219, 1998.
4. K. Nakano, T. Masuzawa, and N. Tokura, A Sub-logarithmic Time Sorting on a Reconfigurable Array, *IEICE Transactions on Information and Systems*, E-74-D, Vol. 11, pp. 3894–3901, 1991.
5. K. Nakano, A Bibliography of Published Papers on Dynamically Reconfigurable Architectures, *Parallel Processing Letters*, Vol. 5, No. 1, pp. 111–124, 1995.
6. C. Steckel, M. Middendorf, H. ElGindy, and H. Schmeck, A Simulator for the Reconfigurable Mesh Architecture, *IPPS/SPDP'98 Workshops, Lecture Notes in Computer Science*, Vol. 1388, *Springer-Verlag*, pp. 99–104, 1998.
7. B. F. Wang, G. H. Chen, and F. C. Lin, Constant Time Sorting on a Processor Array with a Reconfigurable Bus System, *Information Processing Letters*, No. 34, Vol. 4, pp. 187–192, 1990.

A Hardware Implementation of PRAM and its Performance Evaluation *

M. Imai, Y. Hayakawa, H. Kawanaka, W. Chen, K. Wada, C.D. Castanho, Y. Okajima, H. Okamoto

Nagoya Institute of Technology, Gokiso, Showa, Nagoya 466-8555, Japan
** E-mail: (imai,zed,sw20,chen,wada,caca,punio,hiroyuki) @phaser.elcom.nitech.ac.jp

1 Introduction

A PRAM (Parallel Random Access Machine)[4] is the parallel computational model most notable for supporting the parallel algorithmic theory. It consists of a number of processors sharing a common memory. The processors communicate by exchanging data through a shared memory cell. Each processor can access any memory cell at one unit of time and all processors operate synchronously under the control of a common clock. These facts make the model a very advantageous platform for considering the inherent parallelism of problems. However,the development of parallel computers which fit this model has not quite matched the theoretical requests. The researches focusing on the reduction of this gap has been carried out [2, 5, 6, 7, 8, 9]. However, most of them give only theoretical analysis; the implementation of the PRAM on hardware level is seldom seen.

Placing emphasis on the realization of the shared memory, we provide several approaches to implement the PRAM model on hardware level. PRAM algorithms can be executed directly and efficiently in these computers without essential modifications. Variants of the PRAM model differ in their handling of simultaneous reading and writing of the same memory cell. In this paper, our object of study is the CRCW PRAM[4] which is the most powerful variant of the PRAM that allows concurrent reading and concurrent writing to the same cell. In the case of concurrent writing to the same cell, there are several methods to determine the content of writing. Here, our implementation is available for all methods. We present three kinds of PRAM-like computers whose shared memory access mechanisms based on tree structure, tree-bus structure, and bus structure, respectively. We realize the shared memory mechanisms and the synchronous mechanism on hardware level by using VHDL(VHSIC Hardware Description Language)[1] and evaluate their performance. The results obtained from the simulations have shown that our approaches can be realized with a practical amount of hardware, and that the running time decreases in proportional to the amount of hardware.

2 Design of the PRAM-like Computers

2.1 Steps of the PRAM Model and Our PRAM-like Computers

In the PRAM each processor has its own local memory and holds its own ID. All the processors operate synchronously in parallel. Usually, one step of the

---

* This work is supported by the Grant-in-Aid for Scientific Research (B)(2) 10205209 (1999) from the Ministry of Education, Science, Sports and Culture of Japan.
** Corresponding e-mail address is: wada@elcom.nitech.ac.jp

J. Rolim et al. (Eds.): IPDPS 2000 Workshops, LNCS 1800, pp. 143-148, 2000.

PRAM model is considered as one instruction of PRAM algorithms, despite containing multiple access to the shared memory. To avoid this ambiguity, we define one step of our PRAM-like computers, called *PRAM step*, as executing only one shared memory access, or arithmetic and logical operations performed in their local memory. The period of one PRAM step lays between two adjacent synchronous events. Thus, an instruction of a PRAM algorithm such as $a(i) := a(i) + b(i)$ would be changed into four PRAM steps: reading $a(i)$ and reading $b(i)$ from the shared memory, evaluating $a(i) + b(i)$, and writing the value to the shared memory cell used for $a(i)$. Obviously, an instruction of a PRAM algorithm can be easily changed into a sequence of PRAM steps without any essential modification.

2.2   Architecture of the PRAM-like Computers

In the PRAM model, let P denote the number of processors and C denote the number of the shared memory cells. The PRAM parallel computer we design consists of N processor units, M memory units, and one synchronous processing unit (Fig. 1(a)), where each processor unit undertakes P/N PRAM processors, each memory unit undertakes C/M PRAM shared memory cells, and the synchronous processing unit takes the responsibility for synchronous processing. The processor units perform arithmetic and logical operations locally, and the shared memory accesses take place through the communication between the processor units and the memory units. At the end of every PRAM step, the synchronous processing proceeds in the order of the following three steps: (1) Each processor unit transmits a PRAM step termination signal to all the memory units. (2) Each memory unit prepares the data to execute the next PRAM step, and sends the memory update termination signal to the synchronous processing unit. (3) When the termination signal has been received from all memory units, the synchronous processing unit sends the synchronous processing termination signal to all the processor units.

The construction of the above units is based on tree structure which facilitates the tasks of sending messages to or receiving messages from multiple units.

(1) Structure of the Processor Units A processor unit consists of $M - 1$ nodes. The leaves are connected with the M memory units such that the processor unit can communicate with each memory unit (see Fig.1 (b)). There are two kinds of nodes: the root and internal nodes. As mentioned above, one processor unit undertakes the work of P/N processors of the PRAM model. The processor-root executes the arithmetic and logical operations, sends memory accessing requests, receives the results from memory reading requests, and sends the synchronous processing request. The processor-nodes route the memory accessing requests to the desired memory unit according to the cell ID, and pass the results of memory reading requests to the processor-root.

(2) Structure of the Memory Units We design three different kinds of memory units. *(i) Tree model:* It consists of $N - 1$ nodes which form a binary tree. The leaves are connected with the N processor units such that the memory unit can treat the access requests from each processor unit (Fig. 1(c)(i)). There are three different kinds of nodes: the leaves, the root and the internal nodes which are denoted as *memory-leaves (ML)*, *memory-root (MR)* and *memory-nodes (MN)*, respectively. As stated above, one memory unit corresponds to C/M shared memory cells of the PRAM model. Each memory-leaf holds the same data, that is, it holds the data corresponding to C/M PRAM shared memory cells. Therefore, a memory-leaf can immediately return the result when receiving a memory reading request. The memory-leaves and memory-nodes undertake the work of resolving the competition of concurrent writing requests which have been sent from its children. When receiving a writing request, the memory-root sends the update information to all the memory-leaves since all of them store the same data. *(ii) Tree-bus Model:* This model uses a bus to transmit the updated information directly from the memory-root to the memory-leaves without passing through the memory-nodes (Fig. 1(c)(ii)). *(iii) Bus Model:* The memory unit of the bus model has the same structure of that in the tree-bus model except that the memory-root, memory-nodes, the edges between them, and the edges from them to memory-leaves are removed. Here, the bus is used for updating the data of the memory-leaves (Fig. 1(c)(iii)).

(3) Structure of the Synchronous Processing Unit The synchronous processing unit consists of $M - 1$ nodes which form a binary tree. Besides, its leaves are connected with the M memory units, and its root is connected with the N processor units by a bus which is used to send the synchronous processing termination signal.

2.3   Memory Accessing and Synchronous Processing

In one PRAM step, the processing is accomplished in the following order: the shared memory access, the arithmetic and logical operations, and the synchronous processing. In order to decrease the number of the requests, each processor unit executes a preprocessing for all the memory access requests.

Memory Accessing: (1) The preprocessing in each processor unit: (a) Change the sequence of access requests such that all the reading requests are listed before the writing requests. (b) (i) Compress the requests which access the same cell into one request by resolving the competition of concurrent accessing, and (ii) compress the reading requests which access the contiguous memory cells into one request such that it holds the information of the number of the contiguous cells and also the smallest address of the cells. The rates of compressibility are different from each algorithm, therefore, considering a trade-off between the preprocessing time and the memory accessing time, we may not execute (b). (c) In order to avoid too many processor units accessing the same memory unit at the same time, rearrange the order of access requests as follows: assuming that the processor ID is i ($0 \leq i \leq N$), revise the kth request into the $((k + i - 1) \, mod(M))$th request. (2) Processing flow between processor and memory units: The processor units send access requests to and receive the results of the accesses from the memory units. The memory units process the access requests from the processor units and send the access results to them. After finishing the access requests of one PRAM step, the processor units send the access request termination signal to all the memory units.

Synchronous Processing: After receiving the access request termination signal from all processor units, each memory unit updates the cells of the writing access requests, and prepares to process the next PRAM step. Then they send the synchronous processing request to the synchronous processing unit. After receiving the synchronous processing request from all the memory units, the synchronous processing unit sends the command to start the next PRAM step.

2.4   The Internal Processing in the Nodes of the Units

The nodes of each kind of units consist of basic elements which are: the sending buffer, the receiving buffer, and the processing part. The processing part executes the processes like routing, and eliminating the competition of concurrent access requests, which are different for each kind of node. Each node communicates with its two children and one parent, however, since the communication between nodes takes more time than the internal processing, in the implementation we allow one node to communicate with two children in parallel.

(1) The Internal Processing in the Nodes of the Processor Unit The internal structures of the processor-root and processor-nodes are shown in Fig. 1(d), where PR-S and PR-R are the processing parts which correspond to the sending and receiving parts. The processor-root has a special part called CPU which undertakes the work of its corresponding P/N processors of the PRAM model.

(2) The Internal Processing in the Nodes of the Memory Unit *(i) Tree Model:* We let TL, TN and TR represent the leaves, internal nodes and the root of the tree. Each of TL-S, TN-S and TR-S executes the process of eliminating the competition of concurrent writing requests, therefore, they have their own memory in order to check the sequences writing requests. One memory unit corresponds to C/M memory cells of the PRAM model. When receiving a reading request, TL-R immediately returns the data to corresponding processor unit. Therefore, each TL-R holds the data of the C/M PRAM memory cells. The other nodes are used for routing (Fig. 1(e)). *(ii) Tree-Bus Model:* It has the same structure as the tree model memory unit, except that a bus is used to transmit the update information directly from the memory-root to memory-leaves. *(iii) Bus Model:* In this case, each node has the same structure shown as (Fig.1(f)), where a part called BN-PROC has a memory used for recording the history of writing requests and a memory used for the shared memory which corresponds to the C/M memory cells of the PRAM model.

(3) The Internal Processing in the Nodes of the Synchronous Processing Unit Each node outputs a signal after receiving the signals from two inputs, therefore, it consists of an AND-gate.

Readers can find the internal processing algorithms of each kind of the nodes and more details in [3].

2.5   Amount of Hardware and Theoretical Processing Time

The processor-root of a processor unit and the memory-leaves of a memory unit perform much more work than the other nodes, therefore, they are considered to take the same amount of hardware as an usual physical processor. The other nodes are mainly used for routing, therefore we call them routing nodes. We treat them differently. We show the required amount of hardware and the theoretical processing time for the memory in Fig 1(g).

3   Evaluation of the Implementation Method

We developed a simulator which measures the time of shared memory accessing and has the following input, output and parameters: [Input:] algorithms of the PRAM model, the number of shared memory cells (C) and the number of processors (P) which are necessary for the algorithms in the PRAM model, [Output:] the time for the shared memory accessing when the algorithms are executed in the PRAM-like computers, and [Parameter:] number of memory units (M) and number of processor units (N).

The PRAM algorithms we use for evaluation are prefix sum, list ranking, and matrix multiplication. This simulator is composed of two parts. One generates all the memory access patterns caused by each PRAM step which is determined by the input algorithm, and then preprocesses the access requests to decrease the number of the access requests. The other part evaluates the processing time of memory access requests derived by the first part. We used VHDL for devising the simulator. VHDL is a hardware description language used to describe the hardware behavior, as well as to design, analyze and simulate complex digital systems [1]. The part of the simulator that generates the access pattern was entirely written in C++. The following is a summary of our experiment of result.

(1) The effect of the number of processor units: Whether there is a compression of access requests or not, the memory access processing time is almost inverse proportional to the number of processor units. Especially, when $x = 2, 4, 8, 16, 32, 64$ and 128, the rate of speed-up are 1.25, 1.5, 2, 3.1, 6.9, 13.7, 24.4, respectively. Therefore, we have obtained an ideal effect.

(2) The effect of the number of memory units: The effect of the number of memory units is very small. In the algorithms we have used for evaluation, the frequency of memory reading accesses is larger than that of memory writing accesses. This is the reason why the effect of the number of memory units is small in our experiment. We believe that regardless the number of processor units, a small number of memory units is sufficient.

(3) The effect of compression of access requests: The effect of the compression of access requests is very small for the list ranking and the prefix sum algorithms if the number of processor units is large; but it is large for the matrix multiplication algorithm. The reason is because the matrix multiplication algorithm performs many reading requests to contiguous memory cells.

(4) The comparison of memory models: Tree model and tree-bus model have almost the same memory access processing time, thus, tree-bus model is better because the structures of the memory-root and the memory-nodes are simpler. The processing time of the bus model is the same as the other two models when the algorithms have few memory writing accesses, however for those ones that perform frequent writing accesses bus model showed to be inefficient.

References

1. K. C. Chang: "Digital Design and Modeling with VHDL and Synthesis", IEEE Computer Society Press (1996).
2. D. Culler, et al: "Towards a realistic model of parallel computation", in Proc.4th ACM SIGPLAN Sym.on PPPP (1993).
3. M. Imai, Y. Hayakawa, H. Kawanaka, W. Chen, K. Wada:" Design and implementation of PRAM on Hardware Level", Technical report TR-99-01 in Wada Lab. of Dept. of Elec. & Comput. Eng., Nagoya Institute of Technology, Japan (1999).
4. Joseph JáJá: "An Introduction to Parallel Algorithms", Addison-Wesley (1992).
5. V. Leppänen, M. Penttonen: "Work-Optimal Simulation of PRAM Models on Meshes", Nordic Journal on Computing, 2(1):51-69(1995).
6. F. Luccio, A. Pietracaprina, G. Pucci: "A Probabilistic Simulation of PRAMs on a Bounded Degree Networks", Information Processing Letters, 28(3):141-147 (1988).
7. F. Luccio, A. Pietracaprina, G. Pucciappanen: "A New Scheme for the Deterministic Simulation of PRAMs in VLSI", Algorithmica, 5(4):529-544 (1990).
8. K.Sato, T. Kurozawa, K. Honda, K. Nakano, T. Hayashi: "Implementing the PRAM Algorithms in the Multithread Architecture and Evaluating the Performance", IPSJ AL, 61(6):39-46 (1988).
9. L. G. Valiant: "A bridging model for parallel computation", CACM 33,8 (1990).

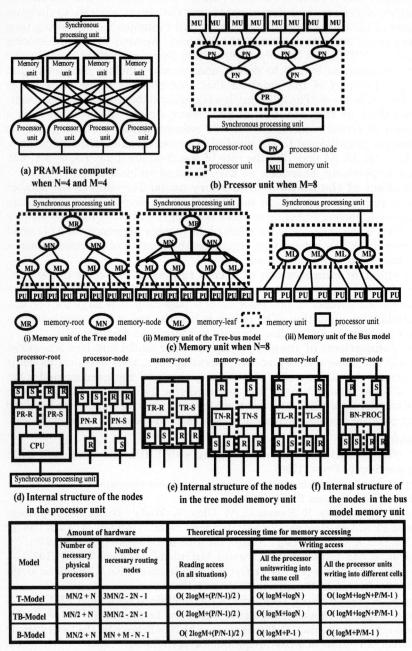

Fig. 1. PRAM-like computer

A Non-Binary Parallel Arithmetic Architecture

Rong Lin[1] and James L. Schwing[2]

1. Department of Computer Science, SUNY at Geneseo
   Geneseo, NY 14454 {lin@cs.geneseo.edu}
2. Department of Computer Science, Central Washington University
   Ellensburg, WA 98926 {schwing@cwu.edu}

Abstract. In this paper we present a novel parallel arithmetic architecture using an efficient non-binary logic scheme. We show that by using parallel broadcasting (or domino propagating) state signals, on short reconfigurable buses equipped with a type of switches, called GP (generate-propagate) shift switches, several arithmetic operations can be carried out efficiently. We extend a recently proposed shift switching mechanism by letting the switch array automatically generate a semaphore to indicate the end of each domino process. This reduces the complexity of the architecture and improves the performance significantly.

1 Introduction

Recently proposed shift switching mechanisms [4-9] allows modulo w (w≥2) arithmetic computations to be efficiently carried out through broadcasting a class of non-binary signals, called state signals, along short buses. The bus structure is closely related to reconfigurable bus systems (REBS, RMESH, Bit Model, RMESH, REBSIS) which have been extensively investigated (refer to [2, 8, 10]). This novel switching technique could lead to VLSI-efficient architectures for high-speed application-specific processor and arithmetic device designs, including parallel comparators, counters, multipliers, and matrix multiplication processors [4-9].

In this paper, we will review the concept of state signals , define (in general) shift switches, and introduce a specific type of shift switches, called the GP (generate and propagate) switches. We will then show that broadcasting state signals on buses equipped with GP shift switches can be used to efficiently construct fast asynchronous adders. In particular, we illustrate a CLA (carry-lookahead) reconfigurable bus-based adder featuring: 1. the combination of state signals, GP switches, and the traditional CMOS domino logic technique on short reconfigurable buses (of width 2 and 3) to achieve a fast VLSI arithmetic design; 2. asynchronous process with a semaphore produced in parallel with the desired arithmetic outputs; 3. full utilization of the inherent speed of each arithmetic opertation; 4. high performance in terms of speed and VLSI area; 5. asynchronous domino process driven by a few control bits; 6. a fast, area compact, stable design and suitable for current CMOS VLSI implementation.

2 The shift switch logic

The shift switch logic is an arithmetic logic which manipulates small integers (as logic operands) represented by state signals by using switc hes (as logic operators) called shift switches to carry out certain basic arithmetic operations (refer to [4-9]). In this section we first introduce the state signals in general and then a GP (generate/propagate) shift switch in particular. A state signal (or S-signal) with integer value I ($0 \leq I \leq w-1$; $w \geq 2$) is represented by bit sequence b_0, b_1, ..b_{w-1} (in this paper we assume the order is either right to left or top to bottom), with the unique bit u (either 0 or 1) in the I-th position. An S-signal is said n- (p-) type denoted by $I_{(w)}$ ($I_{(\overline{w})}$) if u=0 (1). An S-signal may also be denoted by $I_{(w)}$ or simply I if only the number of bits and/or its value is of interest (Figure 1). It may be convenient to represent an S-signal just by the sequence of its bits (variables). For example, (g1, g0, p) is a 3-bit S-signal, for which g1 is

---

The work was supported by National Science Foundation under grant MIP-9630870.

J. Rolim et al. (Eds.): IPDPS 2000 Workshops, LNCS 1800, pp. 149-154, 2000.

carry generate-1 bit, g0 is carry generate-0 bit, and p is carry propagate bit. In this paper we consider only S-signals with small values, say from 0 to 3.

A degenerate state signal, I ($0 \leq I \leq w-1, w \geq 2$), is a state signal with the 0-th bit removed. Thus it is represented by bit sequence $b_{w-1}, b_2,..,b_1$ and its value is 0 if there is no unique bit in the sequence (i.e. all bits are identical for $w>2$), otherwise it is equal to the corresponding S-signal (with the 0-th bit added). A degenerate S-signal is denoted by $I_{(w)}'$ ($I_{(\overline{w})}'$) or $I_{(w)}'$, or simply I (Figure 2), and we may just call it an S-signal if no confusion arises. Again it may be convenient to represent a degenerate S-signal just by the sequence of its bits (variables; e.g. $X_{(\overline{3})}' =(A,B)$). The type of an S-signal with $w>2$ is implied by its representation, but for $w=2$ it is not, and it must be indicated explicitly (when we need to identify the unique bit). Often we use $I_{(w)}$ for an arbitrary-value w-bit S-signal.

<table>
<tr><td>↑↑↑
0 0 1</td><td>↑↑↑
0 1 0</td><td>↑↑↑
1 0 0</td><td>1 →
0 →</td><td>0 →
1 →</td><td>0 →
1 →</td><td>1 →
0 →</td><td>← 0
← 0</td><td>← 1
← 0</td><td>← 0
← 1</td><td>← 1
← 1</td></tr>
<tr><td>$0_{(\overline{3})}$</td><td>$1_{(\overline{3})}$</td><td>$2_{(\overline{3})}$</td><td>$0_{(\overline{2})}$</td><td>$1_{(\overline{2})}$</td><td>$0_{(2)}$</td><td>$1_{(2)}$</td><td>$X_{(\overline{3})}'=0$</td><td>$Y_{(\overline{3})}'=1$</td><td>$Z_{(\overline{3})}'=2$</td><td>$R_{(2)}'=0$</td></tr>
</table>

Figure 1. Some state signals.   **Figure 2**. Degenerate state signals.

Given a function $F(X, Y)=U$ where X and Y and U are (small) S-signals, a pass-transistor (or transmission gate) based digital filter switch which implements function F is called a shift switch with respect to function F. A shift switch which implements function $GP(X, Y)=U$, such that $U=Y$ if $Y> 0$ and $U=X$ if $Y=0$, is called GP (generate/propagate) shift switch (Figure 3).

If a GP switch receives an additional input P/E (precharge/evaluation enable) and produces an additional output V, such that, the functional relations between inputs, P/E, X, Y, and outputs, U and V can be described by Table 1 or 2. The corresponding GP switch is called GP1 or GP2. The block symbols of GP1 and GP2 are shown in Figures 4.a and 4.b. Switch GP2 has a 3-bit input X and a 3-bit output U, while GP1 has a 2-bit input X and a 2-bit output U. The switches are assumed to work on two phases (precharge and evaluation). During the precharge phase (P/E=0), both switches have the same corresponding input/output values (all are 0, represented by appropriate S-signals). During the evaluation phase (P/E=1), both switches produce horizontal output U=GP (X, Y) denoted by S-signals $GP(X,Y)_{(\overline{3})}'$ and $GP(X,Y)_{(\overline{3})}'$ respectively, but they produce different vertical outputs for GP1 $V= GP(X,Y)_{(\overline{3})}$; while for GP2, $V=Y_{(\overline{3})}$.

$U_{(\overline{3})}'=GP(X,Y)$ ← GP ← X $U_{(\overline{3})}'=2$ $^{0}_{1}$ ← GP ← $^{0}_{1}$ $X_{(\overline{3})}'=2$ $U_{(\overline{3})}'=1$ $^{1}_{0}$ ← GP ← $^{0}_{1}$ $X_{(\overline{3})}'=2$ $U_{(\overline{3})}'=2$ $^{0}_{1}$ ← GP ← $^{0}_{0}$ $X_{(\overline{3})}'=0$

↑↑↑ Y   ↑↑↑ 0 0 1 $V_{(\overline{3})}=0$ ↑↑↑ 0 1 0 $V_{(\overline{3})}=1$ ↑↑↑ 1 0 0 $V_{(\overline{3})}=2$

Figure 3. (a) The functionality of a GP shift switch.

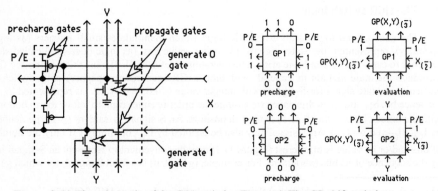

Figure 3.(b) The schematic of the GP1 switch.   **Figure 4.** The GP shift switches.

Table 1. The function of GP1 switch

precharge/evaluation (P/E)	input		output	
	X	Y	U	V
0 (precharge)	$O_{(3)}'$ or 11	$O_{(\bar{4})}'$ 000	$O_{(3)}'$ 11	$O_{(3)}$ 110
1 (evaluation)	$X_{(\bar{3})}'$	$Y_{(3)}$	$GP(X,Y)_{(\bar{3})}'$	$GP(X,Y)_{(3)}$

Table 2. The function of GP2 switch

precharge/evaluation (P/E)	input		output	
	X	Y	U	V
0 (precharge)	$O_{(4)}'$ or 111	$O_{(\bar{4})}'$ 000	$O_{(4)}'$ 111	$O_{(\bar{4})}'$ 000
1 (evaluation)	$X_{(\bar{3})}$	$Y_{(3)}$	$GP(X,Y)_{(\bar{3})}$	$Y_{(\bar{3})}$

3   The small shift switch adder architecture

We first define precharged carry and sum domino logic units, based on the traditional designs [1, 3, 12] and then use them as local components to construct a shift switch (7-bit) adder.

Figures 5, 6 and 7 show block symbols for carry unit 0, carry unit j and sum unit j respectively (for $1 \leq j \leq 31$, note that the units will also be used for large adders). The computations for carries and sums consist of two phases: precharge and evaluation. The logic results are produced during the evaluation phase. During the precharge phase the control signal P/E is set to 0, while $\overline{P/E}$ is set to 1. It is easy to verify that (1) the carry unit 0 are precharged as follows: $g1_j$ (carry generate-1), $g0_j$ (carry generate-0), and p_j (carry propagate) are all high (Figure 5.a); the outputs of carry unit j are all low (Figure 6.a); and the outputs of all sum units are all high (Figure 7.a). When inputs are received P/E is set to 1, the evaluation phase begins. The logic units are discharged as follows: For carry units j: $g1_j$ ($g0_j$) is discharged to high if a carry of 1 (0) is generated; p_j is discharged to high if a carry-propagate is produced, i.e. a_j and b_j have different values since $g1_j = a_j \cdot b_j$; $g0_j = \overline{a_j + b_j}$; and $p_j = a_j \oplus b_j$. Note that S-signal $cb(j)_{(\bar{3})}$ =(g1, g0, p) is called the j-th carry-bit S-signal for $1 \leq j \leq 31$. For carry unit 0: c_0 is discharged to high if both a_0 and b_0 are 1, and to low otherwise, while S_0 is discharged to low if a_0 and b_0 have the same value. For sum unit j ($1 \leq j \leq 31$): S_j is discharged to low if p_j and c_{j-1} have the same value, since $S_j = p_j \oplus c_{j-1}$ (note that c_j denotes the j-th carry-out).

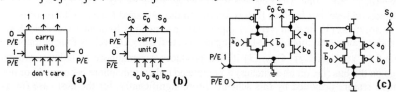

Figure 5. Carry unit 0 (a) Precharge phase; (b) Evaluation phase; (c) the schematic.

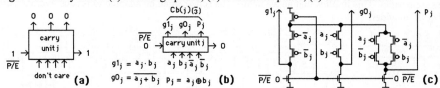

Figure 6. Carry unit j ($1 \leq j \leq 31$) (a) Precharge; (b) Evaluation; (c) the schematic.

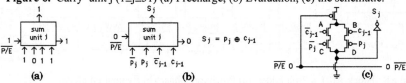

Figure 7. Sum unit j ($1 \leq j \leq 31$) (a) Precharge; (b) Evaluation; (c) the schematic.

Figure8 shows a 7-bit adder, called block A(0). It consists of six GP1 switches connected with seven carry units and six sum units. When signal P/E is set to 0, all GP1 switches, carry units and sum units are precharged as described above.

Next let us discuss the evaluation phase: First, the discharge of each carry unit j generates a carry-bit S-signal $cb(j)_{(\overline{3})}=(g1j, g0j, pj)$, which then discharges the corresponding GP1 switch immediately if $cb(j)_{(\overline{3})}\neq 0$, otherwise it will open the propagation gates of the GP1 switch for the carry propagation. The bus (with the switches) is then partitioned (by the propagation gates). In the worst case carry propagation (discharging) will start from port A or B passing through all 6 pass transistors on the bus. Next, the top outputs of each GP1 (except the last one) go to gate the sum units. Finally the discharge occurs on each sum unit in parallel and simply produces each sum bit of the adder. The worst case delay of the adder is T_c (discharge a carry unit) + 7TGP (tdischarge all seven cascaded pass-transistors on the bus) + T_S (discharge a sum unit).

4   The larger shift switch adder architecture

A larger adder can be constructed by several blocks that contain GP switches organized in three levels. We now illustrate a 32-bit adder (refer to Figure 9). The first level consists of six blocks: blockA(0), the 7-bit adder (Figure 8), blockA(i) (for 1≤i≤4) (Figures 10) and blockA(5) (Figure 11). The second level is a single block, block B (Figure 12). The third level consists of five blocks: blockC(i) (for 1≤i≤4, Figure 13) and blockC(5) (Figure 14).

BlockA(i) (for 1≤i≤4,) contains five cascaded GP2s and carry units. The switches directly send the input carry-bit S-signals $cb(7+j)_{(\overline{3})}=(g1j, g0j, pj)$ (for j=5(i-1)) to the sum units and produce an output S-signal at the left end of the bus. In other words, in the vertical direction the bus simply produces carry-bit S-signals for each bit; in horizontal direction, it produces signal cg(i) =(gg1, gg0, gp) which is called (the i-th) group-carry S-signal, and gp=1 if all p_j of the group are 1, otherwise gp=0 and gg1=1 (gg0=1) if the sum of the group's two bit segments definitely generates a carry 1 (0). BlockA(5) produces only carry-bit signal for each bit.

The worst case delay of the first level (except blockA(0)) can be specified as T_c (discharge a carry unit) + 7TGP(discharge seven cascaded pass-transistors) + T_{inv} (inverter delay, low to high), i.e. $T_c + 7TGP + T_{inv}$. The best case delay can specified as: $T_c + TGP + T_{inv}$.

The second level bus is block B. It is similar to the cascaded (four) switches of blockA(0), except as follows: (1) its four vertical inputs are group-carry S-signals from blockA(i) (for 1≤i≤4) and its horizontal input is a 2-bit S-signal of actual carry C_6; (2) its four outputs are 2-bit S-signals for C_{11}, C_{16}, C_{21}, C_{26}; (3) it also produces a semaphore to indicate whether the discharge on the bus (i.e. on all switches) is completed (when semaphore becomes 0) or not. This is a unique feature provided by this scheme. More significantly, for the best case (or average case) of the adder inputs the domino discharge involves only no more than about 3 cascaded pass-transistors. This implies that an asynchronous scheme can be adapted to reduce average case delay of an adder.

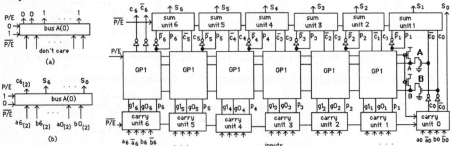

Figure 8. Block A(0), a 7-bit adder, (a) Precharge phase (b) Evaluation phase.

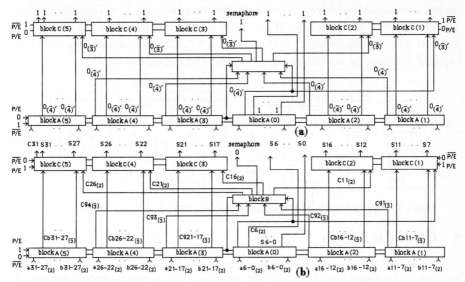

Figure 9. The 32-bit asynchronous adder scheme: (a) precharge; (b) evaluation.

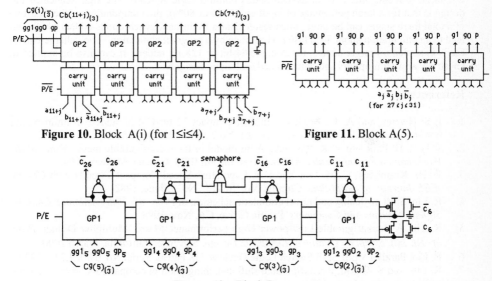

Figure 10. Block A(i) (for $1 \leq i \leq 4$).   **Figure 11.** Block A(5).

Figure 12. Block B.

The worst case delay of the second level (excluding the generation of semaphore) can be specified as: $5T_{GP}$ (time to discharge five cascaded pass-transistors) + T_{inv} (an inverter delay, low to high). The best case delay can be specified as: $T_{GP} + T_{inv}$.

The third level bus is blockC(i) (for $1 \leq i \leq 5$). It is similar to the cascaded (four) switches with sum units in blockA(0), except that during evaluation phase the actual group carry-in (horizontal input) is available now and is represented by S-signal ($C_{6+j}, \overline{C}_{6+j}$) for $j=5(i-1)$ and $1 \leq i \leq 4$). The worst case delay of the third level can be specified as: $6T_{GP}$ (discharging six cascaded pass-transistors) + T_s (discharging a sum unit), i.e. $6T_{GP}+T_s$. The best case of the first level delay +

the best case second level delay + the worst case third level delay is.(T_c + T_{GP}+ T_{inv})+ (T_{GP} + T_{inv}) + ($6T_{GP}$ + T_s.) = T_c + $8T_{GP}$ +$2T_{inv}$ + T_s. By contrast, if the adder is a synchronous adder, the adder delay is estimated as: T_c + $18T_{GP}$ + $2T_{inv}$+ T_s. This indicates the adder will run an expected 50% faster for best case inputs, or 30% faster for average case inputs.

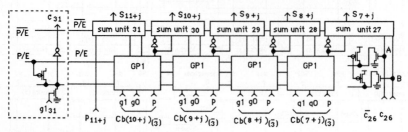

Figure 13. Block C(i) for $1 \leq i \leq 5$ (note that the dotted box is attached only for i=5).

5   Concluding remarks

A novel non-binary arithmetic architecture using the combination of state signals [4-10], GP switches and traditional CMOS domino logic techniques [1, 3] on short reconfigurable buses has been presented. The design produces a semaphore as a by-product to indicate the end of domino discharge process, thus it is an asynchronous domino logic scheme. The significance of the design is that for a large percentage of input cases (about 80%), the computation in the first two of a total three stages can be done much faster than a traditional synchronous adder, which can lead to a computation about 30% faster on average by a program simulation. The scheme can be easily extended for larger adders, for example 64-bit or 128-bit adders. It can also be applied to construct array multipliers and other related arithmetic designs.

References

1. I. S. Hwang, and A. L.. Fischer, Ultrafast compact 32-bit CMOS adders in multi-output domino logic, IEEE J. Solid-State Circ., 24, No 2 April 1989, pp. 358-369.
2. J. Jang, H. Park and V. K. Prasanna, A bit model of the reconfigurable mesh, *Proc. of the Workshop on Reconfigurable Architectures*, the 8th IPPS, Cancun, Maxico, 1994.
3. R. H. Krambeck, C. M. Lee, and H.S. Law, High-Speed Compact Circuits with CMOS, *IEEE Journal of Solid-State Circuits,* Vol SC-17, No. 3, June, 1982.
4. R. Lin, Shift switching and novel arithmetic schemes, in *Proc. of 29th Asilomar Conf. on Signals, Systems and Computers*, Pacific Grove, CA, Nov. 1995.
5. R. Lin, A Reconfigurable Low-power High performance Matrix Multiplier Design, *Proc. of 1th Intl. Symp. on Quality of Electronic Design,* San Jose, California. March 2000.
6. R. Lin, Parallel VLSI Shift Switch Logic Devices, US Patent, Serial No. 09/022,248, 1999.
7. R. Lin and S. Olariu, Reconfigurable shift switching parallel comparators, in *Intl. Journal of VLSI Design*, March, 1999.
8. R. Lin and S. Olariu, Efficient VLSI architecture for Columnsort, *IEEE Transactions on Very Large Scale Integration (VLSI) systems,* VOL. 7, No. 1. March, 1999.
9. R. Lin, and S. Olariu, Reconfigurable buses with shift switching --concept and architectures, in *IEEE Trans. on Parallel And Distributed System.* January 1995.
10. R. Lin, K. Nakano, S. Olariu, M.C. Pintoti, J.L. Schwing, A.Y. Zomaya Scalable hardware-algorithms for binary prefix sums, in Proc. of 6th International Workshop on Reconfigurable Architecture(RAW),1999.
11. R. Miller, V. K. P. Kumar, D. Reisis, and Q. F. Stout, Parallel Computations on Reconfigurable Meshes, *IEEE Transactions on Computers*, 42 (1993), 678--692.
12. N. Weste and K. Eshraghian, *PRINCIPLES OF CMOS VLSI DESIGN, A systems Perspective* (Second Edition), A,ddison-Wesley Publishing Company, 1993.

Multithreaded Parallel Computer Model with Performance Evaluation *

J. Cui[1], J. L. Bordim[1], K. Nakano[1], T. Hayashi[1], N. Ishii[2]

[1] Department of Electrical and Computer Engineering
Nagoya Institute of Technology
[2] Department of Intelligence and Computer Engineering
Nagoya Institute of Technology
Showa-ku, Nagoya 466-8555, Japan

Abstract. The main contribution of this work is to introduce a multithreaded parallel computer model (MPCM), which has a number of multithreaded processors connected with an interconnection network. We have implemented some fundamental PRAM algorithms, such as prefix sums and list ranking algorithms, and evaluated their performance. These algorithms achieved optimal speedup up to at least 16 processors.

1   Introduction

The Parallel Random Access Machine (PRAM) is a standard parallel computer model with a shared memory [2, 5]. The PRAM has a number of processors (PE for short) working synchronously and communicating through the shared memory. Each processor is a uniform-cost Random Access Machine (RAM) with standard instruction set. This model essentially neglects any hardware constraints which a highly specified architecture would impose. In this respect, the model gives free rein in the presentation of algorithms by not admitting limitations on parallelism which might be imposed by specific hardware. However, the PRAM is an unrealistic parallel computer model, because it has a shared memory uniformly accessed by processors.

Parallel computers based on the principle of multithreaded have been developed, such as HEP [12], Monsoon [9, 10], Horizon [7], and MASA [3]. In this paper, we introduce the *multithreaded parallel computer model* (MPCM for short), which has a number of *multithreaded processors* (MP for short) connected with an interconnection network. This model is based on the multithreaded computer architecture [1, 8, 6, 11, 12, 13] which allows fast context-switching, and communicates through message passing. Each multithreaded processor is highly pipelined, and runs several threads in time-sharing manner. Furthermore, processors switch threads in every clock cycle.

---

* This work is in part supported by the Grant-in-Aid for Scientific Research (B)(2) 10205209 (1999) from the Ministry of Education, Science, Sports and Culture of Japan.

J. Rolim et al. (Eds.): IPDPS 2000 Workshops, LNCS 1800, pp. 155–160, 2000.

The main contribution of this work is to develop an MPCM simulator, implement fundamental PRAM algorithms on it, and evaluate their performance. The PRAM algorithms we have implemented include prefix sums algorithm and list ranking algorithm. We coded these algorithms using MPCM machine instructions, and evaluated the number of execution cycles performed. The algorithms we have implemented achieved linear speedup up to at least 16 processors.

2   Multithreaded parallel computer model

An MP has several register sets, each of which has a program counter (PC) and several (say, 32) registers. Each register can store several (say, 32) bits. Although the multithreaded processor has several register sets, it has a single control unit that fetches, decodes, executes an instruction, and writes the result. Thus, an instruction specified by the PC of each register set is processed by the control unit in turn. The instruction may read/write the registers and the local memory of the MPs. However, it cannot directly access the registers in the other register sets. Each execution flow performed by a register set is regarded as a thread. All register sets (threads) share the same program and each register set executes these machine instructions sequentially.

Let $R_1, R_2, \ldots, R_m$ denote m register sets of the MPs. Suppose that each register set executes T instructions of a program. Let I_i^t ($1 \leq i \leq m$, $1 \leq t \leq T$) denote the instruction that R_i executes at time t. Note that a sequence $I_i^1, I_i^2, \ldots, I_i^T$ is not a program code for PE_i; this is a resulting sequence of the instructions executed by PE_i. The behavior of MP is described as follows:

for $t = 1$ **to** T **do**
    for $i = 1$ **to** m **do**
        R_i executes I_i^t

A *phase* is a single iteration of the internal loop, that is, in a single phase the MP executes m instructions $I_1^t, I_2^t, \ldots, I_m^t$.

In most commercial microprocessors, the pipeline mechanism is employed to decrease the average instruction execution time. An instruction process is partitioned into several stages. The pipeline may stall due to the dependency of executed instructions. On the other hand, in the multithreaded processor, instructions executed in a phase have no dependency. This fact allows us to have a large number of pipeline stages.

3   Multithreaded parallel computer model simulator

Our MPCM simulator uses the *DLX* RISC architecture instruction set [4]. We added several instructions for communication and synchronization between processors to the instruction set. These instructions include *network write*, *network read* and *barrier synchronization*. We have implemented MPs having 40 pipeline stages, which is reasonable for highly pipelined arithmetic computation unit [4].

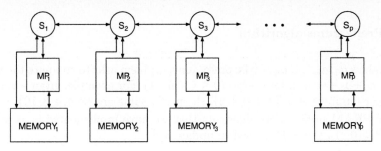

Fig. 1. The multithreaded parallel computer model.

We use one-dimensional linear array interconnection network model as shown in Figure 1. We chose this network topology because it has the weakest communication ability, and also because it is widely used. This weak ability will focus on the power of the MPCM. Each processor is connected to a switch and each neighboring switch is connected by a bidirectional link. A packet can transfer only 32 bit(1 word) to the left or to the right direction. We assume that it takes 4 cycles to transfer a packet to the neighboring switch. This is reasonable because the frequency of the clock of VLSI chips is 3-5 times faster than that of the mother board. From this assumption, a packet transfer from MP_i to MP_j needs at least $4|i - j|$ cycles. Furthermore, we need to consider the case that one processor and its right neighbor will transfer to its left processor simultaneously. Since each communication link transfers one packet at the same time, one of these two packets can not be transferred, in this case, the packet with further destination is transferred first. Here, we assume that each switch has only an unbounded buffer and each switch only process the packet at the first position of buffer at any time.

When we implement PRAM algorithms in our MPCM simulator, n PRAM processors $PE_0, PE_1, PE_2, \cdots, PE_{n-1}$ are equally assigned to p multithreaded processors $MP_0, MP_1, MP_2, \cdots, MP_{p-1}$. Thus, each processor $MP_i(0 \leq i \leq p - 1)$ has n/p register sets $R_{i,0}, R_{i,1}, \cdots, R_{i,n/p-1}$, and these register sets perform tasks that the PRAM processors $PE_{(i-1)\cdot n/p}, PE_{(i-1)\cdot n/p+1}, \cdots, PE_{i\cdot n/p-1}$ would execute.

Furthermore, memory assignment is important in multithreaded architecture model. In our research, it is known which memory location each processor should process. For example, for an array $a[0, n - 1]$ of input data, processor PE_i processes $a[i]$, so shared memory module on the PRAM can be simply divided into p local memory modules. We assume that the input data is allocated to corresponding local memory in advance and ignore the time for input and output.

4   PRAM algorithms implemented in the MPCM

In this section, we briefly describe PRAM algorithms that we have implemented in our simulator.

4.1   Prefix sums algorithm

For n values $a_0, a_1, \cdots, a_{n-1}$, the prefix sums problem asks to compute the values of $p_i = a_0 + a_1 + \cdots + a_i$ for every i ($0 \leq i \leq n-1$). For example, given a sequence of integer numbers $A = \{3, 1, 0, 4, 2\}$, the prefix sums are $\{3, 4, 4, 8, 10\}$.

The PRAM prefix sums algorithm that we have implemented is as follows: Each $PE_i (0 \leq i \leq n - 1)$ is used to update $a[i]$.

for $j = 0$ to $\lceil \log n \rceil - 1$ **do**
    for $i = 1$ to $n - 1$ **do in parallel**
        if $i - 2^j \geq 0$ **then** $a[i] = a[i] + a[i - 2^j]$

The above prefix sums algorithm runs in $O(\log n)$ time using n processors [5]. However, this algorithm is not work optimal because the product of the computing time and the number of processors is $O(n \log n)$. By assigning two or more data to each processor, cost optimization can be achieved. The input data of size n is equally partitioned into s groups such that the ith ($1 \leq i \leq s$) group is $A_i = \{a((i - 1) \cdot \frac{n}{s} + 1), a(i \cdot \frac{n}{s} + 2), \ldots, a(i \cdot \frac{n}{s})\}$. Then, the (local) prefix sums within each group are computed in $O(\frac{n}{s})$ time using a single processor. After that, the prefix sums of the sums of $A_0, A_1, \ldots A_{s-1}$ are computed in $O(\log s)$ time using the prefix sums algorithm described above. Finally, the prefix sums are added to the local prefix sums in obvious way. Clearly, this algorithm runs in $O(n/s + \log s)$ time using s processors.

4.2   List ranking algorithm

Consider a linked list of n nodes whose order is specified by an array p such that $p[i]$ contains a pointer to the next node i in the list, for $1 \leq i \leq n$. We assume that $p[i] = i$ when i is the tail of the list. The list ranking problem is to determine the distance of each node from the tail of the list.

The PRAM algorithm to determine the position of each node on a linked list is as follows.

for $i = 0$ to $n - 1$ do in parallel
    if $p[i] = i$ then $r[i] = 0$ else $r[i] = 1$
for $j = 1$ to $\lceil \log n \rceil$ do
    for $i = 1$ to $n - 1$ do in parallel
        begin
            $r[i] = r[i] + r[p[i]]$
            $p[i] = p[p[i]]$
        end

This algorithm runs in $O(\log n)$ time using n processors [5].

5   Performance evaluation

This section shows the performance evaluation of the above PRAM algorithms using our simulator.

Table 1 shows the number of cycles for n input data, p processors and n/p register sets/processor. We can verify that with $n = 512$, the speedup is almost linear up to 16 processors. This is because the number of threads in each processor is larger than the pipeline stages, and hence, enough instructions can be provided to the MP control unit. On the other hand, when $p \geq 32$, the number of threads range from 1 to 16, and the algorithm runs in 8000 to 9000 cycles regardless of the number of processors. When $p = 512$, the speedup turns down due to a large amount of communication among processors. In the case of $n = 8k(8192)$, linear speedup is nearly achieved up to 256 processors. If we increase input data, and use the same number of processors, the number of register sets n/p increase, and the number of threads put into the pipeline also increase.

Table 1. The number of cycles of the prefix sums algorithm

n \ p	1	2	4	8	16	32	64	128	256	512
512	95250	51712	26175	13471	9424	8996	8698	8510	8318	8910
1k	211986	114176	57727	29631	15711	11196	10760	10470	10350	10244
2k	466962	249856	126207	64639	34111	18975	14040	13564	13330	13338
4k	1042411	542720	273919	140031	73599	40639	24287	18932	18488	18366
8k	2252780	1193936	590847	301567	157951	94847	59455	33759	27968	27636
16k	4841452	2555857	1292235	646143	337407	184063	116095	78271	51423	45340

Table 2 shows the number of cycles of the optimal prefix sums algorithm for 64k input data, p processors and m register sets/processor. To each register set, $64k/mp$ input data is assigned. The smallest number of cycles is achieved with $m = 64$ and p ranging from 1 to 16. That is because not only 64 is larger than the pipeline stages of 40, but also the input data assigned to each register set for $m = 64$ is larger than with $m > 64$. If the number of processors is larger than 16, the smallest number of cycles varies according to the amount of threads, input data assigned to each register set and communication overhead.

Table 2. The number of cycles of the optimal prefix sums algorithm: 64k input

m \ p	1	2	4	8	16	32	64	128	256	512	1024	2048
1	16256573	16259137	8135003	4073865	2044111	1029941	523467	270796	145053	82942	53103	40432
2	16259380	8135462	4074372	2044666	1030548	524094	271392	1455588	83346	53248	40062	37132
4	8135700	4074540	2044852	1030744	524300	271608	145804	83536	53352	39948	36544	41480
8	4078820	2045028	1030912	524500	271768	145956	83656	53404	39872	36212	40632	55569
16	2045352	1031152	524708	272052	146241	83956	53681	40116	36337	40516	54960	86856
32	1031532	525092	272432	146640	84380	54144	40612	36808	40892	55064	94656	159416
64	806591	421440	226655	130015	82655	60511	52319	54047	66655	96927	160286	–
128	840063	452992	259263	163519	117183	96703	100031	111964	136255	196956	–	–
256	904959	517888	325503	230783	209772	188580	177023	204974	254828	–	–	–
512	1036031	649728	458495	364287	320255	318828	325375	369919	–	–	–	–

Table 3 shows the number of cycles of the list ranking algorithm for n−node linked list and p processors. Linear speedup is nearly achieved up to 16 proces-

Table 3. The number of cycles of the list ranking algorithm

p \ n	1	2	4	8	16	32	64	128	256	512
512	129024	64205	38509	24726	19324	18635	18527	18860	19480	25834
1k	283648	141515	84865	53660	37403	31412	30812	31076	32679	33890
2k	618496	309805	185579	117971	82408	64463	57732	57181	58340	62123
4k	1361887	665855	402404	258380	181745	143661	124289	117539	117564	120900
8k	2924511	1450445	868441	551296	386144	303014	261414	240420	233131	235939
16k	6250463	3108902	1865968	1201185	845634	668873	578670	533555	512162	505642

sors. However, for $p \geq 32$, the speedup raises very slowly, because communication among processors is so random that it spends most execution time. In order to decrease the communication overhead we need to employ higher performance networks, such as mesh and hypercube.

References

1. M. Amamiya, H. Tomiyasu, S. Kusakabe, Datarol: a parallel machine architecture for fine-grain multithreading, *Proc. 3rd Working Conference on Massively Parallel Programming Models*, 151–162, 1998.
2. A. Gibbons and W. Rytter, *Efficient Parallel Algorithm*, Cambridge University Press, 1998.
3. R. H. Halstead and T. Fujita, MASA: A multithreaded processor architecture for parallel symbolic computing, *Proc. 15th International Symposium on Computer Architecture*, 443–451, 1988.
4. John L. Hennessy, and David A. Patterson, *Computer Architecture–A Quantitative Approach*, Morgan Kaufmann, 1990.
5. J. JáJá. *An Introduction to Parallel Algorithms*. Addison-Wesley, 1992.
6. Robert A. Iannucci ed., *Multithreaded Computer Architecture: A Summary of the state of the Art*, Kluwer Academic, 1990.
7. J. T. Kuehn and B. J. Smith, The Horizon supercomputing system: architecture and software, Proc. Supercomputing 88, 28–34, 1988.
8. M. Loikkanen and N. Bagherzadeh, A fine-grain multithreading superscalar architecture, *Proc. of Conference on Parallel Architectures and Compilation Techniques*, 1996.
9. G. M. Papadopoulos and D. E. Culler, Monsoon: an explicit token-store architecture, *Proc 17th International Symposium on Computer Architecture*, 82–91, 1990.
10. G. M. Papadopoulos and K. R. Traub Multithreading: A revisionist view of dataflow architecture, *Proc 18th International Symposium on Computer Architecture*, 342–351, 1991.
11. R. G. Prasadh and C.-L Wu, A Benchmark Evaluation of a Multi-Threaded RISC Processor Architecture, *Proc. of International Conference on Parallel Processing*, pp. 84–91, 1991.
12. B. J. Smith, Architecture and applications of the HEP multiprocessor system, *Proc. of SPIE –Real-Time Signal Processing IV*, Vol. 298, Aug, 1981
13. J.-Y. Tsai and P. C. Yew, The Superthreaded Architecture: Thread Pipelining with Run-time Data Dependence Checking and Control Speculation *Proc. of Conference on Parallel Architectures and Compilation Techniques*, 1996.

Workshop on Parallel and Distributed Computing in Image Processing, Video Processing, and Multimedia (PDIVM 2000)

Organizers

Sethuraman Panchanathan, Arizona State University, USA
Andreas Uhl, Salzburg University, Austria

Preface

In the recent years, computing with visual and multimedial data has emerged as a key technology in many areas. However, the creation, processing, and management of these data types require an enormous computational effort, often too high for single processor architectures. Therefore, this fact taken together with the inherent data parallelism in these data types makes image processing, video processing, and multimedia natural application areas for parallel and distributed computing.

The Workshop on Parallel and Distributed Computing for Image Processing, Video Processing, and Multimedia (PDIVM 2000) brings together practitioners and researchers working in all aspects of parallel and distributed computing in these fields. It may be seen as a continuation of the workshops on Parallel Processing and Multimedia held at IPPS'97 and IPPS/SPDP'98 with extended scope. The meeting serves as a forum for exchange of novel ideas on corresponding hardware developments, software tools, algorithms, system solutions, and all types of applications.

PDIVM 2000 aims to act as a platform for topics related, but not limited, to

- Parallel and distributed architectures and algorithms
- Dynamically reconfigurable architectures
- Parallel DSP systems and Media processors
- Application specific parallel architectures
- Languages, software environments and programming tools
- Parallel and distributed video and multimedia servers
- Networked multimedia systems, QoS techniques
- Applications, e.g. remote sensing, medical imaging, satellite image processing, set-top boxes, HDTV, mobile multimedia, cameras

J. Rolim et al. (Eds.): IPDPS 2000 Workshops, LNCS 1800, pp. 161–162, 2000.
© Springer-Verlag Berlin Heidelberg 2000

Committees

Workshop Co-Chairs

Sethuraman Panchanathan, Arizona State University, USA
Andreas Uhl, Salzburg University, Austria

Program Committee

Laszlo Boezoermenyi, Univ. Klagenfurt, Austria
Michael Bove Jr., MIT Media Lab, USA
Larry S. Davis, Univ. of Maryland, College Park, USA
Edward J. Delp, Purdue University, USA
Divyesh Jadav, IBM Research Center, Almaden, USA
Ashfaq A. Khokhar, University of Delaware, USA
Sami Levi, Motorola Corporate Research Labs
Ming L. Liou, Hong Kong University of Science and Technology, China
Reinhard Lueling, Univ. Paderborn, Germany
Peter Pirsch, Univ. of Hannover, Germany
Edwige Pissaloux, Univ. Rouen, France
Viktor K. Prasanna, Univ. Southern California, USA
Subramania Sudharsanan, SUN Microelectronics
Ming-Ting Sun, Univ. of Washington, USA
Wayne Wolf, Princeton Univ., USA

MAJC-5200: A High Performance Microprocessor for Multimedia Computing

Subramania Sudharsanan

Sun Microsystems, Inc., Palo Alto, CA 94303, USA

Abstract. The newly introduced Microprocessor Architecture for Java Computing (MAJC) supports parallelism in a hierarchy of levels: multiprocessors on chip, vertical micro threading, instruction level parallelism via a very long instruction word architecture (VLIW) and SIMD. The first implementation, MAJC-5200, includes some key features of MAJC to realize a high performance multimedia processor. Two CPUs running at 500 MHz are integrated into the chip to provide 6.16 GFLOPS and 12.33 GOPS with high speed interfaces providing a peak input-output (I/O) data rate of more than 4.8 G Bytes/second. The chip is suitable for a number of applications including graphics/multimedia processing for high-end set-top boxes, digital voice processing for telecommunications, and advanced imaging.

1 Introduction

MAJC-5200 is a high performance general purpose microprocessor (based on MAJC [1]) exceptionally suitable for multimedia computing. The processor targets networked and communication devices, client platforms, and application servers delivering digital content and Java applications. It is a multiprocessor system on a chip integrating two CPUs, a memory controller, a PCI controller, two high bandwidth I/O controllers, a data transfer engine, and a crossbar interfacing all the blocks. This article describes the MAJC-5200 microprocessor and points out its capabilities in multimedia computing. The descriptions start with an overview of the MAJC architecture in Section 2. Section 3 details the MAJC-5200 microprocessor. The CPU, its pipeline, various functional blocks, memory subsystem, and external interfaces are described. Instruction set architecture as implemented in MAJC-5200 is given in Section 4. In Section 5, we list benchmark numbers for some multimedia and signal processing applications. Concluding remarks are in Section 6.

2 Architecture

The design of MAJC has been primarily towards exploiting the advances in both hardware and software technologies to address the new computational challenges. These challenges come from the increase in communication bandwidth and the techniques to communicate and process the data that is increasingly of auditory and visual type. The advances in the semiconductor technology boost the number

J. Rolim et al. (Eds.): IPDPS 2000 Workshops, LNCS 1800, pp. 163-170, 2000.

of transistors in a chip and the frequency they can be switched at. Software technology improvements in compilers, Java, and multi-threaded applications make it possible for high level language development of multimedia and telecommunication infrastructure applications to match the *internet-speed* time-to-market requirements. MAJC is scalable, exploits parallelism at a hierarchy of levels and modular for ease of implementation. At the highest level of parallelism, MAJC provides inherent support for multiple processors on a chip. The next level is the ability of *vertical micro-threading* which is attained through hardware support for rapid, low overhead context switching. The context switches can be triggered through either due to a long latency memory fetch or other events. The next hierarchy of parallelism comes from an improved very long instruction word (VLIW) architecture. The instruction packets can vary in length, up to 128 bits, with a maximum of four instructions each of 32 bits per packet. The lowest level of parallelism comes from single instruction multiple data (SIMD) or sub-word parallelism.

3   MAJC-5200 Microprocessor

3.1   Building Blocks

The first implementation of MAJC, MAJC-5200, is a 500 MHz, dual-CPU multimedia processor with a high I/O bandwidth. It implements several key parallelism features of MAJC. The two CPUs share a coherent four-way set-associative 16-KB data cache and common external interfaces. Each of these CPU is a four-issue MAJC VLIW engine. Each CPU contains its own two-way set-associative instruction cache of 16 KB. A high throughput bandwidth requirement is addressed by a multitude of interfaces with built-in controllers. The main memory is a direct Rambus DRAM (DRDRAM) with an interface supporting a peak transfer rate of 1.6 GB/s. A direct interface to 32-bit/66 MHz PCI provides DMA and programmed I/O (PIO) capabilities to transfer up to 264 MB/s. There are two other interfaces that support up to 4.0 GB/s that could be used for high speed parallel (64-bits at 250 MHz) interfaces: North and South UPA. (Universal Port Architecture or UPA has been an interface for graphics and multi-processor configurations of UltraSparc- based systems [3].) The NUPA block contains a 4 KB input FIFO buffer that can also be accessed by both CPUs. The other specialized block in the chip is a graphics preprocessor (GPP). The GPP has built-in support for real-time 3D geometry decompressing, data parsing, and load balancing between the two processors. An on-chip Data Transfer Engine (DTE) provides DMA capabilities amongst these various memory and i/o devices, with the bus interface unit acting as a central crossbar. The chip block diagram is depicted in Figure 1.

3.2   MAJC CPU

The core of each CPU as depicted in Fig. 2 has four functional units, FU0 through FU3. The VLIW instruction packet can have one, two, three, or four

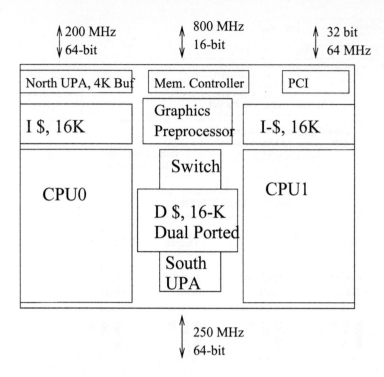

Fig. 1. MAJC-5200 Block Diagram

instructions in it. A two-bit header indicates the issue width, reducing unnecessary **nops** in the instruction stream. The first instruction in a packet must be an FU0 instruction which can be a memory operation (loads and stores), control flow operation (branch, jump and link, etc.) or an ALU (shift, add, etc.) type. FU0 interfaces with the LSU to perform the memory operations. Some special compute instructions are also executed in the FU0. The instructions for FU1-3 are of compute types.

Processor Pipeline During the fetch stage, a 32-byte aligned data is brought from the instruction cache unit. The next stage aligns an instruction packet of 16 bytes based on the header bits. The aligned instructions are then placed in an instruction buffer to feed the decode stage. The branch prediction information bits are looked at this stage to prepare for both static and dynamic predictions. The instruction decode and operand read from the register file occur in the next stage. Instructions then enter the execution pipelines in the functional units. Since the instruction scheduling is a compiler driven task in a VLIW machine, only the non-deterministic loads and long latency instructions are interlocked through a score-boarding mechanism. All the other instructions have a deterministic delay. MAJC-5200 provides precise exception handling capabilities for most instructions.

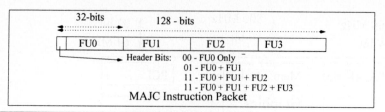

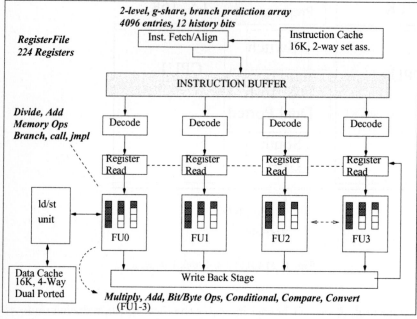

Fig. 2. MAJC CPU

The results are committed to the register file via the write back (WB) stage. In general, computational results become visible to other functional units from Trap/WB stage onwards, with some optimizations for results of FU0 and FU1. Majority of the instructions take a single cycle, with some taking two and four cycles. Within a functional unit, the results are bypassed to younger instructions as soon as available. The results of FU0 are forwarded to FU2 and FU3 with one cycle delay and are visible to FU1 in the next cycle. Similarly, the results of FU1 are forwarded to FU0 without any delay. This complete bypass between FU0 and FU1 enables a simple two-scalar performance for programs that exhibit lower levels of ILP.

Each CPU has a total of 224 logical registers partitioned into 96 global, accessible by all units, and 128 local, 32 for each FU, local registers. The Load/Store unit (LSU) on each CPU handles all memory operations to and from the registers. The LSU aggressively implements a non-blocking memory subsystem to allow multiple outstanding loads and stores. It provides buffering for up to five loads and eight stores. It allows a maximum of four cache misses without blocking the execution and handles out- of-order data returns. Non-faulting prefetch instructions that prefetch a block of data (32-byte) from the main memory to the data cache are also queued in LSU. Support for memory barrier and *atomic* instructions for synchronization of threads is also part of the LSU unit. The data cache access is brisk, allowing a two cycle load-to-use performance. Coupled with the synchronization instructions, this shared data cache provides a powerful, very low overhead communication between the two CPUs thus allowing a wide variety of compiler and run-time optimizations.

4 Instruction Set

MAJC-5200 provides support for 64-, 32-, 16-, and 8-bit integers in addition to 32- and 16-bit fixed point numbers. Both single (32-bit) and double (64-bit) IEEE 754-1985 floating point numbers are supported as well. A variety of load/store operations supporting byte, short, word, long, and group (1,2,4,8, and 32 bytes) can be issued as cached, non-cached, or non-allocating ones. The prefetch instruction is useful in programs with predictable data access patterns common in multimedia and image processing/graphics and other signal processing applications. MAJC-5200 implements branch (on conditions), call, and jump-and-link as FU0 instructions for program flow control. Predication instructions that conditionally pick (FU1-3), move (all FUs), and store (FU0) facilitate compiler to generate codes with fewer conditional branches.

Standard logical operations, AND, OR, XOR; logical and arithmetic shifts; and arithmetic operations, ADD and SUB, on 32-bit integers can be executed on all the functional units. All the units are capable of setting arbitrary constants. FU1-3 additionally provide saturated addition and subtraction of 32-bit integers. The FU1-3s further provide two cycle 32-bit integer multiplication and fused multiply-add (or multiply-sub) instructions that can be fully pipelined. Instructions that produce the high 32-bit part of a 64-bit product of two 32-bit integers facilitate obtaining 64-bit multiplies. Integer (32-bit) divide instruction is non-pipelined and available as an FU0 instruction.

SIMD versions of the arithmetic and logical operations available on FU1-3 take just one cycle. These operate on 16-bit short integer pairs or S.15 or S2.13 fixed point formats (Sign.integer.fraction). Furthermore, four different saturation modes can be enabled to automatically saturate the results. Fully pipelined SIMD counterpart of the multiply-add fused operation again has both regular and saturated versions operating on the same data types. In addition a dot product instruction that preserves the full 32-bits of precision multiplies and accumulates two pairs of 16-bit values. Saturated S.31 product of two S.15 quantities can also be obtained. Another set of unique FU0 SIMD instructions are six-cycle parallel divide and reciprocal square-root for S2.13 fixed point format data. These instructions form a basis for very powerful DSP, media, and graphics capabilities: multiply-add fused and dot product instructions aiding the filtering and transforming operations; and the FU0 SIMD helping the graphics lighting routines. At 500 MHz clock frequency, the peak performance becomes more than 12.33 GOPs for the 16-bit quantities.

The single precision floating point instructions are equally powerful. Pipelined addition, subtraction, multiply and fused multiply-add are available in the FU1-3s with just four cycle latency. Divide and reciprocal square-root instructions implemented in the FU0 have just six cycle latency. These make the processor capable of 6.16 GFLOPS at 500 MHz. The result is a very powerful single precision floating point FFT and graphics transform routines. Functional units FU1-3 provide double precision floating point addition, subtraction, and multiply operations. These instructions are partially pipelined for optimal performance and simpler scheduling by the compiler. Other floating point (single and double precision) instructions include minimum and maximum of two numbers; and a negate operation.

The bit manipulation instructions in the FU1-3s (bit field extract and leading zero detect) are quite useful in parsing compressed bit streams and in handling data communications. The bit extract is also a general purpose alignment instruction since the field extracted can span two registers. The byte-shuffle instruction in the FU1-3s provides a very versatile permutation function that can be used for alignment, table look-up, and zeroing byte-fields in a register. The pixel distance instruction computes the L_1-norm distance between two vectors of four packed bytes and accumulates them to a register. This instruction, combined with byte-shuffle provides excellent motion estimation performance for video coding applications.

5   Performance in Multimedia Applications

MAJC-5200 is slated to become the primary microprocessor in a variety of products that include high-end graphics systems, telecommunication infrastructure and document processing. For 3D graphics processing, MAJC-5200 has two features that significantly enhance its performance: the graphics pre-processor (GPP) and the two CPUs. The GPP decompresses compressed polygon information and distributes the uncompressed information to the CPUs using a load balancing mechanism. The geometry transformation and lighting are then performed using the CPUs. This pipelined architecture delivers a performance of between 60 and 90 million triangles per second. With such performance, MAJC-5200 can be used in a variety of graphics system architecture including multichip configurations to realize high-end graphic systems [4]. All the performance numbers provided are estimated using instruction accurate and cycle accurate simulators.

The performance in video and image processing applications is mainly attained using the instruction set that is suitable for several key kernels. Table 1 lists a few video signal processing benchmarks that indicate the performance of MAJC using a single CPU. Of course, in several of the applications it is possible to obtain thread level parallelism to effectively use both the CPUs. The versatile bit and byte manipulation operations help the variable length decoding common for image and video decompression. Combining this with static ILP and software pipelining, one can decode a variable length symbol and perform inverse zip-zag transform and inverse quantization [5] within 18 cycles. Motion estimation for a video encoder is significantly sped up via the byte permutation and pixel distance operations. Using a logarithmic search mechanism [5], a motion vector with a ±16 range can be found within about 3000 cycles. Large register file aids in convolution operations since the filter coefficients, image data, and the intermediate values can be easily stored in registers avoiding memory operations even when software pipelining techniques that typically require many registers are employed.

Table 1. Video/Image Processing Benchmarks (From Simulators)

Benchmark Kernel	Per Single MAJC CPU
8x8 IDCT	304 cycles
8x8 DCT + Quantization	200 cycles
MPEG-2 VLD+IZZ+IQ	27 MSymbols/sec
Motion Est./ ±16 MV range	3000 cycles
5x5 Convolution (512x512)	1.65 Mcycles
512x512 Color Conversion	0.9 MCycles

MAJC performance for one dimensional signal processing benchmarks also stands out as given in Table 2. The performance primarily stems from fused multiply-add instructions and corresponding memory bandwidth to actively keep the functional units computing. The benchmark kernels are commonly used ones in comparing digital signal processors and are heavily used in communication and speech processing applications. One interesting point to note is that unlike traditional DSPs that have smaller register files, MAJC-5200 is capable of using the compute efficient Radix-4 FFT algorithms. Bit reversal for FFT is however required to be performed using table look-up since no bit-reversed addressing is available. All the given benchmarks are for floating point numbers.

Using similar kernels, several applications have been developed. A small set is given here in Table 3. Again, these performances are estimated using simulators for a single CPU of the MAJC-5200 microprocessor running at 500 MHz.

6   Conclusion

MAJC-5200, the first implementation of the MAJC architecture, has been briefly described. Covered are the major architectural features and the the multimedia capabilities of the microprocessor. At 500 MHz, the processor delivers more than

Table 2. Signal Processing Benchmarks (From Simulators)

Benchmark Kernel	Per Single MAJC CPU
Cascade of eight 2nd order Biquads	63 cycles
64-sample, 64-tap FIR	2757 cycles
64-sample, 16th order IIR	2021 cycles
64-sample, 64-tap Complex FIR	8643 cycles
Single Sample, 16th order LMS	64 cycles
Max Search, maximum value in array of 40	126 cycles
Radix-2, 1024-point complex FFT	25196 cycles
Radix-4, 1024-point complex FFT	16996 cycles
Bit reversal, 1024-point	2484 cycles

Table 3. Application Performance (From Simulators)

Application	Single MAJC-5200 CPU Utilization
G.723.1 (encode) - float	1.6 % (1% without memory effects)
G.729.A (encode) - float	2.0 % (1% without memory effects)
MPEG-2 Video Decode (5Mbps, MP@ML)	75% (43 % without memory effects)
AC-3, MP2 Audio Decode	3-5 %
JPEG Baseline Encode	40 MB/s
Proprietary Lossless Coding	40 MB/s
H.263 Codec (128 kbps, 15 fps, CIF)	50 %

6 GFLOPS and 12 GOPS of raw performance. The performance of MAJC-5200 in graphics, image/video processing, and signal processing has been shown to show the suitability of it in a number of high performance applications.

References

1. M.Tremblay "A microprocessor architecture for the new millennium," Hot Chips 11, Palo Alto, CA, August 1999.
2. L.Kohn, *et al* "Visual Instruction Set (VIS) in UltraSparc," *Proceedings of Compcon*, IEEE CS Press, pp. 462-469,1995.
3. Sun Microsystems, Inc. http://www.sun.com/microelectronics/UltraSPARC/
4. M. Tremblay, "A VLIW convergent multiprocessor system on a chip," Microprocessor Forum, San Jose, CA, Oct. 1999.
5. V. Bhaskaran and K. Konstantinides, *Image and Video Compression Standards: Algorithms and Architecture*, Second Edition, Kluwer, Boston, 1997.

A Novel Superscalar Architecture for Fast DCT Implementation

Zhang Yong[1] and Min Zhang[2]

[1]DSP Lab, S2, School of Electronic and Electrical Engineering,
Nanyang Technological University, Singapore, 639798
eyzhang@ntu.edu.sg

[2]Reseach Institute of Planning & Design
Huaihe River Water Resources Commission of MWR,
41, Fengyang West Rd., Bengbu Anhui, P. R. China, 233001
zmxw@bb.ah.cninfo.net

Abstract. This paper presents a new superscalar architecture for fast discrete cosine transform (DCT). Comparing with the general SIMD architecture, it speeds up the DCT computation by a factor of two at the cost of small additional hardware overheads.

1. Introduction

The Discrete Cosine Transform (DCT) is currently the most popular and effective transform coding scheme to reduce the redundancy of the signals in visual communication and multimedia applications. The broader use of DCT in the wider application areas underlines the requirement for a more efficient and systematic approach to DCT implementation.

The SIMD (Single Instruction Multiple Data) processor has been widely employed for DCT implementation in video encoders because of two reasons. Firstly, more flexibility can be achieved by processing the task under software control. Moreover, since all of the low level tasks (e.g., DCT, motion estimation and quantization) in the whole video compression system have similar computing characteristics, a common processor architecture can be shared for these computing requirement so as to simplify the design problem and save the hardware cost. Fig. 1 illustrates a typical SIMD architecture used in both video signal processors (VSP) [1] and generalized multimedia processors [2]. It properly exploits temporal parallelism and spatial parallelism to process DCT as well as other low level computation-intensive tasks.

However, comparing with the functional specific DCT implementations, the SIMD model cannot achieve best performance for Fast DCT (FDCT) algorithms, consequently lowering the whole system's throughput. In this paper, we will present a superscalar modification of the general SIMD structure shown in Fig. 1. It completes the DCT computation twice as fast as the original SIMD does, at the cost of small hardware overheads.

J. Rolim et al. (Eds.): IPDPS 2000 Workshops, LNCS 1800, pp. 171-177, 2000.

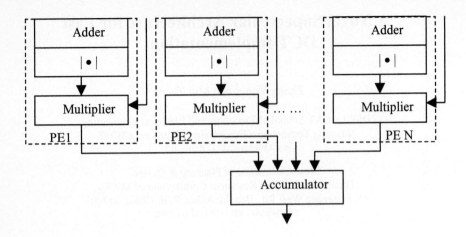

|•| -- Absolute operator

Fig. 1. A typical SIMD architecture

2. Modified SIMD Architecture for Fast DCT

A typical 8-point DCT transform is computed in the following form:

$$X_i = c_i \sum_{k=0}^{7} x_k \cos(\frac{(2k+1)i\pi}{16}), \ 0 \le i \le 7 \tag{1}$$

where x_i are the original data, X_i are the transformed values and c_i are weight parameters defined as

$$c_i = \begin{cases} \sqrt{2} & i = 0 \\ 1 & otherwise \end{cases} \tag{2}$$

In order to accelerate the DCT computation, many fast DCT algorithms have been developed. One of the most efficient methods is performing DCT with the aid of the fast Fourier transformation (FFT) [3]. Fig. 2 illustrates a computing flow chart for such FDCT algorithm, in which d_i denote the cosine coefficients and y_i, z_i are the intermediate results during transform.

It can be seen that there are many features in the FDCT algorithm that impede the SIMD architecture working effectively:

- There are many data transmissions among PEs, therefore a complex interconnection network is required for the FDCT implementation.
- The operation components in different datapaths (x_i to X_i) are various, this irregularity results in that the computation units in PEs cannot be fully utilized.
- Eight PEs are required in order to perform eight-point DCT concurrently. Since the datawidth of one datapath in DCT is normally 16 bits, the total bus width reaches

128bits, which is difficult for physical implementation in terms of the current semiconductor technology.

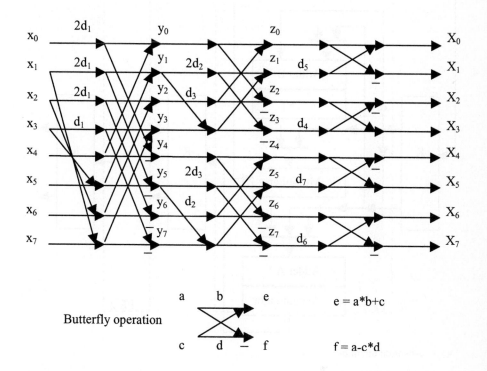

Fig. 2. Fast DCT flow chart

For these reasons, the FDCT algorithm is seldom employed in previous SIMD processors despite of its merit on computational efficiency. In order to solve this problem, some modifications have to be made for the SIMD architecture shown in Fig. 1.

Firstly, the PEs are connected each other to form a PE chain by fetching the result of one PE to its neighboring PE. With this manner, the whole computation flow is mapped onto the SIMD architecture. It can be deduced from Fig. 2 that the 3-PE Chain can fulfill the butterfly computations in 8-point FDCT.

Secondly, a register file **R** is added to each PE for storing the intermediate computation results. The complex interconnection of the PEs is consequently avoided by addressing the register properly.

In addition, an additional adder is introduced to the PE for assisting the butterfly computations.

The new architecture is shown in Fig. 3.

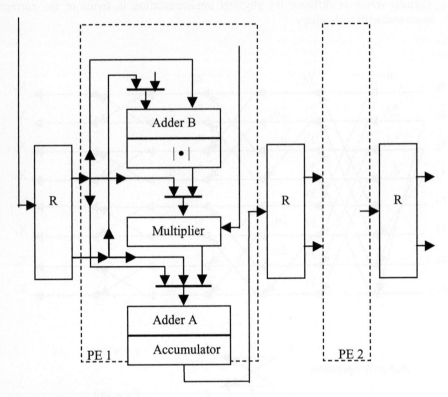

| • | -- Absolute operator

Fig. 3. Modified SIMD architecture

3. Superscalar Execution of FDCT

In a typical superscalar processor [5], the incoming instruction stream is fetched and decoded several instructions at a time, the instructions are then initiated for execution in parallel based primarily on the availability of operand data, rather than their original program sequence. Borrowing this important feature of the superscalar processor, we schedule the operations in the FDCT algorithm onto the architecture of Fig. 3 in terms of whether the operand data is ready. The timing chart of one 8-point FDCT execution is illustrated in Fig. 4, where:

the labels on the t axis are the cycle numbers of the execution;

'R' denotes reading the operand from register files;

'W' denotes writing the result to register files;

'*' denotes the multiplication;

'A+', 'A-', 'B+', 'B-' denote the addition and the subtraction of adder A and B respectively.

4. Comparison and Conclusion

As for the SIMD architecture shown in Fig. 1, when N=4, in order to compute one element in 8-point DCT, two cycles are required (see equation (1)). By contrast, as shown in Fig. 4, the new architecture completes each in one cycle at the pipeline output when the pipeline is full. Notice that in most DCT applications, the data to be transformed are input continuously, thus the preload overhead of the pipeline can be ignored. Therefore, the throughput of the modified superscalar architecture is twice faster than that of the original one.

The modified superscalar architecture and the original SIMD as well as their control circuits have been implemented by the logic synthesis procedure, which are followed by an estimation for the related gate numbers [4]. The results show that the hardware resources of the former are only 14% more than that of the latter.

In conclusion, the proposed architecture gains speedup of two over the previous SIMD one at the cost of small hardware overheads.

REFERENCES

[1] J. Hilgenstock, et al., "A video signal processor for MIMD multiprocessing", Proceedings of Design Automation Conference, 1998, pp. 50 –55.
[2] R. B. Lee, "multimedia extensions for general-purpose processors", IEEE Workshop on Signal Processing Systems, 1997, pp. 9 –23
[3] A. N. Netravali and B. G. Haskell, "Digital Pictures, -Representation, Compression, and Standards", Plenum Press, 1995.
[4] Zhang Yong, "Research on Video Encoder Design", Ph. D. Dissertation, Zhejiang University, 1999.
[5] D. A. Patterson, J. L. Hennessy, "Computer Architecture a Quantitative Approach", Morgan Kaufmann Publishers Inc., 1996.

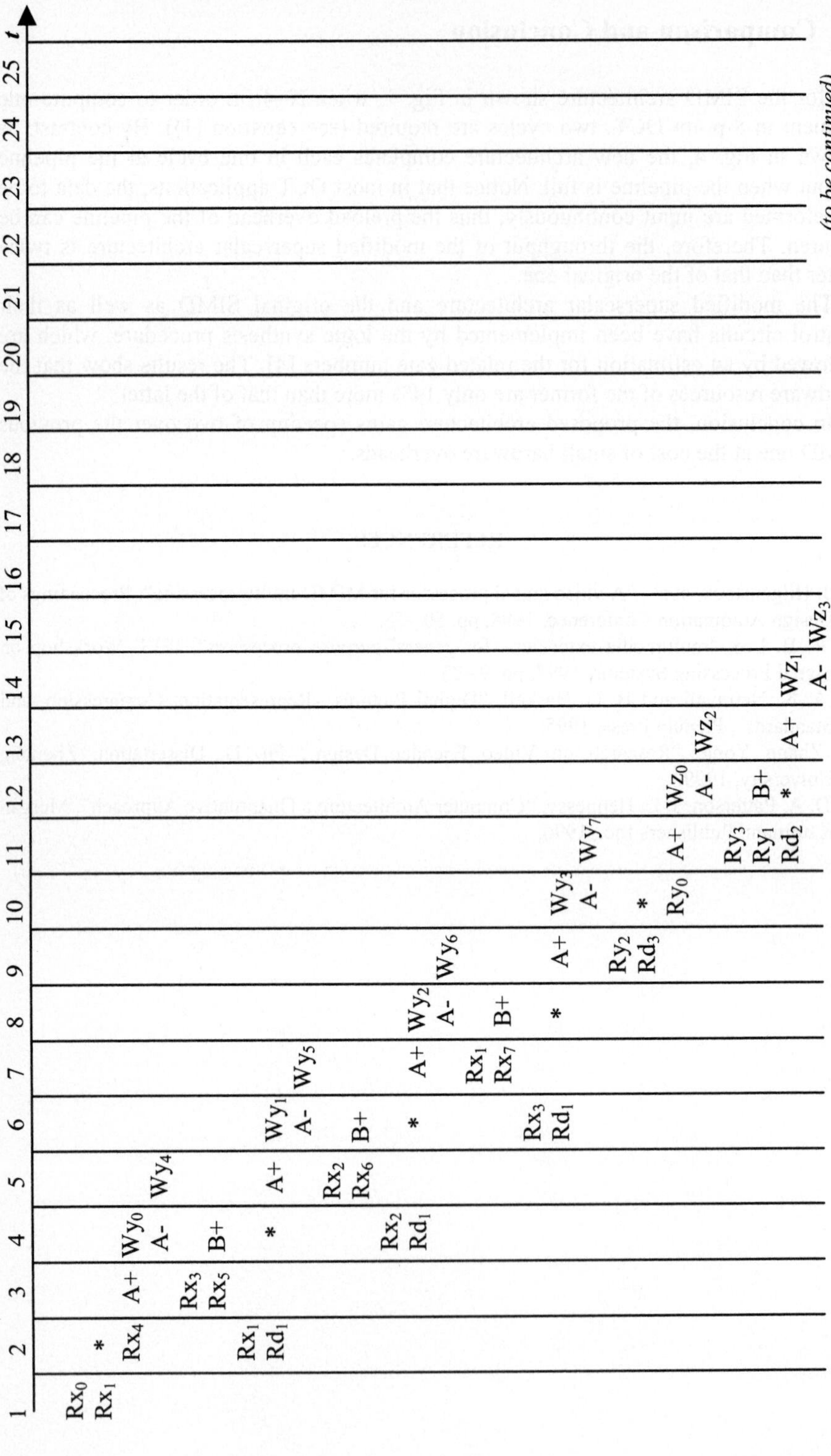

t

25 24 23 22 21 20 19 18 17 16 15 14 13 12 11 10 9 8 7 6 5 4 3 2 1

Rx_0
Rx_1

$*$
Rx_4
Rx_1
Rd_1

A+
Rx_3
Rx_5

Wy_0
A-
B+
$*$
Rx_2
Rd_1

Wy_4
A+
Rx_2
Rx_6

Wy_1
A-
B+
$*$
Rx_3
Rd_1

Wy_5
A+
Rx_1
Rx_7

Wy_2
A-
B+
$*$

Wy_6
A+
Ry_2
Rd_3

Wy_3
A-
$*$
Ry_0

Wy_7
A+
Ry_3
Ry_1
Rd_2

Wz_0
A-
B+
$*$

Wz_2
A-

Wz_1
A-

Wz_3

(to be continued)

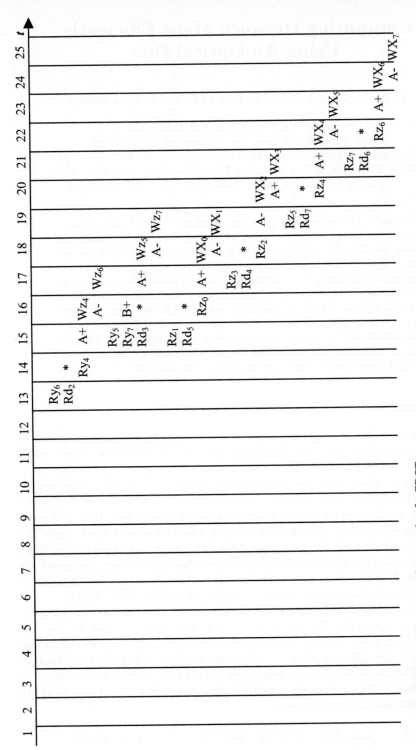

Fig. 4. Timing chart of the superscalar execution for FDCT

Computing Distance Maps Efficiently Using An Optical Bus

Yi Pan[1], Yamin Li[2], Jie Li[3], Keqin Li[4], and Si-Qing Zheng[5]

[1] University of Dayton, Dayton, OH 45469-2160, USA
[2] The University of Aizu Aizu-Wakamatsu 965-80 Japan
[3] University of Tsukuba Tsukuba Science City, Ibaraki 305-8573, Japan
[4] State University of New York New Paltz, New York 12561-2499, USA
[5] University of Texas at Dallas Richardson, TX 75083-0688, USA

Abstract. This paper discusses an algorithm for finding a distance map for an image efficiently using an optical bus. The computational model considered is the arrays with a reconfigurable pipelined bus system (LARPBS), which is introduced recently based on current electronic and optical technologies. It is shown that the problem for an $n \times n$ image can be implemented in $O(\log n \log \log n)$ time deterministically or in $O(\log n)$ time with high probability on an LARPBS with n^2 processors. We also show that the problem can be solved in $O(\log \log n)$ time deterministically or in $O(1)$ time with high probability on an LARPBS with n^3 processors. The algorithm compares favorably to the best known parallel algorithms for the same problem in the literature.

1 Introduction

In the areas of image processing and computer vision, it is usually required to extract information about the shape and the position of the foreground pixels relative to each other for a digital image. There are many computational techniques which can be used for such an information retrieval. One technique to accomplish this task is the distance map (or distance transform). The distance transform is to convert a binary image to another image, such that each pixel has a value to represent the distance between it and its nearest black pixel. The new image is called the distance map of the old image. Note that this problem is different from the problem of finding nearest neighbors, where the search for nearest black neighbors is only performed for black pixels, not for all pixels in an image. Hence, finding the distance map for an image is more difficult.

Consider a black and white $n \times n$ binary image: i.e., a 2-dimensional array where $a_{ij} = 0$ or 1, for $i, j = 0, 1, 2, ..., n - 1$. The index i stands for the row and the index j for column. Pixel $(0,0)$ is the upper left point in the image. There are many different distance transforms available using different distance metrics. The Euclidean distance transform (EDT) is to find for each point (i, j) its Euclidean distance from the set of all black pixels $B = \{(i, j) : a_{ij} = 1\}$. In other words, we compute the array

$$d_{ij} = \min_{(x,y) \in B} \{((i - x)^2 + (j - y)^2)^{1/2}\}, \quad 0 <= i, j <= n - 1.$$

J. Rolim et al. (Eds.): IPDPS 2000 Workshops, LNCS 1800, pp. 178-185, 2000.

The EDT is a basic operation in computer vision, pattern recognition, and robotics. For instance, if the black pixels represent obstacles, then d_{ij} tells us how far the point (i, j) is from these obstacles. This information is useful when one tries to move a robot in the free space (white pixels of the image) and to keep it away from the obstacles (black pixels).

Finding the distance transform with respect to the Euclidean metric is rather time consuming. Yamada [11] presented an EDT algorithm that runs in $O(n)$ time using n^2 processors on an 8-neighbor connected mesh. Kolountzakis and Kutulakos presented an $O(n^2 \log n)$ sequential algorithm for the EDT problem [5]. They also showed that the time complexity of their algorithm is $O((n^2 \log n)/r)$ on an EREW PRAM model with $r <= n$ processors. Chen and Chuang improved the algorithm in [2] by reducing the time complexity to $O(n^2/r + n \log r)$ on an EREW PRAM model. Bossomaier $et\ al.$ studied the speedup and efficiency of Yamada's algorithm on the CM-2 Connection machine, and introduced new techniques to improve its performance [1]. More recently, Chen and Chuang proposed an algorithm for computing the EDT on a mesh-connected SIMD computer [3]. For an $n \times n$ image, their algorithm runs in $O(n)$ time on a 2-dimensional $n \times n$ torus-connected processor array. In this paper, we propose more efficient parallel algorithms for computing the Euclidean distance transform on the LARPBS model. For the same problem, one of our algorithm runs in $O(\log n \log \log n)$ time or in $O(\log n)$ time with high probability on an LARPBS with n^2 processors. We also show that the time complexity can be further reduced if n^3 processors are used. The results presented in this paper improve on all previous results in the literature.

The computational model used is reviewed in the next section. Our algorithm for computing the distance map is described in section 3. Section 4 discuss the possibility of further reducing the time complexity of the algorithm by using more processors. We conclude our paper in section 5.

2   The LARPBS Model

Massively parallel processing using optical interconnections poses new challenges. In general, it does not worth the effort to improve performance by simply replacing the metal interconnections of a parallel system with optical interconnections in a one-to-one fashion. This is due to the fact that in doing so the inherent large bandwidth and high parallelism of optical interconnects are underutilized. The characteristics of optical interconnects have significant implications. New system configurations need to be designed due to the changes in architectural freedom and constraints when optical interconnects are incorporated. To fully utilize the bandwidth offered by the optical interconnections, scheduling and data communication schemes based on new resource metrics need to be investigated. Algorithms for a wide variety of applications need to be developed under novel computation models that are based on optical interconnections. Computations under these new models, which can be drastically different from existing theoretical PRAMs and parallel systems with electrical interconnections, may require new algorithm design techniques and performance measures.

A linear array with a reconfigurable pipelined bus system (LARPBS) consists of N processors $P_1, P_2, ..., P_N$ connected by an optical bus. In addition to the tremendous communication capabilities, a LARPBS can also be partitioned into $k \geq 2$ independent subarrays $LARPBS_1, LARPBS_2, ..., LARPBS_k$, such that $LARPBS_j$ contains processors $P_{i_{j-1}+1}, P_{i_{j-1}+2}, ..., P_{i_j}$, where $0 = i_0 < i_1 < i_2 \cdots < i_k = N$. The subarrays can operate as regular linear arrays with pipelined optical bus systems, and all subarrays can be used independently for different computations without interference (see ([7, 8]) for an elaborated exposition, and ([9]) for similar reconfigurable pipelined optical bus architectures).

For ease of algorithm development and specification, a number of basic communication, data movement, and global operations on the LARPBS model implemented using the coincident pulse processor addressing technique ([4, 6]) have been developed ([7, 8, 9, 10]). They include one-to-one routing, multicast, broadcast, segmented broadcast, compression, minimum finding, binary sum and image transpose. All of them except minimum finding can be performed in a constant number of bus cycles [7, 10]. Minimum finding can be done either in $O(\log \log N)$ time deterministically or in $O(1)$ time with high probability [8]. These powerful primitives that support massive parallel communications, plus the reconfigurability of the LARPBS model, make the LARPBS very attractive in solving problems that are both computation and communication intensive, such as image processing. Our algorithms are developed using these operations as building blocks. For further implementation details on these operations, the reader is referred to [7, 8, 10].

3   Algorithm Using n^2 Processors

In our algorithm description, the following terms are used. We divide the image into two parts for each pixel. Pixels to the left of pixel (i, j) are referred to pixels in the left plane of pixel (i, j). Similarly, Pixels to the right of pixel (i, j) are referred to pixels in the right plane of pixel (i, j).

If we denote the nearest black pixel to pixel (i, j) as $M(i, j)$, then the distance from (i, j) to $M(i, j)$ is the EDT at (i, j). We can reduce the search time for $M(i, j)$ by dividing the image into sections and search the sections separately.

In order to compute (1), it is sufficient to give an algorithm for the computation of

$$l_{ij} = \min_{(x,y) \in B} \{((i-x)^2 + (j-y)^2)^{1/2}\}, 0 \leq i, j \leq n-1.$$

Here, l_{ij} is the distance of the point (i, j) from the part of B that is to the left of (i, j). Computing the distance from the right and from the left and then taking the minimum of the two gives d_{ij}. So only the computation of l_{ij} is described here.

For a given j, we define W_j as the set of all black pixels to the left of column j. Denote the pixel in W_j nearest to (i, j) as $W(i, j)$. The following lemma is due to Kolountzakis and Kutulakos [5]:

Lemma 1 *Let $A = (a, j)$ and $B(b, j)$, $b < a$, be two pixels on the same column with A above B. Let $W(a, j) = (x, y)$ and $W(b, j) = (z, w)$. Then $z \leq x$; namely, $W(a, j)$ is above or on the same row as $W(b, j)$. Similarly, for two pixels on the same row, if $W(i, a) = (s, t)$ and $W(i, b) = (u, v)$ with $b < a$, then $v \leq t$; namely, $W(i, a)$ is left to or on the same column as $W(i, b)$.*

The search spaces for the nearest black pixels can be reduced by applying this lemma. Suppose it is known that $W(i, j) = (z, w)$. By lemma 1, we know $W(i, k)$ must be between columns 0 and $w - 1$ if $k < j$, and between columns w and $n - 1$ if $k > j$. Thus, we do not need to search all the columns.

Now, we describe the algorithm using the basic data movement operations described in the preceding sections. The basic idea of the algorithm is: 1) to find the nearest black pixel in each column for all pixels (including white pixels). 2) to search the nearest black pixel of a pixel in its left region followed by finding the nearest black pixel of a pixel in its right region. 3) we obtain the nearest black pixel for a pixel in the whole image by selecting the nearest black pixel in both regions.

Assume that initially each pixel a_{ij} is stored in the LARPBS with n^2 processors in row-major order. That is, $a(i*n+j) = a_{ij}$, for $0 \leq i, j < n$. Each processor has several local variables. In the following discussion, we use $D(i*n+j)$ and $D(i, j)$ interchangeably, both representing a variable D in processor $k = i*n+j$. The algorithm consists of the following steps:

Step 1: In this step, we calculate for pixel (including white pixels) at (i, j) the row index of the closest black pixel in the j-th column on or above the i-th row of the image. Initially, if $a_{ij} = 1$, set $UPNEAREST(i * n + j) = i$. Otherwise, set $UPNEAREST(i*n+j) = \infty$. In order to do this, the image is first transposed into column-major order, and the array is segmented into n subsystems. Each subsystem contains a column of the image. Then, a right segmented broadcast is performed using $UPNEAREST(i*n+j)$ as the local data value and the pixel value a_{ij} as the local logical value. The results are stored in $UPNEAREST(i * n + j)$.

Step 2: Calculate for pixel at (i, j) the row index of the closest black pixel in the j-th column on or below the i-th row of the image. Initially, if $a_{ij} = 1$, $DOWNNEAREST(i * n + j) = i$. Otherwise, $DOWNNEAREST(i * n + j) = \infty$. At the beginning of this step, the image is already in column-major order, and the processor array is already segmented into n subsystems. A left segmented broadcast is performed using that $DOWNNEAREST(i*n+j)$ as the local data value and a_{ij} the local logical value. The final results are stored in $DOWNNEAREST(i * n + j)$.

Step 3: Calculate the EDT values in their left regions for all the points in the image. In order to do this, the image is first transposed back into row-major order, and the array is segmented into n subsystems. Denote these subsystems as $ARR(i)$, $0 \leq i \leq n - 1$. Each subsystem $ARR(i)$ contains the i-th row of the image. We use subarray $ARR(i)$ to calculates EDT values in their left regions of row i for $0 \leq i \leq n - 1$. The computation is carried out through bisectioning each row until all EDT values in their left regions are calculated in a given row. Since the EDT values in their left regions for all rows are calculated similarly, in the following discussion, we concentrate on row i. First, we use subarray $ARR(i)$ to calculate the EDT value in the left region for point $(i, n/2,)$. This is done as follows. Assume that $DOWNNEAREST(i, j) = i_1$, $UPNEAREST(i, j) = i_2$, $PE(i, j)$ first calculates $DOWNDIST(i, j) =((i_1 - i)^2 + (j - n/2))^{1/2}$ and $UPDIST(i, j) =((i_2 - i)^2 + (j - n/2))^{1/2}$. These are the distances of the nearest black pixels in its column and the point $(i, n/2)$, whose EDT value is being computed. Then, $PE(i, j)$ compares $DOWNDIST(i, j)$ and $UPDIST(i, j)$, and put the smaller in $DIST(i, j)$. Finally, use the linear subarray $ARR(i, *)$ of n processors to find the minimum of the n distances $DIST(i, j)$, $0 \leq j \leq n - 1$. Clearly, this can be done in $O(\log\log n)$ time deterministically.

Suppose that $W(i, n/2) = (z, w)$. Column w divides row i into two subsections. Use lemma 1, to find the EDT value in the left reqion for another point, we need only to search the minimum values among the $DIST(i, j)$ values in its corresponding subsection depending on its position. Hence, we divide the subarray $ARR(i, *)$ into 2 subarrays using w as the partition column, and use the first subarray to calculate the EDT value in the left reqion at point $(i, n/4)$ and the second subarray to calculate the EDT value in the left reqion at point $(i, 3n/4)$. Notice that the sizes of the two subarrays may be different in a row and different rows have different partitions. The reconfiguration is done via disconnecting the bus at $PE(i, w)$ for $0 \leq i \leq n - 1$. This process is carried out for all rows in the system. To calculate the EDT value in the left reqion at point $(i, n/4)$, we only need to find the minimum of $DIST(i, j)$, for $0 \leq j < w$, in the first subarray. This is done as follows. Assume that $DOWNNEAREST(i, j) = i_1$, and $UPNEAREST(i, j) = i_2$. $PE(i, j)$ in the left subarray first calculates $DOWNDIST(i, j) =((i_1 - i)^2 + (j - n/4))^{1/2}$ and $UPDIST(i, j) =((i_2 - i)^2 + (j - n/4))^{1/2}$. Then, $PE(i, j)$ compares $DOWNDIST(i, j)$ and $UPDIST(i, j)$, and put the smaller in $DIST(i, j)$. Finally, use the first subarray in row i to find the minimum of the distances $DIST(i, j)$, $0 \leq j < w$. Since the size of the subarray is at most n, this can be done in $O(\log\log n)$ time using the deterministic minimum finding algorithm [8]. Similarly, to calculate the EDT value in the left reqion at point $(i, 3n/4)$ in the second subarray also takes at most $O(\log\log n)$ steps. Now, we can further partition a subsection into two parts and we reconfigure the subarray $ARR(i)$ into four subarrays according to the values $W(i, 3n/4)$, $W(i, n/2)$, and $W(n/4)$. We can find the EDT values in their left regions of the middle point in each of the four subsections using a similar procedure. Again, we only need to search each subarray to perform the minimum finding of the $DIST$ values base on lemma 1. Hence, at most $O(\log\log n)$ time is needed in this iteration since the minimum findings can be performed concurrently in these four subarrays. This process continues until the size of each subsection is 1.

Step 4: Once the distance of the point (i, j) the left region of (i, j) is computed, a method similar to the one used in step 3 can be employed to calculate the distance of the point (i, j) in the right region.

Step 5: Comparing the distances to a pixel in its left region and right region and then taking the minimum of the two gives d_{ij}. This step involves only a local operation.

At the end of the algorithm, the EDT result is in $EDT(i, j)$ for $0 \leq i, j \leq n - 1$. The time can be calculated as follows. Step 1 calculates the indices of the closest black pixel in the k-th column on or above the i-th row of the image. Similarly, step 2 finds the index of the closest black pixel in the k-th column on or below the i-th row of the image. Since both a transpose operation and a segmented broadcast operation require $O(1)$ time, step 1 takes $O(1)$ time. Similarly, step 2 uses $O(1)$ time. In step 3, several operations such as transpose operation, segmented broadcast, and minimum finding, and several local operations are employed. All operations use $O(1)$ time, except the minimum finding operation which uses $O(\log \log n)$ time. Clearly, we need $\log n$ iterations to find all the EDT valuesv in row i. Since all the n rows are calculated concurrently and $\log n$ iterations are needed, the time taken in step 3 is $O(\log n \log \log n)$. Therefore, the total time used in the algorithm is $O(\log n \log \log n)$.

As pointed before, minimum finding can also be performed in $O(1)$ with high probability using a randomised algorithm [8]. Since each row is partitioned into many segments in each iteration, the probability of using $O(1)$ time in step 3 depends on the number of segments and the probability of the minimum finding algorithm used in each segment. In our algorithm, if the size of a sub-array is smaller than 256, we can simply use the deterministic $O(\log \log n)$ time minimum-finding algorithm given above. For $n = 256$, $\log_2 \log_2 256 = 3$, and $O(\log \log 256) = O(1)$. If the size of a subarray is larger than 256, the above randomised minimum finding algorithm is used. Clearly, during the first iteration, there is only one segment. During the second iteration, we have two segments. In general, during the i-th iteration, we have 2^i segments, where $1 \leq i \leq \log n$. The total number of minimum findings in each row is less than n since each minimum finding corresponds to the calculation of the EDT value for one pixel. Hence, the total number of minimum findings in step 3 is $X = n^2/256$. Let P_i be the probability of $O(1)$ running time for a particular minimum finding i during the algorithm. The probability of the algorithm running in $O(1)$ time is the product of all X P_i's. The total number of randomized minimum finding operations is no larger than $X = n^2/256$, and each has a probability of at least $1 - 256^{-\alpha}$ (α is a function of C used in the minimum finding algorithm) of requiring $O(1)$ time. Thus, the probability of all the minimum finding operations running in $O(1)$ time during step 3 is $(1 - 256^{-\alpha})^{n^2/256}$. For images with practical sizes, this probability is very close to 1. To see this, consider $\alpha = 4$ and an LARPBS of 1,000,000 processors. The probability of all the minimum finding operations running in $O(1)$ time during step 3 is at least 0.9999991. Hence, step 3 uses $O(\log n)$ time with high probability. Similarly, step 4 requires $O(\log n)$ time with high probability. Step 5 performs some local comparisons and hence uses $O(1)$ time. Since each step in the algorithm requires either $O(1)$ time or

$O(\log n)$ time, the total time of the algorithm is $O(\log n)$.

In the above analysis for steps 3 and 4, we require α to be no less than 4. Fortunately, this can be easily achieved because a Monte Carlo algorithm which runs in $O(T(N))$ time with probability of success $1 - O(1/N^{\alpha})$ for some constant $\alpha > 0$ can be turned into a Monte Carlo algorithm which runs in $O(T(N))$ time with probability of success $1 - O(1/N^{\beta})$ for any large constant $\beta > 0$ by running the algorithm for $\lceil \beta/\alpha \rceil$ consecutive times and choosing one that succeeds without increasing the time complexity. Summarizing the above discussions, we obtain the following results:

Theorem 1 *The distance map problem defined on an $n \times n$ images can be solved using an LARPBS of n^2 processors in $O(\log n \log \log n)$ time deterministically or in $O(\log n)$ time with high probability for any practical size of n.*

4   Algorithm Using n^3 Processors

We can further reduce the time in the above algorithm through using more processors. In this section, we describe an algorithm which works on an LARPBS with n^3 processors. The n^3 algorithm is similar to the n^2 algorithm described in the preceding section. The only difference is that we use n^2 to find the EDT values in a row instead of using n processors, thus reducing the time used.

Initially, the n^2 pixels are stored in the first n^2 processors. Actually, all the steps except steps 3 and 4 use the first n^2 processors only, and hence have the same steps as in the n^2 algorithm described in the preceding section.

Now we describe step 3 in the new algorithm in detail. Notice that all values such as $DIST$'s have been computed and stored in local processors. We divide the LARPBS into n subsystems with each having n^2 processors. Denote these subsystems as LARPBS-i, $0 \le i \le n - 1$. Distribute the n rows of pixels along with the computed values such as $DIST$'s in the previous steps to the first n processors of the n subsystems. Thus, each subsystem is responsible for a row of pixels. There are n EDT values to be computed in a row and each subsystem has n^2 processors. Hence, we can let n processors calculate an EDT value and all the EDT values can be computed concurrently. An EDT value can be computed using the deterministic minimum finding algorithm or the randomized minimum finding algorithm on the $DIST$ values computed in step 2. Obviously, this step involves only broadcast, multicast, array reconfiguration, and minimum finding operations. All these operations takes $O(1)$ time except the minimum finding algorithm. Using a similar analysis described previously, it is easily obtained that step 3 can be computed in $O(\log \log n)$ time deterministically or in $O(1)$ time with high probability. Hence, we have the following results:

Theorem 2 *The distance map problem defined on an $n \times n$ images can be solved using an LARPBS of n^3 processors in $O(\log \log n)$ time deterministically or in $O(1)$ time with high probability for any practical size of n.*

5   Conclusions

Due to the high bandwidth of an optical bus and several efficiently implemented data movement operations, the distance map problem is solved efficiently on the LARPBS model. It should be noted that algorithms on plain a mesh or a reconfigurable mesh cannot reach the time bounds described in this paper.

References

1. T. Bossomaier, N. Isidoro, and A. Loeff, "Data parallel computation of Euclidean distance transforms," Parallel Processing Letters, vol. 2, no. 4, pp. 331-339, 1992.
2. L. Chen and H. Y. H. Chuang, "A fast algorithm for Euclidean distance maps of a 2-D binary image," Information Processing Letters, vol. 51, pp. 25-29, 1994.
3. L. Chen and H. Y. H. Chuang, "An efficient algorithm for complete Euclidean distance transform on mesh-connected SIMD," Parallel Computing, vol. 21, pp. 841-852, 1995.
4. D. Chiarulli, R. Melhem, and S. Levitan, "Using Coincident Optical Pulses for Parallel Memory Addressing," IEEE Computer, vol. 20, no. 12, pp. 48-58, 1987.
5. M.N. Kolountzakis and K.N. Kutulakos, "Fast computation of Euclidean distance maps for binary images," Information Processing Letters, vol. 43, pp. 181-184, 1992.
6. R. Melhem, D. Chiarulli, and S. Levitan, "Space Multiplexing of Waveguides in Optically Interconnected Multiprocessor Systems," The Computer Journal, vol. 32, no. 4, pp. 362-369, 1989.
7. Yi Pan and Keqin Li, "Linear array with a reconfigurable pipelined bus system: concepts and applications," Special Issue on "Parallel and Distributed Processing" of Information Sciences, vol. 106, no. 3/4, pp. 237- 258, May 1998. (Also appeared in International conference on Parallel and Distributed Processing Techniques and Applications, Sunnyvale, CA, August 9-11, 1996, 1431-1442)
8. Y. Pan, K. Li, and S.Q. Zheng, "Fast nearest neighbor algorithms on a linear array with a reconfigurable pipelined bus system," Parallel Algorithms and Applications, vol. 13, pp. 1-25, 1998.
9. S. Rajasekaran and S. Sahni, "Sorting, selection and routing on the arrays with reconfigurable optical buses," IEEE Transactions on Parallel and Distributed Systems, vol. 8, no. 11, pp. 1123-1132, Nov. 1997.
10. J. L. Trahan, A. G. Bourgeois, Y. Pan, and R. Vaidyanathan, "Optimally scaling permutation routing on reconfigurable linear arrays with optical buses," Proc. of the Second Merged IEEE Symposium IPPS/SPDP '99, San Juan, Puerto Rico, pp. 233-237, April 12-16, 1999.
11. H. Yamada, "Complete Euclidean distance transformation by parallel operation," Proc. 7th International Conference on Pattern Recognition, pp. 69-71, 1984.

Advanced Data Layout Optimization for Multimedia Applications

Chidamber Kulkarni[1,2] Francky Catthoor[1,3] Hugo De Man[1,3]

[1] IMEC, Kapeldreef 75, B3001 Leuven, Belgium.
[2] Also Ph.D. student at the Dept of EE, Kath Univ Leuven
[3] Also Professor at the Dept of EE, Kath Univ Leuven
<kulkarni,catthoor,deman@imec.be>

1 Introduction and Related Work

Increasing disparity between processor and memory speeds has been a motivation for designing systems with deep memory hierarchies. Most data-dominated multimedia applications do not use their cache efficiently and spend much of their time waiting for memory accesses [1]. This also implies a significant additional cost in increased memory bandwidth, in the system bus load and the associated power consumption apart from increasing the average memory access time.

In this work, we are mainly targeting the embedded (parallel) real-time multimedia processing (RMP) application domain since algorithms in there lend themselves to very good compile-time analysis. Although embedded RMP applications are relatively regular, but certainly not perfectly linear/affine in the loop and index expressions, the simultaneous presence of complex accesses to large working sets makes most of the existing approaches largely to fail in taking advantage of the locality. Earlier studies have shown that the majority of the execution time is spent in cache stalls due to cache misses for image processing applications [1] as well as scientific applications [12]. Hence the reduction of such cache misses is of crucial importance.

Source-level program transformations to modify the execution order can improve the cache performance of these applications to a large extent [3, 6–9] but still a significant amount of cache misses are present. Storage order optimizations [3, 4] are very helpful in reducing the capacity misses. So in the end mostly conflict cache misses related to the sub-optimal data layout remain. Array padding has been proposed earlier to reduce the latter [11, 14, 15]. These approaches are useful for reducing the cross-conflict misses. However existing approaches do not eliminate the majority of the conflict misses. Besides [2, 6, 14], very little has been done to measure the impact of data organization (or layout) on the cache performance. Thus there is a need to investigate additional data layout or organization techniques to reduce these cache misses.

The fundamental relation which governs the mapping of data from the main memory to a cache is given as below :

$$(Block\ Address)\ MOD\ (Number\ of\ Sets\ in\ Cache) \qquad (1)$$

J. Rolim et al. (Eds.): IPDPS 2000 Workshops, LNCS 1800, pp. 186–193, 2000.

Based on the number of lines in a set we define direct mapped, n-way associative and fully associative cache [16]. It is clear that, if we arrange the data in the main memory so that they are placed at particular block addresses depending on their lifetimes and sizes, we can control the mapping of data to the cache and hence (largely) remove the influence of associativity on the mapping of data to the cache. The problem is however that trade-offs normally need to be made between many different variables. This requires a global data layout approach which to our knowledge has not yet been published before. This has been the motivation for us to come up with a new formalized and automated methodology for optimized data organization in the higher levels of memory. Our approach is called main memory data layout organization (MDO). This is our main contribution, which will be demonstrated on real-life applications.

The remaining paper is organized as follows : Section 2 presents an example illustration of the proposed main memory data layout organization methodology. This is followed by the introduction of the general memory data layout organization problem and the potential solutions in section 3. Experimental results on three real-life test vehicles are presented in section 4. Some conclusions from this work are given in section 5.

2   Example Illustration

In this section we will briefly introduce the basic principle behind main memory data layout organization (MDO) using an example illustration.

Consider the example in figure 1. The initial algorithm in figure 1(a) needs three arrays to execute the complete program. Note that the initial main memory data layout in figure 1(b) is single contiguous irrespective of the array and cache sizes. The initial algorithm can have $3N$ (cross-) conflict cache misses for a direct mapped cache, in the worst case i.e. when each of the arrays are placed at an (initial) address, which is a multiple of the cache size. Thus to eliminate all the conflict cache misses, it is necessary that none of the three arrays gets mapped to the same cache locations.

The MDO optimized algorithm, as shown in figure 1(c), will have no (cross-) conflict cache misses at all. This is because, in the MDO optimized algorithm the arrays always get mapped to fixed and non-overlapping locations in the cache. This happens because of the way the data is stored in the main memory, as shown in figure 1(d). To obtain this modified data layout, the following steps are carried out :

1. the initial arrays are split into sub-arrays of equal size. The size of each sub-array is called the *tile size.*
2. different arrays are merged so that the sum of their tile-sizes equals the cache size. Now store the merged arrays recursively till all the concerned arrays are completely mapped in the main memory. Thus we now have a new array which comprises all the arrays but the constituent arrays are stored in such a way that they get mapped into cache so as to remove the conflict misses. This new array is represented by "x[]" in figure 1(c).

In figure 1(c) and (d), two important observations need to be made : (1) there is a recursive allocation of different array data, with each recursion equal to the cache size and (2) the generated addressing, which is used to impose the modified data layout on the linker, contains modulo operations. These can be removed afterwards through a seperate optimization stage [5].

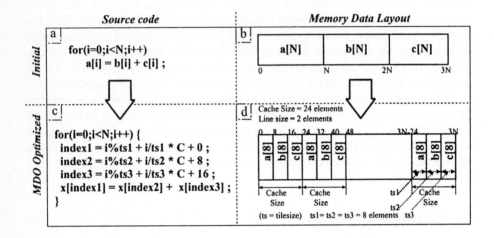

Fig. 1. Example illustration of MDO optimization on a simple case.

3   Main Memory Data Layout Organization (MDO)

In this section, we will first present a complete problem formulation invovling the two stages namely the tile size evaluation and the array merging as introduced in section 2. To deal with complex realistic applications the optimal solution would require too much CPU time. So we have also developed a heuristic, which has been automated in a source-to-source precompiler step.

3.1   The General Problem

The general main memory data layout organization problem for efficient cache utilization (DOECU) can be stated as, "For a given program with m-loop nests and n-variables (arrays), obtain a data layout which has the least possible conflict misses". This problem has two sub problems. First, the tile size evaluation problem and secondly the array merging/clustering problem.

Tile Size Evaluation Problem : Let x_i be the tile size of the array i and C be the cache size. For a given program we need to solve the m equations as below to obtain the needed (optimal) tile sizes. This is required because of two reasons. Firstly, an array can have different effective size in different loop nests. We define effective size as "the number of elements of an array accessed in a loop nest". This number can thus represent either the complete array size or a partial size and is represented as *effsize*. The second reason being that different loop nests have different number of arrays which are simultaneously alive.

$$L_1 = x_1 + x_2 + x_3 + ... + x_n = C \qquad (2)$$

$$L_2 = x_1{}^1 + x_2{}^1 + x_3{}^1 + ... + x_n{}^1 = C \qquad (3)$$

$$\cdot \quad \cdot \quad \cdot \quad \cdot$$

$$L_m = x_1{}^{(m-1)} + x_2{}^{(m-1)} + x_3{}^{(m-1)} + ... + x_n{}^{(m-1)} = C \qquad (4)$$

The above equations need to be solved so as to minimize the number of conflict misses. In this paper, we assume that all the arrays which are simultaneously alive have an equal probability to conflict (in the cache). The optimal solution to this problem comprises solving ILP problem [10, 13], which requires large CPU time. Hence we have developed heuristics which provide good results in a reasonable CPU time.

Array Merging/Clustering Problem : We now further formulate the general problem using the loop weights. The weight in this context is the probability of conflict misses calculated based on the simultaneous existence of arrays for a particular loop-nest i.e. sum of effective sizes of all the arrays as given below :

$$L_{wk} = \sum_{i=1}^{n} effsize_i \qquad (5)$$

Hence, now the problem to be solved is, which variables to be clustered or merged and in what order i.e. from which loop-nest onwards so as minimize the cost function. Note that we have to formulate the array merging problem this way because, we have many tile sizes for each array[1] and there are different number of arrays alive in different loop nest. Thus, using above loop weights we can identify loop nests which can potentially have more conflict misses and focus on clustering arrays in these loop nests.

3.2   The Pragmatic Solution

We now discuss some pragmatic solutions for the above problem. These solutions comprise heuristics, which are less complex and faster from the point of view of automation. First, we briefly discuss how the two stages of the problem are solved as below :

1. The first step involves evaluation of the effective sizes for each array instances in the program. Next, we perform a proportionate allocation based on the effective size of every array in every loop nest. This means that arrays with

---

[1] In the worst case, one tile size for every loop nest in which the array is alive.

larger effective sizes get larger tile sizes and vice-versa. Thus the remaining problem is the merging of different arrays.

2. The second step involves the merging/clustering of different arrays with their tile sizes. To achieve this we first arrange all the loop nests (in our internal model), in ascending order of their loop weights as calculated earlier. Next, we start merging arrays from the loop nest with highest loop weight and go on till the last remaining array has been merged. Note that once the total tile size is equal to the cache size, we start a second cluster and so on. This is done in a relatively greedy way, since we do not explore for the best possible solution extensively.

We have automated two heuristics in a prototype tool, which is a source-to-source (C-to-C) pre-compiler step. The basic principle of these two heuristics are given below :

1. *DOECU I :* In the first heuristic, the tile size is evaluated individually for each loop nest i.e. the proportionate allocation is performed based on the effective sizes of each array in the particular loop nest itself. Thus we have many alternatives[2] for choosing the tile size for an array. In the next step, we start merging the arrays from the loop nest with the highest weight, as calculated earlier, and move to the loop nest with the next highest weight and so on till all the arrays are merged. In summary, we evaluate the tile sizes locally but perform the merging globally based on loop weights.

2. *DOECU II :* In the second heuristic, the tile sizes are evaluated by a more global method. Here we first accumulate the effective sizes for every array over all the loop nests. Next we perform the proportionate allocation for every loop nest based on the accumulated effective sizes. This results in lesser difference between tile size evaluated for an array in one loop nest compared to the one in another loop nest. This is necessary since sub-optimal tile sizes can result in larger self conflict misses. The merging of different arrays is done in a similar was as in the first heuristic.

4   Experimental Results

This section presents the experimental results of applying MDO, using the prototype DOECU tool, on three different real-life test-vehicles namely a cavity detection algorithm used in medical imaging, a voice coder algorithm which is widely used in speech processing and a motion estimation algorithm used commonly in video processing applications. Note that all three algorithms are quite large and due to limitations in space we will not explain the algorithmic details of these applications[3].

---

[2] We could have in the worst case, a different tile size for every array in every loop nest for the given program.

[3] In brief, cavity detection algorithm is 8 pages and has 10 loopnests, voice coder algorithm is 12 pages and has 22 loopnests and motion estimation is 2 pages and has one loopnest with a depth of six.

The initial C source code is transformed using the prototype DOECU tool, which also generates back the transformed C code. These two C codes, initial and MDO optimized, are then compiled and executed on the Origin 2000 machine and the performance monitoring tool "perfex" is used to read the hardware counters on the MIPS R10000 processor.

Table 1, table 2 and table 3 show the obtained results for the different measures for all the three applications. Note that table 3 has same result for both the heuristics since the motion estimation algorithm has only one (large) loop nest with a depth of six namely six nested loops with one body.

	Initial	DOECU (I)	DOECU (II)
Avg memory access time	0.482180	0.203100	0.187943
L1 Cache Line Reuse	423.219241	481.172092	471.098536
L2 Cache Line Reuse	4.960771	16.655451	23.198864
L1 Data Cache Hit Rate	0.997643	0.997926	0.997882
L2 Data Cache Hit Rate	0.832236	0.943360	0.958676
L1–L2 bandwidth (MB/s)	13.580039	4.828789	4.697513
Memory bandwidth (MB/s)	8.781437	1.017692	0.776886
Actual Data transferred L1–L2 (in MB)	6.94	4.02	3.70
Actual Data transferred L2–Memory (in MB)	4.48	0.84	0.61

Table 1. Experimental Results for the cavity detection algorithm using the MIPS R10000 Processor.

The main observations from the results are : MDO optimized code has a larger spatial reuse of data both in the L1 and L2 cache. This increase in spatial reuse is due to the recursive allocation of simultaneously alive data for a particular cache size. This is observed from the L1 and L2 cache line reuse values. The L1 and L2 cache hit rates are consistently greater too, which indicates that the tile sizes evaluated by the tool were nearly optimal, since sub-optimal tile sizes will generate more self conflict cache misses.

	Initial	DOECU (I)	DOECU (II)
Avg memory access time	0.458275	0.293109	0.244632
L1 Cache Line Reuse	37.305497	72.854248	50.883783
L2 Cache Line Reuse	48.514644	253.450867	564.584270
L1 Data Cache Hit Rate	0.973894	0.986460	0.980726
L2 Data Cache Hit Rate	0.979804	0.996070	0.998232
L1–L2 bandwidth (MB/s)	115.431450	43.473854	49.821937
Memory bandwidth (MB/s)	10.130045	0.707163	0.315990
Actual Data transferred L1–L2 (in MB)	17.03	10.18	9.77
Actual Data transferred L2–Memory (in MB)	1.52	0.16	0.06

Table 2. Experimental Results for the voice coder algorithm using the MIPS R10000 Processor.

Since the spatial reuse of data is increased, the memory access time is reduced by an average factor 2 all the time. Similarly the bandwidth used between L1-L2 cache is reduced by a factor 0.7 to 2.5 and the bandwidth between L2 cache - main memory is reduced by factor 2-20. This indicates that though the initial algorithm had larger hit rates, the hardware was still performing many redundant data transfers between different levels of the memory hierarchy. These redundant transfers are removed by the modified data layout and heavily decrease the system bus loading. This has a large impact on the global system performance, since most (embedded) multimedia applications require to operate with peripheral devices connected using the off-chip bus. In addition also the system power consumption goes down.

	Initial	DOECU (I/II)
Avg memory access time	0.782636	0.289850
L1 Cache Line Reuse	9132.917055	13106.610419
L2 Cache Line Reuse	13.500000	24.228571
L1 Data Cache Hit Rate	0.999891	0.999924
L2 Data Cache Hit Rate	0.931034	0.960362
L1–L2 bandwidth (MB/s)	0.991855	0.299435
Memory bandwidth (MB/s)	0.311270	0.113689
Actual Data transferred L1–L2 (in MB)	0.62	0.22
Actual Data transferred L2–Memory (in MB)	0.20	0.08

Table 3. Experimental Results for the motion estimation algorithm using the MIPS R10000 Processor.

Since we generate complex addressing, we also perform address optimizations to remove the addressing overhead [5]. Our studies have shown that we are able to not only remove the complete overhead in addressing but gain by upto 20% in the final execution time measure on R10000 and PA-8000 processors, compared to the initial algorithm.

5 Conclusions

The main contributions of this paper are : (1) MDO is an effective approach at reducing the conflict cache misses to a large extent. We have presented a complete problem formulation and possible solutions. (2) This technique has been automated as part of a source-to-source precompiler for multimedia applications, called ACROPOLIS. (3) The results indicate a consistent gain in the data cache hit rate and a reduction in the memory access time and the memory bandwidth. Hence the system bus load and the energy consumption improve (reduce) significantly.

Acknowledgements : We thank Cedric Ghez and Miguel Miranda at IMEC for implementing the address transformations. Also we thank the ATOMIUM group for providing the C code parser and dumper.

References

1. P.Baglietto, M.Maresca and M.Migliardi, "Image processing on high-performance RISC systems", *Proc. of the IEEE*, vol. 84, no. 7, pp.917-929, july 1996.
2. D.C.Burger, J.R.Goodman and A.Kagi, "The declining effectiveness of dynamic caching for general purpose multiprocessor", *Technical Report*, University of Wisconsin, Madison, no. 1261,1995.
3. E.De Greef, "Storage size reduction for multimedia applications", *Doctoral Dissertation*, Dept. of EE, K.U.Leuven, January 1998.
4. F.Catthoor, S.Wuytack, E.De Greef, F.Balasa, L.Nachtergaele, A.Vandecappelle, "Custom Memory Management Methodology – Exploration of Memory Organization for Embedded Multimedia System Design", ISBN 0-7923-8288-9, Kluwer Acad. Publ., Boston, 1998.
5. S.Gupta, M.Miranda, F.Catthoor, R.Gupta, "Analysis of high-level address code transformations", In proc. of design automation and test in europe (DATE) conference, Paris, March 2000.
6. M.Kandemir, J.Ramanujam, A.Choudhary, "Improving cache locality by a combination of loop and data transformations", *IEEE trans. on computers*, vol. 48, no. 2, pp. 159-167, 1999.
7. C.Kulkarni, F.Catthoor, H.De Man, "Code transformations for low power caching in embedded multimedia processors,", *Intnl. Parallel Proc. Symp.(IPPS/SPDP)*, Orlando FL, pp.292-297,April 1998.
8. D.Kulkarni and M.Stumm, "Linear loop transformations in optimizing compilers for parallel machines", *The Australian computer journal*, pp.41-50, may 1995.
9. M.Lam, E.Rothberg and M.Wolf, "The cache performance and optimizations of blocked algorithms", *In Proc. 6th Intnl. Conference on Architectural Support for Programming Languages and Operating Systems (ASPLOS-IV)*, pp.63-74, Santa Clara, Ca., 1991.
10. C.L.Lawson and R.J.Hanson, "Solving least squares problems", *Classics in applied mathematics*, SIAM, Philadelphia, 1995.
11. N.Manjikian and T.Abdelrahman, "Array data layout for reduction of cache conflicts", *Intl. Conference on Parallel and Distributed Computing Systems*, 1995.
12. K.S.McKinley and O.Temam, "A quantitative analysis of loop nest locality", *Proc. of 8th Intnl. Conference on Architectural Support for Programming Languages and Operating Systems (ASPLOS-VIII)*, Boston, MA, October 1996.
13. G.L.Nemhauser, L.A.Wolsey, "Integer and Combinatorial Optimization", J.Wiley&Sons, New York, N.Y., 1988.
14. P.R.Panda, N.D.Dutt and A.Nicolau, " Memory data organization for improved cache performance in embedded processor applications", *In Proc. ISSS-96*, pp.90-95, La Jolla, Ca., Nov 1996.
15. P.R.Panda, H.Nakamura, N.D.Dutt and A.Nicolau, "Augmented loop tiling with data alignment for improved cache performance", *IEEE trans. on computers*, vol. 48, no. 2, pp. 142-149, 1999.
16. D.A.Patterson and J.L.Hennessy, "Computer architecture A quantitative approach", *Morgan Kaufmann Publishers Inc.*, San Francisco,1996.

Parallel Parsing of MPEG Video in a Multi-threaded Multiprocessor Environment

Suchendra M. Bhandarkar   Shankar R. Chandrasekaran

Department of Computer Science, The University of Georgia,
Athens, Georgia 30602-7404, USA

Abstract. Video parsing refers to the detection of scene changes and special effects in the video stream and is used to extract key frames from a video stream. In this paper, we propose parallel algorithms for the detection of scene changes and special effects in MPEG video in a multi-threaded multiprocessor environment. The parallel video parsing algorithms are capable of detecting abrupt scene changes (cuts), gradual scene changes (dissolves) and dominant camera motion in the form of pans and zooms in an MPEG1-coded video stream while entailing minimal decompression of the MPEG1 video stream. Experimental results on real video clips in a multi-threaded multiprocessor environment are presented. These algorithms are useful when real-time parsing of streaming video or high-throughput parsing of archival video are desired.

1 Introduction

One of the greatest challenges in modern multimedia systems is the rapid and accurate extraction of *key* information from images, audio and video. Video parsing is thus a very important preprocessing step in the extraction of *key* information from the video which can then used for indexing, browsing and navigation of the video [7]. Video parsing refers to the detection of scene breaks, gradual scene changes such as fades and dissolves, and camera motion such as pans and zooms. Most video streams are compressed for storage and transmission. In this paper, we propose parallel algorithms for the detection of abrupt and gradual scene changes and special effects arising from camera motion, in MPEG1-coded video. The parallel video parsing algorithms entail minimal decompression of the video stream and thus result in considerable savings in terms of execution time and memory usage. The parallel video parsing algorithms are designed and implemented in a multi-threaded multiprocessor environment and are useful when real-time parsing of streaming video or high-throughput parsing of archival MPEG1 video are desired.

2 Scene change detection in MPEG1 video

We will discuss briefly the two different approaches for detecting scene changes that are incorporated in our parallel video parsing algorithms. We will call them the *motion-luminance approach* [1, 2] and the *spatio-temporal approach* [5]. Both

J. Rolim et al. (Eds.): IPDPS 2000 Workshops, LNCS 1800, pp. 194-201, 2000.

approaches use the MPEG1 video stream as input. We present a brief discussion of the MPEG1 video format in this paper and refer the interested reader to [3] for a detailed exposition.

2.1   Description of the MPEG1 video format

MPEG1 video compression [3] relies on two basic techniques: block-based motion compensation for reduction of temporal redundancy and Discrete Cosine Transform (DCT)-based compression for the reduction of spatial redundancy. The motion information is computed using 16×16 pixel blocks (called macroblocks) and is transmitted with the spatial information. The MPEG1 video stream consists of three types of frames: intra-coded (I) frames, predictive-coded (P) frames and bidirectionally-coded (B) frames. I frames use only DCT-based compression with no motion compensation. P frames use previous I frames for motion encoding whereas B frames may use both previous and future I or P frames for motion encoding.

An I frame in MPEG1-coded video is decomposed into 8×8 pixel blocks and the DCT computed for each block. The DC term of the block DCT, termed as $DCT(0,0)$ is given by $\frac{1}{8} \sum_{i=0}^{7} \sum_{j=0}^{7} I_B(i,j)$ where I_B is the image block. The DC image represents the average luminance and chrominance value of each macroblock in each frame. Operations of a global nature performed on the original image can be performed on the DC image [6, 7] without significant deterioration in the final results.

For P and B frames in MPEG1-coded video, motion vectors (MVs) are defined for each 16×16 pixel region of the image, called a *macroblock*. P frames have macroblocks that are *motion compensated* with respect to an I frame in the immediate past. These macroblocks are deemed to have a *forward predicted* MV (FPMV). B frames have macroblocks that are motion compensated with respect to, either a reference frame in the immediate past, immediate future or both. Such macroblocks are said to have an FPMV, *backward predicted* MV (BPMV) or both, respectively. The prediction error signal is compressed using the DCT and transmitted in the form of 16×16 pixel macroblocks.

2.2   The motion-luminance approach

The motion-luminance approach exploits the motion, luminance and chrominance information embedded in the MPEG1 video stream. The motion information is encoded in the form of MVs at the macroblock level for each frame whereas the luminance and chrominance information is encoded as a DC image. **Scene cut detection using DC images:** Let X_i, $i = 1, 2, \ldots, N$ be a sequence of DC images. The difference sequence D_i, $i = 1, 2, \ldots, N-1$, is generated where $D_i = d(X_i, X_{i+1})$. Each element of D_i is the difference of the cumulative DC values of two successive images. Based on luminance and chrominance information, a scene cut is deemed to occur between frames X_l and X_{l+1} if and only if (1) The difference D_l is the maximum within a symmetric sliding window of size $2m-1$, i.e. $D_l \geq D_j$, $j = l-m+1, \ldots, l-1, l+1, \ldots, l+m-1$, and (2) D_l is at least n

times the magnitude of the second largest maximum in the sliding window. The parameter m is set to be smaller than the minimum expected duration between the scene changes [6]. Since the DC images have a *luminance* component and two *chrominance* components the difference image is the weighted sum of the absolute values of the individual component differences.

Scene cut detection using motion information: Since a *cut* in a video stream is characterized by an abrupt scene change, a typical motion estimator will need to intracode almost all the macroblocks at a cut. Thus, by computing the extent of motion or motion distance (MD) within each macroblock in the current frame relative to the corresponding macroblock in the previous frame and setting it to a very high value (theoretically ∞) if such correspondence cannot be determined, scene cuts can be detected. The computation of the MD is complicated by the fact that the corresponding macroblocks of successive frames may have MVs based on different reference frames. In [1] we have outlined an algorithm to compute the MD under these circumstances. The MD between two frames is computed as the sum of the MDs of the individual macroblocks. A MD sequence $MD_i, i = 1, 2, ..., N - 1$, is generated where MD_l is the MD between frames X_l and X_{l+1}. Analogous to the DC image-based approach, a scene cut is deemed to occur between frames X_l and X_{l+1} based on motion information if and only if: (1) The difference MD_l is the maximum within a symmetric sliding window of size $2m-1$, i.e. $MD_l \geq MD_j, j = l-m+1, ..., l-1, l+1, ..., l+m-1$, and (2) MD_l is at least n times the magnitude of the second largest maximum in the sliding window.

In order to exploit both, the motion information and the luminance and chrominance information, we designed an integrated motion-luminance approach to video parsing [2]. In the integrated motion-luminance approach, scene cuts are detected by computing the weighted sum W_i of the values of D_i and MD_i. A scene cut is deemed to occur between frames X_l and X_{l+1} iff the difference W_l is a maximum within a symmetric sliding window and W_l is at least n times the magnitude of the second largest maximum in the sliding window [2]. We have experimentally determined that the integrated motion-luminance approach performs better than either the motion-based approach or the luminance and chrominance-based approach used in isolation [2].

Gradual scene change detection using DC images

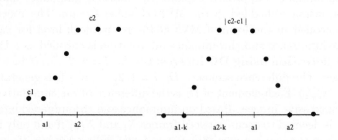

Fig. 1. Gradual Transition and Difference Sequence

A gradual scene transition is modeled as a linear transition from c_1 to c_2, in the time interval $[a_1, a_2]$ (Fig. 1) and modeled as [6]:

$$g_n = \begin{cases} c_1, & n < a_1, \\ \frac{c_2 - c_1}{a_2 - a_1}(n - a_2) + c_2, & a_1 \leq n < a_2, \\ c_2, & n \geq a_2 \end{cases} \tag{1}$$

Assuming that $k > a_2 - a_1$, the difference sequence $G_i^k(g_n) = d(X_i, X_{i+k})$ is given by [6]:

$$G_i^k(g_n) = \begin{cases} 0, & n < a_1 - k, \\ \frac{|c_2 - c_1|}{a_2 - a_1}[n - (a_1 - k)], & a_1 - k \leq n < a_2 - k, \\ |c_2 - c_1|, & a_2 - k \leq n < a_1, \\ -\frac{|c_2 - c_1|}{a_2 - a_1}(n - a_2), & a_1 \leq n < a_2, \\ 0, & n \geq a_2 \end{cases} \tag{2}$$

The plots of g_n and $G_i^k(g_n)$ are shown in Fig. 1. The plateau between $a_2 - k$ and a_1 has a maximum constant height of $|c_2 - c_1|$ if $k > a_2 - a_1$. In order to detect the plateau width, it is required that for fixed k:
(1) $|G_i^k - G_j^k| < \epsilon$, where $j = i - s, \ldots i - 1, i + 1, \ldots, i + s$, and
(2) $G_i^k \geq l \times G_{i-\lfloor k/2 \rfloor -1}^k$ or $G_i^k \geq l \times G_{i+\lfloor k/2 \rfloor +1}^k$, for some large value of l.

Since the width of the plateau is $k - (a_2 - a_1) + 1$, the value of k should be chosen to be $\approx 2(a_2 - a_1)$ where $a_2 - a_1$ is the (expected) length of the transition.

Gradual scene change detection using motion information

During a dissolve, the motion estimation algorithm typically finds the best matching blocks in the reference frame(s) for blocks in the current frame but at the cost of higher prediction error. Also, in a typical dissolve, the error is uniformly distributed in space and value over all of the macroblocks. These observations are encapsulated in the following metrics (i) The average error for fram l $E_{l,avg} = \frac{1}{MN} \sum_{i=1}^{M} \sum_{j=1}^{N} E_l(i,j)$ over the $M \cdot N$ macroblocks should be high, (ii) The error variance $\sigma_{E,l}^2 = \frac{1}{MN} \sum_{i=1}^{M} \sum_{j=1}^{N} [E_{l,avg} - E_l(i,j)]^2$ should be high and (iii) The error cross covariance $\sigma_{l,ij}^2 = \frac{\sum_{i=1}^{M} \sum_{j=1}^{N} ij E_l(i,j)}{\sum_{i=1}^{M} \sum_{j=1}^{N} E_l(i,j)} - i_{avg} j_{avg}$,

where $i_{avg} = \frac{\sum_{i=1}^{M} \sum_{j=1}^{N} i E_l(i,j)}{\sum_{i=1}^{M} \sum_{j=1}^{N} E_l(i,j)}$ and $j_{avg} = \frac{\sum_{i=1}^{M} \sum_{j=1}^{N} j E_l(i,j)}{\sum_{i=1}^{M} \sum_{j=1}^{N} E_l(i,j)}$, should be low. A

motion-based decision criterion of the form $E_l = \frac{E_{l,avg} + \sigma_{E,l}^2}{\sigma_{l,ij}^2}$ is formulated based on the above observations. As in the case of scene cut detection, an integrated decision criterion V_l in the form of a weighted sum of G_l and E_l is computed for each frame l. A gradual scene change is deemed to occur between frames X_l and X_{l+1} iff the difference V_l is a maximum within a symmetric sliding window and V_l is at least n times the magnitude of the second largest maximum in the sliding window [2]. Unlike a scene cut, a gradual scene change is expected to persist over several frames.

Camera pan and zoom detection: For a typical pan frame, the motion vectors for all the macroblocks are similarly oriented. The average motion vector direction is computed for a frame. The sum D of the deviations of the motion vector direction in each macroblock in the frame from the average motion vector direction is computed. A deviation sequence D_i is generated and thresholded to detect the pan. Values of D less than the threshold signify a pan. For a zoom frame, the motion vectors are directed either inward or outward. The orientation of the motion vectors in a majority of the macroblocks is used to detect zoom-ins and zoom-outs [2]. Both, pans and zooms are expected to persist over several frames.

2.3   The spatio-temporal approach

Three 1-D slices are extracted from each frame in the MPEG-1 video stream. The three slices are the strips of the image in vertical, horizontal and diagonal direction. A spatio-temporal image is the collection of slices in the sequence at the same positions [5].

Scene cut and camera pan detection: The Markov Random Field (MRF) energy model is used to locate cuts and wipes (cuts) based on color and texture discontinuities at the boundaries of regions. The video stream is segmented at the scene cut boundary [5].

Gradual scene change detection: In the 2-D spatio-temporal image, a gradual scene change (i.e., dissolve) is characterized by a smooth transition from one region to another. The image is segmented into three portions, two regions representing successive shots and one narrow region representing the dissolve duration.

3   Parallel video parsing

Two parallel implementations corresponding to the motion-luminance approach and the spatio-temporal approach are presented. The input is an MPEG1 video stream, either from over a network or a storage device. The units of parallel work are slice, picture and a group of pictures (GOP). There are three types of processes:

Scan process which reads the streaming video, parses the header information and puts the units of work in a task queue [4]. The scan rate is varied to cope up with the speed of worker processes described below.

Worker process is the main process which does the computation on the units of work. The system consists of several worker processes which communicate and coordinate with other processes.

Display process is used for the collection and display of the results at the end of computation.

In both approaches to parallel video parsing, the slice-level has the finest work granularity followed by the frame-level which is coarser-grained and the GOP-level which is the coarsest-grained.

3.1 Parallel video parsing using the motion-luminance approach

The parallelization involves the worker processes working on the units of work and arriving at an intermediate result. The intermediate result, which is usually an array of values, is further divided among the processes or worked on by one process depending on the load, to yield the final result.

Scene cut detection using DC images: The difference sequence of $2m$ DC images is obtained using one of the following.

Slice-level implementation: The unit of parallel work is a slice. The scan process puts the slices in a task queue. The worker processes grab the slices, decode them and compute the sum of the DC values in the macroblocks. All processes working on the same frame coordinate/communicate with each other. The process that finishes last in a frame computes the aggregate sum. Once a process finishes with a slice, it takes another slice from the task queue. The process that finishes last in a frame also computes the difference between the current sum and the previous sum of DC values. This entails process coordination and data transfers.

Frame-level implementation: The parallel unit of work is a frame. A scene cut is detected by all the worker processes. The $2m$ frames are divided among the worker processes such that each process gets n adjacent frames. Inter-frame differences are computed using a tree-structured computation.

GOP-level implementation: Each process handles a GOP. If the GOP is closed, then inter-process communication is not needed while computing the DC sum. Each process grabs a GOP from the task queue and computes the DC sum for each frame in the group. The results are fed to another process which computes the difference and detects the scene cut.

Scene cut detection using motion vectors: The parallel unit of work is a closed set of frames. The scan process determines a closed set and enqueues it in the task queue. A GOP can be a unit of work if it is always closed. This entails minimum communication among the worker processes. At the frame and slice level, the unit of work is a pair of frames or a pair of slices. All the worker processes work on a pair of frames or on all the slices in a single frame. Both, involve high synchronization and communication overhead.

Gradual scene change detection using motion vectors: In the slice-level approach, each process computes the prediction error-based metrics described earlier for all the macroblocks in its slice. A single process collects all the slice-level values and computes the prediction error-based metrics for a frame. This is followed by the process that detects gradual scene changes. The coarse-grained implementation at the GOP level or frame level, is similar to the slice-level implementation but the resulting work granularity is large entailing less frequent inter-process communication.

Gradual scene change detection using DC images: The worker processes grab two frames that are k sampling intervals apart and compute the difference of their DC values. There is an initial delay for the arrival of the kth frame.

Detection of camera pans and zooms: These typically involve computation on a single frame. The worker processes can work with single frames (at the frame level) or a closed set of frames (at the GOP level). Analysis at the frame level requires communication between the worker processes.

In all the above approaches, groups of processes working on a closed set of frames restrict the inter-process communication only to that group.

3.2   Parallel video parsing using the spatio-temporal approach

This approach is parallelized in 4 distinct stages: (1) Extraction of 1-D slices from the images, (2) Formation of the 2-D spatio-temporal image, (3) Segmentation of the spatio-temporal image, and (4) Detection of the scene change. Each of the stages is divided among the worker processes but stages 3 and 4 could also be accomplished by a single process. All three levels of parallelism slice-level, frame-level and GOP-level are implemented.

4   Experimental Results and Conclusions

The multi-threaded implementation of the parallel video parsing algorithms is done on an SMP and a network of workstations. The MPEG1 video clips, used for testing were either generated in our laboratory or downloaded over the Web. The scene cuts in the *Tennis* sequence at frames 90 and 150 are successfully detected using the motion-luminance approach (Fig. 4) and the MV approach (Fig. 2) but not the DC image-based approach (Fig. 3). The *Spacewalk1* sequence which contains 3 different dissolve sequences (between frames 74 and 85, 155 and 166, and 229 and 242) is successfully parsed using the motion-luminance approach (Fig. 7) and the DC image-based approach (Fig. 5) but not the MV approach (Fig. 6). An example video clip with zooms between frames 77 and 90, and frames 145 and 167 is also successfully parsed (Fig. 8). The pan (between frames 90 and 153) in another example video clip is also successfully detected (Fig. 9).

All three levels of parallelism (slice, frame and GOP level) exhibited significant speedup. The GOP-level exhibited the highest speedup whereas the slice-level exhibited the lowest speedup. This is accounted for by the greater inter-processor communication overhead associated with slice-level parallelism (which has the finest granularity) when compared to GOP-level parallelism (which has the coarsest granularity). We are in the process of testing our parallel implementation on a larger set of video data. We also intend to extend this work to more recent video encoding standards (MPEG4 and MPEG7).

References

1. S.M. Bhandarkar and A.A. Khombadia, Motion-based parsing of compressed video, *Proc IEEE Intl. Wkshp. Multimedia Database Mgmt. Sys.*, Dayton, Ohio, August 5-7, 1998, pp 80-87.
2. S.M. Bhandarkar, Y.S. Warke and A.A. Khombadia, Integrated parsing of compressed video, *Proc. Intl. Conf. Visual Inf. Mgmt. Sys.*, Amsterdam, The Netherlands, June 2-4, 1999, pp 269-276.

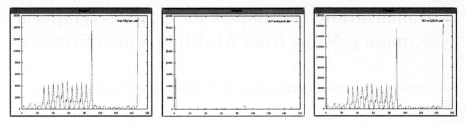

Fig. 2. *Tennis*: Motion edge plot

Fig. 3. *Tennis*: DC difference plot

Fig. 4. *Tennis*: Integrated approach plot

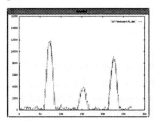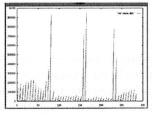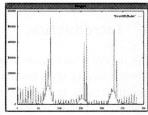

Fig. 5. *Spacewalk1*: DC *k*-difference plot

Fig. 6. *Spacewalk1*: Error variance plot

Fig. 7. *Spacewalk1*: Integrated approach plot

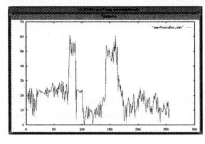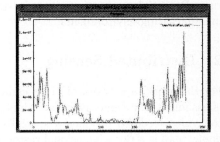

Fig. 8. *Pan7*: Plot of % of pixels satisfying zoom criteria

Fig. 9. *Pan5*: Plot of MV angle variance

3. V. Bhaskaran and K. Konstantinides, *Image and Video Compression Standards:Algorithms and Architectures*, Kluwer Academic Publishers, 1995, pp 161-194.

4. A. Bilas, J. Fritts and J.P. Singh, Real-time parallel MPEG-2 decoding in software, Technical Report 516-96, Department of Computer Science, Princeton University, March 1996.

5. C.W. Ngo, T.C. Pong and R.T. Chin, Detection of gradual transitions through temporal slice analysis, *Proc. IEEE Conf. Computer Vision and Pattern Recognition*, Fort Collins, Colorado, June 23-25, 1999, pp 36-41.

6. B.L. Yeo and B. Liu, Rapid scene analysis on compressed video, *IEEE Trans. Cir. and Sys. for Video Tech.*, Vol. 5(6), 1995, pp 533-544.

7. H.J. Zhang, C.Y. Low, and S.W. Smoliar, Video parsing and browsing using compressed data, *Jour. Multimedia Tools Appl.*, Vol. 1(1), 1995, pp. 89-111.

Parallelization Techniques for Spatial-Temporal Occupancy Maps from Multiple Video Streams

Nathan DeBardeleben, Adam Hoover, William Jones and Walter Ligon

Parallel Architecture Research Laboratory
Clemson University
{ndebard, ahoover, wjones, walt}@parl.clemson.edu

1 Introduction

We describe and analyze several techniques to parallelize a novel algorithm that fuses intensity data from multiple video cameras to create a spatial-temporal occupancy map. Instead of tracking objects, the algorithm operates by recognizing freespace. The brevity of operations in the algorithm allows a dense spatial occupancy map to be temporally computed at real-time video rates. Since each input image pixel is processed independently, we demonstrate parallel implementations that achieve nearly ideal speedup on a four processor shared memory architecture. Potential applications include surveillance, robotics, virtual reality, and manufacturing environments.

2 Distributed Sensing

For this work, a network of video cameras resembling a security video network is assumed. The cameras are all connected to a single computer that processes the video feeds to produce a spatial-temporal occupancy map [1]. The occupancy map is a two-dimensional raster image, uniformly distributed in the floorplane. Each map pixel contains a binary value, signifying whether the designated floorspace is empty or occupied. Figure 1 shows an example occupancy map where grey cells indicate the space is occupied and white cells indicate the space is empty. A spatial frame of the occupancy map is computed from a set of intensity images, one per camera, captured simultaneously. Temporally, a new map frame can be computed on each new video frame sync signal. Thus in effect, the map is itself a video signal, where the pixel values denote spatial-temporal occupancy. A previous implementation of such a network has shown that a frame rate of 5 Hz is feasible [1]. Our goal is to improve the temporal resolution by providing a frame rate approaching 30 Hz through the use of a parallelized implementation of the algorithm.

3 Algorithms

All the calculations necessary to create the mapping from the *camera space* to the *occupancy map space* are independent of image content. Therefore it can be

J. Rolim et al. (Eds.): IPDPS 2000 Workshops, LNCS 1800, pp. 202–209, 2000.

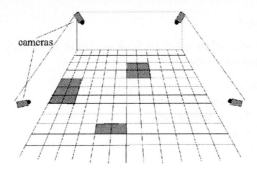

Fig. 1. A spatial occupancy map.

computed off-line and stored as a look-up table. The mapping provides a two-way relation, so that it may be applied in two different manners. The look-up table $L_1[n, c, r]$ relates each image pixel for each camera to a unique occupancy map cell. The look-up table $L_2[x, y]$ relates each occupancy map cell to a set of image pixels, where each set may include any number of pixels (including zero) from each camera. The use of $L_1[n, c, r]$ and $L_2[x, y]$ lead to different algorithms, which we will refer to as image-based and map-based.

Both the image-based and map-based algorithms show great potential for parallelism on a multiprocessor architecture. We describe three different divisions of the processing workload, and the corresponding parallel algorithms. We measure the performance of all the algorithms in Section 4, in terms of speed of execution.

In the following descriptions we maintain the following notation: $O[x, y]$ is the occupancy map, $I[n, c, r]$ is a set of live images from N cameras, and $B[n, c, r]$ is a set of background images acquired during system initialization. The indices x and y refer to map coordinates, c and r refer to image coordinates, and n refers to camera number. $L_1[n, c, r]$ and $L_2[x, y]$ refer to look-up tables storing the mappings described by $\mathcal{F}$ (Equation 1). The threshold T controls the sensitivity of the algorithm, i.e. as the threshold decreases, the system becomes more sensitive to denoting space as occupied. This is demonstrated and discussed further in Section 4.

$$\mathcal{F} : I[n, c, r] \leftrightarrow O[x, y] \tag{1}$$

The arrays $O[x, y]$, $I[n, c, r]$, $B[n, c, r]$, $L_1[n, c, r]$ and $L_2[x, y]$ are multi-dimensional, yet they can be accessed in one-dimensional order because they have discrete boundaries. For the sake of clarity, in the following algorithm descriptions we maintain the multi-dimensional notation. However, loops on (x, y), on (c, r), and on (n, c, r), can be written using a single-index loop. This reduction in loop overhead yields faster executions.

3.1 Image-based

The image-based algorithm uses the look-up table $L_1[n, c, r]$, and is described by the following pseudo-code:

```
loop ... time ...
  loop x = 0 ... map columns
    loop y = 0 ... map rows
      O[x,y] = 1
    end loop
  end loop
  loop n = 0 ... number of cameras
    loop c = 0 ... image columns
      loop r = 0 ... image rows
        if (|I[n,c,r]-B[n,c,r]| < T)
          O[L1[n,c,r]] = 0
        end if
      end loop
    end loop
  end loop
end loop
```

The arrays $I[n, c, r]$, $B[n, c, r]$, and $L_1[n, c, r]$ are accessed in sequential order, which can be exploited by a cache memory. The array $O[x, y]$ is accessed in non-sequential order.

Entries in $L_1[n, c, r]$ that are unused (entries for image pixels which do not map to ground plane points) are given a sentinel value that points to a harmless memory location outside the occupancy map. For instance, the occupancy map array is allocated as $X \times Y + 1$ cells, and the address of the extra cell becomes the sentinel. An alternative is to add a second conditional statement testing a mask. For each camera, a mask is initially generated that distinguishes available floorspace from non-floorspace. In the code given above, the inner-most loop is modified as follows to test for occupation only if the mask states that this space is floor.

```
if (M[n,c,r] == 0)
  if (|I[n,c,r]-B[n,c,r]| < T)
    O[L1[n,c,r]] = 0
  end if
end if
```

In this case an extra conditional statement is executed for every pixel, whereas in the original code non-useful assignment statements may be executed for some pixels. The relative performance of these variations is described in Section 4.

3.2 Map-based

The map-based algorithm uses the look-up table $L_2[x, y]$. Entries in $L_2[x, y]$ are sets of image pixel identities. The size of each set varies depending on how many image pixels view the occupancy map cell. This detail can be simplified by placing a maximum on set size, so that $L_2[x,y]$ may be implemented as a three-dimensional array. The constant set size S is selected so that at least 95% of the mappings in Equation 1 may be found in $L_2[x,y,s]$. Once the pixel has been identified as unoccupied, the algorithm need not further traverse $L_2[x,y,s]$ in the s dimension. This is a form of short-circuit evaluation. The map-based algorithm is described by the following pseudo-code:

```
loop ... time ...
  loop x = 0 ... map columns
    loop y = 0 ... map rows
      O[x,y] = 1
      loop s = 0 ... S
        if (|I[L2[x,y,s]]-B[L2[x,y,s]]| < T)
          O[x,y] = 0
          exit loop s
        end if
      end loop
    end loop
  end loop
end loop
```

In the map-based algorithm, the arrays $L_2[x,y,s]$ and $O[x,y]$ are accessed in sequential order, while the arrays $I[n,c,r]$ and $B[n,c,r]$ are accessed in non-sequential order.

As with the image-based algorithm, unused entries in $L_2[x,y,s]$ may be handled using sentinel addressing or masking. The sentinel version of the code is shown above. In this case entries in $L_2[x,y,s]$ which do not map to image pixels are given a sentinel value that points to memory locations outside the image and background image spaces that cause the conditional statement to fail.

3.3   Image-level parallelism

The image-based algorithm can be split into equal numbers of iterations on the camera loop. In this case, given P processors and N cameras, each processor works on the images provided by $\frac{N}{P}$ cameras. Figure 2 illustrates the workload. In the pseudo-code for the image-based algorithm given above, the camera loop is modified as follows:

```
loop n = (N/P)p ... (N/P)(p+1)
```

where $0 \leq p < P$ identifies a particular processor. This algorithm provides contiguous blocks of memory for the live and background images to each processor, but requires $\frac{N}{P}$ to be an integral number in order to maintain a balanced workload. This algorithm also produces write hazards, because multiple processors may write to the same occupancy map cell at the same time.

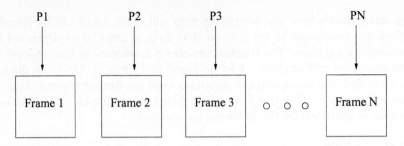

Fig. 2. The processor workload using image-level parallelism.

3.4   Pixel-level parallelism

The image-based algorithm can be split into equal numbers of iterations on the image pixels. In this case, given P processors and N cameras producing $R \times C$ size images, each processor works on $\frac{RC}{P}$ pixels of each image. Figure 3 illustrates the workload. In the pseudo-code for the image-based algorithm given above, the

Frame 1	Frame 2	Frame 3		Frame N
P1	P1	P1		P1
P2	P2	P2		P2
P3	P3	P3	ooo	P3
o o o	o o o	o o o		o o o
PN	PN	PN		PN

Fig. 3. The processor workload using pixel-level parallelism.

image rows loop is modified as follows:

```
loop r = (R/P)p ... (R/P)(p+1)
```

where $0 \le p < P$ identifies a particular processor. This algorithm does not provide contiguous blocks of memory for the live and background images to each processor, but maintains a more balanced workload in the case $\frac{N}{P}$ is not an integral number. This algorithm also produces write hazards, because multiple processors may write to the same occupancy map cell at the same time.

3.5   Map-level parallelism

The map-based algorithm can be split into equal numbers of iterations on the map cells. In this case, given P processors and an $X \times Y$ size occupancy map, each processor works on all the image data for $\frac{XY}{P}$ cells. Figure 4 illustrates the workload. In the pseudo-code for the map-based algorithm given above, the map rows loop is modified as follows:

```
loop y = (Y/P)p ... (Y/P)(p+1)
```

where $0 \leq p < P$ identifies a particular processor. This algorithm has no write hazards, because only one processor may write to each map cell. However, the workload balance is directly related to the uniformity of distribution of mappings in $L2[x, y, s]$. If some areas of the map are scarcely covered by image data while other areas are densely covered, then the workload will be correspondingly unbalanced.

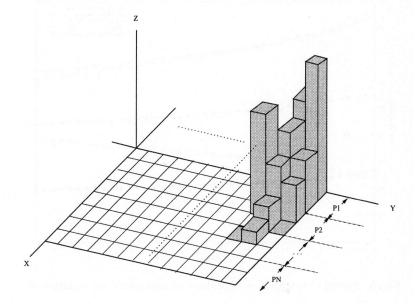

Fig. 4. The processor workload using map-level parallelism.

4   Results

The frame rate of our system depends on the number of cameras, the size of the camera images, the size of the occupancy map, and the algorithm and computer architecture. The frame rate is also upper-bounded by the frame rate of the cameras. In our case, we are using NTSC cameras (video signals), which fixes the camera image size to 640×480 and upper-bounds the frame rate at 30 Hz. We are using an NTSC signal to output the map, fixing the map size to 640×480. The remaining variables are the number of cameras, and the algorithm and computer architecture.

Fixing the number of cameras at four, we examined the performance of the sequential and parallel algorithms on a multi-processor architecture. Simulations were conducted on a Sun HPC 450 with four UltraSparc II processors operating at 300 MHz. A set of real look-up tables used in the sequential prototype were re-used for these experiments. Live images were simulated using a set of randomly valued arrays. The images were replaced on each iteration of the time-loop, to simulate real system operation, so that the 1 MB cache on each processor would have to re-load. Figure 5 plots the frame rates of each algorithm as a function of the threshold T, which is varied across the reasonable range of operation.

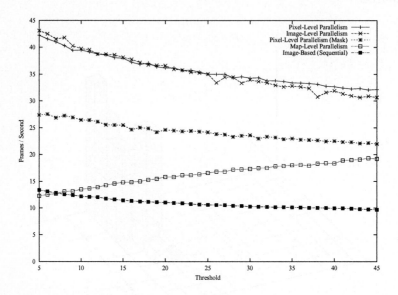

Fig. 5. System throughput of algorithms on multiprocessor architecture.

Based on Figure 5, we observe six results:

1. Both the map-based and image-based parallel algorithms achieved almost linear speedup in the number of processors compared to the sequential algorithms. For instance, between thresholds of 5 and 35, the pixel-level parallel algorithm showed the best average speedup of 3.3 over the image-based sequential algorithm (the theoretical maximum is 4.0, the number of processors).

2. As in our prototype system, the simulations showed a greater performance for the image-based algorithms compared to the map-based algorithms (we show only the fastest map-based algorithm in Figure 5). We suppose this is due to the fact that three out of the four arrays are accessed in sequential order in the image-based algorithms (see Section 3.1), while only two out of four arrays are accessed in sequential order in the map-based algorithms (see Section 3.2). The benefit provided by the increased hit rate in the cache memory (in the image-based parallel algorithms) outweighs the benefit provided by the avoidance of write hazards (in the map-based parallel algorithm).

3. Both the image-based parallel algorithms (pixel-level and image-level) performed equally. This suggests that the small penalty incurred by having a few (in our case four) non contiguous blocks of memory for each processor is relatively insignificant (see Section 3.4). Therefore the pixel-level algorithm is to be preferred, specifically in cases where the number of cameras is not an integral multiple of the number of processors.

4. Using an image mask decreased performance, as compared to sentinel (out-of-map or out-of-images) addressing for unused lookup table entries. The execution of an extra conditional statement for every pixel, along with the cost of loading an additional large array into memory, was more costly than executing the relatively small number of superfluous assignment statements.

5. The performance of each of the algorithms appears to degrade as the threshold increases, with the exception of the map-level algorithm. The map-level algorithm provides a short-circuit mechanism in the inner-most loop as discussed in Section 3.2 while the image-based algorithms do not.

6. It should be noted that, while using simulated I/O, frame rates exceeding the NTSC upper-bound of 30 Hz are indicative of being able to process incoming data at a rate faster than it becomes available. In a physical implementation, this would translate into one or more of the processors being idle waiting for the next frame to arrive from the video capture device.

The sequential prototype described above was constructed in 1997. The multiprocessor hardware described above was constructed in 1998. In 1999, we are constructing a second prototype using a Dell workstation with two Intel processors operating at 450 MHz. Based on the above experiments, we expect this system to operate at approximately 20 Hz. Based on projections of computer architecture performance [2], we expect that an average computer will be able to operate our system at 30 Hz for twenty cameras in the year 2004.

5   Conclusion

We describe and analyze several techniques to parallelize a novel algorithm that fuses intensity data from multiple video cameras to create a spatial-temporal occupancy map. This work provides a foundation to explore distributed sensing on a much larger scale. Future work will include increasing both the number of input data streams as well as the size of the output occupancy map to provide enhanced spatial resolution and coverage.

References

1. A. Hoover and B. Olsen, "A Real-Time Occupancy Map from Multiple Video Streams", in *IEEE ICRA*, 1999, pp. 2261-2266.
2. D. Patterson and J. Hennessy, Computer Architecture: A Quantitative Approach, second edition, Morgan Kaufmann, 1996.

Heuristic Solutions for a Mapping Problem in a TV-Anytime Server Network*

Xiaobo Zhou[1], Reinhard Lüling[1], Li Xie[2]

[1] Paderborn Center for Parallel Computing, University of Paderborn
Fürstenallee 11, D-33102 Paderborn, Germany
Email: {zbo, rl}@uni-paderborn.de

[2] Department of Computer Science, Nanjing University, P.R.China
Email: {xieli}@netra.nju.edu.cn

Abstract

This paper presents a novel broadband multimedia service called TV-Anytime. The basic idea of this service is to store broadcast media assets onto media server systems and allow clients to access these streams at any time. We propose a hierarchical structure of a distributed server network to support a high quality TV-anytime service. A key issue, how to map the media assets onto such a hierarchical server network is addressed and formalized as a combinatorial optimization problem. In order to solve this optimization problem, a set of heuristic solutions by use of a parallel simulated annealing library is proposed and verified by a set of benchmark instances. Finally, the TV Cache is presented as a prototype of a scalable TV-Anytime system.

1 Introduction

Distributed multimedia systems are constantly growing in popularity thanks also to the presence of the Internet [1] [2] [3] [4]. Whereas an Internet newspaper can be accessed at any time, the access to high bandwidth audio/video information (all of which we shall henceforth refer to in this paper as media assets) provided by broadcasting companies is strictly time dependent and synchronous. In order to access broadband media interactively and time-independently, one has to make intensive use of mirroring mechanisms. The problem with todays VCRs is that recording is not very comfortable and especially the amount of content that can be recorded is usually limited. Therefore, one alternative is to install server systems that store a huge amount of digitally broadcasted programs and make these content accessible for the clients.

---

*This research was partly supported by a grant from Siemens AG, München and Axcent Media AG, Paderborn in the framework of the project HiQoS - High Performance Multimedia Services with Quality of Service Guarantees.

J. Rolim et al. (Eds.): IPDPS 2000 Workshops, LNCS 1800, pp. 210-217, 2000.

Advances in high-speed broadband network technologies and the cascading ratio of price to storage capacity make it feasible to provide the content of todays broadcasters in a similar way as the content stored on the Internet: This means the content can be accessed interactively and time-independently. A service that aims to realize such features is called *TV-Anytime* [2].

2   A Hierarchical TV-Anytime Server Network

It is the idea of the TV-Anytime service to record broadcast media assets and to make them available independently from the time they are broadcasted. The full potential of a TV-Anytime service may be reached if the recording of media coming from broadcasters is combined with indexing information provided by the broadcasters and with profile information coming from the clients. If the broadcasters deliver the metadata for each broadcast event, and these metadata are matched with information about the clients preferences, a personal TV program can be set up for the client that delivers the favorite TV contents to the client in a time-independent and interactive way. In this form broadcast media and Internet media can be consumed in the same fashion.

First two commercial implementations of the TV-Anytime service are based on the online-encoding of analog TV signals and are available for the consumer-market in form of digital VCRs [3] [4]. However, the presented architectures have two main drawbacks:

- In comparison with the large amount of TV content that is broadcasted every second, the storage capacity of the systems is rather small. In a scenario where the client describes his personal profile for an automatic recording of media assets, it can be expected that the automatic mechanism will not perfectly match the client's preferences because the systems allow to store only a few media assets.
- Clearly, some clients' preference program may be the same, e.g., popular news clips, sports. Therefore, it would be a good strategy to store media assets onto a server in a way that these content are offered over the network to some clients simultaneously, minimizing the overall storage space consumption and required communication bandwidth in the multicasting way [5].

To be successful, the scalability and robustness of the server network are likely to become predominant issues. The problems above lead directly to our proposal of a hierarchical structure of a server network for a TV-Anytime system. As depicted in Figure 1, we connect a number of media servers by a backbone network and some nodes of the network are connected to a number of clients. Such a hierarchical server network can include a small system that is installed close to the clients as well as larger systems that are installed within a local network or a public broadband network to provide one media asset to a large number of clients. This hierarchical structure can scale to a very large network dynamically and is suitable for Internet/Intranet working. A client connected to a media server has direct and consequently cheap access to all media assets stored on this local server. It is also possible for the client to access media assets stored on remote servers via the backbone network, although this remote access is more expensive.

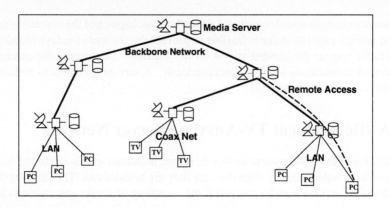

Figure 1: A hierarchical structure for TV-Anytime system.

3   The Media Mapping Problem

To implement a metropolitan TV-Anytime service on such a hierarchical server network, a new set of challenging research problems have been set forth. One of the key problems is how to map the media assets onto the server network and provide each requested media asset to the clients by use of the highest possible quality. The problem takes the access patterns and restrictions of the underlying hardware into account.

3.1   A Feature of the Mapping Problem

A very special feature of audio/video streams is that a media asset can be provided in different quality of service (QoS) which is determined by the encoding bit-rate of the media stream. Clearly, by use of a high encoding bit-rate, the media stream can be presented in good quality but also takes a large amount of storage capacity on the server and perhaps the communication load on the backbone network. Thus, the tradeoff that has to be solved by the mapping problem is to provide each client the selected media assets in the highest possible quality, taking the restrictions of the storage capacity of the media servers, the communication bandwidth of the backbone network, and the access patterns of the clients into account. This combinatorial problem can be formulated mathematically as a NP-complete optimization problem and can therefore be solved by use of some heuristic optimization solutions.

The mapping problem extends the File Allocation Problem (FAP) [6], a well known problem in parallel and distributed processing, using the nice feature of media assets that the assets are scalable in size as they are encoded in different bit-rates.

3.2   Formalizing the Mapping Problem

We assume the hierarchy of a server network is the binary tree with ring, i.e, all nodes in the leaf level of the tree are connected by a ring. All clients are connected to the leaf servers. The servers on upper levels are used as backup servers. Such a hierarchical structure for media server network can be modeled as a graph $N = (H, E)$, $H =$

$\{H_1, H_2\}$. H_1 is the set of backup servers which store assets with different bit-rates to provide them to leaf servers. H_2 is the set of leaf servers which all clients are connected with directly. The access patterns of clients connected to the leaf servers in H_2 is modeled as a graph $A = (V_A, E_A)$, and $V_A = H_2$, $E_A \subseteq \{(v, a) \mid v \in V_A, a \in M\}$. For a given server network N, an access pattern A, a storage capacity function c and a set of bandwidth B, now the question is how to map the media assets in such a way onto the server network that each access request can be fulfilled and totally the QoS is the best. An access request came from a client that is connecting with a leaf server v is fulfilled, if server v stores the requested media asset directly, or there is a path in the communication network to a backup server v' hosting the requested media asset and the communication path provides sufficient bandwidth to stream the asset to the requesting leaf v. Thus, the mapping problem can be formalized as follows:

Given: A server network $N = (H, E)$, $H = \{H_1, H_2\}$, access structure $A = (V_A, E_A)$ and a set of encoding bit-rates B.

Question: Is there a mapping $\pi_a \colon A \to P(E)$ and $\pi_b \colon A \to B$ with $\pi_a((v, a)) = \{(v, v_1), (v_1, v_2), \dots, (v_{l-1}, v_l)\}$ for a $l \in \mathcal{N}$ and $\pi_a((v, a)) = v_l$ and $\pi_b((v, a)) = b$ such that:

$$\sum\nolimits_{(v,a) \in A, \pi_a((v,a)) = h} \pi_b((v, a)) \le c(h), \forall h \in H \text{ and}$$
$$\sum\nolimits_{(v,a) \in A, (u_1, u_2) \in \pi_a((v,a))} \pi_b((v, a)) \le w_e(e), \forall e \in E, e = (u_1, u_2).$$

The optimization target is formalized as follows:

$$QoS := \sum\nolimits_{(v,a) \in E_A} \pi_b((v, a)) \to Max.$$

4 Parallel Simulated Annealing Algorithms

Some heuristic methods, such as Simulated Annealing (SA) [7], have been shown to be very efficient for the solution of combinatorial optimization problems, e.g., TSP, but also for very specific problems in industrial applications, such as the vehicle routing problem. SA is based on a local search procedure. It defines a neighborhood relation on the set of all feasible solutions of a given optimization problem. Starting from an initial solution, a search process is performed on the network of feasible solutions, identifying one node of this graph as the current solution in each iteration.

For large instances of real-world optimization problem, even fast heuristics require a considerable amount of computational time. ParSA [8], a parallel library for simulated annealing, has been developed at University of Paderborn in order to speed up the computation. The generic decisions, such as the choice of the initial temperature, the cooling schedule, and stopping condition, are connected with parameters of the parSA algorithm itself. We need to think over the class of decisions which is problem specific and involved the space of feasible solutions, the cost function and the neighborhood function. Our intention is to parallelize our heuristic solutions with the parSA and therefore are able to reduce the computational time significantly.

4.1 Initial Solution

To compute an initial solution, each media asset a that is requested by clients connected with leaf server v has to be placed at least once on a server that lies on a path from v to the root of server network. Here, we propose a three-step initial solution algorithm.

According to the access pattern A, the first step of the initial solution algorithm places each requested media asset a, $(v, a) \in A$, with the minimal bit-rate on a server that is as close as possible to the leaf node v, because it is possible that no feasible solution will be found if the media assets can only be encoded with relatively high bit-rate. This makes sense for practical applications, since in this case each media asset can be provided to the clients at least in a low bit-rate quality. Then, in the second step the bit-rate of assets that have been mapped on the leaf servers is increased, if there is available storage capacity in leaf servers. In the third step of the algorithm, the bit-rate of media assets that have been mapped on the backup server is increased. The process starts at the backup servers that are nearest to the leaf servers.

4.2   Neighborhood Structure

4.2.1   Neighborhood Structure - Phase I

For a given feasible assignment of the media assets onto the hierarchical server network we compute a neighboring assignment in two phases.

In the first phase, a node in the hierarchical network of servers is identified. If the selected server v is a leaf server, an asset $a \in M$ and $(v, a) \in A$ will be chosen randomly. If the selected server v is a backup server, an asset a according to its *leaf children*'s access patterns will be chosen randomly. Then, the bit-rate of asset a is increased if it has been mapped onto the selected server, otherwise, it will be mapped onto the server. If above operations induce that the storage capacity of server v exceeds its limitation, the bit-rate of one or more media assets that have been mapped onto v will be decreased, which also perhaps induce the deletion of one or more media assets if they are encoded with the minimal bit-rate. We consider the assets that will decrease their encoding bit-rates or be deleted from the server v are chosen with *exponential distribution*, giving a higher priority to those assets which are mapped only with small bit-rates onto the server network. Therefore, the redundant copies which only have smaller bit-rates are deleted with higher probability. In experiments, we found these intelligent neighborhood structures leaded to better results.

4.2.2   Neighborhood Structure - Phase II

The perturbation of the current solution also effects other servers which access the media assets from the selected server. Thus, in the second phase, for all $(v, a) \in A$, if an asset a has not been mapped onto server v and therefore can not be accessed directly by the clients connected with the v, we have to find a path in the hierarchical network from v to a backup server v' which has been mapped a copy of a.

Ideally, the routing path is established to a backup server that stores a copy of requested asset a. This copy is encoded with a highest possible bit-rate. However, this greedy algorithm will get easily blocked for later requests because of violating the backbone bandwidth restrictions. On one hand each client hopes to get highest possible quality service. On the other hand, it is the case that each client hopes it can be given the minimal bit-rate quality if there is no enough bandwidth for its request. Therefore, the policy of selecting suitable backup servers might have to be based on the tradeoff between highest possible bit-rate and the total overhead onto communication bandwidth of backbone network.

Originally, we assumed asset migration only take place between servers that have a common path from the leaf server to the root which means there is only one path from selected backup server to the leaf server. However, in this case we found in some leaf servers there are some requests that can not be satisfied due to the communication load in their local routing trees while there are some idle communication bandwidth in other local routing trees. Since the network structure is the tree with ring, we allow asset migration to take place via the ring thus there are alternative routing paths. We proposed a backtracking routing algorithm to make the communication load in the tree evener and to induce more satisfied requests.

5 Performance Evaluation

To investigate in detail the performance of the algorithms for the mapping problem, we define a set of benchmark instances that reflect the implementation of large scale distributed media archives and their typical access patterns. Then we compare the Simulated Annealing solutions of the algorithms and the Upper bound of the solution.

Let M be the set of available media assets. The access pattern described by the graph $A = (V_A, E_A)$ is determined by a random process, which means each leaf's clients request a media asset $a \in M$ with a given probability p. In this way, we can identify a number of benchmark classes $R\_n\_m\_c_{Atom}\_w_e\_B_p$ given in table 1. The n represents the number of servers. The m represents the number of total media assets, thus $m = | M |$. The c_{Atom} (MB) represents the base storage capacity quantity. For a tree with k levels, we assume that the total storage capacity is $2^{k+1} \cdot c_{Atom}$. Each leaf node is given $3 \cdot c_{Atom}$, each backup node is given $1 \cdot c_{Atom}$ and the root is given $2 \cdot c_{Atom}$. The w_e represents the same communication bandwidth (MB) of each link in the tree network. Many studies in the literature dealing with service quality estimation of digitally coded video/audio sequences use a five-level scale for quality rating. For the set B, we assign the bit-rates from a minimum of 5MB/s (bad quality), 10MB/s (poor), 15MB/s (fair), 25MB/s (good), to a maximum of 40MB/s (excellent).

Benchmark Class	n	m	c_{Atom}	w_e	B	access pattern
$R\_7\_256\_500\_100\_\frac{1}{2}$	7	256	500	100	$\{5,10,15,25,40\}$	random $p = \frac{1}{2}$
$R\_15\_256\_750\_100\_\frac{1}{2}$	15	256	750	100	$\{5,10,15,25,40\}$	random $p = \frac{1}{2}$
$R\_15\_512\_750\_100\_\frac{1}{3}$	15	512	750	100	$\{5,10,15,25,40\}$	random $p = \frac{1}{3}$
$R\_15\_512\_1000\_150\_\frac{1}{3}$	15	512	1000	150	$\{5,10,15,25,40\}$	random $p = \frac{1}{3}$
$R\_31\_512\_750\_100\_\frac{1}{4}$	31	512	750	100	$\{5,10,15,25,40\}$	random $p = \frac{1}{4}$
$R\_31\_512\_1000\_150\_\frac{1}{4}$	31	512	1000	150	$\{5,10,15,25,40\}$	random $p = \frac{1}{4}$
$R\_31\_1024\_1000\_150\_\frac{1}{6}$	31	1024	1000	150	$\{5,10,15,25,40\}$	random $p = \frac{1}{6}$
$R\_31\_1024\_1250\_250\_\frac{1}{6}$	31	1024	1250	250	$\{5,10,15,25,40\}$	random $p = \frac{1}{6}$
$R\_63\_1024\_1000\_100\_\frac{1}{8}$	63	1024	1000	100	$\{5,10,15,25,40\}$	random $p = \frac{1}{8}$
$R\_63\_1024\_1250\_150\_\frac{1}{8}$	63	1024	1250	150	$\{5,10,15,25,40\}$	random $p = \frac{1}{8}$

Table 1: Definition of benchmark instances in a hierarchical server network.

We use parallel simulated annealing library (parSA) to test the set of benchmark instances defined above. The optimization target in cost function of parSA is the QoS defined in section 3.3. Thus, we compare the gap of parSA solution with Upper bound of QoS to verify the proposed heuristic solutions.

As simulated annealing is a stochastic method for the solution of the combinatorial optimization problem, we performed each run of the algorithm 10 times and took the average result. Table 2 shows the gap between the resulting QoS computed by the simulated annealing algorithms as well as the Upper bound of QoS for the benchmark instances. The measurement shows that the differences between the upper bound and the results gained by the parallel simulated annealing algorithms are very small, ranging from about 1.4 percent down to about 0.2 percent. It can be concluded that the algorithms can find good heuristic solutions.

Benchmark Class	$\frac{Upperbound - SA solution}{Upperbound}$	$Average\ bit - rate$
$R\_7\_256\_750\_100\_\frac{1}{2}$	0.21%	17.57
$R\_15\_256\_750\_100\_\frac{1}{2}$	0.80%	18.81
$R\_15\_512\_750\_100\_\frac{1}{3}$	0.24%	13.86
$R\_15\_512\_1000\_150\_\frac{1}{3}$	1.41%	18.42
$R\_31\_512\_750\_100\_\frac{1}{4}$	1.08%	17.82
$R\_31\_512\_1000\_150\_\frac{1}{4}$	0.58%	24.05
$R\_31\_1024\_1000\_150\_\frac{1}{6}$	1.42%	18.21
$R\_31\_1024\_1200\_250\_\frac{1}{6}$	0.76%	22.35
$R\_63\_1024\_1000\_100\_\frac{1}{8}$	0.78%	26.60
$R\_63\_1024\_1250\_150\_\frac{1}{8}$	1.31%	33.21

Table 3: Performance of Neighbor_2 + backtracking routing Algorithm.

It can be seen from the average bit-rate of satisfied requests as depicted in Table 2, that most of the requests can be responded with *good* quality of service. It is expected that the average bit-rate of satisfied requests can be increased if extra storage capacity and communication bandwidth is added into the server network. For instance, the average bit-rate of benchmark $R\_15\_512\_750\_100\_\frac{1}{3}$ is 33% higher after 33% c_{Atom} and 50% w_e was added into the benchmark $R\_15\_512\_1000\_150\_\frac{1}{3}$. The same conclusion can be made from the comparison of $R\_31\_512\_750\_100\_\frac{1}{4}$ and $R\_31\_512\_1000\_150\_\frac{1}{4}$, $R\_31\_1024\_1000\_150\_\frac{1}{6}$ and $R\_31\_1024\_1200\_250\_\frac{1}{6}$, etc.

We also found that the difference between Upper bound of QoS and SA solution is mostly due to the wasted bandwidth in the backbone network. With a good bandwidth distribution strategy, we expect that the gap of SA solution and Upper bound of QoS will converge to 0.

6   TV Cache - A TV-Anytime System

In this paper, we proposed a set of heuristic solutions to solve the media asset mapping problem, a combinatorial optimization problem that arises in a hierarchical TV-Anytime system. The presented algorithms are combined with a parallel simulated annealing library (parSA) to test a set of benchmark instances. It is verified that the formalized optimization problem can be solved efficiently achieving near to optimal solutions in short time. In the optimal case, this parallel algorithms can be performed on the network of servers using the computational power that is available there.

The problem studied in this paper has a lot of practical relevance for the design and development of the prototype of a commercial TV-Anytime system, TV-Cache, which integrates Web technologies and the delivery of media streams in a seamless way.

The basis of the TV-Cache system is a commercial server system that performs MPEG streaming on the basis of a clustered PC architecture. The PCs run the Linux operating system. The media server can be used on a single PC, but also on a closely connected cluster of PCs, if a larger number of clients has to be supported. In the smallest configuration the system is used within the living room of a client. The client is connected to the Internet and to the inhouse antenna providing digital broadcast audio/ video. Larger configurations are based on PC systems and are used as inhouse systems that feed the coax network of an apartment complex with TV-Anytime services or within a company or an ADSL network. Control information is transmitted via the Internet while media assets are transmitted by use of broadband technologies within the server network. This backbone connection allows to mirror media streams from one server to the others. Thus, the model and algorithms discussed in this paper are applied here.

References

[1] F.Cortes Gomez, Reinhard Lüling. *A Parallel Continuous Media Server for Internet Environments*. Proc. of International Conference on High-Performance Computing and Networking(HPCN Europe'98), Lecture Notes in Computer Science, 1998, pp.78-86.

[2] Reinhard Lüling. *Hierarchical Video-on-Demand Servers for TV-Anytime Services*, Proc. of 8th International Conference on Computer Communications and Networks (IC^3N), Boston, Massachusetts, IEEE Press, 1999, pp.110-117.

[3] ReplayTV: http://www.replay.com.

[4] Tivo: http://www.tivo.com.

[5] D.L.Eager, M.K.Vernon, J.Zahorjan. *Minimizing Bandwidth Requirements for On-Demand Data Delivery*. Tech.Report #4105, Computer Science Dept., University of Wisconsin - Madison, Aug. 1999.

[6] L.W.Dowdy, D.V.Foster. *Comparative Models of the File Assignment Problem*. Computing Surveys, Vol.14, No.2, 1982.

[7] S.Kirkpatrick, C.D.Gelatt, M.P.Vecchi. *Optimization by Simulated Annealing*. Science, Vol.220, No.4598, May 1983, pp.671-680.

[8] S.Tschoeke, G.Kliewer. *The parSA Parallel Simulated Annealing Library*. Technical Report, Department of Mathematics and Computer Science, University of Paderborn, http://www.uni-paderborn.de/~parsa.

RPV: A Programming Environment for Real-time Parallel Vision —Specification and programming methodology—

Daisaku Arita, Yoshio Hamada, Satoshi Yonemoto and Rin-ichiro Taniguchi

Department of Intelligent Systems, Kyushu University
6-1 Kasuga-koen, Kasuga, Fukuoka 816-8580 Japan
{arita,yhamada,yonemoto,rin}@limu.is.kyushu-u.ac.jp

Abstract. A real-time distributed image processing system requires data transfer, synchronization and error recovery. However, it is difficult for a programmer to describe these mechanisms. To solve this problem, we are developing a programming tool for real-time image processing on a distributed system. Using the programming tool, a programmer indicates only data flow between computers and image processing algorithms on each computer. In this paper, we outline specifications of the programming tool and show sample programs on the programming tool.

1 Introduction

Recently, the technology of computer vision is applied to more various fields. CDV (Cooperative Distributed Vision) project[1, 2] in Japan aims to establish scientific and technological foundations to realize real world oriented practical computer vision systems. One of the research issue of the CDV project is observation of objects/environments with multiple sensors. When we use multiple sensors, or cameras, a distributed system with multiple computers is more suitable than than a centered system with only one computer because the performance of a distributed system can be easily increased to adapt the number of sensors by increasing the number of computers. To construct such a high-performance distributed system with low cost, we are developing a PC cluster, a set of PCs connected via high speed network, for real-time image processing[3, 4].

Though PC-based distributed systems have many merits, they have also some problems. One of them is that it is difficult for a user to make a system with high performance and stability, because when a user writes programs for a real-time distributed vision system, he or she must make attentions to data transfer, synchronization and error recovery. Their description requires a lot of knowledge about both hardware and software such as network, interruption, process communication and so on, and it is not quite easy. In this paper, we will propose PRV(Real-time Parallel Vision) programming tool for real-time image processing on a distributed system. Using RPV programming tool, a user does not have to write programs of data transfer mechanism, synchronization mechanism and error recovery functions, but only need to write programs of data flow between computers and processing algorithms on each computer.

J. Rolim et al. (Eds.): IPDPS 2000 Workshops, LNCS 1800, pp. 218-225, 2000.

2   System Overview

2.1   Hardware Configuration

Our PC cluster system consists of 14 Pentium-III×2 based PCs. All the PCs are connected via Myrinet, a crossbar-switch based gigabit network, and six of them have real-time image capture cards, ICPCI, which can capture uncompressed images from a CCD camera in real-time. Six CCD cameras are synchronized by a sync-generator, and, therefore, image sequences captured by those cameras are completely synchronized.

In addition, the internal clocks of all the PCs are synchronized by *Network Time Protocol*[5], and the time stamp when each image frame is captured is added to each image frame. Comparing the time stamps of image frames captured by different capturing components with each other, the system identifies image frames taken at the same time.

2.2   Software Architecture

On our PC cluster we consider that the following parallel processing schemes and their combinations are executed. From the viewpoint of program structure, each PC corresponds to a component of a structured program of image processing.

Data gathering Images captured by multiple cameras are processed by PCs and integrated on the succeeding processing stage.

Pipeline parallel processing The whole procedure is divided into sub-functions, and each sub-function is executed on a different PC sequentially.

Data parallel processing Image is divided into sub-images, and each sub-image is processed on a different PC in parallel.

Function parallel processing Images are multicast to multiple PCs, on which different procedures are executed in parallel, and their results are unified in the succeeding processing stage.

2.3   Modules

In each PC, the following four modules are running to handle real-time image data(See Figure 1). Each of them is implemented as a UNIX process.

Data Processing Module(DPM) This module is the main part of the image processing algorithms, and is to process data input to the PC. It receives data from a DRM and sends data to a DSM via UNIX shared memory.

In DPM, any programs should consist of three elements: a main loop to process input stream, in which one iteration is executed in one frame time; pre-processing before entering the loop; post-processing after quitting the loop (Figure 1). The main loop is executed according to the following procedure to process image sequence continuously.

1. Wait for a signal from FSM to start processing. If a signal arrives before starting to wait, an error recovery function is invoked.

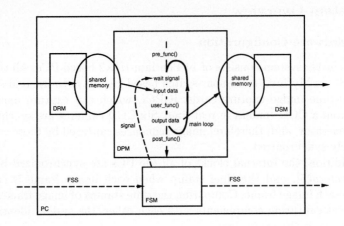

Fig. 1. Modules and Functions

2. Get input data. If input data has not been received, an error recovery function is invoked.
3. Execute a user-defined function representing one iteration of the main loop, which is named user_func here. Function user_func receives synchronous input data I and asynchronous input data A and sends output data O. Synchronous input data I are main streams of data, which originates from image capture cards and are transferred between PCs synchronously. They are synchronized at the beginning of function user_func (described at previous step). Asynchronous input data A can be used for feedback and cooperative processing. They are not synchronized at the beginning of function user_func.
4. Put output data. Because output data are directly written to shared memory in order to avoid data copy, only a notification of write-done is sent to DSM.

Before entering the main loop, a pre-processing function, which is named pre_func here, is executed. Function pre_func is a user-defined function, which is used to initialize DPM and to prepare for the main processing. After exiting the main loop, a post-processing function, which is named post_func here, is executed.

Data Receiving Module(DRM) This module is to receive data from other PCs via messages[1], and has buffers for queuing data. When a data request demand arrives from its succeeding DPM, it returns pointers to data.

Data Sending Module(DSM) This module is to send data to other PCs via messages, and has buffers for queuing data. When processed data arrives from its preceding DPM, it sends the data to the succeeding PCs.

---

[1] Message passing mechanism is developed using PM library[6].

Frame Synchronization Module(FSM) This module is introduced to make executions of different DPM synchronize with each other[4]. FSM sends FSSs to the succeeding FSM, and/or receives FSSs from the preceding FSM. FSM also sends start signals to activate the DPM in the PC.

3   RPV Programming Tool

Describing an entire program of real-time image processing on the PC cluster is not simple, because we have to describe real-time data transfer and synchronization mentioned above. To make the programming simple, we are developing RPV, as a C++ library, a programming environment for real-time image processing on the PC cluster. With RPV, users only have to describe essential structures of the programs, which are image processing algorithms on each PC and connection, i.e., data flow among PCs.

3.1   Class RPV_Connection

Data flow among PCs is described in Class RPV_Connection. Each PC sends and receives data according to the value of Class RPV_Connection. The specification of class RPV_Connection is shown in Fig 2. Member **keyword** indicates which function should be invoked in the PC (See examples in Figure 5 and Figure 6). The value of RPV_Connection varies with the PC, and it should be careful to define the values consistently in the programs of the PCs. To avoid this difficulty, here, we have designed a method with which the value of RPV_Connection can be defined by referring to a unique "connection file." The information is stored in a table with the following column headings:

#PCno   keyword   i_PC   i_size   i_num   o_PC   o_size   a_PC   a_size   a_num

Each row describes connections on one PC. The columns are space-separated and show the PC number, keyword, IPC_m, sizes of $I_{m,t}$, S, OPC_n, sizes of $O_{n,t}$, APC_l, sizes of $A_{l,r}$ and R. Multiple specifications in one column are separated by commas. A '-', a 'c' in column i_PC and a '<' in column o_size are used to indicate that there is no entry in a column, the PC captures images from a camera and the PC broadcasts the left-neighbor data respectively.

3.2   Function RPV_Invoke

Function RPV_Invoke generates all modules (DRM, DPM, DSM and FSM). Its arguments specify the connection relation among PCs (in RPV_Connection), the synchronization mode and parameters required to execute DPM including function names corresponding to **user_func**, **pre_func** and **post_func**. The specification of Function RPV_Invoke is shown in Figure 3.

A user programs only these functions, **pre_func**, **user_func** and **post_func**, without concerning data transfer and synchronization. These functions can pass data via their last arguments.

```
class RPV_Connection{
    int myPC_no;          // PC number
    char* keyword;        // keyword indicating functions invoked in a PC
    int input_PC_num;     // the number of PCs from which data are received: M
    int* input_PC;        // PC numbers data are received from: IPC_m
                          //   (m = 0, ···, M − 1)
    int* input_data_size; // sizes of received data: sizes of I_{m,t}(m = 0, ···, M − 1)
    int input_frame_num;  // the number of frames processed one time: S
    int output_PC_num;    // the number of PCs to which data are sent: N
    int* output_PC;       // PC numbers data are sent from: OPC_n(n = 0, ···, N − 1)
    int* output_data_size;// sizes of sent data: sizes of O_n, t(n = 0, ···, N − 1)
    int asynch_PC_num;    // the number of PCs from which asynchronous data are
                          //    received: L
    int* asynch_PC;       // PC numbers asynchronous data are received from: APC_l
                          //   (l = 0, ···, L − 1)
    int* asynch_data_size;// sizes of asynchronous data: sizes of A_l, r(l = 0, ···, L − 1)
    int asynch_data_num;  // the number of asynchronous data processed one time: R
};
```

Fig. 2. Definition of Class `RPV_Connection`

```
void RPV_Invoke(
    RPV_Connection* connect,                        // Connection informations
    RPV_SynchMode sync_mode,                        // synchronization mode
    int frame_num,                                  // the number of processed frame
    void* (*pre_func)(void*),                       // Function executed before loop
    void* pre_func_arg,                             // Argument for pre_func
    void* (*user_func)(RPV_Input*, RPV_Output*,     // Function executed in loop
                       RPV_Asynch*, void*),
    void* user_func_arg,                            // Argument for user_func
    void* (*post_func)(void*),                      // Function executed after loop
    void* post_func_arg                             // Argument for post_func
);
```

Fig. 3. Function `RPV_Invoke`

3.3   Sample Programs

Since programming style with RPV is based on SPMD (Single Program Multiple Data) paradigm, the same main program is executed on all the PCs. Of course, image processing functions to be executed vary with the PC, depending on the keyword described in the connection file. Figure 4 is an outline of a sample system, which captures video images of a human with 12 color markers, calculates 3D positions of color markers and animates an avatar in a display

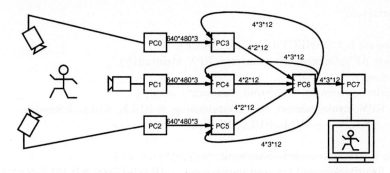

Numeral values are data sizes of data flow. For example 640*480*3
means image width * image height * byte per pixel and
4*2*12 means sizeof(int) * 2D coordinates * 12 markers.

Fig. 4. Sample System

PC	keyword	i_PC	i_size	num	o_PC	o_size	a_PC	a_size	a_num
0	smooth	c	640*480*3	1	3	640*480*3	-	-	-
1	smooth	c	640*480*3	1	4	640*480*3	-	-	-
2	smooth	c	640*480*3	1	5	640*480*3	-	-	-
3	calc2D	0	640*480*3	1	6	4*2*12	6	4*3*12	1
4	calc2D	1	640*480*3	1	6	4*2*12	6	4*3*12	1
5	calc2D	2	640*480*3	1	6	4*2*12	6	4*3*12	1
6	calc3D	3,4,5	4*2*12,4*2*12,4*2*12	1	7,3,4,5	4*3*12,<,<,<	-	-	-
7	display	6	4*3*12	1	-	-	-	-	-

Fig. 5. Sample Connection File

in real-time. PC0, PC1 and PC2 capture and smooth video images. PC3, PC4
and PC5 extract color markers and calculate 2D positions of color markers. PC6
calculates 3D positions of color markers. PC7 make an animation of an avatar.

Figure 5 is a connection file that initializes class RPV_Connection. In this
sample, functions with keyword "smooth", functions with keyword "calc2D", a
function with keyword "calc3D" and a function with keyword "display" are
invoked in PC0, PC1 and PC2, in PC3, PC4 and PC5, in PC6 and in PC7
respectively. PC0, PC1 and PC2 capture images, and PC6 broadcasts output
data to PC7, PC3, PC4 and PC5.

Figure 6 is function main of all PCs, which calls function RPV_Invoke. Figure 7
is a program of function user_func executed in PC3 – PC5. Because, as shown in
these sample programs, a user only programs static image processing algorithms
and describes information about data flow, he or she can easily construct a
real-time distributed image processing system.

```
int main(void)
{
   ifstream fs(CONNECTION_FILE_NAME);
   const RPV_Connection* connect = RPV_MainInit(fs);
   fs.close();
   if (strcmp(connect->keyword, "smooth") == 0) {
     RPV_Invoke(connect, RPV_DataMissing, 0, NULL, NULL, &Smooth,
                 NULL, NULL, NULL);
   }
   else if (strcmp(connect->keyword, "calc2D") == 0) {
     ReadBackgroundArg read_background_arg(BACKGROUND_FILE_NAME);
     RPV_Invoke(connect, RPV_DataMissing, 0,
                 &ReadBackground, &read_background_arg, &Calculate2D,
                 &read_background_arg.background, NULL, NULL);
     ...............
}
```

Fig. 6. Sample Program (main)

```
#include <fstream.h>
#include <rpv.h>

void* Calculate2D(const RPV_Input* id, RPV_Output* od,
                 const RPV_Asynch* ad, void* a)
{
   const RPV_RGB24<IMAGE_WIDTH,IMAGE_HEIGHT>* i_data
     = (const RPV_RGB24<IMAGE_WIDTH,IMAGE_HEIGHT>*)
                    id->data_ptr[0][0];
   Positions2D<MARKER_NUM>* o_data
     = (Positions2D<MARKER_NUM>*)od->data_ptr[0];
   const Positions3D<MARKER_NUM>* a_data
     = (const Positions3D<MARKER_NUM>*)ad->data_ptr[0][0];
   const RPV_RGB24<IMAGE_WIDTH,IMAGE_HEIGHT>* background
     = (RPV_RGB24<IMAGE_WIDTH,IMAGE_HEIGHT>*)a;

   RPV_RGB24<IMAGE_WIDTH,IMAGE_HEIGHT>* sub_data
     = Subtraction(i_data, background);

   // Searching for markers from the subtracted image
   SearchMarker(sub_data, a_data, o_data);

   return NULL;
}
```

Fig. 7. Sample Program (Calculate2D)

4   Conclusion

In this paper, in order to makes it easy to program real-time distributed image processing systems, we proposed RPV programming tool, which is implemented as a C++ library. Because a user have to only describe data flow and processing algorithms on PCs, he or she does not have to concern such problems of the systems like data transfer, synchronization and error recovery.

Our future works are as follows:

- performance, program simplicity and applicability of RPV programming tool should be carefully evaluated.
- sophisticated job scheduling scheme should be considered.
- error recovery process should be improved.

Acknowledgement

This work has been supported by "Cooperative Distributed Vision for Dynamic Three Dimensional Scene Understanding (CDV)" project (JSPS-RFTF96P00501, Research for the Future Program, the Japan Society for the Promotion of Science).

References

1. Takashi Matsuyama: "Cooperative Distributed Vision", *Proceedings of 1st International Workshop on Cooperative Distributed Vision*, pp.1–28, 1997.
2. Takashi Matsuyama: "Cooperative Distributed Vision – Integration of Visual Perception, Action, and Communication –", *Proceedings of Image Understanding Workshop*, 1999.
3. D. Arita, N. Tsuruta and R. Taniguchi: Real-time parallel video image processing on PC-cluster, *Parallel and Distributed Methods for Image Processing II, Proceedings of SPIE*, Vol.3452, pp.23–32, 1998.
4. D. Arita, Y. Hamada and R. Taniguchi. A Real-time Distributed Video Image Processing System on PC-cluster, *Proceedings of International Conference of the Austrian Center for Parallel Computation(ACPC)*, pp.296–305, 1999.
5. D. L. Mills: Improved algorithms for synchronizing computer network clocks, *IEEE/ACM Trans. Networks*, Vol.3, No.3, pp.245–254, 1995.
6. H. Tezuka, A. Hori, Y. Ishikawa and M. Sato: PM: An Operating System Coordinated High Performance Communication Library, *High-Performance Computing and Networking* (eds. P. Sloot and B. Hertzberger), 1225 of Lecture Notes in Computer Science, Springer-Verlag, pp.708–717, 1997.

Parallel Low-Level Image Processing
on a Distributed-Memory System

Cristina Nicolescu and Pieter Jonker

Delft University of Technology
Faculty of Applied Physics
Pattern Recognition Group
Lorentzweg 1, 2628CJ Delft, The Netherlands
email: cristina,pieter@ph.tn.tudelft.nl

Abstract. The paper presents a method to integrate parallelism in the DIPLIB sequential image processing library. The library contains several framework functions for different types of operations. We parallelize the filter framework function (contains the neighborhood image processing operators). We validate our method by testing it with the geometric mean filter. Experiments on a cluster of workstations show linear speedup.

1 Introduction

For effective processing of digital images it is essential to compute the data using a variety of techniques such as filtering, enhancement, feature extraction, and classification. Thus, there is a great need for a collection of image processing routines which can easily and effectively be used on a variety of data. We used in our research an image processing library called DIPLIB (Delft Image Processing LIBrary) [11] developed in the Pattern Recognition Group, Delft University of Technology. It provides a basic set of image handling and processing routines and a framework for expanding the set of image processing routines.

While the library provides the necessary functionality and flexibility required for image processing applications tasks, it is clear that for real-time processing, many important image processing tasks are too slow. For example, a filter operation which removes noise from a 1024×1024 pixel image requires several minutes to complete on a common desktop workstation. This is unreasonable in real-time image processing. A method to speedup the execution is to use existing workstation cluster for parallelism [1, 2, 3, 5]. In [3] the authors present the design and implementation of a parallel image processing toolkit (PIPT), using a model of parallelism designed around MPI. Currently, we are developing a parallel/distributed extension to the DIPLIB. We developed 2 parallel versions of the library on top of MPI [7] and CRL [6], respectively. As a consequence, the code remains portable on several platforms.

The paper is organized as follows. Section 2 presents a classification of low-level image processing operators. Section 3 describes an approach of integrating parallelism in the sequential image processing library. Execution times obtained for the geometric mean filter are presented and interpreted in Section 4. Section 5 concludes the paper and Section 6 presents future work.

J. Rolim et al. (Eds.): IPDPS 2000 Workshops, LNCS 1800, pp. 226-233, 2000.

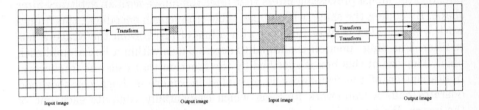

Fig. 1. Point low-level operator   **Fig. 2.** Neighborhood low-level operator

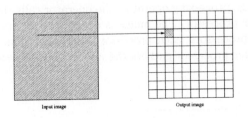

Fig. 3. Global low-level operator

2 Low-level image processing operators

Low-level image processing operators can be classified as *point operators, neighborhood operators* and *global operators,* with respect to the way the output pixels are determined from the input pixels [4].

The simplest case are the *point operators* where a pixel from the output image depends only on the value of the same pixel from the input image, see Figure 1. More generally, *neighborhood operators* compute the value of a pixel from the output image as an operation on the pixels of a neighborhood around a corresponding pixel from the input image, possibly using a kernel mask. The values of the output pixels can be computed independently, see Figure 2. The most complex case consists of *global operators* where the value of a pixel from the output image depends on all pixels from the input image, see Figure 3. It can be observed that most of the image processing operators exhibit natural parallelism in the sense that the input image data required to compute a given area of the output is spatially localized. This high degree of natural parallelism exhibited by many image processing algorithms can be easily exploited by using parallel computing and parallel algorithms.

3 Integrating parallelism in an image processing library

To quicken the development of image processing operators, the DIPLIB library supplies several framework functions.

One of the available frameworks is responsible for processing different types of image filters (neighborhood image processing operators). The framework is intended for image processing filters that filter the image with an arbitrary filter shape. By coding the shape with a pixel table (run length encoding), this framework will provide the filter function with a line of pixels from the image it has to filter. The filter function is allowed to access pixels within a box around each pixel. The size of this box is specified by the function that calls the framework. A description of the framework is presented in Figure 4. Each neighborhood operator calls a framework function which sequentially compute each line of the image. We parallelize the framework approach by data decomposition on a distributed-memory system and in this way we obtain parallelization on all image processing operators (i.e. filters) that are using the framework. So, given a neighborhood image processing operator, the operator calls the framework function on the master processor. The image is distributed by the master processor row-stripe across processors, each processor is computing its part of the image and then the master processor gathers the image back, see Figure 5.

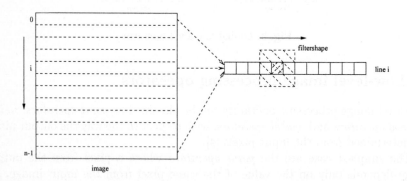

Fig. 4. Library framework function

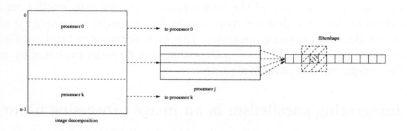

Fig. 5. Library framework function after parallelization

4   Experimental results

We measure the execution time of applying a $k \times k$ window size geometric mean filter on different image sizes. Geometric mean filter is used to remove the gausssian noise in an image. The definition of the geometric mean filter is as follows:

$$GeometricMean = \prod_{(r,c) \in W} [I(r,c)]^{\frac{1}{k^2}}$$

where (r,c) are the image pixel coordinates in window W, $I(r,c)$ is the pixel value and k is the width of the window, measured in pixels.

The filter has been implemented under CRL (C Region Library) [6] and MPI (Message Passing Interface) [7] on a distributed memory system. Two artificial images sizes 256×256 and 1024×1024 are tested on up to 24 processors. The run times in seconds using CRL are tabulated in Table 1. The run times in seconds using MPI are tabulated in Table 2.

The relative speedups computed as $SP(N) = \frac{T(1)}{T(N)}$ are plotted in Figures 6,7,8 and 9 for each image and window sizes of 3×3, 9×9 and 15×15. One may note the sharp increase of the speedup with increasing number of processors for both images. One may also observe that better performance is obtained with larger image sizes and larger window sizes. Thus, the lowest speedup corresponds to the 256×256 image and 3×3 window while the highest speedup to the 1024×1024 image and 15×15 window. The reason is the image operator granularity increases with the image size and the window size. As a consequence, communication time is less predominant compared to computation time and better performance is obtained. Some differences can be noted in the run times between CRL and MPI. This is because on our distributed memory system CRL runs on LFC (Link-level Flow Control) [8] directly while our MPI port uses Panda [9] as a message passing layer and Panda runs on top of LFC, so some overhead appears.

Table 1

Parallel geometric mean filtering execution time (in seconds) with CRL

N	256x256 3x3	256x256 9x9	256x256 15x15	1024x1024 3x3	1024x1024 9x9	1024x1024 15x15
1	1.10	9.23	24.30	16.93	145.89	398.05
2	0.61	4.68	12.21	8.79	73.19	199.35
4	0.31	2.24	5.86	4.46	36.71	99.99
8	0.17	1.13	2.99	2.34	18.46	50.31
16	0.12	0.61	1.57	1.29	9.42	25.47
24	0.10	0.47	1.13	0.94	6.40	17.19

Table 2

Parallel geometric mean filtering execution time (in seconds) with MPI

N	256x256 3x3	256x256 9x9	256x256 15x15	1024x1024 3x3	1024x1024 9x9	1024x1024 15x15
1	1.10	9.23	24.30	16.93	145.89	398.05
2	0.69	5.44	14.72	9.12	74.89	202.35
4	0.35	2.72	7.35	5.35	37.25	100.89
8	0.19	1.36	3.66	2.79	19.84	51.08
16	0.13	1.01	1.81	1.48	11.45	26.42
24	0.82	0.97	1.2	1.12	8.23	18.58

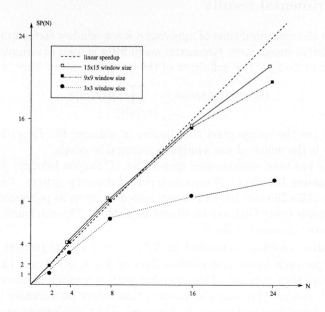

Fig. 6. Speedup of geometric mean filtering on a 256×256 image size for different window sizes, with CRL

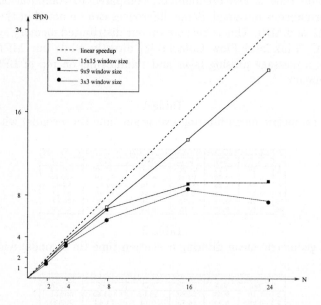

Fig. 7. Speedup of geometric mean filtering on a 256×256 image size for different window sizes, with MPI

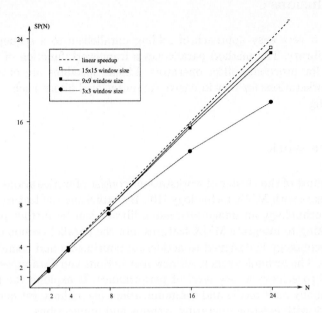

Fig. 8. Speedup of geometric mean filtering on a 1024 × 1024 image size for different window sizes, with CRL

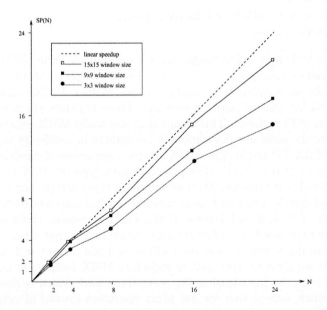

Fig. 9. Speedup of geometric mean filtering on a 1024 × 1024 image size for different window sizes, with MPI

5   Conclusions

We present a very easy approach of adding parallelism to a sequential image processing library. The method parallelizes a framework function of the library responsible for processing filter operators. Linear speedups are obtained on a cluster of workstations for very intensive computational filters, such as geometric mean filtering.

6   Future work

Nowadays, most of the cluster of workstations consist of workstations containing Intel processors with MMX technology [10]. By exploiting the features added by the MMX technology, an image processing library can be further parallelized. We are working to integrate MMX features into the parallel version of DIPLIB.

MMX technology is designed to accelerate multimedia and communications applications. The technology includes new instructions and data types that allow applications to achieve a new level of performance. It exploits the parallelism inherent in many multimedia and communications algorithms, yet maintains full compatibility with existing operating systems and applications.

The highlights of the technology are:

- Single Instruction, Multiple Data (SIMD) technique
- 57 new instructions
- Eight 64-bit wide MMX technology registers
- Four new data types

A process technique called Single Instruction Multiple Data (SIMD) behaves as a SIMD parallel architecture but at a lower level. Special MMX instructions to perform the same function on multiple pieces of data. MMX technology provides eight 64-bit general purpose registers. These registers are aliased on the floating point (FP) registers. This means that physically MMX registers and FP mantissas are the same but the content is interpreted in a different way depending on the MMX/FP mode. The MMX registers are accessed directly using the register names MM0 to MM7. The principal data type of MMX technology is the packed fixed-point integer. The four new data types are: packed byte, packed word, packed double word and quad word all packed into one 64-bit quantity in quantities of 8, 4, 2 and 1 respectively. For this reason, given an array of element type byte, word or double word the processing of that array with simple operations (addition, subtraction, etc.) will be 8, 4 and respective 2 times faster. In Figure 10 we show an approach of including MMX features in our DIPLIB library. We begin by parallelizing the point framework which is similar to the filter framework, except that we use point operators instead of neighborhood operators. A master processor is distributing the image in a row-stripe way to the slave processors and each slave processor is computing its part of the image by applying the point operator to each line of that part of image. If the slave processor is enabled with MMX technology we exploit the MMX features of processing in parallel more elements of a line. This part has to be coded using MMX instructions.

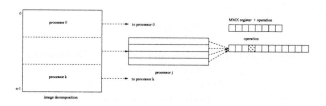

Fig. 10. Adding MMX features to DIPLIB image processing library

References

1. P.Challermvat, N. Alexandridis, P.Piamsa-Niga, M.O'Connell: *Parallel image processing in heterogenous computing network systems*, Proceedings of IEEE International Conference on Image Processing 16-19 sept. 1996,Lausanne,vol.3,pp.161-164
2. J.G.E. Olk, P.P. Jonker: *Parallel Image Processing Using Distributed Arrays of Buckets*, Pattern recognition and Image Analysis, vol. 7, no. 1,pp.114-121,1997
3. J.M. Squyres, A. Lumsdaine, R. Stevenson: *A toolkit for parallel image processing*, Proceedings of the SPIE Conference on Parallel and Distributed Methods for Image processing, San Diego, 1998
4. S.E.Umbaugh: *Computer Vision and Image Processing - a practical approach using CVIPtools*, Prentice Hall International Inc.,1998
5. I.Pitas: *Parallel Algorithms for Digital Image Processing, Computer Vision and Neural Networks*, John Wiley&Sons, 1993
6. K.L.Johnson, M.F.Kaashoek and D.A.Wallach: *CRL: High-Performance All-Software Distributed Shared Memory*, Proccedings of the Fifteenth Symposium on Operating Systems Principles, 1995
7. M.Snir, S.Otto, S.Huss, D.Walker and J.Dongarra: *MPI - The Complete Reference, vol.1, The MPI Core*, The MIT Press, 1998
8. R.A.F. Bhoedjang, T. Ruhl and H.E. Bal: *Efficient Multicast on Myrinet Using Link-level Flow Control*, Proceedings of International Conference on Parallel Processing, pp. 381-390, Minneapolis MN, 1998
9. T.Ruhl, H, Bal, R. Bhoedjang, K. Langendoen and G. Benson: *Experience with a portability layer for implementing parallel programming systems*, Proceedings of International Conference on Parallel and Distributed Processing Techniques and Applications, pp. 1477-1488, Sunnyvale CA, 1996
10. J.E. Lecky: *How to optimize a machine vision application for MMX*, Image Processing Europe, March Issue, pp. 16-20, 1999.
11. *http://www.ph.tn.tudelft.nl/Internal/PHServices/onlineManuals.html*

Congestion-free Routing of Streaming Multimedia Content in BMIN-based Parallel Systems

Harish Sethu

Department of Electrical and Computer Engineering
Drexel University
Philadelphia, PA 19104, USA
sethu@ece.drexel.edu

Abstract. Multimedia servers are increasingly employing parallel systems for the retrieval, scheduling and delivery of streaming multimedia content. However, given non-zero blocking probabilities in interconnection networks of most real parallel systems, jitter-free constant-rate delivery of streaming data cannot often be guaranteed. Such a guarantee can be best accomplished through the elimination of all congestion in the network. In this paper, we focus on folded Benes BMIN networks and achieve a fast convergence to congestion-free transmissions using an approach that combines flow-based adaptive routing with a distributed version of Opferman's looping algorithm. This combined approach significantly reduces the buffering requirements at the receiving stations. In addition, it also achieves a low start-up delay, important for interactive web-based multimedia applications.

1   Introduction

A popular trend in the architecture of multimedia servers and other networked multimedia systems is the use of commercial parallel computing systems such as the IBM RS/6000 SP as distribution engines, co-ordinating the retrieval, scheduling and delivery of streaming multimedia content. New, popular and commercially promising applications such as web-based interactive communications have further fueled this trend. This is driven in part by the fact that parallel system interconnects and the associated hardware interfaces and communication subsystem software services offer high-bandwidth network-level access to system I/O at very low latencies. In many networked multimedia systems, aggregation and/or striping is achieved before the data reaches the network, and the primary issue is the guaranteed-rate jitter-free delivery of continuous data streams between sets of source and destination end-points of the network, with no source paired with more than one destination or vice versa. Guaranteed-rate jitter-free delivery is best accomplished through elimination of all congestion in the network. This provides our motivation to consider the problem of congestion-free routing of multiple streaming multimedia traffic across packet switched multistage interconnection networks popular in parallel systems.

J. Rolim et al. (Eds.): IPDPS 2000 Workshops, LNCS 1800, pp. 234–241, 2000.

In this paper, we focus on a class of Bidirectional Multistage Interconnection Networks (BMINs) used in several offerings of commercial parallel systems [1,2]. Within this class of topologies, we specifically consider those topologies which possess the rearrangeable non-blocking property. A rearrangeable non-blocking topology offers the potential for congestion-free routing; that is, given the knowledge of all exclusive source-destination pairs in the network, one can always use a set of rules by which a congestion-free set of paths can be established between each source-destination pair. The looping algorithm by Opferman and Tsao-Wu for Benes networks is a good example of such a rule [3]. Centralized control routines can easily implement such algorithms in software. However, large delays associated with software involvement are not suitable for interactive web-based multimedia servers. This provides a motivation for a technique that allows the start of transmissions immediately upon request, and which subsequently seeks a fast convergence of the routing to a congestion-free state. Such a technique serves an additional goal of minimizing the disruptions and the associated jitter and/or buffer overflows that occur when there are changes in the pairings between traffic sources and destinations. In this work, we assume that the transient congestions that may occur due to control packets are inconsequential.

In this paper, we present an approach that can be implemented in hardware in the switching elements of parallel systems and which achieves congestion-free, and thus, jitter-free transmission within a short period of time after the start of transmissions. Our method combines flow-based adaptive routing (as opposed to adaptively routing each individual packet) and a distributed implementation of Opferman's looping algorithm modified for the class of networks under consideration. This technique enables a convergence to congestion-free routing in less time than it takes to even complete the execution of the modified looping algorithm. This is because, as the algorithm establishes routes for source-destination pairs, the flow-based adaptive routing begins to achieve greater stability. The two algorithms in tandem, achieve fast convergence to congestion-free routing, thus reducing the buffering requirements at the receiving portals of the multimedia streams, while also achieving a very small delay in starting up the transmissions.

In Section 2, we describe the topology of folded Benes networks. Sections 3 and 4 describe flow-based adaptive routing and the distributed version of the looping algorithm, respectively. Section 5 briefly discusses simulation results, and Section 6 concludes the paper.

2   Folded Benes networks

BMINs may be thought of as Benes networks or symmetrically extended Banyan networks which are folded in the middle to create bidirectional links between switches. Some of IBM's RS/6000 SP system topologies [1,5] and the CM-5 data router [2], are examples of commercial realizations of this family of network topologies. Recognizing that many real BMINs are *folded Benes* networks, we refer to them as such and limit discussion, in this paper, to this class of BMINs.

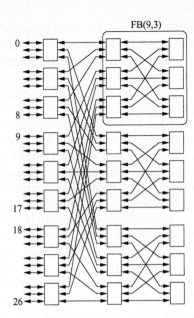

Figure 2. An FB(27, 3) network

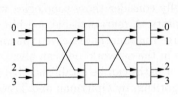

Figure 1(a). A 4-by-4 Benes network

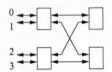

Figure 1(b). An FB(4, 2) network

Denote by $FB(N, b)$, an $N \times N$ folded Benes network using $2b \times 2b$ switches, where N is a power of b. Figure 1(a) shows a 4×4 Benes network using 2×2 switches, which when folded yields an $FB(4,2)$ network shown in Figure 1(b), i.e., a 4×4 folded Benes network using 4×4 switches. This is shown only for the sake of illustration—it may be noted that an $FB(4,2)$ network does not quite make sense since one may just as well use a single 4×4 switch. Figure 2 shows an $FB(27, 3)$ network.

3   Flow-based adaptive routing

In this work, we define a flow as the string of packets that originate from the same source and are headed to the same destination. This concept of flows is similar to that used in some implementations of IP over ATM to achieve connection-oriented behavior in a connectionless protocol [4].

The technique of flow-based adaptive routing merges the two concepts of flow switching and adaptive routing. In our implementation of this technique, the switches maintain a table of active flows, henceforth referred to as the *flow table*. Each entry in this table associates an output port with a flow. Packets belonging to a given flow are all queued for transmission at the output port specified by the entry in this table. As far as possible, the switch attempts to ensure that no output port is associated with more than one flow in the flow table. This is in contrast to traditional packet switching in which each packet is independently routed. Unlike in the popular use of the concept of flow switching in the Internet, our technique allows adaptivity in the path taken by the flow. For example, when a given flow is experiencing congestion, the switch that observes this congestion changes the entries in the flow table seeking to avoid further congestion. This triggers corresponding changes in the tables for other switches downstream as well.

Whenever a switch detects that one of its output queues corresponding to a certain flow, f, has more than a certain threshold number of packets in it, it seeks to change the output port entry corresponding to this flow. In folded Benes networks, the first half of the hops in the path of a flow can be without congestion, since there are as many valid output ports as there are flows seeking to access these output ports [6]. On the return path, i.e., during the second half of the hops in the path of a flow, destination-based contention causes congestion. In this return path, often, two flows have to share a link to reach their respective destinations. Changing the flow table entries in such a way that f's output port is now associated with another flow that is headed to a destination maximally distant from that of f, has the potential to reduce downstream destination-based contention. This is based on the premise that paths to distant nodes will likely diverge in subsequent hops of the packets, and therefore will not cause congestion. The probability of downstream congestion, therefore, can be reduced by switching the output port entry of the flow associated with the congested output port with the output port entry of a flow with a destination as distant as possible from that of the flow experiencing the congestion.

The following algorithm captures the core of our implementation of flow-based adaptive routing. This algorithm determines the destination output port of each new packet that arrives at the switch. In the following, F is defined as the set of all flows that pass through the switch executing the algorithm. T is the threshold number of packets necessary in an output queue, in order for the algorithm to consider switching the output ports in the flow table entries.

1. $f =$ flow the packet belongs to;
2. if flow table entry of f is empty, then,
   Set a random idle output port as the output port entry of f;
3. if output queue of output port entry for f has more than T packets, then,
   Find $g \in F$ such that $|destination(f) - destination(g)|$ is maximum;
   Pick a random number, r, between 0 and 1;
   If $r < \alpha$, exchange the output port numbers in entries for f and g;
4. Output port of new packet $=$ output port in flow table entry of f;

An important aspect of adaptive routing is the potential for oscillations in the paths taken without achieving the desired convergence to a congestion-free set of paths. This phenomenon requires that we use a dampening factor, $\alpha \leq 1$, as in Step 3 of the above algorithm, so that flow table entries are not switched during each cycle that the corresponding queue length is above the threshold number. Instead, a switch in the entries is made only with a probability α when the threshold is reached.

The above routing technique can guarantee a very high rate but not the 100% of the input rate necessary for jitter-free delivery of streaming data. Besides, the routing using this algorithm is not stable since it does not necessarily converge to a congestion-free set of paths for any given permutation of source-destination pairs. The distributed algorithm described in the next section, in conjunction with flow-based adaptive routing achieves the desired convergence.

4 The distributed modified looping algorithm

The looping algorithm, as first proposed by Opferman [3], was for Benes networks with 2×2 switches. In this paper, we present a generalization of the algorithm to $b \times b$ switches, modify it to work in conjunction with the flow-based adaptive routing algorithm, and apply it to folded Benes networks. The algorithm presented here assumes that flow-based adaptive routing is in effect during the time that the looping algorithm is executing. This is a crucial aspect of the algorithm, which helps to simplify the distributed algorithm while also allowing a low start-up delay. To simplify the presentation of the looping algorithm, we assume the worst-case scenario where each node has multimedia streams to receive as well as to send; recall that, in a folded Benes topology, each end-point of the network can be both a receiving as well as a sending portal.

We define an entry in the flow table in a switch as a *forward* entry if the source node of the flow is lesser number of hops away from the switch than the destination node of the flow. Similarly, define an entry in the flow table in a switch as a *backward* entry if the destination node of the flow is lesser number of hops away from the switch than the source node of the flow. Thus, given a flow with a 5-hop path from node 0 to node 26, the entries corresponding to this flow in the first two switches in the path are forward entries while the entries in the last two switches in the path are backward entries. It is possible that an entry is neither a forward nor a backward entry. In a folded Benes network, the same switch can be both a first switch in the path of some flow, and the last switch in the path of some other flow. Note also that a first-stage switch merely means a switch directly connected to the nodes. A second-stage switch is one that has at least one other switch between itself and a node. In general, a stage-s switch has $s - 1$ switches between itself and a node.

As in the original looping algorithm, congestion-free paths are established in a recursive fashion. In the first application of the algorithm, entries are locked only in the first-stage switches of the network. The algorithm is next applied within each of the FB subnetworks interconnecting the first-stage switches; this results in the locking of entries in the second stage switches of the overall network. This recursive application of the algorithm continues on smaller and smaller subnetworks, until the entries in all stages are locked. The three steps enumerated below describe the algorithm executed to lock the entries in the first stage of the network or subnetwork under consideration.

1. Select a first-stage switch that already has a forward locked entry in the flow table, and with at least one unlocked forward entry. If no such switch is

found, select a first-stage switch with no locked forward entries. If no such switch is found either, the algorithm ends.

2. Lock the only unlocked forward entry, or a randomly chosen unlocked forward entry in the selected switch. Lock the corresponding backward entry of the flow in the last switch in the path of the flow. Now, select an unlocked backward entry in this switch, and go to Step 3. If there are no unlocked backward entries in this switch, go to Step 1.

3. Lock the selected unlocked backward entry in the switch. Now, in the first switch in the path of the corresponding flow, lock the forward entry. If there is an unlocked forward entry in this switch, select this switch and go to Step 2; otherwise, go to Step 1.

As shown in Figure 2, an $FB(N,b)$ network may be thought of as being divided into a first stage of $2b \times 2b$ switches, and a second (or last) stage of b $FB(N/b, b)$ subnetworks. In the first application of the algorithm, only the entries in the first-stage switches are locked. Consider a flow that has its corresponding first-stage switch entries locked in this first recursion. This does not lock the path of the flow within the $FB(N/b, b)$ subnetworks. However, from the nature of FB networks, the entry and exit points of the flow within the $FB(N/b, b)$ network are locked and thus the algorithm can now be applied within the b $FB(N/b, b)$ subnetworks. In the subsequent applications of the algorithm within the $FB(N/b, b)$ subnetworks, the entries in the first stage of the $N/b \times N/b$ subnetworks are then locked. When an entry is locked, no flow other than the one in the locked entry may use the output port associated with the locked entry.

We now describe the core aspects of the distributed modified looping algorithm. Assume the model of the $FB(N, b)$ network as shown in Figure 2, with the first stage of switches connected to the sources and destinations, and interconnected via b separate $FB(N/b, b)$ subnetworks. The algorithm begins in the first stage switches, which use short control packets to exchange information. The distributed algorithm begins in a designated switch in the first stage which generates a control packet; randomly chooses a flow that has a destination in another switch; and sends the control packet to this destination switch which is also a first-stage switch. Let the source of this flow be i, and the destination be $d(i)$. A lock is placed in the first-stage switches on the path from i to its destination, $d(i)$. Placing a lock on a path and associating it with a flow is equivalent to establishing a deterministic non-adaptive path for the flow through the corresponding switch. The switch connected to $d(i)$, upon receiving the control packet, responds with a new control packet headed to the switch connected to the source of packets arriving at a destination within this switch connected to $d(i)$. This control packet uses an $FB(N/b, b)$ subnetwork different from the one used by the previous control packet. This happens automatically because no more than one flow can use the output port in a locked entry, and by the nature of FB networks, a switch in the first stage is connected to each of the subnetworks by exactly one cable. Once again, the first stage switches in the path of the control packet establish locks associating the path with the flow. This pattern of choosing different subnetworks to reach different destinations connected to the same switch, is continued until all inputs and outputs are connected via paths that are locked. As far as possible, an attempt is made to merely lock entries, rather than rewrite entries as well as lock them.

Exceptions occur when, for example, in Step 1, a first-stage switch receives a control packet instructing it to lock an entry, which is the last unlocked entry in the switch. The algorithm now needs to continue at a new first-stage switch, as described in Step 1 of the recursive algorithm. We ensure this through using a bit-map of sources, within each control packet, to indicate which flows have already had entries locked in the first stage. For example, if a flow with source address S already has its entries locked, bit S in the length-N bit-map would be set to a 1. A switch can now determine which flow entries are not yet locked, even though all of its own flow entries are locked. An informational control packet can now be sent to the first-stage switch with an unlocked entry. When locks are being placed in the second-stage switches, i.e., within the first-stage switches of the $FB(N/b, b)$ subnetworks, bit $\lfloor S/b \rfloor$ in the length-(N/b) bit-map is set to a 1.

The three-step algorithm described above is a constructive proof that non-overlapping paths can be established in the first stage switches of an FB network. Now, note that an $FB(N, b)$ network is constructed out of a first stage of switches connected to b $FB(N/b, b)$ subnetworks. For example, Figure 2 shows that an $FB(27, 3)$ network has a first stage of switches and its subsequent stages are a set of 3 $FB(9, 3)$ subnetworks. Thus, after non-overlapping paths are established in the first stage switches of a $FB(N, b)$ network, non-overlapping paths can be similarly established in the first stage switches of the $FB(N/b, b)$ subnetworks. These first stage switches of the subnetworks are really the second stage switches of the overall network. Recursively applying the algorithm and thus locking flow table entries in each of the stages of the network leads to congestion-free routing of any given permutation of exclusive source-destination pairs.

5   Simulation results

Our simulation environment consists of output-queued switches, with an architecture that allows one switch hop in a minimum of 4 cycles in the absence of blocking. All input and output queues are set to a size of 4 packets. All links between switches can be traversed in exactly one cycle.

Our simulations indicate a fast convergence to congestion-free routing in the vast majority of randomly chosen permutations, even before all the flow entries are locked. However, the worst-case convergence time is what is relevant in the design of the system, especially with regard to buffering requirements. In a 2-stage $FB(16, 4)$ network, for example, the worst-case convergence time is equal to (i) the time taken to lock 15 paths (the last path is automatically determined by elimination) plus (ii) the time taken for 3 control packets to be sent to other first-stage switches when the sending switch has all its forward entries locked (part of Step 1 of the looping algorithm). This time depends on the input rate since the number of cycles it takes to traverse a path depends on the probability of blocking, which depends on the input rate. However, even at a 100% input load, our simulation of flow-based adaptive routing indicates that the time taken to communicate between two first-stage switches is never more than 12 cycles. Assuming a 4 nanosecond cycle time, this translates to an observed maximum convergence time of less than 1 μs. Note that, during this time, packets are being delivered to the destinations although at a slightly slower rate than the rate at which the cables in the network can send or receive data. The effective output rate varies with the chosen permutation; however, at a 100% input load, an effective output rate of at least 95% was achieved for all the permutations attempted in our simulations. Assuming that the data absorption rate is the same as the rate at which data is being sent from the senders, it should be expected that the receivers should buffer some data before beginning the absorption so as to achieve jitter-free reception. At 1 GB/s, using the earlier example of an $FB(16, 4)$ network, this implies a buffering requirement of just 50 bytes. On the other hand, if one does not attempt convergence to congestion-free routing, the buffering required is unbounded since flow-based adaptive routing alone cannot guarantee stable congestion-free routing. The self-stable nature of this technique can similarly minimize the buffering requirements when disruptions occur due to the introduction of a new set of data streams replacing an older set.

6 Conclusion

In this paper, we have presented a mechanism that combines the concept of flow-based routing, which has shown to have worked very well in the Internet, with Opferman's looping algorithm, originally designed for circuit-switched telecommunication networks. We have added adaptivity in the model of flow-based routing, and used a distributed version of the looping algorithm modified for folded Benes networks and for use in conjunction with flow-based adaptive routing. This combined approach allows a fast convergence to congestion-free routing and therefore, jitter-free constant-rate delivery of multimedia streams, while reducing buffering requirement as well as the start-up delay.

References

1. Sethu, H., Stunkel, C. B., Stucke, R. F.: IBM RS/6000 SP Large-System Interconnection Network Topologies, *Proceedings of the 27th Int'l. Conf. Parallel Processing*, Minneapolis, August 1998.
2. Heller, S.: Congestion-Free Routing on the CM-5 Data Router, *Lecture Notes in Computer Science* **853** (1994) 176–184.
3. Opferman, D. C., Tsao-Wu, N. T.: On a class of rearrangeable switching networks—Part I: Control algorithms, *Bell Syst. Tech. J.* **50** (1982) 1579–1618.
4. Newman, P., Minshall, G., Lyon, T. L.: IP Switching—ATM under IP, *IEEE/ACM Transactions on Networking*, **6**(2) (1998) 117–129.
5. C. B. Stunkel, *et al.*:, "The SP2 High-Performance Switch," *IBM Systems Journal*, **34**(2) (1995) 185–204.
6. Y. Aydogan, *et al.*:, "Adaptive Source Routing in Multicomputer Interconnection Networks," *Proceedings of the 10th Int'l Parallel Processing Symp.*, April 1996.

Performance of On-Chip Multiprocessors
for Vision Tasks*
(Summary)

Y. Chung[1], K. Park[1], W. Hahn[1], N. Park[2], and V. K. Prasanna[2]

[1] Hardware Architecture Team, Electronics and Telecommunications Research Institute,
P.O.Box 8, Daeduk Science Town, Daejeon, 305-350, Korea
{yongwha, kyong, wjhan}@computer.etri.re.kr
[2] Department of EE-Systems, EEB-200C, University of Southern California,
Los Angeles, CA90089-2562, USA
{neungsoo, prasanna}@halcyon.usc.edu

Abstract. Computer vision is a challenging data intensive application. Currently, superscalar architectures dominate the processor marketplace. As more transistors become available on a single chip, the "on-chip multiprocessor" has been proposed as a promising alternative to processors based on the superscalar architecture. This paper examines the performance of vision benchmark tasks on an on-chip multiprocessor. To evaluate the performance, a program-driven simulator and its programming environment were developed. DARPA IU benchmarks were used for evaluation purposes. The benchmark includes integer, floating point, and extensive data movement operations. The simulation results show that the proposed on-chip multiprocessor can exploit thread-level parallelism effectively.

1 Introduction

High performance computing is needed to satisfy the computational requirements of vision tasks. Several efforts have been directed towards providing parallel processing support for vision. A brief summary of leading research efforts in parallel processing for vision can be found in [1,2]. These efforts can be grouped into three categories, based on the nature of computing platforms they utilize: *special-purpose VLSI processors*[3], *specialized vision systems*[4], and *general-purpose parallel machines*[2].

In this paper, we examine the performance of a collection of vision tasks on an **on-chip multiprocessor**. The performance of general-purpose **Commodity-Off-The-Shelf(COTS)** microprocessors has been improving at a phenomenal rate over the last decade. Currently, the most COTS processors such as *Intel Pentium Series, Compaq*

---

* The work at USC was supported by the DARPA Data Intensive Systems program under contract F33615-99-1-1483 monitored by Wright Patterson Airforce Base.

J. Rolim et al. (Eds.): IPDPS 2000 Workshops, LNCS 1800, pp. 242-249, 2000.

Alpha21264, IBM PowerPC620, Sun UltraSparc-2, HP PA8000, and *MIPS R10000* use the **superscalar** design technique[5]. Such superscalar processors execute multiple instructions in a single cycle by exploiting **Instruction-Level Parallelism(ILP)**[5]. However, significant speed-ups may not be achieved by using this technique because of the limitation imposed by the instruction window size and available ILP in a typical sequential program[6]. Moreover, considerable design effort is required to develop such high performance processors. Researchers have been studying alternatives to the superscalar architecture[5]. The "on-chip multiprocessor"[7] is one of the alternatives considered as a next generation processor. The key feature of such an architecture is a multiprocessor in a single chip that shares an off-chip 2nd level cache to exploit **Thread-Level Parallelism(TLP)**[7], in addition to ILP. For scientific workloads, the on-chip multiprocessor has been shown to overcome the ILP limitation[8]. However, the performance of the on-chip multiprocessor on vision tasks (that have varying computational characteristics) is not known.

To investigate the suitability of the on-chip multiprocessor for vision tasks, we conducted performance studies using a dedicated architectural simulator. It is a program-driven, cycle-level simulator consisting of a Pre-Processing Unit as an instruction simulator and a Post-Processing Unit as a performance simulator. Also, a programming environment was developed to support multithreaded programming on multiple processor cores. Our simulations focused on the performance characteristics of the on-chip multiprocessor including the **Instructions Per Cycle(IPC)** and the total number of **execution cycles** as the number of processor cores is increased. Vision tasks were chosen from the widely studied **DARPA Image Understanding Benchmark**[9]. The simulation results show that "on-chip multiprocessor" is an attractive candidate architecture for vision tasks.

The organization of the paper is as follows. Overview of the vision tasks and the on-chip multiprocessor architecture considered in this paper are given in Section 2 and 3, respectively. In Section 4, the architectural simulator and its programming environment are explained. Simulation results are shown in Section 5, and concluding remarks are made in Section 6.

2 Selected Vision Tasks

The vision tasks considered in this paper are selected from the Image Understanding Benchmark[9]. This benchmark performs the recognition of a "mobile" sculpture, given the input images from intensity and range sensors. The benchmark performs low-level operations such as *convolution, thresholding, connected components labeling, edge tracking, median filter, Hough transform, convex hull*, and *corner detection*. It also performs *grouping* operations and *graph matching* which are representative examples of intermediate-level and high-level processing, respectively. The benchmark utilizes information from two sensors in order to complete the interpretation process. It makes use of both integer and floating-point representations.

The benchmark performs both bottom-up (data-directed) and top-down (knowledge or model-directed) processing. The top-down processing can involve processing of low and intermediate-level data to extract additional features from the data, or can involve control of low and intermediate-level processes to reduce the total amount of computation required.

In the benchmark, the processing begins with low-level operations on the intensity and depth images, followed by grouping operations on the intensity data to extract candidate rectangles. These candidates are used to form partial matches with the stored models. For each of these models multiple hypothetical poses may be established. For each of the pose, stored information is used to probe the depth and intensity images in a top-down manner. Each probe tests a hypothesis for the existence of a rectangle in a given location in the images. Rejection of a hypothesis, which only occurs when there is strong evidence that a rectangle is actually absent, results in the elimination of the corresponding model pose. Confirmation of the hypothesis results in the computation of a match strength for the rectangle, and it also results in the updating of its representation in the model pose with new size, orientation, and position information. After a probe has been performed for every unmatched rectangle in the list of model poses, an average match strength is computed for each pose that has not been eliminated. The model pose with the highest average is selected as the best match. More details of the benchmark can be found in [9].

3 On-Chip Multiprocessor

In this paper, we use *Raptor*[10], an on-chip multiprocessor, consisting of four independent processor cores, called General Processor Units(GPUs), and one graphic co-processor, called Graphic Co-processor Unit(GCU). Due to the limited die size, we have chosen four GPUs that are integrated into a single chip. The GCU is shared by four GPUs. Also, in order to control GPUs/GCU and to provide an interface to outside world, additional four component units are included in Raptor namely: Inter-processor Bus Unit(IBU), External Cache Control Unit(ECU), Multiprocessor Control Unit(MCU), and Port Interface Unit(PIU). The IBU is a shared bus connecting the GPUs and the ECU. The MCU distributes the interrupts across the GPUs and provides synchronization resources among the GPUs. The PIU is a mutiprocessor-ready bus interface to communicate with the exterior of the Raptor. The four GPUs execute all instructions except extended graphic instructions with their own register files and program counters, but share the ECU through the IBU. A GPU performs graphic instructions with Single Instruction Stream Multiple Data Stream(SIMD) style and pixel processing hardware. The salient features of Raptor can be summarized as follows:

- Single chip 4-way multiprocessor sharing off-chip 2nd level cache
- 64-bit data and 64-bit virtual address
- SPARC V9 Instruction Set Architecture(ISA)
- Extension of graphic instruction set
- Multiple cache structure consisting of on-chip 1st level cache and off-chip 2nd

level cache
- Harvard structure of 1st level cache consisting of 16 Kbyte instruction cache and 16 Kbyte of data cache
- On-chip 2nd level cache controller handling 4 Mbyte of unified off-chip 2nd level cache

4 Simulation Environment

To evaluate the Raptor quantitatively, we developed a dedicated simulator, called *RapSim*. Also, a programming environment, called *MMOS*(Multithreaded Mini-OS), was developed to support a multithreaded programming on the multiple GPUs. The overall environment of the RapSim and the MMOS is shown in Fig. 1.

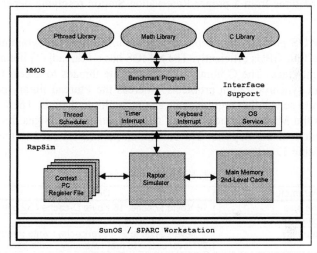

Fig. 1. Simulation Environment

The **RapSim** is a program-driven micro architecture simulator that models the four GPUs and a memory hierarchy shared by the four GPUs. The RamSim consists of a Pre-Processing Unit and a Post-Processing Unit. The Pre-Processing Unit of the RapSim is an instruction set simulator while the Post-Processing Unit is a performance simulator.

The Pre-Processing Unit consists of four components: a processor model for executing instructions, data structures for register files, proxy model for processing system calls, and a model of 1st level cache. The Pre-Processing Unit fetches the instructions and the data from the shared memory hierarchy including 2nd level cache and executes the instructions, and generates an on-the-fly trace consumed by the Post-Processing Unit. The Pre-Processing Unit starts the simulation by loading a benchmark binary file compiled and statically linked with the MMOS library into memory model. During the loading of the benchmark binary, a proper starting program counter is set in the processor model. Trap table and trap handlers are initialized in the memory model and a stack is constructed in the memory model. Then the processor model executes the instructions using the internal resources like the execution units, register files, and 1st level cache.

As the Pre-Processing Unit runs its instruction streams, it generates an on-the-fly trace, a sequence of executed instructions. Each entry of the trace contains enough information so that the Post-Processing Unit can perform the performance simulation using the trace as inputs. The Post-Processing Unit is a RISC pipeline model conducting performance simulation by using the instruction traces generated from the Pre-Processing Unit. It is modeled as a 2-issue **superscalar** including Reservation Stations(RS) and a Reorder Buffer(ROB) to support out-of-order executions. Two instructions in a Trace Buffer are fetched and pre-decoded in a cycle. The pre-decoded instructions in an Instruction Buffer are decoded and issued into proper Reservation Stations(RS), and the Reorder Buffer is updated simultaneously. Each execution unit runs safe instructions from a proper Reservation Station resolving dependency problems.

The **MMOS** provides the RapSim with a multithreaded programming environment to utilize four GPUs efficiently. The MMOS has a Pthread[11] library, C library, and RapSim interface. The C library allows multiple threads to access the shared C library without synchronization problems, whereas the Pthread library provides synchronization and scheduling requirements among multiple threads. The RapSim interface connects the MMOS to the RapSim, and schedules and assigns threads into the processor models of the RapSim. The simulation parameters used in the experiment are listed in Table 1.

Table 1. Simulation Parameters

Parameter	Default Value
1st level cache size	16 Kbyte I-cache, 16 Kbyte D-cache / 32 bytes per line
2nd level cache size	4 Mbyte / 32 bytes per line
Write update policy	1st level cache to 2nd level cache : write through 2nd level cache to main memory : write back
1st level cache access latency	1 cycle
2nd level cache access latency	4 cycles
Main memory access latency	10 cycles

5   Simulation Results and Analysis

Three sets of simulations were conducted for each vision task described in Section 2. The image size was 512X512. The three sets of simulations were:

- Sequential: a non-multithreading running on a 1-GPU configuration.
- 2-Threads: 2-way multithreading running on a 2-GPU configuration.
- 4-Threads: 4-way multithreading on a 4-GPU configuration.

The *Instructions Per Cycle(IPC)* and the total number of *execution cycles* were measured as our performance metrics.

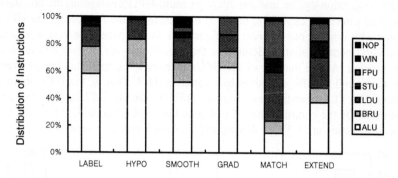

Fig. 2. Distribution of Instructions Executed on a 1-GPU Configuration

To characterize the computational requirement of each vision task in the object recognition system, we break down the instructions executed on a 1-GPU configuration into seven components, as shown in Fig. 2. In this Fig. 2, ALU, BRU, LDU, STU, FPU, WIN represent ALU, branch, load, store, FPU, window register instructions, respectively. The low/intermediate level tasks(LABEL, HYPO) perform operations on the integer representation of the intensity image. The low level tasks(SMOOTH, GRAD) perform operations on the floating-point representation of the depth image thus the FPU is used in these tasks. However, the high level tasks(MATCH, EXTEND) involve less ALU operations and involve more LDU/STU operations due to the graph traversal (used in object recognition). In addition they also contain more FPU operations due to the probe operations on the depth image data.

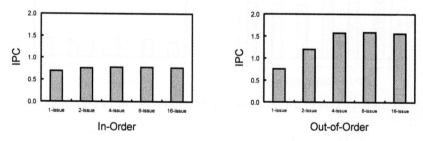

Fig. 3. Effects of Issue Width and Out-of-Order Execution

Fig. 3 shows the limitation of ILP on which a typical superscalar architecture relies. It shows the effects of the issue width on IPC with in-order and out-of-order execution. In this experiment, we used the SimpleScalar[12], which is a well-known simulator for superscalar microprocessors. The improvement by increasing the issue width with the in-order execution was negligible, whereas the IPC was increased up to the 4-issue configuration with the out-of-order execution. Overall, the absolute IPC values with out-of-order were larger than those with in-order. However, the IPC obtained by exploiting ILP only, on which typical superscalar architecture relies, was limited to 1.6 even with the 16-issue configuration. Therefore, an additional level of parallelism needs to be considered to satisfy the time performance required by the selected vision tasks.

The execution cycles and the IPCs on multi-GPU configurations are shown in Fig. 4. The effects of exploiting the TLP by increasing the number of the GPUs are analyzed with 2-issue, out-of-order configuration. In this paper, the execution cycle of a multi-GPU configuration is defined as the execution cycle of the GPU running the parent process. As the number of the GPUs increases, the total number of execution cycles drops and the IPC improves by exploiting TLP. That is, the IPCs obtained by exploiting the TLP from the base 2-issue configuration are 2.37 for 2-thread and 4.7 for 4-thread. On the contrary, the IPCs obtained by exploiting the ILP from the base 2-issue configuration are 1.57 for 4-issue and 1.58 for 8-issue, as shown in Fig. 3.

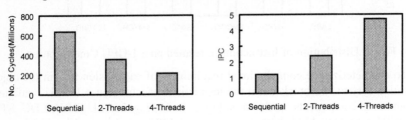

Fig. 4. Execution Cycle and IPC of Each Configuration

The IPC of each vision task in the object recognition system is shown in Fig. 5. Even with very different computational characteristics, each task shows similar IPC by exploiting TLP as well as ILP. Also, compared to the IPC on a 8-issue superscalar, TLP can provide 4 times improvement over the ILP-only solution on the average.

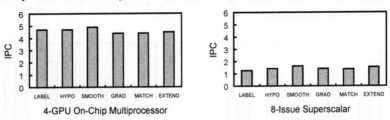

Fig. 5. IPC of Each Task on a 4-GPU On-Chip Multiprocessor and a 8-issue Superscalar

6 Concluding Remarks

The on-chip multiprocessor has been proposed as a promising candidate for a billion-transistor architecture. By integrating simple processor cores, it can exploit TLP(Thread-Level Parallelism) as well as ILP(Instruction-Level Parallelism). In this paper, we have evaluated the suitability of the on-chip multiprocessor for vision tasks. For this evaluation, first we developed a program-driven dedicated architecture simulator and its programming environment. Using simulations, we analyzed the IPC and the total number of execution cycles of the architecture in performing benchmark vision tasks.

Our simulation results showed that more than 4.7 IPC was achieved on the 2-issue, 4-thread on-chip multiprocessor by exploiting TLP as well as ILP. Using ILP only, less than 1.6 IPC was achieved on an 8-issue superscalar architecture. We are currently conducting further evaluation of advanced memory architectures using DARPA DIS benchmarks[13,14].

Acknowledgement

We would like to thank Puneet Goel for his assistance in preparing the final version of this manuscript.

References

1. Prasanna Kumar, V.: Parallel Architectures and Algorithms for Image Understanding. Academic Press (1991)
2. Wang, C., Bhat, P., and Prasanna, V.: High Performance Computing for Vision. IEEE Proceedings, Vol. 84, No. 7 (1996) 931-946
3. Annaratone, M., et al.: The Warp Computer: Architecture, Implementation, and Performance. IEEE Tr. Computers, Vol. 36, No. 12 (1987) 1523-1538
4. Weems, C., Riseman, E., and Hanson, A.: Image Understanding Architecture: Exploiting Potential Parallelism in Machine Vision. IEEE Computer, Vol. 25, No. 2 (1992) 65-68
5. Wilson, J.: Challenges and Trends in Processor Design. IEEE Computer, Vol. 30, No. 1 (1997) 39-50
6. Wall, D.: Limits of Instruction Level Parallelism. WRL Research Report, Digital Western Research Laboratory (1993)
7. Hammond, L., et al.: A Single-Chip Multiprocessor. IEEE Computer, Vol. 30, No. 9 (1997) 79-85
8. Singh, J., Weber, W., and Gupta, A.: SPLASH: Stanford Parallel Applications for Shared Memory. Computer Architecture News, Vol. 20, No. 1 (1992) 5-44
9. Weems, C., et al.: The DARPA Image Understanding Benchmark for Parallel Computers. Journal of Parallel and Distributed Computing, Vol. 11, No. 1 (1991) 1-24
10. Park, K., et al.: On-Chip Multiprocessing with Simultaneous Multithreading. Technical Report, ETRI (1999)
11. POSIX P1003.4a: Threads Extension for Portable Operating Systems, IEEE (1994)
12. Burger, D. and Austin, T.: The SimpleScalar Tool Set, Version 2.0. Technical Report, University of Wisconsin (1997)
13. Bondalapati, K., Dutta, D., Narayanan, S., Prasanna, V. K., Ragahavendra, C., and Seshadri, A.: Optimizing DRAM-based Memory System Performance. Submitted to the 27th Annual International Symposium on Computer Architecture
14. Musmanno, J. F.: DARPA DIS Benchmarks. Atlantic Aerospace Electronics Corp. (1999)

Parallel Hardware-Software Architecture for computation of Discrete Wavelet Transform using the Recursive Merge Filtering algorithm

Piyush Jamkhandi, Amar Mukherjee, Kunal Mukherjee and Robert Franceschini

School of Electrical Engineering and Computer Science,
University of Central Florida,
Orlando, Florida,
USA
E-mail: {piyush, amar, mukherje, rfrances}@cs.ucf.edu

Abstract. We present an FPGA -based parallel hardware-software architecture for the computation of the Discrete Wavelet Transform (DWT), using the Recursive Merge Filtering (RMF) algorithm. The DWT is built in a bottom-up fashion in logN steps, successively building complete DWTs by "merging" two smaller DWTs and applying the wavelet filter to only the "smooth" or DC coefficient from the smaller DWTs. The main bottleneck of this algorithm is the data routing process, which can be reduced by separating the computations into two types to introduce parallelism. This is achieved by using a virtual mapping structure to map the input. The data routing bottleneck has been transformed into simple arithmetic computations on the mapping structure. Due to the use of the FPGA -RAM for the mapping structure, the total number of data accesses to the main memory are reduced. This architecture shows how data routing in this problem can be transformed into a series of index computations

1 Introduction

In this paper, we present a multi-threaded architecture for the computation of Discrete Wavelet Transform (DWT) using the Recursive Merge Filtering (RMF) algorithm. The RMF algorithm[1] overcomes the main disadvantage of conventional wavelet coders-decoders of not being able to generate the code for the image until the complete image has been transformed. The use of the RMF algorithm overcomes the performance, functionality and reliability drawbacks of the current DWT methods. This paper builds upon the RMF algorithm by introducing an efficient data routing technique, based on the transformation of data routing operations to arithmetic computations. We show how the architecture can be used to compute the DWT of an image bottom-up and then carry out hierarchical merging of the sub-blocks to obtain the wavelet transform. To implement this multi- threaded architecture we use the

J. Rolim et al. (Eds.): IPDPS 2000 Workshops, LNCS 1800, pp. 250-256, 2000.
© Springer-Verlag Berlin Heidelberg 2000

XC6200 FPGA, which suits the given application ideally. The essential idea is to separate the actual computation for the transform from the data routing. This can be achieved by using virtual position mapping for each image position. Using virtual mapping, the data computation can be performed on the actual data inputs, while the data routing is performed on the virtual mapping. Using a virtual coordinate system to define the position of the data being routed, the data routing is transformed into a set of coordinate additions/subtractions. We give a set of generalized equations, which can be applied to any block of the input data to perform the wavelet transform of the entire image.

We first describe the RMF algorithm formally. This is followed by an analysis of the various sub-tasks to be carried out for the DWT using RMF. We then present the formal notation for the transformation from data routing to simple coordinate addition and subtraction. The analysis is further developed to define the multi-threaded architecture for computation of the DWT using RMF. An implementation strategy using FPGAs is then described. Results of the simulation of this architecture showing the reduction in the main memory accesses are finally presented.

2 Formal Description of the Recursive Merge Filtering Algorithm

In this section we briefly describe the Recursive Merge Filter (RMF) algorithm to compute the DWT for 1-dimensional. The RMF algorithm computes the DWT of a 1-dimensional array of length N (where $N=2^k$ for some positive integer k) in a bottom-up fashion, by successively "merging" two smaller DWTs (four in 2-D), and applying the wavelet filter to only the "smooth" or DC coefficients.

2.1 RMF Operator

We will first formally define our primitive, the RMF operator, *in terms of the array indices of two DWT arrays* being merged. This takes as inputs two DWTs, DWT_1 and DWT_2, each of length 2^k, and outputs a DWT of length 2^{k+1}:

$$RMF[DWT_1(0:2^k-1), DWT_2(0:2^k-1)]$$

$$= RMF[DWT_1(0:2^{k-1}-1), DWT_2(0:2^{k-1}-1)] \bullet DWT_1(2^{k-1}:2^k-1) \bullet DWT_2(2^{k-1}:2^k-1)$$
$$\text{if } k > 0$$

$$= RMF[DWT_1(0:0), DWT_2(0:0)] = h(DWT_1(0), DWT_2(0)) \bullet g(DWT_1(0), DWT_2(0))$$
$$\text{if } k=0$$

The RMF operator is defined recursively on sub-arrays of the original DWTs. The first half of DWT_1 and DWT_2 are recursively passed to the RMF operator, and the remaining coefficients of the two DWTs are concatenated (symbolized by '$\bullet$') at the end, as shown. The recursion terminates when the length of the DWTs being merged becomes equal to one - at this point, the RMF uses the Haar filters h and g[2], to generate the low pass and high pass coefficients.

3 DWT in terms of the RMF operator

A recursive notation for the discrete wavelet transform (DWT) of an array $x(n)$ of length $N=2^k$, which directly leads to a recursive procedure to compute the DWT is given below.

$$DWT[x(0:2^k-1)] = RMF[DWT[x(0:2^{k-1}-1), DWT[x(2^{k-1}:2^k-1)]] \quad \text{if } k > 1$$

$$DWT[x(0:1)] = [h(x(0),x(1)), g(x(0),x(1))] \quad \text{if } k = 1$$

The recursion terminates when the length of the array becomes two. At this point, the Haar filters h and g are applied to generate the low pass and high pass coefficients. For proof of the equivalence of the RMF algorithm and the FWT algorithm, refer to Mukherjee et al.[1]

4 RMF Algorithm Computations and Data Shifting

For a 2D image, the RMF process begins by computing the 2x2 transform of the 2D input data. This is done by selecting a 2x2-block row wise and computing its transform. This process is continued until all the 2x2 blocks in the image are transformed. Once all the blocks are transformed, sets of four adjacent 2x2 blocks are selected for the merging process. The process of merging the 2x2 blocks into 4x4 blocks continues until all the 2x2 blocks are merged into a set of 4x4 blocks. After the completion of this process, a set of four adjacent 4x4 blocks is merged to obtain the next level 8x8 block. This process of merging continues until all the blocks are merged to obtain a single merged block, at which time we are done with the process. A detailed extension of the RMF for 2D is given in Kunal et al. The merging process is inherently an exchange of blocks of data with the adjacent same-sized blocks representing four DWTs of four sub-images. If the size of the four quadrants of the merged data block is 2x2 then we apply the basic RMF-2D operation to the top-left matrix quadrant. In case the size of the quadrants is greater than 2x2 we recursively apply the Merge operation, until the size of the quadrants becomes 2x2 at which time we apply the RMF-2D operator to the top left quadrant. Along with the recursive Merge operations and the final RMF-2D operation, after a set of four blocks have been merged, we apply a 1D RMF on the bottom left quadrant in row wise manner. The 1D RMF operator is also applied to the upper right quadrant in column wise manner. One of the main disadvantages of the RMF algorithm, in its current form, is the large amount of data movement that has to be performed at every merge step.

5 Transformation of Data Routing to Address Computation

As seen above, the data shifting constitutes a major part of the overall process of DWT computation using RMF. We propose the use of a virtual mapping index called the *rMap* to reduce the total number of data routings. The *rMap* index is a pointer from every image position to a data value. The *rMap* index can be visualized as a structure with two components: x and y. The x and y values store the current co-ordinate position of a pixel on the transformed image. Consider a position (i, j) on an image. During the computation of the DWT the pixel value at (i, j) is shifted to a different position depending upon the computation. Instead of shifting the actual data around the storage array, we add/subtract co-ordinate values from the *rMap.x* and *rMap.y* values thereby shifting the data. Thus the new value of the pixel at (i, j) is given by *(rMap(i, j).x, rMap(i, j).y)*. By the use of the *rMap* index we can compute the data transform independent of the data shifting process. At every instance during the computation, there is direct positional correlation between the initial image locations and the *rMap* index.

6 Equations for Data Shifting

In this section, we develop a generalized set of data movement equations, which can be applied to any block size. These set of equations constitute the complete DWT using the RMF algorithm. In Figure 1.1 we have a generalized block at any stage in the *log N* steps taken to complete the process. We know from the discussion of the *Merge* operator the type of data block movement that is to be carried out. For example, consider block 1,shown in Figure 1.1.

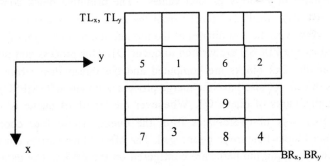

Figure 1.1 Block coordinate system and quadrants.

We know from the definition of the RMF algorithm that the data in block 1 has to be shifted to the position in block 9. Given any data item in block 1 in position *(x, y)*, it is moved to position :

$$x+\frac{(BR_x-TL_x)}{4}, y+\frac{(BR_y-TL_y)}{4}$$

Where : (BR_x, BR_y) and (TL_x, TL_y) are the bottom right and top left coordinates of the block whose four quadrants are to be merged. Thus, we have managed to transform the data routing process into a simple arithmetic compute process. Similarly we define a set of primitives for the computation of the of the DWT using the RMF algorithm. These primitives define the operations to be carried out during the computation such as 1D-RMF, data shifting, and recursive operations.

7 Hardware-Software Architecture

The main aim of this architecture is to exploit the inherent parallelism between the data transforms and the data shifting process. This is done by a multi-threaded architecture using an FPGA. The overall architecture of the system is shown in Figure 1.2. We use two thread for process computations: one for the data compute and the other for the data routing. The data routing is carried out on the FPGA and the *rMap* index is stored on the RAM provided on the FPGA board. This allows faster access to the data for the data routing process carried out on the FPGA using an adder/subtractor circuit configured on the FPGA during the setup of the application. The main components of the architecture are :

- Primitive Block Computation Software Unit (PBCSU)

- Hardware-Software based Merge Process (MPS and the FPGA)

- Main memory based Queuing structure.

We briefly outline the working of the architecture. Initially the input image is stored in the main memory in an array of data values. The primitive block computation software unit (PBCSU) , reads 2x2 blocks from the main memory and computes the basic 2D RMF. However, the coordinates of the top left and bottom right of each 2x2 block computed by the PBCSU are written to queue Q1. This process continues as a thread until all the 2x2 blocks are computed and the thread then terminates. In addition to this thread, we also have the merge process software unit (MPSU) thread, which checks the status of queue Q1. Whenever the length of queue is > 4, the MPSU reads four sets of coordinates from the queue. These four coordinates represent the four blocks that are to be merged. The MPSU then carries out the data movement operation using the hardware configured on the FPGA. The data for the addition process is read from the FPGA on-board RAM. Once the merging process is completed, the coordinates of the merged block are written to queue Q2. The MPSU repeats the process for all 2x2 blocks read from Q1. The process continues until all the blocks in the queue Q1 are processed at which time all blocks of size 2x2 have been merged into 4x4 blocks. The MPSU then begins to read the block coordinates from the queue Q2, merges the blocks and writes the resulting coordinates to queue Q1. This process of switching between queue Q1 and queue Q2 is repeated until all the blocks are processed and we have a single entry in one of the queues.

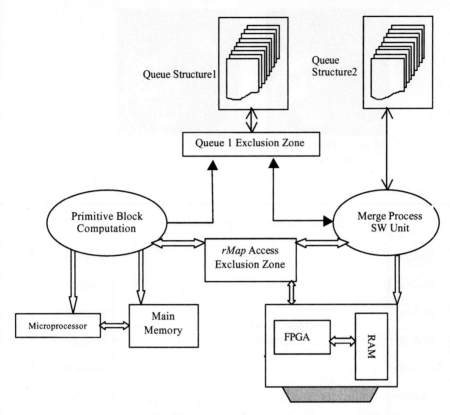

Figure 1.2 Hardware Software Architecture for DWT

8 FPGA Implementation and Resource Use

The H.O.T Works board from VCC[3] has been provided with onboard RAM, which can be used to store the *rMap* index. This technique of implementation of the *rMap* index use is efficient as the data shifting process can be carried out by the means of an addition/subtraction circuit configured on the FPGA. The *rMap* index contains a series of (x,y) pairs which point to a specific location in the original data matrix.

Figure 1.3 shows the comparative number of data accesses for the conventional RMF and the hardware-software implementation of RMF. We see a substantial decrease in the total number of direct accesses. Although we need to reset the data array to the correct positions after the completion of all blocks of a certain level, we can do so by using the block access mechanism rather than singular data accesses. This blocks access mechanism is a fraction of the initial data accesses. The figure below shows the original gray map image along with the reverse-transformed image (.PGM format).

Original Image <img-face.pgm>   *Re-constructed Image*

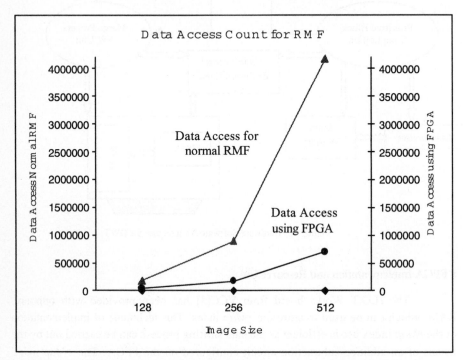

Fig. 1.3 Chart showing the reduction in the main memory data accesses. The data accesses are transformed into FPGA board RAM accesses.

References

[1] K. Mukherjee, "Image Compression and Transmission using Wavelets and Vector Quantization, Ph.D. Dissertation, University of Central Florida, 1999.
[2] S. G. Mallat, "A Theory for Multiresolution Signal Decomposition: The Wavelet Representation", IEEE Transactions on Pattern Analysis and Machine Intelligence, vol.11, no.7, pp. 674- 693, July 1989
[3] VCC H.O.T Works Board Manual

Fifth International Workshop on
High–level Parallel Programming Models
and Supportive Environments
HIPS 2000

Chair: Martin Schulz

schulzm@in.tum.de

Lehrstuhl für Rechnertechnik und Rechnerorganisation, LRR–TUM
Institut für Informatik, Technische Universität München
80290 München, Germany

May 1st, 2000
Cancun/Mexico

Preface

Following the long tradition of this well established event, the 5th International Workshop on High-Level Parallel Programming Models and Supportive Environments (HIPS 2000) provides a forum for researchers and developers from both academia and industry to meet and discuss the newest approaches and results in this active research area. It is again held in conjunction with IPDPS (formerly known as IPPS/SPDP), one of the premier events in the area of parallel and distributed processing.

Despite this long tradition of the HIPS workshop series, the topic — how to efficiently exploit parallel and distributed architectures with respect to performance and ease-of-use — has not lost any of its importance. On the contrary, with the rise of cluster architectures built from commodity–off–the–shelf components and possibly even interconnected with high–performance networking technologies such as SCI, Myrinet, or GigaNet, high-performance parallel computing has become available to a larger base of users than ever before.

Nevertheless, programming is still mostly restricted to the message passing paradigm, which with the advent of the de facto standards Parallel Virtual Machine (PVM) and Message Passing Interface (MPI) has reached a certain level of maturity. However, in terms of convenience and productivity, this parallel programming style is often considered to correspond to assembler-level programming of sequential computers.

J. Rolim et al. (Eds.): IPDPS 2000 Workshops, LNCS 1800, pp. 257-260, 2000.
© Springer-Verlag Berlin Heidelberg 2000

One of the keys for a (commercial) breakthrough of parallel processing are therefore easy-to-use high-level programming models that allow to produce truly efficient code. In this respect, languages and packages have been established that are more convenient than explicit message passing and allow higher productivity in software development; examples are High Performance Fortran (HPF), OpenMP, thread packages for shared memory-based programming, and Distributed Shared Memory (DSM) environments.

Yet, current implementations of high-level programming models often suffer from low performance of the generated code, from the lack of corresponding high-level development tools, e.g. for performance analysis, and from restricted applicability, e.g. to the data parallel programming style. This situation requires strong research efforts in the design of parallel programming models and languages that are both at a high conceptual level and implemented efficiently. In addition, it is necessary to design and implement supportive tools and integrated programming environments to assist in the development of parallel and distributed applications. Hardware and operating system support for high-level programming, e.g. distributed shared memory and monitoring interfaces, are further areas of interest contributing to this goal.

The HIPS workshop series provides a forum to discuss these issues and present new approaches providing users with high-level models and tools that allow the easy and efficient exploitation of parallel and distributed architectures. It addresses a wide range of topics form different areas including, but not limited to:

- Concepts and languages for high-level parallel programming.
- Concurrent object-oriented programming.
- Distributed objects and components.
- Structured parallel programming (skeletons, patterns, ...).
- Software engineering principles.
- Automatic parallelization and optimization.
- High-level programming environments.
- Performance analysis, debugging techniques and tools.
- Distributed shared memory.
- Implementation techniques for high-level programming models.
- Operating system support for runtime systems.
- Architectural and high-speed communication support for high-level programming models.

These topics attracted numerous submissions to this year's HIPS 2000. Each submitted paper was assigned to at least 3 program committee members and underwent a rigorous review process. The reviewers spent a lot of time and gave very detailed comments and recommendations, which lead to an initial list of accepted papers. The program committee members were then given the opportunity to resolve differences in opinion. At the end, after a short second reviewing process for some disputed papers, 10 high-quality papers (listed further below) have been selected for publication. These are presented, together with one invited talk, during the workshop on May 1st, 2000.

Workshop Chair

- Martin Schulz, Technische Universität München, Germany

Steering Committee

- Michael Gerndt, Forschungszentrum Jülich, Germany
- Hermann Hellwagner, Universität Klagenfurt, Austria
- Frank Müller, Humboldt Universität Berlin, Germany

Program Committee

- Arndt Bode, Technische Universität München, Germany
- Helmar Burkhart, Universität Basel, Switzerland
- John Carter, University of Utah, USA
- Karsten Decker, Swiss Center for Scientific Compting, Switzerland
- Michael Gerndt, Forschungszentrum Jülich, Germany
- Hermann Hellwagner, Universität Klagenfurt, Austria
- Francois Irigoin, Ecole des Mines de Paris, France
- Vijay Karamcheti, New York University, USA
- Peter Keleher, University of Maryland, USA
- Gabriele Keller, University of Technology, Sydney, Australia
- Piyush Mehrotra, ICASE / NASA Lanley Research Center, USA
- Frank Müller, Humboldt Universität Berlin, Germany
- Susanna Pelagatti, Universita di Pisa, Italy
- Thierry Priol, IRISA, France
- Martin Schulz, Technische Universität München, Germany
- Xian-He Sun, Louisianna State University, USA
- Domenico Talia, Universita della Calabria, Italy
- George Thiruvathukal, DePaul University, USA

Acknowledgments

I would like to thank the following people for their significant contribution to the success of this workshop: the steering committee members listed above for their assistance and support, Jose Rolim and Danuta Sosnowska from the IPDPS team for their organizational support, the many programmers of the START software package used for the electronic submission process, especially Rich Gerber, and last but not least the members of the program committee and the anonymous external reviewers for their excellent work during the review process.

May 2000   Martin Schulz
                                                              Chair of HIPS 2000

Accepted Papers for HIPS 2000

Session 1

- Pipelining Wavefront Computations: Experiences and Performance
  E Christopher Lewis and Lawrence Snyder, University of Washington

- Specification Techniques for Automatic Performance Analysis Tools
  Michael Gerndt and Hans-Georg Eßer, Research Centre Juelich

- PDRS: A Performance Data Representation System
  Xian-He Sun, Louisiana State University and Illinois Institute of Technology
  Xingfu Wu, Northwestern University and Louisiana State University

Session 2

- Clix - A Hybrid Programming Environment for Distributed Objects and Distributed Shared Memory
  Frank Müller, Humboldt University Berlin
  Jörg Nolte, GMD FIRST
  Alexander Schlaefer, University of Washington

- Controlling Distributed Shared Memory Consistency from High Level Programming Languages
  Yvon Jegou, IRISA/INRIA

- Online Computation of Critical Paths for Multithreaded Languages
  Yoshihiro Oyama, Kenjiro Taura, and Akinori Yonezawa, University of Tokyo

Session 3

- Problem Solving Environment Infrastructure for High Performance Computer Systems
  Daniel C. Stanzione, Jr. and Walter B. Ligon III, Clemson University

Session 4

- Combining Fusion Optimizations and Piecewise Execution of Nested Data-Parallel Programs
  Wolf Pfannenstiel, Technische Universität Berlin

- Declarative concurrency in Java
  Rafael Ramirez and Andrew E. Santosa, National University of Singapore

- Scalable Monitoring Technique for Detecting Races in Parallel Programs
  Yong-Kee Jun, Gyeongsang National University
  Charles E. McDowell, University of California at Santa Cruz

Pipelining Wavefront Computations: Experiences and Performance*

E Christopher Lewis and Lawrence Snyder

University of Washington
Department of Computer Science and Engineering
Box 352350, Seattle, WA 98195-2350 USA
{*echris,snyder*}@*cs.washington.edu*

Abstract. Wavefront computations are common in scientific applications. Although it is well understood how wavefronts are pipelined for parallel execution, the question remains: How are they best presented to the compiler for the effective generation of pipelined code? We address this question through a quantitative and qualitative study of three approaches to expressing pipelining: programmer implemented via message passing, compiler discovered via automatic parallelization, and programmer defined via explicit parallel language features for pipelining. This work is the first assessment of the efficacy of these approaches in solving wavefront computations, and in the process, we reveal surprising characteristics of commercial compilers. We also demonstrate that a parallel language-level solution simplifies development and consistently performs well.

1 Introduction

Wavefront computations are characterized by a data dependent flow of computation across a data space, as in Fig. 1. Wavefronts are common, and the scientific applications in which they appear demand parallel execution [9, 14]. Although the dependences they contain imply serialization, it is well known that wavefront computations admit efficient, parallel implementation via pipelining [5, 15], *i.e.*, processors only partially compute local data before sending data to dependent neighbors. The question of how wavefront computations are best presented to the compiler for the effective generation of pipelined parallel code remains open.

In this paper, we address this question in a study of three approaches for expressing wavefront computations: programmer implemented via message passing, compiler discovered via automatic parallelization, and programmer defined via explicit parallel language features for pipelining. The general parallel programming implications of these approaches are well known, but not in the context of wavefront computations. They are all potentially efficient, but each has a downside. Message passing programming requires considerable time to develop, debug, and tune. The benefits of automatic parallelization are only realized when a program is written in terms that the compiler is able to parallelize. And high-level parallel languages only benefit those who are willing to learn them. We assess these issues in the context of pipelining wavefront computations.

* This research was supported in part by DARPA Grant F30602-97-1-0152

J. Rolim et al. (Eds.): IPDPS 2000 Workshops, LNCS 1800, pp. 261-268, 2000.
© Springer-Verlag Berlin Heidelberg 2000

```
do j = 2, m
   do i = 2, n
      a(i,j) = (a(i,j)+
               a(i-1,j))/2.0
   enddo
enddo
```

(a) WF/1D/VERT

```
do j = 2, m
   do i = 2, n
      a(i,j) = (a(i,j)+
               a(i,j-1))/2.0
   enddo
enddo
```

(b) WF/1D/HOR

```
do j = 2, m
   do i = 2, n
      a(i,j) = (a(i,j)+
               a(i-1,j)+
               a(i,j-1))/3.0
   enddo
enddo
```

(c) WF/2D

```
do j = 2, m
   do i = 2, n
      a(i,j) = (a(i,j)+
               a(i-1,j))/2.0
   enddo
enddo
do j = 2, m
   do i = 2, n
      a(i,j) = (a(i,j)+
               a(i,j-1))/2.0
   enddo
enddo
```

(d) WF/1D/BOTH

Fig. 1. Wavefront kernel computations.

We evaluate the three approaches by developing each of the four wavefront kernels in Fig. 1 on two dissimilar parallel machines (IBM SP-2 and Cray T3E). The kernels are representative of a large class of wavefronts (*e.g.*, those in SWEEP3D, SIMPLE, and Tomcatv), and they are sufficiently simple that they allow us to focus on the first order implications of their parallelization. We use the Message Passing Interface (MPI) [12] as an illustration of message passing, High Performance Fortran (HPF) [7][1] of automatic parallelization, and the ZPL parallel programming language [13] of language-level representation.

This work is the first quantitative and qualitative assessment of developing parallel wavefront computations by the three approaches. Furthermore, we compare the development experience and performance of these approaches via a common set of kernels. The evidence we gather both confirms widely held beliefs about these representations and challenges conventional wisdom. We find that a language-level representation is both simple to develop and consistently efficient. In addition, our study reveals surprising characteristics of commercial HPF compilers.

This paper is organized as follows. The next section describes the representations that we consider. Sect. 3 relates our experiences parallelizing wavefront computations, and Sect. 4 presents performance data for each representation. The final section gives conclusions.

---

[1] HPF is not strictly an automatically parallelized language, but it lacks intrinsic or annotational support for pipelining, relegating pipelining to an automatic parallelization/optimization task.

2 Representations: MPI, HPF, and ZPL

This section summarizes the three representations we consider. MPI and HPF are well known, so we only address ZPL in any detail.

Although often efficient, message passing—in this case MPI—programs are laborious to develop, for the programmer must manage every detail of parallel implementation. This is illustrated by the 626 line kernel of the ASCI SWEEP3D benchmark [1], only 179 lines of which are fundamental to the computation. The remainder manage the complexities of implementing pipelining via message passing. Furthermore, by obscuring the true logic of a program, complexity hinders maintenance and modification. Conceptually small changes may result in substantially different implementations.

An HPF program is a sequential Fortran 77/90 program annotated by the programmer to guide data distribution (via DISTRIBUTE) and parallelization (via INDEPEN-DENT) decisions [7]. The HPF standard does not include annotations to identify computations that may be pipelined, but Gupta *et al.* indicate that the IBM xlHPF compiler for the IBM SP-2 automatically recognizes and optimizes them [6]. Some forms of task-level pipelining are supported by HPF2, but no commercial compilers support the new standard. Furthermore, a representation of this form would look more like an MPI code, thus sacrificing the benefits of HPF.

ZPL is a data-parallel array programming language [13].[2] It supports all the usual scalar data types (*e.g.*, integer and float), operators (*e.g.*, math and logical), and control structures (*e.g.*, if and while). As an array language, it also offers array data types and operators. ZPL is distinguished from other array languages by its use of *regions* [3]. A region represents an index set and may precede a statement, specifying the extent of the array references within its dynamic scope. For example, the following Fortran 90 (slice-based) and ZPL (region-based) array statements are equivalent.

```
F90: a(n/2:n,n/2:n) = b(n/2:n,n/2:n) + c(n/2:n,n/2:n)
ZPL: [n/2..n,n/2..n] a = b + c;
```

When all array references in a statement do not refer to exactly the same set of indices, array operators are applied to individual references, selecting elements from the operands according to some function of the enclosing region's indices. ZPL provides a number of array operators (*e.g.*, shifts, reductions, parallel prefix, broadcasts, and permutations), but this discussion requires only the shift operator, represented by the @ symbol. It shifts the indices of the enclosing region by some offset vector, called a *direction*, to determine the indices of its argument array that are involved in the computation. For example, the following ZPL statement performs a four point stencil computation. Let the directions north, south, west, and east represent the programmer defined vectors $(-1,0)$, $(1,0)$, $(0,-1)$, and $(0,1)$, respectively.

```
[1..n,1..n] a := (b@north + b@south + b@west + b@east) / 4.0;
```

In a scalar language, wavefront computations are implemented by loop nests that contain non-lexically forward loop carried true data dependences. Traditional array language semantics do not allow the programmer to express such a dependence at the array

---

[2] This ZPL summary is sufficient for this discussion. Details appear in the literature [13, 4].

```
[2..m,2..n] a := (a+a'@north)/2.0;        [2..m,2..n] a := (a+a'@west)/2.0;
```

(a)   (b)

```
[2..m,2..n] a := (a+a'@west              [2..m,2..n] a := (a+a'@north)/2.0;
                +a'@north)/3.0;           [2..m,2..n] a := (a+a'@west)/2.0;
```

(c)   (d)

Fig. 2. ZPL wavefront kernels corresponding to those in Fig. 1.

level, so ZPL provides the *prime* operator for this purpose. Primed array references refer to values written in previous iterations of the loop nest that implements it.[3] For example, the ZPL statements in Fig. 2 are semantically equivalent to the corresponding loop nests in Fig 1. The prime operator permits the array-level representation of arbitrary loop-carried flow data dependences, but here we only describe its use in wavefronts.

ZPL may further be distinguished from other parallel languages by its what-you-see-is-what-you-get (WYSIWYG) performance model [2]. The language shields programmers from most of the tedious details of parallel programming, yet the parallel implications of a code are readily apparent in the source text. Naturally, the prime operator—indicating that pipelined parallelism is available—supports this model [4].

3   Parallelization Experiences: MPI, HPF, and ZPL

In this section we describe our experiences writing and tuning wavefront computations.

3.1   MPI

A message passing implementation of pipelining is conceptually simple, but it is surprisingly complex in practice. As an illustration, the WF/2D kernel is 40 lines long, which is large when compared to the single loop nest that is represents. In addition, its development, debugging, and performance tuning consumed three hours, a long time for such a trivial computation. Naturally, with message passing even moderate computations will be slow to develop and lengthy.

Furthermore, despite the conceptual similarity between the four kernels, the four MPI implementations differ in significant ways, such as location of communication, allocation of ghost cells, and indexing. The structure of each code is closely tied to the distribution of data and the dependences that define the wavefront, so there is little code reuse between the four implementations. In addition, we are faced with the problem of finding the best tile size (*i.e.*, the granularity of the pipeline). In order to contain development time, we forgo a dynamic scheme [10] in favor of direct experimentation for each kernel on each machine. Naturally, the results will not extend to other machines and different problem sizes.

---

[3] This discussion excludes the mechanism for enlarging the scope of the primed reference.

3.2   HPF

HPF programmers need not manage per processor details and explicit communication. Nevertheless, they direct the compiler's parallelization via annotations. HPF lacks annotations to identify wavefront computations, so the compiler is solely responsible for recognizing and optimizing them from their scalar representations [8, 11]. We consider the Portland Group, Inc. PGHPF and IBM xlHPF compilers separately, below.

PGHPF. The HPF compiler from Portland Group, Inc. (PGHPF) does not perform pipelining. We determine this by examining the intermediate message passing Fortran code produced by the -Mftn compiler flag. The performance data will confirm this.

PGHPF strictly obeys the INDEPENDENT annotations, redistributing arrays before and after loop nests so that all the annotation specified parallelism is exploited. An implication of this is that parallel loops exploit parallelism—at the cost of data redistribution—even when the user specified data distribution precludes parallelism. In this way PGHPF extracts some parallelism from two of the kernels—as we will see in the next section—but it is not competitive with a pipelined implementation.

Another implication of strictly respecting annotations is that they must be placed very carefully. If an INDEPENDENT annotation is placed on the inner loop in WF/1D/HOR, the compiler will redistribute the array inside the j loop, resulting in performance three orders of magnitude worse than that of the loop nest in WF/1D/VERT. The programmer may interchange the two loops, making the outer loop INDEPENDENT, but the resulting array traversal will have poor cache performance.

While the loop nests in WF/1D/VERT, HOR, and BOTH can use redistribution to exploit parallelism, that in WF/2D can not, for it contains dependence in both dimensions. Only pipelining will extract parallelism from this code. We found that because the loop contains no INDEPENDENT annotations, every array element read is potentially transmitted in the inner loop. It appears that only the source and destination processors of each scalar communication block while the communication takes place, thus other processors are permitted to compute ahead, limited only by data dependences. This realizes a crude form of fine grain pipelining when arrays happen to be traversed in the right way. Despite this, the inner loop communication prevent this code from being competitive with a true pipelined implementation.

XLHPF. A published report indicates that IBM xlHPF performs pipelining [6]. The compiler does not provide an option for viewing the intermediate message passing code and the parallelization summary excludes this information, so we experimentally confirm that the compiler does indeed perform pipelining. Specifically, we observe that an HPF wavefront computation has single node performance comparable to the equivalent Fortran 77 program and that it achieves speedup beyond this for multiple processors.

Unlike PGHPF, xlHPF only exploits parallelism on INDEPENDENT loops that iterate over a distributed dimension. This fact and the pipelining optimization result in good parallel performance for all of the kernels. Despite this, we find that the pipelining optimization fails on even modestly more complex wavefront. For example, loops that iterate from high to low indices or contain non-perfectly nested loops are not pipelined. Certainly, they could be. But the lesson is that when optimizing arbitrary code, certain cases or idioms may easily be over looked. Conversely, a language-level solution makes explicit both the semantic and performance implications of a computation.

3.3   ZPL

The ZPL representations of the kernels are trivial to express (Fig. 2). It is apparent to both the programmer and the compiler how parallelism may be derived from them, and tuning was unnecessary.

4   Performance

We gather performance data on two parallel machines: a 272 processor Cray T3E-900 (450MHz DEC Alpha 21164 nodes) and 192 processor IBM SP-2 (160MHz Power2 Super Chip nodes). We use a number of compilers in this evaluation: on the T3E, we use the Cray CF90 Version 3.2.0.1 and Portland Group, Inc. PGHPF v2.4-4; on the SP-2, we use IBM xlf Fortran v4.1.0.6, IBM xlHPF v1.4, IBM xlc v3.1.4.0, and PGHPF v2.1. On both machines we use the University of Washington zc v1.15 ZPL compiler [16]. All compilers are used with the highest optimization level that guarantees the preservation of semantics.

We study four different representations: C+MPI, ZPL, xlHPF, PGHPF. The C+MPI code is a well tuned pipelined message-passing program. It represents the best that can be achieved on these machines using the C programming language. Because the xlHPF compiler is only available on the SP-2, we do not have results for it on the T3E.

For all the experiments, the a array is distributed across a dimension that gives rise to a loop carried dependence (*e.g.*, the first dimension in WF/1D/VERT) so as to isolate the impact of pipelining. Although it appears that WF/1D/VERT and WF/1D/HOR do not require pipelining (*i.e.*, there exists a distribution that permits complete parallel execution), these kernels may appear in a context that requires a different distribution.

We find that the single processor execution times for C+MPI and ZPL—which generates C code—are comparable to that of a sequential C program. Similarly, on a single processor, xlHPF and PGHPF match sequential Fortran. On the T3E, sequential C code typically executes in twice the time of comparable Fortran codes, while on the SP-2 this ratio varies with character of the kernel. Such disparities between C and Fortran implementations of the same computation are common. In any case, that single processor execution times match sequential languages (within each language domain) indicate observed scaling behavior is relative to an efficient baseline.

All the performance data is summarized by the graphs in Fig. 3. These graphs depict performance (*i.e.*, inverse execution time), so higher bars indicate faster execution. Furthermore, the performance is scaled relative to C+MPI. First, observe that the ZPL performance keeps pace with that of C+MPI. This indicates that ZPL is both performing as well and scaling as well as the hand coded program. At times the ZPL code even surpasses C+MPI, because it performs low level optimizations for more efficient array access. Consider the PGHPF performance. It is competitive on a single processor for the WF/1D/VERT and WF/1D/BOTH kernels, but it quickly trails off as the number of processors increases. This is because, PGHPF redistributes the data to achieve parallelism, which does not scale. The SP-2 exaggerates this effect, because its high communication costs outweigh the benefits of redistribution. Furthermore, for WF/1D/HOR and WF/2D significant communication appears in the inner loop, resulting in abysmal performance (the bars are not even visible!).

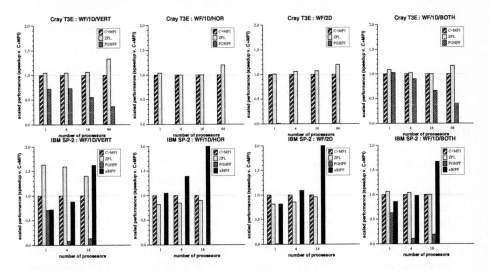

Fig. 3. Performance summary. Kernel names are from Fig. 1. Note that all PGHPF bars are present, but they are very small for WF/1D/HOR and WF/2D.

XlHPF is competitive with the C+MPI and ZPL, because it performs pipelining. The single processor bars highlight disparities in local computation performance. ZPL performs considerably better than any of the others for WF/1D/VERT. We hypothesize that the dependences in this kernel thwart proper array access optimization by the xl optimizer (used by both the Fortran and C compilers). The ZPL code does not suffer from this, because its compiler generates direct pointer references rather than using C arrays. When the C+MPI code is modified in this way, its performance matches ZPL. Conversely, ZPL is worse for WF/1D/HOR. Again, we believe this is an optimization issue. When the ZPL code is modified to use C arrays rather than pointer manipulation, it matches HPF. The summary is that when we ignore the differences that arise from using C versus Fortran, the C+MPI, xlHPF, and ZPL kernel performance are comparable. Nevertheless, as stated in the previous section, we found a number of wavefronts that even the xlHPF compiler failed to optimize.

5   Conclusion

We have evaluated the experience and performance of expressing wavefront computations by three different approaches: programmer implemented via message passing, compiler discovered via automatic parallelization, and programmer defined via explicit parallel language features for pipelining. Our study reveals that in developing wavefronts, each approach can produce an efficient solution, but at a cost. The message passing codes took considerably longer to develop and debug than the other approaches. The HPF codes did not reliably perform well. Although one compiler produced efficient code, the other was three orders of magnitude worse. Even the better compiler failed to pipeline some very simple cases. We find that the language-level approach embodied in ZPL simplifies program development and results in good performance that is consistently achieved.

Acknowledgements. This research was supported in part by a grant of HPC time from the Arctic Region Supercomputing Center.

References

1. Accelerated Strategic Computing Initiative.   ASCI SWEEP3D homepage. *http://www.llnl.gov/asci_benchmarks/asci/limited/sweep3d/sweep3d_readme.html.*
2. Bradford L. Chamberlain, Sung-Eun Choi, E Christopher Lewis, Calvin Lin, Lawrence Snyder, and W. Derrick Weathersby. ZPL's WYSIWYG performance model. In *Third IEEE International Workshop on High-Level Parallel Programming Models and Supportive Environments*, pages 50–61, March 1998.
3. Bradford L. Chamberlain, E Christopher Lewis, Calvin Lin, and Lawrence Snyder. Regions: An abstraction for expressing array computation. In *ACM SIGAPL/SIGPLAN International Conference on Array Programming Languages*, pages 41–49, August 1999.
4. Bradford L. Chamberlain, E Christopher Lewis, and Lawrence Snyder. Language support for pipelining wavefront computations. In *Proceedings of the Workshop on Languages and Compilers for Parallel Computing*, 1999.
5. Ron Cytron. Doacross: Beyond vectorization for multiprocessors. In *International Conference on Parallel Processing*, pages 836–844, 1986.
6. Manish Gupta, Sam Midkiff, Edith Schonberg, Ven Seshadri, David Shields, Ko-Yang Wang, Wai-Mee Ching, and Ton Ngo. An HPF compiler for the IBM SP2. In *Proceedings of the 1995 ACM/IEEE Supercomputing Conference (CD-ROM)*, 1995.
7. High Performance Fortran Forum. *HPF Language Specification, Version 2.0.* January 1997.
8. Seema Hiranandani, Ken Kennedy, and Chau-Wen Tseng. Compiler optimizations for Fortran D on MIMD distributed-memory machines. In *Supercomputing '91*, pages 96–100, Albuquerque, NM, November 1991.
9. K. R. Koch, R. S. Baker, and R. E. Alcouffe. Solution of the first-order form of three-dimensional discrete ordinates equations on a massively parallel machine. *Transactions of the American Nuclear Society*, 65:198–9, 1992.
10. David K. Lowenthal and Michael James. Run-time selection of block size in pipelined parallel programs. In *Proceedings 13th International Parallel Processing Symposium and 10th Symposium on Parallel and Distributed Processing*, pages 82–7, 1999.
11. Anne Rogers and Keshav Pingali. Process decomposition through locality of reference. In *ACM SIGPLAN PLDI '89*, pages 69–80, June 1989.
12. Marc Snir, Steve Otto, Steven Huss-Lederman, David Walker, and Jack Dongarra. *MPI—The Complete Reference.* The MIT Press, Cambridge, Massachusetts, 2nd edition, 1998.
13. Lawrence Snyder. *The ZPL Programmer's Guide.* The MIT Press, 1999.
14. David Sundaram-Stukel and Mark K. Vernon. Predictive analysis of a wavefront application using LogGP. In *Seventh ACM SIGPLAN Symposium on Principles and Practice of Parallel Programming*, May 1999.
15. Michael Wolfe. *High Performance Compilers for Parallel Computing.* Addison-Wesley, Redwood City, CA, 1996.
16. ZPL Project. ZPL project homepage. *http:/www.cs.washington.edu/research/zpl.*

Specification Techniques for Automatic Performance Analysis Tools

Michael Gerndt, Hans-Georg Eßer

Central Institute for Applied Mathematics
Research Centre Juelich
{m.gerndt, h.g.esser}@fz-juelich.de

Abstract. Performance analysis of parallel programs is a time-consuming task and requires a lot of experience. It is the goal of the KOJAK project at the Research Centre Juelich to develop an automatic performance analysis environment. A key requirement for the success of this new environment is its easy integration with already existing tools on the target platform. The design should lead to tools that can be easily retargeted to different parallel machines based on specification documents. This article outlines the features of the APART Specification Language designed for that purpose and demonstrates its applicability in the context of the KOJAK Cost Analyzer, a first prototype tool of KOJAK.

1 Introduction

Current performance analysis tools for parallel programs assist the application programmer in measuring and interpreting performance data. But, the application of these tools to real programs is a time-consuming task which requires a lot of experience, and frequently, the revealed performance bottlenecks belong to a small number of well-defined performance problems, such as load balancing and excessive message passing overhead. It is the goal of the KOJAK project (*Kit for Objective Judgement and Automatic Knowledge-based detection of bottlenecks*) at the Research Centre Juelich to develop an environment that automatically reveals well-defined typical bottlenecks [www.fz-juelich.de/zam/kojak].

We designed KOJAK [6] such that it is not implemented for a single target environment only, e.g. the Cray T3E currently installed at our center, but can easily be ported to other target platforms as well. KOJAK will use specification documents to interface to existing performance data supply tools and to specify potential performance problems of the target programming paradigm.

In parallel with the development of KOJAK automatic performance analysis techniques are investigated in the ESPRIT IV *Working Group on Automatic Performance Analysis: Resources and Tools* (APART) [www.fz-juelich.de/apart]. This article demonstrates the main features of the *APART Specification Language* (ASL) [3] within the context of the *KOJAK Cost Analyzer* (COSY) (Section 3). The performance data analyzed in COSY are specified as an ASL object model (Section 4.1) and represented at runtime via a relational database scheme.

J. Rolim et al. (Eds.): IPDPS 2000 Workshops, LNCS 1800, pp. 269-276, 2000.

The performance problems COSY is aiming at are specified as ASL performance properties (Section 4.2) based on the performance data model and are implemented via SQL queries (Section 5).

2   Related work

The use of specification languages in the context of automatic performance analysis tools is a new approach. Paradyn [8] performs an automatic online analysis and is based on dynamic monitoring. While the underlying metrics can be defined via the *Metric Description Language* (MDL) [9], the set of searched bottlenecks is fixed. It includes *CPUbound*, *ExcessiveSyncWaitingTime*, *ExcessiveIOBlockingTime*, and *TooManySmallIOOps*.

A rule-based specification of performance bottlenecks and of the analysis process was developed for the performance analysis tool OPAL [5] in the SVM-Fortran project. The rule base consists of a set of parameterized hypothesis with proof rules and refinement rules. The proof rules determine whether a hypothesis is valid based on the measured performance data. The refinement rules specify which new hypotheses are generated from a proven hypothesis [4].

Another approach is to define a performance bottleneck as an event pattern in program traces. EDL [1] allows the definition of compound events based on extended regular expressions. EARL [10] describes event patterns in a more procedural fashion as scripts in a high-level event trace analysis language which is implemented as an extension of common scripting languages like Tcl, Perl or Python.

3   Overall Design of the KOJAK Cost Analyzer

COSY [7] analyzes the performance of parallel programs based on performance data of multiple test runs. It identifies program regions, i.e. subprograms, loops, if-blocks, subroutine calls, and arbitrary basic blocks, with high parallelization overhead based on the region's speedup. It explains the parallelization overhead by identifying performance problems and ranking those problems according to their severity.

COSY is integrated into the CRAY T3E performance analysis environment. The performance data measured by Apprentice [2] are transferred into a relational database. The implementation of the interface between COSY and Apprentice via the database facilitates the integration with other performance data supply tools on CRAY T3E as well as the integration with other environments.

The database includes static program information, such as the region structure and the program source code, as well as dynamic information, such as execution time, number of floating point, integer and load/store operations, and instrumentation overhead. For each subroutine call the execution time as well as the pass count with the mean value and standard deviation, as well as the minimum and maximum values are stored.

After program execution Apprentice is started. Apprentice then computes summary data for program regions taking into account compiler optimizations. The resulting information is written to a file and transferred into the database. The database includes multiple applications with different versions and multiple test runs per program version. The data model is outlined in Section 4.1.

The user interface of COSY allows to select a program version and a specific test run. The tool analyzes the dynamic data and evaluates a set of performance properties (Section 4.2). The main property is the total cost of the test run, i.e. the cycles lost in comparison to optimal speedup, other properties explain these costs in more detail. The basis for this computation is the test run with the smallest number of processors. The performance properties are ranked according to their severity and presented to the application programmer.

4   Performance Property Specification

COSY is based on specifications of the performance data and performance properties. The specifications are presented in ASL in the next two subsections. ASL supports the following concepts:

Performance property: A performance property characterizes one aspect of the performance behavior of an application. A property has one or more conditions that can be applied to identify this property. It has a confidence expression that returns a value between zero and one depending on the strength of the indicating condition. Finally it has a severity expression that returns a numerical value. If the severity of a property is greater than zero, this property has some negative effect on the program's performance.

Performance problem: A performance property is a performance problem, iff its severity is greater than a user- or tool-defined threshold.

Bottleneck: A program has a unique bottleneck, which is its most severe performance property. If this bottleneck is not a performance problem, the program does not need any further tuning.

The entire specification consists of two sections. The first section models performance data while the second section specifies performance properties based on the data model.

4.1   Data Model

The performance data can be easily modeled via an object-oriented approach. ASL provides constructs to specify classes similar to Java with single-inheritance only. Classes in the data model have attributes but no methods, since the specification will not be executed. The ASL syntax is not formally introduced in this article due to space limitations, instead, we present the performance data model used in COSY.

```
class Program {                    class ProgVersion {
  String Name;                       DateTime Compilation;
  setof ProgVersion Versions;        setof Function Functions;
}                                    setof TestRun Runs;
                                     SourceCode Code;
                                   }
```

The *Program* class represents a single application which is identified by its name. COSY can store multiple programs in its database. An object of that class contains a set of *ProgVersion* objects, each with the compilation timestamp, the source code, the set of functions (static information) and the executed test runs (dynamic information).

```
class TestRun {                    class Function {
  DateTime Start;                    String Name;
  int NoPe;                          setof FunctionCall Calls;
  int Clockspeed;                    setof Region Regions;
}                                  }
```

A *TestRun* object determines the start time and the processor configuration. A *Function* object specifies the function name, the call sites, and the program regions in this function. All this information is static.

```
class Region {                     class TotalTiming {
  Region ParentRegion;               TestRun Run;
  setof TotalTiming TotTimes;        float Excl;
  setof TypedTiming TypTimes;        float Incl;
}                                    float Ovhd;
                                   }
```

The *Region* class models a program region with its parent region and its performance data gathered during execution. Performance data are modeled by two classes, according to the internal structure of Apprentice. The *TotalTiming* class contains the summed up exclusive and inclusive computing time as well as the overhead time. As there may be several test runs, there are also possibly several *TotalTiming* objects for a region.

The *TypedTiming* class determines the execution time for special types of overhead such as I/O, message passing and barrier synchronization – Apprentice knows 25 such types. As with the *TotalTiming* objects, there is a set of *TypedTiming* objects for every test run, but for each region there is at most one object per timing type and per test run.

```
class TypedTiming {                class FunctionCall {
  TestRun Run;                       Function Caller;
  TimingType Type;                   Region CallingReg;
  float Time;                        setof CallTiming Sums;
}                                  }
```

TypedTiming objects have three attributes: The *TestRun* attribute *Run* codes the specific test run of the program, *Type* (an enumeration type) is the work type that is being considered in this object and *Time* is the time spent doing work of this type.

Call sites of functions are modeled by the *FunctionCall* class. A function call has a set of *CallTiming* objects which store the differences of the individual processes. A *CallTiming* object is composed of the *TestRun* it belongs to, the minimum, maximum, mean value, and standard deviation over a) the number of calls and b) the time spent in the function. For the four extremal values the processor that was first or last in the respective category is memorized.

Due to the design of Apprentice, the data model does not make use of inheritence. More complex data models can be found in [3].

4.2   Performance Properties

property	**is** PROPERTY *pp-name* '(' *arg-list* ')' '{'
	[LET *def* ∗ IN]
	pp-condition
	pp-confidence
	pp-severity
	'};'
arg	**is** *type ident*
pp-condition	**is** CONDITION ':' *conditions* ';'
conditions	**is** *condition*
	or *condition* OR *conditions*
condition	**is** ['(' *cond-id* ')']*bool-expr*
pp-confidence	**is** CONFIDENCE ':' MAX '(' *confidence-list* ')' ';'
	or CONFIDENCE ':' *confidence* ';'
confidence	**is** ['(' *cond-id* ')' '->'] *arith-expr*
pp-severity	**is** SEVERITY ':' MAX '(' *severity-list* ')' ';'
	or SEVERITY ':' *severity* ';'
severity	**is** ['(' *cond-id* ')' '->'] *arith-expr*

Fig. 1. ASL property specification syntax.

The property specification (Figure 1) defines the name of the property, its context via a list of parameters, and the condition, confidence, and severity expressions. The property specification is based on a set of parameters. These parameters specify the property's context and parameterize the expressions. The context specifies the environment in which the property is evaluated, e.g. the program region and the test run.

The condition specification consists of a list of conditions. A condition is a predicate that can be prefixed by a condition identifier. The identifiers have to be unique in respect to the property since the confidence and severity specifications can refer to the conditions via those condition identifiers.

The confidence specification is an expression that computes the maximum of a list of confidence values. Each confidence value is computed via an arithmetic expression resulting in a value in the interval of zero and one. The value can be guarded by a condition identifier introduced in the condition specification. The condition identifier represents the value of the condition. The severity specification has the same structure as the confidence specification. It computes the maximum of the individual severity values of the conditions.

The following example properties are checked by COSY. They demonstrate the ASL language features. Most of the property specifications make use of the following two functions:

```
TotalTiming Summary(Region r, TestRun t) = UNIQUE({s IN r.TotTimes
                                                  WITH s.Run==t});
float Duration(Region r, TestRun t) = Summary(r,t).Incl;
```

The first function *Summary* takes a *Region r* and a *TestRun* object and returns the unique *TotalTiming* object which is a member of *r.TotTimes* belonging to that test run. The second function *Duration* uses *Summary* to extract the total execution time of the specified region in the specified test run. Note that all timings in the database are summed up values of all processes.

The first property *SublinearSpeedup* determines the lost cycles in relation to the test run with the minimal number of processors.

```
Property SublinearSpeedup(Region r, TestRun t, Region Basis) {
LET TotTimes MinPeSum = UNIQUE({sum IN r.TotTimes WITH sum.Run.NoPe ==
                               MIN(s.Run.NoPe WHERE s IN r.TotTimes)});
    float TotalCost = Duration(r,t) - Duration(r,MinPeSum.Run)
IN
    CONDITION:  TotalCost>0;    CONFIDENCE: 1;
    SEVERITY:   TotalCost/Duration(Basis,t);
}
```

The property is based on the total costs, i.e. the lost cycles compared to a reference run with the smallest number of processors. If *TotalCost* is greater than zero, the region has the *SublinearSpeedup* property. The confidence value, which is one in all examples here, might be lower than one if the condition is only an indication for that property. The severity of the *SublinearSpeedup* property is determined as the fraction of the total costs compared to the duration of *Basis* in that test run.

```
Property MeasuredCost (Region r, TestRun t, Region Basis) {
    LET float Cost = Summary(r,t).Ovhd;
    IN  CONDITION:  Cost > 0;    CONFIDENCE: 1;
        SEVERITY:   Cost / Duration(Basis,t);
}
```

The total costs can be split up into measured and unmeasured costs. The *MeasuredCost* property determines that more detailed information might be available (*Summary(r,t).Ovhd* is the overhead measured by Apprentice). If the severity of its counterpart, the *UnmeasuredCost*, is much higher, the reason cannot be found with the available data.

```
Property SyncCost(Region r, TestRun t, Region Basis) {
   LET float Barrier = SUM(tt.Time WHERE tt IN r.TypTimes AND tt.Run==t
                                       AND tt.Type == Barrier);
   IN CONDITION:  Barrier > 0;    CONFIDENCE: 1;
      SEVERITY:   Barrier / Duration(Basis,t);
}
```

The *SyncCost* property determines that barrier synchronization is a reason for overhead in that region. Its severity depends on the time spent for barrier synchronization in relation to the execution time of the ranking basis.

```
Property LoadImbalance(FunctionCall Call, TestRun t, Region Basis) {
   LET CallTiming ct = UNIQUE ({c IN Call.Sums WITH c.Run == t});
       float Dev  = ct.StdevTime;
       float Mean = ct.MeanTime;
   IN  CONDITION:  Dev > ImbalanceThreshold * Mean;  CONFIDENCE: 1;
       SEVERITY:   Mean / Duration(Basis,t);
}
```

The *LoadImbalance* property is a refinement of the *SyncCost* property. It is evaluated only for calls to the barrier routine. If the deviation is significant, the barrier costs result from load imbalance.

5   Implementation

The design and implementation of COSY ensures portability and extensibility. The design requires that the performance data supply tools are extended such that the information can be inserted into the database. This extension was implemented for Apprentice with the help of Cray Research. The database interface is based on standard SQL and therefore, any relational database can be utilized. We ran experiments with four different databases: Oracle 7, MS Access, MS SQL server, and Postgres. For all those databases, except MS Access, the setup was in a distributed fashion. The data were transferred over the network to the database server. While Oracle was a factor of 2 slower than MS SQL server and Postgres, MS Access outperformed all those systems. Insertion of performance information was a factor of 20 faster than with the Oracle server.

COSY is implemented in Java and is thus portable to any Java environment. It uses the standard JDBC interface to access the database. Although accessing the database via JDBC is a factor of two to four slower than C-based implementations, fetching a record from the Oracle server takes about 1 ms, the portability of the implementation outweighs the performance drawbacks. The overall performance depends very much on the work distribution between the client and the database. It is a significant advantage to translate the conditions of performance properties entirely into SQL queries instead of first accessing the data components and evaluating the expressions in the analysis tool.

6   Conclusion and Future Work

This article presented a novel design for performance analysis tools. As an example, COSY, a prototype component of the KOJAK environment, was presented. The design enables excellent portability and integration into existing performance environments. The performance data and the performance properties are described in ASL and can therefore easily be adapted to other environments. For this prototype, the specification is manually translated into a relational database scheme and the evaluation of the conditions and the severity expressions of the performance properties is transformed into appropriate SQL queries and ranking code by the tool developer. In the future, we will investigate techiques for the automatic generation of the database design from the performance property specification and the automatic translation of the property description into executable code.

References

1. P. Bates, J.C. Wileden: *High-Level Debugging of Distributed Systems: The Behavioral Abstraction Approach*, The Journal of Systems and Software, Vol. 3, pp. 255-264, 1983
2. CRAY Research: *Introducing the MPP Apprentice Tool*, Cray Manual IN-2511, 1994, 1994
3. Th. Fahringer, M. Gerndt, G. Riley, J.L. Träff: *Knowledge Specification for Automatic Performance Analysis*, to appear: APART Technical Report, Forschungszentrum Jülich, FZJ-ZAM-IB-9918, 1999
4. M. Gerndt, A. Krumme: *A Rule-based Approach for Automatic Bottleneck Detection in Programs on Shared Virtual Memory Systems*, Second Workshop on High-Level Programming Models and Supportive Environments (HIPS '97), in combination with IPPS '97, IEEE, 1997
5. M. Gerndt, A. Krumme, S. Özmen: *Performance Analysis for SVM-Fortran with OPAL*, Proceedings Int. Conf. on Parallel and Distributed Processing Techniques and Applications (PDPTA'95), Athens, Georgia, pp. 561-570, 1995
6. M. Gerndt, B. Mohr, F. Wolf, M. Pantano: *Performance Analysis on CRAY T3E*, Euromicro Workshop on Parallel and Distributed Processing (PDP '99), IEEE Computer Society, pp. 241-248, 1999
7. A. Lucas: *Basiswerkzeuge zur automatischen Auswertung von Apprentice-Leistungsdaten*, Diploma Thesis, RWTH Aachen, Internal Report Forschungszentrums Jülich Jül-3652, 1999
8. B.P. Miller, M.D. Callaghan, J.M. Cargille, J.K. Hollingsworth, R.B. Irvin, K.L. Karavanic, K. Kunchithapadam, T. Newhall: *The Paradyn Parallel Performance Measurement Tool*, IEEE Computer, Vol. 28, No. 11, pp. 37-46, 1995
9. Paradyn Project: *Paradyn Parallel Performance Tools: User's Guide*, Paradyn Project, University of Wisconsin Madison, Computer Sciences Department, 1998
10. F. Wolf, B. Mohr: *EARL - A Programmable and Extensible Toolkit for Analyzing Event Traces of Message Passing Programs*, 7th International Conference on High-Performance Computing and Networking (HPCN'99), A. Hoekstra, B. Hertzberger (Eds.), Lecture Notes in Computer Science, Vol. 1593, pp. 503-512, 1999

PDRS: A Performance Data Representation System[*]

Xian-He Sun [1,2]   Xingfu Wu [3,1]

[1] Dept. of Computer Science, Louisiana State University, Baton Rouge, LA 70803

[2] Dept. of Computer Science, Illinois Institute of Technology, Chicago, IL 60616

[3] Dept. of Electrical and Computer Engineering, Northwestern University, Evanston, IL 60208
sun@cs.iit.edu   wuxf@ece.nwu.edu

Abstract. We present the design and development of a Performance Data Representation System (PDRS) for scalable parallel computing. PDRS provides decision support that helps users find the right data to understand their programs' performance and to select appropriate ways to display and analyze it. PDRS is an attempt to provide appropriate assistant to help programmers identifying performance bottlenecks and optimizing their programs.

1 Introduction

Many performance measurement systems have been developed in recent years. While these systems are important, their practical usefulness relies on an appropriate understanding of the measured data. When monitoring a complex parallel program, the amount of performance data collected may be very huge. This huge amount of performance data needs to be processed for further performance evaluation and analysis. A general performance measurement system always provides a facility that assists manipulation of this performance data. Data manipulation functions are often dependent on performance data organization and representation. The difficulty in providing an adequate performance environment for high performance computing is the lack of appropriate models, representations and associated evaluation methods to understand measured data and locate performance bottlenecks. Performance Data Representation System (PDRS) proposed in this paper is designed to attack this difficulty. PDRS is a general-purpose integrated system supported by performance database representation and the combination of performance visualization and auralization. It is based on our recent success in automatic performance evaluation and prediction.

Many performance measurement systems exist right now [3, 4, 5]. While these performance systems have made their contribution to the state-of-the-art of performance

---

[*] This work was supported in part by National Science Foundation under NSF grant ASC-9720215 and CCR-9972251.

J. Rolim et al. (Eds.): IPDPS 2000 Workshops, LNCS 1800, pp. 277-284, 2000.

evaluation, none of them has addressed the data presentation and understanding issue adequately. With the advance in performance measurement and visualization techniques, and increased use of large, scalable computing systems, data presentation and management becomes increasingly important. The PDRS is a post-execution performance data representation system designed for scalable computing, and is distinct from existing performance systems. First, while it supports conventional visualization views, it is designed based on the most recent analytical results in scalability and statistical analysis to reveal the scaling properties of a scalable computing system. Second, the system uses relational database, SQL and Java JDBC techniques such that performance information is easily retrieved, compared and displayed. Because of the complexity and volume of the data involved in a performance database, it is natural to exploit a database management system (DBMS) to archive and retrieve performance data. A DBMS will help not only in managing the performance data, but also in assuring that the various performance information can be presented in some reasonable format for users. Third, the system is implemented based on the combination of performance visualization and auralization techniques and object-oriented Java techniques such that it is easy for users to understand and use. Finally, the system supports the SDDF data format. It can be either used as a stand-alone application or easily integrated into other existing performance environments.

2 Design and Implementation of PDRS

Figure 2.1 depicts the design framework of PDRS. The technical approaches used to develop these components are discussed below section by section.

2.1 Trace Data Module

This module is in charge of collecting original performance data of parallel programs, and stores them with SDDF [1].

The large volume of data involved in parallel computations requires that instrumentation to collect the data selectively and intelligently. One way to collect data of a parallel program is to instrument the program executable so that when the program runs, it generates the desired information. PDRS is designed to use the Scala Instrumentation System (SIS) [11] to get the SDDF trace data file. PDRS also provides a general interface that can be used under any system, which provides the SDDF trace data interface.

2.2 Data Management Module

This module is in charge of performance data filtering and mapping.

Event histories of parallel programs are valuable information sources for performance analysis but the problem is how to extract the useful information from massive amounts of low-level event traces. Our system performs the data filtering as a preparation to store the event history into a relational database. The SDDF is a trace description language that specifies both data record structures and data record instances. We are building a performance database based on the SDDF specification. Our data management module is being implemented in Oracle DBMS.

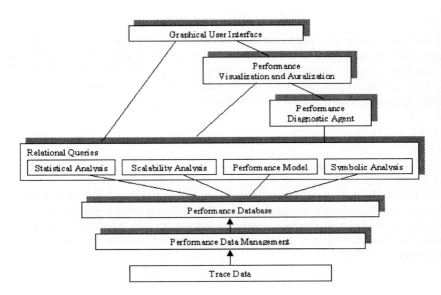

Figure 2.1 Design framework of PDRS

2.3 Performance Database

We classify the performance data saved in the SDDF tracefiles into five groups: processor information, memory information, program information, communication information and I/O information. Each group is represented as an entity relation in the performance database. An individual event in a relation is treated as a tuple with a given unique identifier.

The information retrieval is achieved by the relational database queries. The example below shows how objects can be retrieved using JDBC [13]. For instance, suppose that we want to get the communication events that occurred in processor 0, the query

select sourcePE, destinationPE, messageLength, event_startTimestamp,
event_endTimestamp from Communication Information where processor = 0.

We may make the following SQL query by JDBC:

ResultSet rs = stmt.executeQuery("select sourcePE, destinationPE,
messageLength, event_startTimestamp, event_endTimestamp
    ***from** Communication Information **where** processor = 0 ");*
while (rs.next()) {
        Object i1 = rs.getObject("sourcePE");
        Object i2 = rs.getObject("destinationPE");
        Object r1 = rs.getObject("messageLength");
        Object r2 = rs.getObject("event_startTimestamp");
        Object r3 = rs.getObject("event_endTimestamp");
}

Multiple versions of performance data are handled by specifying a version attribute in each tuple. By specifying a version number in each database query, we can get multiple versions of program performance for comparison. In addition to the default PDRS performance parameters, new performance parameters such as sound files can also be added by users and be supported by the database.

2.4 Relational Queries Module

This module includes four parts: Symbolic Analysis, Statistical Analysis, Scalability Analysis, and Performance Model Generator. The module is being implemented in JDBC. Its structure is shown in Figure 2.2. Java applications include the PDA, PVA, and GUI module implemented by Java. The JDBC provides a bridge between Java applications and performance database.

Figure 2.2 Relational Queries Module

We use symbolic evaluation [2, 6] that combines both data and control flow analysis to determine variable values, assumptions about and constraints between variable values, and conditions under which control flow reaches a program statement. Computations are represented as symbolic expressions defined over the program's problem and machine size. Each program variable is associated with a symbolic expression describing its value at a specific program point

Statistical Analysis determines code and/or machine effects, finds the correlation between program phases, identifies the scaling behavior of "difficult-segments", and provides statistical performance data [12] for the PDA (Performance Diagnostic Agent)

module and GUI. The development of the scalability analysis is based on newly developed algorithms for predicting performance in terms of execution time and scalability of a code-machine combination [8, 9, 11, 15]. Analytical and experimental results show that scalability combined with initial execution time can provide good performance prediction, in terms of execution times. In addition, crossing-point analysis [9] finds fast/slow performance crossing points of parallel programs and machines. In contrast with execution time, which is measured for a particular pair of problem and system size, range comparison compares performance over a wide range of ensemble and problem size via scalability and crossing-point analysis.

In addition to high-level performance prediction, PDRS also supports low-level performance analysis to identify performance bottlenecks and hardware constrains based on performance models chosen by the user. For example, we have proposed an empirical memory model based on a simplified mean value parameterization [14] to separate CPU execution time from stall time due to memory loads/stores. From traced information or information from the analysis modules, performance models can be generated to predict the performance at the component level, as well as over-all performance.

2.5 Performance Diagnostic Agent (PDA) Module

This module provides performance advice in order to help users find performance bottlenecks in their application programs. It also provides performance comparison and suggestions based on real performance results and predicted performance ranges. The PDA is based on our approaches to statistical analysis, scalability analysis and performance model generator. The function operation algorithm for this module is as follows.

Algorithm (Performance diagnosis):
Performance analysis requests; .
switch (analysis type) {
Statistical:
Retrieve the performance information required;
Get or compute the predicted performance range;
Compute the real result of requested performance parameter;
Compare the result with the performance range;
If (the result is not in the performance range)
Give an explanation (using graphics and sound);
break;
Scalability:
Retrieve the performance information required;
Get or compute the predicted scalability results;
Compute the real scalability results;
Compare the real result with the predicted results;
Explain the compared results (using graphics and sound);

break;
Models:
Retrieve the performance information required;
Get the predicted performance range;
Compute the real result of requested performance parameter;
Compare the result with the performance range;
If (the result is not in the performance range)
Give an explanation (using graphics and sound);
break;
Default: *printf("No such analysis type");*
break;
}

In the algorithm, the PDA can provide suggestions and explanations when performance bottlenecks occur. Based on the statistical analysis, the PDA can retrieve the performance information from the performance database, then may provide the advice about program performance.

2.6 Performance Visualization and Auralization (PVA) Module and Graphical User Interface Module

This PVA module provides some graphical display of performance information about users' application programs and platforms. It is natural to use different visual objects to represent various performance data and use visualization techniques to gain insight into the execution of parallel programs so that their performance may be understood and improved. The basic goal of this module is to use graphics and sound (Java 2D, Java 3D and JavaSound) to display some advice and performance views about application programs. For example, based on performance comparison, some performance bottlenecks can be found in graphics. Some suggestions can be given in graphics, such as what applications are suitable for the platforms, what platforms are suitable for solving the applications, and how to modify the application program to be suitable for the platforms. When performance bottlenecks occur, sound is used to inform users about some performance problem in their application programs. The sound files are stored in a performance database.

The Graphical User Interface module is an integrated user-friendly graphical interface. It integrates the whole functions of the PVA module, and directly displays the performance data requested by users.

Figures 2.3 and 2.4 are two views of PDRS GUI. Figure 2.3 shows speed comparison of PDD and PT algorithms [7]. Figure 2.4 shows the Kiviat graph for performance comparison.

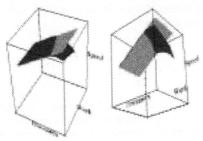

Figure 2.3 Speed comparison of PDD and PT algorithms

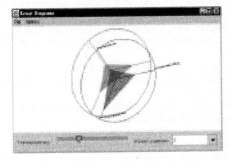

Figure 2.4 Kiviat Graph for Performance Comparison

3 Summary

We have presented the design of a Performance Data Representation System (PDRS) based on our current success of the development of the SCALA [10, 11] performance system for scalable parallel processing. While the PDRS has not been fully implemented at this time, some of its key components have been implemented and tested. Implementation results are very encouraging. PDRS highlights the performance data representation using relational database. Integrated into advanced restructuring compilation and performance analysis system, the proposed PDRS attempts to lift performance evaluation system to a new level. It is designed to provide developers a guideline on performance optimization, to assist the purchasers selecting systems best suited to their needs, and to give valuable feedback to vendors on bottlenecks that can be alleviated in future products. It has the potential to provide users with much more useful information than current existing performance systems. Many advanced technologies, such as database management, object-oriented programming, visualization and auralization are used in the PDRS. The integration of these technologies into compilation and performance analysis system is new, and very challenging. It can motivate many new research and development issues. PDRS is only a first step toward the automatic performance analysis and optimization.

References

1. R. Aydt, *The Pablo Self-Defining Data Format*, Department of Computer Science, University of Illinois, April 1995, ftp://bugle.cs.uiuc.edu/pub/Release/Documentation/SDDF.ps.
2. T. Fahringer and B. Scholz, Symbolic evaluation for parallelizing compilers, in *Proc. of the 11th ACM International Conference on Supercomputing*, Vienna, Austria, ACM Press, July 1997, 261-268.
3. J. Kohn and W. Williams, ATExpert, *Journal of Parallel and Distributed Computing* 18, 1993, 205-222.
4. A.D. Malony and G.V. Wilson, Future directions in parallel performance environment, *Performance Measurement and Visualization of Parallel Systems*, Eds: G. Haring and G. Kotsis, Elsevier Science Publishers B.V., 1993, 331-351.
5. B. P. Miller, M.D. Callaghan, J.M. Cargille, J.K. Hollingsworth, R.B. Irvin, K.L. Karavanic, K. Kunchithapadam, and T. Newhall, The Paradyn parallel performance measurement tools, *IEEE Computer* 28, 11, 1995.
6. M. Scheibl, A. Celic, and T. Fahringer, *Interfacing Mathematica from the Vienna Fortran Compilation System*, Technical Report, Institute for Software Technology and Parallel Systems, Univ. of Vienna, December 1996.
7. X.-H. Sun, H. Zhang, and L. Ni, Efficient tridiagonal solvers on multicomputers, *IEEE Transactions on Computers* 41, 3 (1992), 286-296.
8. X.-H. Sun and D. Rover, Scalability of parallel algorithm-machine combinations, *IEEE Transactions on Parallel and Distributed Systems*, June 1994, 599-613.
9. X.-H. Sun, Performance range comparison via crossing point analysis, *Lecture Notes in Computer Science* 1388 (J. Rolim, ed.), Springer, March 1998.
10. X.-H. Sun, T. Fahringer, M. Pantano, and Z. Zhan, SCALA: A performance system for scalable computing, in *Proc. of the Workshop on High-Level Parallel Programming Models & Supportive Environments, Lecture Notes in Computer Science* 1586, Springer, April 1999.
11. X.-H. Sun, M. Pantano, and Thomas Fahringer, Integrated range comparison for data-parallel compilation systems, *IEEE Transactions on Parallel and Distributed Systems*, Vol. 10, May, 1999, 448-458.
12. X.-H. Sun, D. He, K. Cameron, and Y. Luo, A Factorial Performance Evaluation for Hierarchical Memory Systems, in *Proc. of the IEEE Int'l Parallel Processing Symposium '99*, April 1999.
13. Sun Microsystems Inc., *JDBC: A Java SQL API*, Version 1.20, http://www.javasoft.com/products/jdbc/index.html, January 1997.
14. M. V. Vernon, E. D. Lazowska, and J. Zahorjan, An accurate and efficient performance analysis technique for multi-processor snooping cache-consistency protocols, in *Proc. 15th Annual Symp. Computer Architecture*, Honolulu, HI, June 1988, 308-315.
15. Xingfu Wu, *Performance Evaluation, Prediction, and Visualization of Parallel Systems*, Kluwer Academic Publishers, Boston, ISBN 0-7923-8462-8, 1999.

Clix* – A Hybrid Programming Environment for Distributed Objects and Distributed Shared Memory

Frank Mueller[1], Jörg Nolte[2], and Alexander Schlaefer[3]

[1] Humboldt University Berlin, Institut f. Informatik, 10099 Berlin, Germany
[2] GMD FIRST, Rudower Chaussee 5, D-12489 Berlin, Germany
[3] University of Washington, CSE, Box 352350, Seattle, WA 98195-2350, USA
`mueller@informatik.hu-berlin.de`, phone: (+49) (30) 2093-3011, fax: -3010

Abstract. Parallel programming with distributed object technology becomes increasingly popular but shared-memory programming is still a common way of utilizing parallel machines. In fact, both models can coexist fairly well and software DSM systems can be constructed easily using distributed object systems. In this paper, we describe the construction of a hybrid programming platform based on the ARTS distributed object system. We describe how an object-oriented design approach provides a compact and flexible description of the system components. A sample implementation demonstrates that three classes of less than 100 lines of code each suffice to implement sequential consistency.

1 Introduction

Object-oriented programming and distributed object technology are considered to be state of the art in distributed and as well as parallel computing. However, typical numerical data-structures like huge arrays or matrices are hard to represent in a distributed object paradigm. Such data structures usually cannot be represented as single objects because this leads to extremely coarse-grained programs thus limiting parallel execution. On the other hand, it is not feasible to represent, e.g., each array element as a remote object because remote object invocation mechanisms are typically not targeted for fine-grained computation.

Distributed shared memory can, in principle, cope better with typical numerical data structures: The idea of shared memory is orthogonal to objects and coarse-grained data-structures can transparently be distributed across multiple machines. This scenario is specifically attractive to networks of SMPs and legacy software in general since the shared memory paradigm is not tied to a specific programming language. However, for performance reasons it is usually necessary to relax memory models and the programmer potentially has to cope with different memory consistency models. Object-oriented programming helps to structure memory models and a distributed object platform eases the implementation of consistency models and associated synchronization patterns significantly.

In this paper, we describe the design and implementation of an experimental integrated platform for distributed shared memory and distributed objects. This

---

* Clustered Linux

J. Rolim et al. (Eds.): IPDPS 2000 Workshops, LNCS 1800, pp. 285–292, 2000.

platform is based on a guest-level implementation of ARTS, a distributed object platform for parallel machines and the VAST framework for volatile and non-volatile user-level memory management [5].

2   The ARTS Platform

Originally, ARTS was a remote object invocation (ROI) system designed as an integral part of the parallel PEACE operating system family [11]. Today ARTS is an open object-oriented framework, which provides the basic services for distributed and parallel processing within a global object space. The ARTS platform provides language-level support for proxy-based distributed and parallel programming. Standard C++-classes are extended by annotations to specify remote invocation semantics as well as parameter passing modes similar to an IDL. These annotations are preprocessed by a generator tool called DOG [5], that generates proxy classes and stubs for synchronous and asynchronous ROI. In addition, collective operations on distributed object groups are also supported.

All annotations are specified as pseudo comments to retain a strong backward compatibility with standard C++. Figure 1 shows an example from the DSM system. A /*!dual!*/ annotation is used to mark classes whose instances

```
class SeqConsistency {
 public: ...
   void setObjectControl (int node,
     /*!copy!*//*!dual!*/SeqObjectControl* ctrl);
   ...
   void setHome (int objectId);
   void acquireR (int objectId, int sender);
   void acquireW (int objectId, int sender);
   void ackR (int objectId, int sender) /*!async!*/;
   void ackW (int objectId) /*!async!*/;
}/*!dual!*/;
```
Fig. 1. The Dual Sequential Consistency Class

need to be accessed remotely. Those classes are referred to as *dual* classes. Note that the notation /*!dual!*/SeqObjectControl is treated as a type specifier that actually refers to two objects: a local proxy object in the client's address space and a remote object encapsulated in some remote server's address space. Any time the client performs a method on the proxy, the corresponding operation is executed on the remote object using ROI techniques. Thus, the global sharing of objects is possible through passing of proxy objects as parameters to remote methods. Parameters to remote methods are passed by value by default. Pointer or reference parameters can be annotated to be treated as input parameters (/*!in*!/), output parameters (/*!out!*/) or both (/*!inout!*/). Permanent copies of passed parameters on the server's heap can be created using a /*!copy!*/ annotation as in the setObjectControl() method (Fig. 1).

Remote method invocation is synchronous if not specified otherwise. An /*!async!*/ annotation as in ackR() can be applied to mark methods that shall be executed asynchronously.

Object groups are described by group classes that define the topology of a distributed object group. Groups are implemented as distributed linked object sets that can be collectively addressed by multicast operations. In principle, a multicast group can have any user-defined topology such as n-dimensional grids, lists or trees. We provide a default tree-based implementation of a distributed object group called `GroupMember` and apply inheritance mechanisms to specify a group membership. When a `GroupMember` class is inherited by a dual class it is possible to create and address whole object groups using a `GroupOf` template.

ARTS provides numerous synchronization mechanisms. However, for space considerations, we will only describe the continuation-based approach here. Whenever a method detects that it cannot be executed immediately because the respective object is, e.g., not yet in a suitable state, the method can block the actual caller and generate a `Continuation` instance for the call. These continuation objects can be stored and used later to reply a result to the client, thus deblocking the client when appropriate. Continuation objects can also be passed as parameters to remote methods. Consequently, it is very easy to forward invocations asynchronously to other objects that might perform activities in parallel and finally reply the result to the initial client at completion of an action.

3   Distributed Shared Memory Abstractions

For performance reasons, we need to support several memory consistency models. Therefore, we adopt a NUMA approach and divide the logical address space into separate memory regions that are each associated with a specific shared memory semantics. Thus, the memory semantics of each shared variable is determined by the coarse-grained memory container in which the variable is located.

The easiest way to achieve this is to mark memory blocks of variables declared in the file scope (global variables):

```
BEGIN(strict); int x; double m[512][512]; END(strict);
```

`BEGIN` and `END` declare the beginning as well as the end of a shared memory area and the argument of the macros specify the memory consistency semantics for this area. Both macros implicitly declare an array that has exactly the size of a physical page. Therefore, we are able to align the address of the area in which the variables are declared to a page boundary. This is necessary to use the MMU support. Likewise, we need to declare an object of page size at the end of such a region to ensure that regions do not overlap.

Memory control objects refer to the beginning and the end of the region they control and set up memory protection mechanisms accordingly. Since we assume SPMD programs, each region is locally mapped to the same address. On Linux systems, we apply the `mprotect()` system call to protect and unprotect pages, page faults are propagated to the standard bus error or segmentation fault signals. These basic mechanisms are implemented by a `ControlledArea` class provided by the VAST framework. This class emulates a complete page table and provides the basis to all segment types that need to catch page faults and handle them according to a specific strategy.

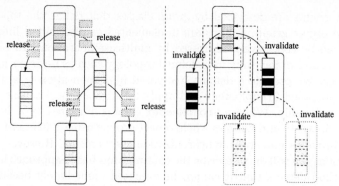

Fig. 2. Collective Operations for Different Modification Modes

All locally mapped regions can be joined to a distributed group that can be collectively addressed. Tasks like global page invalidation or update protocols are, therefore, very easy to implement (Fig. 2).

4   Object-Oriented Design of the Clix-DSM

The properties of a DSM protocol typically include the following aspects: The **consistency model** (sequential, release, entry, scope, etc.), the **granularity** of an object (byte, word, page, any), the **sharing resolution** that still guarantees a consistent view (bit, byte, word), the **directory** to determine who handles requests for accesses (home-base or migration-base), the **modification mode** for propagating changes (update or invalidate) and the **communication mode** for request handling (synchronous or asynchronous).

Each of these aspects has its own abstraction: Memory objects provide the proper sharing resolution, memory controls to oversee access to an object at the granularity level and directory items record the internal state of an object. In addition, aspects of the implementation of a concrete consistency model are isolated into separate classes to allow their reuse in the context of other consistency models, e.g., copy sets record nodes who have copies of a memory object and a scheduler queues and issues incoming requests for synchronous communication. Finally, the abstractions are combined to realize a concrete consistency model by defining each of the aspects above. A concrete consistency model also provides the operational semantics for message exchanges, which then determine the modification and communication mode. The design also supports multi-threading for each node both on the implementation level and the user level. The next section provides more details for an actual implementation of a DSM protocol.

5   Sample Implementation

In the following, several aspects of a sample implementation of sequential consistency of a page-based protocol with invalidation, a static home node per page and synchronous communication shall be given. Sequential consistency is implemented by two classes that are tightly coupled: `SeqConsistency` (SC) (Fig. 1) and `SeqObjectControl` (SOC) (Fig. 3).

```
class SeqObjectControl {
 public: ...
  void protect(int objectId);
  void unprotect(int objectId);
  void invalidate(int objectId);
  void setMemObjectW(int objectId,/*!in!*/MemPage* pMemObject)/*!async!*/;
  void setMemObjectR(int objectId,/*!in!*/MemPage* pMemObject,
    /*!in!*/ Continuation* pContinuation) /*!async!*/;
  void copyMemObjectW(int objectId, int node) /*!async!*/;
  void copyMemObjectR(int objectId, int node,
    /*!in!*/ Continuation* pContinuation ) /*!async!*/;
  void unprotectW(int objectId) /*!async!*/;
  void invalidateW(int objectId) /*!async!*/;
}/*!dual!*/;
```
Fig. 3. The Sequential Object Control Class

The two classes are separated due to their communication structure. SC implements an external server, i.e., it accepts requests from other nodes and handles them. Requests are handled in mutual exclusion since it is a dual class whose instances resemble a monitor on the current node. The actions for handling a request may include modifications of the access right for a memory objects, which are handed off to the corresponding SOC instance. The SOC delegates the task of modifying the MMU access rights to the MemorySegment (MS), an instance of a non-dual class derived from the VAST framework (Fig. 4).

MS overloads the handler of page faults defined by VAST. The handler is an integral part of the consistency model but cannot be transfered into SC and SOC since it is triggered by a page fault controlled through VAST. Upon activation of the handler, it simply distinguishes read and write accesses and issues a remote object call to the corresponding method of the owner's SC instance. In the case of a home-based system, the SC instance of the memory object is selected that corresponds to the home node to delegate the request. The called method is acquireR() or acquireW() for read and write faults, respectively.

```
class MemSegment: public ControlledArea {
 public: ...
  void initSegment(caddr_t area, size_t size, size_t psize);
  void protect(int objectId);
  void unprotect(int objectId);
  void invalidate(int objectId);
  void setConsistency(int objectId, /*!dual!*/SeqConsistency* pSeqCons);
};
```
Fig. 4. The Memory Segment Class

The example in Fig. 5 describes the operations for a read fault on node R. The fault handler of the MS (a regular class) synchronously calls acquireR() of the SC (a dual class, depicted as a shaded box) on the page's home node H in step (1). The request is queued on the home node and a continuation is generated, which blocks the synchronous caller for now but allows further requests to be received on H. If no prior request was active, the next request

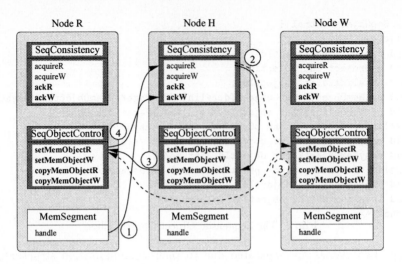

Fig. 5. Read Fault Handling

is dequeued and handled, which would be R's request in this case. As a result, R is added to the copy set. If no writer was active, the SOC of H is requested to hand over the page to node R via `copyMemObjectR()` in step (2). Notice that this request involves an asynchronous remote object invocation (depicted as a bold face method) to prevent the SC from blocking since it may have to accept further requests. The page is transfered from H's SOC to R's SOC via `setMemObjectR()` through another asynchronous call in step (3) and the faulting thread on node R is reactivated through a continuation. Finally, an acknowledgement is sent asynchronously from R's SOC to H's SC in step (4), which would result in handling the next request if H's queue is not empty. Had some other node W had write access to the page, then W would have been instructed by H's `acquireR()` method to reduce page rights to read-only and pass a copy to node R via `copyMemObjectR()`, depicted by the dashed arcs of steps (2) and (3), respectively. Write faults are handled in a similar fashion by the orthogonal methods for write access.

The advantages of the object-oriented design become apparent when the size of the implementation is regarded. SC, SOC and MS consist of slightly under 100 lines of code (excluding comments), out of which 10 lines comprise the page fault handler. This remarkably small implementation facilitates its validation considerably during the software development cycle since testing and debugging are restricted to a small amount of code. The prerequisite for such abstractions are underlying components that have been tested independently and can be reused for other protocols. In fact, this implementation of sequential consistency also illustrates how other protocols can be implemented.

Clix also provides synchronization in a distributed environment. The mechanisms range from FIFO-Locks and condition variables in a POSIX-like style to semaphores and barriers. All synchronization mechanisms are implemented as a separate class over ARTS' communication mechanisms or on top of each other and comprise about 20 lines of code each.

Other consistency models can be supported in much the same way as sequential consistency. Weak consistency models of pages can already fall back onto a class Diff (see Fig. 6) that creates a run-length difference between two memory pages or applies a difference to a page by updating it with the new values. This

```
template<class Granularity, class MemObj>
class Diff {
 public:
  Diff(MemObj& obj1, MemObj& obj2) { ... }  // create diff of 2 objects
  Diff(char* buf,size_t buf_size) { ... }   // byte stream constructor
  Diff(const Diff& source) { ... }          // copy constr. (for Arts ROI)
  ~Diff() { ... }                           // destructur (for Arts ROI)
  void merge(MemObj& obj) { ... }           // apply diff object locally
  char* addr() { ... }                      // address of byte stream
  size_t size() const { ... }               // length of byte stream
}
```

Fig. 6. Diffs with Arbitrary Sharing Resolutions

class is provided as a template class and can be instantiated to support arbitrary sharing resolutions (called granularity in the figure) down to the level of a byte. Thus, different sharing resolutions may be supported for each memory object under the same consistency model. The user has the privilege to decide whether a coarse resolution is sufficient for certain tasks of the application, thereby yielding better performance, or if a smaller granularity is required due to the choice of data structures within other parts of the application.

6   Related Work

Software DSM system have been realized for a number of consistency models, sometimes even with an object-oriented approach. Implementations of sequential consistency, such as by Li and Hudak [10], were non-object based. The models of release and scope consistency typically do not build on objects either [9, 8]. Entry consistency, on the other hand, is often associated with the object-oriented paradigm due to the granularity of the protocol [3, 6]. The Panda system [2] used similar design goals but focused on transaction-based DSMs and persistent objects based on C++ language extensions that were not as powerful as those of ARTS. None of these approaches take advantage of facilities such as remote object invocation on the ARTS environment or VAST's memory management. Consequently, the implementation of their protocols tend to be more complex than in our CLIX approach. The ARTS platform incorporates many features similar to those found in common middleware layers like CORBA as well as parallel programming languages such as CC++ [7], pC++ [4] and Mentat [1]. However, system programming still needs significantly more control over runtime issues than languages designed for application level programming usually provide. ARTS originated in the area of parallel operating systems and, therefore, it does not rely on complex compiler technology but extensible OO-frameworks. Thus most remote object invocation mechanisms as well as high-level collective operations can be controlled by standard C++ inheritance mechanisms. The

VAST framework originated as a platform for persistent containers to ease check-pointing and incremental parallel computations. While many OO-databases and object-stores like Texas [12] exist, the organization of VAST as an OO framework made it possible to reuse many parts of VAST without change for environments that VAST was originally not designed for.

7   Conclusions

We have shown that parallel programming with distributed object technology can be combined with shared-memory programming by constructing a software DSM using distributed object systems. A sample implementation demonstrates that three classes of less than 100 lines of code each suffice to implement sequential consistency. In addition, per node multi-threading, distributed synchronization facilities and facilities for weak consistency protocols are provided, which are also comparably small. The description of such compact protocols facilitates the development, maintenance and validation of DSM systems.

References

1. A.Grimshaw. Easy-to-use parallel processing with Mentat. *IEEE Computer*, 26(5), May 1993.
2. H. Assenmacher, T. Breitbach, P. Buhler, V. Hubsch, and R. Schwarz. PANDA - supporting distributed programming in C++. In O. Nierstrasz, editor, *Object-Oriented Programming*, number 707 in LNCS, pages 361–383. Springer, 1993.
3. B. Bershad, M. Zekauskas, and W. Sawdon. The midway distributed shared memory system. In *COMPCON Conference Proceedings*, pages 528–537, 1993.
4. Francois Bordin, Peter Beckman, Dennis Gannon, Srinivas Narayana, and Shelby X. Yang. Distributed pC++: Basic Ideas for an Object Parallel Language. *Scientific Programming*, 2(3), Fall 1993.
5. L. Büttner, J. Nolte, and W. Schröder-Preikschat. ARTS of PEACE – A High-Performance Middleware Layer for Parallel Distributed Computing. *Special Issue of the Journal of Parallel and Distributed Computing on Software Support for Distributed Computing*, 1999.
6. J. B. Carter. Design of the Munin Distributed Shared Memory System. *J. Parallel Distrib. Comput.*, 29(2):219–227, September 1995.
7. K. M. Chandy and C. Kesselman. CC++: A Declarative Concurrent Object-Oriented Programming Notation. In *Research Directions in Concurrent Object-Oriented Programming*. MIT Press, 1993.
8. Liviu Iftode, Jaswinder Pal Singh, and Kai Li. Scope consistency: a bridge between release consistency and entry consistency. In *Symposium on Parallel Algorithms and Architectures*, pages 277–287, June 1996.
9. Pete Keleher, Alan L. Cox, Sandhya Dwarkadas, and Willy Zwaenepoel. An evaluation of software-based release consistent protocols. *J. Parallel Distrib. Comput.*, 29:126–141, September 1995.
10. K. Li and P. Hudak. Memory coherence in shared virtual memory systems. *ACM Trans. Comput. Systems*, 7(4):321–359, November 1989.
11. W. Schröder-Preikschat. *The Logical Design of Parallel Operating Systems*. Prentice Hall International, 1994. ISBN 0-13-183369-3.
12. V. Singhal, S. V. Kakkad, and P. R. Wilson. Texas: An Efficient Portable Persistent Store. In *Proceedings of the 5th International Workshop on Persistent Object Systems*, San Miniato, Italy, September 1992.

Controlling Distributed Shared Memory Consistency from High Level Programming Languages

Yvon JÉGOU

IRISA / INRIA
Campus de Beaulieu,
35042, RENNES CEDEX, FRANCE,
Yvon.Jegou@irisa.fr

Abstract One of the keys for the success of parallel processing is the availability of high-level programming languages for on-the-shelf parallel architectures. Using explicit message passing models allows efficient executions. However, direct programming on these execution models does not give all benefits of high-level programming in terms of software productivity or portability. HPF avoids the need for explicit message passing but still suffers from low performance when the data accesses cannot be predicted with enough precision at compile-time. OpenMP is defined on a shared memory model. The use of a distributed shared memory (DSM) has been shown to facilitate high-level programming languages in terms of productivity and debugging. But the cost of managing the consistency of the distributed memories limits the performance. In this paper, we show that it is possible to control the consistency constraints on a DSM from compile-time analysis of the programs and so, to increase the efficiency of this execution model.

1 Introduction

The classical compilation scheme for High Performance Fortran (HPF) is based on the application of the owner-compute rule which, after distributing the ownership of array elements to the processors, distributes the charge of executing each instruction to the processor owning the variable modified by this instruction. It is possible to map HPF distributed arrays on a distributed shared memory and to avoid the burden of explicitly updating the local memories through message passing. OpenMP was primarily defined for shared memory multi-processors. But it is possible to transform OpenMP programs for an SPMD computation model if the accesses to the non-private data can be managed, for instance, through a software DSM.

However many authors have reported poor performance on sequentially consistent distributed shared memories for HPF and for OpenMP mainly because the high latencies in inter-processor communications penalize the management of sequential consistency protocols. The situation is even worse in the presence of false sharing.

The independence of the iterations of a parallelized loop guarantees that the modifications to the data during one iteration need not be visible for the execution of the other iterations of the same loop. The propagation of the modifications can be delayed until the next synchronization point. Many projects such as TreadMarks [1] or Munin

J. Rolim et al. (Eds.): IPDPS 2000 Workshops, LNCS 1800, pp. 293-300, 2000.
© Springer-Verlag Berlin Heidelberg 2000

[3] have shown that the use of a relaxed consistency software DSM along with some consistency control on synchronizations can avoid the cost of maintaining the memory consistency during the execution of a parallel loop.

Mome (**Mo**dify-**me**rge) is a software relaxed consistency, multiple writers DSM. Mome mainly targets the execution of programs from the High Performance Computing domain which exhibit loop-level parallelism. This DSM allows to generate consistency requests on sections of the shared space and so to take advantage of the analysis capacity of modern compilers.

Next section compares implicit and explicit consistency management and their relations with HPF and OpenMP compilation models. Section 3 presents the Mome DSM consistency model and Sect. 4 gives some details on the possible controls of the DSM behavior from the programs. Some experimental results are presented in Sect. 5.

2   Implicit versus Explicit Consistency Management

Using an implicit consistency management, the modifications to the shared memories become automatically visible after synchronization operations such as locks or barriers. Race-free parallel programs should produce the same results as with a sequential consistency management when run with an implicit consistency management. TreadMarks [1] follows this model and implements a *release consistency* model with a *lazy invalidate* protocol. Using an explicit consistency management scheme, the modifications from other processors are integrated in a memory section only upon explicit request from the application. Entry consistency [2] implements a form of explicit consistency management through the association of shared objects with synchronization variables. The Mome DSM implements another form of explicit management where a processor can specify the memory section which must be updated without making reference to synchronization variables.

The choice between these schemes depends mainly on the compilation technique in use. The execution flow is controlled in HPF by the owner-compute rule. This rule states that each data element must always be updated by the same processor. As long as this rule can be applied, a distributed data element is always up-to-date on the owning processor. For the execution of a parallel loop, it is necessary to check for the consistency of the non-owned accessed elements only. The explicit consistency management scheme seems to be well adapted to the HPF model. The HPF compilation strategy for Mome is identical to the compilation strategy for an explicit message passing model. The main difference is that the message passing calls in the generated code are replaced by consistency requests to the DSM on the same memory sections. In many HPF implementations, the generated code communicates with the run-time system only through object descriptors. The adaption of these implementations to the Mome DSM is then limited to replacing the calls to the message passing layer by calls to the DSM interface inside the run-time system of the language. No modification of the source program is necessary.

In the OpenMP computation model, the parallelism is specified through program directives. The compilation scheme is not based on the owner-compute rule. Each data element can potentially be modified by any processor. It is still possible to use an explicit consistency scheme if a static data access analysis can be performed at compile-time. An implicit consistency management scheme seems to be the only choice in many cases for the OpenMP programming model.

But real applications do not always fit exactly in one of these models. A strict application of the owner-compute rule on HPF programs does not always produce efficient results, for instance in the presence of indirect array accesses or with complex loop bodies. Moreover, poor data access analysis can lead to update requests on the whole shared data using the explicit consistency scheme. Although the OpenMP computation model is not based on data access analysis, the performance of the resulting codes can often be optimized if the data access patterns are considered in the loop distributions.

3   Mome DSM Consistency Model

Basically, Mome implements a simple relaxed consistency model for the DSM page management. A processor must send a consistency request to the DSM specifying a shared memory address range each time the view of this address range must integrate modifications from other processors. Mome maintains a global clock and most of the consistency requests make reference to this clock. For each memory page, the DSM keeps track of the date of the oldest modification which has not yet been propagated. If this modification date precedes the reference date of a consistency request, all known modifications are merged (exclusive or $\oplus$ of the modified pages) and a new version of the page is created. In the other case, the current version of the page is validated and no modification needs to be integrated. The date associated to a consistency request can be the current date, the date of the last synchronization barrier, the last release date of an acquired mutex lock, or the date of any other event detected by the application.

All the consistency requests are implemented in Mome using a two-step sequence. The first step, at the time of the call, is local to the requesting processor: the consistency constraints (mainly the date) are recorded in the local page descriptor and, if necessary, the memory section is unmapped from the application. The second step is applied when the application tries to read or write inside the memory section and generates a page fault. This second step involves a transaction with a global page manager which can confirm the local copy of the page if the current version has been validated. If modifications to the memory section need to be made visible, the processor receives the new version of the page after all known modifications have been merged.

This two-step procedure limits the updates to the pages which are really used by the application and implements a form of *lazy updates*. Using a combination of consistency requests and prefetch requests, the DSM interface allows to update the local memory before the application generates a page fault.

4   Consistency Management Optimizations

In order to fully exploit the analysis capacity of modern compilers, the Mome interface offers some level of control on the consistency management strategy through prefetching, through on-the-fly consistency checks and through page manager redistribution.

It is possible to combine the consistency requests with non blocking read or write prefetch requests on the shared data. This possibility especially targets the case where the compile-time data access analysis can predict with enough precision which parallel variables are to be accessed in the near future. Data prefetching reduces the page fault latency.

The current implementation of Mome uses a directory-based page management. All processors share a global page manager for each page. The Mome DSM allows for the migration of a page manager, for instance, on the processors making the most frequent requests. This possibility reduces the latency of the consistency requests when enough affinity exists between the data distribution and the distribution of the control. This is the case when the owner-compute rule is applied on HPF programs: the managers can be distributed according to the data distribution.

The basic consistency management strategy of Mome uses a two-step procedure where all access rights are removed from the application during the first step and a transaction with the global manager is generated on the resulting page faults. Although the cost of the consistency management can be reduced using data prefetches, at least two system calls and one page fault need to be generated for each page even if no update of the current copy is needed. This situation comes from the fact that, using Mome, all consistency requests must be generated by the consumers. At the opposite of TreadMarks, Mome does not automatically propagate the write notices. The *on-the-fly consistency checks* we are currently integrating in the Mome DSM partly fulfills this weakness. In on-the-fly mode, which can be selected at the page level, the global page manager broadcasts some information to all processor holding a copy of a page as soon as some other processor is allowed to modify the page. Using a simple test based on the local state of the page and on the visible date of the manager, it is possible to know if pending modifications are present on the page. This on-the-fly consistency management avoids unnecessary system calls when the program exhibits enough data locality on the processors. This situation often happens on OpenMP codes because the affinity between distributed data and distributed iteration space are not exploited by the compilers as it is the case in HPF using the owner-compute rule. This system is close to the write-notice propagation of TreadMarks, although in Mome the propagation of this information is not tied to synchronization requests.

5 Experiments

5.1 Simulated Code

This section shows the usefulness of some Mome features on the tomcatv parallel application. The program was transformed manually because no HPF compilation system is currently available for the Mome DSM. The modifications were limited to the mapping of the arrays on the DSM shared space, to the distribution of the iteration spaces on the processors and to the insertion of consistency requests before the loops.

During an external iteration of tomcatv, each data element of a two-dimensional grid is updated from the values of its neighbors in the previous iteration. Using HPF, all arrays of tomcatv are column-wise block distributed. Each processor reads one column updated by each of its two neighbors during each iteration.

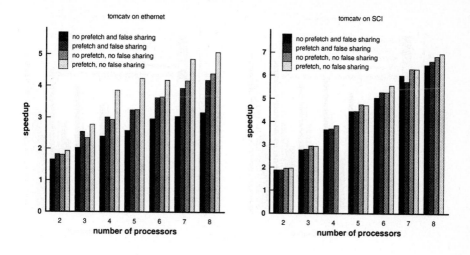

Figure1. Data prefetch on tomcatv

tomcatv was run on a 1023×1023 problem for the false sharing case and on a 1024×1024 problem when false sharing is avoided. The experimentation platform is a group of PCs connected through an SCI ring as well as a 100Mbits Ethernet network. Each node is a dual Pentium II processor running SMP Linux 2.2.7.

5.2 Data Prefetch

The experiments of Fig. 1 consider the application of the HPF owner-compute rule (consistency requests only on the pages containing shared data) combined with read prefetches. The prefetch operations are asynchronous and the computation is started before the requested data is present. Prefetching the left column of a block in tomcatv does not speedup the computation. But the prefetch request loads the right column of the block asynchronously during the block computation. With the SCI communication layer, the latency of page-fault resolution is low enough and the improvement is small. This is not the case for the Ethernet layer, as shown in Fig. 1. False sharing increases the latency of page-fault resolution because the modifications to the page must be gathered and merged. Prefetching in the presence of false sharing improves the speedup, mainly when the latency of the communications is high. The decision to insert prefetch requests is straightforward using the HPF compilation model. In fact, it is possible to systematically insert a read prefetch request everywhere the message passing version inserts a message receive. In general, prefetching pages used in the near future optimizes the execution. But prefetching unused pages can overload the system and delay the treatment of useful requests. Prefetching should be avoided in case of weak data analysis. In [6], M. Karlsson and P. Stenström propose a technique where prefetching is decided after analyzing the recent requests from the application. Such a pure run-time technique does not depend on the analysis ability of the compilers.

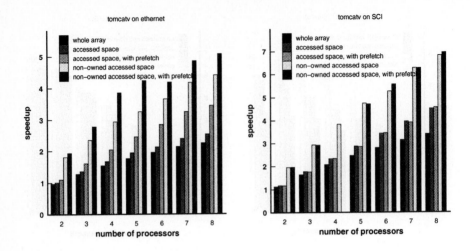

Figure2. Data synchronization on tomcatv

5.3 Consistency Management Strategy

In the experiments of Fig. 1, the consistency requests were generated for non-owned accessed data only. Such optimizations are not possible on OpenMP without a global analysis of the loops in the program. Figure 2 compares the execution speedups when the whole shared space, or the accessed space, or only the non-owned accessed space are synchronized. Using the current implementation of Mome, a page transaction costs a local thread synchronization and can generate a system call if the page must be unmapped. The on-the-fly optimization should avoid the page transactions on unused pages. During the loop execution, each accessed page produces a page-fault and generates communication with its page manager. In the experiments of Fig. 2, only the non-owned columns necessitate processor communication. Without prefetching, the performance for the case where only the accessed parts are synchronized is close to the performance of the case where the whole shared space is synchronized. In fact, the extra cost is due to the thread synchronizations only because the pages which are never accessed are not mapped in the application. Prefetching increases the efficiency of the execution on Ethernet. The performance difference between the accessed case and the non-owned case comes from the extra pages faults generated for the owned part of the array. This difference shows the benefits of a strict application of the owner-compute rule. This difference also shows the speedup expected from the integration of the on-the-fly consistency check: this check should avoid all transactions on the pages which are not modified by other processors.

5.4 Manager Distribution

Mome uses a directory-based management of the shared pages. When the owner-compute rule is applied, managing a page on the processor in charge of the data contained in this

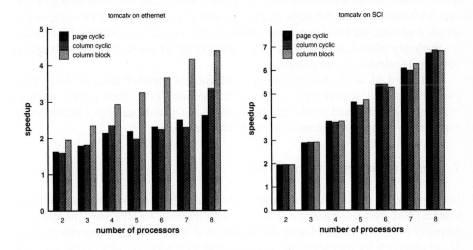

Figure3. Managers distribution on `tomcatv`

page should optimize the communications. OpenMP does not propose any directive for loop distribution control. However, for many parallel loops, it is possible to control the distribution of the iteration space through static scheduling. In this case, it is also possible to insert a call to the DSM interface and to redistribute the page managers according to this scheduling. The technique proposed by F. Mueller in [8] could also be applied at run-time for the case where the affinity between the iteration space and the data cannot be analyzed at compile-time. Figure 3 shows the performance of program executions using different manager distributions on Ethernet and on SCI. The page cyclic distribution is the initial page distribution of the Mome DSM. The latency using SCI is low enough and the performance does not depend on the distribution. But this is not the case for Ethernet. The latency of inter-processor communication directly affects the efficiency of page-fault resolution.

6   Related Work

Several papers describe the integration of run-time and compile-time approaches for the execution of parallelized codes on a software DSM. Dwarkadas et al. [5] augment the DSM interface with entry points which can exploit the message passing capabilities of the underlying hardware. Using these entry points the compiler can request for direct data transfers and reduce the overhead of run-time consistency maintenance. In our system, we have restricted the use of the information received from the compiler to consistency management. But all data transfers go through the DSM page migrations.

In [4], Chandra and Larus also consider explicit management of the shared data between the producers (writers) and the consumers (readers). Using the owner-compute rule, the owners explicitly update the local memories of the readers. This optimization make the non-owned blocks available before the parallel loop executes, so that no access fault occurs during the loop. At the opposite of Chandra and Larus's proposition which deactivate the DSM consistency management in order to optimize, we consider a completely relaxed consistency model and the compiler inserts consistency requests only on the relevant memory sections.

H. Lu and Al. in [7] describe an implementation of OpenMP using the SUIF compiler on networks of workstation upon the TreadMarks DSM. Their implementations considers all parallel and synchronization features of OpenMP and is not limited to the exploitation of loop parallelism as in our experimentations.

7   Conclusion and Future Work

In this paper, we showed that it is possible to optimize the comportment of applications using a DSM from information extracted at compile-time. An HPF run-time system is currently being implemented upon the Mome DSM and should show the effectiveness of our approach for the execution of parallel codes on networks of workstations. The integration of on-the-fly consistency checks is also in progress and will reduce the cost of making consistent large amounts of shared data, as it is the case with OpenMP when no data access analysis is performed.

References

[1] C. Amza, A. L. Cox, S. Dwarkadas, P. Keleher, H. Lu, R. Rajamony, and W. Zwaenepoel. Treadmarks: Shared memory computing on networks of workstations. *IEEE Computer*, 29(2):18–28, February 1996.

[2] B. N. Bershad, M. J. Zekauskas, and W. A. Sawdon. The midway distributed shared memory system. In *Proc. of the 38th IEEE Int'l Computer Conf. (COMPCON Spring '93)*, pages 528–537, February 1993.

[3] J. B. Carter, J. K. Bennett, and W. Zwaenepoel. Techniques for reducing consistency-related communication in distributed shared memory systems. *ACM Transactions on Computer Systems*, 13(3):205–243, August 1995.

[4] Satish Chandra and Larus James R. Optimizing communication in hpf programs on fine-grain distributed shared memory. In *6th ACM SIGPLAN Symposiun on Principles and Practice of Parallel Programming*, June 1977. Las Vegas, June 18-21.

[5] Sandhya Dwarkadas, Honghui Lu, Alan L. Cox, Ramakrishnan Rajomony, and Willy Zwaenepoel. Combining compile-time and run-time support for efficient software distributed shared memory. In *Proceedings of IEEE, Special Issue on Distributed Shared Memory*, volume 87, No 3, pages 467–475, March 1999.

[6] M. Karlsson and P. Stenström. Effectiveness of dynamic prefetching in multiple-writer distributed virtual shared memory system. *Journal of Parallel and Distributed Computing*, 43(2):79–93, July 1997.

[7] H. Lu, Y. C. Hu, and W. Zwaenepoel. Openmp on networks of workstations. In *Proceedings of Supercomputing'98*, 1998.

[8] F. Mueller. Adaptative dsm-runtime behavior via speculative data distribution. In J. Jose and al., editors, *Parallel and Distributed Processing – Workshop on Run-Time Systems for Parallel Programming*, volume 1586 of *LNCS*, pages 553–567. Springer, April 1999.

Online Computation of Critical Paths
for Multithreaded Languages

Yoshihiro Oyama, Kenjiro Taura, and Akinori Yonezawa

Department of Information Science, Faculty of Science, University of Tokyo
E-mail: {oyama,tau,yonezawa}@is.s.u-tokyo.ac.jp

Abstract. We have developed an instrumentation scheme that enables programs written in multithreaded languages to compute a critical path at runtime. Our scheme gives not only the length (execution time) of the critical path but also the lengths and locations of all the subpaths making up the critical path. Although the scheme is like Cilk's algorithm in that it uses a "longest path" computation, it allows more flexible synchronization. We implemented our scheme on top of the concurrent object-oriented language Schematic and confirmed its effectiveness through experiments on a 64-processor symmetric multiprocessor.

1 Introduction

The scalability expected in parallel programming is often not obtained in the first run, and then performance tuning is necessary. In the early stages of this tuning it is very useful to know what the *critical path* and how long it is. The length of an execution path is defined as the amount of time needed to execute it, and the critical path is the longest one from the beginning of a program to the end. One of the advantages of getting critical path information is that we can use it to identify the cause of low scalability and to pinpoint the "bottleneck" program parts in need of tuning.

This paper describes a simple scheme computing critical paths on-the-fly, one that when a program is compiled inserts instrumentation code for computing the critical paths. An instrumented program computes its critical path while it is running, and just after termination it displays critical-path information on a terminal as shown in Fig. 1. Our instrumentation scheme is constructed on top of a high-level parallel language supporting *thread creation* and *synchronization via first-class data structures (communication channels)*. Because the instrumentation is expressed as rules for source-to-source translation from the parallel language to the same language, our scheme is independent of the implementation details of parallel languages such as the management of activation frames. Although our primary targets are shared-memory programs, our scheme can also be applied to message-passing programs. We implemented the scheme on top of the concurrent object-oriented language Schematic [7, 8, 12] and confirmed its effectiveness experimentally.

This work makes the following contributions.

J. Rolim et al. (Eds.): IPDPS 2000 Workshops, LNCS 1800, pp. 301-313, 2000.
© Springer-Verlag Berlin Heidelberg 2000

```
frame entry point                         frame exit point                   elapsed time    share
====================================================================================================
main()                              ---   move_molecules(mols,100)              741 usec      9.2%
spawn                                                                            10 usec      0.1%
move_molecules(mols, n)             ---   spawn move_one_mol(mol[i])             39 usec      0.5%
spawn                                                                            10 usec      0.1%
move_one_mol(molp)                  ---   calc_force(molp, &f)                  366 usec      4.6%
spawn                                                                            10 usec      0.1%
calc_force(molp, fp)                ---   return                               4982 usec     61.9%
communication                                                                    15 usec      0.2%
sumf += f  (in move_one_mol)        ---   send(r, sumf)                         504 usec      6.3%
communication                                                                    15 usec      0.2%
v = recv(r)  (in move_molecules)    ---   send(s, v*2)                          128 usec      1.6%
communication                                                                    15 usec      0.2%
u = recv(s)  (in main)              ---   die                                  1207 usec     15.0%
====================================================================================================
Critical path length:                                                          8042 usec    100.0%
```

Fig. 1. An example of critical-path information displayed on a terminal. All subpaths in a critical path are shown.

- It provides an instrumentation scheme for languages where threads synchronize via first-class synchronization data. As far as we know, no scheme for computing critical paths for this kind of parallel languages has yet been developed.
- Our instrumentation scheme also gives the length of each subpath in a critical path. As far as we know, previous schemes either provide the length of only the critical path [1, 6] or provide a list of procedures and the amount of time each contributes to the critical path [3].
- The usefulness of our scheme has been demonstrated through realistic applications running on symmetric multiprocessors.

Instrumentation code has usually been inserted into the synchronization parts and entry/exit points of procedures by programmers, but our scheme is implemented by the compiler of a high-level multithreaded language. The compiler inserts instrumentation code automatically, thus freeing programmers from the need to rewrite any source code. In an approach using low-level languages such as C and parallel libraries such as MPI or Pthread, on the other hand, a programmer has to modify a program manually. The source modification approach [10] may require much effort by programmers and result in human errors.

The rest of this paper is organized as follows. Section 2 clarifies the advantages of obtaining critical path information. Section 3 describes our instrumentation scheme and Sect. 4 gives our experimental results. Section 5 describes related work and Sect. 6 gives our conclusion and mentions future work.

2 Benefits of Getting Critical Path Information

Computing critical path information brings us the following benefits because it helps us understand parallel performance.[1]

---

[1] The usefulness of critical paths for understanding performance is described in [1].

- A critical path length indicates an *upper bound* on the degree to which performance can be improved by increasing the number of processors. When the execution time is already close to critical path length, the use of more processors is probably futile and may even be harmful.
- Critical path information is essential for performance prediction [1].
- Critical path information helps identify the cause of low scalability. If the critical path is very short, for example, low scalability is likely to result from the increase of overhead or workload. If it is close to the actual execution time, it should be shortened.
- Programmers can pinpoint the program parts whose improvement will affect the overall performance and thus avoid tuning unimportant parts.
- A compiler may be able to optimize a program adaptively by using critical path information. This topic is revisited in Sect. 6.

3   Our Scheme: Online Longest Path Computation

3.1   Target Language

Our target is the C language extended by the addition of thread creation and channel communication.[2] *Channels* are data structures through which threads can communicate values. Channels can contain multiple values. If a channel has multiple values and multiple receiving threads at the same time, which pair communicates is unspecified. Channels can express a wide range of synchronization data including locks, barriers, and monitors. The target language has the following primitives.

spawn $f(x_1, \ldots, x_n)$: It creates a thread to calculate $f(x_1, \ldots, x_n)$. A spawned function f must have a **void** type.

send(r, v): It sends a value v to a channel r.

recv(r): It attempts to receive a value from a channel r. If r has a value, this expression receives that value and returns it. Otherwise, the execution of the current thread is suspended until a value arrives at r.

die: It terminates the whole program after displaying the critical path from the beginning of the program up to this statement.

Since we can encode sequential function calls and return statements straightforwardly with thread creation and channel communication, we do not consider these calls and statements separately.

Figure 2 shows a sample program in the target language. The function **fib** creates threads for computing the values of recursive calls. The values are communicated through channels. l's are labels, which are used later for representing program points. Labels are added automatically by the compiler in a preprocessing phase.

---

[2] Our scheme was originally designed for process calculus languages. For readability, however, this paper describes the scheme in an extension to C.

```
fib(r, x) {
  l1 : if (x < 2) {
        l2 : send(r, 1);
      } else {
            x1 = x - 1;
            r1 = newChannel();
      l3 : spawn fib(r1, x1);          main() {
            x2 = x - 2;                   s = newChannel();
            r2 = newChannel();            spawn fib(s, 2);
      l4 : spawn fib(r2, x2);            u = recv(s);
      l5 : v2 = recv(r2);                print(u);
      l6 : v1 = recv(r1);                die;
            v = v1 + v2;                }
      l7 : send(r, v);
      }
}
```

Fig. 2. A multithreaded program to compute the second Fibonacci number.

Data communicated by threads must be contained in channels and accessed through send and receive operations even in shared-memory programs because our scheme ignores the producer-consumer dependency of the data not communicated through channels.

3.2   Computed Critical Paths

A *dynamic* structure of a parallel program can be represented by a directed acyclic graph (DAG). Figure 3 shows a DAG for the program in Fig. 2. A node in a DAG represents the beginning of a thread or one of the parallel primitives spawn, send, recv, or die. A DAG has three kinds of directed edges:

Arithmetic edges They represent intraframe dependency between nodes. An arithmetic edge from a node X to a node Y is weighted with a value representing the amount of time that elapses between the leaving of X and the reaching of Y.

Spawn edges They represent dependency from a spawn node to a node for the beginning of a spawned thread. Spawn edges are weighted with a predefined constant value representing the difference between the time the spawn operation starts and the time a newly-created thread becomes runnable.

Communication edges They represent dependency from a sender node to its receiver node. Communication edges are weighted with a predefined constant value representing communication delay between the time the send operation starts and the time the sent value becomes available to the receiver.

A path is weighted with the sum of the weights of included edges, and the critical path we compute is the one from the node for the beginning of a program to the die node with the largest weight. The critical path *length* we compute is represented in a DAG as the weight of a critical path.

Unfortunately, the shape of the DAG representing program execution may vary between different runs. For example, communication edges may connect different pairs of nodes in different runs because of the nondeterministic behavior

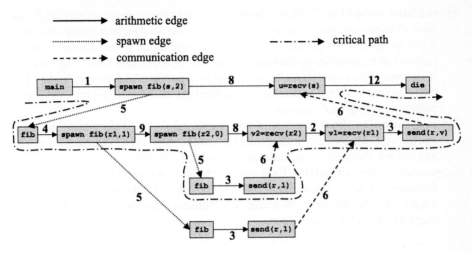

Fig. 3. A DAG representing a dynamic structure of the program in Fig. 2.

of communication. Furthermore, the execution time for each program part will change for various reasons (e.g., cache effects). Therefore, we compute the critical path of *a DAG that is created in an actual run.*

The DAG model described above assumes the following.

- Thread creation cost is constant (all the spawn edges in Fig. 3 have the same weight).
- Communication cost is constant (all the communication edges in Fig. 3 have the same weight).
- Sends, receives, and spawns themselves are completed instantaneously (the nodes in Fig. 3 have no weight).

The first and second assumptions do not take data locality into account. The cost of creating a thread, for example, depends on the processor the thread is placed on; the cost will be small when the thread gets placed on the processor its creator is on. Similarly, interprocessor communication is more costly than intraprocessor communication; two threads on the same processor can communicate through the cache but those on different processors cannot. Our current implementation assumes that a newly created thread and its creator are *not* on the same processor and that all communication is interprocessor communication. Improving our scheme by taking data locality into account would be an interesting thing to do.

The third assumption is reasonable if sends, receives, and spawns are cheap enough compared with the other program parts represented by DAG edges. Otherwise the cost of the operations should be taken into account by giving DAG nodes nonzero weights.

A critical path is shown to programmers in the form of a sequence of *subpaths*, of which there are the following three kinds:

Intraframe subpaths Each is a sequence of arithmetic edges representing the part of execution that begins when control enters a function frame and ends when control leaves the frame. The information about an intraframe subpath consists of its location in the source code and its execution time. The location is expressed by specifying the entry and exit points of a function frame. The entry points are the beginnings of a function body and receives. The exit points are sends and spawns.

Spawn subpaths Each consists of one spawn edge. The information about a spawn subpath consists of the constant weight only, so the information about any spawn subpath is the same as that about any other.

Communication subpaths Each consists of one communication edge. The information about a communication subpath also consists only of the constant weight, so the information about any communication subpath is also the same as that about any other.

3.3   Instrumentation

Our scheme computes critical paths through a *longest path computation* (Fig. 4). Programs instrumented with our scheme maintain the information on the critical path from the beginning of a program to the currently executing program point. In a spawn operation the information about the critical path to the spawn operation is passed to the newly created thread. In a send operation the information about the critical path to the operation is attached to a sent value. A receive operation compares the length of the critical path to the receive with the length of the critical path to the sender of the received value *plus* the weight of a communication edge. The larger of the compared values is used as the length of the critical path to the program part following the receive.

Figure 5 shows an instrumented version of the function fib in Fig. 2. An instrumented program maintains in the three local variables (represented in Fig. 5 as *el*, *et*, and *cp*) the following three values:

Entry label (*el*): It represents a program point that indicates the beginning of the currently executing subpath.

Entry time (*et*): It represents the time when the currently executing subpath started.

Critical path (*cp*): It represents a sequence of subpaths that makes up the longest path up to the beginning of the currently executing subpath.

The function currTime() returns the current value of a timer provided by the operating system.[3] $[v,cp]$ is a data structure comprising a communicated value v and the information on a critical path cp. length(cp) returns the length of a critical path cp. $\{b, e, t\}$ is a data structure representing an intraframe subpath that begins at a label b, ends at a label e, and takes the amount t of time to execute. addInFrSubPath(cp, $\{b, e, c\}$) appends an intraframe subpath $\{b, e, c\}$ to

---

[3] Our scheme does not assume a *global* timer shared among processors. Threads communicate critical path information only, not the value of a timer.

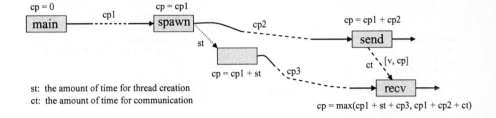

Fig. 4. Computation of the longest path.

a critical path cp and returns the extended critical path. $\texttt{addSpawnSubPath}(cp)$ and $\texttt{addCommSubPath}(cp)$ respectively append a spawn subpath and a communication subpath to a critical path cp and return the extended critical path.

The length of the critical path to the current program point is equal to

$\texttt{length}(cp) +$ (the current time $- \, et$)

while the control is in arithmetic edges. The value (the current time $- \, et$) is called a *lapse* and is the time between the entry label and the current program point. The relations between et, el, and cp are illustrated in Fig. 6, and the lapses kept by the variables t_1, t_2, t_3, t_4, and t_5 in the program in Fig 5 are illustrated in Fig. 7.

As noted earlier, our DAG model assumes that sends, receives, and spawns take no time and thus that there is no time difference between the end of one edge and the start of the edge executed next. In actual execution, however, there are delays between two DAG edges. The delays are descheduling time and suspension time. An underlying thread system may delay the start of a runnable thread because it is not always possible to assign a processor to all runnable threads. A thread may be suspended from the time it starts trying to receive a value until the time it receives the value.

Our instrumentation excludes these kinds of delays from a critical path. We "stop the timer" when a thread reaches a node and "restart the timer" when it leaves the node. When reaching a node, a thread calculates the current lapse and sets it to a variable t. When leaving the node, the thread "adjusts" the value of the variable et to make the current lapse still equal to t.[4]

Figure 8 shows source-to-source translation rules for our instrumentation.

Function Definition A function obtains an additional argument cp, through which a critical path is passed between function frames. S' comes up out of S by applying all other transformation rules.

Spawn A critical path cp is extended with an intraframe subpath and a spawn subpath. The resulting extended critical path is passed to a spawned function.

---

[4] The processing enables our scheme to be used in languages with garbage collection (GC). The GC time can be excluded from computed critical paths by allowing threads to jump to a GC routine only when they are in a DAG node.

```
fib(r, x, cp) {                                 t₃ = currTime() - et;
  el = l₁;                                       [v2, cp'] = recv(r2);
  et = currTime();                               cp'' = addCommSubPath(cp');
  if (x < 2) {                                   if (t₃ + length(cp) < length(cp'')) {
    t₀ = currTime() - et;                          cp = cp'';
    cp' = addInFrSubPath(cp, {el, l₂, t₀});        el = l₅;
    send(r, [1, cp']);                             et = currTime();
    et = currTime() - t₀;                        } else {
  } else {                                         et = currTime() - t₃;
    x1 = x - 1;                                   }
    r1 = newChannel();
                                                  t₄ = currTime() - et;
    t₁ = currTime() - et;                         [v1, cp'] = recv(r1);
    cp' = addInFrSubPath(cp, {el, l₃, t₁});       cp'' = addCommSubPath(cp');
    cp'' = addSpawnSubPath(cp');                  if (t₄ + length(cp) < length(cp'')) {
    spawn fib(r1, x1, cp'');                        cp = cp'';
    et = currTime() - t₁;                          el = l₆;
                                                    et = currTime();
    x2 = x - 2;                                   } else {
    r2 = newChannel();                             et = currTime() - t₄;
                                                  }
    t₂ = currTime() - et;
    cp' = addInFrSubPath(cp, {el, l₄, t₂});       v = v1 + v2;
    cp'' = addSpawnSubPath(cp');
    spawn fib(r2, x2, cp'');                      t₅ = currTime() - et;
    et = currTime() - t₂;                         cp' = addInFrSubPath(cp, {el, l₇, t₅});
                                                  send(r, [v, cp']);
                                                  et = currTime() - t₅;
                                                }
                                              }
```

Fig. 5. Instrumented version of the function `fib` in Fig. 2.

Send A critical path to a send statement is computed and attached to a value.
Receive The length of the critical path to a receive is compared with the length
of the received critical path extended with a communication subpath. If the
former is shorter, the extended critical path is set to the variable cp and a
new subpath starts. Otherwise the thread simply adjusts the entry time.

If critical paths are to be computed accurately, the underlying system should
have the following characteristics.

- Fast and scalable execution of `currTime()` (i.e., negligible overhead for its
  execution)
- Threads that are not preempted during the execution of an arithmetic edge.
  In other words, once a thread acquires a processor and starts executing
  an arithmetic edge, it will not lose the processor until it reaches a node.
  Otherwise, the amount of time that elapses while the thread is descheduled
  would be included in the weight of the arithmetic edge.

In our experimental platform the first assumption holds but the second does not.
The experimental results were therefore affected by the measurement perturbation. It seems difficult to solve the perturbation problem without the cooperation
of a thread scheduler.

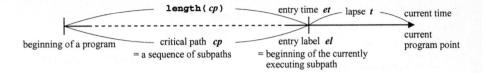

Fig. 6. The relation between the times kept by programs instrumented with our scheme.

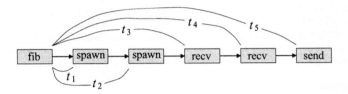

Fig. 7. The lapses kept by the variables t_1, t_2, t_3, t_4, and t_5 in the program in Fig. 5.

3.4 Potential Problems and Possible Solutions

One potential problem with our scheme is the instrumentation overhead, which increases the overall execution time. We can reduce overall execution time by not instrumenting selected functions. Calls to those functions would then be regarded as arithmetic operations. On the other hand, the critical path length computed by our scheme does not include most of the instrumentation overhead: it does not include the amount of time needed for appending a subpath, for attaching critical path information to sent values, or for comparing two critical paths in receives. We thus expect the critical path length computed with an instrumented program is close to that of the uninstrumented program.

Another potential problem is that storing critical path information may require a large amount of memory, even though our scheme requires less memory than do the tracefile-based schemes that record the times of all inter-thread events. Memory usage can be reduced with the technique that *compresses* path information by finding repeated patterns [5]. It would also be possible to, when the memory usage exceeds a threshold, discard the information on subpaths and begin keeping only critical path length.

4 Experiments

We implemented our scheme on top of the concurrent object-oriented language Schematic [7, 8, 12] and tested its usefulness through experiments on a symmetric multiprocessor, the Sun Ultra Enterprise 10000 (UltraSPARC 250MHz × 64, Solaris 2.6). To get the current time we used the **gethrtime** function in the Solaris OS. We tested our scheme on the following application programs.

Function Definition

$f(x_1,\ldots,x_n)$ { l : S } $\longrightarrow$
    $f(x_1,\ldots,x_n,cp)$ {
        el = l;
        et = currTime();
        S'
    }

Send

l : send(r,v); $\longrightarrow$
    t = currTime() - et;
    cp' = addInFrSubPath($cp,\{el,l,t\}$);
    send($r,[v,cp']$);
    et = currTime() - t;

Spawn

l : spawn $f(x_1,\ldots,x_n)$; $\longrightarrow$
    t = currTime() - et;
    cp' = addInFrSubPath($cp,\{el,l,t\}$);
    cp'' = addSpawnSubPath(cp');
    spawn $f(x_1,\ldots,x_n,cp'')$;
    et = currTime() - t;

Receive

l : v = recv(r); $\longrightarrow$
    t = currTime() - et;
    $[v,cp']$ = recv(r);
    cp'' = addCommSubPath(cp');
    if (t + length(cp) < length(cp'')) {
        /* Sender is later */
        cp = cp'';
        el = l;
        et = currTime();
    } else {
        /* Receiver is later */
        et = currTime() - t;
    }

Termination

l : die; $\longrightarrow$
    t = currTime() - et;
    cp' = addInFrSubPath($cp,\{el,l,t\}$);
    printCriticalPath(cp');
    exit();

Fig. 8. Translation rules for instrumentation to compute critical paths.

Prime A prime generator using the Sieve of Eratosthenes. This program consists of two parts, one that divides prime candidates by smaller primes and another that gathers primes into a list. It is unclear in what ratio the two parts contribute to a critical path.

CKY A parallel context-free grammar parser. It is an irregular program in which a number of fine-grain threads synchronize via channels. Like **Prime**, it consists of two parts. One part for lexical rule application and another for production rule application. It is not obvious in what ratio the two parts contribute to a critical path.

Raytrace A raytracer that calculates RGB values of pixels in parallel. After calculating a pixel value, a processor sends it to a buffer object which manages file I/O. The object caches pixel values and flushes them when 64 values are accumulated. Accesses to the buffer object are mutually excluded and hence the object becomes a bottleneck.

The experimental results are shown in Fig. 9, where each line graph has one line showing the execution time of the instrumented program (Inst'ed), one showing the execution time of the uninstrumented program (Original), and one showing the critical path length computed by the instrumented program (CPL). The bar chart compares the execution time, on a single processor, between uninstrumented and instrumented programs. Garbage collection did not happen in any of the experiments.

In all the applications, as the number of processors increased, the execution time of uninstrumented programs got very close to the computed critical path

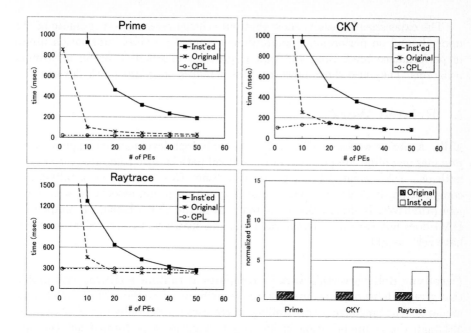

Fig. 9. Measured execution times and computed critical path lengths.

length. The critical path lengths computed with the instrumented programs were extremely close to the best runtimes for the uninstrumented programs. On a single processor the uninstrumented programs were from four to ten times faster than the instrumented ones.

The critical path length we compute can also be used for performance prediction: the performance on large-scale multiprocessors can be predicted from that on a single processor. Cilk's work [1] showed that a program's execution time on P processors is approximately $T_1/P + T_\infty$, where T_1 is the execution time on a single processor and T_∞ is the critical path length.[5] For **Prime** there was almost no difference (5% or less) between the actual execution time and the predicted execution time.

The execution time for each part of **CKY** depends on thread scheduling order. This order changed as the number of processors was increased, and thus the length of the critical path also changed. The critical path length computed for **Raytrace** was greater than the execution time. We do not know why.

We acquired information useful for future tuning. More than 95 percent of the critical path length for **Prime** was due to the time needed for gathering primes into a list. The application of lexical rules made up only four percent of the critical path length for **CKY**.

---

[5] $T_1/P + T_\infty$ is the simplest expression of all the ones they proposed for performance prediction. More sophisticated expressions can be found in their papers.

The experimental results show that our scheme gives a good estimate of the upper bound on performance improvement, that it provides information useful in performance prediction, and that it provides information useful for tuning programs. They also show, however, that it makes programs much slower.

5   Related Work

Cilk. As far as we know, Cilk [1, 6] is the only high-level parallel language that provides an online computation of the critical path. It also computes the critical path of a DAG created in an actual run. Since Cilk is based on a fully-strict computation, it deals only with implicit synchronization associated with fork-join primitives. It should also be noted that in the Cilk scheme, the shape of a DAG does not vary in different runs and the spawn and communication edges have zero weight.

Paradyn. Hollingsworth's work [3] computes at runtime a critical path profile that gives us a list of procedures whose "contributions" to the critical path length are large. The language model Hollingsworth uses is essentially the same as ours, although the communication in his target language is performed not through channels but in the form of message-passing designating a peer. His scheme, like ours, attaches critical path information to messages. But our scheme, unlike his, displays information about the subpaths in a critical path.

Tracefile-Based Offline Schemes. In Dimemas [10] parallel programs call the functions in the instrumented communication library and generate a tracefile after execution. The tracefile contains the parameters and timings of *all* communication operations. A critical path is constructed a posteriori in a tracefile-based *simulation* to which a programmer gives architecture parameters and task-mapping directions. ParaGraph [2] is a tool for visualization of the tracefiles generated by the instrumented communication library PICL. ParaGraph can visualize critical path information. In our scheme the instrumented programs maintain information only on the subpaths that may be included in a critical path. Our on-the-fly computation therefore requires much less memory than do tracefile-based computations.

6   Conclusion and Future Work

We developed an instrumentation scheme that computes critical paths on-the-fly for multithreaded languages supporting thread creation and channel communication. We implemented the scheme and confirmed that the critical path information computed makes it easier to understand the behavior of parallel programs and gives a useful guide to improving the performance of the programs. If our scheme is incorporated into parallel languages, performance tuning in those languages will become more effective.

One area for future work is adaptive optimization utilizing critical path information. A critical path can be shortened if a compiler recompiles with powerful time-consuming optimizations a set of performance-bottleneck procedures whose share of the critical path length is large. We may be able to extend the framework of HotSpot [11] and Self [4], which compiles frequently executed bytecode dynamically, by giving higher compilation priority to bottleneck procedures. Another adaptive optimization exploiting critical path information would be to have a runtime give higher scheduling priority to threads executing a critical path. Our previous work [9] gives heuristics for efficient thread scheduling in synchronization bottlenecks, on which multiple operations are blocked and are likely to be in a critical path. Critical path information makes more adaptive and more reasonable scheduling possible without the help of heuristics. The use of this information might thus become an essential part of adaptive computation.

References

1. M. Frigo, C. E. Leiserson, and K. H. Randall. The Implementation of the Cilk-5 Multithreaded Language. In *Proceedings of the ACM SIGPLAN 1998 Conference on Programming Language Design and Implementation*, pages 212–223, 1998. See also The Cilk Project Home Page http://supertech.lcs.mit.edu/cilk/
2. M. T. Heath and J. E. Finger. Visualizing the Performance of Parallel Programs. *IEEE Software*, 8(5):29–39, 1991.
3. J. K. Hollingsworth. Critical Path Profiling of Message Passing and Shared-memory Programs. *IEEE Transactions on Parallel and Distributed Systems*, pages 1029–1040, 1998.
4. U. Hölzle and D. Ungar. A Third-Generation SELF Implementation: Reconciling Responsiveness with Performance. In *Proceedings of the Ninth Annual Conference on Object-Oriented Programming Systems, Languages, and Applications (OOPSLA '94)*, pages 229–243, 1994.
5. J. R. Larus. Whole Program Paths. In *Proceedings of the ACM SIGPLAN 1999 Conference on Programming Language Design and Implementation (PLDI '99)*, pages 259–269, 1999.
6. C. E. Leiserson, 1999. personal communication.
7. Y. Oyama, K. Taura, T. Endo, and A. Yonezawa. An Implementation and Performance Evaluation of Language with Fine-Grain Thread Creation on Shared Memory Parallel Computer. In *Proceedings of 1998 International Conference on Parallel and Distributed Computing and Systems (PDCS '98)*, pages 672–675, 1998.
8. Y. Oyama, K. Taura, and A. Yonezawa. An Efficient Compilation Framework for Languages Based on a Concurrent Process Calculus. In *Proceedings of Euro-Par '97 Parallel Processing*, volume 1300 of *LNCS*, pages 546–553, 1997.
9. Y. Oyama, K. Taura, and A. Yonezawa. Executing Parallel Programs with Synchronization Bottlenecks Efficiently. In *Proceedings of International Workshop on Parallel and Distributed Computing for Symbolic and Irregular Applications (PDSIA '99)*. World Scientific, 1999.
10. Pallas GmbH. *Dimemas*. http://www.pallas.de/.
11. Sun Microsystems. *The Java HotSpot^{TM} Performance Engine*.
12. K. Taura and A. Yonezawa. Schematic: A Concurrent Object-Oriented Extension to Scheme. In *Proceedings of Workshop on Object-Based Parallel and Distributed Computation (OBPDC '95)*, volume 1107 of *LNCS*, pages 59–82, 1996.

Problem Solving Environment Infrastructure for High Performance Computer Systems

Daniel C. Stanzione, Jr. and Walter B. Ligon III

Parallel Architecture Research Lab
Clemson University
dstanzi@clemson.edu
http://www.parl.clemson.edu/

Abstract. This paper presents the status of an ongoing project in constructing a framework to create problem solving environments (PSEs). The framework is independent of any particular architecture, programming model, or problem domain. The framework makes use of compiler technology, but identifies and addresses several key differences between compilers and PSE. The validity of this model is being tested through the creation of several prototype PSEs, which apply to significantly different domains, and target both parallel computers and reconfigurable computers.

1 Introduction

A number of approaches have been taken to make it simpler to create applications for High Performance Computing (HPC) systems. Existing programming languages have been extended with parallel constructs, notably PVM and implementations of the Message Passing Interface(MPI). The resulting systems remained more difficult to program than sequential computers, and have been referred to as the "assembly language" of parallel computing. New languages have been created in which parallelism is inherent, but to date none of these languages have seen widespread adoption.

In recent years, more and more interest has been paid to the idea of creating problem solving environments (PSEs) for high performance computers. While a number of prototype PSEs have been created, the construction of these PSEs remains largely an ad hoc procedure. The PSE community has repeatedly made calls for infrastructure to be developed which supports the creation of PSEs.

This paper describes ongoing research in creating just such an infrastructure. Presented here is an architecture for the creation of PSEs. The architecture uses a layered model which provides abstractions for the hardware, programming model, mathematical model, and whatever science models are used by a particular problem solving environment. The layered architecture employs a model somewhat similar to an open compiler. The goal is to produce PSEs that provide abstraction bridges, i.e. the PSEs can be used by both computer scientists and users in the domain of the PSE, but each sees the environment through a

J. Rolim et al. (Eds.): IPDPS 2000 Workshops, LNCS 1800, pp. 314–323, 2000.

different level of abstraction. Another goal is to decouple the specification of the application from the target architecture in order to make applications at least somewhat portable between widely varying types of high performance computing systems.

The following sections describe the proposed PSE architecture in more detail. An implementation of the infrastructure proposed by the model is described, followed by a brief look at two PSEs under construction employing this infrastructure which are used to validate the model.

2   The Proposed Model for Problem Solving Environments

This section proposes a model for PSE construction which draws heavily upon compiler technology and concepts from software engineering and object-oriented programming to create a framework from which effective PSEs can be constructed. The fundamental structure of this model is a number of independent tools which interact through a shared open representation of the design in question. A simple block diagram of this model is shown in fig. 1. The tools which access and act on the design are known as *agents*. The design itself is stored in the *Algorithm Description Format* (ADF). The role of the *manager* is to coordinate the actions of the agents, and load and store ADF designs from the *library*.

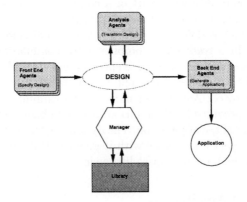

Fig. 1. Basic Structure of an Environment

The basic idea is similar to that of a traditional compiler. A compiler is in essence a set of tools that make a series of transformations on a user's design to transform it into a running application, typically using one or more internal formats to store the design. Agents in this new environment perform many of the same functions. They are involved with specification, analysis, and transformation of designs, and for generating application code from the design. A shortcoming of the compiler model is that once constructed, compilers are difficult to extend. A primary cause if this is that the compiler's internal formats are almost always mysterious. This model attempts to remedy this problem by making the internal format used by all the agents to store, analyze, and transform the design a well-documented open format, ADF. In this sense this architecture can be considered an open compiler for PSEs.

In addition to the open format making the PSE extensible, this model differs from a compiler in three significant ways. A traditional compiler represents an individual program at one level of abstraction; the source language. While some compilers support more than one language for the same back end, this model differs substantially in that it can support multiple *simultaneous* abstractions to represent the same problem. Each agent can present a different abstraction of the same design, and allow the user to interact with only the aspects of the design important to that user or to the task being performed by that particular agent. A second significant difference between PSEs and traditional compilers is that the internal format of compilers represents only information about the computation itself. In the PSE model, the internal format can be used to store many other types of additional information other than what is considered the traditional source program. ADF can be used to represent the data flow within the computation, information about the architecture the application can or has been run on, performance information and results from past runs, monitoring information, higher level functional descriptions of the problem, etc. This flexibility allows the internal format to be used by many of the tools expected in a PSE that are not traditionally part of a compiler, such as experimental and runtime management, steering and monitoring tools.

The third significant difference with the traditional compiler model is that ADF is dynamic. In a traditional compiler, once compilation takes place, the executable is independent of the source specification. In PSEs constructed with this model, the job of the environment is not necessarily complete once the application is generated. Agents can monitor the running design, and use the design description provided by the internal format to back annotate information about the execution of the application. The dynamic nature of the design and the way agents can interact and iterate with it give the PSE the ability to manage the application throughout it's entire life cycle.

2.1   The Layered Architecture

If the model above is accepted as the way in which PSEs are to be constructed, what remains is to provide the integration method through which the various agents in the PSE can interact. This problem is addressed through the creation of a layered architecture, outlined in fig. 2. Agents that exist within a layer share a set of common attributes which exist on the designs on which they operate. Definition of a set of attributes for a particular layer defines that layer of the environment. These agreed upon attributes allow the agents within a layer to interact. The clear definition of attribute sets at each layer also provides the facility for reuse, as agents written for a particular environment can be reused in another as long as that agent's subset of the design attributes exist in each environment.

There are 5 different layers which provide 5 levels of abstraction. In multi-physics environments, or environments targeted at multiple architectures, there may be multiple implementations of particular layers existing concurrently within a single environment. The abstraction hierarchy provides a two-way abstraction, with details about the computation and the computer hidden as you move up the hierarchy, and details about the problem domain and mathematics hidden as you move down.

Fig. 2. PSE Architecture

2.2   Level 0 - Infrastructure

Level 0 is the common or kernel layer, and provides the infrastructure necessary for the creation of any PSE targeted at any domain. Level 0 should provide the following:

- An open, shared, attributed graph format
- A database system for managing and maintaining libraries of designs
- A model for agents (both interactive and autonomous) to interact with the designs and with each other
- A mechanism for creating attribute sets and components

The internal format represents a directed attributed graph. Each graph is composed of an attribute list, and a set of nodes connected by links. Each node consists of an attribute list, and lists of input and output ports. Each of the ports has its own attribute list as well. There are no restrictions placed on attributes in any of the lists; there can be any number of attributes, and they can contain any kind of information. This provides tremendous flexibility in what ADF can be used for. Support for hierarchy need not be explicitly provided, as it can be done by specific environments via the attribute mechanism.

In addition to the design format, an agent API is provided. The API provides mechanisms to create and modify all portions of the graph, and to locate, store and sort the graphs in an environment. The API provides a method for searching designs via any attribute or attribute/value pair within the design. Furthermore, the API allows agents to share an open design, and provide a method for agents synchronize their changes to designs.

A core set of attributes are assigned in level 0 that make it possible to create named attribute sets for use in higher levels, and to check designs and pieces of designs for conformance with particular sets. The attributes on the ports function as an interface description for the design. If two designs comply to the same attribute set, and the port specifications match, then those two designs may be connected. The attribute set mechanism in conjunction with ports and the layered architecture form an extremely versatile component system. When two components are connected using this layered approach, the semantic correctness, mathematical correctness, and if the connection makes sense in terms of the problem domain are all checked by examining the attribute set at each layer.

The attribute set and layered architecture also serve to provide a bridge between abstractions. Users working on the design at any level can work independently of those on other layers. Level 0 provides services similar to many of the distributed object frameworks such as CORBA [2], the Common Component Architecture [3], or DCOM [4]. Level 0 attempts to provide the same support for PSEs equivalent to that which SUIF [1] provides for compilers

2.3   Level 1 - Hardware Abstractions

Level 1 is the "back-end" layer. At this level, abstractions are provided to describe the hardware in a particular HPC target system, and the basic constructs for implementing designs on that system. Level 1 seeks to provide abstractions at the level of an "assembly language" for the architecture in question. Level 1 provides the following services and abstractions:

- A representation of any distinctive characteristics of the compute platform
- Attributes suitable to describe applications on the target system
- A component model for composing code fragments
- Basic data types
- Low-level optimization, analysis, debugging, and code generation agents
- Resource management services
- A library of design constructs

A level 1 implementation for a cluster computer would provide the ability to describe and generate message-passing programs (or, at the very least, collections of processes capable of communication through the cluster's network), the ability to schedule the nodes on the cluster and dispatch jobs, program debugging services, and performance visualization (Data visualization would not be included here, as that would entail knowledge of the context of the problem; however, visualization of processor or memory utilization would be appropriate). A level 1 implementation for a configurable computing target might include a set of hardware macros for basic operations, a set of attributes which described the interconnection of these modules, and agents to perform code generation ,optimization, placement, and routing.

Many existing pieces of software provide services at an abstraction layer suitable for level 1, including message passing or shared memory services such as those provided by PVM [5], or implementations of MPI or OpenMP. The components at this level operate with a level of abstraction similar to that presented in the Interface Description Language [6]. Level 1 would also be the appropriate place to employ metacomputing software.

2.4   Level 2 - Programming Model

Level 2 is the programming model layer. At this level, abstractions are provided which operate at approximately the level of most parallel programming languages. This is accomplished by enforcing some programming style in a somewhat machine-neutral way, for example data parallelism, task parallelism, or object parallelism. An attribute set is required to represent more sophisticated data structures and concepts such as parallel tasks, communication patterns, *Forall* loops, and data distributions.

Agents should be created at this level which provide an interface suitable for use by a good parallel programmer. This level is also the appropriate place for source level optimizations performed by many source-to-source compilers to take place.

The abstraction presented at this level is roughly equivalent to that presented by many parallel languages and programming environments, such as HPF [13], Mentat [7], Data Parallel C [13], Enterprise [10],or Hence [11].

2.5   Level 3 - Mathematics

Level 3 is the layer at which mathematical abstractions are provided. It moves an abstraction level above that of dealing with programs to the level of dealing with a particular class of problems. Typical level 3 abstractions would include geometry, discretization, data range information, and linear algebra constructs such as vectors and matrices.

At level 3, the environments begin to become more customized to a particular problem domain, so not every environment will necessarily include all the abstractions listed above. Services provided at level 3 via agents might include an equation-based interface to the user, or an agent which presents a choice of linear system solvers to apply to the problem. The separation of the mathematical specification of the problem from the science specification also makes it simpler to apply techniques like adaptive mesh refinement to the problem without having to deal with the specifics of the science and unfamiliar terminology. It also allows for some simple sanity checking (i.e. matrix conformity) to be done in a straightforward way.

Level 3 provides a level of abstraction roughly equivalent to that provided by some of the ultra high level languages or general purpose PSEs currently in existence, such as Matlab, SciVis, or Scientific IDL [3]. At this level, existing libraries that are appropriate for the domain can also be integrated with the environment.

2.6   Level 4 - Domain Specific Interface

Level 4 turns the PSE into a domain specific solver. Level 4 provides domain specific abstractions, and a user interface that operates in domain terms. The user interface (simply a collection of agents) translates the user's requirements into a specification usable by the underlying layers. It is at this level that many of the artifacts of creating programs are hidden in favor of solving problems. For instance, an unknown could be represented as an electric current in level 3, a column vector in level 2, and an array in levels 1 and 0. The concept of data flow can be hidden beneath graph templates that present only the steps necessary to solve the problem, in the order that the user would normally solve the problem.

3   Implementation

In this section a quick look is provided at an implementation of the level 0 toolkit, and at two PSEs currently being constructed to use this toolkit, one which supports electromagnetics application on clusters of workstations, and a second which supports image processing applications on reconfigurable computers. Due to space constraints, details of the attribute sets used at each level in each environment have been omitted. However, more detailed information about the toolkit and each of the environments, including examples and other environments not mentioned here, can be found at [15] and [16].

3.1   CECAAD

The Clemson Environment for Computer Aided Application Design (CECAAD) is a prototype implementation of level 0 as described in the previous section. CECAAD is an environment toolkit, and is the basis for the construction of the PSEs described at the end of this paper.

CECAAD consists of the ADF, the manager, the launcher, and a set of core agents. At the core of CECAAD is the ADF. ADF is the internal format for representing a directed attributed graph, as described in the previous chapter. The ADF manager provides synchronization of the actions of the agents on ADF designs and all I/O functions related to the library of designs. A small set of attributes are included in level 0 that implement the attribute set mechanism previously described.

Level 0 also contains several agents which are either useful for all environments or serve as a basis for the creation of more complicated domain specific agents. The ADF editor is a level 0 agent which provides a graphical view of ADF designs and allows the creation or changing of any ADF construct. The ADF text translator is another level 0 agent that translates ADF designs to and from a simple text-based language. The translator provides a basis for the creation of agents which could integrate existing languages for expressing parallel or scientific computation into CECAAD based environments. The Partition Agent attaches a "task" attribute to each node, and provides a graphical interface to group nodes into particular tasks.

3.2   An Electromagnetics Environment for Cluster Computers

A CECAAD-based PSE is currently being created to allow for parallel solution of integral equation method of moment problems in electromagnetics using Beowulf-class cluster computers. Problems of this type typically have roughly the same parallel structure, though numerically they can be very different.

Level 1 abstracts the cluster computer itself, and provides an abstraction for message passing between the nodes in the cluster. The hardware abstraction includes attributes for representing a heterogeneous cluster with various network topologies. A code generation agent uses the level 1 specification of the application along with the specification of the hardware to generate a PVM or MPI-based application. Another agent gathers information from the running program and adds performance information to the specification of both the hardware and the application, which is used by static and dynamic load balancing agents. The level 2 abstraction is of a data parallel programming model. Since the target problems have a fairly static data flow structure, an editor agent using the level 2 abstraction is employed by a user with computer science expertise to create ADF templates of the target problems.

Level 3 provides mathematical abstractions. At this level, a discretization agent allows the user to select basis functions to represent the geometry of the problem to be solved, and a linear system solver agent allows the user to select an appropriate solution method and convergence criteria. Level 4 provides the domain specific interface. At this level, the user employs a geometry editor to graphically define the problem's geometry in terms of insulators, conductors and dielectric materials. The user uses another agent to choose the quantity to solve for, and either select the equation to be used from a library, or to provide custom code (usually Fortran) that is to be used in conjunction with the geometry to fill the matrix (The user supplied code is wrapped in an ADF node that is then bound to the template supplied by the computer scientist).

This environment allows the parallelism and the details of the cluster to be hidden from the electromagnetics user, and allows the computer science user to collaborate without ever examining any of the electromagnetics involved. Applications could be ported to new architectures, or underlying models (such as shared memory) without any changes to the electromagnetics users specification of the problem.

3.3   An Image Processing Environment for Reconfigurable Computers

Reconfigurable Computers based on Field Programmable Gate Array (FPGA) technology are an emerging class of HPC system with the potential for providing enormous performance. Unfortunately, the problems associated with generating applications for this type of platform are even more daunting than those in parallel computing, as "programming" a reconfigurable system using conventional methods is akin to ASIC design in complexity. RCADE [14](the Reconfigurable Computing Application Development Environment) is a CECAAD based environment for FPGA based computing systems, which allows the user to work at roughly the level of a visual high level language to generate image processing applications. The user connects components which represent arithmetic and image filtering functions and basic loop constructs together in a data flow graph. Each component represents a pre-placed logic macro for the target platform.

At level 1, RCADE abstracts the computing platform, and uses attributes to represent the performance and interconnections between the FPGAs, as well as the routing resources between the logic blocks within the devices. At level 2, a data flow programming model is imposed which introduces the components and basic data types. Level 3 views the components from an equation based perspective, showing the mathematical transformations on the data, and level 4 adds an image processing layer by representing components in terms of the filtering operation they perform on the data.

A small library of RCADE components has been implemented on Xilinx 4000 series FPGAs. Currently, existing agents include a VHDL code generator and a partitioning tool based on level 1 specifications, a throughput analysis agent and pipeline balancing agent based on level 2 with latency information drawn from level 1, a simple component selection agent to match level 1 macros with level 2 specifications, and a data range analysis tool which examines level 3 information in order to allow the user to adjust the precision throughout the application in order to generate chip area savings.

Among the agents currently being developed for RCADE include automatic spatial and temporal partitioning, a macro generator which will use level 1 information to resize the hardware macros for components to specific geometries and bit precisions, and a graphical user interface to the level 4 specification. Eventually, the RCADE back end will be fused with a web-based interface for doing remote-sensing/image processing applications on parallel computers which currently exists at Clemson.

4   Conclusion

As PSEs become more prevalent to perform steadily larger and more complex scientific computation on steadily more advanced computing platforms, the need to better understand the construction of these PSEs will also increase. This paper has presented a proposal for a PSE architecture and infrastructure, which leverages compiler technology and current approaches to creating high-performance applications. This infrastructure provides multiple abstractions to multiple groups of users, and allows these users to collaborate via these abstractions. Several different environments using different types of computers and different problem domains are being created to show the utility and versatility of this infrastructure.

Future work includes the completion of the prototype environments to more thoroughly test the infrastructure, and a public release of the CECAAD implementation with a more robust agent collaboration model.

References

1. Robert P. Wilson, Monica S. Lam, and John L. Hennessy et al. Suif: An infrastructure for research on parallelizing and optimizing compilers. Technical report, Computer Systems Laboratory, Stanford University, 1996.
2. OMG et al.   Corba components: Joint revised submission.   Technical report, Dept. of Computer Science,Rice University, December 21 1998. ftp://ftp.omg.org/pub/docs/orbos.98-12-02.pdf.
3. Rob Armstrong, Dennis Gannon, Al Geist, and et al. Toward a common component architecture for high-performance scientific computing. In *Proceedings of the 8th IEEE Int'l Symposium on HPDC*, pages pp.115–132. IEEE Computer Society, IEEE Computer Society Press, Nov. 1999.
4. R. Sessions. *COM and DCOM: Microsoft's Vision for Distributed Objects*. John Wiley & Sons, 1997.
5. A. Beguelin, JJ Dongarra, G.A. Geist, R.Manchek, and V.S. Sundaram. A users' guide to the pvm parallel virtual machine. Technical Report ORNL/TM-11826, Oak Ridge National Laboratory, July 1991.
6. Richard Snodgrass. *The Interface Description Language:Definition and Use*. Computer Science Press, 1989.
7. A. S. Grimshaw. Easy to use object-oriented parallel programming with mentat. *IEEE Computer*, pages 39–51, May 1993.
8. HyperParallel Technologies.   Hyper c parallel programming language. http://www.meridian-marketing.com/HYPER_C/index.html, June 1999.
9. Jagannathan Dodd Agi. Glu: A high level system for granular data-parallel programming. *Concurrency: Practice and Experience*, 1995.
10. Schaeffer, Szafron, and Duane Lobe an Ian Parsons. The enterprise model for developing distributed applications. *Parallel and Distributed Technology*, 1995.
11. A. Beguelin, J.J. Dongarra, G.A. Geist, R. Manchek, and V. S. Sunderam. Visualization and debugging in a heterogeneous environment. *Computer*, 26(6):88–95, June 1993.
12. P. Bellows and B. Hutchings, "JHDL - An HDL for Reconfigurable Systems", *Proceedings of FCCM '98*, April, 1998.
13. Francois Bodin, Thierry Priol, Piyush Mehotra, and Dennis Gannon, "Directions in Parallel Programming: HPF, Shared Virtual Memory, and Object Parallelism in pC++", *Journal of Scientific Computing*, Vol. 2, no. 3, pp 7-22, June, 1993.
14. Ligon, Stanzione, et al, "Developing Applications in RCADE", *Proc of the IEEE Aerospace Conf*, March 1999.
15. The Clemson PSE web site,URL: *http://www.parl.clemson.edu/pse/*, 2000.
16. The RCADE web site, URL: *http://www.parl.clemson.edu/pse/rcade*, 2000.

Combining Fusion Optimizations and Piecewise Execution of Nested Data-Parallel Programs

Wolf Pfannenstiel

Technische Universität Berlin
wolfp@cs.tu-berlin.de

Abstract. Nested data-parallel programs often have large memory requirements due to their high degree of parallelism. Piecewise execution is an implementation technique used to minimize the space needed. In this paper, we present a combinination of piecewise execution and loop-fusion techniques. Both a formal framework and the execution model based on threads are presented. We give some experimental results, which demonstrate the good performance in memory consumption and execution time.

1 Introduction

Nested data-parallelism is a generalization of flat data-parallelism that allows arbitrary nesting of both aggregate data types and parallel computations. Nested data-parallel languages allow the expression of large amounts of data parallelism. The flattening transformation introduced by Blelloch & Sabot [2] exposes the maximum possible data parallelism of nested parallel programs. As parallelism is expressed by computations on vectors, the memory requirements of flattened programs are proportional to the degree of parallelism. Most implementations, e.g. NESL [1], use libraries of vector operations which are targeted by the compiler. While this approach encapsulates machine-specific details in the library and facilitates code generation, it has at least two drawbacks [4]. First, it further increases memory consumption by introducing many temporary vectors. Second, code-optimizations cannot be applied across library functions. Thus, high memory consumption is one of the most serious problems of nested data-parallel languages.

Palmer, Prins, Chatterjee & Faith [5] introduced an implementation technique known as *Piecewise Execution*. Here, the vector operations work on vector pieces of constant size only. In this way, low memory bounds for a certain class of programs are achieved, but the management of the pieces requires an interpreter.

To tackle both drawbacks of the library approach, Keller & Chakravarty [4] have proposed to abandon the library and to use a new intermediate compiler language $\mathcal{L}_{DT}$ instead, whose main feature is the separation of local computations from global operations (communication and synchronization) using the idea of *Distributed Types*. Local computations can be optimized, the most important optimization being the fusion of consecutive loops.

In this paper, we propose combining piecewise execution and loop-fusion into a single framework. We extend the compiler language $\mathcal{L}_{DT}$ by *Piecewise Types*.

J. Rolim et al. (Eds.): IPDPS 2000 Workshops, LNCS 1800, pp. 324–331, 2000.

All optimizations possible in $\mathcal{L}_{DT}$ can still be applied to our extended language. Our implementation uses a self-scheduled thread model, which was introduced in [6].

The rest of the paper is organized as follows. Sect. 2 gives a brief summary of related work. Sect. 3 describes the combination of fusion and piecewise execution. The multi-threaded execution model is looked at briefly in Sect. 4 and we present results of some experiments. Finally, Sect. 5 gives an overview of future work.

2   Related Work

Our work was inspired by two key ideas: *Piecewise Execution*, described in [5], and *Distributed Types*, as proposed in [3].

Piecewise Execution. Piecewise execution is based on the observation that many nested data-parallel programs contain pairs of generators and accumulators. Generators produce vectors that are much larger than their arguments. Accumulators return results that are smaller than their input. The NESL programs in Fig. 1 are examples of matching generator/accumulator pairs. The operation [s:e] enumerates all integers from s to e, plus_scan calculates all prefix sums of a vector, {x * x : x in b} denotes the elementwise squaring of of b, and sum adds up all vector elements in parallel. SumSq(1,n) returns the sum of all squares from 1 to n. The excess parallelism must be sequentialized to

```
function SumSq (s,e) =
  let
      a = [s:e];
      c = {x * x : x in a}
  in
      sum(c);
```

```
function SumSqScan (s,e) =
  let
      a = [s:e];
      b = plus_scan (a);
      c = {x * x : x in b}
  in
      sum(c);
```

Fig. 1: Example programs in NESL

match the size of the parallel machine. However, if the generator executes in one go, it produces a vector whose size is proportional to the degree of parallelism. Then, memory consumption may be so high that the program cannot execute.

In piecewise execution, the computation exposes a consumer/producer pattern. Each operation receives only a piece, of constant size, of its arguments at a time. The operation consumes the piece to produce (a part of) its result. After the current piece is completely consumed, the next piece is requested from the producer. Once a full piece of output has been produced, it is passed to the consumer as input. Large vectors never exist in their full length, so piecewise execution enables larger inputs to be handled. However, some means of control are needed to keep track of the computation. In [5], an interpreter is employed to manage control flow. Piecewise execution is a tradeoff between space and time.

The memory consumption can be reduced dramatically, while the computation times are likely to increase owing to the overhead associated with serializing flattened programs. Indeed, interpretation involves a significant overhead.

Limits of Piecewise Execution. Some operations are not well suited for piecewise execution, e.g. `permute`. Potentially the complete data vector must be known to produce the first part of the result, because the first index may refer to the last element of the input. Whenever one of these operations occurs in a program, piecewise execution is not possible without buffering more than one piece at a time. A more detailed discussion of limitations can be found in [5].

Distributed Types. In the library-approach, the compilation of nested data-parallel programs is finished after the flattening transformation. The parallel work is delegated to a set of library functions. Optimizations across functions are not possible owing to the rigid interfaces of the library. Keller & Chakravarty [4] decompose library functions into two fractions. First, computations with purely local meaning are extracted, i.e. sequential code that references only local memory. Second, all other computations, such as interprocessor communication or synchronization, are declared as global operations. Adjacent local computations build a block of sequential code to which processor-local code optimizations can be applied. Blocks of global operations can be optimized as well, e.g. two send operations may be combined to form just one.

The intermediate language $\mathcal{L}_{DT}$ is introduced featuring *Distributed Types*, a class of types used to distinguish local from global values. A local value consists of one value per processor and is denoted by $\langle\langle.\rangle\rangle$. The components have only local meaning – their layout and size are known only on the local processor. The function $split\_scalar : \alpha \to \langle\langle\alpha\rangle\rangle$ transforms a scalar of type α into a local value whose components all have the specified value (e.g. by broadcasting it). The function $split\_agg : [\alpha] \to \langle\langle[\alpha]\rangle\rangle$ takes a global vector and splits it into chunks. To transform a local vector into a global one, $join\_agg : \langle\langle[\alpha]\rangle\rangle \to [\alpha]$ is used. The higher-order operation $\langle\langle.\rangle\rangle$ takes a function f and applies it to all components of a local value. To transform a program, the function boundaries of library operations are removed by decomposing and inlining their code. All local computations are represented by instances of two canonical higher-order functions.

1. **Loop** : $(\alpha_1 \times \alpha_2 \to \beta) \times (\alpha_1 \times \alpha_2 \to \alpha_2) \times (\alpha_1 \times \alpha_2 \to Bool) \to [\alpha_1] \times \alpha_2 \to [\beta] \times \alpha_2$
2. **Gen** : $(\alpha_2 \to \beta) \times (\alpha_2 \to \alpha_2) \times (\alpha_2 \to Bool) \to Int \times \alpha_2 \to [\beta] \times \alpha_2$

Loop represents calculations on vectors. It receives a computation function, an accumulating and a filter function as input. The filter can be used to restrict the output to elements for which the function returns *True*. **Gen** is similar to **Loop** except that it loops over an integer rather than a vector, i.e. it can be used to generate vectors. The *split* and *join* operations are used to embed the values into the distributed types. The interface of the vector operations remains the same. When two consecutive operations are split up, often the final *join* of the first operation and the initial *split* of the second one can be eliminated, leaving larger blocks of local computations.

Fusion Optimizations. Deforestation is a well-known technique for fusing sequential computations on aggregate data structures. As local computation-blocks in $\mathcal{L}_{DT}$ form sequential code, these techniques can be applied here, too. A number of transformations to fuse adjacent **Loop** and **Gen** constructs are presented in [3]. The main benefits are fewer vector traversals and fewer (or maybe no) intermediate vectors. Consider the function SumSq given in Fig. 1. In the library approach, three functions are used to implement the program, namely enumerate (corresponding to [s:e]), vector_mult (elementwise multiplication) and sum. Transforming the functions into $\mathcal{L}_{DT}$, three blocks of code are formed. In the first part, parameters are split into local values. The last part is the global reduction of all partial sums. All computations in between, which realize the computations inside the three vector operations, form one local block and can be fused into a single **Gen** construct. Thus, fusion combines generation and reduction of the vectors such that no vector is actually created, rendering piecewise execution superfluous here.

Limits of Fusion. Loop and **Gen** constructs cannot be fused across global operations, because the purely local semantics of argument values is destroyed by the global operation. Consider SumSqScan in Fig. 1. Here, plus_scan is split into three parts, the middle part being a global operation originating from plus_scan, which propagates partial sums across processors. Unlike in the previous example, not all local code blocks can be fused, because the propagation forms a barrier between the blocks. Whenever a global operation occurs between pairs of **Loop** or **Gen** constructs, fusion is not possible.

3   Combining Fusion and Piecewise Execution

To combine the benefits of piecewise execution and fusion, we extend $\mathcal{L}_{DT}$ by *Piecewise Types*. We call the extended language $\mathcal{L}_{PW}$. It still contains all features of $\mathcal{L}_{DT}$, as we want to support all its optimizations. We use <.> to denote the type constructor for piecewise types. We provide a higher-order function <.>, which embeds a computation into a piecewise execution context. If we have $f : \alpha_1 \times \ldots \times \alpha_n \to \beta_1 \times \ldots \times \beta_m$, then $<f>$ has type $<\alpha_1> \times \ldots \times <\alpha_n> \to <\beta_1> \times \ldots \times <\beta_m>$. This means $<f> (v_1, \ldots, v_n)$ denotes the piecewise execution of f, where all the arguments v_i must be of a piecewise type. To make a value ready for piecewise execution, we supply a function $pw\_in : \alpha \to <\alpha>$. As only certain parts of programs are to be executed in a piecewise manner we provide $pw\_out : <\alpha> \to \alpha$, which transforms a value of piecewise type into a value of its original type.

Fig. 2 (left) shows an abstract and already fused definition of SumSqScan. (The definitions of the instance functions f, g, k, f', g' and k' are not given since they are not needed for understanding the transformation.) The **Gen** construct implements the `enumerate` generator plus the first local part of `plus_scan` (formerly a **Loop**). The function $propagate_+$ is the global operation that hinders full fusion. The **Loop** construct realizes the second part of `plus_scan`, the elementwise squaring plus the local part of `sum`. Finally, the local values are combined to give the global result by $join_+$, which globally sums up all values. The value $accg$ is the initial accumulating value needed for **Gen**. (Its definition is omitted for simplicity's sake.)

$function\ SumSqScan'(sg, eg) =$
  $\textbf{let}$
    $n\quad\ = split\_scalar(sg - eg)$
    $acc\ = split\_scalar(accg)$
    $(v, b) = \langle\!\langle \textbf{Gen}\ (f, g, k)\rangle\!\rangle(n, acc)$
    $c\quad\ = propagate\_ + (b)$
    $d\quad\ = \langle\!\langle \pi_2 \circ \textbf{Loop}\ (f', g', k')\rangle\!\rangle(v, c)$
  $\textbf{in}$
  $join\_ + (d)$

$function\ SumSqScan'_{pw}(sg, eg) =$
  $\textbf{let}$
    $n\quad\ = pw\_in(sg - eg)$
    $np\quad = \ll split\_scalar \gg (n)$
    $acc\quad = pw\_in(accg)$
    $accp\ = \ll split\_scalar \gg (acc)$
    $(vp, bp) = \ll \langle\!\langle \textbf{Gen}\ (f, g, k)\rangle\!\rangle \gg (np, accp)$
    $cp\quad = \ll propagate\_ + \gg (bp)$
    $dp\quad = \ll \langle\!\langle \pi_2 \circ \textbf{Loop}\ (f', g', k')\rangle\!\rangle \gg (vp, cp)$
    $d\quad\ = \ll join\_ + \gg (dp)$
  $\textbf{in}$
  $pw\_out(d)$

Fig. 2: Fused (left) and piecewise (right) versions of `SumSqScan`

To execute the function in a piecewise fashion, we lift the operations to piecewise types, convert the arguments into piecewise values first, and finally transform the piecewise output into ordinary values again. The transformed program in $\mathcal{L}_{PW}$ notation is shown in Fig. 2 (right). The input and output have the same type as in the original version, i.e. the caller is not affected.

The canonical representation of vector computations using **Loop** and **Gen** is well-suited for an automatic analysis of memory-critical program patterns. A generator is always a **Gen** with filter function that is not unconditionally false ($\lambda x.\lambda a.False$). (Furthermore, it must not be followed by a projection of only the accumulating value (π_1).) An accumulator is a **Loop** that has a truely restricting filter function (i.e. not $\lambda x.\lambda a.True$). An automatic analysis and transformation of such program fragments remains to be developed.

4   Implementation and Benchmarks

The piecewise behavior of program fragments is realized by employing threads, which simulate the producer/consumer behavior in a couroutine-like fashion. Viewing the program as a DAG, the sink node calculates the overall result. To do so, it requests a piece of input from the node(s) on which it depends. The demand is propagated until a node is reached that still has input to produce its next piece of output. Initially, only the source nodes hold the program input. Whenever a thread completes a piece of output, it suspends and control switches to the consumer node. If a thread runs out of input before a full piece of output has been built, it restarts (one of) its producers. Control moves up and down in the DAG self-scheduled until the last node has consumed all its input by producing the last piece of the overall result. A detailed description of the execution model can be found in [7]. To enable piecewise execution, the underlying thread model needs to provide only a processor-local switching protocol including control-flow mechanisms and local data-exchange among threads. The model itself allows only sequential execution as there are no means of communication or synchronization among different processors. However, the model does not pose restrictions on the operations executed inside threads, so e.g. the use of message passing libraries like MPI on distributed memory machines is possible. (On SMP machines, semaphores or signals may be employed to coordinate and synchronize threads.)

We can adopt the library approach by realizing each library function as one thread (on each processor). These threads run synchronously working on the same data-parallel operation. Of course, the code must be enriched by control statements to realize the switching behavior for piecewise execution. Communication and synchronization among processors are realized by means of a message passing library. If consecutive function calls have the same data production/consumption rates, they can be encapsulated into one thread. It is also possible to fuse the original program as far as possible first and then embed the remaining code into threads, attaching the piecewise control-structures.

We transformed a number of examples manually and implemented them on a Cray T3E. The programs are written in C in SPMD style using the Stack-Threads library [8] to realize our thread model. The StackThreads mechanisms have a purely local semantics. Communication and synchronization among processors is realized using the Cray shmem communication library. We implemented combinations of library and fused code with piecewise execution, exploiting the generality of our thread model.

One of the examples implemented is the Line-of-Sight algorithm, which determines all objects that are visible from a specified observation point, given the altitudes of the objects lying in the observer's viewing direction. The altitude of the objects might be given by a height function of X and Y coordinates. The algorithm would then take two points in the 2D plane as the observation point and focus, respectively, and calculate which objects on the connecting line are visible using a specified resolution between points. Defined in this way, the algorithm exhibits the typical generator/accumulator pattern. In Fig. 3, running times are shown for two piecewise program versions par_pw_norm and par_pw_fusion, the first of which being an adaption of the library version and the latter a fused variant. The performance depends heavily on the piece size chosen. If the output is small, we find a pattern that is very similar to results we observed for other programs (e.g. SumSqScan), too: a piece size (per processor) below 500 is so small that the overhead for switching dominates computation times resulting in high execution times. The range between 1K and 3K gives the best running times. The sweet spot at 1K elements corresponds to the size of the second-level cache on the Alpha processors. Above 3K, execution times rise significantly because the vector elements begin stepping on one another in the cache. Beyond 6K,

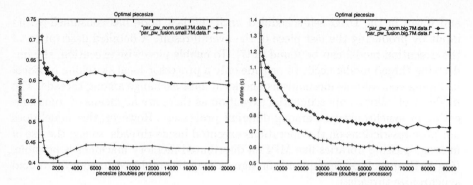

Fig. 3: Optimal piece size for Line-of-Sight with small and large output

the performance improves again slowly owing to the decreased thread overhead. However, the time never again drops below the minimum time attained for small piece sizes. If the output is large, pw_out (necessary for assembling the piecewise generated result) has a big communication overhead, which is worse if the piece size is small. This overhead offsets the improved cache usage and dominates the running times of the piecewise program versions. Here, the bigger the piece size, the better the performance. We measured the absolute speedups of four different versions. The piece sizes for the piecewise programs were set to the best values determined in the previously. The results are shown in Fig. 4. The com-

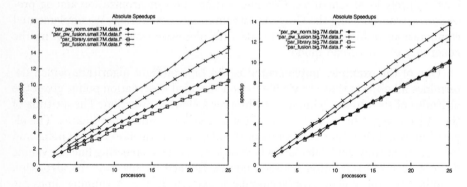

Fig. 4: Absolute speedups of Line-of-Sight with small and large output

bination of fusion and piecewise techniques yields the fastest results if there are few visible points. The fused version (par_fusion) is slightly slower. The plain library-approach (par_library) achieves the worst runtime. Using piecewise execution with one thread per library function slightly increases the speedups. The optimized use of the cache working on small pieces of the memory would appear to compensate the multithreading overhead. The improvements of fusion and piecewise execution seem to mix well. Both reduce memory requirements in an orthogonal way. For large output sizes, owing to the communication overhead of pw_out, par_fusion performs better than par_pw_fusion, and par_pw_norm is not faster than par_library anymore. However, the essential benefit from piecewise execution can be seen in both cases. The piecewise versions can run on any number of processors. For small output, both par_library and par_fusion need at least four processors to handle 7 million objects. For many output values, par_library needs six processors and par_fusion five to execute.

The performance results for SumSqScan are similar to those for Line-Of-Sight with small output. Piecewise execution combined with fusion gives the best results [6]. Experimental results of further examples and more detailed analysis can be found in [7].

5   Conclusion and Future Work

We have combined piecewise execution with fusion optimization expressed in a special intermediate language. We use an improved implementation technique for piecewise execution based on cost-effective multithreading. Piecewise execution does not necessarily mean increasing runtime. On the contrary, the combination of fusion and piecewise execution resulted in the best performance for typical examples. Piecewise execution allows us to execute a large class of nested data-parallel programs that could not normally run owing to insufficient memory.

We intend to develop transformation rules that automatically find and transform program fragments suitable for piecewise execution to implement them in a compiler.

References

1. G. E. Blelloch. NESL: A nested data-parallel language. Technical report, School of Computer Science, Carnegie Mellon University, 1995.
2. G. E. Blelloch and G. Sabot. Compiling collection-oriented languages onto massively parallel computers. *Journal of Parallel and Distributed Computing*, 8(2):119–134, 1990.
3. G. Keller. *Transformation-based Implementation of Nested Data Parallelism for Distributed Memory Machines*. PhD thesis, Technische Universität Berlin, 1999.
4. G. Keller and M. M. T. Chakravarty. On the distributed implementation of aggregate data structures by program transformation. In *HIPS '99*. IEEE CS, 1999.
5. D. Palmer, J. Prins, S. Chatterjee, and R. Faith. Piecewise execution of nested data-parallel programs. In *LCPC '95*. Springer, 1996.
6. W. Pfannenstiel. Piecewise execution of nested parallel programs — a thread-based approach. In P. Amestoy, P. Berger, M. Daydé, I. Duff, V. Frayssé, L. Giraud, and D. Ruiz, editors, *EuroPar'99*, LNCS 1685, pages 445–449. Springer, 1999.
7. W. Pfannenstiel. Thread-based piecewise execution of nested data-parallel programs: Implementation and case studies. Technical Report 99-12, TU Berlin, 1999.
8. K. Taura and A. Yonezawa. Fine-grain multithreading with minimal compiler support - a cost effective approach to implementing efficient multithreading languages. In *PLDI '97*. ACM, 1997.

Declarative concurrency in Java

Rafael Ramirez and Andrew E. Santosa

National University of Singapore
School of Computing
S16, 3 Science Drive 2, Singapore 117543
{rafael,andrews}@comp.nus.edu.sg
Tel. +65 8742909, Fax +65 7794580

Abstract. We propose a high-level language based on first order logic
for expressing synchronization in concurrent object-oriented programs.
The language allows the programmer to *declaratively* state the system
safety properties as temporal constraints on specific program points of
interest. Higher-level synchronization constraints on methods in a class
may be defined using these temporal constraints. The constraints are
enforced by the run-time environment. We illustrate our language by expressing synchronization of Java programs. However, the general underlying synchronization model we present is *language independent* in that
it allows the programmer to glue together separate concurrent threads
regardless of their implementation language and application code.

1   Introduction

The task of programming concurrent systems is substantially more difficult than
the task of programming sequential systems in respect of both correctness (to
achieve correct synchronization) and efficiency. One of the reasons is that it is
very difficult to separate the concurrency issues in a program from the rest of
the code. Synchronization concerns cannot be neatly encapsulated into a single unit which results in their implementation being scattered throughout the
source code. This harms the readability of programs and severely complicates
the development and maintenance of concurrent systems. Furthermore, the fact
that program synchronization concerns are intertwined with the rest of the code,
also complicates the formal treatment of the concurrency issues of the program
which directly affects the possibility of formal verification, synthesis and transformation of concurrent programs. In Java, for instance, the problem of writing
multi-threaded applications is that synchronization code ensuring data integrity
tends to dominate the source code completely. This produces code difficult to
understand and modify. To illustrate the problem we will consider in Section 4
the implementation of a *bounded stack* data type. A basic specification without
synchronization requires only a few lines of code whereas the specification with
synchronization code added is even too complicated to fit onto a single page, its
readability is very poor and it is extremely difficult to reason formally about its
correctness.

J. Rolim et al. (Eds.): IPDPS 2000 Workshops, LNCS 1800, pp. 332-339, 2000.

We believe that the system concurrency issues are best treated as orthogonal to the system base functionality. In this paper we propose a first order language in which the safety properties of concurrent programs are declaratively stated as constraints. Programs are annotated at points of interest so that the run-time environment enforces specific temporal constraints between the visit times of these points. The declarative nature of the constraints provides great advantages in reasoning, deriving and verifying concurrent programs [4]. Higher-level constraints on class methods may be defined using the temporal constraints on program points. The constraints are *language independent* in that the application program can be specified in any object-oriented language. The model has a procedural interpretation which is based on the *incremental* and *lazy* generation of constraints, i.e. constraints are considered only when needed to reason about the execution order of current events.

Section 2 describes some related work. Section 3 presents the language that we propose for specifying the synchronization constraints of concurrent object-oriented programs. In Section 4 we illustrate the use of the language by implementing a bounded stack data type. Section 5 mentions some implementation issues and finally Section 6 summarizes the contributions and indicates some areas of future research.

2   Related work

2.1   Concurrent object-oriented programming

Various attempts have been done by the object-oriented community to separate concurrency issues from functionality. Recently, some researchers have proposed *aspect-oriented programming* (*AOP*) [7] which encompasses the separation of various program "aspects" (which include synchronization) into codes written in "aspect languages" specific for the aspects. The aspect programs are later combined with the base language program using an *aspect weaver*. In this area, the closest related work is the work by De Volder and D'Hondt [15]. Their proposal utilizes a full-fledged logic programming language as the aspect language. In order to specify concurrency issues in the aspect language, basic synchronization declarations are provided which increase program readability. Unfortunately, the declarations have no formal foundation. This reduces considerably the declarativeness of the approach since correctness of the program concurrency issues directly depend on the implementation of the the declarations.

Closer to our work are *path expressions* (e.g., PROCOL [2]) and constructs similar to *synchronization counters* (e.g., Guide [9] and DRAGOON [1]). These proposals, as ours, differ from the AOP approach in that the specification of the system concurrent issues are part of the final program. Unfortunately, synchronization counters have limited expressiveness since it is not possible to order methods explicitly. Path expressions are more expressive in this respect but it cannot express some important synchronization (e.g., producers-consumers) without embedding guards that increases complexity.

An important issue is that in all of the other proposals mentioned above, method execution is the smallest unit of concurrency. This is impractical in actual concurrent programming where we often need finer-grained concurrency.

2.2   Temporal constraints

Most of the previous work on temporal constraints formalisms has concentrated on the specification and verification of real-time systems, e.g. RTL [6] and Hoare logic with time [13]. In addition, there has been research on language notations that consider the synthesis of real-time programs (e.g. TCEL [5], CRL [14], TimeC [10] and Tempo [3]). Tempo is the closest related work. It is a declarative concurrent programming language in which processes are explicitly described as partially ordered set of events. Here, we consider a language in which events are associated with programs points of interest in order to specify the safety properties of a collection of objects. This is done by constraining the execution order of the events by imposing temporal constraints on them.

3   Logic programs for concurrent programming

In this section we describe the logic language which we propose for stating the safety properties of concurrent object-oriented systems. The main emphasis in the language is for it to be declarative on the one hand, and amenable to execution on the other. Unlike other approaches to concurrent programming, our proposal is concerned only with the specification of the synchronization issues of a system. This is, the application functionality is abstracted away and hence can be written in any conventional object-oriented programming language.

3.1   Events and constraints

Many researchers, e.g. [8,11], have proposed methods for reasoning about temporal phenomena using partially ordered sets of events. Our approach to concurrent programming is based on the same general idea. The basic idea here is to use a constraint logic program to represent the (usually infinite) set of constraints of interest. The constraints themselves are of the form $X < Y$, read as "X precedes Y" or "the execution time of X is less than the execution time of Y", where X and Y are events, and $<$ is a partial order.

The constraint logic program is defined as follows[1]. Constants range over events classes $E, F, \ldots$ and there is a distinguished (postfixed) functor $+$. Thus the terms of interest, apart from variables, are $e, e+, e++, \ldots, f, f+, f++, \ldots$. The idea is that e represents the first event in the class E, $e+$ the next event, etc. Thus, for any event X, $X+$ is implicitly preceded by X, i.e. $X < X+$. We denote by $e(+N)$ the N-th event in the class E. Programs facts are of the form $p(t_1, \ldots, t_n)$ where p is a user defined predicate and the t_i are ground terms. Program rules are of the form $p(X_1, \ldots, X_n) \leftarrow B$ where the X_i are distinct variables and B a rule body whose variables are in $\{X_1, \ldots, X_n\}$. A program is a finite collection of rules and is used to define a family of partial

---

[1] For a complete description, see [12]

orders over events. Intuitively, this family is obtained by unfolding the rules with facts indefinitely, and collecting the (ground) constraints of the form $e < f$. Multiple rules for a given predicate symbol give rise to different partial orders. For example, since the following program has only one rule for p:

$p(e, f)$.
$p(E, F) \leftarrow E < F, p(E+, F+)$.

it defines just one partial order $e < f$, $e+ < f+$, $e++ < f++$, ... In contrast,

$p(e, f)$.
$p(E, F) \leftarrow E < F, p(E+, F+)$.
$p(E, F) \leftarrow F < E, p(E+, F+)$.

defines a family of partial orders over $\{e, f, e+, f+, e++, f++, e+++ \ldots\}$. We will abbreviate the set of clauses $H \leftarrow Cs_1$, ..., $H \leftarrow Cs_n$ by the clause $H \leftarrow Cs_1; \ldots; Cs_n$ (disjunction is specified by the disjunction operator ';').

Example 1. An example discussed in almost every textbook on concurrent programming is the producer and consumer problem. The problem considers two types of processes: producers and consumers. Producers create data items one at a time which then must be appended (represented by event *append*) to a buffer. Consumers remove (represented by event *remove*) items from the buffer and consume them. If we assume an infinite buffer, the only safety property needed is that the consumer never attempts to remove an item from an empty buffer. This property can be expressed by

$p(append, remove)$.
$p(X, Y) \leftarrow X < Y, p(X+, Y+)$.

A more practical *bounded buffer* can store only a finite number of data elements. Thus, an extra safety property is that the producer attempts to append items to the buffer only when the buffer is not full. For instance, this safety property for a system with a buffer of size 5 can be expressed by

$p(remove, append(+5))$.
$p(X, Y) \leftarrow X < Y, p(X+, Y+)$.

3.2   Markers and events

In order to refer to the visit times at points of interest in the program we introduce markers. A marker declaration consists of an event name enclosed by angle brackets, e.g. `<e>`. Markers annotations can be seen simply as program comments (i.e. they can be ignored) if only the functional semantics of an application is considered. Markers are associated with programs points between instructions, possibly in different threads. Constraints may be specified between program points delineated by these markers. For a marker M, $time(M)$ (read as "the visit time at M") denotes the time at which the instruction immediately preceding M has just been completed. In the following, we will refer to $time(M)$ simply by M whenever confusion is unlikely. Given a pair of markers, constraints can be stated to specify their relative order of execution in all executions of the program. If the execution of a thread T_1 reaches a program point whose execution time is constrained to be greater than the execution time of a not yet executed program point in a different thread T_2, thread T_1 is forced to suspend execution. In the presence of loops and procedure calls a marker is typically visited several times during program execution. Thus, in general, a marker M associated with a program point p represents an event class E where each of its instances $e, e+, e + + \ldots$ corresponds to a visit to p during program execution (e represents the first visit, $e+$ the second, etc.).

3.3   Constraints and methods

Sometimes it is more convenient to express the synchronization aspects of an object-oriented program in terms of its methods rather that in terms of specific program points in its code. Thus, we allow in the language higher-level constraints of the form $p(c, m_1, m_2) \Leftarrow q(e_1, e_2, \ldots e_k)$, where m_1 and m_2 are methods in class c and $e_1, e_2 \ldots e_k$ program points (i.e. markers) in the code of either m_1 or m_2. This may be seen as adding some syntactic sugar on top of the base language previously defined.

4   Synchronization constraints

In this section we illustrate how the proposed language is used to specify the concurrency issues (safety properties) of concurrent object-oriented systems by presenting an example: a bounded stack data type.

Consider implementation in Java of a *bounded stack* data type. A basic specification without synchronization code is as follows (ignore for the moment markers <ai>, i.e. treat them as comments):

```
class BoundedStack  {
  static final int MAX = 10;
  int pos = 0;
  Object[] contents = new Object [MAX];

  public Object peek () { <a1>
    return contents [pos]; <a2> }
  public Object pop () { <a3>
    return contents [--pos]; <a4> }
  public void push (Object e) { <a5>
    contents [pos++]=e ; <a6> }
}
```

The specification of the class *BoundedStack* with synchronization code added is shown in the program listing of the next page.

Let us consider, for instance, the declaration of the *peek* method with synchronization as shown in the listing. It specifies the safety property that the *peek* method waits until there are no more threads currently executing a *push* or a *pop* method, i.e. mutual exclusion between *peek* and *push* and between *peek* and *pop*. It is clear that the synchronization code completely dominates the source code: almost all of the code for the *peek* method is synchronization code. Furthermore, it is very difficult to formally reason about the correctness of the code.

```
public class BoundedStack {
  private static final int MAX = 10;
  private int pos = 0;
  private Object[] contents = new Object[MAX];
  private int activeReaders_ = 0;
  private int activeWriters_ = 0;
  private int waitingReaders_ = 0;
  private int waitingWriters_ = 0;

  private boolean empty() {
    return pos == 0; }
  private boolean full() {
    return pos == MAX; }

  public Object peek() {
    synchronized(this) {
      ++waitingReaders_;
      while (waitingWriters_ == 0 &&
             activeWriters_ == 0) {
        try { wait(); }
        catch (InterruptedException ex) {}
      }
      --waitingReaders_;
      ++activeReaders_;
    }
    try {
      return contents[pos];
    } finally {
      synchronized(this) {
        --activeReaders_;
        notifyAll();
      }
    }
  }
  public Object pop() {
    synchronized(this) {
      ++waitingWriters_;
          while (activeReaders_ == 0 &&
                 activeWriters_ == 0 &&
                 !empty()) {
            try { wait(); }
            catch (InterruptedException ex) {}
          }
          --waitingWriters_;
          ++activeWriters_;
    }
    try {
      return contents[--pos];
    } finally {
      synchronized(this) {
        --activeWriters_;
        notifyAll();
      }
    }
  }
  public void push(Object e) {
    synchronized(this) {
      ++waitingWriters_;
      while (activeReaders_ == 0 &&
             activeWriters_ == 0 &&
             !full()) {
        try { wait(); }
        catch (InterruptedException ex) {}
      }
      --waitingWriters_;
      ++activeWriters_;
    }
    contents[pos++] = e;
    synchronized(this) {
      --activeWriters_;
      notifyAll();
    }
  }
}
```

Similarly, the safety properties that no thread attempts to remove (*pop*) an item from an empty stack and no thread attempts to append (*push*) into a full stack require coding of additional synchronization code in the *pop* and *push* methods.

These safety properties can be elegantly and formally expressed by using temporal constraints as follows. The requirement that the *peek* method waits until there are no more threads currently executing a *push* or a *pop* method may be implemented by

$mutex(a1, a2, a3, a4).$
$mutex(a1, a2, a5, a6).$
$mutex(X1, X2, Y1, Y2) \leftarrow X2 < Y1, mutex(X1+, X2+, Y1, Y2);$
$\qquad\qquad Y2 < X1, mutex(X1, X2, Y1+, Y2+).$

where $a1, a2 \ldots a6$ are the markers on our initial Java program. We may define equivalent higher-level constraints restricting the execution of the *peek*, *pop* and *push* methods by defining:

$mutex(Stack, peek, push) \Leftarrow mutex(a1, a2, a3, a4)$
$mutex(Stack, peek, pop) \Leftarrow mutex(a1, a2, a5, a6)$

The requirement that no thread attempts to remove an item from an empty stack and no thread attempts to append into a full stack may be respectively implemented by

$p(a8, a5)$.   and $p(a6, a7(+MAX))$.

$p(A, B) \leftarrow A < B, p(A+, B+)$. $p(A, B) \leftarrow A < B, p(A+, B+)$.

5   Implementation

The constraint logic programs have a procedural interpretation that allows a correct specification to be executed in the sense that events are only executed as permitted by the constraints represented by the program. This procedural interpretation is based on an incremental execution of the program and a *lazy* generation of the corresponding partial orders. Constraints are generated by the constraint logic program only when needed to reason about the execution times of current events. A description of how this procedural interpretation of constraint logic programs is implemented can be found in [12].

Fairness is implicitly guaranteed by our implementation. Every event that becomes enabled will eventually be executed (provided that the program point associated with it is reached). This is implemented by dealing with event execution requests in a first-in-first-out basis. Although fairness is provided as the default, users, however, may intervene by specifying priority events using temporal constraints (on how to do this, see [12]). It is therefore possible to specify unfair scheduling.

A prototype implementation of the ideas presented here has been written using the language Java. Java was used both to implement the constraint language and to write the code of a number of applications. The discussion of these applications and further details of implementation can be found in a companion paper.

6   Conclusion

We have presented a high-level language for expressing synchronization constraints in concurrent object-oriented applications. In the language, the safety properties of the system are *explicitly* stated as temporal constraints. Programs are annotated at points of interest so that the run-time environment enforces specific temporal relationships between the visit times of these points. Higher-lever constraints on class methods may also be defined. Constraints are *language independent* in that the application program can be specified in any conventional concurrent object-oriented language. The constraints have a procedural interpretation that allows the specification to be executed. The procedural interpretation is based on the incremental and *lazy* generation of constraints, i.e. constraints are considered only when needed to reason about the execution time of current events.

This paper presents work in progress so several important issues are still to be considered. Our implementation is still in a prototype stage, thus several efficiency issues have still to be addressed. In particular, we will focus on how the two key features of incrementality and laziness may be most efficiently achieved. Another important issue is how to deal with progress properties. Currently, constraints explicitly state all safety and timing properties of programs. However, the progress (liveness) properties of programs remain implicit. It would be desirable to be able to express these properties explicitly as additional constraints, but so far we have not devised a way to do that. Future versions may also include deadlock detection feature. We are considering a mechanism that checks user constraints for cycles (e.g., $A < B, B < A$) whenever a timeout occurred.

References

1. Atkinson, C. 1991. *Object-Oriented Reuse, Concurrency and Distribution: An Ada-Based Approach.* Addison-Wesley.
2. Van den Bos, J. and Laffra, C. 1989. *PROCOL: A parallel object language with protocols.* ACM SIGPLAN Notices 24(10):95–112, October 1989. Proc. of OOPSLA '89.
3. Gregory, S. and Ramirez, R. 1995. *Tempo: a declarative concurrent programming language.* Proc. of the ICLP (Tokyo, June), MIT Press, 1995.
4. Gregory, S. 1995. *Derivation of concurrent algorithms in Tempo.* In LOPSTR95: Fifth International Workshop on Logic Program Synthesis and Transformation.
5. Hong, S. and Gerber, R. 1995. *Compiling real-time programs with timing constraint refinement and structural code motion,* IEEE Transactions on Software Engineering, 21.
6. Jahnaian F. and Mok A. K. 1987. *A graph theoretic approach for timing analysis and its implementation,* IEEE Transactions on Computers, C36(8).
7. Kiczales, G., Lamping, J., Mendhekar, A., Maeda, C., Lopes, C., Loingtier, J.-M. and Irwin, J. 1997. Aspect-oriented programming. In *ECOOP '97—Object-Oriented Programming,* Lecture Notes in Computer Science, number 1241, pp. 220–242, Springer-Verlag.
8. Kowalski, R.A. and Sergot, M.J. 1986. A logic-based calculus of events. New Generation Computing 4, pp. 67–95.
9. Krakowiak, S., Meysembourg, M., Nguyen Van, H., Riveill, M., Roisin, C. and Rousset de Pina, X. 1990. *Design and implementation of an object-oriented strongly typed language for distributed applications.* Journal of Object-Oriented Programming 3(3):11–22.
10. Leung, A., Palem, K. and Pnueli, A. 1998. *Time C: A Time Constraint Language for ILP Processor Compilation,* Technical Report TR1998-764, New York University.
11. Pratt, V. 1986. Modeling concurrency with partial orders. International Journal of Parallel Programming 15, 1, pp. 33–71.
12. Ramirez, R. 1996. *A logic-based concurrent object-oriented programming language,* PhD thesis, Bristol University.
13. Shaw, A. 1989. *Reasoning about time in higher-level language software,* IEEE Transactions on Software Engineering, 15(7).
14. Stoyenko, A. D., Marlowe, T. J. and Younis, M. F. 1996. *A language for complex real-time systems,* Technical Report cis9521, New Jersey Institute of Technology.
15. De Volder, K. and D'Hondt, T. 1999. Aspect-oriented logic meta programming. In *Meta-Level Architectures and Reflection,* Lecture Notes in Computer Science number 1616, pp. 250–272. Springer-Verlag.

Scalable Monitoring Technique for Detecting Races in Parallel Programs*

Yong-Kee Jun[1]** and *Charles E. McDowell*[2]

[1] Dept. of Computer Science, Gyeongsang National University
Chinju, 660-701 South Korea
jun@nongae.gsnu.ac.kr
[2] Computer Science Department, University of California
Santa Cruz, CA 95064 USA
charlie@cs.ucsc.edu

Abstract. Detecting races is important for debugging shared-memory parallel programs, because the races result in unintended nondeterministic executions of the programs. Previous on-the-fly techniques to detect races have a bottleneck caused by the need to check or serialize *all accesses* to each shared variable in a program that may have nested parallelism with barrier synchronization. The new scalable monitoring technique in this paper reduces the bottleneck significantly by checking or serializing *at most* $2(B+1)$ *non-nested accesses in an iteration* for each shared variable, where B is the number of barrier operations in the iteration. This technique, therefore, makes on-the-fly race detection more scalable.

1 Introduction

A race is a pair of unsynchronized instructions, in a set of parallel threads, accessing a shared variable where at least one is a write access. Detecting races is important for debugging shared-memory parallel programs, because the races result in unintended nondeterministic executions of the programs. Traditional cyclical debugging with breakpoints is often not effective in the presence of races. Breakpoints can change the execution timing causing the erroneous behavior to disappear.

On-the-fly race detection instruments either the program to be debugged [1, 3, 5], or the underlying system [8, 12, 13], and monitors an execution of the program to report races which occur during the monitored execution. One drawback of existing on-the-fly techniques is the run-time overhead which is incurred from the need to check or serialize all accesses to the same shared-memory location. Every access must be compared with the previous accesses stored in a shared data

---

* University Research Program supported by Ministry of Information and Communication in South Korea.

** In Gyeongsang National University, he is also involved in both Institute of Computer Research and Development, and Information and Communication Research Center, as a research professor.

J. Rolim et al. (Eds.): IPDPS 2000 Workshops, LNCS 1800, pp. 340-347, 2000.

structure, often called the access history. In addition, the access history must be updated. This overhead has limited the usefulness of on-the-fly techniques.

The overhead can be reduced by detecting only the first races [3, 5, 8, 9, 12], which, intuitively, occur between two accesses that are not causally preceded by any other accesses also involved in races. It is important to detect the first races efficiently, because the removal of the first races can make other races disappear. It is even possible that all races reported by other on-the-fly algorithms would disappear once the first races were removed. A previous paper [5] presents a scalable on-the-fly technique to detect the first races, in which at most two accesses to a shared variable in each thread must be checked. However, this technique is restricted to parallel programs which have neither nested parallelism nor inter-thread synchronization.

In this paper, we introduce a new scalable technique for programs which may have nested parallelism and barrier synchronization. After first describing the background information on this work, in Section 3 we introduce a set of accesses, called *filtered accesses*. The set includes any accesses involved in first races. We then introduce two filtering procedures which examine if the current access is a filtered access during the execution of a program. Checking only filtered accesses is sufficient for detecting first races in the execution instance and reduces the number of non-nested accesses in an iteration that must be checked or serialized to *at most* $2(B + 1)$ for each shared variable, where B is the number of barrier operations in the iteration. Before concluding the paper, we briefly mention some related work in section 5.

2 Background

This work applies to shared-memory parallel programs [10, 11] with nested fork-join parallelism using parallel sections[1] or parallel loops. In this paper we use PARALLEL DO and END DO as in PCF Fortran [11]. The program may have inter-thread coordination in the loops using barriers. The *nesting level* of an individual loop is equal to one plus the number of the enclosing outer loops, and each loop may enclose zero or more disjoint loops at the same level. For example, Figure 1 shows a parallel loop of nesting depth two, which has two loops in the second nesting level.

In an execution of the program, multiple threads of control are created at a PARALLEL DO and terminated at the corresponding END DO statement. These fork and join operations are called *thread operations*. The concurrency relationship among threads is represented by a directed acyclic graph, called a *Partial Order Execution Graph (POEG)* [1]. A vertex of a POEG represents a thread operation, and an arc originating from a vertex represents a thread starting from the corresponding thread operation. Figure 1 shows a POEG that is an execution instance of the program shown in the same figure, where a small filled circle on a thread represents an access executed by the thread to shared variable X. If the program contains barrier synchronization, the POEG will contain additional edges to reflect the induced ordering.

---

[1] The work in this paper also can be applied to parallel sections without difficulty.

...

```
PARALLEL DO I = 1, 2
    ... = X                      {r0, r8}
    IF ... THEN
       PARALLEL DO I = 1, 2
          ... = X                {r1, r2}
          ... = X                {r3, r4}
       END DO
       ... = X                   {r5, r9}
    IF ... THEN
       PARALLEL DO I = 1, 2
          IF ... THEN ... = X {r6}
          IF ... THEN X = ...  {w10}
          X = ...              {w7, w13}
       END DO
       X = ...                   {w11, w14}
    IF ... THEN ... = X          {r12}
END DO
```

...

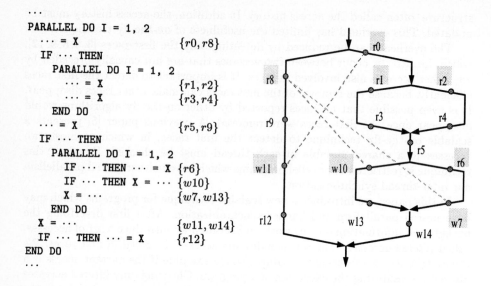

Fig. 1. A Parallel Program and Partial Order Execution Graph

Because the graph captures the *happened-before* relationship [6], it represents a partial order over the set of events executed by the program. Concurrency determination is not dependent on the number or relative execution speeds of processors executing the program. An event e_i *happened before* another event e_j if there exists a path from e_i to e_j in the POEG, and e_i is *concurrent* with e_j if neither one happened before the other. For example, consider the accesses in Figure 1, where $r0$ happened before $w7$ because there exists a path from $r0$ to $w7$, and $r0$ is concurrent with $w11$, because there is no path between them.

Definition 1. *A dynamic iteration of a parallel loop in the i-th nesting level, called a* level i *iteration consists of three components: (1) a thread, t_f, in the i-th nesting level that starts immediately at a fork operation, (2) a thread, t_j, in the i-th nesting level that terminates at the corresponding join operation, and (3) a set of threads, T, such that t_f happened before T which happened before t_j.*

Two accesses to a shared variable are *conflicting* if at least one of them is a write. If two accesses, a_i and a_j, are conflicting and concurrent then the two accesses constitute a *race* denoted a_i-a_j. An access a_j is *affected* by another access a_i, if a_i happened before a_j and a_i is involved in a race. A race a_i-a_j is *unaffected*, if neither a_i nor a_j are affected by any other accesses. The race is *partially affected*, if only one of a_i or a_j is affected by another access. A *tangle* T is a set of partially affected races such that if a_i-a_j is a race in T then exactly one of a_i or a_j is affected by a_k such that a_k-a_l is also in T. A *tangled race* is a partially affected race that is in a tangle.

Definition 2. *A first race is either an unaffected race or a tangled race.*

There are twenty seven races in the POEG shown in Figure 1; all of the accesses in the POEG are involved in races. Among these, only three races, $\{r0\text{-}w11, w7\text{-}r8, r8\text{-}w10\}$, are first races which are tangled races. Eliminating the three tangled races may make the other seven affected races disappear. The term tangled race was introduced by Netzer and Miller [9] describing the situation when no single race from a set of tangled races is unaffected by the others. Note that there can never be exactly one tangled race in an execution. They also introduce a tighter notion of first race, called *non-artifact race*, which uses the event-control dependences to define how accesses affect each other.

3   Scalable Monitoring Technique

On-the-fly race detection performs a relatively expensive check each time a monitored access is executed. In the worst case, this check must be done for every access. If we can determine a smaller set of accesses which includes all the accesses involved in first races, then the number of expensive access history operations can be reduced. In this section, we first define such a set, called *filtered accesses*, and then introduce two filtering procedures which examine if the current access is a filtered access during execution of the program. Checking only filtered accesses is sufficient for detecting first races in the execution instance, and reduces the number of expensive checks or serializing accesses to *at most* $2(B+1)$ *nonnested accesses in an iteration* for each shared variable of a parallel loop, where B is the number of barrier operations in the iteration.

Detecting first races in programs with nested parallelism requires detecting the happened-before relationship between nested iterations. We first exploit the nested iterations to indicate a set of filtered accesses.

Definition 3. *A read (write) access, a_i, is a level k filtered read (write), if and only if (1) a_i is in level k iteration I_k, and (2) there does not exist any other access, a_j, such that a_j is in I_k, a_j happened before a_i, and there are no barrier operations on the path from a_j to a_i in the POEG.*

Definition 4. *A write access, w_i, is a level k filtered r-write, if and only if (1) a_i is in level k iteration I_k, (2) there exists a level k filtered read, r_i, such that r_i happened before w_i, and (3) there does not exist any other write access, w_j, such that w_j is in I_k, w_j happened before w_i, and there are no barrier operations on the path from w_j to w_i in the POEG.*

For example, consider the accesses in Figure 1. There are five filtered accesses in the two level 1 iterations of the execution instance: two filtered reads $\{r0, r8\}$ and three filtered r-writes $\{w7, w10, w11\}$. Among these accesses, $\{r8, w11\}$ are in the same level 1 iteration which is in the left in the POEG, and $\{r0, w7, w10\}$ are in the other level 1 iteration. The reads, $\{r0, r8\}$, are level 1 filtered reads because there does not exist any other access that happened before $r0$ or $r8$ in their level 1 iterations. The access $w7$ is a level 1 filtered r-write, because there exists a filtered read $r0$ in the first level that happened before $w7$, and there does not exist any other write access that happened before $w7$ in the level 1 iteration. In the second level, we have four iterations in the figure, which have five reads

```
                              0  CheckWrite(X, k, bv, c_w)
                              1  if ¬P(X, k, bv_k) ∧ ¬F(X, k, bv_k) then
0  CheckRead(X, k, bv, c_r)   2    for i := 1 upto k do
1  if ¬P(X, k, bv_k) ∧ ¬F(X, k, bv_k) then  3      F(X, i, bv_i) := true;
2    for i := 1 upto k do     4    endfor
3      P(X, i, bv_i) := true; 5    CheckWriteFiltered(X, k, bv, c_w);
4    endfor                   6  elseif P(X, k, bv_k) ∧ ¬F(X, k, bv_k) then
5    CheckReadFiltered(X, k, bv, c_r);  7    for i := 1 upto k do
6  endif                      8      F(X, i, bv_i) := true;
7  EndCheckRead               9    endfor
                              10    CheckR-writeFiltered(X, k, bv, c_w);
                              11  endif
                              12  EndCheckWrite
```

Fig. 2. Checking to filter read or write access

and three writes. Among the eight accesses, five are level 2 filtered accesses: three filtered reads $\{r1, r2, r6\}$, one filtered write $\{w10\}$, and one filtered r-write $\{w7\}$. The shaded access names in the POEG distinguish the filtered accesses from the other accesses.

Note that the write access $w7$ is a filtered r-write not only in a level 1 iteration but also in a level 2 iteration. It is necessary to record the fact that some accesses, such as $w7$, are filtered accesses in iterations at multiple levels, because these accesses can be involved in races at different levels.

Definition 5. *A nested filtered access in a level i iteration is a filtered access, a_j, in a level k iteration $(i < k)$ such that a_j is also a level i filtered access.*

In Figure 1, the filtered write $w10$ in a level 2 iteration is a nested filtered r-write in a level 1 iteration. On the other hand, the filtered read, $r6$, in a level 2 iteration is not a nested filtered access in a level 1 iteration.

In summary of the accesses in Figure 1, we have fifteen accesses to a shared variable X, which include eight filtered accesses, of which five are involved in three first races. The following theorem shows that monitoring the filtered accesses is a sufficient condition for detecting first races in an execution instance. The proof can be found elsewhere [2].

Theorem 1. *If an access a_i is involved in a first race, a_i is a filtered access.*

Determining if each access is a filtered access is solely a function of comparing the current access with other previous accesses to the same shared variable in the iterations at multiple nesting levels. For this purpose, we define two states of nested iteration for each shared variable, which are checked in every access to determine if it is a filtered access.

Definition 6. *A partially-filtered iteration for a shared variable X in the i-th nesting level is an iteration in which there exists a level i filtered read to X. A fully-filtered iteration for a shared variable X in the i-th nesting level is an iteration in which there exists a level i filtered write or a level i filtered r-write to X.*

Benchmarks		Static		Dynamic			
Kernel	Input Set	Accesses		Access Checks		Filtered Checks	
		Read	Write	Read	Write	Read	Write
MG	$64 \times 64 \times 64$	39	23	347,223,802	25,971,050	68,496,214	19,127,166
FT	$128 \times 128 \times 32$	16	3	221,448	101,380	113,924	101,380
EP	67108864	4	3	93,463,643	26,485,851	41,950,826	26,483,873

Table 1. Race Instrumentation Statistics

Figure 2 shows two algorithms used to check, in each barrier partition bv, if the current read (c_r) or write access (c_w) to a shared variable X is a level k filtered access. The barrier vector, bv, selects a barrier partition for each nesting level. Each thread and barrier operation determines the value of bv for the corresponding partition and nesting level. $P(X, i, bv_i)$ and $F(X, i, bv_i)$ indicate if a level i iteration I_i is partially-filtered and fully-filtered, respectively, in a partition bv_i in I_i for shared variable X. If there are B non-nested barriers in I_i, then $0 \leq bv_i \leq B$, resulting in $(B+1)$ unique variable pairs for I_i. These *private boolean* variables are initialized to *false* at the start of every barrier partition of level i iteration.

Now, look at the procedure **CheckRead()** (**CheckWrite()**). If the condition in line 1 is true, it sets all the $P(X, i, bv_i)$ ($F(X, i, bv_i)$) to true, where ($i \leq k$) [line 2-4]. If the current read (write) access is the only access in a barrier partition of a level k iteration, it is a filtered read (write) in the iteration and all the level i iterations are guarranteed to be partially-filtered (fully-filtered). And then it performs **CheckReadFiltered()** (**CheckWriteFiltered()**) to invoke the detection protocol incurring the expensive centralized check or serialization. The modifications of $P(X, i, bv_i)$ and $F(X, i, bv_i)$ in the higher level iterations, do not need to be serialized because the value can only change from false to true.

If the condition in line 1 of **CheckRead()** is false, it does nothing, because if an iteration is fully-filtered or partially-filtered in a barrier partition, there exists at least one previous filtered access in the iteration.

In the case of **CheckWrite()**, it tests another condition. If the line 6 condition is true, **CheckWrite()** sets all the $F(X, i, bv_i)$ to true, where ($i \leq k$) [line 7-9]. If there exists a write access in a barrier partition of a level k iteration, all the level i iterations are fully-filtered. **CheckWrite()** then performs **CheckR-writeFiltered()** to invoke the detection protocol incurring the expensive centralized operation. Otherwise, it does nothing, because if an iteration is fully-filtered in a barrier partition there exists at least one previous filtered write or filtered r-write in the iteration.

Some experiments were performed on a set of three serial NAS benchmarks, in which a small set of common shared variables were monitored and filtered. Table 1 shows that filtering reduced the number of expensive checks to less than half in the case of read accesses, although the benchmarks are fine-grained. The following theorem shows that these algorithms reduce the number of required checks significantly in a monitored execution. The proof appears elsewhere [2].

Theorem 2. *If there exists a set of accesses to a shared variable in an iteration, the set involves at most* $2(B + 1)$ *non-nested filtered accesses in an iteration, where B is the number of barrier operations in the iteration.*

4   Related Work

Many approaches for efficiently detecting races on-the-fly for parallel programs have been reported. In this section, we briefly mention some important work to improve the scalability of on-the-fly race detection. This work falls into two groups: compiler support [7] to reduce the number of monitored accesses, and underlying system support [8, 12, 13] using scalable distributed shared memory systems. Our technique is novel in that the scalability is provided with simple but powerful instrumented code which can be applied to most existing techniques.

Mellor-Crummey [7] describes an instrumentation tool for on-the-fly race detection which applies compile-time analysis to identify variable references that need not be monitored at run-time. Using dependence analysis and interprocedural analysis of scalar side effects, the tool was able to reduce the dynamic counts of instrumented operations by 70-100% for the programs tested. Even with the impressive reductions in dynamic counts of monitoring operations, Mellor-Crummey reports that monitoring overhead for run-time detection of data races ran as high as a factor of 5.8.

Min and Choi [8] propose a technique of on-the-fly race detection to minimize the number of times that the monitored program is interrupted for run-time checking of accesses to shared variables. This scheme uses information from the underlying hardware-based, distributed shared-memory, cache coherence protocol and then requires additional hardware support, processor scheduling, cache management and compiler support.

Richards and Larus [13] propose a similar technique to that of Min and Choi in a software-based coherence protocol for a fine-grained data-maintaining distributed shared memory system. To detect data races on-the-fly in programs with barrier-only synchronization, this technique resets access histories at barriers, and monitors only the first read and write after obtaining a copy of a coherence block. They obtain substantial performance improvment but risk missing races. They report an implementation of this technique running on a 32-processor CM-5, and some experiments in which monitored applications had slowdowns ranging from 0-3.

Perković and Keleher [12] implemented on-the-fly race detection in a page-based release-consistent distributed shared memory system, which maintains ordering information that enables the system to make a constant-time determination of whether two accesses are concurrent without compiler support. They extended the system to collect information about the referenced locations and check at barriers for concurrent accesses to shared locations. Although they statically eliminate over 99% of non-shared accesses in applications, an average of 68% of the total overhead in race detection is the run-time overhead to determine whether an access is to shared memory. Nonetheless, the majority of the results are for non-shared accesses. They report that the applications slow down by an average factor of approximately 2.

5   Conclusion

In this paper, we present a new scalable on-the-fly technique for detecting races in parallel programs which may have nested parallelism with barrier synchronization. Our technique reduces the monitoring overhead to require serializing *at most* $2(B + 1)$ *non-nested accesses, in an iteration* for a shared variable, where B is the number of barrier operations in the iteration. It is important to detect races efficiently, because detecting races might require several iterations of monitoring, and the cost of monitoring a particular execution is still expensive. The technique in this paper can be applied to most existing techniques, therefore, making on-the-fly race detection scalable and more practical for debugging shared-memory parallel programs. We have experimented the technique on a prototype system of race debugging, called *RaceStand* [4], and have been extending it for the programs which have more general types of inter-thread coordination than barrier synchronization.

References

1. Dinning, A., and E. Schonberg, *"An Empirical Comparision of Monitoring Algorithms for Access Anomaly Detection,"* 2nd Symp. on Principles and Practice of Parallel Programming, pp. 1-10, ACM, March 1990.
2. Jun, Y., *"Improving Scalablility of On-the-fly Detection for Nested Parallelism,"* TR OS-9905, Dept. of Computer Science, Gyeongsang National Univ., March 1999.
3. Jun, Y., and C. E. McDowell, *"On-the-fly Detection of the First Races in Programs with Nested Parallelism,"* 2nd Int. Conf. on Parallel and Distributed Processing Techniques and Applications, pp. 1549-1560, CSREA, August 1996.
4. Kim, D., and Y. Jun, *"An Effective Tool for Debugging Races in Parallel Programs,"* 3rd Int. Conf. on Parallel and Distributed Processing Techniques and Applications, pp. 117-126, CSREA, July 1997.
5. Kim, J., and Y. Jun, *"Scalable On-the-fly Detection of the First Races in Parallel Programs,"* 12th Int. Conf. on Supercomputing, pp. 345-352, ACM, July 1998.
6. Lamport, L., *"Time, Clocks, amd the Ordering of Events in Distributed System,"* Communications of ACM, 21(7): 558-565, ACM, July 1978.
7. Mellor-Crummey, J., *"Compile-time Support for Efficient Data Race Detection in Shared-Memory Parallel Programs,"* 3rd Workshop on Parallel and Distributed Debugging, pp. 129-139, ACM, May 1993.
8. Min, S. L., and J. D. Choi, *"An Efficient Cache-based Access Anomaly Detection Scheme,"* 4th Int. Conf. on Architectural Support for Programming Language and Operating Systems, pp. 235-244, ACM, April 1991.
9. Netzer, R. H., and B. P. Miller, *"Improving the Accuracy of Data Race Detection,"* 3rd Symp. on Prin. and Practice of Parallel Prog., pp. 133-144, ACM, April 1991.
10. OpenMP Architecture Review Board, *OpenMP Fortran Application Program Interface*, Version 1.0, Oct. 1997.
11. Parallel Computing Forum, *"PCF Parallel Fortran Extensions,"* Fortran Forum, 10(3), ACM, Sept. 1991.
12. Perković, D., and P. Keleher, *"Online Data-Race Detection vis Coherency Guarantees,"* 2nd Usenix Symp. on Operating Systems Design and Implementation, pp. 47-58, ACM/IEEE, Oct. 1996.
13. Richards, B., and J. R. Larus, *"Protocol-Based Data-Race Detection,"* 2nd Sigmetrics Symp. on Parallel and Dist. Tools, pp. 40-47, ACM, August 1998.

3rd IPDPS Workshop on
High Performance Data Mining

Preface

The explosive growth in data collection in business and scientific fields has literally forced upon us the need to analyze and mine useful knowledge from it. Data mining refers to the entire process of extracting useful and novel patterns/models from large datasets. Due to the huge size of data and amount of computation involved in data mining, high-performance computing is an essential component for any successful large-scale data mining application.

This workshop provided a forum for presenting recent results in high performance computing for data mining including applications, algorithms, software, and systems. High-performance was broadly interpreted to include scalable sequential as well as parallel and distributed algorithms and systems. Relevant topics for the workshop included:

1. Scalable and/or parallel/distributed algorithms for various mining tasks like classification, clustering, sequences, associations, trend and deviation detection, etc.
2. Methods for pre/post-processing like feature extraction and selection, discretization, rule pruning, model scoring, etc.
3. Frameworks for KDD systems, and parallel or distributed mining.
4. Integration issues with databases and data-warehouses.

These proceedings contain 9 papers that were accepted for presentation at the workshop. Each paper was reviewed by two members of the program committee. In keeping with the spirit of the workshop some of these papers also represent work-in-progress. In all cases, however, the workshop program highlights avenues of active research in high performance data mining.

We would like to thank all the authors and attendees for contributing to the success of the workshop. Special thanks are due to the program committee and external reviewers for help in reviewing the submissions.

February 2000
Mohammed J. Zaki
Vipin Kumar
David B. Skillicorn
Editors

J. Rolim et al. (Eds.): IPDPS 2000 Workshops, LNCS 1800, pp. 348-349, 2000.
© Springer-Verlag Berlin Heidelberg 2000

Workshop Co-Chairs

Mohammed J. Zaki (Rensselaer Polytechnic Institute, USA)
Vipin Kumar (University of Minnesota, USA)
David B. Skillicorn (Queens University, Canada)

Program Committee

Philip K. Chan (Florida Institute of Technology, USA)
Alok Choudhary (Northwestern University, USA)
Umeshwar Dayal (Hewlett-Packard Labs., USA)
Alex A. Freitas (Pontifical Catholic University of Parana, Brazil)
Ananth Grama (Purdue University, USA)
Robert Grossman (University of Illinois-Chicago, USA)
Yike Guo (Imperial College, UK)
Jiawei Han (Simon Fraser University, Canada)
Howard Ho (IBM Almaden Research Center, USA)
Chandrika Kamath (Lawrence Livermore National Labs., USA)
Masaru Kitsuregawa (University of Tokyo, Japan)
Sanjay Ranka (University of Florida, USA)
Vineet Singh (Hewlett-Packard Labs., USA)
Domenico Talia (ISI-CNR: Institute of Systems Analysis and Information Technology, Italy)
Kathryn Burn-Thornton (Durham University, UK)

External Reviewers

Eui-Hong (Sam) Han (University of Minnesota, USA)
Wen Jin (Simon Fraser University, Canada)
Harsha S. Nagesh (Northwestern University, USA)
Srinivasan Parthasarathy (University of Rochester, USA)

Implementation Issues in the Design of I/O Intensive Data Mining Applications on Clusters of Workstations

R. Baraglia[1], D. Laforenza[1], Salvatore Orlando[2],
P. Palmerini[1] and Raffaele Perego[1]

[1] Istituto CNUCE, Consiglio Nazionale delle Ricerche (CNR), Pisa, Italy
[2] Dipartimento di Informatica, Università Ca' Foscari di Venezia, Italy

Abstract This paper investigates *scalable* implementations of out-of-core I/O-intensive Data Mining algorithms on affordable parallel architectures, such as clusters of workstations. In order to validate our approach, the K-means algorithm, a well known *DM Clustering* algorithm, was used as a test case.

1 Introduction

Data Mining (DM) applications exploit huge amounts of data, stored in files or databases. Such data need to be accessed to discover patterns and correlations useful for various purposes, above all for guiding strategic decision making in the business domain. Many DM applications are strongly I/O intensive since they need to read and process the input dataset several times [1,6,7]. Several techniques have been proposed in order to improve the performance of DM applications. Many of them are based on parallel processing [5]. In general, their main goals are to reduce the computation time and/or reduce the time spent on accessing out-of-memory data.

Since the early 1990s there has been an increasing trend to move away from expensive and specialized proprietary parallel supercomputers towards clusters of workstations (COWs) [3]. Historically, COWs have been used primarily for science and engineering applications, but their low cost, scalability, and generality provide a wide array of opportunities for new domains of application [13]. DM is certainly one of these domains, since DM algorithms generally exhibit large amounts of data parallelism. However, to efficiently exploit COWs, parallel implementations should be adaptive with respect to the specific features of the machine (e.g. they must take into account the memory hierarchies and caching policies adopted by modern hardware/software architectures).

Specific *Out-of-Core* (OoC) techniques (also known as External Memory techniques) [3,14] can be exploited to approach DM problems that require huge amounts of memory. OoC techniques are useful for all applications that do not completely fit into the physical memory. Their main goal is to reduce memory hierarchy overheads by bypassing the OS virtual memory system and explicitly

J. Rolim et al. (Eds.): IPDPS 2000 Workshops, LNCS 1800, pp. 350-357, 2000.

managing I/O. Direct control over data movements between main memory and secondary storage is achieved by splitting the dataset into several small blocks. These blocks are then loaded into data structures which will certainly fit into physical memory. They are processed and, if necessary, written back to disks. The knowledge of the patterns used by the algorithm to access the data can be exploited in an effective way to reduce I/O overheads by overlapping them with useful computations. The access pattern exploited by the DM algorithm discussed in this paper is simple, since read-only datasets are accessed sequentially and iteratively. Note that the general purpose external memory mechanism provided by the operating system – in our case, the Unix `read()` system call – is specifically optimized for this kind of data access.

This paper investigates *scalable* implementations of I/O-intensive DM algorithms on affordable parallel architectures, such as clusters of PCs equipped with main memories of a limited size, which are not sufficiently big to store the whole dataset (or even a partition of it). The test case DM application used to validate our approach is based on the on-line K-means algorithm, a well known *DM Clustering* algorithm [8,10]. The testbed COW was composed of three SMPs, interconnected by a 100BaseT switched Ethernet, where each SMP was equipped with two Pentium II - 233 MHz processors, 128 MB of main memory, and a 4GB UW-SCSI disk. Their OS was Linux, kernel version 2.2.5-15. The paper is organized as follows. Section 2 discusses implementation issues related to the design of I/O-intensive DM applications. Section 3 deals with the K-means algorithm and its parallel implementation. Finally, Section 4 discusses the results of our experiments and draws some conclusions.

2   Implementation of I/O Intensive DM Applications

As mentioned above, we are interested in DM algorithms that access sequentially the same dataset several times. The repeated scanning of the whole dataset entails good spatial locality but scarce temporal locality. The latter can only be exploited if the whole dataset entirely fits into the physical memory. In general, however, this condition cannot be taken for granted because "real life" datasets are generally very large. Moreover, the physical memory is of limited size, and other running processes contend for its usage. The adoption of an OoC algorithm, which takes advantage of possible prefetching policies implemented by both software drivers and disk controllers [11], and which allows to exploit *multitasking* or *multithreading* strategies in order to overlap I/O latencies with useful computations, is thus mandatory.

The best policy might thus appear to be to adopt OoC algorithms only if a dataset does not fit into the physical memory. When the memory is large enough, an in-core approach might seem more efficient, since all the dataset is read once from disk, and is repeatedly accessed without further I/O operations. Clearly such an in-core strategy might fail when other processes use the main memory, thus causing swapping on the disk. We believe that "smart" OoC approaches are always preferable to their in-core counterparts, even when datasets are small with respect to memory size. This assertion is due to the existence of a *buffer cache* for block devices in modern OSs, such as Linux [2]. The available physical memory left unused by the kernel and processes is dynamically enrolled in the buffer cache on demand. When the requirement for primary memory increases, for example because new processes enter the system, the memory allocated to buffers is reduced. We conducted experiments to compare in-core and out-of-core versions of a simple test program that repeatedly scans a dataset which fits into physical memory. We observed that the two versions of the program have similar performances. In fact, if we consider the OoC version of this simple program, at the end of the first scan the buffer cache contains the blocks of the whole dataset. The following scans of the dataset will not actually access the disk at all, since they find all the blocks to be read in the main memory, i.e. in the buffer cache. In other words, due to the mechanisms provided by the OS, the actual behavior of the OoC program becomes in-core.

We also observed another advantage of the OoC program over the in-core solution. During the first scan of the dataset, the OoC program takes advantage of OS prefetching. In fact, during the processing of a block the OS prefetches the next one, thus hiding some I/O time. On the contrary, I/O time of in-core programs cannot be overlapped with useful computations because the whole dataset has to be read before starting the computation.

In summary, the OoC approach not only works well for small datasets, but it also scales-up when the problem size exceeds the physical memory size, i.e., in those cases when in-core algorithms fail due to memory swapping. Moreover, to improve scalability for large datasets, we can also exploit multitasking techniques in conjunction with OoC techniques to hide I/O time. To exploit multitasking, non-overlapping partitions of the whole dataset must be assigned to distinct tasks. The same technique can also be used to parallelize the application, by mapping these tasks onto distinct machines. This kind of data-parallel paradigm is usually very effective for implementing DM algorithms, since computation is generally uniform, data exchange between tasks is limited, and generally involves a global synchronization at the end of each scan of the whole dataset. This synchronization is used to check termination conditions and to restore a consistent global state. Consistency restoration is needed since the tasks start each iteration on the basis of a consistent state, generating new local states that only reflect their partial view of the whole dataset.

Finally, parallel DM algorithms implemented on COWs also have to deal with load imbalance. In fact, workload imbalance may derive either from different capacities of the machines involved or from unexpected arrivals of external jobs. Since the programming paradigm adopted is data parallel, a possible solution to this problem is to dynamically change partition sizes.

3   A Test Case DM Algorithm and its Implementation

There is a variety of applications, ranging from marketing to biology, astrophysics, and so on [8], that need to identify subsets of records (clusters) presenting characteristics of *homogeneity*. In this paper we used a well known clustering algorithm, the **K-means** algorithm [10] as a case study representative of a class of I/O intensive DM algorithms. We deal with the on-line formulation of K-means, which can be considered as a competitive learning formulation of the classical K-means algorithm. K-means considers records in a dataset to be represented as *data-points* in a high dimensional space. Clusters are identified by using the concept of proximity among data-points in this space. The K-means algorithm is known to have some limitations regarding the dependence on the initial conditions and the shape and size of the clusters found [9,10]. Moreover, it is necessary to define *a priori* the number K of clusters that we expect to find, even though it is also possible to start with a small number of clusters (and associated centers), and increase this number when specific conditions are observed. The three main steps of the on-line K-means sequential algorithm are: (1) start with a given number of centers randomly chosen; (2) scan all the data-points of the dataset, and for each point p find the center closest to p, assign p to the cluster associated with this center, and move the center toward p; (3) repeat step 2 until the assignment of data-points to the various clusters remains unchanged. Note that the repetition of step 2 ensures that centers gradually get attracted into the middle of the clusters. In our tests we used synthetic datasets and we fixed a priori K.

Parallel Implementation. We implemented the OoC version of the algorithm mentioned above, where data-points are repeatedly scanned by sequentially reading small blocks of 4 KBytes from the disk. The program was implemented using MPI according to an SPMD paradigm. A non overlapping partition of the input file, univocally identified by a pair of boundaries, is processed by each task of the SPMD program. The number of tasks involved in the execution may be greater than the number of physical processors, thus exploiting multitasking. This parallel formulation of our test case is similar to those described in [12,4], and requires a new consistent global state to be established once each scan of the whole dataset is completed. Our global state corresponds with the new positions reached by the K centers. These positions are determined by summing the vectors corresponding with the centers' movements which were separately computed by the various tasks involved. In our implementation, the new center positions are computed by a single task, the root one, and are broadcast to the others. The root task also checks the termination condition.

The load balancing strategy adopted is simple but effective. It is based on past knowledge of the bandwidths of all concurrent tasks (i.e. number of points computed in a unit of time). If a load imbalance is detected, the size of the partitions is increased for "faster" tasks and decreased for "slower" ones. This requires input datasets to be replicated on all the disks of our testbed. If complete replication is too expensive or not possible, file partitions with overlapping boundaries can be exploited as well. Let NP be the total number of data-points, and $\{p_1, \ldots, p_n\}$ the n tasks of the SPMD program. At the first iteration $np_i^1 = NP/n$ data-points are assigned to each p_i. During iteration j each p_i measures the elapsed time T_i^j spent on elaborating its own block of np_i^j points, so that $t_i^j = T_i^j / np_i^j$ is the time taken by each p_i to elaborate a single point, and $b_i^j = 1/t_i^j = np_i^j / T_i^j$ is its bandwidth. In order to balance the workload, the numbers np_i^{j+1} of data-points which each p_i has to process in the next iteration are then computed on the basis of the various b_i^j ($np_i^{j+1} = \rho_i^j \cdot NP$, where $\rho_i^j = b_i^j / \sum_{i=1}^{n} b_i^j$). Finally, values np_i^{j+1} are easily translated into partition boundaries, i.e. a pair of offsets within the replicated or partially replicated input file.

4   Experimental Results and Conclusions

Several experiments were conducted on the testbed with our parallel implementation of K-means based on MPI. Data-parallelism, OoC techniques, multitasking, and load balancing strategies were exploited. Note that the successful adoption of multitasking mainly depends on (1) the number of disks with respect to the number of processors available on each machine, and (2) the computation granularity (i.e., the time spent on processing each data block) with respect to the I/O bandwidth. In our experiments on synthetic datasets, we tuned this computational granularity by changing the number K of clusters to look for. Another important characteristics of our approach is the size of the partitions assigned to the tasks mapped on a single SMP machine. If the sum of these sizes is less than the size of the physical main memory, we guess that the behavior of the OoC application will be similar to its in-core counterpart, due to a large enough buffer cache. Otherwise, sequential accesses carried out by a task to its dataset partition will entail disk accesses, so that the only possibility of hiding these I/O times is to exploit, besides OS prefetching, some form of moderated multitasking.

Figure 1 shows the effects of the presence of the buffer cache. On a single SMP we ran our test case algorithm with a small dataset (64 MB) and small computational granularity ($K=3$). Bars show the time spent by the tasks in computing (t_comp), in doing I/O and being idle in some OS queue (t_io + t_idle), and in communication and synchronization (t_comm). The two bars on the left hand side represent the first and the second iterations of a sequential implementation of the test case. The four bars on the right hand side regard the parallel implementation (2 tasks mapped on the same SMP). Note that in both cases the t_io and t_idle are high during the first iteration, since the buffer cache is not able to fulfill the read requests (cache misses). On the other hand, these times almost disappear from the the second iteration bars, since the accessed blocks are found in the buffer cache (cache hits).

Figure 2 shows the effects of multitasking on a single SMP when the disk has to be accessed. Although a small dataset was used for these experiments, the bars only refer to the first iteration, during which we certainly need to access the disk. Now recall that our testbed machines are equipped with a single disk each. This represents a strong constraint on the I/O bandwidth of our platform. This is particularly evident when several I/O-bound tasks, running in parallel on an SMP, try to access this single disk. In this regard, we found that our test case has different behaviors depending on the computational granularity.

For a fine granularity (K=8), the computation is completely I/O-bound. In this condition it is better to allocate a single task to each SMP (see Figure 2.(a)). When we allocated more then one task, the performance worsened because of the limited I/O bandwidth and I/O conflicts. For a coarser granularity (K= 32), the performance improved when two tasks were used (see Figure 2.(b)). In the case of higher degrees of parallelism the performance decreases. This is due to the overloading of the single disk, and to noises introduced by multitasking into the OS prefetching policy.

Figure 3 shows some speedup curves. The plots refer to 20 iterations with K=16. We used at most two tasks per SMP. Note the super-linear speedup achieved when 2 or 3 processors were used. These processors belong to distinct SMPs, so that this super-linear speedup is due to the exploitation of multiple disks and to the effects of the buffer cache. In fact, when moderately large datasets were used (64 MB or 128 MB) the data partitions associated with the tasks mapped on each SMP fit into the buffer caches. Overheads due to communications, occurring at the end of each iteration, are very small and do not affect speedup.

In the case of a larger dataset (384 MB), whose size is greater than the whole main memory available, when the number of tasks remains under three, linear speedups were obtained. For larger degrees of parallelism, the speedup decreases. This is still due to the limited I/O bandwidth on each SMP.

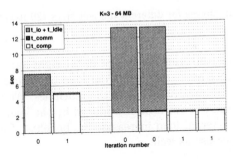

Figure1. Execution times of two iterations of the test case on a single SMP.

Figure 4 shows the effectiveness of the load balancing strategy adopted. Both plots refer to experiments conducted using all the six processors of our testbed with the 64MB dataset and $K = 32$. The plot in the left hand side of the figure shows the number of blocks dynamically assigned to each task by our load balancing algorithm as a function of the iteration index. During time interval $[t1, t3]$ ($[t2, t4]$) we executed a CPU-intensive process on the SMP A (M) running tasks A0 and A1 (M1 and M1). As it can be seen, the load balancing algorithm quickly detects the variation in the capacities of the machines, and correspondingly adjusts the size of the partitions by narrowing partitions assigned to slower

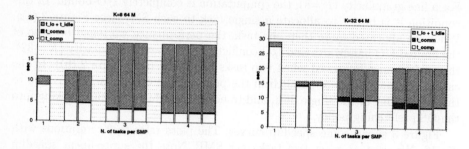

Figure2. Execution times of the first iteration on a single SMP by varying the number of tasks exploited and the computational granularities: (a) $K=8$ and (b) $K=32$.

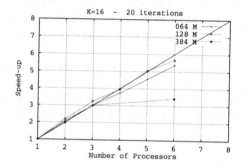

Figure3. Speedup curves for different dataset sizes (20 iterations, $K = 16$).

machines and enlarging the others. The plot in the right hand side compares the execution times obtained exploiting or not our load balancing strategy as a function of the external load present in one of the machines. We can see that in the absence of external load the overhead introduced by the load balancing strategy is negligible. As the external load increases, the benefits of exploiting the load balancing strategy increase as well.

In conclusion, this work has investigated the issues related to the implementation of a test case application, chosen as a representative of a large class of DM I/O-intensive applications, on an inexpensive COW. Effective strategies for managing I/O requests and for overlapping their latencies with useful computations have been devised and implemented. Issues related to data parallelism exploitation, OoC techniques, multitasking, and load balancing strategies have been discussed. To validate our approach we conducted several experiments and discussed the encouraging results achieved. Future work regards the evaluation of the possible advantages of exploiting lightweight threads for intra-SMP parallelism and multitasking. Moreover, other I/O intensive DM algorithms have to be considered in order to define a framework of techniques/functionalities useful for efficiently solving general DM applications on COWs, which, unlike homo-

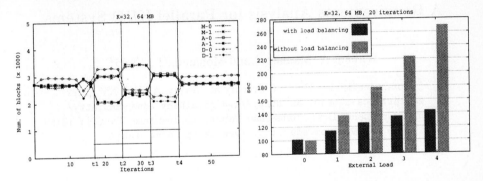

Figure4. Effectiveness of the load balancing strategy.

geneous MPPs, impose additional issues that must be addressed using adaptive strategies.

References

1. Jain A.K. and Dubes R.C. *Algorithms for Clustering Data*. Prentice Hall, 1988.
2. M. Beck et al. *Linux Kernel Internals, 2nd ed*. Addison-Wesley, 1998.
3. Rajkumar Buyya, editor. *High Performance Cluster Computing*. Prentice Hall PTR, 1999.
4. I. S. Dhillon and D. S. Modha. A data clustering algorithm on distributed memory machines. In *ACM SIGKDD Int. Conf. on Knowledge Discovery and Data Mining*, 1999.
5. A. A. Freitas and S. H. Lavington. *Mining Very Large Databases with Parallel Processing*. Kluwer Academin Publishers, 1998.
6. V. Ganti, J. Gehrke, and R. Ramakrishnan. Mining Very Large Databases. *IEEE Computer*, 32(8):38–45, 1999.
7. E. Han, G. Karypis, and V. Kumar. Scalable Parallel Data Mining for Association Rules. *IEEE Transactions on Knowledge and Data Engineering*. To appear.
8. J.A. Hartigan. *Clustering Algorithms*. Wiley & Sons, 1975.
9. G. Karypis, E. Han, and V. Kumar. Chameleon: Hierarchical Clustering Using Dynamic Modeling. *IEEE Computer*, 32:68–75, 1999.
10. Mac Queen, J.B. Some Methods for Classification and Analysis of Multivariate Observation. 5th *Berkeley Symp. on Mathematical Statistics and Probability*, pages 281–297. Univ. of California Press, 1967.
11. Chris Ruemmler and John Wilkes. An Introduction to Disk Drive Modeling. *IEEE Computer*, 27(3):17–28, March 1994.
12. K. Stoffel and A. Belkoniene. Parallel k-means clustering for large datasets. *EuroPar'99 Parallel Processing*, Lecture Notes in Computer Science, No. 1685. Springer-Verlag, 1999.
13. Sterling T.L., Salmon J., Becker D.J., and Savarese D.F. *How to Build a Beowulf. A guide to the Implementation and Application of PC Clusters*. The MIT Press, 1999.
14. J. S. Vitter. External Memory Algorithms and Data Structures. In *External Memory Algorithms (DIMACS Series on Discrete Mathematics and Theoretical Computer Science)*. American Mathematical Society, 1999.

A Requirements Analysis for Parallel KDD Systems

William A. Maniatty[1] and Mohammed J. Zaki[2]

[1] Computer Science Dept., University at Albany, Albany, NY 12222
maniatty@cs.albany.edu, http://www.cs.albany.edu/~maniatty/
[2] Computer Science Dept., Rensselaer Polytechnic Institute, Troy, NY 12180
zaki@cs.rpi.edu, http://www.cs.rpi.edu/~zaki/

Abstract. The current generation of data mining tools have limited capacity and performance, since these tools tend to be sequential. This paper explores a migration path out of this bottleneck by considering an integrated hardware and software approach to parallelize data mining. Our analysis shows that parallel data mining solutions require the following components: parallel data mining algorithms, parallel and distributed data bases, parallel file systems, parallel I/O, tertiary storage, management of online data, support for heterogeneous data representations, security, quality of service and pricing metrics. State of the art technology in these areas is surveyed with an eye towards an integration strategy leading to a complete solution.

1 Introduction

Knowledge discovery in databases (KDD) employs a variety of techniques, collectively called *data mining*, to uncover trends in large volumes of data. Many applications generate (or acquire) data faster than it can be analyzed using existing KDD tools, leading to perpetual data archival without retrieval or analysis. Furthermore, analyzing sufficiently large data sets can exceed the available computational resources of existing computers. In order to reverse the vicious cycle induced by these two problematic trends, the issues of performing KDD faster than the rate of arrival and increasing capacity must simultaneously be dealt with. Fortunately, novel applications of parallel computing techniques should assist in solving these large problems in a timely fashion.

Parallel KDD (PKDD) techniques are not currently that common, though recent algorithmic advances seek to address these problems (Freitas and Lavington 1998; Zaki 1999; Zaki and Ho 2000; Kargupta and Chan 2000). However, there has been no work in designing and implementing large-scale parallel KDD systems, which must not only support the mining algorithms, but also the entire KDD process, including the pre-processing and post-processing steps (in fact, it has been posited that around 80% of the KDD effort is spent in these steps, rather than mining). The picture gets even more complicated when one considers persistent data management of mined patterns and models.

Given the infancy of KDD in general, and PKDD in particular, it is not clear how or where to start, to realize the goal of building a PKDD system

J. Rolim et al. (Eds.): IPDPS 2000 Workshops, LNCS 1800, pp. 358-365, 2000.

that can handle terabyte-sized (or larger) central or distributed datasets. Part of the problem stems from the fact that PKDD draws input from diverse areas that have been traditionally studied in isolation. Typically, the KDD process is supported by a hierarchical architecture consisting of the following layers: (from bottom to top) I/O Support, File System, Data Base, Query Manager, and Data Mining. However, the current incarnations of this architecture tend to be sequential, limiting both problem size and performance. To implement a successful PKDD toolkit, we need to borrow, adapt, and enhance research in fields such as super-, meta- and heterogeneous-computing environments, parallel and distributed databases, parallel and distributed file systems, parallel I/O, mass storage systems, and so on (not to mention the other fields that make up KDD — statistics, machine learning, visualization, etc.).

This paper represents a first step in the process of unifying these diverse technologies and leveraging them within the PKDD system. We do this by discussing the system requirements for PKDD and the extant solutions (or lack thereof), i.e., the *what* and the *how* of PKDD. These requirements follow from: the basic requirements imposed by KDD (Section 2), current KDD algorithmic techniques (Section 3), the trends in commodity hardware design (Section 4) and software requirements (Section 5). One difficulty in making such a survey is that each research community has its own jargon, which we will try to make accessible by describing it within a common PKDD framework.

2 PKDD Requirements

We begin by discussing the wish-list or desirable features of a functional PKDD system, using it to guide the rest of the survey. We mainly concentrate on aspects that have not received wide attention as yet.

Algorithm Evaluation: Algorithmic aspects that need attention are the ability to handle high dimensional datasets, to support terabyte data-stores, to minimize number of data scans, etc. An even more important research area is to provide a rapid development framework to implement and conduct the performance e-valuation of a number of competing parallel methods for a given mining task. Currently this is a very time-consuming process, and there are no guidelines when to use a particular algorithm over another.

Process Support: The toolkit should support all KDD steps, from pre-processing operations for like sampling, discretization, and feature subset selection, to post-processing operations like rule grouping and pruning and model scoring. Other aspects include (persistent) pattern management operations like caching, efficient retrieval, and meta-level mining.

Location Transparency: The PKDD system should be able to seamlessly access and mine datasets regardless of their location, be they centralized or distributed.

Data Type Transparency: The system should be able to cope with heterogeneity (e.g., different database schemas), without having to materialize a join of multiple tables. Other difficult aspects deal with handling unstructured (hyper-)text, spreadsheet, and a variety of other data types.

System Transparency: This refers to the fact that the PKDD system should be able to seamlessly access file systems, databases, or data archives. Databases and data warehouses represent one kind of data repositories, and thus it is crucial to integrate mining with DBMS to avoid extracting data to flat files. On the other hand, a huge amount of data remains outside databases in flat-files, weblogs, etc. The PKDD system must therefore bridge the gap that exists today between the database and file-systems worlds (Choudhary and Kotz 1996). This is required since database systems today offer little functionality to support mining applications (Agrawal *et al.* 1993), and most research on parallel file systems and parallel I/O has looked at scientific applications, while data mining operations have very different workload characteristics.

Security, QoS and Pricing: In an increasingly networked world one constantly needs access to proprietary third-party and other remote datasets. The two main issues that need attention here are security and Quality-of-Service (QoS) issues in data mining. We need to prevent unauthorized mining, and we need to provide cost-sensitive mining to guarantee a level of performance. These issues are paramount in web-mining for e-commerce.

Availability, Fault Tolerance and Mobility: Distributed and parallel systems have more points of failure than centralized systems. Furthermore temporary disconnections (which are frequent in mobile computing environments) and reconnections by users should be tolerated with a minimal penalty to the user. Many real world applications cannot tolerate outages, and in the presence of QoS guarantees and contracts outages, can breach the agreements between providers and users. Little work has been done to address this area as well.

In the discussion below, due to space constraints, we choose to concentrate only on the algorithmic and hardware trends, and system transparency issues (i.e., parallel I/O and parallel and distributed databases), while briefly touching on other aspects (a more detailed paper is forthcoming).

3   Mining Methods

Faster and scalable algorithms for mining will always be required. Parallel and distributed computing seems ideally placed to address these big data performance issues. However, achieving good performance on today's multiprocessor systems is a non-trivial task. The main challenges include synchronization and communication minimization, work-load balancing, finding good data layout and data decomposition, and disk I/O minimization.

The parallel design space spans a number of systems and algorithmic components such as the hardware platform (shared vs. distributed), kind of parallelism (task vs. data), load balancing strategy (static vs. dynamic), data layout (horizontal vs. vertical) and search procedure used (complete vs. greedy).

Recent algorithmic work has been very successful in showing the benefits of parallelism for many of the common data mining tasks including association rules (Agrawal and Shafer 1996; Cheung *et al.* 1996; Han *et al.* 1997; Zaki *et al.* 1997), sequential patterns (Shintani and Kitsuregawa 1998; Zaki 2000), classification (Shafer *et al.* 1996; Joshi *et al.* 1998; Zaki *et al.* 1999; Sreenivas *et al.* 1999), regression (Williams *et al.* 2000) and clustering (Judd *et al.* 1996; Dhillon and Modha 2000; S. Goil and Choudhary 1999).

The typical trend in parallel mining is to start with a sequential method and pose various parallel formulations, implement them, and conduct a performance evaluation. While this is very important, it is a very costly process. After all, the parallel design space is vast and results on the parallelization of one serial method may not be applicable to other methods. The result is that there is a proliferation of parallel algorithms without any standardized benchmarking to compare and provide guidelines on which methods work better under what circumstances. The problem becomes even worse when a new and improved serial algorithm is found, and one is forced to come up with new parallel formulations. Thus, it is crucial that the PKDD system support rapid development and testing of algorithms to facilitate algorithmic performance evaluation.

One recent effort in this direction is discussed by (Skillicorn 1999). He emphasizes the importance of and presents a set of cost measures that can be applied to parallel algorithms to predict their computation, data access, and communication performance. These measures make it possible to compare different parallel implementation strategies for data-mining techniques without benchmarking each one.

A different approach is to build a data mining kernel that supports common data mining operations, and is modular in design so that new algorithms or their "primitive" components can be easily added to increase functionality. An example is the MKS (Anand *et al.* 1997) kernel. Also, generic set-oriented primitive operations were proposed in (Freitas and Lavington 1998) for classification and clustering, which were integrated with a parallel DBMS.

4   Hardware Models and Trends

The current hardware trends are that memory and disk capacity are increasing at a much higher rate than their speed. Furthermore, CPU capacity is roughly obeying Moore's law, which predicts doubling performance approximately every 18 months. To combat bus and memory bandwidth limitations, caching is used to improve the mean access time, giving rise to Non-Uniform Memory Access architectures. To accelerate the rate of computation, modern machines frequently increase the number of processing elements in an architecture. Logically, the memory of such machines is kept consistent, giving rise to a shared memory model, called Symmetric Multiprocessing (SMP) in the architecture community and *shared everything* in the database community (DeWitt and Gray 1992; Valduriez 1993). However the scalability of such architectures is limited, so for higher degrees of parallelism, a cluster of SMP nodes is used. This model, called *shared-nothing* in database literature, is also the preferred architecture for parallel databases (DeWitt and Gray 1992).

Redundant arrays of independent (or inexpensive) disks (RAID) (Chen *et al.* 1994) has gained popularity to increase I/O bandwidth and storage capacity, reduce latency, and (optionally) support fault tolerance. In many systems, since the amount of data exceeds that which can be stored on disk, *tertiary storage* is used, typically consisting of one or more removable media devices with a juke box to swap the loaded media.

In addition to the current trends, there have been other ideas to improve the memory and storage bottlenecks. *Active Disks* (Riedel *et al.* 1997) and *Intelligent Disks* (Keeton *et al.* 1998) have been proposed as a means to exploit the improved processor performance of embedded processors in disk controllers to allow more complex I/O operations and optimizations, while reducing the amount of traffic over a congested I/O bus. Intelligent RAM (IRAM) (Kozyrakis and Patterson 1998) seeks to integrate processing elements in the memory. Active disks and IRAM are not currently prevalent, as the required hardware and systems software are not commonly available.

5   Software Infrastructure

Since our goal is to use commodity hardware, much of the support for our desired functionality is pushed back into the software. In this section we discuss some of the system transparency issues in PKDD systems, i.e., support for seamless access to databases and file systems and parallel I/O. We review selected aspects of these areas.

The most common database constructions currently in use are relational databases, object oriented databases, and object-relational databases. The data base layer ensures referential integrity and provides support for queries and/or transactions on the data set (Oszu and Valduriez 1999). The data base layer is frequently accessed via a query language, such as SQL. We are primarily interested in parallel and distributed database systems (DeWitt and Gray 1992; Valduriez 1993), which have data sets spanning disks. The primary advantages of such systems are that capacity of storage is improved and that parallelizing of disk access improves bandwidth and (for large I/O's) can reduce latency. Early on parallel database research explored special-purpose database machines for performance (Hsiao 1983), but today the consensus is that its better to use available parallel platforms, with shared-nothing paradigm as the architecture of choice. Shared-nothing database systems include Teradata, Gamma (D. DeWitt et al. 1990), Tandem (Tandem Performance Group 1988), Bubba (Boral *et al.* 1990), Arbre (Lorie *et al.* 1989), etc. We refer the reader to (DeWitt and Gray 1992; Valduriez 1993; Khan *et al.* 1999) for excellent survey articles on parallel and distributed databases. Issues within parallel database research of relevance to PKDD include the data partitioning (over disks) methods used, such as simple *round-robin* partitioning, where records are distributed evenly among the disks. *Hash partitioning* is most effective for applications requiring associative access since records are partitioned based on a hash function. Finally, *range partitioning* clusters records with similar attributes together. Most parallel data mining work to-date has used a round-robin approach to data partitioning. Other methods might be more suitable. Exploration of efficient multidimensional indexing structures for PKDD is required (Gaede and Gunther 1998). The vast amount of work on parallel relational query operators, particularly parallel join algorithms, is also of relevance (Pirahesh *et al.* 1990). The use of DBMS *views* (Oszu and Valduriez 1999) to restrict the access of a DBMS user to a subset of the data, can be used to provide security in KDD systems.

Parallel I/O and file systems techniques are geared to handling large data sets in a distributed memory environment, and appear to be a better fit than distributed file systems for managing the large data sets found in KDD applications. Parallel File Systems and Parallel I/O techniques have been widely studied; (Kotz) maintains an archive and bibliography, which has a nice reference guide (Stockinger 1998). Use of parallel I/O and file systems becomes necessary if RAID devices have insufficient capacity (due to scaling limitations) or contention for shared resources (e.g. buses or processors) exceeds the capacity of SMP architectures. The Scalable I/O initiative (SIO) includes many groups, including the *Message Passing Interface* (MPI) forum, which has adopted a MPI-IO API (Thakur *et al.* 1999) for parallel file management. MPI-IO is layered on top of local file systems. MPI uses a run time type definition scheme to define communication and I/O entity types. The ROMIO library (Thakur *et al.* 1999) implements MPI-IO in Argonne's MPICH implementation of MPI. ROMIO automates scheduling of aggregated I/O requests and uses the ADIO middleware layer to provide portability and isolate implementation dependent parts of MPIO. PABLO, another SIO member group, has created the *portable parallel file systems* (PPFS II), designed to support efficient access of large data sets in scientific applications with irregular access patterns. More information on parallel and distributed I/O and file systems appears in (Kotz ; Carretero *et al.* 1996; Gibson *et al.* 1999; Initiative ; Moyer and Sunderam 1994; Nieuwejaar and Kotz 1997; Schikuta *et al.* 1998; Seamons and Winslett 1996).

Users of PKDD systems are interested in maximizing performance. Prefetching is an important performance enhancing technique that can reduce the impact of latency by overlapping computation and I/O (Cortes 1999; Kimbrel *et al.* 1996; Patterson III 1997). In order for prefetching to be effective, the distributed system uses *hints* which indicate what data is likely to be used in the near future. Generation of accurate hints (not surprisingly) tends to be difficult since it relies on predicting a program's flow of control. Many hint generation techniques rely on traces of a program's I/O access patterns. (Kimbrel *et al.* 1996) surveyed a range of trace driven techniques and prefetching strategies, and provided performance comparisons. (Madhyastha and Reed 1997) recently used machine learning tools to analyze I/O traces from the PPFS, relying on artificial neural networks for on-line analysis of the current trace, and hidden markov models to analyze data obtained by profiling. (Chang and Gibson 1999) developed *SpecHint* which generates hints via speculative execution. We conjecture that PKDD techniques can be used to identify reference patterns, to provide hint generation and to address open performance analysis issues (Reed *et al.* 1998).

As we noted earlier, integration of various systems components for effective KDD is lagging. The current state of KDD tools can accurately be captured by the term *flat-file mining*, i.e., prior to mining, all the data is extracted into a flat-file, which is then used for mining, effectively bypassing all database functionality. This is mainly because traditional databases are ill-equipped to handle/optimize the complex query structure of mining methods. However, recent work has recognized the need for integrating of the database, query management and data mining layers (Agrawal and Shim 1996; Sarawagi *et al.* 1998). (Agrawal and Shim 1996) postulated that better integration of the query manager, database and data mining layers would provide a speedup. (Sarawagi *et al.* 1998) confirmed that performance improvements could be attained, with the best performance obtained in *cache-mine* which caches and mines the query results on a local disk. SQL-like operators for mining association rules have also been developed (Meo *et al.* 1996). Further, proposals for data mining query language (Han *et al.* 1996; Imielinski and Mannila 1996; Imielinski *et al.* 1996; Siebes 1995) have emerged. We note that most of this work is targeted for serial environments. PKDD efforts will benefit from this research, but the optimization problems will of course be different in a parallel setting. Some exceptions include the parallel generic primitives proposed in (Freitas and Lavington 1998), and Data Surveyor (Holsheimer *et al.* 1996), a mining tool that uses the Monet database server for parallel classification rule induction. We further argue that we need a wider integration of parallel and distributed databases and file systems, to fully mine all available data (only a modest fraction of which actually resides in databases). Integration of PKDD and parallel file systems should enhance performance by improving hint generation in prefetching. Integrated PKDD can use parallel file systems for storing and managing large data sets and use distributed file system as an access point suited to mobile clients for management of query results.

6 Conclusions

We described a list of desirable design features of parallel KDD systems. These requirements motivated a brief survey of existing algorithmic and systems support for building such large-scale mining tools. We focused on the state-of-the-art in databases, and parallel I/O techniques. We observe that implementing a effective PKDD system requires integration of these diverse sub-fields into a coherent and seamless system. Emerging issues in PKDD include benchmarking, security, availability, mobility and QoS, motivating fresh research in these disciplines. Finally, PKDD approaches may be used as a tool in these areas (e.g. hint generation for prefetching in parallel I/O), resulting in a bootstrapping approach to software development.

References

R. Agrawal and J. Shafer. Parallel mining of association rules. *IEEE Trans. on Knowledge and Data Engg.*, 8(6):962–969, December 1996.

R. Agrawal and K. Shim. Developing tightly-coupled data mining applications on a relational DBMS. In *Int'l Conf. on Knowledge Discovery and Data Mining*, 1996.

R. Agrawal, T. Imielinski, and A. Swami. Database mining: A performance perspective. *IEEE Trans. on Knowledge and Data Engg.*, 5(6):914–925, December 1993.

S. Anand, *et al.* Designing a kernel for data mining. *IEEE Expert*, pages 65–74, March 1997.

H. Boral, *et al.* Prototyping Bubba, a highly parallel database system. *IEEE Trans. on Knowledge and Data Engg.*, 2(1), March 1990.

J. Carretero *et al.* ParFiSys: A parallel file system for MPP. *ACM Operating Systems Review*, 30(2):74–80, 1996.

F. Chang and G. Gibson. Automatic i/o hint generation through speculative execution. In *Symp. on Operating Systems Design and Implementation*, February 1999.

P. M. Chen, *et al.* RAID: High-performance, reliable secondary storage. *ACM Computing Surveys*, 26(2):145–185, June 1994.

D. Cheung, *et al.* A fast distributed algorithm for mining association rules. In *4th Int'l Conf. Parallel and Distributed Info. Systems*, December 1996.

A. Choudhary and D. Kotz. Large-scale file systems with the flexibility of databases. *ACM Computing Surveys*, 28A(4), December 1996.

T. Cortes. *High Performance Cluster Computing*, Vol. 1, chapter Software Raid and Parallel File Systems, pages 463–495. Prentice Hall, 1999.

D. DeWitt et al. The GAMMA database machine project. *IEEE Trans. on Knowledge and Data Engg.*, 2(1):44–62, March 1990.

D. DeWitt and J. Gray. Parallel database systems: The future of high-performance database systems. *Communications of the ACM*, 35(6):85–98, June 1992.

I. S. Dhillon and D. S. Modha. A clustering algorithm on distributed memory machines. In *Zaki and Ho*, 2000.

A. Freitas and S. Lavington. *Mining very large databases with parallel processing.* Kluwer Academic Pub., 1998.

V. Gaede and O. Gunther. Multidimensional access methods. *ACM Computing Surveys*, 30(2):170–231, 1998.

G. Gibson, *et al.* NASD scalable storage systems. In *USENIX99, Extreme Linux Workshop*, June 1999.

J. Han, *et al.* DMQL: A data mining query language for relational databases. In *SIGMOD Workshop on Research Issues in Data Mining and Knowledge Discovery*, June 1996.

E-H. Han, G. Karypis, and V. Kumar. Scalable parallel data mining for association rules. In *ACM SIGMOD Conf. Management of Data*, May 1997.

M. Holsheimer, M. L. Kersten, and A. Siebes. Data surveyor: Searching the nuggets in parallel. In U. Fayyad, G. Piatetsky-Shapiro, P. Smyth, and R. Uthurusamy, editors, *Advances in Knowledge Discovery and Data Mining*. AAAI Press, 1996.

D. Hsiao. *Advanced Database Machine Architectures.* Prentice Hall, 1983.

T. Imielinski and H. Mannila. A database perspective on knowledge discovery. *Communications of the ACM*, 39(11), November 1996.

T. Imielinski, A. Virmani, and A. Abdulghani. DATAMINE: Application programming interface and query language for database mining. In *Int'l Conf. Knowledge Discovery and Data Mining*, August 1996.

Scalable I/O Initiative. *http://www.cacr.caltech.edu/SIO.* California Institute of Technology.

M. Joshi, G. Karypis, and V. Kumar. ScalParC: A scalable and parallel classification algorithm for mining large datasets. In *Int'l Parallel Processing Symposium*, 1998.

D. Judd, P. McKinley, and A. Jain. Large-scale parallel data clustering. In *Int'l Conf. Pattern Recognition*, 1996.

H. Kargupta and P. Chan, editors. *Advances in Distributed Data Mining.* AAAI Press, 2000.

K. Keeton, D. Patterson, and J.M. Hellerstein. The case for intelligent disks. *SIGMOD Record*, 27(3):42–52, September 1998.

M.F. Khan, *et al.* Intensive data management in parallel systems: A survey. *Distributed and Parallel Databases*, 7:383–414, 1999.

T. Kimbrel, *et al.* A trace-driven comparison of algorithms for parallel prefetching and caching. In *USENIX Symp. on Operating Systems Design and Implementation*, pages 19–34, October 1996.

D. Kotz. The parallel i/o archive. Includes pointers to his Parallel I/O Bibliography, can be found at *http://www.cs.dartmouth.edu/pario/*.

C. E. Kozyrakis and D. A. Patterson. New direction in computer architecture research. *IEEE Computer*, pages 24–32, November 1998.

R. Lorie, *et al.* Adding inter-transaction parallelism to existing DBMS: Early experience. *IEEE Data Engineering Newsletter*, 12(1), March 1989.

T. M. Madhyastha and D. A. Reed. Exploiting global input/output access pattern classification. In *Proceedings of SC'97*, 1997. On CDROM.

R. Meo, G. Psaila, and S. Ceri. A new SQL-like operator for mining association rules. In *Int'l Conf. Very Large Databases*, 1996.

S. A. Moyer and V. S. Sunderam. PIOUS: a scalable parallel I/O system for distributed computing environments. In *Scalable High-Performance Computing Conf.*, 1994.

N. Nieuwejaar and D. Kotz. The galley parallel file system. *Parallel Computing*, 23(4), June 1997.

M. T. Oszu and P. Valduriez. *Principles of Distributed Database Systems.* Prentice Hall, 1999.

R. H. Patterson III. *Informed Prefetching and Caching.* PhD thesis, Carnegie Mellon University, December 1997.

Pirahesh *et al. Parallelism in Relational Data Base Systems.* In *nt'l Symp. on Parallel and Distributed Systems*, 1990.

D. A. Reed, *et al.* Performance analysis of parallel systems: Approaches and open problems. In *Joint Symposium on Parallel Processing (JSPP)*, June 1998.

E. Riedel, G. A. Gibson, and C. Faloutsos. Active storage for large-scale data mining and multimedia. In *Int'l Conf. on Very Large Databases*, August 1997.

H. Nagesh S. Goil and A. Choudhary. MAFIA: Efficient and scalable subspace clustering for very large data sets. Technical Report 9906-010, Northwestern University, June 1999.

S. Sarawagi, S. Thomas, and R. Agrawal. Integrating association rule mining with databases: alternatives and implications. In *ACM SIGMOD Conf. on Management of Data*, June 1998.

E. Schikuta, T. Fuerle, and H. Wanek. ViPIOS: The vienna parallel input/output system. In *Euro-Par'98*, September 1998.

K. E. Seamons and M. Winslett. Multidimensional array I/O in Panda 1.0. *Journal of Supercomputing*, 10(2):191–211, 1996.

J. Shafer, R. Agrawal, and M. Mehta. Sprint: A scalable parallel classifier for data mining. In *Int'l Conf. on Very Large Databases*, March 1996.

T. Shintani and M. Kitsuregawa. Mining algorithms for sequential patterns in parallel: Hash based approach. In *2nd Pacific-Asia Conf. on Knowledge Discovery and Data Mining*, April 1998.

A. Siebes. Foundations of an inductive query language. In *Int'l Conf. on Knowledge Discovery and Data Mining*, August 1995.

D. Skillicorn. Strategies for parallel data mining. *IEEE Concurrency*, 7(4):26–35, October-December 1999.

M. Sreenivas, K. Alsabti, and S. Ranka. Parallel out-of-core divide and conquer techniques with application to classification trees. In *Int'l Parallel Processing Symposium*, April 1999.

H. Stockinger. Dictionary on parallel input/output. Master's thesis, Dept. of Data Engineering, University of Vienna, February 1998.

Tandem Performance Group. A benchmark of non-stop SQL on the debit credit transaction. In *SIGMOD Conference*, June 1988.

R. Thakur, W. Gropp, and E. Lusk. On implementing mpi-io portably and with high performance. In *Workshop on I/O in Parallel and Distributed Systems*, May 1999.

P. Valduriez. Parallel database systems: Open problems and new issues. *Distributed and Parallel Databases*, 1:137–165, 1993.

G. Williams, *et al.* The integrated delivery of large-scale data mining: The ACSys data mining project. In *Zaki and Ho*, 2000.

M. J. Zaki and C-T. Ho, editors. *Large-Scale Parallel Data Mining*, LNCS Vol. 1759. Springer-Verlag, 2000.

M. J. Zaki, *et al.* Parallel algorithms for fast discovery of association rules. *Data Mining and Knowledge Discovery: An International Journal*, 1(4):343–373, December 1997.

M. J. Zaki, C.-T. Ho, and R. Agrawal. Parallel classification for data mining on shared-memory multiprocessors. In *Int'l Conf. on Data Engineering*, March 1999.

M. J. Zaki. Parallel and distributed association mining: A survey. *IEEE Concurrency*, 7(4):14–25, 1999.

M. J. Zaki. Parallel sequence mining on SMP machines. In *Zaki and Ho*, 2000.

Parallel Data Mining on ATM-Connected PC Cluster and Optimization of its Execution Environments

Masato OGUCHI[12] and Masaru KITSUREGAWA[1]

[1] Institute of Industrial Science, The University of Tokyo
7-22-1 Roppongi, Minato-ku Tokyo 106-8558, Japan
[2] Informatik4, Aachen University of Technology
Ahornstr.55, D-52056 Aachen, Germany
oguchi@tkl.iis.u-tokyo.ac.jp

Abstract. In this paper, we have constructed a large scale ATM-connected PC cluster consists of 100 PCs, implemented a data mining application, and optimized its execution environment. Default parameters of TCP retransmission mechanism cannot provide good performance for data mining application, since a lot of collisions occur in the case of all-to-all multicasting in the large scale PC cluster. Using a TCP retransmission parameters according to the proposed parameter optimization, reasonably good performance improvement is achieved for parallel data mining on 100 PCs.

Association rule mining, one of the best-known problems in data mining, differs from conventional scientific calculations in its usage of main memory. We have investigated the feasibility of using available memory on remote nodes as a swap area when working nodes need to swap out their real memory contents. According to the experimental results on our PC cluster, the proposed method is expected to be considerably better than using hard disks as a swapping device.

1   Introduction

Looking over the recent technology trends, PC/WS clusters connected with high speed networks such as ATM are considered to be a principal platform for future high performance parallel computers. Applications which formerly could only be implemented on expensive massively parallel processors can now be executed on inexpensive clusters of PCs. Various research projects to develop and examine PC/WS clusters have been performed until now[1][2][3]. Most of them however, only measured basic characteristics of PCs and networks, and/or some small benchmark programs were examined. We believe that data intensive applications such as data mining and ad-hoc query processing in databases are quite important for future high performance computers, in addition to the conventional scientific applications[4].

Data mining has attracted a lot of attention recently from both the research and commercial community, for finding interesting trends hidden in large transaction logs. Since data mining is a very computation and I/O intensive process,

J. Rolim et al. (Eds.): IPDPS 2000 Workshops, LNCS 1800, pp. 366-373, 2000.

parallel processing is required to supply the necessary computational power for very large mining operations. In this paper, we report the results on parallel data mining on ATM-connected PC clusters, consists of 100 Pentium Pro PCs.

2   Our ATM-connected PC cluster and its communication characteristics

We have constructed a PC cluster pilot system which consists of 100 nodes of 200MHz Pentium Pro PCs connected with an ATM switch. An overview of the PC cluster is shown in Figure 1. Each node of the cluster is equipped with 64Mbytes main memory, 2.5Gbytes IDE hard disk, and 4.3Gbytes SCSI hard disk. Solaris ver.2.5.1 is used as an operating system.

All nodes of the cluster are connected by a 155Mbps ATM LAN as well as an Ethernet. HITACHI's AN1000-20, which has 128 port 155Mbps UTP-5, is used as an ATM switch. Interphase 5515 PCI ATM adapter and RFC-1483 PVC driver, which support LLC/SNAP encapsulation for IP over ATM, are used. Only UBR traffic class is supported in this driver.

TCP/IP is used as a communication protocol. TCP is not only a very popular reliable protocol for computer communication, but also contains all functions as a general transport layer. Thus the results of our experiments must be valid even if other transport protocol is used, for investigating reliable communication protocols on a large scale cluster.

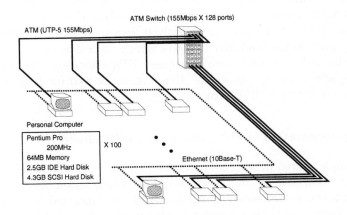

Fig. 1. An overview of the PC cluster

3 Parallel data mining application and its implementation on the cluster

3.1 Association rule mining

Data mining is a method of the efficient discovery of useful information such as rules and previously unknown patterns existing among data items embedded in large databases, which allows more effective utilization of existing data. One of the best known problems in data mining is mining of the association rules from a database, so called "basket analysis"[5][6]. Basket type transactions typically consist of transaction id and items bought per-transaction. An example of an association rule is "if customers buy A and B then 90% of them also buy C". The best known algorithm for association rule mining is Apriori algorithm proposed by R. Agrawal of IBM Almaden Research[7].

In order to improve the quality of the rule, we have to analyze very large amounts of transaction data, which requires considerably long computation time. We have studied several parallel algorithms for mining association rules until now[8], based on Apriori. One of these algorithms, called HPA(Hash Partitioned Apriori), is implemented and evaluated.

Apriori first generates candidate itemsets, then scans the transaction database to determine whether the candidates satisfy the user specified minimum support. At first pass (pass 1), a support for each item is counted by scanning the transaction database, and all items which satisfy the minimum support are picked out. These items are called large 1-itemsets. In the second pass (pass 2), 2-itemsets (length 2) are generated using the large 1-itemsets. These 2-itemsets are called candidate 2-itemsets. Then supports for the candidate 2-itemsets are counted by scanning the transaction database, large 2-itemsets which satisfy the minimum support are determined. This iterative procedure terminates when large itemset or candidate itemset becomes empty. Association rules which satisfy user specified minimum confidence can be derived from these large itemsets.

HPA partitions the candidate itemsets among processors using a hash function, like the hash join in relational databases. HPA effectively utilizes the whole memory space of all the processors. Hence it works well for large scale data mining. In the detail of HPA, please refer to [8][9].

3.2 Implementation of HPA program on PC cluster

We have implemented HPA program on our PC cluster. Each node of the cluster has a transaction data file on its own hard disk. Solaris socket library is used for the inter-process communication. All processes are connected with each other by socket connections, thus forming mesh topology. In the ATM level, PVC (Permanent Virtual Channel) switching is used since the data is transferred continuously among all the processes.

Transaction data is produced using data generation program developed by Agrawal, designating some parameters such as the number of transaction, the

Table 1. The number of candidate and large itemsets

C	the number of candidate itemsets
L	the number of large itemsets
T	the execution time of each pass [sec]

pass	C	L	T
pass 1		1023	11.2
pass 2	522753	32	69.8
pass 3	19	19	3.2
pass 4	7	7	6.2
pass 5	1	0	12.1

number of different items, and so on. The produced data is divided by the number of nodes, and copied to each node's hard disk.

The parameters used in the evaluation is as follows: The number of transaction is 10,000,000, the number of different items is 5000 and minimum support is 0.7%. The size of the transaction data is about 800Mbytes in total. The message block size is set to be 8Kbytes and the disk I/O block size is 64Kbytes.

The numbers of candidate itemsets and large itemsets, and the execution time of each pass executed on 100 nodes PC cluster are shown in Table 1. Note that the number of candidate itemsets in pass 2 is extremely larger than other passes, which often happens in association rules mining.

4   Optimization of transport layer protocol parameters

4.1   Broadcasting on the cluster and TCP retransmission

The execution times of pass $3 - 5$ are relatively long in Table 1, although they do not have large number of itemsets. At the end of each pass, a barrier synchronization and exchange of data are needed among all nodes, that is, all-to-all broadcasting takes place. Even if the amount of broadcasting data is not large, cells must be discarded at the ATM switch if timing of the broadcasting is the same at all nodes. Since pass $3 - 5$ have little data to process, actual execution time is quite short, thus broadcasting is performed almost simultaneously in all nodes, which tend to cause network congestion and TCP retransmission as a result. We have executed several experiments to find the better retransmission parameters setting suitable for such cases.

We use TCP protocol implemented in Solaris OS, whose parameters can be changed with user level commands. Two parameters changed here are 'maximum interval of TCP retransmission' and 'minimum interval of TCP retransmission', which we call 'MAX' and 'MIN' respectively. The default setting is MAX = 60000 [msec] and MIN = 200 [msec] in the current version of Solaris. The interval of

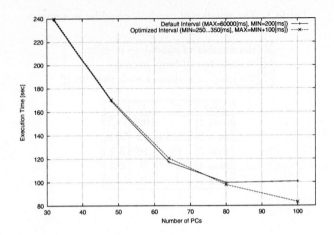

Fig. 2. Execution time of HPA program on PC cluster

TCP retransmission is dynamically changed in the protocol, within the limits between MAX and MIN.

As a result of experiments, we have found that the default value of MAX is not suitable for the cluster, which might cause the unnecessary long retransmission interval. MAX should be set to be smaller than the default value, such as MAX = MIN + 100[msec]. Moreover, MIN is better to be set as random value, which can prevent the collision of the cells at ATM switch.

4.2   Total performance of HPA program using proposed method

HPA program is executed using the proposed parameter setting of TCP on the PC cluster pilot system. The execution time of HPA program is shown in Figure 2. In this Figure, one line indicates the case using default TCP retransmission parameters, i.e. MAX = 60000[msec] and MIN = 200[msec], and the other line indicates the case using random parameters (MIN = 250 ... 350[ms], MAX = MIN + 100[ms]).

Reasonably good speedup is achieved up to 100 PCs using proposed optimized parameters. Since the application itself is not changed, the difference only comes from TCP retransmission, occurred along with barrier synchronization and all-to-all data broadcasting.

5   Dynamic remote memory acquisition

5.1   Dynamic remote memory acquisition and its experiments

As shown in section 3, the number of candidate itemsets in pass 2 is very much larger than in other passes in association rule mining. The number of itemsets is strongly dependent on user-specified conditions, such as the minimum support value, and it is difficult to predict how large the number will be before execution. Therefore, it may happen that the number of candidate itemsets increases dramatically in this step so that the memory requirement becomes extremely large. When the required memory is larger than the real memory size, part of the contents of memory must be swapped out. However, because the size of each data item is rather small and all the data is accessed almost at random, swapping out to a storage device is expected to degrade the total performance severely.

We have executed several experiments in which available memory in remote nodes is used as a swap area when huge memory is dynamically required. In the experiments, a limit value for memory usage of candidate itemsets is set at each node. When the amount of memory used exceeds this value during the execution of the HPA program(in Pass 2), part of the contents is swapped out to available memory in remote nodes, that is, application execution nodes acquire remote memory dynamically. Although such available remote nodes could be found dynamically in a real system, we selected them statically in these experiments. On the other hand, when an application execution node trys to access an item that had been swapped out, a pagefault occurs.

The basic behavior of this approach has something in common with distributed shared memory systems[10], memory management system in distributed operating systems[11], or cache mechanism in client-server database systems[12]. For example, if data structures inside applications are considered in distributed shared memory, almost the same effect can be expected. That is to say, it is possible to program almost the same mechanism using some types of distributed shared memory systems. Thus, our mechanism might be regarded as equivalent to a case of distributed shared memory optimized for a particular application.

We have executed experiments of the proposed mechanism on the PC cluster. The parameters used in the experiment are as follows. The number of transactions is 1,000,000, the number of different items is 5,000, and the minimum support is 0.1%. The number of application execution nodes is 8 in this evaluation. The number of memory available nodes is varied from 1 to 16. With these conditions, the total number of candidate itemsets in pass 2 is 4,871,881. Since each candidate itemset occupied 24 bytes in total(structure area + data area), approximately 14-15Mbytes of memory were filled with these candidate itemsets at each node.

5.2   Remote update method

When memory usage is limited, the execution time is much longer than when there is no memory limit. This is because the number of swapouts is extremely large. In Table 2 the numbers of pagefaults on each application execution node are shown. Because most of the memory contents are accessed repeatedly, a kind of thrashing seems to happen in these cases. In order to prevent this phenomenon, a method for restricting swapping operations is proposed.

When usage of memory reaches the limit value at a node, it acquires remote memory and swaps out part of its memory contents. The contents will be

Table 2. The numbers of pagefaults on each application node

Usage limit	node 1	node 2	node 3	node 4	node 5	node 6	node 7	node 8
12[MB]	1606258	2925254	1306521	2361756	1671840	1723410	2166277	2545003
13[MB]	885798	1896226	593000	1374688	932374	896150	1326941	1375398
14[MB]	254094	1003757		512984	286945	191102	601657	407628
15[MB]		268039						

swapped in again if this data is accessed later. Instead of swapping, it is sometimes better to send update information to the remote memory when a pagefault occurs. That is to say, once some contents are swapped out to memory in a distant node, they are fixed there and accessed only through a remote memory access interface provided by library functions. This remote update method has been applied only to the itemsets counting phase, for simplicity.

The access interface function has been developed to realize the remote update operations. The execution time using this method is shown in Figure 3. This figure shows the execution time of pass 2 of the HPA program, when the number of memory available nodes is 16. The execution times for dynamic remote memory acquisition, according to this method and the previous simple swapping case, are compared in the Figure. The execution time using hard disks as a swapping device is also shown, for comparison. Seagate Barracuda 7,200[rpm] SCSI hard disks have been used for this purpose. Other conditions are the same as the case of dynamic remote memory acquisition.

The execution time using hard disks as swapping devices is very long, especially when the memory usage limit is small, because each access time to a hard disk is much longer than that for remote memory through the network. The execution time of dynamic remote memory acquisition with simple swapping is better than for swapping out to hard disks. It increases, however, when the memory usage limit is small, since the number of pagefaults becomes extremely large in such a case.

Compared to these results, the execution time of dynamic remote memory acquisition with remote update operations is quite short, even when the memory usage limit is small. It seems to be effective to provide a simple remote access interface for the itemsets counting phase, because the number of swapping operations during this phase is very large. These results indicate that, performance of the proposed remote memory acquisition with remote update operations is considerably better than other methods.

References

1. C. Huang and P. K. McKinley: "Communication Issues in Parallel Computing Across ATM Networks", *IEEE Parallel and Distributed Technology*, Vol.2, No.4, pp.73-86, 1994.

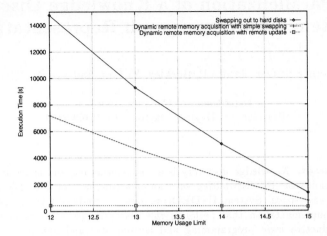

Fig. 3. Comparison of proposed methods

2. R. Carter and J. Laroco: "Commodity Clusters: Performance Comparison Between PC's and Workstations", *Proceedings of the Fifth IEEE International Symposium on High Performance Distributed Computing*, pp.292-304, August 1996.

3. D. E. Culler et al.: "Parallel Computing on the Berkeley NOW", *Proceedings of the 1997 Joint Symposium on Parallel Processing(JSPP '97)*, pp.237-247, May 1997.

4. T. Tamura, M. Oguchi, and M. Kitsuregawa: "Parallel Database Processing on a 100 Node PC Cluster: Cases for Decision Support Query Processing and Data Mining", *Proceedings of SuperComputing '97*, November 1997.

5. U. M. Fayyad et al.: "Advances in Knowledge Discovery and Data Mining", *The MIT Press*, 1996.

6. V. Ganti, J. Gehrke, and R. Ramakrishnan: "Mining Very Large Databases", *IEEE Computer*, Vol.32, No.8, pp.38-45, August 1999.

7. R. Agrawal, T. Imielinski, and A. Swami: "Mining Association Rules between Sets of Items in Large Databases", *Proceedings of the ACM International Conference on Management of Data*, pp.207-216, May 1993.

8. T. Shintani and M. Kitsuregawa: "Hash Based Parallel Algorithms for Mining Association Rules", *Proceedings of the Fourth IEEE International Conference on Parallel and Distributed Information Systems*, pp.19-30, December 1996.

9. M. J. Zaki: "Parallel and Distributed Association Mining: A Survey", *IEEE Concurrency*, Vol.7, No.4, pp.14-25, 1999.

10. C. Amza et al.: "TreadMarks: Shared Memory Computing on Networks of Workstations", *IEEE Computer*, Vol.29, No.2, pp.18-28, February 1996.

11. M. J. Feeley et al.: "Implementing Global Memory Management in a Workstation Cluster", *Proceedings of the ACM Symposium on Operating Systems Principles*, pp.201-212, December 1995.

12. S. Dar et al.: "Semantic Data Caching and Replacement", *Proceedings of 22nd VLDB Conference*, September 1996.

The Parallelization of a Knowledge Discovery System with Hypergraph Representation*

Jennifer Seitzer, James P. Buckley, Yi Pan, and Lee A. Adams

Department of Computer Science
University of Dayton, Dayton, OH 45469-2160

Abstract. Knowledge discovery is a time-consuming and space inten-sive endeavor. By distributing such an endeavor, we can diminish both time and space. System **INDED**(pronounced "indeed") is an induc-tive implementation that performs rule discovery using the techniques of inductive logic programming and accumulates and handles knowl-edge using a deductive nonmonotonic reasoning engine. We present four schemes of transforming this large serial inductive logic programming (ILP) knowledge-based discovery system into a distributed ILP discov-ery system running on a Beowulf cluster. We also present our data parti-tioning algorithm based on locality used to accomplish the data decom-position used in the scenarios.

1 Introduction

Knowledge discovery in databases has been defined as the non-trivial process of identifying valid, novel, potentially useful, and understandable patterns in data [PSF91]. Data mining is a commonly used knowledge discovery technique that attempts to reveal patterns within a database in order to exploit implicit information that was previously unknown [CHY96]. One of the more useful ap-plications of data mining is to generate all significant associations between items in a data set [AIS93]. A discovered pattern is often denoted in the form of an IF-THEN rule (IF *antecedent* THEN *consequent*), where the antecedent and con-sequent are logical conjunctions of predicates (first order logic) or propositions (propositional logic) [Qui86]. Graphs and hypergraphs are used extensively as knowledge representation constructs because of their ability to depict causal chains or networks of implications by interconnecting the consequent of one rule to the antecedent of another.

In this work, using the language of logic programming, we use a hypergraph to represent the knowledge base from which rules are mined. Because the hyper-graph gets inordinantly large in the serial version of our system [Sei99], we have devised a parallel implementation where, on each node, a smaller sub-hypergraph is created. Consequently, because there is a memory limit to the size of a storable hypergraph, by using this parallel version, we are able to grapple with problems

* This work is partially supported under Grant 9806184 of the National Science Foundation.

J. Rolim et al. (Eds.): IPDPS 2000 Workshops, LNCS 1800, pp. 374-381, 2000.

involving larger knowledge bases than those workable on the serial system. A great deal of work has been done in parallelizing unguided discovery of association rules originally in [ZPO97] and recently refined in [SSC99]. The novel aspects of this work include the parallelization of both a nonmonotonic reasoning system and an ILP learner. In this paper, we present the schemes we have explored and are currently exploring in this pursuit.

2   Serial System INDED

System **INDED** is a knowledge discovery system that uses inductive logic programming (ILP) [LD94] as its discovery technique. To maintain a database of background knowledge, **INDED** houses a deduction engine that uses deductive logic programming to compute the current state (current set of true facts) as new rules and facts are procured.

2.1   Inductive Logic Programming

Inductive logic programming (ILP) is a new research area in artificial intelligence which attempts to attain some of the goals of machine learning while using the techniques, language, and methodologies of logic programming. Some of the areas to which ILP has been applied are data mining, knowledge acquisition, and scientific discovery [LD94]. The goal of an inductive logic programming system is to output a rule which *covers* (entails) an entire set of positive observations, or examples, and *excludes* or *does not cover* a set of negative examples [Mug92]. This rule is constructed using a set of known facts and rules, knowledge, called domain or *background* knowledge. In essence, the ILP objective is to synthesize a logic program, or at least part of a logic program using examples, background knowledge, and an entailment relation. The following definitions are from [LD94].

Definition 2.1 (coverage, completeness, consistency) *Given background knowledge $\mathcal{B}$, hypothesis $\mathcal{H}$, and example set $\mathcal{E}$, hypothesis $\mathcal{H}$ **covers** example $e \in \mathcal{E}$ with respect to $\mathcal{B}$ if $\mathcal{B} \cup \mathcal{H} \models e$. A hypothesis $\mathcal{H}$ is **complete** with respect to background $\mathcal{B}$ and examples $\mathcal{E}$ if all positive examples are covered, i.e., if for all $e \in \mathcal{E}^+$, $\mathcal{B} \cup \mathcal{H} \models e$. A hypothesis $\mathcal{H}$ is **consistent** with respect to background $\mathcal{B}$ and examples $\mathcal{E}$ if no negative examples are covered, i.e., if for all $e \in \mathcal{E}^-$, $\mathcal{B} \cup \mathcal{H} \not\models e$.*

Definition 2.2 (Formal Problem Statement) *Let $\mathcal{E}$ be a set of training examples consisting of **true** $\mathcal{E}^+$ and **false** $\mathcal{E}^-$ ground facts of an unknown (target) predicate T. Let $\mathcal{L}$ be a description language specifying syntactic restrictions on the definition of predicate T. Let $\mathcal{B}$ be background knowledge defining predicates q_i which may be used in the definition of T and which provide additional information about the arguments of the examples of predicate T. The ILP problem is to produce a definition $\mathcal{H}$ for T, expressed in $\mathcal{L}$, such that $\mathcal{H}$ is complete and consistent with respect to the examples $\mathcal{E}$ and background knowledge $\mathcal{B}$. [LD94]*

2.2   Serial Arichitecture

System **INDED** (pronounced "indeed") is comprised of two main computation engines. The deduction engine is a bottom-up reasoning system that computes the current state by generating a stable model [2], if there is one, of the current ground instantiation represented internally as a hypergraph, and by generating the well-founded model [VRS91], if there is no stable model[GL90]. This deduction engine is, in essence, a justification truth maintenance system which accommodates non-monotonic updates in the forms of positive or negative facts.

The induction engine, using the current state created by the deduction engine as the background knowledge base, along with positive examples $\mathcal{E}^+$ and negative examples $\mathcal{E}^-$, induces a rule(s) which is then used to augment the deductive engine's hypergraph. We use a standard top-down hypothesis construction algorithm (learning algorithm) in **INDED**[LD94]. This algorithm uses two nested programming loops. The outer (covering) loop attempts to cover all positive examples, while the inner loop (specialization) attempts to exclude all negative examples. Termination is dictated by two user-input values to indicate sufficiency and necessity stopping criterea. The following diagram illustrates the discovery constituents of **INDED** and their symbiotic interaction.

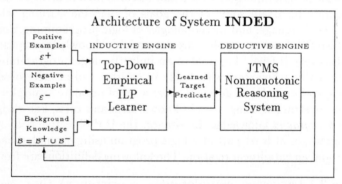

The input files to **INDED** that initialize the system are the extensional database (EDB) and the intensional database (IDB). The EDB is made up of initial ground facts (facts with no variables, only constants). The IDB is made up of universally quantified rules with no constants, only variables. Together, these form the internal ground instantiation represented internally as the deduction engine hypergraph.

3   Parallelizing INDED

Our main goals in parallelizing **INDED** are to obtain reasonably accurate rules faster and to decrease the size of the internal deduction hypergraph so that the

---

[2] Although the formal definitions of these semantics are cited above, for this paper, we can intuitively accept stable and well-founded models as those sets of facts that are generated by transitively applying modus ponens to rules.

problem space is increased. The serial version is very limited in what problems it can solve because of memory limitations. Work has been done in the direct partitioning of a hypergraph [KAK99]. In our pursuit to parallelize **INDED**, however, we are exploring the following schemes where indirect hypergraph reductions are performed. Each of the following scenarios has been devised to be implemented on a Beowulf cluster [Buy99] using MPI [GLS99]:

1. large grained control parallel decomposition where one node runs the induction engine while another node runs the deduction engine
2. large grained control parallel decomposition where a pipeline of processors are established each operating on a different current state as created in previous (or subsequent) pipelined iterations
3. data parallel decomposition where each node runs the same program with smaller input files (hence smaller internal hypergraphs).
4. speculative parallel approach where each node attempts to learn the same rule using a different predicate ranking alogrithm in the induction engine.

4   Naive Decomposition

In this decomposition, we create a very coarse grain system in which two nodes share the execution. One node houses the deduction engine; the other houses the induction. Our strategy lets the induction engine initially discover a target predicate from positive and negative examples and an initial background knowledge base. Meanwhile, the deduction engine computes the current state using initial input files. This current state is sent to the induction engine as its background knowledge base in the subsequent iteration. The learned predicate from the induction engine from one iteration is then fed into the deductive engine to be used during the next iteration in its computation of the current state. This is then used as the background knowledge for the induction engine during the subsequent iteration.

In general, during iteration i, the induction engine computes new intensional rules for the deduction engine to use in its computation of the current state in iteration $i + 1$. Simultaneously, during iteration i, the deduction engine computes a new current state for the induction engine to use as its background knowledge in iteration $i + 1$. The above process is repeated until all target predicates specified have been discovered. As we extend this implementation, we expect to acquire a pipe-lined system where the deduction engine is computing state S_{i+1} while the induction engine is using S_i to induce new rules (where i is the current iteration number).

5   Data Parallel Decomposition with Data Partitioning

In this method, each worker node runs **INDED** when invoked by a master MPI node [GLS99]; each worker executes by running a partial background knowledge base which, as in the serial version, is spawned by its deduction engine. In particular, each worker receives the full serial intensional knowledge base (IDB) but only a partial extensional knowledge base (EDB). The use of a partial EDB creates a significantly smaller (and different) hypergraph on each Beowulf worker node. This decomposition led to a faster execution due to a significantly smaller internal hypergraph being built. The challenge was to determine the best way to decompose the serial large EDB into smaller EDB's so that the rules obtained were as accurate as those learned by the serial version.

5.1 Data Partitioning and Locality

In this data parallel method, our attention centered on decomposition of the input files to reduce the size of any node's deduction hypergraph. We found that in many cases data transactions exhibited a form of locality of reference. Locality of reference is a phenomenon ardently exploited by cache systems where the general area of memory referenced by sequential instructions tends to be repeatedly accessed. Locality of reference in the context of knowledge discovery also exists and should be exploited to increase the efficiency of rule mining. A precept of knowledge discovery is that data in a knowledge base system are nonrandom and tend to cluster in a somewhat predictable manner. This tendency mimics locality of reference. There are three types of locality of reference which may coexist in a knowledge base system: spatial, temporal, and functional. In spatial locality of reference, certain data items appear together in a physical section of a database. In temporal locality of reference, the data items that are used in the recent past appear in the near future. For example, if there is a sale in a supermarket for a particular brand of toothpaste on Monday, we will see a lot of sales for this brand of toothpaste on that day. In functional locality of reference, we appeal to a semantic relationship between entities that have a strong semantic tie that affects data transactions relating to them. For example, cereal and milk are two semantically related objects. Although they are typically located in different areas of a store, many purchase transactions of one, include the other. All three of these localities can be exploited in distributed knowledge mining, and help justify the schemes adpoted in our implementations discussed in the following sections.

5.2 Partitioning Algorithm

To retain all global dependencies among the predicates in the current state, all Beowulf nodes receive a full copy of the serial IDB. The serial EDB, the initial large set of facts, therefore, is decomposed and partitioned among the nodes. The following algorithm transforms a large serial extensional database (EDB) into p smaller EDB's to be placed on p Beowulf nodes. It systematically creates sets based on constants appearing in the positive example set $\mathcal{E}^+$. Some facts from the serial EDB could appear on more than one processor. The algorithm is of linear complexity requiring only one scan through the serial EDB and positive example set E+.

Algorithm 5.1 (EDB Partitioning Algorithm) This algorithm is $O(n)$, where n is the number of facts in the EDB.

Input: Number of processors p in Beowulf
          Serial extensional database (EDB)
          Positive and negative example set $\mathcal{E}^+$, $\mathcal{E}^-$
Output: p individual worker node EDB's

BEGIN ALGORITHM 5.1
For each example $e \in \mathcal{E}^+ \cup \mathcal{E}^-$ Do
     For each constant $c \in e$ Do
          create an initially empty set S_c of facts
Create one (initially empty) set S_{none} for facts that have
          no constants in any example $e \in \mathcal{E}^+ \cup \mathcal{E}^-$
For each fact $f \in$ EDB Do
     For each constant $c' \in f$ Do
          $S_{c'} = S_{c'} \cup f$
          If no set exists for c then
               $S_{none} = S_{none} \cup f$
Distribute the contents of S_{none} among all constant sets
Determine load balance by summing all set cardinalities
     to reflect total parallel EDB entries K
Define $min\_local\_load = \lceil K/p \rceil$
Distribute all sets S_{c_i} evenly among the processors
     so that each processor has an EDB of roughly
     equal cardinality such that each node has EDB of
     cardinality $\geq min\_local\_load$ as defined above.
END ALGORITHM 5.1

6   Global Hypergraph using Speculative Parallelism

In this parallelization, each Beowulf node searches the space of all possible rules independently and differently. All input files are the same on all machines. Therefore, each worker is discovering from the same background knowledge base. Every rule discovered by **INDED** is constructed by systematically appending chosen predicate expressions to an originally empty rule body. The predicate expressions are ranked by employing various algorithms, each of which designates a different search strategy. The highest ranked expressions are chosen to constitute the rule body under construction. In this parallelization of **INDED**, each node of the Beowulf employs a different ranking procedure, and hence, may construct very different rules.

We are considering two possibilities for handling the rules generated by each worker. In the first, as soon as a process converges (finds a valid set of rules), it broadcasts a message to announce the end of the procedure. When the message is received by other processes, they are terminated. The other possibility we are considering is to combine the rules of each worker. Different processes may generate different rules due to the use of different ranking algorithms. These rules may be combined after all the processes are terminated, and only good rules are retained. In this way, not only can we speed up the mining process, but we can also achieve a better and richer quality of solutions.

7   Current Status and Results

The current status of our work in these parallelization schemes is as follows. We have implemented the naive decomposition and enjoyed a 50 per cent reduction in execution time. Thus far, however, the bulk of our efforts have centered on implementing and testing the data parallel implementation on an eleven node Beowulf cluster. Here, we also experienced a great reduction in execution time. Figure 1 illustrates the consistent reduction of time as the number of nodes increased. The problem domain with which we are currently experimenting relates to the diagnosis of diabetes. The accuracy of the discovered rules by the cluster has varied. The rule learned by serial **INDED** is

```
inject_insulin(A) <-- insulin_test4(A) .
inject_insulin(A) <-- iddm(A) .
```

We attribute the variance of rule accuracy by the clusters to our partitioning algorithm. We anticipate extensive refinement of this algorithm as we continue this work.

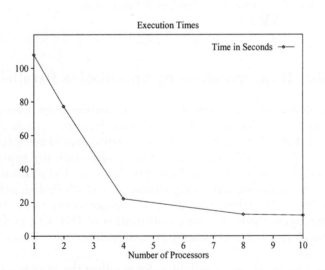

Fig. 1. Performance of Rule Mining on a Cluster.

8   Current and Future Work

We have delineated four parallelization schemes of transforming a large serial reasoning system that both deduces new facts as well as discovers new rules. In successfully implementing two of these schemes,we have found that one of the most interesting problems of parallel rule discovery is effective partitioning of the data. We have performed experimentation with one algorithm and antic- ipate extensive experimentation with new partitioning algorithms of the EDB and background knowlege. Additionally, we are currently implementing the spec- ulative parallelization scheme discussed above, and are enhancing and devising new predicate ranking algorithms used by the induction engine.

References

[AIS93] R. Agrawal, T. Imielinski, and A. Swami. Mining association rules between sets of items in large databases. *SIGMOD Bulletin*, pages 207–216, May 1993.

[Buy99] Rajkumar Buyya. *High Performance Cluster Computing Programming and Applications*. Prentice-Hall, Inc, 1999.

[CHY96] M.S. Chen, J. Han, and P.S. Yu. Data mining: An overview from a database perspective. *IEEE Transactions on Knowledge and Data Engineering*, 8(6), 1996.

[GL90] Michael Gelfond and Vladimir Lifschitz. The stable model semantics for logic programming. In *Proceedings of the Fifth Logic Programming Symposium*, pages 1070–1080, 1990.

[GLS99] William Gropp, Ewing Lusk, and Anthony Skjellum. *Using MPI; Portable Parallel Programming with Message Passing Interface*. The MIT Press, 1999.

[KAK99] G Karypis,R. Agrawal, V. Kumar. Multilevel hypergraph partitioning: Ap- plications in VLSI domain. *IEEE Transactions of VLSI Systems*, 7(1),pages 69–79, Mar 1999.

[LD94] Nada Lavrac and Saso Dzeroski. *Inductive Logic Programming*. Ellis Horwood, Inc., 1994.

[Mug92] Stephen Muggleton, editor. *Inductive Logic Programming*. Academic Press, Inc, 1992.

[PSF91] Piatetsky-Shapiro and Frawley, editors. *Knowledge Discovery in Databases*, chapter Knowledge Discovery in Databases: An Overview. AAAI Press/ The MIT Press, 1991.

[Qui86] J. R. Quindlan. Induction of decision trees. *Machine Learning*, 1:81–106, 1986.

[Sei99] Jennifer Seitzer. **INDED**: A symbiotic system of induction and deduction. In *MAICS-99 Proceedings Tenth Midwest Artificial Intelligence and Cognitive Science Conference*, pages 93–99. AAAI, 1999.

[SSC99] L. Shen, H. Shen, and L. Chen. New algorithms for efficient mining of associa- tion rules. In *7th IEEE Symp. on the Frontiers of Massively Parallel Computation, Annapolis, Maryland*, pages 234–241, Feb 1999.

[VRS91] A. VanGelder, K. Ross, and J. Schlipf. The well-founded semantics for gen- eral logic programs. *Journal of the ACM*, 38(3):620–650, July 1991.

[ZPO97] M. J. Zaki, S. Parthasarathy, and M. Ogihara. Parallel algorithms for dis- covery of association rules. *Data Mining and Knowledge Discovery*, 1:5–35, 1997.

Parallelisation of C4.5 as a Particular Divide and Conquer Computation

Primo Becuzzi, Massimo Coppola, Salvatore Ruggieri, and Marco Vanneschi

Dipartimento di Informatica, Universita' di Pisa

Abstract In this work we show the research track and the current results about the application of structured parallel programming tools to develop scalable data-mining applications. We discuss the exploitation of the divide and conquer nature of the well known C4.5 classification algorithm in spite of its in-core memory requirements. The opportunity of applying external memory techniques to manage the data is advocated. Current results of the experiments are reported.

The main research goal of our group in the past years has been the design and implementation of parallel programming environments to ease the engineering of High Performance applications. To provide a test bed for the current environment, as well as to investigate the theoretical problems involved in the parallelisation of largely used algorithms, we are developing parallel versions of Data-Mining (DM) applications.

Work is ongoing [2,3] to develop DM computational kernels that exhibit both code and performance portability over various parallel architectures, where performance means parallel speed-up and scalability to large databases.

Here we present the current results about the C4.5 algorithm, focusing on two main issues:

- the parallel scalability achievable using structured programming tools (a software engineering perspective)
- the evaluation of strategies to improve the support for tree-structured irregular computations, regarding C4.5 as an algorithm of the Divide and Conquer class.

Dealing with data which exceeds in size the local memory is the most common issue in High Performance Data Mining. We plan to enhance C4.5, which is an in-core classifier, to efficiently manage huge data sets.

We will introduce in the SkIE programming environment some external memory operations (like scan and sort), that exploit at first the sum of all the local memories. We will use those primitives to turn the C4.5 algorithm into a kind of out-of-core classifier. We will investigate if the same approach can be applied to the lower level of the memory hierarchy, disk-resident data.

The the paper is organized as follows. In section 1 we define the problem. Section 2 describes our programming environment. We compare our approach with previous ones in section 3, together with the explanation of the first results. Section 4 explains in detail the current results. The aim and the improvements expected from future work are the subject of the last section.

J. Rolim et al. (Eds.): IPDPS 2000 Workshops, LNCS 1800, pp. 382-389, 2000.

1   Problem statement

We are interested in the parallelisation of the C4.5 core, that is the building of the decision tree, as described in [5]. We'll leave out the evaluation and simplification phases, and we won't discuss the replication of cases with unspecified attributes, nor the windowing and trials techniques, even if these can be conveniently applied.

General divide-and-conquer ($D\&C$) algorithms split (*divide*) large problems into smaller and smaller ones of the same form; the smallest ones are solved, then a recomposition (*conquer*) phase combines the solutions of larger and larger subproblems, up to the initial one. We give here a $D\&C$ description of the core of C4.5.

Input: a database D of n elements, each one a k-tuple of attributes $(a_1, \ldots a_k)$. Each $a_i \in A_i$, where A_i can be an interval (continuous attributes) or a finite set of values (categorical ones). One of the categorical attributes is distinguished as the class of each tuple.

Output: a decision tree T, i.e. a tree which predicts the class of tuples in terms of the values of the other attributes. Leaves of the tree are homogeneous subsets of the data. Each interior node (decision node) defines a split in the data determined by the test of the values of a single attribute, one branch is made for each outcome of the test.

D1: $\forall i \in \{1, k\}$ evaluate for attribute i the information gain $g(D, i)$; let j be the index that maximizes it.

D2cat: if A_j is categorical, let $c = |A_j|$. Split D into c classes according to the value of a_c of each tuple.

D2cont: if A_j is continuous, let $c = 2$ and split D in half such that $D_1 \leq D_2$. (*) Find *in all the data* the threshold value $t \in A_c$ that is the best approximation of the split point.

D3:   build the root node for T and register the choices made so far.

Rec: each $D_1, \ldots D_c$ that does satisfy the stopping criterion becomes a leaf. Process recursively all the others.

Conquer: add each returned leaf or tree $T_1, \ldots T_c$ as the corresponding son of the root of T. Return T as answer.

This is the *growth* phase of the tree, that is followed by a *prune* phase which is less hard to accomplish and we do not describe here.

Steps D1–D3 are the divide phase. The cost of step D1 is $O(n)$ operations for each categorical attribute, $O(n \log n)$ for continuous ones. Categorical attributes already used by an ancestor node are never checked again, since their information gain is zero. Both D2 steps, which are exclusive, require $O(n)$ operations, but finding the threshold (*) in step D2cont has an additional cost $O(N \log N)$ in the size N of the whole initial database.

This behaviour, more deeply analyzed in [6], can impair load predictability and balancing of parallel implementations, and prevents them from being truly $D\&C$ computations. As already noted in [4], the threshold information is not needed until the pruning phase, so the threshold selection can be delayed and computed in an amortized way at the end of the classification phase.

We will assume now that the Conquer step in C4.5 has nearly no cost, and we will come back later to this point.

From an I/O operational point of view, all the steps require linear scans, or sorting and searching through the data in the current partition. Other parallel algorithms like [7] and [8] have been devised to deal with huge data partitions through special data structures and scheduling policies. We want to design a small set of general primitives that allow us to express the algorithm and can be implemented in a standard, user-friendly way in a high-level language. The research in the field of external memory algorithms [10] has produced theoretical analysis and scalable solutions for such a class of basic operations, that are suitable for application to the memory hierarchy of a parallel architecture.

We wanted to maintain the C4.5 sequential results as a reference point[1], and exploit the *D&C* aspect of the computation. Thus solutions that require to change the split criterion to binary, like in [7], or using a different splitting method [8], were regarded as not fully satisfying.

2   The Programming Environment

The SkIE environment [9] we are using and developing is a parallel coordination language based on the concept of *skeleton* . A skeleton is a basic form of parallelism that builds blocks of parallel code by composing simpler blocks (eventually sequential code) in an abstract, modular way. The aim of this approach is at reaching both source code and performance portability across different parallel architectures.

The skeleton run-time support deals with almost all the low level details of parallelism and concurrency (i.e. process mappings, communications, scheduling and load balancing). Application development is thus enhanced by software reuse, rapid prototyping, and a lesser need for performance debugging.

The SkIE user has to pick up a conceptual parallelisation that can be expressed using only the available skeletons. He is relieved of most of the low-level details of the parallelisation, but on the other hand he has little intervention on these aspects, in a trade-off between expressive power and efficiency.

The SkIE semantics is data-flow oriented, with an explicit vision of the streams of tasks. Among the used skeletons, the farm exploits task parallelism over a stream, providing automatic load distribution and balancing. The pipe is the functional composition, with pipeline parallelism exploited. A loop skeleton allows repeated processing of (part of) a stream by another parallel module. Other skeletons like map, reduce, deal with the basic forms of data-parallelism.

All communication set-up is transparently handled by the compiler at the interfaces between the modules, but this imposes some constraints on having fixed-size data structures as parameters. To overcome this limitation and allow

---

[1] Even if it is argued in the literature that a split criterion that requires sorting uses too much computation w.r. to the accuracy of the results.

more expressiveness, a virtual shared memory support is being integrated into the high-level interface. The abstraction is defined of dynamic, out-of-core shared data objects (SO), that are stored in the aggregate memory of the computer.

3 Related work and first experiments

With respect to the parallelisation, following [4] we can classify most of the previous approaches into *attribute parallel*, which assign each A_i to a processor to execute step D1 in parallel, *data parallel* ones, which split the database among the processors and handle most operations collectively, and *task parallel* ones, which try to exploit the recursive definition to start separate computations at each branch.

The tree structure can be highly irregular, and this reflects in the computation. Since the classification workload cannot be foreseen for any given subtree, pure data parallel solutions in the literature exhibit no good parallel scalability, like the synchronous approach in [8]. As discussed in the same paper, even partitioned, more task-oriented solutions with static load-balancing cannot properly handle the irregular load, and a hybrid solutions is proposed.

We choose, instead, to explore task parallelism at first. A similar choice was made in [11], where data parallelism is combined with pipelining. Our experiments are a valuable research alternative, since we take advantage of the automatic load balancing and task pipelining in the SkIE skeletons.

Our prototypes exploits task parallelism, with each worker processor expanding a node of the C4.5 tree into a subtree, up to a certain depth l.

We refer to [1] for a detailed presentation of the first parallel version, which uses farm parallelism in a structure close to that in Fig.1.

Since the user has no control on the task distribution done by the farm skeleton, in our pure task approach there are only two parameters to tune: the depth l of each expansion, and the selection order of waiting tasks.

Using a fixed, on-demand scheduling of tasks requires some property to hold for the given task stream. This first version of the program, which we call MP, was impaired by an excessive communication load and computation variance. The communications were all the same size, comparable to the database size.

The computation of thresholds in step D2cont (*), and the use of too high values for the l parameter led to a highly variable worker load, which resulted in poor load balancing. Only minor improvements were obtained by using more complex or adaptive strategies to adjust the value of l at run-time.

To improve upon the MP solution, we have started a new research path by introducing in the environment the abstraction of shared objects (SO), to be used to remove data replication and unwanted centralization points. This strategy has lead to (1) use the SO to improve communications, (2) switch to a full *D&C* computation, delaying threshold calculation. Next steps are (3) distribute the database among the workers, providing remote access methods, (4) turn the decision tree into a SO itself, thus removing the centralization point.

It is correctly pointed out in [8] that task parallelism alone for classification is not scalable because of large nodes. A further step will be to introduce data-parallel collective operations on the external data structures. Once distributed operation are provided, a first phase of the computation could proceed in a data parallel fashion (either doing attribute or data partitioning).

We argue that it is possible to achieve an efficient implementation of these operations. We will take advantage of the fact that, in the SkIE environment, this implementation is completely independent from the details of the DM algorithms.

4   Current results

Up to now, the path described has been followed to its 2nd step, with work ongoing to reach step 3. We call the prototype at step 1 MP+SM, and the current one DT (it is the same with the Delayed Threshold calculation).

Fig. 1 shows the parallel structure we used for both. The data are still replicated. The task parallelism is applied in the Divide phase through a farm skeleton, each task contains a single node which has to be expanded into a subtree. As we have said, the depth l of the computed subtrees is used to tune the amount of expansion. The expansion politics is explained later.

A single process owns the decision tree and does the Conquer step; it collects and joins subtrees. Nodes still needing to be expanded are sent back to the input of the parallel loop. The Conquer phase is executed in pipeline with the Divide phase inside the loop, to hide its latency.

All the test results are measured on a QSW CS-2 with 10 SPARC processors and a high performance communication network, using the data set Adult from the UCI repository.

With the farm dynamic load balancing, we don't need to evaluate in advance the computation needed by the subtree, which is repeatedly expanded no more than l levels each time. It is enough that the variance of the node workload is bounded. Our work has been oriented at reducing this variance.

Each task requires an amount of communication proportional to the size of its partition, so using the shared objects to store the data keeps communication and computation costs closer. The gain is clear in Fig.2a from the comparison between MP and MP+SM execution times.

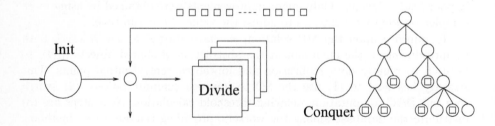

Figure1. Parallel structure of the prototype

The communication latency becomes almost independent from the size of the task, see in Fig.3b the communication cost for one worker. The communication overhead exceeds the computation only for very small tasks.

The MP+SM solution has another source of load imbalancing in the threshold calculation, which is $O(N \log N)$ operations. Delaying the calculation of thresholds until the end of the growth phase minimizes the load variance, allows a $D\&C$ formalisation and reduces the computational effort [4,6].

The comparison of MP+SM and TD in fig.3a underlines the positive effect of delaying the threshold calculation. The overall effect on the completion time is even greater, as reported in fig.2a.

Now the behaviour of the application is easier to analyze than with MP and MP+SM, with different parameters having more understandable effects.

Using a simple depth-first visit does not allow to exploit task parallelism in full. Moreover, since the average expansion cost increases with node size, giving priority to the bigger nodes results in a regular, decreasing computational load, which can be easily balanced.

The depth first selection in the first versions of MP made unsuccessful both the standard scheduling policies of the support and any attempt to use adaptive expansion at the workers.

If task computation time is too low w.r. to communication, parallelism is useless. Varying the amount of node expansion in the sequential code now succeeds in controlling the computation to communication ratio. In fig.3b it is easy to see the task size where communication (and idle) time and computation time on average balance. Here a second parameter controlling the expansion comes into play, the maximum amount of nodes for a subtree. In fig.2b we can see that raising this parameter from 512 to 2048 allows most of the small nodes to be computed immediatly instead of becoming new smaller tasks, this way improving the computation to communication ratio.

In fig.3b the communication time is higher for the same code when run with 6 workers instead of just one. The Conquer process is too busy to keep up with the output of the farm: it definitely becomes a bottleneck. This is partly due to the amount of communication, and mostly to the fact that the sequential code inside the module spends most of the time inside recursive visits the tree.

Theoretically the conquer operation should be fast, but the amount of work required to update the centralized data structure, as well as some inefficiency in its implementation, are no longer negligible. The effect shows up clearly with more and more worker nodes, and gets worse when there is more useful parallelism, as the arrival rate of new tasks increases.

5   Expected Results and Future work

The next change to the application will be needed to remove the Conquer bottleneck. We could change the tree visits into heap accesses, lowering the overhead, but we see as a long term solution to store the tree in the shared memory space.

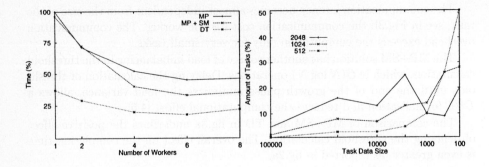

Figure2. (a) Results of the three different parallel implementations. (b) Distribution of task size with a varying subtree buffer size.

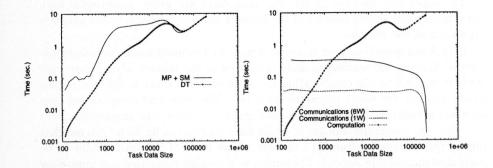

Figure3. (a) Computation time of task w.r. to size, with and without threshold calculation. (b) Computation vs communication and idle time per task, 1 and 6 workers.

Handling the merge operations in a decentralized way will lead to higher parallel scalability.

The utilization of the shared space is also needed to overcome the current assumption that the database fits in memory. A separate development path has already started about maintaining the database and the tree as linked structures in the shared memory space.

The current results, obtained through the integration of the shared objects abstraction into the SkIE skeleton programming model, strongly suggest that the approach is effective in solving the communication/computation problems for a class of irregular computations involving large data sets.

SkIE makes also easy to use the current parallel application as a building block for bigger ones. It is straightforward to set up a control loop that uses a set of C4.5 parallel blocks to scan multiple windows of a database at the same time.

We have not yet addressed the first few iterations in the task parallelisation: the expansion time for the first nodes accounts for 10% to 20% of the computation time, and things would get worse with bigger databases. The solution will be to compute the huge tasks exploiting data parallelism, and to switch to task parallelism as soon as no big tasks are still waiting. This can be done by using a new skeleton of the SkIE environment, which can switch from on demand scheduling to data-driven one. Secondary memory algorithms could be used to efficiently implement global operations like distributed sorting of the data.

Summing up we could move from a main-memory implementation of C4.5 to a skeleton structured implementation, where the main layer of the memory is the aggregate memory of the parallel architecture, whether a massively parallel architecture or a cluster of workstation with virtual shared memory support. The solutions devised should also be applied to the design of a generic skeleton composition that implements, at least, those D&C computations in which the conquer function is basically a merge operation.

References

1. P. Becuzzi, M. Coppola, D. Laforenza, S. Ruggieri, D. Talia, and M. Vanneschi. Data analysis and data mining with parallel architectures: Techniques and experiments. Technical report, Consorzio Pisa Ricerche, project "Parallel Intelligent Systems for Tax Fraud Detection", December 1998.
2. P. Becuzzi, M. Coppola, and M. Vanneschi. Association rules in large databases, additional results. http://www.di.unipi.it/~coppola/ep99talk.ps, Aug 1999.
3. P. Becuzzi, M. Coppola, and M. Vanneschi. Mining of Association Rules in Very Large Databases: a Structured Parallel Approach. In *Euro-Par'99 Parallel Processing*, volume 1685 of *LNCS*. Springer, 1999.
4. John Darlington, Yike Guo, Janjao Sutiwaraphun, and Hing Wing To. Parallel Induction Algorithms for Data Mining. In *Advances in intelligent data analysis: reasoning about data IDA'97*, volume 1280 of *LNCS*, 1997.
5. J.R. Quinlan. *C 4.5: Programs for Machine Learning*. Morgan Kaufmann, San Mateo, 1993.
6. S. Ruggieri. Efficient C4.5. Draft, http://www-kdd.di.unipi.it/software.
7. John Shafer, Rakesh Agrawal, and Manish Mehta. SPRINT: A Scalable Parallel Classifier for Data Mining. In *Proceedings of the 22nd VLDB Conference*, 1996.
8. A. Srivastava, E.H. Han, V. Kumar, and V. Singh. Parallel Formulations of Decision-Tree Classification Algorithms. *Data Mining and Knowledge Discovery*, 3(3), 1999.
9. M. Vanneschi. PQE2000: HPC Tools for Industrial Applications. *IEEE Concurrency: Parallel, Distributed & Mobile Computing*, 6(4):68–73, Oct-Dec 1998.
10. Jeffrey Scott Vitter. External Memory Algorithms and Data Structures: Dealing with MASSIVE DATA. Draft, http://www.cs.duke.edu/~jsv, January 2000.
11. Mohammed J. Zaki, Ching-Tien Ho, and Rakesh Agrawal. Scalable Parallel Classification for Data Mining on Shared-Memory Multiprocessors. In *Proc. of the IEEE Int'l Conference on Data Engineering*, March 1999.

Scalable Parallel Clustering for Data Mining on Multicomputers

D. Foti, D. Lipari, C. Pizzuti and D. Talia

ISI-CNR
c/o DEIS, UNICAL
87036 Rende (CS), Italy
{pizzuti,talia}@si.deis.unical.it

Abstract. This paper describes the design and implementation on MIMD parallel machines of *P-AutoClass*, a parallel version of the *AutoClass* system based upon the Bayesian method for determining optimal classes in large datasets. The *P-AutoClass* implementation divides the clustering task among the processors of a multicomputer so that they work on their own partition and exchange their intermediate results. The system architecture, its implementation and experimental performance results on different processor numbers and dataset sizes are presented and discussed. In particular, efficiency and scalability of *P-AutoClass* versus the sequential *AutoClass* system are evaluated and compared.

1 Introduction

Clustering algorithms arranges data items into several groups so that similar items fall into the same group. This is done without any suggestion from an external supervisor, classes and training examples are not given a priori. Most of the early cluster analysis algorithms come from the area of statistics and have been originally designed for relatively small data sets. In the recent years, clustering algorithms have been extended to efficiently work for knowledge discovery in large databases and some of them are able to deal with high-dimensional feature items. When used to classify large data sets, clustering algorithms are very computing demanding and require high-performance machines to get results in reasonable time. Experiences of clustering algorithms taking from one week to about 20 days of computation time on sequential machines are not rare. Thus, scalable parallel computers can provide the appropriate setting where to execute clustering algorithms for extracting knowledge from large-scale data repositories.

Recently there has been an increasing interest in parallel implementations of data clustering algorithms. Parallel approaches to clustering can be found in [8, 4, 9, 5, 10]. In this paper we consider a parallel clustering algorithm based on Bayesian classification for distributed memory multicomputers. We propose a parallel implementation of the AutoClass algorithm, called *P-AutoClass*, and validate by experimental measurements the scalability of our parallelization strategy. The *P-AutoClass* algorithm divides the clustering task among the processors of a parallel

J. Rolim et al. (Eds.): IPDPS 2000 Workshops, LNCS 1800, pp. 390-398, 2000.

machine that work on their own partition and exchange their intermediate results. It is based on the message passing model for shared-nothing MIMD computers.

This paper describes the design and implementation of *P-AutoClass*. Furthermore, experimental performance results on different processor numbers and dataset sizes are presented and discussed. The rest of the paper is organized as follows. Section 2 provides a very short overview of Bayesian classification and sequential AutoClass. Section 3 describes the design and implementation of *P-Autoclass* for multicomputers. Section 4 presents the main experimental performance results of the algorithm. Section 5 describes related work and section 6 contains conclusions.

2 Bayesian Classification and AutoClass

The Bayesian approach to unsupervised classification provides a probabilistic approach to induction. Given a set $X = \{X_1,...,X_I\}$ of data instances X_i, with unknown classes, the goal of Bayesian classification is to search for the best class description that predict the data. Instances X_i are represented as ordered vectors of attribute values $\overrightarrow{X_i} = \{X_{i1},...,X_{ik}\}$.

In this approach class membership is expressed probabilistically, that is an instance is not assigned to a unique class, instead it has a probability of belonging to each of the possible classes. The classes provides probabilities for all attribute values of each instance. Class membership probabilities are then determined by combining all these probabilities. Class membership probabilities of each instance must sum to 1, thus there not precise boundaries for classes: every instance must be a member of some class, even though we do not know which one. When every instance has a probability of about 0.5 in any class, the classification is not well defined because it means that classes are abundantly overlapped. On the contrary, when the probability of each instance is about 0.99 in its most probable class, the classes are well separated.

A Bayesian classification model consists of two sets of parameters: a set of discrete parameters T which describes the functional form of the model, such as number of classes and whether attributes are correlated, and a set of continuos parameters $\overrightarrow{V}$ that specifies values for the variables appearing in T, needed to complete the general form of the model. The probability of observing an instance having particular attribute value vector is referred to as *probability distribution or density function* (p.d.f.). Given a set of data X, AutoClass searches for the most probable pair $\overrightarrow{V}$, T which classifies X. This is done in two steps:

- For a given T, AutoClass seeks the maximum posterior (MAP) parameter values $\overrightarrow{V}$.

- Regardless of any $\overrightarrow{V}$, AutoClass searches for the most probable T, from a set of possible Ts with different attribute dependencies and class structure.

Thus there are two levels of search: *parameter level search* and *model level search*. Fixed the number classes and their class model, the space of allowed parameter values is searched for finding the most probable $\overrightarrow{V}$. Given the parameter values, AutoClass calculates the likelihood of each case belonging to each class L and then calculates a set of weights $w_{ij}=(L_i/\Sigma_j L_j)$ for each case. Given these weigths, weighted statistics relevant to each term of the class likelihood are calculated. These statistics are then used to generate new MAP values for the parameters and the cycle is repeated.

Based on this theory, Cheeseman and colleagues at NASA Ames Research Center developed AutoClass [1] originally in Lisp. Then the system has been ported from Lisp to C. The C version of AutoClass improved the performance of the system of about ten times and has provided a version of the system that can be easily accessed and used by researchers at a variety of universities and research laboratories.

3 P-AutoClass

In spite of the significant improvement of the C version performance, because of the computational needs of the algorithm, the execution of AutoClass with large datasets requires times that in many cases are very high. For instance, the sequential AutoClass runs on a dataset of 14K tuples, each one composed of a few hundreds bytes, have taken more the 3 hours on Pentium-based PC. Considering that the execution time increases linearly with the size of dataset, more than 1 day is necessary to analyze a dataset composed of about 140K tuples, that is not a very large dataset. For the clustering of a satellite image AutoClass took more than 130 hours [6] and the analysis of protein sequences the discovery process required from 300 to 400 hours [3].

These considerations and experiences suggest that it is necessary to implement faster versions of AutoClass to handle very large data set in reasonable time. This can be done by exploiting the inherent parallelism present in the AutoClass algorithm implementing it in parallel on MIMD multicomputers. Among the different parallelization strategies, we selected the SPMD approach. In AutoClass, the SPMD approach can be exploited by dividing up the dataset among the processors and by the parallel updating on different processors of the weights and parameters of classifications. This strategy does not require to replicate the entire dataset on each processor. Furthermore, it also does not have load balancing problems because each processor execute the same code on data of equal size. Finally, the amount of data exchanged among the processors is not so large since most operations are performed locally at each processor.

3.1 Design of the parallel algorithm

The main steps of the structure of the AutoClass program are described in figure 1. After program starting and structure initialization, the main part of the algorithm is devoted to classification generation and evaluation (*step 3*). This loop is composed of a set of substeps specified in figure 2. Among those substeps, the *new try of classification* step is the most computationally intensive. It computes the weights of each items for each class and computes the parameters of the classification. These operations are executed by the function `base_cycle` which calls the three functions `update_wts`, `update_parameters` and `update_approximations` as shown in figure 3.

1. Program Start and Files Reading ;

2. Data Structures Initialized ;

3. Classification Generation and Evaluation (called also *BIG_LOOP*) ;

4. Check the Stopping Conditions, if they are not verified go to step 3, else go to step 5;

5. Store Results on the Output Files.

Fig. 1. Scheme of the sequential AutoClass algorithm.

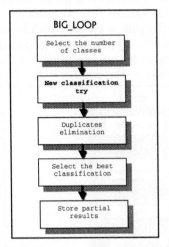

Fig. 2. Main steps of the *BIG_LOOP*.

We analyzed the time spent in the **base_cycle** function and it resulted about the 99,5% of the total time, therefore we identified this function as that one where parallelism must be exploited to speed up the AutoClass performance.

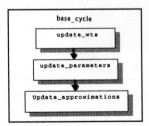

Fig. 3. The structure of the *base_cycle* function.

In particular, analyzing the time spent in each of the three functions called by **base_cycle**, it appears, as observed in other experiences [7], that the **update_wts** and **update_parameters** functions are the time consuming functions whereas the time spent in the **update_approximation** is negligible. Therefore, we studied the parallelization of these two functions using the SPMD approach. To maintain the same semantics of the sequential algorithm of AutoClass, we designed the parallel version by partitioning data and local computation on each of **P** processors of a distributed memory MIMD computer and by exchanging among the processors all the local variables that contribute to form global values of a classification.

3.1.1 Parallel update_wts

In AutoClass the class membership of each data item is expressed probabilistically. Thus every item has a probability that it belongs to each of the possible classes. In the AutoClass algorithm, class membership is expressed by weights. The function **update_wts** calculates the weights w_{ij} for each item i of the active classes to be the normalized class probabilities with respect to the current parameterizations.

The parallel version of this function first calculates on each processing element the weights w_{ij} for each item belonging to the local partition of the data set and sum the weights w_j of each class j $(w_j = \Sigma_i w_{ij})$ relatively to its own data. Then all the partial w_j values are exchanged among all the processors and summed in each of them to have the same value in every processor. This strategy is described in figure 4.

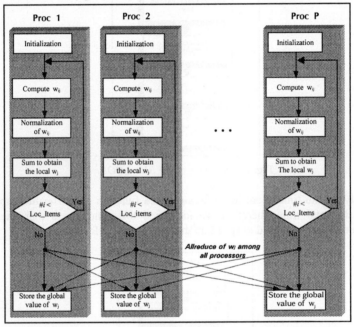

Fig. 4. The parallel version of the *update_wts* function.

To implement the total exchange of the w_j values in the **update_wts** function we used a *global reduction* operation (Allreduce) that sums all the local copies in the all processes (*reduction* operation) and places the results on all the processors (*broadcast* operation).

3.1.2 Parallel update_parameters

The **update_parameters** function computes for each class a set of class posterior parameter values, which specify how the class is distributed along the various attributes. To do this, the function is composed of three nested loops, the external loop scans all the classes, then for each class all the attributes are analyzed and in the inner loop all the items are read and their values are used to compute the class parameters.

In parallelizing this function, we executed the partial computation of parameters in parallel on all the processors, then all the local values are collected on each processor before to utilize them for computing the global values of the classification parameters. Figure 5 shows the scheme of the parallel version of the function. To implement the total exchange of the parameter values in the `update_parameters` function we used a *global reduction* operation that sums all the local copies in all the processes and places the results on every processor.

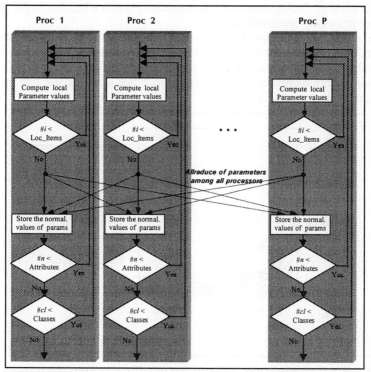

Fig. 5. The parallel version of the *update_parameters* function.

P-AutoClass has been implemented using the Message Passing Interface (MPI) toolkit on a Meiko Computing Surface 2 using the version 3.3 of sequential AutoClass C. Because of the large availability of MPI, *P-AutoClass* is portable practically on every parallel machine from supercomputers to PC clusters.

4   Experimental results

We run our experiments on a Meiko CS 2 with up to 10 SPARC processors connected by a fat tree topology with a communication bandwidth of 50 Mbytes/s in both directions. We used a synthetic dataset composed of 100000 tuples each one composed of two real attributes. To perform our experiments we used different partitions of data from 5000 tuples to the entire dataset, and asked the system to find the best clustering starting with different number of clusters (*start_j_list=2, 4, 8, 16, 24, 50, 64*). Each classification has been repeated 10 times and results presented here represent the mean values obtained after these classifications. We measured the elapsed time, the speedup and the scaleup of P-AutoClass. *Speedup* gives the efficiency of the parallel algorithm when the number of processors varies. It is defined as the ratio of the execution time for clustering a dataset on *1* processor to the execution time for clustering the same dataset on *P* processors. Another interesting measure is scaleup. *Scaleup* captures how a parallel algorithm handles larger datasets when more processors are available.

Figure 6 shows the elapsed times of P-AutoClass on different numbers of processors. We can see that the total execution time substantially decreases as the number of used processors increases. In particular, for the largest datasets the time decreases in a more significant way. We can observe that as the dataset size increases the time gain increases as well.

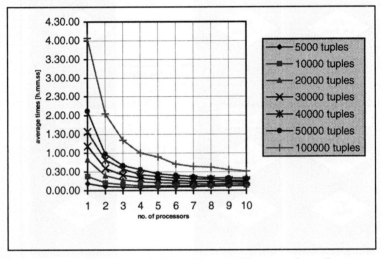

Fig. 6. Average elapsed times of *P-AutoClass* on different numbers of processors.

In figure 7 the speedup results obtained for different datasets are given. We can observe that the *P-AutoClass* algorithm scales well up to 10 processors for the largest datasets, whereas for small datasets the speedup increases until the optimal number of processors are used for the given problem (e.g., 4 procs for 5000 tuples or 8 procs for 20000 tuples). When more processes are used we observe that the algorithm does not scale because the processors are not effectively used and the communication costs increases.

For scaleup measures we evaluated the execution time of a single iteration of the *base_cycle* function by keeping the number of data items per processor fixed while increasing the number of processors. To obtain more stable results we asked *P-AutoClass* to group data into 8 and 16 clusters. Figure 8 shows the scaleup results. For all the experiments we have 10000 tuples per processor. We started with 10000 tuples on 1 processor up to 100000 tuples on 10 processors. It can be seen that the parallel AutoClass algorithm, for a given classification, shows a nearly stable pattern. Thus it delivers nearly constant execution times in number of processors showing good scaleup.

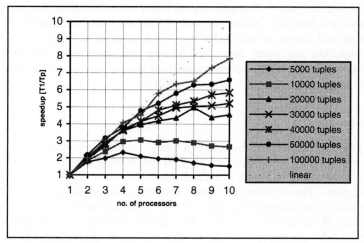

Fig. 7. Speedup of *P-AutoClass* on different numbers of processors.

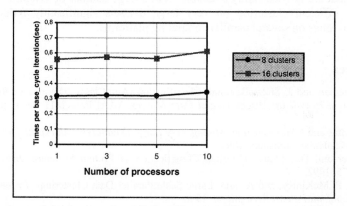

Fig. 8. Scaleup of the *base_cycle* of *P-AutoClass* on different sets of tuples scaled by the number of processors.

5   Related work

In the past few years there has been an increasing interest in parallel implementations of data mining algorithms [2]. The first approaches to the parallelization of AutoClass have been done on SIMD parallel machines by using compilers that automatically generated data-parallel code starting from the sequential program to which only few instructions have been added. The first data-parallel version of AutoClass has been developed in *Lisp to run on a Connection Machine-2 [1], and the second one has been developed adding C* code to the C source to run it on a CM-5 [9]. These approaches are simple from the programming point-of-view but they do not exploit all the potential parallelism. In fact, it is not known how well compilers extract parallelism from sequential programs.

The only experience we know about the implementation of AutoClass on a MIMD computer is described in [7]. This prototype is based on the exploitation of parallelism in the `update_wts` function. Concerning to the SIMD approach, our implementation is more general and allows to exploit parallelism in a more complete, flexible, and portable way. On the other hand, considering the mentioned MIMD parallel implementation, *P-AutoClass* exploits parallelism also in the parameters computing phase, with a further improvement of performance.

6   Conclusion

In this paper we proposed *P-AutoClass*, a parallel implementation of the AutoClass algorithm based upon the Bayesian method for determining optimal classes in large datasets. We have described and evaluated the *P-AutoClass* algorithm on a MIMD parallel computer. The experimental results show that *P-AutoClass* is scalable both in terms of speedup and scaleup. This means that for a given dataset, the execution times can be reduced as the number of processors increases, and the execution times do not increase if, while increasing the size of datasets, more processors are available. Finally, our algorithm is easily portable to various MIMD distributed-memory parallel computers that are now currently available from a large number of vendors. It allows to perform efficient clustering on very large datasets significantly reducing the computation times on several parallel computing platforms.

References

1. P. Cheeseman and J. Stutz. Bayesian Classification (AutoClass): Theory and Results. In *Advances in Knowledge Discovery and Data Mining*, AAAI Press/MIT Press, pp. 61-83, 1996.
2. A.A. Freitas and S.H. Lavington. *Mining Very Large Databases with Parallel Processing*. Kluwer Academic Publishers, 1998.
3. L. Hunter and D.J. States. Bayesian Classification of Protein Structure. *IEEE Expert*, 7(4):67-75, 1992.
4. D. Judd, P. McKinley, and A. Jain. Large-Scale Parallel Data Clustering. *Proceedings of the Int. Conf. on Pattern Recognition*, 1996.
5. D. Judd, P. McKinley, A. Jain. Performance Evaluation on Large-Scale Parallel Clustering in NOW Environments. *Proceedings of the Eight SIAM Conf. on Parallel Processing for Scientific Computing*, Minneapolis, March 1997.
6. B. Kanefsky, J. Stutz, P. Cheeseman, and W. Taylor. An Improved Automatic Classification of a Landsat/TM Image from Kansas (FIFE). Technical Report FIA-94-01, NASA Ames Research Center, May 1994.
7. R. Miller and Y. Guo. Parallelisation of AutoClass. *Proceedings of the Parallel Computing Workshop (PCW'97)*, Canberra, Australia, 1997.
8. C.F. Olson. Parallel Algorithms for Hierarchical Clustering. *Parallel Computing*, 21:1313-1325, 1995.
9. J.T. Potts. Seeking Parallelism in Discovery Programs. Master Thesis, University of Texas at Arlington, 1996.
10. K. Stoffel and A. Belkoniene. Parallel K-Means Clustering for Large Data Sets. *Proceedings Euro-Par '99*, LNCS 1685, pp. 1451-1454, 1999.

Exploiting Dataset Similarity for Distributed Mining *

Srinivasan Parthasarathy and Mitsunori Ogihara

Department of Computer Science
University of Rochester
Rochester, NY 14627-0226

{srini,ogihara}@cs.rochester.edu

Abstract. The notion of similarity is an important one in data mining. It can be used to provide useful structural information on data as well as enable clustering. In this paper we present an elegant method for measuring the similarity between homogeneous datasets. The algorithm presented is efficient in storage and scale, has the ability to adjust to time constraints. and can provide the user with likely causes of similarity or dis-similarity.

One potential application of our similarity measure is in the distributed data mining domain. Using the notion of similarity across databases as a distance metric one can generate clusters of similar datasets. Once similar datasets are clustered, each cluster can be independently mined to generate the appropriate rules for a given cluster. The similarity measure is evaluated on a dataset from the Census Bureau, and synthetic datasets from IBM.

1 Introduction

Similarity is a central concept in data mining. Research in this area has primarily progressed along two fronts: object similarity [2, 8, 7] and attribute similarity [5, 9]. The former quantifies how far from each other two objects in the database are while the latter refers to the distance between attributes. Discovering the similarity between objects and attributes enables reduction in dimensions of object profiles as well as provides useful structural information on the hierarchy of attributes.

In this paper we extend this notion of similarity to homogeneous distributed datasets. Discovering the similarity between datasets enables us to perform "meaningful" distributed datamining. Large business organizations with nation-wide and international interests usually rely on a homogeneous distributed database to store their transaction data. This leads to multiple data sources with a common structure. In order to analyze such collection of databases it seems important to cluster them into small number of groups to contrast global trends with local trends rather than apply traditional methods which simply combine them into a single logical resource. A limitation of traditional methods is that the joining is not based on the database characteristics, such as the demographic, economic conditions, and geo-thermal conditions. Mining each database individually is unacceptable as it is likely to generate too many spurious patterns (outliers). We argue for a hybrid solution. First cluster

* This work was supported in part by NSF grants CDA-9401142, CCR-9702466, CCR-9701911, CCR-9725021, INT-9726724, and CCR-9705594; and an external research grant from Digital Equipment Corporation.

J. Rolim et al. (Eds.): IPDPS 2000 Workshops, LNCS 1800, pp. 399-406, 2000.

the datasets, and then apply the *traditional distributed mining* approach to generate a set of rules for each resulting cluster.

The primary problem with clustering such homogenous datasets is to identify a suitable distance (similarity) metric. The similarity metric depends not only on the kind of mining task being performed but also on the data. Therefore, any measure of similarity should be flexible to both the needs of the task, and data. In this paper we present and evaluate such a similarity metric for distributed association mining. We believe that this metric can be naturally extended to handle other mining tasks such as discretization and sequence mining as well. We then show how one can cluster the database sources based on our similarity metric in an I/O and communication efficient manner. A novelty of our approach to clustering, other than the similarity measure is how we merge datasets without communicating the raw data itself.

The rest of this paper is organized as follows: Section 2 formally defines the problem, and describes our proposed similarity measure. We then present our method for clustering distributed datasets using the aforementioned similarity metric in Section 3. We experimentally validate our approach on real and synthetic datasets in Section 4. Finally, we conclude in Section 5.

2   Similarity Measure

Our similarity measure adopts an idea recently proposed by Das et al [5] for measuring attribute similarity in transaction databases. They propose comparing the attributes in terms of how they are individually correlated with other attributes in the database. The choice of the other attributes (called the probe set) reflects the examiner's viewpoint of relevant attributes to the two. A crucial issue in using this similarity metric is the selection of the probe set. Das *et al.* [5] observed that this choice strongly affects the outcome. However, they do not provide any insight to automating this choice when no apriori knowledge about the data is available. Furthermore, while the approach itself does not limit probe elements to singleton attributes, allowing for complex (boolean) probe elements and computing the similarities across such elements can quickly lead to problems of scale.

We propose to extend this notion of similarity to datasets in the following manner. Our similarity measure compares the datasets in terms of how they are correlated with the attributes in the database. By restricting ourselves to frequently occurring patterns, as probe elements, we can leverage existing solutions (Apriori [3]) for such problems to generate and interactively prune the probe set. This allows us to leverage certain powerful features of associations to handle the limitations described above. First, by using associations as the initial probe set we are able to obtain a "first guess" as to the similarity between two attributes. Second, since efficient solutions for the association problem exist, similarities can be computed rapidly. Third, once this "first guess" is obtained we are able to leverage and extend [1] existing work in interactive (online) association mining [1] to quickly compute similarities under boolean constraints, provide insights into the causes of similarity and dis-similarity, as well as to allow the user to interact and prune the probe space. Finally, we can leverage existing work on sampling to compute the similarity metric accurately and efficiently in a distributed setting.

---

[1] In addition to the interactions supported in [1] we also support **influential attribute** identification. This interaction basically identifies the (set of) probe attribute(s) that contribute most to the similarity metric.

2.1   Association Mining Concepts

We first provide basic concepts for association mining, following the work of Agrawal *et al.* [3]. Let $\mathcal{I} = \{i_1, i_2, \cdots, i_m\}$ be a set of m distinct *attributes*, also called *items*. A set of items is called an *itemset* where for each nonnegative integer k, an itemset with exactly k items is called a *k-itemset*. A *transaction* is a set of items that has a unique identifier *TID*. The *support* of an itemset A in database $\mathcal{D}$, denoted $\sup_{\mathcal{D}}(A)$, is the percentage of the transactions in in $\mathcal{D}$ containing A as the subset. The itemsets that meet a user specified *minimum support* are referred to as *frequent* itemsets or as *associations*. An *association rule* is an expression of the form $A \Rightarrow B$, where A and B are disjoint itemsets. The *confidence* of an association rule $A \Rightarrow B$ is $\frac{\sup_{\mathcal{D}}(A \cup B)}{\sup_{\mathcal{D}}(A)}$, i.e., the fraction of the datasets containing B over those containing A.

The data mining task for discovering association rules consists of two steps: finding all frequent itemsets (i.e., all associations) and finding all rules whose confidence levels are at least a certain value, the *minimum confidence*. We use our group's ECLAT [11] association mining algorithm to compute the associations.

2.2   Similarity Metric

A measure of similarity between two entities reflects how close they are to one another. Let X and Y be two entities whose similarity we want to measure. We denote $Sim(X, Y)$ to mean the similarity measure between X and Y. Ideally we would like Sim to satisfy the following three properties:

- Identity: $Sim(X, Y) = 1$ corresponds to the fact that the two entities are identical in all respects.
- Distinction: $Sim(X, Y) = 0$ corresponds to the fact that the two entities are distinct in all respects.
- Relative Ordinality: If $Sim(X, Y) > Sim(X, Z)$, then it should imply that X is more similar to Y than it is to Z.

The first two properties bound the range of the measure while the third property ensures that similarities across objects can be meaningfully compared. This last property is particularly useful for clustering purposes.

Now we define our metric. Let A and B respectively be the association sets for a database $\mathcal{D}$ and that for a database $\mathcal{E}$. For an element $x \in A$ (respectively in B), let $\sup_{\mathcal{D}}(x)$ (respectively $\sup_{\mathcal{E}}(x)$) be the frequency of x in $\mathcal{D}$ (respectively in $\mathcal{E}$). Define

$$Sim(A, B) = \frac{\sum_{x \in A \cap B} \max\{0, 1 - \alpha| \sup_{\mathcal{D}}(x) - \sup_{\mathcal{E}}(x)|\}}{\|A \cup B\|}$$

where α is a scaling parameter. The parameter α has the default value of 1 and is to reflect how significance the user view variations in supports are (the higher α is the more influential variations are). For $\alpha = 0$ the similarity measure is identical to $\frac{\|A \cap B\|}{\|A \cup B\|}$, i.e., support variance carries no significance.

2.3   Sampling and Association Rules

The use of sampling for approximate, quick computation of associations has been studied in the literature [10]. While computing the similarity measure, sampling can be used at two levels. First, if generating the associations is expensive (for large datasets) one can sample the dataset and subsequently generate the association set from the sample, resulting in huge I/O savings. Second, if the association sets are large one can estimate the distance between them by sampling, appropriately modifying the similarity measure presented above. Sampling at this level is particularly useful in a distributed setting when the association sets, which have to be communicated to a common location, are very large.

3   Clustering Datasets

Clustering is commonly used for partitioning data [6]. The clustering technique we adopt is the simple tree clustering. We use the similarity metric of databases defined in Section 2 for as the distance metric for our clustering algorithm. Input to the algorithm is simply the number of clusters in the final result. At the start of the clustering process each database constitutes a unique cluster. Then we repeatedly merge the pair of clusters with the highest similarity and merge the pair into one cluster until there are the desired number of clusters left.

As our similarity metric is based on associations, there is an issue of how to merge their association lattices when two clusters are merged. A solution would be to combine all the datasets and recompute the associations, but this would be time-consuming and involve heavy communication overheads (all the datasets will have to be re-accessed). Another solution would be to intersect the two association lattices and use the intersection as the lattice for the new cluster, but this would be very inaccurate. We take the half-way point of these two extremes.

Suppose we are merging two clusters $\mathcal{D}$ and $\mathcal{E}$, whose association sets are respectively A and B. The value of $\sup_{\mathcal{D}}(x)$ is known only for all $x \in A$ and that of $\sup_{\mathcal{E}}(x)$ is known only for all $x \in B$. The actual support of x in the join of $\mathcal{D}$ and $\mathcal{E}$ is given as

$$\frac{\sup_{\mathcal{D}}(x) \cdot \|\mathcal{D}\| + \sup_{\mathcal{E}}(x) \cdot \|\mathcal{E}\|}{\|\mathcal{D}\| + \|\mathcal{E}\|}.$$

When x does not belong to A or B, we will approximate the unknown sup-value by a "guess" θ [2], which can be specific to the cluster as well as to the association x.

4   Experimental Analysis

In this section we experimentally evaluate our similarity metric[3]. We evaluate the performance and sensitivity of computing this metric using sampling in a distributed setting. We then apply our dataset clustering technique to synthetic datasets from IBM and on a real dataset from the Census Bureau, and evaluate the results obtained.

4.1   Setup

All the experiments (association generation, similarity computation) were performed on a single processor of a DECStation 4100 containing four 600MHz Alpha 21164 processors, with 256MB of memory per processor.

---

[2] We are evaluating two methods to estimate θ. The strawman is to randomly guess a value between 0 and the minimum support. The second approach is to estimate the support of an item based on the available supports of its subsets.

[3] Due to lack of space we do not detail our experimentation on choice of α.

We used different synthetic databases with size ranging from 3MB to 30MB, which are generated using the procedure described in [3]. These databases mimic the transactions in a retailing environment. Table 1 shows the databases used and their properties. The number of transactions is denoted as $numT$, the average transaction size as T_l, the average maximal potentially frequent itemset size as I, the number of maximal potentially frequent itemsets as $\|L\|$, and the number of items as Size. We refer the reader to [3] for more detail on the database generation.

Database	$numT$	T_l	I	$\|L\|$	Size
D100	100000	8	2000	4	5MB
D200	200000	12	6000	2	12MB
D300	300000	10	4000	3	16MB
D400	400000	10	10000	6	25MB

Table 1. Database properties

The Census data used in this work was derived from the County Business Patterns (State) database from the Census Bureau. Each dataset we derive (dataset per state) from this database contains one transaction per county. Each transaction contains items which highlight information on subnational economic data by industry. Each industry is divided into small, medium and large scale concerns. The original data has numeric data corresponding to number of such concerns occurring in the county. We discretize these numeric values into three categories: high, middle and low. So an item "high-small-agriculture" would correspond to a high number of small agricultural concerns. The resulting set of datasets have as many transactions as counties in the state and a high degree of associativity.

4.2 Sensitivity to Sampling Rate

In Section 2 we mentioned that sampling can be used at two levels to estimate the similarity efficiently in a distributed setting. If association generation proves to be expensive, one can sample the transactions to generate the associations and subsequently use these associations to estimate the similarity accurately. Alternatively, if the number of associations in the lattice are large, one can sample the associations to directly estimate the similarity. We evaluate the impact of using sampling to compute the approximate the similarity metric below.

For this experiment we breakdown the execution time of computing the similarity between two of our databases D300 and D400 under varying sampling rates. The two datasets were located in physically separate locations. We measured the total time to generate the associations for a minimum support of 0.05% (Computing Associations) for both datasets (run in parallel), the time to communicate the associations from one machine (Communication Overhead) to another and the time to compute the similarity metric (Computing Similarity) from these association sets. Transactional sampling influences the computing the associations while association sampling influences the latter two aspects of this experiment. Under association sampling, each processor computes a sample of its association set and sends it to the other, both then compute a part of similarity metric (in parallel). These two values are then merged appropriately, accounting for duplicates in the samples used. While both these sampling levels (transaction and association) could have different

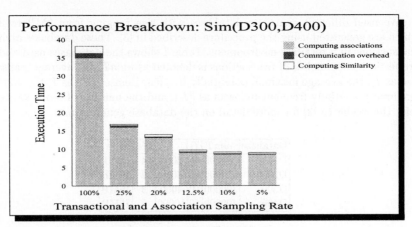

Fig. 1. Sampling Performance

sampling rates, for expository simplicity we chose to set both at a common value. We evaluate the performance under the following sampling rates, 5%, 10%, 12.5%, 20%, and 25%. Figure 1 shows the results from this experiment.

Breaking down the performance it is clear that by using sampling at both levels the performance improves dramatically. For a sampling rate of 10% the time to compute associations goes down by a factor of 4. The communication overhead goes down by a factor of 6 and the time to compute the similarity goes down by a factor of 7. This yields an overall speedup of close to 5. Clearly, the dominant factor in this experiment is computing the associations (85% of total execution time). However, with more traffic in the system, as will be the case when computing the similarity across several datasets (such as in clustering), and when one is modifying the probe set interactively, the communication overhead will play a more dominant role.

The above experiment affirms the performance gains from association and transactional sampling. Next, we evaluate the quality of the similarity metric estimated using such approximation techniques for two minimum support values (0.05% and 0.1%). From Table 2 it is clear that using sampling for estimating the similarity metric can be very accurate (within 2% of the ideal (**Sampling Rate** 100%)) for all sampling rates above 5%. We have observed similar results (speedup and accuracy) for the other dataset pairs as well.

Support	SR-100%	SR-25%	SR-20%	SR-10%	SR-5%
0.05%	0.135	0.134	0.136	0.133	0.140
0.1%	0.12	0.12	0.12	0.12	0.115

Table 2. Sampling Accuracy: Sim(D300,D400)

4.3 Synthetic Dataset Clustering

We evaluated the efficacy of clustering homogeneous distributed datasets based on similarity. We used the synthetic datasets described earlier as a start point. We randomly split each of the datasets D100, D200, D300, and D400 into 10 datasets of roughly equal size. For the sake of simplicity in exposition we describe only the experiment that used only first three subsets from each. We ran a simple tree-based clustering algorithm on these twelve datasets. Figure 2 shows the result. The numbers attached to the joins are the *Sim* metric with $\alpha = 1.0$. Clearly the datasets from the same origin are merged first. Given four as the desired number of clusters (or a merge cutoff of 0.2), the algorithm stops right after executing all the merges depicted by full lines, combining all the children from the same parents into single clusters and leaving apart those from different parents. This experiment illustrates two key points. First, the similarity metric coupled with our merging technique seem to be an efficient yet effective way to cluster datasets. Second, hypothetically speaking, if these 12 datasets were representative of a distributed database, combining all 12 and mining for rules would have destroyed any potentially useful structural rules that could have been found if each cluster were mined independently (our approach).

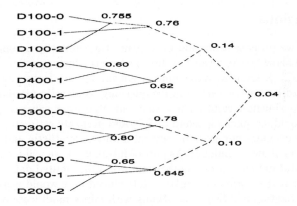

Fig. 2. Dataset Clustering

4.4   Census Dataset Evaluation

Table 3 shows the *Sim* values (with $\alpha = 1.0$) for a subset of the Census data for the year of 1988. As mentioned earlier each dataset corresponds to a state in the US. When asked to break the eight states into four clusters the clustering algorithm returned the clusters [IL, IA, TX], [NY, PA], [FL], and [OR,WA]. On looking at the actual *Sim* values it is clear that NY and PA have a closeted preference for one another IL, IA, and TX have strong preference for one another. OR has a stronger preference for IL, IA and TX, but once IL, IA, and TX were merged it preferred being merged with WA. Interestingly three pairs of neighboring states, i.e., (OR,WA), (IL,IA), and (NY,PA), are found in the same cluster.

An interesting by-play of discretization of the number of industrial concerns into three categories (high, middle and low) is that states with larger counties (area-wise), such as PA, NY and FL tend to have higher associativity (since each county has many items) and thereby tend to have less affinity to states with lower associativity. By probing the similarity between IA and IL further the most influential attribute is found to be agricultural concerns (no surprise there). The reason TX was found to be similar to these states was again due to agricultural concerns, a somewhat surprising result. However, this made sense, when we realized that cattle farming is also grouped under agricultural concerns! Interestingly, we found that the Census data benefited, performance-wise, from association sampling due its high associativity.

State	IL	NY	PA	FL	TX	OR	WA
IA	0.54	0.01	0.01	0.16	0.44	0.26	0.1
IL		0.02	0.02	0.24	0.52	0.30	0.16
NY			0.31	0.14	0.01	0.04	0.08
PA				0.05	0.01	0.03	0.04
FL					0.24	0.21	0.21
TX						0.32	0.16
OR							0.25

Table 3. Census Dataset: *Sim* Values (support = 20%)

5 Conclusions

In this paper we propose a method to measure the similarity among homogeneous databases and show how one can use this measure to cluster similar datasets to perform meaningful distributed data mining. An interesting feature of our algorithm is the ability to interact via informative querying to identify attributes influencing similarity. Experimental results show that our algorithm can adapt to time constraints by providing quick (speedup of 5-7) and accurate estimates (within 2%) of similarity. We evaluate our work on several datasets, synthetic and real, and show the effectiveness of our techniques. As part of future work we will focus on evaluating and applying dataset clustering to other real world distributed data mining tasks. It seems likely that the notion of similarity introduced here would work well for tasks such as Discretization and Sequence Mining with minor modifications if any. We are also evaluating the effectiveness of the merging criteria described in Section 3.

References

1. C. Aggarwal and P. Yu. Online generation of association rules. In *ICDE'98*.
2. R. Agrawal, C. Faloutsos, and A. Swami. Efficient similarity search in sequence databases. In *Foundations of Data Organization and Algorithms*, 1993.
3. R. Agrawal, H. Mannila, R. Srikant, H. Toivonen, and A. Inkeri Verkamo. Fast discovery of association rules. In U. Fayyad and et al, editors, *Advances in Knowledge Discovery and Data Mining*, pages 307–328. AAAI Press, Menlo Park, CA, 1996.
4. R. Agrawal and R. Srikant. Fast algorithms for mining association rules. In *20th VLDB Conf.*, 1994.
5. G. Das, H. Mannila, and P. Ronkainen. Similarity of attributes by external probes. In *KDD 1998*.
6. U. M. Fayyad, G. Piatetsky-Shapiro, and P. Smyth. The KDD process of rextracing useful information from volumes of data. *Communications of ACM*, 39(11):27–34, 1996.
7. R. Goldman, N.Shivakumar, V. Suresh, and H. Garcia-Molina. Proximity search in databases. In *VLDB Conf.*, 1998.
8. H. Jagadish, A. Mendelzon, and T. Milo. Similarity based queries. In *PODS*, 1995.
9. R. Subramonian. Defining *diff* as a data mining primitive. In *KDD 1998*.
10. H. Toivonen. Sampling large databases for association rules. In *VLDB Conf.*, 1996.
11. M. J. Zaki, S. Parthasarathy, M. Ogihara, and W. Li. New algorithms for fast discovery of association rules. In *KDD*, 1997.

Scalable Model for Extensional and Intensional Descriptions of Unclassified Data

Hércules A. Prado[1,2], *, Stephen C. Hirtle[2], **, Paulo M. Engel[1], ***

[1] Universidade Federal do Rio Grande do Sul
Instituto de Informática
Av. Bento Gonçalves, 9500 - Bairro Agronomia
Porto Alegre / RS - Brasil
Caixa Postal 15.064 - CEP 91.501-970
Fone: +55(051)316-6829
Fax: +55(051)319-1576
[2] University of Pittsburgh
Department of Information Sciences and Telecommunications
135 North Bellefield Ave. Pittsburgh, PA 15.260
Phone: +1(412)624-9434
Fax: +1(412)624-2788
{prado, engel}@inf.ufrgs.br and hirtle+@pitt.edu

Abstract. Knowledge discovery from unlabeled data comprises two main tasks: identification of "natural groups" and analysis of these groups in order to interpret their meaning. These tasks are accomplished by unsupervised and supervised learning, respectively, and correspond to the taxonomy and explanation phases of the discovery process described by Langley [9]. The efforts of Knowledge Discovery from Databases (KDD) research field has addressed these two processes into two main dimensions: (1) scaling up the learning algorithms to very large databases, and (2) improving the efficiency of the knowledge discovery process. In this paper we argue that the advances achieved in scaling up supervised and unsupervised learning algorithms allow us to combine these two processes in just one model, providing extensional (who belongs to each group) and intensional (what features best describe each group) descriptions of unlabeled data. To explore this idea we present an artificial neural network (ANN) architecture, using as building blocks two well-know models: the ART1 network, from the Adaptive Resonance Theory family of ANNs [4], and the Combinatorial Neural Model (CNM), proposed by Machado ([11] and [12])). Both models satisfy one important desiderata for data mining, learning in just one pass of the database. Moreover, CNM, the intensional part of the architecture, allows one to obtain rules directly from its structure. These rules represent the insights on the groups. The architecture can be extended to other supervised/unsupervised learning algorithms that comply with the same desiderata.

---

* Researcher at EMBRAPA — Brazilian Enterprise for Agricultural Research and lecturer at Catholic University of Brasília (Supported by CAPES - Coordenaçao de Aperfeiçoamento de Pessoal de Nível Superior, grant nr. BEX1041/98-3)
** Professor at University of Pittsburgh
*** Professor at Federal University of Rio Grande do Sul

J. Rolim et al. (Eds.): IPDPS 2000 Workshops, LNCS 1800, pp. 407-414, 2000.
© Springer-Verlag Berlin Heidelberg 2000

1   Introduction

Research in Knowledge Discovery from Databases (KDD) has developed along two main dimensions: (1) improving the knowledge discovery process, and (2) scaling up this same process to very large databases. Machine Learning, as an important field related to KDD, is founded on three principles [19]:

1. Modeling of cognitive processes, aiming to select characteristics of interest to be formalized as knowledge;
2. Computer science, which offers a formalism to support the descriptions of those characteristics, as well as providing approaches to evaluate the degree of computational difficulty of the issues involved; and
3. Applications, where one departs from practical needs to the implementation of systems.

   In this article, we depart from a characterization of the concept formation activity as a cognitive process, proposing a computational approach to support this activity, from the point of view of KDD. We take the concept of performance as given by the relation functionality/resources applied. By this way, we present a model where funcionality is increased while the resources applied are just slightly changed.

2   Motivation

According to Wrobel [19], a concept is "a generalized description of sets of objects". In this sense, Easterlin and Langley [5] analyse the concept formation process as follows:

1. Given a set of objects (instances, events, cases) descriptions, usually presented incrementally;
2. find sets of objects that can be grouped together (aggregation), and
3. find intensional description of these sets of objects (characterization).

   Murphy and Medin [14] discuss two hypothesis that constrain the way objects are grouped in concepts:

1. Similarity hypothesis: this hypothesis sustain that what defines a class is that its members are similar to each other and not similar to members of other classes.
2. Correlated attribute hypothesis: this hypothesis states that "natural groups" are described according to clusters of features and that categories reflect the cluster structure of correlations.

   The first hypothesis presents a problem that is: the similarity criteria must be applied to a pre-defined set of features and the definition of this set is affected by the previous knowledge one has over the objects. However, when just a small knowledge about the data exists, this criteria is used as a first approximation. Over this approximation, the correlated attribute hypothesis is applied. In a broad sense, what is desirable in this process is to provoke the mental operations that can lead to a problem solution ([16] and [15]). Actually, since it seems that the discovery process, as a rule, requires the human judgment [9], it is useful to leave available to the analyst all relevant information to evaluate both hypothesis when searching for the classes' structures.

3 Proposed Architecture

The research on the KDD realm has emphasized improvements in the processes of supervised and unsupervised learning. More recently, many unsupervised learning algorithms have been scaled up according to the desiderata proposed by Agrawal *et al.* [1] for this kind of learning algorithm. Considering these advances and the ones in supervised learning, we believe there is enough room to scale up the combined process of unsupervised and supervised learning in order to obtain better descriptions of unclassified data. By "better descriptions" we mean obtaining intensional (what are the main characteristics of each class) descriptions, beyond the extensional (what objects are members of each class) ones, usually provided. We explore our idea with a hybrid architecture, based into two well-know models: ART1 [4], used for cluster binary data, and CNM ([11] and [12]), used to map the input space in the formed classes. Both ANNs present an important characteristic to support data mining: they learn in just one pass of the entire data set. The model is illustrated by Figure 1.

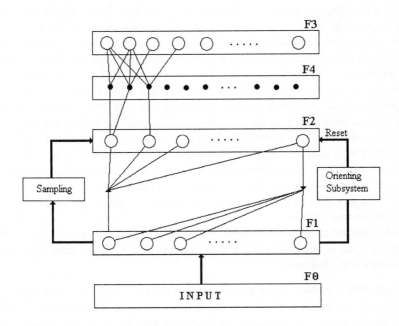

Fig. 1. Describing unclassified data

The architecture is composed by five layers according to the schema: Input layer (F0, where the examples are introduced in the architecure); Aggregation module (F1 and F2, where the classes are defined); Characterization module (F2, F3 , and F4, where the classes are explained).

For the characterization module, it requires the pre-existence of classes that would not be available when the process starts. We overcome this problem by creating the classes by means of a sampling subsystem. The creation of classes through sampling was explored by Guha [8] with consistent results. As a consequence of the sampling process, a complete execution of the system will take more than one pass of the input data set. Considering γ the size of the sample and **D** the size of the input data set, a complete execution of the system will take, precisely, $\mathbf{D} + \gamma$ records. In the next two sections, we describe each model used as building blocks for our architecture.

4   ART1 Neural Network

ART1 (F1 and F2 layers) is a member of the so-called ART family, that stands by Adaptive Resonance Theory [10], [3], [4] and [6].

ART1 is a competitive recurrent network with two layers, the input layer and the clustering layer. This network was developed to overcome the plasticity-stability dilemma [7], allowing an incremental learning, with a continuous updating of the clusters proto-types, and preserving the previously stored patterns. The clustering algorithm proceeds, in general steps, as folows: (a) the first input is selected to be the first cluster; (b) each next input is compared with each existing cluster; the first cluster where the distance to the input is less then a threshold is chosen to cluster the input. Otherwise, the input defines a new cluster. It can be observed that the number of clusters depends on the threshold and the distance metric used to compare the inputs with the clusters. For each input pattern presented to the network, one output unit is declared winner (at the first pattern, the own input pattern defines the cluster). The winner backpropagates a signal that encodes the expected pattern template. If the current input pattern differs more than a defined threshold from the backpropagated signal, the winner are temporarily disabled (by the Orienting System) and the next closest unit is declared winner. The process continues until an output unit become a winner, considering the threshold. If no one of the output units become a winner, a new output unit is defined to cluster the input pattern. Graphically, an ART1 network can be illustrated by Figure 2, where it appears with four input and six ouput neurons. t_{ij} and b_{ij} are, respectively, bottom-up and top-down connections.

One important characteristic of this ANN is that it works in just one pass, what is interesting when we are processing a huge amount of data. The training algorithm for this network is the following:

- **Step1. Initialization**: The bottom-up $b_{ij}(t)$ and top-down $t_{ij}(t)$ weight connection between input node i and output node j at time t are set up. The fraction ρ (vigilance parameter) is defined, indicating how close an input must be to a stored exemplar to match. Initialize N and M, numbers of input and output nodes.
- **Step 2. Apply New Input**
- **Step 3. Compute Matching Scores**: The bottom-up weights are applied to the input pattern, generating the output signal: μ_j.
- **Step 4. Select Best Matching Exemplar**: $\mu_j^* = \max\{\mu_j\}$ is taken as the best exemplar.
- **Step 5. Vigilance Test**: The best exemplar and the input pattern are compared, according to ρ. If the distance is acceptable, the control flows to **Step 7**, otherwise **Step 6** proceeds.

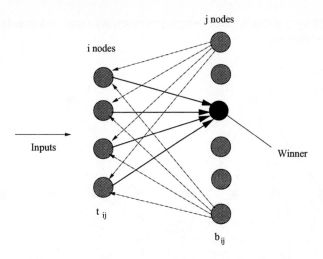

Fig. 2. Architecture of an ART network [3]

- **Step 6. Disable Best Matching Exemplar**: The ouput of the best matching node selected in Step4 is temporarily set to zero and no longer takes part in the maximization of Step4. Then go to Step 3.
- **Step 7. Adapt Best Matching Exemplar**:

$$t_{ij_*}(t+1) = t_{ij_*}(t)t(x_i)$$

$$b_{ij_*}(t+1) = \frac{t_{ij_*}(t)x_i}{\frac{1}{2} + \sum_{i=0}^{N-1} t_{ij_*}(t)x_i}$$

- **Step 8. Repeat by Going to Step 2**: First enable any node disabled in Step 6.

5 Combinatorial Neural Model (CNM)

CNM (F2, F3, and F4 layers) is a hybrid architecture for intelligent systems that integrates symbolic and connectionist computational paradigms. This model is able to recognize regularities from high-dimensional symbolic data, performing mappings from this input space to a lower dimensional output space. Like ART1, this ANN also overcomes the plasticity-stability dilemma [7].

The CNM uses supervised learning and a feedforward topology with: one input layer, one hidden layer - here called combinatorial - and one output layer (Figure 3). Each neuron in the input layer corresponds to a concept - a complete idea about an object of the domain, expressed in an object-attribute-value form. They represent the evidences of the domain application. On the combinatorial layer there are aggregative fuzzy AND neurons, each one connected to one or more neurons of the input layer by arcs with adjustable weights. The output layer contains one aggregative fuzzy OR neuron for each possible class (also called hypothesis), linked to one or more neurons on the combinatorial layer. The synapses may be excitatory or inhibitory and they are characterized by a strength value (weight) between zero (not connected) to one (fully connected synapses).

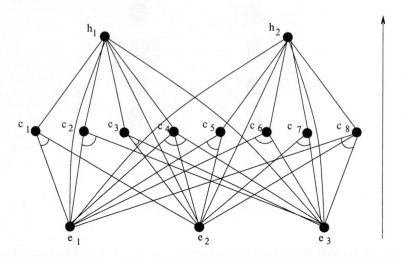

Fig. 3. Complete version of CNM for 3 input evidences and 2 hypotheses [11]

The network is created completely uncommited, according to the following steps: (a) one neuron in the input layer for each evidence in the training set; (b) a neuron in the output layer for each class in the training set; and (c) for each neuron in the output layer, there is a complete set of hidden neurons in the combinatorial layer which corresponds to all possible combinations (length between two and nine) of connections with the input layer. There is no neuron in the combinatorial layer for single connections. In this case, input neurons are connected directly to the hypotheses.

The learning mechanism works in only one iteration, and it is described bellow:

PUNISHMENT_AND_REWARD_LEARNING_RULE
- **Set** to each arc of the network an accumulator with initial value zero;
- **For each** example case from the training data base, **do**:
  - *Propagate* the evidence beliefs from input nodes until the hypotheses layer;
  - **For each** arc reaching a hypothesis node, **do**:
    * **If** the reached hypothesis node corresponds to the correct class of the case
    * **Then** *backpropagate* from this node until input nodes, increasing the accumulator of each traversed arc by its evidencial flow (Reward)
    * **Else** *backpropagate* from the hypothesis node until input nodes, decreasing the accumulator of each traversed arc by its evidencial flow (Punishment).

After training, the value of accumulators associated to each arc arriving to the output layer will be between [-T, T], where T is the number of cases present in the training set. The last step is the prunning of network; it is performed by the following actions: (a) remove all arcs whose accumulator is lower than a threshold (specified by a specialist); (b) remove all neurons from the input and combinatorial layers that became disconnected from all hypotheses in the output layer; and (c) make weights of the arcs arriving at the output layer equal to the value obtained by dividing the arc accumulators by the largest arc accumulator value in the network. After this pruning, the network becomes operational for classification tasks. This ANN has been applied with success in data mining tasks ([2], [17], and [18]).

6 Ongoing Work

This paper presents an architecture to scale up the whole process of concept formation according two main constraints: the identification of groups composed by similar objects and the description of this groups by the higher correlated features. The ongoing work includes the implementation and evaluation of this architecture, instantiated to ART1 and CNM, and its extension to cope with continuous data.

References

1. Agrawal, R., Gehrke, J., Gunopulos, D., Raghanavan, P. Automatic Subspace Clustering of High-Dimensional Data for Data Mining Applicati ons. In: Proceedings of ACM SIGMOD98 International Conference on Management of Data, Seattle, Washington, 1998.
2. Beckenkamp, F. G., Feldens, M. A., Pree, W.: Optimizations of the Combinatorial Neural Model. IN: Vth Brazilian Symposium on Neural Networks. (SBRN'98), Belo Horizonte, Brazil.
3. Bigus, J. P. Data Mining with Neural Networks. [S.l.]: McGraw-Hill, 1996. p.3-42.
4. Carpenter, G. and Grossberg, S. Neural Dynamics of Category Learning and Recognition: Attention, Memory, Consolidation, and Amnesia. In: Joel L. Davis (ed.), Brain structure, learning, and memory. AAAS Symposia Series, Boulder, CO: Westview Press, 1988. p.233-287.
5. Easterlin, J.D., Langley, P.: A Framework for Concept Formation. In: Seventh Annual Conference of the Cognitive Science Society, Irvine, CA, 1985.
6. Engel, P. M. Lecture Notes. Universidade Federal do Rio Grande do Sul. Porto Alegre-RS, Brazil: CPGCC da UFRGS, 1997.
7. Freeman, J. A., Skapura, D. M.: Neural Networks, Algorithms, Applications, and Program Techniques. [S.l.]: Addison-Wesley, 1992. p.292-339.
8. Guha, S., Rastogi, R., Shim, K. Cure: An Efficient Clustering Algorithm for Larg e Databases. In: Proceedings of ACM SIGMOD98 International Conference on Managemen t of Data, Seattle, Washington, 1998.
9. Langley, P. The Computer-Aided Discovery of Scientific Knowledge. In: Proc. of the First International Conference on Discovery Science, Fukuoka, Japan, 1998.
10. Lippmann, D. An Introduction to Computing with Neural Nets, IEEE ASSP Magazine. April, 1987.
11. Machado, R. J., Rocha, A. F.: Handling knowledge in high order neural networks: the combinatorial neural network. Rio de Janeiro: IBM Rio Scientific Center, Brazil, 1989. (Technical Report CCR076).
12. Machado, R. J., Carneiro, W., Neves, P. A.: Learning in the combinatorial neural model, IEEE Transactions on Neural Networks, v.9, p.831-847. Sep.1998.
13. Medin, D., Altom, M.W., Edelson, S.M. and Freko, D. Correlated symptoms and simulated medical classification. Journal of Experimental Psychology: Learning, Memory and Cognition, 8:37-50, 1983.
14. Murphy, G. and Medin, D.: The Role of Theories in Conceptual Coherence. Psychological Review, 92(3):289-316, July, 1985.

15. Pereira, W. C. de A. Resoluçao de Problemas Criativos: Ativaçao da Capacidade de Pensar. Departamento de Informaçao e Documentaçao/EMBRAPA, Brasília-DF, 1980. 54pp.
16. Polya, G. How to Solve It: A New Aspect of Mathematical Method. Princeton: Princeton University Press, 1972. 253pp.
17. Prado, H. A., Frigeri, S. R., Engel, P. M.: A Parsimonious Generation of Combinatorial Neural Model. IN: IV Congreso Argentino de Ciencias de la Computación (CACIC' 98), Neuquén, Argentina, 1998.
18. Prado, H. A. do; Machado, K.F.; Frigeri, S. R.; Engel, P. M. Accuracy Tuning in Combinatorial Neural Model. PAKDD' 99 - Pacific-Asia Conference on Knowledge Discovery and Data Mining. Proceedings ... Beijing, China, 1999
19. Wrobel, S. Concept Formation and Knowledge Revision. Dordrecht, The Netherlands: Kluwer, 1994. 240pp.

Parallel Data Mining
of Bayesian Networks from
Telecommunications Network Data

Roy Sterritt, Kenny Adamson, C. Mary Shapcott, and Edwin P. Curran

University of Ulster,
Faculty of Informatics, Newtownabbey, County Antrim, BT37 0QB, Northern Ireland.
{r.sterritt, k.adamson, cm.shapcott, ep.curran}@ulst.ac.uk
http://www.ulst.ac.uk/

Abstract. Global telecommunication systems are built with extensive redundancy and complex management systems to ensure robustness. Fault identification and management of this complexity is an open research issue with which data mining can greatly assist.
This paper proposes a hybrid data mining architecture and a parallel genetic algorithm (PGA) applied to the mining of Bayesian Belief Networks (BBN) from Telecommunication Management Network (TMN) data.

1 Introduction and the Global Picture

High-speed broadband telecommunication systems are built with extensive redundancy and complex management systems to ensure robustness. The presence of a fault may not only be detected by the offending component and its parent but the consequence of that fault discovered by other components. This can potentially result in a net effect of a large number of alarm events being raised and cascaded to the element controller, possibility flooding it in a testing environment with raw alarm events. In an operational network a flood is prevent by filtering and masking functions on the actual network elements (NE). Yet there can still be a considerable amount of alarms depending on size and configuration of the network, for instance, although unusual to execute, there does exist the facility for the user to disable the filtering/masking on some of the alarm types.

The behaviour of the alarms is so complex it appears non-deterministic. It is very difficult to isolate the true cause of the fault/multiple faults. Data mining aims at the discovery of interesting regularities or exceptions from vast amounts of data and as such can assist greatly in this area.

Failures in the network are unavoidable but quick detection and identification of the fault is essential to ensure robustness. To this end the ability to correlate alarm events becomes very important.

This paper will describe how the authors, in collaboration with NITEC, a Nortel Networks R&D lab, have used parallel techniques for mining bayesian networks from telecommunications network data. The primary purpose being fault management - the induction of a bayesian network by correlating offline

J. Rolim et al. (Eds.): IPDPS 2000 Workshops, LNCS 1800, pp. 415-422, 2000.

event data, and deducing the cause (fault identification) using this bayesian network with live events.

The problems encountered using traditional computing and how these can be overcome using a high performance computing architecture and algorithm are reported along with the results of the architecture and parallel algorithm.

1.1   Telecommunication Fault Management, Fault Correlation and Data Mining BBNs

Artificial intelligence and database techniques are useful tools in the interrogation of databases for hitherto unseen information to support managerial decision making or aid advanced system modelling. Knowledge discovery in databases is a technique for combing through large amounts of information for relationships that may be of interest to a domain expert but have either been obscured by the sheer volume of data involved or are a product of the volume[1].

Bayesian Belief Networks is a technique for representing and reasoning with uncertainty[2]. It represents CAUSEs and EFFECTs as nodes and connects CAUSEs and EFFECTs as networks with a probability distribution. Bayesian Belief Networks have been successfully used to build applications, such as medical diagnosis, where multiple causes bear on a single effect[2]. We proposed to employ both techniques in conjunction to model complex systems that act in a non-deterministic manner. These problems are common to real-life industrial systems that produce large amounts of information. The data-handling requirements, computationally expensive techniques and real-time responsiveness make parallel processing a necessity.

The exemplar under which the architecture is being developed is a telecommunications application, involving the Synchronous Digital Hierarchy (SDH) - the backbone of global communications. It offers flexibility in dealing with existing bandwidth requirements and provide capabilities for increasingly sophisticated telecommunications services of the future[3]. One key area of interest to engineers is the management of events and faults in a network of SDH multiplexers[4]. An event is a change of status within a network that produces a corresponding alarm message or fault. When a network error occurs, each multiplexer determines the nature of the problem and takes steps to minimise any loss in signal or service. To facilitate the process of error recovery, there are many levels of redundancy built into a network, e.g. signal re-routing and the use of self-healing rings. The management of faults is complex because :

- the occurrence of faults is time-variant and non-deterministic;
- faults can produce a cascade of other faults;
- fault-handling must be performed in real-time.

Although the behaviour of individual multiplexers is accurately specified, the complex interactions between a number of multiplexers and the variability of real-life network topologies means that the process of fault occurrence is more complicated than a single specification will imply. This problem is compounded by the growth of networks and the increasing variety of network topology. The application of data mining and in particular parallel data mining can assist greatly.

Data mining aims at the discovery of interesting regularities or exceptions from vast amounts of data. As has been stated, fault management is a critical but difficult area of telecommunication network management since networks produce a vast quantity of data that must be analysed and interpreted before faults can be located. Alarm correlation is a central technique in fault identification yet it is difficult to cope with incomplete data and the sheer complexity involved.

At the heart of alarm event correlation is the determination of the cause. The alarms represent the symptoms and as such are not of general interest[5]. There are two real world concerns, the;

- sheer volume of alarm event traffic when a fault occurs;
- cause not the symptoms.

A technique that can tackle both these concerns would be best, yet this can be difficult to achieve.

1.2   The Architecture

We proposed a parallel mining architecture for the elucidation of an accurate system representation based primarily on Bayesian Belief Networks that are induced using Knowledge Discovery techniques. The architecture has a modular design that can be reconfigured according to application specification. The system was prototyped using the INMOS transputer as the target hardware. Within the telecommunications domain, it is hoped that the application of the system will ultimately assist in fault management but also in the analysis of test data.

The actual realisation of the architecture for the situation is shown in Figure 1. It can be seen that the input data is available in the form of log data from the element controller. The team identified a need for the efficient preparatory processing of data that appeared in the event log format. This led to the design and preliminary implementation of a data cleaner and a data pre-processor. The data cleaner allows a user to specify the format of a text document in a generic way using a template, and to specify filtering conditions on the output. In the telecommunications application the text document is an event log, and the template file can be altered if the structure of the event records changes. The data cleaner parses a log file and passes the resulting events to the pre-processor which time-slices the information and creates an intermediate file for use in the induction module.

The Bayesian net, created as a result of induction, is potentially useful in a fault situation where the faults most likely to be responsible for observed alarms can be computed from the net and relayed to a human operator. For this reason there is a deduction module in the architecture whereby observed conditions in the telecommunications network can be fed into the Bayesian net and changes in the probabilities of underlying fault conditions can be computed[6].

However, the components are able to operate in isolation, provided that they are provided with files in the correct input format. In particular the induction component and the deduction component use data in Bayesian Network Interchange format[7].

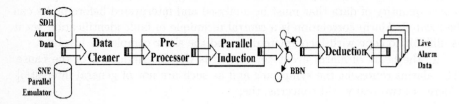

Fig. 1. The Architecture

2   The Parallel Data Mining Algorithm

2.1   The Need for Parallelism

In this case, as in many others, the structure of the graphical model (the Bayesian net) is not known in advance, but there is a database of information concerning the frequencies of occurrence of combinations of different variable values (the alarms). In such a case the problem is that of induction - to induce the structure from the data. Heckerman has a good description of the problem[8]. There has been a lot of work in the literature in the area, including that of Cooper and Herskovits[9]. Unfortunately the general problem is NP-hard [10]. For a given number of variables there is a very large number of potential graphical structures which can be induced. To determine the best structure then in theory one should fit the data to each possible graphical structure, score the structure, and then select the structure with the best score. Consequently algorithms for learning networks from data are usually heuristic, once the number of variables gets to be of reasonable size. There are $2^{k(k-1)/2}$ distinct possible independence graphs for a k-dimensional random vector: this translates to 64 probabilistic models for k= 4, and 32, 768 models for k = 6. Several different algorithms were prototyped and tested using the telecommunications data but since the potential number of graph candidates is so large a genetic algorithm was developed.

2.2   Parallel Cause And Effect Genetic Algorithm (P-CAEGA)

Goldberg describes many ways to view genetic algorithms (GA) [11]: as problems solvers, as a basis for competent machine learning, as a computational model of innovation and creativity and so on. In the work described here the problem is to find the best cause-and-effect network in a very large solution space of all possible cause-and-effect networks since the problem is NP-hard a heuristic search technique must be used. This led to a consideration of genetic algorithms, since they have been shown to work well in many application areas[12] and offers a robust means of searching for the globally optimal solution[13].

The genetic algorithm works on a population of solutions, which change as the algorithm cycles through a sequence of generations, until a satisfactory solution has been found. Initialisation consists of the creation of an initial population, a pool of breeders. In each generation each breeder is scored, and the best breeders are selected (possibly with some uncertainty) to breed and create solutions for the next generation. A solution created by breeding is a genetic mixture of its two parents, and may also have been subject to random mutations which allow new gene sequences to be created. Solutions are directed graphs, viable solutions (those which will be scored and allowed to breed in the next generation) are directed acyclic graphs. The scoring function used was an adaptation of one proposed by Cooper and Herskovits[9], in which the best fit to the experimental data is calculated using Bayesian techniques. High scoring structures have a greater chance of being selected as parents for the next generation.

Due to the sheer volume of data involved in data mining[1], the time required to execute genetic algorithms and the intrinsic parallel nature of genetic algorithms[14], it was decided to implement a parallel version of the CAEGA algorithm (P-CAEGA).

There are a number of approaches to parallelising an algorithm for execution on more than one processor. An architecture with common memory can be used which allows efficient communication and synchronisation. Unfortunately these systems cannot be scaled to a large number of processors because of physical construction difficulties and contention for the memory bus[15].

An alternative is the distributed programming model or message passing model. Two environments widely researched in this area are the Local Area Network (LAN) using Parallel Virtual Machine (PVM) and Transputer Networks using Inmos development languages[16]. In the network architecture the hardware scales easily to a relatively large number of processors but this is eventually limited because of network contention. The Transputer hardware is a point-to-point architecture with dedicated high-speed communications with no contention and no need for addressing. The system can be highly scaleable if the program is constructed accordingly.

The parallel prototype implementation was carried out on T805 INMOS Transputers connected to a Sun Workstation with development performed in parallel C. The sequential prototype of CAEGA had been coded in Pascal. This was converted to C. The C code was then decomposed into processes that needed to be run sequentially and those that could be executed in parallel. Communications channels were used to transfer data between processes.

The first parallel version is a straightforward master-slave (processor farm) implementation. Breeding (reproduction, crossover and mutation) was carried out in parallel. In fact the scoring was also implemented in parallel. The selection had to be implemented sequentially and thus remained on the master (the root processor which is the controller, and is connected to the host). This was necessary, as all of the structures from the new generation needed to be re-mixed to form new parents from the gene pool before distribution to the slaves for breeding. The remaining processors are utilised as slaves, which carry out the breeding in parallel and report the new structures and their scores to the master (root processor).

As was anticipated from preliminary investigations the scaleability achievable is limited because of the overhead of communications. For less than 8 processors the echo n-2 holds (excluding the master; n-1). It is believed, with further work on the efficiency of the algorithm this could be improved, but a linear increase (excluding the master) is not expected because of the sheer amount of communications involved.

This implementation represents a straight forward first prototype. It is a direct parallelisation of an CAEGA which did not change the underlying nature of the algorithm. This has resulted in global communications, which limits the scaleable - speedup ratio. In a LAN implementation this would be even more restrictive due to communications costs. In general an effective concurrent design will require returning to first principles.

2.3   Results

Typical results of the application of the algorithms described are shown below. The data which is shown results from an overnight run of automated testing, but does not show dependencies on underlying faults. About 12,000 individual events were recorded and the graph in Figure 2 shows a generated directed graph in which the width of the edge between two nodes (alarms) is proportional to the strength of connection of the two variables. It can be seen that PPI-AIS and LP-EXC have the strongest relationship, followed by the relationship between PPI-Unexp-Signal and LP-PLM. Note that the directions of the arrows are not important as causal indicators but variables sharing the same parents do form a group.

The graph in Figure 3 shows the edge strengths as strongest if the edge remains in models which become progressively less complex - where there is a penalty for complexity. It can be seen that the broad patterns are the same in the two graphs but that the less strong edges are different in the two graphs. In the second graph the node NE-Unexpected-Card shows links to three other nodes, whereas it has no direct links in the first graph.

The results from an industrial point of view are very encouraging. The case for using a genetic algorithm holds and parallelising it speeds up this process. The algorithm is currently being used by NITEC to analyse their data produced by a SDH network when a fault occurs. From their point of view it has been a worthwhile effort for the speed-up.

It has been established that genetic scoring and scaling could be implemented in parallel but the communications cost in transmitting these back to the master removed any benefit from just having the master perform these functions.

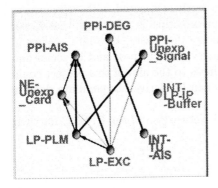

Fig. 2. Example results BBN of TMN data

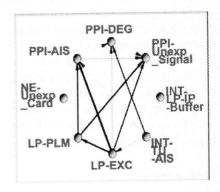

Fig. 3. Another set of results

2.4 Future Potential Research

This study assessed GAs as a solution provider. As Goldberg states "some of us leave the encounter with far more"[11] . Future development will take the basic solution further and remove the limitation on scaleability. What is required is a concurrent algorithm as opposed to a modification of a sequential algorithm. The next planned development is a redesign of the algorithm into the "continental" algorithm (current development name).

P-CAEGA as it stands can be classified as global parallelisation. Every individual has a chance to mate with all the rest (i.e. random breeding), thus the implementation did not affect the behaviour of the original algorithm (CAEGA).

The "continental" version would be a more sophisticated parallel approach where the population is divided into sub-populations , relatively isolated from each other. This model introduces a migration element that would be used to send some individuals from one sub-population to another. This adapted algorithm would yield local parallelism and each processor could be thought of as a continent where the majority of the breeding occurs between residents with limited migration. This modification to the behaviour of the original algorithm would vastly decrease the communications cost, and present a more scaleable implementation.

3   Conclusion

The association between the University of Ulster and Nortel in the area of fault analysis has continued with the Garnet project. Garnet is using the techniques developed in NETEXTRACT to develop useful tools in the area of live testing. The basic idea is to regard the Bayesian nets as abstract views of the test network's response to stimuli. In a further development, the Jigsaw project entails the construction of a data warehouse for the storage of the test data, with a direct link to data mining algorithms.

Although this paper has described the work completed with reference to Nortel's telecommunications network, the architecture is generic in that it can extract cause-and-effect nets from large, noisy databases, the data arising in many areas of science and technology, industry and business and in social and medical domains. The corresponding hypothesis to the aim of this project could be proposed - that cause and effect graphs can be derived to simulate domain experts' knowledge and even extend it.

The authors would like to thanks Nortel Networks and the EPSRC for their support during the project.

References

1. Agrawal, R., Imielinski, T., Swami, A.,"Database Mining: a Performance Perspective",IEEE Trans. KDE, 5(6), Dec, pp914-925, 1993.
2. Guan, J.W., Bell, D.A., Evidence Theory and Its Applications, Vol. 1&2, Studies in Computer Science and AI 7,8, Elsevier, 1992.
3. ITU-T Rec. G.782 Types and General Characteristics of SDH Muliplexing Equipment, 1990.
4. ITU-T Rec. G.784 SDH Management, 1990.
5. Harrison K. "A Novel Approach to Event Correlation", HP Labs., Bristol. HP-94-68, July, 1994, pp. 1-10.
6. McClean, S.I. and Shapcott, C.M., 1997, "Developing BBNs for Telecommunications Alarms", Proc. Conf. Causal Models & Statistical Learning, pp123-128, UNICOM, London.
7. MS Decision Theory and Adaptive Systems Group, 1996. "Proposal for a BBN Interchange Format", MS Document.
8. Heckerman D, 1996. "BBNs for Knowledge Discovery" In Fayyad UM, Piatetsky-Shapiro G, Smyth P and Uthurusamy R (Eds.), Advances in Knowledge Discovery and Data Mining AAAI Press/MIT Press, pp273-305.
9. Cooper, G.F. and Herskovits, E., 1992. "A Bayesian Method for the Induction of Probabilistic Networks from Data". ML, 9, pp309-347
10. Chickering D.M. and D. Heckerman, 1994. "Learning Bayesian networks is NP-hard". MSR-TR-94-17, Microsoft Research, Microsoft Corporation, 1994.
11. Goldberg, D.E., "The Existential Pleasures of Genetic Algorithms". Genetic Algorithms in Engineering and Computer Science , (England: John Wiley and Sons Ltd., 1995) Ed. Winter, G., et al 23-31.
12. Holland, J. H., Adaptation in Natural and Artifical Systems, (Ann Arbor: The University of Michigan Press, 1975)
13. Larranga, P., et al., Genetic Algorithms Applied to Bayesian Networks,Proc. of the Applied Decision Technology Conf., 1995,pp283-302.
14. Bertoni, A., Dorigo, M., Implicit Parallelism in Genetic Algorithms. AI (61) 2, 1993, pp307-314.
15. Ben-Ari, M., Principles of Concurrent and Distributed Programming, (Hertfordshire: Prentice Hall International (UK) Ltd. ,1990)
16. Almeida, F., Garcia, F., Roda, J., Morales, D., Rodriguez C., A Comparative Study of two Distributed Systems: PVM and Transputers, Transputer Applications and Systems, (IOS Press, 1995) Ed. Cook, B., et al 244-258.

IRREGULAR'00

SEVENTH INTERNATIONAL WORKSHOP ON SOLVING IRREGULARLY STRUCTURED PROBLEMS IN PARALLEL

General Chair: Sartaj Sahni, University of Florida

Program Co-chairs:

Timothy Davis, University of Florida
Sanguthevar Rajasekeran, University of Florida
Sanjay Ranka, University of Florida

Steering Committee:

Afonso Ferreira, CNRS-I3S-INRIA Sophia Antipolis
José Rolim, University of Geneva

Program Committee:

Cleve Ashcraft, The Boeing Company
Iain Duff, CLRC Rutherford Appleton Laboratory
Hossam ElGindy, University of Newcastle
Apostolos Gerasolous, Rutgers University, New Brunswick
John Gilbert, Xerox Palo Alto Research Center
Bruce Hendrickson, Sandia Labs
Vipin Kumar, University of Minnesota
Esmond Ng, Lawrence Berkeley National Laboratory
C. Pandurangan, Indian Institute of Technology, Madras
Alex Pothen, Old Dominion University
Padma Raghavan, Univ. of Tennessee
Rajeev Raman, King's College, London
John Reif, Duke University
Joel Saltz, University of Maryland
Horst Simon, Lawrence Berkeley National Laboratory
Jaswinder Pal Singh, Princeton University
Ramesh Sitaraman, University of Massachusetts
R. Vaidyanathan, Louisiana State University
Kathy Yelick, University of California, Berkeley

Invited Speakers:

William Hager, University of Florida
Vipin Kumar, University of Minnesota
Panos Pardalos, University of Florida

J. Rolim et al. (Eds.): IPDPS 2000 Workshops, LNCS 1800, pp. 423-426, 2000.

FOREWORD

The Seventh International Workshop on Solving Irregularly Structured Problems in Parallel (*Irregular '00*) is an annual workshop addressing issues related to deriving efficient parallel solutions for unstructured problems, with particular emphasis on the inter-cooperation between theoreticians and practitioners of the field. Irregular'00 is the seventh in the series, after Geneva, Lyon, Santa Barbara, Paderborn, Berkeley, and Puerto Rico.

Twelve of the submitted papers have been selected for presentation by the Program Committee on the basis of referee reports. The final scientific program of Irregular '00 consists of four sessions and three invited talks.

We wish to thank all of the authors who responded to the call for papers and our invited speakers. Thank the members of the Program Committee and anonymous reviewers for reviewing and selecting papers. We also would like to thank the IPDPS'00 conference co-chair José Rolim and members of the steering committee for helping us in organizing this workshop.

January 2000

Timothy Davis
Sanguthevar Rajasekeran
Sanjay Ranka
Sartaj Sahni

Irregular'00 Final Program

8:30am-9:15am

Invited Talk by William Hager (University of Florida): Load Balancing and Continuous Quadratic Programming.

9:15am-10:25 Finite element methods–applications and algorithms

Parallel Management of Large Dynamic Shared Memory Space: A Hierarchical FEM Application, Xavier Cavin, Institut National Polytechnique de Lorraine and Laurent Alonso, INRIA Lorraine.

Efficient Parallelization of Unstructured Reductions on Shared Memory Parallel Architectures, Siegfried Benkner, University of Vienna and Thomas Brandes, German National Research Center for Information Technology (GMD).

Parallel FEM Simulation of Crack Propagation–Challenges, Status, and Perspectives, Bruce Carter, Chuin-Shan Chen, Gerd Heber, Antony R. Ingraffea, Roland Krause, Chris Myers, and Paul A. Wawrzynek, L. Paul Chew, Keshav Pingali, Paul Stodghill, and Stephen Vavasis, Cornell University; Nikos Chrisochoides and Demian Nave University of Notre Dame; and Guang R. Gao, University of Delaware.

10:25 -10:55 Coffee break

10:55 -12:05pm Architecture and system software support

Support for Irregular Computations in Massively Parallel PIM Arrays, Using an Object-Based Execution Model, Hans P. Zima, University of Vienna and Thomas L. Sterling, California Institute of Technology.

Executing Communication-Intensive Irregular Programs Efficiently, Vara Ramakrishnan and Isaac D. Scherson, University of California, Irvine.

Non-Memory-based and real-time zerotree building for wavelet zerotree coding systems, Dongming Peng and Mi Lu, Texas A & M University.

12:05 -1:20 Lunch break

1:20pm-2:05pm

Invited Talk by Vipin Kumar (University of Minnesota): Graph Partitioning for Dynamic, Adaptive and Multi-phase Computations (joint work with Kirk Schloegel and George Karypis).

2:05pm-3:15 Graph partitioning - algorithms and applications

A Multilevel Algorithm for Spectral Partitioning with Extended Eigen-Models, Suely Oliveira and Takako Soma, University of Iowa.

An Integrated Decomposition and Partitioning Approach, Jarmo Rantakokko, University of California, San Diego

Ordering Unstructured Meshes for Sparse Matrix Computations on Leading Parallel Systems, Leonid Oliker and Xiaoye Li, Lawrence Berkeley Nat. Lab.; Gerd Heber, Cornell University; and Rupak Biswas, MRJ/Nasa Ames Res. Center.

3:15-3:45 Coffee break

3:45-4:30

Invited Talk by Panos Pardalos (University of Florida): A GRASP for computing approximate solutions for the Three-Index Assignment Problem

4:30-5:40pm Graph algorithms and sparse matrix methods

On Identifying Strongly Connected Components in Parallel, Ali Pinar, University of Illinois; Lisa Fleischer, Columbia University; and Bruce Hendrickson, Sandia National Laboratories.

A Parallel, Adaptive Refinement Scheme for Tetrahedral and Triangular Grids, Alan Stagg, Los Alamos National Laboratory; Jackie Hallberg, US Army Eng. Res. and Dev. Center, Coastal and Hydraulics Lab.; and Joseph Schmidt, Reston, Virginia.

PaStiX: A Parallel Sparse Direct Solver Based on a Static Scheduling for Mixed 1D/2D Block Distributions, Pascal Henon, Pierre Ramet, and Jean Roman, Universite Bordeaux I.

Load Balancing and Continuous Quadratic Programming

William W. Hager

Department of Mathematics, University of Florida,
358 Little Hall, Gainesville, FL 32611-8105. hager@math.ufl.edu

Abstract. A quadratic programming approach is described for solving the graph partitioning problem, in which the optimal solution to a continuous problem is the exact solution to the discrete graph partitioning problem. We discuss techniques for approximating the solution to the quadratic program using gradient projections, preconditioned conjugate gradients, and a block exchange method.

1 Extended Abstract

In a parallel computing environment, tasks can be modelled as nodes on a graph and the communication links between tasks as edges on the graph. The problem of assigning the tasks to different processors while minimizing the communication between processors is equivalent to partitioning the nodes of the graph into sets chosen so that the number of edges connecting nodes in different sets is as small as possible. The graph partitioning problem was first exposed by Kernighan and Lin in the seminal paper in 1970. Since Sahni and Gonzales show in 1976 that graph partitioning is an NP hard problem, exact solutions can be computed only when the number of nodes is small. To solve large problems, approximation techniques have been developed that include exchange techniques, spectral methods, geometric methods, and multilevel methods. Our approach to the graph partitioning problem is based a quadratic programming formulation. For the problem of partitioning an n node graph into k sets, we exhibit a quadratic programming problem (quadratic cost function and linear equality and inequality constraints) in nk variables x_{ij}, $1 \leq i \leq n$, $1 \leq j \leq k$, where x_{ij} is a continuous variable taking values on the interval $[0, 1]$. This quadratic program is equivalent to the discrete graph partitioning problem in the sense that there exists an optimal 0/1 solution to the quadratic program and the assignment of node i to set j if $x_{ij} = 1$ is optimal in the graph partitioning problem. Based on this equivalence between graph partitioning and a quadratic problem, we have applied tools from optimization theory to solve the graph partitioning problem. In this talk, we discuss techniques for approximating the solution to the quadratic program using gradient projections, preconditioned conjugate gradients, and a block exchange method. The advantages and disadvantages of the optimization approach compared to other approaches to graph partitioning are also examined. For papers related to this talk, see http://www.math.ufl.edu/~hager.

J. Rolim et al. (Eds.): IPDPS 2000 Workshops, LNCS 1800, pp. 427-427, 2000.

Parallel Management of Large Dynamic Shared Memory Space: A Hierarchical FEM Application

Xavier Cavin* and Laurent Alonso**

ISA research team, LORIA***
Campus Scientifique, BP 239, F–54506 Vandœuvre–lès–Nancy CEDEX, France
Xavier.Cavin@loria.fr

Abstract. We show in this paper the memory management issues raised by a parallel irregular and dynamic hierarchical application, which constantly allocates and deallocates data over an extremely large virtual address space. First, we show that if memory caches data locality is necessary, a lack of virtual pages locality may greatly affect the obtained performance. Second, fragmentation and contention problems associated with the required parallel dynamic memory allocation are presented. We propose practical solutions and discuss experimentation results obtained on a cache–coherent non uniform memory access (ccNUMA) distributed shared memory SGI Origin2000 machine.

1 Introduction

The radiosity equation [Kaj86] is widely used in many physical domains and in computer graphics, for its ability to model the global illumination in a given scene. It looks like a Fredholm integral equation of the second kind, which can be expressed as:

$$f(y) = f_e(y) + \int_\Omega k(x,y) f(x) \, dx \ , \tag{1}$$

where f is the radiosity equation to determine. Generally, this unknown function is defined over an *irregular, non–uniform* physical domain Ω, mainly described in terms of polygonal surfaces (some of them having non–zero initial radiosity f_e).

Finding an analytical solution to (1) is not possible in general. Numerical approximations must be employed, generally leading to very expensive algorithms. The fundamental reason for their high cost is that each surface of the input scene may potentially influence all other surfaces via reflection.

A common resolution technique is the weighted residual method, often referred as "finite element method" (FEM). Early radiosity algorithms can be analyzed as FEM using piecewise constant basis functions. Later, hierarchical algorithms, inspired by adaptive N–body methods, have been introduced

---

* Institut National Polytechnique de Lorraine.
** INRIA Lorraine.
*** UMR 7503, a joint research laboratory between CNRS, Institut National Polytechnique de Lorraine, INRIA, Université Henri Poincaré and Université Nancy 2.

J. Rolim et al. (Eds.): IPDPS 2000 Workshops, LNCS 1800, pp. 428–434, 2000.
© Springer-Verlag Berlin Heidelberg 2000

by [HSA91] to increase the efficiency of the computations. They are based on a multi–level representation of the radiosity function, which is *dynamically* created as the computation proceeds, subdividing surfaces where necessary to increase the accuracy of the solution. Since energy exchanges can occur between any levels of the hierarchies, *sub–computation times are highly variable and change at every step of the resolution.*

This dynamic nature, both in memory and computing resources, combined to the non–uniformity of the physical domain being simulated makes the parallelization of hierarchical radiosity algorithms a challenge, since straightforward parallelizations generally fail to simultaneously provide the load balancing and data locality necessary to efficient parallel execution, even on modern distributed shared memory multiprocessor machines [SHG95].

In a recent paper [Cav99], we have proposed appropriate partitioning and scheduling techniques for a parallel hierarchical wavelet radiosity algorithm, that deliver an optimal load balancing, by minimizing idle time wasted on locks and synchronization barriers, while still exhibiting an excellent data locality. However, our experiments seemed to show that this was still not sufficient to perform extremely large computations with optimal parallel performance. Indeed, dealing in parallel with a dynamically growing huge amount of memory (for a whole building simulation, it is not rare that more than 20 Gbytes may be required) is not free of problems to have it done in an efficient way. This is even more complicated since most of this memory management is generally hidden to the programmer. If this can be a great facility when all works well, it quickly becomes damageable when problems start to occur.

We show in this paper the two main causes of the performance degradation, and experiment practical solutions to overcome them on a 64–processor SGI Origin2000 ccNUMA machine. The first problem concerns the irregular memory access patterns of our application, which have to be handled within an extremely large virtual address space. The issues are discussed in Sect. 2, and we show how some SGI IRIX operating system facilities can help enhancing virtual pages locality and consequently reduce computation times. Parallel dynamic memory allocation is the second problem and comes in two different flavors: *fragmentation* and *contention*. We experiment in Sect. 3 available IRIX and public domain solutions, and propose enhancements leading to an efficient parallel memory allocator. Finally, Sect. 4 concludes and presents future work.

2 Efficient Irregular Memory Accesses within a Large Virtual Address Space

2.1 Understanding ccNUMA Architecture

In order to fully benefit from a computer system performance, it is really important to understand the underlying architecture. The SGI Origin2000 is a scalable multiprocessor with distributed shared memory, based on the Scalable Node 0 (SN0) architecture [Cor98]. The basic building block is the node board, composed of two MIPS R10000 processors, each with separate 32 Kbytes first level

(L1) instruction and data caches on the chip, with 32–byte cache line, and a unified (instruction and data), commonly 4 Mbytes, two–way set associative second level (L2) off–chip cache, with 128–byte cache line. Large SN0 systems are built by connecting the nodes together via a scalable interconnection network.

The SN0 architecture allows the memory to be physically distributed (from 64 Mbytes to 4 Gbytes per node), while making all memory equally accessible from a software point of view, in a ccNUMA approach [LL97]. A given processor only operates on data that are resident in its cache: as long as the requested piece of memory is present in the cache, access times are very short; on the contrary, a delay occurs while a copy of the data is fetched from memory (local or remote) into the cache. The trivial conclusion is that a program shall use these caches effectively in order to get optimal performance.

Obviously, if the shared memory is seen as a contiguous range of virtual memory addresses, the physical memory is actually divided into pages, which are distributed all over the system. For *every* memory access, the given virtual address must be translated into the physical address required by the hardware. A hardware cache mechanism, the translation lookaside buffer (TLB), keeps the 64×2 most recently used page addresses, allowing an instant virtual–to–physical translation for these pages. This allows a 2 Mbytes memory space (for the default 16 Kbytes page size) to be addressed without translation penalty. Programs using larger virtual memory (the common case) may refer to a virtual address that is not cached in the TLB. In this case (TLB miss), the translation is done by the operating system, in the kernel mode, thus adding a non–negligible overhead to the memory access, *even if it is satisfied in a memory cache.*[1]

2.2 Enhancing Virtual Pages Locality

Having an optimal data caches locality appears to be a necessary, but no sufficient, condition to get optimal performance, since a L1 or L2 cache hit may be greatly delayed by a TLB miss. It is really important, however, to understand that data caches locality does not necessarily implies virtual pages (i.e. TLB) locality. Indeed, the data are stored in memory caches with a small size granularity (128 bytes for the 4 Mbytes of L2 cache), and may thus come from a large number (much greater that 64 TLB entries) of different pages. Then, the application may exhibit an optimal data locality through these data and a poor TLB locality at the same time.

Unfortunately, this is the case for our hierarchical application, which by nature exhibits high data locality, but suffers from highly irregular data access patterns. Whatever the input data, our application affords very high cache hits rates of more than 95 % for both L1 and L2 caches, for *any number* of processors used [CAP98]. At the same time, it is clear that the irregular memory accesses towards a large number of virtual pages are responsible for many TLB misses, especially as the size of the input data increases.[2] This is confirmed by the analysis of the sequential run of the application with the default 16 Kbytes page

---

[1] See [Cor98] for the precise read latency times.

[2] Monitoring of L1, L2 and TLB misses is done with the IRIX `perfex` command.

size, where about 44 % of the execution time is lost due to TLB misses. Table 1 shows that when the number of processors used increases, the total number of TLB misses quickly falls down by 33 % with 16 processors, and then more slowly decreases. This seems to be due to the fact that using N processors allows to use $64 * N$ TLB entries at the same time. This suggests that next generations of MIPS processors should contain more TLB entries.

Increasing the number of TLB entries appears to enhance TLB locality. Nevertheless, for a fixed number of available TLB entries, an alternate solution is to increase the total memory space they can address, by simply telling the operating system to increase the size of a memory page, thanks to the IRIX `dplace` command. As shown by Table 1, doing this reduces the execution times, at least with at the small 16–processor scale. Indeed, using a larger number of processors multiplies the number of available TLB entries, thus reducing the TLB misses problem and the benefits of this solution.

Table 1. Impact of page sizes on TLB misses and execution times

	Page size	Processors						
		1	4	8	16	24	32	40
Execution time	16k	13 835	3 741	1 866	1 054	761	653	610
(s)	1m	10 813	–	1 538	891	701	612	590
TLB misses	16k	17 675	14 019	12 487	11 676	10 176	9 888	9 497
($\times 10^6$)	1m	5 479	–	3 939	3 769	3 813	3 842	3 688

3   Efficient Parallel Dynamic Memory Allocation

Since the memory required by our application has to be dynamically allocated all along the execution, an efficient dynamic memory allocator has to be employed, both to reduce fragmentation problems, which may cause severe memory waste, and to allow contention free parallel manipulations. As quoted in [WJNB95], "memory allocation is widely considered to be either a solved problem, or an insoluble one". This appears to be true for the common programmer using the default memory allocation package available on his machine. Once again, most of the time, this solution is not a bad one, but when the memory allocator runs badly, one discovers to be in front of a mysterious "black box".

The role of a memory allocator is to keep track of which parts of the memory are in use, and which parts are free, and to provide the processes an efficient service, minimizing wasted space without undue time cost, or *vice et versa*. Space consideration appears to be of primary interest. Indeed, worst case space behavior may lead to complete failure due to memory exhaustion or virtual memory trashing. Obviously, time performance is also important, especially in parallel, but this is a question rather of implementation than of algorithm, even if considerations can be more complicated.

We believe that it is primordial to rely on existing memory allocators, rather than to develop *ad hoc* storage management techniques, for obvious reasons of software clarity, flexibility, maintainability, portability and reliability. We focus here on the three following available ones:

1. the IRIX C standard memory allocator, which is a complete black box;
2. the IRIX alternative, tunable, "fast main memory allocator", available when linking with the -lmalloc library;
3. the LINUX/GNU libc parallel memory allocator[3], which is an extension of famous Doug Lea's Malloc[4], and is also parameterizable.

3.1   The Fragmentation Problem

Fragmentation is the inability to reuse memory that is free. An application may free blocks in a particular order that creates holes between "live" objects. If these holes are too numerous and small, they cannot be used to satisfy further requests for larger blocks. Note here that the notion of fragmented memory at a given moment is completely relative to further requests. Fragmentation is the central problem of memory allocation and has been widely studied in the field, since the early days of computer science. Wilson et al. present in [WJNB95] a wide coverage of literature (over 150 references) and available strategies and policies. Unfortunately, a unique general optimal solution has not emerged, because it simply does not exist. Our goal here is not to propose a new memory allocation scheme, but rather to report the behavior of the chosen allocators in terms of fragmentation. Algorithmic considerations have been put aside, since it is difficult to know the principles a given allocator is implemented on.

We have chosen to use the experimental methodology proposed in [WJNB95] to illustrate the complex memory usage patterns of our radiosity application: all allocation and deallocation requests done by the program are written to a file during its execution. The obtained trace only reflects the program behavior, and is *independent* of the allocator. Figure 1 shows the profile of memory use for a complete run of our radiosity application. Although it has been done on a small input test scene, it is representative of what happens during the computations: many temporary, short–live objects are continuously allocated and deallocated to progressively build the long–live objects of the solution (here for instance, 120 Mbytes of data are allocated for only 10 Mbytes of "useful" data).

Then, the trace is read by a simulator, which has first been linked with the allocator to be evaluated: this allows to precisely monitor the way it behaves, including the potential fragmentation of the memory. Unfortunately, on such small input test scenes, none of the three allocators suffers from fragmentation (86 % of the allocated memory is actually used). The fragmentation problem only occurs with large input test scenes, the computation of which can not (yet) be traced with the tools we use[5]. We can just report what we have observed

---

[3] See http://www.dent.med.uni-muenchen.de/~wmglo/malloc-slides.html

[4] See http://gee.cs.oswego.edu/dl/html/malloc.html

[5] At ftp://ftp.dent.med.uni-muenchen.de/pub/wmglo/mtrace.tar.gz

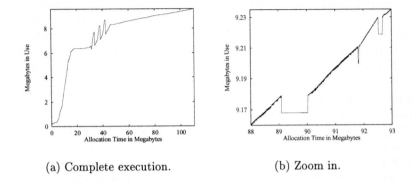

(a) Complete execution.   (b) Zoom in.

Fig. 1. Profile of memory use in a short radiosity computation (534 initial surfaces, 20 iterations, 5 478 final meshes): these curves plot the overall amount of live data for a run of the program, the time scale being the allocation time expressed in bytes

during our numerous experiments: the standard IRIX C allocator appears to be the better one, while the alternative IRIX allocator leads to catastrophic failure with large input data on its default behavior (our experiments with the parameters have not been successful either); LINUX/GNU libc allocator is closer to the IRIX C allocator, although a little bit more space consuming.

3.2   Allocating Memory in Parallel

Few solutions have been proposed for parallel allocators (some are cited at the beginning of Sect. 4 in [WJNB95]). However, none of them appears to be implemented inside the two IRIX allocators, since the only available solution is to serialize memory requests with a single lock mechanism. This is obviously not the right way for our kind of application, which constantly allocates and deallocates memory in parallel. Generally, with this strategy, only one or two processes are running at a given moment, while all remaining ones are idle.

We first considered implementing our own parallel allocator, based on the *arena* facilities provided by IRIX. Basically, each process is assigned to its own arena of memory, and is the only one responsible for memory operations in it: we observed we could achieve optimal time performance, without any contention. Unfortunately, the allocation inside a given arena is based on the alternative IRIX allocator, which gives, as previously said, very poor results in terms of fragmentation. We then considered the LINUX/GNU libc allocator, which exhibits better space performance and provides parallelism facilities, based on similar ideas as ours. Unfortunately, a few implementation details greatly limit the scalability to four or eight threads, which is obviously insufficient for us. We thus fixed these minor problems to have the LINUX/GNU libc allocator more looks like our parallel version, and finally get a parallel allocator which proves to be (for the moment) rather efficient both in space and time performance.

4   Conclusion

We show in this paper the problems raised by the memory management of a large, dynamically evolving, shared memory space within an irregular and non–uniform parallel hierarchical FEM algorithm. These problems are not widely covered in the literature, and there are few available solutions. We first propose practical techniques to enhance memory accesses performance. We then study the characteristics of our application in terms of memory usage, and show that it is greatly suitable to fragmentation problems. Available allocators are considered and experimented to find which one gives the best answer to the request patterns. Finally, we design a parallel allocator, based on the LINUX/GNU libc one, which appears to give good performance in terms of time and space.[6] However, deeper insights, inside both our application and available memory allocators, will be needed to better understand the way they interact, and we believe this is still an open and beautiful problem.

Acknowledgments. We would like to thank the Centre Charles Hermite for providing access to its computing resources. Special thanks to Alain Filbois for the hours spent on the Origin.

References

[CAP98]   Xavier Cavin, Laurent Alonso, and Jean-Claude Paul. Experimentation of Data Locality Performance for a Parallel Hierarchical Algorithm on the Origin2000. In *Fourth European CRAY-SGI MPP Workshop*, Garching/Munich, Germany, September 1998.

[Cav99]   Xavier Cavin. Load Balancing Analysis of a Parallel Hierarchical Algorithm on the Origin2000. In *Fifth European SGI/Cray MPP Workshop*, Bologna, Italy, September 1999.

[Cor98]   David Cortesi. Origin2000 (TM) and Onyx2 (TM) Performance Tuning and Optimization Guide. Tech Pubs Library guide Number 007-3430-002, Silicon Graphics, Inc., 1998.

[HSA91]   Pat Hanrahan, David Salzman, and Larry Aupperle. A Rapid Hierarchical Radiosity Algorithm. In *Computer Graphics (ACM SIGGRAPH '91 Proceedings)*, volume 25, pages 197–206, July 1991.

[Kaj86]   James T. Kajiya. The Rendering Equation. In *Computer Graphics (ACM SIGGRAPH '86 Proceedings)*, volume 20, pages 143–150, August 1986.

[LL97]   James Laudon and Daniel Lenoski. The SGI Origin: A ccNUMA Highly Scalable Server. In *Proceedings of the 24th Annual International Symposium on Computer Architecture*, pages 241–251, Denver, June 1997. ACM Press.

[SHG95]   Jaswinder Pal Singh, John L. Hennessy, and Annoop Gupta. Implications of Hierarchical N-body Methods for Multiprocessor Architectures. *ACM Transactions on Computer Systems*, 13(2):141–202, May 1995.

[WJNB95] Paul R. Wilson, Mark S. Johnstone, Michael Neely, and David Boles. Dynamic storage allocation: A survey and critical review. In *Proceedings of International Workshop on Memory Management*, volume 986 of *Lecture Notes in Computer Science*. Springer-Verlag, September 1995.

---

[6] Feel free to contact us if you want to evaluate this clone.

Efficient Parallelization of Unstructured Reductions on Shared Memory Parallel Architectures[*]

Siegfried Benkner[1] and Thomas Brandes[2]

[1] Institute for Software Technology and Parallel Systems
University of Vienna, Liechtensteinstr. 22, A-1090 Vienna, Austria
sigi@ieee.org
[2] Institute for Algorithms and Scientific Computing (SCAI)
German National Research Center for Information Technology (GMD)
Schloß Birlinghoven, D-53754 St. Augustin, Germany
brandes@gmd.de

Abstract. This paper presents a new parallelization method for an efficient implementation of unstructured array reductions on shared memory parallel machines with OpenMP. This method is strongly related to parallelization techniques for irregular reductions on distributed memory machines as employed in the context of High Performance Fortran. By exploiting data locality, synchronization is minimized without introducing severe memory or computational overheads as observed with most existing shared memory parallelization techniques.

1 Introduction

Reduction operations on unstructured meshes or sparse matrices account for a large fraction of the total computational costs in many advanced scientific and engineering applications. Typical examples of such applications include crash simulations, fluid-dynamics codes, weather forecasting models, electromagnetic problem modeling, and many others. In order to fully exploit the potential of parallel computers with such applications using high-level parallel languages like OpenMP [11] or HPF [8], it is of paramount importance to apply efficient parallelization strategies to reduction operations performed on irregular data structures. Unstructured reductions are usually implemented by means of loops containing indirect (vector-subscripted) array accesses. When parallelizing such loops it is crucial to avoid high synchronization or serious communication overheads, respectively. Several different parallelization techniques targeted either for distributed or for shared memory parallel architectures have been described in the literature [10, 2, 7, 1] and have been integrated in parallelizing compilers.

In this paper we present a new parallelization method for unstructured array reductions on shared memory parallel computers and compare it to existing parallelization strategies. Without loss of generality we discuss these techniques in the context of finite-element methods (FEM). Section 2 describes existing shared memory parallelization techniques for unstructured reductions and their support in OpenMP. Section 3 briefly discusses parallelization for distributed

---

[*] The work described in this paper was supported by NEC Europe Ltd. as part of the ADVICE project in cooperation with the NEC C&C Research Laboratories and by the Special Research Program SFB F011 AURORA of the Austrian Science Fund.

J. Rolim et al. (Eds.): IPDPS 2000 Workshops, LNCS 1800, pp. 435–442, 2000.

memory machines and high-level language support for unstructured reductions as provided in HPF. Section 4 presents an optimized parallelization method for unstructured array reductions on shared memory machines. This method is strongly related to an efficient handling of irregular reductions on distributed memory machines [1]. By exploiting data locality, synchronization is minimized without introducing severe memory or computational overheads as observed with most existing shared memory parallelization techniques. We have implemented this parallelization method in a compilation system [4] that translates high-level data parallel programs into shared memory parallel programs utilizing OpenMP to realize thread parallelism and synchronization. Performance results presented in Section 5 for a typical FEM application kernel verify the effectiveness of our approach and its superiority compared to existing techniques.

Figure 1 shows an example of an unstructured reduction operation on a finite-element mesh. The mesh consists of NELEMS elements and NNODES nodes, whereby each element comprises four nodes. The arrays NODE and ELEM store certain physical quantities[1], e.g. positions for each node or forces for each element, respectively. The integer array IX captures the connectivity of the mesh, i.e. the array section IX(1:4,I) contains the four node numbers of element I.

```
integer, parameter :: NNODES=...,NELEMS=...
real,    dimension (NNODES)  :: NODE
real,    dimension (NELEMS)  :: ELEM
integer, dimension (4, NELEMS) :: IX

! unstructured reduction loop
do I = 1, NELEMS
   VAL = Work(ELEM(I))
   do K = 1, 4
      NODE(IX(K,I)) = NODE(IX(K,I)) + VAL
   end do
end do
```

```
real, dimension (NNODES) :: NTMP
 ...
!$omp parallel, private (VAL, K, NTMP)
   NTMP = 0.0
!$omp do
   do I = 1, NELEMS
      VAL = Work(ELEM(I))
      do K = 1, 4
         NTMP(IX(K,I)) = NTMP(IX(K,I)) + VAL
      end do
   end do
!$omp critical
   NODE = NODE + NTMP
!$omp end critical
!$omp end parallel
```

Fig. 1. Unstructured reduction on a finite-element mesh: sequential version (left) and OpenMP version (right).

In each iteration of the unstructured reduction loop shown in Figure 1 a value VAL (e.g. an elemental force) is computed for an element of the mesh by means of a function (Work) and added to all nodes comprising this element. Since the addition is assumed to be an associative and commutative operation, the order in which the loop iterations are executed does not change the final result (except for possible round-off errors). However, when parallelizing the loop it has to be taken into account that different loop iterations may update the same node. This is caused by the fact that neighboring elements of the mesh may share one or more nodes, and thus IX(:,I) may have the same values for different iterations. If the reduction loop is parallelized on a shared memory machine by partitioning the loop iterations among concurrent threads, synchronization will be required to ensure that two distinct threads do not update one node at the same time. On

---

[1] To simplify presentation, only one physical quantity is stored per node and element.

distributed memory machines, where the mesh is to be partitioned with respect to the local memories of the processors, communication will be required for nodes at processor boundaries.

2 Unstructured Reductions on Shared Memory Machines

In this section we briefly discuss existing techniques for parallelizing unstructured reductions on shared memory machines by means of thread parallelism, as for example offered by OpenMP.

Array Privatization The central idea of the array privatization technique is that every processor gets an own private copy of the reduction array and performs its part of the loop iterations independently of other processors. Subsequently, the private results of all processors are combined to yield the final result. In order to ensure correct results, this ultimate step requires synchronization. Since the reduction clause of OpenMP [11] supports only reductions on scalar variables, parallelization of array reductions based on array expansion has to be programmed explicitly as shown in the right hand side of Figure 1. The original loop is enclosed in a `parallel` region and the temporary array `NODE_TMP` is declared as `private` to enforce that each thread gets its own copy. The I-loop is parallelized by relying on the default work sharing mechanism of OpenMP. As a consequence, a chunk of loop iterations is assigned to each thread, and all threads execute their chunk of iterations independently of each other. After parallel execution of the loop, the array assignment statement, protected by a `critical` section, ensures that each thread adds its local result of the reduction operation stored in `NODE_TMP` to the shared array `NODE`.

Parallelizing compilers for shared memory architectures like POLARIS [2] or SUIF [7], automatically translate sequential reduction loops as outlined in Figure 1 but utilize a thread library instead of OpenMP to implement multithreading. Array privatization is a good solution for reductions on scalar variables and for small arrays. For our FEM example, `NNODES << NELEMS` should be fulfilled in order to keep both memory and execution overhead for the final array assignment reasonably small.

Array Expansion The array expansion technique splits the reduction loop into two separate loops, whereby the first loop only evaluates the function `Work` for each element of the mesh. In order to store the results of `Work` for each element, the variable `VAL` has to be expanded into an array of size `NELEMS`. The second loop solely performs the reduction operation by reading the expanded `VAL` variable. The first part is executed in parallel without requiring any synchronization, whereas the second part has to be executed serially by the master thread only in order to avoid synchronization.

In our example, array expansion is a feasible solution only if `NNODES >> NELEMS`, since the memory overhead and the execution time for the serial part of the computation increase with the number of elements.

Atomic Reductions OpenMP provides the atomic directive for enforcing the atomic updating of a specific memory location, rather than exposing it to multiple, simultaneously writing threads. Using this directive, the reduction loop of

Figure 1 can be parallelized by means of a **parallel do** directive and by inserting an **atomic** directive prior to the assignment of **NODE**. This technique does not require any additional memory but may cause high synchronization overheads.

Although the atomic directive may be replaced by a **critical** directive, the atomic directive permits better optimizations. In contrast to a critical section, more than one thread may execute the assignment at the same time as long as different memory locations are updated.

3   Unstructured Reductions on Distributed Memory Machines

Parallelization of unstructured reductions on distributed memory machines is more complex than on shared memory machines. Since a distributed memory architecture provides no global address space, the reduction arrays have to be distributed to the local memories of the processors and accesses to array elements on other processors require communication. As a consequence of the indirect array accesses, analysis of array access patterns, which is a pre-requisite for communication generation, cannot be performed at compile time and, therefore, runtime parallelization techniques have to be applied. In the following the parallelization of unstructured array reductions for distributed memory machines is discussed in the context of High Performance Fortran (HPF) [8].

```
!hpf$ distribute(block) :: ELEM, NODE
!hpf$ align IX(*,i) with ELEM(i)
      ...
!hpf$ independent, on home (ELEM(I)), new(VAL, K), reduction(NODE)
      do I = 1, NELEMS
         VAL = Work(ELEM(I))
         do K = 1, 4
            NODE(IX(K,I)) = NODE(IX(K,I)) + VAL
         end do
      end do
```

Fig. 2. Unstructured Reductions in High Performance Fortran.

HPF provides high-level directives for specifying the distribution of arrays to abstract processors according to various formats (**block, cyclic**, etc.). The **independent** directive may be used to assert that a loop does not contain loop-carried dependences and therefore may be parallelized. In this context, temporary variables that are (conceptually) private for each loop iteration may be specified by means of a **new** clause. Moreover, a **reduction** clause may be used to indicate that dependences caused by associative and commutative reduction operations can be ignored. As opposed to OpenMP, in HPF also array variables may appear within a reduction clause. With the **on home** clause the loop iteration space may be partitioned according to the distribution of an array.

Inspector/Executor Parallelization Technique In order to parallelize the code shown in Figure 2, an HPF compiler distributes the arrays **ELEM, NODE**, and **IX** as specified by the distribution and alignment directives to the local memories of the processors. Parallel execution of the reduction loop is usually performed in two phases based on the inspector/executor strategy [10]. During the inspector phase each processor determines for its share of iterations the set of non-local elements of NODE it needs to access and derives the corresponding communication schedules (i.e. gather/scatter schedules). In the executor phase, on each processor non-local data to be read are gathered from the respective owner processors according to the gather-schedules by means of message-passing communication and are stored in local buffers. This is followed by a local computation phase where each processor executes, independently of the other processors, its share of loop iterations on its local part of the NODE array or the local buffers, respectively. Finally, a global communication phase based on the scatter-schedules takes place combining all those elements of NODE that have been written by processors not owning them. Since the inspector phase may be very time-consuming, it is essential to amortize the preprocessing overhead over multiple executions of a loop by reusing communication schedules [9, 1, 3] as long as communication patterns do not change. This is possible since unstructured reductions are performed in many codes within a serial time-step loop and the communication patterns are invariant for all (or at least many) time-steps.

Some HPF compilers also employ alternative strategies akin to array privatization or array expansion as discussed in Section 2. However, these techniques usually exhibit a larger memory and/or communication overhead.

Non-Local Access Patterns The main task of the inspector phase when preprocessing a loop with irregular array accesses (vector subscripts) is to determine, on each processor, the set of non-local array elements to be accessed. Once the non-local access pattern has been determined, the required communication can be derived. Recently, the concept of halos [1] has been proposed for HPF, enabling the explicit specification of non local data access patterns for distributed arrays. A *halo*, which in its simplest form comprises a list of global indices, specifies the set of non-local elements to be accessed at runtime for each abstract processor participating in the execution of an HPF program. The information provided by a halo significantly reduces the overheads of the inspector phase and alleviates computation and reuse of communication schedules. Figure 3 sketches a mesh partitioned in a node-based way, and the corresponding halo describing the set of non-local nodes to be accessed on each processor. By making the required communication explicit, the size of the halo area provides an appropriate measure for data locality. In the next section we show how the concepts of data distribution and halos can be utilized in order to parallelize irregular reductions efficiently for shared memory parallel architectures.

4   Exclusive Ownership Technique

In this section we present a new parallelization method for unstructured reductions on shared memory machines. This strategy is an extension of the atomic reduction technique described in Section 2 that avails itself with the concepts of data distribution and halos in order to minimize synchronization overheads. It can be employed for compiling an HPF program for multithreaded execution on shared memory parallel computers or for parallelizing irregular array reductions with OpenMP directly. We outline this technique for the FEM reduction loop shown in Figure 1.

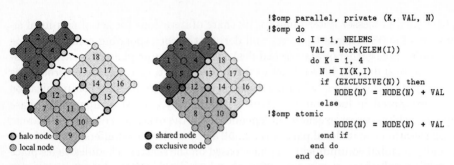

```
!$omp parallel, private (K, VAL, N)
!$omp do
      do I = 1, NELEMS
         VAL = Work(ELEM(I))
         do K = 1, 4
            N = IX(K,I)
            if (EXCLUSIVE(N)) then
               NODE(N) = NODE(N) + VAL
            else
!$omp atomic
               NODE(N) = NODE(N) + VAL
            end if
         end do
      end do
```

Fig. 3. Distributed mesh and halo (left); virtually distributed mesh with exclusive ownership (middle); OpenMP code based on exclusive ownership (right).

As a starting point, a *virtual* distribution is determined for the arrays ELEM, NODE and IX with respect to abstract processors. Data distribution is referred to as virtual, since the arrays are not actually distributed but allocated in an unpartitioned manner in shared memory. Based on the virtual distribution, each array element is associated with a unique abstract processor which becomes the *owner* of this element. Ownership is then used to determine the work sharing for the reduction loop with respect to abstract processors, whereby each abstract processor will be implemented by a separate thread. In our example, the loop iteration space is partitioned such that each iteration I is assigned to the abstract processor owning ELEM(I) of the mesh. Assuming a block distribution for ELEM, the loop iteration space can be partitioned by relying on the standard work sharing mechanism of OpenMP. In order to minimize synchronization overheads for the assignment to NODE, we introduce, based on halos, the concept of *exclusive ownership*. An element of array NODE is exclusively owned by an abstract processor (thread) if it is owned by that processor and not contained in the halo of any other processor. Synchronization via atomic updates is necessary only for loop iterations that access nodes not exclusively owned by the executing thread (shared nodes), while exclusively owned nodes can be handled like private data requiring no synchronization.

In Figure 3 the halo and exclusive ownership information is shown for a simple mesh together with the resulting OpenMP code. In the code, exclusive ownership information is represented by means of a logical array EXCLUSIVE which can be easily derived from the halo of the array NODE. We assume that the halo is either supplied by a domain partitioning tool or explicitly computed before the reduction loop is executed by analyzing the indirection array IX. In the latter case, the analysis of the indirection array is similar to an inspector phase as applied in the context of distributed memory parallelization, yet much simpler, since due to the shared address space, no communication is required. Exclusive ownership information can be reused employing techniques for communication schedule reuse [9, 1, 3], as long as the indirection array is not changed.

Gutiérrez et al. [6] presented a parallelization method for irregular reductions on shared memory machines that exploits locality similar to our method. The iteration space of a reduction loop is partitioned among threads in such a way that conflict-free writing on the reduction array is guaranteed and no synchronization is required. For this purpose, *loop index prefetching* arrays are built before the loop is executed by employing techniques similar to those applied in an inspector/executor strategy. The construction of the loop-index prefetching arrays becomes very complex and the algorithms presented in [6] work only for one or two reductions within one loop iteration. As opposed to our technique, some loop iterations have to be executed by more than one processor, since more than one element of the reduction array may have to be updated during each iteration. In the context of the FEM example presented previously, all iterations that manipulate elements on the distribution boundary of the mesh would have to be executed by all threads that own a node of this element. As a consequence, redundant computations are introduced, with an overhead depending on the computational costs of the function `Work`.

5   Performance Results

For the evaluation of the exclusive ownership technique we used a kernel from an industrial crash simulation code [5]. The kernel is based on a time-marching scheme to perform stress-strain calculations on a finite-element mesh consisting of 4-node shell elements. In each time-step elemental forces are calculated for every element of the mesh (cf. function `Work`) and added back to the forces stored at nodes by means of unstructured reduction operations. Besides the computation of elemental forces, the unstructured reduction operations to obtain the nodal forces represent the most important contribution to the overall computational costs. Table 1 shows the elapsed times measured on an SGI Origin 2000 (MIPSpro Fortran compiler, version 7.30) for different variants of a crash kernel performing 100 iterations on a mesh consisting of 25600 elements and 25760 nodes. In the table the entry *halo (DM)* refers to an HPF version parallelized with the Adaptor compiler [4] for distributed memory according to the inspector/executor strategy, *privatization (SM)*, *expansion (SM)* and *atomic (SM)* refer to the different shared memory parallelization strategies discussed in Section 2, *redundant (SM)* refers to the method based on loop-index prefetching, and *exclusive (SM)* to our exclusive ownership technique. All versions marked with (SM) utilize thread parallelism based on OpenMP, while the HPF version (DM) is based on process parallelism and relies on MPI for communication.

The irregular mesh used in this evaluation exhibits a high locality. There are only 160 non-exclusive (shared) nodes for two processors, 324 for three, and 480 nodes for four processors, respectively. As a consequence, both the distributed memory version and the shared memory versions that exploit data locality (i.e. exclusive ownership technique and loop index prefetching) show very satisfying results. The versions based on array privatization and array expansion achieve some speed-up, yet they exhibit very poor scaling due to the high synchronization overhead or computational overheads introduced by the serial code section, respectively. The version using atomic updates for all assignments to the node array scales but exhibits an overhead of about a factor of two. The best performance is obtained with the exclusive ownership strategy. The version based on

	NP = 1	NP = 2	NP = 4	NP = 8	NP = 16	NP = 32
halo (DM)	6.39	3.58	1.76	0.99	0.61	0.40
privatization (SM)	5.57	3.81	4.33	8.53	16.83	37.12
expansion (SM)	6.43	6.03	5.28	4.91	5.04	5.68
atomic (SM)	11.39	6.51	3.48	2.06	1.41	1.27
redundant (SM)	5.35	2.95	1.53	0.82	0.65	0.38
exclusive (SM)	5.10	2.79	1.47	0.74	0.55	0.34

Table 1. Execution times (secs) for crash simulation kernel on the SGI Origin 2000.

index prefetching is slightly worse since for redundant computations of elemental forces more time is required than for atomic updates with our strategy.

6   Summary and Conclusion

An efficient handling of unstructured reductions is crucial for many scientific applications. The usual methods for implementing reductions on shared memory parallel computers based on privatization, array expansion, or atomic updates, may not yield satisfying results for unstructured array reductions as they do not exploit data locality. The parallelization technique presented in this paper exploits data locality in order to minimize synchronization. The performance results verify that the concept of ownership in a shared memory programming model is essential for an efficient realization of unstructured reductions.

References

1. S. Benkner. Optimizing Irregular HPF Applications Using Halos. In *Proceedings of IPPS/SPDP Workshops.* LNCS 1586, 1999.
2. W. Blume, R. Doallo, and R. Eigenmann et.al. Parallel Programming with Polaris. *IEEE Computer,* 29(12):78–82, 1996.
3. T. Brandes and Germain C. A Tracing Protocol for Optimizing Data Parallel Irregular Computations. LNCS 1470, pages 629–638, September 1998.
4. T. Brandes and F. Zimmermann. Adaptor – A Transformation Tool for HPF Programs. In *Programming environments for massively parallel distributed systems,* pages 91–96. Birkhäuser Boston Inc., April 1994.
5. J. Clinckemaillie, B. Elsner, and G. Lonsdale et al. Performance issues of the parallel PAM-CRASH code. *The International Journal of Supercomputer Applications and High Performance Computing,* 11(1):3–11, Spring 1997.
6. E. Gutiérrez, O. Plata, and E.L. Zapata. On Automatic Parallelization of Irregular Reductions on Scalable Shared Memory Systems. In *Euro-Par'99 Parallel Processing, Toulouse,* pages 422–429. LNCS 1685, Springer-Verlag, September 1999.
7. M. Hall and al. Maximizing Multiprocessor Performance with the SUIF Compiler. *IEEE Computer,* 29(12):84–90, 1996.
8. High Performance Fortran Forum. High Performance Fortran Language Specification. Vers. 2.0, Rice University, January 1997.
9. Ravi Ponnusamy, Joel Saltz, and Alok Choudhary. Runtime-compilation techniques for data partitioning and communication schedule reuse. In *Proceedings Supercomputing '93,* pages 361–370, 1993.
10. J. Saltz, K. Crowley, R. Mirchandaney, and H. Berryman. Run-time scheduling and execution of loops on message passing machines. *Journal of Parallel and Distributed Computing,* 8(2):303–312, 1990.
11. The OpenMP Forum. OpenMP Fortran Application Program Interface. Technical Report Ver 1.0, SGI, October 1997.

Parallel FEM Simulation of Crack Propagation – Challenges, Status, and Perspectives*

Bruce Carter[1], Chuin-Shan Chen[1], L. Paul Chew[2], Nikos Chrisochoides[3], Guang R. Gao[4], Gerd Heber[1], Antony R. Ingraffea[1], Roland Krause[5], Chris Myers[1], Demian Nave[3], Keshav Pingali[2], Paul Stodghill[2], Stephen Vavasis[2], and Paul A. Wawrzynek[1]

[1] Cornell Fracture Group, Rhodes Hall, Cornell University, Ithaca, NY 14853
{bcarter,dchen,heber,myers}@tc.cornell.edu, wash@stout.cfg.cornell.edu, ari1@cornell.edu
[2] CS Department, Upson Hall, Cornell University, Ithaca, NY 14853
{chew,pingali,stodghil,vavasis}@cs.cornell.edu
[3] CS Department, University of Notre Dame, Notre Dame, IN 46556
{nikos,dnave}@cse.nd.edu
[4] EECIS Department, University of Delaware, Newark, DE 19716
ggao@capsl.udel.edu
[5] Center for Comp. Mech., Washington University in Saint Louis, MO 63130
rokrau@pocus.wustl.edu

Abstract. Understanding how fractures develop in materials is crucial to many disciplines, e.g., aeronautical engineering, material sciences, and geophysics. Fast and accurate computer simulation of crack propagation in realistic 3D structures would be a valuable tool for engineers and scientists exploring the fracture process in materials. In the following, we will describe a next generation crack propagation simulation software that aims to make this potential a reality.

1 Introduction

Within the scope of this paper, it is sufficient to think about crack propagation as a dynamic process of creating new surfaces within a solid. During the simulation, crack growth causes changes in the geometry and, sometimes, in the topology of the model. Roughly speaking, with the tools in place before the start of this project, a typical fracture analysis at a resolution of 10^4 degrees of freedom, using boundary elements, would take about 100 hours on a state-of-the-art single processor workstation. The goal of this project is it to create a parallel environment which allows the same analysis to be done, using finite elements, in 1 hour at a resolution of 10^6 degrees of freedom. In order to attain this level of performance, our system will have two features that are not found in current fracture analysis systems:

---

* This work was supported by NSF grants CCR-9720211, EIA-9726388, ACI-9870687, and EIA-9972853.

Parallelism – Current trends in computer hardware suggest that in the near future, high-end engineering workstations will be 8- or 16-way SMP "nodes", and departmental computational servers will be built by combining a number of these nodes using a high-performance network switch. Furthermore, the performance of each processor in these nodes will continue to grow. This will happen not only because of faster clock speeds, but also because finer-grain parallelism will be exploited via multi-way (or superscalar) execution and multi-threading.

Adaptivity – Cracks are (hopefully) very small compared with the dimension of the structure, and their growth is very dynamic in nature. Because of this, it is impossible to know a priori how fine a discretization is required to accurately predict crack growth. While it is possible to over-refine the discretization, this is undesirable, as it tends to dramatically increase the required computational resources. A better approach is to *adaptively* choose the discretization refinement. The dynamic nature of crack growth and the need to do adaptive refinement make crack propagation simulation a highly irregular application. Exploiting parallelism and adaptivity presents us with three major research challenges,

- developing algorithms for parallel mesh generation for unstructured 3D meshes with automatic element size control and provably good element quality,
- implementing fast and robust parallel sparse solvers, and
- determining efficient schemes for automatic, hybrid h-p refinement.

To tackle the challenges of developing this system, we have assembled a multi-disciplinary and multi-institutional team that draws upon a wide-ranging pool of talent and the resources of 4 universities.

2   System Overview

Figure 1 gives an overview of a typical simulation. During pre-processing, a solid model is created, problem specific boundary conditions (displacements, tractions, etc.) are imposed, and flaws (cracks) are introduced. In the next step, a volume mesh is created, and (linear elasticity) equations for the displacements are formulated and solved. An error estimator determines whether the desired accuracy has been reached, or further iterations, after subsequent adaptation, are necessary. Finally, the results are fed back into a fracture analysis tool for post-processing and crack propagation.

Figure 1 presents the simulation loop of our system in its final and most advanced form. Currently, we have sequential and parallel implementations of the outer simulation loop (i.e., not the inner refinement loop) running with the following restrictions: right now, the parallel mesher can handle only polygonal (non-curved) boundaries. Curved boundaries can be handled by the sequential meshers, though (see section 3). We have not yet implemented unstructured h-refinement and adaptive p-refinement, although the parallel formulator can handle arbitrary p-order elements.

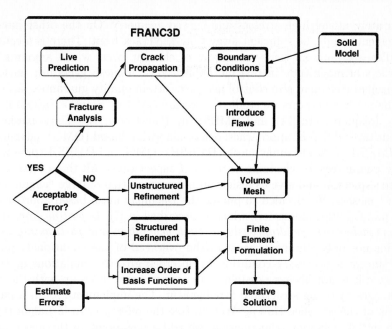

Fig. 1. Simulation loop.

3   Geometric Modeling and Mesh Generation

The solid modeler used in the project is called OSM. OSM, as well as the main pre- and post-processing tool, FRANC3D, is freely available from the Cornell Fracture Group's website [4]. FRANC3D - a workstation based FRacture ANalysis Code for simulating arbitrary non-planar 3D crack growth - has been under development since 1987, with hydraulic fracture and crack growth in aerospace structures as the primary application targets since its inception. While there are a few 3D fracture simulators available and a number of other software packages that can model cracks in 3D structures, these are severely limited by the crack geometries that they can represent (typically planar elliptical or semi-elliptical only). FRANC3D differs by providing a mechanism for representing the geometry and topology of 3D structures with arbitrary non-planar cracks, along with functions for 1) discretizing or meshing the structure, 2) attaching boundary conditions at the geometry level and allowing the mesh to inherit these values, and 3) modifying the geometry to allow crack growth but with only local re-meshing required to complete the model. The simulation process is controlled by the user via a graphic user-interface, which includes windows for the display of the 3D structure and a menu/dialogue-box system for interacting with the program.

The creation of volume meshes for crack growth studies is quite challenging. The geometries tend to be complicated because of internal boundaries (cracks). The simulation requires smaller elements near each crack front in order to accurately model high stresses and curved geometry. On the other hand, larger elements might be sufficient away from the crack front. There is a considerable difference between these two scales of element sizes, which amounts to three orders of magnitude in real life applications. A mesh generator must provide automatic element size control and give certain quality guarantees for elements. The mesh generators we studied so far are QMG by Steve Vavasis [14], JMESH by Joaquim Neto [11], and DMESH by Paul Chew [12]. These meshers represent three different approaches: octree-algorithm based (QMG), advancing front (JMESH), and Delaunay mesh (DMESH). QMG and DMESH come with quality guarantees for elements in terms of aspect ratio. All these mesh generators are sequential and give us insight into the generation of large "engineering quality" meshes. We decided to pursue the Delaunay mesh based approach first for a parallel implementation, which is described in [5]. *Departing from traditional approaches, we simultaneously do mesh generation and partitioning in parallel.* This not only eliminates most of the overhead of the traditional approach, it is almost a necessary condition to do crack growth simulations at this scale, where it is not always possible or too expensive to keep up with the geometry changes by doing structured h-refinement. The implementation is a parallelization of the so-called Bowyer-Watson (see the references in [5]) algorithm: given an initial Delaunay triangulation, we add a new point to the mesh, determine the simplex containing this point and the point's cavity (the union of simplices with non-empty circumspheres), and, finally, retriangulate this cavity. One of the challenges for a parallel implementation is that this cavity might extend across several submeshes (and processors). What looks like a problem, turns out to be the key element in unifying mesh generation and partitioning: the newly created elements, together with an adequate cost function, are the best candidates to do the "partitioning on the fly". We compared our results with Chaco and MeTis in terms of equidistribution of elements, relative quality of mesh separators, data migration, I/O, and total performance. Table 1 shows a runtime comparison be-

Table 1. Total run time in seconds on 16 processors.

Mesh Size	PPGK	SMGP0	SMGP1	SMGP2	SMGP3
200K	90	42	42	42	42
500K	215	65	87	64	62
1000K	439	97	160	91	94
2000K	1232	133	310	110	135

tween ParMeTis with PartGeomKway (PPGK) and, our implementation, called SMGP, on 16 processors of an IBM SP2 for meshes of up to 2000K elements. The numbers behind SMGP refer to different cost functions used in driving the partitioning [5].

4   Equation Solving and Preconditioning

So far, our focus has been on iterative solution methods. It is part of our future work to explore the potential of direct methods. We chose PETSc [13, 1] as the basis for our equation solver subsystem. PETSc provides a number of Krylov space solvers, such as Conjugate Gradient and GMRES, and a number of widely-used preconditioners, such as (S)SOR, and ILU/ICC. We have augmented the basic library with third party packages, including BlockSolve95 [8] and the Barnard's SPAI [2]. In addition, we have implemented a parallel version of the Global Extraction Element-By-Element (GEBE) preconditioner [7] (which is unrelated to the EBE preconditioner of Winget and Hughes [15]), and added it to the collection using PETSc's extension mechanisms. The central idea of GEBE is to extract subblocks of the global stiffness matrix associated with elements and invert them, which is highly parallel.

The ICC preconditioner is frequently used in practice, and is considered to be a good preconditioner for many elasticity problems. However, we were concerned that it would not scale well to the large number of processors required for our final system. We believed that GEBE would provide a more scalable implementation, and we hoped that it would converge nearly as well as ICC. In order to test our hypothesis, we ran several experiments on the Cornell Theory Center SP-2. The preliminary performance results for the gear2 and tee2 models are shown in Tables 2 and 3, respectively. gear2 is a model of a power transmission gear with a crack in one of its teeth. tee2 is a model of a T steel profile. For each model, we ran the Conjugant Gradient solver with both PETSc's IC(0) preconditioner and our own parallel implementation of GEBE on 8 to 64 processors. (**PC** – preconditioner type, p – number of processors, t_{PC} – time for preconditioner setup, t_{it} – time per cg iteration. The iteration counts in Tables 2 and 3 correspond to a 10^{15} reduction of the residual error, which is completely academic at this point.) The experimental results confirm our hypothesis: 1)

Table 2. Gear2 (79,656 unknowns)

PC	p	t_{PC}	t_{it}	Iters.
IC(0)	8	17.08	0.2	416
GEBE	8	9.43	0.19	487
IC(0)	16	15.47	0.27	422
GEBE	16	6.71	0.11	486
IC(0)	32	8.51	0.32	539
GEBE	32	3.73	0.08	485
IC(0)	64	11.00	0.28	417
GEBE	64	4.74	0.07	485

Table 3. Tee2 (319,994 unknowns)

PC	p	t_{PC}	t_{it}	Iters.
IC(0)	32	30.00	0.29	2109
GEBE	32	35.70	0.21	2421
IC(0)	64	23.60	0.29	2317
GEBE	64	7.60	0.12	2418

GEBE converges nearly as quickly as IC(0) for the problems that we tested.

2) Our naive GEBE implementation scales much better than PETSc's IC(0) implementation which uses BlockSolve95.

5 Adaptivity

Understanding the cost and impact of the different adaptivity options is the central point in our current activities. Our implementation follows the approach of Biswas and Oliker [3] and currently handles tetrahedra, while allowing enough flexibility for an extension to non-tetrahedral element types. For relatively simple, two-dimensional problems, stress intensity factors can be computed to an accuracy sufficient for engineering purposes with little mesh refinement by proper use of singularly enriched elements. There are many situations though when functionals other than stress intensity factors are of interest or when the singularity of the solution is not known *a priori*. In any case the engineer should be able to evaluate whether the data of interest have converged to some level of accuracy considered appropriate for the computation. It is generally sufficient to show, that the data of interest are converging sequences with respect to increasing degrees of freedom. Adaptive finite element methods are the most efficient way to achieve this goal and at the same time they are able to provide estimates of the remaining discretization error. We define the error of the finite element solution as $\mathbf{e} = \mathbf{u} - \mathbf{u}^{FE}$ and a possible measure for the discretization error is the energy norm, $\|\mathbf{e}\|_{E(\Omega)}^2 = \frac{1}{2}\mathcal{B}(\mathbf{e}, \mathbf{e})$. The error estimator introduced by Kelly et.al. [9, 6] is derived by inserting the finite element solution into the original differential equation system and calculating a norm of the residual using interpolation estimates. An error indicator computable from local results of one element of the finite element solution is then derived and the corresponding error estimator is computed by summing the contribution of the error indicators over the entire domain. The error indicator is computed with a contribution from the interior residual of the element and a contribution of the *stress jumps* on the faces of an element. Details on the computation of the error estimator from the finite element solution can be found in [10].

6 Future Work

The main focus of our future work will be on improving the performance of the existing system. We have not yet done any specific performance tuning, like locality optimization. This is not only highly platform dependent, but also has to be put in perspective to the forthcoming runtime optimizations, like dynamic load balancing. We are considering introducing special elements at the crack tip, and non-tetrahedral elements (hexes, prisms, pyramids) elsewhere. On the solver side, we explore new preconditioners (e.g., support tree preconditioning), and multigrid, as well as sparse direct solvers, to make our environment more effective and robust. There is a port of our code base to the new 64 4-way SMP node NT cluster at the Cornell Theory Center underway.

7 Conclusions

At present, our project can claim two major contributions. The first is our parallel mesher/partitioner, which is the first practical implementation of its kind with quality guarantees. This technology makes it possible, for the first time, to fully automatically solve problems using unstructured h-refinement in a parallel setting. The second major contribution is to show that GEBE outperforms ICC, at least for our problem class. We have shown that, not only does GEBE converge almost as quickly as ICC, it is much more scalable in a parallel setting than ICC.

References

[1] Satish Balay, William D. Gropp, Lois Curfman McInnes, and Barry F. Smith. Efficient management of parallelism in object-oriented numerical software libaries. In E. Arge, A.M. Bruaset, and H.P. Langtangen, editors, *Modern Software Tools in Scientific Computing*. Birkhauser Press, 1997.

[2] Stephen T. Barnard and Robert Clay. A portable MPI implementation of the SPAI preconditioner in ISIS++. In *Eighth SIAM Conference for Parallel Processing for Scientific Computing*, March 1997.

[3] R. Biswas and L. Oliker. A new procedure for dynamic adaption of three-dimensional unstructured grids. *Applied Numerical Mathematics*, 13:437–452, 1994.

[4] http://www.cfg.cornell.edu/.

[5] Nikos Chrisochoides and Demian Nave. Simultaneous mesh generation and partitioning for Delaunay meshes. In *8th Int'l. Meshing Roundtable*, 1999.

[6] J.P. de S.R. Gago, D.W. Kelly, O.C. Zienkiewicz, and I. Babuška. A posteriori error analysis and adaptive processes in the finite element method: Part II – Adaptive mesh refinement. *International Journal for Numerical Methods in Engineering*, 19:1621–1656, 1983.

[7] I. Hladik, M.B. Reed, and G. Swoboda. Robust preconditioners for linear elasticity FEM analyses. *International Journal for Numerical Methods in Engineering*, 40:2109–2127, 1997.

[8] Mark T. Jones and Paul E. Plassmann. Blocksolve95 users manual: Scalable library software for the parallel solution of sparse linear systems. Technical Report ANL-95/48, Argonne National Laboratory, December 1995.

[9] D.W. Kelly, J.P. de S.R. Gago, O.C. Zienkiewicz, and I. Babuška. A posteriori error analysis and adaptive processes in the finite element method: Part I – Error analysis. *International Journal for Numerical Methods in Engineering*, 19:1593–1619, 1983.

[10] Roland Krause. *Multiscale Computations with a Combined h- and p-Version of the Finite Element Method*. PhD thesis, Universität Dortmund, 1996.

[11] J.B.C. Neto et al. An algorithm for three-dimensional mesh generation for arbitrary regions with cracks. submitted for publication.

[12] http://www.cs.cornell.edu/People/chew/chew.html.

[13] http://www.mcs.anl.gov/petsc/index.html.

[14] http://www.cs.cornell.edu/vavasis/vavasis.html.

[15] J.M. Winget and T.J.R. Hughes. Solution algorithms for nonlinear transient heat conduction analysis employing element-by-element iterative strategies. *Computational Methods in Applied Mechanical Engineering*, 52:711–815, 1985.

Support for Irregular Computations in Massively Parallel PIM Arrays, Using an Object-Based Execution Model

Hans P. Zima[1,2] and Thomas L. Sterling[1]

[1] CACR, California Institute of Technology, Pasadena, CA 91125, U.S.A.

[2] Institute for Software Science, University of Vienna, Austria

E-mail: {zima,tron}@cacr.caltech.edu

Abstract

The emergence of semiconductor fabrication technology allowing a tight coupling between high-density DRAM and CMOS logic on the same chip has led to the important new class of Processor-In-Memory (PIM) architectures. Furthermore, large arrays of PIMs can be arranged into massively parallel architectures. In this paper, we outline the salient features of PIM architectures and discuss macroservers, an object-based model for such machines. Subsequently, we specifically address the support for irregular problems provided by PIM arrays. The discussion concludes with a case study illustrating an approach to the solution of a sparse matrix vector multiplication.

1 Introduction

Processor-in-Memory or PIM architecture couples processor cores and DRAM blocks providing direct access to memory row buffers to increase the memory bandwidth achieved by two orders of magnitude. Current generation PIMs are very basic, treating memory as a physical resource. But future generations of PIM architecture may be well suited to the processing of irregular data structures. PIM based system architecture may take a number of forms from a simple replacement of dumb memory chips with PIM chips to complete systems comprising an array of PIM chips. Manipulating irregular data structures favors specific system architecture characteristics that provide high inter-PIM chip communication bandwidth, efficient address manipulation, and fast response to light-weight service requests.

In this paper, we first characterize some properties of irregular problems (Section 2). The subsequent Section 3 provides a short overview of an object-based execution model for PIM arrays that we are currently developing, in parallel

J. Rolim et al. (Eds.): IPDPS 2000 Workshops, LNCS 1800, pp. 450–456, 2000.
© Springer-Verlag Berlin Heidelberg 2000

with the design of a PIM array architecture. Section 4 will then discuss in more detail some of the features of future generation systems that are designed to support the efficient processing of irregular problems. We finish with concluding remarks in Section 6.

2   Irregular Problems

Many advanced scientific problems are of an irregular nature and contain a large degree of inherent parallelism. This includes sparse matrix computations, sweeps over unstructured grids, tree searches, and particle-in-cell codes; moreover, many relevant problems are of an adaptive nature. In order to execute such codes efficiently on a massively parallel array of PIMs, a viable tradeoff between the exploitation of locality and parallelism has to be found. As a consequence, the performance of irregular codes is largely determined by the decisions made relating to their data and work distributions.

First, a *data distribution* must be determined by partitioning large data structures and distributing them across the memories of the machine. Secondly, the choice of the *work distribution* has to made in conjunction with the data distribution, taking into account dependencies and access patterns in the code.

Regular problems, which are characterized by regular, usually linear, data access patterns, can be efficiently implemented on conventional architectures exploiting compile-time knowledge [7]. In contrast, the data access patterns as well as the data and work distributions typical for irregular algorithms must be largely resolved at runtime. Such algorithms pose major problems for most existing parallel machines, mainly as a result of the memory access latency which makes the runtime translation of indirect references and the associated communication very expensive. The problem is compounded by the preprocessing that is often required to effectively organize bulk communication of data [5]. As will be discussed in more detail in Section 4, PIM arrays can offer significant support for this kind of problems.

Below we summarize some of the typical features of iregular applications.

- **Irregular Data Distributions:** The management of irregular data distributions is one of the key issues to be dealt with. Relevant topics include the generation of the data structure, its representation in memory, the implementation of access mechanisms, and (total as well as incremental) redistribution algorithms.

- **Thread Groups:** Irregular algorithms require sophisticated mechanisms for the generation of groups of cooperating threads. One example is a set of data parallel threads working on a problem in loosely synchronous SPMD mode; another one a thread structure arising from a tree search. Critical operations – all of which may be intra-chip as well as inter-chip include thread generation, communication, prefix and reduction operations, mutual exclusion, and condition synchronization.

- **Address Translation:** Address translation refers to the problem of mapping an indirect reference (such as $A(X(I))$ or a pointer value) to a memory address. For irregular data structures, this translation must, in general, be performed at runtime, since neither the data distributions nor the value of index arrays need to be statically known.

3   Macroservers

We have defined an object-based model that provides an abstract programming paradigm for a massively parallel PIM array. Here we give a short outline of this model; for a more detailed specification see [8].

The central concept of the model is called a *macroserver*. A macroserver is an autonomous, active object that comes into existence by being created based upon a *macroserver class*, which essentially establishes an encapsulation for a set of variables and methods as well as a context for threads and their synchronization.

At any point in time, a macroserver has a well-defined *home*, which is a location in virtual memory where the *metadata* needed for the management of the object can be found. Macroservers are associated with a set of variables whose values define its state. Variables may be distributed across the memories of the PIM; the underlying distribution mechanism can be thought of as a generalization of the corresponding concept in languages such as Vienna Fortran [2] or HPF [4], extended to LISP-type data structures and allowing arbitrary mappings as well as incremental redistribution.

The activation of a method in a macroserver gives rise to the generation of a (synchronous or asynchronous) thread. Threads can directly read and modify the state of the macroserver to which they belong. Each thread can activate methods of its own or of another macroserver accessible to it; as a consequence, the model offers intra-server as well as inter-server concurrency, reflecting the multi-level parallelism of the underlying architecture. Special support is provided for the operation of thread groups.

The model provides a simple mechanism for mutual exclusion (atomic methods) and allows synchronization via condition variables and futures.

Metadata includes information about the state variables (such as types and distributions), the signatures of methods, and representations of the threads currently operating in the macroserver. Most of the metadata will be stored at the home of the object, allowing an efficient centralized management by the associated PIM.

4   PIM Arrays and Their Support for Irregular Computations

PIM combines memory cell blocks and processing logic on the same integrated circuit die. The row buffer of the typical memory block may be on the order of 1K or more bits which are acquired in a single memory cycle delivering data at a typical rate of 4 Gigabytes per second at the sense amps of the memory. A processor designed to directly manipulate this data simultaneously can typically operate at a performance of a Gigaops. Multiple memory-processor pairs can occupy the same die with a possible performance total today of 4 Gigaops, with higher rates for simple arithmetic or small field manipulations. A typical memory system comprising an array of PIM chips could provide a total throughput of 256 Gigaops or, for byte level operations, 1 Teraops peak throughput. While in some designs core processors developed as conventional microprocessors are "dropped" into the PIM to minimize development time and cost as well as exploit existing software such as compilers, in other cases much simpler processors may be designed explicitly for the PIM context, minimizing the die area consumed while optimizing for effective memory bandwidth. A typical PIM chip includes multiple modules of memory blocks and processors, memory bus interface, and shared functional units such as floating point arithmetic units. Also, addresses employed within the PIM by the on-chip processors are usually physical.

A system incorporating PIM chips may differ from the conventional structure outlined above in important ways to take advantage of the capabilities of PIM and employ them in the broader system. While in the most simple structure, the PIM chips may simply replace regular memory retaining the systems processor with its layered cache hierarchy, other structures diverging from this will deliver improved performance. Providing a separate highly parallel network for just inter-PIM communications and a second level of bulk storage as backing store for the PIMs are two examples. Beyond this is the "Gilgamesh" (billion logic gate array mesh) which is a multi-dimensional structure of PIM chips in a standalone system.

The processing and manipulation of irregular data structures imposes additional requirements than can be effectively handled with current generation PIM technology and architecture. However, many of the advantages that PIM has for dense contiguous data computing would also convey to metadata-organized irregular data structures if augmented with advanced mechanisms. An important advance that PIM enables is the exploitation of fine grain parallelism intrinsic to irregular structures. This is because many different parts of the distributed structure can be processed simultaneously by the many PIM processors throughout the memory and because the nodes of the structure can be handled efficiently in the memory itself.

Virtual memory management including address translation is central to the advanced architecture requirements of PIM for irregular data. Such data incorporates virtual user addresses within the data structure itself, which must be managed directly by the PIM. Conventional TLB based methods employed by microprocessors are likely to be poorly suited because the access patterns encountered by the PIM internal processors often will not experience the necessary level of temporal locality required to make them effective. An innovative approach employs "intrinsic address translation" in which the virtual to physical address mapping is incorporated in the data structures metadata. With bi-directional links, any page movement across physical memory can cause automatic update of the translation data in other linked pages. A second method is a "set associative" approach that is a hybrid mapping; partly physical and partly virtual as in cache systems but in main memory instead. This allows efficient address translation between memory chips while permitting random placement and location of pages anywhere within a designated memory chip. Both methods when combined provide relatively general and highly scalable virtual to physical memory address translation.

A second requirement is direct inter-PIM message driven communication without system processor intervention. This is being explored by the DIVA project [3] and HTMT project [1]. When a pointer indicates continuation of a structure on an external chip, the computation must be able to "jump" across this physical gap while retaining logical consistency. The "parcel" model [1] permits such actions to follow distributed data structures across PIM arrays. A parcel is a message packet that not only conveys data, but also specifies the actions to be performed. The arrival of a parcel causes a PIM to respond by instantiating the specified thread, carrying out the action on the designated local data, and providing a necessary response, possibly by returning another parcel or by continuing to move through the data structure.

A third requirement is architecture support for thread management. Because of the potentially random distribution of the data structure elements or substructures, the order of service requests may be unknown until runtime. A PIM chip may be required to service multiple parcels concurrently. A PIM comprises several subsystems that can operate simultaneously. Finally, some PIM threads may require access to remote resources such as other PIM chips and such accesses will impose delays. One mechanism not yet incorporated in past PIM implementations is multithreading, a means of rapidly switching among multiple concurrent thread contexts. Multithreading provides an efficient low level mechanism for managing multiple active threads of execution. This is useful while waiting several cycles for a memory access or the propagation delay for a shared ALU to permit other work to be performed by the remaining resources. Each time a new parcel arrives, the active thread can be suspended until the incident parcel is at least stored, thereby freeing the receiver hardware resources for the next arrival. Multithreaded architecture will greatly facilitate the manipulation and processing of irregular data structures because it provides a runtime adaptive means of applying system resources to computational needs as determined by the structure of the data itself.

5   Case Study: Sparse Matrix Vector Multiply

In this section, we outline an approach for the parallelization of a sparse matrix-vector multiplication using our model, based partly on concepts developed in [6].

We first take a look at the sequential algorithm. Consider the operation $S = A.B$, where $A(1 : N, 1 : M)$ is a sparse matrix with q nonzero elements, and $B(1 : M)$ and $S(1 : N)$ are vectors. We assume that the nonzero elements of A

```
INTEGER :: C(q), R(N+1)
REAL :: D(q), B(M), S(N)
INTEGER :: I, J
  DO I = 1,M
    S(I)=0.0
    DO J = R(I), R(I+1)-1
      S(I) = S(I) + D(J)*B(C(J))
    ENDDO J
  ENDDO I
```

Figure 1: Sparse Matrix-Vector Multiply: Core Loop

are enumerated using row-major order; the k-th element in this order is called the k-th nonzero element of A.

In the **Compressed Row Storage (CRS)** format, A is represented by three vectors, D, C, and R:

- the **data vector**, $D(1 : q)$, stores the sequence of nonzero elements of A, in the order of their enumeration;

- the **column vector**, $C(1 : q)$, contains in position k the column number, in A, of the k-th nonzero element in A; and

- the **row vector**, $R(1 : N + 1)$, contains in position i the number of the first nonzero element of A in that row (if any); $R(N + 1)$ is set to $q + 1$.

Based upon this representation, the core loop of the sequential algorithm can be formulated in Fortran as shown in Figure 1.

The first step in developing a parallel version of the algorithm consists of defining a *distributed sparse representation* of A. This essentially combines a data distribution with a sparse format such as CRS. More specifically, a data distribution [7, 8] is interpreted as if A were a dense array, specifying a local distribution segment for each PIM node. The distributed sparse representation is then obtained by representing the submatrix constituting the local distribution segment in the sparse format (CRS in our case).

A number of data distributions have been used for this purpose, including *Multiple Recursive Decomposition (MRD)* and cyclic distributions [6]. MRD is a method that partitions A into nn rectangular distribution segments, where nn is the number of memory nodes. These segments are the result of a recursive construction algorithm that aims at achieving load balancing by having approximately the same number of nonzero elements in each segment.

Based upon a distributed sparse representation using MRD, a parallel algorithm for the sparse matrix vector product can now be formulated by applying (a slightly modified version of) the algorithm in Fig. 1 in parallel to all distribution segments, and then combining the partial results for each row of the original matrix in a reduction operation. Because of lack of space, we do not discuss further details of the parallel algorithm here (see [8]). However, we outline a number of topics that illustrate the support of the PIM array architecture for this kind of algorithm:

- The CRS representation of the local data segments can be stored and processed locally in each PIM node by microservers.

- The indirect references involving D and B can be resolved in the memory; making the implementation of an inspector/executor scheme [5, 7] much more efficient than for distributed-memory machines.

- The PIM array network offers efficient support for spawning a large number of "similar" parallel threads and for executing reduction operations.

6   Conclusion

In this paper, we have discussed the design of massively parallel PIM arrays, together with an object-based execution model for such architectures. An important focus of this work in progress is the capability to deal effectively with irregular problems.

References

[1] J.B.Brockman,P.M.Kogge,V.W.Freeh,S.K.Kuntz, and T.L.Sterling. Microservers: A New Memory Semantics for Massively Parallel Computing. *Proceedings ACM International Conference on Supercomputing (ICS'99)*, June 1999.

[2] B. Chapman, P. Mehrotra, and H. Zima. Programming in Vienna Fortran. *Scientific Programming*, 1(1):31–50, Fall 1992.

[3] M.Hall,J.Koller,P.Diniz,J.Chame,J.Draper, J.LaCoss, J.Granacki, J.Brockman, A.Srivastava, W.Athas, V.Freeh, J.Shin, and J.Park. Mapping Irregular Applications to DIVA, a PIM-Based Data Intensive Architecture. *Proceedings SC'99*, November 1999.

[4] High Performance Fortran Forum. *High Performance Fortran Language Specification, Version 2.0*, January 1997.

[5] J.Saltz,K.Crowley,R.Mirchandaney, and H.Berryman. Run-Time Scheduling and Execution of Loops on Message-Passing Machines. *Journal of Parallel and Distributed Computing*, 8(2),pp.303-312, 1990.

[6] M.Ujaldon,E.L.Zapata,B.M.Chapman,and H.P.Zima. Vienna Fortran/HPF Extensions for Sparse and Irregular Problems and Their Compilation. *IEEE Transactions on Parallel and Distributed Systems*, Vol.8, No.10, pp.1068-1083 (October 1997).

[7] H. Zima and B. Chapman. Compiling for Distributed Memory Systems. *Proceedings of the IEEE*, Special Section on Languages and Compilers for Parallel Machines, pp. 264-287, February 1993.

[8] H.Zima and T.Sterling. Macroservers. An Object-Based Model for Massively Parallel Processor-in-Memory Arrays. *Caltech CACR Technical Report*, January 2000 (in preparation).

Executing Communication-Intensive Irregular Programs Efficiently

Vara Ramakrishnan*   Isaac D. Scherson

Department of Information and Computer Science
University of California, Irvine, CA 92697
{vara,isaac}@ics.uci.edu

Abstract. We consider the problem of efficiently executing completely irregular, communication-intensive parallel programs. Completely irregular programs are those whose number of parallel threads as well as the amount of computation performed in each thread vary during execution. Our programs run on MIMD computers with some form of space-slicing (partitioning) and time-slicing (scheduling) support. A hardware barrier synchronization mechanism is required to efficiently implement the frequent communications of our programs, and this constrains the computer to a fixed size partitioning policy.

We compare the possible scheduling policies for irregular programs on fixed size partitions: local scheduling and multi-gang scheduling, and prove that local scheduling does better. Then we introduce competitive analysis and formally analyze the online rebalancing algorithms required for efficient local scheduling under two scenarios: with full information and with partial information.

1   Introduction

The universe of parallel programs can be broadly divided into regular and irregular programs. Regular programs have a fixed number of parallel threads, each of which perform about the same amount of computation. Irregular programs have a varying number of parallel threads and/or threads which perform unequal amounts of computation. Since the behavior of regular programs is predictable, scheduling them well is relatively easy. There are several static (job arrival time) and dynamic scheduling methodologies [3] which work very well on regular programs. For this reason, we focus on irregular programs, which waste computing resources if managed poorly.

There are two main classes of parallel computers, MIMD (multiple-instruction multiple-data) and SIMD (single-instruction multiple-data). MIMD computers have full-fledged processors which fetch, decode and execute instructions independently of each other. SIMD computers have a centralized control mechanism which fetches, decodes and broadcasts each instruction to specialized processors, and they all execute the instruction simultaneously.

---

* This research was supported primarily by PMC-Sierra, Inc., San Jose, California. http://www.pmc-sierra.com

J. Rolim et al. (Eds.): IPDPS 2000 Workshops, LNCS 1800, pp. 457-468, 2000.

We assume that the parallel computer is an MIMD computer. MIMD computers have overwhelming advantages in cost and time-to-market, provided by using off-the-shelf processors, whereas SIMD computers need custom-built processors. MIMD computers can also be used more efficiently for the following reasons: 1. An MIMD computer does not force unnecessary synchronization after every instruction, or unnecessary sequentialization of non-interfering branches of computation, as an SIMD computer does. 2. Many jobs can be run simultaneously on different processors (or groups of processors) of an MIMD computer. These factors, among others, justify our assumption and explain the continuing market trend towards MIMD computers, exemplified by the Thinking Machines CM-5, the Cray Research T3D and, more recently, the Sun Microsystems Enterprise 10000 and the Hewlett-Packard HyperPlex. To enable sharing by multiple programs, MIMD computers usually provide some form of time-slicing (scheduling) or space-slicing (partitioning) or both.

Parallel irregularity can be classified as follows: variation in parallelism (number of threads) during execution (termed *X-irregularity*), and variation in the amount of computation performed per thread (termed *Y-irregularity*). Of course, a program may be both X and Y-irregular: *completely irregular*, which is the class of programs we consider. This means our programs definitely exhibit X-irregularity, the behavior of spawning and terminating parallel threads at run-time. We then have to make run-time decisions on where to schedule newly spawned threads, and whether to rebalance remaining threads after threads terminate.

Several examples of communication-intensive irregular parallel programs can be found among parallelized electronic design automation applications, and parallel search algorithms used in internet search engines and games. Besides being irregular, these programs require frequent communications among several or all of the running threads.

In Section 2 of this paper, we discuss why the most efficient way of implementing frequent communications is barrier synchronization, using a hardware tree mechanism. For a detailed discussion of barrier synchronization and implementation methods, see [4]. Hardware barrier synchronization trees impose fixed size processor partitions on the computer. In Section 3, we discuss the scheduling strategies available for fixed size partitions. We show that local scheduling does better than a policy based on gang scheduling for Y-irregular programs, provided we are able to balance threads across processors.

In Section 4, we discuss the problem of balancing threads across processors. Due to frequent communications between threads, it is essential to schedule a newly spawned thread almost immediately. Therefore, the only practical way to schedule a new thread is to temporarily run it on the same processor as its parent thread, and periodically rebalance the threads in a partition to achieve an efficient schedule. We outline an optimal online algorithm which decides when to rebalance, and prove that it does no worse than 2 times the optimal off-line strategy. The optimal online algorithm requires complete information about the scheduling state of the partition, which is prohibitively expensive to gather.

Therefore, we propose a novel way to gather partial state information by using the barrier synchronization tree available on the computer. Then, we propose and analyze an online algorithm which uses only this partial state information. We show that our online algorithm based on partial state information performs no worse than n times the optimal off-line strategy, where n is the number of processors. This implies that our worst case performance is limited to the performance of running the program sequentially. We intend to expand on this work by demonstrating experimentally that the average case performance of our partial state online algorithm is close to the optimal online algorithm.

2 Constraints on Execution

This section describes the constraints placed by our program characteristics on their efficient execution. We outline the most effective way to implement the frequent communications of our programs, and its impact on partitioning and scheduling options.

2.1 Barrier Synchronization

The overhead of implementing frequent communications as multiple pairwise communications across the data network is very high. An efficient alternative is *barrier synchronization*: a form of synchronization where a point in the code is designated as a barrier and no thread is allowed to cross the barrier until all threads involved in the computation have reached it. Usually, threads are barrier synchronized before and after each communication. This ensures that all data values communicated are current, without doing any pairwise synchronizations between threads. Barrier synchronization can be implemented in software or in hardware.

Software barriers are implemented using shared semaphores or message passing protocols based on recursive doubling. Software implementations provide flexible usage, but they suffer from either the sequential bottleneck associated with shared semaphores or the large communication latencies of message passing protocols. The time for a software barrier to complete is measured in data network latencies, and is proportional to n for shared semaphores and to $\Theta(\log n)$ for protocols based on recursive doubling, where n is the number of processors.

Hardware barriers are implemented in their simplest form as a single-bit binary tree network, called a barrier tree. The barrier tree takes single-bit flags from the processors as its inputs, and essentially implements recursive doubling in hardware, using AND gates. The time for a hardware barrier to complete is measured in gate delays, and is proportional to the height of the tree, or equivalently, $\Theta(\log n)$.

Although barrier trees are commonly implemented as complete binary trees, this is not essential. A barrier tree can be any spanning tree of the processors, and in fact, such an implementation facilitates partitioning of the computer for multiple programs [5].

Instead of a single-bit tree, the barrier tree can be constructed with m-bit edges, and any associative logic on m-bit inputs (such as maximum, minimum, sum) can replace the AND gates in the barrier tree. In this case, barriers will complete after a hardware latency no worse than $\Theta(\log n \log m)$. [4] shows that a tree which computes the maximum of its input flags, a *max-tree*, is useful in synchronizing irregular programs efficiently. Later in this paper, we discuss a way to also use such a tree in scheduling irregular programs.

In general, hardware barrier trees are intrinsically parallel and have very low latency, so they are at least an order of magnitude faster than software barriers.

2.2   Fixed Size Processor Partitions

If barrier synchronization is implemented in software, it imposes no constraints on the partitioning policy. For example, a partition may be defined loosely as the set of processors assigned to the threads of a job, and it may be possible to increase the partition size whenever a new thread is spawned. Due to this flexibility, scheduling new threads is an easier problem on such computers.

If barrier synchronization is implemented as a hardware tree, the barrier tree dictates the possible processor partitions on the computer. This is because we need to ensure that a usable portion of the barrier tree is available on each partition. Then, space-slicing can only be done by assigning fixed size partitions (usually of contiguous processors) to each program. Assuming there is only one barrier tree available per partition, partitions must be non-overlapping. It may be possible to resize partitions dynamically, but this is a slow operation since all the threads running on the partition must be quiesced and the barrier tree reconfigured for the new partition size. Therefore, resizing has to be done at infrequent enough intervals that we may assume a fixed size partition for the duration of a program. We treat a run-time partition resizing as a case where a program terminates all its threads simultaneously, and a new program with the same number of threads starts on the new partition. Note that if the computer does not support partitioning at all, then the entire computer can be treated as one fixed size partition for our purposes.

Scheduling in the presence of a hardware barrier tree is a harder problem and more applicable to the class of programs we are interested in. Therefore, in the rest of this paper, we assume that we are executing on computers with a hardware barrier synchronization mechanism, specifically a max-tree, and fixed size, non-overlapping partitions.

3   Scheduling on Fixed Size Partitions

Since barrier synchronization is implemented in hardware, when a new thread is spawned by a program, it must be scheduled on one of the processors within the fixed size partition. Given this fixed size partitioning policy, let us consider what the scheduling options are:

If there is no time-slicing available on the computer, threads cannot be pre-empted, meaning that a job will relinquish its entire partition only upon completion. Without time-slicing, each processor can run no more than one thread, so there is no way to run a job whose number of threads exceeds the number of processors in the largest partition (which may be the entire computer). This also means it is only possible to run X-irregular jobs which can predict the maximum number of threads they may have at any time during execution (this may not be feasible), and there is a large enough partition to accommodate that number. Therefore, to run X-irregular jobs without restrictions, it is essential that time-slicing be available on the computer.

One form of time-slicing called *gang scheduling* is possible on fixed size partitions. In gang scheduling, each processor has no more than one thread assigned to it, and threads never relinquish processors on an individual basis. At the end of a time-slice, all threads in the partition are preempted simultaneously using a centralized mechanism called multi-context-switch. Then, new threads (either of the same or a different job) are scheduled on the partition. (Note that it is possible to schedule different jobs in each time-slice because the state of the barrier tree can be saved along with the job's other context information, effectively allowing the barrier tree to be time-sliced as well.) By gang scheduling threads of the same job in more than one time-slice, called *multi-gang scheduling*, it is possible to run X-irregular jobs. However, this is inefficient because each processor will be idle after its thread reaches a barrier, wasting the rest of the time-slice. To address this, the time-slice can be selected to match the communication frequency, but if the job is Y-irregular, the time-slice has to be large enough to allow the longest thread to reach its barrier. In such a case, it would be helpful to allow the barrier tree to trigger the multi-context-switch hardware when the longest thread reaches the barrier, rather than using fixed length time-slices (this feature is not available on any computers we know of). There would still be some idling on most of the other threads' processors due to Y-irregularity, but this cannot be completely eliminated in multi-gang scheduling.

An alternate form of time-slicing, called *local scheduling*, mitigates the idling caused by Y-irregularity. Local scheduling is possible within fixed size partitions, and requires that each processor is capable of individually time-slicing multiple threads allocated to it. These threads must all belong to the same job, since threads from multiple jobs cannot share a barrier tree simultaneously and there is only one barrier tree per partition. In a local scheduled partition, the processor preempts each thread when it reaches a barrier, giving other threads a chance to run and reach the barrier. The processor only sets its flag on the barrier tree after all its threads have reached the barrier. In other words, the processor locally synchronizes all its threads and places the result on the barrier tree.

3.1   Handling Y-Irregularity

Barriers divide the program execution into *barrier phases*, and any Y-irregularity in the program is fragmented into these barrier phases. There will be some processing resources wasted in each barrier phase because not all threads have the same amount of computation to perform before reaching the next barrier. We show that local scheduling can usually do better than multi-gang scheduling in eliminating some of this waste.

Theorem. *Given a Y-irregular program with a large number of threads distributed evenly across processors, multi-gang scheduling cannot perform any better than local scheduling.*

Proof. Let the total number of threads in the job be N, and the number of processors in the partition be a much smaller number n. The number of threads on each processor is either $m = \lceil \frac{N}{n} \rceil$, or $m - 1$.

Let the time for thread i on processor j to reach the barrier be t_{ij}, with discrete values varying between 0 and M. (If a processor j has only $m - 1$ threads, then $t_{mj} = 0$.) In multi-gang scheduling, the time for one barrier phase to complete is

$$T_m = \sum_{i=1}^{m} \max_{j=1}^{n}(t_{ij})$$

while in local scheduling, the time for one barrier phase to complete is

$$T_l = \max_{j=1}^{n} \sum_{i=1}^{m}(t_{ij})$$

T_m is the sum of the maximum t_{ij} values across all the processors. T_l is the maximum among the sums of the t_{ij} values on each processor. The only way T_l could be as large as T_m is when the largest t_{ij} values all happen to occur on exactly the same processor, whose sum would then be selected as the maximum. For all other cases, T_l would be smaller than T_m. Therefore, $T_m \geq T_l$. This proves the theorem. □

Since the odds of all the largest t_{ij} values occurring on the same processor are very low, local scheduling generally does better than multi-gang scheduling. Intuitively, local scheduling tends to even out differences in barrier phase times across processors by averaging across the local threads.

For the above reason, as well as the fact that barrier tree triggered multi-context-switch mechanisms are not available on any existing computer, we assume that local scheduling as opposed to multi-gang scheduling is used to run X-irregular jobs on our computer. This gives rise to the problem of ensuring that the job's threads are distributed evenly across processors, which is addressed in Section 4.

Pure local scheduling has the disadvantage of not allowing a partition to be shared by more than one job. This means that a decision to run a particular job may have a potentially large, persistent and unpredictable impact on future jobs wanting to run on the computer [2]. To avoid this problem, a combination of both forms of time-slicing called *family scheduling* [1] is possible as well. In family scheduling, it is assumed that the number of threads in a job is larger than the partition size, and they are distributed across the processors as evenly as possible (the number of threads on any two processors may differ by at most 1). Multiple jobs are gang scheduled on the partition. Within its allotted partition and gang scheduled time-slice, multiple threads of the job are local scheduled on each processor. For our purposes, we can treat family scheduling and local scheduling as equivalent, since the gang scheduling time-slice is usually much larger than barrier phase times on our jobs.

3.2   Handling X-Irregularity

Since communications are implemented using barrier synchronizations, they cannot complete unless all threads participate. Therefore, to ensure job progress, all threads must be executed simultaneously or at least given some guarantee of execution within a short time bound. This means newly spawned threads must be scheduled almost immediately to keep processors from idling. Since we have a fixed partition size, the only practical option is to schedule a newly spawned thread on the same processor as its parent thread. This will temporarily cause an imbalance on that processor (and violate the family scheduling rule that the number of threads on any two processors differ by at most 1).

A thread's probability of spawning other threads may be data dependent, causing some processors to become heavily loaded compared to others. Therefore, to ensure efficient execution, spawned threads will have to be migrated to other processors. Similarly, when threads terminate, some processors may be underutilized till the remaining threads are migrated to rebalance the load.

4   Online Rebalancing of Threads

At every barrier synchronization point, we have the opportunity to gather information about the state of the partition. If we find an imbalance in the number of threads across processors, we can make the decision whether to correct the imbalance or to leave the threads where they are and continue running till the imbalance gets worse. In making this decision, we must weigh the cost of processor idling due to the imbalance against the cost of rebalancing. In addition, there is also the cost associated with gathering information about the state of the system to enable our decision making, but we will ignore this for the moment. Note that we use the term *cost* to indicate the time penalty associated with an action.

We must make our decisions *online*: at a given barrier, we have knowledge of the previous imbalances in the system, and the current state. We also know the cost of rebalancing the system, which varies as a function of the imbalance. With this partial knowledge, we must decide whether to rebalance the system or not. In contrast, a theoretical *offline* algorithm knows the entire sequence of imbalances in advance and can make rebalancing decisions with the benefit of foresight.

Consider a fixed sequence of imbalances σ. Let $C_{OPT}(\sigma)$ be the cost incurred by the optimal offline algorithm on this sequence. Let $C_A(\sigma)$ be the cost incurred by an online algorithm A on the same sequence. Algorithm A is said to be r-competitive if for all sequences of imbalances σ, $C_A(\sigma) \leq r \cdot C_{OPT}(\sigma)$. The competitive ratio of algorithm A is r. This technique of evaluating an online algorithm by comparing its performance to the optimal offline algorithm is called *competitive analysis*. It was introduced in [6] and has been used to analyze online algorithms in various fields. The optimal offline algorithm is often referred to as an *adversary*.

Note that a thread's time to reach a barrier may be data dependent, which would mean that all threads cannot be counted as equals when rebalancing decisions are made. However, due to the dynamic nature of thread behavior over the duration of a program, it is very hard to predict a thread's time to reach a barrier. Even if that information were predictable, using it to further improve scheduling decisions is usually not feasible. This is because the variation in times to reach a barrier is limited by the very small time for each barrier phase (due to communication frequency), and it is difficult to make scheduling decisions with low enough overhead to actually recover some of that small time wasted in each barrier phase.

In the rest of this paper, we assume that each thread utilizes its barrier phase fully for computation. In addition, we also assume that the barrier phase time does not vary significantly over the duration of the program. The above two assumptions enable us to treat threads of a job as equals, and rebalancing has the sole objective of equalizing the *number* of threads across processors.

4.1   Costs of Imbalance and Rebalancing

Initially, we assume that complete information about the number of threads on all processors in the job's partition is available at each processor. This information is actually expensive to gather, but we will discuss this expense and alternatives in the next section.

We first define two terms which are used to analyze the costs of imbalance and rebalancing:

- The *system imbalance*, δ is the maximum thread imbalance on any processor. If there are n processors and N threads, let $m = \lfloor \frac{N}{n} \rfloor$ denote the average number of threads in the system. If k_j is the number of threads on processor j, then $\sum_{j=0}^{n-1} k_j = N$. The thread imbalance on processor j is $\delta_j = k_j - m$. Note that δ_j values can be either positive or negative, and their maximum, by definition, has a value of 0 or higher. The system imbalance, $\delta = \max_{j=0}^{n-1} \delta_j$.
- The *aggregate system imbalance*, Δ is the number of threads that need to be moved in order to balance the system.
  Let δ_{Aj} denote the absolute value of the thread imbalance on each processor. In other words, for each processor, j, $\delta_{Aj} = |\delta_j|$. The aggregate system imbalance, $\Delta = \frac{\sum_{j=0}^{n-1} \delta_{Aj}}{2}$.

The cost of running without rebalancing at a barrier is c, where $c = t \cdot \delta$, and t is the time for any thread to reach any barrier, which we have assumed to be a constant for a given program. The cost of rebalancing the tree is C, where $C = x + y \cdot \Delta$, and x and y are constants which depend on the implementation of the data network on the machine. These costs c and C are used to analyze online rebalancing algorithms. For our analysis, we assume that x is negligibly small compared to $y \cdot \Delta$. Note that for communication-intensive programs, t is very small, so C is significantly larger than c on any machine.

When the job is initially scheduled or after the last rebalancing, the threads in the partition are evenly distributed across all processors. At this point, δ may be 0 or 1, depending on whether N is a multiple of n. Therefore, the lowest value of c which represents a correctable system imbalance is generally $2t$. It is reasonable to assume that the system places a limit M on the maximum number of threads that can be run per processor, and this limit is helpful in bounding the values of c and C later on.

An online algorithm would maintain a sum Σ of all the c values encountered (resetting Σ whenever c has a value of 0 or t, corresponding to $\delta = 0$ or 1). When Σ equals some threshold T, a system rebalance is triggered, at a cost of C.

4.2 Optimal Online Algorithm

Case 1: Consider an algorithm which selects a T smaller than C. Therefore, $T = C - \epsilon$. To minimize the performance of this algorithm, the adversary would keep the system at a minimum imbalance at all times, by spawning or terminating threads. Therefore, the algorithm pays the cost C of rebalancing the system, while the adversary never pays a rebalancing cost. However, both the algorithm and the adversary pay the cost $\Sigma = T$ of running with an imbalance. The algorithm's cost is $C + T$, while the adversary's cost is T, making the algorithm's competitive ratio $r = \frac{2C - \epsilon}{C - \epsilon} = 1 + \frac{C}{C - \epsilon}$. Therefore, $r \geq 2$ for all values of ϵ, and has a minimum value of 2 occurring when $\epsilon = 0$.

Case 2: Consider an algorithm which selects a T larger than C. Therefore, $T = C + \epsilon$. To minimize the performance of this algorithm, the adversary would rebalance its system as early as possible, therefore paying no costs for running with an imbalance. Once again, the algorithm's cost is $C + T$, while the adversary's cost is C, making the algorithm's competitive ratio $r = \frac{2C + \epsilon}{C} = 2 + \epsilon/C$. Again, $r \geq 2$ for all values of ϵ, and has a minimum value of 2 occurring when $\epsilon = 0$.

From the above two cases, we see that the optimal online algorithm selects $T = C$, and has a competitive ratio of 2.

4.3 Low Overhead Alternatives

The cost of gathering complete information about the system configuration at each barrier is too high, requiring n phases of communication over the data network, where each processor is allowed to declare how many threads are assigned to it. To enable hardware barrier synchronization, we assumed that a max-tree is available on the computer. This tree can be used to inexpensively compute the maximum and minimum number of threads per processor on our system. This is done in two phases:

1. Each processor j places its k_j in its max-tree flag, and the tree returns their maximum, $\max_{j=1}^{n} k_j$ to all the processors. This value directly corresponds to the maximum number of threads running on any processor.

2. Each processor places $M - k_j$ in its max-tree flag, and the tree returns their maximum, $\max_{j=1}^{n}(M - k_j)$ to all the processors. By subtracting this value from M, each processor calculates the minimum number of threads running on any processor, since $M - \max_{j=1}^{n}(M - k_j) = M - [M - \min_{j=1}^{n} k_j] = \min_{j=1}^{n} k_j$.

We wish to consider algorithms which estimate costs only based on the difference between the minimum and maximum number of threads on the processors, without knowing the actual average number of threads on the system.

The average number of threads on the system has to lie between the minimum and maximum number of threads on any processor. Therefore, the algorithm faces the worst uncertainty in guessing the average when the maximum and minimum are as far apart as possible. This happens in system configurations where at least one processor has 0 threads and at least one has M threads. We refer to the set of system configurations with this property as *0-M* configurations. We need to consider only *0-M* configurations since we are interested in estimating the algorithm's worst case behavior.

The rebalancing period p is the number of barriers run with an imbalance, after which the algorithm chooses to rebalance. The algorithm would have to guess values of c and C, and select a value for p that is as close as possible to $\frac{C}{c}$.

If the algorithm underestimates p, it would rebalance too often, and the adversary's strategy would be to keep the system at a minimum imbalance at all times. The adversary would never pay a rebalancing cost, while the algorithm would pay it more often than it would with complete information. Both pay the cost of running with the minimum imbalance at all times.

If the algorithm overestimates p, it would rebalance too infrequently, and the adversary's strategy would be to cause the maximum system imbalance and rebalance immediately. The adversary pays the rebalancing cost, while the algorithm pays the rebalancing cost as well as the cost of running with the maximum imbalance for longer than it would with complete information.

Now we analyze *0-M* configurations to arrive at the minimum and maximum values that δ and Δ can take. For any *0-M* configuration, without loss of generality, we can also assume that processor 0 has 0 threads, and processor $n - 1$ has M threads.

Note that the system imbalance δ is always attributable to the processor with the largest number of threads running on it. Therefore, we may assume that $\delta_{n-1} = \delta$.

The maximum δ, denoted by δ_{MAX}, has to occur when $\delta_{n-1} = M - m$ is maximized, which is when m is minimized. This happens when processors 0 through $n-2$ all have 0 threads, making $m = \frac{M}{n}$, and $\delta_{n-1} = M - \frac{M}{n} = \frac{M(n-1)}{n}$. (Any other configuration would have more threads on some of the processors 0 through $n - 2$, making the mean m higher and reducing the value of δ_{n-1}.) Therefore, δ_{MAX} is $\frac{M(n-1)}{n}$. For this configuration, referred to as *Configuration α*, $\Delta = \frac{(n-1)\cdot\frac{M}{n}+M-\frac{M}{n}}{2}$, which simplifies to $\frac{M(n-1)}{n}$.

Similarly, the minimum δ, denoted by δ_{MIN}, has to occur when $\delta_{n-1} = M - m$ is minimized, which is when m is maximized. This happens when processors 1 through $n - 1$ all have M threads, making $m = \frac{M(n-1)}{n}$, and $\delta_{MIN} = M - \frac{M(n-1)}{n} = \frac{M}{n}$. For this configuration, referred to as *Configuration β*, $\Delta = \frac{\frac{M(n-1)}{n}+(n-1)\cdot[M-\frac{M(n-1)}{n}]}{2}$, which simplifies to $\frac{M(n-1)}{n}$.

Note that Δ has the same value, $\frac{M(n-1)}{n}$ at δ_{MIN} and at δ_{MAX}. This is also the minimum number of threads that need to be moved to balance any *0-M* configuration. Therefore, the minimum value of Δ, Δ_{MIN} is $\frac{M(n-1)}{n}$.

The maximum value of Δ, Δ_{MAX} occurs when half the processors have M threads and the other half have 0 threads. In this configuration, $m = \frac{M}{2}$. To balance this configuration, $\frac{M}{2}$ threads have to be moved from $\frac{n}{2}$ processors (who have M threads) to the others. Therefore, $\Delta_{MAX} = \frac{Mn}{4}$, and corresponds to $\delta = \frac{M}{2}$. We refer to this as *Configuration γ*.

Assume there are no limits (other than M) on the number of threads that can be spawned or terminated in one barrier phase. Considering only *0-M* configurations, δ can take any value in the range $[\frac{M}{n}, \frac{M(n-1)}{n}]$ at a barrier, regardless of its previous value. Similarly, Δ can take any value in the range $[\frac{M(n-1)}{n}, \frac{Mn}{4}]$ at a barrier, depending only on δ and regardless of its previous value. However, to analyze the worst case behavior of any algorithm, it is sufficient to consider configurations α, β and γ, since these provide the worst case values of δ and Δ.

$c = t \cdot \delta$, with three choices of δ values: $\frac{M}{n}$, $\frac{M}{2}$ and $\frac{M(n-1)}{n}$. $C = y \cdot \Delta$, with just two choices of Δ values: $\frac{M(n-1)}{n}$ and $\frac{Mn}{4}$.

If the algorithm underestimates p, its cost is $p \cdot t \cdot \frac{M}{n} + y \cdot \frac{Mn}{4}$, while the adversary's cost is $p \cdot t \cdot \frac{M}{n}$. This makes the competitive ratio $r = 1 + \frac{yn^2}{4pt}$.

If the algorithm overestimates p, its cost is $p \cdot t \cdot \frac{M(n-1)}{n} + y \cdot \frac{M(n-1)}{n}$, while the adversary's cost is $y \cdot \frac{M(n-1)}{n}$. This makes the competitive ratio $r = \frac{pt+y}{y}$.

We propose an algorithm which selects $p = \frac{y}{t} \cdot \frac{n}{2}$, corresponding to configuration γ. By exhaustively considering all possible combinations of δ and Δ values, one can prove that any algorithm does best by choosing this value, thus showing that our algorithm is optimal among incomplete information alternatives. (The proof is omitted here due to lack of space.) By substituting the value of p in the equations for r above, we see that the competitive ratio of our algorithm is n.

5   Summary and Future Work

In this paper, we classify irregular parallel programs based on the source of their irregularity, into X-irregular, Y-irregular and completely irregular programs. Our programs are communication-intensive besides being completely irregular. This limits us to fixed size partitions, since frequent communication is efficiently implemented using a hardware barrier tree. We compare the possible scheduling algorithms for completely irregular programs on fixed size partitions: multi-gang scheduling and local scheduling, and show that local scheduling does better to mitigate the inefficiencies of Y-irregularity. However, to handle the effects of X-irregularity, threads need to be rebalanced on processors periodically. We propose and analyze online algorithms for rebalancing threads, including an n-competitive algorithm which is efficient due to its low information gathering cost.

We intend to run simulations based on our program characteristics to experimentally show the following: Although our algorithm's worst case behavior is in $\Theta(n)$, its average behavior is fairly close to the performance of the 2-competitive, optimal algorithm which has a far greater information gathering overhead. Based on our experiments, we also intend to propose algorithms for machines where the rebalancing cost $C = x + y\Delta$ has a large value of x, making infrequent rebalancing advantageous.

References

1. R. M. Bryant and R. A. Finkel. A stable distributed scheduling algorithm. In *International Conference on Distributed Computing Systems*, April 1981.
2. D. G. Feitelson and M. A. Jette. Improved utilization and responsiveness with gang scheduling. In D. G. Feitelson and L. Rudolph, editors, *Job Scheduling Strategies for Parallel Processing - Lecture Notes in Computer Science*, volume 1291, pages 238–261. Springer Verlag, 1997.
3. D. G. Feitelson and L. Rudolph. Parallel job scheduling: Issues and approaches. In D. G. Feitelson and L. Rudolph, editors, *Job Scheduling Strategies for Parallel Processing - Lecture Notes in Computer Science*, volume 949, pages 1–18. Springer Verlag, 1995.
4. V. Ramakrishnan, I. D. Scherson, and R. Subramanian. Efficient techniques for fast nested barrier synchronization. In *Symposium on Parallel Algorithms and Architectures*, pages 157–164, July 1995.
5. V. Ramakrishnan, I. D. Scherson, and R. Subramanian. Efficient techniques for nested and disjoint barrier synchronization. *Journal of Parallel and Distributed Computing*, 58:333–356, August 1999.
6. D. D. Sleator and R. E. Tarjan. Amortized efficiency of list update and paging rules. *Communications of the ACM*, 28:202–208, February 1985.

Non-Memory-Based and Real-Time Zerotree Building for Wavelet Zerotree Coding Systems

Dongming Peng and Mi Lu

Electrical Engineering Department, Texas A&M University, College Station, TX77843, USA

1   Introduction

The wavelet zerotree coding systems, including Embedded Zerotree Wavelet (EZW)[1] and its variants Set Partitioning In Hierarchical Trees (SPIHT)[2] and Space Frequency Quantization (SFQ)[3][4], have three common procedures: 1) 2-D Discrete Wavelet Transform (DWT)[5], 2) zerotree building and symbol generation from the wavelet coefficients (illustrated in Figure 1), and 3) quantization of the magnitudes of significant coefficients and entropy coding, where the second procedure is an important one. All recently propsed architectures ([6]-[9]) for wavlelet zerotree coding use memories to build zerotrees. In this paper we contribute to building the zerotrees in a non-memory-based way with real-time performance leading to the decrease of hardware cost and the increasement of processing rate which is especially desirable in video coding. One of our main ideas is to rearrange the DWT calculations taking advantage of parallel and pipelined processing so that *any* parent coefficient and its children coefficients in zerotrees are guaranteed to be calculated and outputted simultaneously.

2   The Architecture for Rearranging 2-Stage 2-D DWT

2.1   Two Preliminary Devices Used in the Architecture for Rearrangement

(1)The Processing Unit(PU) shown as in Figure 2 rearranges the calculation of wavelet filtering so that the filter is cut to half taps based on the symmetry between the negative and positive wavelet filter coefficients. x, a and c are the input sequence, low- and high-pass filtering output sequence respectively. While a datum of sequence x is fed and shifted into the PU per clock cycle, a datum of a is calculated every even clock cycle and a datum of c is calculated every odd clock cycle. The PU in Figure 2(a) can be extended to a parallel format as in Figure 2(b) where if a number of data from sequence x x_{k+8}, x_{k+7}, ..., x_k are fed to the PU in parallel at a clock cycle, then x_{k+9}, x_{k+8}, ..., x_{k+1} are fed at the next cycle. (2)In Figure 3 the TU is a systolic array with $(N_w+3) \times N$ cells, where

J. Rolim et al. (Eds.): IPDPS 2000 Workshops, LNCS 1800, pp. 469-475, 2000.

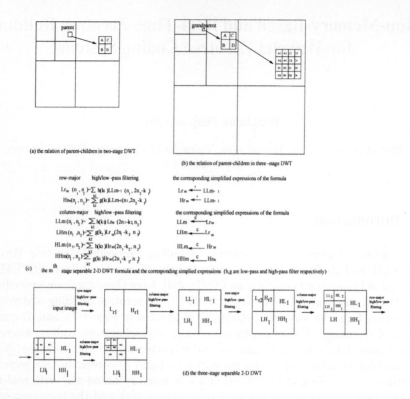

(a) the ralation of parent-children in two-stage DWT

(b) the relation of parent-children in three -stage DWT

(c) the mth stage separable 2-D DWT formula and the corresponding simplied expressions (h,g are low-pass and high-pass filter respectively)

(d) the three-stage separable 2-D DWT

Fig. 1. the algorithms of EZW and 2-D DWT

N_w is the width of the wavelet filter and N is the width or the height of the input (square) image to DWT. N is hundreds or thousands of times greater than N_w for most applications of 2-D wavelet transforms (e.g. image/video systems). A cell transfers its content to the next adjacent cell once it receives a datum from its preceding cell. The leftmost cells in odd rows and rightmost cells in even rows have output ports and copy their data to outside. The upper-left cell has an input port and the TU uses it to recieve the input sequence. An element in matrix X is fed to the TU per clock cycle in the order according to the indices in Figure 3(b). The TU's (N_w+3) outputs and its newly arrived element belong to the same column in X. An example for the positions of the X's elements in TU after 3N clock cyles of inputting X is illustrated in Figure 3(c).

2.2   the Proposed Architecture and the Analysis of Its Operations

The architecture for the rearrangement of DWT is proposed in Figure 4, and the corresponding timing of operations is presented in Figure 5. Every four sibling coefficients in the first decomposition stage are designed to be calculated together (meanwhile their parent is generated by PU_3).

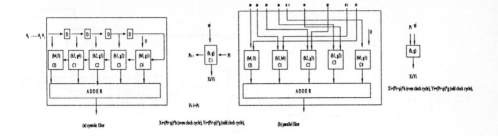

Fig. 2. the structures of PU (Processing Unit)

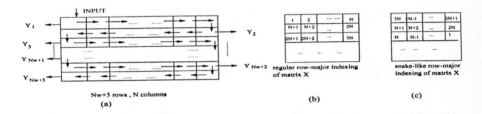

Fig. 3. A new transpose unit (TU)

Due to the row-major dyadic subsampling, the row-major high/low-pass filtering is alternatively executed by the PU_1 in Figure 4 point by point in each row. Based on similar column-major dyadic subsampling, the PU_2 takes turns to execute column-major high/low-pass filtering and selects appropriate inputs from TU_1 to generate four sibling coefficients consecutively. The order to calculate the siblings is as A, then B, C and D in the example of four siblings illustrated in Figure 1(a). After PU_2's calculation of point A by taking Y_1, ..., Y_{Nw} as inputs, PU_2 has to calculate B by taking Y_3, ..., Y_{Nw+2} as inputs, then PU_2 comes back to take Y_1, ..., Y_{Nw} to calculate C, then PU_2 takes Y_3, ..., Y_{Nw+2} again to calculate D.

In Figure 5 it can be seen that PU_2 calculates the same kind of column-major convolution (high-pass or low-pass filtering) during the period when a row of input image is sequentially fed to the system in N clock cycles, and the coefficients in four subbands LL_1, LH_1, HL_1 and HH_1 are calculated in turns in a longer period when four rows of input image are fed. The right column in Figure 5 is to show the operations of PU_3 in Figure 4. In PU_3 the second stage of DWT is completed and the parent coefficients are generated. PU_3 takes sequential input of LL_1 coefficients from PU_2 to perform row-major filtering in the first quarter of the period during which four rows of input image are fed to the system. Then in the next three quarters (i.e., i=4q+1, 4q+2 or 4q+3 in Figure 5), PU_3 performs the column-major convolutions by taking the result from the first quarter's row-major filtering as input.

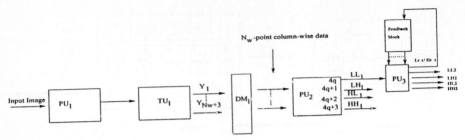

1) DM1 is a demultiplexer that select Nw-point data from Nw+3 outputs of TU1.

2) PU2 is a paralle filter as in Figure 2(b) and has four output ports active at different time.

3) PU3 is the hybrid version of PU that can take either sequential or parallel inputs.

4) Feedback block consists of 2 separate TUs and Demultriplexers to select Lr2 / Hr2 into respective TU and to select outputs from 2 TUs into PU3.

Fig. 4. The architecture for two-stage DWT and the zerotree building

3   The Design Extended to General Stages of DWT

Now we consider what should be done to modify the architecture in Figure 4 for m stages of wavelet decomposition. Because the input image is fed into the system in the same way as before, the first stage row-major high/low-pass filtering is still performed in PU_1 alternatively as designated in Figure 5.

Regarding the first stage column-major high/low-pass fitering performed in PU_2, we note that there are 4^{m-1} coefficients in the first stage decomposition corresponding to the same ancestor in the last stage (stage m) decomposition. To satisfy the restriction of generating parent and children simultaneously, it is required that these 4^{m-1} "kindred" coefficients be calculated together. (Meanwhile these coefficients' parents in the intermediate stages of decompostion should be calculated together too.) Note that these 4^{m-1} coefficients are located in 2^{m-1} adjacent rows and 2^{m-1} adjacent columns in their subband. PU_2 should alternatively select appropriate inputs among 2^{m-1} different groups of parallel column-major data from TU_1, and perform column-major filtering to generate the 4^{m-1} kindred coefficients in turns, where the coefficeints calculated with the same group of input belong to the same row. Accordingly, TU_1 is an extended version in Figure 3(a) and is supposed to have output ports Y_1, ..., Y_{Nw+M} with M equal to 2^m.

PU_3 carries out the rest computation in DWT. The second stage decomposition is achieved as follows. In the first quarter of the period when 2^m rows of input image are fed to the system, PU_3 gets its inputs, i.e., the coefficients in LL_1 subband from TU_1, and alternatively performs the second stage low/high-pass row-major convolution. The calculated results, or the coefficients in Lr_2 and Hr_2 are stored in two TUs hidden in PU_3's feedback block. In the second quarter, the Lr_2 coefficients are fed back to PU_3 to be column-major filtered to get the results in LL_2 and LH_2. In the third and fourth quarter, the Hr_2 points are fed back to PU_3 to be used to calculate out HL_2 and HH_2 respectively. The PU_3 achieves further stage decompositions in the available intervals during its execution of the second stage decomposition.

By reason of limit space in this paper, the operations of processors are described with basic principles and not with many details. To sum up, TU_1 is modified to have (N_w+M) rows; PU3's feedback block has changed a little bigger so that it holds N_w+M rows of coefficients in the results of row-major wavelet filtering; and the switches have become complicated to select appropriate data at diffrent time.

4   Performance Analysis and Conclusion

Since N_w (the width of wavelet filters) is far less than N (the width or length of input image) and the size of boudary effect of wavelet transforms is only dependent on N_w, in this paper we ignore the boudary effect to simplify our expressions knowing that it can be resolved by a little adjustment in either timing or architecture. The area of the proposed architecture is dominated by PUs and TUs. A PU contains pN_w MACs (Multiplyer and Accumulator Cell), where p is the number of precision bits of data, thus three PUs contain $3pN_w$ MACs. Because a TU is necessary for the column-major filtering in every stage decomposition, and the number of cells in TU at the ith stage decomposition is $(N/2^i)(N_w+2^i)$, where the first item is the length of a row and the second item is the number of rows in the TU, the total area for TUs is $O(mN+N_wN)$, where m is the number of stages in DWT. Note that the TUs except TU_1 are hidden in PU3's feedback block. Thus the whole area of the architecture for m stage DWT is $\mathbf{A}=O(pNN_w+pNm)$. The input image is assumed to be fed with one pixel per clcok cycle. The system's latency (execution time) $\mathbf{T}$ is N^2 clock cycles. Thus the product of $\mathbf{A}$ and $\mathbf{T}$ for the system is $O(pN^3(N_w+m))$, where N^2 is the input size of the algorithm. Our proposed architecture is comparable to conventional DWT architectures ([10]-[14]) in the aspect of area, latency, the product of them, or the hardware utilization, even though not only the DWT but also the zerotree building is achieved in this architecture.

We have proposed a non-memory-based design in which the input image is recursively decomposed by DWT and zerotrees are built in real-time. The computation of wavelet-based zerotree coding is strongly featured by the computation locality in that the calculations of coefficients on a certain zerotree (only) depend on the same local sub-area of the 2-D inputs. This desirable feature has been exploited in this paper by calculating children and their parent simultaneously in the rearranged DWT, so that most intermediate data need not be held for future calculations.

References

1. J.M. Shapiro, Embedded image coding using zerotrees of wavelet coefficients, *IEEE Transactions on Signal Processing,* Volume: 41, 1993, Page(s): 3445 -3462.

2. A. Said, W.A. Pearlman, A new, fast, and efficient image codec based on set partitioning in hierarchical trees, *IEEE Transactions on Circuits and Systems for Video Technology*, Volume: 6, June 1996, Page(s): 243 -250.

3. Zixiang Xiong, K. Ramchandran, M.T. Orchard, Wavelet packet image coding using space-frequency quantization, *IEEE Transactions on Image Processing*, Volume: 7, June 1998, Page(s): 892 -898.

4. Zixiang Xiong, K. Ramchandran, M.T. Orchard, Space-frequency quantization for wavelet image coding, *IEEE Transactions on Image Processing*, Volume: 6, May 1997, Page(s): 677 -693.

5. M. Vetterli, J. Kovacevic, Wavelets and Subband Coding, *Prentice Hall*, 1995.

6. Li-Minn Ang, Hon Nin Cheung, K. Eshraghian, VLSI architecture for significance map coding of embedded zerotree wavelet coefficients, *Proceddings of 1998 IEEE Asia-Pacific Conference Circuits and Systems*, 1998. Page(s): 627 -630.

7. Jongwoo Bae, V. K. Prasanna, A fast and area-efficient VLSI architecture for embedded image coding, *Proceedings of International Conference on Image Processing*, Volume: 3, 1995, Page(s): 452 -455.

8. J.M. Shapiro, A fast technique for identifying zerotrees in the EZW algorithm, *Proceedings of 1996 IEEE International Conference on Acoustics, Speech, and Signal Processing*, Volume: 3, 1996, Page(s): 1455-1458.

9. J. Vega-Pineda, M. A. Suriano, V. M. Villalva, S. D. Cabrera, Y.-C. Chang, A VLSI array processor with embedded scalability for hierarchical image compression, *1996 IEEE International Symposium on Circuits and Systems*, Volume: 4, 1996, Page(s): 168 -171.

10. Jer Min Jou, Pei-Yin Chen, Yeu-Horng Shiau, Ming-Shiang Liang, A scalable pipelined architecture for separable 2-D discrete wavelet transform, *Design Automation Conference, Proceedings of the ASP-DAC '99. Asia and South Pacific*, Volume: 1, Page(s): 205 -208.

11. M. Vishwanath, R. M. Owens, M. J. Irwin, VLSI architectures for the discrete wavelet transform, *IEEE Transactions on Circuits and Systems II: Analog and Digital Signal Processing*, Volume: 42, May 1995, Page(s): 305 -316.

12. Chu Yu, Sao-Jie Chen, Design of an efficient VLSI architecture for 2-D discrete wavelet transforms, *IEEE Transactions on Consumer Electronics, Volume: 45, Feb. 1999, Page(s): 135 -140*.

13. V. Sundararajan, K. K. Parhi, Synthesis of folded, pipelined architectures for multidimensional multirate systems, *Proceedings of the 1998 IEEE International Conference on Acoustics, Speech and Signal Processing*, Volume: 5, 1998, Page(s): 3089 -3092.

14. Chu Yu, Sao-Jie Chen, VLSI implementation of 2-D discrete wavelet transform for real-time video signal processing, *IEEE Transactions on Consumer Electronics*, Volume: 43, Nov. 1997, Page(s): 1270-1279.

15. T. Acharya, Po-Yueh Chen, VLSI implementation of a DWT architecture, *Proceedings of the 1998 IEEE International Symposium on Circuits and Systems*, Volume: 2, 1998, Page(s): 272 -275.

```
real-time Alg. DWT amenable to EZW
{
   for i=0 to N-1  /* row */
      for j=0 to N-1 /* column */
         Do DWT(i,j)
}
```

DWT(i,j)   /* q,s are any non-negative integers */
{

	what PU₁ does	what PU₂ does	what PU₃ does
if i=4q			
if j=even	$Ln \xleftarrow{r} x$	$LL_1 \xleftarrow{c} Ln$	
if j=odd	$Lr_1 \xleftarrow{r} x$	$LL_1^1 \xleftarrow{c} Lr_1^1$	
if j=4s			$Lr_2 \xleftarrow{r} LL_1$
if j=4s+1			$Hr_2 \xleftarrow{r} LL_1$
if j=4s+2			$Lr_2^1 \xleftarrow{r} LL_1^1$
if j=4s+3			$Hr_2^1 \xleftarrow{r} LL_1^1$
if i=4q+1			
if j=even	$Ln \xleftarrow{r} x$	$LH_1 \xleftarrow{c} Lr_1^+$	
if j=odd	$Lr_1 \xleftarrow{r} x$	$LH_1^1 \xleftarrow{c} Lr_1^{1+}$	
if j=4s			$LL_2 \xleftarrow{c} Ln$
if j=4s+2			$LH_2 \xleftarrow{c} Lr_2^1$
if i=4q+2			
if j=even	$Ln \xleftarrow{r} x$	$HL_1 \xleftarrow{c} Hr_1$	
if j=odd	$Hr_1 \xleftarrow{r} x$	$HL_1^1 \xleftarrow{c} Hr_1^1$	
if j=4s+2			$HL_2 \xleftarrow{c} Hr_2$
if i=4q+3			
if j=even	$Ln \xleftarrow{r} x$	$HH_1 \xleftarrow{c} Hr_1^+$	
if j=odd	$Hr_1 \xleftarrow{r} x$	$HH_1^1 \xleftarrow{c} Hr_1^{1+}$	
if j=4s+2			$HH_2 \xleftarrow{c} Hr_2^+$

Notation:

(1).The meaning of superscript "k" is explained as the following.(k is an integer)

Using the simplified expression as in figure 1(c),we call the calculation of $\xleftarrow{r}$ as "r" arrow;$\xleftarrow{c}$ as "c" arrow.

Suppose A corresponds to a part of a component in figure 1 (d). If row-wise signal A and A^k are on the right side of "r" arrow, and A is from the ith row,then A is from the kth row.

If column-wise signal A and A^k are on the right side of "c" arrow, and A is from $Y_i,....Y_{Nw+i}$ (See TU's output in figure 3), then A is from $Y_{i+2k},.....Y_{Nw+2k}$ (in the same column)

If A and A^k are on the left side of "c" arrow or "r" arrow,they are corresponding to 2 coefficients in the same column and in ith row and (i+k)th row respectively.

The meaning of "+" is explained as the following:

Column-wise signal A ,A and A^+ are on the right side of "c" arrow. A is from $Y_i,......Y_{Nw+i}$ (see TU's output in figure 3) the A^+ is from $Y_{i+1},....Y_{Nw+i+1}$, A^{k+} is from $Y_{i+2k+1},.......Y_{Nw+i+2k+1}$

(they are in the same column)

(2) Because of PU's feature of alternative high/low -pass filtering,we use A and A alternative to get column-wise low-high pass outputs.

(3) The reason why we use A and A^1 (A standing for Lr_1 , LL_1) alternatively to generate B and B (B standing for LL_1 , LH_1 ,....) is based on the restriction that any siblings be generated consecutively. Because of dyadic downsamplings for column-wise convojution. $B \xleftarrow{c} A^1$ if $B \xleftarrow{c} A$

(4) Lr_2 and Hr_2 are fed in 2 seperate TUs in PU3's feedback block. The TUs can hold Nw+2 rows of data at most. Careful readers may find that $LL_2 \xleftarrow{c} Lr_2$ and $LH_2 \xleftarrow{c} Lr_2$ are executed alternately point by point, however, $HL_2 \xleftarrow{c} Hr_2$ and $HH_2 \xleftarrow{c} Hr_2$ are executed alternately row-by row. This paradox can be resolved by a little manipulation in 2 TUs in feedback block . Anyway, during the time of $4q+1 \le i \le 4q+3$, no new data are generated for Lr_2 or Hr_2 ,so the old data in feedback block can be held until i=4q+4.

(5) This real time algorithm dictates the operations happening in figure 4. The control signals for PU ,TU ,DM ,PU ,PU , feedback block can be easily implemented locally and periodically. Their details are not discussed because of the limited length of this paper.

Fig. 5. The timing for the operations in the architecture for two-stage DWT and the zerotree building

Graph Partitioning for Dynamic, Adaptive and Multi-phase Computations

Vipin Kumar

Joint work with Kirk Schloegel and George Karypis, Computer Science Department, University of Minnesota. http://www.cs.umn.edu/~kumar, email kumar@cs.umn.edu

Abstract. Algorithms that find good partitionings of highly unstructured graphs are critical in developing efficient algorithms for problems in a variety of domains such as scientific simulations that require solution to large sparse linear systems, VLSI design, and data mining. Even though this problem is NP-hard, efficient multi-level algorithms have been developed that can find good partitionings of static irregular meshes. The problem of graph partitioning becomes a lot more challenging when the graph is dynamically evolving (e.g., in adaptive computations), or if computation in multiple phases needs to be balanced simultaneously. This talk will discuss these challenges, and then describe some of our recent research in addressing them.

J. Rolim et al. (Eds.): IPDPS 2000 Workshops, LNCS 1800, pp. 476-476, 2000.
© Springer-Verlag Berlin Heidelberg 2000

A Multilevel Algorithm for Spectral Partitioning with Extended Eigen-Models

Suely Oliveira[1] and Takako Soma

[1] The Department of Computer Science,
The University of Iowa, Iowa City, IA 52242, USA,
oliveira@cs.uiowa.edu,
WWW home page: http://www.cs.uiowa.edu/~oliveira

Abstract. Parallel solution of irregular problems require solving the graph partitioning problem. The extended eigenproblem appears as the solution of some relaxed formulations of the graph partitioning problem. In this paper, a new subspace algorithm for the solving the extended eigenproblem is presented. The structure of this subspace method allows the incorporation of multigrid preconditioners. We numerically compare our new algorithm with a previous algorithm based on Lanczos iteration and show that our subspace algorithm performs better.

1 Introduction

One of the main problems encountered when dealing with irregular problems on parallel architectures is mapping the data into the various processors. Traditionally graph partitioning has been used to achieve this goal.

Kernighan and Lin developed an effective combinatorial method based on swapping vertices [10]. Multilevel extensions of the Kernighan-Lin algorithm have proven effective for graphs with large numbers of vertices [9]. Like combinatorial methods, spectral methods have proven effective for large graphs arising from FEM discretizations [12]. In fact, currently various software packages (METIS [8] and Chaco [4] and others) combine multilevel combinatorial algorithms and spectral algorithms. The spectral algorithms which are used in these packages are based on the models that partition a graph by finding the second smallest eigenvector of its graph Laplacian using an iterative method. This is in fact the model that we used in [7] and [6].

However, as the example below shows, there are number of reasons for modifying the traditional model and spectral heuristics for graph partitioning. Recently, Hendrickson et al. [5] pointed out the problems with traditional models. Consider Figure 1. Assume we have already partitioned the graph into two pieces, left and right halves, and that we have similarly divided the left half graph into top and bottom quadrants. When partitioning the right half graph between processors 3 and 4 we should like messages to travel short distances. The mapping shown in the left-hand figure is better since the total message distance is less than that for the right-hand figure. Note that even though in modern computers

J. Rolim et al. (Eds.): IPDPS 2000 Workshops, LNCS 1800, pp. 477-484, 2000.

the distance between processors may not imply very different timings, smaller distances traveled by the messages will decrease congestion problems. The basic idea is to associate with each vertex in the partitioned subgraph a value which reflects its net desire or preference to be in the bottom quadrant instead of the top quadrant. Note that this preference is a function only of edges that connect the vertex to vertices which are not in the current subgraph. The preferences need to be considered when subgraphs are partitioned. These preferences should be propagated through the recursive partitioning process. In order to to overcome this problem, Hendrickson et al. used an extended eigen-model in [5].

Another way in which extended eigen-models occur is when some nodes are pre-assigned to some partitions, and we wish to assign the remaining nodes to the remaining partitions. In this situation the extended eigen-model can also be used to assigning the remaining nodes. This can be a much smaller problem, and would be very useful for dynamic re-partitioning.

In other words, extended eigen-models can be used to develop more refined models of the true costs of a partition, and to develop algorithms which give more useful and effective partitionings.

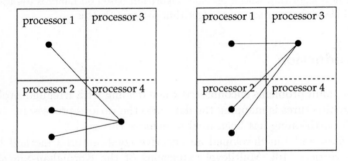

Fig. 1. Partition between the 4 processors.

2   Extended Eigen-Model

Here we describe the motivation given in [5], but notice that there are other applications in which similar problems to extended eigenproblems need to be solved [1]. Let $G = (V, E)$ be a graph with vertices $v \in V$ and edges $e_{ij} \in E$. We also allow either edges or vertices to have positive weights associated with them, which we denote by w_v and $w_{e_{ij}}$ respectively. Assume we want to divide V into two subsets V_1 and V_2, and that we have a vector d of preferences for the vertices of V to be in V_1. The cost associated with a partition now has two components. First, every edge $e_{ij} \in E$ crossing between V_1 and V_2 contributes a value of $w_e(e_{ij})$. Second, for each vertex in V_1 with a negative preference we add the magnitude of the preference to the cost, and similarly for each vertex in

V_2 with a positive preference. Our goal is to find a partition of the vertices into two sets of nearly equal size in which this combined cost function is minimized.

A graph partition can be described by assigning a value of $+1$ to all vertices in one set and a value of -1 to all vertices in the other. If we denote the value assigned to a vertex i by $x(i)$, then the simple function $(x(i) - x(j))^2/4$ is equal to 1 if the vertices i and j are in different partitions and equal to 0 otherwise. If $d(i)$ is the preference for a vertex to be in the set denoted by $+1$ which we will define to be V_1, then the new problem we want to solve is

$$\text{Minimize } f(x) = \tfrac{1}{4} \sum_{e_{ij} \in E} w_e(e_{ij})(x(i) - x(j))^2 - \tfrac{1}{2} \sum_{i \in V} d(i)x(i)$$
Subject to
(a) $\sum_{i \in V} w_v(i)x(i) \approx 0$ \hfill (1)
(b) $x(i) = \pm 1.$

After some transformations and relaxing the integrality constraint that $x(i) = \pm 1$, this model can be expressed as

$$\text{Minimize } f(x) = y^T A y - 2h^T y$$
Subject to
(a) $s^T y = 0$ \hfill (2)
(b) $y^T y = \omega_v$

where A is the graph Laplacian matrix for the edge weights w_e, $\omega_v = |V|$ the number of vertices, $s_i = w_v(i)$, and $h_i = d(i)$ for all i.

We introduce Lagrange multipliers λ and μ, and look for stationary points of the Lagrangian function

$$F(y, \lambda, \mu) = y^T A y - 2h^T y + \lambda(s^T y) + \mu(\omega_v - y^T y) = 0. \tag{3}$$

satisfying (a) and (b).

Taking derivatives with respect to the components of y, we obtain

$$\nabla_y F(y, \lambda, \mu) = 2Ay - 2h + \lambda s - 2\mu y = 0. \tag{4}$$

We can calculate λ by left multiplying Equation 4 by s^T. Since s is orthogonal to y and s is a zero eigenvector of A, we have $\lambda = 2s^T h/\omega_v$. We now define

$$g = h - \frac{s^T h}{\omega_v} s, \tag{5}$$

which allows us to rewrite Equation 4 as

$$Ay = \mu y + g. \tag{6}$$

This extended eigenproblem must be solved subject to the constraints in (2). In fact, we need to use the smallest μ for which a solution of (2, 6) can be found. A proof of this fact can be found using [11, Thm. 4.3, p. 78], for example.

In [5], Hendrickson used a numerical method for solving this extended eigen-problem based on the Lanczos algorithm, which will be described in section 4. Next section we present our theoretical motivation for studying subspace algorithms for the extended eigenproblem. In section 5 we compare our subspace algorithms with the method of [5].

3   Subspace Algorithms for Solving Extended Eigenproblems

Generalized Davidson algorithms are subspace methods for large symmetric eigenvalue problem. They solve the eigenvalue problem $Ay = \mu y$ by constructing an orthonormal basis $V_k = \{v_1 \ldots, v_k\}$ at each kth iteration step and then finding an approximation for the eigenvector y of A by using a vector y_k from the subspace spanned by V_k. Previously, we have developed a Davidson-type algorithm for solving the $Ay = \lambda y$ model where A is the graph Laplacian matrix [6,7]. Now, we present Davidson-type algorithms to solve the extended eigenproblems.

Our problem is to find the smallest μ and the corresponding y such that $Ay = \mu y + g$. The corresponding optimization problem being

$$
\begin{aligned}
&\text{Minimize } f(y) = y^T A y - 2h^T y \\
&\text{Subject to} \\
&\text{(a) } s^T y = 0 \\
&\text{(b) } y^T y = \omega_v.
\end{aligned} \tag{7}
$$

The solution of this problem is given by the solution of Equation 4. Notice that if we represent y in a subspace V with $y = Vz$, we can rewrite (7) as

$$
\begin{aligned}
&\text{Minimize } f(z) = z^T V^T A V z - 2h^T V z \\
&\text{Subject to} \\
&\text{(a) } s^T V z = 0 \\
&\text{(b) } z^T V^T V z = \omega_v.
\end{aligned} \tag{8}
$$

Again, the solution of this problem can be obtained from the Lagrangian

$$
\tilde{F}(z, \lambda, \mu) = z^T V^T A V z - 2h^T V z + \lambda s^T V z - \mu(\omega_v - z^T V^T V z) = 0. \tag{9}
$$

This leads to

$$
\begin{aligned}
\nabla_z \tilde{F}(z, \lambda, \mu) &= V^T 2 A V z - V^T 2h + \lambda V^T s - 2\mu V^T V z \\
&= V^T [2 A V z - 2h + \lambda s - 2\mu V z] \\
&= V^T [2 A y - 2h + \lambda s - 2\mu y] \\
&= V^T \nabla_y F(y, \lambda, \mu).
\end{aligned} \tag{10}
$$

Consequently, The subspace problem (8) can be used to estimate the solution of the original problem (7). That gives us an opportunity to develop Davidson-type algorithms for the extended eigenproblem.

4   Algorithm

The problem is to find the smallest μ and the corresponding y such that

$$
\begin{aligned}
&Ay = \mu y + g \\
&\text{Subject to} \\
&\text{(a) } s^T y = 0 \\
&\text{(b) } y^T y = \omega_v.
\end{aligned} \tag{11}
$$

The problem of finding the smallest μ and the corresponding y is not simply a matter of a linear system eigensolver because the norm constraint on y must also be satisfied. An iterative approach can be used. That is we guess a value of μ, solve for y and check whether $y^T y = \omega_v$ and adjust our guess for μ accordingly.

The existing approach of [5] is described here. We multiply $Ay = \mu y + g$ by $Q_j^T \in \mathbb{R}^{j \times n}$ (the transpose of the Lanczos basis at step j) and look for a solution of form $y = Q_j z$, obtaining

$$
Q_j^T A Q_j z = \mu Q_j^T Q_j z + Q_j^T g. \tag{12}
$$

From the standard Lanczos process we have after j steps that $Q_j^T A Q_j = T_j + Q_j^T r_j e_j^T$ where T_j is a tridiagonal matrix, Q_j is orthogonal, r_j is the residual vector and $e_j = (0, \ldots, 0, 1, 0, \ldots, 0)^T$ where 1 is at the jth position. Hence equation (12) becomes $(T_j + Q_j^T r_j e_j^T)z = \mu z + Q_j^T g$. Since in exact arithmetic $\|Q_j^T r_j e_j^T\|$ is zero, we can solve

$$
(T_j - \mu I)z = Q_j^T g. \tag{13}
$$

Substituting $\tilde{g} = Q_j^T g$ we obtain

$$
(T_j - \mu I)z = \tilde{g}. \tag{14}
$$

where we seek the pair (μ, v) corresponding to the left-most value of μ such that the applicable constraints are satisfied.

Now, the constraint equations are considered. Using $y = Q_j z$ the norm constraint on z becomes $\omega_v = y^T y = z^T Q_j^T Q_j z = z^T z = \omega_v$. Q_j can be constructed in such a way that all of its columns are orthogonal to s. The problem is then to find the smallest μ and the corresponding z

$$
\begin{aligned}
&(T_j - \mu I)z = \tilde{g} \\
&\text{Subject to} \\
&\text{(a) } \|z\|_2^2 = \omega_v,
\end{aligned} \tag{15}
$$

which can be rewritten as

$$
\|(T_j - \mu I)^{-1}\tilde{g}\|_2^2 = \omega_v. \tag{16}
$$

since T_j is tridiagonal, the cost of calculating $(T_j - \mu I)^{-1}\tilde{g}$ is $O(k)$. In [5], equation (16) was solved using a bisection method.

However, there are a number of numerical difficulties with the Lanczos method, which are due to roundoff error [3, §9.2.2]. One of the main difficulties is that the Q_j matrices may become highly non-orthogonal for large j. Usually, some sort of re-orthogonalization scheme is needed [3, §§9.2.3–9.2.4], or else a method of removing "ghost" eigenvalues [3, §9.2.5]. Either approach results in code that is more reliable but less efficient.

We need to choose a method for solving the problem on a subspace. Instead of Lanczos we use an eigendecomposition approach which is described next. We consider the eigen decomposition of $A = Q\Lambda Q^T$, where Q is orthogonal. Substituting this into Equation (12), we have

$$(A - \mu I)^{-1}g = Q(\Lambda - \mu I)^{-1}Q^T g, \tag{17}$$

$$||(A - \mu I)^{-1}g||_2^2 = ||Q(\Lambda - \mu I)^{-1}Q^T g||_2^2 = ||(\Lambda - \mu I)^{-1}\hat{g}||_2^2, \tag{18}$$

where $\hat{g} = Q^T g$. The inversion of $(\Lambda - \mu I)$ has cost $O(k)$, where k is the dimension of the system. For each A only one eigendecomposition is needed, which costs $O(k^3)$. This makes it a competitive approach that is also very reliable. Since our minimization problem is

$$z = (\Lambda - \mu I)^{-1}\hat{g}$$
$$\text{Subject to } ||z||_2^2 = \omega_v, \tag{19}$$

we can rewrite it as

$$||(\Lambda - \mu I)^{-1}\hat{g}||_2^2 = \omega_v. \tag{20}$$

Equation (20) can be solved using a bisection method, but Brent's method gives better convergence rates [2].

Below we present the main steps of the Davidson algorithm for the extended eigenproblem.

1. Define V_1 as $V_1 = [v_1]$.
2. Find the extended eigenproblem solution (μ, y) on the subspace. The projected matrix on the subspace is $S_j = V_j^T A V_j$ where V_j is the current orthogonal basis.
3. Use Ritz values $(u_j = V_j y_j)$ and vectors to estimate residual: $r_j = Au_j - \mu_j u_j - g = W_j y_j - \mu_j V_j y_j - g$, where $W_j = AV_j$.
4. Solve $t_j = M_j r_j$ approximately. (This corresponds to preconditioning the residual.)
5. Orthonormalize t_j against the current orthogonal basis V_j. Append the orthonormalized vector to V_j to give V_{j+1}.

5   Numerical Results

We compared our new subspace algorithm with the Lanczos based algorithm previously used for the extended eigenproblem. Our preconditioner was a multigrid algorithm and research is under development exploring the use of other multilevel preconditioners for this algorithm. The two algorithms have been implemented and run on a HP VISUALIZE Model C240 workstation, with a 236MHz PA-8200 processor and 512MB RAM. Both algorithms were run with square matrices of various sizes. Figure 2 compares the observed running timings. From this graph we can see that the subspace algorithm takes less than the Lanczos based algorithm for all the test cases, specially as the problem sizes get bigger. In addition, there are other problems inherent to the Lanczos algorithm that will be avoided with our new preconditioned subspace algorithm.

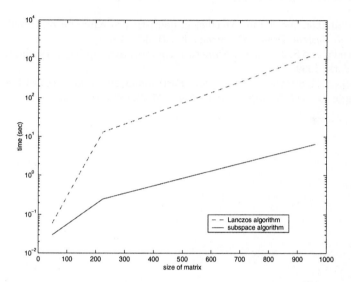

Fig. 2. Timings comparing the subspace implementation of extended eigenproblem against the Lanczos implementation of extended eigenproblem.

References

1. C. J. Alpert, T. F. Chan, D. J.-H. Huang, A. B. Kahng, I. L. Markov, P. Mulet, and K. Yan. Faster Minimization of Linear Wirelength for Global Placement. In *Proc. Intl. Symposium on Physical Design*, 1997.
2. R. P. Brent. *Algorithms for Minimization without Derivatives*. Automatic Computation. Prentice–Hall, Englewood Cliffs, NJ, 1973.
3. G. Golub and C. Van Loan. *Matrix Computations*. Johns Hopkins University Press, Baltimore, Maryland, 3rd edition, 1996.
4. B. Hendrickson and R. Leland. The Chaco user's guide, version 2.0. Technical Report SAND-95-2344, Sandia National Laboratories, 1995.
5. B. Hendrickson, R. Leland, and R. V. Driessche. Enhancing data locality by using terminal propagation. In *Proc. 29th Hawaii Intl. Conf. System Science*, volume 16, 1996.

6. M. Holzrichter and S. Oliveira. A graph based Davidson algorithm for the graph partitioning problem. *International Journal of Foundations of Computer Science*, 10:225–246, 1999.

7. M. Holzrichter and S. Oliveira. A graph based method for generating the Fiedler vector of irregular problems. In *Lecture Notes in Computer Science 1586, Parallel and Distributed Processing, 11 IPPS/SPDP'99*, pages 978–985, 1999.

8. G. Karypis and V. Kumar. METIS: Unstructured graph partitioning and sparse matrix ordering system Version 2.0. Technical report, Department of Computer Science, University of Minnesota, 1995.

9. George Karypis and Vipin Kumar. A fast and high quality multilevel scheme for partitioning irregular graphs. *SIAM J. Sci. Comput.*, 20(1):359–392 (electronic), 1999.

10. B. Kerninghan and S. Lin. An efficient heuristic procedure for partitioning graphs. *The Bell System Technical Journal*, 49:291–307, 1970.

11. J. Nocedal and S. J. Wright. *Numerical Optimization*. Springer, Berlin, Heidelberg, New York, 1999.

12. A. Pothen, H. Simon, and K. P. Liou. Partitioning sparse matrices with eigenvector of graphs. *SIAM J. Matrix. Anal. Appl.*, 11(3):430–452, 1990.

An Integrated Decomposition and Partitioning Approach for Irregular Block-Structured Applications

Jarmo Rantakokko

Dept. of Computer Science and Engineering, University of California, San Diego
9500 Gilman Dr, La Jolla, CA 92093-0114, USA
jarmo@cs.ucsd.edu

Abstract. We present an integrated domain decomposition and data partitioning approach for irregular block-structured methods in scientific computing. We have demonstrated the method for an application related to Ocean modeling. Our approach gives better results than other methods that we have found in the literature. We have compared the domain decomposition part with the Berger-Rigoutsos point clustering algorithm and the partitioning part with a multilevel method used in Metis, a bin packing method, and an inverse space filling curve algorithm. The partitions produced with our approach give lower run times, i.e. higher parallel efficiency, for our application than all the other partitions we have tried. Moreover, the integrated approach gives a possibility to further improve the parallel efficiency compared to doing the decomposition and partitioning in two separate steps.

1 Introduction

Irregular block decompositions are commonly used in scientific applications where partial differential equations are solved numerically. For example, in structured multiblock methods the computational grid is decomposed into blocks and the blocks are fitted around or within an object. The blocks may then be of different sizes and connected to each other in an irregular fashion. Similarly, in structured adaptive mesh refinement techniques we may have irregular regions with high error. The flagged points, i.e. the high error points, are clustered together and a new refined level of grids with an irregular block decomposition is created.

The problem we focus on in this paper originates from Ocean modeling and relates to both application domains discussed above. Here, we have an irregular geometry of water points but we still use a rectangular structured grid covering both land and water. The land points are then masked out in the computations. Still, the inactive points will consume both processor power and memory. It is then necessary to have an irregular block decomposition to cover the active points as efficiently as possible, minimizing the overheads associated with the inactive points. The strategy to use a structured grid and then mask out points to handle irregular boundaries is used, e.g. in HIROMB [12], MICOM [3],

J. Rolim et al. (Eds.): IPDPS 2000 Workshops, LNCS 1800, pp. 485-496, 2000.

and OCCAM[1] [5], but is not limited to ocean modeling applications. We have
seen the same techniques used in oil/ground water flow simulations [15] and
in electro-magnetics computations. It is a simple and yet efficient method to
handle irregular geometries, especially if it is combined with an irregular block
decomposition minimizing the fraction of inactive points in the blocks.

Our emphasis is to create an efficient partitioning and block decomposition
method for the irregularly structured applications discussed above. The method
should minimize the number of inactive points in the blocks, give a good load
balance, a small number of communication dependencies, while keeping the total
number of blocks small. This paper presents an integrated block decomposition
and partitioning approach and compares it with other techniques found in the
literature.

2   The partitioning approach

2.1   Overview

Our algorithm consist of three steps or phases. The idea is to first cluster the
water points in "dense" blocks. We strive to get a block efficiency, i.e. the fraction
of active points, above a given threshold. The next step is to distribute the
blocks onto the processors with a small load imbalance ratio and a low number
of inter-processor dependencies. The final step is to try to merge blocks on the
same processor into larger rectangular blocks. The emphasis with the last step
is to reduce the total number of blocks. There is a small cost associated with
each block, e.g. starting up loops, calling functions, and updating the block
boundaries. The two first steps will be explained in detail below.

2.2   The clustering algorithm

We want to cluster the active points in rectangular blocks with a high block
efficiency. To begin with, we create a minimal bounding box surrounding all
active points. We split this block recursively and create new bounding boxes
around the resulting blocks until the block efficiency is above a given threshold.

In each recursion step, we choose only to split those blocks with a fraction
of active points below a given threshold. This threshold can either be chosen as
a fixed rate or a relative rate. Fixed rate means that we choose blocks with a
fraction of active points below $X\%$ (X is a constant). Relative rate means that
we choose blocks that has relatively many inactive points compared to the other
blocks, i.e. we split only those blocks with the most inactive points. For example,
we can choose to split all blocks with $Y\%$ more inactive points than in average.

In the splitting phase, we have a search window around the center of the
block and do the split across the longest side where we cut through the least
number of active points. In other words, we cut along the grid line, within the
window, with the smallest signature S_i. The signature S_i for grid line i is defined

---

[1] OCCAM can also handle unstructured partition shapes.

as the sum of the active points along the line (surface in 3D), i.e. $S_i = \sum_j f_{ij}$ where $f_{ij} = 1$ for water points and $f_{ij} = 0$ for land points. See Figure 1. The purpose of the window is twofold: it limits the search space and avoids cuts into too thin blocks.

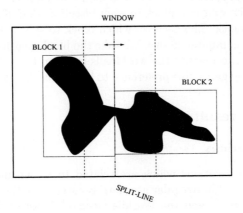

Fig. 1. We define a search window around the center and find the line that cuts through the least number of active points. Then, we set up minimal bounding boxes around the resulting blocks and repeat the algorithm recursively.

2.3   The distribution method

The second phase in our algorithm is to distribute the blocks onto the processors. We treat this in terms of a graph partitioning problem. The blocks constitute the vertices and the data dependencies between the blocks constitute the edges. We choose the vertex-weights proportional to the workload on the blocks and the edge-weights proportional to the communication volume between the corresponding blocks.

The Recursive Spectral Bisection method (RSB) [14] uses a heuristic to minimize the global edge-cut in the partitioning of the graph. The partitions can still be improved in the fine details with a local refinement method, e.g. Greedy, Kernighan-Lin [4]. We have implemented the RSB-method combined with a Greedy-like local refinement method [13]. In each bisection step we sort the vertices according to the Fiedler vector into two parts with approximately equal workloads. Then, we improve this assignment by swapping vertices between the two partitions using the local refinement method. The refinement method exploits gains computed from the edges in the graph. The gain of a vertex v to move from partition A to partition B, (A and B neighbors), is defined as

$$g(v, A, B) = \sum_{w \in B} E_w - \sum_{w \in A} I_w,$$

where E_w (external degree) is the edge-weight between vertex v in partition A and vertex w in partition B. I_w (internal degree) is the edge-weight between vertex v and vertex w, both in partition A.

If the blocks are large or we have a relatively small number of blocks compared to the number of processors, it will be difficult to balance the workload. Our remedy is to further split one block in each bisection step, if necessary. We can not just split any block. The decision which block to split is based on the block sizes and the gains computed above. We take the block that has the highest gain and is large enough to balance the workloads between the two partitions in the bisection step. Large gains give preference to blocks on the partition boundaries.

3   Numerical results

3.1   The applications

We have chosen applications which are related to Ocean modeling. We solve Poisson's equation within irregular geometries covered by the shape of Lake Superior[2], the Baltic Sea[3], and the World's Oceans[4]. Most of the runs are made for the Lake Superior case while the other geometries are only used to complement the results. These applications have a general interest in oceanography even though we solve a 2D equation with a constant forcing function. In Ocean modeling there is a part in the solver (the Barotropic part) where an elliptic equation is solved for all surface points.

We solve the 2D Poisson's Equation with a constant forcing function and with homogeneous Dirichlet boundary conditions.

$$\begin{cases} U_{xx} + U_{yy} = -1 & \text{in } \Omega \\ U = 0 & \text{at } \partial\Omega \end{cases}$$

The equation is discretized with the centered five point finite difference stencil and the corresponding system of linear equations (unsymmetric) is solved with a Krylov subspace method, the Conjugate-Gradient Squared method. Inside the loops, in the computations, we have a condition for each grid point checking if it is land or water. If it is a water point we apply the finite difference stencil otherwise we skip the computations for that grid point. The Baltic sea grid contains initially 25% water points, the Lake Superior grid contains 44%, and the World contains 66% water points.

3.2   Comparison with the Berger-Rigoutsos algorithm

Our first task is to evaluate if the block decomposition in the first step is efficient compared to other related work. To do this, we compare it with the Berger-Rigoutsos point clustering algorithm from structured adaptive mesh refinement

---

[2] The grid was created from boundary data provided by R. Banks, UCSD
[3] The data was provided by the Swedish Meteorological and Hydrological Institute
[4] The computational grid was extracted from an image provided by B. Huffaker, SDSC

applications [2]. This algorithm is very commonly used and can be considered as a standard benchmark in the area. Another possible candidate for a comparison could be the less complex algorithm by James Quirk [11], but it has no explicit criteria for optimizing the block efficiency.

The Berger-Rigoutsos algorithm is similar to ours, it is only the splitting criteria that differs. Here, the first criterion is to look for zero signatures $S_i = 0$ and make the cut along the most central of these. If all signatures are non-zero then a second criterion is to look for zero-crossings where the Laplacian of the signatures, $\Delta_i = S_{i+1} - 2S_i + S_{i-1}$, changes sign. The best zero-crossing is the one whose magnitude $Z_{i+1/2} = |\Delta_{i+1} - \Delta_i|$ is largest, and this index is chosen as the cutting line. In our algorithm we only have one cutting criterion. We find the smallest signature and make the cut at the corresponding grid line. Our algorithm does not include the arithmetic of computing a second derivative and thus is faster and simpler than the Berger-Rigoutsos method.

Unfortunately, the Berger-Rigoutsos algorithm is not well suited for our applications. Their algorithm is based on edge-detection techniques. The boundary (coast) line is too irregular and this produces only noise in the second derivative of the signature. The cutting lines become randomly placed and the algorithm does not maximize the fraction of active points in the resulting blocks. A modification is then to damp out the high frequency oscillations in the signature. This improves the block efficiency somewhat but increases the complexity of the algorithm considerably. Our clustering algorithm still gives a higher fraction of active points with the same number of blocks, or, for a given fraction of active points we get fewer blocks and a lower run-time.

We have made experiments running the solver on one processor with the initial decomposition produced either with our algorithm or with the Berger-Rigoutsos algorithm (see Figure 2). The differences in the run-times, without the decomposition times, are quite small as the overhead in processing a larger number of blocks (for a given fraction of active points) is small. Note, this overhead depends very much on the solver but also on the grid sizes and the memory hierarchy of the computer. In adaptive mesh refinement applications it is more important to reduce the number of "low-error" points in the blocks as all points are treated equally giving a high cost for these "non-necessary" points.

3.3   Parallel performance

Here, we also include the partitioning phase and compare the parallel run-times from the solver with data partitions produced with other techniques. The simplest possible data decomposition is a regular block-block decomposition of the entire domain, not reducing the number of inactive points. This will give a poor load balance as some blocks will cover more water points than others. A remedy is then to use the Recursive Coordinate Bisection method [1] to get blocks with a balanced workload. But, this introduces an irregular and imbalanced communication pattern increasing the total communication time.

Figure 3 shows the results from partitioning and running the Lake Superior application on IBM SP2 and ASCI Blue Pacific. We get similar results from the

Baltic Sea and the World's Oceans applications. For the Baltic Sea geometry the irregular decomposition gives an improvement of almost 50% (half the run-time) compared to the two other partitions and for the World geometry a $10 - 15\%$ improvement. The solver runs faster with our irregular domain decomposition. This is mostly due to the reduced number of inactive (land) points. Another observation is that the run-times using the irregular decomposition scale well with the number of processors.

The comparison above shows that reducing the number inactive points gives a better efficiency. To further explore the properties of our algorithm, we have made another set of experiments where we compare our distribution algorithm by replacing it in step 2 with other partitioning methods (step 1 and 3 are the same for all methods). First, we find the block decomposition that gives the best serial run-time (step 1) and then we distribute the blocks with Metis [7], a bin packing method [6], and an inverse space filling curve algorithm [9], respectively (step 2). Finally, we merge blocks within the same partitions (step 3). In Metis we use their recursive multilevel method *pmetis* with default settings. As a bin-packing method we use the *greedy* algorithm suggested in [6] but ignore all neighbor relations. The space filling curve we have chosen uses Morton ordering of the blocks with the center point as a reference. The load balancing is done with the greedy algorithm above but now following the ordering of the blocks.

Our distribution method outperforms all these other methods (see Figure 4). After step 1, we have 130 blocks to distribute to the processors. The granularity is too coarse and all the other methods fail to give a good load balance above 8 processors. We get an almost perfect load balance for all processor configurations due to the extra split in each recursion step. We can even improve our method by decreasing the number of blocks in the initial decomposition, i.e. allowing for a higher fraction of inactive points (*Our** in Figure 4), without sacrificing the load balance. This enhancement is not directly applicable to the other methods as it would be even more difficult to get a good load balance with a smaller number of blocks.

The problem with the other methods is that we will have some very large blocks where there are wide areas of water and this prohibits the algorithms from producing a well balanced workload. An obvious improvement is to split these larger blocks into smaller pieces. We have then included another splitting step in phase one that takes blocks larger than average size and splits them uniformly into four pieces. We have applied this for Metis (labeled *Metis** in Figure 4) and it improves the parallel efficiency considerably, but falls behind our method. This is due to that it makes more splits and introduces more blocks than our approach, which gives a minimal increase in the number of blocks in the second phase to maintain the load balance.

If we want to improve the parallel efficiency even more, as above with *Our**, by trading off the fraction of active points against the number of blocks, we have to search a 2-dimensional parameter space to find the optimal domain decomposition (x-number of splits to reduce the number of inactive points and y-number of splits to even out the block sizes). The merge step, phase 3 where we

Table 1. Parallel run-times IBM SP2, Lake Superior, 16 processors. The different columns represents the number of split steps increasing the fraction of active points. The rows corresponds to the number of steps to split large blocks, making the load balancing easier. The resulting distribution of the blocks is done with Metis. The best run-time corresponds to a parallel efficiency of 72%. The results show that it is difficult with this strategy to find the optimal partitioning, we have to search a 2-dimensional parameter space.

	Better Block Efficiency →							
	5	6	7	8	9	10	11	12
0					3.294	2.706	1.677	1.500
1				1.300	0.770	0.783	**0.643**	0.773
2			1.363	0.754	0.649	**0.660**	0.653	**0.739**
3		0.864	0.732	0.661	**0.597**	0.668	0.701	0.835
4	0.910	0.823	0.646	0.666	0.604	0.681	0.715	0.813
5	0.751	0.678	**0.619**	0.682	0.649	0.739	0.854	
6	0.682	**0.662**	0.650	**0.625**	0.635	0.753		
7	0.711	0.725	0.623	0.740	0.762			
8	0.700	0.671	0.680	0.711				
9	**0.677**	0.682	0.719					
10	0.709	0.669						
11	0.732							

Better Load Balance ↓

combine small blocks assigned to the same processor, becomes very important for this improvement strategy. The results would otherwise not be competitive at all, the number of blocks would then be too high.

As an example, we have applied this strategy distributing the resulting blocks with Metis onto 16 processors, see Table 1. The table shows that we have to do lots of runs to find the minimal running time, making the partitioning problem very hard. We have managed to increase the efficiency up to 72%, computed for the best run-time in the table. However, our method (Our^*) still gives the best result, with 78% efficiency. We get a better load balance with fewer blocks in our method.

4   Conclusions

We have here developed a new integrated block decomposition and partitioning method for irregularly structured problems arising in Ocean modeling. The approach is not only limited to Ocean modeling but is also suitable for other similar applications, e.g. structured adaptive mesh refinement applications solving time-independent elliptic equations. Here, the parallelism is within a level rather than between the levels, [8]. Partitioning the data of a grid level is then very similar to what we are doing here. It remains to see if our algorithm is competitive for dynamic problems as well. Some preliminary results show that phase 1, the regridding part, dominates the total partitioning and decomposition time. The partitioning time (phase 2) accounts only for a few percent of this.

We have compared our domain decomposition method with the Berger-Rigoutsos grid clustering algorithm and our method gives better results for the applications here. Our approach is simpler and faster but we still get a higher block efficiency, i.e. a higher fraction of active points in the blocks. We have shown that is not necessary to introduce the complexity of the Berger-Rigoutsos algorithm to get comparable or even better results.

We have also compared the distribution method with other algorithms found in the literature, i.e. Metis, an inverse space filling curve, and a bin-packing method. Our method outperforms the other algorithms. We get a better load balance with fewer blocks and then less serial overhead in the solver, resulting in a shorter parallel run-time. In addition, the integrated decomposition and partitioning approach makes it possible[5] to even more increase the parallel efficiency by allowing for a higher fraction of inactive points but less blocks assigned to each processor as we use larger machine configurations.

It is possible to introduce the combined block-splitting and partitioning ideas with other methods than the recursive spectral bisection method. For example, the space filling curve and the bin-packing methods can easily be modified to include an extra step to split blocks to maintain the load balance. The space filling curve can locally be lengthen from splitting a block. A similar approach has been taken in [10]. For a bin-packing method we can take the highest bin, split one block, and redistribute the parts. Then, repeat this with the currently highest bin until we have a balanced workload. For a pure graph partitioning method, like Metis, this enhancement is not directly applicable. Splitting a node introduces new edges and this information is not available in the initial graph, the graph has to be rebuilt from the new block structure.

Finally, we like to mention that our irregular block decomposition and partitioning algorithm has successfully been used by the Swedish Meteorological and Hydrological Institute to partition the Baltic Sea [16, 17]. As a result, it has been possible to increase the grid resolution in the forecast model from 3 Nautic miles on a Cray C90 down to 1 Nautic mile on a Cray T3E and still solve the problem within the same time limits.

Acknowledgments

The work in this paper was financially supported by the *Swedish Foundation for International Cooperation in Research and Higher Education (STINT)* and the *US Department of Energy* by Lawrence Livermore National Laboratory under contract W07405-Eng-48, ONR contract N00014-93-1-0152. We would like to thank Scott Baden (UCSD) and Michael Thuné (Uppsala University) for advice on how to improve this paper.

References

1. M.J. Berger, S. Bokhari, *A partitioning strategy for non-uniform problems on multiprocessors*, ICASE Report No. 85-55, NASA Langely Research Center, Hampton VA, 1985.

---

[5] In adaptive mesh refinement one may not want to change the initial mesh generation as this could affect the solution making it very difficult to ensure consistent behavior.

2. M.J. Berger, I. Rigoutsos, *An algorithm for point clustering and grid generation*, IEEE Transactions on Systems, Man and Cybernetics, 21(1991), pp. 1278-1286.

3. R. Bleck, S. Dean, M. O'Keefe, A. Sawdey, *A comparison of data-parallel and message-passing versions of the Miami Isopycnic Coordinate Ocean Model (MI-COM)*, Parallel Computing, 21:1695-1720, 1995.

4. C.M. Fiduccia, R.M. Mattheyses, *A Linear-Time Heuristic for Improving Network Partitions*, in Proceedings of 19th IEEE Design Automation Conference, IEEE, pp. 175-181, 1982.

5. G.S. Gwilliam, *The OCCAM Global Ocean Model*, in Proceedings of the sixth ECMWF workshop on the use of parallel processors in meteorology, Reading, UK, November 21-25, 1994.

6. M.A. Iqbal, J.H. Saltz, S.H. Bohkari, *Performance tradeoffs in static and dynamic load balancing strategies*, Technical Report 86-13, ICASE, NASA Langley Research Center, Hampton, VA, 1986.

7. G. Karypis, V. Kumar, *Metis: Unstructured Graph Partitioning and Sparse Matrix Ordering System*, Technical Report, University of Minnesota, Department of Computer Science, Minneapolis, 1995.

8. S. Kohn, *A Parallel Software Infrastructure for Dynamic Block-Irregular Scientific Calculations*, UCSD CSE Dept. Tech. Rep. CS95-429, (Ph.D. Dissertation), Jun. 1995.

9. C. Ou, S. Ranka, *Parallel remapping algorithms for adaptive problems*, Technical Report, Center for Research on Parallel Computation, Rice University, 1994.

10. M. Parashar, J. Brown, *On Partitioning Dynamic Adaptive Grid Hierarchies*, Technical Report, Department of Computer Science, University of Texas, Austin, 1996.

11. J. Quirk, *A parallel adaptive grid algorithm for computational shock hydrodynamics*, Applied Numerical Mathematics, 20:427–453, Elsevier, 1996.

12. J. Rantakokko, *A framework for partitioning structured grids with inhomogeneous workload*, Parallel Algorithms and Applications, Vol 13, pp:135-151, 1998.

13. J. Rantakokko, *Data Partitioning Methods and Parallel Block-Oriented PDE Solvers*, Ph.D. thesis, Department of Scientific Computing, Uppsala University, Sweden, 1998.

14. H.D. Simon, *Partitioning of unstructured problems for parallel processing*, Computing Systems in Engineering, 2:135–148, 1991.

15. M.F. Wheeler, T. Arbogast, S. Bryant, J. Eaton, Q. Lu, M. Peszynska, *A Parallel Multiblock/Multidomain Approach for Reservoir Simulation*, in proceedings of the 1999 SPE Reservoir Simulation Symposium, Houston, Texas, 14-17 February, 1999.

16. T. Wilhelmsson, J. Schüle, J. Rantakokko, L. Funkquist, *Increasing Resolution and Forecast Length with a Parallel Ocean Model*, in proceedings of the Second EuroGOOS International Conference, Rome, Italy, March 11-13, 1999.

17. T. Wilhelmsson, J. Schüle, *Running an Operational Baltic Sea Model on the T3E*, in proceedings of the Fifth European SGI/Cray MPP Workshop, 1999.

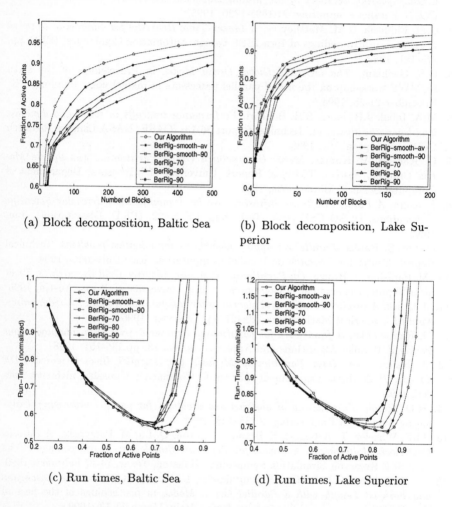

(a) Block decomposition, Baltic Sea

(b) Block decomposition, Lake Superior

(c) Run times, Baltic Sea

(d) Run times, Lake Superior

Fig. 2. Comparison of the Berger-Rigoutsos algorithm with our algorithm. We have made five different runs with the Berger-Rigoutsos algorithm. Three runs where we only split blocks below a fixed fraction of active points (70%, 80%, and 90%) and two runs with the modified damped version (*smooth*), using a fixed threshold (90%) of active points and a relative threshold of active points (*av*). With the relative threshold we only split those blocks that has more than the average number of inactive points per block in each recursion step. This criterion is also used in our algorithm.

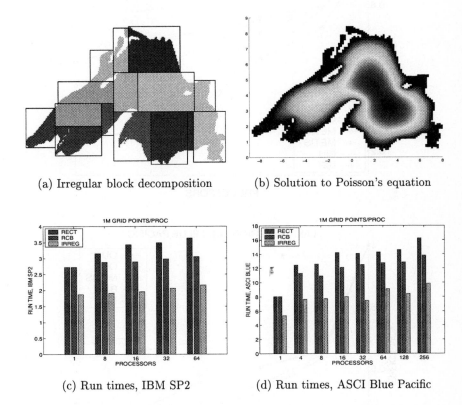

(a) Irregular block decomposition   (b) Solution to Poisson's equation

(c) Run times, IBM SP2   (d) Run times, ASCI Blue Pacific

Fig. 3. Parallel results from the Lake Superior application. (a) An example of an irregular block decomposition produced with our algorithm for 8 processors. The different colors indicate processor assignment. (b) Solution to Poisson's equation with constant forcing function and homogeneous Dirichlet boundary conditions. (c) Parallel run-times on IBM SP2 scaling the problem size linearly with the number of processors. Timings for a fixed number of iterations. We compare three different partitioning strategies, a regular block-block decomposition (*RECT*), the Recursive Coordinate bisection method (*RCB*), and our irregular block decomposition (*IRREG*). (d) The same as (c) but on ASCI Blue Pacific. The irregular block decomposition improves the performance considerably and scales well with the number of processors.

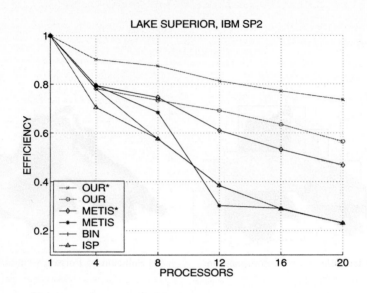

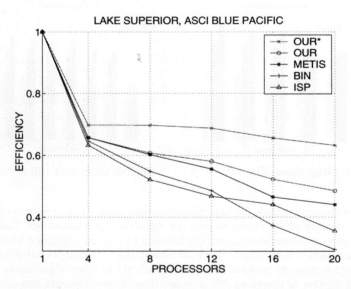

Fig. 4. Parallel Efficiency ($E_p = T_1/pT_p$ where T_1 = serial run time, T_p = parallel run time, and p = number of processors). We solve for the Lake Superior geometry with constant problem size 2000 × 2000 grid points. We take the block decomposition that gives the best run-time T_1 on a single processor and distribute the blocks with *Our* standard method, the *Metis* partitioning software, a *Bin* packing method disregarding neighbor relations, and an inverse space filling curve (*ISP*) method. We also have a second variant of our method, *Our**, where we trade-off the fraction of active points for a smaller number of blocks. The *Metis** method has an extra step to split up large blocks so that the algorithm can produce a better load balance.

Ordering Unstructured Meshes for Sparse Matrix Computations on Leading Parallel Systems

Leonid Oliker[1], Xiaoye Li[1], Gerd Heber[2], and Rupak Biswas[3]

[1] NERSC, Lawrence Berkeley National Laboratory, Berkeley, CA 94720, USA
{oliker,xiaoye}@nersc.gov
[2] Cornell Theory Center, Rhodes Hall, Cornell University, Ithaca, NY 14853, USA
heber@tc.cornell.edu
[3] MRJ/NASA Ames Research Center, Moffett Field, CA 94035, USA
rbiswas@nas.nasa.gov

Abstract. Computer simulations of realistic applications usually require solving a set of non-linear partial differential equations (PDEs) over a finite region. The process of obtaining numerical solutions to the governing PDEs involves solving large sparse linear or eigen systems over the unstructured meshes that model the underlying physical objects. These systems are often solved iteratively, where the sparse matrix-vector multiply (SPMV) is the most expensive operation within each iteration. In this paper, we focus on the efficiency of SPMV using various ordering/partitioning algorithms. We examine different implementations using three leading programming paradigms and architectures. Results show that ordering greatly improves performance, and that cache reuse can be more important than reducing communication. However, a multithreaded implementation indicates that ordering and partitioning are not required on the Tera MTA to obtain an efficient and scalable SPMV.

1 Introduction

The ability of computers to solve hitherto intractable problems and simulate complex processes using mathematical models makes them an indispensable part of modern science and engineering. Computer simulations of large-scale realistic applications usually require solving a set of non-linear partial differential equations (PDEs) over a finite region. For example, one thrust area in the U.S. DoE Grand Challenge projects is to design future accelerators such as the Spallation Neutron Source (SNS). Our colleagues at the Stanford Linear Accelerator Center (SLAC) need to model complex RFQ cavities with large aspect ratios [6]. Unstructured grids are currently used to resolve the small features in a large computational domain; dynamic mesh adaptation will be added in the future for additional efficiency.

The PDEs for electromagnetics are discretized by the FEM method, which leads to a generalized eigenvalue problem $Kx = \lambda Mx$, where K and M are the stiffness and mass matrices, and are very sparse. In a typical cavity model, the

J. Rolim et al. (Eds.): IPDPS 2000 Workshops, LNCS 1800, pp. 497-503, 2000.

number of degrees of freedom is about one million. For such large eigenproblems, direct solution techniques quickly reach the memory limits. Instead, the most widely-used methods are Krylov subspace methods, such as Lanczos or Jacobi-Davidson. In all the Krylov-based algorithms, sparse matrix-vector multiplication (SPMV) must be performed repeatedly. Therefore, the efficiency of SPMV usually determines the eigensolver speed. SPMV is also one of the most heavily used kernels in large-scale numerical simulations.

On uniprocessor machines, numerical solutions of such complex, real-life problems can be extremely time-consuming, a fact driving the development of increasingly powerful parallel (multiprocessor) supercomputers. The unstructured, dynamic nature of many systems worth simulating, however, makes their efficient parallel implementation a daunting task. Furthermore, modern computer architectures, based on deep memory hierarchies, show acceptable performance only if users care about the proper distribution and placement of their data [1]. Single-processor performance crucially depends on the exploitation of locality, and parallel performance degrades significantly if inadequate partitioning of data causes excessive communication and/or data migration. The traditional approach would be to use a partitioning tool like MeTiS [5], and to post-process the resulting partitions with an enumeration strategy for enhanced locality. Although, optimizations for partitioning and locality may then be treated as separate problems, real applications tend to show a rather intricate interplay of both.

In this paper, we investigate the efficiency of SPMV using various ordering/partitioning algorithms. We examine different implementations using different programming paradigms and architectures. Results on state-of-the-art parallel supercomputers show that ordering greatly improves performance, and that cache reuse can be more important than reducing communication. However, a multithreaded implementation indicates that ordering and partitioning are not required on the Tera MTA to obtain an efficient and scalable SPMV.

2   Partitioning and Linearization

Space-filling curves have been demonstrated to be an elegant and unified linearization approach for certain problems in N-body and FEM simulations. The linearization of a higher-dimensional spatial structure, i.e. its mapping onto a one-dimensional hyperspace, is exploited in two ways: First, the "locality preserving" nature of the construction fits elegantly into a given memory hierarchy, and second, the partitioning of a contiguous linear object is trivial. For our experiments, we pursued both strategies with some modifications. In the following, we briefly describe the two enumeration techniques and the general-purpose graph partitioner which were used. In future, we plan to integrate ordering algorithms into the eigensolver at SLAC and evaluate the overall performance gain.

2.1   Cuthill-McKee Algorithms (CM)

The particular enumeration of the vertices in a FEM discretization controls, to a large extent, the sparseness pattern of the resulting stiffness matrix. The bandwidth, or profile, of the matrix, has a significant impact on the efficiency of linear systems and eigensolvers. Cuthill and McKee [2] suggested a simple algorithm based on ideas from graph theory. Starting from a vertex of minimal degree, levels of increasing "distance" from that vertex are first constructed. The enumeration is then performed level-by-level with increasing vertex degree (within each level). Several variations of this method have been suggested, the most popular being reverse Cuthill-McKee (RCM) where the level construction is restarted from a vertex of minimal degree in the final level. The class of CM algorithms are fairly straightforward to implement and largely benefit by operating on a pure graph structure, i.e. the underlying graph is not necessarily derived from a triangular mesh.

2.2   Self-Avoiding Walks (SAW)

These were proposed recently [3] as a mesh-based (as opposed to geometry-based) technique with similar application areas as space-filling curves. A SAW over a triangular mesh is an enumeration of the triangles such that two consecutive triangles (in the SAW) share an edge or a vertex, i.e. there are no jumps in the SAW. It can be shown that walks with more specialized properties exist over arbitrary unstructured meshes, and that there is an algorithm for their construction whose complexity is linear in the number of triangles in the mesh. Furthermore, SAWs are amenable to hierarchical coarsening and refinement, i.e. they have to be rebuilt only in regions where mesh adaptation occurs, and can therefore be easily parallelized. SAW, unlike CM, is not a technique designed specifically for vertex enumeration; thus, it cannot operate on the bare graph structure of a triangular mesh. This implies a higher construction cost for SAWs, but several different vertex enumerations can be derived from a given SAW.

2.3   Graph Partitioning (MeTiS)

Some excellent parallel graph partitioning algorithms have been developed and implemented over the last 5-10 years that are extremely fast while giving good load balance quality and low edge cuts. Perhaps the most popular is MeTiS [5], belonging to the class of multilevel partitioners. MeTiS reduces the size of the graph by collapsing vertices and edges using a heavy edge matching scheme, applies a greedy graph growing algorithm for partitioning the coarsest graph, and then uncoarsens it back using a combination of boundary greedy and Kernighan-Lin refinement to construct a partitioning for the original graph.

3   Experimental Results

To perform a SPMV, $y \leftarrow Ax$, we assume that the nonzeros of matrix A are stored in a compressed sparse row format. The dense vector x is stored sequentially in memory with unit stride. Various numberings of the mesh elements/vertices result in different nonzero patterns of A, which in turn cause different patterns for accessing the entries of x. Moreover, on a distributed-memory machine, they imply different amounts of communication.

Our experimental test mesh consists of a two-dimensional Delaunay triangulation, generated by Triangle [7]. The mesh is shaped like the letter "A", and contains 661,054 vertices and 1,313,099 triangles. The underlying matrix was assembled by assigning a random value to each (row, column) entry corresponding to the vertex endpoints (v_1, v_2) of the edges in the mesh. This simulates a stencil computation where each vertex needs to communicate with its nearest neighbors. The final matrix is extremely sparse containing only 2,635,207 nonzeros. The number of floating point operations required for a SPMV is twice the number of nonzeros (5,270,414 for our test matrix).

3.1   Distributed-Memory Implementation

In our experiments, we use the parallel SPMV routines in Aztec [4] implemented using MPI. The matrix A is partitioned into blocks of rows, with each block assigned to one processor. Two routines are of particular interest: AZ_transform and AZ_matvec_mult. The first initializes the data structures and the communication schedule, while the second performs the matrix-vector multiply. In Table 1, we report the runtimes of these two routines on the 450 MHz Cray T3E at NERSC. The original natural ordering (ORIG) is the slowest and clearly unacceptable on distributed-memory machines. For AZ_matvec_mult, the key kernel routine, RCM is slightly but consistently faster than SAW, while MeTiS requires almost twice the RCM execution time. However, MeTiS, RCM, and SAW, all demonstrate excellent scalability up to the 64 processors that were used for these experiments. The pre-processing times in AZ_transform are more than an order of magnitude larger than the corresponding times for AZ_matvec_mult (except for ORIG where it is two to three orders of magnitude larger).

Table 1. Runtimes (in seconds) for different orderings on the Cray T3E

P	AZ_matvec_mult				AZ_transform			
	ORIG	MeTiS	RCM	SAW	ORIG	MeTiS	RCM	SAW
4	0.228	0.093	0.046	0.054	181.556	0.573	0.419	0.466
8	0.161	0.046	0.024	0.029	207.567	0.296	0.219	0.266
16	0.080	0.022	0.012	0.013	140.269	0.159	0.122	0.177
32	0.046	0.010	0.006	0.006	21.095	0.132	0.179	0.124
64	0.027	0.005	0.003	0.004	6.682	0.142	0.158	0.114

To better understand the various partitioning/ordering algorithms, we have built a simple performance model to predict the parallel runtime. First, using the T3E's hardware performance monitor, we collected the average number of cache misses per processor. This is reported in Table 2. SAW has the fewest number of cache misses. In comparison, RCM, MeTiS, and ORIG, have between two and three times that number. Second, we gathered statistics on the average

Table 2. Locality and communication statistics of `AZ_matvec_mult`

	Avg. Cache Misses (10^6)				Avg. Comm. (64-bit words)				Max. Message Count			
P	ORIG	MeTiS	RCM	SAW	ORIG	MeTiS	RCM	SAW	ORIG	MeTiS	RCM	SAW
4	1.02	0.63	0.71	0.48	85323	115	418	585	3	3	2	3
8	0.62	0.31	0.36	0.26	60425	130	424	524	7	3	2	6
16	0.30	0.16	0.17	0.11	33084	132	432	385	15	4	2	8
32	0.16	0.08	0.10	0.06	17139	112	434	319	31	5	2	9
64	0.10	0.05	0.06	0.03	8717	95	434	241	63	6	2	14

communication volume and the maximum number of messages per processor, both of which are also shown in Table 2. MeTiS transfers the least amount of data, whereas RCM has the fewest number of messages.

In our model, we estimate the total parallel runtime as $T = T_f + T_m + T_c$, where T_f, T_m, and T_c are the estimated times to perform floating-point operations, to service the cache misses, and to communicate the x vector. Given that a floating-point operation requires 1/900 microseconds and that each cache miss latency is 0.08 microseconds (both from documentation), and assuming that the MPI bandwidth and latency are 50 MB/second and 10 microseconds (both from measurement), respectively, we can estimate the total runtime based on the information in Table 2. We found a maximum deviation of 74% from the measured runtimes. The model showed that servicing the cache misses was extremely expensive and required more than 95% of the total time for MeTiS, RCM, and SAW, and almost 80% for ORIG which has relatively more communication.

3.2   Shared-Memory Implementation

The shared-memory version of SPMV was implemented on the Origin2000, which is a SMP cluster of nodes each containing two processors, some local memory, and 4 MB secondary cache per processor. This parallel code was written using SGI's native pragma directives, which create IRIX threads. Each processor is assigned an equal number of rows in the matrix. SPMV proceeds in parallel without the need for any synchronization, since there are no concurrent writes. The two basic implementation approaches described below were taken.

The SHMEM strategy naively assumes that the Origin2000 is a flat shared-memory machine. Arrays are not explicitly distributed among the processors, and nonlocal data requests are handled by the cache coherent hardware. Alternatively, the CC-NUMA strategy addresses the underlying distributed-memory nature of the machine by performing an initial data distribution. Sections of the sparse matrix are appropriately mapped onto the memories of their corresponding processors using the default "first touch" data distribution policy of the Origin2000. The computational kernels of both the SHMEM and CC-NUMA implementations are identical and were simpler to implement than the MPI version. Table 3 shows the runtime of SPMV using both approaches with the ORIG, RCM, and SAW orderings of the mesh.

Table 3. Runtimes (in seconds) for different orderings running in SHMEM and CC-NUMA modes on the SGI Origin2000

P	SHMEM			CC-NUMA		
	ORIG	RCM	SAW	ORIG	RCM	SAW
1	0.348	0.254	0.250	0.341	0.247	0.244
2	0.200	0.150	0.147	0.189	0.152	0.091
4	0.167	0.138	0.134	0.191	0.025	0.028
8	0.169	0.111	0.104	0.088	0.004	0.006
16	0.200	0.184	0.206	0.060	0.002	0.005
32	0.483	0.453	0.359	0.054	0.006	0.006
64	0.947	0.931	0.891	0.065	0.015	0.008

As expected, the CC-NUMA implementation shows significant performance gain over SHMEM. Within the CC-NUMA approach, RCM and SAW dramatically reduce the runtimes as compared to ORIG, indicating that an ordering algorithm is necessary to achieve good performance on distributed shared-memory systems. There is little difference in parallel performance between RCM and SAW because both reduce the number of secondary cache misses and the non-local memory references of the processors. However, there is a slowdown in performance when using more than 16 processors. This is due to the increased surface-to-volume ratio of the mesh partitions, which cause the overhead of cache coherence and false sharing to grow with the numbers of processors.

3.3 Multithreaded Implementation

The Tera MTA is a supercomputer recently installed at SDSC. The MTA has a radically different architecture than current high-performance computer systems. Each processor has support for 128 hardware streams, where each stream includes a program counter and a set of 32 registers. One program thread can be assigned to each stream. The processor switches among the active streams at every clock tick, while executing a pipelined instruction. The uniform shared memory of the MTA is flat, and physically distributed across hundreds of banks. Rather than using data caches to hide latency, the MTA processors use multithreading to tolerate latency. Once a code has been written in the multithreaded model, no additional work is required to run it on multiple processors.

The multithreaded implementation of the SPMV was trivial, requiring only MTA compiler directives. Load balancing is implicitly handled by the operating system which dynamically assigns rows to threads. Special synchronization constructs were not required since there are no possible race conditions in the multithreaded SPMV. For this implementation, no special ordering is required to achieve parallel performance and scalability. Results, shown in Table 4, indicate that there is enough instruction level parallelism in SPMV to tolerate the relatively high overhead of memory access. However, MTA runtimes will generally be slower than traditional cache-based systems for load balanced applications with substantial cache reuse.

Table 4. Runtimes (in seconds) on the Tera MTA using 60 streams per processor

P=1	P=2	P=4	P=8
0.0812	0.0406	0.0203	0.0103

4   Work in Progress

In this paper, we examined different ordering strategies for SPMV, using three leading programming paradigms and architectures. We plan to port the distributed-memory implementation of SPMV onto the newly installed RS/6000 SP machine at NERSC. In addition, we will examine the effects of partitioning the sparse matrix using MeTiS and then performing RCM or SAW orderings on each subdomain. Combining both schemes should minimize interprocessor communication and significantly improve data locality. Future research will focus on evaluating the effectiveness of the parallel Jacobi-Davidson eigensolver, when various orderings are applied to the underlying sparse matrix. Finally, we intend to extend the SAW algorithm to three-dimensional meshes and modify it to efficiently handle adaptively refined meshes in a parallel environment.

Acknowledgements

The first two authors were supported by the Office of Computational and Technology Research, U.S. DoE Division of Mathematical, Information, and Computational Sciences under contract DE-AC03-76SF00098. The third author was supported by NSF under grants CISE-9726388 and MIPS-9707125 while he was at the University of Delaware. The work of the fourth author was supported by NASA under contract NAS 2-14303 with MRJ Technology Solutions.

References

1. Burgess, D.A., Giles, M.B.: Renumbering Unstructured Grids to Improve the Performance of Codes on Hierarchical Memory Machines. Adv. Engrg. Soft. **28** (1997) 189–201
2. Cuthill, E., McKee, J.: Reducing the Bandwidth of Sparse Symmetric Matrices. Proc. ACM Natl. Conf. (1969) 157–172
3. Heber, G., Biswas, R., Gao, G.R.: Self-Avoiding Walks over Adaptive Unstructured Grids. Springer-Verlag LNCS **1586** (1999) 968–977
4. Hutchinson, S.A., Prevost, L.V., Shadid, J.N., Tuminaro, R.S.: Aztec User's Guide, v 2.0 Beta. Sandia National Laboratories TR SAND95-1559 (1998)
5. Karypis, G., Kumar, V.: A Fast and High Quality Multilevel Scheme for Partitioning Irregular Graphs. SIAM J. Sci. Comput. **20** (1998) 359–392
6. McCandless, B., Li, Z., Srinivas, V., Sun, Y., Ko, K.: Omega3P: A Parallel Eigensolver for Modeling Large, Complex Cavities. Proc. Intl. Comput. Accel. Phys. Conf. (1998)
7. Shewchuk, J.R.: Triangle: Engineering a 2D Quality Mesh Generator and Delaunay Triangulator. Springer-Verlag LNCS **1148** (1996) 203–222

A GRASP for Computing Approximate Solutions for the Three-Index Assignment Problem

Invited Talk

Renata M. Aiex
Department of Computer Science, Catholic U. of Rio de Janeiro

Panos M. Pardalos
Department of Industrial and Systems Engineering, U. of Florida

Leonidas S. Pitsoulis
Department of Industrial and Systems Engineering, U. of Florida

Mauricio G. C. Resende
Information Sciences Research Center, AT&T Labs Research

ABSTRACT

In this talk a greedy randomized adaptive search procedure (GRASP) is presented for computing approximate solutions to the NP-hard three-index assignment problem (AP3). A FORTRAN implementation of this GRASP is tested on several problem instances of the AP3. A parallelization strategy that combines GRASP and path relinking is also proposed and tested for the problem. Computational results indicate that the GRASP and the parallel hybrid heuristic provide good approximate solutions to a variety of AP3 instances.

J. Rolim et al. (Eds.): IPDPS 2000 Workshops, LNCS 1800, pp. 504-504, 2000.
© Springer-Verlag Berlin Heidelberg 2000

On Identifying Strongly Connected Components in Parallel*

Lisa K. Fleischer[1], Bruce Hendrickson[2], and Ali Pınar[3]

[1] Industrial Engrg. & Operations Research, Columbia Univ., New York, NY 10027
`lisa@ieor.columbia.edu`
[2] Parallel Computing Sciences, Sandia National Labs, Albuquerque, NM 87185-1110
`bah@cs.sandia.gov`
[3] Dept. Computer Science, University of Illinois, Urbana, IL 61801
`alipinar@cse.uiuc.edu`

Abstract. The standard serial algorithm for strongly connected components is based on depth first search, which is difficult to parallelize. We describe a divide-and-conquer algorithm for this problem which has significantly greater potential for parallelization. For a graph with n vertices in which degrees are bounded by a constant, we show the expected serial running time of our algorithm to be $O(n \log n)$.

1 Introduction

A *strongly connected component* of a directed graph is a maximal subset of vertices containing a directed path from each vertex to all others in the subset. The vertices of any directed graph can be partitioned into a set of disjoint strongly connected components. This decomposition is a fundamental tool in graph theory with applications in compiler analysis, data mining, scientific computing and other areas.

The definitive serial algorithm for identifying strongly connected components is due to Tarjan [15] and is built on a depth first search of the graph. For a graph with m edges, this algorithm runs in the optimal $O(m)$ time, and is widely used in textbooks as an example of the power of depth first search [7].

For large problems, a parallel algorithm for identifying strongly connected components would be useful. One application of particular interest to us is discussed below. Unfortunately, depth first search (DFS) seems to be difficult to parallelize. Reif shows that a restricted version of the problem (lexicographical DFS) is $\mathcal{P}$-Complete [14]. However, Aggarwal and Anderson, and Aggarwal, et al. describe randomized NC algorithms for finding a DFS of undirected and directed graphs, respectively [1, 2]. The expected running time of this latter algorithm is

---

* This work was funded by the Applied Mathematical Sciences program, U.S. Department of Energy, Office of Energy Research and performed at Sandia, a multiprogram laboratory operated by Sandia Corporation, a Lockheed-Martin Company, for the U.S. DOE under contract number DE-AC-94AL85000.

J. Rolim et al. (Eds.): IPDPS 2000 Workshops, LNCS 1800, pp. 505–511, 2000.

$O(\log^7 n)$[1], and it requires an impractical $n^{2.376}$ processors. To our knowledge, the deterministic parallel complexity of DFS for general, directed graphs is an open problem. Chaudhuri and Hagerup studied the problem for acyclic [5], and planar graphs [10], respectively, but our application involves non-planar graphs with cycles, so these results don't help us. More practically, DFS is a difficult operation to parallelize and we are aware of no algorithms or implementations which perform well on large numbers of processors. Consequently, Tarjan's algorithm cannot be used for our problem.

Alternatively, there exist several parallel algorithms for the strongly connected components problem (SCC) that avoid the use of depth first search. Gazit and Miller devised an NC algorithm for SCC, which is based upon matrix-matrix multiplication [9]. This algorithm was improved by Cole and Vishkin [6], but still requires $n^{2.376}$ processors and $O(\log^2 n)$ time. Kao developed a more complicated NC algorithm for planar graphs that requires $O(\log^3 n)$ time and $n/\log n$ processors [11]. More recently, Bader has an efficient parallel implementation of SCC for planar graphs [3] which uses a clever *packed–interval* representation of the boundary of a planar graph. When n is much larger than p the number of processors, Bader's approach scales as $O(n/p)$. But algorithms for planar graphs are insufficient for our needs.

Our interest in the SCC problem is motivated by the discrete ordinates method for modeling radiation transport. Using this methodology, the object to be studied is modeled as a union of polyhedral finite elements. Each element is a vertex in our graph and an edge connects any pair of elements that share a face. The radiation equations are approximated by an angular discretization. For each angle in the discretization, the edges in the graph are directed to align with the angle. The computations associated with an element can be performed if all its predecessors have been completed. Thus, for each angle, the set of computations are sequenced as a topological sort of the directed graph. A problem arises if the topological sort cannot be completed – i.e. the graph has a cycle. If cycles exist, the numerical calculations need to be modified – typically by using old information along one of the edges in the cycle, thereby removing the dependency. So identifying strongly connected components quickly is essential. Since radiation transport calculations are computationally and memory intensive, parallel implementations are necessary for large problems. Also, since the geometry of the grid can change after each timestep for some applications, the SCC problem must be solved in parallel.

Efficient parallel implementations of the topological sort step of the radiation transport problem have been developed for *structured* grids, oriented grids that have no cycles [4, 8]. Some initial attempts to generalize these techniques to unstructured grids are showing promise [12, 13]. It is these latter efforts that motivated our interest in the SCC problem.

---

[1] Function $f(n) = \Theta(g(n))$ if there exist constants $c_2 \geq c_1 > 0$ and N such that for all $n \geq N$, $c_1 g(n) \leq f(n) \leq c_2 g(n)$. If $f(n) = \Omega(g(n))$, then this just implies the existence of c_1 and N. If $f(n) = O(g(n))$, then c_2 and N exist.

In the next section we describe a simple divide-and-conquer algorithm for finding strongly connected components. In §3 we show that for constant degree graphs our algorithm has an expected serial complexity of $O(n \log n)$. Our approach has good potential for parallelism for two reasons. First, the divide-and-conquer paradigm generates a set of small problems which can be solved independently by separate processors. Second, the basic step in our algorithm is a reachability analysis, which is similar to topological sort in its parallelizability. So we expect the current techniques for parallelizing radiation transport calculations to enable our algorithm to perform well too.

2   A Parallelizable Algorithm for Strongly Connected Components

Before describing our algorithm, we introduce some notation. Let $G = (V, E)$ be a directed graph with vertices V and directed edges E. An edge $(i, j) \in E$ is directed from i to j. We denote the set of strongly connected components of G by $SCC(G)$. Thus $SCC(G)$ is a partition of V. We also use $SCC(G, v)$ to denote the (unique) strongly connected component containing vertex v. We denote by $V \setminus X$ the subset of vertices in V which are not in a subset X. The size of vertex set X is denoted $|X|$.

A vertex v is *reachable* from a vertex u if there is a sequence of directed edges $(u, x_1), (x_1, x_2), \ldots, (x_k, v)$ from u to v. We consider a vertex to be reachable from itself. Given a vertex $v \in V$, the *descendants* of v, $Desc(G, v)$, is the subset of vertices in G which are reachable from v. Similarly, the *predecessors* of v, $Pred(G, v)$, is the subset of vertices from which v is reachable. The set of vertices that is neither reachable from v nor reach v is called the *remainder*, denoted by $Rem(G, v) = V \setminus \{Desc(G, v) \cup Pred(G, v)\}$.

Given a graph $G = (V, E)$ and a subset of vertices $V' \subseteq V$, the *induced subgraph* $G' = (V', E')$ contains all edges of G connecting vertices of V', i.e. $E' = \{(u, v) \in E : u, v \in V'\}$. We will use $\langle V' \rangle = G' = (V', E')$ to denote the subgraph of G induced by vertex set V'. The following Lemma is an immediate consequence of the definitions.

Lemma 1. *Let $G = (V, E)$ be a directed graph, with $v \in V$ a vertex in G. Then*

$$Desc(G, v) \cap Pred(G, v) = SCC(G, v).$$

Lemma 2. *Let G be a graph with vertex v. Any strongly connected component of G is a subset of $Desc(G, v)$, a subset of $Pred(G, v)$ or a subset of $Rem(G, v)$.*

Proof. Let u and w be two vertices of the same strongly connected component in G. By definition, u and w are reachable from each other. The proof involves establishing $u \in Desc(G, v) \iff w \in Desc(G, v)$ and $u \in Pred(G, v) \iff w \in Pred(G, v)$, which then implies $u \in Rem(G, v) \iff w \in Rem(G, v)$. Since the proofs of these two statements are symmetric, we give just the first: If $u \in Desc(G, v)$ then u must be reachable from v. But then w must also be reachable from v, so $w \in Desc(G, v)$. $\qquad\square$

With this background, we can present our algorithm which we call DCSC (for Divide-and-Conquer Strong Components). The algorithm is sketched in Fig. 1. The basic idea is to select a random vertex v, which we will call a *pivot* vertex, and find its descendant and predecessor sets. The intersection of these sets is $SCC(G, v)$ by Lemma 1. After this step, the remaining vertices are divided into three sets $Desc(G, v)$, $Pred(G, v)$, and $Rem(G, v)$. By Lemma 2, any additional strongly connected component must be entirely contained within one of these three sets, so we can divide the problem and recurse.

$DCSC(G)$
   If G is empty **then Return.**
   Select v uniformly at random from V.
   $SCC \leftarrow Pred(G, v) \cap Desc(G, v)$
   Output SCC.
   $DCSC(\langle Pred(G, v) \setminus SCC \rangle)$
   $DCSC(\langle Desc(G, v) \setminus SCC \rangle)$
   $DCSC(\langle Rem(G, v) \rangle)$

Fig. 1. A divide-and-conquer algorithm for strongly connected components.

3   Serial Complexity of Algorithm DCSC

To analyze the cost of the recursion, we will need bounds on the expected sizes of the predecessor and descendant sets. The following two results provide such bounds.

Lemma 3. *For a directed graph G, there is a numbering π of the vertices from 1 to n in which the following is true. All elements $u \in Pred(G, v) \setminus Desc(G, v)$ satisfy $\pi(u) < \pi(v)$; and all elements $u \in Desc(G, v) \setminus Pred(G, v)$ satisfy $\pi(u) > \pi(v)$.*

Proof. If G is acyclic, then a topological sort provides a numbering with this property. If G has cycles, then each strongly connected component can be contracted into a single vertex, and the resulting acyclic graph can be numbered via topological sort. Assume a strongly connected component with k vertices was assigned a number j in this ordering. Assign the vertices within the component the numbers $(j, \ldots, j + k - 1)$ arbitrarily and increase all subsequent numbers by $k - 1$.   □

It is important to note that we do not need to construct an ordering with this property; we just need to know that it exists.

Corollary 1. *Given a directed graph G and a vertex numbering π from Lemma 3, then $|Pred(G, v) \setminus SCC(G, v)| < \pi(v)$ and $|Desc(G, v) \setminus SCC(G, v)| \leq n - \pi(v)$ for all vertices v.*

The cost of algorithm DCSC consists of four terms, three from the three recursive invocations and the fourth from the cost of determining the set of predecessors and descendants. For a graph with all degrees bounded by a constant, this last term is linear in the sizes of the predecessor and descendant sets. Let $T(n)$ be the expected runtime of the algorithm on bounded degree graph G with n vertices. For a particular pivot i, let p_i, d_i and r_i represent the sizes of the recursive calls. That is, $p_i = |Pred(G, v) \setminus SCC(G, v)|$, $d_i = |Desc(G, v) \setminus SCC(G, v)|$ and $r_i = |Rem(G, v)|$. If vertex number i is selected as the pivot, then the recursive expression for the run time is

$$T(n) = T(r_i) + T(d_i) + T(p_i) + \Theta(n - r_i). \qquad (1)$$

Clearly, $T(n) = \Omega(n)$, since we eventually must look at all the vertices. Also, in worst case $T(n) = O(n^2)$, since each iteration takes at most linear time and reduces the graph size by at least 1. We show here that the expected behavior of $T(n)$ is $\Theta(n \log n)$. The average case analysis will require summing the cost over all pivot vertices and dividing by n. So for a graph with constant bound on the degrees, the expected runtime, $ET(n)$, of the algorithm is

$$ET(n) = \frac{1}{n} \sum_{i=1}^{n} [T(p_i) + T(d_i) + T(r_i) + \Theta(n - r_i)]. \qquad (2)$$

Theorem 1. *For a graph in which all degrees are bounded by a constant, algorithm DCSC has expected time complexity $O(n \log n)$.*

Proof. We analyze (2) by partitioning the vertices according to their value of r_i. Let $S_1 := \{v | r_{\pi(v)} < n/2\}$ and $S_2 := \{v | r_{\pi(v)} \geq n/2\}$. We analyze each case separately and show that the separate recursions lead to an $O(n \log n)$ expected run time. Thus, the average of these recursions will also.

Case 1: $r_i < \frac{n}{2}$.

Note that Corollary 1 implies the lower and upper bounds: $p_i \leq i-1 \leq p_i + r_i$ and $d_i \leq n - i \leq d_i + r_i$. Since $r_i < n/2$, it follows that $\min\{i, n - i\} \leq r_i + \min\{p_i, d_i\} < \frac{3n}{4}$. By symmetry, it is enough to consider $p_i \leq d_i$. Then we can bound $\min\{d_i, r_i + p_i\}$ from below:

$$\min\{d_i, r_i + p_i\} \geq \min\{i, n/4\}. \qquad (3)$$

By superlinearity of $T(n)$, we have $T(r_i) + T(p_i) + T(d_i) + \Theta(n - r_i) \leq T(r_i + p_i) + T(d_i) + \Theta(n - r_i)$. Using (3), we can bound the contribution of each i by either $T(i - 1) + T(n - i) + \Theta(n)$ or by $T(n/4) + T(3n/4) + \Theta(n)$. This latter case, through a well-known analysis, has a solution of $O(n \log n)$. In the former, at worst, all i contributing here lie at the extremes of the interval $[1, n]$. If there are $2q$ of them, their total contribution to (2) is at most $\frac{2}{n} \sum_{i=1}^{q} [T(i - 1) + T(n - i) + \Theta(n)]$. Then, an analysis similar to that used for quicksort yields a $O(n \log n)$ recursion. We reproduce this below for completeness.

$$\frac{2}{2q} \sum_{i=1}^{q} [T(i-1) + T(n-i) + \Theta(n)]$$

$$= \Theta(n) + \frac{1}{q} [\sum_{i=1}^{q-1} (c_1 i \log i) + \sum_{i=n-q+1}^{n} (c_1 i \log i)]$$

$$\leq \Theta(n) + \frac{1}{q} [c_1 \log q \sum_{i=1}^{q-1} i + c_1 \log n \sum_{i=n-q+1}^{n} i]$$

$$= \Theta(n) + \frac{c_1}{2} [(q-1) \log q + (2n - q + 1) \log n]$$

$$= c_1 n \log n + [\Theta(n) - \frac{c_1}{2}(q-1)(\log n - \log q)].$$

The expression in brackets in the last inequality is < 0 for large enough choice of c_1.

Case 2: $r_i \geq \frac{n}{2}$.

By superlinearity of $T(n)$, we can rewrite equation (1) as $T(n) \leq T(r_i) + T(d_i + p_i) + \Theta(n - r_i)$. If we let $a = n - r_i$, we can rewrite this as a function of a and n:

$$T_2(a, n) \leq T(n - a) + T(a) + \Theta(a).$$

By our assumptions in this case, we have that $1 \leq a \leq n/2$. We show that this recursion is $O(n \log n)$ by first showing that this holds for $a = 1, n/2$, and then showing that T_2, as a function of a in the range $[1, n-1]$, is convex. Thus, its value in an interval is bounded from above by its values at the endpoints of the interval.

It is easy to see that $T_2(1, n) = \Theta(n)$, and $T_2(n/2, n) = \Theta(n \log n)$. We suppose, by induction, that $T(r) = c_1 n \log n$ for an appropriate constant c_1 for $r < n$, and that the constant in the Θ term is c_2. Thus, the first derivative of $T_2(a, n)$ with respect to a is

$$c_1(\log a - \log(n - a)) + c_2.$$

The second derivative is

$$c_1(\frac{1}{a} + \frac{1}{n - a}),$$

which is positive for $a \in [1, n-1]$. Thus $T_2(a, n)$ is convex for $a \in [1, n/2]$, and hence $T_2(n) = \Theta(n \log n)$ in this case.   □

4   Future Work

For the radiation transport application that motivated our interest in this problem, the graphs will often be acyclic, and any strongly connected components will usually be small. For acyclic graphs, a topological sort, using the methodologies being developed for this application, will terminate. By coupling a termination detection protocol to the topological sort, we can use existing parallelization approaches to quickly determine whether or not a cycle exists. Besides quickly excluding graphs which have no cycles, using topological sort as a preprocessing step allows for the discarding of all the visited vertices, reducing the size of the problem that needs to be addressed by our recursive algorithm. With Will McLendon III and Steve Plimpton, we are implementing such a hybrid scheme and will report on its performance in due course.

Acknowledgements

We benefited from general discussions about algorithms for parallel strongly connected components with Steve Plimpton, Will McLendon and David Bader. We are also indebted to Bob Carr, Cindy Phillips, Bill Hart and Sorin Istrail for discussions about the analysis.

References

1. A. AGGARWAL AND R. J. ANDERSON, *A random NC algorithm for depth first search*, Combinatorica, 8 (1988), pp. 1–12.
2. A. AGGARWAL, R. J. ANDERSON, AND M.-Y. KAO, *Parallel depth-first search in general directed graphs*, SIAM J. Comput., 19 (1990), pp. 397–409.
3. D. A. BADER, *A practical parallel algorithm for cycle detection in partitioned digraphs*, Tech. Rep. Technical Report AHPCC-TR-99-013, Electrical & Computer Eng. Dept., Univ. New Mexico, Albuquerque, NM, 1999.
4. R. S. BAKER AND K. R. KOCH, *An S_n algorithm for the massively parallel CM-200 computer*, Nuclear Science and Engineering, 128 (1998), pp. 312–320.
5. P. CHAUDHURI, *Finding and updating depth-first spanning trees of acyclic digraphs in parallel*, The Computer Journal, 33 (1990), pp. 247–251.
6. R. COLE AND U. VISHKIN, *Faster optimal prefix sums and list ranking*, Information and Computation, 81 (1989), pp. 334–352.
7. T. H. CORMEN, C. E. LEISERSON, AND R. L. RIVEST, *Introduction to Algorithms*, MIT Press and McGraw-Hill, Cambridge, MA, 1990.
8. M. R. DORR AND C. H. STILL, *Concurrent source iteration in the solution of 3-dimensional, multigroup discrete ordinates neutron-transport equations*, Nuclear Science and Engineering, 122 (1996), pp. 287–308.
9. H. GAZIT AND G. L. MILLER, *An improved parallel algorithm that computes the BFS numbering of a directed graph*, Inform. Process. Lett., 28 (1988), pp. 61–65.
10. T. HAGERUP, *Planar depth-first search in $O(\log n)$ parallel time*, SIAM J. Comput., 19 (1990), pp. 678–704.
11. M.-Y. KAO, *Linear-processor NC algorithms for planar directed graphs I: Strongly connected components*, SIAM J. Comput., 22 (1993), pp. 431–459.
12. S. PAUTZ. Personal Communication, October 1999.
13. S. PLIMPTON. Personal Communication, May 1999.
14. J. H. REIF, *Depth-first search is inherently sequential*, Inform. Process. Lett., 20 (1985), pp. 229–234.
15. R. E. TARJAN, *Depth first search and linear graph algorithms*, SIAM J. Comput., 1 (1972), pp. 146–160.

A Parallel, Adaptive Refinement Scheme for Tetrahedral and Triangular Grids

Alan Stagg[1], Jackie Hallberg[2], and Joseph Schmidt[3]

[1] Los Alamos National Laboratory, Applied Physics Division
P.O. Box 1663, MS P365
Los Alamos, NM 87545
stagg@lanl.gov
[2] U.S. Army Engineer Research and Development Center
Coastal and Hydraulics Laboratory
3909 Halls Ferry Road
Vicksburg, MS 39180
pettway@juanita.wes.army.mil
[3] 2420 Wanda Way
Reston, VA 20191
roig.and.schmidt@erols.com

Abstract. A grid refinement scheme has been developed for tetrahedral and triangular grid-based calculations in message-passing environments. The element adaption scheme is based on edge bisection of elements marked for refinement by an appropriate error indicator. Hash table/linked list data structures are used to store nodal and element information. The grid along inter-processor boundaries is refined consistently with the update of these data structures via MPI calls. The parallel adaption scheme has been applied to the solution of an unsteady, three-dimensional, nonlinear, groundwater flow problem. Timings indicate efficiency of the grid refinement process relative to the flow solver calculations.

1 Introduction

Adaptive grid methods based on point insertion and removal have been popular for a number of years for achieving greater solution accuracy with relative cost efficiency. However, issues related to implementing such schemes on parallel systems are just now being addressed, and much work is needed to identify the best approaches.

In this paper we present a new approach for the h-refinement of irregular tetrahedral and triangular grids in message-passing environments. Data structures have been selected to simplify implementation and coding complexity as much as possible for refinement, coarsening, and load balancing components. This software has been implemented in the Department of Defense code ADH (ADaptive Hydrology) under development at the U.S. Army Engineer Research

J. Rolim et al. (Eds.): IPDPS 2000 Workshops, LNCS 1800, pp. 512-518, 2000.

and Development Center. ADH is a modular, parallel, finite element code designed to support groundwater, surface water, and free-surface Navier-Stokes modeling [1].

1.1 Serial Element Adaption Scheme

Given an initial grid, the model subdivides grid elements to achieve the desired resolution in regions of interest. The parallel grid adaption scheme developed here is based on the geometric splitting algorithm of Liu and Joe [2]. Elements are refined by edge bisection according to an error indicator, and elements can be merged to increase efficiency where high resolution is not required. A grid closure step is utilized to eliminate non-conforming nodes. Element edges are selected for bisection based on a modified longest-edge bisection approach in which the oldest edge in the element is first flagged for bisection followed by the longest edge. Refinement and coarsening of a tetrahedral element are illustrated in Figure 1. Here a new node is added to an edge, creating two new tetrahedra. The new elements can be merged to recover the original element by removing the inserted node.

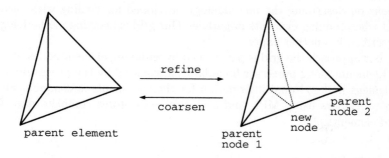

Fig. 1. Tetrahedral Grid Adaption Based on Edge Bisection

The refinement process in ADH begins with determination of elements to be split according to an explicit residual error indicator. The following pseudo-code describes the basic steps in the refinement scheme.

Refinement pseudo-code

```
loop over elements
      refine element via edge bisection if its error > threshold;

conforming_grid = false;
do while conforming_grid == false {
      conforming_grid = true;
```

```
loop over elements
     if element has an edge with a newly inserted node {
          refine element;
          conforming_grid = false;
     }
}
```

2   Parallel Implementation

The parallel implementation of local grid refinement schemes like the edge bisection scheme presents a number of challenges for the message-passing environment. First, in the standard approach where the grid is partitioned and subdomains are assigned to processors, the subdomains must be refined and coarsened consistently along processor boundaries. Also, closure requirements may force element refinement to spread to a processor that has no elements marked for refinement by the error indicator. Finally, the local adaption process will likely lead to load imbalances among the processors, and nodes and elements must be transferred between processors to maintain processing efficiency. In this paper we focus on describing the methodology developed for dealing with such difficulties when refining elements in parallel. Our grid coarsening methodology will be described in another paper.

In our approach the grid is partitioned by assigning element nodes to processors. Elements along processor boundaries are shared by the processors owning the element nodes. Nodal information for these elements is communicated between processors using MPI, and each processor stores complete data for its shared elements [3].

2.1   Data Structures

Data structures have been selected to simplify the parallel implementation of the adaption scheme and to facilitate the coupling of the refinement, coarsening, and load balancing components. During early work, we realized that common techniques like the use of tree structures for refinement could adversely impact other adaption components such as load balancing. In this case, the use of graph partitioners and the resulting grid point movement between processors requires splitting refinement trees between processors. To avoid these difficulties we use hash table/linked list structures [4]. Such structures are naturally suited for grid adaption since they are dynamic in nature and facilitate node and element searches. These structures support all grid refinement, coarsening, and load balancing requirements without complicating the implementation of any single component.

Hash tables are used to store nodes and element edges. Each entry in the node hash table consists of a node number local to the processor and a corresponding node identifier in the global grid. Each entry in the edge hash table consists of the two local node numbers that define the edge, an integer edge rank based on comparative lengths of the edges, and an integer that stores the new node number if a node has been inserted on the edge. These node and edge hash tables are constructed prior to each refinement step, and the memory is freed upon completion of the refinement step.

2.2   Refinement

The grid refinement scheme presented here is primarily a local process and is thus amenable to parallel processing. The principal requirement in the parallel environment is that the grid along inter-processor boundaries be refined consistently. The elements that are shared along these boundaries are duplicated on each of the processors that own the elements' nodes. If one of these duplicated elements is marked for refinement, each processor that stores a copy of the element must bisect that element's edges in the same way.

To insure that shared elements are refined consistently, all element edges are ranked for bisection based on their lengths so that the edges are uniquely and consistently identified throughout the global grid. This ranking also facilitates refinement in elements that are not shared between processors. Integer ranks are utilized rather than computed edge lengths so that processors are easily able to make consistent edge bisection decisions when multiple edges in an element have the same length. The edge hash tables are used to store these ranks.

In our partitioning approach, an edge shared by two processors will appear in each of the processors' hash tables, and a protocol must be established to maintain consistency of the edge hash tables between processors. To support this communication, edge communication lists are constructed which provide a mapping between these duplicated edge storage locations. For each such edge, one of the processors sharing the edge is assigned ownership of it.

The edge ranking follows construction of the edge communication lists. After a parallel odd-even transposition sort, global ranks for edges are returned to the processors owning the edges, and the ranks are stored by these processors in their edge hash tables. These processors then communicate the ranks to the processors sharing the edges using the edge communication lists that have been constructed. Finally, the receiving processors store the ranks in their edge hash tables to complete the edge ranking process.

Elements are selected for refinement based on the error indicator, and edges in these elements are selected for bisection based on their age and rank. The oldest edge is identified first for bisection. However, if all edges in an element have the same age, the longest edge (identified by rank comparisons) is marked for bisection. If the marked edge has not been bisected by an adjacent element, a new node is created and its number is stored in the edge hash table. Edges that have not been bisected have a negative entry in the new node location in the edge data structure. Two new elements are created using the new node on the bisected edge, and the element Jacobians and other data are corrected for these elements. The new node entries for the edges in these new elements are reinitialized to indicate that new nodes are not present.

2.3   Grid Closure

After the initial refinement step, further refinement may be required to obtain a conforming grid. In the serial case each element's edges are checked for the presence of new nodes in the edge hash table. If any element has an edge with a new node, that element is marked for refinement according to the established rules. The refinement process continues iteratively until a conforming grid is obtained.

In the parallel environment this procedure is complicated by the fact that shared edges may be bisected by only one of the processors sharing the edge. To maintain consistency of the edge hash tables, processors owning shared edges must communicate new node information to the other processors sharing the edges. If a message indicates that an edge has a new node, then the receiving processor creates a new node for the edge and updates its hash table.

An example is illustrated in Figure 2 where three elements are distributed over two processors as indicated by the shaded background. The center and right elements are shared by the two processors. In the first step, processor P0 splits the left element because of high error. Next the original center element is split in the closure phase because that element now has a new node on one of its edges. Note that the center element's longest edge is bisected rather than the edge with the new node. Following this second step, the copy of the right element on processor P0 can be refined since the shared edge owned by processor P0 has been updated with the new node number. However, the new node on the shared edge must be communicated to processor P1 using the edge communication lists to inform processor P1 that its copies of the center and right elements must be refined. In this example grid refinement has spread from processor P0 to processor P1 even though processor P1 did not have any elements marked for refinement by the error indicator.

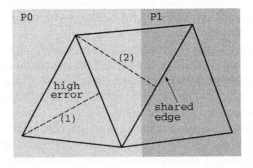

Fig. 2. Grid Closure on Two Processors

Similarly, processors may bisect edges they do not own. To handle this situation, the edge communication lists are utilized in reverse order (the send list becomes a receive list, and vice verse) so that processors owning shared edges can update their hash tables if other processors bisect them. After this communication, the elements with new nodes on edges are refined, and the process is repeated iteratively until a conforming grid is obtained.

3   Groundwater Application

The capabilities of the parallel grid refinement scheme have been investigated for the solution of a draining heterogeneous column. In this problem a column is filled with a mixture of clay, silt, and sand. The column consist primarily of sand with a clay lens near the bottom and silt lenses in several places throughout the column. Initially, the column is completely saturated with water, and then the water is allowed to drain from the bottom of the column. The grid is allowed to refine and coarsen locally as dictated by the explicit error indicator, and dynamic load balancing is utilized to improve processor efficiency.

A snapshot of the adaptively refined grid for the hegerogeneous column is illustrated in Fig. 3. The area shaded black represents the clay material, while

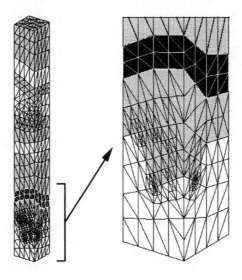

Fig. 3. Adaptively Refined Grid for Heterogeneous Column

the sand and silt are represented by the gray and white regions, respectively. Grid refinement is visible at the sand/silt interface in the lowest points of the sand. This refinement is indicative of the large head gradient from the sand to the silt. Water travels through the sand at a faster rate than through the silt due to the silt's lower conductivity. As a result, the water ponds, or collects, in the low points of the sand until the pressure is great enough to push the flow across the interface.

The grid adaption software is currently being tested, and preliminary timings are only available for the heterogeneous column described above. Due to the coarse grid size for this problem, timings were only obtained on up to 8 processors of the Origin2000. During the refinement step, the number of nodes was increased by 35%, and the total time spent in grid refinement was an order of magnitude less than the time spent in the flow solver for timings on one through eight processors. The time spent in grid closure, including communication, was less than half of the total time in grid refinement for these cases. Though preliminary, these timings indicate efficiency of the grid refinement process relative to the flow solver.

4   Conclusion

A parallel refinement scheme has been developed for tetrahedral and triangular grids. The refinement scheme and data structures described here have been developed to facilitate the parallel implementation of both grid refinement and coarsening. The refinement and coarsening schemes are based on communicating a minimum set of data and reconstructing information locally without the use of tree structures. The goal with this approach is a balanced design between refinement, coarsening, and load balancing in terms of efficiency and ease of implementation. Preliminary application of the adaptive grid scheme to an unsteady groundwater flow problem has demonstrated the capability and efficiency of the method.

Funding for this project was provided by the Department of Defense High Performance Computing Modernization Office, and computer time was provided by the ERDC Major Shared Resource Center in Vicksburg, MS. Permission to publish this paper was given by the Chief of Engineers and by Los Alamos National Laboratory.

References

1. Jenkins, E.W., Berger, R.C., Hallberg, J.P., Howington, S.E., Kelley, C.T., Schmidt, J.H., Stagg, A.K., and Tocci, M.D., "Newton-Krylov-Schwarz Methods for Richards' Equation," submitted to the *SIAM Journal on Scientific Computing*, October 1999.
2. Liu, A. and Joe, B., "Quality Local Refinement of Tetrahedral Meshes Based on Bisection," *SIAM Journal on Scientific Computing*, vol. 16, no. 6, pp. 1269-1291, November 1995.
3. Snir, M., Otto, S., Huss-Lederman, S., Walker, D., and Dongarra, J., *MPI- The Complete Reference, Volume 1, The MPI Core*, The MIT Press, Cambridge, Massachusetts, 1998.
4. Cormen, T., Leiserson, C., and Rivest, R., *Introduction to Algorithms*, The MIT Press, Cambridge, Massachusetts, 1990.

PaStiX : A Parallel Sparse Direct Solver Based on a Static Scheduling for Mixed 1D/2D Block Distributions *

Pascal Hénon, Pierre Ramet, and Jean Roman

LaBRI, UMR CNRS 5800, Université Bordeaux I & ENSERB
351, cours de la Libération, F-33405 Talence, France
{henon|ramet|roman}@labri.u-bordeaux.fr

Abstract. We present and analyze a general algorithm which computes an efficient static scheduling of block computations for a parallel $L.D.L^t$ factorization of sparse symmetric positive definite systems based on a combination of 1D and 2D block distributions. Our solver uses a supernodal fan-in approach and is fully driven by this scheduling. We give an overview of the algorithm and present performance results and comparisons with PSPASES on an IBM-SP2 with 120 MHz Power2SC nodes for a collection of irregular problems.

1 Introduction

Solving large sparse symmetric positive definite systems $Ax = b$ of linear equations is a crucial and time-consuming step, arising in many scientific and engineering applications. Consequently, many parallel formulations for sparse matrix factorization have been studied and implemented; one can refer to [6] for a complete survey on high performance sparse factorization.

In this paper, we focus on the block partitioning and scheduling problem for sparse LDL^T factorization without pivoting on parallel MIMD architectures with distributed memory; we use LDL^T factorization in order to solve sparse systems with complex coefficients. In order to achieve efficient parallel sparse factorization, three pre-processing phases are commonly required:

• The *ordering* phase, which computes a symmetric permutation of the initial matrix A such that factorization will exhibit as much concurrency as possible while incurring low fill-in. In this work, we use a tight coupling of the Nested Dissection and Approximate Minimum Degree algorithms [1, 10]; the partition of the original graph into supernodes is achieved by merging the partition of separators computed by the Nested Dissection algorithm and the supernodes amalgamated for each subgraph ordered by Halo Approximate Minimum Degree.

• The *block symbolic factorization* phase, which determines the block data structure of the factored matrix L associated with the partition resulting from the ordering phase. This structure consists of N column blocks, each of them containing a dense symmetric diagonal block and a set of dense rectangular

---

* This work is supported by the *Commissariat à l'Énergie Atomique* CEA/CESTA under contract No. 7V1555AC, and by the GDR ARP (iHPerf group) of the CNRS.

off-diagonal blocks. One can efficiently perform such a block symbolic factorization in quasi-linear space and time complexities [5]. From the block structure of L, one can deduce the weighted elimination quotient graph that describes all dependencies between column blocks, as well as the supernodal elimination tree.

• The *block repartitioning and scheduling* phase, which refines the previous partition by splitting large supernodes in order to exploit concurrency within dense block computations, and maps the resulting blocks onto the processors of the target architecture.

The authors presented in [8] a preliminary version describing a mapping and scheduling algorithm for 1D distribution of column blocks. This paper extends this previous work by presenting and analysing a mapping and scheduling algorithm based on a combination of 1D and 2D block distributions. This algorithm computes an efficient static scheduling [7, 11] of the block computations for a parallel solver based on a supernodal fan-in approach [3, 12, 13, 14] such that the parallel solver is fully driven by this scheduling. This can be done by very precisely taking into account the computation costs of BLAS 3 primitives, the communication cost and the cost of local aggregations due to the fan-in strategy.

The paper is organized as follows. Section 2 presents our algorithmic framework for parallel sparse symmetric factorization, describes our block repartitioning and scheduling algorithm, and outlines the supernodal fan-in solver driven by this precomputed scheduling. Section 3 provides numerical experiments on an IBM SP2 for a large class of sparse matrices, including performance results and analysis. According to these results, our PASTIX software appears to be a good competitor with the current reference software PSPASES [9]. Then, we conclude with remarks on the benefits of this study and on our future work.

2   Parallel solver and block mapping

Let us consider the block data structure of the factored matrix L computed by the block symbolic factorization. Let us remind that a column block holds one dense diagonal block and some dense off-diagonal blocks. From this block data structure, we can introduce the boolean function $off\_diag(k, j)$, $1 \leq k < j \leq N$, that returns *true* if and only if there exists an off-diagonal block in column block k facing column block j. Then, we can define the two sets $BStruct(L_{k*}) := \{i < k \mid off\_diag(i, k) = true\}$ and $BStruct(L_{*k}) := \{j > k \mid off\_diag(k, j) = true\}$. Thus, $BStruct(L_{k*})$ is the set of column blocks that update column block k, and $BStruct(L_{*k})$ is the set of column blocks updated by column block k.

Let us now consider a parallel supernodal version of sparse LDL^t factorization with total local aggregation: all non-local block contributions are aggregated locally in block structures. This scheme is close to the Fan-In algorithm [4] as processors communicate using only aggregated update blocks. If memory is a critical issue, an aggregated update block can be sent with partial aggregation to free memory space; this is close to the Fan-Both scheme [2]. So, a block j in column block k will receive an aggregated update block only from every processor in set $Procs(L_{jk}) := \{map(j, i) \mid i \in BStruct(L_{k*}) \text{ and } j \in BStruct(L_{*i})\}$, where the $map(,)$ operator is the 2D block mapping function. These aggregated update blocks, denoted in the following AUB_{jk}, are stored by processors in $Procs(L_{jk})$, and can be built from the block symbolic factorization. The pseudo-code of LDL^T factorization can be expressed in terms of dense block computations (see figure 1); these computations are performed, as much as possible, on compacted sets of blocks for BLAS efficiency.

The proposed algorithm can yield 1D or 2D block distributions. The block computations as well as the emissions and receptions of aggregated update block are fully ordered with a compatible priority. Block computations, or tasks, can be classified in four types :

- COMP1D(k) : update and compute contributions for column block k;
- FACTOR(k) : factorize the diagonal block k;
- BDIV(j,k) : update the off-diagonal block j on column block k ($k < j \leq N$);
- BMOD(i,j,k) : compute the contribution for block i in column block k facing column block j ($k < j \leq i \leq N$).

For a given k, $1 \leq k \leq N$, we define :
- N_p : total number of local tasks for processor p;
- $K_p[n]$: n^{th} local task for processor p;
- for the column block k, symbol $\star$ means $\forall j \in BStruct(L_{\star k})$;
- let $j \geq k$; sequence $[j]$ means $\forall i \in BStruct(L_{\star k}) \cup \{k\}$ with $i \geq j$.

```
On processor p :
  1. For n = 1 to N_p Do
  2.    Switch ( Type(K_p[n]) ) Do
  3.       COMP1D(k): Receive all AUB_[k]k for A_[k]k
  4.          Factor A_kk into L_kk D_k L^t_kk
  5.          Solve L_kk F^t_* = A^t_*k and D_k L^t_*k = F^t_*
  6.          For j ∈ BStruct(L_*k) Do
  7.             Compute C_[j] = L_[j]k F^t_j
  8.             If map([j], j) == p Then A_[j]j = A_[j]j − C_[j]
  9.             Else AUB_[j]j = AUB_[j]j + C_[j] , if ready send AUB_[j]j to map([j], j)
 10.       FACTOR(k): Receive all AUB_kk for A_kk
 11.          Factor A_kk into L_kk D_k L^t_kk and send L_kk D_k to map([k], k)
 12.       BDIV(j,k): Receive L_kk D_k and all AUB_jk for A_jk
 13.          Solve L_kk F^t_j = A^t_jk and D_k L^t_jk = F^t_j and send F^t_j to map([j], k)
 14.       BMOD(i,j,k): Receive F^t_j
 15.          Compute C_i = L_ik F^t_j
 16.          If map(i, j) == p Then A_ij = A_ij − C_i
 17.          Else AUB_ij = AUB_ij + C_i , if ready send AUB_ij to map(i, j)
```

Fig. 1. Outline of the parallel factorization algorithm.

Before running the general parallel algorithm we presented above, we must perform a step consisting in the partitioning and mapping of the blocks of the symbolic matrix onto the set of processors. The partitioning and mapping phase aims at computing a static regulation that balances workload and enforces the precedence constraints imposed by the factorization algorithm; the block elimination tree structure must be used there.

The existing approaches for block partitioning and mapping rise several problems, which can be divided into two categories: on one hand, problems due to the measure of workload, and on the other hand those due to the run-time scheduling of block computations in the solver. To be efficient, solver algorithms are block-oriented to take advantage of BLAS subroutines, which efficiencies are far from being linear in terms of number of operations. Moreover, workload encompasses block computations but does not take into account the other phenomena that occur in parallel fan-in factorization, such as extra-workload generated by the fan-in approach and idle-waits due to the latency generated by message passing.

The obtaining of high performances at run-time requires to compute efficient solutions to several scheduling problems that define the orders in which:
- to process tasks that are locally ready what is crucial for minimizing idle time;
- to send and to process the reception of aggregate update blocks (due to fan-in) and blocks used by BDIV or BMOD tasks, which determines what block will be ready next for local computation.
To tackle these problems, the partitioning and mapping step generates a fully ordered schedule used in parallel solving computations. This schedule aims at statically regulating all of the issues that are classically managed at run-time.

To make our scheme very reliable, we estimate the workload and message passing latency by using a BLAS and communication network time model, which is automatically calibrated on the target architecture. Unlike usual algorithms, our partitioning and mapping strategy is divided in two distinct phases.

The partitioning phase is based on a recursive top-down strategy over the block elimination tree. Pothen and Sun presented such a strategy in [11]. For each supernode, starting by the root, we assign it to a set of candidate processors Q for its mapping. Given the number of such candidate processors and the cost (that takes into account BLAS effect) of the supernode, we choose a 1D or 2D distribution strategy; the mapping and scheduling phase will use this choice to distribute this supernode. Then, recursively, each subtree is assigned to a subset of Q proportionally to its workload. Consequently, this strategy leads to a 2D distribution for the uppermost supernodes in the block elimination tree, and to a 1D for the others. Moreover, we allow a candidate processor to be in two sets of candidate processors for two subtrees having the same father in the elimination tree. The mapping phase will choose the better proportion to use such a processor in the two sets. By this way we avoid any problem of rounding to integral numbers. Furthermore, the column blocks corresponding to large supernodes are split using the blocking size suitable to achieve BLAS efficiency.

Once the partitioning phase is over, the task graph is built, such that each task is associated with the same set of candidate processors as the one of the supernode from which the task depends. The scheduling phase maps each task onto one of these processors; it uses a greedy algorithm that consists in mapping each task as it comes during the simulation of the parallel factorization. Thus, for each processor, we define a timer that will hold the current elapsed computation time, and a ready task heap that will contain at a time all tasks that are not yet mapped, that have received all of their contributions, and for which the processor is candidate. The algorithm then starts by mapping the leaves of the elimination tree (which have only one candidate processor). When a task is mapped onto a processor, the mapper: updates the timer of this processor according to our BLAS model; computes the time at which a contribution from this task is computed; puts into the heaps of their candidate processors all tasks all of the contributions of which have been computed.

After a task has been mapped, the next task to be mapped is selected by taking the first task of each ready tasks heap, and by choosing the one that comes from the lowest node in the elimination tree. Then, we compute for each of its candidate processors the time at which it will have completed the task if it is mapped onto it, thanks to: the processor timer; the time at which all contributions to this task have been computed (taking into account the fan-in overcost); the communication cost modeling that gives the time to send the contributions. The task is mapped onto the candidate processor that will be able to compute it the soonest.

After this phase has ended, the computations of each task are ordered with respect to the rank at which the task have been mapped. Thus, for each processor p, we obtain a vector K_p of the N_p local task numbers fully ordered by priority, and the solver described at figure 1 will be fully driven by this scheduling order.

3   Numerical Experiments

In this section, we describe experiments performed on a collection of sparse matrices in the RSA format; the values of the metrics in table 1 come from scalar column symbolic factorization.

Table 1. Description of our test problems. NNZ_A is the number of off-diagonal terms in the triangular part of matrix A, NNZ_L is the number of off-diagonal terms in the factored matrix L and OPC is the number of operations required.

Name	Columns	NNZ_A	$NNZ_{L(SCOTCH)}$	$OPC_{(SCOTCH)}$	$NNZ_{L(METIS)}$	$OPC_{(METIS)}$
B5TUER	162610	3873534	2.542e+07	1.531e+10	2.404e+07	1.237e+10
BMWCRA1	148770	5247616	6.597e+07	5.702e+10	6.981e+07	6.124e+10
MT1	97578	4827996	3.115e+07	2.109e+10	3.455e+07	2.269e+10
OILPAN	73752	1761718	8.912e+06	2.985e+09	9.065e+06	2.751e+09
QUER	59122	1403689	9.119e+06	3.281e+09	9.586e+06	3.448e+09
SHIP001	34920	2304655	1.428e+07	9.034e+09	1.481e+07	9.462e+09
SHIP003	121728	3982153	5.873e+07	8.008e+10	5.910e+07	7.587e+10
SHIPSEC5	179860	4966618	5.650e+07	6.952e+10	5.256e+07	5.509e+10
THREAD	29736	2220156	2.404e+07	3.884e+10	2.430e+07	3.583e+10
X104	108384	5029620	2.634e+07	1.713e+10	2.728e+07	1.412e+10

We compare factorization times performed in double precision real by our PASTIX software and by PSPASES version 1.0.3 based on a multifrontal approach [9]. PASTIX makes use of version 3.4 of the SCOTCH static mapping and sparse ordering software package developed at LaBRI. PSPASES is set to make use of METIS version 4.0 as its default ordering library. In both cases, blocking size is set to 64 and the IBM ESSL library is used.

These parallel experiments were run on an IBM SP2 at CINES (Montpellier, France), which nodes are 120 MHz Power2SC thin nodes (480 MFlops peak performance) having 256 MBytes of physical memory each. Table 2 reports the performances of our parallel block factorization for our distribution strategy and the one of the PSPASES software.

Table 2. Factorization performance results (time in seconds and Gigaflops) on the IBM SP2. For each matrix, the first line gives the PaStiX results and the second line the PSPASES ones.

Name	Number of processors					
	2	4	8	16	32	64
B5TUER	29.54 (0.52)	15.52 (0.99)	8.86 (1.73)	5.25 (2.91)	3.96 (3.87)	2.91 (5.25)
	-	-	14.04 (0.88)	7.32 (1.69)	3.91 (3.16)	2.58 (4.79)
BMWCRA1	-	-	30.97 (1.84)	17.28 (3.30)	9.89 (5.76)	6.94 (8.21)
	-	-	-	-	24.87 (2.49)	13.48 (4.61)
MT1	37.92 (0.56)	20.35 (1.04)	11.29 (1.87)	6.65 (3.17)	4.33 (4.87)	3.51 (6.01)
	-	-	-	10.71 (2.12)	5.70 (3.98)	3.59 (6.32)
OILPAN	7.28 (0.41)	3.81 (0.78)	2.15 (1.39)	1.39 (2.14)	1.00 (3.00)	0.87 (3.42)
	-	5.23 (0.53)	2.79 (0.99)	1.73 (1.59)	1.25 (2.20)	0.93 (2.96)
QUER	8.35 (0.39)	4.46 (0.74)	2.57 (1.28)	1.67 (1.96)	1.16 (2.83)	0.93 (3.53)
	23.80 (0.14)	13.11 (0.26)	3.22 (1.07)	2.01 (1.72)	1.30 (2.65)	0.96 (3.59)
SHIP001	20.98 (0.43)	10.91 (0.83)	6.07 (1.49)	3.63 (2.49)	2.43 (3.72)	1.96 (4.60)
	-	15.32 (0.62)	8.11 (1.17)	4.48 (2.11)	2.98 (3.17)	2.14 (4.42)
SHIP003	-	109.78 (0.73)	45.83 (1.75)	25.75 (3.11)	16.60 (4.83)	12.03 (6.66)
	-	-	-	-	21.28 (3.57)	14.08 (5.39)
SHIPSEC5	-	79.12 (0.88)	35.68 (1.95)	20.51 (3.39)	13.99 (4.97)	11.58 (6.00)
	-	-	-	-	21.80 (2.52)	11.81 (4.66)
THREAD	78.04 (0.50)	41.14 (0.94)	22.85 (1.70)	13.50 (2.88)	10.42 (3.73)	6.70 (5.80)
	-	-	-	21.02 (1.70)	11.24 (3.19)	6.41 (5.59)
X104	31.53 (0.54)	19.69 (0.87)	9.98 (1.72)	7.49 (2.29)	5.09 (3.37)	3.85 (4.44)
	-	-	-	8.32 (1.70)	4.92 (2.87)	3.18 (4.44)

An important remark to help comparisons is that PSPASES uses a Cholesky (LL^T) factorization, intrinsicaly more BLAS efficient and cache-friendly than the LDL^T one used by PaStiX. For instance, for a dense 1024x1024 matrix on one Power2SC node, the ESSL LL^T factorization time is 1.07s whereas the ESSL LDL^T factorization time is 1.27s.

A multi-variable polynomial regression has been used to build an analytical model of these routines. This model and the experimental values obtained for communication startup and bandwidth are used by the partitioning and scheduling algorithm. It is important to note that the number of operations actually performed during factorization is greater than the OPC value because of amalgamation and block computations.

Good scalability is achieved for all of the test problems, at least for moderately sized parallel machines; the measured performances vary between 3.42 and 8.21 Gigaflops on 64 nodes. Results show that PaStiX compares very favorably to PSPASES and achieves better solving times in almost all cases up to 32 processors. The comparison on larger cases still appears favorable to PaStiX almost up to 64 processors. On the other smaller cases, performance is quite equivalent on 64 processors when scalability limit is reached.

4   Conclusion and Perspectives

In this paper, we have presented an efficient combination of 1D and 2D distributions and the induced static scheduling of the block computations for a parallel sparse direct solver. This work is still in progress; we are currently working on a more efficient criterion to switch between the 1D and 2D distributions, to improve scalability. We are also developing a modified version of our strategy to take into account architectures based on SMP nodes.

References

[1] P. Amestoy, T. Davis, and I. Duff. An approximate minimum degree ordering algorithm. *SIAM J. Matrix Anal. and Appl.*, 17:886–905, 1996.

[2] C. Ashcraft. The fan-both family of column-based distributed Cholesky factorization algorithms. *Graph Theory and Sparse Matrix Computation, IMA, Springer-Verlag*, 56:159–190, 1993.

[3] C. Ashcraft, S. C. Eisenstat, and J. W.-H. Liu. A fan-in algorithm for distributed sparse numerical factorization. *SIAM J. Sci. Stat. Comput.*, 11(3):593–599, 1990.

[4] C. Ashcraft, S. C. Eisenstat, J. W.-H. Liu, and A. Sherman. A comparison of three column based distributed sparse factorization schemes. In *Proc. Fifth SIAM Conf. on Parallel Processing for Scientific Computing*, 1991.

[5] P. Charrier and J. Roman. Algorithmique et calculs de complexité pour un solveur de type dissections emboîtées. *Numerische Mathematik*, 55:463–476, 1989.

[6] I. S. Duff. Sparse numerical linear algebra: direct methods and preconditioning. Technical Report TR/PA/96/22, CERFACS, 1996.

[7] G. A. Geist and E. Ng. Task scheduling for parallel sparse Cholesky factorization. *Internat. J. Parallel Programming*, 18(4):291–314, 1989.

[8] P. Hénon, P. Ramet, and J. Roman. A Mapping and Scheduling Algorithm for Parallel Sparse Fan-In Numerical Factorization. In *EuroPAR'99, LNCS 1685*, pages 1059–1067. Springer Verlag, 1999.

[9] M. Joshi, G. Karypis, V. Kumar, A. Gupta, and Gustavson F. PSPASES : Scalable Parallel Direct Solver Library for Sparse Symmetric Positive Definite Linear Systems. Technical report, University of Minnesota and IBM Thomas J. Watson Research Center, May 1999.

[10] F. Pellegrini, J. Roman, and P. Amestoy. Hybridizing nested dissection and halo approximate minimum degree for efficient sparse matrix ordering. In *Proceedings of IRREGULAR'99, Puerto Rico, LNCS 1586*, pages 986–995, April 1999. Extended version to be published in Concurrency: Practice and Experience.

[11] A. Pothen and C. Sun. A mapping algorithm for parallel sparse Cholesky factorization. *SIAM J. Sci. Comput.*, 14(5):1253–1257, September 1993.

[12] E. Rothberg. Performance of panel and block approaches to sparse Cholesky factorization on the iPSC/860 and Paragon multicomputers. *SIAM J. Sci. Comput.*, 17(3):699–713, May 1996.

[13] E. Rothberg and A. Gupta. An efficient block-oriented approach to parallel sparse Cholesky factorization. *SIAM J. Sci. Comput.*, 15(6):1413–1439, November 1994.

[14] E. Rothberg and R. Schreiber. Improved load distribution in parallel sparse Cholesky factorization. In *Proceedings of Supercomputing'94*, pages 783–792. IEEE, 1994.

Workshop on
Java for Parallel and Distributed Computing

Denis Caromel   Serge Chaumette   Geoffrey Fox   Peter Graham

INRIA   LaBRI   NPAC   PDSL and ADSL

Nice Univ.   Bordeaux 1 Univ.   Syracuse Univ.   Univ. of Manitoba

This series of workshops was initially created during the IPPS/SPDP conference that was held in Orlando, Florida, in April 1998. At that time Java was beginning to play a larger and more important role in the domain of parallel and distributed computing as was reflected by the significant number of papers that year that referred to this technology. It was thus decided to set up a forum for the community attending IPPS/SPDP (now merged as IPDPS for this Y2K Cancun edition) to discuss the topic. This workshop continues to provide an opportunity to share experience and views of current trends and activities in the domain.

The 2000 version of this workshop follows last year's workshop that was held in San Juan, Puerto Rico. Its focus is on Java for parallel and distributed computing, supporting environments, and related technologies. This year the organizers broadened the focus and added a new topic "Internet-based Parallel and Distributed Computing" to both cover new areas of interest and also to more fully recognize the impact of some of the subjects covered by papers presented last year in Puerto Rico.

The papers selected for presentation this year focus on efficiency and portability. They present effective solutions on how to cope with the most recent technologies in a transparent manner (papers on cluster computing and IPv6), without losing sight of efficiency (papers on native code integration and asynchronous parallel and distributed computing), portability and standards (paper on Java Message Passing API). The organizers again decided this year to reserve space for invited sessions and for open discussions in addition to the regular papers.

The quality and variety of the topics covered by the submitted papers reflect the vitality of this research area. The work of the referees and the program committee have led to the selection of a collection of papers reflecting current research efforts in the field. The organizers believe that the papers presented here advance the field significantly and that other researchers in the area will want to become familiar with their contents.

January 2000

Denis Caromel
Serge Chaumette
Geoffrey Fox
Peter Graham

J. Rolim et al. (Eds.): IPDPS 2000 Workshops, LNCS 1800, pp. 526–527, 2000.
© Springer-Verlag Berlin Heidelberg 2000

Workshop Organization

Program Co-Chairs

Denis Caromel   INRIA, University of Nice Sophia Antipolis,
                         Nice, FRANCE
Serge Chaumette   LaBRI, University Bordeaux 1,
                         Bordeaux, FRANCE
Geoffrey Fox   NPAC, Syracuse University,
                         Syracuse, NY, USA
Peter Graham   PDSL and ADSL, University of Manitoba,
                         Winipeg, Manitoba, CANADA

Program Committee

Jack Dongarra   University of Tenessee, USA
Ian Foster   Argonne National Laboratory, USA
Doug Lea   State University of New York at Oswego, USA
Vladimir Getov   University of Westminster, London, UK
Dennis Gannon   Indiana University, USA
Siamak Zadeh   Sun Microsystems Computer Corp., USA
George K. Thiruvathukal Loyola University, USA
Michael Philippsen   University of Karlsruhe, GERMANY
David Walker   Cardiff University, UK

An IP Next Generation Compliant *Java*™ *Virtual Machine*

Guillaume Chelius[1] and Éric Fleury[2]

[1] École Normale Supérieure de Lyon, Guillaume.Chelius@ens-lyon.fr
[2] LORIA, INRIA, Eric.Fleury@loria.fr

Abstract This paper discusses the implementation of an IPv6.java.net package allowing distributed applications to benefit from all the new aspects of the IPv6 protocol such as: security, flow control, mobility and multicast. Moreover, we present a full IP Next Generation Compliant *Java*™ *Virtual Machine*, that is a JVM which is able to handle both IPv4 and IPv6 protocols from standard Java APIs and in a fully transparent way for the users/programmers.

1 Introduction

Recently, there has been a strong interest in using heterogeneous, remote resources for high performance distributed computing. The concept of *Computational Grid* envisioned in Foster and Kesselman [11] has emerged to capture the vision of a network computing system that provides broad access not only to massive information resources but also to massive computational resources. At the same time, several other opportunities have also been highlighted, including the potential to realize time-shared, collaborative metacomputing and to exploit portability and code mobility through a language such as Java [4,12,13,16]. As a result, there have been many highly successful projects that focus on particular aspect of distributed computing and the networking part takes a major position, not only in the research fields but also in the deployment of ambitious projects propounding very large-scale metacomputing and the computational grid [10,14,20].

In the early nineties, in this context of technological improvement in both computer science and telecommunication areas, the expansion of the Internet has become such huge that the format of addresses was not adapted to the number of potential hosts any more. This was the main reason which pushed researchers to develop a new version of the IP protocol. In addition, this development was also a good way to introduce advances made in network research during the twenty five last years. These advances concern a wide domain covering security, flow control, mobility and multicast. All this work led to the conception of Internet protocol version 6 (IPv6 for short), also known as *IP next generation*. Much of the current IPv6 architecture has already been ratified, and implementations are emerging [9] even if – as an evolving technology – IPv6 is far from being a production network proposition at present. Several issues still remain.

In this paper, we address the advantages that distributed applications may benefit when using IPv6 and we present our implementation of an IPv6 Compliant *Java*™ *Virtual Machine*. One of the main reasons of the popularity of the Java programming

J. Rolim et al. (Eds.): IPDPS 2000 Workshops, LNCS 1800, pp. 528-535, 2000.

language is its support for distributed computing [3]. Java's API for sockets, URL and other networking facilities is much simpler than what is offered by other programming languages like C or C++. Moreover, the Java™ Remote Method Invocation (RMI) was designed to make distributed application programming as simple as possible, by hiding the remoteness of distributed objects as much as possible. All these APIs that were developed and used with an IPv4 JVM will be fully available on our IPv6 Compliant *Java™ Virtual Machine*. This goal can be obtain by two complementary approaches: providing a `IPv6.java.net` package (API for IPv6) and/or running a JVM above an IPv6 stack and thus inherit all the enhancements of IPv6.

Section 2 contains a brief overview of IP next generation, and points out the main modifications from the IPv4 protocol that may present improvements for distributed applications. Section 3 presents the integration/implementation of IPv6 into Java, *i.e.*, an API supporting IPv6. Section 4 presents some extensions and improvement we add and results to validate our approach. Section 5 concludes with future research and software design directions.

2 A Quick overview of IPv6

The aim is to present briefly the main important differences between IPv4 and IPv6 that should benefit distributed applications in terms of improvement and new facilities. For a more precise description of IPv6, please refer to [6,15].

2.1 Addressing format

The ability to sustain continuous and uninterrupted growth of the Internet could be viewed as the major driving factor behind IPv6. IP address space depletion and the Internet routing system overload are some of the major obstacles that could preclude the growth of the Internet. Even though the current 32 bit IPv4 address structure can enumerate over 4 billion hosts on as many as 16.7 million networks, the actual address assignment efficiency is far less than that, even on a theoretical basis [15]. This inefficiency is exacerbated by the granularity of assignments using Class A, B and C addresses. By extending the size of the address field in the network layer header from 32 to 128 bits, IPv6 raised this theoretical limit to 2^{128} nodes. Therefore, IPv6 could solve the IP address space depletion problem for the foreseeable future.

2.2 Modifications of the IP stack protocols

IP. The modification of the address length is not the only change that occurred in the IP protocol. The aim behind all the modifications is to optimize the treatment inside a router in order to reduce the latency in each intermediate router along a packet's delivery path. The most important optimization is the simplification of the header format which has now a fixed size. Some IPv4 header fields (*e.g.*, checksum, fragmentation) have been dropped or made optional to reduce the common-case processing cost of packet handling and to keep the bandwidth overhead of the IPv6 header as low as possible in spite of the increased size of the addresses. IPv6 options are placed in separate headers that are located in packets between the IPv6 header and the transport-layer header. Since most IPv6 option headers are not examined or processed by any router along a packet's delivery path until it arrives at its final destination, this organization facilitates a major improvement in router performance for packets containing options. A key extensibility feature of IPv6 is the ability to encode, within an option, the action which a router or host should perform if the option is unknown. This permits the incremental deployment of additional functionalities into an operational network with a minimal danger of disruption.

ICMPv6. ICMPv6 is more than a simple evolution of ICMP. It now also supports the functionalities of ARP and IGMP. It still handles control or error messages like *unreachable destination, no route to host, packet too big* but it becomes also responsible for the *neighbor discovery* protocol. This protocol is used to perform different tasks: *Address Resolution, Neighbor Unreachability Detection*, some configurations and *Redirection Indication* and to support autoconfiguration.

TCP and UDP. Modifications made to TCP and UDP concern two main areas. The first one deals with data integrity: a pseudo IP header has been added to the computation of the checksum (it is now mandatory for UDP). The second one concerns packet length. It is now possible to send *jumbograms*, which are packets longer than 65536 bytes. They can notably be useful in distributed computation. For example, an IP over Myrinet will benefit from the introduction of jumbogram. Since there is no physical MTU fixed in Myrinet, packet can be of any size which allow to achieve efficient throughput.

2.3   Other new features

Support for authentication and privacy: IPsec. IPv6 includes the definition of an extension which provides support for authentication and data integrity. This extension is included as a basic element of IPv6 and support for it will be required in all implementations. IPv6 also includes the definition of an extension to support confidentiality by using encryption. Support for this extension will be strongly encouraged in all implementations. The use of IPsec in a computational grid context can be very useful in the deployment of sensitive applications that are potentially *griddable*. The fact that security is pushed to the network layer will increase the performance of such systems and allows an easy creation of VPN (*Virtual Private Network*) that can be deployed over the Internet.

Multicast. The scalability of multicast routing is improved by adding a "scope" field to multicast addresses. Multicast provides more flexibility in a more efficient way and it is heavily used even in the most basic protocol like Neighbor Discovery Protocol (NDP). Therefore, multicast is mandatory in every IPv6 hosts/routers whereas it was optional in IPv4. The principle of multicast opens numerous perspectives in the domains of metacomputing or distributed computation. Indeed, it enables the dispatch of information and communication between several entities in a new and efficient way and appears to be a key point in Distributed Interactive Application (DIS) [17].

Mobility. The principle of mobility is quite new. The aim is that computers can stay connected to the Internet despite their physical moves. It assumes that a computer can automatically obtain an address when it is connected to an IP network but also that it is reachable on a constant address. The result is a great flexibility in network management and architecture. A protocol, based on the use of *extensions*, was written to negotiate such associations (between the *mother address* of a computer and the router where it is actually located). This new capability may allow new interfaces to be easily designed to support a range of scientific activities across distributed, heterogeneous computing platforms [18].

Quality of service capabilities. A new capability is added to enable the labeling of packets belonging to particular traffic "flows" for which the sender has requested special handling, such as non-default quality of service or "real-time" service. This traffic class will be useful to enhance distributed applications that have specific features (interactivity, hard time constraint, ...) like DIS [5]. A new type of address, called a "cluster address" is defined to identify topological regions rather than individual nodes. The use of cluster addresses in conjunction with the IPv6 source route capability allows nodes additional control over the path their traffic takes.

3   Developing an IPv6 package for Java

The first step of our project, bringing IPv6 to Java, was to create a package, called `fr.loria.resedas.net6`, that provides the possibility to deal with IPv6 networking programming. This API has to be coherent with regard to the Java architecture and, in particular, with the existing network programming API as described in [8].

3.1   Architecture

If we look at the different classes of the *java.net* package, three main levels can be discerned. The first one is composed of classes which only deal with Java mechanisms: the *exceptions*. The second one is composed of classes which implement session or application mechanisms. Finally, the third one deals with transport mechanisms and roughly speaking with TCP/IP. We will call it the *networking set* and we will work at this level since this is typically the right place where we need to plug IPv6 functionalities.

In order to provide a good integration ability to our package and to make the development of upper-layer objects easier, we consider the structure of the *networking set* and decide to keep the same structure. The first obvious advantage of this choice is code reuse. Not only for the design (and coding) of the package itself but also for network interacting objects. The transition from already existing IPv4 objects, towards an IPv6 networking ability would be simple, flexible and easy to perform. In most cases, it can be automatically done by syntactic transformation (we used `sed`). A second profit is that we did not introduce any new classes. Thus, the underlying semantic of Java is not changed, the security is not depreciated and in a more practical way, network programming with Java remains the same. The last but not the least advantage is the facility of integrating such a package in a JVM. We will see in details in section 4.2 what that really means.

The first decision we had to take in the translation of the networking set from IPv4 to IPv6 was a purely syntactic one. We have to find new names and then upgrade all the references off the concerned objects. This step was performed automatically by using a simple stream editor. In order to handle IPv6 specifications, like the storage size of addresses, the other major modification was focused on the internals of objects. Concept changes between IPv4 and IPv6 were also responsible for some other modifications. Of course, we did not limit ourselves to the already existing interface but improved it by integrating new IPv6 features.

3.2 Implementation of the underlying mechanisms

Under Java, the development of new features, outside the standart Java class library, requires to deal with mechanisms that are themselves not available from the Java™ Virtual Machine. In our case, these mechanisms deal with operations on IPv6 sockets and addresses. In order to be able to perform such operations, we had to develop a library which would embed the Java™ virtual machine into native network operations.

In order to build our library we use the *Java™ Native Interface* (JNI). JNI allows Java code that runs inside a JVM to interoperate with applications and libraries written in other programming languages, such as C, C++, and assembly. The main problem we encountered was the question of thread safety. Working in Java induces working in a multi-thread environment, and such an environment requires to take some precautions. Without being *Java-thread safe* the library could not be effective. This is the reason why we could not use classical system calls in the library without any particular precautions.

In fact, only a limited number of system calls were concerned. In particular, they are mostly I/O calls like *open*, *read* or *write*. The important point is that all these calls are already used in the JVM, at least to support the *java.net* package. Thus, the *thread safety* problem has been already addressed. In the JVM, the controversial system calls are mapped in what we will call pseudo functions which take all precautions in connection with threads. Since this pseudo functions were already implemented for the system calls in question, we decided to integrate our library into the architecture of the JVM by proceding the same way as the classical network library. Our C functions would call the same pseudo functions. For more information about network programming, please refer to [19].

4   Results and extensions

The last step in the development of the package was the introduction of specific features provided by IPv6.

4.1   A raw level and new options

In a classical Java environment, raw sockets are not accessible. This could represent a lack of flexibility in several situations. For example, one could want to use *emulators* to reproduce particular environments for testing purposes and thus generate specific traffic (ICMP, router alert...). We could also need to monitor network traffic when using or optimizing parallel applications [2]. For these purposes, we wrote a raw socket interface for Java.

New IPv6 options were also added to the network API. If multicast already existed in Java, IPsec had never been included before in a Java package, though it was defined under IPv4. Flow control is another new feature proposed in the package. It relies on the mechanisms described in section 2.3.

4.2   An IPv6 compliant JVM

The good behavior of our IPv6 package opens new perspectives. Instead of just adding a new package to the JVM, we decided to completely replace the IPv4 network mechanisms present inside a classical JVM by the IPv6 one described above, in order to have a transparent IPv6 compliant JVM. To achieve this task, we replaced the original *java.net* objects with ours and integrated the IPv6 library in the internals of the JVM. The consequence is that any Java objects can not only communicate under IPv6 without any modifications, but also manage IPv4 messages. Indeed, IPv6 provides simple and flexible transition from IPv4. The key transition objective is to allow IPv6 and IPv4 hosts to interoperate. The Simple Internet Transition (SIT) specify for example a model of deployment where all hosts and routers upgraded to IPv6 in the early transition phase are "dual" capable (*i.e.*, implement complete IPv4 and IPv6 protocol stacks) In other words, SIT ensures that IPv6 hosts can interoperate with IPv4 hosts anywhere in the Internet. This means that our JVM allows any Java program to work under IPv6 as well as under IPv4. In particular, any Java application using networking will work in a IPv6 environment by using indistinctly IPv4 addresses or IPv6 addresses and becomes able to performed *Remote Method Invocations* (RMI) on a IPv4 JVM or in a fully IPv6 network environment. Another interesting point is that the modifications made to the JVM are not irreversible. Indeed, it is very easy to switch between an IPv4/IPv6 JVM and classical IPv4 JVM.

4.3   Results

To validate the IPv6 package, we translated several Java applications from IPv4 to IPv6. This was done by doing only syntactic transformations. We did it for small applications (secured telnet clients, ping programs) and for larger ones (chat programs based on multicast). We used for example jmrc (a Java multicast chat program) and introduced flow control and security to it.

To validate the IPv6 compliant JVM, we run several unmodified Java applications on an dual stack platform. The experiments show that both protocols are fully supported. This means, for example, that telnet servers are now accepting IPv6 and IPv4 connections, or that a web server like Jigsaw can run without any modification over IPv6. Tests of RMI were also performed and were conclusive since RMI clients and servers now support IPv6 networks.

We are also carried out several experiments and benchmarks, especially concerning the security part and more generally the overhead introduced by IPv6. Unfortunately, at the time of sending the camera ready version of this paper and by a lack of space we can not include graphs but the work will be done soon and we will be able to present our finding in the Workshop in May 2000. First results show that on a local network, IPv4 performs a little better. These results can be easily explained since their is no routers by definition on a local network, and IPv6 was designed to optimize and reduce the latency in each intermediate router.

Our IPv6 API has been developed under several IPv6 stacks for several OS: Linux, FreeBSD, NetBSD and Windows NT. Since the progress of the different stacks are not the same, all the features proposed by the IPv6 package are not always supported in all OS. A new version of the library is under development for Solaris. Note that our package is running under JdK 1.1 and JdK 1.2.

5   Conclusion

We have discussed the design and development of a IPv6 Compliant Java™ Virtual Machine. We have also highlighted the benefits of using IPv6 in distributed applications compare to the IPv4 currently available. Our results have shown that the IPv6 package fulfills the needs of programmers not only in terms of compatibility with IPv4 but also in terms of new functionalities and we hope in terms of performances. The main point is that every Java applications running under our IPv6 JVM are still compatible with IPv4 and thus modification to the application code is not necessary. In order to show that a IPv6 Compliant Java™ Virtual Machine, which is simply a JVM where the networking part has been replaced by our IPv6 package, makes transition from IPv4 to IPv6 free of development/coding, we installed and run the original Jigsaw web server. This server ipv6.loria.fr can be reached from any hosts IPv4 or IPv6 anywhere on the Internet.

In the future, we plan to continue to improve the development of our IPv6 Compliant Java™ Virtual Machine on other IPv6 stacks. We plan also to develop computational distributed application, especially under the Globus metacomputing systems. For this purpose it seems interesting to take advantage of all the new features provided by IPv6: security, flow control and multicast and we plan to develop an IPv6 version over Myrinet in order to be able to send jumbogram data packets which will provide efficient API for developing MPI in Java [1,7].

References

1. M. Baker, B. Carpenter, G. Fox, and S. H. Koo. mpiJava: An object-oriented Java interface to MPI. In *Parallel and Distributed Processing*, volume 1586 of *Lecture Notes in Computer Science*, pages 748–762, San Juan, Puerto Rico, April 1999.
2. A. M. Bakić, M. W. Mutka, and D. T. Rover. An on-line performance visualization technology. In *Heterogeneous Computing Workshop (HCW '99)*, pages 47–59, San Juan, Puerto Rico, April 1999. IEEE.
3. B. Carpenter, Y.-J. Chang, G. Fox, and X. Li. Java as a language for scientific parallel programming. *Lecture Notes in Computer Science*, 1366, 1998.
4. B. Carpenter, G. Zhang, G. Fox, and X. Li. Towards a Java environment for SPMD programming. *Lecture Notes in Computer Science*, 1470, 1998.

5. C. Chassot, A. Loze, F. Garcia, L. Dairaine, and L. R. Cardenas. Specification and realization of the QoS required by a distributed interactive simulation application in a new generation internet. In M. Diaz, P. Owezarski, and P. Sénac, editors, *Interactive Distributed Multimedia Systems and Telecommunication Services (IDMS'99)*, volume 1718 of *Lecture Notes in Computer Science*, pages 75–91, Toulouse, France, October 1999.

6. G. Cizault. *IPv6: Théorie et pratique*. O'Reilly, 1998.

7. K. Dincer. A ubiquitous message passing interface implementation in java: *jmpi*. In *International Parallel Processing Symposium, Symposium on Parallel and Distributed Processing (IPPS / SPDP 1999)*, pages 203–207, San Juan, Puerto Rico, April 1999. IEEE.

8. Java API documentation. home page. http://java.sun.com/docs/index.html.

9. Robert Fink. Network integration — boning up on IPv6 — the 6bone global test bed will become the new Internet. *Byte Magazine*, 23(3):96NA–3–96NA–8, March 1998.

10. I. Foster and C. Kesselman. Globus: A metacomputing infrastructure toolkit. *The International Journal of Supercomputer Applications and High Performance Computing*, 11(2):115–128, Summer 1997.

11. I. Foster and C. Kesselman. *The Grid: Blueprint for a new computing infrastructure*. Morgan Kaufmann, 1998.

12. G. Fox. Editorial: Java for high-performance network computing. *Concurrency: Practice and Experience*, 10(11–13):821–824, September 1998. Special Issue: Java for High-performance Network Computing.

13. Java Grande. home page. http://www.javagrande.org.

14. Andrew S. Grimshaw, William A. Wulf, James C. French, Alfred C. Weaver, and Paul F. Reynolds, Jr. Legion: The next logical step toward a nationwide virtual computer. Technical Report CS-94-21, Department of Computer Science, University of Virginia, June 08 1994. Mon, 28 Aug 1995 21:06:39 GMT.

15. C. Huitema. *IPv6: The New Internet Protocol*. Prentice Hall, 1996.

16. R.van Nieuwpoort, J. Maassen, T. Bal, H. E.and Kielmann, and R. Veldema. Wide-area parallel computing in java. In *Proceedings of the ACM Java Grande Conference*, New York, NY, June 1999. ACM Press.

17. David Powell. Group communication. *Communications of the ACM*, 39(4):50–97, April 1996. (special section Group Communication).

18. M. J Skidmore, M. J. Sottile, and A. D. Cuny, J. E.and Malony. A prototype notebook-based environment for computational tools. In *High Performance Networking And Computing Conference*, Orlando, USA, November 1998.

19. Richard W Stevens. *UNIX Network Programming*. Software Series. Prentice Hall PTR, 1990.

20. Globus Metacomputing Toolkit. home page. http://www.globus.org.

An Approach to Asynchronous Object-Oriented Parallel and Distributed Computing on Wide-Area Systems*

M. Di Santo[1], F. Frattolillo[1], W. Russo[2] and E. Zimeo[1]

[1] University of Sannio, School of Engineering, Benevento, Italy
[2] University of Calabria, DEIS, Rende (CS), Italy

Abstract. This paper presents a flexible and effective model for object-oriented parallel programming in both local and wide area contexts and its implementation as a Java package. Blending *remote evaluation* and *active messages*, our model permits programmers to express asynchronous, complex interactions, so overcoming some of the limitations of the models based on message passing and RPC and reducing communication costs.

1 Introduction

Exploiting geographically distributed systems as high-performance platforms for large-scale problem solving is becoming more and more attractive, owing to the high number of workstations and clusters of computers accessible via Internet and to the spreading in the scientific community of platform-independent languages, such as Java [14]. Unfortunately, the development of efficient, flexible and transparently usable distributed environments is difficult due to the necessity of satisfying new requirements and constraints. (1) *Host heterogeneity*: wide-area systems solve large-scale problems by using large and variable pools of computational resources, where hosts often run different operating systems on different hardware. (2) *Network heterogeneity*: wide-area systems are characterized by the presence of heterogeneous networks that often use a unifying protocol layer, such as TCP/IP. While this allows hosts in different networks to interoperate, it may limit the performance of specialized high-speed networking hardware, such as Myrinet [1]. (3) *Distributed code management*: a great number of hosts complicates the distributed management of both source and binary application code. (4) *Use of non-dedicated resources*: traditional message-passing models are too static in order to support the intrinsic variability of the pool of hosts used in wide-area systems. In this context, the interaction schemes based on one-sided communications are more apt, even if some of them, by adopting synchronous client/server models, do not ensure an efficient parallelization of programs.

Starting from these considerations, we propose a flexible and effective model for object-oriented parallel programming in both local- and wide-area contexts.

---

* Work carried out under the financial support of the M.U.R.S.T. in the framework of the project "Design Methodologies and Tools of High Performance Systems for Distributed Applications" (MOSAICO)

It is based on the *remote evaluation* [13] and *active messages* [3] models and overcomes some of the limitations of the models based on message passing and RPC, thanks to its completely asynchronous communications and to the capability of expressing complex interactions, which permit applications to reduce communication costs. Moreover, its ability of migrating application code on demand avoids the static distribution and management of application software.

The model has been integrated into a minimal, portable, efficient and flexible middleware infrastructure, called *Moka*, implemented as a Java library. *Moka* allows us both to directly write object-oriented, parallel and distributed applications and to implement higher-level programming systems. *Moka* applications are executed by a parallel abstract machine (*PAM*) built on top of a variable collection of heterogeneous computers communicating by means of a transparent, multi-protocol transport layer, able to exploit high-speed, local-area networks. The *PAM* appears as a logically fully-interconnected set of abstract nodes (*AN*), each one wrapping a Java Virtual Machine. Each physical computer may host more than one *AN*.

2   Related work

De facto standard environments for parallel programming on clusters of workstations are doubtless PVM [7] and MPI [12]. Both use an execution model based on processes that communicate by way of message passing, which offers good performances but supports only static communication patterns. Moreover both PVM and MPI are rather complex to be used by non-specialists and, on the wide-area scale, present the distributed code management problem. Java implementations of PVM [4] and MPI [8] simplify a little the use of message passing, especially for object-oriented parallel programming.

A different and more attractive approach to distributed and parallel computing is the one proposed by Nexus [5] and NexusJava [6]. These systems support fully asynchronous communications, multithreading and dynamic management of distributed resources in heterogeneous environments. The communication scheme is based on the use of global pointers (one-sided communication), which allow software to refer memory areas (Nexus) or objects (NexusJava) allocated in different address spaces. NexusJava is rather similar to *Moka* both in the programming model and in the architecture, but it has a too low-level and verbose programming interface and limits distributed interactions to the invocations of methods explicitly registered as handlers.

Commonly used middlewares based on Java RMI [14], generally, do not directly provide asynchronous mechanisms on the client site. Some recent systems, such as ARMI [11], transform synchronous RMI interactions into asynchronous ones by using either an explicit or an implicit multithreading approach. However, in both cases, inefficiencies due to the scheduling of fine grain threads are introduced, especially on commodity hardware. Instead, *Moka* provides more efficient, fully asynchronous communications at system level.

3   The *Moka* programming model

The model is based on the following three main concepts. (1) *Threads*, which represent the active entities of a computation. (2) *Global objects*, which, through a global naming mechanism, allow the applications to build a global space of objects used by the threads to communicate and synchronize themselves. (3) *Active objects* (auto-executable objects), which allow threads to interact with global objects, by using a one-sided asynchronous communication mechanism.

A *Moka* computation is fully asynchronous and evolves through subsequent changes of global objects' states and through their influence on the control of application threads. These have to be explicitly created and managed by the program, which must pay attention to protect objects against simultaneous accesses from the threads running on the same *AN*.

An active object (*AO*) can be asynchronously and reliably communicated, as a message, to an *AN* where it may arrive with an unlimited delay and without preserving the sending order. When an *AO* reaches its destination node, the execution of its handler (a function connected to the *AO*) is automatically carried out. The handler code, when not already present on the node, is dynamically retrieved and loaded by *Moka*. So it is possible to program according to a pure MIMD model, where applications are organized as collections of components, each one implementing a class used to build active or global objects, loaded on the nodes only when necessary (code on demand promoted by servers).

The automatic execution of handlers implies that a message is automatically and asynchronously received without using any explicit primitive. This semantics requires the existence of a specific activity (*network consumer*) devoted to the tasks of extracting active objects from the network and of executing their handlers. A solution is based on the *single-threaded upcall* model, in which a single network consumer serially runs in its execution environment the handlers of the received objects. On the other hand, when a handler may suspend on a condition to be satisfied only by the execution of another *AO*, in order to avoid the deadlock of the system, the program must use the *popup-threads* model and so explicitly ask for the handler to be executed by a separate, dedicated thread (*network consumer assistant*) [9].

Application threads and handlers interact by using a space of global objects (*GOS*). A global object (*GO*) is abstractly defined as a pair (GN, GO_IMPL). GN is the global name of the *GO* and univocally identifies it in the *Moka* system. GO_IMPL is the concrete *GO* representation, an instance of a user defined class physically allocated on an *AN*. A thread can access a GO_IMPL by making a query to the *GOS*, which is organized as a distributed collection of *name spaces*, each one (NS_i) allocated on a different *AN*. Therefore, in order to access a GO_IMPL, it is necessary to know its location (node i). *Moka* offers two ways in order to do this: (1) by means of a static, immutable association (GN, i), established at the GN creation time; (2) by using a dynamic approach, where the GO_IMPL location is explicitly specified at access time. In this latter case, one different implementation per node may be bound to a given GN; so, it

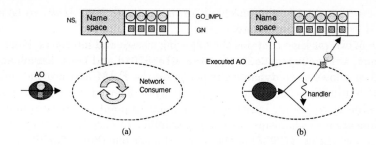

Fig. 1. Creation of a new *GO*. (a) An *AO*, containing a *GO_IMPL* and a *GN*, is arriving on an node. (b) The network consumer runs the *AO* handler, which creates the association (*GN*, *GO_IMPL*) in the NS_i

is possible to realize a replicated implementation of a *GO*, even if the program must explicitly ensure the consistency of replicas.

At the start of computation, only *AO* handlers can refer the local *namespace* (NS_i). Subsequently, the NS_i reference may be passed to other activities running on the same node. When *GO_IMPL*s become shared resources, they must be explicitly protected from simultaneous accesses. The creation of a *GO* on a node i requires three operations executed on the node where the request starts: (1) generate a *GN*; (2) create a specific *AO* containing both the *GN* and the *GO_IMPL*; (3) send the *AO* to the node i (see fig. 1), where its handler execution binds the *GN* to the *GO_IMPL* in the local namespace NS_i. An existent *GO* can be remotely accessed through the following operations: (1) reclaim from the *GN* the identifier of the node where the *GO* resides; (2) send to this node an *AO* which looks up NS_i with *GN* as key, in order to get the *GO_IMPL*; (3) execute on the *GO_IMPL* the operations abstractly requested on the *GO*.

Through a special form of active objects (*AOc*), *Moka* provides a deferred synchronous send primitive that calls for a result produced by the execution of operations tied to a dispatched *AOc*. Sending an *AOc* does not suspend the caller which immediately receives a *promise*, an object that will get the result in the future. A suspension will occur only if the result is reclaimed before it is really available. An *AOc* can be modified and forwarded many times before returning the result (agent like model); so the caller may receive the result of a complex distributed interaction.

4   The Java API of *Moka*

A Java package implements the proposed model, by offering an API that allows programs to create and dynamically configure the *PAM*, to create and send active objects, and to create global objects. Thread management and synchronization on global objects, are instead committed to the Java language default mechanisms. In fact, differently from the proposal in [2], we do not provide

Moka-level synchronization mechanisms, because we want *Moka* to be a minimal and extensible middleware.

The *Moka* package contains the following classes and interfaces: Moka, ActiveObject, ActiveObjectCall, Promise, GlobalUid and LocalNameSpace. The Moka class allows us to dynamically configure the *PAM* and provides programmers with primitives for either asynchronously or deferred synchronously sending active objects to nodes, by using either point-to-point or point-to-multi-point communication mechanisms, with the possibility to specify the single threaded upcall model (Moka.SINGLE) or the popup threads one (Moka.POPUP). It is worth noting that, when a synchronous send primitive is used, the result is to be caught by an instance of Promise. An active object is an instance of a user-defined class implementing one of the Java interfaces ActiveObject and ActiveObjectCall, to be respectively used for asynchronous and deferred synchronous interactions. Instances of GlobalUid and LocalNameSpace classes respectively implement the global name (GN) of a GO and the namespace of a node i (NS_i). In particular, *Moka* creates one LocalNameSpace instance per node, which is automatically passed to all the handlers. Moreover, the *Moka* API provides three classes of active objects: Create, InvokeVoidMethod and InvokeMethod, that, using the reflection Java package, allow programs to respectively create a global object, invoke on a GO a void method or a method returning a value.

For the sake of clarity, in the following, we present a simple program that multiplies in parallel two square matrices, A and B.

```
public class Main implements ActiveObject {
 public void handler(LocalNameSpace ns) {
  float[] a, b; int dim;
  <read matrices a and b as mono-dimensional arrays of dim*dim>;
  GlobalUid gn = new GlobalUid(); ns.bind(gn, new Matrix(b));
  Moka.broadcast(new Create(gn, Matrix.class, b), Moka.SINGLE);
  int numNodes = Moka.size(); int rfn = dim/numNodes;
  float[] subM = new float[dim*rfn];
  Promise[] result = new Promise[numNodes];
  for(int node = 0; node < numNodes; node++) {
   System.arraycopy(a, node*dim*rfn, subM, 0,rfn*dim);
   result[node] = Moka.call(new SubMatrix(gid,subM),node,Moka.SINGLE); }
  for(int node = 0; node < numNodes; node++) {
   float[] r = (float[])result[node].getValue(); <print r>; }
 }
}
public class Matrix implements Serializable {
 private float mat[];
 public Matrix(float[] m) { mat = m; }
 public float[] multiply(float[] a) { return < a x mat >; }
}
public class SubMatrix implements ActiveObjectCall {
 private float[] part; private GlobalUid gn;
 public SubMatrix(GlobalUid g, float[] a) {gid = g; part = a;}
 public Object handler(LocalNameSpace ns) {
```

```
        return ((Matrix)ns.lookUp(gn)).multiply(part);
    }
}
```

The algorithm is organized as follows: B is replicated on each node of the
PAM, whereas A is dissevered into submatrices, each one formed by an equal
number of rows and assigned to a different node of the PAM. The computation
evolves through the parallel multiplication of each submatrix with B. In more
detail, the program creates the PAM and sends a first (Main) AO to one of its
nodes, where the handler replicates B as a global object by generating a global
uid (gn) and wrapping B in an instance of the class Matrix (og_impl); locally,
this is obtained by directly binding, in the local name space, gn to a local instance of Matrix; remotely, by broadcasting an AO of the class Create, which
takes charge to create a remote og_impl and bind it to gn in the name space of
the remote node. Moreover, the handler divides A into dim/numNodes submatrices, where dim is the dimension of A and numNodes is the size of the PAM,
obtained by invoking the primitive Moka.size. Each submatrix is wrapped into
an instance of the class SubMatrix, which implements the ActiveObjectCall
interface, and sent to each node by using the deferred synchronous send primitive
Moka.call; all these invocations immediately return a promise. Remotely, the
handler of SubMatrix gets the local instance of Matrix and invokes its method
multiply. The resulting values are caught by $Moka$ and implicitly sent to the
node that executed the main AO; here, they are extracted from the promise by
using the blocking Promise.getValue primitive.

5   Transparent vs. non-transparent distributed interactions

The $Moka$ model is based on non-transparent, distributed interactions among
threads and global objects, whereas the middlewares based on the RMI model
make possible the transparent invocation of methods on remote objects. Unfortunately, in order to realize transparency, these systems use IDLs (*Interface
Description Language*) and stub generators, which complicate the development
of distributed applications, especially when many remote objects are used. In addition, using RMI, the interactions between different address spaces are limited
to the invocations of remote object methods, explicitly exported by an interface.
So, it is not possible to remotely invoke class methods, to directly access the
public data members of an object and to dynamically create remote objects.
$Moka$ instead, thanks to the possibility of remotely executing all the locally
executable operations, permits us to efficiently realize any kind of interactions
among address spaces, without using stub generators and preserving a non transparent polymorphism between local and remote (global) objects. Moreover, our
model does not require global objects to be instances of particular classes, as
Java RMI does; instead any object can become a global one after its binding in a
NS_i. In addition, the use of distributed interactions as first class objects allows

Moka to minimize communication costs when a distributed interaction implies the execution of methods whose results are used as arguments of other methods on the same *AN*. Therefore, we argue that while transparent distributed interactions allow programs to easily express remote invocations of methods, a non-transparent lower-level approach seems more efficient when object-oriented programming is used for high-performance parallel computing or when the interest is in the development of higher level systems.

6   Performance evaluation

At the present, *Moka* implements three transport modules, two respectively based on TCP and reliable UDP with multicast support, to be used on the Internet and local-area Ethernet networks, and the third, based on Illinois Fast Messages [10], to be used on Myrinet networks. All these protocols can be contemporary used by the *PAM*, because it is possible to specify a different protocol for each pair of nodes.

In the following we present two graphs. The first graph shows the performances of the presented matrix multiplication example on a cluster composed of bi-processor PCs (PentiumII 350MHz) interconnected by a multi-port Fast Ethernet repeater and using the TCP as communication transport. We can observe that, without taking into account the time necessary to transfer the right matrix, a good speedup is achieved for matrices over 400×400.

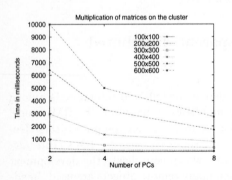 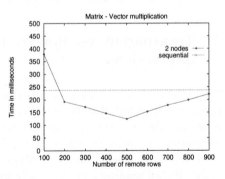

The second graph shows the time needed to compute the product between a 1000×1000 matrix and a vector on two Sun UltraSparcs interconnected by Ethernet, without taking into account the time to transfer the matrix. We can observe that the best performance is reached for a fifty-fifty splitting of the matrix rows on the two machines. Anywhere, the graph shows an absolute speedup even for the other, less favorable splittings of the matrix. The anomaly in the case of 100 remote rows is due to the need of initial remote code loading.

A third experiment was realized on a small wide-area system composed of a PentiumII 400MHz PC interconnected to 2 Sun UltraSparcs by means of a frame relay network (512 Kbps CIR, 2 Mbps CBIR). On this system, the product of two matrices shows a little speedup only for the dimension of 500×500 floats (21 sec. parallel, 33 sec. sequential). In fact, for smaller dimensions the grain is too small, while for larger ones the available bandwith is saturated.

7   Conclusions

We have shown how the integration of the *remote evaluation* model and the *active messages* one permits Java programs to set-up global objects spaces and, with the use of promises, to realize distributed asynchronous interactions that overcome some limitations of message passing and RPC. The proposed model has been integrated into the middleware *Moka* which, thanks to the use of a multi-protocol transport, ensures acceptable performances both in local- and wide-area networks. At this end, we will provide a better integration of default Java serialization with the FM libray in order to improve performances on Myrinet clusters, which currently are only slightly better than the ones provided by TCP clusters.

References

1. N. J. Boden et al.: Myrinet: A Gigabit-per-Second Local Area Network. *IEEE Micro*, 15(1):29-36, 1995.
2. D. Caromel, W. Klauser, and J. Vayssiere: Towards Seamless Computing and Meta-computing in Java. *Concurrency: Pract.&Exp.*, 10(11-13):1043-1061, 1998.
3. T. von Eicken et al.: Active Messages: A Mechanism for Integrated Communication and Computation. *19th Ann. Int'l Symp. Computer Architectures, ACM Press*, NY, 256-266, 1992.
4. A. Ferrari: JPVM: Network Parallel Computing in Java. *ACM Workshop on Java for High-Performance Network Computing*. Palo Alto, 1998.
5. I. Foster, C. Kesselmann, and S. Tuecke: The Nexus Approach to Integrating Multithreading and Communication. *J. of Par. and Distr. Computing*, 37:70-82, 1996.
6. I. Foster, G. K. Thiruvathukal, and S. Tuecke. Technologies for Ubiquitous Super-computing: A Java Interface to the Nexus Communication System. *Concurrency: Pract.&Exp.*, June 1997.
7. A. Geist et al.: *PVM: Parallel Virtual Machine*. The MIT Press, 1994.
8. V. Getov and S. Mintchev: *Towards a Portable Message Passing in Java*. http://perm.scsise.vmin.ac.uk/Publications/javaMPI.abstract.
9. K. Langendoen, R. Bhoedjang, an H. Bal: Models for Asynchronous Message handling. *IEEE Concurrency*, 28-38, April-June 1997.
10. S. Pakin, V. Karamcheti, and A. A. Chien: Fast Messages: Efficient, Portable Communication for Workstation Clusters and MPPs. *IEEE Concurrency*, 60-73, April-June 1997.
11. R. R. Raje, J. I. William, and M. Boyles: An Asynchronous Remote Method Invocation (ARMI) Mechanism for Java. *Concurrency: Pract.&Exp.*. 9(11):1207-1211, 1997.
12. M. Snir et al.: *MPI: The Complete Reference*. The MIT Press, 1996.
13. J. W. Stamos, and D. K. Gifford: Remote Evaluation. *ACM Transactions on Computer Systems*, 12(4):537-565, October 1990.
14. http://www.javasoft.com.

Performance Issues for Multi-language Java Applications

Paul Murray[1], Todd Smith[1], Suresh Srinivas[1], and Matthias Jacob[2]

[1] Silicon Graphics, Inc., Mountain View, CA
{pmurray, tsmith, ssuresh}@sgi.com
http://www.sgi.com
[2] Princeton University, Princeton, NJ
mjacob@cs.princeton.edu
http://www.cs.princeton.edu

Abstract. The Java programming environment is increasingly being used to build large-scale multi-language applications. Whether these applications combine Java with other languages for legacy reasons, to address performance concerns, or to add Java functionality to preexisting server environments, they require correct and efficient native interfaces. This paper examines current native interface implementations, presents performance results, and discusses performance improvements in our IRIX Java Virtual Machine and Just-In-Time Compiler that have sped up native interfacing by significant factors over previous releases.

1 Introduction

The Java programming environment [1] allows fast and convenient development of very reliable and portable software written in 100% Java. However, in recent years, it has increasingly been used to build large-scale multi-language applications, a trend which is due to several factors.

In many circumstances Java has to interface with existing legacy code written in C/C++. Important examples of this include Java bindings for OpenGL [2], Win32, VIA [3], and MPI [4]. In some performance-critical situations, such as oil and gas applications or visual simulations, application developers have to write core portions of their software in higher performance languages such as C, C++, or Fortran. Finally, developers of server environments need to be able to embed a Java Virtual Machine into servers written in other languages in order to provide Java functionality; examples of this include Java Servlets in the Apache Web Server and Java Datablades in the Informix Database. All these types of applications raise issues of correctness and efficiency in native interface implementations that must be addressed.

This paper is a result of our experiences, as VM and JIT compiler implementors working with our customers, with the effects of interfacing Java with other languages such as C, C++, and Fortran. In Section 2 of the paper we go into detail about the various choices that designers of multi-language Java applications have and which ones are suitable for what purposes. In Section 3 we examine

J. Rolim et al. (Eds.): IPDPS 2000 Workshops, LNCS 1800, pp. 544-551, 2000.
© Springer-Verlag Berlin Heidelberg 2000

performance of current Java Native Interface (JNI) [5] implementations. Section 4 discusses issues surrounding memory management and threading in multi-language applications. In Section 5 we present a Fast JNI implementation in the IRIX JVM and JIT that speeds up native interfacing and multi-threaded data accesses by a factor of 3–4x over what is implemented in the standard JavaSoft reference implementation. In Section 6 we discuss related research and in Section 7 we present our conclusions.

2   Choices for Native Interface

Among the many choices that designers of multi-language Java applications have (Fig. 1), the fundamental differences lie in whether Java can coexist and interact directly with C/C++ object models. While j/Direct and KNI tie them together closely, allowing Java objects to be used in C++ contexts and vice versa, JNI takes the approach of separating Java and other languages in a very clear way, which is much more safe and portable, though not very efficient.

As Fig. 1 indicates, JNI is a popular and portable interface, supported by Sun, their licensees, and also other independent software vendors working on clean-room JVM implementations [6][7]. It also allows embedding a Java Virtual Machine within a C/C++ application. This is useful in the context of large servers that need to allow the development of server plugins in Java. The rest of the paper will only look in depth at JNI and its implementation in JVM's derived from JavaSoft's implementation.

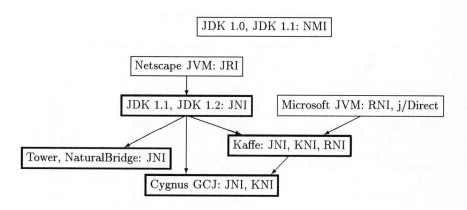

Fig. 1. Choices for interfacing from Java to C/C++. NMI was the original design from Sun and is no longer in active use. Microsoft spawned its own proprietary extensions in RNI (Raw Native Interface) and j/Direct. JNI (Java Native Interface), which evolved from Netscape's work on JRI (Java Runtime Interface), was introduced in JDK 1.1 by Sun and is used in Java2. Clean room implementors such as Tower, NaturalBridge, Kaffe, and GCJ support JNI. Kaffe and GCJ also support KNI (Kaffe Native Interface)

Recommendation 1 *The Java Native Interface (JNI) is a portable and widely supported API for interfacing with C/C++/Fortran codes. Designers should look closely at this before choosing other interfaces.*

3   Java Native Interface Performance

JNI was initially introduced in JavaSoft's JDK 1.1 and the implementations have matured. But there are still wide differences in performance of JNI implementations. This section examines performance results for SGI IRIX and IA-32 Linux and makes recommendations for designers. The SGI tests were performed on an Origin 2000 (300 Mhz R10K) and the Linux test results were obtained on a Dell Pentium II (350 Mhz) running Red Hat Linux. The JVM's used in the tests include SGI's IRIX JDK 1.1.6 [8], SGI's IRIX Java2 [9], Blackdown's [10] Linux JDK 1.1.6, Blackdown's Linux Java2-prev2, and IBM's Linux JDK 1.1.8 [11].

3.1   Cost of a Native Call

One of the most critical pieces of information for developers is the overhead introduced by JNI. One way to determine this is to compare the costs of calling a regular Java method and a JNI method, giving us a rough idea of the overhead introduced for making a native call. In Table 1 we compare the costs of calling an empty Java method and an empty native method one million times.

Table 1. Native call overhead

JVM	Java Method (seconds)	Native Method (seconds)	Slowdown
SGI 1.1.6	41.3	98.0	2.4
SGI 1.1.6 (JIT)	20.8	98.0	4.7
SGI Java2	75.2	76.6	1.0
SGI Java2 (JIT + Fast JNI)	12.7	21.1	1.7
Blackdown 1.1.7	15.0	79.1	5.3
IBM 1.1.8	13.1	81.0	6.2
IBM 1.1.8 (JIT)	0.6*	28.7	47.8*
Blackdown Java2	35.3	49.1	1.4
Blackdown Java2 (Sun JIT)	2.9*	92.8	32.0*
Blackdown Java2 (Inprise JIT)	39.9	49.2	1.2

*The slowdown factors for IBM 1.1.8 and Blackdown Java2 with the Sun JIT are incorrect since the cost of calling a regular Java method is obviously too small; it is likely that JIT optimizations have eliminated the calls of the empty Java method.

From these results we can draw the following conclusions:

- The additional overhead of calling a native method is significant in all cases.
- The cost of calling a native method is lower in Java2 than in JDK 1.1.x.
- If a JIT has specific support for fast JNI calls, they can run significantly faster than under the interpreter (SGI Java2 and IBM 1.1.8); not having such support may cause them to run much slower (Blackdown Java2 with Sun JIT).

Recommendation 2 *For designers of multi-language applications it is impor-tant to choose a JVM with support in the JIT for making fast JNI calls, such as IBM's Linux JDK 1.1.8 and SGI's IRIX Java2. In the absence of a JIT, de-signers should consider newer JVM's such as Java2 over earlier ones like JDK 1.1.x.*

3.2 Cost of Accessing Data from Native Code

Native methods invariably have to access data either in the form of incoming parameters or Java data structures such as arrays. The tests in this section simulate the way that the Java bindings to OpenGL pass around and access data using JNI.

We provide results for several tests that measure data access performance in Table 2. Each one involves calling a native method one million times. In the first one, `scalar`, the method takes 3 float parameters. In the second, `array`, it takes one parameter, a float array of length 3. In the third, `scalar2`, it takes 3 float parameters and then passes them to a helper function. In the fourth, `array-extract`, it takes one parameter, a float array, and extracts it as a C array.

Table 2. Costs of data access from native code. All results given in seconds

JVM	scalar	array	scalar2	array-extract
SGI 1.1.6 (JIT)	128.2	126.7	140.8	410.8
SGI Java2 (JIT)	22.3	31.4	22.3	223.3
IBM 1.1.8	103.7	109.0	105.7	572.7
IBM 1.1.8 (JIT)	29.4	43.7	31.9	511.1
Blackdown 1.1.7	102.3	110.0	107.7	285.8
Blackdown Java2	88.8	53.1	92.2	220.8
Blackdown Java2 (Sun JIT)	106.9	101.5	111.4	286.4
Blackdown Java2 (Inprise JIT)	87.9	53.1	90.7	221.7

When we compare the results to that in the previous section, we see that with support in the JIT, the cost of passing scalar parameters and accessing them is not significantly higher than calling an empty native method. The results also show that in almost all cases, the cost of passing an array is higher than the cost of passing scalar values. The cost of extracting an array is very high for all the JVM's. In general, Java2 implementations perform better than the 1.1.x implementations, and IBM and SGI outperform Blackdown.

Recommendation 3 *Designers of multi-language applications can build native implementations and use scalar parameter passing and expect reasonable performance when there is JIT and JVM support for fast native interface calling. If they need to pass and extract arrays with the current JVM's, they should be aware of significant costs and should try to optimize by caching and reusing data extracted from the arrays.*

4   Embedding the JVM in Servers: Memory Management and Threading Interactions

One of the big advantages of programming in 100% Java is that all the memory management is fully handled by the JVM and is largely a black box as far as programmers are concerned. The programmer has no real control over Java object locations and lifetimes, and garbage collectors can choose to move objects around. In current JVM's the Java heap and the native heap are in separate areas in the process address space. Crossing from one to the other usually requires copying of data. Although JNI provides a mechanism to access data within the Java heap directly through the `GetPrimitiveArrayCritical` call, the VM implementation decides whether or not a copy is required.

JNI currently does not provide a way to allocate the heap at a specific address, which may be desired by large servers with embedded JVM's, where the Java heap is only one segment of the total memory the server must manage. In general, the cost of native heap allocation in the presence of a JVM is higher than in single-threaded C code since it is done in a thread-safe manner. Finally, the safety of the JVM can be compromised by native routines writing to memory areas they do not own, such as the JVM data structures.

Large servers often already have a threading model, necessitating some complex cooperation between the server code and an embedded JVM. JNI currently has no way of knowing any details about threads created by the server code. This may be necessary, if for example these threads need to use JNI, if the garbage collector needs to scan their C stacks for object references, or if the JVM needs to accurately handle and deliver signals to its own threads as well as server threads.

Recommendation 4 *Embedding a Java VM in an industrial-strength server application needs much more support in both JNI and the JVM than is currently present. Designers should exercise caution before embarking on such Grande applications with current JVM technologies.*

5   Implementation of Fast JNI on IRIX

5.1   Fast JNI Calling Optimization in the MIPS JIT

Sun has defined a JIT API which provides the services that a JIT compiler needs from the VM, including calling JNI methods. However, using it introduces two layers of overhead, translating from the JIT calling convention to the JVM's standard calling convention and then to the JNI calling convention, with a fair amount of other machinery along the way. In all, the journey from a JIT-compiled method to a native method contains three or four levels of function calls, setting up two stack frames, and moving the method arguments around among the registers and memory two or three times. Obviously this has a negative impact on performance, particularly for native methods which are small, such as those in the java.lang.Math library.

Since our JIT calling convention and the JNI calling convention are actually very similar, we were able to write an optimized version of this path which handles the vast majority of cases, including all the native methods in the standard libraries. In these cases, the journey from JIT-compiled method to native method now requires one function call, setting up two stack frames, and one fairly quick rearrangement of the method arguments within registers. The benefits of this can be seen in the performance results in Table 3.

Table 3. Speedup from fast JNI calling optimization

Test	Java2 (seconds)	Java2 + Fast JNI (seconds)	Speedup from Fast JNI
Empty Method	69.3	21.1	3.1
scalar	104.2	22.3	4.5
array	74.0	31.4	2.4
scalar2	110.0	23.1	4.8
array-extract	290.7	223.3	1.3

5.2   JNI Pinning Lock Optimization in the JVM

Prowess, a multi-threaded Java oil and gas application from Landmark Graphics, showed poor scalability running on multiprocessor IRIX. Using the Speedshop performance tools and a tracing-enabled POSIX threads library, we found the primary problem to be contention for the JNI pinning lock as the number of threads increased. The JNI pinning lock is held in the reference implementation of the JNI GetPrimitiveArrayCritical call while adding the object to a global hash table. By changing the implementation to a lock-per-bucket strategy this contention was largely avoided, resulting in a very large performance improvement, shown in Table 4.

Table 4. Prowess throughput with JNI pinning lock optimization

Threads	Single Lock Throughput (MBytes/sec)	Lock-per-Bucket Throughput (MBytes/sec)	Speedup
2	10.0	70.0	7.0
8	6.5	35.0	5.4

6   Related Work

IBM's JVM team has also developed JNI optimizations [12] similar to those we have implemented in SGI's Java2 JVM. Researchers have developed a variety of solutions for reducing the overhead of interfacing Java to low-level native code. They include extending the Java runtime with new primitives that are inlined at JIT compilation time [3], providing a new type of Java object whose storage is outside the Java heap [3], and introducing a new buffer class and extra garbage collector features that eliminate the need for extra copies [13].

Dennis Gannon et. al. [14] at Indiana University have studied the problem of object-interoperability in the context of RMI and Java/HPC++. Other researchers [15][16] are building tools to automatically create Java bindings to high-performance standard libraries.

7   Conclusions

The lessons we have learned from experience working with our customers in interfacing Java with other languages include:

- While JVM's are beginning to bridge the gap in cost between native method calls and Java method calls, it still deserves careful attention from developers.
- JNI and current JVM's provide insufficient support for industrial server-side applications that want to embed a JVM.
- Our IRIX/MIPS Fast JNI implementation has given substantial performance improvement over previous releases and has improved the performance of real-world applications such as Magician [2] and Prowess.

Acknowledgements

We would like to thank Alligator Descartes of Arcana Technologies [2] for providing us with the benchmark programs used in the performance section. We would like to thank Brian Sumner, an SGI Apps consulting engineer, for working on the Landmark Graphics Prowess application and identifying various performance issues with it. We would like to thank the engineers within SGI working on the Netscape Enterprise Server and at Informix working on database servers for helping us understand some of the issues surrounding embedding JVM's in servers.

References

1. J. Gosling, B. Joy, and G. Steele. *The Java Language Specification.* Addison-Wesley, 1996.
2. Magician: Java Bindings for OpenGL. http://arcana.symbolstone.org/products/magician/index.html.
3. Matt Welsh and David Culler. Jaguar: Enabling Efficient Communication and I/O from Java. In *Concurrency: Practice and Experience, Special Issue on Java for High-Performance Applications*, December 1999. http://ninja.cs.berkeley.edu/pubs/pubs.html.
4. Glenn Judd, Mark Clement, Quinn Snell, and Vladimir Getov. Design Issues for Efficient Implementation of MPI in Java. In *Proceedings of the ACM 1999 Java Grande Conference*, June 1999.
5. Sun Microsystems, Inc. *Java Native Interface Specification.* http://java.sun.com/products/jdk/1.2/docs/guide/jni/index.html.
6. NaturalBridge. *BulletTrain^{TM} Optimizing Compiler and Runtime for JVM Bytecodes.* http://www.naturalbridge.com/.
7. Tower Technologies. *TowerJ3.0: A New Generation Native Java Compiler and Runtime Environment.* http://www.towerj.com.
8. SGI JDK 3.1.1 (based on Sun's JDK 1.1.6). http://www.sgi.com/Products/Evaluation/#jdk_3.1.1.
9. Java2 Software Development Kit v 1.2.1 for SGI IRIX. http://www.sgi.com/developers/devtools/languages/java2.html.
10. Blackdown JDK port of Sun's Java Developer's Toolkit to Linux. http://www.blackdown.org/.
11. IBM Developer Kit and Runtime Environment for Linux, Java Technology Edition, Version 1.1.8. http://www.ibm.com/java/jdk/118/linux/index.html.
12. IBM's Java Technology Presentations. http://www.developer.ibm.com/java/jbdays.html.
13. Chi-Chao Chang and Thorsten von Eicken. Interfacing Java with the Virtual Interface Architecture. In *Proceedings of the ACM 1999 Java Grande Conference*, June 1999.
14. Dennis Gannon and F. Berg et. al. Java RMI Performance and Object Model Interoperability: Experiments with Java/HPC++. *Concurrency: Practice and Experience*, 10:941–946, 1998.
15. Vladimir Getov, Susan Flynn-Hummel, and Sava Mintchev. High-Performance Parallel Programming in Java: Exploiting Native Libraries. In *Proceedings of the ACM 1998 Java Grande Conference*, June 1998.
16. H. Casanova, J. Dongarra, and D. M. Doolin. Java Access to Numeric Libraries. *Concurrency: Practice and Experience*, 9:1279–1291, Nov 1997.

MPJ: A Proposed Java Message Passing API and Environment for High Performance Computing

Mark Baker
University of Portsmouth
Hants, UK, PO1 2EG
Mark.Baker@port.ac.uk

Bryan Carpenter
NPAC at Syracuse University
Syracuse, NY 13244, USA
dbc@npac.syr.edu

January 24th 2000

Abstract

In this paper we sketch out a proposed reference implementation for message passing in Java (MPJ), an MPI-like API from the Message-Passing Working Group of the Java Grande Forum [1,2]. The proposal relies heavily on RMI and Jini for finding computational resources, creating slave processes, and handling failures. User-level communication is implemented efficiently directly on top of Java sockets.

1. Introduction

The Message-Passing Working Group of the Java Grande Forum was formed late 1998 as a response to the appearance of several prototype Java bindings for MPI-like libraries. An initial draft for a common API specification was distributed at Supercomputing '98. Since then the working group has met in San Francisco and Syracuse. The present API is now called MPJ.

Currently there is no complete implementation of the draft specification. Our own Java message-passing interface, mpiJava, is moving towards the "standard". The new version 1.2 of the software supports direct communication of objects via object serialization, which is an important step towards implementing the specification in [1]. We will release a version 1.3 of mpiJava, implementing the new API.

The mpiJava wrappers [2] rely on the availability of platform-dependent native MPI implementation for the target computer. While this is a reasonable basis in many cases, the approach has some disadvantages.

J. Rolim et al. (Eds.): IPDPS 2000 Workshops, LNCS 1800, pp. 552-559, 2000.

° The two-stage installation procedure – get and build native MPI then install and match the Java wrappers – is tedious and off-putting to new users.

° On several occasions in the development of mpiJava we saw conflicts between the JVM environment and the native MPI runtime behaviour. The situation has improved, and mpiJava now runs on various combinations of JVM and MPI implementation.

° Finally, this strategy simply conflicts with the ethos of Java, where pure-Java, write-once-run-anywhere software is the order of the day.

Ideally, the first two problems would be addressed by the providers of the original native MPI package. We envisage that they could provide a Java interface bundled with their C and Fortran bindings, avoiding the headache of separately installing the native software and Java wrapper. Also they are presumably in the best position to iron-out low-level conflicts and bugs. Ultimately, such packages should represent the fastest, industrial-strength implementations of MPJ.

Meanwhile, to address the last shortcoming listed above, this paper considers production of a pure-Java reference implementation for MPJ. The design goals are that the system should be as easy to install on distributed systems as we can make it, and that it be sufficiently robust to be useable in an Internet environment. Ease of installation and use are special concerns to us. We want a package that will be useable not only by experienced researchers and engineers, but also in, say, an educational context.

We are by no means the first people to consider implementing MPI-like functionality in pure Java, and working systems have already been reported in [3, 4], for example. The goal here is to build on the some lessons learnt in those earlier systems, and produce software that is standalone, easy-to-use, robust, and fully implements the specification of [1].

Section 2 reviews our design goals, and describes some decisions followed from these goals. Section 3 reviews the proposed architecture. Various distributed programming issues posed by computing in an unreliable environment are discussed in Section 4, which covers basic process creation and monitoring. This section assumes free use of RMI and Jini. Implementation of the message-passing primitives on top of Java sockets and threads is covered in section 5.

2. Some design decisions

A MPJ "reference implementation" can be implemented as Java wrappers to a native MPI implementation, or it can be implemented in pure Java. It could also be implemented principally in Java with a few simple native methods to optimize operations (like marshalling arrays of primitive elements) that are difficult to do efficiently in Java. Our proposed system focuses on the latter possibility – essentially pure Java, although experience with DOGMA [3] and other systems strongly suggests that optional native support for marshalling will be desirable. The aim is to provide an implementation of MPJ that is maximally portable and requires the minimum of support for anomalies found in individual systems.

We envisage that a user will download a jar-file of MPJ library classes onto machines that may host parallel jobs. Some installation "script" (preferably a parameterless script) is run on the host machines. This script installs a daemon (perhaps by registering a persistent *activatable* object with an existing rmid daemon). Parallel java codes are compiled on any host. An mpjrun program invoked on that host transparently loads all the user's class files into JVMs created on remote hosts by the MPJ daemons, and the parallel job starts. The only required parameters for the mpjrun program should be the class name for the application and the number of processors the application is to run on. These seem to be an irreducible minimum set of steps; a conscious goal is that the user need do no more than is absolutely necessary before parallel jobs can be compiled and run.

In light of this goal one can sensibly ask if the step of installing a daemon on each host is essential. On networks of UNIX workstations – an important target for us – packages like MPICH avoid the need for special daemons by using the rsh command and its associated system daemon. In the end we decided this is not the best approach for us. Important targets, notably networks of NT workstations, do not provide rsh as standard, and often on UNIX systems the use of rsh is complicated by security considerations. Although neither RMI or Jini provide any magic mechanism for conjuring a process out of nothing on a remote host, RMI does provide a daemon called rmid for restarting *activatable objects*. These need only be installed on a host once, and can be configured to survive reboots of the host. We propose to use this Java-centric mechanism, on the optimistic assumption that rmid will become as widely run across Java-aware platforms as rshd is on current UNIX systems.

In the initial reference implementation it is likely that we will use Jini technology [5, 6] to facilitate location of remote MPJ daemons and to provide a framework for the required fault-tolerance. This choice rests on our guess that in the medium-to-long-term Jini will become a ubiquitous component in Java installations. Hence using Jini paradigms from the start should eventually promote interoperability and compatibility between our software and other systems. In terms of our aim to simplify *using* the system, Jini multicast discovery relieves the user of the need to create a "hosts" file that defines where each process of a parallel job should be run. If the user actually *wants* to restrict the hosts, a unicast discovery method is available. Of course it has not escaped our attention that eventually Jini discovery may provide a basis for much more dynamic access to parallel computing resources.

Less fundamental assumptions bearing on the organization of the MPJ daemon are that standard output (and standard error) streams from all tasks in an MPJ job are merged non-deterministically and copied to the standard output of the process that initiates the job. No guarantees are made about other IO Operations – for now these are system-dependent.

The main role of the MPJ daemons and their associated infrastructure is thus to provide an environment consisting of a group of processes with the user-code loaded and running in a *reliable* way. The process group is reliable in the sense that no *partial failures* should be visible to higher levels of the MPJ implementation or the user code.A partial failure is the situation where some members of a group of cooperating processes are unable to continue because other members of the group have crashed, or the network connection between members of the group has failed. To quote [7]: *partial failure is a central reality of distributed computing*. No software technology can guarantee the absence of *total* failures, in which the whole MPJ job dies at essentially the same time (and all resources allocated by the MPJ system to support the user's job are released). But total failure should be the *only* failure mode visible to the higher levels. Thus a principal role of the base layer is to detect partial failures and cleanly abort the whole parallel program when they occur.

Once a reliable cocoon of user processes has been created through negotiation with the daemons, we have to establish connectivity. In the reference implementation this will be based on Java sockets. Recently there has been interest in producing Java bindings to VIA [8, 9]. Eventually this may provide a better platform on which to implement MPI, but for now sockets are the only realistic, portable option. Between the socket API and the MPJ API there will be an intermediate "MPJ device" level. This is modelled on the abstract device interface of MPICH [10]. Although the role is slightly different here – we do not really anticipate a need for multiple device-specific implementations. The API is actually not modelled in detail on the MPICH device, but the level of operations is similar.

3. Overview of the Architecture

A possible architecture is sketched in Figure 1. The bottom level, process creation and monitoring, incorporates initial negotiation with the MPJ daemon, and low-level services provided by this daemon, including clean termination and routing of output streams. The daemon invokes the MPJSlave class in a new JVM. MPJSlave is responsible for downloading the user's application and starting that application. It may also directly invoke routines to initialize the message-passing layer. Overall, what this bottom layer provides to the next layer is a reliable group of processes with user code installed. It may also provide some mechanisms – presumably RMI-based – for global synchronization and broadcasting simple information like server port numbers.

The next layer manages low-level socket connections. It establishes all-to-all TCP socket connections between the hosts. The idea of an "MPJ device" level is modelled on the Abstract Device Interface (ADI) of MPICH. A minimal API includes non-blocking standard-mode send and receive operations (analogous to MPI_ISEND and MPI_IRECV, and various wait operations – at least operations equivalent to MPI_WAITANY and MPI_TESTANY). All other point-to-point communication modes can be implemented correctly on top of this minimal set. Unlike the MPICH device level, we do not incorporate direct support for groups, communicators or (necessarily) datatypes at this level (but we do assume support for message contexts). Message buffers are likely to be *byte* arrays. The device level is intended to be implemented on socket send and recv operations, using standard Java threads and synchronization methods to achieve its richer semantics.

The next layer is base-level MPJ, which includes point-to-point communications, communicators, groups, datatypes and environmental management. On top of this are higher-level MPJ operations including the collective operations. We anticipate that much of this code can be implemented by fairly direct transcription of the src subdirectories in the MPICH release – the parts of the MPICH implementation above the abstract device level.

3.1 Process creation and monitoring

We assume that an MPJ program will be written as a class that extends MPJApplication. To simplify downloading we assume that the user class also implements the Serializable interface. The default communicator is passed as an argument to main. Note there is no equivalent of MPI_INIT or MPI_FINALIZE. Their functionality is absorbed into code executed before and after the user's main method is called.

High Level MPI	Collective operations
	Process topologies
Base Level MPI	All point-to-point modes
	Groups
	Communicators
	Datatypes
MPJ Device Level	isend, irecv, waitany, ...
	Physical process ids (no groups)
	Contexts and tags (no communicators)
	Byte vector data
Java Socket and Thread APIs	All-to-all TCP connections
	Input handler threads
	synchronized methods, wait notify
Process Creation and Monitoring	MPJ service daemon
	Lookup, leasing, distributed events (Jini)
	exec java MPJSlave
	Serializable objects, RMIClassLoader

Figure 1: Layers of an MPJ reference implementation

3.2 The MPJ daemon

The MPJ daemon must be installed on any machine that can host an MPJ process. It will be realized as an instance of the class MPJService. It is likely to be an *activatable* remote object registered with a system rmid daemon. The MPJ daemon executes the Jini discovery protocols and registers itself with available Jini lookup services, which we assume are accessible as part of the standard system environment (Figure 2). The daemon passes the id of the new slave into the java command that starts the slave running. We assume the daemon is running an RMI registry, in which it publishes itself. The port of this registry is passed to the slave as a second argument. The first actions of the slave object are to look up its master in the registry, then call back to the master and install a remote reference to itself (the slave) in the master's slave table. The net effect is that the client receives a remote reference to a new slave object running in a private JVM. In practice a remote destroySlave method that invokes the Process.destroy method will likely be needed as well.

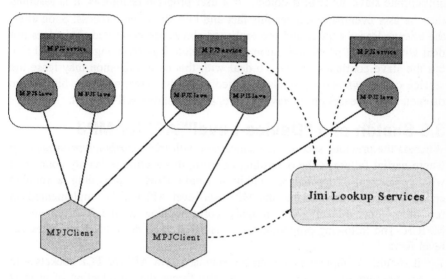

Figure 2: Independent clients may find MPJService daemons through the Jini lookup service. Each daemon may spawn several slaves.

3.3 Handling MPJ aborts – Jini events
If any slave JVM terminates unexpectedly while the runTask method is in progress, a RemoteException will be passed to the thread that started the remote call. The thread should catch the exception, and generate an MPJAbort event. This is a Jini remote event – a subclass of RemoteEvent. Early in the process of creating a slave, the MPJ daemons will have registered themselves with the client as handlers for MPJAbort events. Their notify method will apply the destroy method to the appropriate slave Process object. Hence if any slave aborts (while the network connection stays intact), all remaining slave processes associated with the job are immediately destroyed.

3.4 Other failures – Jini leasing
The distributed event mechanism can rapidly clean up processes in the case where some slaves disappear unexpectedly, but it cannot generally reclaim resources in the case where:

○   the client process is killed during execution of an MPJ job,

○   the daemon process is killed while it has some active slaves,

○   the case of network failures that do not directly affect the client.

There is a danger that orphaned slave processes will be left running in the network. The solution is to use the Jini leasing paradigm. The client leases the services of each daemon for some interval, and continues renewing leases until all slaves terminate, at which point it cancels its leases. If the client process is killed (or it connection to the slave machine fails), its leases will expire. If a client's lease expires the daemon applies the destroy method to the appropriate slave Process object. If a user program deadlocks, it is assumed that the user eventually notices this fact and kills the client process. Soon after, the client's leases expire, and the orphaned slaves are destroyed. This does not deal with the case where a daemon is killed while it is servicing some MPJ job, but the slave continues to run. To deal with this case a daemon may lease the service of its own slave processes immediately after creating them. Should the daemon die, its leases on its slaves expire, and the slaves self-destruct.

3.5 Sketch of a "Device-Level" API for MPJ

Whereas the previous section was concerned with true *distributed programming* where partial failure is the overriding concern, this section is mainly concerned with *concurrent programming* within a single JVM – providing a reliable environment. We assume that the MPJ user-level API will be implemented on top of a "device-level" API, roughly corresponding to the MPID layer in MPICH. The following properties are considered to be desirable for the device-level API:

1. It should be implementable on the standard Java API for TCP sockets – in the absence of select, this essentially forces the introduction of at least one receive thread for each input socket connection.

2. It should be efficiently implementable (and will be implemented) with precisely this minimum required number of threads.

3. It should be efficiently implementable with at least two protocols:
    a)   The naive eager-send protocol, assuming receiver threads have unlimited buffering.
    b)   A ready-to-send/ready-to-receive/rendezvous protocol requiring receiver threads only have enough buffering to queue unserviced "ready" messages.

4. The basic operations will include isend, irecv and waitany (plus some other "wait" and "test" operations). These suffice to build legal implementations of all the MPI communication modes. Optimized entry points for the other modes can be added later.

5. It is probable that all handling of groups and communicators will be outside the device level. The device level only has to correctly interpret absolute process ids and integer contexts from communicators.

6. It may be necessary that all handling of user-buffer datatypes is outside the device level – here the device level only deals with byte vectors.

4. Conclusions and Future Work

In this paper we have discussed the findings from preliminary research into the design and implementation issues for producing a reference message passing interface for Java (MPJ). Our proposed design is based on our experiences from creating and then supporting the widely used mpiJava wrappers, experimental prototypes, and from on-going discussions with the Message-Passing Working Group of the Java Grande Forum. Our preliminary research has highlighted a number of design issues that the emerging Java technology, Jini, can help us address. In particular the areas of application distribution, resource discovery and fault tolerance. Issues that other message passing systems typically fail to address. Apart from a few minor questions about the exact syntax of the MPJ API, the design of our reference MPJ environment is complete and preliminary implementation work is underway. We believe that with our current man power we will be able to report on a limited release by the time that the workshop takes place in early May 2000.

5. References

1. B. Carpenter, et al. MPI for Java: Position Document and Draft API Specification, Java Grande Forum, JGF-TR-3, Nov. 1998, http://www.javagrande.org/
2. M. Baker, D. Carpenter, G. Fox, S. Ko and X. Li, mpiJava: A Java MPI Interface, to appear in the *International Journal Scientific Programming* – ISSN 1058-9244, http://www.cs.cf.ac.uk/hpjworkshop/
3. G. Judd, et. al., DOGMA: Distributed Object Group Management Architecture, ACM 1998 Workshop on Java for High-Performance Network Computing. Palo Alto, February 1998, *Concurrency: Practice and Experience*, 10(11-13), 1998,
4. K. Dincer, jmpi and a Performance Instrumentation Analysis and Visualization Tool for jmpi, *First UK Workshop on Java for High Performance Network Computing, Europar '98*, September 1998, http://www.cs.cf.ac.uk/hpjworkshop/
5. K. Arnold et. al., The Jini Specification, *Addison Wesley*, 1999
6. W. Edwards, Core Jini, *Prentice Hall*, 1999
7. J. Waldo, et. al., A Note on Distributed Computing, *Sun Microsystems Laboratories*, SMLI TR-94-29, 1994
8. M. Welsh, Using Java to Make Servers Scream, Invited talk at ACM 1999 Java Grande Conference, San Francisco, CA, June, 1999
9. C. Chang and T. von Eiken, Interfacing Java to the Virtual Interface Architecture, ACM 1999 Java Grande Conference, June, 1999, ACM Press
10. MPICH – A Portable Implementation of MPI, http://www.mcs.anl.gov/mpi/mpich/
11. MPI Forum, MPI: A Message-Passing Interface Standard, University of Tennessee, Knoxville, TN, USA, June 1995, http://www.mcs.anl.gov/mpi
12. G. Crawford, Y. Dandass and A. Skjellum, The JMPI Commercial Message Passing Environment and Specification: Requirements, Design, Motivations, Strategies, and Target Users, http://www.mpi-softtech.com/publications

Implementing Java Consistency Using a Generic, Multithreaded DSM Runtime System

Gabriel Antoniu[1], Luc Bougé[1], Philip Hatcher[2],
Mark MacBeth[2]*, Keith McGuigan[2], Raymond Namyst[1]

[1] LIP, ENS Lyon, 46 Allée d'Italie, 69364 Lyon Cedex 07, France
[2] Dept. Computer Science, Univ. New Hampshire, Durham, NH 03824, USA

Abstract. This paper describes the implementation of Hyperion, an environment for executing Java programs on clusters of computers. To provide high performance, the environment compiles Java bytecode to native code and supports the concurrent execution of Java threads on multiple nodes of a cluster. The implementation uses the PM2 distributed, multithreaded runtime system. PM2 provides lightweight threads and efficient inter-node communication. It also includes a generic, distributed shared memory layer (DSM-PM2) which allows the efficient and flexible implementation of the Java memory consistency model. This paper includes preliminary performance figures for our implementation of Hyperion/PM2 on clusters of Linux machines connected by SCI and Myrinet.

1 Introduction

The Java programming language is an attractive vehicle for constructing parallel programs to execute on clusters of computers. The Java language design reflects two emerging trends in parallel computing: the widespread acceptance of both a thread programming model and the use of a (possibly virtually) shared memory. While many researchers have endeavored to build Java-based tools for parallel programming, we think most people have failed to appreciate the possibilities inherent in Java's use of threads and a "relaxed" memory model.

There are a large number of parallel Java efforts that connect multiple Java virtual machines by using Java's remote-method-invocation facility (e.g., [2, 3, 13]) or by grafting an existing message-passing library (e.g., [4, 5]) onto Java. In contrast, we view a cluster as executing a *single* Java virtual machine. The separate nodes of the cluster are hidden from the programmer and are simply resources for executing Java threads with true concurrency: the Java threads are mapped onto the native threads available at the nodes. The private memories of the nodes are also hidden from the programmer: our implementation supports the illusion of a shared memory within a distributed cluster. This illusion is consistent with respect to the Java memory model, which is "relaxed" in that it does not require sequential consistency.

---

* Current affiliation: Sanders, A Lockheed Martin Company, PTP02-D001, P.O. Box 868, Nashua, NH, USA

J. Rolim et al. (Eds.): IPDPS 2000 Workshops, LNCS 1800, pp. 560–567, 2000.
© Springer-Verlag Berlin Heidelberg 2000

The difficulty is that the Java memory model does not exactly match one of the "classical" memory consistency models that are supported by existing distributed shared memory (DSM) systems. Moreover, our approach requires a tight integration of a Java specific DSM system with a native thread management system. Designing such an environment from scratch is a huge task, and only very few such projects have succeeded. In contrast, we have built our system on top of a new, generic, multi-protocol, multithreaded DSM runtime platform, which provides all the primitives necessary for a distributed implementation of the Java consistency model. This allowed us to complete this task within a few weeks. Moreover, the genericity allowed us to explore several alternative implementation strategies with an invaluable flexibility. The preliminary performance figures reported are definitely encouraging.

2 Executing Java programs on distributed clusters

2.1 Concurrent programming in Java

It is relatively easy to write parallel programs in Java. Support for threads is part of the Java API, and provides similar functionality to POSIX threads. Threads in Java are represented as objects. The class java.lang.Thread contains all of the methods for initializing, running, suspending, querying and destroying threads. Critical sections of code can be protected by monitors. Monitors in Java are available through the use of the keyword synchronized. The keyword can be used inline in the code as a statement, or as a modifier for a method. As a statement, synchronized must be provided an object reference and a block of code to protect. A method modified with synchronized uses the instance of the object it is being called on and protects the method body. Every object has exactly one lock associated with it and in both cases a lock is acquired on the referenced object, the protected code is executed, and then the lock is released. Note that the class java.lang.Thread also contains the methods wait(), notify(), and notifyAll(), which provide functionality similar to POSIX condition variables. From monitors and the wait/notify methods, other synchronization constructs, such as barriers and semaphores, can easily be built.

Java has a well-defined memory model [6]. All threads share the same central memory, so all objects, static values, and class objects are accessible to every thread. Thus, reads and writes of such values should be protected by monitors when appropriate to prevent race conditions. The Java memory model allows threads to keep locally cached copies of objects. Consistency is provided by requiring that a thread's object cache be flushed upon entry to a monitor and that local modifications made to cached objects be transmitted to the central memory when a thread exits a monitor.

2.2 The Hyperion System

Hyperion is an environment for the high-performance execution of Java programs developed at the University of New Hampshire. Hyperion supports high performance by utilizing a Java-bytecode-to-C translator and by supporting parallel execution via the distribution of Java threads across the multiple processors of a cluster of workstations.

We are interested in computationally intensive programs that can exploit parallel hardware. Therefore we expect that the added cost of compiling to native machine code will be recovered many times over in the course of executing a program. (This focus on compilation distinguishes us from projects investigating the implementation of Java interpreters on top of a distributed shared memory [1, 15].)

To produce an executable for a user program, Java bytecode is first generated from the Java source using a standard Java compiler. (We currently use Sun's javac.) The bytecode in the generated class files is compiled using Hyperion's java2c. The resulting C code is compiled using a native C compiler for the cluster and then linked with the Hyperion runtime library and any necessary external libraries.

Hyperion's Java-bytecode-to-C compiler, java2c, uses a very simple approach to code generation. Each virtual machine instruction is translated directly into a separate C statement or macro invocation, similar to the approaches taken in the Harissa [10] or Toba [14] compilers. Currently, java2c supports all non-wide format instructions as well as exception handling.

The Hyperion run-time system is structured as a collection of subsystems in order to support both easy porting to new target architectures and experimentation with different implementation techniques for individual components.

- The threads subsystem provides support for lightweight threads, on top of which Java threads can be implemented.
- The communication subsystem supports the transmission of messages between the nodes of a cluster.
- The memory subsystem is responsible for the allocation, management (including synchronization mechanisms), and garbage collection of Java objects. Table 1 provides the key primitives for implementing memory consistency. On a distributed implementation these primitives need to be supported by the underlying DSM layer.

loadIntoCache	Load an object into the cache
invalidateCache	Invalidate all entries in the cache
updateMainMemory	Update memory with modifications made to objects in the cache
get	Retrieve a field from an object previously loaded into the cache
put	Modify a field in an object previously loaded into the cache

Table 1. The Hyperion DSM subsystem API

The compiler generates explicit calls to put and get to access shared data. Objects are loaded from the main memory to the local cache using loadIntoCache. The primitives invalidateCache and updateMainMemory are called on entering/exiting monitors to ensure consistency, as described in Section 3.2.

To implement the above primitives, we utilize DSM-PM2, a generic, multi-protocol DSM layer built on top of the PM2 multithreaded runtime system. It provides an easy-to-use API for specifying consistency protocols.

3   Implementing Java consistency

3.1   DSM-PM2: a generic, multi-protocol DSM layer

PM2 (Parallel Multithreaded Machine) [11] is a multithreaded environment for distributed architectures. Its programming interface is based on *Remote Procedure Calls* (RPCs). Using RPCs, the PM2 threads can invoke the remote execution of user defined *services*. Such invocations can either be handled by a pre-existing thread or they can involve the creation of a new thread. Threads running on the same node can freely share data. Threads running on distant nodes can only interact through RPCs.

PM2 provides a *thread migration* mechanism that allows threads to be transparently and preemptively moved from one node to another during their execution. Such functionality is typically useful to implement dynamic load balancing policies. The interactions between thread migration and data sharing are handled through a *distributed shared memory* facility: the *DSM-PM2* layer.

DSM-PM2 provides a programming interface to manage static and dynamic data to be shared by all the threads running within a PM2 session, whatever their location. To declare *static* shared data, one simply brackets the corresponding C declarations by specific DSM-PM2 keywords. *Dynamic* shared data are allocated as needed by calling a specific allocation routine instead of the ordinary malloc primitive.

Since the DSM-PM2 API is intended both for direct use and as a target for compilers, no pre-processing is assumed in the general case and accesses to shared data are detected using page faults. Applications can thus be programmed as if a true physical shared memory was available. Nevertheless, an application can choose to bypass the fault detection mechanism by controlling the accesses with explicit calls to get/put primitives. In some cases, the resulting cost may be smaller than the overhead of the underlying fault handling subsystem. DSM-PM2 copes with this approach as well, as illustrated by the implementation of the Java runtime discussed in Section 3.2.

Since existing DSM applications require different consistency models, DSM-PM2 has been designed to be generic so as to support multiple consistency models. Sequential consistency and Java consistency are currently available. Moreover, new consistency models can be easily implemented using the existing generic DSM library routines.

These primitives may also be used to provide alternative protocols for a given consistency model. For instance, the usual action on a access fault is to bring the data to the accessing thread. Alternatively, one may choose to preemptively migrate the thread to the data: this may be much more efficient in certain cases! One may even build hybrid protocols, which are able to dynamically switch from one policy to another depending of some external load information.

The overall structure of the DSM-PM2 layer is presented in Figure 1. The *DSM page manager* is essentially dedicated to the low-level management of memory pages. It implements a distributed table containing page ownership information and maintains the appropriate access rights on each node. The *DSM communication module* is responsible for providing elementary communication mechanisms such as delivering requests for page copies, sending pages, and invalidating pages. It implements a convenient high-level communication API based on RPCs. On top of these two components, the *DSM protocol library* provides elementary routines that are used as building blocks to implement consistency protocols. For instance, it includes routines to bring a copy of a page to a thread, to migrate a thread to a page, to invalidate all the copies of a page, etc. Finally, the *DSM protocol policy* layer is responsible for implementing consistency models out of a subset of the available library routines and for associating each application data with its own consistency model.

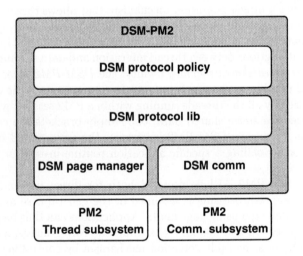

Fig. 1. Overview of the DSM-PM2 software architecture

3.2 Using DSM-PM2 to build a Java consistency protocol

Java consistency requires a MRMW (*Multiple Reader Multiple Writer*) protocol: an object can be replicated and copies may be concurrently modified on different nodes. To guarantee consistency, the accesses to shared data have to be protected by monitors (corresponding to the keyword synchronized at the language level).

On entering a monitor, the primitive invalidateCache is called. Objects are loaded from the main memory to the local cache using loadIntoCache. Finally, on exiting a monitor, the modifications carried out in the local cache are sent to the main memory via the updateMainMemory primitive. We have implemented these protocol primitives using the programming interface of the lower-level DSM-PM2 components: *DSM page manager* and *DSM communication module*.

Main memory and caches. To implement the concept of *main memory* specified by the Java model, the runtime system associates a *home node* to each object. It is in charge of managing the reference copy. Initially, the objects are stored on their home nodes. They can be replicated if accessed on other nodes. Note that at most one copy of an object may exist on a node and this copy is shared by all the threads running on that node. Thus, we avoid wasting memory by associating caches to nodes rather than to threads.

Access detection. Hyperion uses specific access primitives to shared data (get and put), which allows us to use explicit checks to detect if an object is present (i.e., has a copy) on the local node. If the object is locally cached, it is directly accessed, else the loadIntoCache primitive is invoked. The default mechanism for access detection provided by DSM-PM2 is thus bypassed and the cost of page fault handling is saved. An alternative we have planned to investigate would be to allow direct access and call loadIntoCache within the page fault handler.

Access rights. Java objects cannot be read-only, so that a node can either have full *read-write* access to an object, or have no access at all. This feature simplifies the protocol implementation, since only these two cases have to be considered.

Sending modifications to the main memory. Since shared data are modified through a specific access primitive (put), the modifications can be recorded at the moment when they are carried out. For this purpose, a bitmap is created on a node when a copy of the page is received. The put primitive records all writes to the page. The modifications are sent to the home node of the page by the updateMainMemory primitive.

Pages and objects Java objects are implemented on top of pages. Consequently, loading an object into the local cache may generate prefetching, since all objects on the corresponding page are actually brought to the current node. On the other hand, updateMainMemory may generate non-required updates, since the modifications of all objects located on the current page are sent to the home node. These side effects do not affect the validity of our protocol with respect to Java consistency.

4 Preliminary performance evaluation

We first evaluate the performance of our protocol primitives on three different platforms. The first column corresponds to measurements carried out on a cluster of Pentium II, 450 MHz nodes running Linux 2.2.10 interconnected by a SCI network using the SISCI protocol. The figures in the next two columns have been obtained on a cluster of Pentium Pro, 200MHz nodes running Linux 2.2.13 interconnected by a Myrinet network using the BIP and TCP protocols respectively. The cost of the loadIntoCache primitive for a 4 kB object can be broken down as follows (time is given in μs):

Operation/Protocols	SISCI/SCI	BIP/Myrinet	TCP/Myrinet
Preparing a page request	1	2	2
Transmitting the request	17	30	190
Processing the request	1	2	2
Sending back a 4 kB page	85	134	412
Installing the page	12	24	24
Total	**116**	**192**	**630**

The processing overhead of DSM-PM2 with respect to the raw transmission time is 10–15%. The overhead related to page installation includes a call to the mprotect primitive to enable writing and a call to malloc to allocate the page bitmap necessary for recording local modifications. This latter cost could be further improved using a custom malloc-like primitive.

As a preliminary test, we ran a program that estimates π by calculating a Riemann sum of 50 million values. Parallelism can be utilized by assigning sections of the Riemann sum to different threads and then doing one final sum reduction to complete the calculation. Compared to an optimized sequential C program, the Java/Hyperion program achieves the following speedups on the Myrinet/BIP cluster described above (time is given in seconds):

# Nodes	Time	Speedup	Efficiency
C code	9.4	1.0	100%
1	9.6	.98	98%
2	4.9	1.9	95%
4	2.5	3.8	95%
6	1.7	5.5	92%
8	1.3	7.2	90%

Even if this *pi* program exhibits good performance, it is admittedly rather simple! For a more complex evaluation of our approach, we are currently working on a minimal-cost graph-coloring application.

5   Conclusion

We propose utilizing a cluster to execute a single Java Virtual machine. This allows us to run threads completely transparently in a distributed environment. Java threads are mapped to native threads available on the nodes and run with true concurrency. Our implementation supports a globally shared address space via the DSM-PM2 runtime system that we configured to guarantee Java consistency.

We plan to make further evaluation tests using more complex applications. Thanks to the genericity of DSM-PM2, we will be able to study alternative protocols for Java consistency. We also intend to perform comparisons between different access detection techniques (segmentation faults vs. explicit locality checks).

References

1. Y. Aridor, M. Factor, and A. Teperman. cJVM: A single system image of a JVM on a cluster. In *Proceedings of the International Conference on Parallel Processing*, Fukushima, Japan, September 1999.
2. F. Breg, S. Diwan, J. Villacis, et al. Java RMI performance and object model interoperability: Experiments with Java/HPC++. In *Proc. ACM 1998 Workshop on Java for High-Performance Network Computing*, pages 91–100, February 1998.
3. D. Caromel, W. Klauser, and J. Vayssiere. Towards seamless computing and meta-computing in Java. *Concurrency: Practice and Experience*, 10:1125–1242, 1998.
4. A. Ferrari. JPVM: Network parallel computing in Java. In *Proc. ACM 1998 Workshop on Java for High-Performance Network Computing*, pages 245–249, 1998.
5. V. Getov, S. Flynn-Hummell, and S. Mintchev. High-performance parallel programming in Java: Exploiting native libraries. In *Proc. ACM 1998 Workshop on Java for High-Performance Network Computing*, pages 45–54, February 1998.
6. J. Gosling, W. Joy, and G. Steele Jr. *The Java Language Specification*. Addison-Wesley, Reading, MA, 1996.
7. P. Launay and J.-L. Pazat. A framework for parallel programming in Java. In *High-Performance Computing and Networking (HPCN '98)*, volume 1401 of *Lect. Notes in Comp. Science*, pages 628–637. Springer-Verlag, 1998.
8. K. Li and P. Hudak. Memory coherence in shared virtual memory systems. *ACM Trans. Computer Systems*, 7(4):321–359, November 1989.
9. F. Mueller. Distributed shared-memory threads: DSM-Threads. In *Proc. of the Workshop on Run-Time Systems for Parallel Programming (RTSPP '97)*, pages 31–40, Geneva, Switzerland, April 1997. Held in conjonction with IPPS '97.
10. G. Muller, B. Moura, F. Bellard, and C. Consel. Harissa: A flexible and efficient Java environment mixing bytecode and compiled code. In *Third Conference on Object-Oriented Technologies and Systems (COOTS '97)*, pages 1–20, Portland, June 1997.
11. Raymond Namyst and Jean-François Méhaut. PM2: Parallel multithreaded machine. a computing environment for distributed architectures. In *Parallel Computing (ParCo '95)*, pages 279–285. Elsevier Science Publishers, September 1995.
12. B. Nitzberg and V. Lo. Distributed shared memory: A survey of issues and algorithms. *IEEE computer*, 24(8):52–60, September 1991.
13. M. Philippsen and M. Zenger. JavaParty — transparent remote objects in Java. *Concurrency: Practice and Experience*, 9(11):1125–1242, November 1997.
14. T. Proebsting, G. Townsend, P. Bridges, et al. Toba: Java for applications — a way ahead of time (WAT) compiler. In *Third Conference on Object-Oriented Technologies and Systems (COOTS '97)*, Portland, June 1997.
15. W. Yu and A. Cox. Java/DSM: A platform for heterogeneous computing. In *Proceedings of the Workshop on Java for High-Performance Scientific and Engineering Computing*, Las Vegas, Nevada, June 1997.

Third Workshop on Bio-Inspired Solutions to Parallel Processing Problems (BioSP3)

May 1, 2000
Cancun, Mexico

This section contains the papers presented at the Third Workshop on Bio-Inspired Solutions to Parallel Processing Problems (BioSP3) which is held in conjunction with IPDPS'2000 Cancun, Mexico, May 1-5, 2000. The workshop aimed to provide an opportunity for researchers to explore the connection between biology-inspired techniques and paradigms and the development of solutions to problems that arise in parallel processing.

It is well known that techniques inspired by biological phenomena can provide efficient solutions to a wide variety of problems in parallel computing and, more generally, in computer science. A vast literature exists on bio-inspired approaches to solving a rather impressive array on problems and, more recently, a number of studies have reported on the success of such techniques for solving difficult problems in all key areas of parallel processing and computer science. Rather remarkably, most of the biology-inspired techniques and paradigms are inherently parallel. Thus, solutions based on such methods can be conveniently implemented on parallel architectures.

In response to the call for papers for this workshop, we received a reasonably large number of submissions from all over the world, leading to a truly international competition. The papers underwent a thorough review process. The Workshop Chairs ranked the manuscripts on their original contributions and also carefully considered the suitability of the topic for the workshop. Thus were selected twelve manuscripts which went through a round of revisions, before they could be finally included in the workshop proceedings.

The collection of twelve papers that are presented here is a good sampler of the theoretical and practical aspects of the research in biology-inspired techniques and paradigms. The papers accepted span a variety of topics ranging from DNA computing, to agent-based techniques, to collective intelligence modeling, to genetic algorithms and neural nets, among others.

We take this opportunity to thank all the authors for their submissions and the program committee for making BioSP3 a success.

Albert Y. Zomaya
Fikret Ercal
Stephan Olariu

J. Rolim et al. (Eds.): IPDPS 2000 Workshops, LNCS 1800, pp. 568–569, 2000.

Workshop Chairs

Albert Y. Zomaya, The University of Western Australia
Fikret Ercal, University of Missouri-Rolla
Stephan Olariu, Old Dominion University

Steering Committee

Peter Fleming, University of Sheffield, United Kingdom
Frank Hsu, Fordham University
Oscar Ibarra, University of California, Santa Barbara
Viktor Prasanna, University of Southern California
Sartaj Sahni, University of Florida - Gainsville
Hartmut Schmeck, University of Karlsruhe, Germany
H.J. Siegel, Purdue University
Les Valiant, Harvard University

Program Committee

Ishfaq Ahmad, Hong Kong University of Science and Technology
David Andrews, University of Arkansas
Juergen Branke, University of Karlsruhe
Jehoshua (Shuki) Bruck, California Institute of Technology
Jens Clausen, Technical University of Denmark, Denmark
Sajal Das, University of North Texas
Afonso Ferreira, SLOOP, CNRS-INRIA-UNSA, France
Ophir Frieder, Illinois Institute of Technology
Eileen Kraemer, University of Georgia
Mohan Kumar, Curtin University of Technology, Australia
Richard Lipton, Princeton University
Rebecca Parsons, University of Central Florida
John H. Reif, Duke University
Peter M.A. Sloot, University of Amsterdam, The Netherlands
Assaf Schuster, Technion, Israel
Franciszek Seredynski, Polish Academy of Sciences, Poland
Ivan Stojmenovic, Ottawa University, Canada
El-ghazali Talbi, Laboratoire d'Informatique Fondamentale de Lille, France

Take Advantage of the Computing Power of DAN Computers

Z. FRANK QIU and MI LU

Department of Electrical Engineering
Texas A&M University
College Station, Texas 77843-3128, U.S.A.
{zhiquan, mlu}@ee.tamu.edu

Abstract. Ever since Adleman [1] solved the Hamilton Path problem using a combinatorial molecular method, many other hard computational problems have been investigated with the proposed DNA computer [3] [25] [9] [12] [19] [22] [24] [27] [29] [30]. However, these computation methods all work toward one destination through a couple of steps based on the initial conditions. If there is a single change on these given conditions, all the procedures need to be gone through no matter how complicate these procedures are and how simple the change is. The new method we are proposing here in the paper will take care of this problem. Only a few extra steps are necessary to take when the initial condition has been changed. This will provide a lot of savings in terms of time and cost.

1 Introduction

Since Adleman [1] and Lipton [25] presented the ideas of solving difficult combinatorial search problems using DNA molecules, there has been a flood of ideas on how DNA can be used for computations [4] [3] [25] [9] [12] [16] [18] [19] [20] [22] [24] [27] [28] [29] [30]. As one liter of water can hold 10^{22} bases of DNA, these methods all take advantage of the massive parallelism available in DNA computers. This also raises the hope to solve the problems intractable for electronic computers. However, the brute force approaches usually take long time to complete due to the long process of the problem and for those existing method, they all need to restart the whole process when the initial condition changes.

In this paper, we propose a new class of algorithms to be implemented on a DNA computer. The algorithms we are going to introduce will not be affected much by the initial condition change. This will give DNA computers great flexibility. Knapsack problems are classical problems solvable by this method. It is unrealistic to solve these problems using conventional electronic computers when the size of them get large due to the NP-complete property of these problems. DNA computers using our method can solve substantially large size problems because of their massive parallelism.

Throughout the paper, we make one assumption that all molecular biological procedures are error free. This is not true for the real world, but there is

J. Rolim et al. (Eds.): IPDPS 2000 Workshops, LNCS 1800, pp. 570–577, 2000.
© Springer-Verlag Berlin Heidelberg 2000

a large amount of finished research work attacking this error-resisting problem [5] [10] [11] [14] [35] [36]. These research work also showed many fault tolerant techniques. We hope that errors which arise during our DNA computer operations can be dealt with by the given techniques.

The rest of the paper are organized as follows: the next section will explain the methodology. The description of how NP-complete problems are solved will be presented in section 3. Section 4 will explain how we can avoid going through all the procedures when a little change is made to the initial condition. The last section will conclude this paper and point out future work.

2   DNA Computation Model

In this section, we include the fundamental model for basic operations to be used in our DNA computation. All DNA operation models necessary for solving those NP-complete problems in the next section are introduced and explained as follows.

2.1   Operations

1. **reset(S)**: This may also be called initialization. It will generate all the strands for the following operations. These strands can be generated either to represent the same value or to represent different values according to the requirement.
2. **addin(x, S)**: This step adds a value x to all the numbers inside the set **S**.
3. **sub(x, S)**: This operation will subtract a value x from all the numbers inside the set **S**.
4. **divide(S, S1, S2, C)**: This step will separate the set **S** into two different sets based on the criteria C. If no criterion is given, then components in these two sets are randomly picked from set **S** and **S** will be evenly distributed into two sets **S1** and **S2**.
5. **Union(S1, S2, S)**: The operation combines, in parallel, all the components of sets **S1** and **S2** into set **S**.
6. **copy(S, S1)**: This will produce a copy of S: **S1**.
7. **select(S, C)**: This operation will select an element of **S** following criteria **C**. If no C is given, then an element is selected randomly.

2.2   Biological Implementation

In this section, we include the fundamental biological operations for our DNA computation model.

1. **reset(S)**: This initialization operation can be accomplished using mature biological DNA operations [1] [4] [6] [13] [15] [30]. It will generate a tube of DNA strands representing the same number, e.g., these strands will consist of exactly the same nucleotides with same order. It may also generate different strands to represent different numbers according to the requirement.

2. **addin(x, S)**: There are some existing arithmetic operations for DNA computers that have been developed [30] [19]. We are going to use the method introduced by [30]. This operation will add a number to all the strands in the tube using the method first introduced by [8]. After the addition is finished, all the new strands inside the tube will represent the sum of the value represented by the original strand and the number we add in no matter what was in the tube before the operation. Readers may refer to [30] if any detailed information about how to perform this addition is needed.

3. **sub(x, S)**: The operation can be accomplished similarly as the **addin** operation shown above. It will simply subtract the value x from all the strands inside the tube.

4. **divide(S, S1, S2, C)**: The necessary operation for this step of DNA computing is to separate one tube of strands into two tubes. Each resultant tube will have approximately half of the strands of the original tube. The criteria C can be containing or not containing a certain segment, e.g. ATTCG, and we may use the metal bead method to extract them [1] [23].

5. **union(S1, S2, S)**: This operation will simply pour two tubes of strands into one.

6. **copy(S, S1)**: We need to make copies of DNA strands of the original tube and double the number of strands we have for this copy operation. The best and easiest method for this will be PCR (Polymerase Chain Reaction). Because PCR is counted as non-stable by [31] [34] [33] [32], we will try to use this operation as few times as possible.

7. **select(S, C)**: This procedure will actually extract out the strand we are looking for. So, it will extract strands from tube **S** following certain criteria C. We may use existing methods introduced in [2] [21] [25].

3   NP-complete Problem Solving

3.1   One Simplified NP-complete Problem

We will show how to solve a simplified knapsack problem [7], one of the NP-complete problems which is unsolvable by currently electronic computers.

Problem: Given an integer K and n items of different sizes such that the ith item has an integer size k_i, find a subset of the items whose sizes sum to exactly K, or determine that no such subset exists [26] [17].

In solving the knapsack problem using DNA computers, we intend to use the methodology presented in the previous section. It can be accomplished as follows:

a-1 **reset(S)**: We will generate a large amount of strands in a tube **S** and all strands in the tube will represent 0, i.e. they are exactly the same. This will assume that all the potential "bags" are empty because each strand is counted as a bag to hold items.

a-2 **divide(S, S1, S2)**: This will give us two almost identical sets that are close to the exact copies of **S** when we have a large amount of strands in the original set **S**.

a-3 **addin(x_i, S1):** We will add integer x_i which represents the size of the ith item to set **S1**.

a-4 **union(S1, S2, S):** Now we have a mixed set with about half of the potential "bags" containing the ith item while others do not.

a-5 Repeat the above steps 2, 3 and 4 until all items have been added in.

a-6 **select(S, C):** The criteria **C** we are using here is whether number **K** exists or not. If we find an integer **K** in the tube, that means we have a subset of the items with sizes sum exactly equals to **K**. If there is no such kind of strand in the tube, then the answer will be no.

The number of steps the algorithm requires is 3n+2, where n is the total number of items. Unlike another possible solution for this kind of knapsack problem introduced by Baum [7], we do not have any restriction like balanced size of items required by Baum. That means, we can solve all simplified version of knapsack problems within polynomial time as long as the total size of these n items can be represented by the method [30] we use.

3.2   An Advanced Problem

In the previous section, we gave the DNA algorithm on one of the NP-complete problems: simplified knapsack problem. Here, we are going to advance the method and try to solve the complete version of the knapsack problem, a more computation intense problem.

Problem: The input is a set X such that each element $x \in X$ has an associated size k(x) and value v(x). The problem is to determine whether there is a subset $B \in X$ whose total size is $\leq K$ and whose total value is $\geq V$.

Algorithm: The advanced algorithm based on the one for the simplified knapsack problem is shown as follows:

b-1 **reset(S):** We will generate a lot of "empty" strands in the set. Here each strands in the tube will be treated as two parts while both are zeros initially. The first part X will be used to represent the size of the items inside the tube and the second part U is for the value of these items. At the very beginning, because the bags are empty, so both the size and the value are zeros. Each strand will be like $5' - X - U - 3'$. We make X and V large enough to hold the total item sizes and total item values so no matter how we operate on them, X and U will not intervene with each other.

b-2 **divide(S, S1, S2):** This will give us two almost identical sets that are close to the exact copies of S when we have a large amount of strands in the original set **S**.

b-3 **addin(x_i, S1):** We will operate on the first set S1. The integer that represents the size of the item:k_i and the integer that represents the value of the item:v_i will be added into different part of all strands in the set. x_i is added to X and v_i is added to U. Set S2 will be untouched. The technique that adding different numbers to their expected locations are based on the locators L_i and R_i shown in [30] b-4 **union(S1, S2, S)**: Now we have a mixed set with about half of the potential "bags" containing the ith item while others do not.

b-5 Repeat the above steps 2, 3 and 4 until all items have been added in.

Because we are not looking for a particular number but for the numbers smaller than K while the associate value U is larger than V, we can not easily pick one strand out. So we go through one extra step before the final result extraction.

b-6 **divide(S, S1, S2, C)**: The criteria here is X≤K or X>K. Let's assume that S1 contains all strands with X≤K while S2 holds the rest.

b-7 **select(S1, C)**: We are going to extract the answer from S1 as S1 is the set that contains those "bags" with items less than full, i.e., X≤ K. As the value of each strand is represented by a certain number of digits, we only need to go through these digits one by one and find the answer larger than V.

4 Problem Reconsideration

In the previous section, we introduced new algorithms for solving NP complete problems: Knapsack problems. Here we are going to show that the advantage of our algorithm, i.e., unlike other existing algorithms [1] [2] [3] [12] [18] [21] [25] that need to restart the whole computation process when there are changes on the initial condition, our algorithm will only need a few extra operations and the new problem will be solved. This will greatly save time and cost for our DNA computer because usually DNA computing needs a lot of expensive materials and takes very long time, e.g., months, to complete.

We first work on the simplified knapsack problem. The initial condition is an integer K and n items of different sizes. After the procedures we showed in section 3.1, we will obtain a bag with size K and have m items inside where m<n. Suppose we want to make a minor change at the initial condition. Let the change be: instead of having n items at the beginning, we lost one of the items. So totally we have n-1 items. If the item is not contributing to the "bag", then nothing will change. If the item is in the "bag" of size K, then we need to generate a totally different new solution. Instead of going through all the steps above, we just add a few new steps to the existing algorithm and it is much easier to obtain the result.

The following are the extra steps we need to add:

A-1 **divide(S, S1, S2, C)**: Seperate the set S into two sets where S1 contains strands with item Y and S2 contains strands without Y. Y is the item we do not want to count.

A-2 **sub(S1, Y)**: S1 is the set left over after the extracting of the previous result.

A-3 **union(S1, S2, S)**: Now S will have no strand containing item Y. So Y has been removed.

A-4 **select(S, C)**: This is the exact same procedure as step a-6 shown above in section 3.1. Still, condition C is regarding whether we have a strand representing number K or not. If at least one such strand exists, then we have a solution. Otherwise, there is no combination of K with these n-1 items.

If more than one item have been removed from the initial list, we need to repeat the above extra steps A-2 and A-4 a few times.

For the complete knapsack problem, similar procedures can be used after the initial condition is changed. If the same modification is performed as the example of simplified knapsack problem showed above: one item is removed from the initial list, the following operations are necessary in order to obtain the solution.

B-1 **union(S1, S2, S)**: This will put the remaining potential answers together.
B-2 **divide(S, S1, S2, C)**: It will sperate the set S into two sets following the criteria C so that set S1 will contain item i which should be removed from the bag and S2 do not have item i in it.
B-3 **sub(y_i, S1)**: Subtract item i from set S1. The detailed operation is that y_i is subtracted from X and the corresponding value v_i is subtracted from U.

Then we only need to perform steps b-6 and b-7 above to see if we have the answer we expected.

5   Conclusion

In this paper, we attempted to solve a set of problems to which DNA computers can apply. As an example, we demonstrate that simplified knapsack problem can be solved efficiently on our DNA computer. We also extend the algorithm to solve the complete version of knapsack problem. These examples have illustrated the advantage of DNA computers.

We note that knapsack problems are NP complete, whether simplified version or complete version. Our method can solve these problems with different complexities within polynomial time. The biggest advantage of using our method to solve these NP-complete problems comparing with other existing methods is that it not only gives out the correct answer, but also saves a lot of computing time and resouces when there are minor changes to the conditions given. We may also extend our algorithm to what is under investigation: the graph connectivity problem. We may also consider cases when the condition change are huge. The future work will include implementing our algorithms in the biological lab and make it more robust.

References

1. Len Adleman. Molecular computation of solutions to combinatorial problems. *Science*, November 1994.

2. Leonard M. Adleman, Paul W.K. Rothemund, Sam Roweis, and Erik Winfree. On applying molecular computation to the data encryption standard. In *Second Annual Meeting on DNA Based Computers*, pages 28–48, June 1996.

3. Joh-Thomes Amenyo. Mesoscopic computer engineering: Automating dna-based molecular computing via traditional practices of parallel computer architecture design. In *Second Annual Meeting on DNA Based Computers*, pages 217–235, June 1996.

4. Martyn Amos. *DNA Computation*. PhD thesis, University of Warwick, UK, September 1997.

5. Martyn Amos, Alan Gibbons, and David Hodgson. Error-resistant implementation of DNA computations. In *Second Annual Meeting on DNA Based Computers*, pages 87–101, June 1996.

6. Eric B. Baum. DNA sequences useful for computation. In *Second Annual Meeting on DNA Based Computers*, pages 122–127, June 1996.

7. Eric B. Baum and Dan Boneh. Running dynamic programming algorithms on a dna computer. In *Second Annual Meeting on DNA Based Computers*, pages 141–147, June 1996.

8. D. Beaver. Molecular computing. Technical report, Penn State University Technical Report CSE-95-001, 1995.

9. Andrew J. Blumberg. Parallel computation on a dna substrate. In *3rd DIMACS Workshop on DNA Based Computers*, pages 275–289, June 1997.

10. Dan Boneh, Christopher Dunworth, Jeri Sgall, and Richard J. Lipton. Making DNA computers error resistant. In *Second Annual Meeting on DNA Based Computers*, pages 102–110, June 1996.

11. Junghuei Chen and David Wood. A new DNA separation technique with low error rate. In *3rd DIMACS Workshop on DNA Based Computers*, pages 43–58, June 1997.

12. Michael Conrad and Klaus-Peter Zauner. Design for a DNA conformational processor. In *3rd DIMACS Workshop on DNA Based Computers*, pages 290–295, June 1997.

13. R. Deaton, R. C. Murphy, M. Garzon, D. R. Franceschetti, and Jr. S. E. Stevens. Good encodings for DNA-based solutions to combinatorial problems. In *Second Annual Meeting on DNA Based Computers*, pages 131–140, June 1996.

14. Myron Deputat, George Hajduczok, and Erich Schmitt. On error-correcting structures derived from DNA. In *3rd DIMACS Workshop on DNA Based Computers*, pages 223–229, June 1997.

15. Dirk Faulhammer, Richard Lipton, and Laura Landweber. Counting DNA: Estimating the complexity of a test tube of DNA. In *Fourth Internation Meeting on DNA Based Computers*, pages 249–252, June 1998.

16. Y. Gao, M. Garzon, R.C. Murphy, J.A. Rose, R. Deaton, D.R. Franceschetti, and S.E. Stevens Jr. DNA implementattion of nondeterminism. In *3rd DIMACS Workshop on DNA Based Computers*, pages 204–211, June 1997.

17. Michael R. Garey and David S. Johnson. *Computers and Intractability*. W. H. Freeman And Company, 1979.

18. Gre Gloor, Lila Kari, Michelle Gaasenbeek, and Sheng Yu. Towards a DNA solution to the shortest common superstring problem. In *Fourth Internation Meeting on DNA Based Computers*, pages 111–116, June 1998.

19. Vineet Gupta, Srinivasan Parthasarathy, and Mohammed J. Zaki. Arithmetic and logic operation with dna. In *3rd DIMACS Workshop on DNA Based Computers*, pages 212–222, June 1997.

20. Masami Hagiya and Masanori Arita. Towards parallel evaluation and learning of boolean mu-formulas with molecules. In *3rd DIMACS Workshop on DNA Based Computers*, pages 105–114, June 1997.

21. Natasa Jonoska and Stephen A. Karl. A molecular computation of the road coloring problem. In *Second Annual Meeting on DNA Based Computers*, pages 148–158, June 1996.

22. Peter Kaplan, David Thaler, and Albert Libchaber. Paralle overlap assembly of paths through a directed graph. In *3rd DIMACS Workshop on DNA Based Computers*, pages 127–141, June 1997.

23. Julia Khodor and David K. Gifford. The efficiency of sequence-specific separation of DNA mixtures for biological computing. In *3rd DIMACS Workshop on DNA Based Computers*, pages 26–34, June 1997.

24. Thomas H. Leete, Matthew D. Schwartz, Robert M. Williams, David H. Wood, Jerome S. Salem, and Harvey Rubin. Massively parallel dna computation: Expansion of symbolic determinants. In *Second Annual Meeting on DNA Based Computers*, pages 49–66, June 1996.

25. Richard Lipton. Using DNA to solve SAT. Unpulished Draft, 1995.

26. Udi Manber. *Introduction To Algorithms*. Addison-Wesley Publishing company, 1989.

27. Nobuhiko Morimoto and Masanori Arita Akira Suyama. Solid phase DNA solution to the hamiltonian path problem. In *3rd DIMACS Workshop on DNA Based Computers*, pages 83–92, June 1997.

28. Mitsunori Ogihara and Animesh Ray. DNA-based parallel computation by 'counting'. In *3rd DIMACS Workshop on DNA Based Computers*, pages 265–274, June 1997.

29. John S. Oliver. Computation with DNA-matrix multiplication. In *Second Annual Meeting on DNA Based Computers*, pages 236–248, June 1996.

30. Z. Frank Qiu and Mi Lu. Arithmetic and logic operations for DNA computers. In *Parallel and Distributed Computing and Networks (PDCN'98)*, pages 481–486. IASTED, December 1998.

31. Sam Roweis, Erik Winfree, Richard Burgoyne, Nickolas Chelyapov, Myron Goodman, Paul Rothemund, and Leonard Adleman. A sticker based architecture for DNA computation. In *Second Annual Meeting on DNA Based Computers*, pages 1–27, June 1996.

32. Erik Winfree. *Algorithmic Self-Assembly of DNA*. PhD thesis, California Institute of Technology, June 1998.

33. Erik Winfree. Proposed techniques. In *Fourth Internation Meeting on DNA Based Computers*, pages 175–188, June 1998.

34. Erik Winfree, Xiaoping Yang, and Nadrian C. Seeman. Universal computation via self-assembly of DNA: Some theory and experiments. In *Second Annual Meeting on DNA Based Computers*, pages 172–190, June 1996.

35. David Harlan Wood. applying error correcting codes to DNA computing. In *Fourth Internation Meeting on DNA Based Computers*, pages 109–110, June 1998.

36. Tatsuo Yoshinobu, Yohei Aoi, Katsuyuki Tanizawa, and Hiroshi Iwasaki. Ligation errors in DNA computing. In *Fourth Internation Meeting on DNA Based Computers*, pages 245–246, June 1998.

Agent Surgery: The Case for Mutable Agents

Ladislau Bölöni and Dan C. Marinescu

Computer Sciences Department
Purdue University
West Lafayette, IN 47907

Abstract. We argue that mutable programs are an important class of future applications. The field of software agents is an important beneficiary of mutability. We evaluate existing mutable applications and discuss basic requirements to legitimize mutability techniques. Agent surgery supports runtime mutability in the agent framework of the Bond distributed object system.

1 Introduction

Introductory computer architecture courses often present an example of self-modifying assembly code to drive home the message that executable code and data are undistinguishable from one another once loaded in the internal memory of a computer. Later on, computer science students learn that writing self-modifying programs or programs which modify other programs is not an acceptable software engineering practice but a rebellious approach with unpredictable and undesirable results. A typical recommendation is *"Although occasionally employed to overcome the restrictions of first-generation microprocessors, self-modifying code should never ever be used today - it is difficult to read and nearly impossible to debug."* [1]

We argue that self-modifying or mutable programs are already around, and they will be an important component of the computing culture in coming years. In particular, this approach can open new possibilities for the field of autonomous agents.

Software agents play an increasingly important role in today's computing landscape due to their versatility, autonomy, and mobility. For example the Bond agent framework, [2], is currently used for a variety of applications including an workflow enactment engine for commercial and business applications [5], resource discovery and management in a wide-area distributed object-system, [7], adaptive video services [6], parallel and distributed computing [4]. Some of these applications require that the functionality be changed while the agent is running. But to change the functionality of an agent we have to modify its structure. This motivates our interest for agent surgery and by extension for program mutability.

J. Rolim et al. (Eds.): IPDPS 2000 Workshops, LNCS 1800, pp. 578-585, 2000.

2   Mutable programs

We propose the term *mutable programs* for cases when the program executable is modified by the program itself or by an external entity. We use the term "mutable" instead of "mutating" to indicate that changes in the program structure and functionality are directed, well specified, non-random. The word mutation is used in genetic algorithms and evolutionary computing to indicate random changes in a population when the "best" mutant in some sense is selected, using a survival of the fittest algorithm. In these cases the source of the program is modified. The case we consider is when a program is modified at run time to change its functionality. We distinguish *weak* and *strong* mutability.

Weak mutability is the technique to *extend* the functionality of the application using external components.The application still keeps its essential characteristics and functionality. The trademark of weak mutability is a well defined API or interface between the program to be extended and the entity providing the additional functionality. The mechanism of extension is either the explicit call of an external application or loading of a library, either in the classical dynamic library sense or in the Java applet / ActiveX control sense.

This is the currently most accepted form of mutability in applications. Examples include: data format plugins, active plugins, skins and themes, applets and embedded applications, automatic upgrades etc.

In case of **strong mutability**, applications change their behavior in a radical manner. The APIs for the modification is loosely defined, or implicit.

Strong mutability can be implemented at any level of granularity. At the machine-code level, typical example are viruses and anti-virus programs. Classical executable viruses are attaching themselves to programs and modify the loader code to add their additional, usually malign, functionality. Certain viruses perform more complex procedures on the executable code, e.g. compression, redirection of function calls etc. Anti-virus programs modify the executable program, by removing the virus and restoring the loader. A special kind of anti-virus programs, called *vaccines* modify an executable to provide a self-checksum facility to prevent further modifications. Another example of strong mutability is *code mangling* , performed to prevent the reverse engineering of the code. A somewhat peculiar example is the case of self-building neural networks.

Component-based systems are most promising for strong mutability because the larger granularity of the components makes the problem more tractable. At the same time the cleaner interface of components and the fact that they are usually well specified entities allow for easier modification. The most promising directions are the custom assembly of agents based on specifications, and runtime modification of component-based agents, called in this paper "agent surgery". An example of application-level strong mutability is the case of dynamically assembled and modified workflows.

Several requirements must be met before mutabilility could become an accepted programming technique, they are: well-defined scope, self-sufficiency, seamless transformation, persistency, and reversibility.

First, the modification should have a **well-defined scope** and lead to a predictable change of program behavior. For example a user downloading a new plugin expects that the only modification of the program behavior is to accept a new data type.

Self-sufficiency requires that the change described by the process should be defined without knowing the internal structure of the application. Thus multiple consecutive changes can applied to an application. It also allows one change to be applied to multiple applications.

Seamless transformation of program behavior, the change should be performed on a running program without the need to stop its execution. The alternative to modify the source code, compile it, and restart the program is not acceptable in many cases. As our dependency upon computers becomes more and more pronounced, it will be increasingly difficult to shut down a complex system e.g. the air-traffic control system, a banking system, or even a site for electronic commerce to add, delete, or modify some of its components. Even today it is not acceptable to restart the operating system of a computer or recompile the browser to add a new plugin necessary to view a webpage. A similar hardware requirement is called hot-plug compatibility, hardware components can be added and sometimes removed from a computer system without the need to shut down the system.

Another requirement is to make the change **persistent**. This requirement is automatically satisfied when the file containing the executable program is modified. If the image of a running program is modified then a mechanism to propagate this change to the file containing the executable must be provided.

The reverse side of the coin is the ability to make the change **reversible**. In a stronger version of this requirement we expect to revert a change while the program is running. A weaker requirement is to restart the agent or the application in order to revert the change. This requirement can be easily satisfied for individual changes, by keeping backup copies of the original program. If we allow multiple changes and then revert only some of them the problem becomes very complex as the experience with **install programs/scripts** shows.

We emphasize that not all these requirements should be satisfied simultaneously. Actually only the predictability is a critical property. Satisfying the additional requirements however, broadens the range of applicability of the technique. Thus, when designing the **mechanisms of mutability** we should attempt to satisfy as many of them as possible.

The common feature of every mechanism of mutability is that we treat code as a data structure. This is an immediate consequence of the von Neumann's concept of stored-program computers, and thus applicable to virtually any program. Yet, in practice, modifying running programs is very difficult, unless they are described by a simple and well-defined defined data structure accesible at run-time. For example, a program written in C++ or C has a structure given by the flow of function calls in the original source code. After the compilation and optimization process however, this structure is very difficult to reconstruct.

The object code of compiled languages or the code generated from the assembly language, while it can be viewed as a data structure, it does not allow us to easily discover its properties. This complexity justifies the point made at the beginning of this paper, that self-modifying machine code should not be used as a programming technique.

In conclusion, programs can be successfully modified if there is a high level, well documented data structure [1] that in some sense is analogous with the genetic information of biological structures. If the program designer chooses to have only part of the code described by this structure (like in the case of plugin API-s) weak mutability is possible. If the entire program is described by the data structure then strong mutability is possible. This is the case of Bond agents whose behavior is based on the multiplan state machine and the agent surgery enables strong mutability.

3   Mutability in Bond

Agent surgery is a modification technique employed in the Bond system to change the behavior of a *running* agent by modifying the data structure describing the multi-plane state machine of the agent.

There are several advantages the multiplane state machines offer over other data structures:

-The behavior of the agent in any moment is determined by a well defined subset of the multiplan state machine (the current state vector).

-The multiplane state machine exhibits *enforcable locality of reference*. While most programs exhibit temporal and spatial locality of reference, this can not be used in mutability procedures, because we don't have a guarantee that the program will not perform suddenly a long jump, which can be intercepted only at a very low level - by the operating system or the hardware. The semantic equivalent of a long jump, the transitions, however, are executed through the messaging function of the agent, thus they can be captured and queued for the duration of the agent surgery.

-The multiplane state machine is *self-describing*. Its structure can be iterated on by an external component. This allows to make the changes persistent by allowing to write out the runtime modifications to a new blueprint script. On the other hand allows for specifying operations independently from the structure of the agent.

Bond agents can be modified using "surgical" `blueprint` scripts. In contrast with the scripts used to create agents which define the structure of a multiplan finite state machine, surgical blueprints are adding, deleting or changing nodes, transitions and/or planes in an existing multiplan finite state machine.

The sequence of actions in this process is:

---

[1] Of course, strong mutability is possible even without this high level data structure, if the external entity is able to figure out the low level data structure manifests itself at the object code level - viruses are doing exactly this. However, this cannot form the basis of a generally viable technique.

(1)A *transition freeze* is installed on the agent. The agent continues to execute normally, but if a transition occurs the corresponding plane is frozen. The transition will be enqueued.

(2) The agent factory interprets the blueprint script and modifies the multiplane state machine accordingly. There are some special cases to be considered: (a) If a whole plane is deleted, the plane is brought first to a soft stop - i.e. the last action completes. (b) If the current node in a plane is deleted, a *failure* message is sent to the current plane.

(3) The transition freeze is lifted, the pending transitions performed, and the modified agent continues its existence.

4   Surgery techniques

The framework presented previously permits almost arbitrary modifications in the structure of the agent. Without some self-discipline however, the modified agent will quickly become a chaotic assembly of active components. A successfull surgical operation is composed on a number of more disciplined elementary operations, with a well specified semantics. In these section we enumerate those techniques which we consider as being the most useful.

4.1   Simple surgical operations

The simplest surgical operations are referring to the adding and removal of individual states and transitions. These operations are executed unconditionally. In the case of direct specification of operations is that the writer of the surgical blueprint must have a good knowledge of the existing agent structure. If new states are added to an existing plane, but not linked to existing states with transitions, they will never be executed. In order to remove existing states, transitions we need to know the name and function of the given states and transitions.

4.2   Replacing the strategy of a state

In this operation the strategy of a state is replaced with a different strategy. The reason of doing this is to improve or adapt the functionality to the specific condition of an agent. For example if an agent is running on, or migrated to a computer which doesn't have a graphical user interface, the strategies implementing the graphic user interface should be replaced with strategies adapted to the specific host, for example with command line programs.

These operations keep the old strategy namespace in which the strategy operates. In certain cases we should replace a group of interrelated strategies at once. For example, running and controlling an external application locally in Bond is done using the Exec strategy group. These strategies allow starting, supervising, terminating local applications, but can not run applications remotely. Now if the application requires remote run, we can replace all these strategies with the corresponding strategies from the RExec strategy group, which run applications using the Unix rexec call.

4.3   Splitting a transition with a state

Transitions in the Bond system represent a change in the strategy. One important way of changing the functionality of an agent is by inserting a new state in a transition. In effect, the existing transition is redirected to the new state, while from the new state the "success" transition is generated to the original target. The name of the new state is generated automatically. This can be later used to add new transitions to the state.

Typical application of these techniques are: logging and monitoring, confirmation checks and interagent synchronization.

4.4   Bypassing states

Bypassing a state is the semantic equivalent of replacing its strategy with a strategy which immediately, unconditionally succeeds. Practically the state is deleted, while all the incoming transitions are redirected to the target of the "success" transition originating from the given state.

The bypass operation can be used to revert the effect of the split operation discussed previously. For example, one application of the bypass operation is to remove the debugging or logging states from an agent.

4.5   Adding and removing planes

The planes of the multiplane finite state machine are expressing parallel activities. Thus, new functionality can be easily added to an agent by creating a new plane which implements that functionality. The new functionality can use the model of the world in the agent (the knowledge accumulated by other planes) and can directly communicate with other planes by triggering transitions in them. Analogously, we can remove functionality by deleting planes. There is direct support in *blueprint* for adding and removing planes.

Of course, like any surgical operation adding and removing planes is a delicate operation. Generally, we can safely add and remove planes which represent an independent functionality. Usually, the planes added to an already working agent can be safely removed.

Examples of using this technique are: adding a visualization plane, adding remote control or negotiation planes and replacing the reasoning plane to include a better reasoning mechanism.

4.6   Joining and splitting agents

Two of the simplest surgical operations on agents are the joining and splitting. When **joining** two agents, the multi-plane state machine of the new agent contains all the planes of the two agents and the model of the resulting agent is created by merging the models of the two agents. The safest way is to separate the two models (for example through use of namespaces), but a more elaborate merging algorithm may be considered. As our design does not specify the knowledge representation method, the best approach should be determined from case to case. The agenda of the new agent is a logical function (usually an "and" or an "or") on the agendas of the individual agents. It is tempting to consider the joining of agents as a boolean operation on agents, and maybe to envision an algebra of agents. While the subject definitely justifies further investigation, the design presented in this paper do not qualify for such an algebra. The more difficult problem is handling the "not" operator, which applied to the agenda would render the current multi-plane state machine useless.

In case of agent **splitting** we obtain two agents, the union of their planes gives us the planes of the original agent The splitting need not be disjoint, some planes (e.g an error handling or housekeeping) may be replicated in both agents. Both agents inherit the full model of the original agent, but the models may be reduced using the techniques presented in section 4.7. The agendas of the new agents are derived from the agenda of the original agent. The conjunction or disjunction of the two agendas gives the agenda of the original agent.

There are several cases when joining or splitting agents are useful: (a) Joining control agents from several sources, to provide a unified control, (b) Joining agents to reduce the memory footprint by eliminating replicated planes, (c) Joining agents to speed up communication, (d) Migrating only part of an agent, (e) Splitting to apply different priorities to parts of the agent.

Joining and splitting of agents is used by our implementation of agents implementing workflow computing.

4.7   Trimming agents

The state machines describing the planes of an agent may contain states and transitions unreachable from the current state. These states may represent execution branches not chosen for the current run, or states already traversed and not to be entered again. The semantics of the agent does not allow some states to be entered again, e.g. the initialization code of an agent is entered only once.

If the implementation allows us to identify the set of model variables which can be accessed by strategies associated with states, we can further identify parts of the model, which can not be accessed by the strategies reachable from the current state. The Bond system uses *namespaces* to perform a mapping of the model variables to strategy variables, thus we can identify the namespaces which are not accessed by the strategies reachable from the current state vector.

If the agenda of the agent can be expressed as a logical function on model variables and this is usually the case, we can simplify the agenda function for any given state, by eliminating the "or" branches that cannot be satisfied from the current state of the agent.

All these considerations allow us to perform the "trimming" of agents, for any given state to replace the agent with a different, smaller agent.

While stopping an agent to "trim" it is not justified for every situation there are several cases when we consider it to be especially useful, like before migration and before checkpointing.

The default migration implementation in the Bond system is using trimming to reduce the amount of data transferred in the migration process.

5   Conclusions

A number of factors, some of them technical, others legal and market-driven motivate an increasing interest in self-modifying programs. Some of the applications of software agents require that the functionality be changed while the agent is running. To support efficiently agent mobility we propose to trim out all the unnecessary components of an agent before migrating it. But to change the functionality of an agent we have to modify its structure. This motivates our interest for agent surgery and by extension for program mutability.

Several requirements must be met before mutabilility becomes an accepted programming technique, mutability should have a well-defined scope, be self-sufficient, seamless, persistent, and reversible. For example we should be able to modify running applications because it is conceivable to stop a critical system to modify its components.

The Bond agent framework is distributed under an open source license (LGPL) and the second beta release of version 2.0 can be downloaded from http://bond.cs.purdue.edu.

Acknowledgments

The work reported in this paper was partially supported by a grant from the National Science Foundation, MCB-9527131, by the Scalable I/O Initiative, and by a grant from the Intel Corporation.

References

1. A. Clements *Glossary of computer architecture terms. Self-modifying code.* http://www-scm.tees.ac.uk/users/a.clements/Gloss1.htm
2. L. Bölöni and D.C. Marinescu, *Biological Metaphors in the Design of Complex Software Systtems, Journal of Future Computer Systems*, Elsevier, (in press)
3. L. Bölöni and D. C. Marinescu, *An Object-Oriented Framework for Building Collaborative Network Agents,* in *Intelligent Systems and Interfaces,* (A. Kandel, K. Hoffmann, D. Mlynek, and N.H. Teodorescu, eds). Kluewer Publising, 2000, (in press).
4. P. Tsompanopoulou, L. Bölöni and D.C. Marinescu *The Design of Software Agents for a Network of PDE Solvers Agents For Problem Solving Applications Workshop, Agents '99,* IEEE Press, pp. 57-68, 1999.
5. K. Palacz and D. C. Marinescu, *An Agent-Based Workflow Management System Proc. AAAI Spring Symposium Workshop "Bringing Knowledge to Business Processes"* Standford University, IEEE Press, (to appear)).
6. K.K. Jun, L. Bölöni, D. K.Y. Yau, and D. C. Marinescu, *Intelligent QoS Support for an Adaptive Video Service, Proc. IRMA 2000,* IEEE Press, (to appear).
7. K.K. Jun, L. Bölöni, K. Palacz and D. C. Marinescu, *Agent-Based Resource Discovery, Proc. Heterogeneous Computing Workshop, HCW 2000,* IEEE Press (to appear).

Was Collective Intelligence[1] before Life on Earth?

Tadeusz SZUBA szuba@mcs.sci.kuniv.edu.kw
Mohammed ALMULLA almulla@mcs.sci.kuniv.edu.kw

Department of Mathematics and Computer Science, Science College,
Kuwait University, P.O. Box 5969 Safat, 13060 KUWAIT.

Abstract Collective Intelligence (*CI*) is formalized through a molecular model of computations and mathematical logic in terms of information_molecules quasi-chaotically displacing and running natural-based inference processes in the environment. *CI* abstracts from definitions of a communication system and Life. The formalization of *CI* is valid for social structures of humans, ants, and bacterial colonies. A simple extrapolation of the definition of CI suggests that a basic form of *CI* emerged on Earth in the "chemical soup of primeval molecules", before well-defined Life did, since *CI* is defined with fewer and weaker conditions than Life is. Perhaps that early, elementary *CI* provided basic momentum to build primitive Life. This successful action boosted a further self-propagating cycle of growth of *CI* and Life. The *CI* of ants, wolves, humans, etc. today is only a higher level of *CI* development. In this paper we provide formalization and a proposed partial proof for this hypothesis.

1. Introduction

We can ask, "What is the relationship between *Life* and *Intelligence?*" We can now build computers able to win chess matches against top masters and to emulate intelligent behavior of animal/human problem solving, but we do not attribute *Life* to such an artificially *intelligent* computer. There are research efforts tackling this problem, e.g. analysis of *Life* through its complexity e.g. [12], attempts to define and simulate *Artificial Life* e.g. [11], etc. Such attempts increase the depth of our knowledge, but still the question remains "How are *Life* and *Intelligence* related?" Formalization of *Collective Intelligence* [15], [16] sheds some light on this problem. The formal definition of *CI* has only three requirements. Information_molecules must emerge in a certain computational space (CS), and such CS can be almost anything: chemical molecules, software agents, ants, humans, or even social structures like cooperating villages. Later on some interaction must emerge between CS, which in a given environment results in the ability to solve specific problems. The emergence of *CI* is viewed in terms of the probability that specific inferences will result. As a result of this restricted set of requirements, we can analyze the problem of the *CI* of human social structures and we can as well go down to the edges of Life, to biofilms of

---

[1] This research was supported by Grant No. SM 174 funded by Kuwait University.

interacting bacteria in bacterial colonies [4] considered as collectively intelligent [3], [5], [6]. We cannot at this moment provide direct evidence that viruses or prions cooperate and create any CI; however - indirectly we can claim it by referring to the DNA computer [1]. Molecules in such a system are able to run inferences (computations) like a digital computer running an Expert System [10]. They are not alive at all. Moreover, the "conclusions" in such a DNA computer are active chemical molecules, thus able to output, i.e. implement a solution to the problem discovered by this CS. If we look at our present perception of *Life* [13], [14] and *Intelligence* through the complexity of individuals and the social structures they create, it is obvious that it is a dynamic system, and that *Life* and *Intelligence* are interleaved. The question is how does it happen? This paper poses for further discussion the following hypotheses:

- PRECEDENCE HYPOTHESIS: *CI* first emerged (perhaps accidentally) as a result of interacting chemical molecules on Earth.
- HYPOTHESIS ON ORIGIN: *Life* emerged later on, probably from *Collectively Intelligent* activity "looking" for stabilization and how to develop/propagate itself.
- HYPOTHESIS ON CYCLES: Dependency between *Life* (at different levels of complexity) and *Intelligence* (individual and collective) is the consecutive result of a spiral (development cycle) of evolution fired at that time and still active.

In our paper the computational model of *CI* will be briefly given and a definition of *Life* presented. On this basis we will try to demonstrate that *CI* is less complicated, and therefore could emerge more easily on a primeval Earth. A draft of the proof for the PRECEDENCE HYPOTHESIS will be also given.

2. Computational Collective Intelligence

CI can be easier to formalize and measure than the intelligence of a single being. Individual intelligence has only been evaluated on the basis of external results of behavior during processes in real life or during IQ tests. As a result, it is necessary to create abstract model activity based on neuropsychological hypotheses e.g. [7], or to use models like Artificial Intelligence. In contrast, many more elements of *CI* activity can be observed, measured, and evaluated. We can observe displacements, actions of beings, exchange of information between them (e.g. language, the ant's pheromone communication system, the language of dance of the honeybees, the crossover of genes between bacteria resulting in spreading a specific resistance to antibiotics, etc. Moreover, individual intelligence and behavior is scaled down as a factor. Underlying the presented *CI* formalization and modeling are these basic observations:

- In a socially cooperating structure it is difficult to differentiate thinking from non-thinking beings (abstract logical beings must be introduced, e.g. messages).
- Observing a being in a social structure, we can extract, label, and define rules of social behavior, e.g. use of pheromones. However, real goals, methods, and interpretations are mainly hidden until a detailed study is done. Thus use of mathematical logic is suggested to allow describing and simulating *CI*, postponing an interpretation of the clauses that label given elements of social behavior.
- The individuals inside a social structure usually cooperate in chaotic, yet non-continuous ways. Even hostile behavior between some beings can increase, to

some extent, the global *CI* of the social structure to which those hostile beings belong. In the social structure, beings move randomly because needs and opportunities of real life force them to do so. Inference processes are made randomly, most of which are not finished at all. This suggests using quasi-Brownian movements [18] for modeling social structure behavior. As a result, the probability of whether a problem can be solved over a certain domain of problems must be used as a measure for the *CI* of a social structure.

- Resources for inference are distributed in space, time, and among beings. Facts, rules, and goals may create inconsistent interleaving systems, multiple copies of facts, rules, and goals are allowed.
- Most of the concepts of human IQ tests are matched to culture, perception, communication, problem solving techniques, and methods of synthesizing the answer. Thus it is necessary to propose a new concept for testing *CI*, which is absolutely independent from points of view.
- The conditions given above are fulfilled by the efficiency of the N-element inference, with separately given interpretations for all formal elements of the test into real inferences or into a production process. *With this concept, in the same uniform way we can model inferring processes within a social structure, as well as production processes. This is very important because some inference processes can be observed only through resultant production processes, e.g. ants gathering to transport a heavy prey.* Separating N-element inferences from interpretation allows us, among other things, to test the intelligence of beings through building a *test environment* for them, where the sole solution is known to us as a given N-element inference.
- Humans infer in all directions: forward, backward, and through generalization. The N-element inference simulated in this model reflects all these cases clearly.

2.1 Computational model of Collective Intelligence

The 1st level computational space *CS* with inside quasi-random traveling facts, rules, and goals c_i is denoted as the multiset $CS^1 = \{c_1,...,c_n\}$. The clauses of facts, rules, and goals are themselves 0-level *CS*. For a given *CS*, we define a membrane (similar to [2]) denoted by $|\cdot|$ which encloses inherent clauses. It is obvious that $CS^1 = \{c_1,...,c_n\} \equiv \{|c_1,...,c_n|\}$. For a certain kind of membrane $|\cdot|$ its type p_i is given, which will be denoted $|\cdot|^{p_i}$ to define which *information_molecules* can pass through it. Such an act is considered Input/Output for the given *CS* with a given $|\cdot|$. It is also allowable to define degenerated membranes marked with $\cdot|$ or $|\cdot$ i.e. the collision-free (with membrane) path can be found going from exterior to interior of an area enclosed by such a membrane, for all types of *information_molecules*. The simplest possible application of degenerated membranes is to make, e.g. streets or other boundaries. If the *CS* also contains other *CSs*, then it is considered a higher order one, depending on the level of the internal CSs. Such internal CS will also be labeled with $\hat{v}_j$ e.g.

$$CS^2 = \left\{ \left| c_1, \ldots CS^1_{v_i}, \ldots c_n \right| \right\} \textit{ iff } CS^1_{v_i} \equiv \left\{ \left| b_1, \ldots, b_n \right| \right\} \textit{ where } b_i \; i = 1 \ldots m, \; c_j \; j = 1 \ldots n \textit{ are clauses}$$

Every c_i can be labeled with $\hat{v}_j$ to denote characteristics of its individual quasi-random displacements. Usually higher level CSs will take fixed positions, i.e. will create structures, and lower level CSs will perform displacements. For a given CS there is a defined position function pos:

$$pos: \; O_i \rightarrow \left\langle position \; description \right\rangle \cup undefined \quad where \quad O_i \in CS$$

If there are any two internal CS objects O_i, O_j in the given CS, then there is a defined distance function $D\left(pos\left(O_i \right), pos\left(O_j \right) \right) \rightarrow \Re$ and a rendezvous distance d. We say that during the computational process, at any time t or time period Δt, two objects O_i, O_j come to rendezvous iff $D\left(pos\left(O_i \right), pos\left(O_j \right) \right) \le d$. The rendezvous act will be denoted by the rendezvous relation $®$ e.g. $O_i \; ® \; O_j$ which is reflexive and symmetric, but not transitive. For another definition of rendezvous, see [8]. The computational process for the given CS is defined as the sequence of frames F labeled by t or Δt, interpreted as the time (given in standard time units or simulation cycles) with a well-defined *start* and *end*, e.g. $F_{t_0}, \ldots, F_{t_e}$. For every frame its multiset $F_j \equiv \left(\left| c_1, \ldots, c_m \right| \right)$ is explicitly given, with all related specifications: $pos(.)$, membrane types p, and movement specifications v if available. The simplest case of CS used in our simulations is the 3-D cube with randomly traveling clauses inside. The process is initialized to start the inference process after the set of clauses is injected into this CS. More advanced examples of the CS include a single main CS^2 with a set of internal CS^1 which take fixed positions inside CS^2, and a number of CS^0 who are either local for a given CS^1_i (because the membrane is not transparent for them) or global for any subset of $CS^1_j \in CS^2$. Modeling the CI of social structures, interpretations in the structure will be given for all CS^m_n, i.e. "this CS is a message"; "this is a single human"; "this is a city", etc. As has been mentioned, the higher level CS^i_j will take a fixed position to model substructures like villages or cities. If we model a single human as CS^1_j, then $\hat{v}_j$ will reflect displacement of the human. Characteristics of the given $\hat{v}_j$ can be purely Brownian or can be quasi-random, e.g. in lattice, but it is profitable to subject it to the present form of CS^i_j. When $\hat{v}_j$ has the proper characteristics, there are the following essential tools: • The goal clause, when it reaches the final form, can migrate toward the defined *Output* location, e.g. membrane of the main CS or even local CS. Thus the appearance of a solution of a problem in the CS can be observable. • Temporarily, the density of some *information_molecules* can be increased in the given area of the CS in such a way that after the given low-level

CS_j^i reaches the necessary form, it migrates to specific area(s) to increase the speed of selected inferences in some areas.

2.2. The Inference Model for Collective Intelligence and its measure

The pattern of inference generalized for any CS has the form:

DEFINITION 1. GENERALIZED INFERENCE IN CS^N

Assuming that $CS = \left\{ ...CS_j^i ...CS_l^k ... \right\}$, we can define:

CS_j^i ® CS_l^k and $U(CS_j^i, CS_l^k)$ and C(one or more CS_n^m of conclusions)

$\vdash$

one or more CS_n^m of conclusions, $R(CS_j^i$ or $CS_l^k)$ ■

The above description should be interpreted as follows: CS_j^i ® CS_l^k denotes rendezvous relation; $U(CS_j^i, CS_l^k)$ denotes that unification of the necessary type can be successfully applied; C(*one or more CS_n^m of conclusions*) denotes that CS_n^m are satisfiable. $R(CS_j^i$ or $CS_l^k)$ denotes that any parent *information_molecules* are retracted if necessary. The standard, e.g. PROLOG inferences are simple cases of the above definition. The above diagram will be abbreviated as $CS_j^i ; CS_l^k \xrightarrow{\ RPP\ } \sum_n CS_n^m$ without mentioning the retracted *information_molecules* given by $R(CS_j^i$ or $CS_l^k)$. In general, successful rendezvous can result in the "birth" of one or more child *information_molecules*. All of them must then fulfill a $C(...)$ condition; otherwise they are aborted. It is difficult to find examples of direct rendezvous and inference between two CS_i^m and CS_j^n if $m, n \geq 1$ without an intermediary involved CS_k^0 $k = 1, 2...$ (messages, pheromones, observation of behavior, etc.). Single beings like humans or ants can be represented as $CS_{individual}^1$. Such beings perform internal inferences (in their brains), independently of higher level, cooperative inferences inside CS_{main} and exchange of messages of the type CS^0. It will be allowable to have internal CS^k inside the main CS, but only as static ones (taking fixed positions) to define sub-structures such as streets, companies, villages, cities, etc. For simplicity, however, we will try to approximate beings as CS^0; otherwise, even statistical analysis would be too complicated. It is also important to assume that the results of inference are not allowed to infer between themselves after they are created. Products of inference must immediately disperse; however, later on, inferences between them are allowed (in [10] this is *refraction*). The two basic definitions for modeling and evaluating Collective Intelligence have the form:

DEFINITION 2: N-ELEMENT INFERENCE IN CS^N

There is a given CS at any level $CS = \left\{ CS_1^{a_1}, ..., CS_m^{a_m} \right\}$, and an allowed Set of Inferences SI of the form $\{$set of premises $CS\} \xrightarrow{I_j} \{$set of conclusions $CS\}$, and one or more CS_{goal} of a goal. We say that $\left\{ I_{a_0}, ..., I_{a_{N-1}} \right\} \subseteq SI$ is an N-element inference in CS^n, if for all $I \in \left\{ I_{a_0}, ..., I_{a_{n-1}} \right\}$ the premises $\in$ present state of CS^n at the moment of firing this inference, all $\left\{ I_{a_0}, ..., I_{a_{n-1}} \right\}$ can be connected into one tree by common conclusions and premises, and $CS_{goal} \in \left\{ set \text{ of conclusions for } I_{a_{N-1}} \right\}$.   ■

DEFINITION 3: *COLLECTIVE INTELLIGENCE* QUOTIENT (IQS)

IQS is measured by the probability P that after time t, the conclusion CM_{goal} will be reached from the starting *state of CS^n*, as a result of the assumed N-element inference. This is denoted $IQS = P(t, N)$.   ■

The above two definitions fulfill all the basic observations underlying the *CI* formalization. The proposed theory of Collective Intelligence allows us surprisingly easily to give formal definitions of the properties of social structures, which are obvious in real life (see [15], [16]).

3. Comprehension and definition of life

Traditionally *Life* has been defined as a material organization, which fulfills certain lists of properties or requirements [13], [14]. The distinction between *alive* and *non-alive* has emerged as the result of "Cartesian "thought in science. Before that, all processes were considered as alive, e.g. clouds. This way of thinking is still observable in so-called "personification" processes, attributing them with human-like behavior. Despite progress in science, we have problems defining *Life* in an efficacious way. For a process to be attributed with *Life,* it must have the following basic properties: *I) metabolism:* a complex of physical and chemical processes occurring within a living cell or organism that is necessary for the maintenance of *Life*; *II) adaptability:* becoming suitable for a new use or situation; *III) self-maintenance:* autonomy; *IV) self-repair*; *V) growth:* development from lower/simpler to a higher (more complex form); or an increase, as in size/number/value/strength, extension or expansion; *VI) replicability:* ability to reproduce or make an exact copy(s) of, e.g. genetic material, a cell, or an organism; *VII) evolution:* the theory that groups of organisms change in time, mainly as a result of natural selection, so that descendants differ morphologically and physiologically from their ancestors **or** as the historical development of a related group of organisms, i.e. phylogeny. Other properties can also be sub-attributed to be properties of *Life*. Most living organisms adhere to these requirements; however, there are material systems which obey only a subset of these rules, e.g. viruses. There are also processes like candle flames, which fulfill most of them, but scientists do not attribute life to. As a result, we can say that *Life* is still a fuzzy concept. Even the properties listed above closely overlap. To help us understand how *Life* developing we should look at theory, e.g. the Proliferation Theory [5].

4. Ordering Collective Intelligence and Life

The hypothesis that *Collective Intelligence* emerged on Earth before *Life* will be demonstrated in the following way. First we will formally define the complexity order for computational spaces CS_i^j. Later, on a complexity axis, (see Fig. 1) we will order all possible CS_i^j which are interesting or could be turning points from a *CI* and/or *Life* point of view. Finally, looking back at properties I-VII required for attributing *Life*, we will point out that it is not probable that *Life* emerged directly, skipping various simple evolutionary steps where *CI* was present.

<u>Definition 4: Strong[2] ordering of *computational spaces* CS^j.</u>

Assume that there are given CS_i^j and CS_k^l composed of elements:

$$CS_i^j = \left\{ CS_{(...)}^{a_1},...,CS_{(...)}^{a_n}, \middle| \cdot \middle|_{(...)}^{p_1},...,\middle| \cdot \middle|_{(...)}^{p_m} \right\} \quad CS_k^l = \left\{ CS_{(...)}^{b_1},...,CS_{(...)}^{b_r}, \middle| \cdot \middle|_{(...)}^{q_1},...,\middle| \cdot \middle|_{(...)}^{q_s} \right\}$$

*where (...) denotes an unspecified identification number for a given object
and $a_i,...,b_i,...,p_i,...,q_i$ are types*

The key problem is how to locate the point where the property *VI (Replicability)* could emerge. It is well known that many even simple chemical molecules can self-replicate in a favorable environment [9]. <u>However we should remember that self-replicating molecules also take "building components" from their environment, absorbing and processing other molecules. Pure self-replication cannot happen in the real world; otherwise the fundamental principle of "constant mass" will be violated.</u> Here we may find a turning point, i.e. computational space of the structure:

$$CS^1 = \left\{ \middle| c_1, c_2,... \middle| \right\} \quad \textit{i.e. information\_molecules inside membrane.}$$

because according to inference processes, various types CS^l can be also ordered by Definition 4 as to how they affect CS^l. An inference process can <u>either</u> reduce the number of c_i which automatically moves CS^l down on the complexity scale (analog of natural selection), <u>or</u> stabilize <u>or</u> even expand it. *CI* as a computational process can also exist as a reduction process, but *Life* with *VI Replicability* cannot go beyond this point. **This validates our PRECEDENCE HYPOTHESIS given earlier.** However, the real *Replicability* can start as early as:

$$CS^2 = \left\{ \middle| c_1, c_2,...,CS_1^1, CS_2^1,... \middle| \right\} \quad \textit{i.e. first local computational spaces } CS_{(...)}^1 \textit{ emerge}$$

From this point on the complexity scale we can speak about gemmation and gamogenesis. At this time draft proofs of the HYPOTHESIS ON ORIGIN and the

---

[2] The condition for ordering can be weakened, e.g. if we require only that elements of the same type must be used, not necessarily in the same quantity.

HYPOTHESIS ON CYCLES HAVE NOT BEEN COMPLETED. We consider them both as highly probable and as subjects for future discussion.

5. Conclusions

We have made an attempt to use the model of *CI* to formalize the relationship between the concepts of *Life* and *Intelligence*. We have proposed a draft proof on the basis of complexity of computational processes that a simple *CI* emerged before *Life* did on Earth. We have also proposed two hypotheses that *Life* is a logical consequence of of emerged *CI* in the environment of Earth. The second hypothesis states that since that time a cyclic development process has run where more and more complex forms of *Intelligence* and *Life* have propagated each other.

6. References

1.   Adleman L. M.: Molecular computations of solutions to combinatorial problems. Science. 11. 1994.
2.   Berry G., Boudol G.: The chemical abstract machine. Theoretical Comp. Science 96. 1992.
3.   Caldwell D. E., et al.: Germ theory versus community theory in understanding and controlling the proliferation of biofilms. Adv. Dental Research 11, 1996.
4.   Caldwell D. E., Costerton J. W.: Are bacterial biofilms constrained to Darwin's concept of evolution through natural selection? Microbiología SEM 12. 1996.
5.   Caldwell D. E. et al.: Do bacterial communities transcend Darwinism? Adv. Microb. Ecol. 15, 1996.
6.   Caldwell D. E., et al.: Cultivation of microbial communities and consortia. In Stahl, D. et al. (ed.), Manual of Environmental Microbiology, Am. Soc. for Microbiology Press. 1996.
7.   Das J. P., Naglieri J. A., Kirby J. R.: Assessment of cognitive processes. (the PASS theory of Intelligence). Allyn and Bacon. 1994.
8.   Fontana W., Buss L. W.: The arrival of the fittest. Bull. Math. Biol. 56, 1994.
9.   Freifelder D.: Molecular biology. Jones and Bartlett Pub. 1987.
10.   Giarratano J., Riley G.: Expert Systems. PWS Pub. II ed. 1994.
11.   Langton C. G.: Life at the edge of chaos". in Artificial Life II. Addison-Wesley. 1992.
12.   Lewin R.: Complexity - Life at the edge of Chaos. Macmillan. 1992.
13.   Pattee H.: Simulations, realizations, and theories of Life". In Artificial Life. C. Langton (Ed.). Addison-Wesley. 1989.
14.   Rosen R.: Life itself - a comprehensive inquiry into the nature, origin, and fabrication of Life. Columbia University Press, 1991.
15.   Szuba T.: Evaluation measures for the collective intelligence of closed social structures. ISAS'97 Gaithersburg/Washington D.C. Sept. 1997.
16.   Szuba T.: A formal definition of the phenomenon of collective intelligence and its IQ measure. Future Generation Computing Journal. Elsevier. 1999.
17.   Szuba T.: Computational Collective Intelligence. Wiley & Sons. To be published Dec. 2000.
18.   Whitney C. A.: Random processes in physical systems. An introduction to probability-based computer simulations. Wiley & Sons. 1990.

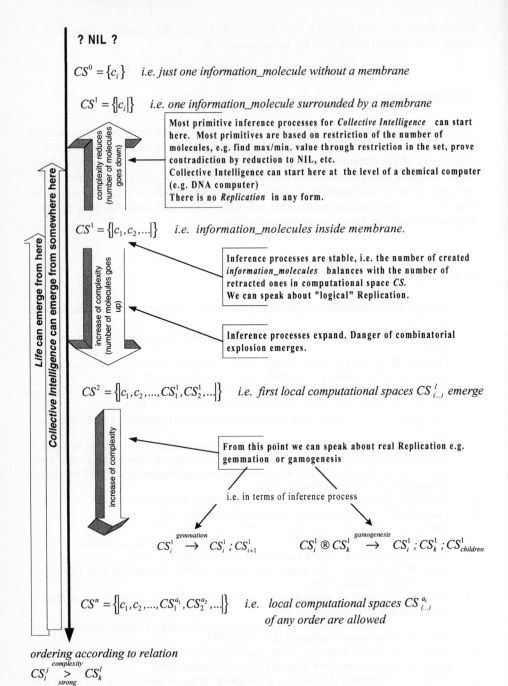

? NIL ?

$CS^0 = \{c_i\}$ *i.e. just one information_molecule without a membrane*

$CS^1 = \{|c_i|\}$ *i.e. one information_molecule surrounded by a membrane*

> Most primitive inference processes for *Collective Intelligence* can start here. Most primitives are based on restriction of the number of molecules, e.g. find max/min. value through restriction in the set, prove contradiction by reduction to NIL, etc.
> Collective Intelligence can start here at the level of a chemical computer (e.g. DNA computer)
> There is no *Replication* in any form.

complexity reduces (number of molecules goes down)

$CS^1 = \{|c_1, c_2, ...|\}$ *i.e. information_molecules inside membrane.*

> Inference processes are stable, i.e. the number of created *information_molecules* balances with the number of retracted ones in computational space *CS*.
> We can speak about "logical" Replication.

increase of complexity (number of molecules goes up)

> Inference processes expand. Danger of combinatorial explosion emerges.

$CS^2 = \{|c_1, c_2, ..., CS_1^1, CS_2^1, ...|\}$ *i.e. first local computational spaces $CS_{(...)}^1$ emerge*

increase of complexity

> From this point we can speak about real Replication e.g. gemmation or gamogenesis

i.e. in terms of inference process

$$CS_i^1 \xrightarrow{\text{gemmation}} CS_i^1 ; CS_{i+1}^1 \qquad CS_i^1 ® CS_k^1 \xrightarrow{\text{gamogenesis}} CS_i^1 ; CS_k^1 ; CS_{children}^1$$

$CS^n = \{|c_1, c_2, ..., CS_1^{a_1}, CS_2^{a_2}, ...|\}$ *i.e. local computational spaces $CS_{(...)}^{a_i}$ of any order are allowed*

Life can emerge from here

Collective Intelligence can emerge from somewhere here

ordering according to relation

$$CS_i^j \overset{\text{complexity}}{\underset{\text{strong}}{>}} CS_k^l$$

Fig. 1. Ordering *Computational SpacesCS* according to their strong complexity.

Solving Problems on Parallel Computers by Cellular Programming

Domenico Talia

ISI-CNR
c/o DEIS, UNICAL,
87036 Rende (CS), Italy
Email : talia@si.deis.unical.it

Abstract. Cellular automata can be used to design high-performance *natural solvers* on parallel computers. This paper describes the development of applications using CARPET, a high-level programming language based on the biology-inspired cellular automata theory. CARPET is a programming language designed for supporting the development of parallel high-performance software abstracting from the parallel architecture on which programs run. We introduce the main constructs of CARPET and discuss how the language can be effectively utilized to implement *natural solvers* of real-world complex problems such as forest fire and circuitry simulations. Performance figures of the experiments carried out on a MIMD parallel computer show the effectiveness of our approach both in terms of execution time and speedup.

1. Introduction

Cellular processing languages based on the cellular automata (CA) model [10] represent a significant class of restricted-computation models [8] inspired to a biological paradigm. They are used to solve problems on parallel computing systems in a wide range of application areas such as biology, physics, geophysics, chemistry, economics, artificial life, and engineering.

CA provide an abstract setting for the development of *natural solvers* of dynamic complex phenomena and systems. Natural solvers are algorithms, models and applications that are inspired by processes from nature. Besides CA, typical examples of natural solvers methods are neural nets, genetic algorithms, and Lindenmayer systems. CA represent a basic framework for *parallel natural solvers* because their computation is based on a massive number of *cells* with local interactions that use discrete time, discrete space and a discrete set of state variable values.

A cellular automaton consists of one-dimensional or multi-dimensional lattice of *cells*, each of which is connected to a finite neighborhood of cells that are nearby in the lattice. Each cell in the regular spatial lattice can take any of a finite number of discrete state values. Time is discrete, as well, and at each time step all the cells in the lattice are updated by means of a local rule called *transition function*, which determines the cell's next state based upon the states of its neighbors. That is, the state

J. Rolim et al. (Eds.): IPDPS 2000 Workshops, LNCS 1800, pp. 595-603, 2000.
© Springer-Verlag Berlin Heidelberg 2000

of a cell at a given time depends only on its own state and the states of its nearby neighbors at the previous time step. Different neighborhoods can be defined for the cells. All cells of the automaton are updated synchronously. The global behavior of the system is determined by the evolution of the states of all cells as a result of multiple interactions. An interesting extension of the CA standard model is represented by *continuous CA* that allow a cell to contain a real, not only an integer value. This class of automata is very useful for simulation of complex phenomena where physical quantities such as temperature or density must be taken into account.

CA are intrinsically parallel and they can be mapped onto parallel computers with high efficiency, because the communication flow between processors can be kept low due to locality and regularity. We implemented the CA features in a high-level parallel programming language, called CARPET [9], that assists cellular algorithms design. Unlike early cellular approaches, in which cell state was defined as a single bit or a set of bits, we define the state of a cell as a set of typed substates. This extends the range of applications to be programmed by cellular algorithms. CARPET has been used for programming cellular algorithms in the CAMEL environment [2, 4].

The goal of this paper is to discuss how the language can be effectively utilized to design and implement scientific applications as parallel natural solvers. The rest of the paper is organized as follows. Sections 2 and 3 introduce the constructs of CARPET and the main architectural issues of the CAMEL system. Section 4 presents a simple CARPET example and describes how the language can be utilized to model the forest fire problem. Finally, performance figures that show the scalability of CARPET programs on a multicomputer are given.

2. Cellular Programming

The rationale for CARPET (*CellulAR Programming EnvironmenT*) is to make parallel computers available to application-oriented users hiding the implementation issues resulting from architectural complexity. CARPET is a high-level language based on C with additional constructs to define the rules of the transition function of a single cell of a cellular automaton. A CARPET user can program complex problems that may be represented as discrete cells across 1D, 2D, and 3D lattices.

CARPET implements a cellular automaton as a SPMD program. CA are implemented as a number of processes each one mapped on a distinct processing element (PE) that executes the same code on different data. However, parallelism inherent to its programming model is not apparent to the programmer. According to this approach, a user defines the main features of a CA and specifies the operations of the transition function of a single cell of the system to be simulated. So using CARPET, a wide variety of cellular algorithms can be described in a simple but very expressive way.

The language utilizes the control structures, the types, the operators and the expressions of the C language. A CARPET program is composed of a declaration part that appears only once in the program and must precede any statement (except those of C pre-processor) and of a program body. The program body has the usual C statements and a set of special statements defined to access and modify the state of a cell and its neighborhood. Furthermore, CARPET permits the use of C functions and procedures to improve the structure of programs.

The declaration section includes constructs that allow a user to specify the dimensions of the automaton (**dimension**), the radius of the neighborhood (**radius**), the pattern of the neighborhood (**neighbor**), and to describe the state of a cell (**state**) as a set of typed substates that can be: *shorts, integers, floats, doubles* and arrays of these basic types. The use of *float* and *double* substates allows a user to define *continuous CA* for modeling complex systems or phenomena. At the same time, formal compliance with the standard CA definition can be easily assured by resorting to a discretized set of values.

In CARPET, the state of a cell is composed of a set of typed substates, unlike classical cellular automata where the cell state is represented by a few bits. The typification of the substates allows us to extend the range of the applications that can be coded in CARPET simplifying writing the programs and improving their readability. Most systems and languages (for example CELLANG [6]) define the cell substates only as integers. In this case, for instance, if a user must store a real value in a substate then she/he must write some procedures for the data retyping. The writing of these procedures makes the program longer and difficult to read or change. The CARPET language frees the user of this tedious task and offers her/him a high level in state declaration. A type identifier must be included for each substate. In the following example the state is constituted of three substates:

```
state (int particles, float temperature, density);
```

A substate of the current cell can be referenced by the variable cell substate (e.g., cell_speed). To guarantee the semantics of cell updating in cellular automata the value of one substate of a cell can be modified only by the **update** operation. After an **update** statement the value of the substate, in the current iteration, is unchanged. The new value takes effect at the beginning of the next iteration.

CARPET allows a user to define a logic neighborhood that can represent a wide range of different neighborhoods inside the same radius. Neighborhoods can be asymmetrical or have any other special topological properties (e.g., hexagonal neighborhood). The neighbor declaration assigns a name to specified neighboring cells of the current cell and a vector name that can be used as an alias in referring to a neighbor cell. For instance, the von Neumann and Moore neighborhoods shown in figure 1, can be defined as follows:

```
neighbor Neumann[4]([0,-1]North,[-1,0]West, [0,1]South, [1,0]East);

neighbor Moore[8] ([1,-1]NEast, [0,-1]North, [-1,-1]NWest, [-1,0] West,
                   [1,0]East ,[-1,1]SWest, [0,1]South [1,1]SEast);
```

A substate of a neighbor cell is referred to, for instance, as NEast_speed. By the vector name the same substate can be referred to also as Moore[0]_speed. This way of referencing simplifies writing loops in CARPET programs.

CARPET permits the definition of global parameters that can be initialized to specific values (e.g., **parameter** (viscosity 0.25)). The value of a parameter is the same in each cell of the automaton. For this reason, the value of each parameter cannot be changed in the program but it can only be modified, during the simulation, by the user interface (UI). CARPET defines also a mechanism for programming non-deterministic rules by a random function. Finally, a user can define cells with different transition functions by means of the **Getx, Gety, Getz** functions that return the value of the coordinates X, Y, and Z of the cell in the automaton.

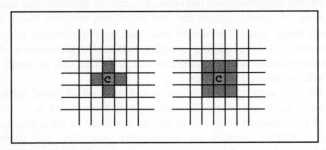

Fig. 1. The von Neumann and Moore neighborhoods in a two-dimensional cellular automaton .

CARPET does not include constructs for configuration and visualization of the data, unlike other cellular languages. As a result, the same CARPET program can be executed with different configurations. The size of lattice, as other details, of a cellular automaton are defined by the UI of the CARPET environment. The UI allows, by menus, to define the size of a cellular automaton, the number of processors on which the automaton will be executed, and to choose colors to be assigned to the cell substates to support the graphical visualization of their values.

3. A Parallel Environment for CARPET

Parallel computers represent the most natural architecture where CA programming environments can be implemented. In fact, when a sequential computer is used to support the simulation, the execution time might become very high since such computer has to perform the transition function for each cell of the automaton in a sequential way. Thus, parallel computers are necessary as a practical support for the effective implementation of high-performance CA [1].

The approach previously mentioned motivated the development of CAMEL (Cellular Automata environMent for systEms ModeLing), a parallel software architecture based on the cellular automata model that constitutes the parallel run-time system of CARPET. The latest version of CAMEL named CAMELot (*CAMEL open technology*) is a portable implementation based on the MPI communication library. It is available on MIMD parallel computers and PC clusters.

The CAMEL run-time system is composed of a set of *macrocell* processes, each one running on a single processing element of the parallel machine, and by a *controller* process running on a processor that is identified as the *Master* processor.

The CAMEL system uses the SPMD approach for executing the CA transition function. Because of the number of cells that compose an automaton is generally greater than the number of available processors, several elementary cells are mapped on each *macrocell* process. The whole set of the *macrocells* implement a cellular automaton and it is called the CAMEL *Parallel Engine*. As mentioned before,

CAMEL provides also a user interface to configure a CARPET program, to monitor the parameters of a simulation and dynamically change them at run time.

CAMEL implements a form of block-cyclic data decomposition for mapping cells on the processors that aims to address the problem of load imbalance experienced when the areas of active cells are restricted to one or few domains and the rest of lattice may be inactive for a certain number of steps [2]. This load balancing strategy divides the computation of the next state of the active cells among all the processors of the parallel machine avoiding to compute the next state of cells that belongs to a stationary region. This is *a domain decomposition* strategy similar to the *scattered decomposition* technique.

4. Programming Examples

To describe practically cellular programming in CARPET, this section shows two cellular programs. They are simple but representative examples of complex systems and phenomena and can explain how the natural solver approach can be exploited by the CARPET language.

4.1. The wireworld program

This section shows the simple wireworld program written by CARPET. This program should familiarize the reader with the language approach. In fact, figure 2 shows how the CARPET constructs can be used to implement the *wireworld* model proposed in the 1990 by A. K. Dewdney [3] to build and simulate a wide variety of circuitry.

In this simple CA model each cell has 4 possible states: space, wire, electron head or electron tail. This simple automaton models electrical pulses with heads and tails, giving them a direction of travel. Cells interact with their 8 neighbours by the following rules: space cells forever remain space cells, electron tails turn into wire cells, electron heads turn into electron tails, wire cells remain wire cells unless bordered by 1 or 2 electron heads.

By taking special care in the arrangement of the wire (initial configuration of the lattice), with these basic rules electrons composed of heads and tails can move along wires and you can build and test diodes, OR gates, NOT gates, memory cells, wire crossings and much more complex circuitry.

4.2. A forest fire model

We show here the basic algorithm of a CARPET implementation of a real life complex application. Preventing and controlling forest fires plays an important role in forest management. Fast and accurate models can aid in managing the forests as well as controlling fires. This programming example concerns a simulation of the propagation of a forest fire that has been modeled as a two-dimensional space partitioned into square cells of uniform size (figure 3).

```
#define space    0
#define wire     1
#define electhead 2
#define electail  3

cadef
{
  dimension 2;         /*bidimensional lattice */
  radius 1;
  state (short content);
  neighbor moore[8] ([0,-1]North,[-1,-1]NorthWest, [-1,0]West,
                     [-1,1]SouthWest,[0,1]South, [1,1] SouthEast,
                     [1,0]East, [1,-1]NorthEast);
}
int i; short count;
{
  count = 0;
  for (i = 0; i<8; i++)
     if (moore[i]_content == electhead)
        count = count + 1;
  switch (cell_content)
  {
    case electail : update(cell_content, wire); break;
    case electhead: update(cell_content, electail); break;
    case wire     : if (count == 1 || count == 2)
                       update(cell_content, electhead);
  }
}
```

Fig. 2. The wireworld program written in CARPET.

Each cell can represents a portion of land of 10x10 meters. Cells in the lattice can have the values included between '0' and '3'. The fire is represented by '0' value, the ground is represented by '1' value, and trees are represented by '2' value if they are alive or by '3' value if they are dead. Fire spreads from a cell which is on fire to a Moore neighbor that is treed, but not on fire.

Each tree follows a simple rule: if it catches fire from one of its neighbors cells, it will burn and spread the fire to any neighboring trees. The fire spreads in all directions. The fire's chance of diffusing towards all the forest depends critically on the density (dens) of trees in the forest.

The most interesting aspect of this simulation is that using a very simple rule it is possible to observe a real complex behavior in the fire evolution. The fire spreads from tree to tree and in some places the fire reaches a dead end surrounded by an empty region. Running the simulation with different densities of trees, by changing the percentage of tree-cells in the lattice, has been observed different spreading of the fire across the forest.

This example shows as using a high-level language designed for programming cellular algorithms can strongly simplify the algorithms design process and reduce the program code. Moreover, the programming environment allows a user to observe the dynamic evolution of the fire spreading on a computer display where the application is visualized as in figure 4.

```
#define tree 2
#define fire 0
#define dead 3
#define land 1

cadef
{ dimension 2;
  radius 1;
  state (short ground);
  neighbor moore[8] ( [0,-1]North,[-1,-1]NorthWest, [-1,0]West,
                      [-1,1]SouthWest,[0,1]South, [1,1] SouthEast,
                      [1,0]East, [1,-1]NorthEast);
  parameter (dens 0.6);
}
  float px;
{ if (step == 0)
  {
    px = ((float) rand())/RAND_MAX;
    if (px < dens)
       update(cell_ground, tree);
    else
       update(cell_ground, land);
  }
  else
  if((cell_ground == tree) && (North_ground == fire ||
     South_ground == fire || East_ground == fire ||
     West_ground == fire || NorthWest_ground == fire ||
     SouthWest_ground == fire ||
     SouthEast_ground == fire || NorthEast_ground == fire))
          update(cell_ground, fire);
  else
    if (cell_ground == fire)
       update(cell_ground, dead);
}
```

Fig. 3. A simple forest fire program written in CARPET.

The simple forest fire model discussed here could be extended considering forests with roads, rivers, and houses, by defining a state value for each cell that represents one of these objects. The transition function will take into account the behavior of fire when these cells will be encountered. It can also be extended considering different weather conditions, wind speed, fuel types, and terrain conditions.

4.3 Performance results

Here are presented some performance figures obtained from the implementation of the forest fire model in CARPET on a Meiko CS2 parallel computer. Table 1 shows the elapsed time (in seconds) for the execution of 1000 steps of the simulation using different lattice sizes on different processor sets.

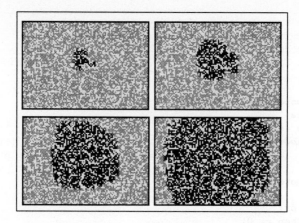

Fig. 4. Four snapshots of visualization of the forest fire simulation by CARPET.

The results are satisfying both in terms of execution time and speedup. In particular, figure 5 shows the speedup measures. On 8 processors, the measured speedup goes from 5.6 when a small lattice is used to 7.9 when we simulated a larger lattice that exploits the computational power of the parallel machine in a more efficient way.

Table 1. Execution times (in sec.) for 1000 steps for three lattice sizes on different processors.

Processors	Lattice sizes		
	56 x 56	112 x 112	224 x 224
1	7.46	27.77	118.93
2	3.66	14.18	59.47
4	1.98	7.03	29.64
8	1.32	3.95	14.91

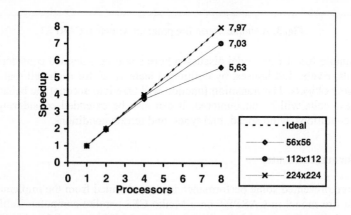

Fig. 5. Speedup of the CARPET forest fire program using different lattice sizes.

These values represent a notable result. In fact, speedup shows how much faster the simulation runs by increasing the number of PEs of a parallel computer. Thus, this

CARPET program appears to be scalable as occurred with other parallel cellular applications developed by the language [4, 5].

6. Conclusion

CARPET has been successfully used to implement several real life applications such as landslide simulation, lava flow models, freeway traffic simulation, image processing, and genetic algorithms [4, 5, 7]. It can also be used to solve parallel computing problems such as routing strategies, task scheduling and load balancing, parallel computer graphics and cryptography.

A portable MPI-based implementation of CARPET and CAMEL (called *CAMELot*) has been implemented recently on MIMD parallel computers such as the Meiko CS-2, SGI machines and PC-Linux clusters. In our opinion and on the basis of our experience, high-level languages such as CARPET can enlarge the practical use of cellular automata in solving complex problems according to the natural solvers approach while preserve high performance and expressiveness.

Acknowledgements

This research has been partially funded by CEC ESPRIT project n° 24,907.

References

[1]   B. P. Brinch Hansen, Parallel cellular automata: a model for computational science, *Concurrency: Practice and Experience*, **5**:425-448, 1993.

[2]   M. Cannataro, S. Di Gregorio, R. Rongo, W. Spataro, G. Spezzano, and D. Talia, A parallel cellular automata environment on multicomputers for computational science, *Parallel Computing*, **21**:803-824, 1995.

[3]   A. K Dewdney, Cellular automata programs that create wireworld, rugworld and other diversions, *Scientific American*, **262**: 146-149, 1990.

[4]   S. Di Gregorio, R. Rongo, W. Spataro, G. Spezzano, and D. Talia, A parallel cellular tool for interactive modeling and simulation, *IEEE Computational Science & Engineering* **3**:33-43, 1996.

[5]   S. Di Gregorio, R. Rongo, W. Spataro, G. Spezzano, and D. Talia, High performance scientific computing by a parallel cellular environment, *Future Generation Computer Systems*, North-Holland, Amsterdam, **12**:357-369, 1997.

[6]   J. D. Eckart, Cellang 2.0: reference manual, *ACM Sigplan Notices*, **27**: 107-112, 1992.

[7]   G. Folino, C. Pizzuti, and G. Spezzano, Solving the satisfiability problem by a parallel cellular genetic algorithm, *Proc. of Euromicro Workshop on Computational Intelligence*, IEEE Computer Society Press, pp. 715-722, 1998.

[8]   D.B. Skillicorn and D. Talia, "Models and languages for parallel computation", *ACM Computing Survey*, **30**:123-169, 1998.

[9]   G. Spezzano and D. Talia, A high-level cellular programming model for massively parallel processing, in: *Proc. 2nd Int. Workshop on High-Level Programming Models and Supportive Environments (HIPS97)*, IEEE Computer Society, pp. 55-63, 1997.

[10] J. von Neumann, *Theory of Self Reproducing Automata*, University of Illinois Press, 1966.

Multiprocessor Scheduling with Support by Genetic Algorithms - based Learning Classifier System

Jerzy P. Nowacki[1], Grzegorz Pycka[2] and Franciszek Seredyński[1,3]

[1] Polish -Japanese Institute of Information Technologies
Koszykowa 86, 02-008 Warsaw, Poland
[2] Polish Telecom
Sw. Barbary 2, 00-686 Warsaw, Poland
[3] Institute of Computer Science, Polish Academy of Sciences
Ordona 21, 01-237 Warsaw, Poland
e-mail: jpnowacki@pjwstk.waw.pl, grzes@crt.tpsa.pl, sered@ipipan.waw.pl

Abstract. The paper proposes using genetic algorithms - based learning classifier system (CS) to solve multiprocessor scheduling problem. After initial mapping tasks of a parallel program into processors of a parallel system, the agents associated with tasks perform migration to find an allocation providing the minimal execution time of the program. Decisions concerning agents' actions are produced by the CS, upon a presentation by an agent information about its current situation. Results of experimental study of the scheduler are presented.

1 Introduction

The problem of multiprocessor scheduling formulated as designing effective algorithms to minimize the total execution time of a parallel program in a parallel architecture remains a challenge for practitioners and researchers. Recently proposed scheduling algorithms show the importance and effectiveness of such issues as *clustering* [1], *clustering* and *task-ordering* [5] or *dynamic critical-path* [3]. The increasing number of algorithms applies methodologies of natural computation, such as *mean-field annealing* [6], *genetic algorithms* [4] or *cellular automata* [7].

In the paper we propose to use a CS [2] to solve the scheduling problem. CSs constitute the most popular approach to genetic algorithms-based machine learning. We investigate a possibility to create a scheduling algoritm based on multi-agent interpretation of the scheduling problem and controlling behavior of the system with use of a single CS.

Section 2 of the paper briefly describes a concept of CSs. Section 3 discusses accepted models of parallel programs and systems, proposes a multi-agent interpretation of the scheduling problem, and presents a concept of a scheduler. Section 4 describes an architecture of a CS used for scheduling. Results of experimental study of the proposed scheduler are presented in Section 5.

J. Rolim et al. (Eds.): IPDPS 2000 Workshops, LNCS 1800, pp. 604-611, 2000.

2   Genetic Algorithms-based Learning Classifier System

A learning CS operating in some environment is a system which is able to learn simple syntactic rules to coordinate its actions in the environment. Three components of a CS can be pointed: CLASSIFIERS - a population of decision rules, called *classifiers*, CREDIT ASSIGNMENT - a system to evaluate rules, and DISCOVERY OF RULES - a system to modify or discover new rules.

A classifier c is a condition-action pair $c = <$ **condition** $>:<$ **action** $>$ with the interpretation of the following decision rule: if a current observed state of the environment matches the **condition**, then execute the **action**. The conditional part of a classifier contains some description of the environment, expressed with use of symbols $\{0,1\}$, and additionally a *don't-care* symbol $\#$. The action part of a classifier contains an action of the CS, associated with the condition.

A usefulness of a classifier c, applied in a given situation, is measured by its *strength str*. A real-valued strength of a classifier is estimated in terms of rewards for its action obtained from the environment. If a measurement of the environment matches a conditional part of a classifier then the classifier is activated and becomes a candidate to send its action to the environment. Action selection is implemented by a competition mechanism based on *auction* [2], where the winner is a classifier with the highest strength.

To modify classifier strengths the simplified *credit assignment* algorithm [2] is used. The algorithm consists in subtracting a tax of the winning classifier from its strength, and then dividing equally the reward received after executing an action, among all classifiers matching the observed state.

A strength of a classifier has the same meaning as a *fitness function* of an individual in genetic algorithm (GA) (see, e.g. [2]). Therefore, a standard GA with three basic genetic operators: selection, crossover and mutation is applied to create new, better classifiers.

3   Multi-agent Approach to Multiprocessor Scheduling

A multiprocessor system is represented by an undirected unweighted graph $G_s = (V_s, E_s)$ called a *system graph*. V_s is the set of N_s nodes representing processors and E_s is the set of edges representing channels between processors. A parallel program is represented by a weighted directed acyclic graph $G_p = < V_p, E_p >$, called a *precedence task graph* or a *program graph*. V_p is the set of N_p nodes of the graph, representing elementary tasks. The weight b_k of the node k describes the processing time needed to execute task k on any processor of the system.

E_p is the set of edges of the precedence task graph describing the communication patterns between the tasks. The weight a_{kl}, associated with the edge (k, l), defines the communication time between the ordered pair of tasks k and l when they are located in neighboring processors. If the tasks k and l are located in processors corresponding to vertices u and v in G_s, then the communication delay between them will be defined as $a_{kl} * d(u, v)$, where $d(u, v)$ is the length of the shortest path in G_s, between u and v.

The purpose of *scheduling* is to distribute the tasks among the processors in such a way that the precedence constraints are preserved, and the *response time T* is minimized. *T* depends on the *allocation* of tasks in the multiprocessor topology and *scheduling policy* applied in individual processors:

$$T = f(allocation, scheduling\_policy). \tag{1}$$

We assume that the scheduling policy is fixed for a given run. The scheduling policy accepted at this work assumes that the highest priority among tasks ready to run in a given processor will have the task with the greatest number of successors. The priority p_k of a task k is calculated using the following recurrent formula:

$$p_k = s_k + \sum_{n_k=1}^{s_k} p_{k_{n_k}}, \tag{2}$$

where, s_k is the number of immediate successors of a task k, and $p_{k_{n_k}}$ is a priority of the n_k immediate successor of the task k.

For the purpose of the scheduling algorithm we specify two additional parameters of a task k mapped into a system graph: a Message Ready Time (MRT) predecessor of the task k, and the MRT successor of the k. A MRT predecessor of a task k is its predecessor which is the last one from which the task k receives data. The task can be processed only if data from all predecessors arrived. A MRT successor of the task k is a successor for which the task is the MRT predecessor.

We propose an approach to multiprocessor scheduling based on a multi-agent interpretation of the parallel program. We assume that an agent associated with a given task can perform a migration in a system graph. The purpose of migration is searching for an optimal allocation of program tasks into the processors, according to (1). We assume that decision about migration of a given agent will be taken by a CS, after presentation by the agent a local information about its location in the system graph.

4   An Architecture of a Classifier System to Support Scheduling

To adjust the CS to use it for scheduling we need to interpret the notion of an environment of the CS. The environment of the CS is represented by some information concerning a position of a given task located in a system graph. A message containing such a information will consist of 7 bits:

- bit 0: value 0 - task does not have any predecessors;
  value 1 - the task has at least one predecessor
- bit 1: value 0 - the task does not have any successors;
  value 1 - the task has at least one successor

- bit 2: value 0 - the task does not have any brothers; value 1 - the task has brothers
- bits 3 and 4:
  values 00 - none MRT successor of the task is alocated on the processor where the task is allocated; values 01 - some MRT successors are allocated on the same processsor where the task is allocated; values 11 - all MRT successors are alocated on the same processor where the task is allocated; values 10 - the task does not has any MRT successors
- bits 5 and 6:
  values 00 - none MRT predecessor of the task is alocated on the processor where the task is allocated; values 01 - some MRT predecessors are allocated on the same processsor where the task is allocated; values 11 - all MRT predecessors are alocated on the same processor where the task is allocated; values 10 - the task does not has any MRT predecessors.

The list of actions of a CS contains 8 actions:

- *action 0*: **do_nothing** - the task does not migrate from a given location (processor) to any other processor of the system
- *action 1*: **random_action** - randomly chosen action from the set of all actions, except the *action 1*, will be performed
- *action 2*: **random_node** - the task migrates to one of randomly chosen processors of the system
- *action 3*: **pred_rnd** - the task migrates to a processor where randomly selected predecessor of the task is located
- *action 4*: **succ_rnd** - the task migrates to a processor where randomly selected successor of the task is located
- *action 5*: **less_neighbours** - the task migrates to a processor where the smallest number of neighbours of the task is located
- *action 6*: **succ_MRT** - the task migrates to a processor where its MRT successor is located
- *action 7*: **pred_MRT** - the task migrates to a processor where its MRT predecessor is located.

Conditional part of a classifier contains information about specific situation of a given task which must be satisfied to execute the action of the classifier. For example, a classifier < #1 #0 0 #0 >:< 6 > can be interpreted in the following way: **IF**: it does not matter whether the task has predecessors or not (symbol: #) **AND IF**: the task has successors (symbol: 1) **AND IF**: it does not matter whether the task has brothers or not (symbol: #) **AND IF**: none among MRT successors of the task is located on the processor where the task is located (symbols: 00) **AND IF**: none among MRT predecessors of the task is located on the processor where the task is located or the task does not has MRT predecessors (symbols: # 0) **THEN**: move the task to the processor where is located a MRT successor of the task (symbol: 6).

	classifier	strength of classifiers after execution of action by subsequent agents						
		initial	0	1	2	3	4	5
0	<#1#00##>:<7>	300.00	298.35	298.20	**395.07**	**490.94**	**485.79**	485.55
1	<#1#0110>:<6>	300.00	299.85	299.70	299.55	299.40	299.25	299.10
2	<10#10#0>:<1>	300.00	299.85	**297.70**	297.56	297.41	297.26	297.11
3	<#1##101>:<1>	300.00	299.85	299.70	299.55	299.40	299.25	299.10
4	<110##1#>:<7>	300.00	299.85	299.70	299.55	299.40	299.25	299.10
5	<11#1##0>:<5>	300.00	299.85	299.70	299.55	299.40	299.25	**297.11**
6	<00##1##>:<6>	300.00	299.85	299.70	299.55	299.40	299.25	299.10
7	<##00###>:<5>	300.00	298.35	298.20	298.05	297.90	297.75	297.60
8	<#0#0011>:<1>	300.00	299.85	299.70	299.55	299.40	299.25	299.10
9	<####11#>:<3>	300.00	299.85	299.70	299.55	299.40	299.25	299.10
10	<#0##1#0>:<7>	300.00	299.85	299.70	299.55	299.40	299.25	299.10
11	<###1101>:<1>	300.00	299.85	299.70	299.55	299.40	299.25	299.10
12	<1011#00>:<1>	300.00	299.85	298.20	298.05	297.90	297.75	297.60
13	<##11###>:<5>	300.00	299.85	298.20	298.05	297.90	297.75	296.11
14	<01###10>:<3>	300.00	**297.85**	297.70	297.56	297.41	297.26	297.11
15	<0###111>:<2>	300.00	299.85	299.70	299.55	299.40	299.25	299.10
16	<0#1#1##>:<6>	300.00	299.85	299.70	299.55	299.40	299.25	299.10
17	<#1010#0>:<6>	300.00	299.85	299.70	299.55	299.40	299.25	299.10
18	<001100#>:<0>	300.00	299.85	299.70	299.55	299.40	299.25	299.10
19	<01110#0>:<3>	300.00	299.85	299.70	299.55	299.40	299.25	299.10

Fig. 1. Initial population of classifiers in few first steps of working the scheduler.

5   Experimental Results

Experiment #1: Step by step simulation (problem: $gauss18 > full2$)

We will analyze some initial steps of the work of the scheduler, solving the scheduling problem for a program graph $gauss18$ ([3], see Fig. 2a) processed in the 2-processor system $full2$. The program contains 18 tasks, and is initially allocated as shown in Fig. 2b with response time $T = 74$. Fig. 1 shows an initial population of the CS containing 20 classifiers with initial values of the strenght of each equal to 300.

The agent A_0 sends first its message $< 0100010 >$ to the CS. The message describes the actual situation of the task 0 (as shown in Fig. 2a, b) and contains the following information: the task 0 does not have any predecessor; it has successors; it does not have any brothers; all MRT successors are located on a processor different than the processor where is located the task 0, and the task does not have any MRT predecessors.

The message matches three classifiers of the CS: the classifier 0, 7 and 14. A winner of the competition between them is the classifier 14, and its action $< 3 >$ is passed to the agent A_0. The action says: migrate to a processor where your randomly chosen predecessor is located. The agent A_0 can not execute this action, because the task 0 does not have any predecessors. So, the allocation of tasks and corresponding value of T is not changed. The CS receives a reward

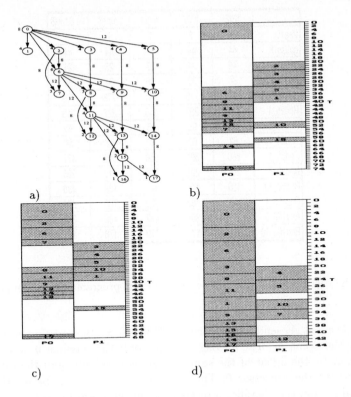

Fig. 2. Program graph *gauss*18 (a) and Gantt charts (b), (c), (d) for allocations of tasks in *full*2, corresponding to actions of classifiers shown in Fig. 1.

equal to 1 for this action, because it does not change the value of T (the value of the reward - a user defined parameter). The reward increases the strength of the classifier 14. New strengths of classifiers, as shown in Fig. 1, in the column corresponding to the agent 0, are the result of applying taxes.

The next agent which sends a message to the CS is the agent A_1, and the message is $< 1011000 >$. The message matches classifiers 2, 12 and 13. A winner of the competition is the classifier 2, which sends the action $< 1 >$ to the A_1. As the result of this action (random_action), the action 0 is chosen (do_nothing). The allocation of tasks remains the same, and the classifier 2 receives the reward equal to 1. All classifiers pay the *life* tax, the classifiers 2, 12, and 13 pay *bid* tax, also the winner pays a tax, what results in new values of strength of classifiers.

The agent A_2 sends to the CS the message $< 1110000 >$. It matches only the classifier 0 and the action $< 7 >$ is executed by the agent. The execution of the action results in the migration of the task 2 from the processor 1 to the processor 0, where the task MRT predecessor is located. Changing allocation of tasks reduces T to the value 68 (see, Fig. 2c). The classifier 0 receives the reward equal to 100 for improvement (the user defined value for improvement) of T.

	full2	full8	ring8	cube8	de Bruijn8
tree15	9	7	7	7	7
gauss18	44	44	44	44	44
g18	46	24	27	25	25
g40	80	32	36	34	34
fft64	2055	710	841	778	779
Rnd25_1	495	289	346	327	313
Rnd25_5	95	95	95	95	95
Rnd25_10	62	62	62	62	62
Rnd50_1	890	394	550	502	477
Rnd50_5	207	201	209	216	205
Rnd50_10	138	141	141	138	138
Rnd100_1	1481	582	789	703	671
Rnd100_5	404	364	432	422	389
Rnd100_10	175	173	179	178	172

Fig. 3. The best response time received for different program and system graphs.

Next, the agent A_3 sends the message $< 1110000 >$ - the same message as the one sent by the agent A_2. The message matches only the classifier 0, what causes the execution by the agent of the same action as previouly, and the migration of the task 3 to the processor 0. The new value of $T = 61$ is better than the previous value, and the classifier 0 receives again the reward equal to 100.

The agent A_4 sends the same message to the CS as agents A_3 and A_2 sent. However, an attempt to execute the action $< 7 >$ by the agent, i.e. migration of the task 4 from the processor 1 to the processor 0, increases T to the value 62, so the execution of the action is cancelled and the classifier 0 receives the reward equal to 0 (the user defined value for causing the result worse).

The action executed by the agent A_5 does not change the value of T. The message $< 1111000 >$ of the agent matches classifiers 5 and 13, and the classifier 5 with the action $< 5 >$ is the winner. The execution of the action, i.e. the migration of the task 5 from the processor 1 to the same processor, obviously does not change tasks' allocation.

Agents execute their actions sequentially, in the order of their numbering in the program graph. After the execution of an action by the agent A_{17}, the sequence of actions is repeated again starting from the agent A_0. In the considered experiment, actions of next several agents do not improve the value of T. The next improvement of T appears as the result of the execution of an action by the agent A_{15} ($T = 46$). Last migration of a task, which causes decreasing T to $T = 44$ takes place in the iteration 38. Found value of T (see, Fig. 2d) is optimal and can not be improved.

Experiment #2: Response time for different scheduling problems

The scheduling algorithm was used to find response time T for deterministic program graphs such as $tree15, gauss18, g18, g40, fft64$ known from literature, processed in different topologies of multiprocessor systems, such as $full2, full5, ring8, cube8, deBruin$. Also a number of random graphs were used, with $25, 50$ and 100 (in the average) tasks ($Rnd25\_x$, $Rnd50\_x$, $Rnd100\_x$), where x denotes ratio of the average communication time a_{kl} in a program graph to the average processing time b_k of tasks in the program graph. Fig. 3 summerizes results. Results obtained for deterministic graphs are the same as known in literature. Results obtained for random graphs were compared with results (not shown) obtained with use of GA-based algorithms, such as parallel GAs of *island* and *diffusion* models. Results obtained with use of the scheduler are significantly better than with use of parallel GAs.

6   Conclusions

We have presented results of our research on development scheduling algorithms with support of the scheduling process by genetic algorithm-based learning classifier system. Results of experimental study of the system are very promising. They show that the CS is able to develop effective rules for scheduling during its operation, and solutions found with use of the CS outperform ones obtained by applying non-learning GA-based algorithms.

Acknowledgement
The work has been partially supported by the State Committee for Scientific Research (KBN) under Grant 8 T11A 009 13.

References

1. S. Chingchit, M. Kumar and L. N. Bhuyan, A Flexible Clustering and Scheduling Scheme for Efficient Parallel Computation, in *Proc. of the IPPS/SPDP 1999*, April 12-16, 1999, San Juan, Puerto Rico, USA, pp. 500-505.

2. D. E. Goldberg, *Genetic Algorithms in Search, Optimization and Machine Learning*, Addison-Wesley, Reading, MA, 1989

3. Y. K. Kwok and I. Ahmad, Dynamic Critical-Path Scheduling: An Effective Technique for Allocating Task Graphs to Multiprocessors, *IEEE Trans. on Parallel and Distributed Systems*. 7, N5, May 1996, pp. 506-521.

4. S. Mounir Alaoui, O. Frieder and T. El-Ghazawi, A Parallel Genetic Algorithm for Task Mapping on Parallel Machines, in J. Rolim et al. (Eds.), *Parallel and Distributed Processing*, LNCS 1586, Springer, 1999, pp. 201-209.

5. A Radulescu, A. J. C. van Gemund and H. -X. Lin, LLB: A Fast and Effective Scheduling for Distributed-Memory Systems, in *Proc. of the IPPS/SPDP 1999*, April 12-16, 1999, San Juan, Puerto Rico, USA, pp. 525-530.

6. S. Salleh and A. Y. Zomaya, Multiprocessor Scheduling Using Mean-Field Annealing, in J. Rolim (Ed.), *Parallel and Distributed Processing*, LNCS 1388, Springer, 1998, pp. 288-296.

7. F. Seredynski, Scheduling tasks of a parallel program in two-processor systems with use of cellular automata, *Future Generation Computer Systems* 14, 1998, pp. 351-364.

Viewing Scheduling Problems through Genetic and Evolutionary Algorithms

Miguel Rocha, Carla Vilela, Paulo Cortez, and José Neves

Dep.Informática - Universidade do Minho - Braga - PORTUGAL
{mrocha,cvilela, pcortez,jneves}@di.uminho.pt

Abstract. In every system, where the resources to be allocated to a given set of tasks are limited, one is faced with scheduling problems, that heavily constrain the enterprise's productivity. The scheduling tasks are typically very complex, and although there has been a growing flow of work in the area, the solutions are not yet at the desired level of quality and efficiency. The *Genetic and Evolutionary Algorithms (GEAs)* offer, in this scenario, a promising approach to problem solving, considering the good results obtained so far in complex combinatorial optimization problems. The goal of this work is, therefore, to apply *GEAs* to the scheduling processes, giving a special attention to indirect representations of the data. One will consider the case of the *Job Shop Scheduling Problem*, the most challenging and common in industrial environments. A specific application, developed for a *Small and Medium Enterprise*, the *Tipografia Tadinense, Lda*, will be presented.
Keywords: Genetic and Evolutionary Algorithms, Job Shop Scheduling.

1 Introduction

In every industrial environment one is faced with a diversity of scheduling problems which can be difficult to solve. Once a good solution is found it produces very tangible results, in terms of the way the resources are used to maximize the profits. The scheduling problems are typically NP-Complete, thus not having the warranty of solvability in polynomial time. Indeed, although there has been a steady evolution in the areas of *Artificial Intelligence (AI)* and *Operational Research (OR)* aiming at the development of techniques to give solution to this type of problems, the basic question has not yet been solved.

The *Genetic and Evolutionary Algorithms (GEAs)* mimic the process of natural selection, and have been used to address complex combinatorial optimization problems. Using an evolutionary strategy, the *GEAs* objective is to maximize/minimize an objective function $f : S(R) \mapsto \Re$, where S is the solution's space. *GEAs* belong to a class of processes designated by *black-box*, once they optimize a function using a strategy independent of the problem under consideration; i.e., require little knowledge of the structure of the universe of discourse.

In this work one aims at studying the application of *GEAs* to address the *Job Shop Scheduling Problem (JSSP)*, a common well-known, scheduling task.

J. Rolim et al. (Eds.): IPDPS 2000 Workshops, LNCS 1800, pp. 612-619, 2000.

This work is part of a larger project on *Genetic and Evolutionary Computation (GEC)*, where the problem of knowledge representation is addressed in terms of indirect representations of the genotype (the genetic constitution of an organism), that are decoded into a solution. This approach allows the same representation, and its genetic operators to be used in distinct problems. The disadvantage relies on the extra work one is forced to at the decoder's level.

This work was structured into three slopes. The former one gives an introduction to *GEAs*, to the scheduling tasks and to the *JSSP*. In the second slope, one devises the way GEAs can be used in this case and present some of the results obtained. Finally, one comes to the conclusions and prospective work.

2   Genetic and Evolutionary Algorithms

2.1   Basic Concepts

The *Genetic and Evolutionary Algorithms (GEAs)* are stochastic adaptive systems, whose search methods model natural genetic inheritance and the *Darwinian* struggle for survival. *GEAs* have been used successfully in applications involving parameter optimization of unknown, non-smooth and discontinuous functions, where the traditional algorithmic approaches have been failing (e.g., in industrial drawing, scheduling, planning, financial calculation, data mining). In a *GEA* one begins with a set of individuals, possible solutions to the problem, and proceeds towards finding the best solutions. They operate on an evolving population by applying operators modeled in accordance to the natural phenomena of selection and reproduction. Each solution is represented by an individual, whose genetic constitution, its chromosome, is a sequence of genes, taken from a fixed alphabet, with a well-established meaning. A quality measure is defined in terms of a fitness function, which measures the quality of each chromosome. A new population is formed from stochastically best samples of the previous generations and some offspring coming from the application of genetic operators. The process goes on throughout time until a proper goal is reached (Figure 1). Typical genetic operators include crossover, where the genetic information of two or more individuals are combined, to generate one or two new individuals, and mutation where the genetic information of one individual is slightly altered.

2.2   GEPE: The Genetic and Evolutionary Programming Environment

In recent years, there has been a remarkable growth on the *Genetic and Evolutionary Computation (GEC)* field. The models, methodologies and techniques that were developed share common features, overlooked in the software design process. In one's approach, the *GEC's* applications are implemented as reusable software modules, providing an incremental, manageable and user friendly environment for software development for beginners, being also a good tool for *experts*, who can develop complex applications with increased reusability. To

Begin
    Population initialization
    Evaluation of the initial population
    WHILE (termination criteria is not met)
        Select the population's ancestors to reproduction
        Create new individuals, through the application of genetic operators
        Evaluate the new individuals (offspring)
        Select the survivors and add the offsprings to the next generation
End

Fig. 1. Pseudocode for a Genetic and Evolutionary Algorithm

achieve all these capabilities, one subscribes to the Object-Oriented Programming Paradigm, which allows modularity, encapsulation and reusability of software, by using template fields, abstract classes, virtual methods, default behaviors and values, and re-definition of methods and operators. The proposed framework, named *Genetic and Evolutionary Programming Environment (GEPE)*[3] has three major conceptual layers (Figure 2), namely the *Individuals*, the *Populations* and the *GEA* ones. Each of these layers is defined by a class hierarchy, whose root is responsible for defining the basic attributes and methods. The derived classes can augment or redefine program's behaviors.

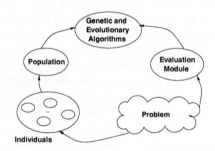

Fig. 2. The GEPE Archetype

At the *Individual*'s level one defines the representation and the genetic operators to be used. The genotype of an individual is a set of genes of a given type, implemented as a template, allowing for different types of representational schemes. At the *Population*'s level one defines the initial population creation, the selection and the reinsertion procedures. At the higher level, the *GEA* one, are defined the general structure of the process and the termination criteria. Apart from these three levels, that make the core of the *GEPE*, one needs an extra module to allow for the definition of the problem dependent procedures; i.e., the decoding and the evaluation ones.

3   Analysis of the Scheduling Processes

The scheduling problem is concerned with the allocation of scarce resources to tasks throughout time; i.e., it is a decision-making process that produces a plan of procedure for a project, allotting the work to be done and the time for it. The resources and tasks can take varied forms. The resources can be machines in a factory or tracks in an airport. The tasks can be operations in a production process, landings and lift-offs in an airport or executions of computer programs. Each task can have a different level of priority, or a finalization time. The problem may be concerned either with the minimization of the last task finalization time, or of the number of tasks to be completed at an earlier time.

Most part of the scheduling problems can be described in terms of the *Job Shop Scheduling Problem*. Consider an industrial atmosphere where n tasks (operations) will be processed by m machines. Each task is associated with a set of restrictions with respect to the machine's sequence, as well as to the processing time in each one of those machines. The tasks can have different duration times and involve different subsets of machines. One's ordeal aims at an ideal sequence of tasks for each machine in order to minimize a pre-defined objective function, in accordance to the restrictions stating that the order's sequence of processing must be followed, a machine cannot process two (or more) tasks at the same time, and that different tasks for the same order cannot be processed simultaneously by different machines.

More formally, one has a set J of n tasks, a set M of m machines, and a set O of operations. For each operation $op \in O$, there is a task $j_{op} \in J$ to which a machine $m_{op} \in M$ is conjuncted, where task j_{op} will be processed, in a given time $t_{op} \in \Re$. There is also a temporary binary ordering relation that decomposes the set O on a group of partially ordered sets, according to the tasks; i.e., if $x \to y$ then $j_x \to j_y$ and there is not a z different from x or y, such that $x \to z$ or $z \to y$. Electing as objective the minimization of the time elapsed, with the processing of all tasks (making the spanned of the objective function), the problem consists on seeking an initial time s_{op}, for each operation op, in such a way that the function: $max(s_{op} + t_{op})$ and $op \in O$, is minimized, taking into attention the restrictions (where t_{op} stands for the operation processing time):

(i) $t_{op} \in O, \forall op \in O$
(ii) $s_x - s_y \geq t_y$ if $y \to x$, and $x, y \in O$
(iii) $(s_i - s_j \geq t_j) \lor (s_j - s_i \geq t_i)$ if $m_i = m_j$, and $i, j \in O$

4   Approaching the JSSP with GEAs

A number of the scheduling applications use an indirect representation; i.e., the *GEAs* act on a population of solutions coded for the scheduling problem, and not directly on the schedule. Thus, a transition on such a representation has to be done by a decoder before the evaluation process. This decoder has to seek the information that is not supplied by the chromosome. Its activity will be concerned with the amount of information coded onto the chromosome; i.e., the least information represented, the more will be done by the decoder, and vice-versa. The domain's knowledge is here used only on the evaluation phase, just to fix the chromosome's fitness (Figure 3).

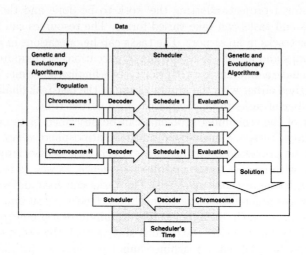

Fig. 3. An Indirect Representation

The representation that is used in one's approach is an *Order- Based (OBR)* one; i.e., the genotype is a permutation of symbols, taken form a given alphabet, with no repetitions. In the *JSSP*, the alphabet will be a set of orders to be scheduled, numbered sequentially. Each chromosome will then be a list of orders (Figure 4). Under this setting, the arrangement of elements on the list represent the order's priorities on the schedule. The scheduling problem is reduced to a sequentially ordered set problem, like the *Traveling Salesman Problem (TSP)*. Per each chromosome, the decoder generates the respective production schedule in accordance to the order's sequence - the first order in the list is scheduled in first place, and so forth.

Order 7	Order 9	Order 5	...

Fig. 4. A chromosome for the JSSP

There are a number of genetic operators developed to work with *OBRs* and the following were used in this work:

– *Order Preserving Crossover (OPX) - OPX* emphasizes the relative order of the genes from both parents, working by selecting cutting points (1 or 2) and then building the several segments, in the offspring, by taking genes from alternating parents, maintaining their relative order.

- *Uniform Order Preserving Crossover (UOPX)* - It has some similarities with the previous one, since it maintains the relative order of the genes given by the ancestors. It works with a randomly generated binary mask, that determines which parent is used to fill a given position.
- *Partially Matched Crossover (PMX)* - Under the *PMX* [1] two crossing points are chosen, defining a *matching section*, used to effect a cross between the two parents, through position-to-position exchange operations.
- *Cycle Crossover (CYCX)* - Cycle crossover [4] performs recombination under the constraint that each gene in a certain position must come from one parent or the other.
- *Edge Crossover (EDGX)* - The edge family of crossover operators is based on maintaining all possible pairs of adjacent genes (edges) in the string. It was specially designed for the *TSP* [6].
- *Maximum Preservative Crossover (MPX)* - The *MPX* operator was designed by Mühlenbein [2] with the purpose to tackle the *TSP* by preserving, in the offspring, subtours contained in both parents.
- *Schleuter Crossover (SCHX)* - The *SCHX* [5] is a variation of the previous operator, with some features similar to the *OPX* ones, that also contemplates the process of inversion of partial tours.
- *Adjacent mutation(ADJ)* - It swaps the positions of two adjacent genes.
- *Non-adjacent mutation (NADJ)* - It swaps the positions of two random genes.
- *K-permutation mutation (KPERM)* - It scrambles a sub-list of genes.
- *Inversion(INV)* - It inverts a partial sub-list of genes.

5   Results

In this section one describes the results obtained with *GEAs* using different configurations of the genetic operators. In all the experiments were used two operators: a crossover and a mutation one. Table 1 shows the results, for all possible combinations of the operators given above. Each result was computed as the average of 20 independent runs. The instance of the *JSSP* used was generated randomly, considering 5 machines, 2 production plans and 50 orders.

6   A practical example

It is one's goal to produce a production scheduling of a typography, such as the Tipografia Tadinense, Lda, a small enterprise located near Oporto, in the north of Portugal. It is intended to produce a solution to give answer to three fundamental questions:

- the priority orders are not handled with the precedence that they deserve;
- there is subutilization of resources (e.g. raw materials, machinery);
- there are too many inactive times due to changes in machine's configuration.

Table 1. Results obtained for a JSSP with different genetic operators

Crossover / Mutation	ADJ	NADJ	INV	KPERM
OPX1	5019.33	1095.41	2929.91	1607.77
OPX2	3024.11	1214.8	1554.36	1797.5
UOPX	1127.49	1133.99	1139.51	1141.4
MPX	1109.83	1116.9	1235.17	1199.22
PMX	1098.78	1165.12	1172.46	1235.15
CYCX	5956.23	**1092.33**	2083.03	3022.5
SCHX	1213.54	1464.34	1358.09	1220.9
EDGX	2428.82	2956.41	2100.06	2763.16

The intended scheduling is a typical *JSSP* one. Indeed, given a production order, it goes through a set of machines, sequentially It is one's aim to minimize the execution time of an orders portfolio, to give orders the right of precedence over others and to reduce dead times. The evaluation function receives as input a chromosome and computes its fitness. Whenever in it there are non-priority orders, preceding priority ones, it computes a penalty, returning the foreseen time for the scheduling process plus the penalty of the orders portfolio.

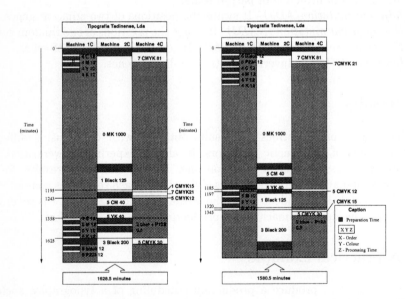

Fig. 5. A Snapshot of a Proper Scheduling and a possible solution to the problem via GEAs

A typical order's portfolio, with their associated priorities, the quantities involved, the executing times, the colors and the machines needed was provided.

In Figure 5(left) one can find the schedule used to process it, resulting in a total time of 1628.5 minutes. One applied *GEAs*, considering a population of 100 individuals, and the best solution found was the following list of orders: 6 4 0 7 5 1 3 2. Based on this chromosome, was prepared a scheduling (Figure 5). The time needed to process the orders portfolio is now of 1580.5 minutes.

The scheduling that falls back upon the skills of the practitioner, leads to a deficient use of resources but also to an endemic customer's non-satisfaction. Indeed, machine 2C hangs out during 13 minutes, while in the scheduling made by GEAs there are no hang out times. In this case there is also a reduction of the of machine's preparation times. On the other hand, if two or more orders use the same color, the preparation time for color changeover is zero. Under the *GEA* approach the processing time for the orders portfolio had a decreased of 48 minutes, approximately 4%, which is very important in terms of productivity gains.

7   Conclusions

It was presented a programming environment that allows one to address complex scheduling tasks in terms of an evolutionary approach. The ongoing work is being directed to the introduction of direct representations, the dynamic scheduling, by considering events such as orders arriving, machines breaking down, raw materials failing, or orders changing priority, the application of Constraint Logic Programming in order to reduce the solutions space and the development of a Client/Server Scheduling Model with Mobil Agents support.

References

1. D. Goldberg and R. Lingle. Alleles, Loci and the Traveling Salesman Problem. In J.Grenfenstette, editor, *Proc. of the 1st Intern. Conf. on Genetic Algorithms and their Applications*, Hillsdale, New Jersey, 1985. Lawrence Erlbaum Assoc.
2. H. Mühlenbein. Evolution in time and space - the parallel genetic algorithm. In G. Rawlins, editor, *Foundations of Genetic Algorithms*, pages 316–337. Morgan-Kaufman, 1991.
3. J. Neves, M. Rocha, H. Rodrigues, M. Biscaia, and J. Alves. Adaptive Strategies and the Design of Evolutionary Applications. In *Proc. of the Genetic and Evolutionary Computation Conference (GECCO99)*, Orlando, Florida, USA, 1999.
4. I.M. Oliver, D.J. Smith, and J. Holland. A Study of Permutation Crossover Operators on the Travelling Salesman Problem. In J.Grenfenstette, editor, *Proc. of the 2nd Intern. Conf. on Genetic Algorithms and their Applications*. Lawrence Erlbaum Assoc., July 1987.
5. M.G. Schleuter. ASPARAGOS - An Asynchronous Parallel Genetic Optimization Strategy. In J.D.Schafer, editor, *Proc. of the 3rd ICGA*. George-Mason Univ., Morgan Kaufman, 1989.
6. T. Starkweather, S. McDaniel, K. Mathias, D. Whitley, and C. Whitley. A Comparison of Genetic Sequencing Algorithms. In R.Belew and L.Booker, editors, *Proc. of the 4th ICGA*, San Diego, July 1991. Morgan-Kaufmann Publishers.

Dynamic Load Balancing Model: Preliminary Assessment of a Biological Model for a Pseudo-Search Engine

Reginald L. Walker

Computer Science Department
University of California at Los Angeles
Los Angeles, California 90095-1596
rwalker@cs.ucla.edu

Abstract. Emulation of the current World Wide Web (WWW) search engines using methodologies derived from Genetic Programming (GP) and Knowledge Discovery in Databases (KDD) were used for the Pseudo-Search Engine's initial parallel implementation of an indexer simulator. The indexer was implemented to follow some of the characteristics currently implemented by AltaVista and Inktomi search engines who index each word in a Web document. This approach has provided very thorough and comprehensive search engine results that have led to the development of a Pseudo-Search Engine Indexer which has in turn provided insight into the computational effort needed to develop and implement an integrated search engine - information crucial to the adaptation of a biological model. The initial implementation of the Pseudo-Search Engine Indexer simulator used the Message Passing Interface (MPI) on a network of SUN workstations and an IBM SP2 computer system.

1 Introduction

Improvements to the fitness measure associated with Genetic Programming (GP) applications have taken various approaches. The evolution of species within a

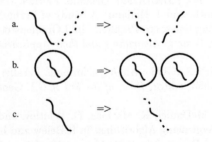

Fig. 1. Genetic programming operators a) crossover, b) reproduction, and c) mutation.

J. Rolim et al. (Eds.): IPDPS 2000 Workshops, LNCS 1800, pp. 620-627, 2000.

GP/Genetic Algorithms (GAs) search space results from an application of cluster analysis techniques that utilize the evaluation of each individual's (of a population) fitness measure. The search space associated with GP/GAs applications is obtained via a re-combination of partially-fit individuals who form potentially fitter offsprings. Thus, the re-combination operators provides the algorithm with a means to intelligently search [1] the entire search space with greater rewards. The similarity measures [8] used to compare and/or cluster individuals within the population are derived from the dimensionless ratio of the distance between two individuals and their maximum possible distance which is the search space's boundaries.

The use of a single population leads to panmictic selection [5] in which the individuals selected to participate in a genetic operation can be from anywhere in the population. The use of subpopulations (evolution of species) is an additional method used to avoid local optima [9]. Also, the GP operators adhere to the closure property [1],[9] because the primary and secondary operators generate a sets of functions and terminals that provide input to other applications of the genetic operators.

2   Methodologies of Genetic Programming

2.1   Overview

Genetic Programming is an evolutionary methodology [2],[4],[14] that extends and expands upon the techniques associated with Genetic Algorithms. The evolutionary force of these methodologies reflects a population's fitness. The foundation for GAs is an artificial chromosome of a fixed length and size designed to map the points in the problem search space. The artificial chromosome is derived by assigning the variables of the problem to specific locations (genes/alleles). The gene value denotes the value of a particular set of gene variables (memes) [1]. Such GAs provide an efficient mechanism for generating/displaying multidimensional search spaces that are highly complex and/or nonlinear.

The hybrid chromosome structure associated with the Pseudo-Search Engine's indexer [13] as shown in Figure 2, follows the methodologies of GP and GAs. Here, a single structure was used to represent subsets (subpopulations) of Web pages that reside at each node (Web site) and that will be eventually be expanded to an allocation of two chromosomes per node. In the chromosome each horizontal member structure represents a Web page that would translate into a meme - genetic component that varies with each allele. The bracket to the left of the Web pages refers to the pages having similar characteristics that comprise a allele and its memes - which represents/corresponds to the most fundamental features contained at each Web site. The addition of new Web pages at a given Web site creates new allele which can in turn grow in size via the addition of new memes (a process that does not alter the chromosome's length). The application of GP crossover operator - which transmits and contains the parents' genetic makeup - results in two new chromosomes. The bracket mechanism here

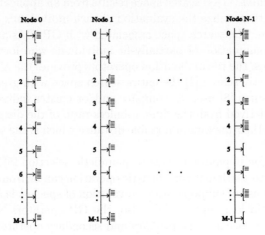

Fig. 2. Distribution of Web Pages.

constitutes an organizational device that numerically orders the structure's Web pages - a phenomenon that facilitates the evolution of diverse nodes.

2.2   Application of the Genetic Operators

The improvement of the solutions generated in the GP applications [9] is due to their evolution over a series of generations. Each successive generation is derived from an application of some combination of the primary genetic operators: *reproduction* and *crossover* (see Figure 1). And while the application of the former results in copying an individual into a new generation and/or subpopulation, the utilization of the latter results in a random selection of distinct points from two selected parents. Based on the values indicated by the fitness measure for two offspring - two potential members of a new generation - certain individuals of the population are stochastically selected as parents. Thus, the resulting children replace their parents in the existing population of solutions associated with a specific GAs/GP application. The migration operator in GP purges an existing sub-population of the least desirable members and in some cases the best member. This process provides GP with another mechanism to avoid local optimals.

Secondary operators, such operators are the *mutation* and *editing* operators, utilize a probability model that modifies individuals prior to their inclusion/addition to the new generation. By randomly selecting an individual and replacing a sub-tree with a new, randomly generated sub-tree the mutation operator avoids local optima in the search space. The editing operator in turn, chooses sets of individuals only when they are in possession of some specified generational frequency. After the application of the editing operator, the identities of the individuals are modified with simpler statements

Table 1. Genetic Programming and its corresponding Evolutionary Operators.

cause	Honeybee Colony	Genetic Programming Operators
bees fly out	migration/swarming	migration
queen and drones mate	reproduction	crossover/reproduction
worker bee lays eggs	mutation	mutation/editing

3   Adapting the Biological Model

3.1   Overview of the Biological Model

The basic characteristics exhibited by the inhabitants of a honeybee colony [3][12] are those that will be utilized to adequately search the Web for valuable information. These basic characteristics will be translated so as to develop a model of an integrated search engine using GP (see Table 1). The evolutionary processes exhibited by the bee colony thus provide a biological model not only for the storage, processing, and retrieval of valuable information but also for Web crawlers, as well as for an advanced communication system.

The biological storage mechanism for Web documents will simulate the interior components of the honeycomb, while the crawler mechanisms will be simulators that emulate the behavior of forager bees (communicating agents). Since the queen is the centralized control and co-ordination mechanism in a bee colony, queen simulators will be developed to control and coordinate all the process interactions within the simulated hives (Web sites). For mating purposes drone simulators will also be introduced in the model. Some of the functionality of the worker bees are integrated as components of a traditional operating system. Worker bees perform various tasks such as: 1) cell cleaning, 2) feeding larvae, 3) comb building, 4) nectar reception within the hive, 5) pollen packing, 6) removing debris, and 7) guarding. Also, they perform routine inspections and give extensive care to the bee larvae that range from one to thousands of visits. Current operating systems (OS) mechansims perform such rudimentary worker bees tasks as the allocation/deallocation of memory, garbage collection, and the basic form of computer security for user data as well the OS itself.

The characteristics provided by this model remove the implementation limitations inherent in a methodology when developing a comprehensible model. The worker bees' conversion of pollen and nectar into honey will be simulated so as to provide a model for the indexing mechanisms that are essential in the organization of various sets of existing Web documents.

The current Pseudo-Search Engine Indexer, capable of organizing limited subsets of Web documents, provides a foundation for the first bee hive simulators. Adaptation of the honeybee model for the refinement of the Pseudo-Search Engine establishes order in the inherent interactions between the indexer, crawler and browser mechanisms by including the social (hierarchical) structure and simulated behavior of this complex system [6]. The simulation of behavior will engender mechanisms that are controlled and co-ordinated in their various levels of complexity.

3.2   Description of the Computer Model

The components of the computer system for the Pseudo-Search Engine can be compared to those of the honeybee colonies' social organization. The model in Table 2 shows the correlation between the honeycomb cells, queens, worker bees, drones, forager bees, guard bees, and such components as propolis, water, pollen, and nectar that are needed to sustain life in the colony. Here, the honeycomb cells, used to store pollen, nectar, and honey as well as brood (eggs and larvae), correspond to the memory cells in a computer. Each queen, responsible for controlling and coordinating most of the activities within a bee colony by releasing over 32 distinct pheromones, is here equivalent to a Web site (each indexer incorporates the functionality of the computer operating system) which manages and coordinates all of the activities associated with the efficient operation of a computer system. The implementation of drone Web sites will also be modeled after that of queen Web sites. Since the forager bees while searching for propolis, pollen, nectar, and sometimes water still serve the queen, the pheromones also regulate the activities that occur around the bee colony.

Most processes associated with the computer model for the Pseudo-Search Engine will have short life spans that are similar to those of the bee colony members. When submitted via the browser interface, the validity of a users requests will be determined by the guard bee emulators that inspect all the bees entering the colony. The forager bees here represent the Web crawlers which will retrieve Web pages from a diverse set of locations; and the diverse quantity of pollen, nectar, honey, and possible water that resides in a bee colony will here represent the diverse set of Web pages that comprised the WWW.

Table 2. Computer Model for the Evolutionary Pseudo-Search Engine.

Honeybee Colony	Computer Model
queens and drones	Web sites
propolis, pollen, nectar, and sometimes water	Web pages
honey	useful/organized information resulting from processed Web pages
honey comb cells	memory locations
brood and worker bees	processes
guard bees	computer security system
possible intruder	possible user
- checked by guard bees	- restricted internal access
- restricted internal access	

4   Computation Measures for the Pseudo-Search Engine

4.1   Overview

Computation measures [5] for GP applications were developed to measure the extent of the difficulty associated with the size of the population. The effort needed to apply genetic operators such as reproduction, crossover, and mutation is minute when compared to the computational effort needed to evaluate the fitness of each individual in a population. Also, as the population size increases, memory constraints become the next factor that requires consideration.

The creation of subpopulations requires the computation of localized fitness measures - a process that improves the overall performance to a degree that makes it directly proportional to the number of nodes (Web sites). One side effect of using GP and/or GAs-based approach is the disorderly sequence of conditions and other operations associated with individuals in the population. The model under development consists of a network of workstations and a network of single-board computer (such as the IBM SP2). The limitations of such system configurations include ongoing management and maintenance of the independent processes on an unruly collection of machines. The configuration for this type of environment consists of a host computer that acts as the file server, a Program Manager, and the network of processing nodes (queens and drones).

4.2   Computational Measures

By incorporating several different computational measures [5] the performance of the genetically enhanced Information Retrieval (IR) system for the Pseudo-Search Engine Indexer (the simulator of the internals of the bee colony's internal mechanisms) can be measured. The components of these measure include 1) the cumulative probability of success by generation i, $P(M, i)$, 2) the population size, M, and 3) a target probability, z.

The number of fitness evaluation for the sequential version with panmictic selection is

$$x = M(i + 1),\tag{1}$$

while the number of fitness evaluation for the parallel version is

$$x = \sum Q_d(i(d) + 1).\tag{2}$$

Here, i represents the generation number for the solution, d is the summation index that runs over the n nodes, Q_d is the subpopulation (deme) size, and $i(d)$ is the number of the last reporting generation from node d at the time when the sequential version satisfied the success criterion. The cumulative probability is computed via the following equation:

$$P(M, i) = \frac{\sum_i \ successful \ runs}{total \ number \ of \ runs}.\tag{3}$$

The number of independent runs needed to achieve a required probability, z, following x fitness evaluations is given by

$$R(z) = \left\lceil \frac{log(1-z)}{log(1-P(M,i))} \right\rceil . \tag{4}$$

The problem size [7] can be computed by using the equation

$$k = p \times g \times e, \tag{5}$$

where p is the population size, g is the number of generations, and e is the number of fitness cases. The number of Web pages to be classified is represented by the number of fitness cases for the Pseudo-Search Engine.

Additional fitness computations are associated with the maintenance of individuals of different species. The number of individuals [10] needed to produce a solution by generation i with probability greater than $z \approx 99\%$ is

$$I(M,i,z) = x \times (i+1) \times \left\lceil \frac{log(1-z)}{log(1-P(M,i))} \right\rceil . \tag{6}$$

This computation also indicates the number of individuals needed for the partial control mechanism [11] - a mechanism essential in the determination of the number of required Web crawlers. Another computational measure [5] used to determine the "computational effort" required in solving the given genetic programming system is computed via the following equation:

$$computational \ effort \ = \ min(I(M,i,z)). \tag{7}$$

5   Conclusion

In supplying its IR system current search engines incorporate some form of a crawler mechanism for the retrieval of Web documents. The implementation issues associated with the Pseudo-Search Engines' indexer provide insight into the caliber of adaptive crawler mechanisms essential in accurately parsing, indexing, and retrieving Web documents. These insights gained from the initial implementation of this simulator have been utilized in the initial phase of the biological model adaption. To facilitate communication among the diverse process mechanisms (the indexer(s), Web crawler(s), and the browser interface(s)), components indispensable in the formation of a fully integrated Pseudo-Search Engine, this biological model provides built-in mechanisms. In addition to representing a benchmark in the determination of the implemented scheme's efficiency, the adopted model also serves as a foundation for future evolutionary expansions of this search engine as World Wide Web documents continue to proliferate.

6   Acknowledgements

The author wishes to thank Walter Karplus and Zhen-Su She for their direction and suggestions. Support for this work came from the Raytheon Fellowship Program. Special thanks to Martha Lovette for her assistance with the figure in this paper.

References

1. Abramson, M.Z., Hunter, L.: Classification using Cultural Co-evolution and Genetic Programming. In: Koza, J.R., Goldberg, D.E., Fogel, D.B., Riolo, R.L. (eds.): Proc. of the 1996 Genetic Programming Conf. MIT Press, Cambridge, MA (1996) 249–254
2. Chapman, C.D., Jakiela, M.J.: Genetic Algorithm-Based Structural Topology Design with Compliance and Topology Simplification Considerations. J. of Mech. Design **118** (1996) 89–98
3. Free,J.B.: The Social Organization of Honeybees (Studies in Biology no. 81). The Camelot Press Ltd, Southampton (1970)
4. Koza, J.R.: Survey of Genetic Algorithms and Genetic Programming. In: Proc. of WESCON '95. IEEE Press, New York (1995) 589–594
5. Koza, J.R., Andre, D.: Parallel Genetic Programming on a Network of Transputers. Technical Report STAN-CS-TR-95-1542. Stanford University, Department of Computer Science, Palo Alto (1995)
6. Marenbach, P., Bettenhausen, K.D., Freyer, S., U., Rettenmaier, H.: Data-Driven Structured Modeling of a Biotechnological Fed-Batch Fermentation by Means of Genetic Programming. J. of Systems and Control Engineering **211 no. I5** (1997) 325–332
7. Oussaidène, M., Chopard, B., Pictet O.V., Tomassini, M.: Parallel Genetic Programming: An Application to Trading Models Evolution. In: Koza, J.R., Goldberg, D.E., Fogel, D.B., Riolo, R.L. (eds.): Proc. of the 1996 Genetic Programming Conf. MIT Press, Cambridge, MA (1996) 357–362
8. Senin, N., Wallace, D.R., Borland, N.: Object-based Design Modeling and Optimization with Genetic Algorithms. In: Banshaf, W., Daida, J., Eiben, A.E., Garzon, M.H., Honavar, V., Jakiela, M., Smith,R.E. (eds.): GECCO-99: Proc. of the Genetic and Evolutionary Computation Conf. Morgan Kaufman Publishers, Inc., San Francisco (1999) 1715–1721
9. Sherrah, J., Bogner, R.E., Bouzerdoum, B.: Automatic Selection of Features for Classification using Genetic Programming. In:Narasimhan, V.L. Jain, L.C. (eds.): Proc. of the 1996 Australian New Zealand Conf. on Intelligent Information Systems. IEEE Press, New York (1996) 284–287
10. Spector, L., Luke, S.: Cultural Transmission of Information in Genetic Programming. In: Koza, J.R., Goldberg, D.E., Fogel, D.B., Riolo, R.L. (eds.): Proc. of the 1996 Genetic Programming Conf. MIT Press, Cambridge, MA (1996) 209–214
11. Sinclair,M.C. , Shami, S.H.: Evolving Simple Software Agents: Comparing Genetic Algorithm and Genetic Programming Performance. In: Proc. of the 2nd Intl. Conf. on Genetic Algorithms in Engineering Systems: Innovations and Applications. IEE Press, London (1997) 421–426
12. von Frisch, K.: Bees: Their Vision, Chemical Senses, and Languages. Cornell University Press, Ithaca, New York (1964)
13. Walker, R.L.: Implementation Issues for a Parallel Pseudo-Search Engine Indexer using MPI and Genetic Programming. In: Proc. of the Sixth International Conf. on Applications of High-Performance Computers in Engineering. WIT Press, Ashurst, Southampton, UK (January 2000). To appear
14. Willis, M.J., Hiden, H.G., Marenbach, P., McKay, B. Montague, G.A.: Genetic Programming: An Introduction and Survey of Applications. In: Proc. of the 2nd Int. Conf. on Genetic Algorithms in Engineering Systems: Innovations and Applications. IEE Press, London (1997) 314–319

A Parallel Co-evolutionary Metaheuristic

Vincent Bachelet and El-Ghazali Talbi

Université des Sciences et Technologies de Lille
LIFL - UPRESA 8022 CNRS
Bâtiment M3 Cité Scientifique,
59655 Villeneuve d'Ascq CEDEX, France
{bachelet, talbi}@lifl.fr

Abstract. In order to show that the parallel co-evolution of different heuristic methods may lead to an efficient search strategy, we have hybridized three heuristic agents of complementary behaviours: A Tabu Search is used as the main search algorithm, a Genetic Algorithm is in charge with the diversification and a Kick Operator is applied to intensify the search. The three agents run simultaneously, they communicate and cooperate via an adaptive memory which contains a history of the search already done, focusing on high quality regions of the search space. This paper presents CO-SEARCH, the co-evolving heuristic we have designed, and its application on large scale instances of the quadratic assignment problem. The evaluations have been executed on large scale network of workstations via a parallel environment which supports fault tolerance and adaptive dynamic scheduling of tasks.

1 Introduction

Usually, the parallelism is considered as a mean to increase the computational power in running several tasks simultaneously. However, a completely different approach has been appeared recently. The objective is no more to speed up the processing, and the implementation may be sequential or parallel. The main interest is the co-evolution of activities. According to this approach, the metaheuristic we present in this paper combines the two advantages of the parallelism: the computational power and the co-evolution. Our objective is to design a robust and efficient search method involving the co-evolution of well known basic agents, having complementary behaviours.

In the design of a efficient heuristic search strategy, two opposite criteria have to be balanced: the exploration (diversification task) of the search space and the exploitation (intensification task) of the solutions that have already been found. Moreover, some algorithms are rather well fitted in diversification and others are rather good in intensification. Instead of trying to improve diversifying heuristics and intensifying ones, a common idea is to hybridize both approaches in order to make both methods compensate themselves. In order to design a well-balanced metaheuristic, we propose the co-evolution of a search agent (a local search), a diversifying agent and an intensifying agent. The three agents exchange information via a passive coordinator called the adaptive memory.

J. Rolim et al. (Eds.): IPDPS 2000 Workshops, LNCS 1800, pp. 628-635, 2000.

In the remainder of the paper, we present an instance of the model of the parallel co-evolutionary metaheuristic we have describe above. In section 2, a NP-hard problem, the quadratic assignment problem (QAP), which we use as a testbed for this study, is presented. In section 3, the idiosyncrasies of the co-evolutionary metaheuristic are described, the three agents and the adaptive memory are detailed without loss of generality with regard to the search problem to be tackled. Then, we give the specifications of the co-operative hybrid involved in the application on the QAP. In section 4, we present the experiments carried out on a parallel environment which supports the adaptive dynamic scheduling of tasks and fault tolerance. Finally, we comment the results obtained on standard large scale instances which reveals the interest in the co-evolution.

2 The Quadratic Assignment Problem

The Quadratic Assignment Problem (QAP) is used in this study for the co-operative hybrid to be evaluated. However, our study is more general and the co-evolutionary metaheuristic we present here is not especially designed for the QAP.

The QAP is a NP-hard problem and exact algorithms (such branch and bound) are restricted to small instances ($n < 25$). Therefore, many heuristics have been proposed in the literature: an extensive survey and recent developments can be found in [3].

The QAP can be stated as follows. Given the size of the instance n, a set of n objects $O = \{O_1, O_2, ..., O_n\}$, a set of n locations $L = \{L_1, L_2, ..., L_n\}$, a flow matrix c, where each element c_{ij} denotes a flow cost between the objects O_i and O_j, a distance matrix d, where each element d_{kl} denotes a distance between location L_k and L_l, find an object-location bijective mapping $U : O \longrightarrow L$, which minimises the function ψ:

$$\psi(U) = \sum_{i=1}^{n} \sum_{j=1}^{n} c_{ij} . d_{U(i)U(j)}$$

To represent a solution of the QAP, we use the same representation for all agents which is based on a permutation of n integers: $u = (u_1, u_2, ..., u_n)$ where $u_i = U(i)$ denotes the location of the object O_i. The distance between two permutations can be defined with regards to the pair-exchange operator (swap) usually used for the QAP. The distance $dist(u, v)$ between u and v is the minimum number of swap operations needed to move from u to v.

3 CO-SEARCH: A parallel co-evolutionary metaheuristic

In the metaheuristic CO-SEARCH (CO-evolutionary Search), three heuristic agents run simultaneously and exchange information via an adaptive memory (AM). As the main search agent, we choose the tabu search (TS) because it is one of the most efficient local search methods now available in terms of solution quality. For the diversification task, we use a genetic algorithm (GA) as it is powerful at the exploration task due to its genetic operators. The intensification task is handled by a kick operator (KO) for it is a uncomplicated way to scramble, more or less, a solution.

3.1   MTS: the search agent

The search agent we propose is a multiple tabu search (MTS). The TS approach has been proposed by Glover [4]. Sometimes, The TS involves some techniques to encourage the exploitation of a region in the search space, or to encourage the exploration of new regions. For this study, the TS does not need any advanced technique for the diversification or the intensification because of the other agents.

We use multiple independent TS tasks running in parallel without any direct cooperation (between the different TS). We choose a multiple short TS run rather than a long run because the TS does not need to evolve far from the initial solution (in other regions). In CO-SEARCH, the TS has been extended with a frequency memory for storing an information about all the solutions it visits along a walk. Actually, the frequency memory stores the frequency of occurency of some events during the search [4]. Let H, the set of events to deal with, and let F_e, the frequency (number of occurency) of the event $e \in H$.

For solving the QAP, the following events are considered: "the object O_i is assigned to the location L_j". The frequency memory is a n-sized squared matrix (n is the size of the instance to be solved). This matrix is computed by adding all the solutions (encoded as matrices) along the TS walk. In this frequency matrix, the sum of any column or any row is equal to length of the TS walk (the number of iterations). Before starting a TS walk, the frequency memory is initialised to zero. Along the walk, it is updated by adding the visited solutions. At the end of the TS search, the frequency memory provides an information about the areas the TS has visited. Then, the TS sends the frequency memory and its best found solution to the adaptive memory.

3.2   The Adaptive Memory

The adaptive memory (AM) is the coordinator of the co-evolutionary meta-heuristic CO-SEARCH. At run time, the AM maintains a history of the search performed by the MTS. The AM is composed of three main parts: the global frequency memory, the set of the initial solutions of the TS and the set of the elite solutions (see Fig. 1).

The global frequency memory and the set of starting solutions are used by the GA for the diversification task. The global frequency memory is the sum of the local frequency memory yielded by the TS. For the intensification task, the set of elite solutions is used. This set gathers the best solutions the TS walks have found. It stores an information about the regions of best quality in the explored regions of the search space. The number of elite solutions is a parameter of the method. When a TS finishes, it communicates its best found solution to the adaptive memory. If the set of elite is not full, then the solution is added; else, and if the new solution is better than the worst elite, then the new solution is added and the worst is discarded. In any case, if the set of elite contains a solution its quality is equal to the quality of the new solution, no replacement is done. According to this last rule, the diversity of the elites is insured, and the adaptive memory represents a greater number of promising solutions.

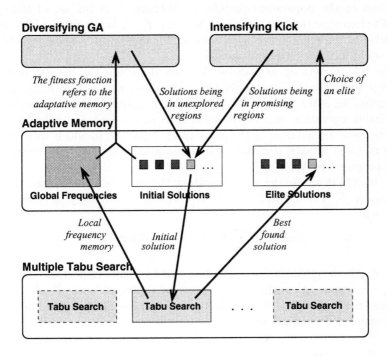

Fig. 1. The Genetic Algorithm cooperate with the Multiple Tabu Search via an Adaptive Memory which ensures the diversity of exchanged solutions.

3.3   The Diversifying GA

The diversifying agent, used in the CO-SEARCH, is a genetic algorithm. Genetic algorithms (GA) are meta-heuristics based on the natural evolution of living beings [6]. In the co-operative hybrid, the aim of the GA is to generate individuals in unexplored regions with regards to the AM. The GA is intrinsically efficient at the diversification task because the genetic operators generate scattered solutions. According to our experience [7], the GA is also able to adapt in a dynamic environment. It is helpful here because the AM, the GA refers to, is continuously being updated by the TS while the hybrid is running. When the GA has evolved a satisfying individual, that solution is given to a TS for being improved.

For the diversifying GA, there is a bi-objective function φ, composed of f and g to be minimised: f is relative to the frequency-based memory and g is relative to the set of initial solutions of the TS:

$$\varphi(u) = \alpha f(u) - g(u) \qquad f(u) = \sum_{e \in E} 1(e, u).F_e \qquad g(u) = \min_{v \in S} dist(u, v)$$

in which α is a coefficient which balances the influence of f and g, u is any individual in the population the GA is evolving, E is the set of the events considered to maintain the frequency memory F, F_e is the frequency stored in F relative to the event e, $1(e, u)$ is equal to 1 if the event e occurs in u else 0, S is the set of initial solutions of the TS.

The GA is running as follows: At each iteration, the genetic operators, crossover and mutation, are applied in competition on the best solution of the population. An other point of the population is randomly chosen, then used for the crossing operation. If a better solution than the best one is build, then it replaces the worst solution of the population. This GA, in which only one solution is replaced at each iteration, is in accordance with the steady-state model of GA proposed by Whitley [9].

For the QAP, we use the usual operators: the swap as mutation and the PMX as crossover [5]. The fitness function optimised by the diversifying GA is defined as follows:

$$\varphi(u) = \frac{1}{N} \sum_{i=1}^{n} F_{u(i)i} \quad - \quad \min_{v \in D}\left(\text{dist}(u, v)\right)$$

where (F_{ij}) is the global frequency matrix, i and j denote respectively the objects and the locations, N is the total number of solutions "stored" in the global frequency memory. By dividing by N, the fitness is independent from the number of iterations of TS already done.

3.4   The intensifying kick

The diversifying agent yields starting solutions to the MTS. The intensifying agent we use in the CO-SEARCH is the Kick operator (KO). To generate a new solution, the KO refers to the set of elite solutions. The KO randomly chooses an elite and scrambles it, the objective being to do a small jump in the search space. If the KO scrambles too much the elite, it may build a solution out of the promising area in which the elite is. So, care should be taken not to scramble too much the elite to avoid the kick to act as a diversifying operator, or it leads to the opposite behaviour the KO is used for. For the QAP, we use a KO which consists in applying several times the swap operator.

4   Experiments

In order to show the influence of the diversifying agent and the intensifying agent, the metaheuristic CO-SEARCH is compared to a MTS. In this comparison, the MTS is considered as a CO-SEARCH in which the GA and the KO are discarded. However, in the MTS, the starting solutions are randomly generated and the

AM contains only the global best found solution. For this comparison, the same amount of CPU effort is used for both methods because the computational effort the GA and the KO are unsignificant with regard to the need of the MTS.

For the evaluation of each method, 1000 TS are done using quite small walks. In the MTS, the 1000 TS walks run simultaneously. In the CO-SEARCH, only a part of the TS walks are launched at the starting in order to provide partial results to update the AM. Actually, 100 TS are started at the beginning of the CO-SEARCH using uniformly distributed initial solutions. Then, when a TS achieves, the AM is updated and the GA or the KO provides an interesting solution for a new TS to restart with. The process is iterated until the 1000 TS are completed. So, the 900 last starting solutions are yielded by the GA or the KO alternately.

The TS walks are $100n$ iterations long, where n is the size of the instance to be solved. Before starting, a TS randomly sets the size of its tabu list between $\frac{n}{2}$ et $\frac{3n}{2}$. The population of the GA is randomly generated at the starting, its size is n. The number of generations used to build a diversified solution is n. The maximum size of the set of elite solutions is $\frac{n}{3}$, and the number of applications of the swap operation in the KO is $\frac{n}{3}$.

We have developed the co-evolutionary metaheuristic CO-SEARCH using the dynamic scheduling system MARS (Multi-user Adaptive Resource Scheduling) [8]. The MARS system harnesses idle cycles of time of a network of workstations. It supports the adaptive parallelism: It reconfigures dynamically the set of processors hosting the application with respect to the idle status of the workstations. A workstation is stated as idle when it is not loaded or its user is away (no one is using the console). The implementation of the parallel metaheuristic CO-SEARCH uses the master / workers paradigm. The master task, as usually done, do the management of the workers, however it also hosts the AM, the GA and the KO. The workers are sequential TS tasks. At run time, the master run on a single computer while the workers are dynamically distributed by the system MARS on the execution platform. So, some workers are running on the idle workstations while the other TS tasks are stored in the master until they are unfolded. When a node (workstation) becomes idle, the master unfolds a TS to it; When a node becomes busy, for it is loaded or owned, the TS task which is running on it is stopped (after the current iteration is completed) and retreated to the master. A checkpoint mechanism is available on the system MARS to prevent from failures or reboot during long runs. The platform we used to evaluate the co-evolutionary metaheuristic CO-SEARCH and the MTS is a network of heterogeneous computers: More than 150 workstations (PC-linux, Sun4-SunOS, Alpha-OSF, Sun-Solaris) and a cluster composed of 16 processors DEC-Alpha. A great deal of these workstations lies in the offices, the laboratories or the teaching rooms of the university.

All the testbed problems are extracted from the QAP library [2] which is a reference for people working on the QAP. The instances of QAP can be divided into several groups depending on their natures, from fully randomly generated instances to real world problems, from strongly structured instances to uniform instances. We propose a taxonomy using three types (type I, II and III). This taxonomy, based on the distribution of the local optima in the search space is the results of a study of the landscapes of the QAP we have carried out [1]. As the results of a heuristic search algorithm depend on this intrinsic nature [1], we have used QAP instances of different types to evaluate the parallel co-evolutionary metaheuristic CO-SEARCH.

Instance	Type	ψ(QAPlib)	Best		Average	
			MTS	CO-SEARCH	MTS	CO-SEARCH
Tai25a	I	1 167 256	0.7361	0.7361	0.7361	0.7361
lipa50a	I	62 093	0.0000	0.0000	0.0000	0.0000
Els19	II	17 212 548	0.0000	0.0000	0.0000	0.0000
Bur26d	II	3 821 255	0.0000	0.0000	0.0000	0.0000
Nug30	II	6 124	0.0000	0.0000	0.0000	0.0000
Tai35b	II	283 315 445	0.0000	0.0000	0.0000	0.0000
Ste36c	II	8 239 110	0.0000	0.0000	0.0000	0.0000
Sko64	III	48 498	0.0000	0.0000	0.0037	0.0033
Tai64c	-	1 855 928	0.0000	0.0000	0.0000	0.0000
Tai100a	I	21 125 314	0.6867	**0.3754**	0.7834	**0.5131**
Tai256c	I	44 759 294	0.2359	**0.1201**	0.2708	**0.1701**
Sko100a	III	152 002	0.0289	**0.0000**	0.0730	**0.0544**
Tai100b	III	1 185 996 137	0.2445	**0.0000**	0.3969	**0.1348**
Wil100	III	273 038	0.0205	**0.0029**	0.0352	**0.0085**
Tai150b	III	498 896 643	0.9246	**0.1903**	1.1276	**0.4393**
Tho150	III	8 133 484	0.0824	**0.0530**	0.1243	**0.0650**

Table 1. Results of the evaluations of the metaheuristics MTS and CO-SEARCH on instances of the QAP. The results are given using the format: $100 \times \frac{\psi(s) - \psi(QAPlib)}{\psi(QAPlib)}$ where s is the best found solution of the method and ψ is the cost function. The column **Best** shows the best result over 3 runs of CO-SEARCH or 10 runs of MTS. The column **Average** shows the average result over the runs. One can observe that the CO-SEARCH is more efficient then the MTS on almost all instances.

The evaluation compare the metaheuristics MTS and CO-SEARCH using quite the same amount of CPU time. To present the results, we distinguish between small instances ($n < 100$) and others (see Tab. 1). For small instances, both MTS and MTS provide the best known solution on almost all instances. Hence, the GA and the KO are no so useful for these instances, or the number of iterations should be reduced. On the opposite, for large instances, one can observe that the CO-SEARCH is very more efficient than the MTS. For 6 / 7 instances the CO-SEARCH provides a better solution; and on average, the CO-SEARCH is always better with a significant improvement for any type of instance. Moreover, the solutions found by the CO-SEARCH are high quality solutions in the sense that they are very close to the best known solutions. Hence, the CO-SEARCH is an efficient method, according to the CPU effort used. The CPU time used for the evaluation correspond to 1000 n iterations of TS, that is 1000 $\frac{n(n-1)}{2}$ times the evaluation of a solution. For example, for $n = 100$ (Tai100a, Wil100, etc, ...), the CO-SEARCH evaluates approximatively 500 millions solutions and the mean running time is 4 hours (depending on the load of the platform).

This evaluation shows the benefits of the co-evolution to solve large instances: The adding of the GA and the KO on the MTS lead to a better equilibrium between the intensification and the diversification which increase the efficiency of the metaheuristic.

5 Conclusion

In this paper, we have proposed CO-SEARCH, an efficient metaheuristic which combines the two advantages of parallelism: the computational power and the co-evolution. The computational power is used at the implementation level to reduce the running time; the co-evolution is used to well-balance the diversification and the intensification in the metaheuristic. The evaluations of the CO-SEARCH we have carried out and the comparison to the MTS (Multiple Tabu Search) shows that the co-evolution increases significantly the efficiency of the method. So, this study has reveals the interest of our approach based on the co-evolution paradigm to solve combinatorial problems.

References

1. V. Bachelet. *Métaheuristiques parallèles hybrides : application au problème d'affectation quadratique.* PhD thesis, Université des Sciences et Technologies de Lille, Villeneuve d'Ascq, France, December 1999.
2. R.E. Burkard, S. Karisch, and F. Rendl. Qaplib: A quadratic assignment problem library. *European Journal of Operational Research*, 55:115–119, 1991.
3. E. Çela. *The Quadratic Assignment Problem Theory and Algorithms.* Kluwer Academic Publishers, 1998.
4. F. Glover and M. Laguna. *Modern Heuristic Techniques For Combinatorial Problems*, chapter 3, pages 70–150. Blackwell Scientific Publications, 1992.
5. D. Goldberg and R. Lingle. Alleles, loci, and the traveling salesman problem. In *International Conference on Genetic Algorithms and their Applications*, 1985.
6. J.H. Holland. *Adaptation in natural and artificial systems.* The University of Michigan Press, Ann Arbor, MI, USA, 1975.
7. Bessiere P., Ahuactzin J.M., Talbi E-G., and Mazer E. The ariadne's clew algorithm: global planning with local methods. *IEEE International Conference on Intelligent Robots Systems IROS, Yokohama, Japan*, July 1993.
8. E-G. Talbi, J-M. Geib, Z. Hafidi, and D. Kebbal. Mars: An adaptive parallel programming environment. In R. Buyya, editor, *High Performance cluster computing*, volume 1, chapter 4. Prentice Hall PTR, 1999.
9. D. Whitley. GENITOR: A different genetic algorithm. In *Proc. of the Rocky Mountain Conference on Artificial Intelligence*, Denver, CO, USA, 1988.

Neural Fraud Detection in Mobile Phone Operations

[1]Azzedine Boukerche* and [2]Mirela Sechi Moretti Annoni Notare

[1]University of North Texas, Denton, TX, USA.
[2]Federal University of Santa Catarina, Brazil

Abstract. With the increasing popularity of wireless and mobile networks, and the needs in the rapidly changing telecommunications industry, the security issue for mobile users could be even more serious than we expect. In this paper, we present an on-line security system for fraud detection of impostors and improper use of mobile phone operations based on a neural network classifier, It acts solely on the recent information and past history of the mobile phone owner activities, and classifies the telephone users into classes according to their usage logs. Such logs contain the relevant characteristics for every call made by the user. As soon as the system identifies a fraud, it notifies both the carrier telecom and the victim about it immediately and not at the end of the monthly bill cycle. In our implementation, we make use of Kohonen model because of its simplicity and its flexibility to adapt to pattern changes. Our results indicate that our system reduces significantly the telecom carrier's profit losses as well as the damage that might be passed to the clients. This might help the carriers to reduce significantly the cost of phone calls and will turn to the users' advantage.

1 Introduction

This paper is motivated by the needs in the rapidly changing wireless and mobile telecommunications industry. Mobile phones will change many aspects of our lives forever, but not until potential users become convinced of the security of the mobile networks [1, 6,7]. Emerging requirements for higher data services such as news-on-demand, web browsing, E-commerce, stock quotes, video-conferencing and better spectrum efficiency are the main drivers identified for the next generation mobile radio systems, and the coming decade of the next millennium.

Before the mobile phones become widely popular, the greatest threat to the network security in most organizations was dial-up lines. While dial-up lines still merit attention, the risks they pose are minor when compared to wireless and mobile connections. To break the system, one needs only to buy a piece of portable radio equipment, such as a scanner for instance, to program a genuine mobile cloned to debit calls from genuine mobile phone, and register the frequencies where mobile phones operate in a surrounding areas. Then, the person committing the fraud may, for example, parks his car around a shopping mall, jots down various frequencies, transfers the data to clones, and then passes them to whomever may be interested in these cloned mobiles. Frauders are also turning to subscription fraud[2] as technical fraud becomes more difficult.

---

* Supported by UNT Faculty Research Grant

[2] Carriers want to make it easier for potential subscribers to sign up for service, so subscription fraud is becoming easier.

J. Rolim et al. (Eds.): IPDPS 2000 Workshops, LNCS 1800, pp. 636-644, 2000.

One might argue that although it is rather easy to clone an AMPS phone, it is much trickier to clone a D-AMPS, a GSM, or an IS-95 phone. Interestingly enough, these systems are designed to provide open access across a vast networked environment. Today's technologies usually are network operation intrusive, i.e., they often limit the connectivity and inhibit easier access to data and services. The traditional analogue cellular phones are very insecure. The 32-bit serial number, the 34-bit phone number and the conversation in a cell can be scanned easily by an all-band receiver. The widely used advanced mobile Phone System (AMPS) is an analogue phone system. Therefore, sending a password or a host name through this system can be a serious security issue. Thus, the security remains, and needs to be resolved in the next wireless network generation. In addition to make the mobile phones hard to clone from a hardware point of view, many prevention methods have to be installed. Thus, new fraud detection tools, neural or other are weapons that should be used in a war that has already started in many countries. we believe that the one-line security system would be an efficient solution to this problem. Currently, Telecom companies are losing a lot of money due to impostors and frauders using mobile phones they don't own. Therefore, rather than ignoring the security concerns of potential users, merchants and telecommunication companies need to acknowledge these concerns and deal with them in a straightforward manner [6, 7].

In this paper, we will show how neural network technology [2, 4] contributes as a new tool to tackle this challenging problem. The remainder of the paper is organized as follows. In Section 2, we present the neural network classifier based upon Kohonen model, , we discuss the implementation of our system. Section 3 reports the experimental results we obtained. In Section 4, we present the conclusion and some of our future work.

2   Neural Fraud Model Construction

2.1   Classification of Mobile Phone Users

Fraud detection of mobile phones falls neatly in principal within the scope of pattern recognition paradigm [4, 6, 7], and its solution can be sought through the construction of appropriate classifier functions. However, the imbalance of good and fraudulent operations made by impostors make it rather a difficult problem. One should also not forget the huge amount of profit losses the telecom carriers have to prevent. Our main approach to identify fraud calls is to classify the mobile phone users into a set of groups according to their log files. We assume that all relevant characteristics that identify the users will be stored in these files; i.e., *where*, at *what* time, and from *where* the calls were made etc. There are several types of impostors that our system will identify: (*i*) those who had changed the mobile phone's owner call patterns; (*ii*) those who bought a mobile phone only for one month (convinced that thy won't pay after the end of the month); and (*iii*) and those who bought mobile phones using other names. Classifying the users into groups will help our system to identify easily if a specific call was made by the mobile phone's owner. Thus, when the call made using a genuine/cloned phone is terminated, the system will check if the characteristics of the call is within the client patterns saved in the file. A warning message could be sent to the client if a fraud was detected. This immediate notification, as opposed to waiting till the end of the monthly bill cycle, will not only help the telecom carriers to reduce their profit's loses, but also the damage that might be passed on to their clients. In our system, we propose to use neural network algorithm to partition the users into classes and create the log files.

2.2   Neural Network Model

Neural networks, when seen as an adaptive machine, can be viewed as a distributed memory machine that is naturally able to store experimental knowledge and make it available for use [2, 3, 4, 5]. Neural networks are similar to the mind in two aspects: (i) the knowledge is acquired by the network by means of a learning process; and (ii) the weights of the connections between neurons, known as synapses, are used to store the knowledge. The procedure used to represent the learning process, commonly called learning algorithm, has the function of modifying the weights of the connections of the network in seeking to reach an initial designed objective. The details of synaptic adjustment and inter-neuronal competition are not known in biology, and may not be the same as that of artificial self-organizing maps.

In our neural fraud detection system, we use Kohonen's model [3, 5] because of its simplicity, its adaptive organization with useful topological maps. and its flexibility to adapt to pattern changes. In Kohonen's model, neurons compete among themselves to be the winner in each iteration. The neuron with a weight vector that generates the minimum Euclidean distance with regard to the input vector is identified as a winner. In this model, both the winner neuron and its neighbor weights are adjusted in each iteration during the learning phase. The synaptic weights begin with an off states (i.e., low values) and an X-input value is given without output specification, i.e., non supervised network. Based upon the input value, an output neuron Y must provide a good response to the respective input value. Y will be then identified as the winner neuron. Both, the Y winner neuron and its neighbors will update their weights accordingly, in order to provide the best response possible to the input value.

The basic model of Kohonen [3] consists of a mapping function ϕ from an input space V into an output space A; where A represents parameters of unit d, V represents the input or stimulus to units d, and W represents the network interconnection ("weights"). In this case, training becomes the process of finding one or more solutions (approximate or exact) for W. The output space consists of n_j nodes that are organized into a topological order. Each node in the output (layer) has a weight vector (or prototype) attached to it. The prototypes $V = \{v_1, v_2, ..., v_c\}$ are essentially a network array of clusters centers $v_i \in R^p$ for $1 \leq i \leq c$. In this context "learning" refers to finding values for $\{v_{ij}\}$. When an input vector x is submitted to this network, distances are computed between each v_r and x The output nodes compete, a (minimum distance) winner node, say v_i, is found; and it is updated using several rules.

During each iteration, the values of the w_i are adjusted. A first iteration step begins with the generation of a stimulus $u \in V$ according to a probability density function $p(u)$. Then, the values w_i and w_j of the nodes in the vicinity of n_i are shifted to a small step toward u: $\delta w_j = \epsilon . h_{j,i}^0 (d^A(n_j, n_i))(u - w_j) n_j \in A$. The extension of the vicinity of n_i is given by the function $h_{j,i}^0$,[3] which depends on $d^A(n_j, n_i)$, i.e., the distance between n_j and n_i in the output space. The step-size ϵ and the (vicinity) radius σ have a profound impact on the convergence of the algorithm[4].

The training phase of the neural network consists of processing the training data set, and satisfy the trainee criteria. The next phase (or testing phase) is similar to the training phase, except that in this case, neuron weights will not be updated anymore. If the network recognizes successfully the input, then the network is considered tested with success. Independently of the inputs, if a network has x neurons in the output

---

[3] A typical choice for $h_{j,i}^0$ is $h^0(d) = e^{-d^2/2\sigma^2}$.

[4] i.e., they should decrease while the number of learning steps increases. In the most cases, they decrease exponentially.

layer, then will be have x possible outputs. The function of competitive transference accepts an input vector of the network for a layer and returns an output of zeros for all neurons with the exception of the winner, i.e., the neuron related with the largest positive element of the network input. The winner output is equal 1.

The learning phase begins with the initialisation[5] of all (w_j) values. The number of learning steps can be defined by the user, or using some termination criterions (e.g., a shorter radius σ_{min}).

Once the "learning" phase is finished, the network posses enough knowledge to start making use of the system[6]. In our case, our neural network fraud detection system must be able to identify and distinguish the calls[7] made by regular users and impostors using different clients' mobile phones - independently even if they were not included in the learning phase.

3   Experimental Study

The goal of our experiments is to show how neural networks can be applied as a tool to identify frauds and impostors that use mobile phones improperly. Thus, our primary study aims at understanding and developing the required neural network model while providing the best performance (efficiency) for our specific application (fraud detection in mobile phone operations). Furthermore, we wish to investigate the impact of Kohonen's model on the performance of our neural fraud detection system.

During the course of our experiments, we split the weekdays and weekends into several groups as follow:

$$Week_{Set} = \{wdays_{00-08h}, wdays_{08-12h}, wdays_{12-14h}, wdays_{14-18h}, wdays_{18-24h}\} \cup$$
$$\{wkend_{00-08Sat}, wkend_{08-12Sat}, wkend_{12-24Sat}, wkend_{00-12Sun}, wkend_{12-24Sun}\}.$$

We wish to determine the following items:

(a) the characteristics of each phone call that must be considered and the size of the networks one should use. In our experiments we investigate the following samples:

- 1 network using the characteristics: called_number, duration, time and day.
- 2 networks (weekdays, weekend) using called_number, duration, and time.
- 10 networks using WeekSet, and the 2 characteristics: called_number and duration.

(b) the quantity of data (i.e., phone calls) necessary to train our system and be able to identify the impostors. We choose to groups the data into the following categories:

- One day (1 file with all 24hrs calls or 5 files representing the differents tariff calls)
- Two days (Monday/Sunday)
- Three days (Monday/Sunday and one Holiday)
- A week (5 days of a weekdays and 2 days of a weekend)
- A month (many users make more calls in the first week of the month, for instance)

(c) the significant/optimal output size and the maximum number of clusters.

- 16 - sufficient only for the combination of called_number ({local, interurban, international, special}) and duration sets {short, median, long, extra long})
- 32 - sufficient to provide the combination of called_number, duration and date (weekdays, weekend) sets

---

[5] either randomly or based on apriori information about the map.

[6] we refer to this phase as the utilization phase.

[7] i.e., mobile utilization.

- 512 - sufficient to provide the combination of called_number, duration, date/time, and WeekSet.
- 1024 - sufficient to provide a refined clusterization (i.e., more sensibility)

The motivation behind investigating all these issues is to observe which configurations successfully train the network most for our application. Once, we have answered all these issues, we have all components we need to build and train the neural network for our specific application. The resulting network is then evaluated for performance. Note that neural networks do not have theoretically proven performance levels; hence, one must do testing to study the performance of the neural network obtained.

3.1   Environment and Implementation Issues

In our experiments, we made use of a Pentium II with 64Mb of RAM and 4Gb of disk space. The data files (i.e., users calls) were collected from a telecom carrier. However, since most users have different calling patterns on the weekends and weekdays, using different discount rate levels, we choose to consider only a representative data set in our experiments that can be supported by the available computer resources (hardware/software limitations).

A good (sub-optimal) classification requires a good selection of criteria to identify the frauders. In our experiments, we choose the following characteristics:

- caller_number (used only for testing/usage; not for training);
- called_number (used to identify the type of calls, i.e., local, international, etc.);
- time of the call
- duration of the call (used to identify long calls that do not match with the user's pattern); and
- date of the call (used to identify the different users' patterns during the week, and holidays).

During our experiments, first, we choose to use only 4 of these characteristics, then we use only 2 of them to train our system. We also vary the number of neural networks from 1 to 10. Initially, the input data set was scaled into values between 0 and 1. In the case of called_number, we use a binary transformation, where one column is transformed to four columns. Each column corresponds to one specific type of call, i.e., local, interurban, international, and special calls (800, 888, 900, 911, 411 etc.).

Using the transformed input data set, our system was trained, then tested to evaluate its performance. Note that in order to train the network successfully, a training schedule is needed: this includes the number of training iterations to be done and the changes that are made during the training phase to get the learning rate parameters. When a network is trained, a learning curve can be observed. Thus, the neural network can be built and trained for our specific application.

• Training the network

Using the data set that contains the called_number and duration characteristics and the 8830 phone calls (representing all calls on Sunday August 1st, from 12:00 to 14:00), we have trained the network choosing 4 points in the (linear) output (providing a maximum of $2^4 = 16$ clusters). Note that we choose not to consider the day/time characteristics since the data set corresponds to the same day (i.e., Sunday, August 1st) and the same time/discount rates (i.e., from 12:00 to 14:00).

Table 1 shows the trainee output, i.e., the iterations and weights updated. We can observe that 4 different clusters were generated, i.e., 1111, 1011, 1101 and 1100. Each cluster groups a set of users of two hours of the day (Sunday in this case).

Column 1 through 24   .... Column 73 through 90

1 1 1 1 1 1 1 1 1 1 1 1 1 1 1 1 1 1 1 1 1 1 1 1 ..	1 1 1 1 1 1 1 1 1 1 1 1 1 1 1 1 1 1
1 1 1 0 1 1 1 1 1 1 1 1 1 1 1 1 1 1 1 1 1 1 1 1 ..	1 1 1 1 1 1 1 1 1 1 1 1 1 1 1 1 1 1
1 1 1 1 1 1 1 1 1 1 1 1 1 1 1 1 1 1 0 1 0 0 0 1 ..	0 1 1 1 0 0 0 0 1 0 0 1 1 1 1 1 0 0
1 1 1 1 1 1 1 1 1 1 1 1 1 1 1 1 1 1 1 1 1 1 1 1 ..	1 1 1 1 1 1 1 1 1 1 1 1 1 1 1 1 0 0

Table 1. Clusters generated by the neural network

Figure 1 shows the clusterization of the phone calls, i.e., the classification of these calls into groups according to their similarities based upon the characteristics chosen.

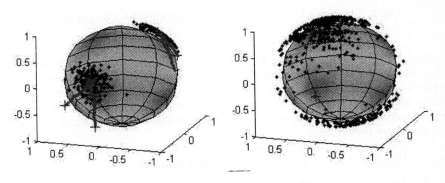

Fig. 1. Clustered Phone calls.

• Testing the network

In order to test our network, we submitted a set of data (after scaling them into [0,1]). We keept track of the generated output. This output represents the user's pattern that will be used in our security system.

Note that the more input data we submit, the more scaled vectors (of 0s and 1s) are generated. Then, the classifier takes the vector (which corresponds to a cluster) that occurs the most as the most probable user's pattern in our security system. For example, if we submitted 10 input vectors, 6 of them were distinct and the others 4 are equals, then the later vector will be chosen as the most probable user's pattern. Small quantities of input test data are not sufficient and they won't provide an efficient security system. Our experiments indicate that significant/good results were obtained when we submitted 10 phone calls of each user. Several test-runs were made in order to observe the classification performance of the networks and the data used. Our initial results were very encouraging.

In our next set of experiments, we wish to study the users' patterns during the weekends and weekdays. Due to limited computational resources, we have settled with the following set of experiments: (i) the use of two networks, one for the weekdays and

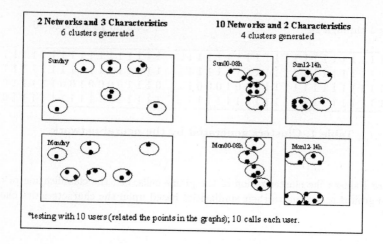

Fig. 2. Clustering Using 2 and 10 Neural networks: A Comparison.

another one for weekends&holiday. In this case, we choose not to consider "date" characteristic; and (2) the use of ten networks, in addition to the weekday/weekend/holiday networks, we choose to consider different networks for each tariff rate during the day. Hence, we included the "time" characteristic in this set of experiments. This second set of experiments allows us to use a large number of input data to train the networks and get better performance.

In order to investigate the different patterns during the weekends and weekdays, we used the (input) files that contain the following information[8]:

(1) Sunday calls, August 1st, from 12:00 to 14:00 - totaling 8.830 calls; and
(2) Monday calls, August 2nd, from 12:00 to 14:00 - totaling 13.456 calls.

Furthermore, in order to investigate the changes of users' patterns during the weekends and weekdays, the two networks (weekends/weekdays) need to be trained using the same weights. Hence, the weights that were automatically generated, and adjusted in the first network were saved. and then were used in the second network.

As we expected, our results indicate that the output vector generated to identify users using the Sunday input file calls was different from the output vector generated using the Monday input file calls of the same user. Thus, the results obtained indicate clearly the the significant difference in the users' patterns during the weekends (Sunday) and weekdays (Monday).

Similarly, we have examined the users' patterns during the day within the different discount periods. To do so, we used the following set of data:

(3) Sunday calls, August 1st, from 12:00 to 14:00 - totaling 8.830 calls; and
(4) Sunday calls, August 1st, from 00:00 to 08:00 - totaling 7.030 calls.

Here again, the results obtained indicate the differences in the user's patterns during discounts period calls. However, the user's patterns were not significantly differents than these obtained during the weekday (Mondays) and weekend rates (Sundays).

---

[8] These data were obtained from a Telecom Carrier.

Space limitations preclude us to describe a complete, factorial testing of all possible combinations of experiments. Thus, in Figure 2, we summarize the results obtained when we used several size of networks with a different number of characteristics, using 10 sets of calls of 10 users. Each one of these 10 tested users was properly classified into their corresponding cluster.

As we can see, reasonable (good) results were obtained using 10 networks, 2 characteristics where 4 clusters were generated, phone calls for two days; and 9 points in the output space[9]. This is mainly due to (i) the fact that "days" (i.e., weekend/weekday) and "time" (00:00-08:00, 12:00-14:00) have different user calls' patterns; and (ii) the increased sensibility of the trainee data for a particular data set that exhibits high similarity, i.e., all calls already belong to a specific day (e.g., Sunday/Monday) and a specific discount rate (e.g., from 12:00 to 14:00, and from 00:00 to 08:00).

Recall that the "testing" phase is done in order to observe the sensibility of the classification obtained. Short alterations inserted in the data set must generate the same output (i.e., cluster). This should help to avoid false alarms to be sent to the clients. Similarly, large alterations must result in a different cluster for the user - in order to guarantee that the client will be aware about the existence of the fraud.

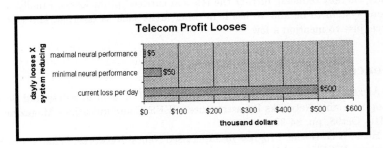

Fig. 3. Telecom Carrier's Saving Using our System

We have tested our system, on real data provided by a telecom carrier, and the results obtained (see Fig. 3) indicate clearly how much saving (in dollars) our security system might provide to the carrier telecom if our system was used. For instance, with minimal neural performance, we obtain a reduction of losses from $500K to $50K, that represents a reduction of profit losses of 10%. While with a maximal neural performance, we obtain a reduction of profit losses to 1%. As we all know it is very hard to have a system to identify all frauds. Indeed, there are several parameters one should not forget, which in most cases are independent to our system. For completeness, we will mention just a few of them:

- Availability of the on-line calls might take minutes, or even hours and depend on the hardware/equipments used by the telecom carrier.
- The time it takes to train and test the neural network (hence our system) depends on the quantity of data, the output points quantity and the available computational resources; and
- The sensibility of the classifier using the on-line data.

---

[9] i.e., that represents $2^9 = 512$ possible clusters.

Nevertheless, our system indicates a significant reduction of the telecom profit losses, and we hope to reduce these losses further in the future .

4   Conclusion

In this paper, we described how neural networks and pattern recognition tools can be applied against frauds in analogue mobile telecommunication networks- A growing security concerns these days for most telecom carriers. To the best of our knowledge there is a lot of work done to secure the usage of mobile phones at the hardware level, but very little work has done at the software level.

In this paper, we present a neural fraud system to identify frauds, impostors and untrusted users of mobile phones, followed by its performance using real data from a telecom carrier. Our results showed clearly the effectiveness of our system to identify the neural frauds. Thereby reducing the profit losses of the telecom carriers and the damages that might be passed to their clients. A reduction of the profit losses to 10% and 1% were obtained using minimal and maximal neural performance respectively.

As a future work; we see several directions. First, we plan to investigate further the performance of our system using several types of characteristics. Next, we plan to parallelize the Kohonen's algorithm to speedup our security system to identify faster the frauds, thereby reducing further the telecom carriers' profit losses. Finally, we will extend our system to be able to use it in other areas such as Internet, and electronic commerce, just to mention a few.

References

1. D.S. Alexander, W.A. Arbaugh, A.D. Keromytis, J.M. Smith, Safety and Security of Programmable Networks Infrastructures, IEEE Communications Magazine, Vol. 36, N10, Oct98, pp. 84-92.
2. D.R. Hush, B.G. Horne, Progress in Supervised Neural Networks: What's New Since Lippmann, IEEE Signal Processing Magazine, January, 1993, pp. 8-39.
3. T. Kohonen, Self-organized Formation of Topologically Correct Feature Maps, Biological Cybernetics 43, 1982, pp. 59-69.
4. R.P Lippmann, Pattern Classification Using Neural Networks. IEEE Communications Magazine, Nov1989, pp. 47-64.
5. J.W. Meyer, Self-Organizing Processes, CONPAR'94 - VAPP VI, Lecture Notes in Computer Science 824, Spring-Verlag, Berlin, 1994, pp.842-853.
6. M.S.M.A.Notare. "A Methodology to Security Management of Distributed Systems Validated in Telecommunication Networks". Phd Thesis. Federal University of Santa Catarina, Computer Science Dept., March 2000.
7. M.S.M.A. Notare, A. Boukerche, F. A. S. Cruz, B. G. Riso, C. B. Westphall. Security Management Against Cloning Mobile Phones, Proc. IEEE Globecom, December 1999.
8. C. Pfleeger, D. Cooper, Security and Privacy: Promising Advances, IEEE Software, Sep/Oct97, pp.110-111.
9. M. Stillerman, C. Marceau, M. Stillman, Intrusion Detection for Distributed Applications, Communications of the ACM, Vol.42, No. 7, July 99, pp.62-69.

Information Exchange in Multi Colony Ant Algorithms

Martin Middendorf[1], Frank Reischle[2], Hartmut Schmeck[3]

Institut für Angewandte Informatik
und Formale Beschreibungsverfahren, Universität Karlsruhe,
D-76128 Karlsruhe, Germany
{[1]middendorf,[3]schmeck}@aifb.uni-karlsruhe.de
[2]Frank.Reischle@gmx.net

Abstract. Multi colony ant algorithms are evolutionary optimization heuristics that are well suited for parallel execution. Information exchange between the colonies is an important topic that not only influences the parallel execution time but also the optimization behaviour. In this paper different kinds of information exchange strategies in multi colony ant algorithms are investigated. It is shown that the exchange of only a small amount of information can be advantageous not only for a short running time but also to obtain solutions of high quality. This allows the colonies to profit from the good solutions found by other colonies and also to search in different regions of the search space by using different pheromone matrices.

1 Introduction

Ant Colony Optimization (ACO) is a metaheuristic to solve combinatorial optimization problems by using principles of communicative behaviour occurring in ant colonies. Ants can communicate information about the paths they found to food sources by marking these paths with pheromone. The pheromone trails can lead other ants to the food sources.

ACO was introduced by Dorigo et al. [4,6]. It is an evolutionary approach where several generations of artificial ants search for good solutions. Every ant of a generation builds up a solution step by step thereby going through several decisions until a solution is found. Ants that found a good solution mark their paths through the decision space by putting some amount of pheromone on the edges of the path. The following ants of the next generation are attracted by the pheromone so that they will search in the solution space near good solutions.

Ant algorithms are good candidates for parallelization but not much research has been done in parallel ant algorithms so far. One reason might be that it seems quite obvious how to parallelize them. Every processor can hold a colony of ants and after every generation the colonies exchange information about their solutions. Then, every colony computes the new pheromone information which is usually stored in some pheromone matrix. Most parallel implementations of ant algorithms follow this approach and differ only in granularity and whether the

J. Rolim et al. (Eds.): IPDPS 2000 Workshops, LNCS 1800, pp. 645-652, 2000.

computations for the new pheromone matrix are done locally in all colonies or centrally by a master processor which distributes the new matrix to the colonies. Exceptions are [3, 9, 10] where the colonies exchange information only after several generations (Similar principles are also used in the island model of genetic algorithms, e.g. [8]). The effect is that the pheromone matrices of the colonies may now differ from each other.

In this paper we investigate different kinds of information exchange between colonies of ants. We show that it can be advantageous for the colonies to exchange not too much information not too often so that their pheromone matrices can develop independently to some extent. This allows the colonies to exploit different regions in the solution space.

2   Ant Algorithm for TSP

The Traveling Salesperson problem (TSP) is to find for n given cities a shortest closed tour that contains every city exactly once. Several ant algorithms have been proposed for the TSP (e.g. [2, 6, 12]). The basic approach of this algorithms is described in the following.

In each generation each of m ants constructs one solution. An ant starts from a random city and iteratively moves to another city until the tour is complete and the ant is back at its starting point. Let d_{ij} be the distance between the cities i and j. The probability that the ant chooses j as the next city after it has arrived at city i where j is in the set S of cities that have not been visited is

$$p_{ij} = \frac{[\tau_{ij}]^{\alpha} [\eta_{ij}]^{\beta}}{\sum_{k \in S} [\tau_{ik}]^{\alpha} [\eta_{ik}]^{\beta}}$$

Here τ_{ij} is the amount fo pheromone on the edge (ij), $\eta_{ij} = 1/d_{ij}$ is a heuristic value, and α and β are constants that determine the relative influence of the pheromone values and the heuristic values on the decision of the ant.

For some constant $m_{best} \leq m$ the m_{best} ants having found the best solutions are allowed to update the pheromone matrix. But before that is done some of the old pheromone is evaporated by multiplying it with a factor $\rho < 1$

$$\tau_{ij} = \rho \cdot \tau_{ij}$$

The reason for this is that old pheromone should not have a too strong influence on the future. Then every ant that is allowed to update adds pheromone to every edge (ij) which is on its tour. The amount of pheromone added to such an edge (ij) is Q/L where L is the length of the tour that was found and Q is a constant:

$$\tau_{ij} = \tau_{ij} + \frac{Q}{L}$$

To prevent that too much of the pheromone along the edges of the best found solution evaporates an elitist strategy is used. In every generation e additional ants - called elitist ants - add pheromone along the edges of the best known solution.

The algorithm stops when some stopping criterion is met, e.g. a certain number of generations has been done or the best found solution has not changed for several generations.

3   Parallel Ant Algorithms

Only a few parallel implementations of ant algorithms have been described in the literature. A very fine-grained parallelization where every processor holds only a single ant was implemented by Bolondi and Bondaza [1]. Due to the high overhead for communication this implementation did not scale very well with a growing number of processors. Better results have been obtained by [1,5] with a more coarse grained variant.

Bullnheimer et al. [3] propose a parallelization where an information exchange between several colonies of ants is done every k generations for some fixed k. They show by using simulations how much the running time of the algorithm decreases with an increasing interval between the information exchange. But it is not discussed how this influences the quality of the solutions.

Stützle [11] compares the solution quality obtained by the execution of several independent short runs of an ant algorithm with the solution quality of the execution of one long run whose running time equals the sum of the running times of the short runs. Under some conditions the short runs prooved to give better results. Also, they have the advantage to be easily run in parallel and it is possible to use different sets of parameters for the runs.

Talbi et al. [13] implemented a parallel ant algorithm for the Quadratic Assignment problem. They used a fine-grained master-worker approach, were every worker holds a single ant that produces one solution. Every worker then sends its solution to the master. The master computes the new pheromone matrix and sends it to the workers.

An island model approach that uses ideas from Genetic Algorithms was proposed in Michels et al. [10]. Here, every processor holds a colony of ants exchanging the locally best solution after every fixed number of iterations. When a colony receives a solution that is better than the best solution found so far by this colony, the received solution becomes the new best found solution. It influences the colony because during trail update some pheromone is always put on the trail that corresponds to the best found solution.

The results of Krüger et al. [9] indicate that it is better to exchange only the best solutions found so far than to exchange whole pheromone matrices and and add the received matrices — multiplied by some small factor — to the local pheromone matrix.

4   Strategies for Information Exchange

We investigate four strategies for information exchange differing in the degree of coupling that is enforced between the colonies through this exchange. Since the results of [9] indicate that exchange of complete pheromone matrices is not advantageous all our methods are based on the exchange of single solutions.

1. Exchange of globally best solution: In every information exchange step the globally best solution is computed and sent to all colonies where it becomes the new locally best solution.
2. Circular exchange of locally best solutions: A virtual neighbourhood is established between the colonies so that they form a directed ring. In every information exchange step every colony sends its locally best solution to its successor colony in the ring. The variable that stores the best found solution is updated accordingly.
3. Circular exchange of migrants: As in (2) the processors form a virtual directed ring. In an information exchange step every colony compares its m_{best} best ants with the m_{best} best ants of its successor colony in the ring. The m_{best} best of these $2m_{best}$ ants are allowed to update the pheromone matrix.
4. Circular exchange of locally best solutions plus migrants: Combination of (2) and (3).

5   Results

We tested our information exchange strategies on the TSP instance eil101 with 101 cities from the TSPLIB [14]. The smallest tourlength for this instance is known to be 629. The parameter values that were used are: $\alpha = 1$, $\beta = 5$, $\rho = 0.95$, $Q = 100$, $e = 10$. All runs were done with $m = 100$ ants that are split into $N \in \{1, 5, 10, 20\}$ colonies. The number m_{best} of ants that are allowed to update the pheromone matrix in a colony was varied with the size of the colony. For colonies of size 100 three ants update, for colonies of size 10 or 20 two ants update, and for colonies of size 5 ants only one ant updates. For strategies (3) and (4) m_{best} migrants where used. Every run was stopped after 500 generations. All given results are averaged over 20 runs.

Some test runs were performed to show the influence of the information exchange strategy on the differences between the pheromone matrices of the colonies. Let $T^{(k)}$ be the pheromone matrix of the kth colony. Then, the average pheromone matrix M is defined by

$$M_{ij} = \frac{1}{N} \sum_{k=1}^{N} T_{ij}^{(k)}$$

To measure the average difference between the pheromone matrices we took the average σ of the variances between the single elements of the matrices, i.e.,

$$\sigma = \frac{1}{n^2} \sum_{i=1}^{n} \sum_{j=1}^{n} S_{ij} \quad \text{where} \quad S_{ij} = \sqrt{\frac{1}{N-1} \sum_{k=1}^{N} (T_{ij}^{(k)} - M_{ij}^{(k)})}$$

The test runs were done with $N = 10$ colonies of 10 ants each. Figure 1 shows the results when an information exchange is done after every $I = 10$, respectively $I = 50$ generations. The figure shows that during the first generations the pheromone matrices in the colonies evolve into different directions which results in a growing average difference σ. One extreme is method (1) which grows up to a maximum at $\sigma = 0.9$ (0.23) after 10 generations for $I = 10$ (respectively 50 generations for $I = 50$). An exchange of the global best solution has the effect that the average difference becomes smaller very fast before it starts to grow slowly until the next information exchange occurs. After about 125 (275) generations σ is smaller than 0.01 in the case $I = 10$ (respectively $I = 50$). The other extreme is method (3) where the exchange of migrants has only a small effect on σ. It increases up to about 0.25 and then degrades very slowly to 0.24 in generation 500 for $I = 10$. The curve for $I = 50$ is nearly the same with only slightly larger σ. The curve of method (2) lies between the curves of methods (1) and (3) but there is a big difference between the cases $I = 10$ and $I = 50$. For $I = 50$ the difference σ goes up to 0.24 and then drops slowly down to 0.12 in generation 500. For $I = 10$ the difference σ goes only up to 0.17 and falls below 0.01 at generation 350. Method (4) behaves similarly to method (2) which means the circular exchange of the locally best solution is the dominating factor.

In order to show how strict the ant algorithm converges to one solution the average number of alternatives was measured that an ant has for the choice of the next city. We are interested only in cities that have at least some minimal chance to be chosen next. For the kth ant that is placed on city i let $D^{(k)}(i) = |\{j \mid p_{ij} > \lambda, j \in [1:n]$ was not visited$\}|$ be the number of possible successors that have a probability $> \lambda$ to be chosen. Then, the average number of alternatives with probability $> \lambda$ during a generation is

$$D = \frac{1}{mn} \sum_{k=1}^{m} \sum_{i=1}^{n} D^{(k)}(i)$$

Clearly, $D \geq 1$ always holds for $\lambda < 1/(n-1)$. Note that a similar measure — the λ-branching factor — was used in [7] to measure dimension of the search space. In contrast to D the λ-branching factor considers all other cities as possible successors not only those cities that have not yet been visited by the ant. Hence, D takes into account only the alternatives that the ants really meet, whereas the λ-branching factor is a more abstract measure and problem dependent.

Figure 2 shows the influence of the information exchange strategies on D for $\lambda = 0.01$ when information exchange is done every $I = 10$, respectively $I = 50$ generations. The figure shows that after every information exchange step the D value becomes larger. But in all cases it falls down below 2 before the 80th generation. After generation 150 the D values for method (1) are always lower than for the other methods. They are below 1.1 after generation 270 for $I = 10$, respectively after generation 290 for $I = 50$. It is interesting that during the first 100-150 generations D falls fastest for method (3) but in generation 500 the D value of 1.08 in case $I = 10$ is the largest. The D values of methods (2) and (4) with circular information exchange of local best solutions are quite similar. They are always larger than those of method (1). Compared to method (3) they are smaller after generation 300 in case $I = 10$ but are always larger in case $I = 50$.

Table 1 shows the lengths of the best found tours after 500 generations with methods (1)-(4) and for the case that no information exchange takes place when $I = 50$. In the case of no information exchange it is better to have one large colony than several smaller ones (see also Figure 3). It was observed that the length of the solution found by one colony does not change any more after generation 250. For methods (1) and (3) there is no advantage to have several colonies over just one colony. It seems that the exchange of only a few migrants in method (1) is so weak that the colonies can not really profit from it. It should be noted that the picture changes when information exchange is done more often. E.g., for $I = 10$ we found that 5 colonies are better than one (best found solution was 638.65 in this case).

Table 1. Different strategies of information exchange: best found solution after 500 generations, $I = 50$

	No information exchange	Exchange of globally best solution	Circular exch. of locally best solutions	Circular exchange of migrants
N=1	640.15	—	—	—
N=5	642.85	640.70	637.10	643.15
N=10	642.85	641.65	637.10	642.75
N=20	648.00	642.90	640.45	645.45

Methods (2) and (4) where local best solutions are exchanged between neighbouring colonies in the ring perform well. Figure 3 shows that the solutions found with method (2) by 5 or 10 colonies are always better than those of one colony after 250 generations, respectively 350 generations, for $I = 50$. In generation 500 the length of the best solution found by 10 colonies is about the same as that found by 5 colonies. Moreover, the curves show that there is still potential for further improvement after 500 generations for the 10 colonies and the 20 colonies. The curves for method (4) are quite similar to those in Figure 3 and are omitted. Table 2 shows the behaviour of method (2) when the information exchange is done more often, i.e., every 10 or 5 generations. For an exchange after every 5 generations the solution quality found in generation 500 is not or only slightly better for the multi colonies compared to the case with one colony. It seems that in this case the information exchange is too much in the sense that the colonies can not evolve into different directions.

Table 2. Circular exchange of locally best solutions

	I=5	I=10	I=50
N=5	642.30	638.90	637.10
N=10	642.90	638.55	637.10
N=20	639.35	638.20	640.45

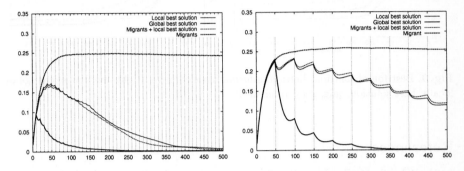

Fig. 1. Difference σ between matrices , Left: migration interval 10, Right: migration interval 50

6   Conclusion

Different methods for information exchange in multi colony ant algorithms were studied. Clearly, ant algorithms with several colonies that exchange not too much information can effectively be parallelized. It was shown that even the solution quality can improve when the colonies exchange not too much information. Instead of exchanging the local best solution very often and between all colonies it is better to exchange the local best solution only with the neighbour in a directed ring and not too often.

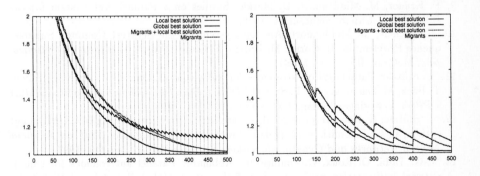

Fig. 2. Average number of alternatives D, Left: migration interval 10, Right: migration interval 50

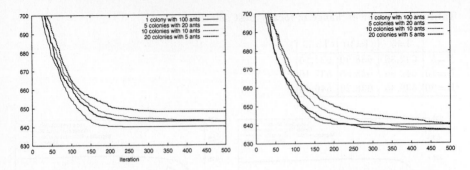

Fig. 3. Best found solution, Left: no information exchange, Right: circular exchange of locally best solution, migration interval 50

References

1. M. Bolondi, M. Bondaza: Parallelizzazione di un algoritmo per la risoluzione del problema del comesso viaggiatore; Master's thesis, Politecnico di Milano, 1993.
2. B. Bullnheimer, R.F. Hartl, C. Strauss: A New Rank Based Version of the Ant System - A Computational Study; CEJOR, Vol 7, 25-38, 1999.
3. B. Bullnheimer, G. Kotsis, C. Strauss: Parallelization Strategies for the Ant System; in: R. De Leone et al. (Eds.), *High Performance Algorithms and Software in Non-linear Optimization*; series: Applied Optimization, Vol. 24, Kluwer, 87-100, 1998.
4. M. Dorigo: Optimization, Learning and Natural Algorithms (in Italian). PhD thesis, Dipartimento di Elettronica, Politecnico di Milano, 1992.
5. M. Dorigo: Parallel ant system: An experimental study; Unpub. manuscript, 1993.
6. M. Dorigo, V. Maniezzo, A. Colorni: The Ant System: Optimization by a Colony of Cooperating Agents; *IEEE Trans. Sys., Man, Cybernetics – B*, 26, 29-41, 1996.
7. L. M. Gambardella, M. Dorigo: Ant-Q: A Reinforcement Learning approach to the traveling salesman problem; *Proceedings of ML-95, Twelfth Intern. Conf. on Machine Learning*, Morgan Kaufmann, 252-260, 1995.
8. U. Kohlmorgen, H. Schmeck, K. Haase: Experiences with fine-grained parallel genetic algorithms; *Ann. Oper. Res.*, 90, 203-219, 1999.
9. F. Krüger, M. Middendorf, D. Merkle: Studies on a Parallel Ant System for the BSP Model; Unpub. manuscript, 1998.
10. R. Michels, M. Middendorf: An Ant System for the Shortest Common Supersequence Problem; in: D. Corne, M. Dorigo, F. Glover (Eds.), *New Ideas in Optimization*, McGraw-Hill, 1999, 51–61.
11. T. Stützle: Parallelization strategies for ant colony optimization; in: A. E. Eiben, T. Bäck, M. Schonauer, H.-P. Schwefel (Eds.), *Parallel Problem Solving from Nature - PPSN V*, Springer-Verlag, LNCS 1498, 722-731, 1998.
12. T. Stützle, H. Hoos: Improvements on the ant system: Introducing MAX(MIN) ant system; in G. D. Smith et al. (Eds.), *Proc. of the International Conf. on Artificial Neutral Networks and Genetic Algorithms*, Springer-Verlag, 245-249, 1997.
13. E-G. Talbi, O. Roux, C. Fonlupt, D. Robillard: Parallel ant colonies for combinatorial optimization problems; in J. Rolim et al. (Eds.) *Parallel and Distributed Processing, 11 IPPS/SPDP'99 Workshops*, LNCS 1586, Springer, 239-247, 1999.
14. http://www.iwr.uni-heidelberg.de/iwr/comopt/soft/TSPLIB/TSPLIB.html

A Surface-Based DNA Algorithm for the Expansion of Symbolic Determinants

Z. FRANK QIU and MI LU

Department of Electrical Engineering
Texas A&M University
College Station, Texas 77843-3128, U.S.A.
{zhiquan, mlu}@ee.tamu.edu

Abstract. In the past few years since Adleman's pioneering work on solving the HPP(Hamiltonian Path Problem) with a DNA-based computer [1], many algorithms have been designed on solving NP problems. Most of them are in the solution bases and need some error correction or tolerance technique in order to get good and correct results [3] [7] [9] [11] [21] [22]. The advantage of surface-based DNA computing technique, with very low error rate, has been shown many times [12] [18] [17] [20] over the solution based DNA computing, but this technique has not been widely used in the DNA computer algorithms design. This is mainly due to the restriction of the surface-based technique comparing with those methods using the DNA strands in solutions. In this paper, we introduce a surface-based DNA computing algorithm for solving a hard computation problem: expansion of symbolic determinants given their patterns of zero entries. This problem is well-known for its exponential difficulty. It is even more difficult than evaluating determinants whose entries are merely numerical [15]. We will show how this problem can be solved with the low error rate surface-based DNA computer using our naive algorithm.

1 Introduction

Although there are a flood of ideas about using DNA computers to solve difficult computing problems [2] [16] [19] [15] since Adleman [1] and Lipton [16] presented their ideas, most of them are using DNA strands in solution. They all take advantage of the massive parallelism available in DNA computers as one liter of water can hold 10^{22} bases of DNA strands. Because they all let DNA strands float in solution, it is difficult to handle samples and strands may get lost during some bio-operations.

A well developed method, in which the DNA strands are immobilized on a surface before any other operations, is introduced to DNA computing area by Liu [18]. This method, which is called surface-based DNA computing, first attaches a set of oligos to a surface (glass, silicon, gold, etc). They are then subjected to operations such as hybridization from solution or exonuclease degradation, in order to extract the desired solution. This method greatly reduces losses

J. Rolim et al. (Eds.): IPDPS 2000 Workshops, LNCS 1800, pp. 653-659, 2000.

of DNA molecules during purification steps [18]. The surface-based chemistries have become the standard for complex chemical syntheses and many other chemistries.

Although the surface-based DNA computer has been demonstrated as more reliable with low error rate and easier to handle [8] [12] [18] [20], only a little research work about utilizing these properties of this kind of computer has been presented [12]. This happens mainly because when the oligos are attached to a surface, we lose flexibility due to the restriction that the oligos can not grow in the direction of the attachment on the surface. In order to take advantage of the new mature method, algorithms of surface-based computing need to be developed.

In this paper, we present a new algorithm to be implemented on a surface-based DNA computer that will take fully advantage of these special properties of low error rate. We will use the expanding symbolic determinants problem as an example to show the advantage of our algorithm comparing with an existing algorithm based on general DNA computer in solution. Both algorithms will be able to solve some intractable problems that are unrealistic to be solved by current conventional electronic computers because of the intense computing power requirement. These problems are harder to solve than the problem in NP-Complete. Our algorithm has all the advantages of surface-based computers over an existing algorithm introduced in [15].

The rest of the paper are organized as follows: the next section will explain the methodology, including the logical and biological operations of surface-based DNA computers. The problem of expansion of symbolic determinants and our algorithm to solve it will be presented in section 3. Section 4 will analyze our new surface-based algorithm and the last section will conclude this paper.

2   Surface-Based Operations

In this section, we show the logical operations of DNA computers and then explain how these operations can be implemented on surface-based DNA computers. All these operations are necessary for solving the computational hard problem given in the next section.

A simple version of surface-based DNA computer uses three basic operations, mark, unmark, and destroy [17] plus the initialization and append operations introduced in [8]. The explanation of these operations are clearly shown as follows.

2.1   Abstract Model

1. **reset(S)**: It can also be called initialization. This step will generate all the strands for the following operations. These strands in set **S** can be generated to represent either the same value or different values according to the requirement.
2. **mark(C, S)**: All strands in set **S** satisfying the constraint **C** are identified as marked. A strand satisfies this constraint if and only if there is a number

represented by a strand with bit i agrees with the bit value specified in the constraint. If no constraint is given, all strands are marked [8].

3. **unmark()**: Unmark all the marked strands.

4. **delete(C)**: All strands satisfying condition **C** are removed from set **S** where $C \in \{marked, unmarked\}$.

5. **append(C, X)**: A word **X** represented by a strand segment is appended to all strands satisfying constraint **C**. **C** can be defined as marked or unmarked. If the constraint is marked strands, a word **X** is appended to all marked strands. Otherwise, a word **X** will be appended to all unmarked strands.

6. **readout(C, S)**: This operation will select an element in **S** following criteria **C**. If no C is given, then an element is selected randomly. We will use this step to obtain the expected answer.

2.2 Biological Implementation

In this section, we include the fundamental biological operations for our surface-based DNA computation model.

1. **reset(S)**: The initialization operation used here is different from those widely used biological DNA operations described in [1] [2] [4] [10] [19]. All the strands generated are attached to a surface instead of floating in the solution. In order to prepare all these necessary strands on the surface, both the surface and one end of the oligonucleotides are specially prepared to enable this attachment. A good attachment chemistry is necessary to ensure that the properly prepared oligonucleotides can be immobilized to the surface at a high density and unwanted binding will not happen on the surface [8] [18] [17].

2. **mark(C, S)**: Strands are marked simply by making them double-strands at the free end as all the strands on the surface are single strands at the beginning. These single strands being added in to the container will anneal with the strand segments that need to be marked. Partial double strands will be formed according to the Watson-Crick(WC) complement rule [1] [16] [6].

3. **unmark()**: This biological operation can be implemented using the method introduced in [8]. Simply washing the surface in distilled water and raising the temperature if necessary will obtain the resultant container with only single strands attaching to the surface. Because with the absence of salt which stabilizes the double strand bond, the complementary strands will denature from the oligonucleotides on the surface and will be washed away.

4. **delete(C)**: This operation can be achieved using some enzymes known as exonucleases which chew up DNA molecules from the end. Detail of this operation is introduced in [8]. Exonucleases exist with specificity for either the single or double stranded form. By picking different enzymes, marked (double strands) or unmarked (single strands) can be destroyed selectively.

5. **append(C, X)**: Different operations are used depending on whether marked or unmarked strands are going to be appended. If X is going to be appended

to all marked strands, the following bio-operations will be used for appending. Since marked strands are double stranded at the free terminus, the append operation can be implemented using the ligation at the free terminus. The method introduced in [8] can be used here. More details may be found in [8]. To append to unmarked strands, simple hybridization of a splint oligonucleotide followed by ligation as explained in [1] [16] may be used.

6. **readout(C, S)**: This procedure will actually extract out the strand we are looking for. There are many existing methods developed for solution based DNA computing readout [1] [6] [20]. In order to use these methods, we have to detach the strands from the surface first. Some enzymes can recognize short sequences of bases called restriction sites and cut the strand at that site when the sequence is double-stranded [8]. When the segment which is attaching to the surface contains this particular sequence, they can all be detached from the surface when the enzyme is added in.

3   Hard Computation Problem Solving

3.1   Expansion of Symbolic Determinants Problem

We will use the expansion of symbolic determinants problem as an example to show how our surface-based DNA computer can be used to solve hard problems that are unsolvable by currently electronic computers.

Problem: Assuming the matrix is n×n:

$$
\begin{array}{cccccc}
a_{11} & a_{12} & a_{13} & \cdot & \cdot & \cdot & a_{1n} \\
a_{21} & a_{22} & & & & \\
a_{31} & & & & & \cdot \\
& \cdot & & & & \\
& \cdot & & & \cdot & \\
& \cdot & & & & \cdot & \cdot \\
a_{n1} & \cdot & \cdot & & \cdot & a_{nn}
\end{array}
$$

Generally, the determinant of a matrix is:

$$det(A) = \sum_{\sigma \in S_n} (-1)^\sigma A_{i_{\sigma_1} 1} \cdots a_{i_{\sigma_n} n} \tag{1}$$

where $S_n = (\sigma_1, \ldots, \sigma_n)$ is a permutation space [13] [5] [14]. A complete matrix expansion has n! items. When there are many zero entries inside, the expansion will be greatly simplified. We are going to solve this kind of problem–to obtain the expansion of matrices with many zero entries in them.

3.2   Surface-Based Algorithm

In order to make the process easy, we encode each item in the matrix a_{ij} by two parts: $(a_{ij})\_L$ and $(a_{ij})\_R$ while all the $(a_{kj})\_L's$ are with the same k but different j and all the $(a_{ik})\_R's$ are with the same k but different i. Using this coding method, all items from the same row will have the same left half code, and all the items from the same column will have the same right code. It seems like that we construct a_{ij} by combining a_i and a_j. So, for example, a_{13} and a_{19} will be represented by the same left half segment but different right halves because they are in the same row but different columns. For another example, a_{14} and a_{84} will have the same right half but different left halves because they are in the same column but different rows. The following is an algorithm using the methodology of the previous section. It can be accomplished as follows:

a-1 **reset(S)**: A large amount of strands will be generated on the surface. All the strands are empty initially, they only have the basic header to be annealed to the surface.

a-2 **append(X, S)**: This will make the strands on the surface grow with X. The X here is $a_{ij} \neq 0$ while i is initially set as one and $j \in (1:n)$. All the strands will grow by one unit and each will contain one item in the first row. After the append operation finishes, wash the surface to get rid of all unnecessary strand segment remained on the surface.

a-3 Repeat the above steps a-2 with i incremented by one until i reaches n. Now we have each strand should represent n units while each unit is an item from one row. So, each strand should have n items from n different rows.

a-4 **mark(X, S)**: We mark all strands containing X and X is initially set as a_i, the code for left half of each item representing the row number, with $i = 0$.

a-5 **delete(UM)**: Destroy all strands that are unmarked. This will eliminate those strands containing less than n rows because no matter what i is, it represents a row and every strand should contain it.

a-6 Repeat the above steps a-4 and a-5 n times with different i's while $i \in (1:n)$. This will guarantee that one item from each row is contained in each strand.

a-7 Repeat the above steps a-4 and a-5 and a-6 with different a_j's, the codes for the right half of each item representing the column number, while $j \in (1:n)$. This is used to keep only those strands that have items from each column and eliminate those that do not satisfy.

a-8 **readout(S)**: Readout all the remaining strands on the surface and they will be the answer for the expansion of our symbolic determinant. Each strand will contain one item from each row and one item from each column.

4   Analysis of the Algorithm

The complexity of this new algorithm is O(n) where n is the size of the matrix. In order to show the advantage of our surface-based DNA computer, we need to analysis the traditional method for expanding the symbolic determinants. The computing complexity of the traditional method is O(n!). Compare with the traditional method, we have solved a problem harder than NP within linear steps. The advantage of using DNA computer to solve the expansion of symbolic determinants problem is huge. Because the surface-based DNA technology is used, the DNA computer will be more reliable with low error-rate.

5   Conclusion

In this paper, we have proposed an algorithm to solve the expansion of symbolic determinants using surface-based model of DNA computer. Compare with other given applications of DNA computers, our problem is a more computation intensive one and our surface-based DNA computer will also reduce the possible errors due to the loss of DNA strands.

Further research includes expanding the application of surface-based DNA computing in order to make DNA computers more robust. With the goal of even lower error rate, we may combine the existing error-resistant methods [3] [7] [9] [11] [21] [22] and the surface-based technology to achieve better results.

References

1. Len Adleman. Molecular computation of solutions to combinatorial problems. *Science*, November 1994.
2. Martyn Amos. *DNA Computation*. PhD thesis, University of Warwick, UK, September 1997.
3. Martyn Amos, Alan Gibbons, and David Hodgson. Error-resistant implementation of DNA computations. In *Second Annual Meeting on DNA Based Computers*, pages 87–101, June 1996.
4. Eric B. Baum. DNA sequences useful for computation. In *Second Annual Meeting on DNA Based Computers*, pages 122–127, June 1996.
5. Fraleigh Beauregard. *Linear Algebra 3rd Edition*. Addison-Wesley Publishing Company, 1995.
6. D. Beaver. Molecular computing. Technical report, Penn State University Technical Report CSE-95-001, 1995.
7. Dan Boneh, Christopher Dunworth, Jeri Sgall, and Richard J. Lipton. Making DNA computers error resistant. In *Second Annual Meeting on DNA Based Computers*, pages 102–110, June 1996.
8. Weiping Cai, Anne E. Condon, Robert M. Corn, Elton Glaser, Tony Frutos Zhengdong Fei, Zhen Guo, Max G. Lagally, Qinghua Liu, Lloyd M. Smith, and Andrew Thiel. The power of surface-based DNA computation. In *RECOMB'97. Proceedings of the first annual international conference on Computational modecular biology*, pages 67–74, 1997.
9. Junghuei Chen and David Wood. A new DNA separation technique with low error rate. In *3rd DIMACS Workshop on DNA Based Computers*, pages 43–58, June 1997.
10. R. Deaton, R. C. Murphy, M. Garzon, D. R. Franceschetti, and Jr. S. E. Stevens. Good encodings for DNA-based solutions to combinatorial problems. In *Second Annual Meeting on DNA Based Computers*, pages 131–140, June 1996.
11. Myron Deputat, George Hajduczok, and Erich Schmitt. On error-correcting structures derived from DNA. In *3rd DIMACS Workshop on DNA Based Computers*, pages 223–229, June 1997.
12. Tony L. Eng and Benjamin M. Serridge. A surface-based DNA algorithm for minimal set cover. In *3rd DIMACS Workshop on DNA Based Computers*, pages 74–82, June 1997.

13. Paul A. Fuhrmann. *A Polynomial Approach To Linear Algebra*. Springer, 1996.

14. Klaus Jänich. *Linear Algebra*. Springer-Verlag, 1994.

15. Thomas H. Leete, Matthew D. Schwartz, Robert M. Williams, David H. Wood, Jerome S. Salem, and Harvey Rubin. Massively parallel dna computation: Expansion of symbolic determinants. In *Second Annual Meeting on DNA Based Computers*, pages 49–66, June 1996.

16. Richard Lipton. Using DNA to solve SAT. Unpulished Draft, 1995.

17. Qinghua Liu, Anthony Frutos, Liman Wang, Andrew Thiel, Susan Gillmor, Todd Strother, Anne Condon, Robert Corn, Max Lagally, and Lloyd Smith. Progress towards demonstration of a surface based DNA computation: A one word approach to solve a model satisfiability problem. In *Fourth Internation Meeting on DNA Based Computers*, pages 15–26, June 1998.

18. Qinghua Liu, Zhen Guo, Anne E. Condon, Robert M. Corn, Max G. Lagally, and Lloyd M. Smith. A surface-based approach to DNA computation. In *Second Annual Meeting on DNA Based Computers*, pages 206–216, June 1996.

19. Z. Frank Qiu and Mi Lu. Arithmetic and logic operations for DNA computer. In *Parallel and Distributed Computing and Networks (PDCN'98)*, pages 481–486. IASTED, December 1998.

20. Liman Wang, Qinghua Liu, Anthony Frutos, Susan Gillmor, Andrew Thiel, Todd Strother, Anne, Condon, Robert Corn, Max Lagally, and Lloyd Smith. Surface-based DNA computing operations: Destroy and readout. In *Fourth Internation Meeting on DNA Based Computers*, pages 247–248, June 1998.

21. David Harlan Wood. applying error correcting codes to DNA computing. In *Fourth Internation Meeting on DNA Based Computers*, pages 109–110, June 1998.

22. Tatsuo Yoshinobu, Yohei Aoi, Katsuyuki Tanizawa, and Hiroshi Iwasaki. Ligation errors in DNA computing. In *Fourth Internation Meeting on DNA Based Computers*, pages 245–246, June 1998.

Hardware Support for
Simulated Annealing and Tabu Search

Reinhard Schneider and Reinhold Weiss

[schneider | weiss]@iti.tu-graz.ac.at
Institute for Technical Informatics
Technical University of Graz, AUSTRIA

Abstract. In this paper, we present a concept of a CPU kernel with hardware support for local-search based optimization algorithms like Simulated Annealing (SA) and Tabu-Search (TS). The special hardware modules are: (i) A linked-list memory representing the problem space. (ii) CPU instruction set extensions supporting fast moves within the neighborhood of a solution. (iii) Support for the generation of moves for both algorithms, SA and TS. (iv) A solution mover managing several solution memories according to the optimization progress. (v) Hardware addressing support for the calculation of cost functions. (vi) Support for nonlinear functions in the acceptance procedure of SA. (vii) A status module providing on-line information about the solution quality. (v) An acceptance prediction module supporting parallel SA algorithms.
Simulations of a VHDL implementation show a speedup of up to 260 in comparison to an existing implementation without hardware support.

1   Introduction

Simulated Annealing (SA)[1] and Tabu-Search (TS)[2][3] are algorithms that are well suited to solving general combinatorial optimization problems which are common in the area of real-time multiprocessor systems. Tindell et al.[4] solved a standard real-time mapping task with several requirements using SA. Axelsson[5] applied SA, TS and genetic algorithms, all three based on the local search concept[6], to the problem of HW/SW Codesign. In [7] the authors introduced a complete tool for handling parallel digital signal processing systems based on parallel SA.

All research projects mentioned use, like many others, SA, TS or other algorithms based on local search to find solutions for partitioning, mapping and scheduling problems in parallel systems. The results show that these algorithms are able to solve even difficult problems with a good solutions quality. The main drawback is the slow optimization speed. This is particularly true for SA. Many researchers have tried to reduce execution time in different ways. One way is to optimize the algorithm itself, which depends strongly on the application and has a limited possible speedup[8]. Another approach is to parallelize SA[9]. With parallel simulated annealing (PSA) it is possible to achieve greater speedup, independent of the problem, without compromising solution quality[10]. PSA is

J. Rolim et al. (Eds.): IPDPS 2000 Workshops, LNCS 1800, pp. 660-667, 2000.
© Springer-Verlag Berlin Heidelberg 2000

already successfully applied to multiprocessor scheduling and mapping[11]. But even with PSA on up-to-date processor hardware, it takes a very long time to compute a multiprocessor schedule for realistic system complexity. This still prevents the on-line use of SA in dynamic systems, and it is also the main reason why our research focuses on supporting SA and TS by dedicated processor hardware. It is evident that a processor supporting local search also simplifies non real-time applications using SA and TS.

Abramson[12] showed that with a custom computing machine (CCM) it is possible to outperform software implementation by several orders of magnitude. Other hardware- implementations ([13][14]) also showed a significant speedup in comparison with a software-implementation. CCMs are very efficient for the problem they are designed for. Unfortunately they can not solve other problems. Even a small change in the characteristic of the problem or an unexpected increase of the problem size means that the CCM itself has to be re-designed. Eschermann et al.[15] tried to build a more flexible processor for SA where fewer parts of the algorithm are implemented in hardware so that different problems can be solved. Unfortunately, this processor has not been developed any further. Up to now, there is no processor available that explicitly supports local search or other nature-inspired algorithms.

Our solution combines the flexibility of a programmable CPU with the speed of dedicated hardware in a flexible, modular concept.

2   Local Search

Optimization algorithms based on local search (LS) have in common that they start from any arbitrary solution and try to find better solutions by stepping through a *neighborhood* of solutions. The neighborhood of a solution i is defined by the neighborhood function $N(i)$. A real cost value c can be mapped to each solution i by a *cost function* $c(i)$. The problem is to find a globally optimal solution $i*$, such that $c* = c(i*) \leq c(i)$ for all solutions i.

Iterative Improvement. A basic version of LS is *iterative improvement*. With this technique, the neighborhood $N(i)$ is searched starting from the current solution i. Then, either the first better solution (*first improvement*) or the solution with the lowest costs within the neighborhood (*best improvement*) is chosen as the new solution. Improved techniques like SA or TS use different strategies to overcome the problem of getting caught in a local minimum. But all these techniques are based on the same few basic functions. The following pseudo-code describes the basic structure of a local-search based algorithm:

```
i=Generate-Initial-Solution
REPEAT
      Move=Select-a-Move-within-Neighborhood-N(i)        (1)
      i'=Apply-Move(i, Move)                              (2)
      dC=Compute-Change-in-Cost(i, i')                    (3)
      IF accept THEN i=i'                                 (4)
UNTIL Stopping-Condition-is-true
```

Key functions of this algorithm are: (1) The selection of a *move*, which means the selection of a transition from one solution to another. (2) Performing the move to obtain the new solution. (3) Computing the difference in costs, and (4) deciding whether to accept the new solution or not. The definition of the neighborhood and the way of computing the costs depend on the problem that has to be solved. The way of selecting a solution from the neighborhood and the criteria for accepting a new state depend on the algorithm:

Simulated Annealing. The selection of a move in SA is based on a stochastic process. This means that one move is chosen at random out of all possible moves within the neighborhood. Therefore, the quality of the pseudo random number generator is important in order not to omit any solution. In SA, a move which leads to an improvement of cost is always accepted, deteriorations of costs are accepted if they fulfill the *Metropolis Criterion* - in analogy to the annealing procedure of metals.

Tabu Search. TS always searches the whole neighborhood. The best solution within the neighborhood is taken as a new solution. In order to avoid getting trapped in a local minimum, TS works with the search history: Solutions that have already been selected some time before are forbidden (taboo). These solutions or the moves that lead to these solutions, respectively, are stored in a *tabu list*. Solutions in the tabu list may still be accepted if they are extraordinary (e.g., if they are significantly better than all other solutions in the neighborhood). These solutions are also stored in a list called the *aspiration list*.

3   Hardware Support

The analysis of possible hardware support for LS-based algorithms was governed by the following objectives:

- The flexibility of the final system should be maximized.
- Hardware support should be modular so that more than one optimization algorithm can be accelerated by the same hardware.
- As many parts of the algorithms as possible should be realized in hardware.
- The final processor should support parallelization.
- The employment in real-time systems should be supported.

The first goal was achieved by designing the hardware support as a CPU kernel extension. Thus, the flexibility of a fully programmable CPU remained. Additionally, data transfer time drops out in this concept because the hardware modules directly interact with the CPU, the bus system and the main memory.

The second goal was achieved by a strict modular design. Different optimization algorithms could be supported by different combinations of the modules. This concept also satisfied the third objective, namely the realization of special, algorithm-dependent functions as modules and their integration in the system.

Parallelization techniques were analyzed only for SA. An acceptance prediction module was introduced which efficiently supports the *decision tree decomposition* algorithm[16], where the processors work in a pipelined, overlapped mode.

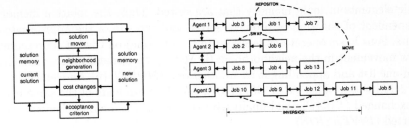

Fig. 1. Left: *Structure of a typical local search based algorithm.* **Right:** *Linked-list representation of the generalized assignment problem with fundamental instructions to generate a neighbor solution.*

In real-time systems, it is most important to *know* about the status of the optimization process. As all optimization algorithms *approach* the optimal solution, knowledge about the quality of the solution reached so far must exist in order to make a decision at a crucial time. Therefore, a statistical module was designed that continuously checks the current status of the optimization process.

Modular View. A modular view of local search based algorithms is depicted in Figure 1, left side. The basic modules are: (i) Two solution memories, (ii) a neighborhood generator, (iii) a solution mover, (iv) support for the calculation of the cost function, and (v), support for the calculation of the acceptance criterion. Modules (i)-(iii) are closely coupled, as they work on the same data structure. Module (iv) strongly depends on the problem, module (v) depends on the algorithm used. Still, there are possibilities to build hardware support for these two modules. Additional modules (an acceptance prediction module and a status module) are implemented to support parallelization and real-time systems.

Solution Memory. A fast solution memory is fundamental as the movement within the solution space is the most frequent operation in local search. In order to speed up moves, it is necessary to find an appropriate problem representation in memory. Generally, a combinatorial optimization can be described by (i) elements of different type and (ii) relations between them. E.g., the generalized assignment problem (GAP) could look as depicted in Figure 1, right side. The problem of mapping tasks to processors is a special case of the GAP problem, where the jobs represent tasks and the agents represent the processors. In this model, the elements and their content form the static description of the problem. All possible combinations of relations represent the solution space. One set of relations represents a special solution. Moving in the solution space means to change relations. Based on this definition, it is easy to define a neighborhood: A solution $i*$ is defined to be within the neighborhood $N(i)$ if the differences in relations is small (e.g. one different relation).

Linked lists[17] and matrices are efficient ways of representing relations between two types of elements. We decided to implement both a memory based on a linked-list representation, and a memory based on a matrix representation. In the special list/matrix memory (*solution memory*), it is only pointers to the static elements in main memory that are stored. Thus, the solution memory is independent of the kind of problem solved and the size of the appropriate elements. Four basic operations (see Figure 1, right side) on the list representation allow movement in the neighborhood: (i) Moving an element means removing it from one list and appending it to an other one *(MOVE)*. (ii) An element can be reordered within a list *(REPOSITION)*. (iii) Two elements of the same list can be exchanged *(SWAP)*. And finally, (iv) the order of a chain of elements can be inverted *(INVERSION)*.

Neighborhood Generation. The generation of the neighborhood depends on the algorithm used. In SA, a new solution is generated at random. Therefore, a set of hardware pseudo random number generators *(PRNG)* is proposed. One of them has to chose the move, another one has to select the source element (job), the third one has to chose the destination agent and/or position in the list according to the selected move. In TS, all possible moves within the neighborhood have to be searched. The neighborhood generator has to check if a selected move is forbidden *(tabu)* or not. This is done by comparing the move with the tabu list and the aspiration list. This search is accelerated by managing the lists in hardware .

Solution Mover. As the current solution and the new solution must be stored until the acceptance decision has been taken, the linked-list memory is duplicated. The result of the acceptance decision determines which memory has to be synchronized to the other. TS also needs to store the best solution reached within the neighborhood. Additionally, the best solution reached so far is stored. This is important for real-time systems, where the optimization has to stop after a fixed time, and the best solution so far should be available. The solution mover module applies the list operation suggested by the neighborhood generator and manages all solution memories and transactions between them.

Cost Function. The cost function strongly depends on the problem solved. A function completely realized in hardware decreases flexibility dramatically. Hence, we suggest to implement only *addressing support* for the cost function: Providing an easy way to access the elements that have been affected by the move, e.g., by a list of these elements, supports in particular cost functions that can be computed incrementally. As the order and size of this list depends on the problem, we suggest to provide a *user-programmable hardware module* (e.g., an FPGA-based module) which is tightly coupled to the linked-list memory. This allows the adaptation of the sequence of elements to any individual problem before the optimization process is started.

Acceptance Criterion. In TS, the best solution found so far is always accepted. Thus, no additional hardware is needed. In SA, moves with a cost improvement are always accepted. If costs rise, SA decides on the acceptance of the move by evaluating the *Metropolis criterion* $e^{\frac{E_i - E_j}{T}} > random(0,1)$. Negative cost differences $(E_i - E_j)$ are weighted by a control parameter T and transformed by non-linear operation. The move is then accepted if the result of $e^{\frac{E_i - E_j}{T}}$ (always between 0 and 1) is greater than a random number between 0 and 1. A hardware pseudo random number generator improves performance significantly, as the random number is provided without CPU interaction. Additionally, as the result of the exponential function is compared with a random number, no high accuracy is needed. Therefore, a hardware lookup table with pre-calculated values for each value of T is sufficient. By means of these tables, the evaluation of the exponential function is done in one cycle.

Status Information. The absolute value of the cost function can not be used as status information, because only its relation to the optimal solution is linked to the quality of the solution. But the optimal solution is not known to the system. Therefore, we use a statistical status information based on the relative cost changes.

Acceptance Prediction. The acceptance prediction module is used to support parallelization in SA. The output value corresponds to the probability of accepting a new move. With this value, a good prediction of the acceptance is available before the actual result of the acceptance criterion is available.

4   Implementation

All modules are implemented using VHDL and are synthesized for emulation on a programmable FPGA chip using Xilinx Foundation software and tools. All modules use parametric problem sizes to be easily adaptable to different systems.

Solution Memory. The module consists of four solution memories, realized as both linked list memories and matrix memories: the current solution, the new solution, the best solution found so far and the best solution in the neighborhood. The latter is used only in TS. Memory synchronization works very fast as all memories are arranged physically side by side and connected by a high speed internal bus.

Move Generator. Moves are generated in two ways: For SA, a set of pseudo random number (PRN) generators, based on cellular automata[18], automatically generates a move. These automata provide excellent PRNs every cycle with a maximum perod of 2^n. For TS, all possible moves have to be considered. These moves are generated sequentially. Each move has to be checked by a move checker . The move checker decides, with the help of the content of the tabu list and the aspiration list, if a move is accepted or not. The search within the lists is realized by parallel comparators.

Status Module. A good estimation of the current status of the optimization can be made by averaging the cost changes over the absolute costs. This only works for problems with a smooth cost function without singular minima, which is the case for mapping tasks in multiprocessor systems.

Acceptance Prediction. The acceptance prediction unit (for SA) uses an averaged cost value, the last cost differences and the last acceptance decision as input values. The output is a prediction value that indicates if the next new solution will be accepted or not. With the help of this value, the network topology of the parallelized processors can change dynamically.

5   Results

Timing results were obtained in two ways: Firstly by simulation with a VHDL simulator, and secondly by calculating the cycles needed per instruction. The time needed for one iteration strongly depends on the time to perform a move in the neighborhood (move generation, solution mover and acceptance decision) and the time needed to calculate the cost difference. The latter strongly depends on the problem and is therefore not discussed any further. The use of our hardware modules shortens the time for move generation and the acceptance decision to one cycle each. The solution mover is more critical. The time needed to perform a particular move depends on the type of memory (linked-list based or matrix-based) and the problem size, which is indicated by parameter n in Table 1.

Table 1. Timing requirements for the solution mover module.

instruction	cycles: matrix memory	cycles: list memory
swap	10	≤ 26
inversion	$n * 8 + 3$	$\leq n * 28 + 6$
remove	$n * 8 + 10$	22
reposition	$n * 8 + 2$	$\leq n * 26 + 2$

In order to assess our solution, a system was designed to solve the travelling salesman problem with SA. Simulations needed 13 cycles for one iteration. With an FPGA running at 13 MHz, the time for one instruction is $1\mu s$. A software implementation on a digital signal processor with a clock speed of 40 MHz needs $86\mu s$. The speedup of the hardware-supported solution is therefore 86 or, assuming that the hardware modules will run with the same speed if directly implemented in a CPU, the speedup will be over 260. The acceptance prediction module showed a hit rate of 90% when suspended for only 10% of the time.

6   Discussion

Nature-inspired algorithms are a fast growing field. New and improved algorithms are developed rapidly. But there is a lack of appropriate computer architectures to support these algorithms. The system described in this paper shows that with an extended CPU it is possible to speed up significantly local-search based algorithms. Even though an ASIC–prototype has to be realized first in order to verify the speedup, the simulation results are respectable. These modules are an attempt to show which functions could be supported by new, intelligent CPU cores. The costs of integrating these modules in a CPU core are small compared to the speedup they provide. The modular concept is very flexible and allows, e.g., support for parallelization. Based on this concept, a lot of new modules can be imagined: Support for other algorithms like genetic algorithm, neuronal networks, qualitative algorithms, etc. A CPU extended by such modules will probably make expensive special solutions dispensable.

References

[1] S Kirkpatrick, C D Gelatt, and M P Vecchi. Optimisation by simulated annealing. *Science*, 220:671–680, 1983.

[2] Fred Glover. Tabu search: 1. *ORSA Journal on Computing*, 1(3):190–206, 1989.

[3] Fred Glover. Tabu search: 2. *ORSA Journal on Computing*, 2(1):4–32, 1990.

[4] K. W. Tindell, A. Burns, and A. J. Wellings. Allocating Hard Real–Time Tasks: An NP-Hard Problem Made Easy. *The Journal of Real–Time Systems*, (4):145–165, 1992.

[5] Jakob Axelsson. Architecture Synthesis an Partitioning of Real-Time Systems: A Comparison of Three Heuristic Search Strategies. In *5th International Workshop on Hardware/Software Codesign*, pages 161–165, March, 24-26 1997.

[6] E. Aarts and K. Lenstra. *Local Search in Combinatorial Optimization*. Interscience Series in Discrete Mathematics and Optimization. John Wiley & Sons, 1997.

[7] Claudia Mathis, Martin Schmid and Reinhard Schneider. A Flexible Tool for Mapping and Scheduling Real-Time Applications on Parallel Systems. In *Proceedings of the Third International Conference on Parallel Processing and Applied Mathematics*, Kazimierz Dolny, Poland, September, 5-7 1999.

[8] E. H. L. Aarts and J. H. M Korst. *Simulated Annealing and Boltzmann Machines*. Interscience Series in Discrete Mathematics and Optimization. John Wiley & Sons, Chichester, U.K., 1989.

[9] Tarek M. Nabhan and Albert Y. Zomaya. Parallel simulated annealing algorithm with low communication overhead. *IEEE Transactions on Parallel and Distributed Systems*, 6(12):1226–1233, December 1995.

[10] Soo-Young Lee and Kyung Geun Lee. Synchronous and asynchronous parallel simulated annealing with multiple Markov chains:. *IEEE Transactions on Parallel and Distributed Systems*, 7(10):993–1008, October 1996.

[11] Martin Schmid and Reinhard Schneider. A Model for Scheduling and Mapping DSP Applications onto Multi-DSP Platforms. In *Proceedings of the International Conference on Signal Processing Applications and Technology*. Miller Freeman, 1999.

[12] David Abramson. A very high speed architecture for simulated annealing. *j-COMPUTER*, 25(5):27–36, May 1992.

[13] J. Niittylahti. Simulated Annealing Hardware Tool. In *The 2nd International Conference on Expert Systems for Development*, pages 187–191, 1994.

[14] Bang W. Lee and Bing J. Sheu. Paralleled hardware annealing for optimal solutions on electronic neural networks. *IEEE Transactions on Neural Networks*, 4(4):588–599, July 1993.

[15] B. Eschermann, O. Haberl, O. Bringmann, and O. Seitzr. COSIMA: A Self-Testable Simulated Annealing Processor for Universal Cost Functions. In *EuroASIC*, pages 374–377, Los Alamitos, CA, 1992. IEEE Computer Society Press.

[16] Daniel R. Greening. Parallel Simulated Annealing Techniques. In *In Emergent Computation*, pages 293–306. MIT Press, Cambridge, MA, 1991.

[17] A. Postula, D.A. Abramson, and P. Logothetis. A Tail of 2 by n Cities: Performing Combinatorial Optimization Using Linked Lists on Special Purpose Computers. In *The International Conference on Computational Intelligence and Multimedia Applications (ICCIMA)*, Feb, 9-11 1998.

[18] P.D. Hortensius, R.D. McLeod, and H.C. Card. "parallel random number generation for vlsi systems using cellular automata". *IEEE Transactions on Computers*, 38(10):1466–1473, October 1989.

Eighth International Workshop on Parallel and Distributed Real-Time Systems

held in conjunction with
International Parallel and Distributed Processing Symposium

May 1-2, 2000
Cancun, Mexico

General Chair

Kenji Toda, Electrotechnical Laboratory, Japan

Program Chairs

Sang Hyuk Son, University of Virginia, USA
Maarten Boasson, University of Amsterdam, The Netherlands
Yoshiaki Kakuda, Hiroshima City University, Japan

Publicity Chair

Amy Apon, University of Arkansas, USA

Steering Committee

David Andrews (**Chair**), University of Arkansas, USA
Dieter K. Hammer, Eindhoven University of Technology, The Netherlands
E. Douglas Jensen, MITRE Corporation, USA
Guenter Hommel, Technische Universitaet Berlin, Germany
Kinji Mori, Tokyo Institute of Technology, Japan
Viktor K. Prasanna, University of Southern California, USA
Behrooz A. Shirazi, The University of Texas at Arlington, USA
Lonnie R. Welch, Ohio University, USA

J. Rolim et al. (Eds.): IPDPS 2000 Workshops, LNCS 1800, pp. 668-670, 2000.
© Springer-Verlag Berlin Heidelberg 2000

Program Committee

Tarek Abdelzaher, University of Virginia, USA
Giorgio Buttazzo, University of Pavia, Italy
Max Geerling, Chess IT, Haarlem, The Netherlands
Jorgen Hansson, University of Skovde, Sweden
Kenji Ishida, Hiroshima City University, Japan
Michael B. Jones, Microsoft Research, USA
Tei-Wei Kuo, National Chung Cheng University, Taiwan
Insup Lee, University of Pennsylvania, USA
Victor Lee, City University of Hong Kong, Hong Kong
Jane Liu, University of Illinois, USA
Doug Locke, Lockheed Martin, USA
G. Manimaran, Iowa State University, USA
Tim Martin, Compaq Computer Corporation, USA
Sang Lyul Min, Seoul National University, Korea
Al Mok, UT Austin, USA
C. Siva Ram Murthy, IIT Madras, India
Hidenori Nakazato, OKI, Japan
Joseph Kee-Yin Ng, Hong Kong Baptist University, Hong Kong
Isabelle Puaut, INSA/IRISA, France
Ragunathan Rajkumar, Carnegie Mellon University, USA
Franklin Reynolds, Nokia Research Center, USA
Wilhelm Rossak, FSU Jena, Informatik, Germany
Shiro Sakata, NEC, Japan
Manas Saksena, University of Pittsburgh, USA
Lui Sha, University of Illinois, USA
Kang Shin, University of Michigan, USA
Hiroaki Takada, Toyohashi University of Technology, Japan
Nalini Venkatasubramanian, University of California at Irvine, USA
Wei Zhao, Texas A&M University, USA

Message from the Program Chairs

The Eighth International Workshop on Parallel and Distributed Real-Time Systems (WPDRTS'00) is a forum that covers recent advances in real-time systems – a field that is becoming an important area in the field of computer science and engineering. It brings together practitioners and researchers from academia, industry, and government, to explore the best current ideas on real-time systems, and to evaluate the maturity and directions of real-time system technology. As the demand for advanced functionalities and timely management of real-time systems continue to grow, our intellectual and engineering abilities are being challenged to come up with practical solutions to the problems faced in design and development of complex real-time systems. The workshop presents the papers that demonstrate recent advances in research pertaining to real-time systems. Topics addressed in WPDRTS'00 include:

Communication and Coordination
Real-Time and Fault-Tolerance
Real-Time Databases
Scheduling and Resource Management
QoS and Simulation

In addition to the regular paper presentation, the workshop also features a Keynote Speech, "Real-Time Application Specific Operating Systems: Towards a Componet Based Solution," by Jack Stankovic, University of Virginia, an invited papers session, and a panel discussion.

We would like to thank all who have helped to make WPDRTS'00 a success. In particular, the Program Committee members carefully reviewed the submitted papers. We also would like to thank the authors of all the submitted papers. The efforts of the Steering Committee chair and the Publicity chair are also greatly appreciated. Finally, we thank the IPDPS organizers for providing an ideal environment in Cancun.

Sang H. Son   **Maarten Boasson**   **Yoshiaki Kakuda**
son@virginia.edu   boasson@signaal.nl   kakuda@ce.hiroshima-cu.ac.jp

Program Chairs
8th International Workshop on Parallel and Distributed Real-Time Systems

A Distributed Real Time Coordination Protocol

Lui Sha[1] and Danbing Seto[2]
[1]CS, University of Illinois at Urbana-Champaign
[2]United Technology Research Center

Abstract: When the communication channels are subject to interruptions such as jamming, the coordination of the real time motions of distributed autonomous vehicles becomes a challenging problem, that differs significantly with fault tolerance communication problems such as reliable broadcast. In this paper, we investigate the issues on the maintenance of the coordination in spite of arbitrarily long interruptions to the communication.

1 Introduction

Internet based instrumentation and controls are an attractive avenue for the development and evolution of distributed real-time systems [1, 2]. However, one of the challenges is the real-time coordination problem in the presence of communication interruptions. In distributed control, coordination concerns with how to synchronize the states of distributed control subsystems in real-time. A prototypical problem is to command a group of unmanned air vehicles, where each vehicle must closely follow a desired trajectory which is planned in real-time.

To synchronize the states of distributed control systems, a reference trajectory[1] is given in real time to each distributed node, a local system. A reference setpoint[2] moves along the reference trajectory according to the specified speed profile. The reference trajectories are designed in such a way that the movements of the reference setpoints represent the synchronized changes of distributed states. The difference between the actual system state and the state represented by the reference setpoint is called tracking error. A tracking error bound specifies the acceptable tracking error on each reference trajectory. A local controller is designed to force the local system's state to follow the reference setpoint closely within the tracking error bound.

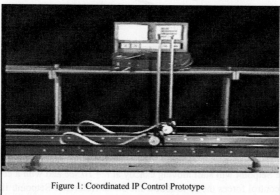

Figure 1: Coordinated IP Control Prototype

---

[1] A reference trajectory is a specification of how a system's state should change as a function of some independent variables. For example, the glide slope that guides aircraft landing specifies both the path and the speed along the path.

J. Rolim et al. (Eds.): IPDPS 2000 Workshops, LNCS 1800, pp. 671-677, 2000.
© Springer-Verlag Berlin Heidelberg 2000

This paper addresses the abstract problem of designing reliable communication protocols for distributed real-time coordination with a concrete example. We will use a simplified Inverted Pendulum (IP) control prototype to introduce the basic concepts in real-time coordination.

The coordinated IP control system has two IPs with a nearly massless plastic rod which ties the tips of the two IPs together as shown in Figure 1. The rod does not affect the inverted pendulum control until the slack is consumed by the difference in IPs' positions. Each IP consists of a metal rod mounted vertically to a hinge on a motorized cart controlled by a computer. The metal rod rotates freely. It will fall down from its upright position if the cart's movement is not properly controlled. The mission of the overall system is to get the two IPs moving in synchrony to a desired position on the track with the IPs standing upright. Apparently, if the two IPs are significantly out of steps with each, they can pull each other down. Therefore, the two carts must keep the pendulums at upright position, and maintain their positions synchronized within a small tolerance of, e.g., 5 cm, to prevent the plastic rod from falling. The tolerance is a function of how tightly the two tips are tied together.

In this experiment, each IP is controlled locally by a computing node on an Ethernet switch. An operator uses a Command Node on the network to send messages commanding the two IPs where to go. A "communication jamming station" is also connected to the same network, so that we can experimentally test the robustness of communication protocol designed for coordinated control.

Example 1: As illustrated in Figure 2, suppose that in a coordinated IP control experiment, the initial positions for the two IPs are at the middle of the two parallel tracks, i.e., $x_1 = 0$ cm, $x_2 = 0$ cm at time $t = 0$. We may command the IPs to move to positions near one end of the parallel tracks at 70 cm with a motion start time $t = 10$ sec, and with a constant speed of 2cm/sec. The system coordination is carried out by sending commands to the IPs. A command specifies both the start time and the reference trajectory. A reference trajectory specifies the path of the motion and

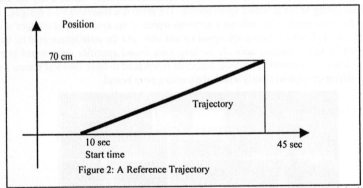

Figure 2: A Reference Trajectory

the speed along the specified path. In this example, paths for the IPs are two straight lines, each connecting track position 0 to track position 70 cm.

Suppose that both IPs receive their commands in time, that is, before the start time $t = 10$ sec. The Local Reference Setpoint will start moving exactly at $t = 10$ sec and with a constant speed of 2 cm/sec. The local control forces the IP to follow the Local Reference Setpoint. If both IPs' con-

---

[2] Giving a reference trajectory, the reference setpoint specifies where the controlled system's state ought to be along the reference trajectory. That is, it is a specification of the desired system state as a function of time. Feedback control is used to force the physical system's state to converge to the reference setpoint.

trols are functioning correctly, the tracking error between the IP position and the Local Reference Setpoint will tend to zero. In the experiment, the error between the two IPs positions is allowed to be as large as *5 cm*, which is called as the global coordination error bound. Typically, in distributed control systems, the global coordination error bound is translated into sufficient conditions that can be observed and controlled locally. For example, if each IP is within *±2.5 cm* of its Local Reference Setpoint, then the global coordination error bound is satisfied. This localized condition is referred to as the local tracking error bound.

In Example 1, both IPs start their motions at the same time. In practice, it is quite common to command different objects starting motions at different specified times.   However, distributed real-time coordination with synchronized start times is the key problem. The problem of using different start times can always be decomposed into two problems: 1) a coordination problem with synchronized start times and 2) a stand alone control problem. For example, suppose that initially one of the IPs, IP_1 is at *– 5cm* while IP_2 is at *0 cm*. We would like them to line up first and then move in synchrony. This problem can be decomposed into two problems: 1) command IP_1 to first move to $x_1 = 0$ *cm,* and 2) command them to move in synchrony as illustrated in Example 1. Obviously, the hard problem is coordinated control with synchronized start times. In the following, we will focus on problems that require synchronous start times.

We have so far assumed that both IPs receive their commands on time. This assumption is unrealistic in an open network. Obviously, if one IP receives its command on time, and the other receives its command much later, the IP that moves first will pull down the other IP. This is an example of coordinated control failure. The design of the real-time coordination communication protocol concerns with the problem of how to send the trajectories to distributed nodes quickly and reliably. That is, in spite of arbitrary long interruptions to any or all of the communication channels, the protocol must guarantee that distributed nodes will never receive a set of inconsistent commands that will lead to coordination failure.

This problem is related to the synchronization of distributed objects [3] in the sense that the states of distributed objects cannot be diverged arbitrarily. However, in the synchronization of distributed objects, the problem is how to force distributed executions to enter a set of prescribed states when certain condition is met, not how to quickly and reliably communicate the trajectories that constrain state transitions.

The communication protocol design for real-time coordination is similar to the design of fault tolerant communication protocols, such as reliable broadcast [4, 5], in the sense that we need to find a way to reliably provide distributed objects with consistent information. However, in real-time coordination, we have a weaker form of consistency constraints due to the existence of tracking error bound. On the other hand, we are faced with a hard constraint on the relative delays between the messages received by coordinating nodes. We will revisit this point after we specify the real-time coordination problem. In Section 2, we will define the problem and show some of the pitfalls in protocol design. In Section 3 we present the solutions. Section 4 is the conclusion and summary.

2   Problem Formulation

In this section, we will define the communication protocol design problem for real-time coordination. Our assumptions are as follows:

Assumption 1: Communication delays change widely and unpredictably. They are normally short with respect to application needs. However, very long delays can happen suddenly without warning.

Assumption 2: Messages are encrypted. Adversaries are unable to forge or alter the content of messages without being detected.

Assumption 3: The clocks of distributed nodes are synchronized.

Assumption 4: The control of objects is precise. Errors due to control algorithms, environments or mechanical problems are negligible.

Assumptions 3 and 4 allow us to ignore tracking errors due to control or clock synchronization inaccuracies. That is, they allow us to focus on the specific problem of tracking errors caused by communication delays. The local tracking error bound in the following discussion is used only to constrain the tracking error due to communication delays. From a system engineering perspective, this is the portion of the tracking error bound that is allocated for tracking errors due to communication delays. It is important to note that the bound for control errors is in the form of $\pm$ **B**, because the object being controlled can either undershoot or overshoot the reference setpoint. The tracking error due to late start is always positive. It measures how much the object lags behind the reference setpoint. We will use the symbol **B** to denote the tracking error bound in the rest of this paper.

Assumption 5: A reliable point-to-point communication protocol is used in all the communications.

In coordinated control (with synchronous starts), we could specify fixed start times as in Example 1. However, we cannot guarantee the coordination to work using fixed start times, if the duration between start time and current time is shorter than the worst case communication delay. Observing this constraint causes long delays to the communication of the trajectories. We are interested in protocols using start times that are set dynamically to take advantage of the window of opportunities in communication – moments at which bandwidth is available.

The simplest coordination protocol using dynamic start time is to let each node start its motion immediately after it has received its command. Although this simple-minded protocol will not work in the presence of arbitrary delays, it helps us to pin down a number of useful concepts. The idea of dynamic start times is to use some communication protocols to dynamically start the motions within a narrow time window. To analyze the worst case relative delay in start times, the System-Start-Time (SST) is defined as the leading edge of the time window. That is, the time at which one of the coordinating nodes makes the first move.

To compute the local tracking error due to late start, we imagine that the Local Reference Setpoint starts at SST, independent of the time at which the local coordination command is received. This is, what the Local Reference Setpoint should have done if there were no delay. We call this idealized setpoint the "SST-Reference Setpoint", and call the trajectory "SST-Reference Trajectory". The tracking error due to late start is then computed as the difference between SST-Reference Setpoint and the actual position of the object due to late start.

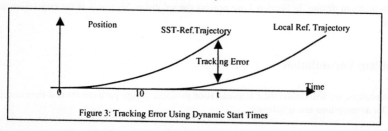

Figure 3: Tracking Error Using Dynamic Start Times

Example 2: Let the system start time SST = 0. But the node starts its motion 10 sec later due to the delay in receiving its coordination command. The SST-Reference Trajectory and the Local Reference Trajectory are illustrated in Figure 3. Note that the physical object follows the Local

Reference Setpoint to move along the Local Reference Trajectory. At time t, the local tracking error due to late start is the difference between the positions of SST-Reference Setpoint and the Local Reference Setpoint. Note that the position of Local Reference Setpoint and the position of the object is identical under Assumptions 1 to 5. That is, the object follows the local reference setpoint perfectly. The tracking error in Figure 3 is caused by the late start of the object. Next, we illustrate some of the difficulties in the design of real-time coordination communication protocol using dynamic start times.

> **Example 3:** The Command Node sends a command to nodes N_1 and N_2. When N_1 (N_2) receives its command, it immediately sends N_2 (N_1) a confirmation message that it has received its command with the given command identifier. When N_1 (N_2) receives its command and the confirmation from N_2 (N_1) that N_2 has also received the corresponding command, N_1 (N_2) starts its motion immediately.

> Unfortunately, the protocol in Example 3 does not work. Suppose that N_1 receives its command and the confirmation from N_2 at time t, and therefore starts its motion at time t. Unfortunately, N_2 receives N_1's confirmation message long after time t. Therefore, the large lag leads to coordination failure.

A moment's reflection tells us that the consensus based dynamic start time protocol is not better than fixed start time protocol. It is not possible for distributed nodes to reach a strictly consistent view of a given command within a duration shorter than the maximal communication delay D. To see this point, note that in any consensus protocol, there will be a decision function, which will return the value *True*, if a certain set of messages are received. All the adversary needs to do is to let one of the coordinating nodes receive all the required messages, and start its motion. However, the adversary jams the required messages to other nodes for duration D. Indeed, if we insist on finding a way to ensure that all the nodes receive the same set of commands, it becomes a reliable broadcast problem. It is not possible to guarantee reliable broadcast within a time window that is less than the worst case communication delay. Fortunately, the real-time coordination problem permits a weaker form of consistency due to the existence of tracking error bounds.

3 Protocol Design

There are two requirements for the design of communication protocols for real-time coordination. First, under a given communication protocol, the tracking error due to late start must always stay within its bound no matter how long is the communication delay. Second, it is desirable to shorten the time that the protocol needs to send the reference trajectories to coordinating nodes.

As a first step to exploit the weaker form of consistency, we develop the Constant Distance Grid (CDG) – Iterative Command (IC) protocol.

A CDG partitions each reference trajectory into a series of k equal distance short segments. Each segment on a trajectory should be no longer than the tracking error bound. Figure 4 illustrates a simple Constant Distance Grid with two parallel reference trajectories. Position 0 on each trajectory marks the starting point of the first segment, Segment 1 of the trajectory.

Given a trajectory in the form of CDG, the Iterative Command (IC) protocol works as follows. The Command Node sends messages to each of the N Maneuver Control Nodes and asks them to move to Position 1 first and wait for further commands. Once the Command Node receives messages that all N Maneuver Control Nodes have reached Position 1, it commands them to move to Position 2 and wait, and so on. The IC protocol is outlined by the following pseudo-code.

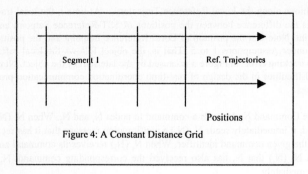

Figure 4: A Constant Distance Grid

Definition 1: The IC Protocol

Initialization:
Each of the N objects will be at its starting position (Position 0) of its trajectory.

Command Node
for j = 1 to k   // k is the last position, the final destination for a trajectory.
  Send Message j to each of N Maneuver Control Nodes to go to Position j;
  Wait for confirmation of reaching Position j from all the N Maneuver Control Nodes;
end

Each Maneuver Control Node
Loop
  Wait for command;
  Move to the commanded Position j;
  Send confirmation to the Command Node
      immediately after reaching the commanded Position j;
end Loop

The CDG-IC protocol is just the IC protocol that uses CDG. We now analyze the tracking error. Recall that the worst case tracking error is computed with respect to the SST-Reference Setpoint, which starts to move as soon as an object makes the first move. Let the segment length be d and the tracking error bound be B and $d \leq B$.

Theorem 1: Under CDG-IC protocol, the local tracking error, e, on Trajectory i is bounded by the tracking error bound, B.

Proof: Suppose that Theorem 1 is false, i.e., $e > B$. Since the segment size $d \leq B$, we have $e > d$. For $e > d$, SST-Reference Setpoint and the object must be in two different segments. Let the object be in Segment i and the SST-Reference Setpoint at Segment j, and $j > i$. For SST-Reference Setpoint in Segment j, a command it must receive a command to go to Position $j+1$. Under CDG-IC protocol, a command to go to Position $j+1$ will be issued only if all the objects have reached Position j, the starting position of Segment j. This contradicts the assumption that the object is in Segment i and the SST-Reference Setpoint in Segment j. Theorem 1 follows.

CDG-IC has the drawback of waiting for all the objects to complete the current command before issuing the next one. However, the movement of electronic messages is much faster than the movement of physical objects. Waiting for the movement of physical objects could waste the window of opportunities in communication.

To speed up the process of sending the trajectories, we have developed Constant Time Grid - Fast Iterative Command (CTG-FIC) protocol. There are two key ideas in CTG-FIC protocol.

- To replace the constant distance grid with a constant time grid. In a constant time grid, the distance of a segment is adjusted in such a way that each segment will take the same time to finish with respect to a giving reference trajectory.
- To send commands to objects to go to Position $(j+1)$ without actually waiting for all the objects actually reaching Position j.

As soon as the Command Node receives all the acknowledgements from all the objects that they have received the command to go to Position j, it sends messages to command them to move to Position $(j+1)$ until the command for the final destination position is successfully received. In order words, we allow an arbitrary number of outstanding, yet to execute commands. This allows us to capitalize on the windows of opportunity in communication. Due to the lack of space, we are unable to show the proof of correctness of the GTG-FIC protocol. Readers who are interested in knowing some of the potential pitfalls in designing fast protocols that allow outstanding commands or interested in the proof of CTG-FIC may send emails to the authors to request for a copy of the report: "Communication Protocols for Distributed Real-Time Coordination in The Presence of Communication Interruptions."

4 Summary and Conclusion

Internet based instrumentation and controls are an attractive avenue for the development and evolution of distributed control systems. However, one of the challenges is the design of communication protocols for real-time coordination in the presence of communication interruptions.

Two protocols were developed to solve the real time coordination problem in the presence of communication interruptions, the Constant Distance Grid - Iterative Command protocol (CDG-IC) and the Constant Time Grid - Fast Iterative Command protocol (CTG - FIC). Both of them can tolerate arbitrary long communication delays without causing coordination failures. However, the completion time of sending a trajectory to a node under CDG-IC depends on the speed of physical systems. CTG-FIC can send successive commands to distributed nodes without waiting for the completion of the earlier commands. Thus, the completion time of CTG-FIC is independent of the speed of the physical systems. It can better exploit the window of opportunities in communication. Due to the limitation of space, only the simpler CDG-IC is described and the CTG-FIC was briefly outlined.

Acknowledgement
This work was sponsored in part by the Office of Naval Research, by EPRI and by the Software Engineering Institute, CMU. The authors want to thank Michael Gagliardi, Ted Marz and Neal Altman for their contributions to the design and implementation of the demonstration software and to thank John Walker for the design and implementation of the hardware. Finally, we want to thank Jane Liu for her helpful comments to an earlier draft.

References:
1. The Proceedings of NSF/CSS Workshop on New Directions in Control Engineering Education, October, 1998. pp.15-16.
2. The Proceedings of Workshop on Automated Control of Distributed Instrumentation, April, 1999.
3. J. P. Briot, R. Cuerraoui and K. P. Lohr, "Concurrency and Distribution in Object-Oriented Programming", ACM Computing Survey, Vol. 30, No. 3, September, 1998.
4. P. M. Melliar-Smith, L. E. Moser and V. Agrawala, "Broadcast Protocols for Distributed Systems", IEEE Transaction on Parallel and Distributed Systems, January, 1990.
5. P. Jalote, "Fault Tolerance in Distributed Systems", Prentice Hall, 1994.

A Segmented Backup Scheme for Dependable Real Time Communication in Multihop Networks

Gummadi P. Krishna   M. Jnana Pradeep and C. Siva Ram Murthy

Department of Computer Science and Engineering
Indian Institute of Technology, Madras - 600 036, INDIA
gphanikrishna@hotmail.com, mjpradeep@yahoo.com, murthy@iitm.ernet.in

Abstract. Several distributed real time applications require fault tolerance apart from guaranteed timeliness. It is essential to provide hard guarantees on recovery delays, due to component failures, which cannot be ensured in traditional datagram services. Several schemes exist which attempt to guarantee recovery in a timely and resource efficient manner. These methods center around a priori reservation of network resources called spare resources along a backup route. In this paper we propose a method of segmented backups which improves upon the existing methods in terms of resource utilisation, call acceptance rate and bounded failure recovery time. We demonstrate the efficiency of our method using simulation studies.

1   Introduction

Any communication network is prone to faults due to hardware failure or software bugs. It is essential to incorporate fault tolerance into QoS requirements for distributed real time multimedia communications such as video conferencing, scientific visualisation, virtual reality and distributed real time control. Conventional applications which use multihop packet switching easily overcome a local fault but experience varying delays in the process. However, real time applications with QoS guaranteed bounded message delays require a priori reservation of resources (link bandwidth, buffer space) along some path from source to destination. All the messages of a *real time session* are routed through over this static path. In this way the QoS guarantee on timeliness is realised but it brings in the problem of fault tolerance for failure of components along its predetermined path. Two *proactive* approaches are in vogue to overcome this problem. The first approach is forward recovery method [1,2], in which multiple copies of the same message are sent along disjoint paths. The second approach is to reserve resources along a path, called *backup path* [3,4], which is disjoint with the primary, in anticipation of a fault in the primary path. The second approach is far more inexpensive than the first if infrequent transient packet losses are tolerable. We focus on the second proactive scheme. Establishment of backup channels saves the time required for reestablishing the channel in reactive methods.

Two different schemes have been widely analysed for the establishment of backup channels. In the first, the spare resources in the vicinity of failed component are used to reroute the channel. This method of *local detouring* [3,4] leads to inefficient resource utilisation as after recovery, the channel path lengths usually get extended significantly. The second method *end to end detouring* was proposed to solve the problem in a resource efficient manner. But end to end detouring has the additional requirement that the primary and backup paths be totally disjoint except the source and destination. This might lead to rejection of a call even when there is considerable bandwidth available in the network. Further, this method of establishing backups might be very inefficient for delay critical applications if the delay of the backup is not within the required limits. In this paper we address these problems by proposing

J. Rolim et al. (Eds.): IPDPS 2000 Workshops, LNCS 1800, pp. 678–684, 2000.

to have *segmented backups* rather than a single continuous backup path from source to destination and show that the proposed method not only solves these problems but also is more resource efficient than the end to end detouring methods with *resource aggregation* through *backup multiplexing* [5-7].

We now explain our concept of segmented backups. Earlier schemes have used end to end backups, i.e., backups which run from source to destination of a dependable connection, with the restriction that the primary and the backup channels do not share any components other than the source and destination. In our approach of *segmented backups*, we find backups for only parts of the primary path. The primary path is viewed as made up of smaller contiguous paths, which we call *primary segments* as shown in Figure 1. We find a backup path, which we call *backup segment*, for *each* segment independently. Note that successive primary segments of a primary path overlap on at least one link and that any two non consecutive segments are disjoint. The primary channel with 9 links shown, has 3 primary segments: the 1st segment spanning the first 3 links, the 2nd spanning link 3 to link 6 and the 3rd the last 4 links, segments overlapping on the 3rd and 6th links. The backup segments established are also shown. In case of a failure in a component along a primary segment the message packets are routed through the corresponding backup segment rather than through the original path, *only* for the length of this primary segment as illustrated. In case of a fault in any component of a primary path, we give the following method of backup segment activation. If only one primary seg-

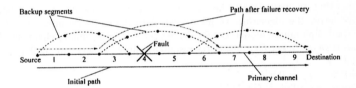

Fig. 1. Illustration of Segmented Backups

ment contains the failed component activate the backup segment corresponding to that primary segment as shown for the failure of link 4. If two successive primary segments contain the failed component activate any one of the two backup segments corresponding to the primary segments. Now we illustrate one of the advantages of

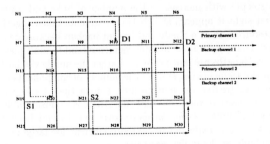

Fig. 2. Establishment of Segmented Backup Channels

the segmented approach over end to end backup approach with a simple example of a 5 X 6 mesh in Figure 2. Suppose the capacity of each link on the mesh is only 1 unit. There are 2 dependable connections to be established : S1 to D1 and S2 to D2. The primary paths (shortest paths) of these connections are shown in the figure. It is not possible to establish end to end backups for both the connections as both the backups contend for the unit resource along the link between N15 to N16. However, segmented backups can be established as shown in the figure.

2   Spare Resource Allocation

It is very important to address the issue of minimizing the amount of spare resources reserved. The idea is to reduce the amount of backup resources reserved by *multiplexing* the backups passing through the same link. We explain the method very briefly below. Refer to [5-7] for more detailed discussion. We note that the resources reserved for backup channels are used only during component failures in their primary channels. We consider *single link* failure model for our analysis, under the assumption that channel failure recovery time i.e., time taken for the fault to be rectified, is much smaller than the network's mean time to failure (MTTF). If primary channels of two connections share no common components and their backup channels with bandwidths b1 and b2 pass through link L, it is sufficient to reserve max(b1, b2) for both the backup channels on the link L in this failure model, as we know that both the backup channels can never be activated simultaneously. This is the idea of multiplexing. We discuss how *deterministic multiplexing* [5,6] applies to our scheme in comparison to earlier schemes.

We use deterministic failure model and calculate the minimum amount of extra resources that are necessary to be reserved to handle all possible cases of failure. We give below the algorithm we use to calculate the spare resources $S_L at link L$ under single link failure model. Let Φ_L denote the set of all primary channels whose backups traverse L. Let R_{P_S} denote the resource required at each link by the primary segment P_S.

Initialise $S_{I,L} = 0 \ \forall \ I, L$
loop for each link $I, I \neq L$
        loop for each primary channel segment $P_S \in \Phi$
            if P_S contains link I **then**
$$S_{I,L} = S_{I,L} + R_{P_S}$$
        endif
        endloop
endloop
$S_L = max\{ \ S_{I,L} \ \} \ \forall \ I \neq L$

It is worth noting the complexity of this multiplexing algorithm. Its execution time increases steeply with increase in the number of links and connections in the network. At first sight it appears as if backup segments taken together, require to reserve more resources than a single end to end backup because segments overlap over the primary channel. But the backup segments tend to multiplex more as their primary segments' lengths are much shorter. Larger the number of backup segments, shorter the primary segments i.e., smaller the number of components in each primary segment and hence, greater the multiplexing among their backup segments. Our method tends to be more resource efficient since there is a considerable improvement in backup segments' *multiplexing capability* over end to end backup's capability. Therefore, our scheme is expected to be more efficient for large networks when a large number of calls are long distance calls.

3   Backup Route Selection

Several elaborate routing methods have been developed which search for routes using various QoS metrics. Optimal routing problem of minimizing the amount of spare resources while providing the guaranteed fault tolerance level is known to be NP-hard. So we resort to heuristics. Several greedy heuristics for selecting end to end backup paths are discussed in [5]. A shortest path search algorithm like Dijkstra's is enough to find the minimum cost path where the cost value for a link can be made a function of delay, spare resource reservation needed etc. The complexity of our problem of selecting segmented backups is far greater as we have to address additional constraints due to our following design goals.

Improving Call Acceptance Rate: Our scheme tends to improve the call acceptance rate over end to end backups due to two main reasons. Firstly, it tends to improve the call acceptance in situations where there exists a primary path but the call gets rejected due to the lack of an end to end disjoint backup path. We have already shown this through a simple example in Figure 2. Secondly, by reserving lesser amount of resources it allows for more calls to be accepted. This method however, has the problem of choosing the appropriate intermediate nodes (the nodes chosen should not only allow backup segments but should also economize on the resource reservation).

Improving Resource Reservation: This sets up two opposing constraints. First, longer the primary segment of a backup segment, lesser will be the number of backup segments required. Too short primary segments can lead to a requirement of large amounts of resources for the large number of backup segments (Note that each of the backup segments requires more resource than the primary segment which it spans). On the contrary shorter primary segments lead to more multiplexing among their backup segments as described before. So we have to choose primary segments which are neither too short nor too long.

Increase in the Delay Due to Backup: We are interested only in backup segments which do not lead to an unreasonable increment in delay in case of a failure in their primary segment, which constrains the choice of intermediary nodes.

Even in case of end to end detouring we face these constraints but we have a very simple way out. The shortest path algorithm run on the network with the nodes of the primary path removed should give a very good solution and if it fails there does not exist any solution. In contrast, for our scheme we do not have the intermediate destinations fixed and we have to choose among the many possible solutions. In our heuristic we run Dijkstra's shortest path algorithm from source to destination removing all links in the primary path. If in the process, Dijkstra's search algorithm comes to any node in the primary path, we mark it as an intermediate node. Then, we take the node previous to it in the primary path (in the order of increasing distance from the source) and using it as the new source try to find a shortest path to the destination recursively. In order to ensure that the primary segment is not too small we use a parameter MINLEAPLEN which indicates the minimum number of nodes in any primary segment. Thus, we remove the first MINLEAPLEN nodes starting from the new source along the primary path every time before beginning the search for the shortest path to the destination. It is also important that the delay increment for any backup segment is below a threshold Δ for the backup to be of use. This tends to prevent lengthy backups for very small primary segments. In case the destination cannot be reached or the Δ condition is violated, we start Dijkstra's algorithm again from the first segment, this time avoiding the nodes which were chosen as the end of first segment, in previous attempts. The number of times we go back and try again (number of retries) is constant and can be set as a parameter. It is to be noted that our scheme tends to perform better in comparison to the scheme in [6] for large networks, with moderate congestion and for long distance calls. Further, it is important to note that for small networks with short distance calls this scheme mimics the end to end backup scheme in [6] as we do allow a backup to be made of just one segment. In case of connections with very short primary path our heuristic chooses the backup with a single segment.

4   Failure Recovery

When a fault occurs in a component in the network, all dependable connections passing through it have to be rerouted through their backup paths. This process is called failure recovery. This has three phases: *fault detection, failure reporting* and *backup activation*. The restoration time, called failure recovery delay, is crucial to many real time applications, and has to be minimized.

In our model, we assume that when a link fails, its end nodes can detect the failure. For failure detection techniques and their evaluation refer to [8]. After fault detection, the nodes which have detected the fault, report it to the concerned nodes for recovering from the failure. This is called failure reporting. After the failure report reaches certain nodes, the backup is activated by those nodes. Failure reporting and backup activation need to use control messages. For this purpose, we assume a real time control channel (RCC) [6] for sending control messages. In RCC, separate channels are established for sending control messages, and it guarantees a minimum rate of sending messages.

Failure Reporting and Backup Activation: The nodes adjacent to a failed component in the primary path of a dependable connection will detect the failure and send failure reports both towards the source and the destination. In the end to end backup scheme, these messages have to reach the source and destination before they can activate the backup path. In our scheme, this is not necessary. Failures can be handled more locally. The end nodes of the primary segment containing the faulty component on receiving the failure reports initiate the recovery process. These two nodes send the activation message along the backup segment, and the dependable connection service is resumed. This process is illustrated in Figure 3. If there are k segments in the backup, then this gives about $O(k)$ improvement in the time for failure reporting. When a fault occurs, not only do we experience a

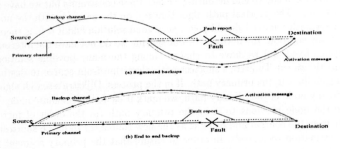

Fig. 3. Illustration of Failure Recovery

disruption of service for some time, but also packets transmitted during the failure reporting time are lost. Most real time applications cannot tolerate much message loss. In our scheme the message loss is reduced to a considerable extent. When a fault occurs in one segment of the primary, only the packets which have entered that segment from the time of the occurrence of the fault till the backup segment activation are lost. Other packets in the segments before and after the failed segment are not affected and will be delivered normally. This is in contrast to the end to end backup case, where *all* packets in transit in the primary path before the failed component, between occurrence of failure and backup activation, are lost.

5   Delay and Scalability

Delay: In Real Time Communication, it is essential to have the delays along both the primary and the backup channels to be as low as possible. Hence, we might have a restriction on the amount by which the delay along the backup exceeds that along the primary path. Let the total delay along the backup path not exceed the delay along the primary by Δ, a specified QoS parameter.

Thus, the constraint for choosing an end to end backup is,

delay(backup path) - delay(primary path) $\leq \Delta$.

In the case of segmented backups, this constraint is,

(delay(backup segment i) - delay(primary segment i)) $\leq \Delta, \forall$ i.

We see that in our case we have to minimize the delay increase for each segment independently. Hence call acceptance rate will be better since it is easier to find small segments than to find big end to end backups satisfying the Δ constraint.

Scalability: The segmented backup scheme scales well since it does not demand global knowledge and does not involve any kind of broadcast. There is no necessity for a network manager and this scheme works well in a distributed network. For Backup Multiplexing each node needs to know the primary paths of the channels whose backups pass through it. This is easily accomplished if the information is sent along with the packet requesting the establishment of backup channel. Upon encountering faults, control messages are not broadcast, but sent only to a limited part of the network affected by the fault.

In large networks, the effectiveness of the segmentation increases as the mean path length of connections increases. Since the calculation of spare resources using multiplexing has to be done per segment independently, this scheme scales better than the earlier end to end methods.

6   Performance Evaluation

We evaluated the proposed scheme by carrying out simulation experiments similar to those in [6], on a 12 X 12 mesh. We also implemented the end to end backup scheme [6] for comparative study.

In the simulated network, neighbour nodes are connected by two simplex links, one in each direction, and all links have identical bandwidth. For simplicity, the bandwidth requirement of all connections was put equal to 1 unit. The delay of each link was set to 1, thereby making delay along any path equal to its path length. Primary channels were routed using a sequential shortest-path search algorithm. The end to end backups were also routed using the same shortest-path search algorithm, with the condition that it does not contain any component of the primary other than source and destination. The amount by which backup path delay can exceed primary path delay was used as a parameter, Δ.

We find the backup segments as described in Section 3. The number of retries was set to 9 in our simulation experiments. The MINLEAPLEN parameter was set to 4. Connections were requested incrementally, between a source and destination chosen randomly, with the condition that no (source, destination) pair is repeated, and the length of the shortest path between them is at least MINPATHLEN. In our simulation studies, connections were only established but not torn down since (i) the computational time required for release of connections is considerably high, and (ii) earlier studies with end to end backups [5,6] also do the same.

The results are shown in Table 1. In this table, we show the statistics at different instants of time as in the simulation. The number of connections requested is proportional to the time. The network load at the time is also shown. Table 1 shows the average amount of spare bandwidth reserved per connection, both for segmented (seg) and end to end (end) backups, for different values of Δ. We show the results for MINPATHLEN=6, and for MINPATHLEN=8. The average path lengths in the two cases was 10.8 and 12.3. The bandwidth of the links was chosen as 100 units for MINPATHLEN=6 and 90 units for MINPATHLEN=8. As expected, the spare bandwidth reserved was much lower for segmented backups. Also, the improvement is seen to be more in the second case. This illustrates that as the average length of connections increases, the effectiveness of segmented backups increases.

The cumulative number of requests rejected till an instant of time was also noted. The number rejected by the segmented backup scheme was seen to be much lesser than that of the end to end scheme.

Table 1. Average amount of spare bandwidth reserved per connection

MINPATHLEN = 6		$\Delta = 2$		$\Delta = 4$		MINPATHLEN = 8		$\Delta = 2$		$\Delta = 4$	
Time	n/w load	end	seg	end	seg	Time	n/w load	end	seg	end	seg
1245	42%	7.55	7.06	7.50	7.07	1284	53%	8.72	8.16	8.71	8.16
1559	51%	7.40	6.88	7.42	6.91	1488	59%	8.76	8.12	8.74	8.13
1846	57%	7.49	6.93	7.52	6.99	1708	63%	8.74	8.09	8.74	8.11
2148	63%	7.51	6.98	7.52	7.11	1911	65%	8.69	8.08	8.68	8.18
2442	67%	7.48	6.99	7.49	7.15	2131	66%	8.64	8.10	8.63	8.20

7 Conclusions

In this paper, we have proposed segmented backups: a failure recovery scheme for dependable real-time communication in distributed networks. This mechanism not only improves resource utilisation and call acceptance rate but also provides for faster failure recovery. We evaluated the proposed scheme through simulations and demonstrated the superior performance of the scheme compared to earlier end to end backup schemes [5-7]. In order to realise the full potential of the method of segmented backups, better routing strategies have to be developed for choosing intermediate nodes optimally. We also need faster algorithms for backup multiplexing.

References

1. P. Ramanathan and K. G. Shin, "Delivery of time-critical messages using a multiple copy approach," *ACM Trans. Computer Systems*, vol. 10, no. 2, pp. 144-166, May 1992.
2. B. Kao, H. Garcia-Molina, and D. Barbara, "Aggressive transmissions of short messages over redundant paths," *IEEE Trans. Parallel and Distributed Systems*, vol. 5, no. 1, pp. 102-109, January 1994.
3. Q. Zheng and K. G. Shin, "Fault-tolerant real-time communication in distributed computing systems," in *Proc. IEEE FTCS*, pp. 86-93, 1992.
4. W. Grover. "The self-healing network: A fast distributed restoration technique for networks using digital crossconnect machines," in *Proc. IEEE GLOBECOM*, pp. 1090-1095, 1987.
5. S. Han and K. G. Shin, "Efficient spare-resource allocation for fast restoration of real-time channels from network component failures," in *Proc. IEEE RTSS*, pp. 99-108, 1997.
6. S. Han and K. G. Shin, "A primary-backup channel approach to dependable real-time communication in multihop networks," *IEEE Trans. on Computers*, vol. 47, no. 1, pp. 46-61, January 1998
7. C. Dovrolis and P. Ramanathan, "Resource aggregation for fault tolerance in integrated services networks," *ACM SIGCOMM Computer Communication Review*, 1999.
8. S. Han and K. G. Shin, "Experimental evaluation of failure detection schemes in real-time communication networks," in *Proc. IEEE FTCS*, pp. 122-131, 1997.

Real-Time Coordination in Distributed Multimedia Systems

Theophilos A. Limniotes and George A. Papadopoulos

Department of Computer Science
University of Cyprus
75 Kallipoleos Str, P.O.B. 20537
CY-1678 Nicosia
Cyprus
E-mail: {theo,george}@cs.ucy.ac.cy

Abstract. The coordination paradigm has been used extensively as a mechanism for software composition and integration. However, little work has been done for the cases where the software components involved have real-time requirements. The paper presents an extension to a state-of-the-art control- or event-driven coordination language with real-time capabilities. It then shows the capability of the proposed model in modelling distributed multimedia environments

1   Introduction

The concept of coordinating a number of activities, possibly created independently from each other, such that they can run concurrently in a parallel and/or distributed fashion has received wide attention and a number of coordination models and associated languages ([4]) have been developed for many application areas such as high-performance computing or distributed systems.

Nevertheless, most of the proposed coordination frameworks are suited for environments where the sub-components comprising an application are conventional ones in the sense that they do not adhere to any real-time constraints. Those few that are addressing this issue of real-time coordination either rely on the ability of the underlying architecture apparatus to provide real-time support ([3]) and/or are confined to using a specific real-time language ([5]).

In this paper we address the issue of real-time coordination but with a number of self imposed constraints, which we feel, if satisfied, will render the proposed model suitable for a wide variety of applications. These constraints are:
• The coordination model should not rely on any specific architecture configuration supporting real-time response.
• The real-time capabilities of the coordination framework should be able to be met in a variety of systems including distributed ones.
• Language interoperability should not be sacrificed and the real-time framework should not be based on the use of specific language formalisms.

J. Rolim et al. (Eds.): IPDPS 2000 Workshops, LNCS 1800, pp. 685-691, 2000.
© Springer-Verlag Berlin Heidelberg 2000

We attempt to meet the above-mentioned targets by extending a state-of-the-art coordination language with real-time capabilities. In particular, we concentrate on the so-called control- or event-driven coordination languages ([4]) which we feel they are particularly suited for this purpose, and more to the point the language Manifold ([1]). We show that it is quite natural to extend such a language with primitives enforcing real-time coordination and we apply the proposed model to the area of distributed multimedia systems.

2   The Coordination Language Manifold

Manifold ([1]) is a control- or event-driven coordination language, and is a realisation of a rather recent type of coordination models, namely the Ideal Worker Ideal Manager (IWIM) one. In Manifold there exist two different types of processes: *managers* (or *coordinators*) and *workers*. A manager is responsible for setting up and taking care of the communication needs of the group of worker processes it controls (non-exclusively). A worker on the other hand is completely unaware of who (if anyone) needs the results it computes or from where it itself receives the data to process. Manifold possess the following characteristics:

• *Processes*. A process is a *black box* with well-defined *ports* of connection through which it exchanges *units* of information with the rest of the world.

• *Ports*. These are named openings in the boundary walls of a process through which units of information are exchanged using standard I/O type primitives analogous to read and write. Without loss of generality, we assume that each port is used for the exchange of information in only one direction: either into (*input* port) or out of (*output* port) a process. We use the notation p.i to refer to the port i of a process instance p.

• *Streams*. These are the means by which interconnections between the ports of processes are realised. A stream connects a (port of a) producer (process) to a (port of a) consumer (process). We write p.o -> q.i to denote a stream connecting the port o of a producer process p to the port i of a consumer process q.

• *Events*. Independent of streams, there is also an event mechanism for information exchange. Events are broadcast by their sources in the environment, yielding *event occurrences*. In principle, any process in the environment can pick up a broadcast event; in practice though, usually only a subset of the potential receivers is interested in an event occurrence. We say that these processes are *tuned in* to the sources of the events they receive. We write e.p to refer to the event e raised by a source p.

Activity in a Manifold configuration is *event driven*. A coordinator process waits to observe an occurrence of some specific event (usually raised by a worker process it coordinates) which triggers it to enter a certain *state* and perform some actions. These actions typically consist of setting up or breaking off connections of ports and streams. It then remains in that state until it observes the occurrence of some other event, which causes the *preemption* of the current state in favour of a new one corresponding to that event. Once an event has been raised, its source generally continues with its activities, while the event occurrence propagates through the environment independently and is observed (if at all) by the other processes according to each observer's own sense of priorities.

More information on Manifold can be found in [1]; the language has already been implemented on top of PVM and has been successfully ported to a number of platforms including Sun, Silicon Graphics, Linux, and IBM AIX, SP1 and SP2.

3   Extending Manifold with a Real-Time Event Manager

The IWIM coordination model and its associated language Manifold have some inherent characteristics, which are particularly suited to the modelling of real-time software systems. Probably the most important of these is the fact that the coordination formalism has no concern about the *nature* of the data being transmitted between input and output ports since they play no role at all in setting up coordination patterns. More to the point, a stream connection between a pair of input-output ports, simply passes anything that flows within it from the output to the input port. Furthermore, the processes involved in some coordination or cooperation scenario are treated by the coordination formalism (and in return treat each other) as black boxes without any concern being raised as to their very nature or what exactly they do. Thus, for all practical purposes, some of those black boxes may well be devices (rather than software modules) and the information being sent or received by their output and input ports respectively may well be signals (rather than ordinary data). Note also that the notion of stream connections as a communication metaphor, captures both the case of transmitting discrete signals (from some device) but also continuous signals (from, say, a media player). Thus, IWIM and Manifold are ideal starting points for developing a real-time coordination framework.

In fact, a natural way to enhance the model with real-time capabilities is by extending its event manager. More to the point, we enhance the event manager with the ability to express real-time constraints associated with the raising of events but also reacting in bound time to observing them. Thus, while in the ordinary Manifold system the raising of some event e by a process p and its subsequent observation by some other process q are done completely asynchronously, in our extended framework timing constraints can be imposed regarding when p will raise e but also when q should react to observing it. Effectively, an event is not any more a pair <e,p>, but a triple <e,p,t> where t denotes the moment in time at which the event occurs. With events that can be raised and detected respecting timing constraints, we essentially have a real-time coordination framework, since we can now guarantee that changes in the configuration of some system's infrastructure will be done in bounded time. Thus, our real-time Manifold system goes beyond ordinary coordination to providing temporal synchronization.

3.1   Recording Time

A number of primitives exist for capturing the notion of time, either relative to world time, the occurrence of some event, etc. during the execution of a multimedia application which we refer to below as presentation. These primitives have been implemented as atomic (i.e. not Manifold) processes in C and Unix. In particular:

- **AP_CurrTime(int timemode)**

returns the current time according to the parameter timemode. It could be world time or relative.

- **AP_OccTime(AP_Event anevent, int timemode)**

returns the time point (in world or relative mode) of an event. Time points represent single instance in time; two time points form a basic interval of time.

- **AP_PutEventTimeAssociation(AP_Event anevent)**

creates a record for every event that is to be used in the presentation and inserts it in the events table mentioned above.

- **AP_PutEventTimeAssociation_W(AP_Event anevent)**

is a similar primitive which additionally marks the world time when a presentation starts, so that the rest of the events can relate their time points to it.

3.2   Expressing Temporal Relationships

There are two primitives for expressing temporal constraints among events raised and/or observed. The first is used to specify when an event must be triggered while the second is used to specify when the triggering of an event must be delayed for some time period.

- **AP_Cause(AP_Event anevent, AP_Event another, AP_Port delay, AP_Port timemode)**

enables the triggering of the event another based on the time point of anevent.

- **AP_Defer(AP_Event eventa, AP_Event eventb, AP_Event eventc, AP_Port delay)**

inhibits the triggering of the event eventc for the time interval specified by the events eventa and eventb. This inhibition of eventc may be delayed for a period of time specified by the parameter delay.

4   Coordination of RT Components in a Multimedia Presentation

We show the applicability of our proposed model by modelling an interactive multimedia example with video, sound, and music. A video accompanied by some music is played at the beginning. Then, three successive slides appear with a question. For every slide, if the answer given by the user is correct the next slide appears; otherwise the part of the presentation that contains the correct answer is re-played before the next question is asked. There are two sound streams, one for English and another one for German.

For each such medium, there exists a separate manifold process. Each such manifold process is a "building block". The coordination set up with the stream connections between the involved processes is shown below (the functionality of some of these boxes is explained later on):

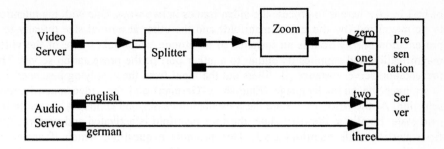

We now show in more detail some of the most important components of our set up. We start with the manifold that coordinates the execution of atomics that take a video from the media object server and transfer it to a presentation server.

```
manifold tv1()
{

begin:(activate(cause1,cause2,mosvideo,splitter,zoom),c
ause1,WAIT).
  start_tv1:(cause2,mosvideo -> ( -> splitter),
             splitter.zoom ->zoom,
             zoom-> (->ps.zero),ps.out1->stdout,WAIT).
  end_tv1:post(end).
  end:(activate(ts1),ts1).
}
```

In addition to the `begin` and `end` states which apply at the beginning and the end of the manifold's execution respectively, two more states are invoked by the AP_Cause commands, namely `start_tv1` and `end_tv1`. At the `begin` state the instances of the atomics `cause1`, `cause2`, `mosvideo`, `splitter`, and `zoom` are activated. These activations introduce them as observable sources of events. This state is *synchronized* to preempt to `start_tv1` with the execution of `cause1`. More to the point, the declaration of the instance `cause1`

```
process cause1 is
AP_Cause(eventPS,start_tv1,3,CLOCK_P_REL)
```

indicates that the preemption to `start_tv1` should occur 3 seconds (relative time) after the raise of the presentation start event `eventPS`.

Within `start_tv1` the other three instances, `cause2`, `mosvideo`, and `splitter`, are executed in parallel. `cause2` synchronizes the preemption to `end_tv1` and its declaration

```
process cause2 is
AP_Cause(eventPS,end_tv1,13,CLOCK_P_REL)
```

indicates that the currently running state must execute the other two atomic instances within 13 seconds. So the process for the media object `mosvideo` keeps sending its data to `splitter` until the state is preempted to `end_tv1`. The `mosvideo` coordinating instance supplies the video frames to the `splitter` manifold. The role of splitter here is to process the video frames in two ways. One with the intention to be magnified (by the zoom manifold) and the other at normal size directly to a presentation port. zoom is an instance of an atomic which takes care of the video magnification and supplies its output to another port of the presentation server. The presentation server instance ps filters out the input from the supplying instances, i.e. it arranges the audio language (English or German) and the video magnification selection. At the end_tv1 state the presentation ceases and control is passed to the end state. Finally at the end state, the tv1 manifold is activated and performs the first question slide manifold ts1. This prompts a question, which if answered correctly prompts in return the next question slide. A wrong answer leads to the replaying of the presentation that relates to the correct answer, before going on with the next question slide. The code for a slide manifold is given below.

```
manifold tslide1()
{
  begin:(activate(cause7),cause7,WAIT).
  start_tslide1:(activate(testslide),testslide,WAIT).
  tslide1_correct: "your answer is correct"->stdout;
                  (activate(cause8),cause8,WAIT).
  tslide1_wrong:"your answer is wrong"->stdout;
                  (activate(cause9),cause9,WAIT).
  end_tslide1:(post(end),WAIT).
  start_replay1:
(activate(replay1,cause10),replay1,cause10,WAIT).
  end_replay1: (activate(cause11),cause11,WAIT).
  end:(activate(ts2),ts2).
}
```

The instance cause7 is responsible for invoking the start_tslide state. The declaration for the cause7 instance is

```
process cause7 is
AP_Cause(end_tv1,start_slide1,3,CLOCK_P_REL)
```

Here we specify that start_slide1 will start 3 seconds after the occurrence of end_tv1. Inside that, the testslide instance is activated and eventually causes preemption to either tslide_correct or tslide1_wrong, depending on the reply.

The tslide_wrong instance causes transition to the start_replay1 state which causes the replay of the required part of the presentation and then preempts through cause10, to end_replay1. That in turn preempts through cause11, to end_replay1, after replaying the relevant presentation. The end_replay marks the end of the repeated presentation and preempts to end_tslide1. The tslide_correct state, also causes the end_tslide1 event through the instance cause8. The end_tslide1, simply preempts to the end state that contains the execution of the next slide's instance.

The main program begins with the declaration of the events used in the program.

```
AP_PutEventTimeAssociation_W(eventPS)
```

is the first event of the presentation and puts the current time as its time point. For the rest of the events the function

```
AP_PutEventTimeAssociation(event)
```

is used which leaves the time point empty. Then the implicit instances of the media manifolds, are executed in parallel at the end of the block. These are

```
(tv1,eng_tv1,ger_tv1,music_tv1)
```

tv1 is the manifold for the video transmission, eng_tv1 is the manifold for the English narration transmission, ger_tv1 is the manifold for the German narration transmission and music_tv1 is the manifold for the music transmission.

5   Conclusions

In this paper we have addressed the issue of real-time coordination in parallel and distributed systems. In particular, we have extended a control- or event-driven coordination language with a real-time event manager that allows expressing timing constraints in the raising, observing, and reacting to events. Thus, state transitions are done in a temporal sequence and affect accordingly the real-time behaviour of the system. We have tested our model with a scenario from the area of multimedia systems where recently issues of coordination and temporal synchronization at the middleware level have been of much interest to researchers ([2]).

References

1.   F. Arbab, "The IWIM Model for Coordination of Concurrent Activities", *First International Conference on Coordination Models, Languages and Applications (Coordination'96)*, Cesena, Italy, 15-17 April, 1996, LNCS 1061, Springer Verlag, pp. 34-56.
2.   G, Blair, J-B. Stefani, *Open Distributed Processing and Multimedia*, Addison-Wesley, 1998.
3.   IEEE Inc., "Another Look at Real-Time Programming", Special Section of the *Proceedings of the IEEE* **79(9)**, September, 1991.
4.   G. A. Papadopoulos and F. Arbab, "Coordination Models and Languages", *Advances in Computers*, Marvin V. Zelkowitz (ed.), Academic Press, Vol. 46, August, 1998, 329-400.
5.   M. Papathomas. G. S. Blair and G. Coulson, "A Model for Active Object Coordination and its Use for Distributed Multimedia Applications", LNCS, Springer Verlag, 1995, pp. 162-175.
6.   S. Ren and G. A. Agha, "RTsynchronizer: Language Support for Real-Time Specifications in Distributed Systems", *ACM SIGPLAN Workshop on Languages, Compilers and Tools for Real-Time Systems*, La Jolla, California, 21-22 June, 1995.

Supporting Fault-Tolerant Real-Time Applications using the RED-Linux General Scheduling Framework [*]

Kwei-Jay Lin and Yu-Chung Wang

Department of Electrical and Computer Engineering
University of California, Irvine, CA 92697-2625
{klin, wangy}@ece.uci.edu

Abstract. In this paper, we study the fault-tolerant support for real-time applications. In particular, we study the scheduling issues and kernel support for fault monitors and the primary-backup task model. Using the powerful scheduling framework in RED-Linux, we can support a jitterless fault monitoring. We can also provide the task execution isolation so that an erroneous runaway task will not take away additional CPU budget from other concurrently running tasks. Finally, we provide a group mechanism to allow the primary and backup jobs of a fault-tolerant task to share both the CPU budget as well as other resources. All these mechanisms make the implementation of fault-tolerant real-time systems easier.

1 Introduction

As more computer-based systems are now used in our daily life, many applications must be designed to meet real-time or response-time requirements, or human safety may be jeopardized. Real-time applications must be fault-tolerant both to timing faults as well as logical faults. Timing faults occur when an application cannot produce a result before its expected deadline. Logical faults occur when an application produce a wrong result before or after the deadline. Both types of faults must be handled in a fault-tolerant real-time system. Supporting fault-tolerant mechanisms in real-time systems therefore is a complex issue. Finding a powerful real-time OS to support fault-tolerant applications is even more difficult.

We have been working on a real-time kernel project based on Linux. Our real-time kernel project is called RED-Linux (*Real-time and Embedded Linux*). For efficiency, we have implemented a mechanism that provides a short task dispatch time [18]. To enhance the flexibility, we provide a general scheduling framework (GSF) in RED-Linux [19]. In addition to the priority-driven scheduling, RED-Linux supports the *time-driven* [7–9] and the *share-driven* (such as the proportional sharing [14] and approximations [2, 17]) scheduling paradigms. In this paper, we investigate how GSF in RED-Linux may support fault-tolerant real-time systems. We review the primitives for many fault-tolerant real-time

---

[*] This research was supported in part by UC/MICRO 98-085, 99-073 and 99-074, Raytheon and GeoSpatial Technologies, and by NSF CCR-9901697.

J. Rolim et al. (Eds.): IPDPS 2000 Workshops, LNCS 1800, pp. 692-698, 2000.

system models and study how to support (or enforce) them in the framework. By adjusting scheduling attribute values and selection criteria in the scheduler, it is possible to implement many fault-tolerant scheduling algorithms in our framework efficiently.

In particular, we study the scheduling issues and kernel support for fault monitors and the primary-backup task model. Using the powerful scheduling framework in RED-Linux, we can support a jitterless fault monitoring. We can also easily specify the CPU budget for each computation so that an erroneous runaway task will not take away the CPU budget reserved for other concurrently running tasks. Finally, we provide a group mechanism to allow the primary and backup jobs of a fault-tolerant task to share both the CPU budget as well as other resources. All these mechanisms make the implementation of fault-tolerant real-time systems easier.

The rest of this paper is organized as follows. Section 2 reviews popular scheduling paradigms used in real-time systems and other real-time OS projects. Section 3 briefly introduces the RED-Linux general scheduling framework. We then study the fault monitoring issues for real-time system in Section 4. Section 5 presents the design of the task group mechanism in RED-Linux. The paper is concluded in Section 6.

2   Related Work on Fault-Tolerant and Real-Time Support

Several previous work has studied the fault-tolerant real-time scheduling issues. Liestman and Campbell [11] propose a scheduling algorithm for frame based, simply periodic uniprocessor systems. Each task has two versions: primary and backup. Task schedules are dynamically selected from a pre-defined schedule tree depending on the completion status of the primary tasks. Chetto and Chetto [5] present an optimal scheduling strategy based on a variant of the EDF algorithm, called EDL, to generate fault-tolerant schedules for tasks that are composed of primary and alternate jobs. Their method provides the ability to dynamically change the schedule and accounting for runtime situations such as successes or failures of primaries. Caccamo and Buttazzo [4] propose a fault-tolerant scheduling model using the primary and backup task model for a hybrid task set consisting of firm and hard periodic tasks on a uniprocessor system. The primary version of a hard task is always scheduled first if it is possible to finish it and the backup task before the deadline. If not, only the backup task is scheduled. Another interesting work related to real-time fault-tolerance is the Simplex architecture [15]. The Simplex architecture is designed for on-line upgrade of real-time software applications by using redundant software components. By allowing different versions of a software component to be executed in sequence or in parallel, real-time application software can be dynamically replaced with negligible down-time. The architecture can also be used for fault tolerance.

Our goal in this paper is not to propose a new fault-tolerant model but to study the OS support for those proposed earlier. Using RED-Linux's general scheduling framework, we hope to be able to support many existing fault-tolerant mechanisms effectively and efficiently. To support the fault-tolerant mechanisms mentioned above, at least two mechanisms are necessary. The first is a way to define the group relationship between related tasks (primary and backup, old and new versions etc.) to allow them to share the budget for CPU or other resources. The other is a predictable monitoring facility. These fault tolerance supports from RED-Linux will be discussed in this paper.

3   The RED-Linux General Scheduling Framework

The goal of the RED-Linux general scheduling framework (GSF) is to support most well-known scheduling paradigms, including the *priority-driven*, the *time-driven* [7–9] and the *share-driven* [14, 2, 17], so that any application can use RED-Linux for real-time support. Two features have been introduced: the general scheduling attributes used in the framework and the scheduler components used to make scheduling decisions.

In our model, the smallest schedulable unit is called a *job*. For systems with periodic activities, we call a job stream as a periodic *task*. Different scheduling paradigms use different attributes to make scheduling decisions. In order for all paradigms to be supported in GSF, it is important for all useful timing information to be included in the framework so that they can be used by the scheduler. We denote four scheduling attributes for each job in GSF: *priority*, *start_time*, *finish_time*, *budget*. Among the four, the start_time and the finish_time together define the *eligible interval* of a job execution. The priority specifies the relative order for job execution. The budget specifies the total execution time assigned to a job. These attributes can be used as constraints. However, these timing attributes can also be used as the selection factors when a scheduler needs to select a job to be executed next.

RED-Linux uses a two-component scheduling framework in . The framework separates the low level scheduler, or *dispatcher*, from the QOS parameter translator, or *allocator*. We also design a simple interface to exchange information between these two components. It is the allocator's responsibility to set up the four scheduling attributes associated with each real-time job according to the current scheduling policy. The dispatcher inspects each job's scheduling attribute values, chooses one job from the ready queue and dispatches it to execution. In addition to assigning attribute values, the allocator also determines the evaluation function of scheduling attributes, since each job has multiple scheduling attributes. This is done by producing an *effective priority* for each job. The allocator uses one or more attributes to produce the effective priority so that the dispatcher will follow a specific scheduling discipline.

More details on the GSF implementation and the performance measurement can be found in [19].

4   The Design of Fault Monitors

To provide fault tolerance, three facilities can be supported: fault detection, fault avoidance, and fault recovery. Fault-tolerant systems must be able to monitor the system and application status closely and predictably. The earlier a fault can be detected and identified, the easier it may be fixed.

Depending on the type and the likelihood of faults to be monitored, cyclic monitoring is often used in systems with safety properties that must always be maintained. For example, many system components send "heartbeat" messages to each other or to a central controller to let them know that the component is still alive and well. Another example is a temperature monitoring facility that constantly reads the temperature sensor and produce a warning if the temperature is too high. Cyclic monitors are scheduled independently from any user applications. Depending on their importance, they must be executed predictably and without jitters so that they do not miss a critical warning window for an important fault. However, the traditional priority driven scheduler may not provide the kind of predictability required by fault-tolerance monitors. There is no guarantee on the execution jitters since the temporal distance between two consecutive executions of a monitor task may be as long as twice the period length [7, 8].

One effective scheduling paradigm in RED-Linux for cyclic monitors is the time-driven (TD) (or clock-driven) paradigm. For embedded systems with steady and well-known input data streams, TD schedulers have been used to provide a very predictable processing time for each data stream [7–9]. Using this scheduling paradigm, the time instances when each task starts, preempts, resumes and finishes are pre-defined and enforced by the scheduler. User applications may specify the exact time and cycle when a monitor should be activated; the Dispatcher will activate the monitor accordingly. However, using the general scheduling framework in RED-Linux, other tasks may use their own schedulers independent of the TD scheduler for monitors. The integration of TD schedules with these application schedulers has many interesting issues. For example, if an application uses the fixed priority driven scheduling such as rate monotonic scheduling in the presence of TD schedulers, can we still guarantee that all periodic jobs will meet their deadlines using the schedulability condition for the RM model [12]?

Suppose a fault-tolerant system has monitor jobs and priority driven (PD) jobs. The monitor jobs are scheduled using TD at exact times. Therefore the priority-driven jobs are scheduled after the TD jobs are executed. If a PD job is running when a TD job is scheduled to start, the PD job will be preempted. In other words, all TD jobs are considered to have a higher effective priority than all PD jobs.

Using the RM scheduling, a system of n tasks are guaranteed to meet their deadlines if the total utilization satisfies the condition:

$$U = \sum_{i=1}^{n} \frac{c_i}{p_i} \le n(2^{1/n} - 1)$$

where a task τ_i must be executed for c_i time units per p_i time interval. However, when a RM system is scheduled after a time-driven scheduler, the execution of a periodic task may be delayed or interrupted by a TD job. To handle this problem, we can treat all TD jobs as "blocking" for PD jobs just like PD jobs are blocked on accessing critical sections. We can model all TD jobs as critical sections for PD jobs. As long as all TD jobs are short enough, the schedulability of all PD jobs can be guaranteed using this approach. In other words, all PD jobs can meet their deadlines as long as:

$$\left(\sum_{j<i} \frac{c_j}{p_j}\right) + \frac{c_i + b_i}{p_i} < i(2^{1/i} - 1)$$

where b_i is the blocking time by all TD jobs for task τ_i. To reduce b_i, we need to make sure that all monitoring jobs have short execution times and are not clustered together.

Another useful facility for fault-tolerant real-time systems is to monitor the timing events when executing user programs. Timing faults, i.e. results produced too late or an application uses more than its share of resources, may cause the system to produce erroneous responses. The OS should provide a powerful yet efficient mechanism to detect timing faults.

In the original GSF reported in [19], the Dispatcher reports only those events when jobs are terminated (voluntarily or involuntarily). Some fault monitoring algorithms need to know the exact time when a job is executed, suspended or terminated. To support these algorithms, we have extended the original GSF feedback mechanism so that the Dispatcher sends these information about job executions to the Allocator.

5   The Implementation of Task Group in RED-Linux

Many fault-tolerant functions are implemented using the primary-backup model or the N-version model so that a specific functionality can be provided by multiple jobs. For real-time scheduling, it is important for all these jobs to share a common CPU budget so that other tasks in the system will have enough CPU time for their execution. A *group* structure is thus introduced to distinguish a set of jobs from others. Every job in the set is assigned the same group number. For example, the primary-backup model [11, 5, 4] have two jobs. The two jobs may have different start time and finish time, as well as different priorities. However, the two jobs must share a total budget. Another example is the imprecise computation model [13] where a task may have many optional jobs that can be used to enhance the result quality. The optional jobs should share a common CPU budget for the group.

To support the task group, we have implemented a hierarchical task group mechanism in RED-Linux. Similar to the concept of the hierarchical file directory structure, a task group may have other task sub-group as a member. For scheduling purposes, each task group is assigned some specific resource budget to be shared by all tasks and task sub-groups in the group. Moreover, each task group may use a different scheduling policy to assign its resources. Therefore, the Dispatcher in GSF needs the capability to adopt different scheduling policies to select the next running job from a set of jobs.

In RED-Linux, each job is in a group and has a group number. Since each sub-group is like a job in a group, each sub-group has a group number as well. Another parameter, the *server* number, is used by each sub-group to identify itself as a server (for scheduling). Each server job is associated with a job queue which holds all jobs that will be scheduled using the server job's budget. Each server can use a different policy to schedule the jobs on its queue. For normal real-time jobs, they do not have a server number. Therefore the server number of a normal real-time job is set to be the same as the group number which it belongs. In other words, a job can be identified as a *server job* if its server number is different from its group number. A server job is a scheduling unit with an allocated budget but no application job. The Dispatcher will not execute the server job. Instead, it will select another job to be served by the server. The group algorithm implemented in the RED-Linux Dispatcher is shown as follows.

1. The Dispatcher selects a job K from all eligible jobs in group 0 according to the scheduling policy of group 0.
2. The server job list J is initialized to NULL.
3. If K is a real-time job, execute it, else if K is a server job with server number i, select a job L from all eligible jobs in group i.

   (a) Set a new timer to be the minimum of the following values: the current time plus the budget of the server job K, the current time plus the budget of L, the finish time of the server job K, the finish time of L.
   (b) Append the server job K to the job list J.
   (c) Set $K = L$ and repeat Step 3.

4. When K finishes or is interrupted, reduce the actual execution time used from the budget of all jobs in the server job list J.

The maintenance of the server list is necessary so that the execution of a job will consume the budget in all groups it belongs to under the whole group hierarchy.

6 Conclusions

In this paper, we present the support for fault-tolerant real-time applications using the general scheduling framework in RED-Linux. The scheduling framework is able to accommodate a variety of scheduling models used in fault-tolerant real-time applications. By using the group mechanism, fault-tolerant primary-backup tasks may share the CPU and other resource budget efficiently. By using the time-driven scheduling, fault monitors can be executed efficiently and without jitter. By using the budget mechanism, tasks are always given their guaranteed share regardless of the possible ill behavior of other tasks.

References

1. L. Abeni and G.C. Buttazzo. Integrating multimedia applications in hard real-time systems. In *Proc. IEEE Real-Time Systems Symposium*, Dec 1998.
2. J.C.R. Bennett and H. Zhang. WF2Q: Worst-case fair weighted fair queueing. In *Proc. of IEEE INFOCOMM'96*, San Francisco, CA, pp. 120-128, March 1996.
3. S. Punnekkat and A. Burns. Analysis of checkpointing for schedulability of real-time systems. In Proceedings of IEEE Real-Time Systems Symposium, pages 198–205, San Fran- cisco, December 1997.
4. M. Caccamo and G. Buttazzo. Optimal Scheduling for Fault-Tolerant and Firm Real-Time Systems. Proceedings of IEEE Conference on Real-Time Computing Systems and Applications, Hiroshima, Japan, Oct 1998.
5. H. Chetto and M. Chetto. An adaptive scheduling algorithm for fault-tolerant real-time systems. Software Engineering Journal, May 1991.
6. A. Demers, S. Keshav, and S. Shenker. Analysis and Simulation of a Fair Queueing Algorithm. In *Journal of Internetworking Research and Experience*, pp.3-26, October 1990.
7. Ching-Chih Han, Kwei-Jay Lin and Chao-Ju Hou. Distance-constrained scheduling and its applications to real-time systems. In *IEEE Trans. Computers*, Vol. 45, No. 7, pp. 814-826, December 1996.
8. Chih-wen Hsueh and Kwei-Jay Lin. An optimal Pinwheel Scheduler Using the Single-Number Reduction Techniques. In *Proc. of IEEE Real-Time Systems Symposium*, December 1996, pp.196-205.
9. Chih-wen Hsueh and Kwei-Jay Lin. On-line Schedulers for Pinwheel Tasks Using the Time-Driven Approach. In *Proc. of the 10th Euromicro Workshop on Real-Time Systems*, Berlin, Germany, June 1998, pp. 180-187.
10. K. Jeffay et al. Proportional Share Scheduling of Operating System Service for Real-Time Applications. In *Proc. IEEE Real-Time Systems Symposium*, Madrid, Spain, pp. 480-491, Dec 1998.
11. A. L. Liestman and R. H. Campbell. A fault-tolerant scheduling problem. IEEE Transactions on Software En- gineering, 12(11):1089–95, November 1986.
12. C.L. Liu and J. Layland. Scheduling Algorithms for Multiprogramming in a Hard Real-Time Environment. *Journal of the ACM*, 20(1):46-61, 1973.
13. J.W.-S. Liu, K.J. Lin, W.-K. Shih, A.C. Yu, J.Y. Chung and W. Zhao. Imprecise Computation. In *Proc. of IEEE* 82:83-94, 1994
14. A. K. Parekh and R. G. Gallager. A Generalized Processor Sharing Approach to Flow Control in Integrated Services Networks: The Single-Node Case. In *IEEE/ACM Trans. Networking*, Vol. 1, No. 3, pp. 344-357, June 1993.
15. L. Sha. Dependable system upgrade. In *Proc. IEEE Real-Time Systems Symposium*, pp. 440-448, Dec 1998.
16. M. Spuri and G.C. Buttazzo. Efficient aperiodic service under the earliest deadline scheduling. In *Proc. IEEE Real-Time Systems Symposium*, Dec 1994.
17. I. Stoica, H. Zhang, and T.S.E. Ng. A Hierarchical Fair Service Curve Algorithm for Link-Sharing, Real-Time and Priority Services. In *Proc. of ACM SIGCOMM'97*, Cannes, France, 1997.
18. Y.C. Wang and K.J. Lin. Enhancing the real-time capability of the Linux kernel. In *Proc. of 5th RTCSA'98*, Hiroshima, Japan, Oct 1998.
19. Y.C. Wang and K.J. Lin. Implementing a general real-time framework in the RED-Linux real-time kernel. In *Proc. of RTSS'99*, Phoenix, Arizona, Dec 1999.

Are COTS Suitable for Building Distributed Fault-Tolerant Hard Real-Time Systems*?

Pascal Chevochot, Antoine Colin, David Decotigny, and Isabelle Puaut

IRISA, Campus de beaulieu, 35042 Rennes, France

Abstract For economic reasons, a new trend in the development of distributed hard real-time systems is to rely on the use of Commercial-Off-The-Shelf (COTS) hardware and operating systems. As such systems often support critical applications, they must comply with stringent real-time and fault-tolerance requirements. The use of COTS components in distributed critical systems is subject to two fundamental questions: are COTS components compatible with *hard* real-time constraints? are they compatible with fault-tolerance constraints? This paper gives the current status of the HADES project, aiming at building a distributed run-time support for hard real-time fault-tolerant applications on top of COTS components. Thanks to our experience in the design of HADES, we can give some information on the compatibility between COTS components and hard real-time and fault-tolerance constraints.

1 Introduction

Real-time systems differ from other systems by a stricter criterion of correctness of their applications. The correctness of a real-time application not only depends on the delivered result but also on the time when it is produced. Critical applications (*e.g.* flight control systems, automotive applications, industrial automation systems) often have *hard* real-time constraints: missing a task deadline may cause catastrophic consequences on the environment under control. It is thus crucial for such systems to use *schedulability analysis* in order to prove, before the system execution, that all deadlines will be met. Moreover, critical applications exhibit fault-tolerance requirements: the failure of a system component should not cause a global system failure.

Mainly due to economic reasons, a new trend in the development of distributed real-time systems is to rely on the use of Commercial-Off-The-Shelf (COTS) hardware and operating systems. As such systems often support critical applications (*e.g.* aircraft control systems), they must comply with stringent real-time and fault-tolerance requirements. The use of COTS components in distributed critical systems is subject to two fundamental questions: are COTS components compatible with *hard* real-time constraints? are they compatible with fault-tolerance constraints?

---

* This work is partially supported by the French Department of Defense (DGA/DSP), #98.34.375.00.470.75.65. Extra information concerning this work can be found in the Web page *http://www.irisa.fr/solidor/work/hades.html*.

J. Rolim et al. (Eds.): IPDPS 2000 Workshops, LNCS 1800, pp. 699-705, 2000.

The objective of our work is to build a run-time support for distributed fault-tolerant hard real-time applications from COTS components (processor and operating system). The run-time support is built as a *middleware* layer running on top of COTS components, and implements scheduling and fault-tolerance mechanisms. This work is achieved in cooperation with the French department of defense (DGA) and Dassault-Aviation. Our work is named hereafter HADES, for *Highly Available Distributed Embedded System*.

This paper gives the current status of the HADES project. Thanks to the experience gained in designing HADES, we can give some information on the compatibility between COTS components and hard real-time and fault-tolerance constraints. The remainder of this paper is organized as follows. Section 2 gives our preliminary results concerning the compatibility between COTS components and hard real-time constraints. Section 3 then deals with the issue of providing fault-tolerance facilities on COTS components. An overview of the prototype of the run-time support of HADES is given in Section 4.

2   COTS and hard real-time constraints

2.1   Methodology

Schedulability analysis is used in hard real-time systems to prove, before the system execution, that all deadlines will be met. Schedulability analysis must have static knowledge about the tasks (*e.g.* worst-case arrival pattern, worst-case execution time, deadline, synchronizations and communications between tasks). *Worst-case execution time analysis (*WCET *analysis)*, through the analysis of a piece of code, returns an upper bound for the time required to execute it on a given hardware. Our approach to be able to apply system-wide (application and run-time support) schedulability analysis to is to combine the use of a *static task model* and WCET *analysis*.

Static task model. Applications must be structured according to a task model that defines off-line:

- The internal structure of tasks: every task is described by a directed acyclic graph whose nodes model synchronization-free computations and edges model precedence constraints and data transfers between nodes.
- A set of attributes related to task execution. *Synchronization attributes* serve at expressing exclusion constraints between nodes. *Timing attributes* express temporal properties for tasks and nodes (task arrival law, deadline, earliest and latest start time). *Distribution attributes* define the site where each node of a task graph executes. *Fault-tolerance attributes* specify for each task or portion of task which replication strategy must be applied, and the requested replication degree.

Figure 1 illustrates the task model by depicting two tasks A and B, distributed on two sites named $Site_1$ and $Site_2$. Distribution attributes indicate that nodes a_1 to a_4, as well as nodes b_1 and b_2 execute on $Site_1$; nodes a_5 and b_3 execute on $Site_2$. Synchronization attributes state that a shared resource R is modified by nodes a_5 and b_3, and thus that there is an exclusion constraint

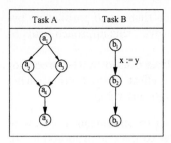

Attributes		
Task name	A	B
Synchronization	$Use(a_5, R, modified)$	$Use(b_3, R, modified)$
Distribution	$Location(a_1..a_4, Site_1)$	$Location(b_1..b_2, Site_1)$
	$Location(a_5, Site_2)$	$Location(b_3, Site_2)$
Timing	$Periodic(A, 50)$	$Periodic(B, 40)$
	$Deadline(A, 40)$	$Deadline(B, 40)$
Fault tolerance	$Active(a_2, 2, Site_1, Site_2)$	$Active(b_1, 2, Site_1, Site_2)$

Figure1. Illustration of the task model

between these two nodes. Timing attributes specify that both A and B are periodic with respective periods 40 and 50, and give the tasks deadlines. Finally, fault-tolerance attributes specify that nodes a_2 and b_1 must be made reliable by using active replication of degree 2 and that the node replicas must be located on $Site_1$ and $Site_2$ (see § 3).

The interest of using the above static task model is that it forces the application developer to provide all informations needed to achieve schedulability analysis.

Use of WCET analysis. To achieve schedulability analysis, the WCETs of tasks have to be known. They can be obtained through the actual execution of tasks, or through the analysis of their code (WCET analysis). We propose to use WCET analysis because it produces a *safe* estimation of the tasks WCETs. A WCET analysis tool named HEPTANE (*Hades Embedded Processor Timing ANalyzEr*) has been developed. It analyses C code and produce timing information for the Pentium processor. HEPTANE operates on two program representations: the program syntax tree, obtained through the analysis of the program source code, and the program control flow graph, obtained through the analysis of the program assembly code generated by the C compiler. User-provided annotations are used to identify the worst-case execution path for loops. HEPTANE is used to obtain the WCET of nodes of application graphs, which, by construction are blocking-free, and to obtain the WCET of the run-time support itself (operating system, and the middleware layer we have developed to provide services for fault-tolerance). In order to be suited to WCET analysis, the middleware layer has been structured into two layers: (i) a low layer, written directly in C and designed so that it never blocks; (ii) a high layer, made of tasks structured using the static task model described above, so that blocking points are statically identified.

Applying WCET analysis to systems that use COTS components raises a number of issues, which are described in the two following paragraphs.

2.2 WCET analysis and COTS hardware

COTS processor include architectural features, such as instruction caches, pipelines and branch prediction. These mechanisms, while permitting performance improvements, are sources of complexity in terms of timing analysis. A trivial approach to deal with them is to act if they were not present (*i.e.* to suppose all memory accesses lead to cache misses and assume there is no parallelism between the execution of successive instructions). However, this approach leads to largely overestimated WCETs.

In order to reduce the pessimism of WCET analysis caused by the processor microarchitecture, HEPTANE takes into account the effect of instruction cache, pipeline and branch prediction when computing programs WCETs.

- *Pipeline.* The presence of pipelines is considered by simulating the flow of instructions in the pipelines, in a method similar to the one proposed in [1].
- *Instruction cache.* Consideration of instruction cache uses static cache simulation [2]: every instruction is classified according to its worst-case behavior with respect to the instruction cache. The instruction classification process uses both the program syntax tree and control flow graph.
- *Branch prediction.* An approach similar to the one used for the instruction cache was used to integrate the effect of branch prediction (see [3] for details). Experimental results show that the timing penalty due to wrong branch predictions estimated by the proposed technique is close to the real one, which demonstrates the practical applicability of our method: from 98% to 100% of results of predictions can be known statically on a set of small benchmark programs. To our knowledge, this work is the first attempt to incorporate the effect of branch prediction on WCET analysis.

2.3   WCET analysis and COTS real-time operating systems

The first obstacle to the use WCET analysis on COTS real-time operating systems is to obtain their source code. However, at least for small kernels targeted for embedded applications, more and more operating systems come with their source code at reasonable cost.

We have undertaken a study aiming at using WCET analysis to obtain the WCETs of the system calls of the RTEMS real-time kernel, restricted for the sake of the experimentation to act on a monoprocessor architecture. We summarize below the results of the study, showing that, to some extent, the structure of the source code of RTEMS is suited to WCET analysis (for details, see [4]).

The first conclusion of this study is that using WCET analysis to obtain the WCET of the system calls of the RTEMS real-time kernel is feasible. The central part of RTEMS was analyzed in less than two months by a student having no a priori knowledge of the internals of RTEMS and WCET analysis. During the study, we discovered interesting properties of the code of RTEMS, that hopefully exist in other real-time operating systems, and that can have an influence of the construction of WCET analyzers suited to the analysis of operating systems: small number of loops, absence of recursion, small number of function calls performed using function pointers.

Most dynamic function calls (*i.e.* function calls through function pointers) could have been replaced by static ones.

We observed that finding the maximum number of iterations for 75% of the loops required an in-depth study of the source code of RTEMS. We found rather pessimistic bounds for a number of loops (dynamic memory allocation routine, scheduler). Bounds for these loops depend on the operational conditions, such as the arrival law of interrupts or the actual number of active tasks at a given priority.

3   COTS and fault tolerance constraints

3.1   Methodology

Several issues have to be dealt with in order to use COTS components (processor and operating systems) to support applications with fault-tolerance requirements. First, COTS hardware generally do not include any fault-tolerance mechanism: messages may get lost on the network and processors may crash or produce incorrect results. Thus, some form of redundancy (spatial and/or temporal) must be used to support a faulty hardware component. Second, most COTS operating systems themselves are not designed to mask the failure of hardware components (*e.g.* machine crashes, network omissions). Consequently, any error-masking mechanism must be provided as a software layer outside the operating system.

Our approach to support applications with fault tolerance constraints is to combine the use of *off-line task replication* and *basic fault-tolerance mechanisms*.

- Off-line task replication (see § 3.2) transforms the task graphs (see § 2.1) to make portions of tasks fault-tolerant through the use of replication. Off-line replication relies on a number of properties that are verified by the run-time support (fail-silence assumption, bounded and reliable communications).
- Basic fault-tolerance mechanisms are provided by the run-time support (see § 3.3) in order to verify these assumptions. Error detection mechanisms have been designed to provide the highest coverage as possible of the fail-silence assumption. A set of mechanisms (group membership, clock synchronization, multicast) altogether guarantee bounded and reliable communications.

3.2   Off-line task replication

Application tasks are made fault-tolerant through the replication of parts of their code (transformation of their structure, defined as a graph, see § 2.1). Task replication is achieved *off-line*. An extensible set of error treatment strategies based on task replication (currently active, passive and semi-active replication) is provided. Task replication is achieved thanks to a tool we have developed, which is named HYDRA [5]. It takes as input parameters the portions of code to be replicated, the replication strategy to be applied and the replication degree.

Figure 2 illustrates the application of active replication on task A taken as example in Figure 1, where node a_2 is made reliable through the use of active replication of degree 2 on $Site_1$ and $Site2$. The replication tool transforms the graph of task A through the addition of: *(i)* two replicas for node a_2 that must be made fault-tolerant (nodes a_2^a and a_2^b); *(ii)* a node that computes a consensus value for the outputs of a_2^a and a_2^b; *(iii)* new edges in the graph.

The main interest of off-line replication is that replicas can be taken into account easily in schedulability analysis, even if only portions of tasks are replicated (see [6]).

Off-line replication is correct as far as the underlying run-time support in charge of executing tasks ensures a set of properties (see [5]) like for instance the

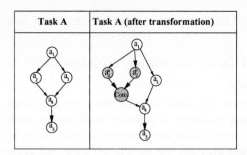

Figure2. Example of application of active replication

exe-fail-silence property, stating that the run-time support of a given site either sends correct results in the time and value domains, or remains silent forever.

3.3 Basic fault-tolerance mechanisms and COTS components

Most COTS operating systems are not designed to mask faulty hardware components. Thus, we have designed a set of fault-tolerance mechanisms, embedded in a middleware layer running on top of a COTS real-time kernel (in our prototype Chorus and RTEMS). Mechanisms fall into two categories:

- *error detection mechanisms:* these mechanisms detect (as far as possible) value and temporal errors and transform them into machine stops in order to reach the fail-silence assumption (property *exe-fail-silence:* given above).
- *fault masking mechanisms:* these mechanisms allow to mask communications faults and site crashes through the provision of fault-tolerant services: group membership service, reliable time-bounded multicast service and clock synchronization service (see [7] for details on these services). The properties of these services are of great importance to ensure *exe-timeliness* and *exe-agreement* properties. These services have been designed using the static task model of § 2.1 to ease their integration in schedulability analysis.

The design of services such as group membership, multicast and clock synchronization on COTS components did not cause any intractable difficulty. However, the problem is that all these services are entirely implemented in software on top of an existing operating system, which leads to modest worst-case performances (approximately two orders of magnitude worse than hardware-implemented solutions like the TTP/C communication chips). An estimation of the influence of these worst-case performance on the range of deadlines that can be supported has still to be achieved.

A set of error detection mechanisms has been integrated into the run-time support. To detect timing errors, the run-time support includes monitoring code to *(i)* detect deadline and WCET exceeding; *(ii)* check that tasks arrival laws conform to their expected arrival laws. These checks are possible because the run-time support is aware of semantic information about the tasks it executes, thanks to the use of a static task model. To detect value errors, the run-time support *(i)* catches every hardware exception (*e.g.* parity error, alignment error, protection violation, division by zero); *(ii)* checks if a set of implementation-dependent invariants hold. We are currently estimating the coverage of the fail-silence assumption provided by these mechanisms.

4   Experimental platform

A prototype of the run-time support of HADES has been developed (see [7] for details). It runs on a network of Pentium PCs, and has been ported on Chorus and RTEMS. It is divided into two layers: (i) a kernel, developed in C, in charge of executing tasks that are developed according to the task model of § 2.1, as well as detecting errors; (ii) a set of services, most of them implementing fault-masking mechanisms. Services are developed according to the task model of § 2.1 to be able to apply system-wide schedulability analysis.

We are currently porting an avionic application provided by our industrial partner Dassault-Aviation on top of our prototype.

5   Concluding remarks

This paper has given current status of the HADES project, aiming at building a distributed run-time support for applications with hard real-time and fault-tolerance constraints. Concerning the support of hard real-time constraints, we highly rely on the use of WCET analysis to provide information for schedulability analysis. We have shown that using this class of techniques on COTS hardware and operating system is feasible. We are currently studying the pessimism induced by the analysis, and are studying the use of WCET analysis on larger (and therefore more realistic) operating systems. Concerning fault-tolerance constraints, we have built on top of a COTS kernel predictable fault tolerance mechanisms (fault masking mechanisms such as reliable multicast and clock synchronization, error detection mechanisms to enforce the fail-silence assumption). Designing fault masking mechanisms on top of a real-time kernel did not cause any particular problem, but is not efficient due to the fact that all these mechanisms are implemented entirely in software. The impact of software-implemented mechanisms on the range of deadlines that can be supported is currently under study. We are currently evaluating the efficiency of the error detection mechanisms that have been integrated into the run-time support.

References

[1] N. Zhang, A. Burns, and M. Nicholson. Pipelined processors and worst case execution times. *Real-Time Systems*, 5(4):319–343, October 1993.

[2] F. Mueller. *Static Cache Simulation and its Application*. PhD thesis, Departement of Computer Sciences, Florida State University, July 1994.

[3] A. Colin and I. Puaut. Worst case execution time analysis for a processor with branch prediction. *Real-Time Systems*, 2000. To appear.

[4] A. Colin and I. Puaut. Worst-case timing analysis of the RTEMS real-time operating system. Technical Report 1277, IRISA, November 1999.

[5] P. Chevochot and I. Puaut. An approach for fault-tolerance in hard real-time distributed systems. Technical Report 1257, IRISA, July 1999. A short version of this paper can be found in the WIP session of SRDS'18 p. 292–293.

[6] P. Chevochot and I. Puaut. Scheduling fault-tolerant distributed hard real-time tasks independently of the replication strategies. In *Proc. of RTCSA'99*, Hong-Kong, China, December 1999.

[7] E. Anceaume, G. Cabillic, P. Chevochot, and I. Puaut. A flexible run-time support for distributed dependable hard real-time applications. In *Proc. of ISORC'99*, pages 310–319, St Malo, France, May 1999.

Autonomous Consistency Technique in Distributed Database with Heterogeneous Requirements

Hideo Hanamura, Isao Kaji and Kinji Mori

Tokyo Institute of Technology, 2-12-1 Ookayama, Meguro, Tokyo 125, Japan

Abstract. Recently, the diversified types of companies have been trying to cooperate among them to cope with the dynamic market and thus integration of their DBs with heterogeneous requirements is needed. But the conventional centralized approach has problems of fault-tolerance, real-time property and flexibility.

Here, the autonomous consistency technique in distributed DB and its architecture are proposed. In this architecture, each site has the autonomy to determine Allowable Volume and to update the DB independently using it. In addition, this volume can be managed dynamically and successfully through autonomous communication among sites, and the system can achieve the adaptation to unpredictable user requirements. As an experimental result, it is shown that this mechanism can adaptively achieves users heterogeneous requirements.

1 Background

As the Information Technology advances, the information sharing among companies is essential for business and the integration of their databases(DBs) becomes more important. Though such kinds of DBs have been integrated by centralized approach [2], it has problems of fault-tolerance, real-time property and flexibility. Therefore, the autonomous approach is proposed to solve these problems. In this paper, the stock management system in Supply Chain Management(SCM) is discussed to explain the new approach.

1.1 Needs in SCM

Only the makers and retailers are considered as the constituents in the SCM. The characteristics of heterogeneous requirements of makers and retailers are described as below.

Retailers Retailers are considered to be dealing in two kinds of products, regular products and non-regular products.

Regular products are usually in stock at retailers in enough quantity. When some customers order this kind of products, the retailers ship them from their own stock. If they do not have enough stock, they order them to makers. On the

J. Rolim et al. (Eds.): IPDPS 2000 Workshops, LNCS 1800, pp. 706-712, 2000.

other hand, non-regular products are not usually in stock at retailers. When the retailers receive the order from customers, they order them to the makers and the makers manufacture them at that time.

Makers Makers deal in both regular products and non-regular products in the same way, namely, check the current stock and manufacture them, if necessary.

2 Approach

2.1 Assurance

The **Assurance** is defined as the achievement of the user satisfaction with heterogeneous requirements in the integrated system.

Assurance Usually, a system is constructed to achieve a single user requirement. But the integrated system has heterogeneous requirements, which are realized in each system before the integration. So, These requirements must be realized in the integrated one, even if they are contradictory. When these heterogeneous requirements are realized in the integrated system – if they are contradictory, the system satisfies them fairly –, it is defined that the integrated system realizes the assurance.

2.2 Goal

The following properties are required in the integrated DB system in SCM.

- Real-Time Property
- Fault Tolerance
- Assurance

As mentioned above, the system must realize the assurance. As for the non-regular products, the requirements are same between maker and retailer and the system realizes the **Immediate Update**, which propagates the result of update operation to all the system immediately.

While, in case of the regular products, the requirements are contradictory between maker and retailer. In this case, the real-time property of update at retailers site is given the priority and the result is propagated to all the system at the earliest. which is called as **Delay Update**.

3 Accelerator

3.1 Allowable Volume

The centralized approach is difficult to achieve the assurance when conflicting requirements coexist in a system or user requirements are changing rapidly. Because it does not provide the operational autonomy at each site which is essential in these cases. So, the autonomous consistency mechanism is introduced.

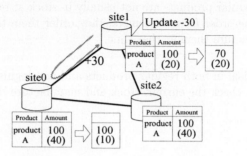

Fig. 1. Example of Allowable Volume

The attribute called as the Allowable Volume(AV) is introduced at each site. The AV is defined on each numeric data in each local DBs. Each site can update the numeric data within it autonomously – without any communication with others. The AV is not fixed volume allocated by some master site but is flexibly managed by communication among sites.

The update of DB with AV is illustrated in Fig. 1, in which each site has 40, 20 and 40 of AV for the product A, totally 100. The stock data of the product A is also 100. If some user updates -30 for product A at site 1, the AV is checked at the site 1 and both the AV and stock data are updated without any communication if the AV is enough. Otherwise, the site requests other sites to transfer the AV. In this figure, since the AV at site 1 is not enough to update, the site 1 requests the AV to site 0 and get +30, then the AV and data at site 1 are updated into 20 and 70.

3.2   System Model in SCM

The structure of the proposed system is shown in Fig. 2.

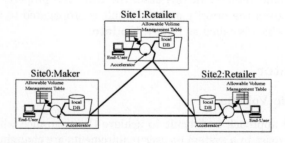

Fig. 2. System Model

In the proposed system, each site has a local DB and they participate in a distributed DB. One of local DBs is called as the base DB which is usually located at makers and works as the primary copy in the case of Immediate Update. The others are located at retailers. The content of all local DBs are the same, which include product names and amount of their stock. In addition, the classification between regular and non-regular products is known. All data are assumed to be delivered to all the sites initially from the base DB.

3.3 Accelerator

The accelerator is proposed to achieve both Immediate Update and Delay Update for realizing the assurance for makers and retailers. It is located at each site with Allowable Volume management table(AV table) and provides DBMS function and AV management function. The AV management function consists of three functions, which is **checking**, **selecting** and **deciding**.

The **checking** function is the function to check the type of user update request with AV table, Immediate Update and Delay Update, when the accelerator receives user update requests. The **selecting** function selects the site to request the transfer of the AV from/to. The **deciding** function decides the volume of AV needed to transfer from/to the other sites, depending on the situation to realize effective AV reallocation.

The accelerator realizes both Immediate Update and Delay Update by using these functions. In following, the behavior of the accelerator is explained.

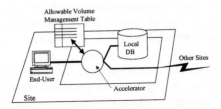

Fig. 3. Accelerator

Delay Update When the accelerator receives the update request, it checks using AV table whether it is Immediate Update or Delay Update. If the AV is defined on it, the accelerator distinguishes it as Delay Update and holds the necessary amount of AV in advance of the update of local DB. In this case, it is not necessary to lock the AV exclusively until the completion of whole transaction. Because the AV is not seen from end-users and if rollback of transaction occurs, the recovery of operation can be done by updating with opposite of update volume, which was used for the transaction at first. Then extra AV can be used by other process while one process accesses the same data.

If the AV is sufficient at the local site, the accelerator updates the AV and completes the update at the local site. But if the AV is not sufficient for update, the accelerator holds all the AV at the site and requests other sites for extra AV. The target site for the request is determined according to the strategy of the accelerator such as the order of the volume the other sites keep. On the other hand, the site receiving the request provides some amount of AV according to its strategy. When the requesting accelerator receives the AV, it checks whether it is sufficient for the update. It requests again to other sites if it is not sufficient. After it gets the enough volume, it updates both data and AV and remaining AV is stored at the local AV table. Otherwise, all accumulated AV is stored in the local AV table.

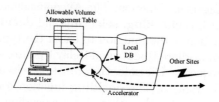

Fig. 4. Delay Update with AV

Immediate Update When the accelerator checks the user requests and finds the AV is not defined on the AV table, it deals with the request as Immediate Update based on primary copy scheme. The requesting accelerator works as the coordinator and the update is processed. At first, it locks the data at the local DB and it also sends the lock request to the other accelerators simultaneously. Then the operations for update are processed at all the sites and ready and commitment messages are exchanged. The requesting accelerator judges the completion of the update with the message from the accelerator at the base DB.

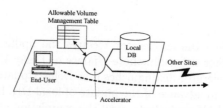

Fig. 5. Immediate Update

3.4   AV management

Each site has its own strategy to manage AV to realize the real-time property and the assurance. But the optimal AV allocation algorithm using global information is not suitable for AV management. Because it is more important to adapt to changing user requirements rapidly. While determining the mechanism of AV management, it is essential to calculate the volume of AV transfer using local information and to make AV circulate among the sites. Then the system achieves the real-time property and the assurance.

4   Simulation

The simulation for the Delay Update is done with the proposed mechanism. A maker and two retailers are modeled as Fig. 2. For AV management, the algorithm which is proposed in the research for electronic commerce money distributed system [1] is utilized. That is, the AV management is occurred in the case when the Allowable Volume is insufficient for update. The requested site is selected according to the amount of AV the site keeps, which information is collected at the necessary communication for AV management and may not be current data. In addition, the requested AV is the amount of shortage needed for the completion of update. And allocated AV is half of the amount of AV that the site keeps.

In the simulation, the number of data items in local DB is 100. In site 0, data is updated to increase the volume by at most 20% of the initial amount of data randomly. On the other hand, at site 1 and site 2, it is updated to decrease at most 10% randomly. The result is shown in the Fig. 6 and Table. 1.

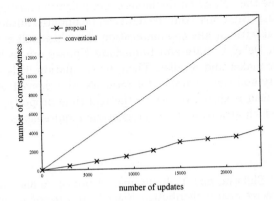

Fig. 6. Number of Updates vs Number of Correspondences

Site	1000	2000	3000	4000	5000	6000	7000	8000
Site1	195	416	669	973	1441	1516	1614	1997
Site2	195	422	634	905	1335	1473	1591	1996

Table 1. Number of Correspondences for Update

In the Fig. 6, the total number of updates in the system is described as the horizontal axis and the number of correspondences for update is the vertical axle (2 messages are counted as 1 correspondence). The line "conventional" shows the number of correspondences for update by the conventional centralized way. The figure shows that the proposed way decreases the correspondences by 75% and most of the update is completed within the local site. Thus the real-time property is attained.

In the Table. 1, it shows the number of correspondences for update in each site. In proposed way, the numbers are almost same between site 1 and site 2 and increases very slowly. That is, it is shown that the real-time property is fairly achieved at the retailer sites – site 1 and site 2.

As the result, in the case of Delay Update, the proposed mechanism is shown to improve the real-time property by decreasing the number of correspondences and to realize the assurance for retailers.

5 Conclusion

In this paper, it is focused on the problem of the heterogeneity of the requirements according to the needs of cooperation among companies in SCM. Then the autonomous approach is proposed to solve this problem. In the proposal, each site can determine the AV, by autonomous communication among them, according to the changing situations. And the data can be updated autonomously at the local site within it without any communication to realize fault tolerance. By the accelerator, both Delay Update and Immediate Update are realized to achieve the assurance for maker and retailer. Then, by simulation, it is shown that the real-time property and the assurance for retailers are achieved. As a result, the proposed mechanism is shown to realize the real-time property, fault-tolerance and assurance, which attains users heterogeneous requirements.

References

1. H. Kawazoe, T. Shibuya, and T. Tokuyama. "optimal on-line algorithms for an electronic commerce money distribution system". In *Proceedings of the 10th ACM SIAM Symposium on Discrete Algorithms (SODA 99)*, pages 527–536, 1999.
2. Amit P. Sheth and James A.Larson. "federated database systems for managing distributed, heterogeneous, and autonomous databases". *ACM Comupting Surveys*, 22(3):183–236, 1990.

Real-time Transaction Processing
Using Two-stage Validation in Broadcast Disks*

Kwok-wa Lam[1], Victor C. S. Lee[1], and Sang H. Son[2]

[1] Department of Computer Science, City University of Hong Kong
csvlee@cityu.edu.hk
[2] Department of Computer Science, University of Virginia
son@cs.virginia.edu

Abstract. Conventional concurrency control protocols are inapplicable in mobile computing environments due to a number of constraints of wireless communications. In this paper, we design a protocol for processing mobile transactions. The protocol fits the environments such that no synchronization is necessary among the mobile clients. Data conflicts can be detected between transactions at the server and mobile transactions by a two-stage validation mechanism. In addition to relieving the server from excessive validation workload, no to-be-restarted mobile transactions will be submitted to the server for final validation. Such early data conflict detection can save processing and communication resources. Moreover, the protocol allows more schedules of transaction executions such that unnecessary transaction aborts can be avoided. These desirable features help mobile transactions to meet their deadlines by removing any avoidable delays due to the asymmetric property.

1 Introduction

Broadcast-based data dissemination becomes a widely accepted approach to communication in mobile computing environments [1], [2], [6], [8]. The distinguishing feature of this broadcast mode is the communication bandwidth asymmetry, where the "downstream" (server to client) communication capacity is relatively much greater than the "upstream" (client to server) communication capacity. The limited amount of bandwidth available for the clients to communicate with the broadcast server in such environments places a new challenge to implementing transaction processing efficiently. Acharya et al [1], [2] introduced the concept of Broadcast Disks (Bdisks), which uses communication bandwidth to emulate a storage device or a memory hierarchy in general for mobile clients of a database system. It exploits the abundant bandwidth capacity available from a server to its clients by broadcasting data to its clients periodically.

Although there is a number of related research work in Bdisks environments [1], [3], [4], only a few of them support transactional semantics [5], [8]. In this

* Supported in part by Direct Allocation Grant no. 7100094 from City University of Hong Kong

J. Rolim et al. (Eds.): IPDPS 2000 Workshops, LNCS 1800, pp. 713-719, 2000.
© Springer-Verlag Berlin Heidelberg 2000

paper, we propose a new approach to processing transactions in Bdisks environments where updates can be originated from the clients. In the new protocol, mobile clients share a part of the validation function with the server and are able to detect data conflicts earlier such that transactions are more likely to meet their deadlines [9].

2   Issues of Transaction Processing in Broadcast Environments

Since the bandwidth from mobile clients to server is limited, concurrency control protocols, which require continuous synchronization with the server to detect data conflicts during transaction execution such as two phase locking, become handicapped in these environments. This is the reason why almost all the protocols proposed for such environments are based on optimistic approach. The eventual termination (commit or abort) of mobile transactions submitted to the server will be broadcast to the clients in the following broadcast cycles. If the mobile transaction submitted to the server could not pass the validation, it will take a long time for the client to be acknowledged and to restart the failed transaction. For a huge number of clients, this strategy will certainly cause intolerable delays and clutter the server. Consequently, it will have a negative impact on the system performance in terms of response time and throughput.

In addition, there are two major problems using conventional optimistic approach in Bdisks environments. First, any "serious" conflicts which leads to a transaction abort can only be detected in the validation phase at the server. Therefore, some transactions, which are destined to abort when submitted to the server, are allowed to execute to the end. Such continuation of execution of these to- be-aborted transactions wastes the processing resources of the mobile clients as well as the communication bandwidth. Second, the ineffectiveness of the validation process adopted by the protocols at the server leads to many unnecessary transaction aborts and restarts because they have implicitly assumed that committed transactions must precede the validating transaction in the serialization order [7].

3   Protocol Design

We assume a central data server that stores and manages all the data objects at the server. Updates submitted from the clients are subject to final validation so that they can be ultimately installed in the central database. Data objects are broadcast by the server and the clients listen to the broadcast channel to perform the read operations. When a client performs a write operation, it pre-writes the value of the data object in its private workspace.

3.1   Broadcasting of Validation Information

We use the timestamp intervals of transactions to reduce the unnecessary aborts and exploit the capability of the server to broadcast validation information to the clients so that the clients can adjust the timestamp intervals of their respective active transactions. Clearly, this method is not supposed to guarantee that the mobile transactions can be terminated (committed or aborted) locally at the clients. It is because the clients do not have a complete and up-to-date view of all conflicting transactions. Thus, all transactions have to be submitted to the server for final validation. Our new strategy places part of the validation function to the clients. In this way, the validation is implemented in a truly distributed fashion with the validation burden shared by the clients. One important issue is that the server and the clients should avoid repeating same part of the validation function. In other words, they should complement each other.

3.2 Timestamp Ordering

We assume that there are no blind write operations. For each data object, a read timestamp (RTS) and a write timestamp (WTS) are maintained. The values of $RTS(x)$ and $WTS(x)$ represent the timestamps of the youngest committed transactions that have read and written data object x respectively. Each active transaction, T_a, at the clients is associated with a timestamp interval, $TI(T_a)$, which is initialised as $[0, \infty)$. The $TI(T_a)$ reflects the dependency of T_a on the committed transactions and is dynamically adjusted, if possible and necessary, when T_a reads a data object or after a transaction is successfully committed at the server. If $TI(T_a)$ shuts out after the intervals are adjusted, T_a is aborted because a non-serializable schedule is detected. Otherwise, when T_a passes the validation at the server, a final timestamp, $TS(T_a)$, selecting between the current bounds of $TI(T_a)$, is assigned to T_a. Let us denote $TI_{lb}(T_a)$ and $TI_{ub}(T_a)$ the lower bound and the upper bound of $TI(T_a)$ respectively. Whenever T_a is about to read (pre- write) a data object written (read) by a committed transaction T_c, $TI(T_a)$ should be adjusted such that T_a is serialized after T_c. Let's examine the implication of data conflict resolution between a committed transaction, T_c, and an active transaction, T_a, on the dependency in the serialization order. There are two possible types of data conflicts that can induce the serialization order between T_c and T_a such that $TI(T_a)$ has to be adjusted.

1) $RS(T_c) \cap WS(T_a) \neq \{\}$ (read-write conflict)
This type of conflict can be resolved by adjusting the serialization order between T_c and T_a such that $T_c \rightarrow T_a$. That is, T_c precedes T_a in the serialization order so that the read of T_c is not affected by T_a's write. Therefore, the adjustment of $TI(T_a)$ should be : $TI_{lb}(T_a) > TS(T_c)$.

2) $WS(T_c) \cap RS(T_a) \neq \{\}$ (write-read conflict)
In this case, the serialization order between T_c and T_a is induced as $T_a \rightarrow T_c$. That is, T_a precedes T_c in the serialization order. It implies that the read of T_a is placed before the write of T_c though T_c is committed before T_a. The adjustment of $TI(T_a)$ should be : $TI_{ub}(T_a) < TS(T_c)$. Thus, this resolution makes it possible for a transaction, which precedes some committed transactions in the serialization order, to be validated and committed after them.

4   The New Protocol

4.1   Transaction Processing at Mobile Clients

The clients carry three basic functions:- (1) to process the read/write requests of active transactions, (2) to validate the active transactions using the validation information broadcast in the current cycle, (3) to submit the active transactions to the server for final validation. These three functions are described by the algorithms **Process**, **Validate**, and **Submit** as shown below and the validation information consists of the following components.

- The *Accepted* and *Rejected* sets contain the identifiers of transactions successfully validated or rejected at the server in the last broadcast cycle.
- The *CT_ReadSet* and *CT_WriteSet* contain data objects that are in the read set and the write set of those committed transactions in the *Accepted* set.
- The $RTS(x)$, a read timestamp and, $FWTS(x)$ and $WTS(x)$, the first and the last write timestamps in the last broadcast cycle, are associated with each data object x in *CT_ReadSet* and *CT_WriteSet*. $FWTS(x)$ is used to adjust $TI_{ub}(T_a)$ of an active transaction T_a for the read-write dependency while $WTS(x)$ is used to adjust $TI_{lb}(T_a)$ for the write-read dependency.

Functions: Process, Validate, and Submit at the Clients
Process (T_a, x, op)
{ if (op = READ)
  { $TI(T_a) := TI(T_a) \cap [WTS(x), \infty)$;
    if $TI(T_a) = []$ then abort T_a;
    else
    { Read(x);
      $TOR(T_a, x) := WTS(x)$;
      $Final\_Validate(T_a) := Final\_Validate(T_a) \cup \{x\}$; }
  }
  if (op = WRITE)
  { $TI(T_a) := TI(T_a) \cap [RTS(x), \infty)$;
    if $TI(T_a) = []$ then abort T_a;
    else
    { Pre-write(x);
      remove x from $Final\_Validate(T_a)$; }
  }
}

Validate
{ // results of previously submitted transactions
  for each T_v in *Submitted*
  { if $T_v \in Accepted$ then
    { mark T_v as committed;
      $Submitted := Submitted - \{T_v\}$; }
    else
    { if $T_v \in Rejected$ then

      { mark T_v as aborted;
        restart T_v;
        $Submitted := Submitted - \{T_v\}$; }
    }
  }
  for each active transaction (T_a)
  { if $x \in CT\_WriteSet$ and $x \in CWS(T_a)$ then
     abort T_a;
    if $x \in CT\_WriteSet$ and $x \in Final\_Validate(T_a)$ then
    { $TI(T_a) := TI(T_a) \cap [0, FWTS(x)]$;
     if $TI(T_a) = []$ then abort T_a;
     else remove x from $Final\_Validate(T_a)$; }
    if $x \in CT\_ReadSet$ and $x \in CWS(T_a)$ then
    { $TI(T_a) := TI(T_a) \cap [RTS(x), \infty)$;
     if $TI(T_a) = []$ then abort T_a; }
  }
}

Submit (T_a)
{ $Submitted := Submitted \cup \{T_a\}$;
  Submit to the server for global final validation
    with $TI(T_a), RS(T_a), WS(T_a), New\_Value(T_a, x)$,
      $Final\_Validate(T_a), TOR(T_a, x)$
    // x of $TOR(T_a, x) \in (WS(T_a) \cup Final\_Validate(T_a))$;
}

4.2   The Server Functionality

The server continuously performs the following algorithm until it is time to broadcast the next cycle. In essence, the server performs two basic functions: (1) to broadcast the latest committed values of all data objects and the validation information and (2) to validate the submitted transactions to ensure the serializability. One objective of the validation scheme at the server is to complement the local validation at clients to determine whether the execution of transactions is globally serializable. Note that the server does not need to perform the validation for those read operations of the validating transactions that have already done at the clients. Only the part of validation that cannot be guaranteed by the clients is required to be performed. At the server, we maintain a validating transaction list that enqueues the validating transactions submitted from the clients, but not yet processed.

    The server maintains the following information: a read timestamp $RTS(x)$ and a write timestamp $WTS(x)$ for each data object x. Each data object x is associated with a list of k write timestamp versions, which are the timestamps of the k most recently committed transactions that wrote x. For any two versions, $WTS(x, i)$ and $WTS(x, j)$, if $i < j$, then $WTS(x, i) < WTS(x, j)$. The latest version is equal to $WTS(x)$. Note that this is not a multiversion protocol as only one version of the data object is maintained.

Validation at the Server
Global_Validate (T_v)
{ Dequeue a transaction in the validating transaction list.
  for each x in $WS(T_v)$
  { if $WTS(x) > TOR(T_v, x)$ then
    { abort T_v;
      $Rejected := Rejected \cup \{T_v\};$ }
    else
    { $TI(T_v) := TI(T_v) \cap [RTS(x), \infty);$
      if $TI(T_v) = []$ then
      { abort T_v;
        $Rejected := Rejected \cup \{T_v\};$ }
    }
  }
  for each x in $Final\_Validated(T_v)$
  { Locate $WTS(x, i) = TOR(T_v, x)$
    if FOUND then
    { if $WTS(x, i + 1)$ exists then
        $TI(T_v) := TI(T_v) \cap [0, WTS(x, i + 1)];$
      if $TI(T_v) = []$ then
      { abort T_v;
        $Rejected := Rejected \cup \{T_v\};$ }
    }
    else
    { abort T_v;
      $Rejected := Rejected \cup \{T_v\};$ }
  }
  // transaction passes the final validation
  $TS(T_v) :=$ lower bound of $TI(T_v) + \epsilon$ // ϵ is a sufficient small value
  for each x in $RS(T_v)$
    if $TS(T_v) > RTS(x)$ then
      $RTS(x) := TS(T_v);$
  for each x in $WS(T_v)$
    $WTS(x) := TS(T_v);$
  $Accepted := Accepted \cup \{T_v\};$
  $CT\_WriteSet := CT\_WriteSet \cup WS(T_v);$
  $CT\_ReadSet := CT\_ReadSet \cup \{RS(T_v) - WS(T_v)\};$
}

5   Conclusions and Future Work

In this paper, we first discuss the issues of transaction processing in broadcast environments. No one conventional concurrency control protocol fits well in these environments due to a number of constraints in the current technology in wireless communication and mobile computing equipment. Recent related research on this area is mainly focused on the processing of read-only transactions. Update mobile transactions are submitted to the server for single round validation. This strategy suffers from several deficiencies such as high overhead, wastage of resources on to-be-restarted transactions, and many unnecessary transaction restarts. These deficiencies are detrimental to transactions meeting their deadlines.

To address these deficiencies, we have designed a concurrency control protocol in broadcast environments with three objectives. Firstly, data conflicts should be detected as soon as possible (at the mobile clients side) such that both processing and communication resources can be saved. Secondly, more schedules of transaction executions should be allowed to avoid unnecessary transaction aborts and restarts since the cost of transaction restarts in mobile environments is particularly high. Finally, any synchronization or communication among the mobile clients or between the mobile clients and the server should be avoided or minimized due to the asymmetric property of wireless communication. These are very desirable features in real-time applications where transactions are associated with timing constraints.

References

1. Acharya S., M. Franklin and S. Zdonik, "Disseminating Updates on Broadcast Disks," Proc. of 22nd VLDB Conference, India, 1996.
2. Acharya S., R. Alonso, M. Franklin and S. Zdonik, "Broadcast Disks: Data Management for Asymmetric Communication Environments," Proc. of the ACM SIGMOD Conference, U.S.A., 1995.
3. Baruah S. and A. Bestavros, "Pinwheel Scheduling for Fault-Tolerant Broadcast Disks in Real-Time Database Systems," Technical Report TR-1996-023, Computer Science Department, Boston University, 1996.
4. Bestavros A., "AIDA-Based Real-Time Fault-Tolerant Broadcast Disks," Proc. of the IEEE Real-Time Technology and Applications Symposium, U.S.A. 1996.
5. Herman G., G. Gopal, K. C. Lee and A. Weinreb, "The Datacycle Architecture for Very High Throughput Database Systems," Proc. of the ACM SIGMOD Conference, U.S.A. 1987.
6. Imielinski T and B. R. Badrinath, "Mobile Wireless Computing: Challenges in Data Management," Communication of the ACM, vol. 37, no. 10, 1994.
7. Lam K. W., K. Y. Lam and S. L. Hung, "Real-time Optimistic Concurrency Control Protocol with Dynamic Adjustment of Serialization Order," Proc. of the IEEE Real-Time Technology and Applications Symposium, pp. 174-179, Illinois, 1995.
8. Shanmugasundaram J., A. Nithrakashyap, R. Sivasankaran, K. Ramamritham, "Efficient Concurrency Control for Bdisks environments," ACM SIGMOD International Conference on Management of Data, 1999.
9. Stankovic, J. A., Son, S. H., and Hansson, J., "Misconceptions about Real-Time Databases," Computer, vol. 32, no. 6, pp. 29-37, 1999.

Using Logs to Increase Availability in Real-Time Main-Memory Database

Tiina Niklander and Kimmo Raatikainen

University of Helsinki, Department of Computer Science
P.O. Box 26 (Teollisuuskatu 23), FIN-00014 University of Helsinki,Finland
{tiina.niklander,kimmo.raatikainen}@cs.Helsinki.FI

Abstract. Real-time main-memory databases are useful in real-time environments. They are often faster and provide more predictable execution of transactions than disk-based databases do. The most reprehensible feature is the volatility of the memory. In the RODAIN Database Architecture we solve this problem by maintaining a remote copy of the database in a stand-by node. We use logs to update the database copy on the hot stand-by. The log writing is often the most dominating factor in the transaction commit phase. With hot stand-by we can completely omit the disk update from the critical path of the transaction, thus providing more predictable commit phase execution, which is important when the transactions need to be finished within their deadlines.

1 Introduction

Real-time databases will be an important part of the future telecommunications infrastructure. They will hold the information needed in operations and management of telecommunication services and networks. The performance, reliability, and availability requirements of data access operations are demanding. Thousands of retrievals must be executed in a second. The allowed unscheduled down time is only a few minutes per year. The requirements originate in the following areas: real-time access to data, fault tolerance, distribution, object orientation, efficiency, flexibility, multiple interfaces, and compatibility [13,14]. Telecommunication requirements and real-time database concepts are studied in the literature [1–3, 7].

The RODAIN[1] database architecture is a *real-time, object-oriented, fault-tolerant*, and *distributed* database management system, which is designed to fulfill the requirements of a modern telecommunications database system. It offers simultaneous execution of firm and soft deadline transactions as well as transactions that do not have deadlines at all. It supports high availability of the data using a hot stand-by, which maintains a copy of the operational database. The hot stand-by is ready to switch to the database server at any time, if the primary server fails. Related systems include ClustRa [4], Dalí[5], and StarBase [6].

---

[1] RODAIN is the acronym of the project name *Real-Time Object-Oriented Database Architecture for Intelligent Networks* funded by Nokia Networks, Solid Information Technology, and the National Technology Agency of Finland.

J. Rolim et al. (Eds.): IPDPS 2000 Workshops, LNCS 1800, pp. 720-726, 2000.

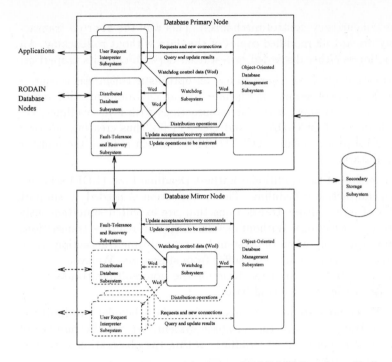

Fig. 1. The Architecture of RODAIN Database Node.

The rest of the paper is organized as follows. The architecture of the RODAIN Database Management System is presented in section 2. The logging mechanism is presented in detail in section 3. Finally, in section 4 we will summarize the results of our experiments, based on a prototype implementation of the RODAIN database system.

2   RODAIN Database

A database designed to be used as a part of telecommunication services must give quick and timely responses to requests. In the RODAIN Database System (see Fig. 1) this is achieved by keeping time-critical data in the main-memory database and using real-time transactions. Real-time transactions have attributes like criticality and deadline that are used in their scheduling. Data availability is increased using a hot stand-by node to maintain a copy of the main-memory database. The hot stand-by, which we call the Database Mirror Node, can replace the main database server, called the Database Primary Node, in the case of failure.

Our main goal in the database design was to avoid as much of the overhead of rollbacks during transaction abort as possible. This was achieved using the deferred write mechanism. In a deferred write mechanism the transaction is allowed to write the modified data to the database area only after it is accepted to commit by the concurrency control mechanism. This way the aborted transaction can simply discard its modified copies of the data without rollbacking. An aborted transaction is either discarded or restarted depending on its properties.

For concurrency control, we chose to use an optimistic concurrency control protocol. Such a protocol seems appropriate to our environment with main-memory data and mainly short, read-only transactions with firm deadlines. We combined the features of OCC-DA [8] and OCC-TI [9], thus creating our own concurrency control protocol called OCC-DATI [11] which reduces the number of unnecessary restarts.

A modified version of the traditional Earliest Deadline First (EDF) scheduling is used for transaction scheduling. The modification is needed to support a small number of non-realtime transactions that are executed simultaneously with the real-time transactions. Without deadlines the non-realtime transactions get the execution turn only when the system has no real-time transaction ready for execution. Hence, they are likely to suffer from starvation. We avoid this by reserving a fixed fraction of execution time for the non-realtime transactions. The reservation is made on a demand basis.

To handle occasional system overload situations the scheduler can limit the number of active transactions in the database system. We use the number of transactions that have missed their deadlines within the observation period as the indication of the current system load level.

The synchronization between Primary and Mirror Nodes within the RO-DAIN Database Node is done by transaction logs and it is the base for the high availability of the main-memory database. Transactions are executed only on the Primary Node. For each write and commit operation a transaction redo log record is created. This log is passed to the Mirror Node before the transaction is committed. The Mirror Node updates its database copy accordingly and stores the log records to the disk. The transaction is allowed to commit as soon as the log records are on the Mirror Node removing the actual disk write from the critical path. It is like the log handling done in [10], except that our processors do not share memory. Thus, the commit time needed for a transaction contains one message round-trip time instead of a disk write.

The database durability is trusted on the assumption that both nodes do not fail simultaneously. If this fails, our model might loose some committed data. This data loss comes from the main idea of using the Mirror Node as the stable storage for the transactions. The data storing to the disk is not synchronized with the transaction commits. Instead, the disk updates are made after the transaction is committed. A sequential failure of both nodes does not lose data, if the time difference between the failures is large enough for the Mirror Node to store the buffered logs to the disk.

The risk of loosing committed data decreases when the time between node failures increases. As soon as the remaining node has had enough time to store the remaining logs to the disk, no data will be lost. In telecommunication the minor risk of loosing committed data seems to be acceptable, since most updates handle data that has some temporal nature. The loss of temporal data is not catastrophic, it will be updated again at a later time.

During node failure the remaining node, called the Transient Node, will function as Primary Node, but it must store the transaction logs directly to the disk before allowing the transaction to commit. The failed node will always become a Mirror Node when it recovers. This solution avoids the need to switch the database processing responsibilities from the currently running node to another. The switch is only done when the current server fails and can no longer serve any requests.

3   Log Handling in the RODAIN Database Node

Log records are used for two different purposes in the RODAIN Database Node. Firstly, they are used to maintain an up-to-date copy of the main-memory database on a separate Mirror Node in order to recover quickly from failures of the Primary Node. Secondly, the logs are stored in a secondary media in the same way as in a traditional database system. These logs are used to maintain the database content even if both nodes fail simultaneously, but they can be also used for, for example, off-line analysis of the database usage.

The log records containing database updates, the after images of the updated data items, are generated during the transaction's write phase. At the write phase the transaction is already accepted for commit and it just updates the data items it has modified during its execution. Each update also generates a log record containing transaction identification, data item identification and an after image of the data item. All transactions that have entered their write phases will eventually commit, unless the primary database system fails. When the Primary Node fails, all transactions that are not yet committed are considered aborted, and their modifications to the database are not performed on the database copy in the Mirror Node.

The communication between the committing transaction and the Log writer is synchronous. The Log Writer on the Primary Node sends the log records to the Mirror Node as soon as they are generated. When the Mirror Node receives a commit record, it immediately sends an acknowledgment back. This acknowledgment is used as an indication that the logs of this specific transaction have arrived to the Mirror Node. The Log writer then allows the transaction to proceed to the final commit step. If a Mirror Node does not exist, then the Log writer (on Transient Node) must store the logs directly to the disk.

The logs are reordered based on transactions before the Mirror Node updates its database copy and stores the logs on disk. The true validation order of the transactions is used for the reordering. This reordering simplifies the recovery process. With logs already ordered, the recovery can simply pass the log once from the beginning to the end omitting only the transactions that do not have a commit record in the log. Likewise, The Mirror Node performs the logged updates to its database only when it has also received the commit record. This way it can be sure that it never needs to undo any changes based on logs.

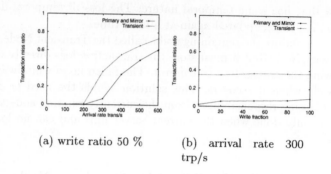

(a) write ratio 50 %   (b)   arrival   rate   300
                            trp/s

Fig. 2. Comparison of normal mode, both Primary and Mirror Node present, and transient mode, only Transient Node, using true log writes.

4   Experimental Study

The current implementation of the RODAIN Database Prototype runs on a Chorus/ClassiX operating system [12]. The measurements were done on computers with a Pentium Pro 200MHz processor and 64 MB of main memory. All transactions arrive at the RODAIN Database Prototype through a specific interface process, that reads the load descriptions from an off-line generated test file. Every test session contains 10 000 transactions and is repeated at least 20 times. The reported values are the means of the repetitions. The test database, containing 30 000 data objects, represents a number translation service. The number of concurrently running transactions is limited to 50. If the limit is reached, an arriving lower priority transaction is aborted. Transactions are validated atomically. If the deadline of a transaction expires, the transaction is always aborted.

The workload in a test session consists of a variable mix of two transactions, one simple read-only transaction and the other a simple write transaction. The read-only service provision transaction reads a few objects and commits. The write transaction is an update service provision transaction that reads a few objects, updates them and then commits. The relative firm deadline of all real-time transactions is 50ms and the deadline of all write transactions is 150ms.

We measured the transaction miss ratio, which represents the fraction of transactions that were aborted. The aborts can be either due to the exceeding of a transaction deadline, a concurrency control conflict, or an acceptance denial due to the load limit. In the experiments, the failures in transaction executions were mainly due to system overload. Occasionally a transaction also exceeded its deadline and was, therefore, aborted.

We compared the performance of our logging mechanism in its normal use with both the Primary and the Mirror Node to a situation where only a single Transient node is running (see Fig. 2). When both nodes are up and running the logs are passed from the Primary to the Mirror Node. When the Transient Node is running alone, it stores the logs directly to the log storage. The experiment

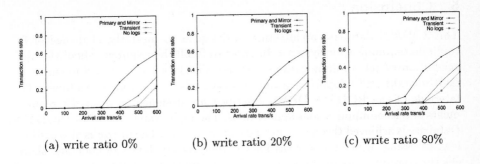

(a) write ratio 0%   (b) write ratio 20%   (c) write ratio 80%

Fig. 3. Comparison of optimal (marked as No logs), single node (Transient), and two node systems (Primary and Mirror).

shows clearly that the use of a remote node instead of direct disk writes increases the system performance.

Since our experiments with disk writing showed that the log storing to the disk can easily become the bottleneck in the log handling, we ran more tests with the disk writing turned off. This scenario is feasible, if the probability of simultaneous failure of both nodes is acceptable and the system can be trusted to run without any other backups. The omission of the disk writes also emphasizes the overhead from our log handling algorithms with the two nodes. If the log storing to the disk system is slower than the median log generation rate, then the system gets trashed from the buffered logs and must reduce the incoming ratio of the transactions to the pace of disk storing. This would then remove most of the benefit of the Mirror Node use. For comparison, we also ran tests on a Transient Node where the logging feature was completely turned off. The results from this optimal situation do not differ much from the results of Transient Node with logging turned off.

From Fig. 3 we can see that the most effective feature of the system performance is the transaction arrival time. At the arrival rate of 200 to 300 transactions per second depending on the ratio of update transactions, the system becomes saturated and most of the unsuccessfully executed (=missed) transactions are due to abortions by overload manager. The effect of the ratio of update transactions is relatively small. There are two reasons for this behavior. First, the update transactions modify only a few items. Thus, the number of log records per transaction is not large either. Secondly, the system generates a commit log record also for read-only transactions, thus forcing the commit times of both transaction types to be quite close.

The benefits of the use of the hot stand-by are actually seen when the primary database system fails. When that happens, the Mirror Node can almost instantaneously serve incoming requests. If, however, the Primary Node was alone and had to recover from the backup on the disk or in the stable memory, like Flash, the database would be down much longer. Such down-times are not allowed in certain application areas such as telecommunication.

5   Conclusion

The RODAIN database architecture is designed to meet the challenge of future telecommunication systems. In order to fulfill the requirements of the next generation of telecommunications systems, the database architecture must be fault-tolerant and support real-time transactions with explicit deadlines. The internals of the RODAIN DBMS described are designed to meet the requirements of telecommunications applications. The high availability of the RODAIN Database is achieved through using a database mirror. The mirror is also used for log processing, which reduces the load at the primary database node and shortens the commit times of transactions allowing more transactions to be executed within their deadlines.

References

1. I. Ahn. Database issues in telecommunications network management. *ACM SIG-MOD Record*, 23(2):37–43, 1994.
2. R. Aranha et al. Implementation of a real-time database system. *Information Systems*, 21(1):55–74, 1996.
3. T. Bowen et al. A scale database architecture for network services. *IEEE Communications Magazine*, 29(1):52–59, January 1991.
4. S. Hvasshovd et al. The ClustRa telecom database: High availability, high throughput, and real-time response. In *Proc. of the 21th VLDB Conf.*, pp. 469–477, 1995.
5. H. Jagadish et al. Dalí: A high performance main memory storage manager. In *Proc. of the 20th VLDB Conf.*, pp. 48–59, 1994.
6. Y. Kim and S. Son. Developing a real-time database: The StarBase experience. In A. Bestavros, K. Lin, and S. Son, editors, *Real-Time Database Systems: Issues and Applications*, pp. 305–324. Kluwer, 1997.
7. Y. Kiriha. Real-time database experiences in network management application. Tech. Report CS-TR-95-1555, Stanford University, USA, 1995.
8. K. Lam, K. Lam, and S. Hung. An efficient real-time optimistic concurrency control protocol. In *Proc. of the 1st Int. Workshop on Active and Real-Time Database Systems*, pp. 209–225. Springer, 1995.
9. J. Lee and S. Son. Performance of concurrency control algorithms for real-time database systems. In V. Kumar, editor, *Performance of Concurrency Control Mechanisms in Centralized Database Systems*, pp. 429–460. Prentice-Hall, 1996.
10. T. Lehman and M. Carey. A recovery algorithm for a high-performance memory-resident database system. In U. Dayal and I. Trager, editors, *Proc. of ACM SIG-MOD 1987 Ann. Conf.*, pp. 104–117, 1987.
11. J. Lindström and K. Raatikainen. Dynamic adjustment of serialization order using timestamp intervals in real-time databases. In *Proc. of 6th Int. Conf. on Real-Time Computing Systems and Applications*, 1999.
12. D. Pountain. The Chorus microkernel. *Byte*, pp. 131–138, January 1994.
13. K. Raatikainen. Real-time databases in telecommunications. In A. Bestavros, K. Lin, and S. Son, editors, *Real-Time Database Systems: Issues and Applications*, pp. 93–98. Kluwer, 1997.
14. J. Taina and K. Raatikainen. Experimental real-time object-oriented database architecture for intelligent networks. *Engineering Intelligent Systems*, 4(3):57–63, September 1996.

Components are from Mars

M.R.V. Chaudron[1] and E. de Jong[1,2]

[1]Technische Universiteit Eindhoven, Dept. of Computer Science
P.O. Box 513, 5600 MB Eindhoven, The Netherlands
m.r.v.chaudron@tue.nl
[2]Hollandse Signaalapparaten B.V., P.O. Box 42, 7550 GD Hengelo, The Netherlands
edejong@signaal.nl

Abstract. We advocate an approach towards the characterisation of components where their qualifications are deduced systematically from a small set of elementary assumptions. Using the characteristics that we find, we discuss some implications for components of real-time and distributed systems. Also we touch upon implications for design-paradigms and some disputed issues about components.

1   Introduction

From different perspectives on software engineering, it is considered highly desirable to build flexible systems through the composition of components. However, no method of design exists that is tailored towards this component-oriented style of system development. Before such a method can emerge, we need a clear notion of what components should be. However, although the component-oriented approach can be dated back to the late 1960's (see [McI68]), recent publications list many different opinions about what components should be [Br98], [Sa97], [Sz98]. This abundance of definitions indicates that we do not yet understand what components and component-oriented software engineering are about.

The discussion on what components should be is complicated by the absence of an explicit statement of (and agreement on) the fundamental starting points. As a result, the motivations behind opinions are often unknown, implicit or unclear. Also, presuppositions are implicitly made that are unnecessarily limiting.

The goal of this paper is twofold: firstly, to make explicit the fundamental starting points of component-based engineering, and secondly, to systematically deduce characteristics of the ideal component.

2   Basic Component Model and Qualification

First we shall introduce a basic model and discuss its consequences for components in general. Next, we consider some implications for components for real-time and distributed systems.

J. Rolim et al. (Eds.): IPDPS 2000 Workshops, LNCS 1800, pp. 727-733, 2000.

Basic component model

In this section we introduce our basic model for reasoning about components. Our aim is to introduce concepts only when necessary. As a result, a lot of possible aspects of components are intentionally not present in our model.

The model we consider consists of the following:
- There are things called *components*.
- Components may be composed by some *composition mechanism*.

We use the following terminology:
- A configuration of a number of composed components is called a *composition*.
- Everything outside a component is called its *environment*.

A pitfall in reasoning about components is that we presuppose they have features that we are familiar with from programming methodology to such a degree that we cannot imagine that the issues addressed by these features can be approached in another way. Typically, many people endow components with features from the object-oriented paradigm. In order to prevent us from doing so, we will adhere to a strict regime for reasoning about components. We fit our reasoning in the form of a logical theory that has axioms and corollaries. We postulate our basic assumptions about components as axioms. From these axioms we aim to deduce corollaries that qualify components and their composition mechanism.

Next, we present our first axioms.

A1   A component is capable of performing a task in isolation; i.e. without being   (1)
        composed with other components.

A2   Components may be developed independently from each other.   (2)

A3   The purpose of composition is to enable cooperation between the   (3)
        constituent components.

Axioms A1 and A2 are generally agreed upon. Already in [Pa72], axiom A2 appears explicitly and A1 is close in spirit to Parnas' observation ".. we are able to cut off the upper levels [of the system] and still have a usable and useful product." The intention of axiom A1 is more explicitly present in recent formulations such as "[a component is an] independent unit of deployment" [Sz98].

To build larger systems out of smaller ones, we want to combine the effects of components. In order to be able to do so, we need a composition mechanism (axiom A3). Note that axiom A3 does not imply that it is a component's purpose to cooperate. In fact, for the functioning of a component it should be immaterial whether it is cooperating with other components (cf. A1). It is the designer (composer) of a composition who attributes meaning to the combined effect of the components. (Meaning [of a composition] is "in the eye of the composer.")

Next, we present a first corollary.

C1   A component is capable of acquiring input from its environment and/or of   (4)
      presenting output to its environment.

This corollary can be motivated in two ways. The first is that performing some task (axiom A1) would be futile without some means to observe its effect. The second can be inferred from A3: in order to achieve cooperation between components, there must be some mechanism that facilitates their interaction.

We proceed by deducing some more qualifications of components.

C2   A component should be independent from its environment.   (5)

This corollary follows from axiom A1: In order for a component to fulfill its task in isolation, it should have no dependencies on this environment. Put more constructively, a design principle for components is to optimize their autonomy.

C3   The addition or removal of a component should not require modification   (6)
      of other components in the composition.

Corollary C3 follows from C2. Suppose that the opposite of C3 was true; i.e. the addition (or removal) of a component does require modification of other components in the composition. Then, clearly, there is a dependency of the components that require modification on the one that is added to (or removed from) the composition.

Corollary C3 expresses the flexibility or openness generally required of component-based systems.

Implications for distributed real-time systems

From the preceding general observations we next shift attention to the design of components for real-time and distributed systems. The corollaries that we present follow straightforwardly from C2. To start with timeliness, C2 leads to the following corollary.

C4   Timeliness of output of a component should be independent from   (7)
      timeliness of input.

Again this is a qualification towards the autonomy of components. One possible means to make the timeliness of output independent of timeliness of input is to build in a mechanism that enables a component to generate output when stimuli do not arrive as anticipated. Typically, such an output can be generated only at the cost of a decrease in the quality of the output.

The next corollary, C5, is the justification of a principle that is known in the area of parallel and distributed systems as *location transparency*. Clearly, C5 follows from corollary C2.

C5   The functioning of a component should be independent of its location in a   (8)
       composition.

Corollary C5 is a constraint on the internals of a component (*internal location transparency*). The counterpart of C5, *external location transparency* (corollary C6) is a qualification of the composition mechanism. Its justification is analogous to that of C3 (by contradiction of the opposite).

C6   The change of location of a component should not require modifications   (9)
       to other components in the composition.

Next, we present our final corrolary of this paper.

C7   A component should be a unit of fault-containment.   (10)

The justification of Corollary C7 is as follows: a component cannot assume that some input is normal and some other is faulty, since this implies a dependency on its environment. Hence, a component has to cater for all possible input.

Corollary C7 entails the following guideline for the design of components: components should shield their output from any anomalies[1] at their input.

3   On Disputed Issues in Component Design

In this section we will discuss some issues in the design of components based on the qualifications that we found in the preceding sections. When this has unexpected

---

[1] Actually, the term "anomaly" is indicative for an assumption about, and hence a dependency on, the environment.

implications we may refer to existing composition systems (e.g. pipe-and-filter [Ri80], or shared tuple-spaces [CG89], [FHA99]) to illustrate that there are systems that do not violate these implications.

Do components have state?

Let us assume that, in some composition, the task of a component is to store some state. The openness or flexibility corollary C3 asserts that the removal of a component should not require modifications to other components in the composition. This suggests that using a component to store data that is to be used by other components is a bad idea, since this storage component may be removed arbitrarily and the data it stored will no longer be available for other components in the system. In other words, a storage component induces dependencies on other components. This reasoning suggests that stacks and queues should not be considered good examples of components.

Although this is a surprising consequence, we see that neither the pipe-and-filter- nor the shared dataspace model require components that store data. In these cases the composition mechanism deals with the storage of data.

The fundamental issue seems to be that of openness versus encapsulation (in the style of abstract data types as encouraged by the object orientation paradigm). Giving priority to openness (as we do here) seems to prohibit encapsulation of storage.

However, a component is free to build up a "state" as long as the effect of this state cannot be observed by the environment. For example, a filter that performs a word-count on input text clearly computes the output by incrementing some local word-counter. However, this local state does not induce a dependency on other components.

Are objects components?

Components are often seen as the next logical step in the evolution of software engineering after objects. Be that as it may, this does not mean that components should be an extension of objects. It may turn out that some features of objects that were introduced to facilitate programming may not be suitable for the purpose of composition.

The following are examples of features of the object-oriented paradigm that seem to hinder composition:

- *The mechanism for cooperation:* The object orientation paradigm uses method invocation (based on message passing) as a mechanism for cooperation. This mechanism requires agreement between the invoking and the invoked object on the order in which methods are executed. Such an order is built into the definition of objects. As a result, addition or removal of an object requires modification to other objects in the system (methods may cease to exist or new methods may need to be introduced), contradicting corollary C3.

In the area of coordination models and languages [PA98], this style of interaction is called *endogenous*. In contrast, in *exogeneous* languages, the interaction between parties is specified outside (textually separate) from the computational code. An example of an exogenous composition language is the pipe-and-filter mechanism from Unix. The specification of the pattern of interaction outside of the components involved in it allows modification of the interaction pattern without requiring modifications to the components.

Also, with method invokation, the initiative for invoking a method may not reside with the object that the method is part of, but with some object in the environment. This is a violation of the independence of components (corollary C2).

- *Encapsulation of data*: One argument is given in the previous subsection as answer to the issue of components and state. Another is given by [HO93]. The essence of that argument is that in an evolving system, the future uses of data cannot be predicted; hence an object that encapsulates data cannot provide the methods for which a need may arise in the future.

The above, however, does not imply that object oriented programming should not be used for implementing components – only that this paradigm does not provide the right abstractions for designing component based systems.

4   Concluding Remarks

The fact that currently many different definitions for components are proposed, suggests that we do not yet fully understand the implications of the requirements for component based engineering. In this paper we pursued the implications of these requirements further than is often done. To this end, we presented a rigorous approach to the qualification of components that makes the fundamental assumptions explicit. In this way, we aim to incrementally develop a model for component-based engineering.

Our investigations suggest that object-orientation has some features that hamper the composability of software needed for component-based software development. Hence, we should investigate alternative composition mechanisms.

We welcome comments and additions to our framework.

Acknowledgements

The authors would like to thank Tim Willemse for his critical comments.

References

[Br98] Broy M., Deimel A., Henn J., Koskimies K., Plasil F., Pomberger G., Pree W., Szyperski C.: What characterizes a (software) component?, Software Concepts & Tools (vol. 19, no. 1), 1998.

[CG89] Carriero, N. and Gelernter, D., Linda in context, Communications of the ACM, vol 32(4), pp. 444-458, April 1989.

[FHA99] Freeman, E., Hupfer, S. and Arnold, K., JavaSpaces(TM) Principles, Patterns and Practice (The Jini Technology Series), Addison-Wesley, 1999.

[HO93] Harrison, W. and Osher, H., Subject-oriented Programming (a critique of pure objects), in: Proceedings of OOPSLA 1993, pp. 411-428.

[McI68] McIlroy, D., Mass Produced Software Components, in "Software Engineering, Report on a conference sponsored by the NATO Science Committee, Garmisch, Germany, 7th to 11th October 1968", P. Naur and B. Randell (eds), Scientific Affairs Division, NATO, Brussels, 1969, 138-155.

[Pa72] Parnas, D.L., On the Criteria to be used in Decomposing Systems into Modules, Communications of the ACM, Vol. 15, No. 12, Dec. 1972.

[Pa98] Papadopoulos, G.A. and Arbab, F., Coordination Models and Languages. In M. Zelkowitz, editor, Advances in Computers, The Engineering of Large Systems, volume 46. Academic Press, August 1998.

[Ri80] Ritchie, D.M., The Evolution of the Unix Time-sharing System, Proceedings of the Conference on Language Design and Programming Methodology, Sydney, 1979, Lecture Notes in Computer Science 79: Language Design and Programming Methodology, Springer-Verlag, 1980 (also at http://cm.bell-labs.com/cm/cs/who/dmr/hist.html).

[Sa97] Sametinger, J., Software Engineering with Reusable Components, Springer, 1997.

[SG96] Shaw, M. and Garlan, D., Software Architecture: Perspectives on an Emerging Discipline, Prentice Hall, 1996.

[Sz98] Szyperski, C., Component Software: Beyond Object-Oriented Programming, Addison-Wesley, 1998.

$2 + 10 \; \succ \; 1 + 50 \; !$

Hans Hansson, Christer Norström, and Sasikumar Punnekkat

Mälardalen Real-Time Research Centre
Department of Computer Engineering
Mälardalen University, Västerås, SWEDEN
han@idt.mdh.se, cen@mdh.se, spt@idt.mdh.se
WWW home page: http://www.mrtc.mdh.se

Abstract. In traditional design of computer based systems some effort, say 1, is spent on the early modeling phases, and some very high effort, say 50, is spent on the later implementation and testing phases. It is the conjecture of this paper that the total effort can be substantially reduced if an increased effort, say 2, is spent on the early modeling phases. Such a shift in focus of efforts will also greatly improve the overall effects (both quality and cost-wise) of the systems developed, thereby leading to a better (denoted by "$\succ$") design process. In this paper, we specifically consider the design of safety-critical distributed real-time systems.

1 Introduction

Designing safety-critical real-time systems involves assessment of functionality, timing and reliability of the designed system. Though several design methods have been proposed in literature (such as HRT-HOOD, DARTS, UPPAAL, UML-RT), none of them have been able to gain widespread acceptance due to the range and magnitude of the issues involved and probably due the restricted focus of these methods.

In Figure 1 we present a generic design model for the development of safety critical distributed real-time systems.

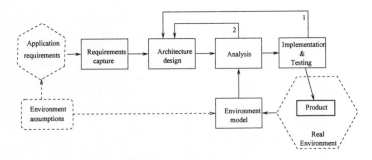

Fig. 1. A generic real-time design model

J. Rolim et al. (Eds.): IPDPS 2000 Workshops, LNCS 1800, pp. 734-737, 2000.

The architecture design is the highest abstraction level for the design and construction of the system. Here the system is partitioned into components, processes for realisation of them are identified, and boundaries for desired quality levels are set. For real-time systems, timing budgets are typically allocated to individual components at this stage. The analysis part in the design process contains both functional analysis (such as, temporal behaviour, reliability modelling, safety, performance) as well as non-functional analysis (such as, testability, maintainability, portability, cost, extensibility). To be able to make these analyses, the architecture has to be described by a language that provides a precise syntax and semantics. Such a language should define the computational model with possible extensions for hierarchical and functional decompositions.

Experiences from an industrial co-operation [1] have further convinced us of the benefits of performing architecture analysis on temporal requirements, communication and synchronisation. Based on these insights we will be focusing on architecture analysis rather than analysis of the implementation. It is apparent that such a shift in focus from the implementation & testing phase to the architecture design and analysis phases, by adding more resources and efforts in these earlier phases, is absolutely necessary to detect many critical issues before they manifest in the product and necessitate a costly product re-design. In terms of figure 1, this amounts to iterating more on the inner loop (marked 2) rather than on the outer loop (marked 1) to improve the quality at a lower cost.

Using such an approach, one of the major issues, i.e., timing compliance of the system was achieved in our project at Volvo [1] by applying a time-budgeting and negotiations strategy for individual tasks. We now present briefly two other major issues, viz. fault modelling and testability, representing a functional and non-functional issue, respectively. An accurate fault modelling and analysis will assist the designer in incorporating sufficient fault-tolerance capabilities into the system, whereas testability analysis can greatly reduce the final testing efforts. It should be noted that, both these issues are addressed in conjunction with their effects on the temporal requirements and properties.

2   Fault Modelling and Analysis

Though there has been sizable amount of research efforts in both the fault tolerance and the real-time realms, these two fields have been more or less treading along parallel paths. These two research domains are all the more relevant in the case of safety-critical systems and their mutual dependencies and interactions need to be analysed for achieving predictable performance. There are very few studies in literature, aimed at bridging the gap between these two areas and many issues remain open that need to be further investigated. One such important issue is the effect of faults on schedulability analysis and on the timing guarantees provided.

The major stumbling block in having an intergrated approach is the orthogonal nature of the two factors, viz., the stochastic nature of faults and the deterministic requirements on schedulability analysis. This calls for development of more realistic fault models which capture the nuances of the environment as well as methods for incorporating such models into the timing analysis with ease. In applications such as automobiles, the systems are often subjected to high degrees of Electro Magnetic Interference (EMI). The common causes for such interferences include cellular phones and other radio equipments inside the vehicle and electrical devices like switches and relays as well as as radars and radio transmissions from external sources and lightning in the environment. These interferences may cause errors in the transmitted data.

In this context we have recently [3] developed a model for calculating worst-case latencies of messages on the Controller Area Network (CAN) under errors. CAN is a predictable communication network widely used in the automotive and automation industries. The basic CAN analysis assumes an error free communication bus, which is not always true. To reduce the risk due to errors, CAN designers have provided elaborate error checking and confinement features, which identify the errors and retransmit the affected messages, thus increasing the message latencies and potentially leading to timing violations.

Tindell and Burns [2] have proposed a model for calculating worst-case latencies of CAN messages under errors. They define an error overhead function $E(t)$, as the maximum time required for error signaling and recovery in any time interval of length t. Their model is relatively simplistic and assumes an initial error-burst followed by sporadic error occurrences (i.e., errors separated by a known minimum time). Our new fault model [3] is more general, in that it

- models intervals of interference as periods in which the bus is not available
- allows more general patterns of interferences to be specified and from that description derive the effect on message transmissions
- allows the combined effects of multiple sources of interference to be modeled.
- considers the potential delay induced by the interference durations

With this fault model it is possible to build parameterised models of different types of interferences originating from different sources. Using these models, realistic worst-case scenarios can be characterised and analysed. We believe that this kind of analysis will be a step towards future design of adaptive scheduling strategies which takes in to account the error occurrences and decides on-line issues such as graceful degradation and choosing different policies for different classes of messages.

3   Testability Analysis

A large part of the effort, time and cost in developing safety-critical real-time (and most other) systems is related to testing. Consequently, one of the most important non-functional quality attributes of a design is its testability, i.e. the effort required to obtain a specific coverage in the testing process. High testability means that relatively few tests have to be exercised. The design with highest testability may however not be the preferred one, since testability typically is in conflict with other desired qualities, such as performance and maintainability.

Using testability measures in choosing between alternative designs that are similar in other respects is however highly desirable, and sacrificing other qualities for increased testability may be a good compromise in many situations.

An intuitive metric for the testability of a system is its number of distinguishable computations. For a sequential program this is proportional to the number of program paths. For concurrent and distributed systems we must additionally consider the possible interleavings of the program executions (the tasks). Clearly, by limiting the freedom in scheduling and by making synchronization between distributed nodes tighter, we can substantially reduce the number of interleavings, thus increasing testability. Testability is further increased if the variations (jitter) in release and execution times of individual tasks can be reduced.

In [4], we introduce a method for identifying the set of task interleavings of a distributed real-time system with a task set having recurring release patterns. We propose a testing strategy which essentially amounts to regarding each of the identified interleavings as a sequential program, and then use sequential techniques for testing it. Due to the large number of interleavings, this in general is a formidable task. We are however convinced that for a sufficiently large class of safety-critical real-time systems this approach is both feasible and desirable.

4   Conclusion and future challenges

In this paper, we have described some important issues in the design of safety-critical distributed real-time systems. We emphasize the potential gain of shifting the focus from implementation & testing phase to the architectural design phase, by obtaining a high effects-efforts ratio. In this context, we also highlighted two of our latest research contributions.

The vision and objective of current research in the Systems Design Laboratory at Mälardalen Real-Time research Centre is to provide engineers with scientific methods and tools for designing safety-critical real-time systems, such that the state-of-art and practice for developing such systems is advanced to a mature engineering discipline. This amounts to developing, adopting and applying theory with industrial applications in mind, as well as designing appropriate engineering tools and methods.

References

1. Christer Norström, Kristian Sandström, and Jukka Mäki-Turja: Experiences and findings from the usage of real-time technology in an industrial project, *MRTC-Technical report*, January 2000.
2. Ken W. Tindell, Alan Burns, and Andy J. Wellings: Calculating Controller Area Network (CAN) Message Response Times. *Control Engineering Practice*, 3(8), 1995.
3. Sasikumar Punnekkat, Hans Hansson, and Christer Norström: Response time analysis of CAN message sets under errors, *MRTC-Technical report*, December 1999.
4. Henrik Thane and Hans Hansson: Towards Systematic Testing of Distributed Real-Time Systems, *20th IEEE Real-Time Systems Symposium*, Phoenix, December 1999.

A Framework for Embedded Real-time System Design [*]

Jin-Young Choi[1], Hee-Hwan Kwak[2], and Insup Lee[2]

[1] Department of Computer Science and Engineering, Korea Univerity
`choi@formal.korea.ac.kr`
[2] Department of Computer and Information Science, University of Pennsylvania
`heekwak@saul.cis.upenn.edu`, `lee@cis.upenn.edu`

Abstract. This paper describes a framework for parametric analysis of real-time systems based on process algebra. The Algebra of Communicating Shared Resources (ACSR) has been extended to ACSR with Value-passing (ACSR-VP) in order to model the systems that pass values between processes and change the priorities of events and timed actions dynamically. The analysis is performed by means of bisimulation or reachability analysis. The result of the analysis is predicate equations. A solution to them yields the values of the parameters that satisfy the design specification. We briefly describe the proposed framework in which this approach is fully automated and identify future work.

1 Introduction

There have been active research on formal methods for the specification and analysis of real-time systems [4, 5] to meet increasing demands on the correctness of embedded real-time systems. However, most of the work assumes that various real-time system attributes, such as execution time, release time, priorities, etc., are fixed *a priori*, and the goal is to determine whether a system with all these known attributes would meet required timing properties. That is to determine whether or not a given set of real-time tasks under a particular scheduling discipline can meet all of its timing constraints.

Recently, parametric approaches which do not require to guess the values of unknown parameters *a priori* have been proposed as general frameworks for the design analysis of real-time systems. Gupta and Pontelli [3] proposed a unified framework where timed automata has been used as a front-end, and the constraint logic programming (CLP) languages as a back-end. We [7] proposed a parametric approach based on real-time process algebra ACSR-VP (Algebra of Communicating Shared Resources with Value Passing). The scheduling problem is modeled as a set of ACSR-VP terms which contain the unknown variables as parameters. As shown in [7], a system is schedulable when it is bisimilar to a non-blocking process. Hence, to obtain the values for these parameters we

---

[*] This research was supported in part by NSF CCR-9619910, ARO DAAG55-98-1-0393, ARO DAAG55-98-1-0466, and ONR N00014-97-1-0505.

J. Rolim et al. (Eds.): IPDPS 2000 Workshops, LNCS 1800, pp. 738–742, 2000.

check a symbolic bisimulation relation between a system and a non-blocking process described both in ACSR-VP terms. The result of the bisimulation relation checking with the non-blocking process is a set of predicate equations of which solutions are the values for parameters that make the system schedulable. In this way, our approach reduces the analysis of scheduling problems into finding solutions of a recursive predicate equation system. We have demonstrated in [7] that CLP techniques can be used to solve predicate equations.

Before we explain an extension of our approach [7], we briefly present some background material below. Due to the space limitation we omit the formal definition of ACSR-VP. Instead, we illustrate the syntax and semantics using the following example process P.

$$P(t) = (t > 0) \rightarrow (a!t + 1, 1).P'(t)$$

The process P has a free variable t. The instantaneous action $(a!t+1, 1)$ outputs a value $t + 1$ on a channel a with a priority 1. The behavior of the process P is as follows. It checks the value of t. If t is greater than 0, then it performs the instantaneous action $(a!t + 1, 1)$ and becomes P' process. Otherwise it becomes NIL. For more information on ACSR-VP a reader refers to [7].

To capture the semantics of an ACSR-VP term, we proposed a Symbolic Graph with Assignment (SGA). SGA is a rooted directed graph where each node has an associated ACSR-VP term and each edge is labeled by boolean, action, assignment, (b, α, θ). Given an ACSR-VP term, an SGA can be generated using the rules shown in [7].

The notion of bisimulation is used to capture the semantics of schedulability of real-time systems. The scheduling problem is to determine if a real-time system with a particular scheduling discipline meets all of its deadlines and timing constraints. In ACSR-VP, if no deadline and constraints are missed along with any computation of the system, then the process that models the system always executes an infinite sequence of timed action. Thus by checking the bisimulation equivalence between the process that models the system and the process that idles infinitely, the analysis of schedulability for the real-time systems can be achieved.

2   A Fully Automatic Approach for the Analysis of Real-time Systems.

In the approach published in [7] a bisimulation relation plays a key role to find solutions for parameters. However, the disadvantage with a bisimulation relation checking method is that it requires to add new τ edges. These new edges will increase the size of a set of predicate equations and the complexity to solve them. To reduce the size of a set of predicate equations, we introduced a parametric reachability analysis techniques.

As noted in [7], finding conditions that make system schedulable is equivalent to finding symbolic bisimulation relation with an infinite idle process. Furthermore, checking the symbolic bisimulation relation with an infinite idle process

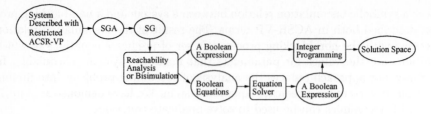

Fig. 1. Our Approach for the Real-time System Analysis

is equivalent to finding conditions that guarantee there is always a cycle in an SGA regardless of a path taken. That is, there is no deadlock in the system under analysis. Hence, we can obtain a condition that guarantees there is no deadlock in the system under analysis by checking possible cycles in an SGA for the system under analysis. We illustrate that this reachability analysis can replace a bisimulation relation checking procedure. With a reachability analysis we can avoid adding new τ edges and reduce the complexity of solving predicate equations.

Utilizing existing CLP techniques seems to be a natural way of solving predicate equations. However, it is not possible to determine if a CLP program terminates. This leads us to identify a decidable subset of ACSR-VP terms. This subset can be classified by defining variables in ACSR-VP terms into two types: control variable and data variable. Control variable is a variable with finite range. The value of a control variable can be modified while a process proceeds. Data variable is the variable that does not change its value. That is, it just hold values "passively" without modification to them. Data variables may assume values from infinite domains. A detailed explanation on a decidable subset of ACSR-VP is given in [6]. We use the term "restricted ACSR-VP" to denote a decidable subset of ACSR-VP.

With a restricted ACSR-VP terms we can reduce a real-time system analysis into solving either a boolean expression or boolean equations with free variables. A decidable subset of ACSR-VP allow us to generate a boolean expression or boolean equations with free variables (BESfv) as a result of reachability analysis or symbolic bisimulation checking. We have developed a BESfv solving algorithm, which is based on maximal fixpoint calculation.

Here we explain the overview of our fully automatic approach, which is a refined version of our previous one [7]. A simplified version of the overall structure of our approach is shown in Figure 1. We describe a system with restricted ACSR-VP terms. With a given set of restricted ACSR-VP processes, an SGA is generated from a restricted ACSR-VP term in order to capture the behavior of a system. Once an SGA is generated, we instantiate all the control variables in each SGA node to form an Symbolic Graph (SG). An SG is a directed graph in which every edge is labeled by (b, α), where b is a boolean expression and α is an action. As an analysis either the symbolic bisimilarity is checked on an SG with an SG of infinite idle process or reachability analysis can be directly performed on an SG of the system. The result is a set of boolean equations or a boolean expression. In the case that a boolean expression with free variables is produced, it can be solved by existing integer programming tools such as Omega Calculator [8]. In the other case that boolean equations with free variables are generated, an algorithm presented in [6] can be applied.

We have applied our framework into several real-time scheduling problems. For real-time scheduling problems, the solution to boolean expression or a set of boolean equations with free variables identifies, if it exists, under what values of unknown parameters the system becomes schedulable. For instance, in the shortest job first scheduling, we may want to know the period of certain jobs that guarantee the scheduling of the system. We let those periods be unknown parameters and describe the system in ACSR-VP process terms. Those unknown parameters are embedded into the derived boolean expression or boolean equations, and consequently the solutions of them represent the values of unknown parameters that make them satisfiable. These solutions represent the valid ranges of periods (i.e., unknown parameters) of the jobs that make the system schedulable.

Our method is expressive to model complex real-time systems in general. Furthermore, the resulting boolean-formulas can be solved efficiently. For instance, there has been active research [2] to solve a boolean expression efficiently, and there are existing tools such as Omega Calculator [8] for a Presburger formulas. Another significant advantage of our method is the size of graphs. Due to the abstract nature of SGA, the size of an SGA constructed from an ACSR-VP term is significantly smaller than that of Labeled Transition Systems (LTS) which requires all the parameters to be known *a priori*. Consequently, this greatly reduces the state explosion problem, and thus, we can model larger systems and solve problems which were not possible by the previous approaches due to state explosions.

Furthermore, our approach is decidable whereas other general framework as [3] is not, and thus, it is possible to make our approach fully automatic when we generate a set of boolean equations or a boolean expression. Since our approach is fully automatic, it can also be used to check other properties as long as they can be verified by reachability analysis.

3 Conclusion

We have overviewed a formal framework for the specification and analysis of real-time systems design. Our framework is based on ACSR-VP, symbolic bisimulation, and reachability analysis. The major advantage of our approach is that the same framework can be used for scheduling problems with different assumptions and parameters. In other real-time system analysis techniques, new analysis algorithms need to be devised for problems with different assumptions since applicability of a particular algorithm is limited to specific system characteristics.

We believe that restricted ACSR-VP is expressive enough to model any real-time system. In particular, our method is appropriate to model many complex real-time systems and can be used to solve the *priority assignment problem, execution synchronization problem,* and *schedulability analysis problem* [9]. We are currently investing how to adapt the proposed frame for embedded hybrid systems, that is, systems with both continous and discrete components.

The novel aspect of our approach is that schedulability of real-time systems can be described formally and analyzed automatically, all within a process-algebraic framework. It has often been noted that scheduling work is not adequately integrated with other aspects of real-time system development [1]. Our work is a step toward such an integration, which helps to meet our goal of making the timed process algebra ACSR a useful formalism for supporting the development of reliable real-time systems. Our approach allows the same specification to be subjected to the analysis of both schedulability and functional correctness.

There are several issues that we need to address to make our approach practical. The complexity of an algorithm to solve a set of boolean equations with free variables grows exponentially with respect to the number of free variables. We are currently augmenting PARAGON, the toolset for ACSR, to support the full syntax of ACSR-VP directly and implementing a symbolic bisimulation algorithm. This toolset will allow us to experimentally evaluate the effectiveness of our approach with a number of large scale real-time systems.

References

1. A. Burns. Preemptive priority-based scheduling: An appropriate engineering approach. In Sang H. Song, editor, *Advances in Real-Time Systems*, chapter 10, pages 225–248. Prentice Hall, 1995.
2. Uffe Engberg and Kim S. Larsen. Efficient Simplification of Bisimulation Formulas. In *Proceedings of the Workshop on Tools and Algorithms for the Construction and Analysis of Systems*, pages 111–132. LNCS 1019, Springer-Verlag, 1995.
3. G. Gupta and E. Pontelli. A constraint-based approach for specification and verification of real-time systems. In *Proceedings IEEE Real-Time Systems Symposium*, December 1997.
4. Constance Heitmeyer and Dino Mandrioli. *Formal Methods for Real-Time Computing*. Jonh Wiley and Sons, 1996.
5. Mathai Joseph. *Real-Time Systems: Specification, Verification and Analysis*. Prentice Hall Intl., 1996.
6. Hee Hwan Kwak. *Process Algebraic Approach to the Parametric Analysis of Real-time Scheduling Problems*. PhD thesis, University of Pennsylvania, 2000.
7. Hee-Hwan Kwak, Jin-Young Choi, Insup Lee, Anna Philippou, and Oleg Sokolsky. Symbolic Schedulability Analysis of Real-time Systems. In *Proceedings IEEE Real-Time Systems Symposium*, December 1998.
8. William Pugh. The Omega test: a fast and practical integer programming algorithm for dependence analysis. *Communications of the ACM*, 8:102–114, August 1992.
9. Jun Sun. *Fixed-priority End-to-end Scheduling in Distributed Real-time Systems*. PhD thesis, University of Illinois at Urbana-Champaign, 1997.

Best-effort Scheduling of (m,k)-firm Real-time Streams in Multihop Networks

A. Striegel and G. Manimaran

Dept. of Electrical and Computer Engineering
Iowa State University, USA
{adstrieg,gmani}@iastate.edu

Abstract. In this paper, we address the problem of best-effort scheduling of (m,k)-firm real-time streams in multihop networks. The existing solutions for the problem ignore scalability considerations because the solutions maintain a separate queue for each stream. In this context, we propose a scheduling algorithm, EDBP, which is scalable (fixed scheduling cost) with little degradation in performance. The proposed EDBP algorithm achieves this by allowing multiplexing of streams onto a fixed number of queues and by using the notion of a look-ahead window. In the EDBP algorithm, at any point of time, the best packet for transmission is selected based on the *state* of the stream combined together with the laxity of the packet. Our simulation studies show that the performance of EDBP is very close to that of DBP-M (a known algorithm for the problem) with a significant reduction in scheduling cost.

1 Introduction

Packet switched networks are increasingly being utilized for carrying real-time traffic which often require quality of service (QoS) in terms of end-to-end delay, jitter, and loss. A particular type of real-time traffic is a real-time stream, in which a sequence of related packets arrive at a regular interval with certain common timing constraints [1]. Real-time streams occur in many applications such as real-time video conferencing, remote medical imaging, and distributed real-time applications. Unlike non-real-time streams, packets in a real-time stream have deadlines by which they are expected to reach their destination.

Packets that do not reach the destination on time contain stale information that cannot be used. There have been many schemes in the literature to deterministically guarantee the meeting of deadlines of all packets in a stream [2,3]. The main limitation of these schemes is that they do not exploit the ability of streams that can tolerate occasional deadline misses. For example, in teleconferencing, occasional misses of audio packets can be tolerated by using interpolation techniques to estimate the information contained in tardy/dropped packets.

On the other hand, there are schemes that try to exploit the ability of streams to tolerate occasional deadline misses by bounding the steady-state fraction of packets that miss their deadlines [4]. The main problem with these approaches is that the deadline misses are not adequately spaced which is often better than encountering spurts of deadline misses. For example, if a few consecutive audio packets miss their deadlines, a vital portion of the talkspurt may be missing and

J. Rolim et al. (Eds.): IPDPS 2000 Workshops, LNCS 1800, pp. 743-749, 2000.
© Springer-Verlag Berlin Heidelberg 2000

the quality of the reconstructed audio signal may not be satisfactory. However, if the misses are adequately spaced, then interpolation techniques can be used to satisfactorily reconstruct the signal [5].

To address this problem, the (m, k)-firm guarantee model was proposed in [1]. A real-time stream with an (m, k)-firm guarantee requirement states that m out of any k consecutive packets in the stream must meet their respective deadlines. When a stream fails to meet this (m, k)-firm guarantee, a condition known as *dynamic end-to-end failure* occurs. The probability of dynamic end-to-end failure is then used as a measure of the QoS perceived by a (m, k) firm real-time stream.

Related Work: The message scheduling algorithms, such as Earliest Deadline First (EDF) and its variants [2, 3] that have been proposed for real-time streams are not adequate for (m, k)-firm streams because they do not exploit the m and k parameters of a stream. For scheduling of (m, k)-firm streams, a best-effort scheme has been proposed in [1] for single hop and has been extended to multihop in [6], with the objective of minimizing the dynamic end-to-end failure.

DBP Algorithms: A scheduling algorithm, Distance Based Priority (DBP), has been proposed in [1] in which each stream is associated with a state machine and a DBP value which depends on the current state of the stream. The state of stream captures the meeting and missing of deadlines for a certain number of previous packets of the stream. The DBP value of a stream is the number of transitions required to reach a failing state, where failing states are those states in which the number of meets is less than m. The lower the DBP value of a stream, the higher its priority. The packet from the stream with the highest priority is selected for transmission. Figure 1 shows the state diagram for a stream with a $(2,3)$-firm guarantee wherein M and m are used to represent meeting a deadline and missing a deadline, respectively. Each state is represented by a three-letter (k-letter) string. For example, MMm denotes the state where the most recent packet missed its deadline and the two previous packets met their deadlines. The edges represent the possible state transitions. Starting from a state, the stream makes a transition to one of two states, depending on whether its next packet meets (denoted by M) or misses (denoted by m) its deadline. For example, if a stream is in state MMm and its next packet meets the deadline, then the stream transits to state MmM. In Figure 1, the failure states are mMm, Mmm, mmM, and mmm.

The Modified DBP (DBP-M) [6] is a multihop version of the original DBP algorithm. In DBP-M, for each stream, the end-to-end deadline is split into link (local) deadlines, along the path from source to destination of the stream, such that the sum of the local deadlines is equal to the end-to-end deadline. DBP-M confronts the problem introduced by multihop networks by having packets transmitted onward until they have missed their respective end-to-end deadlines. Thus, although a packet may miss its local deadline, it is still given a chance to meet its end-to-end deadline.

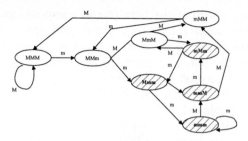

Fig. 1. DBP state diagram of a (2,3) stream

Motivation for Our Work: DBP and DBP-M use a separate queue for each stream at every node along the path of a stream (connection). That is, for each stream that is flowing across the network, a separate queue is created and per-stream state information is maintained at each node along the path of the stream. This solution is not scalable as the number of queues increases with the number of streams which results in high scheduling cost in terms of computational requirements. Similarly, the per-stream state information incurs overhead in terms of computational and memory requirements. The second aspect has been addressed by the Differentiated Services model [9]. In this paper, we address the first aspect by proposing an algorithm that reduces the scheduling cost by maintaining a fixed number of queues.

There exists a tradeoff between dynamic failure performance and the scheduling cost involved in achieving that performance. With the DBP and DBP-M extreme, a significant amount of scheduling cost is required to maintain the one queue per one stream ratio. Given a link that has N streams flowing across it, a DBP-M implementation requires N queues and requires $O(N)$ scheduling cost. However, this queue to stream ratio does deliver the best dynamic end-to-end failure performance for a given set of (m, k) streams.

In contrast, classical EDF scheduling and its variations require only one (or a fixed number of) queue(s) per link, i.e. the streams are multiplexed onto the queue(s), hence requiring a scheduling cost of $O(1)$. These methods incur the least scheduling cost but deliver the poorest end-to-end dynamic failure performance for (m, k) streams. Therefore, a better algorithm would require less scheduling cost than DBP-M but would provide better dynamic failure performance than classic EDF scheduling. This is the principal motivation for our work; in it, an integrated heuristic is proposed that allows multiplexing of streams while still providing adequate dynamic failure performance.

2   EDBP Scheduling Algorithm

The proposed EDBP algorithm aims at providing the same dynamic failure performance as that of DBP-M with a minimal scheduling cost by allowing queues to have more than one stream multiplexed. EDBP meets this goal by its integrated heuristic (EDBP value) that incorporates the DBP state of a stream together with the laxity of the packet. The EDBP algorithm has two key parts.

The first part deals with selecting the best (highest priority) packet from a window of packets in each queue (Steps 1-4). The second part selects the best packet from those packets chosen in the first part and transmits it (Steps 5-6). For the EDBP algorithm, the following notations are used:

Q_i: i^{th} queue; P_j: j^{th} packet in a queue

S_x: stream that produced P_j; w: window size

$EDBP(P_j)$: EDBP value of packet P_j

$EDBPS(S_x)$: EDBP state of stream S_x

The packets in a queue are stored in FIFO order. The cost of algorithm has two parts: queue insertion cost and scheduling cost. The insertion cost is high for EDF because it uses a priority queue and is unit cost for DBP and EDBP. EDF has a unit scheduling cost whereas the scheduling costs of DBP and EDBP are N and $w*Q$, respectively, where N is the number of streams and Q is the number of queues. The EDBP algorithm for transmitting a packet is given in Figure 2 below. Following it, the steps of the algorithm are discussed in detail.

Begin

For each queue Q_i perform Steps 1-4

1) For each P_j from P_0 to P_{w-1}, determine if the packet has missed its end-to-end deadline, such packets are then dropped.

2) Local Deadline $(P_j) = \frac{End-to-EndDeadline(P_j)}{Number\ of\ Hops\ in\ the\ path\ of\ stream\ S_x}$

   Laxity (P_j) = Local Deadline(P_j) - *current time*

   BucketWidth = Max $(|(Laxity(P_0)|, |(Laxity(P_1)|,..., |(Laxity(P_{w-1})|) + 1$

3) Calculate the EDBP value for each packet P_j.

   $EDBP(P_j)$ = BucketWidth * $EDBPS(S_x)$ + $Laxity(P_j)$

4) Select P_j that has the lowest EDBP value, called best packet.

5) Repeat steps 2-4, treating the best packet from each queue Q_i as a packet in an overall queue and with a window size (w) equal to the number of queues available.

6) Schedule the packet with the lowest EDBP value.

End

Fig. 2. EDBP scheduling algorithm for transmitting a packet

Step 1: The EDBP algorithm examines a window of w packets from each queue starting from P_0 (head packet in queue) up to P_{w-1} to determine if the packet has missed its end-to-end deadline. Thus, if a packet cannot meet its end-to-end deadline, the packet is dropped and the EDBP state of the corresponding stream for the node is adjusted accordingly. As with DBP-M, a packet is not dropped based on its local deadline. The use of the end-to-end deadline as a dropping mechanism is to give the packet a chance to meet its end-to-end deadline by scheduling the packet ahead of time in the downstream nodes across its path.

Step 2: In order to combine the EDBP state of a given packet P_j with the packet's laxity, the EDBP state must be converted to a meaningful value. Therefore, the EDBP algorithm uses the notion of buckets and offsets. The idea of a bucket is to group together the streams that have similar DBP states and the laxity is used as an offset inside the group (bucket). The local deadline cannot be used for the calculation of the bucket width as it is a relative value. However, the laxity of a packet is an absolute value related to the maximum end-to-end deadline in the network. In this step, for each queue, a window of packets is examined to determine the packet with the largest absolute laxity value.

However, the maximum laxity value itself cannot simply be used to determine the bucket width. Consider the case where all of the packets in the window have missed their local deadline and the maximum laxity value is negative. Because the maximum laxity value is negative, priority inversion would occur as a lower EDBP heuristic value means a higher priority. To handle this case, the EDBP heuristic uses the maximum absolute laxity value. Thus, the value is always positive and priority inversion cannot occur.

Consider a second case where all of the packets have a local deadline of zero. Thus, without further modification, the EDBP state of the respective streams would essentially drop out of the EDBP heuristic. To handle this case, the maximum laxity value is further modified by adding one. This ensures that the modified laxity value will always be greater than or equal to one, thus eliminating the possibility of priority inversion or the elimination of the term corresponding to the EDBP state.

Steps 3, 4: Following the bucket width calculation, the best packet for the queue must be selected. The EDBP heuristic itself is divided into two parts, the bucket calculation and the bucket offset calculation. Each packet is placed into its appropriate bucket by multiplying the value of the EDBP state with the bucket width. After the bucket calculation is complete, each packet is appropriately offset into its bucket by adding the laxity value for that packet.

For the EDBP algorithm, a modification of the DBP state calculation is proposed. As with the initial DBP algorithm, the DBP value of a stream in the non-failing state is the number of transitions required to reach a failing state. Consider a (2,3)-firm stream where with a previous history of MMM. The DBP value would be 2, representing the two transitions required to reach a failing state. In the EDBP heuristic, the DBP state is expanded to allow negative values, thus allowing the EDBP state to discern between levels of dynamic failure between different streams. When the stream has reached a failing state, EDBP expands upon the initial DBP algorithm by setting the EDBP value equal to one minus the number of transitions to return to a non-failing state. Under the initial DBP algorithm, a (2,3) stream with a history of Mmm would yield a DBP value of 0. However, when one examines the state diagram for the (2,3) stream, it is discovered that two transitions are required to return to a non-failing state. Under the EDBP algorithm, this stream would receive an EDBP value of -1, thus appropriately placing the packet at a priority level denoting its level of dynamic failure.

Best Packet Selection - Steps 5, 6: Once the best packet has been selected from each queue Q_i, the overall best packet is selected among these packets for transmission. To accomplish this, Steps 2-4 are repeated again with the following modifications. First, the queue being examined is now a queue of the best packets from each queue Q_i. Second, the window size for the EDBP algorithm is equal to the number of queues available. The best overall packet thus obtained will have the lowest EDBP value and is transmitted.

3   Performance Study

A network simulator was developed to evaluate and compare the performance of the EDBP algorithm with that of the DBP-M and EDF algorithms. The simulator uses a single queue for EDF, one queue per stream for DBP-M, and a fixed number of queues (which is an input parameter to the simulator) for EDBP. For our simulation studies, we have selected ARPANET as the representative topology. The algorithms were evaluated using the probability of dynamic failure as the performance metric. In our simulation, one millisecond (ms) is represented by one simulation clock tick.

Source and destination nodes for a stream were chosen uniformly from the node set. The local deadline for each stream was fixed with the end-to-end deadline equal to the fixed local deadline times the number of hops in the stream's path. The m and k values of a stream in the network are exponentially distributed with the condition that $m < k$. The mean inter-arrival time of streams in the network follow a Poisson distribution and stay active for an exponentially distributed duration of time. Packets are assumed to be of fixed size and each link has a transmission delay of one millisecond.

Effect of Number of Queues: In Figure 3, the effect of the number of queues on the probability of dynamic failures is examined in the EDBP algorithm. Thus, in the best case, the number of queues is equal to the number of streams. This is exemplified by the DBP-M algorithm. The EDBP algorithm has been split into two versions, one with $N/2$ queues and the other with $N/4$ queues ($N = 16$). Each increase in the number of queues results in an appropriate increase in the dynamic failure performance of the EDBP algorithm. For this figure, the performance of the EDF and DBP-M algorithms remain unchanged as the queue parameter has no effect on these algorithms. From Figure 3, one can deduce that an increase of the number of queues reduces the multiplexing degree that in turn increases the performance of the EDBP algorithm. The performance of the EDBP algorithm at N/2 queues is extremely close to the performance of the DBP-M algorithm while requiring only half of the scheduling cost of DBP-M.

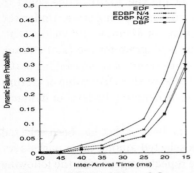
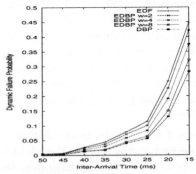

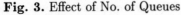

Fig. 3. Effect of No. of Queues   **Fig. 4.** Effect of Window Size

Effect of Window Size: However, in a given setting, it may not be practical or even possible to increase the number of queues available. Figure 3 repeats the same settings used in Figure 3, except that the window size is varied instead of the number of queues. Three versions of the EDBP algorithm are examined with $w = 2, 4, 8$. As the window size increases, the dynamic failure performance of the EDBP algorithm increases because the window size offsets the penalty imposed by the multiplexing of streams onto a given queue.

When the effect of window size is compared to the effect of additional queues in the EDBP algorithm, our experiments show that the increase in queues produces a more profound effect than an increase in window size. The underlying cause is due to the multiplexing of streams onto queues. Consider a scenario in which a stream (S_x) with a small period (high rate) and another stream (S_y) with a large period (low rate) are multiplexed onto the same queue. In this case, S_x will have a higher chance of having its packets inside the window than S_y. This results in more dynamic failure for S_y. However, as the number of available queues increases, the chance of these streams being separated into different queues increases as well, thus explaining the difference in performance. Therefore, to obtain the best performance from the EDBP algorithm, the window size must be appropriately tuned to the degree of multiplexing.

4   Conclusions

In this paper, we have addressed the problem of best-effort scheduling of (m, k)-firm real-time streams in multihop networks. The proposed algorithm, EDBP, allows multiplexing of streams onto a fixed number of queues and aims at maximizing the dynamic failure performance with minimal scheduling cost. Our simulation studies have shown that the performance is close to that of the DBP-M algorithm with a significantly lower scheduling cost.

References

1. M. Hamdaoui and P. Ramanathan, "A dynamic priority assignment technique for streams with (m,k)-firm guarantees," *IEEE Trans. Computers*, vol.44, no.12, pp.1443-1451, Dec. 1995.
2. D. Ferrari and D.C. Verma, "A scheme for real-time channel establishment in wide-area networks," *IEEE JSAC*, vol.8, no.3, pp.368-379, Apr. 1990.
3. H. Zhang, "Service disciplines for guaranteed performance service in packet-switching networks," *Proc. IEEE*, vol.83, no.10, pp. 1374-1396, Oct. 1995.
4. D. Yates, D.T.J. Krouse, and M.G. Hluchyj, "On per-session end-to-end delay distributions and call admission problem for real-time applications with QoS requirements," in *Proc. ACM SIGCOMM*, pp.2-12, 1993.
5. Y.-J. Cho and C.-K. Un, "Performance analysis of reconstruction algorithms for packet voice communications,", *Computer Networks and ISDN Systems*, vol. 26, pp. 1385-1408, 1994.
6. W. Lindsay and P. Ramanathan, "DBP-M: A technique for meeting end-to-end (m,k)-firm guarantee requirements in point-to-point networks," in *Proc. IEEE Conference on Local Computer Networks*, pp. 294-303, Nov. 1997
7. S.S. Panwar, D. Towsley, and J.K. Wolf, "Optimal scheduling policies for a class of queues with customer deadlines to the beginning of service," *Journal of the ACM*, vol.35, no.4, pp.832-844, Oct. 1988.
8. S. Shenker and L. Breslau, "Two issues in reservation establishment," in *Proc. ACM SIGCOMM*, pp.14-26, 1995.
9. W. Weiss, "QoS with Differentiated Services," *Bell Labs Technical Journal*, pp. 44-62, Oct.-Dec 1998.

Predictability and Resource Management in Distributed Multimedia Presentations

Costas Mourlas

Department of Computer Science, University of Cyprus,
75 Kallipoleos str., CY-1678 Nicosia, Cyprus
mourlas@ucy.ac.cy

Abstract. The continuous media applications have an implied temporal dimension, i.e. they are presented at a particular rate for a particular
length of time and if the required rate of presentation is not met the
integrity of these media is destroyed. We present a set of language constructs suitable for the definition of the required QoS and a new real-time
environment that provides low-level support to these constructs. The emphasis of the proposed strategy is given on deterministic guarantees and
can be considered as a next step for the design and the implementation
of *predictable* continuous media applications over a network.

1   Introduction

The current interest in network and multimedia technology is focused on the
development of distributed multi-media applications. This is motivated by the
wide range of potential applications such as distributed multi-media information
systems, desktop conferencing and video-on-demand services. Each such application needs Quality of Service (QoS) guarantees, otherwise users may not accept
them as these applications are expected to be judged against the quality of traditional services (e.g. radio, television, telephone services). The traditional network
environments although they perform well in static information spaces they are
inadequated for continuous media presentations, such as video and audio.

In a distributed multimedia information system (see figure 1) there is a set of
Web-based applications where each application is allocated on a different node
of the network and can require the access of media servers for continuous media
data retrieval. These continuous media servers can be used by any application
running in parallel on a different node of the network. Each such presentation
has specific timing and QoS requirements for its continuous media playback.
This paper presents a new set of language constructs suitable for the definition
of the required QoS and the real-time dimension of the media that participate
in multimedia presentations as well as a runtime environment that provides low-
level support to these constructs during execution.

2   The Proposed Language Extensions for QoS definition

Playing a set of multimedia presentations in a traditional network architecture
two main problems are met. Firstly, the best-effort service model provided by

J. Rolim et al. (Eds.): IPDPS 2000 Workshops, LNCS 1800, pp. 750-756, 2000.

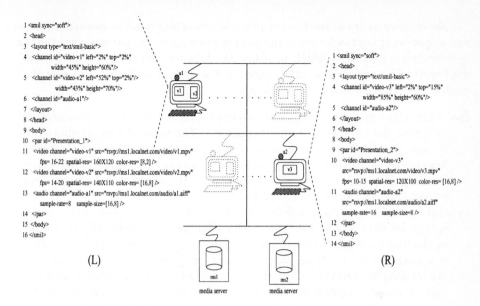

```
1 <smil sync="soft">
2 <head>
3   <layout type="text/smil-basic">
4    <channel id="video-v1" left="2%" top="2%"
             width="45%" height="60%"/>
5    <channel id="video-v2" left="52%" top="2%"/>
             width="43%" height="70%"/>
6    <channel id="audio-a1"/>
7   </layout>
8 </head>
9 <body>
10 <par id="Presentation_1">
11   <video channel="video-v1" src="rsvp://ms1.localnet.com/video/v1.mpv"
          fps= 16-22 spatial-res= 160X120 color-res= [8,2] />
12   <video channel="video-v2" src="rsvp://ms1.localnet.com/video/v2.mpv"
          fps= 14-20 spatial-res= 140X110 color-res= [16,8] />
13   <audio channel="audio-a1" src="rsvp://ms1.localnet.com/audio/a1.aiff"
          sample-rate=8   sample-size=[16,8] />
14 </par>
15 </body>
16 </smil>
```
(L)

```
1 <smil sync="soft">
2 <head>
3   <layout type="text/smil-basic">
4    <channel id="video-v3" left="2%" top="15%"
             width="85%" height="60%"/>
5    <channel id="audio-a2"/>
6   </layout>
7  </head>
8  <body>
9  <par id="Presentation_2">
10   <video channel="video-v3"
          src="rsvp://ms1.localnet.com/video/v3.mpv"
          fps= 10-15 spatial-res= 120X100 color-res= [16,8] />
11   <audio channel="audio-a2"
          src="rsvp://ms1.localnet.com/audio/a2.aiff"
          sample-rate=16   sample-size=8 />
12 </par>
13 </body>
14 </smil>
```
(R)

media server   media server

Fig. 1. *A Distributed Multimedia Information System*

the existing systems does not address the temporal dimension of the continuous media data during their retrieval and transmission phase. Resource reservation even if it is required, it is not the final answer to the end-users. The end-users actually care on how to exploit all the available (and reserved) resources in a best way such that the multimedia application will be presented according to the expected quality requirements. For example, a 10% reservation of the total bandwidth to a video presentation means that the video can be played either colored with a rate of 10 frames per second or grey-scaled with a rate of 18 frames per second. The decision has to be taken by the end-users and the multimedia authors, providing high-level language primitives and special annotation for the definition of any quality requirement. This new set of high-level language constructs will be presented in the following paragraphs and comes as a continuation of our previous work described in [6].

The language that will be extended is SMIL [9], a language for Web-based multimedia presentations which has been developed by the W3C working group on synchronized multimedia. These extensions are introduced along the lines of SMIL, and there is an attempt to reuse terminology wherever feasible. SMIL describes four fundamental aspects of a multimedia presentation: temporal specifications, spatial specifications, alternative behaviour specifications and hypermedia support. In this section we introduce and define a fifth aspect of a multimedia presentation, called *quality specifications*. In our extended SMIL language, the two continuous media objects can be described together with their quality requirements within a document via the following syntax:

- ⟨video cmo-attributes v-qos-attributes⟩, and
- ⟨audio cmo-attributes a-qos-attributes⟩.

The extensions are defined by the two new sets of attributes `v-qos-attributes` and `a-qos-attributes` for video and audio respectively. The set `cmo-attributes` is curenlty supported by SMIL to define the location and duration of the media object. The new `v-qos-attributes` and `a-qos-attributes` lists describe quality requirements using the attributes:

fps : The value of *fps* defines the temporal resolution of a video presentation by giving the number of frames per second. The value of this attribute can be any positive integer or a range of positive integers. For example giving `fps=14-18` as attribute to a video object, it means that the accepted values for this video presentation can be any rate between 14 and 18 frames per second (Figure 1 lines: L-11,L-12,R-10).

spatial-res : The *spatial-res* definition of a video presentation specifies the spatial resolution in pixels required for displaying the video object. In our model, the concepts of *layout* and *resolution* are separated. The *resolution* is a quality concept. If an ordered list of resolutions is given (e.g. *spatial-res=[180X130, 120X70]*) then the video object will be presented with the highest possible spatial resolution according to the availability of system resources and can be altered at run time (lines: L-11,L-12,R-10).

color-res : This attribute specifies the color resolution in bits required for displaying the video object. Typical values are 2, 8, 24 …. If an ordered list of integer values is given (e.g. *color-res=[8,2]*) then the video object will be presented with the highest possible color resolution. (lines: L-11,L-12,R-10).

sample-rate : The value of *sample-rate* for an audio object defines in KHz the rate that the analog signal is sampled. If we need, for example, telephone quality the analog signal should be sampled 8000 times per second (i.e. *sample-rate = 8*), (lines: L-13,R-11).

sample-size : This attributes of an audio object specifies the sample size in bits of each sample. If an ordered list of integer values is given (e.g. *sample-size=[16,8]*) then each sample will be represented with a number of bits equal with one of the values given. For telephone quality, each sample of the signal is coded with 8 bits whereas for CD quality it is coded with 16 bits. The highest value that can be used for every sample it is decided at run time according to the availability of the resources (lines: L-13,R-11).

The above language primitives form a complete set for QoS definition of every distinct continuous media that participate in a multimedia presentation. If several media streams have to be combined then inter-media synchronization is another important factor of quality specification but this subject has been extensively studied and completely supported by the standard SMIL language.

3   The Proposed Runtime Environment

We view every different multimedia presentation s_i as a *periodic* task τ_i with period T_i. Every periodic task τ_i is allocated on a different node of the distributed system and requires in each period the retrieval of a number of media blocks from the remote disk of the server. CS^i_j is the deterministic disk access time that task τ_i requires in every period to retrieve data for all of its streams from the server S_j (communication delays can be included in the evaluation of every CS^i_j). Every data retrieval section on a remote shared server S is guarded by a `lock(S)` statement. These locks are released after the data retrieval using the `unlock(S)` statement. The term "critical section" will be used to denote any data retrieval section of a task defined between a `lock(S)` and the corresponding `unlock(S)` statement.

We follow a rate monotonic strategy for priority assignments. Periodic tasks are assigned priorities inversely to tasks periods (ties are broken arbitrarily). Hence, task τ_i with period T_i receives higher priority than τ_j with period T_j if $T_i < T_j$.

The period T_i and the computational requirements CS^i_j of every task are determined by the desired QoS of the stream that the task represents as well as system resources (processor speed, disk access time). The formal procedure of transforming the set of distributed multimedia presentations with quality of service expectations to a set of periodic tasks is described in our previous work [7, 6]. We have to notice here that the scheduling analysis that follows does not consider ranges of QoS values and this task is left as future work.

A periodic task τ can have multiple non-overlapping critical sections, e.g.
$$\tau = \{ \ldots \texttt{lock(S}_1)\ldots\texttt{unlock(S}_1)\ldots\ldots\texttt{lock(S}_2)\ldots\texttt{unlock(S}_2)\ldots \}$$
but not any nested critical section. Each task is characterized by two components (CS^i, T_i), $1 \le i \le n$, where CS^i is the set $\{CS^i_j \mid j \ge 1\}$ that includes all the critical sections of the task τ_i. CS^i_j is the critical section of task τ_i guarded by statement `lock(S`$_j$`)`. We define as C_i the total deterministic computation requirement of all data retrieval sections of task τ_i, i.e $C_i = \sum_{x \in CS^i} x$.

Each server S_j can be either *locked* by a task τ_i if τ_i is within its critical section CS^i_j or *free* otherwise. Suppose that a task τ_i requires to lock server S_j and enter its critical section CS^i_j issuing the operation `lock(S`$_j$`)`. Then the following cases can occur:

1. The server S_j is *free*. Then, the server S_j is allocated to the task τ_i, the task τ_i proceeds to its critical section and the state of S_j becomes *locked*. A server S_j *locked* by task τ_i can not be accessed by any other task.
2. If case 1 does not hold, i.e. server S_j is currently *locked*, then after its release it is allocated to the highest priority task that is asking for its use. The task τ_i will proceed to its critical section if and only if server S_j has been allocated to τ_i.

By the definition of the protocol, a task τ_i can be blocked by a lower priority task τ_j, only if τ_j is executing within its critical section CS^j_l when τ_i asked for the use of the shared server S_l. Note also that the proposed synchronization protocol prevents deadlocks due to the fact that for any task τ_i there is no nested critical section. Thus, τ_i will never ask in its critical section for the use of any other server and so a blocking cycle (deadlock) cannot be formed.

We can easily conclude that a set of n periodic tasks, each one bound to a different node $\wp$ of a network can be scheduled using the proposed synchronization protocol if the following conditions are satisfied:

$$\forall i, 1 \le i \le n, \quad C_i + B_i \le T_i \tag{1}$$

The term B_i represents the total worst case blocking time that task τ_i has to wait for the allocation of the required media servers in every period T_i. Once B_is have been computed for all i, the conditions (1) can then be used to determine the schedulability of the set of tasks.

3.1   Determination of Task Blocking Time

Here, we shall compute the worst-case blocking time B_l^i that a task τ_i has to wait the allocation of server S_l, following a response-time-analysis type formulation [3]. This longest blocking time occurs at the *critical instance* for τ_i.

Definition 3.1 A *critical instance* for task τ_i occurs whenever a request from τ_i to lock a server occurs simultaneously with the requests of all higher-priority tasks to lock this server. At that instance also, the lower priority task with the longest critical section executes its critical section holding the lock of that server.

Theorem 3.1 Consider a set of n tasks $\tau_1, \ldots, \tau_n$ arranged in descending order of priority. Each task is bound to a different node $\wp_i$ of the network and the proposed synchronization protocol is used for the allocation of the servers. Let

$$H_l^i = \{CS_l^j \mid 1 \le j < i\}, \qquad$$ - set of critical sections used by tasks with higher priorities than τ_i accessing the same server S_l

$$L_l^i = \{CS_l^j \mid i < j \le n\}, \qquad$$ - set of critical sections used by tasks with lower priorities than τ_i accessing the same server S_l

$$\beta_l^i = \max(L_l^i). \qquad$$ - blocking time due to lower priority tasks

Then, the worst case blocking time B_l^i each time task τ_i attempts to allocate server S_l and execute its critical section is equal to:

$$B_l^i = \sum_{CS_l^j \in H_l^i} \lceil \frac{B_l^i + \Delta t}{T_j} \rceil * CS_l^j + \beta_l^i, \quad 0 < \Delta t < 1 \quad \text{if} \sum_{CS_l^j \in H_l^i} \frac{CS_l^j}{T_j} < 1 \tag{2}$$

Proof: The smallest integer value that satisfies equation 2 above represents the longest blocking time B_l^i for a task τ_i trying to enter its critical section CS_l^i at its worst-case task set phasing, i.e. at its critical instance.

If the worst-case task set phasing occurs at time $t_0 = 0$ then the right-hand side of the equation represents the sum of the computational requirements for server S_l for all inputs from higher levels at the time interval $[0, B_l^i + \Delta t)$ as well as the duration of one (actually the maximum) critical section of the lower priority tasks in L_l^i namely β_l^i. Task τ_i will enter its critical section at time B_l^i

when the server S_l becomes *free*, i.e. after its consecutive use from tasks during the worst-case phasing. At that time and during the interval $[B_l^i, B_l^i + 1)$, server S_l becomes *free* for first time after t_0 and thus task τ_i will have the opportunity to lock S_l. The fact that server S_l is idle at time $t \in [B_l^i, B_l^i + 1)$ leads to the result that the sum of the computational requirements for server S_l over the interval $[0, t)$ equals B_l^i. Notice that an arbitrary value lying between zero and one is actually needed to check the load of the server at the interval $[B_l^i, B_l^i + 1)$, and this value is represented by the term Δt.

In all cases, the sum $\sum_{CS_l^j \in H_l^i} \frac{CS_l^j}{T_j}$ should be less than one. This sum represents the work load of server S_l or the utilization factor of the server due to higher priority tasks and should be less than one otherwise all these higher priority tasks could block repeatedly the task τ_i and in this case B_l^i will be unbounded (condition of formula 2). Hence the Theorem follows.   □

Equations of the form 2 above do not lend themselves easily to analytical solution. However, a solution to this equation can be found by iteration. The total worst-case blocking duration B_i experienced by task τ_i is the sum of all these blocking durations, i.e. $B_i = \sum_{CS_j^i \in CS^i} B_j^i$. Once these blocking terms B_i, $1 \leq i \leq n$, have been determined, conditions (1) give a complete solution for the real-time task synchronization and scheduling in the distributed environment.

4   Related Work

A significant amount of work has been carried out for making resource allocations to satisfy specific application-level requirements. The Rialto operating system [2] was designed to support simultaneous execution of independent real-time and non-real-time applications. The RT-Mach microkernel [4] supports a processor reserve abstraction which permits threads to specify their CPU resource requirements. If admitted by the kernel, it guarantees that the requested CPU demand is available to the requestor.

The Lancaster QoS Architecture [1] provides extensions to existing microkernel environments for the support of continuous media. The QoS Broker [8] model addresses also the requirements for resource guarantees, QoS translation and admission control, so a new system architecture is proposed which provides all these issues. The Nemesis operating system is described in [5] as part of the Pegasus Project, whose goal is to support both traditional and multimedia applications.

We have to notice at this point that few of the above efforts address the problem of distributed multimedia applications and very few of all the current multimedia architectures provide any synchronization strategy and a theory for the analysis and the predictability of a set of multimedia applications executed in a distributed environment. Many CPU allocation schemes have been presented for multimedia applications based on the restrictive assumption that the applications are independent of one another and do not have access to multiple resources simultaneously.

5 Conclusions

In this paper, we studied a set of language extensions and a runtime environment suitable for creating and playing distributed multimedia information systems with QoS requirements. At the language level a set of language extensions for SMIL was presented suitable for the definition of the required QoS and the real-time dimension of the media that participate in a multimedia presentation. The runtime part is mainly focused on the maintenance of real-time constraints accross continuous media streams. It is based on a task oriented model that employs a *periodic-based service discipline* which provides the required service rate to a continuous media presentation independent of traffic characteristics of other presentations.

One direction of our future work will be on the ability of the runtime environment to support the required quality of service when the required quality lies within a range, by giving the minimal and the upper bound for the expected quality (e.g. *fps=18-22*). The runtime system will try to provide the best value in the range and it will be also authorised to modify this value at run-time towards the upper or the lower bound value according to the availability of the resources. This adaptation of quality of service will make the best use of the resources currently available to distributed applications and will give a fair solution to the presentation of continuous media applications over a network without sacrificing the ability to execute these applications *predictably* in time.

References

1. G.Coulson, G.S. Blair, P. Robin, and D. Shepherd. Supporting Continuous Media Applications in a Micro-Kernel Environment. In Otto Spaniol, editor, *Architectures and Protocols for High-Speed Networks*. Kluwer Academic Publishers, 1994.
2. M. B. Jones, D. Rosu, and M. Rosu. CPU Reservations and Time Constraints: Efficient, Predictable Scheduling of Independent Activities. In *Proceedings of the 16th ACM Symposium on Operating Systems Principles*, October 1997.
3. M. Joseph and P. Pandya. Finding Response Times in a Real-Time System. *The Computer Journal*, 29(5):390–395, 1986.
4. C. Lee, R. Rajkumar, and C. Mercer. Experiences with Processor Reservation and Dynamic QOS in Real-Time Mach. In *Proceedings of the Multimedia Japan 96*.
5. I. Leslie, D. McAuley, R. Black, T. Roscoe, P. Barham, D. Evers, R. Fairbairns, and E. Hyden. The Design and Implementation of an Operating System to Support Distributed Multimedia Applications. *IEEE Journal on Selected Areas in Communications*, 14(7):1280–1297, September 1996.
6. C. Mourlas. A Framework for Creating and Playing Distributed Multimedia Information Systems with QoS Requirements. In *Proceedings of the 2000 ACM Symposium on Applied Computing, SAC 2000* (accepted for publication) .
7. C. Mourlas, David Duce, and Michael Wilson. On Satisfying Timing and Resource Constraints in Distributed Multimedia Systems. In *Proceedings of the IEEE ICMCS'99 Conference*, volume 2, pages 16–20. IEEE Computer Society, 1999.
8. Klara Nahrstedt and Jonathan M. Smith. The QoS Broker. *IEEE Multimedia*, 2(1):53–67, Spring 1995.
9. W3C. SMIL Draft Specification. *See:* http://www.w3.org/TR/WD-smil.

Quality of Service Negotiation for Distributed, Dynamic Real-time Systems

Charles D. Cavanaugh[1], Lonnie R. Welch[2], Behrooz A. Shirazi[1], Eui-nam Huh[2], and Shafqat Anwar[1]

[1]Computer Science and Engineering Dept. The University of Texas at Arlington Box 19015, Arlington, TX 76019-0015
{cavan|shirazi|anwar}@cse.uta.edu
[2]School of Electrical Engineering and Computer Science Ohio University Athens, OH 45701-2979
{welch|ehuh}@ace.cs.ohiou.edu

Abstract. Dynamic, distributed, real-time systems control an environment that varies widely without any time-invariant statistical or deterministic characteristic, are spread across multiple loosely-coupled computers, and must control the environment in a timely manner. In order to ensure that such a system meets its timeliness guarantees, there must be a means to monitor and maintain the quality of service in the system. The QoS manager is a monitoring and diagnosis system for real-time paths, collections of time-constrained and precedence-constrained applications. These applications may be distributed across multiple, heterogeneous computers and networks. This paper addresses the QoS negotiation features of the QoS manager and its interaction with the middleware resource manager. The major contributions of the paper are the negotiation algorithms and protocol that minimize the impact on the other paths' QoS while maximizing the unhealthy path's QoS. The approach and algorithms for QoS negotiation are presented.

1 Introduction

Dynamic, distributed, real-time systems possess three characteristics. First, the environment that they control is not deterministic and cannot be characterized by time-invariant statistical distributions. Second, the system is spread across multiple loosely coupled computers. Third, the system must control the environment in a timely manner. Existing solutions for monitoring real-time systems [1] and for real-time scheduling are usually based on the assumption that the processes have worst-case execution times. In dynamic environments, such as air traffic control [2], robotics, and automotive safety, this assumption does not hold [3].

The dynamic real-time path [4][5] (Fig. 1) is a collection of time-constrained and precedence-constrained applications. These applications may be distributed across multiple, heterogeneous computers and networks. The QoS manager's tasks are to monitor path health, diagnose the causes of poor health, and request computation and communication resources to maintain and restore health.

J. Rolim et al. (Eds.): IPDPS 2000 Workshops, LNCS 1800, pp. 757-765, 2000.

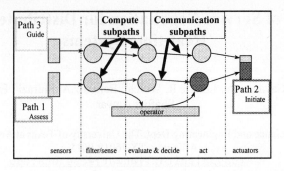

Fig. 1. Path composition

The problem of mapping applications to resources is to assign resources to consumers such that the delivered QoS meets or exceeds the QoS requirement (if possible). If this is not possible, some of the resources that are in use by a low criticality real-time application may need to be diverted to a high criticality real-time application. The QoS manager and resource manager must negotiate a solution that is mutually acceptable. QoS negotiation is the process of the QoS manager and the resource manager trading off resources for some applications while improving the QoS of the applications having higher criticality.

The rest of this paper is organized as follows: the QoS negotiation architecture and approach are explained in Section 2, the negotiation algorithms and protocol are presented in Section 3, a sample experiment using manual techniques to illustrate QoS negotiation is shown in Section 4, related work is summarized in Section 5, and a summary and statement of future work is in Section 6.

2 QoS Negotiation Architecture and Approach

The QoS negotiation architecture is presented in Fig. 2. The QoS monitor's job is to combine the monitored data into QoS metrics for the path and applications and to translate and pass along relevant application load and resource usage information. The analyzer's function is to detect QoS violations and calculate trends for QoS metrics, load, and resource usage. The diagnosis component determines the causes of the QoS violations by recognizing conditions that indicate a particular malfunction. The negotiator has two functions. First, it selects actions that will remedy the malfunctions and requests resources for applications if necessary. Second, it negotiates the highest possible QoS with the resource manager when the resource manager indicates that resource availability does not allow a certain action or resource request to be carried out. Negotiation involves trading off some actions for alternative actions that provide the highest possible QoS assurance under the resource availability constraints. The resource manager obtains current utilization levels for communication and computation resources from host monitors. Moreover, resource unification is required to map heterogeneous resource requirements into available target host. Then, the RM finds resources that meet (unified) resource requirements. If the hosts are feasible, then it predicts queuing delays to analyze schedulability.

QoS prediction will result from candidate reallocation actions. Finally, resource allocation selects and performs a reallocation action (through program control and startup daemons) based on predicted QoS. A new selection technique is used to guarantee the customer's QoS.

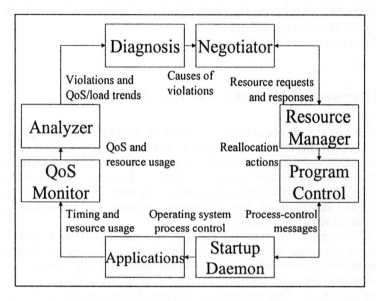

Fig. 2. QoS Negotiation Architecture

The three phases of QoS negotiation correspond to the three phases of diagnosing poor path health: path-local, resource-local, and global diagnosis. During phase I, path-local diagnosis, the QoS manager requests for allocation actions involving the unhealthy subpaths that it identifies. For example, one application within a path may be unhealthy; and the QoS manager would request that it be scaled up. During phase II, resource-local diagnosis, the QoS manager requests actions involving any software that is sharing resources with the unhealthy path. For example, the QoS manager may request that some competing application program be moved off a host. The QoS manager does not need to know the specific application program that is involved, as it is the resource manager's responsibility to maintain the system resources. During phase III, global diagnosis, the QoS manager requests actions that involve any resource. For example, the QoS manager may request that a less critical application be moved in order to free up space on a host that is not currently in use by the unhealthy subpath. The resource manager is responsible for finding the best host for the application or path while balancing the load among other paths and applications.

The three phases of negotiation are illustrated in the following scenario:

QM: Application x on host A is unhealthy and using 20% of CPU. Phase I: can you migrate it to another host? (QM adds action to list of attempted actions.)

RM: No. No combination of host idle times adds up to 20%. Provide QoS information, ranked application actions, and resource usage.

QM: (Marks previous action as unsuccessful.)   Phase II: can you move competing application y (also on host A), which uses 15% of CPU to another host (to free up 15% of CPU on host A)? (QM adds action to list of attempted actions.)

RM: No. No combination of host idle times adds up to 15%. Phase III: I can free up resources on host A by moving a less-critical application to a host with the lowest utility. (RM carries out action.)

QM: (Marks previous action as successful.)

3 QoS Negotiation Algorithm and Protocol

The QoS manager and resource manager maintain high QoS and manage resources, respectively. Whenever there is a conflict between obtaining enough resources to ensure high QoS and providing enough resources to the rest of the software, the QoS manager and resource manager negotiate a solution. To do this, both need algorithms to work toward their goals as well as a protocol for them to communicate with each other. The flowcharts of the algorithms illustrate the algorithms, and the communication steps show the protocol.

The following are the steps that the QoS manager takes once it detects a QoS violation. A flowchart of the process is shown in Fig. 3. First, the QoS manager identifies unhealthy computation and communication subpaths. Depending on the phase of negotiation (path-local, resource-local, or global), and the constraints on allowable actions, the QoS manager then selects actions to remedy the unhealthy subpaths. Each subpath has a resource requirement that is proportional to the slowdown that the subpath is experiencing. The slowdown is the ratio of the current subpath latency to the subpath's minimum latency for the same data stream size while on the same resource. For example, if the current subpath latency is 0.4 seconds at a data stream size of 1,000 on a particular resource, and the lowest latency that it has experienced in that same situation is 0.3 seconds, then the subpath's slowdown is 0.4 / 0.3, or approximately 1.333. This implies that it requires (133% - 100%) or 33% more resources to run at its best. The slowdown is due to contention, so moving it to another resource is a likely solution. The QoS manager ranks the actions based on their resource requirements in descending order and groups actions that involve moving subpaths off a particular host and actions that involve replicating a particular subpath (if it is replicable). The groups are automatically ranked, since the groups are made from the sorted list.

Once action selection is complete, the QoS manager requests resources by sending to the resource manager the ranked action requests (one from each group) along with the criticalities, current latencies, and resource usage information. If the resource manager responds that it can carry out the action, then the QoS manager monitors the stability of the system once the actions are carried out to ensure that QoS is indeed improved.

However, if the RM cannot do the action, then the RM sends a negotiation request to the QM. The QoS manager responds by sending out the next ranked action in each group, or it goes to the next phase of negotiation. The RM responds with a level of degradation in the QoS that is to be expected by the QM. The QoS manager calculates the slowdown that would be associated with the degradation and derives a benefit value for the path from it. If the benefit is at least as favorable as the QM requires, then the QoS manager responds to the RM with an acknowledgement; otherwise, the QoS manager proceeds to the next phase of negotiation.

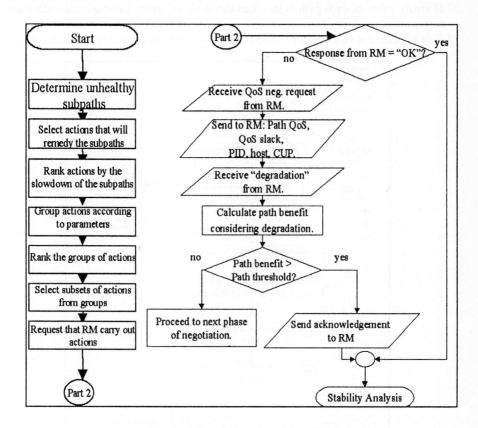

Fig. 3. QoS manager QoS negotiation algorithm

The steps that the resource manager takes to negotiate with the QoS manager and to allocate resources are listed below. A flowchart of the process is shown in Fig. 4.
1. Find a feasible host corresponding to resource needs
2. If a host is feasible, then do step 4
3. Else do step 8
4. Predict queuing delay and execution time on feasible hosts
5. If the task with predicted response time is schedulable, then do step 7
6. Else do step 8
7. Predict QoS and allocate it best host and exit
8. Send "QoS negotiation requests" to all QoS managers
9. Receive path QoS information and ranked list of actions and applications' resource usage from each QM
10. Calculate current utility value of each path
11. Select negotiable paths based on the minimum utility value
12. Calculate host utility values and find the host, H_j, with the minimum utility

13. Select application, a_i, in the ranked list of recommended actions
14. Test feasibility of allocating the application, a_i, on the host, H_j.
15. If not feasible, then pick next path and do step 11
16. If feasible, then recalculate utility value of path
17. If utility value of each path is less than threshold of utility value of each path then do step 4
18. Else allocate the violated application to the host that has the minimum utility value

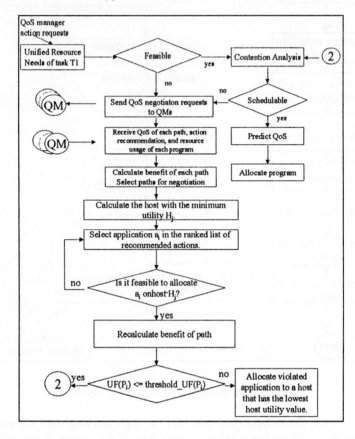

Fig. 4. Resource Manager QoS Negotiation Algorithm

4 Experimental Results

Sample experimental results were obtained by specifying two DynBench [6] periodic paths in the spec language: a higher criticality sensing path, *D:H:Higher_Sensing*, and a lower criticality sensing path, *D:L:Lower_Sensing*.   These paths were started simultaneously, and the experiment generator was used to bring the data stream size (the load) to 1600 tracks for each path.   Then, the filter and ED applications of each path were manually replicated to simulate the QoS manager's requesting that they be scaled up. The latency was brought down at that point. However, 500 more tracks were added to each path in order to overload the paths again. When no action was taken, the system became unstable, despite the fact that the loaded applications in both paths were already replicated. All four available hosts were in use. This instability is evident on the left-hand side of Fig. 5. Negotiation was simulated by manually moving the higher criticality path's filter and ED replicas to a more powerful host, named *texas*. In addition, resources were taken away from the lower criticality path by terminating the additional replicas of the lower criticality path's filter and ED applications, resulting in a normal QoS for the higher-criticality path (*C*) and a degraded QoS for the lower-criticality path (*D*), as shown on the right-hand side of Fig. 5. This combination of manual actions simulates the behavior of QoS negotiation and thus serves as a prototypical experiment. The scenario is a case by which an implementation of the negotiation algorithm should be tested.

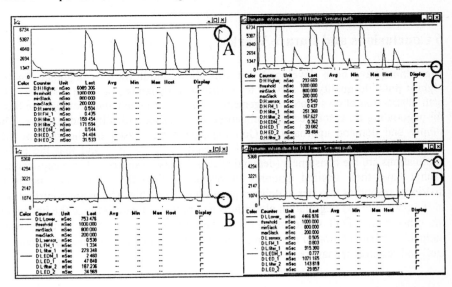

Fig. 5. Instability caused by overload (2100 tracks per path), without negotiation (*left*). The higher (*A*) and lower (*B*) criticality paths are fluctuating. Stability restored after negotiation (*right*). The higher criticality path experiences normal QoS (*C*); the lower criticality path experiences degraded QoS (*D*)

5 Previous Work in QoS Negotiation

To summarize the related work, the related work in negotiation is narrowly defined. The DeSiDeRaTa project promotes a broader view of negotiation: maximizing the quality of service provided to the most critical applications while minimizing the impact on other applications. The QuO project [7][8] terms the adaptation of object methods to the load as negotiation. Adaptation is only one aspect of QoS/resource management in DeSiDeRaTa, with dynamic optimization of system resource utilization and application QoS being other capabilities of DeSiDeRaTa's QoS negotiation. The University of Colorado DQM's [9][10] negotiation concept is a means of raising and lowering the operating level (the algorithm's complexity) based on current CPU usage conditions. This use of the term "negotiation" is similar to QuO's use of the term. EPIQ's [11] description of negotiation falls under this description as well, with the switching of regions of feasible quality being done in response to current conditions. The RTPOOL project [12] describes negotiation as the client's specifying a static deadline for a task with a reward for scheduling the task. The server does a preliminary static schedulability analysis of worst-case timing characteristics, and its algorithm shuffles the tasks to maximize the reward. DeSiDeRaTa is a dynamic system that maintains the required quality of service under dynamic workloads, where worst-case execution times and time-invariant statistical timing characteristics are unknown. Furthermore, it uses the path abstraction.

6 Conclusions and Future Work

Algorithms have been developed that will allow middleware to negotiate for the highest possible quality of service for distributed, dynamic real-time systems. The path abstraction allows QoS management to be decentralized and provides the basis for negotiating for resources for applications of differing criticality and purpose. The supply and demand approach to QoS management is based on the concept that resources (supply space) are limited in quantity and capacity and that the paths' applications are the consumers (demand space) of these resources. If the applications cannot have their desired amounts of resources, then the middleware needs to distribute resources in order to deliver the best QoS possible. The major contributions of the paper are the negotiation algorithms and protocol that minimize the impact on the other paths' QoS while maximizing the unhealthy path's QoS. Future work includes implementation of the negotiation algorithms and integration into the current QoS and resource managers.

References

1. Tsai, J. J. P., and S. J. H. Yang. *Monitoring and Debugging of Distributed Real-Time Systems*. Los Alamitos, CA: IEEE Computer Society Press, 1995.
2. Cavanaugh, C. D., L. R. Welch, and C. Bruggeman. *A Path-Based Design for the Air Traffic Control Problem*. Arlington, TX: The University of Texas at Arlington Department of Computer Science and Engineering, 1999. Technical Report, TR-CSE-99-001.
3. Harrison, R. D. "Combat System Prerequisites on Supercomputer Performance Analysis." Proceedings of the NATO Advanced Study Institute on Real Time Computing, 1994.
4. Welch, L. R., B. Ravindran, B. Shirazi, and C. Bruggeman. "Specification and Analysis of Dynamic, Distributed Real-Time Systems." Proceedings of the 19th IEEE Real-Time Systems Symposium, Madrid, Spain, December 2-4, 1998.
5. Welch, L. R., P. V. Werme, B. Ravindran, L. A. Fontenot, M. W. Masters, D. W. Mills, and B. A. Shirazi. "Adaptive QoS and Resource Management Using A Posteriori Workload

Characterizations." Proceedings of the 5th IEEE Real-Time Technology and Applications Symposium (RTAS '99), May 1999.

6. Welch, L. R., and B. A. Shirazi. "A Dynamic Real-time Benchmark for Assessment of QoS and Resource Management Technology." Proceedings of the 5th IEEE Real-Time Technology and Applications Symposium (RTAS '99), May 1999.

7. Loyall, J. P., R. E. Schantz, J. A. Zinky, and D. E. Bakken. "Specifying and Measuring Quality of Service in Distributed Object Systems." Proceedings of the 1st International Symposium on Object-Oriented Real-Time Distributed Computing (ISORC '98), Kyoto, Japan, April 1998.

8. Zinky, J. A., D. E. Bakken, and R. E. Schantz. "Architectural Support for Quality of Service for CORBA Objects", *Theory and Practice of Object Systems,* 3(1) 1997.

9. Brandt, S., G. Nutt, T. Berk, and J. Mankovich, "A Dynamic Quality of Service Middleware Agent for Mediating Application Resource Usage", *Proceedings of the 19th IEEE Real-Time Systems Symposium (RTSS '98),* December 1998.

10. Brandt., S., G. Nutt, T. Berk, and M. Humphrey, "Soft Real-Time Application Execution with Dynamic Quality of Service Assurance", *Proceedings of the 6th IEEE/IFIP International Workshop on Quality of Service (IWQoS '98),* pp. 154-163, May 1998.

11. Liu, J. W. S., K. Nahrstedt, D. Hull, S. Chen, and B. Li. "EPIQ QoS Characterization Draft Version." *http://epiq.cs.uiuc.edu/qo-970722.pdf*

12. Abdelzaher, T. F., E. M. Atkins, and K. Shin, "QoS Negotiation in Real-Time Systems and its Application to Automated Flight Control" accepted to *IEEE Transactions on Software Engineering,* 1999. (Earlier version appeared in *IEEE Real-Time Technology and Applications Symposium,* Montreal, Canada, June 9-11, 1997.

An Open Framework for Real-Time Scheduling Simulation

Thorsten Kramp, Matthias Adrian, and Rainer Koster

Distributed Systems Group, Dept. of Computer Science
University of Kaiserslautern, P.O. Box 3049, 67653 Kaiserslautern, Germany
{kramp,koster}@informatik.uni-kl.de

Abstract. Real-time systems seek to guarantee predictable run-time behaviour to ensure that tasks will meet their deadlines. Optimal scheduling decisions, however, easily impose unacceptable run-time costs for many but the most basic scheduling problems, specifically in the context of multiprocessors and distributed systems. Deriving suitable heuristics then usually requires extensive simulations to gain confidence in the chosen approach. In this paper we therefore present FORTISSIMO, an open framework that facilitates the development of taylor-made real-time scheduling simulators for multiprocessor systems.

1 Introduction

Real-time systems are defined as those systems in which correctness of the system depends not only on the logical result of computation, but also on the time at which the results are produced. Predictability is therefore of paramount concern with the scheduling algorithm being responsible for deciding which activity is allowed to execute at some instant of time so that the maximum number of tasks meet their deadlines. Unfortunately, optimal scheduling decisions easily become prohibitively expensive at run time or even computationally intractable, specifically for multiprocessors and distributed systems [15]. In these cases, heuristics may serve as viable alternatives, providing 'good enough' behaviour at acceptable run-time overhead. While certain properties of sophisticated heuristics can be derived analytically, it is often desirable to verify these results or even to find new approaches empirically. Thus, a customisable and extensible testbed is needed for observing the behaviour of a scheduling algorithm under well-controlled conditions. Such a scheduling simulator must provide enough infrastructure to let the real-time researcher concentrate on the details of the scheduling algorithm and yet must be open to new requirements. That is, in addition to a powerful dispatching core flexible load generators and statistics gathering facilities are needed.

By now, however, real-time scheduling simulators have been commonly build with a particular scheduling problem or execution environment in mind [11, 16]. In this paper we therefore present FORTISSIMO, an open object-oriented framework not exclusively aimed at simulating a particular class of scheduling algorithms but to serve as a starting point for the development of taylor-made real-time scheduling simulators for multiprocessor architectures [8]. Consequently,

J. Rolim et al. (Eds.): IPDPS 2000 Workshops, LNCS 1800, pp. 766–772, 2000.

FORTISSIMO is not a ready-to-run application, yet offers a frame of ideas to work in. Short of the concrete scheduling policy the framework consists of a number of ready-to-use components for workload creation, integration with dispatchers, and collecting run-time statistics. These components are realised as well-documented C++ classes and serve as the base from which the adaptation of FORTISSIMO to specific simulation requirements evolves. Thus, FORTISSIMO tries to support the real-time architect by coping with various scheduling paradigms rather than forcing him or her into a single notion.

Among the scheduling paradigms explicitly considered for hard real-time systems are static table-driven approaches such as cyclic executives [12], static priority-driven and dynamic best-effort policies such as rate monotonic scheduling or earliest deadline first [9], and dynamic planning-based strategies such as the SPRING scheduling-algorithm [14]. Task semantics, however, is not limited to hard real-time environments. Support for aperiodic and sporadic real-time activities [5], reasoning with value functions [6, 17] as well as requirements derived from techniques such as skip-over scheduling [7], imprecise computation [10], and task pair scheduling [4] have been included.

The remainder of this paper is organised as follows. Section 2 discusses related work that partially has influenced some of our design decisions. Then, in Section 3, the architecture of FORTISSIMO as well as the communication between the components are described. Section 4 finally summarises our experience and briefly outlines future work on FORTISSIMO.

2   Related Work

Naturally, concepts of other real-time scheduling simulation projects found their way into FORTISSIMO. Among the projects that have influenced our design, SPRING and STRESS come closest.

SPRING [14] is a research real-time operating system supporting multiprocessors and distributed systems. A project spin-off [3], the SPRING simulation testbed, has influenced the design of workload generation and scheduling components of FORTISSIMO. However, the primary focus of the SPRING simulator seemingly was to evaluate the planning-based dynamic-priority assignment policy used in SPRING. As a consequence, the simulator provides strong support for this kind of scheduling in a distributed environment, yet falls short when it comes to basically different scheduling strategies.

STRESS [1], in contrast, is a simulation environment for hard real-time systems consisting of a simulation core that is supplemented by a graphical front-end for control and display. The approach chosen comprises a full-featured simulation language to specify both the system environment and task semantics. The simulation engine is quite elaborate, including some feasibility tests and support for multiprocessing as well as networking; tasks may synchronise via critical sections or message-passing. Since STRESS is targeted at hard real-time systems there is no build-in support for soft-deadline or value-function scheduling and it is unclear whether the simulation language is rich enough to cope with imprecise

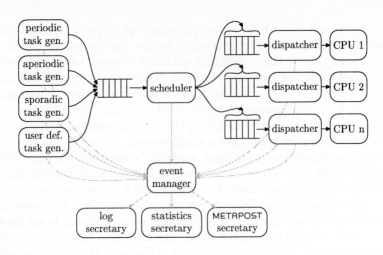

Fig. 1. Architecture of the framework

computing or task-pair scheduling, for instance. Task creation in FORTISSIMO, however, works in a similar way to STRESS.

3   Theory of Operation

As mentioned before, FORTISSIMO not only provides the basic infrastructure to build a real-time scheduling simulator suited for particular needs, but supports a number of scheduling paradigms right out of the box. Consequently, in order to add a new scheduler or task model, in most cases the real-time architect should need to refine or add only some specific classes rather than to redo everything from scratch. In FORTISSIMO, each class logically belongs to one of three independent modules, namely *workload generation, scheduling and dispatching*, and *gathering statistics*, with *tasks* and *events* serving as glue between these modules. The overall architecture is sketched in Fig. 1.

The workload component creates tasks according to user-defined patterns. Firstly, as part of the initialisation, the scheduler is allowed to check the feasibility of the specified task set as it will be generated from so-called *task generators*. Then, during simulation, the task generators create jobs for these tasks (e. g., instances of a periodic task) and place them into a global FIFO arrival queue. After removing a job from this queue, the scheduler can reject the job based on some feasibility test, accept and integrate it into its schedule, or react in a completely different way implemented by the user. An example would be putting the job aside and executing it only if additional execution time becomes available due to jobs that temporarily require less execution time than planned for.

As soon as a new job has been successfully scheduled, it is assigned to a *system dispatcher*, each dispatcher being exclusively responsible for one CPU. Again, scheduler and dispatchers communicate via queues with one ready queue per dispatcher in which tasks are placed by the scheduler. The scheduler, however, retains full access to the ready queues to simply add or remove some task, or to perform complete reschedules if necessary. It is therefore the responsibility of the scheduler to sort the jobs in the ready queues to reflect its policy — the dispatchers simply execute the job that is currently at the front of their queue, automatically performing a context switch if a different task moves to the front at any time. Since CPUs are simply abstractions and time passes by as ticks from a logical clock, the execution of a task merely consists of decrementing an execution counter and updating internal busy/idle statistics.

When some job has completed execution, it is handed to the *statistics facilities*. Because some information of interest is often spread out over the complete lifetime of jobs and tasks, the statistics module also processes events from other components of the framework.

Based on this overview, the following sections give a closer look at each component. A longer version of this paper describes in more detail how schedulers can be implemented in FORTISSIMO and how the framework can be configured [8].

3.1   Taks Model and Workload Generation

Workload generation in FORTISSIMO is split among independent *task generators*, each one responsible for the generation of a single class of tasks. Readily available are generator classes for periodic tasks whose jobs re-arrive by some fixed amount of time, sporadic tasks whose frequency is limited by some minimum inter-arrival time, and aperiodic tasks whose arrival pattern is modelled by some stochastic assumptions. In addition, a user can create completely new task generators or customise the available ones via inheritance to produce workload patterns currently not explicitly supported.

Timing parameters of a task include its average-case computation time, its worst-case computation time, and its deadline; the first invocation of a task may be delayed by some initial offset to construct arbitrary task phasings, in order to prevent or enforce critical instants, for example. Furthermore, a directed precedence graph without cycles may be used to explicitely define predecessor/successor relationships. The basic classes of hard, firm, and soft constraints are employed categorising a deadline miss as resulting in a catastrophy, in the computation being useless, or in a degraded quality of service, respectively. Whenever this scheme is insufficient, two value functions per task may be used to describe the value of finishing the task up to and after its deadline. Each task may be assigned a base priority during setup while at run time an additional temporary priority per task can be used to support dual-priority scheduling [2] and priority-inheritance protocols [13], for instance. Besides these fundamental paradigms, skip-over scheduling, the notion of imprecise computations, and task-pair scheduling are also readily supported. While in FORTISSIMO skip-over scheduling is limited to periodic tasks, support for imprecise computations and task-pair scheduling is available for both periodic and sporadic tasks.

To assess the behaviour of scheduling algorithms many simulation runs with varying load patterns are needed. Hence, virtually all task characteristics may be chosen randomly by FORTISSIMO according to given stochastic distributions. Parameters such as arrival patterns and actual computation time may vary for each job. Additionally, for a sequence of simulation runs changing task set characteristics may be specified.

3.2 Scheduling and Dispatching

Scheduling Algorithms are not built into FORTISSIMO, but have to be implemented and linked by the user. Schedulers, however, can be derived from a base class `Schedule` providing some default behaviour that can be customised selectively. We believe that this approach, besides promising some additional flexibility, allows analysing the computation time of the scheduler itself, already at the simulation stage.

A typical scheduler might work as follows within FORTISSIMO. The scheduler is invoked every tick of the logical clock and, provided it implements a preemptive algorithm, may perform a reschedule in response. If no new jobs have become ready since the last tick, the scheduler then falls asleep again until its next invocation. If new jobs have arrived, it removes these jobs one by one from the global arrival queue. For some algorithms providing guarantees, then a run-time admission test is performed. If the new job cannot be executed without jeopardizing the deadlines of either the new job itself or already guaranteed tasks, it is rejected and usually removed from the system. Otherwise, a new schedule must be constructed comprising the jobs already scheduled as well as the new job. For this, the scheduler typically has to retrieve the jobs already accepted and scheduled from the dispatchers' ready queues. Then, the jobs are sorted and re-inserted into the individual ready queues, possibly causing context switches.

Like the scheduler, dispatchers are invoked every tick of the logical clock. At any time, the dispatcher will run the job that is currently at the front of its ready queue, which subsequently becomes the active job until it terminates normally, the scheduler aborts the job for some reason, or the dispatcher's ready queue has changed.

Finally, jobs are run on *virtual processors*. Execution is simulated simply by decrementing the remaining execution time of the running job. In future versions, a more powerful processing model may, for instance, take interrupts and context-switching overhead into account.

3.3 Logging and Statistics

Whenever an important action is executed within the framework, this is signalled by an event. Each event carries the relevant information about the time and cause that lead to its creation, supplemented by additional data as needed. An *event manager* uniformly collects and distributes these events to so-called *secretaries*, which are registered with the event manager for certain types of events. Various types of action can be taken by a secretary upon arrival of a new event. Simple log secretaries just write a formatted line onto some output device, other secretaries

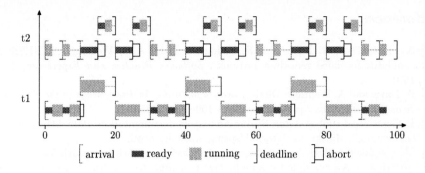

Fig. 2. Skip-over scheduling simulation run

may update some kind of statistical analysis data, and even more sophisticated ones may act as a gateway transforming the event into messages for a graphical user display. At the time of writing, secretaries for logging events, for collecting statistical data, and for visualizing a simulation run as a METAPOST figure are implemented.

Fig. 2 shows an example run of a skip-over scheduler that tolerates missed deadlines to a certain degree provided 'most' of a tasks deadlines are met [7]; a skip parameter s per task denotes the tolerance of that task to missing deadlines such that at least $s - 1$ task instances must meet their deadlines after missing a deadline. The skip parameter of tasks t_1 and t_2 is set to 3 and 2, respectively; that is, after one aborted job of t_1, two jobs of t_1 must be executed in time, and no two successive jobs of t_2 may be aborted.

4   Conclusions

In this paper we have presented FORTISSIMO, an open object-oriented framework to simulate the scheduling of real-time tasks. The versatility of FORTISSIMO has been verified by implementing a wide range of fundamentally different scheduling policies such as rate-monotonic scheduling, earliest deadline first, the sporadic server algorithm, an imprecise computation policy, skip-over scheduling, and task-pair scheduling. Although the task model already provides a sound basis, we intend to add support for critical sections, resource reservation, task semantics including inter-task communication, and more elaborate precedence relations to the scheduling core. Furthermore, in addition to the multiprocessor support already implemented, an infrastructure to simulate real-time scheduling in distributed systems is under development. A graphical user interface, finally, will increase the ease the use of FORTISSIMO and illustrate behaviour of scheduling policies at run time; for the latter, the event mechanism already provides the necessary internal hooks. Despite these loose ends, however, we believe that even the scheduling core as described in this paper might already serve real-time architects to develop taylor-made simulators based on FORTISSIMO to evaluate their algorithms and heuristics.

References

[1] N. C. Audsley, A. Burns, M. F. Richardson, and A. J. Wellings. STRESS: A simulator for hard real-time systems. *Software Practice and Experience*, July 1994.

[2] R. Davis and A. Wellings. Dual-priority scheduling. In *Proceedings of the Sixteenth Real-Time Systems Symposium*, pages 100–109, 1995.

[3] E. Gene. Real-time systems: Spring simulators documentation, 1990. http://www-ccs.cs.umass.edu/spring/internal/spring_sim_docs.html.

[4] M. Gergeleit and H. Streich. Task-pair scheduling with optimistic case execution times—An example for an adaptive real-time system. In *Proceedings of the Second Workshop on Object-Oriented Real-Time Dependable Systems (WORDS)*, February 1996.

[5] T. M. Ghazalie and T. P. Baker. Aperiodic servers in a deadline scheduling environment. *Journal of Real-Time Systems*, 7(9):31–67, 1995.

[6] E. D. Jensen, C. D. Locke, and H. Tokuda. A time-driven scheduling model for real-time operating systems. In *Proceedings of the Sixth IEEE Real-Time Systems Symposium*, December 1985.

[7] G. Koren and D. Shasha. Skip-over: Algorithms and complexity for overloaded systems that allow skips. In *Proceedings of Sixteenth IEEE Real-Time Systems Symposium*. IEEE, 1995.

[8] T. Kramp, M. Adrian, and R. Koster. An open framework for real-time scheduling simulation. SFB 501 Report 01/00, Department of Computer Science, University of Kaiserslautern, Germany, January 2000.

[9] C. L. Liu and J. W. Layland. Scheduling algorithms for multiprogramming in a hard-real-time environment. *Journal of the ACM*, 20(1):46–61, 1973.

[10] J. W. S. Liu, K.-J. Lin, W.-K. Shih, A. C. Yu, J.-Y. Chung, and W. Zhao. Algorithms for scheduling imprecise computations. *IEEE Computer*, 24(5):58–68, May 1991.

[11] J. W. S. Liu, J. L. Redondo, Z. Deng, T. S. Tia, R. Bettati, A. Silberman, M. Storch, R. Ha, and W. K. Shih. PERTS: A prototyping environment for real-time systems. In *Proceedings of the Fourteenth Real-Time Systems Symposium*, pages 184–188. IEEE, December 1993.

[12] C. D. Locke. Software architectures for hard real-time applications: Cyclic executives vs. fixed-priority executives. *Journal of Real-Time Systems*, 4(1):37–53, 1992.

[13] L. Sha, R. Rajkumar, and J. P. Lehoczky. Priority inheritance protocols: An approach to real-time synchronisation. Technical Report CMU-CS-87-181, Computer Science Department, Carnegie Mellon University, 1987.

[14] J. A. Stankovic and K. Ramamritham. The Spring kernel: A new paradigm for hard real-time operating systems. *IEEE Software*, 8(3):62–72, May 1991.

[15] J. A. Stankovic, M. Spuri, M. Di Natale, and G. Buttazzo. Implications of classical scheduling results for real-time systems. *IEEE Computer*, 28(6):16–25, June 1995.

[16] A. D. Stoyenko. A schedulability analyzer for Real-time Euclid. In *Proceedings of the Eighth Real-Time Systems Symposium*, pages 218–227. IEEE, December 1987.

[17] H. Tokuda, J. W. Wendorf, and H.-Y. Wang. Implementation of a time-driven scheduler for real-time operating systems. In *Proceedings of the Eighth IEEE Real-Time Systems Symposium*, December 1987.

5th International Workshop on Embedded/Distributed HPC Systems and Applications (EHPC 2000)

Workshop Co-Chairs

Devesh Bhatt

Honeywell Technology Center
3660 Technology Drive
Minneapolis, MN 55418, USA
devesh.bhatt@honeywell.com

Lonnie R. Welch

Ohio University
School of Engineering and Computer Science
Athens, OH 45701-2979, USA
welch@ohio.edu

Preface

The International Workshop on Embedded/Distributed HPC Systems and Applications (EHPC) is a forum for the presentation and discussion of approaches, research findings, and experiences in the applications of High Performance Computing (HPC) technology for embedded/distributed systems. Of interest are both the development of relevant technology (e.g.: hardware, middleware, tools) as well as the embedded HPC applications built using such technology.

We hope to bring together industry, academia, and government researchers/users to explore the special needs and issues in applying HPC technologies to defense and commercial applications.

Topics of Interest

- **Algorithms and Applications:** addressing parallel computing needs of embedded military and commercial applications areas such as signal/image processing, advanced vision/robotic systems, smart-sensor based systems, industrial automation/optimization, vehicle guidance.

- **Networking Multiple HPC Systems:** in-the-large application programming models/API's, partitioning/mapping, system integration, debugging and testing tools.

- **Programming Environments:** software design, programming, and parallelization methods/tools for DSP-based, reconfigurable, and mixed-computation-paradigm architectures.

- **Operating Systems and Middleware:** distributed middleware service needs (e.g. QoS, object distribution) of high-performance embedded applications, configurable/optimal OS features needs, static/dynamic resource management needs.

J. Rolim et al. (Eds.): IPDPS 2000 Workshops, LNCS 1800, pp. 773-775, 2000.
© Springer-Verlag Berlin Heidelberg 2000

- **Architectures:** special-purpose processors, packaging, mixed-computation-paradigm architectures, size/weight/power modeling and management using hardware and software techniques.

EHPC 2000 Contents

The EHPC 2000 workshop will feature technical paper presentations, and an open discussion session. This year, we have papers covering several topic areas of interest. The following is a highlight of the papers.

In the algorithm and applications area, Yang et al. present a reconfigurable, dynamic load balancing parallel sorting algorithm applicable to information fusion. Hadden et al. present system health management application domain which would benefit from embedded HPC architectures.

In the programming environments area, Janka and Wills present a specification and design methodology for signal-processing systems using high-performance middleware and front-end tools. Patel et al. present performance comparison of high-performance real-time benchmarks using hand-crafted design versus automated glue-code generation from data-flow specification using their design tool.

In the operating systems and middleware area, we have several papers ranging from network load monitoring to communication scheduling for high-performance applications. Islam et al. present a technique for evaluating network load based upon dynamic paths using embedded application benchmarks. Pierce et al. present an architecture for mining of performance data for HPC systems, extending the capabilities of current instrumentation tools. Huh et al. present an approach for predicting the real-time QoS in dynamic heterogeneous resource management systems. VanVoorst and Seidel present the use of a real-time parallel communication benchmark to compare several MPI implementations. West and Antonio present an approach for optimizing the communication scheduling in parallel Space-Time Adaptive Processing (STAP) applications.

In the architecture area, we have papers on software and hardware perspectives on power management, as well as a new architecture for embedded applications. Osmulski et al. present a probabilistic power-prediction tool for Xilinx 4000-series reconfigurable computing devices. Unsal et al. present an energy consumption model addressing task assignment and network toplogy/routing, using replication of shared data structures. Schulman et al. present a system-on-chip architecture containing an array of VLIW processing elements, with reconfiguration times much smaller than FPGA-based architectures.

Program Committee

Ashok Agrawala, Univ. of Maryland, USA
Bonnie Bennett, Univ. of St. Thomas, USA
Bob Bernecky, NUWC, USA
Alberto Broggi, Universita' di Pavia, Italy
Hakon O. Bugge, Scali Computer, Norway
Richard Games, MITRE, USA

A Probabilistic Power Prediction Tool for the Xilinx 4000-Series FPGA

Timothy Osmulski, Jeffrey T. Muehring, Brian Veale, Jack M. West, Hongping Li,
Sirirut Vanichayobon, Seok-Hyun Ko, John K. Antonio, and Sudarshan K. Dhall

School of Computer Science
University of Oklahoma
200 Felgar Street
Norman, OK 73019
Phone: (405) 325-7859
antonio@ou.edu

Abstract. The work described here introduces a practical and accurate tool for predicting power consumption for FPGA circuits. The utility of the tool is that it enables FPGA circuit designers to evaluate the power consumption of their designs without resorting to the laborious and expensive empirical approach of instrumenting an FPGA board/chip and taking actual power consumption measurements. Preliminary results of the tool presented here indicate that an error of less than 5% is usually achieved when compared with actual physical measurements of power consumption.

1 Introduction and Background

Reconfigurable computing devices, such as field programmable gate arrays (FPGAs), have become a popular choice for the implementation of custom computing systems. For special purpose computing environments, reconfigurable devices can offer a cost-effective and more flexible alternative than the use of application specific integrated circuits (ASICs). They are especially cost-effective compared to ASICs when only a few copies of the chip(s) are needed [1]. Another major advantage of FPGAs over ASICs is that they can be reconfigured to change their functionality while still resident in the system, which allows hardware designs to be changed as easily as software and dynamically reconfigured to perform different functions at different times [6].

Often a device's performance (i.e., speed) is a main design consideration; however, power consumption is of growing concern as the logic density and speed of ICs increase. Some research has been undertaken in the area of power consumption in CMOS (complimentary metal-oxide semiconductor) devices, e.g., see [4, 5]. However, most of this past work assumes design and implementation based on the use of standard (basic cell) VLSI techniques, which is typically not a valid assumption for application circuits designed for implementation on an FPGA.

J. Rolim et al. (Eds.): IPDPS 2000 Workshops, LNCS 1800, pp. 776-783, 2000.

2   Overview of the Tool

A probabilistic power prediction tool for the Xilinx 4000-series FPGA is overviewed in this section. The tool, which is implemented in Java, takes as input two files: (1) a *configuration file* associated with an FPGA design and (2) a *pin file* that characterizes the signal activities of the input data pins to the FPGA. The configuration file defines how each CLB (configurable logic block) is programmed and defines signal connections among the programmed CLBs. The configuration file is an ASCII file that is generated using a Xilinx M1 Foundation Series utility called *ncdread*. The pin file is also an ASCII file, but is generated by the user. It contains a listing of pins that are associated with the input data for the configured FPGA circuit. For each pin number listed, probabilistic parameters are provided which characterize the signal activity for that pin.

Based on the two input files, the tool propagates the probabilistic information associated with the pins through a model of the FPGA configuration and calculates the activity of every internal signal associated with the configuration [1]. The activity of an internal signal s, denoted a_s, is a value between zero and one and represents the signal's relative frequency with respect to the frequency of the system clock, f. Thus, the average frequency of signal s is given by $a_s f$.

Computing the activities of the internal signals represents the bulk of computations performed by the tool [1]. Given the probabilistic parameters for all input signals of a configured CLB, the probabilistic parameters of that CLB's output signals are determined using a well-defined mathematical transformation [2].   Thus, the probabilistic information for the pin signals is transformed as it passes through the configured logic defined by the configuration file. However, the probabilistic parameters of some CLB inputs may not be initially known because they are not directly connected to pin signals, but instead are connected to the output of another CLB for which the output probabilistic parameters have not yet been computed (i.e., there is a feedback loop). For this reason, the tool applies an iterative approach to update the values for unknown signal parameters. The iteration process continues until convergence is reached, which means that the determined signal parameters are consistent based on the mathematical transformation that relates input and output signal parameter values, for every CLB.

The average power dissipation due to a signal s is modeled by $\frac{1}{2} C_{d(s)} V^2 a_s f$, where $d(s)$ is the Manhattan distance the signal s spans across the array of CLBs, $C_{d(s)}$ is the equivalent capacitance seen by the signal s, and V is the voltage level of the FPGA device. The overall power consumption of the configured device is the sum of the power dissipated by all signals. For an $N \times N$ array of CLBs, Manhattan signal distances can range from 0 to $2N$. Therefore, the values of $2N + 1$ equivalent capacitance values must be known, in general, to calculate the overall power consumption. Letting S denote the set of all internal signals for a given configuration, the overall power consumption of the FPGA is given by:

$$P_{\text{avg}} = \sum_{s \in S} \frac{1}{2} C_{d(s)} V^2 a_s f$$

$$= \frac{1}{2} V^2 f \sum_{s \in S} C_{d(s)} a_s. \tag{1}$$

The values of the activities (i.e., the a_s's) are dependent upon the parameter values of the pin signals defined in the pin file. Thus, although a given configuration file defines the set S of internal signals present, the parameter values in the pin file impact the activity values of these internal signals.

3   Calibration of the Tool

Let S_i denote the set of signals of length i, i.e., $S_i = \{s \in S \mid d(s) = i\}$. So, the set of signals S can be partitioned into $2N + 1$ subsets based on the length associated with each signal. Using this partitioning, Eq. 1 can be expressed as follows:

$$P_{\text{avg}} = \frac{1}{2}V^2 f \left(C_0 \sum_{s \in S_0} a_s + C_1 \sum_{s \in S_1} a_s + \cdots + C_{2N} \sum_{s \in S_{2N}} a_s \right). \tag{2}$$

To determine the values of the tool's capacitance parameters, actual power consumption measurements are taken from an instrumented FPGA using different configuration files and pin input parameters. Specifically, $2N + 1$ distinct measurements are made and equated to the above equation using the activity values (i.e., the a_s's) computed by the tool. For the j-th design/data set combination, let P_j denote the measured power and let $A_{j,k}$ denote the aggregate activity of all signals of length k. The resulting set of equations is then solved to determine the $2N + 1$ unknown capacitance parameter values:

$$\frac{1}{2}V^2 f \begin{pmatrix} A_{0,0} & A_{0,1} & \cdots & A_{0,2N} \\ A_{1,0} & A_{1,1} & \cdots & A_{1,2N} \\ \vdots & & \ddots & \vdots \\ A_{2N,0} & A_{2N,1} & & A_{2N,2N} \end{pmatrix} \begin{pmatrix} C_0 \\ C_1 \\ \vdots \\ C_{2N} \end{pmatrix} = \begin{pmatrix} P_0 \\ P_1 \\ \vdots \\ P_{2N} \end{pmatrix}. \tag{3}$$

Solving the above equation for the vector of unknown capacitance values is how the tool is calibrated.

4   Power Measurements

For this study, a total of 70 power measurements were made using 5 different configuration files and 14 different data sets. Descriptions of these configuration files and data sets are given in Tables 1 and 2, respectively. All of the configuration files listed in Table 1 each take a total of 32-bits of data as input. The first three configurations (fp_mult, fp_add, int_mult) each take two 16-bit operands on each clock cycle, and the last two (serial_fir and parallel_fir) each take one 32-bit complex operand on each clock cycle. The 32 bits of input data are numbered as 0 through 31 in Table 2, and two key parameters are used to characterize these bits: an *activity factor, a* and a *probability factor, p*. The activity factor of an input bit is a value between zero and one and represents the signal's relative frequency with respect to the frequency of the system clock, f. The probability factor of a bit represents the fraction of time that the bit has a value of one.

Fig. 1 shows plots of the measured power for all combinations of the configuration files and data sets described in Tables 1 and 2. For all cases, the clock was run at $f =$ 30 MHz. With the exception of the fp_mult configuration file, the most active data set file (number 6) is associated with the highest power consumption. Also, the least active data set file (number 5) is associated with the lowest power consumption across all configuration files. There is somewhat of a correlation between the number of components utilized by each configuration and the power consumption; however, note that even though the serial_fir implementation is slightly larger than parallel_fir, it consumes less power. This is likely due to the fact that the parallel_fir design requires a high fan-out (and thus high routing capacitance) to drive the parallel multipliers.

Table 1. Characteristics of the configuration files.

Configuration File Name	Description	Component Utilization of Xilinx 4036xla
fp_mult	Custom 16-bit floating point multiplier with 11-bit mantissa, 4-bit exponent, and a sign bit [3].	368
fp_add	Custom 16-bit floating point adder with 11-bit mantissa, 4-bit exponent, and a sign bit [3].	339
int_mult	16-bit integer array multiplier; produces 32-bit product [3].	509
serial_fir	FIR filter implementation using a serial-multiply with a parallel reduction add tree. Input data is 32-bit integer complex. Constant coefficient multipliers and adders from core generator.	1060
parallel_fir	FIR filter implementation using a parallel-multiply with a series of delayed adders. Input data is 32-bit integer complex. Constant coefficient multipliers and adders from core generator.	1055

Table 2. Characteristics of the data sets.

Data Set Number	Description
1	Pins 0 through 15 $\Rightarrow p = 0.0$ and $a = 0.0$. Pins 16 through 31 $\Rightarrow p = 0.5$ and $a = 1.0$
2	Pins 0 through 15 $\Rightarrow p = 0.0$ and $a = 0.0$ Pins 16 through 31 $\Rightarrow p = 0.75$ and $a = 0.4$
3	Pins 0 through 15 $\Rightarrow p = 0.25$ and $a = 0.45$ Pins 16 through 31 $\Rightarrow p = 0.0$ and $a = 0.0$
4	Pins 0 through 15 $\Rightarrow p = 0.\,5$ and $a = 1.0$ Pins 16 through 31 $\Rightarrow p = 0.0$ and $a = 0.0$
5	Pins 0 through 31 $\Rightarrow p = 0.0$ and $a = 0.0$
6	Pins 0 through 31 $\Rightarrow p = 0.5$ and $a = 1.0$
7	Even numbered pins $\Rightarrow p = 0.0$ and $a = 0.0$ Odd numbered pins $\Rightarrow p = 0.5$ and $a = 1.0$
8	Even numbered pins $\Rightarrow p = 0.3$ and $a = 0.5$ Odd numbered pins $\Rightarrow p = 0.7$ and $a = 0.5$
9	Even numbered pins $\Rightarrow p = 0.5$ and $a = 1.0$ Odd numbered pins $\Rightarrow p = 0.0$ and $a = 0.0$
10	Even numbered pins $\Rightarrow p = 0.8$ and $a = 0.1$ Odd numbered pins $\Rightarrow p = 0.2$ and $a = 0.15$
11	For all pins, p and a selected at random (different from data set 12).
12	For all pins, p and a selected at random (different from data set 11).
13	Pins 0 through 2, $p = 0.1$ and $a = 0.1$ Pins 3 through 5, $p = 0.2$ and $a = 0.2$, etc., p's continue to increase in steps of 0.1 and a's increase to 0.5 in steps of 0.1 and then decrease back down to 0.0.
14	Pin 0, $p = 0.1$ and $a = 0.2$ Pin 1, $p = 0.2$ and $a = 0.4$ Pin 2, $p = 0.3$ and $a = 0.6$, etc., p's continue to increase to 1.0 in steps of 0.1 (and then decrease) and a's increase to 1.0 in steps of 0.2 (and then decrease).

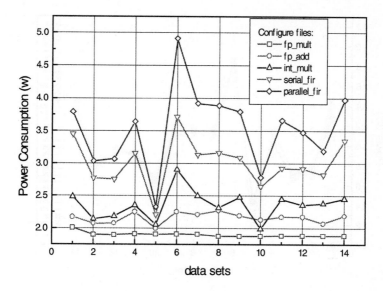

Fig. 1. Measured power consumption for the configuration files and data sets described in Tables 1 and 2.

5   Experimental Evaluation of the Tool

Because 73 values are used to model all of the internal capacitances of the device used in this study, at least three more measurement scenarios are required to calibrate all capacitance values (by solving the complete set of linear equations defined by Eq. 3). Fortunately, however, we were able to calibrate a subset of capacitance values by considering the power consumption of the two FIR filters (serial_fir and parallel_fir). This was because there turned out to be a total of only 28 non-zero entries for the rows of the matrix of Eq. 3, corresponding to aggregate activities for the two FIR filter designs.

Fig. 2 shows the measured power consumption curve along with 29 different prediction curves generated by the tool for the serial FIR filter design. One of the prediction curves corresponds to predicted values based on using all 28 measured values to calibrate the tool's capacitance values (this curve is labeled "all" in the legend of the figure). This curve naturally has excellent accuracy; predicted power consumption values match measured values nearly perfectly.[1] The remaining 28 prediction curves are associated with capacitance values determined by using all but one of the measured data values to calibrate the tool (the data set not used is indicated in the legend of the figure). For each of these curves, the data set not used in the

---

[1] The reason the predicted values do not match measured values exactly is because the equations used to determine capacitance values did not have full rank, and thus a least-squares solution was determined.

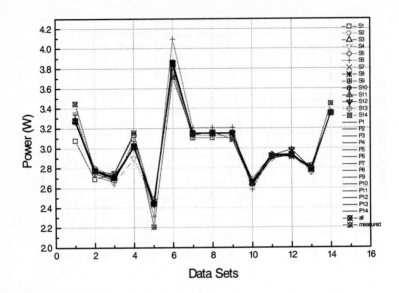

Fig. 2. Measured and predicted power consumption curves using various calibration scenarios for the serial FIR filter implementation.

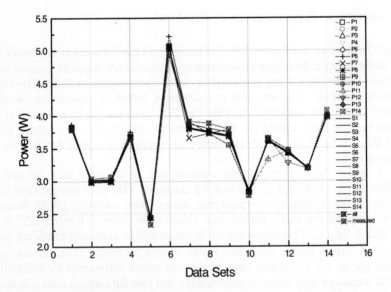

Fig. 3. Measured and predicted power consumption curves using various calibration scenarios for the parallel FIR filter implementation.

calibration of the tool's capacitance values generally associates with the highest error in the predicted value for that data point. For example, note that when data set number six for the serial FIR (labeled S6 in the figure's legend) was not used in the calibration process, the resulting prediction for that value was highest (around 10% error). When data sets associated with the parallel FIR design were not included, the prediction curves did not change, thus those curves are all drawn as solid lines with no symbols. Fig. 3 shows the same type of results as Fig. 2, except for the parallel FIR instead of the serial FIR.

6   Summary

To summarize the results for both filter designs, when all 28 sets of measurements are used to calibrate the tool, the maximum error in predicted versus measured power is typically less than about 5%. With one data set removed, the maximum error increases to about 10%, and the predicted value with this highest error is typically associated with the data set not used in calibrating the tool. This level of error is acceptable for most design environments, and represents a considerable accomplishment in the area of power prediction for FPGA circuits. Thus, these preliminary results indicate that the tool is able to adequately predict power consumption (i.e., for data sets not used in calibrating the tool). By using more data sets to calibrate the tool in the future, it is expected that even greater prediction accuracy and robustness will be achieved.

Acknowledgements

This work was supported by DARPA under contract no. F30602-97-2-0297. Special thanks go to Annapolis Micro Systems, Inc. for their support and for providing the instrumented FPGA board that was used to take power measurements.

References

1. T. Osmulski, *Implementation and Evaluation of a Power Prediction Model for Field Programmable Gate Array*, Master's Thesis, Computer Science, Texas Tech University, 1998.
2. K. P. Parker and E. J. McClusky, "Probabilistic Treatment of General Combinatorial Networks," *IEEE Trans. Computers*, vol. C-24, pp. 668-670, June 1975.
3. B. Veale, *Study of Power Consumption for High-Performance Reconfigurable Computing Architectures*, Master's Thesis, Computer Science, Texas Tech University, 1999.
4. T. L. Chou, K. Roy, and S. Prasad, "Estimation of Circuit Activity Considering Signal Correlations and Simultaneous Switching," *Proc. IEEE Int'l Conf. Comput. Aided Design*, pp. 300-303, Nov. 1994.
5. A. Nannarelli and T. Yang, "Low-Power Divider," *IEEE Trans. Computers*, Vol. 48, No. 1, Jan. 1999, pp. 2-14.
6. *Xilinx XC4000E and XC4000X Series Field Programmable Gate arrays, Product Specification*, Xilinx Inc., v1.5, http://www.xilinx.com/partinfo/databook.htm#xc4000, 1999.

Application Challenges: System Health Management for Complex Systems

George D. Hadden [1], Peter Bergstrom[1], Tariq Samad[1] ,Bonnie Holte Bennett[2] ,
George J. Vachtsevanos[3] , and Joe Van Dyke[4]

[1]Honeywell Technology Center, 3660 Technology Drive, Minneapolis, MN 55418
george.d.hadden@htc.honeywell.com
[2]Knowledge Partners of Minnesota, Inc., 9 Salem Lane, Suite 100,
St. Paul, MN 55118-4700
bbennett@kpmi.com
[3]The Georgia Institute of Technology, School of Electrical and Computer Engineering,
Atlanta, Georgia 30332-0250
gjv@ece.gatech.edu
[4]Systems Analysis and Software Engineering, 253 Winslow Way West, Bainbridge Island,
Washington, 98110
joevandyke@Predict-DLI.com

Abstract. System Health Management (SHM) is an example of the types of challenging applications facing embedding high-performance computing environments. SHM systems monitor real-time sensors to determine system health and performance. Performance, economics, and safety are all at stake in SHM, and the emphasis on health management technology is motivated by all these considerations. This paper describes a project focusing on condition-based maintenance (CBM) for naval ships. *Condition-based maintenance* refers to the identification of maintenance needs based on current operational conditions. In this project, system architectures and diagnostic and prognostic algorithms are being developed that can efficiently undertake real-time data analysis from appropriately instrumented machinery aboard naval ships and, based on the analysis, provide feedback to human users regarding the state of the machinery – such as its expected time to failure, the criticality of the equipment for current operation.

1   Introduction

Although some aspects of system operation, such as feedback control, are by now widely automated, others such as the broad area of system health management (SHM) still rely heavily on human operators, engineers, and supervisors. In many industries, SHM is viewed as the next frontier in automation.

System health management has always been a topic of significant interest to industry. Only relatively recently, however, have the numerous aspects of health

J. Rolim et al. (Eds.): IPDPS 2000 Workshops, LNCS 1800, pp. 784-791, 2000.

management begun to be viewed as facets of one overall problem. The term itself has gained currency only recently. We now understand SHM as encompassing all issues related to off-nominal operations of systems – including equipment, process/plant, and enterprise. As for the capabilities that fall under the SHM label, the following are particularly notable:

- Fault detection: identifying that some element or component of a system has failed.
- Fault identification: identifying *which* element has failed.
- Failure prediction: identifying elements for which failure may be imminent and estimating their time to failure.
- Modelling and tracking degradation: quantifying gradual degradation in a component or the system.
- Maintenance scheduling: determining appropriate times for preventive or corrective operations on components.
- Error correction: estimating 'correct' values for parameters, the measurements of which have been corrupted.

Technologists are seeking to exploit advances in diverse fields for developing SHM solutions. As might be expected, the variety and complexity of problems that SHM encompasses preclude any single-technology answers. Hardware, software, and algorithmic technologies are all required and are being explored. An SHM solution can require a hardware architecture design, integrating sensors, actuators, computational processors, and communication networks. Different algorithmic techniques may be needed for signal processing, including Fourier and wavelet transforms and time series models. Artificial intelligence methods such as expert systems and fuzzy logic can be helpful in allowing human expertise and intuition to be captured. There is also increasing interest in fundamental modelling, especially in failure mode effects analysis (FMEA), a systematic approach for identifying what problems can potentially occur with products and processes. Finally, software architectures are required to manage the multiple devices, data streams, and algorithms. With Internet-enabled architectures, an SHM system can be physically distributed across large distances.

1.1   Challenges in system health management

Our successes in capturing common failure mechanisms has resulted in safer, more reliable, and more available systems. An interesting corollary is that we are now seeing failure modes that were rarely seen before. The lack of empirical data or experiential knowledge in such cases renders many methods unusable. Other types of knowledge must be relied upon in such cases, generally based on a human expert's understanding of system operation.

Another failing with many conventional methods for fault identification is that they assume that faults occur singly. Surprising relationships can occur among various failure modes. A fault in one device may cause problems in otherwise unrelated machines that depend on it for their input (perhaps separated by several intervening devices). Compound faults often do not have independent symptoms, and predicting or diagnosing multiple faults is not simply a matter of dealing with each separately.

Even when there is a single fault, its symptoms will be masked by any number of additional symptoms generated by logically upstream and downstream subsystems. Also, SHM must deal with the large differences in the time scales. Vibration data from a motor may need to be collected at nearly a megahertz for shaft balance problems to be detectable, whereas flooding in a distillation column is a phenomenon that occurs on a time scale of many minutes. System architectures and algorithms, that can deal with these extremes of sampling rates, are needed and not readily available.

1.2   Condition-Based Maintenance for Naval Ships

This project, supported by the Office of Naval Research of the U. S. Department of Defence, is focusing on condition-based maintenance (CBM) for naval ships. *Condition-based maintenance* refers to the identification of maintenance needs based on current operational conditions. In this project, system architectures and diagnostic and prognostic algorithms are being developed that can efficiently undertake real-time data analysis from appropriately instrumented machinery aboard naval ships and, based on the analysis, provide feedback to human users regarding the state of the machinery – such as its expected time to failure. Using these analyses, ship maintenance officers can determine which equipment is critical to repair before embarking on their next mission – a mission that could take the better part of a year.

1.2.1 MPROS Architecture

The development of the CBM system, called MPROS (for Machinery Prognostic and Diagnostic System), had two phases. The first phase had MPROS installed and running in the lab. During the second phase, we extended MPROS's capability somewhat and installed it on the Navy hospital ship *Mercy* in San Diego.

MPROS is a distributed, open, extensible architecture for hosting multiple on-line diagnostic and prognostic algorithms. Additionally, our prototype contains four sets of algorithms aimed specifically at centrifugal chilled water plants. These are:

1. PredictDLI's (a company in Bainbridge Island, Washington, that has a Navy contract to do CBM on shipboard machinery) vibration-based expert system adapted to run in a continuous mode.
2. State-based feature recognition (SBFR), an Honeywell Technology Center (HTC)-developed embeddable technique that facilitates recognition of time-correlated events in multiple data streams.
3. Wavelet Neural Network (WNN) diagnostics and prognostics developed by Professor George Vachtsevanos and his colleagues at Georgia Tech. This technique

is aimed at vibration data; however, unlike PredictDLI's, their algorithm excels at drawing conclusions from transitory phenomena.

4. Fuzzy logic diagnostics and prognostics also developed by Georgia Tech that draws diagnostic and prognostic conclusions from nonvibrational data.

Since these algorithms (and others we may add later) have overlapping areas of expertise, they may sometimes disagree about what is ailing the machine. They may also reinforce each other by reaching the same conclusions from similar data. In these cases, another subsystem, called *Knowledge Fusion* (KF), is invoked to make some sense of these conclusions. We use a technique called *Dempster-Shafer Rules of Evidence* to combine conclusions reached by the various algorithms. It can be extended to handle any number of inputs.

MPROS is distributed in the following sense: Devices called *Data Concentrators* (DCs) are placed near the ship's machinery. Each of these is a computer in its own right and has the major responsibility for diagnostics and prognostics. Except for Knowledge Fusion, the algorithms described above run on the DC. Conclusions reached by these algorithms are then sent over the ship's network to a centrally located machine containing the other part of our system – the *Prognostic/Diagnostic/Monitoring Engine* (PDME). KF is located in the PDME. Also in the PDME is the *Object-Oriented Ship Model* (OOSM). The OOSM represents parts of the ship (e.g., compressor, chiller, pump, deck, machinery space) and a number of relationships among them (e.g., part-of, proximity, kind-of). It also serves as a repository of diagnostic conclusions – both those of the individual algorithms and those reached by KF. Communication among the DCs and the PDME is done using Distributed Common Object Module (DCOM), a standard developed by Microsoft.

1.2.2 Data Concentrator hardware

The DC hardware (Figure 1 shows the HTC-installed DC) consists of a PC104 single-board Pentium PC (about 6 in. x 6 in.) with a flat-screen LCD display monitor, a PCMCIA host board, a four-channel PCMCIA DSP card, two multiplexer (MUX) cards, and a terminal bus for sensor cable connections. The operating system is Windows 95™, and there are connections for keyboard and mouse. Data is stored via DRAM. The DC is housed in a NEMA enclosure with a transparent front door and fans for cooling. Overall dimensions are 10 in. x 12 in. x 4 in. The system was built entirely with commercial off-the-shelf components with the exception of the MUX cards, which are a PredictDLI hardware subcomponent, and the PCMCIA card, which was modified from a commercial two-channel unit to meet the needs of the project.

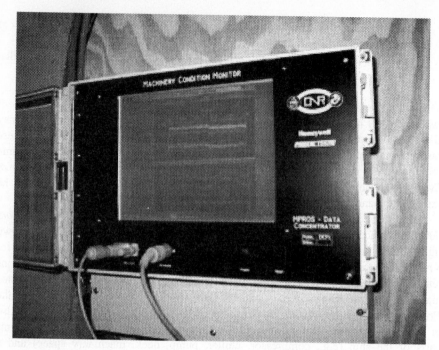

Figure 1 **Data concentrator installed at HTC**

2   MPROS Software

Figure 2 shows a diagram of the MPROS system. The PDME consists entirely of software and runs on any sufficiently powerful Windows NT machine. A potentially large number (on the order of a thousand) DCs are installed on the ship and report diagnostic and prognostic conclusions to the PDME over the ship's network. In the following, we describe the various software parts of the system.

2.1   PDME

The PDME is the logical center of the MPROS system. Diagnostic and prognostic conclusions are collected from DC-resident as well as PDME-resident algorithms. Fusion of conflicting and reinforcing source conclusions is performed to form a prioritized list for use by maintenance personnel.

The PDME is implemented on a Windows NT platform as a set of communicating servers built using Microsoft's Component Object Model (COM) libraries and services. Choosing COM as the interface design technique has allowed us to build some components in C++ and others in Visual Basic, with an expected improvement in development productivity as the outcome. Some components were prototyped using Microsoft Excel, and we continue to use Excel worksheets and macros to drive some testing of the system. Communications between DC and PDME components depend on Distributed COM (DCOM) services built into Microsoft's operating systems.

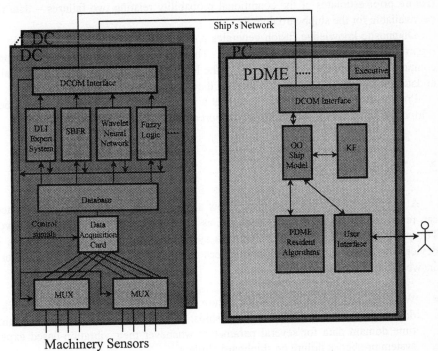

Machinery Sensors
Figure 2 The MPROS system

2.2   Knowledge fusion

Knowledge fusion is the co-ordination of individual data reports from a variety of sensors. It is higher level than pure 'data fusion,' which generally seeks to correlate common-platform data. Knowledge fusion, for example, seeks to integrate reports from acoustic, vibration, oil analysis, and other sources, and eventually to incorporate trend data, histories, and other components necessary for true prognostics.

Implementation To date, two levels of knowledge fusion have been implemented: one for diagnostics and one for prognostics.

Our approach for implementing knowledge fusion for diagnostics uses Dempster-Shafer belief maintenance for correlating incoming reports. This is facilitated by use of a heuristic that groups similar failures into logical groups.

Dempster-Shafer theory is a calculus for qualifying beliefs using numerical expressions. For example, given a belief of 40% that A will occur and another belief of 75% that B or C will occur, it will conclude that A is 14% likely, B or C is 64% likely, and assign 22% of belief to unknown possibilities. This maintenance of the likelihood of unknown possibilities is both a differentiator and a strength of Dempster-Shafer theory. It was chosen over other approaches (e.g., Bayes nets) because the others require prior estimates of the conditional probability relating two failures – data not yet available for the shipboard domain.

Diagnostic knowledge fusion generates a new fused belief whenever a diagnostic report arrives for a suspect component. This updates the belief for that suspect component and for every other failure in the logical group for that component. It also updates the belief of 'unknown' failure for the logical group for that component.

Prognostic knowledge fusion generates a new prognostic vector for each suspect component whenever a new prognostic report arrives.

3   Validation

A question we are often asked is, 'How are you going to prove that your system can really predict failures?' This question, as it turns out, is quite difficult to answer. The problem is that we are developing a system we claim will predict failures in devices, and that in real life, these devices fail relatively rarely. We have several answers to this question:

- We are still going to look for the failure modes. We have a number of installed data collectors both on land and on ships. In addition, PredictDLI is collecting time domain data for several parameters whenever their vibration-based expert system predicts a failure on shipboard chillers.
- As Honeywell upgrades its air conditioning systems to be compliant with new nonpolluting refrigerant regulations, older chillers become obsolete and are replaced. We have managed to acquire one of these chillers and are now constructing a test plan to collect data from this chiller.
- Seeded faults are worth doing. Our partners in the Mechanical Engineering Department of Georgia Tech are seeding faults in bearings and collecting the data. These tests have the drawback that they might not exhibit the same precursors as real-world failures, especially in the case of accelerated tests.
- Honeywell, York, PredictDLI, the Naval Research Laboratory, and WM Engineering, have archived maintenance data that we will take advantage of.

Although persuasive, these answers are far from conclusive. The authors would welcome any input on how to validate a failure prediction system.

4   Conclusions

In the not too distant past, automation was employed largely to manage systems under nominal operating conditions. The realm of automation rarely extended to abnormal conditions – people were expected to handle these. Whether it was equipment failure, severe environmental disturbances, or other sorts of disruptions, the responsibility for predicting and diagnosing faults and returning the system to normal operation rested squarely on human staff. Developers of control systems and their applications were concerned about these issues only to the extent that they needed to provide the appropriate information and decision support to operators, engineers, and supervisors. The actual prognosis, diagnosis, and remedial actions were generally outside the scope of automation.

We have succeeded in our original mission almost too well, and this success has led to a broadening of our ambitions for automation and control systems. This has happened even as the scale and complexity of the physical systems – whether naval ships or commercial buildings or factories – have dramatically increased.

As might be expected, problem complexity translates to solution complexity. For instance, the more time we have to plan our response before a failure occurs, the better off we are – catastrophic failures can be avoided, human safety can be maximized, repair actions can be combined, and so on. To increase this time, we must find new ways to access data that we have not sensed before. In addition, we have to construct software that derives prognostic and diagnostic conclusions from increasingly subtle correlations among the sensed data.

5   Acknowledgment

The authors gratefully acknowledge the support of the Office of Naval Research, grant number N00014-96-C-0373. Joe Van Dyke participated in this project while employed at Predict DLI.

References

Bennett, B.H. and Hadden, G.D. (1999) Condition-based maintenance: algorithms and applications for embedded high performance computing. *Proceedings of the 4th International Workshop on Embedded HPC Systems and Applications (EHPC'99).*

Bristow, J., Hadden, G.D., Busch, D., Wrest, D., Kramer, K., Schoess, J., Menon, S., Lewis, S. and Gibson, P. (1999) Integrated diagnostics and prognostics systems. *Proceedings of the 53rd Meeting of the Society for Machinery Failure Prevention Technology* (invited).

Hadden, G.D., Bennett, B.H., Bergstrom, P., Vachtsevanos, G. and Van Dyke, J. (1999) Machinery diagnostics and prognostics/condition based maintenance: a progress report. *Proceedings of the 53rd Meeting of the Society for Machinery Failure Prevention Technology.*

Hadden, G.D., Bennett, B.H., Bergstrom, P., Vachtsevanos, G. and Van Dyke, J. (1999) Shipboard machinery diagnostics and prognostics/condition based maintenance: a progress report. *Proceedings of the 1999 Maintenance and Reliability Conference (MARCON99).*

Accommodating QoS Prediction in an Adaptive Resource Management Framework

E. Huh[1], L. R. Welch[1], B. A. Shirazi[2], B. Tjaden[1], and C. D. Cavanaugh[2]

[1] 339 Stocker Center, School of Electrical Engineering and Computer Science,
Ohio University, Athens, OH 45701
[2] Department of Computer Science Engineering,
The University of Texas at Arlington, Arlington, TX 76019
[1]{ehuh|welch|tjden@ace.cs.ohiou.edu}
[2]{shirzai|cavan@cse.uta.edu}

Abstract. *Resource management for dynamic, distributed real-time systems requires handling of unknown arrival rates for data and events; additional desiderata include: accommodation of heterogeneous resources, high resource utilization, and guarantees of real-time quality-of-service (QoS). This paper describes the techniques employed by a resource manager that addresses these issues. The specific contributions of this paper are: QoS monitoring and resource usage profiling; prediction of real-time QoS (via interpolation and extrapolation of execution times) for heterogeneous resource platforms and dynamic real-time environments; and resource contention analysis.*

1 Introduction

In [1], real-time systems are categorized into three classes: (1) deterministic systems, which have a priori known worst case arrival rates for events and data, and are accommodated by the Rate Monotonic Analysis (RMA) approach (see [2]); (2) stochastic systems, which have probabilistic arrival rates for events and data, and can be handled using statistical RMA [3] and real-time queuing theory [4]; and (3) dynamic systems, which operate in highly variable environments and therefore have arrival rates that cannot be known a priori.

This paper presents a resource management approach for dynamic allocation to handle execution times represented using a time-variant stochastic model. Additionally, we show how to accommodate heterogeneity of resources and QoS prediction.

Section 2 provides an overview of the resource manager (RM) approach. Sections 3-5 explain each component used in our RM approach. Section 6 presents experimental assessments of our techniques.

J. Rolim et al. (Eds.): IPDPS 2000 Workshops, LNCS 1800, pp. 792–799, 2000.

2   Overview of RM approach

Our approach to resource management is based on the dynamic path model of the *demand space* [5], [8], [9]. This demand space model is a collection of dynamic real-time paths, each of which consists of a set of communicating programs with end-to-end QoS requirements. The demand space system model is described in Table 1.

Table 1. Demand space system model

Symbol	Description
P_i	a name of path "i"
a_{ij}	name of application j in path "i"
H_k	a name of host "k"
$|P_i.DS| = tl$	data stream sizes of path "i" (or workload or tactical load)
$T(a_{ij}, tl)$	period of a_j in P_i with workload tl
$C_{obs}(a_{ij}, tl, H_k)$	observed execution time of a_j at cycle c with tl in path "i" on H_k
$C_{req}(a_{ij}, tl, H_k)$	required execution time of a_j at cycle c with tl in path "i" on H_k
$C_{prof}(a_{ij}, tl, H_p)$	profiled execution time of a_j at cycle c with tl in path "i" on H_p
$C_{pred}(a_{ij}, tl, H_k)$	predicted execution time of a_j at cycle c with tl in path "i" on H_k
$D_{obs}(a_{ij}, tl, H_k)$	observed queuing delay of a_j at cycle c with tl in path "i" on H_k
$CUP_{obs}(a_{ij}, tl, H_k)$	observed CPU usage on H_k for the a_j in P_i with tl
$CUP_{req}(a_{ij}, tl, H_k)$	required minimum CPU usage on H_k for a_j in P_i with tl
$CUP_{ureq}(a_{ij}, tl, H_k)$	required, unified minimum CPU usage on the target H_k for the a_j in P_i with tl
$MEM_{req}(a_{ij}, tl, H_k)$	memory usage of a_j in path "i" on H_k with tl
$\lambda_{req}(P_i)$	required latency of P_i (=QoS)
$\lambda_{pred}(c+1, P_i)$	predicted latency of path P_i at cycle c+1
$\psi(P_i)$	required slack interval for each QoS requirement = $[\psi_{min}(P_i), \psi_{max}(P_i)]$

Table 2. Supply space system model

Symbol	Description
H_k	host name "k"
$SPECint95(H_i)$	the fixed point operation performance of SPEC CPU95 of H_i
$SPECfp95(H_i)$	the floating point operation performance of SPEC CPU95 of H_i
$SPEC\_RATE(H_i)$	the relative host rating of H_i
$Threshold\_CPU(H_i)$	the CPU utilization threshold of H_i
$Threshold\_MEM(H_i)$	the memory utilization threshold of H_i
$CUP(H_i, t)$	the CPU usage (user + kernel) percentage of H_i at time t
$CIP(H_i, t)$	the idle-percentage of H_i at time t
$FAM(H_i, t)$	the free-available-memory of H_i at time t
$MF(H_i, t)$	the number of page faults on H_i at time t
$INT(H_i, t)$	the number of interrupts on H_i at time t
$CALL(H_i, t)$	the number of system calls on H_i at time t
$CTX(H_i, t)$	the process context switching rate on H_i at time t
$CMI(H_i, t)$	the number of packet-in received on H_i at time t
$CMO(H_i, t)$	the number of packet-out transferred on H_i at time t
$COL(H_i, t)$	the number of collisions occurred on H_i at time t
$LM_i(H_j, t)$	the ith load metrics in host j at time; $LM_i(H_i, t) \in \{FAM(H_i,t), MF(H_i,t), INT(H_i, t), CIP(H_i, t), CUP(H_i,t), CALL(H_i, t), CTX(H_i, t)\}$

We also model the resources or the *supply space* (described in Table 2), which consists of host features, host resources, and host load metrics.

The resource management problem is to map the set of all paths P_i onto the set of hardware resources, such that all $\lambda_{req}(P_i)$ are satisfied. Since the workloads of the P_i

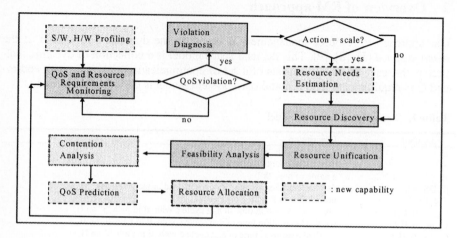

Fig. 1. Overview of resource manager

vary, the mapping needs to be adapted dynamically. The flow of our adaptive resource management approach shown in Fig. 1.

Each step is described in detail in the subsequent sections of this document.

3   Software and Hardware Profiling

In order to manage resources in an efficient manner, it is necessary to understand the resource usage characteristics of the members of the demand space and the relative resource capabilities of the members of the supply space.

S/W profiling measures an application's execution time, period, CPU usage, and memory usage that are collected passively by an external process (a monitor) that reads the *proc* table periodically to obtain process data. Three different techniques are tested as follows: (1) the process calls *getrusage* once per period, (2) an external monitor reads *ps_info* block in the *proc* table once per second, and (3) an external monitor reads *ps_usage* block in the *proc* table once per second.

An exponential moving average is applied to measurements for all techniques for filtering. Initial profiling is done during application development and profiles are refined through dynamic profiling. The accuracy of exponential moving average of *ps_usage* block in the *proc* table is almost as good as *getrusage* shown in Fig. 2.

H/W profiling measures capabilities of hosts relative to a reference host using the Standard Performance Evaluation Corporation (*SPEC*). *SPEC* is a standardized set of relevant benchmarks that can be applied to the newest generation of high-performance computers (see [10]). To achieve overall, relative system performance, the mean throughput is compared to a reference machine, a Sun-Sparc-10/40Mhz.

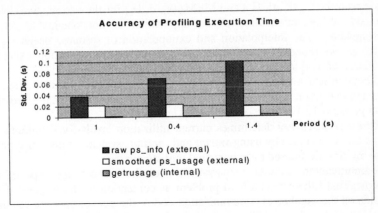

Fig. 2. Comparison of profiling techniques

We use *SPECfp95* (a measure of a host's floating point performance) and *SPECint95* (a measure of a host's fixed point performance) to derive the relative, *normalized* host rating as follows:

$$SPEC\_RATE(H)=AVG(SPECint95(H)/Max(SPECint95(H),SPECfp95(H)/Max(SPECfp95(H))),$$
where $\forall_j$.

4   QoS and Resource Utilization Monitoring

This section discusses our approach to QoS and resource monitoring, resource needs estimation, and resource discovery.

This module observes end-to-end real-time QoS of dynamic, distributed paths, and monitors resource requirements for dynamic software profiling to determine execution time, period, and memory usage. The memory usage (of main memory for allocation of workloads) is observed by taking the process residence set size from the *proc* table. The execution time of an application consists of the user- and the kernel-time, each of which corresponds to accurate computation of CPU utilization measured for a "move" action as follows:

$$CUP_{req}(a_{ij},tl, H_k)= C_{obs}(a_{ij}, tl, H_k) / T(a_{ij},tl).$$

Also, the *cycle time of the QoS monitor* called validity interval is used for the period $(T(a_{ij},tl))$ of an application to calculate the CPU resource requirement, while conventional approaches use the arrival time of workload for the period, which causes poor utilization on the dynamic environment.

Interpolation and extrapolation uses profiles to *estimate resource needs* of a new replica of a scalable application. When the current path QoS is greater than minimum slack of the QoS requirement, and QoS Manager (QM) recommends a "scale up" action decision based on the rate of workload trend, the resource requirements for the new workload tl (*tl = current tl / (current replicas+1)*) that will be distributed equally among replicas, need to be modified at run-time to request resource needs to the supply space. Hence, initial profiles of the violated application are the only way to decide required $C_{req}(a_{ij}, tl, H_k)$ and $MEM_{req}(a_{ij}, tl, H_k)$ for the various workloads as the boundary of execution time of an application is not obtainable on dynamic environments. The interpolation and extrapolation of resource needs for a "scale down" action proceeds exactly the same as a "scale up" action except for the calculation of workload tl ($tl = current\ tl / (replicas - 1)$).

The examined average of errors between the observed execution times and the estimated execution times that are examined by the piecewise linear regression using 2 data points is 12.1 milliseconds (1% CPU usage).

Resource discovery determines current utilization levels for communication and computation resources by using *vmstat* and *netstat* system calls once per second. These metrics are filtered by exponential moving average.

Communication resource management over the broadcasting type of networks (Ethernet/Fast Ethernet) is a hard problem as contention of those types of networks depends on the number of communication nodes, the size of packets, retransmission strategies, and collisions. The network load in terms of delay clearly has a strong relationship with collisions. Hence, our approach considers network load of hosts that are part of a real-time path is computed using the number of packet received / transmitted, and collisions. For a single host, the network load (net_load_of_host) is computed as follows:

$$net\_load\_of\_host = (1 + COL(H_i, t))*(CMI(H_i, t) + CMO(H_i, t))$$

5 Resource Selection

This section explains techniques for resource unification (mapping heterogeneous resource requirements into a canonical form), feasibility analysis, contention analysis, QoS prediction, and resource allocation analysis and enactment.

The role of **resource unification** is to map heterogeneous resource requirements into a canonical form of each resource metric. To allocate delivered $CUP_{req}(a_{ij}, tl, H_k)$, RM needs to determine the relative amount of the resources available on the target host. There might exist two approaches, which are static, and dynamic for resource management considering heterogeneity of resources. The static approach uses stable system information like benchmarks and CPU clock rate. In Globus project (see [6]), benchmark rates are used as the resource requirement (e.g. 100Gflops). The dynamic approach uses low level system parameters (see [7]). In Windows NT, a popular operating system, it is very complicated to access the dynamic system parameters that the operating system provides. Eventually, as a host-level, the global scheduler that will handle any types of systems, RM needs to use general system characteristics instead of dynamic, specific system parameters in the operating system layer. In our approach, using a static system information, *SPEC_RATE* shown in section 3, resources are unified into a canonical form as follows:

$$CUP_{ureq}(a_{ij}, tl, H_t) = C_{pred}(a_{ij}, tl, H_t) / T(tl, a_{ij})$$
$$C_{pred}(a_{ij}, tl, H_t) = C_{req}(a_{ij}, tl, H_p) * SPEC\_RATE(H_k) / SPEC\_RATE(H_t),$$
where H_t: a target host, H_k: host that the resource requirements are measured.

Feasibility analysis finds resources, which will meet $CUP_{ureq}(a_{ij}, tl, H_t)$. The *thresholds* are used for adaptable resource supply to tolerate the difference between unified and actual resources. For example, if the available CPU resource is greater than the unified CPU resource requirement plus the threshold, then the host becomes a candidate.

Contention analysis phase predicts queuing delays of applications among candidates. The queuing delay of an application in a path based real-time system is one of the critical elements for the RM to examine schedulability, when periods (of applications or paths) overlap each other. Currently, in our approach, observed system load metrics of hosts ($LM_i(H_i,t)$) are applied to get the delay on heterogeneous hosts.

First, we predict the queuing delay of the application in the target host by the observed queuing delay multiplied by the ratio of monitored load metrics between current host and target host ($D_{pred}(a_{ij},tl,H_i) = D_{obs}(a_{ij}, tl, H_k) * LM_i(H_p,t) / LM_i(H_i,t)$). Second, we use the execution time and current CPU usage on the target host ($D_{pred}(a_{ij},tl,H_i) = C_{pred}(a_{ij}, tl, H_i) * CUP(H_p,t)$). If one of our approaches can approximately uncover observed queuing delay, it becomes a generic solution in point of the host-level, global RM. An experiment is assessed for these approaches in section 6.

RM predicts a real-time QoS (considering contention) that will result from candidate reallocation actions. In general, when a customer requests QoS, this step tells the next QoS ($\lambda_{pred}(c+1,P_i)$) to the customer in addition to resource supply. If a single application in a path is violated, the path QoS is easily computed by adding predicted latency of the application instead of the current latency. Otherwise, the path QoS is accumulated until the last application's latency is predicted.

Resource allocation selects and performs a reallocation action based on the predicted QoS. Using the predicted QoS, RM can guarantee new reallocation. By testing predicted QoS path latency $\lambda_{pred}(c+1,P_i)$) by $\psi_{max}(P_i) < \lambda_{pred}(c+1,P_i) < \psi_{min}(P_i)$ called *pre-violation test*, QoS violation of the path at the next cycle can be detected by RM. Therefore, RM now can see QoS in addition to an amount of resources being supplied.

An allocation schemes for the violated application are considered based on QoS slack called "QoS Allocation (QA)", where QoS slack = $\lambda_{req}(P_i)$ - $\lambda_{pred}(c+1, P_i)$; which has passed the above *pre-violation test*. A *greedy, heuristic QA* scheme finds a host H_i which has minimum $\lambda_{pred}(c+1,P_i)$; and it is in top 50th percentile of the average network load among all candidate hosts.

6 Experiments

We have used DynBench (see [8]) and D-SPEC (see [9]) as an assessment tool and specification language for dynamic resource management. DynBench uses an identical scenario for experiments, respectively. A CPU load generator has been developed to allow the user to adjust CPU usage. The profiled execution times are measured on a sun-ultra-1 (140Mhz). Prediction is performed on sun-ultra-10 (333Mhz) as a source node and sun-ultra-10 (300Mhz) as a target node, respectively.

Experiment 1 shows predicted latency of a filter application a target host. Initially, 30% of CPU usage is used. In Fig. 3, using general system load metrics, several

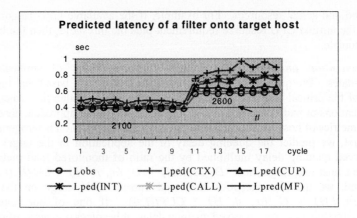

Fig. 3. Predicting latency of an application

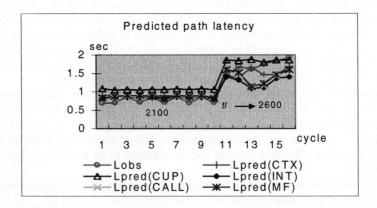

Fig. 4. Path prediction

methods of predicting latency are tested, and they are compared to the observed latency, which is measured at offline on the host with the same scenario

Each method specified in parenthesis predicts latency of the filter application as follows:

- Lobs : observed application latency
- Lpred(CTX) = $C_{pred}(a_{ij},tl,H_i)+D_{obs}(a_{ij},tl,H_k) * CTX(H_i,t)/CTX(H_k,t)$
- Lpred(CUP) = $C_{pred}(a_{ij},tl,H_i)+C_{pred}(a_{ij},tl,H_i) * CUP(H_i,t)$
- Lpred(INT) = $C_{pred}(a_{ij},tl,H_i) + D_{obs}(a_{ij},tl,H_k) * INT(H_i,t)/INT(H_k,t)$
- Lpred(CALL) = $C_{pred}(a_{ij},tl,H_i)+D_{obs}(a_{ij},tl,H_k) * CALL(H_i,t)/CALL(H_k,t)$
- Lpred(MF) = $C_{pred}(a_{ij},tl,H_i) + D_{obs}(a_{ij},tl,H_k) * MF(H_i,t)/MF(H_k,t)$

The results of experiment show that the queuing delay is a more important issue than the execution time to predict accurate latency of an application, when a host is overloaded. Overall average error (Lobs() - Lpred(CUP)) is *0.031 seconds.*

Experiment 2 shows the path latency comparison between the predicted (Lpred) and the observed (Lobs) latency. Note that predicted path latency is the sum of each application's predicted latency on to the target host. Fig. 4 explains that the Lpred(CUP) approach is most accurate approach, when the system is overloaded $(CUP(H_x, t)) > 70$ at workload 2600). To fully utilize CPU resource, the measurement of the queuing delay, when CPU usage is high, is very important for the distributed real-time system. The average error between the predicted and the observed latency is *0.084 seconds.*

7   Conclusions and Ongoing Work

The experimental results show that our approach achieves good CPU utilization by analyzing system contention and by predicting QoS accurately. The accuracy of the techniques is shown by noting that the predicted CPU resource needs differ from observed ones by no more than 4.5%. Ongoing work includes proactive RM and dynamic QoS negotiation.

References

1. Welch, L.R., Masters, M.W.,: Toward a Taxonomy for Real-Time Mission-Critical Systems. Proceedings of the First International Workshop on Real-Time Mission-Critical Systems 1999)
2. Liu, C.L., Layland, H.W.,: Scheduling Algorithm for multiprogramming in hard real-time environment. JACM, Vol. 20. (1973) 46-61
3. Atlas, A., Bestavros, A.,: Statistical Rate Monotonic Scheduling. Proceedings of Real-Time Systems Symposium (1998)
4. Lehoczky, J.P.: Real-Time Queueing Theory. Proceedings of IEEE Real-Time Systems Symposium, IEEE CS Press (1996) 186-195
5. Welch, L.R., Ravindran, B., Harrison, R., Madden, L., Masters, M., Mills, W.,: Challenges in Engineering Distributed Shipboard Control Systems. The IEEE Real-Time Systems Symposium. (1996)
6. Czajkowski, K., Foster, I., Kesselman, C., Martin, S., Smith, W., Tuecke, S.,: A Resource Management Architecture for Metacomputing Systems. Proceedings in IPPS/SPDP '98 Workshop on Job Scheduling Strategies for Parallel Processing (1998)
7. Chatterjee, S., Strosnider, J.,: Distributed Pipeline Scheduling: A Framework for Distributed, Heterogeneous Real-Time System Design. In the Computer Journal, British Computer Society, Vol. 38. No. 4. (1995)
8. Welch, L.R., Shirazi, B.A.,: A Dynamic Real-Time Benchmark for Assessment of QoS and Resource Management Technology. IEEE Real-time Application System (1999)
9. Welch, L.R., Ravindran,B., Shirazi, B.A., Bruggeman,C., : Specification and Analysis of Dynamic, Distributed Real-Time Systems. Proceedings of the 19th IEEE Real-Time Systems Symposium, IEEE Computer Society Press (1998) 72-81
10. OSG Group: SPEC CPU95. http://www.spec.org.

Network Load Monitoring in Distributed Systems

Kazi M Jahirul Islam[*], Behrooz A. Shirazi[*], Lonnie R. Welch[+], Brett C. Tjaden[+], Charles Cavanaugh[*], Shafqat Anwar[*]

[*]University of Texas at Arlington Department of CSE Box 19015 Arlington, TX 76019-0015

{islam|shirazi|cavan}@cse.uta.edu, anwar@swbell.net

['] Ohio University School of Electrical Engineering and Computer Science Athens, OH 45701-2979

{welch|tjaden}@ohio.edu

Abstract. Monitoring the performance of a network by which a real-time distributed system is connected is very important. If the system is adaptive or dynamic, the resource manager can use this information to create or use new processes. We may be interested to determine how much load a host is placing on the network, or what the network load index is. In this paper, a simple technique for evaluating the current load of network is proposed. If a computer is connected to several networks, then we can get the load index of that host for each network. We can also measure the load index of the network applied by all the hosts. The dynamic resource manager of DeSiDeRaTa should use this technique to achieve its requirements. We have verified the technique with two benchmarks – LoadSim and DynBench.

1 Introduction

The DeSiDeRaTa project is providing innovative resource management technology that incorporates knowledge of resource demands in the distributed, real-time computer control systems domain. This project involves building middleware services for the next generation of ship-board air defense systems being developed by the U.S. Navy. DeSiDeRaTa technology differs from related work in its incorporation of novel features of dynamic real-time systems. The specification language, mathematical model and dynamic resource management middleware support the dynamic path paradigm, which has evolved from studying distributed, real-time application systems. The dynamic path is a convenient abstraction for expressing end-to-end system objectives, and for analyzing timeliness, dependability and scalability. Novel aspects of the dynamic path paradigm include its large granularity and its ability to accommodate systems that have dynamic variability[1].

The resource manager is responsible for making all resource allocation decisions. The resource manager component computes allocation decisions by interacting with the system data repository and obtaining software and hardware system profiles. The allocation decision may involve migrating programs to different hosts, starting additional copies of programs (for scalability), or restarting failed programs (for survivability). The system data repository component is responsible for collecting and maintaining all system information.

J. Rolim et al. (Eds.): IPDPS 2000 Workshops, LNCS 1800, pp. 800-807, 2000.

The resource management architecture (attached at end of this document) consists of components for adaptive resource management and QoS negotiation, data broker, path monitoring and diagnosis, resource monitoring, and resource management consoles. The adaptive resource management and QoS negotiation component is responsible for making resource management decisions. This component computes the allocation decision by interacting with the data broker and obtaining software and hardware system profiles. The allocation decision may involve migrating programs to different host nodes, starting additional copies of programs (for scalability), or restarting failed programs (for survivability). The resource management component carries out its decisions by communicating with a daemon program (on each host) to start up and control programs on each host[4].

The data broker component is responsible for collecting and maintaining all system information. The data broker reads the system description and requirements expressed using the specification language and builds the data structures that model the system. Dynamically measured software performance metrics, such as path latency and throughput, and resource usage characteristics, such as program page faults and resident size, are collected and maintained by the path monitoring and diagnosis component. The data broker obtains measurements of the dynamic attributes of the software from the monitoring component. Hardware resource profiles are collected and maintained by the resource monitoring component, and fed to the data broker on demand as well as periodically. The data broker thus provides a single interface for all system data. The path monitoring and diagnosis component monitors the performance of software systems at the path-level. This component determines the changing requirements of the software by interacting with the data broker. When a path fails to meet the requirements, this component performs diagnosis of the path, and determines the ``bottleneck" node of the path. Resource management consoles display system and path status, and allow dynamic attributes, such as deadlines, to be modified. All communication for such consoles is through the data broker[6].

As mentioned earlier, the resource manager utilizes software and hardware system profiles to make allocation decisions. To obtain the system profiles, the resource manager continuously monitors the whole system and calculates various metrics. These metrics provide the guidelines for choosing among different allocation possibilities and optimizing resource usage. One of the components that is monitored by the resource manager is the network which connects the different computers that form the distributed real-time system. Several network parameters are of interest including host-to-host delay, network load index, host load index, etc. Host-to-host delay measures the time required to transmit a message from a specific host to another. Host load index measures the load applied to the network by a specific host. Network load index measures the total amount of load applied to the network by all the hosts that are currently connected or communicating through the network. Furthermore, since computers may be connected to multiple networks, they may have multiple IP addresses. Therefore, we can measure the host load index of a host on a specific network; or we may be interested in the load index of each network[7].

Depending on these parameters, the resource manager might select a different host to initiate a new process; or it might send data to a different host to get the result in an acceptable time using the least busy network or route. In a multi-homed network, it might also route data through an alternate network where load index indicates low traffic.

In this paper, we will formulate a host load index and network load index. We will also explain the experimental procedure that was followed to get the results. At the end we will also discuss the limitations of our approach and present some ideas for future work that will allow us to improve our method.

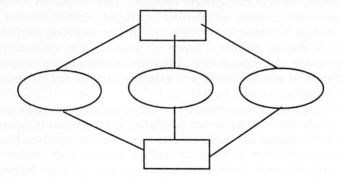

Fig. 1. Formal definition of the problem

We will use the above model for the illustration of the problem. Let us assume that there are n hosts $Host_1$, $Host_2$, ... $Host_n$ and m networks Net_1, Net_2, ...Net_m in the system (Figure 1). We are interested in finding the host load index for each host, $Host_i$. We are also interested in finding the load index on all the networks through which the hosts are connected. That means, if they are connected through k different networks, we are interested in measuring the load index on all k different networks. We also want to measure the load index of each network Net_i. That will help us to select the least loaded network for transmission.

2 Load Simulator

The LoadSimulation[1], hereafter referred as LoadSim, is able to compose and simulate the resource utilization (CPU cycles, network bandwidth, latency, etc.) of a large-scale distributed system that may consists of many interacting processes executing on many computers networked together. Here, simulation of distributed system load is achieved by means of replicated copies of a configurable LoadSim computer program. Each replicated copy must be capable of being initialized to a potentially different host computer and network resource utilization profile. LoadSim replicates are mapped onto a heterogeneous network of computers by a set of support services that allows the user to specify and control the topology and characteristics of the LoadSim configuration under test. The LoadSim further provides the ability to collect metrics on the performance of the simulated large-scale system.

---

[1] Some parts of LoadSimulation have been taken from "Requirements for a Real-time Distributed LoadSimulation" written by Timothy S. Drake. He may be contacted at drakets@nswc.navy.mil

The primary goals of the benchmark are to provide the ability to objectively assess the network communication protocols (e.g. TCP/IP, UDP, etc.), network bandwidth, network latency characteristics and to place additional load on partial implementations of real systems in order to assess the impact of the load that would be placed on the computing resource base by missing components if those components were present.

As LoadSim can place additional load on a partial implementations of real systems, we used this tool to apply load on the network. The DynBench benchmark application is modeled after typical distributed real-time military applications such as an air defense subsystem. Figure 2 shows the three *dynamic paths* from the DynBench benchmark application. The *detect* path (path 1) is a continuous path that performs the role of examining radar sensor data (radar tracks) and detecting potential threats to a defended entity. The sensor data are filtered by software and are passed to two evaluation components, one is software and the other is a human operator. The detection may be performed manually, automatically, or semi-automatically (automatic detection with manual approval of engagement recommendation). When a threat is detected and confirmed, the transient *engage* path (path 2) is activated, resulting in the firing of a missile to engage the threat. After a missile is in flight, the quasi-continuous *guidance* path (path 3) uses sensor data to track the threat, and issues guidance commands to the missile. The *guidance* path involves sensor hardware, software for filtering/sensing, software for evaluating and deciding, software for acting, and actuator hardware.[5]

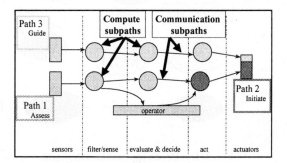

Fig. 2. The DynBench dynamic paths

A (simulated) radar sensor periodically generates a stream of data samples (representing the positions of moving bodies) based on equations of motion defined in a *scenario* file. The data stream is provided to the Filter Manager, which distributes the current workload among replicas of the filter program (Figure 3). Each filter uses a least mean square regression algorithm to filter "noise" and to correlate the data points into three equations that describe the motion of a body. The equations of motion for each of the observed bodies are sent to the evaluate and decide manager, which distributes the workload among the evaluate and decide programs. Evaluate and decide processes determine if the current position of an observed body is within a "critical region" defined by a doctrine file.

When a body first enters the critical region, it is passed to the action manager in the initiation path. Action manager distributes the workload among the action programs, which calculate equations of motion to intercept bodies of interest. A simulated actuator operates the initiation of motion of the intercepting body.

Whenever engaged objects are present in the sensor data, the evaluate and decide programs transmit the equations of motion of those bodies to the monitor and guide manager, which pairs identified bodies of targets with their corresponding interceptors. The corresponding pairs of equations are equally distributed among monitor and guide processes, which monitor the progress of the interceptor relative to the new positions of their intended target. If necessary, a new fight equation is calculated for the interceptor and sends to sensor. If an interception occurs, this process sends a request to remove the target and interceptor from the data stream.

There is also a deconflict path added to DynBench application subsystem. The deconflict path is designed to pre-launch the intercept bodies, and check if there is some conflict for the interceptor flight path with some other tracks or interceptors before the interceptor hit its target track. If there is a conflict, deconflict will send a warning message to Radar Display.

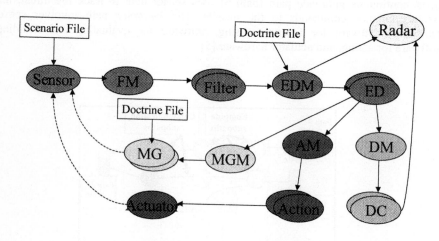

Fig. 3. DynBench application subsystem

We have used the DynBench as the load generator. We can put specific amount of tracks in the sensor and that will be processed by the whole benchmark. The increasing amount of tracks will put increasing amount of load on the network.

3 Previous Work

Philip M. Irey IV, Robert D Harrison and David T. Marlow have previously analyzed LAN performance in [2] and have shown an approach to evaluate the applicability of currently available commercial products in real-time distributed systems. They also had defined a few parameters for this purpose. Depending on these parameters and explained measurement methodologies we can determine the applicability of the commercial products in real-time systems. Andrej Sostaric, Milan Gabor Andreas Gygi had developed Mtool[3]. It can be used in performance monitoring in networked multi-platform systems. They emphasized on the three-tier architecture and used Java technology to achieve platform independence.

4 Experimental Procedure

We used *netstat* to collect our statistics. We are mainly interested in the TCP/IP suite of protocols, and the command '*netstat –i 1*' is used to produce a packet transmission summary once per second. We used this command in different hosts that may be connected through one or more networks to gather the statistics. This command gives number of packets in, number of packets out and number of collision in every second. We tried to formulate the load index using any linear combination of these three parameters.

To generate the load, we used two different tools. One is load simulator. We put the specific amount of load in every 200 millisecond on the network. The other tool is DynBench, a benchmark for DeSiDeRaTa. We were sending a specific amount of tracks or load from Sensor to Filter Manager for a specific amount of time. We tested for each load for approximately three minutes; that means each amount of load was applied on the network for three minutes. Later, we increased the load and run the experiment again. The test continued in this manner.

We performed our experiments in Sun workstations that are connected to a LAN. We collected statistics from different hosts in both cases. We are sure that the load was sent from the specified source host to the specific destination host. Because the mean and standard deviation of packets in, packets out and collision of the source and destination differs significantly from the other hosts; both mean and standard deviation were much higher in the source and destination.

Load generator	DynBench			Load Simulator		
	Host 1	Host 2	Network	Host 1	Host 2	Network
Out	0.53306	0.62177	0.57686	0.90124	0.92911	0.91584
Out+Collision	0.54933	0.62177	0.58216	0.94287	0.92911	0.94481
Out+in	0.53596	0.61334	0.58968	0.91077	0.91070	0.91472
In	0.53698	0.60707	0.59190	0.92778	0.90018	0.91252
Collision	0.55735	N/A	0.56566	0.96815	N/A	0.97252
In+Collision	0.55380	0.60707	0.60402	0.96350	0.90018	0.94345
In+Out+Collisn.	0.54741	0.61334	0.59450	0.93961	0.91070	0.93282

Table 1. Coefficient of correlation

Table 1 shows the coefficient of correlation of different combination of the parameters with the load that was applied on the network. Here data is moving from "Host 1" to "Host 2". Two scenarios of load generation are shown here for comparison: the first load was produced by DynBench; the second load was generated from Load Simulator. Figs. 4 and 5 summarize the results. We have also measured the load index of the whole network using the same approach. In the above graph, network specifies the load index of the whole network. It is the sum of the specified parameters.

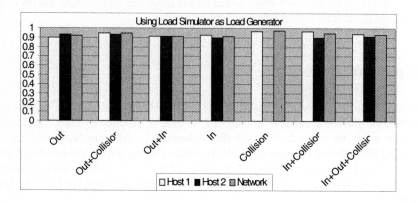

Fig. 4.

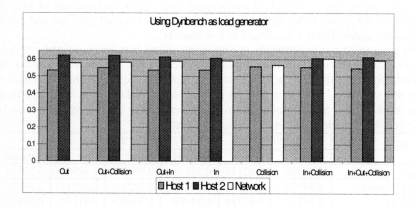

Fig. 5.

Several interesting characteristics can be observed from here. When a host is transmitting data to another host, the number of collisions gives a better approximation of the load index at the source than at the destination. From the receiver's point of view, the number of packets transmitted by the receiver reflects its load index. Here, both "out" and "out+collision" are the same. Experimental data shows that during packet transmission, the receiver does not experience any collision. As TCP/IP uses handshaking to inform the sender about the reception of data, the number of packets out somehow reflects actually about the number of packets transmitted to this host. This is supported by both load simulator and DynBench. For measuring the network load index, two different parameters are chosen by two different load generators. DynBench proposes sum of "In+Collision" of all the hosts connected to the network while according to Load Simulator, it is sum of Collision.

5 Conclusion

In this paper, we have proposed a simple non-intrusive technique for measuring the load applied on a network. We have used a simple tool, *netstat –i*, for that. We generated load using DynBench and LoadSimulator and then measured the load index. We have shown all the combinations of three parameters – packet in, packet out and collision, to determine which one best describes the load index.

Another way of measuring load index is to send time-stamped packets from one host to another host. This can measure the delay in the network, which also gives a fair indication of applied load. We can also measure the delay between two hosts through each network they are connected to. This gives the network load index.

References

1. L. R. Welch, B. A. Shirazi, B. Ravindran and C. Bruggeman, "DeSiDeRaTa: QoS Management Technology for Dynamic, Scalable, Dependable, Real-Time Systems", Proceedings of The 15th IFAC Workshop on Distributed Computer Control Systems, September 1998.
2. Philip M. Irey IV, Robert D Harrison and David T. Marlow, "Techniques for LAN Performance Analysis in a Real-Time Environment", Real-Time Systems, 14 21-44(1998) Kluwer Academic Publishers.
3. Andrej Sostaric, Milan Gabor Andreas Gygi, "Performance Monitoring in Network Systems", 20th Int. Conf. Information Technology Interfaces ITI '98, June 16-19, 1998.
4. L. R. Welch, B. Ravindran, B. Shirazi, and C. Bruggeman, "Specification and analysis of dynamic, distributed real-time systems", in Proceedings of the 19th IEEE Real-Time Systems Symposium, 72-81, IEEE Computer Society Press, 1998
5. L. R. Welch, B. A. Shirazi, "A Dynamic Real-Time Benchmark for Assessment of QoS and Resource Management Technology", RTAS 99
6. B. Ravindran, L. R. Welch, B. A. Shirazi, Carl Bruggeman, Charles Cavanaugh, "A Resource Management Model for Dynamic, Scalable, Dependable, Real - Time Systems"
7. L. R. Welch, P. Shirolkar, Shafqat Anwar, Terry Sergeant, B. A. Shirazi, "Adaptive Resource Management for Scalable, Dependable, Real - Time Systems"

A Novel Specification and Design Methodology Of Embedded Multiprocessor Signal Processing Systems Using High-Performance Middleware

Randall S. Janka[1] and Linda M. Wills[2]

[1] Georgia Institute of Technology, Georgia Tech Research Institute,
Atlanta, GA 30332-0856 USA
randall.janka@gtri.gatech.edu
[2] Georgia Institute of Technology, School of Electrical and Computer Engineering,
Atlanta, GA 30332-0250 USA
linda.wills@ee.gatech.edu

Abstract. Embedded signal processing system designers need to be able to prototype their designs quickly and validate them early. This must be done in a manner that avoids premature commitment to the implementation target, especially when that target includes costly COTS parallel multiprocessing hardware. A new specification and design methodology known as MAGIC enables the designer to move from an executable specification through design exploration and on to implementation with minimal loss of specification and design information by leveraging compuation middleware (VSIPL) and communication middleware (MPI). Maintaining such information is a quality known as "model continuity," which is established using the MAGIC specification and design methodology.

1 Introduction

Embedded signal processing system designers need to be able to prototype their designs quickly and validate them early. This results in quicker time to market as well as early detection of errors, which is less costly. There is tremendous complexity in the specification and design of these systems even when we restrict the technology space to commercial-off-the-shelf (COTS) multiprocessing (MP) hardware and software. We need a way to manage this complexity and accomplish the following goals:

- Enable the designer to quickly evaluate and validate design prototypes.
- Reduce and manage the level of detail that needs to be specified about the system in order to make sound decisions at each stage of the design process.
- Allow the design space to be explored *without* committing too early to a particular technology (hardware platform).
- Enable constraints identified and derived in one stage to be applied consistently in other stages of the design process.

In other words, we need to be able to benchmark and validate in early stages (at the appropriate level of detail and without premature commitment) – a process we call

J. Rolim et al. (Eds.): IPDPS 2000 Workshops, LNCS 1800, pp. 808–815, 2000.
© Springer-Verlag Berlin Heidelberg 2000

"virtual benchmarking" [1]. We also need to carry information gained (constraints and design rationale) through to later stages, a quality known as "model continuity." We have developed a new methodology to do this by exploiting computation and communication middleware that are emerging as standards in the embedded real-time COTS multiprocessing domain.

2   The Need for Model Continuity in Specification & Design Methodologies

The process of designing embedded real-time embedded multiprocessor signal processing systems is plagued by a lack of coherent specification and design methodology. A canonical waterfall design process is commonly used to specify, design, and implement these systems with COTS MP hardware and software. Powerful frameworks exist for each individual phase of this canonical design process, but no single methodology exists which enables these frameworks to work together coherently, i.e., allowing the output of a framework used in one phase to be consumed by a different framework used in the next phase.

This lack of coherence usually leads to design errors that are not caught until well into the implementation phase. Since the cost of redesign increases as the design moves through these three stages, redesign is the most expensive if not performed until the implementation phase. We have developed design rules and integrated commercial tools in such a way that designs targeting COTS MP technologies can be improved by providing a coherent coupling between these frameworks, a quality known as model continuity.

The basic information flow of a COTS MP specification and design (SDM) methodology is shown in **Fig. 1**. To appreciate how our SDM establishes model continuity, we first illustrate how model continuity is missing in today's COTS MP methodologies, as shown in **Fig. 2**. Currently, constants such as filter coefficients can be passed from MATLAB .m files into a CASE SDM or a simpler vendor software development environment, but that is the only link from the requirements specification and design specification to the implementation phase in the whole design process. Not having an executable requirements model and a channel for passing it to the design analysis phase leads to *model discontinuity*, which is the total absence or minimal presence of model continuity.

3   The MAGIC Specification and Design Methodology

We have developed and prototyped a new SDM which we call the MAGIC[1] SDM [2]. The means of accomplishing model continuity using the frameworks we chose for the MAGIC SDM is illustrated in **Error! Reference source not found.**. Solid boxes are

---

[1] MAGIC–Methodology Applying Generation, Integration, and Continuity.

documents or frameworks. Dashed boxes are aggregates of frameworks that contain executable specifications or the design analysis environment. Solid lines are automated channels, where system model information can be passed between frameworks without manual intervention. Dashed lines are semi-automated channels where some human intervention is required to move system model information between frameworks.

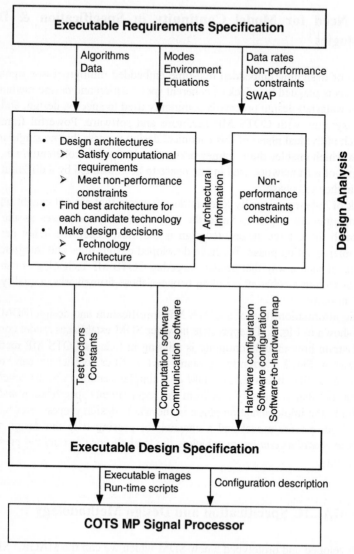

Fig. 1. Basic flow of information needed to support model continuity.

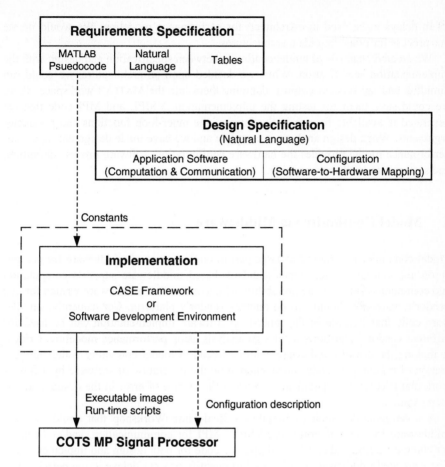

Fig. 2. How model continuity is currently lacking in current COTS MP SDM.

The executable workbook was fundamental in providing model continuity between specification and design. It was created using Excel with links created between worksheets that contained data (benchmarks, reliability statistics, form factor constraints, etc.) and models (benchmark conversions, process estimates, latency estimates, etc.). The data link to Simulink[2] was manual; architectural parameters were computed in Excel and then implemented in Simulink by hand since Simulink does not support scaling for parallelization. VSIPL[3] (computation middleware) and MPI[4] (communication middleware) functions were "generated" using our code generation rules and entered into our executable workbook. Once in our workbook, we could compute

---

[2] Simulation and rapid prototyping framework from The MathWorks.

[3] Vector Scalar Image Processing Library–an open-standards API for computation.

[4] Message-Passing Interface–an open-standard API for multiprocessing and parallel processing communication. Its real-time cousin is "MPI/RT."

token delays to be used in eArchitect[5] for performance modeling. We would iterate this process for other candidate architectures.

We created channels of model continuity between specification and design with the implementation specification. When we decided upon an architecture, we could run Simulink and tap process outputs, dumping them into the MATLAB workspace where we could save them for testing the implementation. VSIPL and MPI code that we generated is available for use in the form of the inner-loop functions and parameter arguments. When design analysis is complete and we have made design decisions, our performance model provides the hardware configuration, software process definition, and software-to-hardware mapping.

4   Model Continuity via Middleware

Model continuity is achieved in large part through the use of middleware for computation and communication. Open standards-based middleware supports computation and communication software portability, which means that middleware written for one vendor's hardware should run on another vendor's platform. Consequently, middleware code that constitutes the inner-loop software implementation can be used for different vendors' platforms for design analysis using performance modeling. Critical to making the use of middleware a strong thread of model continuity is the autogeneration of middleware code, since automating the generation of software by a framework that is correct in specification reduces the chance of error in the design and implementation.

A code generator such as Simulink's Real-Time Workshop that could generate middleware for computation using VSIPL, MPI for communication, and/or MPI/RT for communication and control will produce code for both design and implementation. The generated middleware can be used to quantify process delays in the performance model framework and as the core for signal processing implementation application software.

Our reasons for choosing VSIPL and MPI are very similar to our reasons for choosing the frameworks discussed above. They are stated here in order of importance with the most important reason stated first:

- Acceptable performance–These middlewares deliver high-performance because they are tightly integrated with the vendors' computation and communication libraries.
- Standards-based–Since all the COTS MP vendors in our domain space support these middleware and actively participate in their standardization processes, frameworks that generate VSIPL and MPI code will be consumable by all of the hardware vendors' SDEs considered in the design phase.
- COTS–They are now becoming commercially available and therefore stable and supported.

---

[5] Performance modeling framework from Viewlogic that supports multiprocessing and high-speed interconnections such as RACEway and Myrinet.

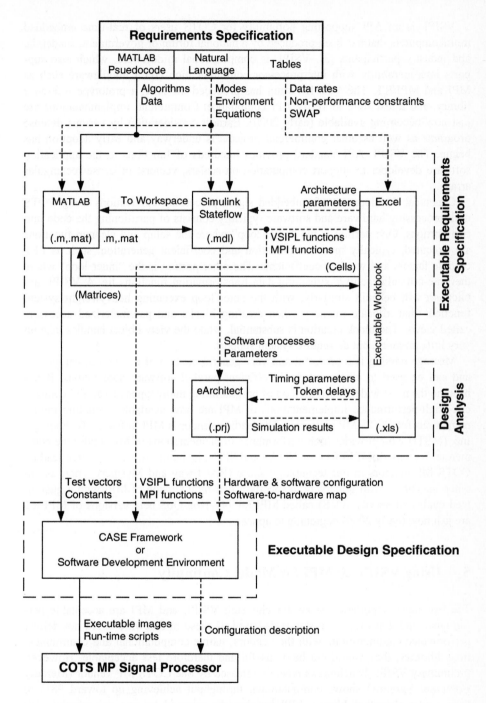

Fig. 3. MAGIC SDM information flow and illustration of model continuity.

VSIPL is an API supporting portability for COTS users of real-time embedded multicomputers that has been produced by a national forum of government, academia, and industry participants [3]. VSIPL is computational middleware, which also supports interoperability with interprocessor communication (IPC) middleware such as MPI and MPI/RT. The VSIPL Forum has produced the API, a prototype reference library, and a test suite to verify API compliance. Commercial implementations are just now becoming available (early 2000). Earnest consideration by various defense programs as well as other commercial projects is underway and early adoption has begun. The VSIPL API standard provides hundreds of functions to the application software developer to support computation on scalars, vectors, or dense rectangular arrays.

Canonical development of embedded signal processing applications using COTS multiprocessing hardware and software typically consists of partitioning the code into two portions. One portion is the "outer loop" where the setup and cleanup functions are executed, typically memory allocation and coefficient generation, such as FFT twiddle factors and window coefficients. The other portion is the "inner loop" where the time-critical repetitive streaming data transformation functions lie. A VSIPL application will be built similarly, with the outer loop executing heavyweight system functions that allocate memory when creating blocks and parameterized accessors called views. The block creation is substantial, while the view object handles take up very little memory, but do require system support.

Message passing is a powerful and very general method of expressing parallelism and can be used to create extremely efficient parallel software applications. It has become the most widely used method of programming many types of parallel computers. High-performance implementations of MPI are now available, including implementations for COTS MP platforms. The leading vendor is MPI Software Technology, Inc. (MSTI) who provides high-performance implementations of MPI under the commercial trademark MPI/PRO for NOWs and SPCs, including two of the three leading COTS MP vendors in our technology space (RACEway and Myrinet). There is another standards effort underway to specify a real-time version of MPI with a guaranteed quality-of-service (QoS) called MPI/RT [4]. Non-QoS beta versions of MPI/RT are just now (early 2000) beginning to appear.

5   Using VSIPL & MPI for Model Continuity

The two most important reasons for choosing VSIPL and MPI are acceptable performance and that they were standards-based. If these middleware could not deliver performance commensurate with the vendors' native computational and communications libraries, they would not be as useful and therefore less acceptable. However, preliminary VSIPL benchmarks recently released by one COTS MP vendor (Mercury Computer Systems) shows computational throughput achieving up toward 98% of their native algorithm library. MPI benchmarks released by one commercial MPI vendor (MSTI) show bandwidths within 5% of the RACE theoretical maximum for large block sizes, which is very close to that achieved by the vendor's own native communication library.

Being standards-based is the other key characteristic of these middleware. The participation of researchers, implementers, and users to form and support these standards goes a long way towards assuring their adoption. It is our opinion that there are two types of standards, official and de facto. Being standard is not a blessing deferred by some official "acronym'd" organization, but something established de facto when companies invest their own resources in products designed to a standard and consumers purchase those products. We are not saying that oversight and management by standards organizations is not worthwhile, we are just saying that real standards are determined by the community. Suffice to say, MPI and VSIPL are currently establishing themselves in the marketplace as standards, and no doubt "official sanctification" will occur sometime later.

Being a genuine de facto standard means that code generated within the MAGIC SDM can be used to estimate communication and computation token delays in performance modeling, as well as for the inner-loop computational code in the implementation. This strengthens the thread of continuity from specification to design (token delays) and implementation (inner-loop code).

6   Conclusion

We have introduced a new specification and design methodology (SDM) in this paper, the MAGIC SDM, that leverages standards-based middleware to achieve model continuity in the specification and design of signal processing systems implemented with COTS hardware and software. This is feasible since middleware generated in the specification and design processes can be used in the physical implementation because of the efficiency of both the VSIPL computation and MPI communication middleware.

References

[1] R. S. Janka and L. M. Wills, "Virtual Benchmarking of Embedded Multiprocessor Signal Processing Systems," in *submitted to IEEE Design and Test of Computers*, 2000, pp. 26.
[2] R. S. Janka, "A Model-Continuous Specification and Design Methodology for Embedded Multiprocessor Signal Processing Systems," a Ph.D. dissertation in the School of Electrical and Computer Engineering. Atlanta, Georgia: Georgia Institute of Technology, 1999, pp. xxiii, 225.
[3] VSIPL Forum, "VSIPL v1.0 API Standard Specification," DARPA and the Navy, Draft http://www.vsipl.org/PubInfo/pubdrftrev.html, 1999.
[4] Real-Time Message Passing Interface (MPI/RT) Forum, "Document for the Real-Time Message Passing Interface (MPI/RT-1.0) Draft Standard," DARPA, Draft http://www.mpirt.org/drafts.html, February 1, 1999.

Auto Source Code Generation and Run-Time Infrastructure and Environment for High Performance, Distributed Computing Systems

Minesh I. Patel Ph.D.[1], Karl Jordan[1], Mattew Clark Ph.D.[1],
and Devesh Bhatt Ph.D.

[1] Honeywell Space Systems-Commercial Systems Operations,
13350 U.S. Highway 19 North,
Clearwater, Florida, USA, 33764
{minesh.patel, karl.l.jordan, mathew.clark}@honeywell.com
[2] Honeywell Technology Center,
Minneapolis, Minnesota,
devesh.bhatt@honeywell.com

Abstract. With the emergence of inexpensive commercial off-the-shelf (COTS) parts, heterogeneous multi-processor HPC platforms have now become more affordable. However, the effort required in developing real-time applications that require high-performance and high input/output bandwidth for the HPC systems is still difficult. Honeywell Inc. has released a suite of tools called the Systems and Applications Genesis Environment (SAGE) which allows an engineer to develop and field applications efficiently on the HPCs. This paper briefly describes the SAGE tool suite, which is followed by a detailed description of the SAGE automatic code generation and run-time components used for COTS based heterogeneous HPC platform. Experiments were conducted and demonstrated to show that the SAGE generated glue (source) code with run-time executes comparably or within 75% efficiency to hand coded version of the Parallel 2D FFT and Distributed Corner Turn benchmarks that were executed on CSPI, Mercury and SKY compute platforms.

1 Introduction

Many Military, Industrial and Commercial systems require real-time, high-performance and high input/output bandwidth performance. Such applications include radar, signal and image processing, computer vision, pattern recognition, real time controls and optimization. The complexities of high performance computing (HPC) resources have made it difficult to port and fully implement the various applications. With the availability of inexpensive HPC systems based on commercial hardware, the high demands of military and industrial applications can be met. However, the potential benefit of using high performance parallel hardware is offset by the effort required to develop the application. Honeywell Inc. has release a set of user friendly tools that

J. Rolim et al. (Eds.): IPDPS 2000 Workshops, LNCS 1800, pp. 816–822, 2000.

offer the application and systems engineer ways to use the computing resources for application development. By tuning processes, improving application efficiency and throughput, and automatic mapping, partitioning and glue (source) code generation, the engineer can improve productivity and turn around time, and lower development cost.

This paper describes the Systems and Applications Genesis Environment (SAGE) and its auto-glue (source) code generation and run-time components. We first provide a brief overview of Honeywell's SAGE tool suite. This is followed by a description of the SAGE's auto glue code generation and run-time components. Finally, the experiments and results describing the comparison between the performance of the auto-generated glue code and hand-coded benchmarking applications, the Parallel 2D-FFT and the distributed corner turn is provided.

1.1 Systems and Applications Genesis Environment (SAGE)

Honeywell has developed an integrated tool suite for system design called the Systems and Applications Genesis Environment (SAGE)[1]. The tool suite provides complete lifecycle development through an integrated combination of tools potentially reducing design and development costs. The SAGE approach to application development is to bring together under a common GUI, a set of collaborating tools designed specifically for each phase of a system's development lifecycle. SAGE consists of the SAGE: Designer, the SAGE: Architecture Trades and Optimization Tool (AToT) and the SAGE: Visualizer.

Typically the design process begins with the Designer. The engineer can use the Designer to describe and capture the hardware and software/application architectures of the system and the mapping between application to hardware, which may be refined or narrowed by AToT. In the Designer, application/system and hardware co-design can be performed using the Designer's three editors, the application editor, data type editor and the hardware editor. The application editor is used to build a graphical view or model of the application by connecting functional or behavioral blocks (hierarchical) in a data flow manner through user defined or COTS functional libraries. The data type editor is used to define the various data types and striping and parallelization relationships for the different functions in the application editor. In the hardware editor, the hardware architecture is built hierarchically from the processor all the way up to the system level. All primitive and hierarchical blocks are stored on software and hardware "shelves" for later reuse. Items on the hardware shelf include workstations, other embedded computers, CPU chips, memory, ASICs, FPGAs, etc. The application and system designs can be refined using the software shelf items such as other COTS functional or user defined blocks. The entire software development environment integrates COTS-supplied components (compilers and run-time system, and libraries), along with custom, user-supplied software and hardware components (application code, libraries, etc.). Combining elements from the hardware shelf, the software shelf, and trade information, the engineer can construct an executable which maps software components onto hardware resources.

Once the performance requirements, application and hardware of the system are captured in the Designer, the information is sent to AToT. AToT will analyze and interpret the captured information, which drives optimization and trade-off activities described in the following section. After the architecture trades process has determined a target hardware architecture, the genetic algorithm based partitioning and mapping capability of AToT assigns the application tasks to the multi-processor, heterogeneous architecture. AToT can be employed for total design optimization, which includes load balancing of CPU resources, optimizing over latency constraints, communication minimization and scheduling of CPUs and busses.

When all the details of the system design have been made, the engineer may instrument and auto-generate the actual application code, which can be compiled and executed on certain supported testbed platforms. The SAGE Visualizer is a configurable instrumentation package that enables the designer to visualize the execution of the application through a variety of graphical displays that are fed by probes placed within the generated code. The Visualizer allows the designer to configure the instrumentation probes to measure application performance, and search for problems in the system, such as bottlenecks or violated latency thresholds.

2   Auto-Glue Code Generation and Run-Time Kernel

The SAGE glue-code generator is implemented in Alter, a programming language similar to Lisp in its syntax and style, which provides a direct interface to the contents of a SAGE model. Alter is designed to enable the tool developer to traverse the objects and arc connections in a model, collect the relevant information from the various attributes and properties, and then output the information in a particular format for the application. In the context of the glue-code generator, Alter traverses through the SAGE model and generates source code that can be compiled with application function libraries and the SAGE run-time as shown in Figure 1. The basic Alter language provides the constructs to perform the traditional programming tasks, such as procedure encapsulation, conditionals, looping, variable declaration, and recursion. The language also includes a set of standard calls to access certain features in SAGE, such as setting or retrieving a property value from an object.

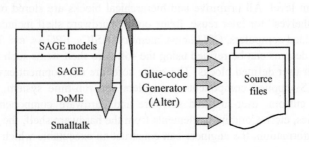

Figure 1.0 The SAGE glue-code generator gains access into the internal SAGE design tool environment, traverses objects in the models to filter relevant information, and then outputs the information in formats particular to the SAGE run-time source files. The SAGE glue-code generator is implemented in Alter, the programming language that facilitates the traversal and manipulation of DoME-based objects and graphs.

The SAGE run-time kernel is responsible for all sequencing of functions, data striping, and buffer management. To better cover the wide range of application domains, it is necessary to capture the notation of complex data distribution between functional software modules. In the data-flow programming model of the SAGE design notation, this requirement is handled by the port striping features. In short, the port striping conventions enable the system designer to define complex data distribution patterns between functions in a multi-threaded environment. A function's port object is the sending and receiving point for all data-flow communication between functions; the striping characteristics of a data-flow connection are defined on the source and destination ports. As mentioned previously, the glue-code generator develops several SAGE run-time source files, using information generated from the application model. For example, the function table is generated from a list of all function instances in the SAGE design. SAGE Designer orders all function instances and assigns them IDs from 0.. N - 1. The SAGE runtime executes functions based on this ID, which is the index of this descriptor into the function table. Similarly, information is extracted from the model that allows the runtime to perform data striping.

A function port can be defined in the model to be of type **replicated** or **striped**. Replicated ports represent data-flow communications in which the data is replicated for each thread of the host function. Striped ports represent data-flow communications in which the data is sliced or divided evenly among the threads of the host function. The port striping type applies to both sending (outgoing) and receiving (incoming) ports.

The runtime is responsible for striping the data based on the model information specified in the glue-code. It performs this operation using data buffers. Located and shared between each port on the sender and receiver functions is the SAGE notion of a *logical buffer*. A logical buffer is a logical representation of the data flow between sender and receiver function threads. It contains the striding information, total buffer size (before striding), thread information (number and type), etc. The logical buffer is defined by the glue-code using the application model's properties. The runtime uses the logical buffer and the striding information to create physical buffers for message transfer.

3   Experiments

In our experiments, we intend to show that SAGE produces executable code that is comparable to hand generated code for the targeted high performance computing platform for a selected set of benchmark applications. It is understood that tools which can auto generate code that can surpass performance wise hand coded application implementations is still work to be done. It is our intention to show that an application or system engineer can develop an application (conceptual, first cut or final version) using SAGE quickly and that the resulting solution is comparable both in performance and code size to hand coded versions. Additionally, the application can be refined for better performance by using the SAGE visualization software and by adding hand tuned functions to the SAGE reuse library for the target hardware platform.

3.1 Benchmark Applications

The benchmark applications chosen are algorithms that have been used by Rome Laboratories and MITRE in their benchmarking efforts of COTS based high performance computing systems such as from Mercury, Sky and CSPI. The applications chosen for our experiments are the parallel 2D FFT and the parallel distributed Corner Turn executing on a 1024x1024 data matrix. The two applications and data set were provided by CSPI. Performance results of the two applications executing on a Mercury, CSPI, SIGI and SKY platforms were obtained from MITRE[2]. For each of the hardware platforms, MITRE performed measurements using several node configurations (node counts). Additionally, high performance-computing vendors developed their own MPI implementation optimized for their hardware. The traditional MPI implementation have a built in function for performing the corner turn operation, namely the MPI_All_to_All function, each vendor implemented their own version tailored to their respective hardware for the most optimal performance.

3.2 Target Machine

The target hardware platform for performing the SAGE glue code and run-time experiments was chosen to be a 200 MHz Power PC 603e based high performance computing system provided by CSPI. The target system contained two quad-Power PC boards with the VxWorks operating system housed within a 21 Slot VME chassis. Each Power PC has 64 Mbytes of DRAM and can communicate through 160 MBytes Myrinet fabric interconnect to each other (intra-board) and to the outside world (inter-board). CSPI also provided all software including the VxWorks operating system, MPI implementation and the CSPI ISSPL functional libraries. As part of the Honeywell IR&D program and corporate alliance with CSPI, the SAGE tool was ported to CSPI target hardware platform. The term "port" corresponds to the capturing of all knowledge associated with programming to the CSPI hardware. Such knowledge that is captures includes the ISSPL function libraries on to the appropriate shelves, the CSPI board specific run-time software and programming methodology. It is expected that within the year, additional hardware platforms will be folded into the SAGE knowledge repository. It should be noted that SAGE hides the complexities of programming to COTS high performance computing hardware from the application developer. Once an application is developed, that application becomes portable to other SAGE supported platforms.

3.3 Experiments and Test Method

The experiments for the SAGE auto glue code generation and run-time components will be conducted in four steps. First, the application will be modeled using the Designer. Second, the different node configurations and mappings will be chosen through the Designer. Third, the glue code will be auto-generated where each node configuration and mapping will be executed ten times where each execution consists of a 100 iterations. The fourth step is the actual execution. The final performance number for that execution will average the 100*10 results into a final average result. When results are reported, a period is defined to be the time between input data sets while latency is the time required to process a single data set. The latency corresponds to the time from when the first data leaves the data source to the time the final result is output to the data sink.

3.4 Results

The results of the experiments are shown in Table 1.0. Table 1.0 shows the actual performance numbers for the two benchmark applications executing on 4 and 8 node configurations with 256, 512 and 1024 data sets. Each entry denotes the average of the 10*100 executions with cumulative averages shown in the last column. The table shows that the SAGE auto-generated code executed within an average 86% of the hand-coded versions on the CSPI hardware. For the distributed corner turn, the SAGE generated code running on the CSPI platform performed as well as the hand-coded CSPI version with an average overhead of 20%. For the 2D FFT, SAGE showed, on average, 17% cost in overhead.

Table 1.0 Comparison of hand-coded and auto-generated code for CSPI

Application	Array Size	Number of Processing Nodes						Average
		CSPI Hand Coded		SAGE AutoGen		% of Hand Coded		
		4	8	4	8	4	8	
2D FFT	256 x 256	14.8	8.496	15.8	9.4	93.7	90.4	92.0
	512 x 512	63.77	33.902	70.22	37.75	90.8	89.8	90.3
	1024 x 1024	267	137	312	169	85.6	81.1	83.3
Corner Turn	256 x 256	6.68	4.27	7.786	4.753	85.8	89.8	87.8
	512 x 512							
	1024 x 1024	86.53	52.2	108.822	65.135	79.5	80.1	79.8
								86.7

For the distributed corner turn, the SAGE generated code running on the CSPI platform performed as well as the hand coded CSPI version with on average 25% cost in overhead. A performance hit was taken on a two-node configuration. Here, the SAGE run-time buffer management scheme assigns unique logical buffers to the data per function which can cause extra data access times when compared to the CSPI implementation. For the 2D FFT, SAGE showed on average 20% cost in overhead.

4 Conclusions

The SAGE tool suite provides a powerful graphical and interactive interface for the creation of executable systems and applications based on customer defined specifications with fewer errors and an order of magnitude reduction in development time. The SAGE auto glue code generation and run-time components delivered and executed the two benchmark applications at 77.5 % of hand code versions. Although the performance of the auto-generated code is not equal to the hand code versions, tools that can generate such code are many years away. Work is currently underway to improve the performance of the glue code generation component that will reach levels of 90% of hand coded performance. The use of SAGE provides the application or systems engineer a way to rapidly develop an application on the target system with reasonable assurances that the performance of the auto-generated code for the application will not be magnitudes different from hand coded versions. And since the current SAGE tool makes the target system transparent to the engineer, the application developed is portable to other SAGE supported hardware platforms. The designer simply needs to re-generate the glue code for the new hardware platform. The time saved by using SAGE can now be more effectively used to perform such tasks as improving the applications performance on the current hardware platform, trading and testing the application on other hardware platforms, and moving on to the next project.

References

1. Honeywell's Systems and Applications Genesis Environment (SAGE™) Product Line, http://www.honeywell.com/sage.
2. Games, Richard, "Cross-Vendor Parallel Performance," Slides Taken from: Real-Time Embedded High Performance Computing State-of-the-Art, MITRE Corporation, Presented at DARPA Embedded Systems PI Meeting, Maui, Hawaii, March 16, 1999.

Developing an Open Architecture for Performance Data Mining

David B. Pierce[1] and Diane T. Rover[2]

[1] MS 1C1, Smiths Industries, 3290 Patterson Ave SE, Grand Rapids, MI, 49512
pierce_david@si.com
[2] Dept. of Elec. and Computer Engineering, Michigan State Univ., E. Lansing, MI, 48824
rover@egr.msu.edu

Abstract. Performance analysis of high performance systems is a difficult task. Current tools have proven successful in analysis tasks but their implementation is limited in several respects. Closed architectures, predefined analysis and views, and specific platforms account for these limitations. Embedded systems are particularly affected by these concerns. This paper presents an open architecture for performance data mining that addresses these limitations. Comparisons of the architecture with current tools show its capabilities address a wider range of system phases and environments.

1 Introduction

Performance analysis of complex systems is a difficult task. As a result, methods and tools to manage and reduce performance data to useable quantities or useful representations are the focus of significant research. Some successful tools include Pablo [1], Paragraph [2], and SPI [3]. These tools receive events from files, embedded instrumentation, or from an Instrumentation System (IS). These tools generally have a predefined set of views selected from a menu, some have options to select data to display, and libraries or executables to compile the tool.

Despite their successes in the lab environment, this class of tools are not an integral part of embedded and high performance systems because:

- The usage environment is limited to a specific OS or target HW,
- The design/source is protected or incomplete, limiting ability for integration,
- The views, processing algorithms, and queries (some tools have no query mechanism) are predefined, limiting flexibility for specific problems,
- The data sources/sinks are limited, limiting the use of the system and its results.

These tools are geared toward a lab environment, but we want to extend performance analysis to other environments. This will support embedded high performance systems, which can utilize performance analysis results for greater efficiency, user directed fault tolerance, and environmental tolerance (the recognition and corrective action operational conditions exceeding worst case design scenarios).

J. Rolim et al. (Eds.): IPDPS 2000 Workshops, LNCS 1800, pp. 823-830, 2000.

A solution to these limitations is to define a Performance Data Mining Architecture (PDMA) that: 1) has an open architecture described in a format consistent with a wide range of system design tools, 2) that addresses the data mining capabilities needed for large quantities of data, and 3) is flexible and extensible (concerning views, algorithms, queries, data exchange, and data storage) allowing for a wide range of systems and interfaces, and further development of the individual pieces as specific systems and applications dictate.

This paper presents the definition of such an architecture, with comparisons to current tools showing the benefits and advantages of such an approach.

2   Unified Modeling Language

To enable the widespread use of a PDMA in system designs, the development of a PDMA must address the system design environment. System designs are documented, reviewed, and analyzed in the early stages through the use of modeling techniques, such as Structured Analysis and UML. Following stages of system design use the model created to generate requirements, test plans and procedures, and in some cases, to generate source code headers and/or source code. While the most effective technique is a subject for debate [4], the great utility of these methods is not.

We have utilized the Unified Modeling Language [5] (UML) for the development of the PDMA. UML is widely accepted within the systems community, and its usage is increasing [6]. By using UML, the design of a PDMA is expressed in the same format as the system design itself, promoting ready implementation into design, analysis, test, and documentation. There are a significant number of tools that can analyze, simulate, and generate code from suitable UML diagrams, securing a spot for data mining at the ground floor of a system, and making it one of the important features of a complete system.

3   A Performance Data Mining Architecture

To begin, we examine current performance analysis tools, which have been successful in at least one phase of a system lifespan. These tools have one or more common tasks: 1) data input (performance events or statistics), 2) computation of statistics or data points, 3) display of data, and 4) user interface. A query function is also present on a few tools.

These common tasks summarize a significant portion of the desired system. However, there are four additional tasks that extend the usage environment of these tools. First, a database function to provide more flexibility to queries and more support for long term or relational computations is needed. Second, an output function for an IS, providing the ability to change instrumentation, based on current data. Third, provision for data exchange with system applications is important, and will support contextual analysis of the system model, requirements, and testing in a wide range of operational environments.

These common tasks then comprise the Use Cases within the Use Case diagram and form the basic requirements. The Use Case diagram is shown in Fig. 1. There is much detail underlying these simple use cases that differentiate the desired architecture from existing tools.

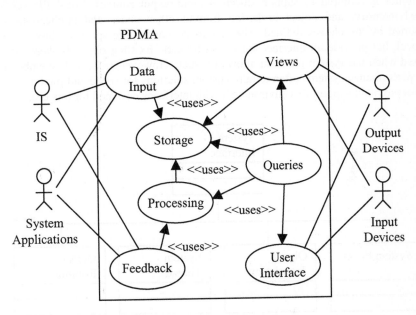

Fig. 1. A Use Case diagram showing the use cases, actors, and protocols for the PDMA. In this diagram, actors on the system represent classes of outside actors, and not individual items. The <<uses>> statement indicates that a use case uses the functionality of another use case (the end with the arrow)

A UML Class Diagram shows the classes that implement the architecture design and provides a vehicle for describing the details of the PDMA. The PDMA consists of three primary classes, System Interface, Analysis Context, and Data Warehouse. These primary classes are separated from each other to preserve data hiding principles and promote independence among system threads. These two principles provide flexibility for many specific system implementations [7].

The System Interface class (shown in Fig. 2) responds to large numbers of data inputs with short processing routines. Data inputs include performance events and statistics from the IS and configuration and loading data from system applications. This demands a relatively high priority thread to prevent queue overruns. It accepts data, converts to internal format as necessary, and routes the data. These items must be done quickly to prevent stealing too much time from other system threads.

It is also responsible for the output of data to the IS and systems applications. ISs accept feedback during operation to control the amount of instrumentation collected from specific instrumentation points. In this case, the data is formatted for output and routed to the IS using the appropriate interface. System applications can also accept feedback to control message routing and priority, system thread priority, and other features that may be determined by current and future research.

Flexibility is required to support different input/output sources. Local file access, shared memory, and object brokering from/to other nodes and applications are supported by the classes defined. The classes do not require specific object broker protocol, but provide an interface for object brokers. Existing object brokers can be utilized when the system platform allows for such. However, systems like embedded high performance systems often require custom solutions for speed and platform. The classes provided support this environment with interfaces designed for this task.

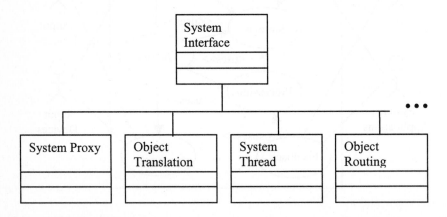

Fig. 2. Class Diagram showing the System Interface primary class

The Data Warehouse class (shown in Fig. 3) handles persistent data and responds to data storage requests and data search requests. It is also responsible for agents or database request to control size, important for embedded solutions with fixed memory. Data storage requests may include the computation of relational information that is stored with performance data. This class is separated to allow the use of a custom database structure or an off-the-shelf database application, which is dictated by the specific application. The processing priority of this task is likely to be low and require more time than the other classes, due to the nature of search requests, which is promoted by separation from the other classes. The interface to this class is tightly controlled through data storage requests and data query requests, enabling the update on either side of the interface without affecting the other.

An important factor for flexibility of the data warehouse implementation is relational information. The interface supports relational data requests and the formation of new relational information. Some key techniques for data mining include the search for new association rules, clustering, classification, sequential patterns, and outlier detection [8]. These techniques are supported in this design, including the use of relational information. The combined use of system application data and performance data also provides new analysis possibilities for environmental tolerance and corrective action.

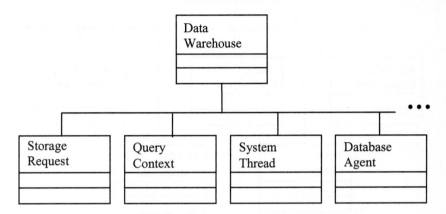

Fig. 3. Class Diagram showing the Data Storage primary class

The Analysis Context class (shown in Fig. 4) contains classes used for a specific analysis problem or context. Each performance analysis request has a specific context and can best be addressed by using an analysis context class instantiation, often running in its own thread. Allowing a separate thread for analysis contexts provides flexibility for assignment and priority of the thread. This supports a wide range of system applications. It is especially important to embedded high performance systems, where changing environmental conditions can be accounted for with dynamic adjustment of analysis threads.

Secondary classes under the Analysis Context class include classes for algorithms, display constructs, interfaces and translations to display hardware that is not high resolution CRT, contextual (display) and operator entered query formation, user interface, and others. The criterion for the definition of these classes is to allow the addition of new algorithm, view, and other objects without affecting the existing objects. Further, the specific system implementation, including hardware and software, must not affect the underlying PDMA, only a few classes defining the interface to such items as hardware or system applications.

The method for separating these analysis context objects is the interface to each of the objects. The Algorithm class can have many possibilities for computation within the instantiated object, but the interface to the View class, the Query class, etc., is maintained. A View object can then be utilized to display computed data, formatting the computed data in anonymous methods (from the algorithm viewpoint). The Display class receives View object data in a standard interface and transforms or translates the data to the specific hardware device involved in the Analysis Context instantiation. This may involve high resolution CRTs, character screen displays, banks of LEDs, alarms, etc.

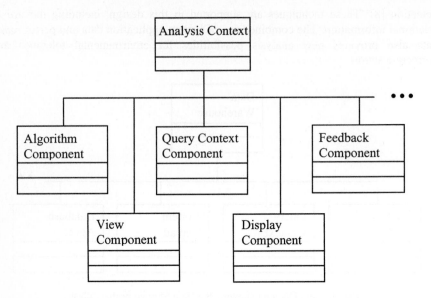

Fig. 4. Class Diagram showing the Analysis Context primary class

The Feedback class performs a similar function to the Display class, as it accepts the outputs from the View class, but it transforms or translates the data for feedback to the IS or system applications. It is separated from the View class because the necessary interface for each is unique enough to warrant it. This is shown by the types of data required by Display objects and Feedback objects.

The previous figures did not show the relationships between the classes. Relationships exist in the form of data objects, including performance data objects, query request objects, query request objects, view data objects, etc. Two of these these relationships are shown in Fig. 5. These objects determine a large part of the interface between the classes. Several more relations have not been shown.

4   Discussion

Current tools handle the presentation of data by providing several displays of data, such as Gantt, histogram, and pie charts, which can be selected. Some tools allow the user to select the data types to be displayed. In the PDMA, this capability is extended in an object-oriented method. A Gantt chart object is a class containing basic parameters such as data orientation, scale, etc. Instantiating a Gantt Chart object accepts the interface parameters and builds a display view (within the objects scope). The internal view of this Gantt Chart is not what is presented to the user however.

Additional modules within the display interface take the display parameters and map them to the display hardware. The display hardware will not always be a high resolution CRT, the common display hardware in the lab. Embedded high performance systems may utilize character displays, banks of LEDs, klaxons, or some other hardware device. The display class handles this responsibility and allows the use of any views with any display technology.

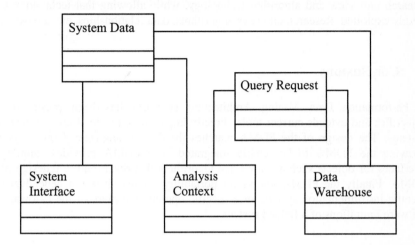

Fig. 5. Class Diagram for the PDMA showing two of the relationships between the classes

Paragraph and other tools limit the user to the predefined selections, since the system does not provide user definition of views, and the source cannot be easily modified. Using the PDMA, a user sets up an analysis context, including an instantiation of the desired view, scaling, data to be displayed in the view and its orientation, etc. Given this interface, the user can define these during operation. Further, the user can define new objects for views, etc., during operation with the user interface. The analysis context can also be instantiated as a perofrmance monitor. In this case, no display is instantiated until an event of interest appears, at which point the display is created.

Additionally, priorities can be assigned to the context, and actions assigned to its results as well. The user can assign priorities or interface to a scheduling algorithm to control the scheduling of tasks to meet the requirements during any specific operating environment. Embedded high performance systems have complex operational environments, which are difficult to accurately predict and design for. Providing capabilities for the operator, coupled with system support, provides a more flexible environment and greater operational success.

The displays allow interactive queries, such as entered queries or button clicks in the context of a display. Each of the query types resolves the display context for mouse clicks, or resolves the textual entry of a query. This provides the interface to the Data Warehouse class, maintaining a simple constant interface to the database.

5   Future PDMA Research

This paper presents the definition of a PDMA considering a wide range of systems from a general point of view. It is purposely designed to promote future analysis research into view and algorithm technology, while allowing that technology to be readily exploited. Research on views, algorithms, data relationships, etc, are expected.

6   Conclusions

A Performance Data Mining Architecture (PDMA) has been presented, that objectifies and extends current tools, directly impacting embedded high performance systems. The design of the PDMA matches the design language of other systems, allowing the PDMA to be readily integrated. The PDMA provides support and interfaces for objects such as views and algorithms that don't require redesign of the PDMA. Finally the PDMA allows for portability because it is not dependent on a specific instrumentation system, or a specific operating system, or the hardware and software limitations of a fielded system.

Acknowledgements

This work was funded in part by DARPA Contract No. DABT63-95-C-0072 and NSG Grant No. ASC-9624149.

References

1.   D. Reed et al., "Virtual Reality and Parallel Systems Performance Analysis", *IEEE Computer*, pp. 57-67, November 1995.
2.   M. Heath and J. Etheridge, "Visualizing the Performance of Parallel Programs", *IEEE Software*, 8(5), September 1991, pp. 29-39.
3.   D. Bhatt, et al., "SPI: An Instrumentation Development Environment for Parallel/Distributed Systems", *Proceedings of the 9th International Parallel Processing Symposium*, April 1995.
4.   R. Agarwal, P. De, and A. Sinha, "Comprehending Object and Process Models: An Empirical Study", *IEEE Transactions of Software Engineering*, Vol. 25, No. 4, July 1999, pp. 541-544.
5.   UML Documentation [Online], available at http://www.rational.com/uml/, April 30, 1999.
6.   B. P. Douglass, *Real-Time UML : Developing Efficient Objects for Embedded Systems*, Addison Wesley Longman, Inc., 1998.
7.   L. Bass, P. Clements, and R. Kazman, *Software Architecture In Practice*, Addison Wesley Longman Inc., 1998.
8.   A. Zomaya, T. El-Ghazawi, and O. Frieder, "Parallel and Distributing Computing for Data Mining", *IEEE Concurrency*, Vol. 7, No. 4, October 1999, pp. 11-13.

A 90k gate "CLB" for Parallel Distributed Computing

Bruce Schulman[1] and Gerald Pechanek[2]

[1] BOPS, Inc. Palo Alto, CA   bruces@bops.com

[2] BOPS, Inc. Chapel Hill, NC   gpechanek@bops.com

Abstract. A reconfigurable architecture using distributed logic block processing elements (PEs) is presented. This distributed processor uses a low-cost interconnection network and local indirect VLIW memories to provide efficient algorithm implementations for portable battery operated products. In order to provide optimal algorithm performance, the VLIWs loaded to each PE configure that PE for processing. By reloading the local VLIW memories, each PE is reconfigured for a new algorithm. Different levels of flexibility are feasible by varying the complexity of the distributed PEs in this architecture.

1 Introduction

As the complexity of portable products has increased, along with the need to support multiple, evolving standards, processor-based solutions have become a requirement at all levels of product architecture.

While a processor provides the needed flexibility, it must do so in an energy efficient and area efficient manner. Since the type of processing required for different products includes communications, video, graphics, and audio functions, multiple data types and algorithmic computational needs must be accommodated.

Due to this wide diversity of requirements, many approaches to providing efficient processing capability in each application have been proposed. These solutions include custom designed ASICs, general-purpose processors with DSP packed-data type instruction extensions, different DSPs in each product, and reconfigurable processor designs using FPGAs. ASICs lack flexibility in the face of changing standards and changing product requirements, measured as their high cost to support changes or multiple similar instances. General-purpose processors for embedded applications are inefficient in energy and area. Reconfigurable processors using FPGAs, even with the latest process improvements, are also inefficient in implementation area and energy use. This is especially true for FPGA implementations of arithmetic units, which are still very large and slow, compared to ASIC or custom arithmetic designs [1].

Even so, there is much work being done to combine the advantages of microprocessors and FPGAs for reconfigurable co-processing units, such as DISC [2] and GARP [3]. These systems may mix general control processors, fixed function ASICs, and FPGAs in a final system, such as Pleides [4]. In addition, companies such

J. Rolim et al. (Eds.): IPDPS 2000 Workshops, LNCS 1800, pp. 831-838, 2000.
© Springer-Verlag Berlin Heidelberg 2000

as Xilinx and Altera provide FPGA's and design solutions for specific reconfigurable algorithmic use [5, 6].

The difficulty with state-of-the-art FPGA designs is that the area, performance, and power cannot compete with standard cell or custom designed logic.   While using FPGAs seems to hold promise, many difficult problems exist that must be solved. Two problems with FPGA designs are the programming model/tools, and consistent and efficient use of silicon area.   It is important that each product has a consistent programming model and a common set of development tools across the numerous applications.   It is equally important to have a programmable design that can efficiently provide high performance and low power in the intended products. Research attempting to improve the implementation efficiency of FPGA-based reconfigurable processors proposes to increase the complexity of the Configurable Logic Blocks (CLBs), to include circuitry better suited for arithmetic use [7]. These additions attempt to provide application specific improvements to the original CLB definition. The goal is still to solve the basic problem of providing processor-level flexibility in a cost and performance efficient manner.

The purpose of a reconfigurable processor is to make effective and efficient use of the available logic for a number of applications by programming the arrangement and interconnection of the logic. We propose to use standard ASIC processes for a set of flexible arithmetic units in a standard PE definition that is programmed through local VLIW memory. Programming the PE can be viewed as a method to optimize the logic make-up of the PE for different algorithms.   With our scalable, parallel distributed processing configuration, the use of the available resources can be configured appropriately, cycle-by-cycle, to meet the requirements of each application.   Further, these features and capabilities are provided in a single architectural definition using a consistent and standardized tool set.

In this paper we present the BOPS® ManArray™ parallel distributed computing architecture and show that by reprogramming the PEs' logic, very high-performance computing can be provided across multiple applications.

2 ManArray Parallel Distributed Computing

The BOPS® iVLIW™ PE is based upon the BOPS ManArray architecture, a parallel-distributed computer architecture targeting System-On-Chip applications. The ManArray architecture supports from 1 to 64 iVLIW PEs and a Sequence Processor (SP) for controlling the array of PEs. The SP is uniquely merged into the PE array for maximum efficiency to provide the SP controller with access to the ManArray network. The ManArray network interconnects clusters of PEs to provide contention-free, scalable, single-cycle communications. The distributed processor uses two basic building blocks, as shown in Figure 1. The PE consists of a register file, a set of execution units, a cluster switch as an interface to the ManArray network, local data memory, and local VLIW memory (VIM). The SP adds an instruction fetch unit and uses the same building block PE elements. Various core processors can be developed from these two reusable IP blocks.

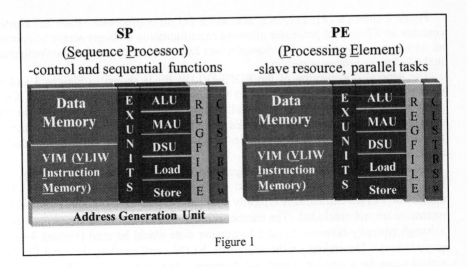

SP
(Sequence Processor)
-control and sequential functions

PE
(Processing Element)
-slave resource, parallel tasks

Figure 1

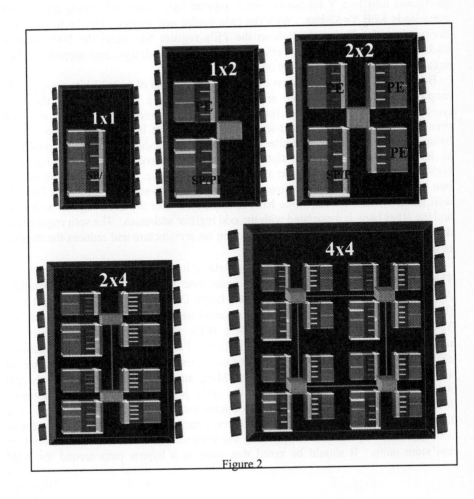

Figure 2

Figure 2 shows a 1x1, 1x2, 2x2, 2x4, and a 4x4 array processor. Each 2x2 cluster contains an SP control processor allowing reconfiguration of larger arrays to operate as subset array processors. For example, two 2x2 array processors can be configured in the 2x4 array processor system.

The general organization of the BOPS iVLIW PE is shown in Figures 3 and 4, which depict the three main interfaces to each PE. These interfaces are an X-bit instruction bus, Y-bit data busses, and a single-port send/receive interface to the cluster switch that interconnects the PEs in the ManArray topology. The instruction format is typically X=32-bits but 16-bit and 24-bit formats are not excluded depending upon an application's needs.

Internal to the PEs are three storage elements, the local PE data memory, the KxS*X iVLIW Memory (VIM), and a multiported NxM-bit register file. The number of VLIWs entries, K is typically less than 128 entries, although larger iVLIW memories are not precluded. The number of instruction slots, S can vary from 1 to 8, although typically between 2 and 5 instruction slots would be used (Figures 3 and 4 respectively). Depending on the arithmetic VLIW configuration, the local PE data memory can be a one- or a two-port memory. The two-port local PE memory is configured into two Y-bit banks which support byte, halfword, word, and double word loads with Y=32-bits. With the twin banks, one memory can be loading and storing data simultaneously to/from the PE's register file while the DMA unit is loading the other bank. This effectively hides DMA delays, and supports a data streaming approach to processing on the array.

Based upon present application evaluations, two banks of 512x32-bits are typically proposed, although there is no architecture limit. In a 5-issue iVLIW, the VIM typically consists of up to 64x160 bits of iVLIW memory with 160-bit read out capability. The VIM is loaded sequentially, one X-bit instruction at a time, after being primed by a LoadV delimiter instruction. The 160-bit field is made up of 5 Y-bit instruction slots, with each slot associated with an execution unit. In addition, a NxM-bit 8-read-port 4-write-port register file is available. This register file is split into two banks of 16x32-bits allowing the architecture to support 64-bit data flows as well as 32-bit data flows. One bank is associated with the even register addresses and the other bank is associated with the odd register addresses. The split register-file design takes full advantage of the instruction set architecture and reduces the number of ports required per register bank.

Finally, the ManArray architecture supports up to eight execution units in each PE. The first release uses five execution units: one Load Unit (LU), one Multiply Accumulate Unit (MAU), one Arithmetic Logic Unit (ALU), one Data Select Unit (DSU), and one Store Unit (SU). The execution units support 1-bit, 8-bit, 16-bit, and 32-bit fixed point data types, and 32-bit IEEE floating point data to meet the requirements of a large number of applications. For high-performance applications, each PE supports 32-bit and 64-bit packed data operations that are interchangeable on a cycle-by-cycle basis. Specifically, the MAU supports quad 16x16 Multiply and Accumulate operations per cycle, and the ALU performs standard adds, subtracts, four 16-bit absolute-value-of-difference operations, and other DSP functions. The DSU performs bit operations, shifts, rotates, permutations, and ManArray network communication operations. Supporting the computational elements are 64-bit load and store units. It should be noted that there is a bypass path around the VIM allowing single 32-bit instructions to be executed separately in classical SIMD mode in each PE and consequently on the array.

We use a linearly scalable switch fabric to connect the PEs with an interconnect maximum length of 2 for large embedded arrays, and length 1 for orthogonal interconnected PEs [8].   This ManArray network is integrated in the architecture of the PEs such that data movement between PEs can be programmed and overlapped with other arithmetic operations and load/store operations. This interconnect is programmable per cycle to allow many different interconnect patterns to match the current processing task.

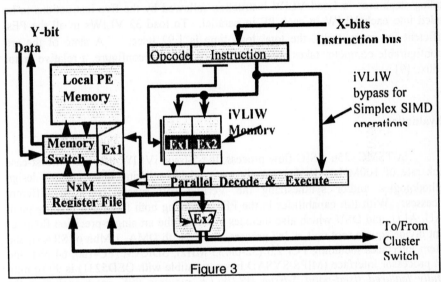

Figure 3

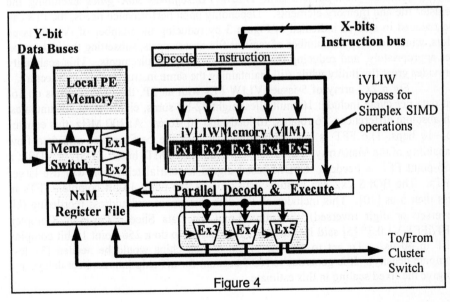

Figure 4

The programmer controls an array of PEs by writing a program for the SP, which includes the personalization of the PEs VLIW memory for the intended algorithm or algorithms to be executed. In addition, the SP controls the DMA unit to move data through the PEs, while controlling the program flow to perform the desired computation. Depending upon the size of the VLIW memory and the number of VLIWs needed for each algorithm in an application, the optimized set of VLIWs for multiple tasks can be resident in the VIM, allowing instantaneous reconfiguration as tasks change. Even with small VIMs that must be reloaded for each task, the loading of a five issue VLIW entry into all the PEs' VIMs @100MHz takes only 60nsec. The loading steps are - a Load VLIW instruction followed by the five instructions to be loaded into each VLIW in each PE in parallel. To load 32 VLIWs in all the PEs, sufficient for many tasks, the total load time is 1.92 μsec. A state of the art, reconfigurable computer takes approximately 100 μs to reconfigure, a relative factor of 50x. [9] .

3 Evaluation

In TSMC .25u ASIC flow process, the 5-issue iVLIW PEs has a worst-case clock rate of 100MHz. Higher speeds are available utilizing more custom design methodologies and/or synthesizing the Verilog soft macro cores to different processes. With full capabilities in the PEs, including both fixed and floating point MAU, ALU, and DSU which also includes a state of the art single-precision floating-point divide/square root unit, a 2x2 processor array with DMA, 1 Mbit of SRAM, and system interfaces including PCI 2.0 (32-bit/33 MHz), SDRAM (PC-100, 64-bit), and host processor interface (MIPS SYSAD bus compatible with QED5231) is 90 sq mm. *A fully featured fixed-point 5-issue iVLIW PE requires 90K gates* excluding the register file and memory elements. Depending upon performance needs, the PEs can be reduced in size as indicated in Figure 3 by reducing the number of iVLIW issue slots, which reduces the number of register file access ports, subsetting the instruction set appropriately, and reducing the local PE memory requirements. This scalability provides great flexibility while still maintaining the same instruction set architecture.

For a 2x2 array of 5-issue iVLIW PEs with an SP, the computations on 16-bit data per cycle include: 16 multiplies, eight 32-bit sums, eight 40-bit accumulates, 16 absolute differences, 16 rotates, 16 loads and 16 stores. At 100 MHz, this equates to ~10 bops. The FFT is one application that uses the performance and data flow capability of the ManArray. For Discrete MultiTone (DMT) based ADSL or VDSL, a 256-point FFT is needed. OFDM-based digital terrestrial television will use larger FFTs. The BOPS 2x2 can continuously process 256-point 16-bit complex FFTs in less than 5 us [10]. This includes all the data movement and address reordering (bit reversed or digit reversed). Xylinx LogiCore Data Sheet "256-Point Complex FFT/iFFT V1.0.3" [5] said it takes 1643 logic slices to do a 256-point 16-bit complex FFT in 40 us. To get to 5us per block, 8 such chips would be needed (8x less compute density). It is generous to NOT account for off-chip interconnect delays and forgive the fixed scaling in this estimate.

The basis of the most popular image compression algorithms (JPEG, MPEG) is the 8x8 iDCT. The algorithm takes in 8-bit data, but needs higher dynamic range to meet the S/N IEEE STD 1190. A 2x2 array can continuously process 8x8 blocks at a rate of 128 MBytes/second [11]. The Altera Discrete Cosine Transform AMPP Datashhet [6] shows an 8x8 iDCT processing rate of 17.5 MBytes/second, so you would need 8 of them to keep up.

4 Conclusions

BOPS offers the highest performance DSP IP in the industry and targets mass-market applications in 3D graphics, multimedia, Internet, wireless communications, VoIP and digital imaging. With this new level of performance and cost/performance, Embedded High Performance Computing can become a reality in consumer products from 3G cell phones with streaming video, to broadband Internet, to higher performance 3D Graphics in Set-top, games, and PCs.

The ManArray Architecture, including PEs, SP, DMA and Cluster Switch, delivers the highest performance, scalable, reusable, reconfigurable DSP IP in the industry. Compared to FPGAs, BOPS delivers more than 8x improvement in performance @100MHz in standard ASIC flow parts. Depending upon the array size, BOPS solutions cover the range from 1 to over 100 billion integer math operations/second.

References

1. O. T. Albaharna, P. Y. K. Cheung, and T. J. Clarke, "Area & Time Limitations of FPGA-based Virtual Hardware," Proceedings of the IEEE International Conference on Computer Design: VLSI in Computers and Processors, pp. 184-189, Cambridge, Mass., October 10-12, 1994, IEEE Computer Society Press.

2. M. J. Wirthlin and B. L. Hutchings, "DISC: The Dynamic Instruction Set computer," Proceedings of the SPIE, Field Programmable Gate Arrays (FPGAs) for Fast Board Development and Reconfigurable Computing, Vol. 2607, pp. 92-102, 1995.

3. J. R. Hauser and J. Wawrzynek, "Garp: A MIPS Processor with a Reconfigurable Coprocessor," Proceedings of IEEE Workshop on FPGAs for custom Computing Machines (FCCM), Napa, CA, April 1997.

4. Marlene Wan, Hui Zhang, Varghese George, Martin Benes, Arthur Abnous, Vandana Prabhu, Jan Rabaey, "Design Methodology of a Low-Energy Reconfigurable Single-Chip DSP System", *Journal of VLSI Signal Processing,* 2000.

5. Xilinx website: http://www.xilinx.com/

6. Altera website: http://www.altera.com/

838   B. Schulman and G. Pechanek

7. Nelson, Brent, "Reconfigurable Computing," HPEC 98 proceedings, September 1998

8. G. G. Pechanek, S. Vassiliadis, and N. Pitsianis, "ManArray Interconnection Network: An Introduction," EuroPar'99, Toulouse, France, Aug. 31-Sept. 3, 1999.

9. National Semiconductor NAPA 1000 – DARPA ITO Sponsored Research 1988. www.darpa.mil/ito/psum1998/e257-0.html

10. N. P. Pitsianis and G. G. Pechanek, "High-Performance FFT Implementation on the BOPS ManArray Parallel DSP," International Symposium on Optical Science, Engineering, and Instrumentation, Denver, Colorado, July 18-23, 1999.

11.   G. G. Pechanek, B. Schulman, and C. Kurak, "Design of MPEG-2 Function with Embedded ManArray Cores," Proceedings DesignCon 2000 IP World Forum section, Jan. 31-Feb. 3, 2000.

Power-Aware Replication of Data Structures in Distributed Embedded Real-Time Systems[*]

Osman S. Unsal, Israel Koren, C. Mani Krishna

Department of Electrical and Computer Engineering
University of Massachusetts, Amherst, MA 01003

Abstract. In this paper, we study the problem of positioning copies of shared data structures to reduce power consumption in real-time systems. Power-constrained real-time systems are of increasing importance in defense, space, and consumer applications. We describe our energy consumption model and present numerical results linking the placement of data structures to energy consumption.

1 System Model

This paper explores the power ramifications of various task assignment heuristics as well as network topology/routing issues. We study distributed real-time systems, with each node having a private memory and each task having a worst-case execution time and deadline. If two tasks reside on different processors then the communication power cost depends on the routing algorithm and topology. The objective is to study the impact of a particular assignment-topology-routing combination on power consumption. To save energy, part of a remote task's data structure may be replicated closer to the consuming node(s). The aim is to find the ideal degree of replication. Increasing the replication increases local memory size and its energy consumption, while decreasing the volume of network transfers and the associated power consumption. Therefore a "sweet spot" may exist, beyond which increasing the degree of replication increases the energy consumption. More formally, the total energy consumed, denoted by E, is:

$$E = \sum_{i=1}^{n\_tasks} E_i \quad \text{where} \tag{1}$$

$$E_i = N_{write_i} \cdot e_{write} + N_{read_i} \cdot e_{read} +$$

$$(\sum_{j=1}^{n\_tasks} N_{net_{ij}}) \cdot e_{net} + Size_{mem_i} \cdot e_{static} \tag{2}$$

[*] This work is supported in part by DARPA through contract No. F30602-96-1-0341. The views and conclusions contained in this document are those of the authors and should not be interpreted as necessarily representing the official policies or endorsements, either expressed or implied, of the Defense Advanced Research Projects Agency, the Air Force or the U.S. Government.

J. Rolim et al. (Eds.): IPDPS 2000 Workshops, LNCS 1800, pp. 839–846, 2000.

Here, E_i is the energy consumed in executing the i-th task, e_{write} and e_{read} are the memory energy consumption per write and read access, e_{static} is the memory static energy consumption, e_{net} is the energy cost of a per-hop data transfer per-bit, N_{write_i} and N_{read_i} are the number of local memory write and read accesses of task i respectively and $N_{net_{ij}}$ is the number of remote accesses from task i to task j.

If the two tasks i and j are assigned to the same node then, $N_{net_{ij}} = 0$.

Memory consistency is preserved by updating all the replicated copies of a data item when a task writes a shared data item to its private memory. For typical programs the writes are at most 15 percent of reads. This characteristic facilitates the usefulness of replication.

All links are assumed to be of the same type, i.e., the link power consumption to transfer one byte is the same for all links. Various routing strategies such as broadcasting or flooding are also implemented in the model. As for the case of multicasting, efficient multicasting algorithms rely on building and trimming a minimum spanning tree [3]. However, this is not optimal from a power point of view since it builds a minimum spanning tree for all the nodes instead of the subset of nodes in the multicast group. To obtain a better solution, we have developed an energy-saving Steiner tree heuristic for systems with multiple multicasting requirements. Given a weighted graph G, the Steiner tree problem is to find a tree that spans a specified subset of nodes of G with minimal total distance on its edges. Various distinct trees that span the same subset of nodes of G can be constructed and one can select the tree that has less total edge cost than that of the other trees in the set. Since the problem is NP-complete, heuristics are needed. We have adapted such a heuristic [5] for our purposes. The heuristic finds a solution with total distance no more than $2(1 - 1/k)$ times that of the optimal tree in time $O(pn^2)$. Here, n is the number of nodes in G, p the number of Steiner points and k the number of leaves in the optimal Steiner tree. A short description of the Steiner heuristic algorithm is given in Figure 1.

For intertask-communication-bound real-time systems, the allocation of tasks to nodes can also have a significant impact on power. We use a steepest-descent heuristic[2] for power-aware task allocation. The heuristic starts from an initial allocation and then reallocates that pair of tasks to the same node which results in the largest decrease in the energy consumption from among the set of candidate task pairs. This reallocation is done iteratively until the energy saving is below a given threshold. Thus, the heuristic tends to assign tasks which communicate heavily to the same node.

2   Numerical Results

For the results in this section, unless otherwise noted, the number of tasks is 10 and the task execution times, periods as well as intertask communication sizes are random. The number of nodes is 4, the write-to-read power ratio is 1.22, 8% of the memory operations are writes and 92% are reads, and per-hop remote

Step 1. For every multicast group repeat steps 2 through 6.

Step 2. Construct the complete graph H from G and S in such a way that the set of nodes in H is equal to S; for every edge (u, v) in H, the distance of (u, v) is set equal to the shortest path between u and v in G.

Step 3. Find a minimum spanning tree T_H of H.

Step 4. Replace each edge (u, v) in T_H by the shortest path between u and v in G; the resulting graph R is a subgraph of G.

Step 5. Find a minimum spanning tree T_R of R.

Step 6. Delete edges in T_R, if necessary, so that all the leaves in T_R are elements of S. The resulting tree is returned as the solution.

Fig. 1. The Steiner heuristic algorithm

access energy cost is three times that of a local access energy cost.

2.1   Effect of Application Write Ratios

We begin by considering a situation in which each node keeps a fraction of the global data structures in its private memory.

Figure 2 illustrates the effect of changing write ratios on power. As can be seen from the figure the optimum energy point shifts towards lower replication as the write percentage gets higher. Another observation is that as the degree of replication gets higher the energy consumption increases sharply for higher write ratios. This stems from the memory consistency constraint and is caused by the need to update all the replicated copies of a data item that has been modified by the local task.

2.2   Impact of Per-hop Transfer Cost

The per-link power cost per bit transferred depends on the interconnection hardware used. For example, a wireless link may consume less power than a twisted-pair link. If the real-time system designer has multiple options for choosing interconnection hardware, he/she can find the optimum degree of memory replication for each option. Figure 3 illustrates this. Here the per-hop energy consumption varies from being equal to memory energy consumption per operation to four times that of the memory energy consumption per operation. We observe that the optimum energy point shifts toward higher degrees of replication as the per-link power consumption is increased. Also, the system energy consumption starts converging at higher degrees of replication. This phenomenon is due to the fact that most of the data is locally replicated, thus decreasing the sensitivity of total energy consumption to the per-link power cost.

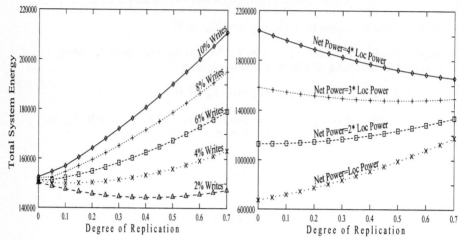

Fig. 2. Effect of write ratios on energy   **Fig. 3.** Effect of per-link power on energy

2.3   Task Allocation and Network Topology

Figure 4 shows the energy consumption impact of the task allocation scheme by comparing the previously mentioned power-aware optimization heuristic with a simple, power-blind round-robin scheme. The resulting saving in energy consumption emphasizes the importance of the task allocation step.

Network topology and routing are also important design considerations in real-time design. Figure 5 shows the energy comparisons of two different choices, a 16-node mesh topology and a 16-node torus topology. The extra wraparound edges of the torus result in lower energy consumption, but the energy difference between the two topologies is not very large.

2.4   Routing Issues

Multicasting has received little attention in real-time systems but is an important problem: sensors providing data to multiple processes and process outputs driving redundant actuators can all benefit from efficient multicasting algorithms. For the baseline case, we implemented the minimal spanning tree truncation scheme [3]. As mentioned in the previous section, we have developed a better multicasting scheme which makes use of a Steiner tree heuristic to find a path with the minimum cost among the multicasting nodes. Figure 6 shows the energy comparison of the two approaches. Here the number of nodes is 16 and the number of tasks is 40.

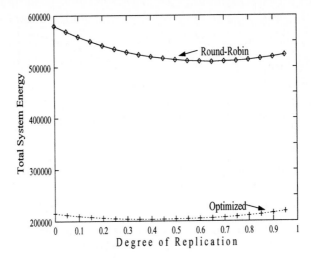

Fig. 4. Impact of task allocation strategy

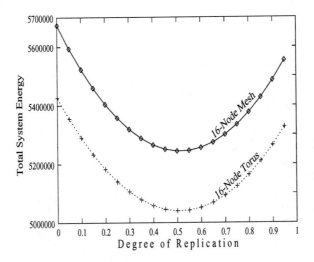

Fig. 5. Topology choice and energy

The routing capabilities of a Real-Time Operating System (RTOS) determines the power impact of multicasting tasks. A minimalist micro-kernel RTOS might just supply a simple flooding model. In this model the multicast message is sent to all the nodes in the system. A slightly more sophisticated RTOS would do a broadcast by sending a unique message to each of the multicast nodes. Broadcasting is considered to be more efficient than flooding [4]. However, as seen in Figure 7, flooding surprisingly does better than broadcasting from an energy point of view. This is because only a single copy of the multicast message is sent in flooding.

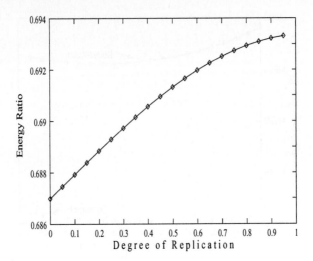

Fig. 6. Energy Ratio of Steiner / Minimal Spanning Tree Truncation

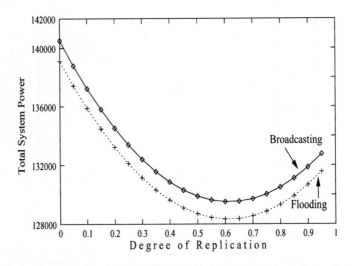

Fig. 7. Flooding versus Broadcasting

2.5   Selective Replication

Up to this point, we have considered the task-to-task communication data structures to be fully replicated. This means that for a multicast group, the data

structure of the multicast source task is replicated at all multicast destination tasks. We now relax this requirement and selectively replicate the data structure of the source task only at some of the destination tasks, thus saving energy. Consider the example of Figure 8. This is a 16-node mesh and part of the task assignment is shown in the figure. Our focus is multicasting group A with task $A.1$ being the source and the other tasks in group A being the destinations. We selectively replicate task $A.1$'s data structure only at task $A.4$'s node. The result is compared against full replication and no replication in Figure 9. Here, the energy is plotted against the per-hop energy cost and it is normalized with respect to the energy consumption of no replication. As can be seen, selective replication results in significant energy savings.

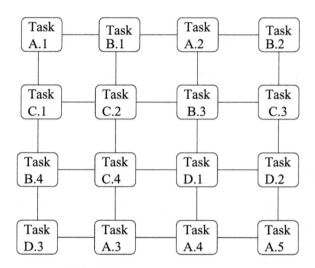

Fig. 8. Example for selective replication

3   Conclusion

We have constructed a model to gauge the power impact of task assignment, network topology and routing strategies within the context of data structure replication to decrease energy. Our results show that substantial energy savings are possible by careful design.

Our model also gives us the ability to calculate the energy impact of new power aware heuristics. We have adapted a Steiner tree heuristic for multicasting and compared its energy consumption with the baseline case of minimal spanning tree truncation.

Currently, we are studying the more general case of heterogeneous data consumption rates at the destination tasks. We are also developing a heuristic which will

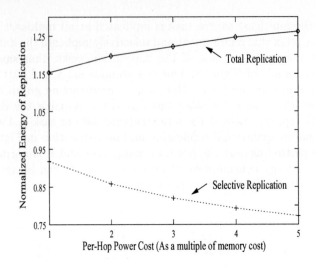

Fig. 9. Advantage of selective replication

optimize the memory replication needs of each task.

References

1. Coumeri, S. L., and Thomas, D. E., Memory Modeling for System Synthesis, www.ece.cmu.edu:80/ thomas/research/List.html
2. Press, W. H., Flannery, B. P., Teukolsky, S. A., Vetterling, W. T., Numerical Recipes, Cambridge University Press, 1989
3. Deering, S. E., and Cheriton, D.R., Multicast Routing in Datagram Internetworks and Extended LANs, ACM Trans. on Computer Systems May 1990
4. Tanenbaum, A. S., Computer Networks, Third Edition, Prentice Hall, 1996
5. Lau, H. T., Combinatorial Heuristic Algorithms with FORTRAN , Springer-Verlag, 1986

Comparison of MPI Implementations on a Shared Memory Machine

Brian VanVoorst[1] and Steven Seidel[2]

[1] Honeywell Technology Center, 3660 Technology Drive, Minneapolis, Minn. 55418
brian_vanvoorst@honeywell.com
[2] Dept. of Computer Science, Michigan Technological Univ., Houghton, Michigan
49931 steve@mtu.edu

Abstract. There are several alternative MPI implementations available
to parallel application developers. LAM MPI and MPICH are the most
common. System vendors also provide their own implementations of
MPI. Each version of MPI has options that can be tuned to best fit
the characteristics of the application and platform. The parallel application developer needs to know which implementation and options are best
suited to the problem and platform at hand. In this study the RTCOMM1
communication benchmark from the Real Time Parallel Benchmark Suite
is used to collect performance data on several MPI implementations for a
Sun Enterprise 4500. This benchmark provides the data needed to create
a refined cost model for each MPI implementation and to produce vi-
sualizations of those models. In addition, this benchmark provides best,
worst, and typical message passing performance data which is of particular interest to real-time parallel programmers.

1   Introduction

Shared memory platforms can support many different versions of the Message
Passing Interface (MPI)[1]. Among the best known MPI implementations are
LAM MPI[3] and MPICH[2]. Vendors also provide MPI implementations particularly suited to their platforms. Each implementation has various options for
tuning its behavior. This creates several choices for an application developer who
is seeking the best possible performance for their application.

The work presented here characterizes several MPI implementations and configurations for the Sun Enterprise 4500. These characterizations are based on
data obtained from the RTCOMM1 communication benchmark, part of the Real
Time Parallel Benchmark Suite [4, 5]. A refined communication cost model for
each implementation is obtained by an iterative process of running RTCOMM1,
examining the output, and adjusting the input to focus on the behavioral features revealed by the most recent data. This process was performed for the MPI
implementations listed in Table 1.

---

[2] This work is partially supported by NSF grant MRI-9871133.

J. Rolim et al. (Eds.): IPDPS 2000 Workshops, LNCS 1800, pp. 847–854, 2000.

Table 1. MPI variations examined

MPI	Mechanism	Option
Sun	SHMEM	
	SHMEM	MPI_SPIN
	SHMEM	MPI_POLLALL
	SHMEM	MPI_EAGER
LAM	TCP/IP	-O -c2c -nger
	SHMEM	-O -c2c -nger
MPICH	SHMEM	

Several system configuration options can also be varied in order to reveal their impact on message passing performance. The configuration options available on the E4500 include locking processes to processors, disabling interrupts, and even disabling individual processors. The effects of these options were also investigated.

2   Approach

Three MPI implementations are studied: Sun's MPI provided with HPC 3.0, LAM MPI[3], and MPICH[2]. LAM MPI was built in both its default TCP/IP version and in its shared memory version. These two builds of LAM MPI are compared to determine the amount of additional overhead created by the TCP/IP version compared to its shared memory version. By default, MPICH builds a shared memory version. No attempt at building a TCP/IP version of MPICH was made. These four implementations of MPI are the subject of the characterization work presented here. The platform used for this work is an 11-processor Sun Microsystems Enterprise 4500 symmetric multiprocessor with 8GB of memory running Solaris 2.7. The processors are 400MHz Sparc II's with 4MB of cache.

The characterization methodology for this work relies heavily on the use of the RTCOMM1 benchmark. RTCOMM1 takes as input a sequence of message size ranges (*e.g.*, 0-128 bytes, 129-4098 bytes, ...) and for each range produces N sample points. The experiments reported here use $N = 20$. A large number of ping-pong operations (sending a message back and forth between two processes) are timed at each sample point. The exact number of ping-pongs is not specified by the input to the benchmark. Instead, a total run time is specified. The benchmark performs a ping-pong measurement for each sample point in a round-robin fashion until the run time expires. The benchmark terminates only after completing a full round of sample points. This ensures that all message size ranges are measured an equal number of times and that any interruption of the benchmark (by, for example, an increase in background load) will not significantly bias the measurement of any one sample point.

For each sample point RTCOMM1 records the fastest (best), slowest (worst), and typical (median) time to complete a ping-pong. At the completion of the benchmark RTCOMM1 fits a line to the typical points of each message size range. This line is the communication cost model for that range of message sizes. RTCOMM1 provides as output these cost models and a series of data files suitable for plotting.

The initial approach to the characterization of each MPI implementation is to oversample with short message ranges and a dense set of sample points. This provides a fine-grained picture of point-to-point communication performance. These measurements reveal interesting regions in the graph of the performance data. It is usually apparent that there are certain message size ranges for which different underlying protocols, buffering schemes, *etc.* are used. Transitions in the graph at the boundaries of these ranges illustrate changes in the performance of the MPI implementation. Based on these observations, the input to the benchmark is adjusted so that the selected ranges match the transition points of the oversampled runs. A few iterations of this approach produces an accurate cost model for each MPI configuration.

3   Results

Due to limited space, only graphs of the most interesting characteristics of the MPI implementations are presented. These features appear at a variety of scales. Those that have the most direct impact on performance are discussed here.

It is important to remember that each point on a graph represents thousands of individually timed messages. When a "best" point takes slightly longer than its neighboring best points it is not due to chance mis-measurement. It is the result of some artifact in the system that did not allow that message to be transmitted faster. Not all such abnormalities can be explained, but they can be measured and their impact on performance can be revealed. All data points shown are actual measurements of point-to-point communication, not averages, computed by halving the best, typical, or worst observed ping-pong measurements.

For all three message passing libraries the typical times are often the same as (or very close to) the best times. This means that out of thousands of trials the median time is usually the same as the best time. Therefore, application developers can be confident that they will usually receive the best possible message passing performance the system has to offer. However, poorer performance sometimes occurs. This is captured by the "worst" observed points. These points often differ by a constant from the best observed times. This constant may be the cost of servicing one interrupt, which might happen only infrequently.

Messages were occasionally observed to be slowed down by orders of magnitude, up to $1/10^{th}$ of a second. This phenomenon can be reproduced by binding the benchmark processes to specific processors, disabling interrupts on those processors, and disabling all other processors except one. This caused many messages delays ranging from 0.02 to 0.1 seconds at consistently spaced intervals of 0.01 seconds. It is unclear why this particular combination of circumstances caused this delay.

3.1   Platform Configuration

Experiments showed that binding a processes to a processor fostered consistent performance. The `processor_bind()` system call prevents the operating system from migrating processes among processors. Under these conditions it appeared that the operating system will not schedule these processors for other work if other processors are available. Using the `psradm` command to disable I/O interrupts on these processors further reduced the possibility of these processors being interrupted while running the benchmark. Experience also indicated that it was necessary to leave more than one processor available for servicing interrupts.

The results presented here were collected from two benchmarking tasks bound to processors 0 and 1 on which interrupts were disabled. The remaining nine processors were available for other purposes but no other user jobs were running on the machine.

3.2   Sun's HPC 3.0 MPI

Sun's MPI implementation delivered the fastest overall point-to-point message passing. However, Sun's MPI was the least consistent and hardest to model for larger-sized messages. Figure 1 shows a plot of messages sizes between the ranges of 210KB and 240KB. No explanation can be offered for the illustrated oscillations in message passing times. While the variance is small (5%), it is large relative to the execution time. This shows that for certain message sizes, a message that is a few bytes longer may be transmitted in less time (as much as 50 microseconds) than the shorter message. This effect is reproducible and starts to occur for messages longer than 64KB. A second observable trend (not illustrated here) is that the difference between the worst and best points increases with message length. This is probably due to an increased chance of being interrupted multiple times while sending a longer message. The cost model for Sun MPI is given in Table 2. Due to the variance in measurements seen in Figure 1 it is not possible to present a precise cost model for messages longer than 64KB.

Table 2. Sun MPI cost model (*Imprecise due to large variance)

Message size (bytes)	Latency (μsec)	Bandwidth (MB/sec)
0 - 256	6	41.53
256 - 512	10	236.7
512- 1K	9	158.2
1K -16K	11	182.0
16K - 32K	29*	197.9*
32K - 1M	35*	208.6*
1M - 2M	277*	219.7*
2M - 4M	594*	225.1*

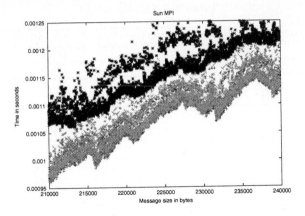

Fig. 1. Sun's MPI performance varies for large message sizes

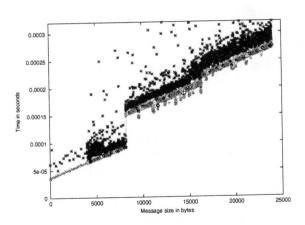

Fig. 2. Oversamples LAM MPI performance

3.3 LAM Shared Memory MPI

The most distinctive feature of LAM MPI is a large jump in latency for messages
of length 8KB, as shown in Figure 2. The magnitude of this increase shows that
it takes almost twice as long to transmit a message of length 8193 as it does to
transmit a message of 8192 bytes. For longer messages LAM message passing
costs are modeled well by a straight line, as given in Table 3. Both the shared
memory version of LAM and the TCP/IP version of LAM were built and tested
in these experiments but the measurements were the same in both cases. Because
the bandwidth is so high in each case it must be that the TCP/IP version is
making use of shared memory.

Table 3. LAM cost model

Message size (bytes)	Latency (μsec)	Bandwidth (MB/sec)
0-256	33	44.48
256 - 8K	35	145.9
8K - 1M	108	141.9
1M - 4M	108	141.7

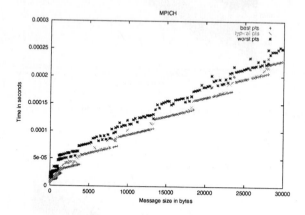

Fig. 3. MPICH cost model is a step function

3.4 MPICH

The performance of MPICH is best characterized by a step function. Figure 3 shows the observed message passing times for MPICH for messages of lengths 0 to 30,000 bytes. The interval of the step shown in Figure 3 is about 4900 bytes and it varies slightly as the message size grows. It then changes to an interval of about 9800 bytes when the message size is greater than 130,000 bytes. This interval also changes slightly as message size grows. The cause for this step function and the variation of interval size is not known but it might be a side effect of padding or buffer allocation and usage. The cost model for MPICH is given in Table 4. For messages longer than 100KB MPICH exhibits two "levels" of message passing times for each message length. Figure 4 shows message passing times for messages between the sizes of 200,000 and 210,000 bytes. Note that about a third of the time message passing times are greater by a fixed amount. This behavior is reproducible but no explanation can be offered here.

4 Conclusions

Sun MPI offers the best performance of the three MPI implementations. Figure 5 summarizes the costs of passing long message using Sun MPI, MPICH, and LAM. The second-order performance characteristics of these message passing interfaces are illustrated in Figures 1-4. Figure 1 shows that for long messages Sun MPI

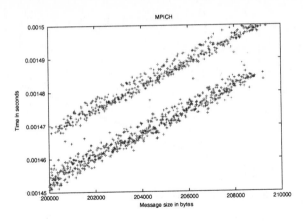

Fig. 4. MPICH bimodal cost behavior

Table 4. MPICH cost model

Message size (bytes)	Latency (μsec)	Bandwidth (MB/sec)
0 - 256	7	48.2
256 - 512	12	195.4
512 - 1K	6	56.5
1K - 130K	$\frac{Size-1K}{4900} * 5$	282.3
130K - 4M	$\frac{Size-130000}{9800} * 5+400$	238.7

exhibits large cost fluctuations. Figure 2 shows that with LAM, messages longer than 8KB have a start up cost three times that of messages shorter than 8KB and that LAM performance is best modeled by one cost function for messages shorter than 8KB and by another for messages longer than 8KB. MPICH is best characterized by a step function whose latency increases with message length, as shown in Figure 3. Figure 4 illustrates MPICH's bimodal cost behavior. The best platform configuration across all MPI implementations required locking processes to processors and disabling interrupts on those processors. These steps helped to ensure that processors remained dedicated to the application.

MPI implementations on the same machine, using the same shared memory message transport mechanism, have very different performance characteristics. The results presented here illustrate significant differences among cost models, scaling behavior, worst-case performance, and other performance characteristics. These differences stem from implementation decisions made by interface developers. LAM and MPICH are portable MPI implementations that are not tuned for specific platforms. The native implementation has a clear advantage in this case. It is also clear that no single implementation of MPI is best for all applications. This suggests that similar studies should be done for other platforms.

It has been shown here that RTCOMM1 can be used to characterize MPI implementations. The communication cost model of a message passing interface and hardware platform is usually described as a linear function determined by

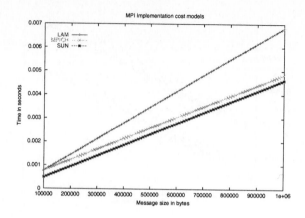

Fig. 5. Comparative message passing performance

a measured startup cost (latency) and bandwidth. RTCOMM1 was used here to show that this is sometimes an oversimplification. This work also demonstrated an approach for using RTCOMM1 to identify and illustrate performance differences between MPI implementations. While this approach does not reveal the causes underlying those differences, the experimental data does admit the construction of more accurate cost models. In addition, RTCOMM1 provides insight into best and worst case message passing performance which is useful for real-time software development.

References

[1] W. Gropp, E. Lusk, N. Doss, and A. Skjellum. A high-performance, portable implementation of the MPI message passing interface standard. *Parallel Computing*, 22(6):789–828, September 1996.

[2] William D. Gropp and Ewing Lusk. *User's Guide for* mpich, *a Portable Implementation of MPI*. Mathematics and Computer Science Division, Argonne National Laboratory, 1996. ANL-96/6.

[3] A. Lumsdaine, *et al. LAM MPI Home page.* http://www.mpi.nd.edu/lam/. University of Notre Dame.

[4] B. VanVoorst, R. Jha, S. Ponnuswammy, C. Nanvati, and L. Pires. DARPA Real Time Parallel Benchmarks: Final report. Technical Report (C013) - Contract Number F30602-94-C-0084, Rome Laboratory, USAF, 1998.

[5] B. VanVoorst, S. Ponnuswammy, R. Jha, and L. Pires. DARPA Real Time Parallel Benchmarks: Low-level benchmark specifications. Technical Report (C006) - Contract Number F30602-94-C-0084, Rome Laboratory, USAF, 1998.

A Genetic Algorithm Approach to Scheduling Communications for a Class of Parallel Space-Time Adaptive Processing Algorithms

Jack M. West and John K. Antonio

School of Computer Science
University of Oklahoma
200 Felgar Street
Norman, OK 73019
Phone: (405) 325-4624
{west, antonio}@ou.edu

Abstract. An important consideration in the maximization of performance in parallel processing systems is scheduling the communication of messages during phases of data movement to reduce network contention and overall communication time. The work presented in this paper focuses on off-line optimization of message schedules for a class of radar signal processing techniques know as space-time adaptive processing on a parallel embedded system. In this work, a genetic-algorithm-based approach for optimizing the scheduling of messages is introduced. Preliminary results indicate that the proposed genetic approach to message scheduling can provide significant decreases in the communication time.

1 Introduction and Background

For an application on a parallel and embedded system to achieve required performance given tight system constraints, it is important to efficiently map the tasks and/or data of the application onto the processors to the reduce inter-processor communication traffic. In addition to mapping tasks efficiently, it is also important to schedule the communication of messages in a manner that minimizes network contention so as to achieve the smallest possible communication time.

Mapping and scheduling can both – either independently or in combination – be cast as optimization problems, and optimizing mapping and scheduling objectives can be critical to the performance of the overall system. For parallel and embedded systems, great significance is placed on minimizing execution time (which includes both computation and communication components) and/or maximizing throughput.

The work outlined in this paper involves optimizing the scheduling of messages for a class of radar signal processing techniques known as space-time adaptive processing (STAP) on a parallel and embedded system. A genetic algorithm (GA) based approach for solving the message-scheduling problem for the class of parallel STAP algorithms is proposed and preliminary results are provided. The GA-based optimization is performed off-line, and the results of this optimization are static

J. Rolim et al. (Eds.): IPDPS 2000 Workshops, LNCS 1800, pp. 855-861, 2000.

schedules for each compute node in the parallel system. These static schedules are then used within the on-line parallel STAP implementation. The results of the study show that significant improvement in communication time performance are possible using the proposed approach for scheduling. Performance of the schedules were evaluated using a RACEway network simulator [6].

2   Overview of Parallel STAP

STAP is an adaptive signal processing method that simultaneously combines the signals received from multiple elements of an antenna array (the spatial domain) and from multiple pulses (the temporal domain) of a coherent processing interval [5]. The focus of this research assumes STAP is implemented using an element-space post-Doppler partially adaptive algorithm, refer to [5, 6] for details. Algorithms belonging to the class of element-space post-Doppler STAP perform filtering on the data along the pulse dimension, referred to as Doppler filtering, for each channel prior to adaptive filtering. After Doppler filtering, an adaptive weight problem is solved for each range and pulse data vector.

The parallel computer under investigation for this work is the Mercury RACE® multicomputer. The RACE® multicomputer consists of a scalable network of compute nodes (CNs), as well as various high-speed I/O devices, all interconnected by Mercury's RACEway interconnection fabric [4]. A high-level diagram of a 16-CN RACEway topology is illustrated in Figure 1. The interconnection fabric is configured in a fat-tree architecture and is a circuit switched network. The RACEway interconnection fabric is composed of a network of crossbar switches and provides high-speed data communication between different CNs. The Mercury multicomputer can support a heterogeneous collection of CNs (e.g., SHARC and PowerPCs), for more details refer to [6].

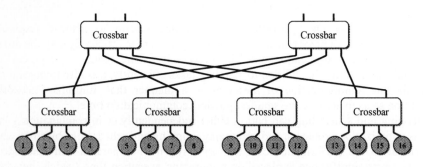

Fig. 1. Mercury RACE® Fat-Tree Architecture configured with 16 CNs.

Achieving real-time performance requirements for STAP algorithms on a parallel embedded system like the Mercury multicomputer largely depends on two major issues. First is determining the best method for distributing the 3-D STAP data cube across CNs of the multiprocessor system (i.e., the mapping strategy). Second is determining the scheduling of communications between phases of computation. In general, STAP algorithms contain three phases of processing, one for each dimension of the data cube (i.e., range, pulse, channel). During each phase of processing, the vectors along the dimension of interest are distributed as equally as possible among the processors for processing in parallel. An approach to data set partitioning in STAP applications is to partition the data cube into sub-cube bars. Each sub-cube bar is composed of partial data samples from two dimensions while preserving one whole dimension for processing. The work here assumes a sub-cube bar partitioning of the STAP data cube, for further details refer to [6]. Figure 2 shows an example of how sub-cube partitioning is applied to partition a 3-D data cube across 12 CNs.

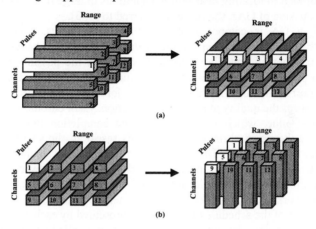

Fig. 2. Illustration of the sub-cube bar mapping technique for the case of 12 CNs. The mapping of the sub-cube bars to CNs defines the required data communications. (a) Example illustration of the communication requirements from CN 1 to the other CNs (2, 3, and 4) after completion of the range processing and prior to Doppler processing. (b) Example illustration of the communication requirements from CN 1 to other CNs (5 and 9) after the completion of Doppler processing and prior to adaptive weight processing.

During phases of data redistribution (i.e., communication) between computational phases, the number of required communications and the communication pattern among the CNs is dependant upon how the data cube is mapped onto the CNs. For example, in Figure 2(a) the mapping of sub-cube bars to CNs dictates that after range processing, CN 1 must transfer portions of it data sub-cube bar to CNs 2, 3, and 4. (Each of the other CNs, likewise, is required to send portions of their sub-cube bar to CNs on the same row.) The scheduling (i.e., ordering) of outgoing messages at each CN impacts the resulting communication time. For example, in Figure 2(a) note CN 1 could order its outgoing messages according to one of 3! = 6 permutations (i.e., [2,3,4], [3,2,4], etc.). Similarly, a scheduling of outgoing messages must be defined for each CN. Improper schedule selection can result in excessive network contention and thereby increase the time to perform all communications between processing phases. The focus in this paper is on optimization of message scheduling, for a fixed mapping, using a genetic algorithm methodology.

3   Genetic Algorithm Methodology

A GA is a population-based model that uses selection and recombination operators to generate new sample points in the solution space [3]. A GA encodes a potential solution to a specific problem on a chromosome-like data structure and applies recombination operators to these structures in a manner that preserves critical information. Reproduction opportunities are applied in such a way that those chromosomes representing a better solution to the target problem are given more chances to reproduce than chromosomes with poorer solutions. GAs are a promising heuristic approach to locating near-optimal solutions in large search spaces [3]. For a complete discussion of GAs, the reader is referred to [1, 3].

Typically, a GA is composed of two main components, which are problem dependent: the *encoding problem* and the *evaluation function*. The *encoding problem* involves generating an encoding scheme to represent the possible solutions to the optimization problem. In this research, a candidate solution (i.e., a chromosome) is encoded to represent valid message schedules for all of the CNs. The *evaluation function* measures the quality of a particular solution. Each chromosome is associated with a fitness value, which in this case is the completion time of the schedule represented by the given chromosome. For this research, the smallest fitness value represents the better solution. The "fitness" of a candidate is calculated here based on its simulated performance. In previous work [6, 7], a software simulator was developed to model the communication traffic for a set of messages on the Mercury RACEway network. The simulation tool is used here to measure the "fitness" (i.e., the completion time) of the schedule of messages represented by each chromosome.

Chromosomes evolve through successive iterations, called generations. To create the next generation, new chromosomes, called offspring, are formed by (a) merging two chromosomes from the current population together using a crossover operator or (b) modifying a chromosome using a mutation operator. Crossover, the main genetic operator, generates valid offspring by combining features of two parent chromosomes. Chromosomes are combined together at a defined crossover rate, which is defined as the ratio of the number of offspring produced in each generation to the population size. Mutation, a background operator, produces spontaneous random changes in various chromosomes. Mutation serves the critical role of either replacing the chromosomes lost from the population during the selection process or introducing new chromosomes that were not present in the initial population. The mutation rate controls the rate at which new chromosomes are introduced into the population. In this paper, results are based on the implementation of a position-based crossover operator and an insertion mutation operator, refer to [1] for details.

Selection is the process of keeping and eliminating chromosomes in the population based on their relative quality or fitness. In most practices, a roulette wheel approach, either rank-based or value-based, is adopted as the selection procedure. In a ranked-based selection scheme, the population is sorted according to the fitness values. Each chromosome is assigned a sector of the roulette wheel based on its ranked-value and not the actual fitness value. In contrast, a value-based selection scheme assigns roulette wheel sectors proportional to the fitness value of the chromosomes. In this paper, a ranked-based selection scheme is used. Advantages of rank-based fitness assignment is it provides uniform scaling across chromosomes in the population and is less sensitive to probability-based selections, refer to [3] for details.

4   Numerical Results

In the experiments reported in this section, it is assumed that the Mercury multicomputer is configured with 32 PowerPC compute nodes. For range processing, Doppler filtering, and adaptive weight computation, the 3-D STAP data cube is mapped onto the 32 processing elements based on an 8×4 process set (i.e., 8 rows and 4 columns), refer to [2, 6]. The strategy implemented for CN assignment in a process set is raster-order from left-to-right starting with row one and column one for all process sets. (The process sets not only define the allocation of the CNs to the data but also the required data transfers during phases of data redistribution.) The STAP data cube consists of 240 range bins, 32 pulses, and 16 antenna elements.

For each genetic-based scenario, 40 random schedules were generated for the initial population. The poorest 20 schedules were eliminated from the initial population, and the GA population size was kept a constant 20. The recombination operators included a position-based crossover algorithm and an insertion mutation algorithm. A ranked-based selection scheme was assumed with the angle ratio of sectors on the roulette wheel for two adjacently ranked chromosomes to be $1 + 1/P$, where P is the population size. The stopping criteria were: (1) the number of generations reached 500; (2) the current population converged (i.e., all the chromosomes have the same fitness value); or (3) the current best solution had not improved in the last 150 generations.

Figure 3 shows the simulated completion time for three genetic-based message scheduling scenarios for the data transfers required between range and Doppler processing phases. Figure 4 illustrates the simulated completion time for the same three genetic-based message scheduling scenarios for the data transfers required between Doppler and adaptive weight processing phases. In the first genetic scenario (GA 1), the crossover rate (P_{xover}) is 20% and the mutation rate (P_{mut}) is 4%. For GA 2, P_{xover} is 50% and P_{mut} is 10%. For GA 3, P_{xover} is 90% and P_{mut} is 50%. Figures 3 and 4 provide preliminary indication that for a fixed mapping the genetic-algorithm-based heuristic is capable of improving the scheduling of messages, thus providing improved performance. All three genetic-based scenarios improve the completion time for both communication phases. In each phase, GA 2 records the best schedule of messages (i.e., the smallest completion time).

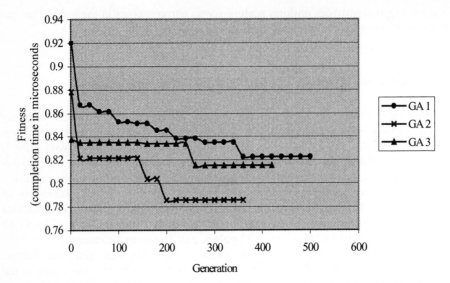

Fig. 3. Simulated completion time of the communication requirements for data redistribution after range processing and prior to Doppler processing for the parameters discussed in Section 4. For GA 1, the crossover rate (P_{xover}) = 20% and the mutation rate (P_{mut}) = 4%. For GA 2, P_{xover} = 50% and P_{mut} = 10%. For GA 3, P_{xover} = 90% and P_{mut} = 50%.

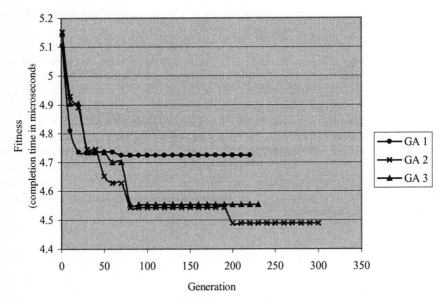

Fig. 4. Simulated completion time of the communication requirements for data redistribution after Doppler processing and prior to adaptive weight computation for the parameters stated in Section 4. For GA 1, the crossover rate (P_{xover}) = 20% and the mutation rate (P_{mut}) = 4%. For GA 2, P_{xover} = 50% and P_{mut} = 10%. For GA 3, P_{xover} = 90% and P_{mut} = 50%.

5.   Conclusion

In conclusion, preliminary data demonstrates that off-line GA-based message scheduling optimization can provide improved performance in a parallel system. Future work will be conducted to more completely study the effect of changing parameters of the GA, including crossover and mutation rates as well as the methods used for crossover and mutation. Finally, future studies will be conducted to determine the performance improvement between a randomly selected scheduling solution and the one determined by the GA. In Figures 3 and 4, the improvements shown are conservative in the sense that the initial generations' performance on the plots represents the best of 40 randomly generated chromosomes (i.e., solutions). It will be interesting to determine improvements of the GA solutions with respect to the "average" and "worst" randomly generated solutions in the initial population.

Acknowledgements

This work was supported by DARPA under contract no. F30602-97-2-0297.

References

1. M. Gen and R. Cheng, *Genetic Algorithms and Engineering Design*, John Wiley & Sons, Inc., New York, NY, 1997.
2. M. F. Skalabrin and T. H. Einstein, "STAP Processing on a Multicomputer: Distribution of 3-D Data Sets and Processor Allocation for Optimum Interprocessor Communication," *Proceedings of the Adaptive Sensor Array Processing (ASAP) Workshop*, March 1996.
3. L. Wang, H. J. Siegel, V. P. Roychowdhury, and A. A. Maciejewski. "Task Matching and Scheduling in Heterogeneous Computing Environments Using a Genetic-Algorithm-Based Approach," *Journal of Parallel and Distributed Computing*, Special Issue on Parallel Evolutionary Computing, Vol. 47, No 1, pp. 8-22, Nov. 25, 1997.
4. The RACE Multicomputer, Hardware Theory of Operation: Processors, I/O Interface, and RACEway Interconnect, Volume I, ver. 1.3.
5. J. Ward, Space-Time Adaptive Processing for Airborne Radar, Technical Report 1015, Massachusetts Institute of Technology, Lincoln Laboratory, Lexington, MA, 1994.
6. J. M. West, *Simulation of Communication Time for a Space-Time Adaptive Processing Algorithm Implemented on a Parallel Embedded System*, Master's Thesis, Computer Science, Texas Tech University, 1998.
7. J. M. West and J. K. Antonio, "Simulation of the Communication Time for a Space-Time Adaptive Processing Algorithm on a Parallel Embedded System," *Proceedings of the International Workshop on Embedded HPC Systems and Applications (EHPC '98)*, in *Lecture Notes in Computer Science 1388: Parallel and Distributed Processing*, edited by Jose Rolim, sponsor: IEEE Computer Society, Orlando, FL, USA, Apr. 1998, pp. 979-986.

Reconfigurable Parallel Sorting and Load Balancing on a Beowulf Cluster: HeteroSort

Pamela Yang[1], Timothy M. Kunau[1], Bonnie Holte Bennett[1], Emmett Davis[1], Bill Wren [2]

[1] University of St. Thomas, Graduate Programs in Software, Mail # OSS 301, 2115 Summit Avenue, Saint Paul, MN 55105
PamelaY@uswest.att.net, kunau@ahc.umn.edu,
bhbennett@stthomas.edu, EmmettDa@aol.com
[2] Honeywell Technology Center, 3660 Technology Drive, Minneapolis, MN 55418
Wren_Bill@htc.honeywell.com

Abstract. HeteroSort load balances and sorts within static or dynamic networks using a conceptual torus mesh. We ported HeteroSort to a 16-node Beowulf cluster with a central switch architecture. By capturing global system knowledge in overlapping microregions of nodes, HeteroSort is useful in data dependent applications such as data information fusion on distributed processors.

1   Introduction

Dynamic adaptability, both within an application's immediate distributed environment as well as future environments to which it will be ported, is a keystone feature for applications implemented on modern networks. Dynamic adaptability is a basis for fault tolerance. A system, which is dynamically adaptive, strives to withstand the assault of hardware glitches, electrical spikes, and component destruction. The research described in this paper set out to develop a high-speed load balancing algorithm, which would balance loads by sorting data across the network of nodes and resulted in developing a reconfigurable system for parallel sorting with dynamic adaptability.

1.1   Dynamic Adaptability

With the increased dependence on distributed and parallel processing to support general as well as safety-critical applications, we must have applications that are fault tolerant. Programs must be able to recognize that current resources are no longer available. Schedulers are employed in the presence of faults to manage resources against program needs using dynamic or fixed priority scheduling for timing correctness of critical application tasks.

We have taken a different approach and refocused on the design of elemental processes such as load balancing and sorting. Instead of depending on schedulers, we design process

J. Rolim et al. (Eds.): IPDPS 2000 Workshops, LNCS 1800, pp. 862–869, 2000.

algorithms where global processes are completed using only local knowledge and recovery resources. This lessens the need for schedulers and eases their workload. [1]

1.2   Beowulf Clusters

Beowulf clusters are one of the most exciting implementations of Linux today. Originating from the Center of Excellence in Space Data and Information Sciences (CESDIS) at the NASA Goddard Space Center in Maryland, the project's mission statement is:

> Beowulf is a project to produce the software for off-the-shelf clustered workstations based on commodity PC-class hardware, a high-bandwidth internal network and the Linux operating system.

The Beowulf project was conceived by Dr. Thomas Sterling, Chief Scientist, CESDIS. One of NASA's imperatives has always been to share technology with universities and industries. With the Beowulf project, NASA has provided the Linux community with the opportunity to spread into scientific areas needing high performance computing power.[2]

1.3   Local Knowledge and Global Processes

An efficient network sort algorithm is highly desirable, but difficult. The problem is that it requires local operations with global knowledge. So, consider a group of data (for example, that of names in a phone directory) which is to be distributed across a number of processors (for example, 26). Then an efficient technique would be for each processor to take a portion of the unsorted data and send each datum to the processor upon which it eventually belongs (A's to processor 1, B's to processor 2, ... Z's to processor 26).

A significant practical feature of HeteroSort is that in our experiments it load balances before it finishes sorting. Since HeteroSort detects when the system is sorted, it also detects termination of load balancing. Chengzhong Xu and Francis Lau in <u>Load Balancing in Parallel Computers: Theory and Practice</u> (Boston: Kluwer Academic Publishers, 1997) state:

---

[1] Examples of these fault tolerant efforts can be found in the work of Jay Strosnider and his colleagues at Department of Electrical and Computing Engineering, Carnegie Mellon University in the Fault-Tolerant Real Time Computing Project.
Katcher, Daniel I., Jay K. Strosnider, and Elizabeth A. Hinzelman-Fortino.
"Dynamic versus Fixed Priority Scheduling: A Case Study"
http://usa.ece.cmu.edu/Jteam/papers/abstracts/tse93.abs.html.

[2] For more information, see The University of St. Thomas Artificial Intelligence and High Performance Parallel Processing Research Laboratory's Beowulf cluster web page: Kunau, Timothy M.

From a practical point of view, the detection of the global termination is by no means a trivial problem because there is a lack of consistent knowledge in every processor about the whole workload distribution as load balancing progresses.[4]

Thus the global knowledge that all names beginning with the same letter belong on a prespecified processor facilitates local operations in sending off each datum. The problem, however, is that this does not adequately balance the load on the system because there may be many A's (Adams, Anderson, Andersen, Allen) and very few Q's or X's. So the optimal loading Aaa-Als on processor 1, Alb-Bix on processor 2, ... Win-Zzz on processor 26) cannot be known until all the data is sorted. Global knowledge (the optimal loading) is unavailable to the local operations (where to send each datum) because it is not determined until all the local operations are finished. HeteroSort combines load balancing within sorting processes.

Traditionally, techniques such as hashing have been used to overcome the non-uniform distribution of data. However, parallel hash tables require expensive computational maintenance to upgrade each sort cycle, thus making them less efficient than HeteroSort, which requires no external tables.

1.4   Related Work

Much of the work in this area deals with linear arrays.[2,3] The general approach is to take linear sort techniques and use either a row major or a snake-like grid overlaid on a regular grid topology of processors.[1] The snake-like grid is used at times with a shear-sort or shuffle sorting program where there is first a row operation and then an alternating column operation. So, either the row or the column connections are ignored in each cycle.

2   Approach

HeteroSort is our load balancing and sorting algorithm. Our initial approach was to use four-connectedness (as an example of N-connectedness) for load balancing and sorting. In traditional linear sorts data is either high or low for the processor it is on, and is sent up or down the sort chain accordingly. Our approach differs in that we defined data to be very high, high, low, or very low. In order to do this we first defined a sort sequence across an array of processors as depicted in Figure 1.

Next we defined the four neighbors. This is easily understood by examining Node 7 in the example of sixteen processors shown in Figure 1. The neighbors for Node 7 are 2, 6, 8, and 10. When Node 7 receives its initial data, it sorts it and splits it into four quarters. The lowest quarter goes to Node 2, the next lowest quarter goes to Node 6, the third quarter goes to Node 8, and the highest quarter goes to Node 10. Thus, the extremely high and low data are shipped on "express pathways" across the coils of the snake network.

8	7	6	5
9	**10**	11	12
16	15	14	13

Fig. 1. The sort sequence is overlaid in a snake-like grid across the array of processors. The lowest valued items in the sort will eventually reside on processor 1 and the highest valued items on processor 16. Node 7's four connected trading partners are in bold: 2, 6, 8, and 10. When Node 7 receives its initial state, it sorts and splits the data into four quarters. The lowest quarter goes to Node 2. The next lowest quarter goes to Node 6, the third quarter to Node 8, and the highest quarter goes to node 10. Thus the extremely high and low data are shipped across the coils of the snake network.

The trading neighbors Node 2 and Node 10 which are not adjacent on the sort sequence (transcoil neighbors) provide a pathway for very low or very high data to pass across the coils of the snake network into another neighborhood of nodes. This provides an express pathway for extremely ill sorted data to move quickly across the network. The concept of four connectedness is easy to understand with an interior node like Node 7, but other remaining nodes in this example are edge nodes, and their implementation differs slightly.

Table 1. Trading partner list. Determining which data is kept at a node depends on how that node falls among the sort order of its neighbors. For example, node 1 falls below all of its neighbors and thus receives the lowest quarter.

Node	Odd Cycle	Even Cycle	Node	Odd Cycle	Even Cycle
1	1 2 4 8 16	1 16 4 8 2	9	8 9 10 12 16	8 9 16 12 10
2	1 2 3 7 15	1 2 15 7 3	10	7 9 10 11 15	9 7 10 15 11
3	2 3 4 6 14	2 3 14 6 4	11	6 10 11 12 14	10 6 11 14 12
4	1 3 4 5 13	3 1 4 13 5	12	5 9 11 12 13	11 9 5 12 13
5	4 5 6 8 12	4 5 12 8 6	13	4 12 13 14 16	12 4 13 16 14
6	3 5 6 7 11	5 3 6 11 7	14	3 11 13 14 15	13 11 3 14 15
7	2 6 7 8 10	6 2 7 10 8	15	2 10 14 15 16	14 10 2 15 16
8	1 5 7 8 9	7 5 1 8 9	16	1 9 13 15 16	15 9 13 1 16

Simply put, we use a torus for full connectivity. So nodes along the "north" edge of the array which have no north neighbors are connected (conceptually) to nodes along the "south" edge and vice versa (transedge neighbors). Similarly, a node along the "east" edge are given nodes along the "west" edge as east neighbors and so forth. The odd cycle column of Table 1 summarizes all the nodes of a sixteen node network.

Thus, the use of the torus for four-connectedness provides full connectivity. The result is a modified shear-sort where both row and column connections are used with each round of sorting. Furthermore, ill-sorted data is quickly moved across the network via torus connections.

The "express pathway" is a conceptual map of the sorting network. Ideally, the operating systems supports express pathways, such as in an Intel Paragon system where we first implemented our algorithm. Where this environmental support is missing, the cost of these non-adjacent operations is higher. In those environments where networks have edges, HeteroSort has three strategies. The first is to still implement the conceptual torus at the higher transmission cost. The second is to re-configure itself to the reality of some nodes having only two or three physical neighbors. A third strategy is particularly useful in heterogeneous environments, where we employ a genetic algorithm to determine the optimal network by minimizing transmission costs.

2.1   Beowulf Clusters

The major portion of this Beowulf background section is abstracted from CESDIS material on their web page: http://www.beowulf.org/

The Beowulf class of computers and its architecture are appropriate to the times. The increasing presence of computers in offices, homes, and schools, has led to an abundance of mass produced cost effective components. The COTS (Commodity Off The Shelf) industry now provides fully assembled subsystems (microprocessors, motherboards, disks and network interface cards). The pressure of the mass market place has driven the prices down and reliability up. In addition, shareware, freeware, and open source development; in particular, the Linux operating system, the GNU compilers and programming tools and the MPI and PVM message passing libraries, provide hardware independent software.

In the taxonomy of parallel computers, Beowulf clusters fall somewhere between MPP (Massively Parallel Processors, like the Convex SPP, Cray T3D, Cray T3E, CM5, etc.) and NOWs (Networks of Workstations). The Beowulf project benefits from developments in both these classes of architecture. MPP's are typically larger and have a lower latency interconnect network than Beowulf clusters. Most programmers develop their programs in message passing style. Such programs can be readily ported to Beowulf clusters. Programming a NOW is usually an attempt to harvest unused cycles on an already installed base of workstations in a lab or on a campus. Programming in this environment requires algorithms that are extremely tolerant of load balancing problems and large communication latency. These programs will directly run on a Beowulf.

A Beowulf class cluster computer differs from a Network of Workstations in that the nodes in the cluster are dedicated to the cluster. This eases load balancing. Also, this allows the Beowulf software provide a global process ID, enabling signals to be sent from one node to another node of the system..

The challenge for our HeteroSort has been to adapt a conceptual mesh torus to a Beowulf cluster architecture. A trade in benefits has been the increased expense of nearest neighbor transactions. In the Beowulf, all transactions pass through a switch. This expense trades for the benefit that all other transactions do not have to traverse a network, passing through intervening nodes.

2.2   Optimization of HeteroSort

HeteroSort's distributed approach can provide an efficient control mechanism for a wide variety of algorithms. It also provides "reconfiguration-on-fault" fault tolerance when a node or network error occurs. HeteroSort automatically reconfigures to account for the failed node(s), and the distributed data is not lost. However, efficient operation requires that major sort axis nodes should reside on near neighbor network physical processors. This minimizes communication costs for efficient operation. And, for a heterogeneous topology, or a homogeneous topology made irregular by failed nodes, automatically achieving this near neighbor configuration for the sort nodes is difficult.

Figure 2 indicates a homogeneous mesh made irregular by two failed nodes. The numbers in the boxes (nodes) indicate the node's position in the sort order.

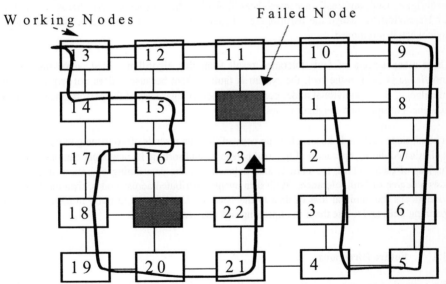

Sort Order is the number in each node

Fig. 2.   Sort order is the number of each node. A homogeneous mesh of 25 nodes made irregular by two failed nodes requires a new sort order for efficient performance. The numbers in the boxes (nodes) indicate the node's position in the new sort order. The lowest valued items in the sort will eventually reside on processor 1 and the highest valued items on processor 23. This new order optimizes near neighbor relations.

We assume that a message cannot be sent across a failed node.   To provide for online reconfiguration of the node sort order, we have developed an adaptive online sort order optimizer named the Scaleable Adaptive Load-balancing (SAL) Online Optimizer (SOO). SOO is performed by using a genetic algorithm which minimizes the total path length of the HeteroSort major sort axis, indicated on the figure by the line from one to 23. Note that other possible minimum path sort orders exist. Also note that for some topologies or failure patterns, strict near neighborness is not achievable. For these cases SOO defines the minimum path that includes store-and-forwards or traversals across other nodes. SOO can optimize given any combination of failed nodes and busses.

3   Fault Tolerance

The most important aspect of our algorithm is that it does not depend on a regular network topology (as, for example, a traditional shear sort does) because the torus can be superimposed on any physical architecture. This yields fault tolerance because our system can dynamically reconfigure itself, and easily accommodates "holes" in the connection. All that is required is for HeteroSort to change the partitioning schema in the data, and to stop sending data to a node when it is removed.

Three other aspects of fault tolerance result from this algorithm. First, since only local knowledge is used in the sort, the system is fault tolerant because it does not require global knowledge. Thus, individual nodes continue to operate regardless of the performance (or even existence) of other non-neighbor nodes.

Second, since each node keeps a backup copy of the data it sends off to its neighbors, if a node is eliminated during operation of the load balancing and sorting, its neighbors can make up for the loss of data. Third, the natural load balancing of the data during operation of the sorts adds a degree of fault tolerance. With data evenly distributed across nodes, then the loss of a node means the minimal loss of data to the system. The intent is to build minimum weight spanning trees and to use them in improving sort efficiency.

3.1   Future Directions

We currently have the concept of near (adjacent) neighbors and far neighbors (which exist with the implementation of the torus structure). This has implications for implementations on heterogeneous and distributed networks. Specifically, the far neighbors are metaphors for nodes on another processor in a distributed system. So, one component of the sort, partition, and send task could be that the data is partitioned not into equal subsets, but into subsets of a size proportional to the speed of the link to that node.

Furthermore, in heterogeneous architectures, the subset size could also be related to the speed of the corresponding neighbor node. Thus, future enhancements will include an applications kernel that will be resident on each node of the heterogeneous network. Upon startup, each kernel will negotiate with its near neighbor kernels to adjust the size of the exchange list (to be load balanced and sorted). The negotiated value will be a function of each node's own capacity in memory, processing, and its number of neighbors. Upon a fault, the kernels will re-negotiate the exchange files with the surviving near neighbors.

Acknowledgments

This research was partially supported by a grant from the Defense Nuclear Agency 93DNA-3. We gratefully acknowledge this support.

Reference

1. Gu, Qian Ping, and Jun Gu: Algorithms and Average Time Bounds of Sorting on a Mesh-Connected Computer. IEEE Transactions on Parallel and Distributed Systems. Vol 5, no 3. (March 1994) 308-315
2. Lin, Yen-Chun: On Balancing Sorting on a Linear Array. IEEE Transactions on Parallel and Distributed Systems. Vol 4, no 5. (May 1993) 566-571
3. Thompson, C.D., and H.T. Kung: Sorting on a Mesh-connected Parallel Computer. Communication of the ACM. Vol 20, no 40,. (April 1977) 263-271
4. Xu, Chengzhong and Francis Lau: Load Balancing in Parallel Computers: Theory and Practice. Kluwer Academic Publishers, Boston (1997)

7th Reconfigurable Architectures Workshop (RAW 2000)

Workshop Chair

Hossam ElGindy, University of New South Wales (Australia)

Steering Chair

Viktor K. Prasanna, University of Southern California at Los Angeles (USA)

Program Chair

Hartmut Schmeck, University of Karlsruhe (Germany)

Publicity Chair

Oliver Diessel, University of South Australia (Australia)

Programme Committee

Jeff Arnold, Independent Consultant (USA)
Peter Athanas, Virginia Tech (USA)
Gordon Brebner, Univ. of Edinburgh (Scotland)
Andre DeHon, Univ. of California at Berkeley (USA)
Carl Ebeling, Univ. of Washington (USA)
Hossam ElGindy, Univ. of New South Wales (Australia)
Reiner Hartenstein, Univ. of Kaiserslautern (Germany)
Brad Hutchings, Brigham Young Univ. (USA)
Mohammed Khalid, Quickturn Design Systems (USA)
Hyoung Joong Kim, Kangwon National Univ. (Korea)
Rainer Kress, Siemens AG (Germany)
Fabrizio Lombardi, Northeastern University (USA)
Wayne Luk, Imperial College (UK)
Patrick Lysaght, Univ. of Strathclyde (Scotland)
William H. Mangione-Smith, Univ. of California, Los Angeles (USA)
Margaret Marek-Sadowska, Univ. of California, Santa Barbara (USA)
William P. Marnane, Univ. College Cork (Ireland)
Margaret Martonosi, Princeton Univ. (USA)
John T. McHenry, National Security Agency (USA)

J. Rolim et al. (Eds.): IPDPS 2000 Workshops, LNCS 1800, pp. 870–872, 2000.

Alessandro Mei, Univ. of Trento (Italy)
Martin Middendorf, Univ. of Karlsruhe (Germany)
George Milne, Univ. of South Australia (Australia)
Koji Nakano, Nagoya Institute of Technology (Japan)
Stephan Olariu, Old Dominion Univ. (USA)
Bernard Pottier, Univ. Bretagne Occidentale (France)
Ralph Kohler, Air Force Research Laboratory (USA)
Mark Shand, Compaq Systems Research Center (USA)
Jerry L. Trahan, Louisiana State Univ. (USA)
Ramachandran Vaidyanathan, Louisiana State Univ. (USA)

Preface

The Reconfigurable Architecture Workshop series provides one of the major international forums for researchers to present ideas, results, and on-going research on both theoretical and industrial/practical advances in Reconfigurable Computing.

The main focus of this year's workshop is " *Run Time Reconfiguration - Foundations, Algorithms, Tools*": Technological advances in the field of fast reconfigurable devices have created new possibilities for the implementation and use of complex systems. Reconfiguration at runtime is one new dimension in computing that blurs the barriers between hardware and software components. Neither the existing processor architectures nor the hardware/software design tools which are available today can fully exploit the possibilities offered by this new computing paradigm. The potential of run time reconfiguration can be achieved through the appropriate combination of knowledge about foundations of dynamic reconfiguration, the various different models of reconfigurable computing, efficient algorithms, and the tools to support the design of run time reconfigurable systems and implementation of efficient algorithms. RAW 2000 provides the chance of creative interaction between these diciplines.

The programme consists of an invited talk by Steven Guccione (Xilinx), technical sessions of refereed papers on various aspects of Run Time Reconfiguration, and a panel discussion on *"The Future of Reconfigurable Computing"*. The 12 paper presentations were selected out of 27 submissions after a careful review process, every paper was reviewed by at least three members of the programme committee. We hope that the workshop will provide again the environment for productive interaction and exchange of ideas.

We would like to thank the organizing committee of IPDPS 2000 for the opportunity to organize this workshop, the authors for their contributed manuscripts, and the programme committee for their effort in assessing the 27 submissions to the workshop.

January 2000   Hartmut Schmeck

Programme of RAW 2000:

Invited Talk - Run-Time Reconfiguration at Xilinx
Steven A. Guccione

JRoute: A Run-Time Routing API for FPGA Hardware
Eric Keller

A Reconfigurable Content Addressable Memory
Steven A. Guccione, Delon Levi, Daniel Downs

ATLANTIS - A Hybrid FPGA/RISC Based Reconfigurable System
O. Brosch, J. Hesser, C. Hinkelbein, K. Kornmesser, T. Kuberka, A. Kugel, R. Männer, H. Singpiel, B. Vettermann

The Cellular Processor Architecture CEPRA-1X and its Configuration by CDL
Christian Hochberger, Rolf Hofmann, Klaus-Peter Volkmann, Stefan Waldschmidt

Loop Pipelining and Optimization for Run Time Reconfiguration
Kiran Bondalapati, Viktor K. Prasanna

Compiling Process Algebraic Descriptions into Reconfigurable Logic
Oliver Diessel, George Milne

Behavioral Partitioning with Synthesis for MultiFPGA Architectures under Interconnect, Area, and Latency Constraints
Preetham Lakshmikanthan, Sriram Govindarajan, Vinoo Srinivasan, Ranga Vemuri

Module Allocation for Dynamically Reconfigurable Systems
Xuejie Zhang, Kamwing Ng

Augmenting Modern Superscalar Architectures with Configurable Extended Instructions
Xianfeng Zhou, Margaret Martonosi

Complexity Bounds for Lookup Table Implementation of Factored Forms in FPGA Technology Mapping
Wenyi Feng, Fred J. Meyer, Fabrizio Lombardi

Optimization of Motion Estimator for RunTimeReconfiguration Implementation
Camel Tanougast, Yves Berviller, Serge Weber

ConstantTime Hough Transform On A 3D Reconfigurable Mesh Using Fewer Processors
Yi Pan

Run-Time Reconfiguration at Xilinx
(invited talk)

Steven A. Guccione

Xilinx Inc.
2100 Logic Drive
San Jose, CA 95124 (USA)
Steven.Guccione@xilinx.com

Abstract. Run-Time Reconfiguration (RTR) provides a powerful, but essentially untapped mode of operation for SRAM-based FPGAs. Research over the last decade has indicated that RTR can provide substantial benefits to system designers, both in terms of overall performance and in terms of design simplicity. While RTR holds great promise for many aspects of system design, it has only recently been considered for commercial application. Two factors seem to be converging to make RTR based system design viable. First, silicon process technology has advanced to the point where million gate FPGA devices are commonplace. This permits larger, more complex algorithms to be directly implemented in FGPAs. This alone has led to a quiet revolution in FPGA design. Today, coprocessing using large FPGA devices coupled to standard microprocessors is becoming commonplace, particularly in Digital Signal Processing (DSP) applications. The second factor is software. Until recently, there was literally no software support available for RTR. Existing ASIC-based design flows based on schematic capture and HDL did not provide the necessary mechanisms to allow implementation of RTR systems. Today, the JBits software tool suite from Xilinx provides the direct support for coprocessing and for RTR. The combination of hardware and software for RTR has already begun to show some impressive results on standard system design methodologies and algorithms. Future plans to enhance both FPGA architectures and tools such as JBits should result in a widening acceptance of this technology.

J. Rolim et al. (Eds.): IPDPS 2000 Workshops, LNCS 1800, pp. 873-873, 2000.
© Springer-Verlag Berlin Heidelberg 2000

JRoute: A Run-Time Routing API for FPGA Hardware

Eric Keller

Xilinx Inc.
2300 55th Street
Boulder, CO 80301
Eric.Keller@xilinx.com

Abstract. JRoute is a set of Java classes that provide an application programming interface (API) for routing of Xilinx FPGA devices. The interface allows various levels of control from connecting two routing resources to automated routing of a net with fanout. This API also handles ports, which are useful when designing object oriented macro circuits or cores. Each core can define its own ports, which can then be used in calls to the router. Debug support for circuits is also available. Finally, the routing API has an option to unroute a circuit. Built on JBits, the JRoute API provides access to routing resources in a Xilinx FPGA architecture. Currently the Virtex™ family is supported.

1   Introduction

JRoute is an API to route Xilinx FPGA devices. The API allows the user to have various levels of control. Using this API along with JBits, the user can create hierarchical and reusable designs through a library of cores. The JRoute API allows a user to perform run-time reconfiguration (RTR) of the routing resources by preserving the elements of RTR that are present in its underlying JBits[1] foundation. RTR systems are different from traditional design flows in that circuit customization and routing are performed at run-time. Since the placement of cores is one of the parameters that can be configured at run-time, the routing is not predefined. This means that auto routing can be very useful, especially when connecting ports from two different cores. Furthering the development of RTR computing designs, JRoute enables the implementation of nontrivial run-time parameterizable designs.

Since JRoute is an API, it allows users to build tools based on it. These can range from debugging tools to extensions that increase functionality. It is important to note that the JRoute API is independent of the algorithms used to implement it. The algorithms discussed in this paper are the initial implementations to further explain the API. This paper is meant to present features and benefits of the API, not the algorithms.

J. Rolim et al. (Eds.): IPDPS 2000 Workshops, LNCS 1800, pp. 874-881, 2000.

2   Overview of the Virtex Routing Architecture

The Virtex architecture has local, general purpose, and global routing resources. Local resources include direct connections between horizontally adjacent configurable logic blocks (CLBs) and feedback to inputs in the same logic block. Each provides high-speed connections bypassing the routing matrix, as seen in Figure 1. General-purpose routing resources include long lines, hex lines, and single lines. Each logic block connects to a general routing matrix (GRM). From the GRM, connections can be made to other GRMs along vertical and horizontal channels. There are 24 single length lines in each of the four directions. There are 96 hex length lines in each of the four directions that connect to a GRM six blocks away. Only 12 in each direction can be accessed by any given logic block. Some hexes are bi-directional, meaning they can be driven from either endpoint. There are also 12 long lines that run horizontal, or vertical for the length of the chip. Long lines are buffered, bi-directional lines that distribute the signals across the chip quickly. Long lines can be accessed every 6 blocks. Each type of general routing resource can only drive certain types of wires. Logic block outputs drive all length interconnects, longs can drive hexes only, hexes drive singles and other hexes, and singles drive logic block inputs, vertical long lines, and other singles. There are also global resources that distribute high-fanout signals with minimal skew. This includes four dedicated global nets with dedicated pins to distribute high-fanout clock signals. The array sizes for Virtex range from 16x24 CLBs to 64x96 CLBs. For a complete description of the Virtex architecture, see [3].

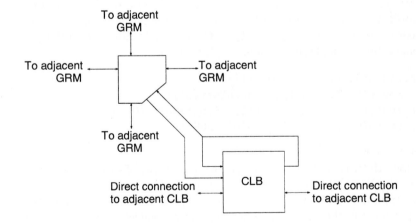

Fig. 1. Virtex routing architecture.

3   JRoute Features

The JRoute API makes routing easier to perform and helps in the development of large systems with reusable libraries. Unlike the standard Xilinx tools, JRoute can perform the routing at run-time. It also provides debugging facilities.

Before describing each of the calls, the architecture description file must first be described. There is a Java class in which all of the architecture information is held. In this class each wire is defined by a unique integer. Also in this class the possible template values are defined, along with which template value each wire can be classified under. A template value is defined as a value describing a direction and a resource type. For example, a template value of NORTH6 describes any hex wire in the north direction, a template value of NORTH1 describes any single wire in the north direction. Similar values are defined for each resource type in each direction that it can go. Also in this Java class is a description of each wire, including how long it is, its direction, which wires can drive it, and which wires it can drive.

3.1   Various Levels of Control

The JRoute API was designed with the goal of providing various levels of control. The calls range from turning on or off a single connection to auto-routing a bus connection.

route (int row, int col, int from_wire, int to_wire):
This call allows the user to make a single connection (i.e. the user decides the path). This can be useful in cases where there is a real time constraint on the amount of time spent configuring the device. However, the user must know what each wire connects to, and which wires are used. This call turns on the connection between from_wire and to_wire in CLB (row,col). The following example shows how to create a route connecting S1_YQ in CLB (5,7) to S0F3 in CLB (6,8) going through Out[1], SingleEast[5], and SingleNorth[0].

```
router.route(5, 7, S1_YQ, Out[1]);
router.route(5, 7, Out[1], SingleEast[5]);
router.route(5, 8, SingleWest[5], SingleNorth[0]);
router.route(6, 8, SingleSouth[0], S0F3);
```

route (Path path):
This call allows the user to define a path. A path is an array of specific resources, for example HexNorth[4], that are to be connected. The path also requires a starting location, defined by a row and column. The router turns on all of the connections defined in the path. The following example shows how to construct and route a path for the same route as in the previous example.

```
int[] p = {S1_YQ, Out[1], SingleEast[5], SingleNorth[0], S0F3};
Path path = new Path(5,7,p);
router.route(path);
```

route (Pin start_pin, int end_wire, Template template):
This call allows the user to specify a template and the router picks the wires. A template is defined as an array of template values, previously defined. The user does not have to know the wire connections and the resources in use. Using a template can also take advantage of regularity which would occur, for example, when connecting each output bit of an adder to an input of another core. The cost is longer execution time, and there is no guarantee that an unused path even exists. For this method a starting pin, defined as a wire at a specific row and column, needs to be defined. As well, the ending wire and the template to follow is specified. The router begins at the start wire, then goes through each wire that it drives, as defined in the architecture class, and checks first if the wire's template value matches the template value specified by the user. If so, then it checks to make sure the wire is not already in use. A recursive call is made with the new wire as the starting point and the first element of the template removed. The call would fail if there is no combination of resources that are available that follow the template. In this case a user action is required. The following example shows how to construct a template and route using it. The source and destination are the same as in the previous two examples. However, the specific resources may differ.

```
int[] t = {OUTMUX, EAST1, NORTH1, CLBIN};
Template template = new Template(t);
Pin src  = new Pin(5, 7, S1_YQ);
router.route(src, SOF3, template);
```

Finally, there are the auto-routing calls. This involves source to sink, source to many sinks, and a bus connection of many sources to an equal number of sinks.

route (EndPoint source, EndPoint sink):
This single source to single sink call allows for auto-routing of point to point connections. An EndPoint is either a Pin, defined by a row, column, and wire, or a Port, which is described in the next subsection. Many algorithms can be used to implement this call. One possibility is to use a maze router [4] [5]. Another possibility that would potentially be faster is to define a set of unique and predefined templates that would get from the source to the sink and try each one. If all of them fail then the router could fall back on a maze algorithm. The benefit of defining the template would be to reduce the search space. The following example shows how to define the end point (Pins) and connect them. The source and sink are the same as in the previous three examples, for the individual connections, path route, and template route. The template followed and the resources used may not necessarily be the same as it would with the other calls.

```
Pin src  = new Pin(5, 7, S1_YQ);
Pin sink = new Pin(6, 8, SOF3);
router.route(src, sink);
```

`route (EndPoint source, EndPoint[] sink):`
This is the method for a source to several sinks. It decides the best path for the entire collection of sinks. This call should be used instead of connecting each sink individually, since it minimizes the routing resources used. Each sink gets routed in order of increasing distance from the source. For each sink, the router attempts to reuse the previous paths as much as possible. Because it is not timing driven, this algorithm is suitable only for non-critical nets. For critical nets, however, the user would need to specify the routes at a lower level. In an RTR environment traditional routing algorithms require too much time. Currently long lines are not supported; only hexes and singles are used. Using long lines would improve the routing of nets with large bounding boxes.

`route (EndPoint[] source, EndPoint[] sink):`
This is a call for bus connections. In a data flow design, the outputs of one stage go to the inputs of the next stage. As a convenience, the user does not need to write a Java loop to connect each one. If used along with cores, this call can be very useful when connecting ports to other ports. For example, the output ports of a multiplier core could be connected to the input ports of an adder core. Using the bus method, the user would not need to connect each bit of the bus.

Each of the auto-routing calls described above use greedy routing algorithms. This was chosen because of the designs that are targeted. Structured and regular designs often have simple and regular routing. Also, in an RTR environment, global routing followed by detailed routing would not be efficient. Furthermore, RTR designs will be changing during the execution. This leads to an undefined definition of what global routing would mean.

3.2   Support for Cores

Another goal when designing the JRoute API was to support a hierarchical and reusable library of run-time parameterizable cores. Before JRoute the user of a core needed to know each pin (an input or output to a logic resource) that needs to be connected. With JRoute, a core can define ports. Ports are virtual pins that provide input or output points to the core. The core can use the ports in calls to the router, instead of specifying the specific pin. To the user there is no distinction between a physical pin, defined as location and wire, and a logical port as they are both derived from the EndPoint class. The core can define a connection from internal pins to ports. It can also specify connections from ports of internal cores to its own ports.

The router knows about ports and when one is encountered, it translates it to the corresponding list of pins. When a port gets routed, the source and sinks connected to the port are saved. This information is useful for the unrouter and the debugging features, which are described later.

There are routing guidelines that need to be followed when designing a core. First, each port needs to be in a group. For example, if there is an adder with an n bit output, each bit is defined as a port and put into the same group.

The group can be of any size greater than zero. Second, the router needs to be called for each port defined. This call defines the connections to the port from pins internal to the core. Finally, a getPorts( ) method must be defined for each group, which returns the array of Ports associated with that group.

3.3   Unrouter

Run-time reconfiguration requires an unrouter. There may be situations when a route is no longer needed, or the net endpoints change. Unrouting the nets free up resources. A core may be replaced with the same type of core having different parameters. In this case the user can unroute the core then replace it. The port connections are removed, but are remembered. If the ports are reused, then they will be automatically connected to the new core. For example, consider a constant multiplier. The system connects it to the circuit and later requires a new constant. The core can be removed, unrouted, and replaced with a new constant multiplier without having to specify connections again. Core relocation is handled in a similar way.

`unroute (EndPoint source);`
An unrouter can work in either the forward or reverse direction. In the forward direction a source pin is specified. The unrouter then follows each of the wires the pin drives and turns it off. This continues until all of the sinks are found.

`reverseUnroute (EndPoint sink);`
In the reverse direction a sink pin is specified. The entire net, starting from the source, is not removed. Only the branch that leads to the specified pin is turned off, and freed up for reuse. The unrouter starts at the sink pin and works backwards, turning off wires along the way, until it comes to a point where a wire is driving multiple wires. It stops there because only the branch to the given sink is to be unrouted.

3.4   Avoiding Contention

`isOn (int row, int col, int wire);`
This call checks to see if the wire in CLB (row,col) is currently in use. The Virtex architecture has bi-directional routing resources. This means that the track can be driven at either end, leading to the possibility of contention. The router makes sure that this situation does not occur, and therefore protects the device. An exception is thrown in cases where the user tries to make connections that create contention. In the auto-routing calls, the router checks to see if a wire is already used, which avoids contention.

3.5   Debugging Features

`trace (EndPoint source);`
A JRoute call traces a source to all of its sinks. The entire net is returned for

the trace. Debugging tools, such as BoardScope [2], can use this to view each sink.

```
reverseTrace (EndPoint sink);
```
A sink is traced back to its source. Only the net that leads to the sink is returned.

4   JRoute versus Routing with JBits

JRoute uses the JBits low-level interface to Xilinx FPGA configuration bit-streams, which only provides manual routing. The JRoute API extensions provide automated routing support, while not prohibiting JBits calls. JRoute facilitates the use of run-time relocatable and parameterizable cores.

Using cores and the JRoute API, a user can create designs without knowledge of the routing architecture by using port to port connections. The user only really needs a small set of architecture-specific cores to start with. For example, a counter can be made from a constant adder with the output fed back to one input ports and the other input set to a value of one.

5   Portability

Currently, JRoute only supports Virtex devices. However, it can be extended to support future Xilinx architectures. The API would not need to change. However, the architecture description class would need to be created for the new architecture. The algorithms as presented in this paper have some architecture dependencies. For example, when routing a single source to a single sink, defining the set of predefined templates is architecture dependent. However, algorithms can be designed that have no architecture dependencies, and could be used with new architectures. These algorithms would use the architecture class to choose wires, check their lengths, and check the connectivity. The path-based router and template-based router have no knowledge of the architecture outside of what the architecture class provides.

6   Future Work

Virtex features such as IOBs and Block RAM will be supported in a future release of JRoute. Also, skew minimization will be addressed. The use of long lines to improve the routing of certain nets will be examined. Finally, different algorithms are being investigated such as [6].

7   Conclusions

JRoute is a powerful abstraction of the Xilinx FPGA routing resources. A routing API facilitates the design of object oriented circuits that are configurable at run-time. There are many options that are made available by JRoute such as connecting two points for which the location is determined dynamically.

Hierarchical core-based design using JRoute permits easier management of design complexity than using only JBits. JRoute automates much of the routing and reduces the need to understand the routing architecture of the device. JRoute also provides support for large designs by allowing cores to define ports. RTR features include the unrouter, which allows cores to be removed or replaced at run-time without having to reconfigure the entire design. Auto-routing calls allow connections to be specified, even if the placement is not known until run-time.

Acknowledgements

Thanks to Cameron Patterson for his guidance and help in understanding routing algorithms. This work was supported by DARPA in the Adaptive Computing Systems (ACS) program under contract DABT63-99-3-0004.

References

1. S. A. Guccione and D. Levi, "XBI: A Java-based interface to FPGA hardware," *Configurable Computing Technology and its uses in High Performance Computing, DSP and Systems Engineering*, Proc. SPIE Photonics East, J. Schewel (Ed.), SPIE - The International Society for Optical Engineering, Bellingham, WA, November 1998.
2. D. Levi and S. A. Guccione, "BoardScope: A Debug Tool for Reconfigurable Systems," *Configurable Computing Technology and its uses in High Performance Computing, DSP and Systems Engineering*, Proc. SPIE Photonics East, J. Schewel (Ed.), SPIE - The International Society for Optical Engineering, Bellingham, WA, November 1998.
3. Xilinx, Inc., *The Programmable Logic Data Book*, 1999.
4. Naveed A Sherwani, *Algorithms for VLSI Physical Design Automation*, Kluwer Academic Publishers, Norwell, Massachusetts, 1993.
5. Stephen D. Brown, Robert J. Francis, Jonathan Rose and Zvonko G. Vranesic, *Field-Programmable Gate Arrays*, Kluwer Academic Publishers, Norwell, Massachusetts, 1992.
6. J. Swartz, V. Betz and J. Rose, "A Fast Routability-Driven Router for FPGAs," *ACM/SIGDA International Symposium on Field Programmable Gate Arrays*, Monterey, CA, 1998.

A Reconfigurable Content Addressable Memory

Steven A. Guccione, Delon Levi and Daniel Downs

Xilinx Inc.
2100 Logic Drive
San Jose, CA 95124 (USA)
Steven.Guccione@xilinx.com
Delon.Levi@xilinx.com
Daniel.Downs@xilinx.com

Abstract. Content Addressable Memories or *CAMs* are popular parallel matching circuits. They provide the capability, in hardware, to search a table of data for a matching entry. This functionality is a high performance alternative to popular software-based searching schemes. CAMs are typically found in embedded circuitry where fast matching is essential. This paper presents a novel approach to CAM implementation using FPGAs and run-time reconfiguration. This approach produces CAM circuits that are smaller, faster and more flexible than traditional approaches.

1 Introduction

Content Addressable Memories or *CAMs* are a class of parallel pattern matching circuits. In one mode, these circuits operate like standard memory circuits and may be used to store binary data. Unlike standard memory circuits, however, a powerful *match mode* is also available. This match mode permits all of the data in the CAM device to be searched in parallel.

While CAM hardware has been available for decades, its use has typically been in niche applications, embedded in custom designs. Perhaps the most popular application has been in cache controllers for central processing units. Here CAMs are often used to search cache tags in parallel to determine if a cache "hit" or "miss" has occurred. Clearly in this application performance is crucial and parallel search hardware such as a CAM can be used to good effect.

A second and more recent use of CAM hardware is in the networking area [3]. As data packets arrive into a network router, processing of these packets typically depends on the network destination address of the packet. Because of the large number of potential addresses, and the increasing performance demands, CAMs are beginning to become popular in processing network address information.

2 A Standard CAM Implementation

CAM circuits are similar in structure to traditional Random Access Memory (RAM) circuits, in that data may be written to and read from the device [5]. In

J. Rolim et al. (Eds.): IPDPS 2000 Workshops, LNCS 1800, pp. 882–889, 2000.

addition to functioning as a standard memory device, CAMs have an additional parallel search or *match* mode. The entire memory array can be searched in parallel using hardware. In this match mode, each memory cell in the array is accessed in parallel and compared to some value. If this value is found in any of the memory locations, a *match* signal is generated.

In some implementations, all that is significant is that a match for the data is found. In other cases, it is desirable to know exactly where in the memory address space this data was located. Rather than producing a simple "match" signal, some CAM implementations also supply the address of the matching data. In some sense, this provides a functionality opposite of a standard RAM. In a standard RAM, addresses are supplied to hardware and data at that address is returned. In a CAM, *data* is presented to the hardware and an *address* returned.

At a lower level, the actual transistor implementation of a CAM circuit is very similar to a standard static RAM. Figure 1 shows transistor level diagrams of both CMOS RAM and CAM circuits. The circuits are almost identical, except for the addition of the *match* transistors to provide the parallel search capability.

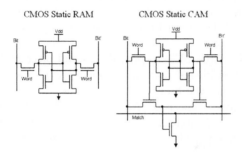

Fig. 1. RAM versus CAM transistor level circuits.

In a CMOS static RAM circuit, as well as in the CAM cell, data is accessed via the *BIT* lines and the cells selected via the *WORD* lines. In the CAM cell, however, the *match* mode is somewhat different. Inverted data is placed on the *BIT* lines. If any cell contains data which does not match, the *MATCH* line is pulled low, indicating that no match has occurred in the array.

Clearly this transistor level implementation is efficient and may be used to produce CAM circuits which are nearly as dense as comparable static RAM circuits. Unfortunately, such transistor level circuits can not be implemented using standard programmable logic devices.

3   An FPGA CAM Implementation

Of course, a content addressable memory is just a digital circuit, and as such may be implemented in an FPGA. The general approach is to provide an array of registers to hold the data, and then use some collection of comparators to see if a match has occurred.

While this is a viable solution, it suffers from the same sort of inefficiencies that plague FPGA-based RAM implementations. Like RAM, the CAM is efficiently implemented at the transistor level. Using gate level logic, particularly programmable or reconfigurable logic, often results in a substantial penalty, primarily in size.

Because the FPGA CAM implementation relies on flip-flops as the data storage elements, the size of the circuit is restricted by the number of flip flops in the device. While this is adequate for smaller CAM designs, larger CAMs quickly deplete the resources of even the largest available FPGA.

4   The Reconfigurable Content Addressable Memory (RCAM)

The *Reconfigurable Content Addressable Memory* or *RCAM* makes use of run-time reconfiguration to efficiently implement a CAM circuit. Rather than using the FPGA flip-flops to store the data to be matched, the RCAM uses the FPGA *Look Up Tables* or *LUTs*. Using LUTs rather than flip-flops results in a smaller, faster CAM.

The approach uses the LUT to provide a small piece of CAM functionality. In Figure 2, a LUT is loaded with data which provides a "match 5" functionality. That is, whenever the binary encoded value "5" is sent to the four LUT inputs, a *match* signal is generated. Note that using a LUT to implement CAM functionality, or any functionality for that matter, is not unique. An N-input LUT can implement any arbitrary function of N inputs, including a CAM.

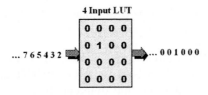

Fig. 2. Using a LUT to match 5.

Because a LUT can be used to implement any function of N variables, it is also possible to provide more flexible matching schemes than the simple match described in the circuit in Figure 2. In Figure 3, the LUT is loaded with values which produce a match on any value but binary "4". This circuit demonstrates the ability to embed a *mask* in the configuration of a LUT, permitting arbitrary disjoint sets of values to be matched, within the LUT. This function is important in many matching applications, particularly networking.

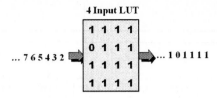

Fig. 3. Using a LUT to match all inputs except 4.

This approach can be used to provide matching circuits such as *match all* or *match none* or any combination of possible LUT values. Note again, that this arbitrary masking only applies to a single LUT. When combining LUTs to make larger CAMs, the ability to perform such masking becomes more restricted.

While using LUTs to perform matching is a powerful approach, it is somewhat limited when used with traditional design tools. With schematics and HDLs, the LUT contents may be specified, albeit with some difficulty. And once specified, modifying these LUTs is difficult or impossible. However, modification of FPGA circuitry at run-time is possible using a run-time reconfiguration tool such as *JBits* [1]. *JBits* permits LUT values, as well as other parts of the FPGA circuit, to be modified arbitrarily at run time and in-system. An *Application Program Interface (API)* into the FPGA configuration permits LUTs, for instance, to be modified with a single function call. This, combined with the partial reconfiguration capabilities of new FPGA devices such as Virtex (tm) permit the LUTs used to build the RCAM to be easily modified under software control, without disturbing the rest of the circuit.

Finally, using run-time reconfiguration software such as *JBits*, RCAM circuits may be dynamically sized, even at run-time. This opens the possibility of not only changing the contents of the RCAM during operation, but actually changing the size and shape of the RCAM circuit itself. This results in a situation analogous to dynamic memory allocation in RAM. It is possible to "allocate" and "free" CAM resources as needed by the application.

5   An RCAM Example

One currently popular use for CAMs is in networking. Here data must be processed under demanding real-time constraints. As packets arrive, their routing information must be processed. In particular, destination addresses, typically in the form of 32-bit *Internet Protocol (IP)* addresses must be classified. This typically involves some type of search.

Current software based approaches rely on standard search schemes such as hashing. While effective, this approach requires a powerful processor to keep up with the real-time demands of the network. Offloading the computationally demanding matching portion of the algorithms to external hardware permits less powerful processors to be used in the system. This results in savings not only

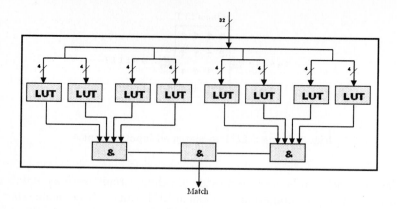

Fig. 4. Matching a 32-bit IP header.

in the cost of the processor itself, but in other areas such as power consumption and overall system cost.

In addition, an external CAM provides networking hardware with the ability to achieve packet processing in essentially constant time. Provided all elements to be matched fit in the CAM circuit, the time taken to match is independent of the number of items being matched. This provides not only good scalability properties, but also permits better real-time analysis. Other software based matching schemes such as hashing are data-dependent and may not meet real-time constraints depending on complex interactions between the hashing algortihm and the data being processed. CAMs suffer no such limitations and permit easy analysis and verification.

Figure 4 shows an example of an IP Match circuit constructed using the RCAM approach. Note that this example assumes a basic 4-input LUT structure for simplicity. Other optimizations, including using special-purpose hardware such as carry chains are possible and may result in substantial circuit area savings and clock speed increases.

This circuit requires one LUT input per matched bit. In the case of a 32-bit IP address, this circuit requires 8 LUTs to provide the matching, and three additional 4-input LUTs to provide the ANDing for the MATCH signal. An array of this basic 32-bit matching block may be replicated in an array to produce the CAM circuit. Again, note that other non-LUT implementations for generating the MATCH circuit are possible.

Since the LUTs can be used to mask the matching data, it is possible to put in "match all" conditions by setting the LUTs to all ones. Other more complicated masking is possible, but typically only using groups of four inputs. While this does not provide for the most general case, it appears to cover the popular modes of matching.

6   System Issues

The use of run-time reconfiguration to construct, program and reprogram the RCAM results in some significant overall system savings. In general, both the hardware and the software are greatly simplified.

Most of the savings accrue from being able to directly reconfigure the LUTs, rather than having to write them directly as in standard RAM circuits. Reconfiguration rather than direct access to the stored CAM data first eliminates all of the read / write access circuitry. This includes the decode logic to decode each address, the wiring necessary to broadcast these addresses, the data busses for reading and writing the data, and the IOBs used to communicate with external hardware.

It should be pointed out that this interface portion of the circuitry is substantial, both its size and complexity. Busses typically consume tri-state lines, which are often scarce. Depending on the addressing scheme, tens of IOBs will necessarily be consumed. These also tend to be valuable resources. The address decoders are also somewhat problematic circuits and often require special purpose logic to be implemented efficiently. In addition, the bus interface is typically the most timing sensitive portion of the circuit and requires careful design and simulation. This is eliminated with the use of run-time reconfiguration.

Finally, the system software is simplified. In a standard bus interface approach, device drivers and libraries must be written, debugged and maintained to access the CAM. And when the system software or processor changes, this software must be ported to the new platform. With the RCAM, all interfacing is performed through the existing configuration port, at no additional overhead.

The cost of using the configuration port rather than direct hardware access is primarily one of setup speed. Direct writes can typically be done in some small number of system cycles. Reconfiguration of the RCAM to update table entries may take substantially longer, depending on the implementation. Partial reconfiguration in devices such as Virtex permit changes to be made more rapidly than in older bulk configuration device, but the speed may be orders of magnitude slower than direct hardware approaches. Clearly the RCAM approach favors applications with slowly changing data sets. Fortunately, many applications appear to fit into this category.

7   Comparison to Other Approaches

While CAM technology has been in widespread use for decades, there has been little interest in producing commercial CAM devices. This recent interest in CAMs, driven primarily by the high-performance networking market, has resulted in commercially available CAM devices. Music Semiconductor [4] and Net Logic [2] are two companies which provide CAM devices tailored specifically for the networking market.

In addition, at least one FPGA manufacturer, Altera, has begun to embed CAM hardware into their Apex(tm) devices. While this circuitry is embedded in an FPGA, it is special purpose and not part of the general configurable fabric. It is included here for comparison, but it should be pointed out that special purpose hardware is readily inserted into FPGAs. The cost here is inflexibility. The special purpose hardware must be used for a specific circuit, at a specific physical location, or not used at all. In this sense, this embedded CAM has more in common with custom solutions than programmable solution. But the specifications are included here for comparison.

CAM (Virtex V1000)	768 x 32	384 x 64
RCAM (Virtex V1000)	3K x 32	1K x 64
Quality Semiconductor	1K x 64	2K x 64
Net Logic	16K x 64	8K x 128
Music Semiconductor	2K/4K/6K x 32	2K/4K/6K x 64
Altera APEX	1K-8K x 32	500-4K x 64

Fig. 5. Some commercially available CAM devices.

Figure 5 gives some sizes for current commercially available devices. While these are custom CAM implementations and can expected to be denser than FPGA implementations, the RCAM sizes are within the general range of that available from custom implementations. In addition, the RCAM circuits are more flexible and may be placed at any location within the FPGA and may be integrated with other logic in the design. Finally, the RCAM approach is approximately 3-4 times denser than attempting to implement a CAM using an FPGA and traditional design approaches. Optimizations using logic such as the Virtex carry chain also indicate improvements of an additional 40%.

8   Associative Processing

Today, advances in circuit technology permit large CAM circuits to be built. However, uses for CAM circuits are not necessarily limited to niche applications like cache controllers or network routers. Any application which relys on the searching of data can benefit from a CAM-based approach. A short list of some potential application areas that can benefit from fast matching are Artificial Intelligence, Database Search, Computer Aided Design, Graphics Acceleration and Computer Vision.

Much of the work in using parallel matching hardware to accelerate algorithms was carried out in the 1960s and 1970s, when several large parallel matching machines were constructed. An excellent survey of so-called *Associative Processors* can be found in Yau and Fung [7].

With the rapid growth both in size and speed of traditional processors in the intervening years, much of the interest in CAMs has faded. However, as real-time constraints in areas such as networking become impossible to meet with traditional processors, solutions such as CAM-based parallel search will almost certainly become more prevalent.

In addition, the use of parallel matching hardware in the form of CAMs can provide another more practical benefit. For many applications, the use of CAM-based parallel search can offload much of the work done by the system processor. This should permit smaller, cheaper and lower power processors to be used in embedded applications which can make use of CAM-based parallel search.

9   Conclusions

The *RCAM* is a flexible, cost-effective alternative to existing CAMs. By using FPGA technology and run-time reconfiguration, fast, dense CAM circuits can be easily constructed, even at run-time. In addition, the size of the *RCAM* may be tailored to a particular hardware design, or even temporary changes in the system. This flexibility is not available in other CAM solutions.

In addition, the *RCAM* need not be a stand-alone implementation. Because the *RCAM* is entire a software solution using state of the art FPGA hardware, it is quite easy to embed *RCAM* functionality in larger FPGA designs.

Finally, we believe that existing applications, primarily in the field of network routing, are just the beginning of *RCAM* usage. Once other applications realize that simple, fast, flexible parallel matching is available, it is likely that other applications and algorithms will be accelerated using this approach.

10   Acknowledgements

Thanks to Kjell Torkellesson and Mario Dugandzic for discussions on networking. And thanks especially to Paul Hardy for early RCAM discussions.

References

1. Steven A. Guccione and Delon Levi. XBI: A Java-based interface to FPGA hardware. In John Schewel, editor, *Configurable Computing Technology and its use in High Performance Computing, DSP and Systems Engineering, Proc. SPIE Photonics East*, pages 97–102, Bellingham, WA, November 1998. SPIE – The International Society for Optical Engineering.
2. Net Logic Microsystems. World Wide Web page http://www.netlogicmicro.com/, 1999.
3. R. Neale. Is content addressable memory (CAM) the key to network success? *Electronic Engineering*, 71(865):9–12, February 1999.
4. Music Semiconductor. World Wide Web page http://www.music-ic.com/, 1999.
5. Neil Weste and Kamram Eshraghian. *Principles of CMOS VLSI Design*. Addison-Wesley Publishing Company, 1985.
6. Xilinx, Inc. *The Programmable Logic Data Book*, 1996.
7. S. S. Yau and H. S. Fung. Associative processor architecture – a survey. *Computing Surveys*, 9(1):3–27, March 1977.

ATLANTIS – A Hybrid FPGA/RISC Based Re-configurable System

O. Brosch, J. Hesser, C. Hinkelbein, K. Kornmesser, T. Kuberka, A. Kugel, R. Männer, H. Singpiel, B. Vettermann

Lehrstuhl für Informatik V, Universität Mannheim, D-68131 Mannheim, Germany
{brosch, hinkelbein, kornmesser, kuberka, kugel, maenner, singpiel}@ti.uni-mannheim.de, jhesser@rumms.uni-mannheim.de, b.vettermann@fh-mannheim.de

Abstract. ATLANTIS is the result of 8 years of experience with large stand-alone and smaller PCI based FPGA processors. Dedicated FPGA boards for computing and I/O plus a private backplane for a data rate of up to 1 GB/s support flexibility and scalability. FPGAs with more than 100k gates and 400 I/O pins per chip are used. CompactPCI provides the basic communication mechanism. Current real-time applications include pattern recognition tasks in high energy physics, 2D image processing, volume rendering, and n-body calculations in astronomy. First measurements and estimations show an acceleration up to a factor of 25 compared to a PC workstation, or commercial volume rendering hardware, respectively. Our CHDL, an object-oriented development environment is used for application programming.

1 Introduction

8 years of experience with FPGA based computing machines show that this new class of computers is an ideal concept for constructing special-purpose processors. As processing unit, I/O unit and bus system are implemented in separate modules, this kind of system provides scalability in computing power as well as I/O bandwidth.

Enable-1 [1] was the first FPGA processor developed at Mannheim University in 1994, tailored for a specific pattern recognition task. More general machines were introduced at about the same time, e.g. DecPeRLe-1 [2] or Splash-2 [3]. Enable-1 was followed by a general-purpose FPGA processor in 1996, the Enable++ [4] system. In addition to the large scale Enable++ system a small PCI based FPGA coprocessor – microEnable [5] – was developed in late 1997. It turned out that the simplicity together with the tight host-coupling of the smaller system was a significant improvement compared to Enable++.

The new FPGA processor ATLANTIS combines advantages of its predecessors Enable-1, Enable++, microEnable and others, and introduces several new features. The first is the ability to combine FPGA and RISC performance. A unique feature is the scalability and the fast data exchange between the different modules due to the CompactPCI and private bus backplane system. Another highlight is the configurable memory system which complements the flexibility of the FPGAs. We use CHDL, an

J. Rolim et al. (Eds.): IPDPS 2000 Workshops, LNCS 1800, pp. 890–897, 2000.

unique object-oriented software tool-set that was at developed our institute, to create and simulate hybrid applications.

2 ATLANTIS System Architecture

A well-tried means to adjust a hybrid system to different applications is modularity. ATLANTIS implements modularity on different levels. First of all there are the main entities host CPU and FPGA processor which allow to partition an application into modules tailored for either target. Next the architecture of the FPGA processor uses one board-type (ACB) to implement mainly computing tasks and another board-type (AIB) to implement mainly I/O oriented tasks. A CompactPCI based backplane (AAB) as interconnect system provides scalability and supports an arbitrary mix of the two board-types, thus providing a high-speed interconnect. Finally modularity is used on the sub-board level by allowing different memory types or different I/O interfaces per board type.

Only FPGA devices with a high I/O pin-count and a complexity in the 100k gate-range are of interest for the ATLANTIS project. Two additional features are important either for our concept or for some applications: support for read-back/test and asynchronous dual ported memory (DP-RAM). In particular the partial reconfiguration is of great interest for co-processing applications involving hardware task switches. These features and a relatively low price guided the decision to use the Lucent ORCA 3T125 in the ATLANTIS system. The latest Xilinx family – the VIRTEX series – is also a good choice but was not available on the market at the time the ACB was designed. However, the AIB carries two VIRTEX XCV600 chips.

The ACB and the AIB both use a PLX9080 as PCI interface. This chip is compatible to the one used with the microEnable FPGA coprocessor. Furthermore the entire on-board support logic – like FPGA configuration and clock control – which is implemented in a large CPLD, is derived from microEnable. This high degree of compatibility ensures that virtually all basic software (WinNT driver, test tools, etc.) are immediately available for ATLANTIS.

Clock generation and distribution is an important issue for large FPGA processors. The basic approach in Atlantis is to provide a central clock from the AAB. Additionally the I/O ports of all FPGAs on either ACB and AIB have their individual clock sources. Finally each ACB and AIB provides a local clock which can be used if the main AAB clock is not available or if the application requires an additional clock. All clocks are programmable in the range of a few MHz up to at least 80 MHz. Programming is done under software control from the CPU module.

2.1 ATLANTIS Computing Board (ACB)

The core of the main processing unit of the ATLANTIS system consists of a 2*2 FPGA matrix. Assuming an average gate count of approximately 186k per chip for the ORCA 3T125 sums up to 744k FPGA gates. Each FPGA has 4 different ports:
- 2 ports @ 72 lines to a neighboring FPGA each in vertical and horizontal direction
- 1 logical I/O port @ 72 lines and

- 1 memory interconnect port @ 206 lines.

Theses 4 ports use a total amount of 422 I/O signals per FPGA. The 72 lines of FPGA interconnect provide for high bandwidth as well as multi-channel communication between chips. The memory interconnect port is built from 2 high-density 124 pin mezzanine connectors per FPGA. Depending on the application, memory modules with different architectures can be used to optimize system performance. E.g. the HEP TRT trigger (see below) will employ memory modules organized as a single bank of 512k * 176 bit of synchronous SRAM per module, leading to a total of 44 MB per ACB. The 3D-rendering algorithm will use a single module of triple width with 512 MB of SDRAM organized in 8 simultaneously accessible banks. A more generalized module – also used for 2D image processing – will take 9 MB of synchronous SRAM organized in 2 banks of 512k * 72 bits.

The I/O port serves different tasks on the 4 FPGAs, depending on the physical connection of the respective chip:

- One FPGA is connected to the PLX9080 PCI interface chip thus providing the host-I/O functionality.
- Two FPGAs are connected to the private backplane bus.
- One FPGA is attached to two parallel LVDS connectors for external I/O.

The connectors can be used to attach I/O modules, e.g. S-Link[1], to set up a down-scaled or test system without the need to add AAB and AIB modules. The 2 backplane ports support high-speed I/O of 1 GB/s @ 66 MHz, 2*64 bits. The host-interface via PCI is compatible to the one used with microEnable, allowing 125 MB/s max. data rate.

2.2 ATLANTIS I/O Board (AIB)

The task of the ATLANTIS I/O units is to connect the ATLANTIS system to its real-world environments via the private backplane bus. To provide a maximum flexibility in connecting to external data sources or destinations a modular design of the I/O boards was selected. Depending on the standard CompactPCI card size every AIB is able to carry up to four mezzanine I/O daughter-boards.

Two Xilinx VIRTEX XCV600 FPGAs control the four I/O ports. Interfacing to the AAB and to the local PCI bridge is done in the same fashion as on the ACB. The default capacity of any of the four channels is 32 + 4 data bits @ 66MHz (or 264 MB/s ignoring the 4 extra bits). Thus the four I/O channels provide the same bandwidth as the 2 backplane ports: 1GB/s. To provide a sustained and high I/O bandwidth even at small block sizes buffering of data can be done in two stages (numbers per I/O channel):

- A 32k * 36 FIFO-style buffer connected directly to the I/O port, implemented with dual-ported memory.
- A 1M * 36 general purpose buffer implemented with synchronous SRAM.

The fact that both FPGAs are connected to the PLX local bus provides a communication means in case channel synchronization, loop-back or the like is needed.

---

[1] S-Link is a FIFO-like CERN internal standard for point-to-point links.

2.3 ATLANTIS Active Backplane (AAB)

ACBs and AIBs share the same I/O-circuit with 160 signal lines. Connections between boards are done using the private bus system of the AAB. The default configuration of the I/O lines will be 4 channels of 32bit plus control, however any granularity from 16 channels of a single byte to 2 channels of 64 bit might be useful.

Different backplanes can be used in order to scale the ATLANTIS system to the respective application. A simple pipelined, passive, i.e. not configurable, backplane is currently used for system and performance tests.

The total bandwidth is 1 GB/s per slot. For example configuring the backplane for two independent pairs of ACBs and AIBs, an integrated bandwidth of 2 GB/s will result for a single ATLANTIS system. Like all other boards, the backplane is controlled by the host CPU via the PCI bus.

2.4 Host CPU

The host computer to be used with ATLANTIS is an industrial version of a standard x86 PC – a CompactPCI computer – that plugs into one of the AAB slots. This industrial computer is equipped with a mobile Intel Pentium-200 MMX or Celeron-450 processor and thus 100% compatible to a standard PC desktop workstation. All standard operating systems can be used, in particular Windows NT and Linux, without the need to adapt drivers or I/O handlers, etc. The compatibility at the device driver level of ATLANTIS with the small scale FPGA processor microEnable allows a quick start using the tools already available.

The CPU module allows to have the complete FPGA development tool-set be run on the target system, as well as the application itself. The ACB and AIB boards act as coprocessors, accelerating time and resource consuming parts of an application, and providing high I/O bandwidth. Moreover, the CPU is needed for control, when task switching and re-configuration of FPGAs is desired. Additionally, high precision floating point operations that are too much resource consuming on FPGAs, may be carried out in the CPU.

2.5 CHDL Development Environment

CHDL (C++ based Hardware Description Language) was designed to support simulation of FPGA coprocessors. The use of commercial VHDL products to simulate FPGA coprocessors shows some insufficiencies:
1. A test bench must be implemented for emulating the FPGA environment using VHDL while the application operating the FPGA is mostly written in C/C++.
2. The test bench has to emulate the behavior of the microprocessor system exactly, including bus system and DMA controllers at the level of bus signals.
3. Implementing the test bench is redundant work because the application already contains the whole algorithm needed for simulation.
CHDL provides a hardware description based on C++ classes for entering structural designs and state machine definitions. A CHDL design description is a traditional C++ program linked to a class library. This enables the developer to implement complex high level software which generates the structural CHDL design automatically.

The developer uses the original application to simulate the designs. No traditional hardware oriented test benches are needed. One single language, C++, is sufficient to manage the whole development process. In both the application and the hardware description the features of this powerful programming language can be used.

More details can be found in [6].

3 Applications

FPGA processors have shown to provide superior performance in a broad range of fields, like encryption, DNA sequencing, image processing, rapid prototyping etc. Very good surveys can be found in [3] and [7]. We are in particular interested in hybrid CPU/FPGA systems for:

- acceleration of computing intensive pattern recognition tasks in High Energy Physics (HEP) and Heavy Ion Physics,
- subsystems for high-speed and high-frequency I/O in HEP,
- 2-dimensional industrial image processing,
- 3-dimensional medical image visualization and
- acceleration of multi-particle interaction (e.g. N-Body [8], SPH) calculations in astronomy.

3.1 High Energy Physics

In the field of HEP many FPGA algorithms have been implemented at our institute during the past 5 years. Results show speedup rates in the range from 10 to 1,000[2] compared to workstation implementations [9]. The most recent HEP pattern matching algorithm tries to find straight or curved tracks in a 2-dimensional input image delivered by a transition radiation tracking detector (TRT) with a repetition rate of up to 100 kHz. The size of the detector image is 80,000 pixels. The number of patterns varies from 240 to more than 2,400 depending on the operating frequency. The working principle of the algorithm is as follows:

- Predefined patterns are stored in a large look-up table (LUT) with every data bit representing one pattern.
- Each pixel in the input image contributes to a number of patterns, defined by the content of the LUT.
- For every pattern a counter increments if its corresponding data bit is set. The total of all counter values builds the track histogram.
- A track is considered valid if its value is above a predefined threshold.

A description of the algorithm and its implementation can be found in [10].

In particular this algorithm is ideally suited for an FPGA implementation because it can be extremely parallelized. Adjustable memory boards allow RAM access with a width of e.g. 4*176 bits. Therefore, 706 straws can be processed simultaneously on a single ACB board equipped with 4 memory modules, thus providing an enormous speed-up compared to other systems, e.g. a state-of-the-art PC.

---

[2] Measured on Enable-1 with parallel histogramming only, no I/O was needed.

3.2 Image processing

Almost all image processing applications involve tasks where image elements (pixels or voxels) have to be processed with local filters. Among others, hardware implementation of algorithmically optimized real-time volume rendering is a current project at our institute in this area.

The following rendering - or ray processing - pipeline is assumed:
- Starting from each pixel of the resulting image rays are cast into the virtual scene.
- At equally distant positions on the rays sample points are generated by tri-linear interpolation of the neighboring voxel values.
- Sample points are classified with opacity or reflectivity according to gray values and gradient magnitude.
- Finally, the absorption for each voxel is determined. The reflected fraction of the light intensity reaching the sample point is calculated and added to the contributions of all other sample points on that ray.

The new architecture uses algorithmic optimizations: regions with no contribution are skipped, and processing is aborted as soon as the remaining intensity drops under an adjustable threshold. To overcome the resulting data and branch hazards in the rendering pipeline multi-threading is introduced. Each ray is considered as a single thread, and after each sample point the context is switched to the next ray. Our implementation has the same speed-up like software implementations of this algorithm, compared to volume rendering without algorithmic optimizations. However, compared to conventional architectures the number of pipeline stalls is reduced from more than 90% to less than 10% of rendering time.

Details of the algorithm and its FPGA implementations can be found in [11].

3.3 Astronomy

Using FPGAs to accelerate complex computations using floating-point algorithms has not been considered a promising enterprise in the past few years. The reason is that general floating-point [12] as well as particular N-Body [13] implementations have shown only poor performance[3] on FPGAs.

Usually N-Body calculations need a computing performance in at least Tera-FLOP range and are accelerated with the help of ASIC based coprocessors [14]. Nonetheless we have recently investigated the performance of a certain sub-task of the N-Body algorithm on the Enable++ system [15]. The results indicate that FPGAs can indeed provide a significant performance increase even in this area.

3.4 Measured and Estimated Performance

HEP. Besides principle parameters like system frequency the DMA performance plays a dominant role for the execution time of the TRT algorithm. Therefore DMA Read/Write access was the main focus of the measurements. Following are some

---

[3] In 1995 approx. 10 MFLOP per Xilinx chip were reported for 18 bit precision, and 40 MFLOP with 32 bit precision on an 8 chip Altera board.

results showing the data throughput over CPCI for various applications, measured with ATLANTIS, microEnable driver, design speed 40 MHz.

Table 1. ATLANTIS DMA performance

Block size (kByte)	1	4	32	256
DMA Read perf. (MB/s)	8.8	24.6	75.3	97.7
DMA Write perf. (MB/s)	7.4	21.6	54.3	65.3

The effect these results suggest for the performance of a distributed system largely depend on the respective application. For the TRT algorithm, the time needed for I/O is indeed the bottle-neck, in case the ATLANTIS sub-systems are employed as co-processors and thus receive their data from the host CPU.

Measurements of histogramming performance were done using a single-memory ACB (176 bit RAM access) [16]. The execution time on the test system (algorithm plus I/O), 19.2 ms compared to 35 ms using a C++ implementation on a Pentium-II/300 standard PC, extrapolates to 2.7 ms using 2 ACB with 4 memory modules each (1408 bit RAM access). This corresponds to a speed-up by a factor of 13.5.

Volume Rendering. The hardware speed is limited by several factors. One is the memory bandwidth. Assuming 100 MHz devices, simulations have shown that 4 Hz frame rates for 1024^3 data sets can be achieved for typical data with hard surfaces and otherwise empty space in between [17]. With our FPGA solution we will achieve a clock rate of >25 MHz that reduces the frame rate accordingly.

For detailed simulation we used a CT data set with 256*256*128 voxels. This data set is viewed from three different viewing directions and three different levels of opacity for soft tissue is applied.

On average one achieves efficiencies of between 90% and 97%. The number of sample points varies between 10-15% of all voxels if the data set consists mainly of empty space and opaque objects and 25-40% for semi transparent opacity levels.

The above results correspond to rendering rates from 20 Hz on semi-transparent data sets to 138 Hz for opaque objects and parallel projection. The results are achieved from images of size 256*128. Perspective views reduce the rendering speed by a factor of about 2.

Comparing these results with the performance of the only commercially available volume rendering hardware, VolumePro [18], simulations suggest a speed-up by a factor of 10 to 25 when using 1024^3 data sets.

4 Summary and Outlook

ATLANTIS is a CompactPCI based computing machine that combines the advantages of FPGA and RISC architectures. Its unique features are scalability, flexibility with respect to memory, configurable high-speed I/O, and it comes with a powerful object-oriented development environment, CHDL.

ATLANTIS has proven its supreme power regarding bandwidth and speed in applications we have investigated so far. An ACB is available since 09/1999 and is currently tested with different memory modules and a simple backplane, with different applications. A second ACB and an AIB will be completed shortly. Though the full system is not available by now (01/2000) it is planned to have an implementation of a HEP trigger application run in a real experiment (FOPI at GSI, Darmstadt, Germany) within this year. Other implementations concern future experiments, or have prototype character.

References

[1] Klefenz F., Zoz R., Noffz K.-H., Männer R., "The ENABLE Machine - A Systolic Second Level Trigger Processor for Track Finding", Proc. Comp. in High Energy Physics, Annecy, France; CERN Rep. 92-07 (1992) 799-802

[2] DECPeRLe-1, an FPGA processor containing 16 Xilinx XC3090 FPGAs, http://pam.devinci.fr/hardware.html#DECPeRLe-1

[3] D. Buell, J. Arnold, W. Kleinfelder, "Splash-2 – FPGAs in a Custom Computing Machine", CS Press, Los Alamitos, CA, 1996

[4] H. Hoegl et al., "Enable++: A Second Generation FPGA Processor", Proc. IEEE Symposium on FPGAs for Custom Computing Machines, pp. 45-53, 1995

[5] microEnable, a PCI based FPGA co-processor by Silicon Software GmbH, http://www.silicon-software.com/

[6] K. Kornmesser et al, "Simulating FPGA-Coprocessors Using the FPGA Development System CHDL", Proc. PACT Workshop on Reconf. Comp., Paris (1998) pp. 78-82

[7] J. Vuillemin et al., "Programmable Active Memories: Reconfigurable Systems Come of Age", Proc. of the 1996 IEEE Trans. On VLSI Systems

[8] R. Spurzem, S.J. Aarseth, "Direct Collisional Simulation of 10,000 Particles Past Core Collapse", Monthly Notices Royal Astron. Soc., Vol. 282, 1996, p. 19

[9] V. Dörsing et al., "Demonstrator Results Architecture – A", ATL-DAQ-98-084, CERN, 26 Mar 1998

[10] A. Kugel et al., "50kHz Pattern Recognition on the Large FPGA Processor Enable++", Proc. IEEE Symp. on FPGAs for Custom Computing Machines, CS Press, Los Alamitos, CA, 1998, pp. 1262-3

[11] J. Hesser, B. Vettermann, "Solving the Hazard Problem for Algorithmically Optimized Real-Time Volume Rendering", Int. Workshop on Vol. Graph. 1999, Swansea, UK

[12] W. Ligon et al, "A Re-evaluation of the Practicality of Floating-Point Operations on FPGAs", Proc. IEEE Symp. on FPGAs for Custom Computing Machines, 1998

[13] H.-R. Kim et al, "Hardware Acceleration of N-Body Simulations for Galactic Dynamics", SPIE Conf. on FPGAs for Fast Board Develop. and Reconf. Comp. 1995, pp. 115-126

[14] J. Makino et al, "GRAPE-4: A Massively Parallel Special-Purpose Computer for Collisional N-Body Simulations", Astrophysical Journal, Vol. 480, 1997, p. 432

[15] T. Kuberka, Diploma Thesis, Universität Mannheim, Germany, 1999

[16] C. Hinkelbein et al, "LVL2 Full TRT Scan FEX Algorithm for B-Physics Performed on the FPGA Processor ATLANTIS", to be publ. as ATL-DAQ-Note, CERN

[17] B. Vettermann et al, "Implementation of Algorithmically Optimized Volume Rendering on FPGA Hardware", IEEE Visualization '99, San Francisco, CA (1999)

[18] VolumePro, a PCI based volume rendering coprocessor by Mitsubishi Electronics America, Inc. RTVIZ, http://www.rtviz.com/

The Cellular Processor Architecture CEPRA–1X and its Configuration by CDL

Christian Hochberger[1], Rolf Hoffmann[2], Klaus–Peter Völkmann[2], and Stefan Waldschmidt[2]

[1] University of Rostock, 18059 Rostock, Germany,
hochberg@informatik.uni-rostock.de
[2] Darmstadt University of Technology, 64283 Darmstadt, Germany,
(hoffmann,voelk,waldsch)@informatik.tu-darmstadt.de

Abstract. The configurable coprocessor CEPRA–1X was developed as a PC plug–in card in order to speed up cellular processing significantly. *Cellular Processing* is an attractive and simple massive parallel processing model. To increase its general acceptance and usability it must be supported by a software environment, an efficient simulator and a special language. For this purpose the cellular description language CDL was defined and implemented. With CDL complex cellular algorithms can be described in a concise and readable form. A CDL program can automatically be transformed into a logical design for the CEPRA–1X. The design is loaded into field programmable gate arrays for the computation of the state transition of the cells. For time dependent or complex rules the design may be reconfigured between consecutive generations. An example is presented to show the generation of logic code.

1 Introduction

Cellular Processing is based on the processing model of Cellular Automata. All cells obey in parallel to the same local rule, which results in a global transformation of the whole generation. The cells are connected to their adjacent cells only. In the two dimensional case 4 neighbours (von Neumann neighbourhood) or 8 neighbours (Moore neighbourhood) are considered. In the three dimensional case up to 26 neighbours can be taken into consideration.

Typical applications are: crystal growth, biological growth, simulation of digital logic, neuronal switching, electrodynamic fields, diffusion, temperature distributions, movement and collision of particles, lattice gas models, liquid flow, wave optics, Ising systems, image processing, pattern recognition and numerical applications.

Cellular algorithms are described in a concise and readable form in the language CDL (Cellular Description Language). CDL has been proved to be very useful for the description of complex cellular algorithms[1]. One version of the compiler generates C or Java code for the software simulator, another version generates a hardware description for field programmable gate array which we use in our coprocessor CEPRA–1X[2].

J. Rolim et al. (Eds.): IPDPS 2000 Workshops, LNCS 1800, pp. 898-905, 2000.

Cellular processing on a conventional computer is time consuming especially for a large number of cells, complex rules and experiments with parameter variations. Special hardware support is necessary to speed up the computation and for realtime visualisation on the fly.

2 Target Architectures

In the course of the cellular processing project at the Technical University of Darmstadt different architectures have been developed, in particular the CEPRA-8L[3], the CEPRA-1X[2], and the CEPRA-3D[4]. A new designed machine CEPRA-S for general purposes is under development. The advantage of the CEPRA processors compared to CAM[5] machines is that complex and probabilistic rules can be computed in one step, whereas the CAM machines must split the problem into cascaded look-up tables.

Coprocessor CEPRA-1X. The CEPRA-1X coprocessor is a plug-in card for the PCI bus. It was designed for 2D cellular processing with visualisation support, but it can be used as a general data stream processor. The cellular field data is stored in the host. For the computation of a new generation the cell states are streamed to the coprocessor, the rule is computed for all the cells in the stream and the new cell states are streamed back to the host.

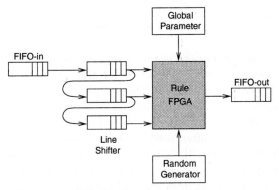

Fig. 1. three line FIFOs

The rule is computed by an FPGA (field programmable gate array) which has to be loaded with a configuration, describing the logic design of the rule. Because three lines are buffered (implemented as FIFOs) each cell has to be read and written exactly once. With the PCI-bus-performance of 133 MByte/sec the performance is 30 million 2D-16-Bit cell operations per second with 9 neighbours. Considering the Belousov-Zhabotinsky reaction described later this is a speed up of about 40 in comparison to a 133 MHz PC. More complex rules will yield higher speed ups, because we use hardware pipelining in the CEPRA-1X. Therefore the computation time is independent of the the rule complexity.

The logic design which has to be loaded into the FPGA is generated by the CDL hardware compiler. The compiler generates intermediate logic code

(*VERILOG*) which is transformed into FPGA configuration data by a tool from XILINX.

The logic description of different rules can be reloaded between the computation of the generations. By this technique time dependent rules can be computed. Complex rules which do not fit into the FPGA can be broken into a sequence of phase rules. The phase rules are loaded between the phases of the generations. The time to reload the FPGA (parallel mode, 8MHz) is 15% of the computation time for a cell field of size 1024×1024.

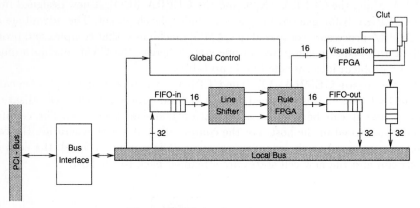

Fig. 2. CEPRA–1X architecture

Software Simulator. For the evaluation of cellular algorithms we have developed a simulator software. Experiments for this simulator consist of three basic parts: the description of the rule, the initial state of the cells in the array and some information about the visualisation.

The simulator allows the user to store the cell state in a structured datatype. One of the easiest and most often used visualisation concepts is the assignment of colours to cell states. Thus the simulator provides a visualisation tool that uses one of the cells components as an index into a colourmap.

The rule is written in C or Java and is linked with a kernel which controls the simulation. The kernel provides a **neighbour** function for the access to the neighbours. The kernel is capable of calling different rules depending on the position within the cellular field. By this technique special rules for borders and corners can be defined.

3   CDL, a Language for Cellular Processing

Until now cellular algorithms are programmed in simulator dependent special languages and data structures. Thus the programmer needs special knowledge of the target architecture, which makes programming a tedious task. The CEPRA–1X processor is programmed in VERILOG, whereas the software simulator is programmed in C or Java. Neither of those languages is convenient and adequate for the programmer to describe cellular algorithms. Also both languages contain elements that are not required for this purpose (e.g. pointer and dynamic memory allocation in C).

The new language CDL was defined with respect to readability, conciseness and portability. While developing a cellular algorithm it is desired to have short turn–around cycles. Thus the usage of a highly interactive software simulator is recommended during the development process. After having tested the algorithm on the software simulator it can be transferred to the CEPRA-1X for fast execution and realtime visualisation.

Features of the Language. The language CDL is intended to serve as an architecture independent language for cellular algorithms. The programmer's benefit is obvious: Switching the target architecture does not require more than just a new compiler run. Moreover CDL contains special elements that make the description of complex conditions very easy (groups, special loop constructs). These elements allow the description of situations like:

- Is there any neighbour that fulfils a certain condition? (one())
- Do all neighbours fulfil a certain condition? (all())
- How many neighbours are in a certain state? (num())

CDL does not contain conditional loops, which has two positive side effects. (1) It enforces the termination of the rule because it is impossible to write endless loops and (2) it enables the compiler to unroll all statements which is extremely important for the synthesis of hardware. CDL allows the user to describe the cell state as a record of arbitrary types. All common data types are available in CDL (integer, boolean, float, etc.). In addition the user can define new types (enumerations and subranges of integers or enumerations).

Example. To give an impression of a CDL program we present the Belousov–Zhabotinsky reaction[6]. It does not show all the special features of CDL, but demonstrates some of the problems that have to be handled quite differently on hardware and software simulators.

```
(1) cellular automaton Belousov_Zhabotinsky ;
(2)
(3) const dimension = 2 ;   // a two-dimensional grid
(4)       distance  = 1 ;   // allow/restrict Moore-neighbourhood
(5)       maxtimer  = 7 ;   // a local constant
(6)       cell      = [0,0]; // relative address of actual cell
(7)                  // *[0,0]  means the contents of the cell
(8) type celltype = record   // celltype defines possible states
(9)       active : boolean;
(10)      alarm  : boolean;
(11)      timer  : 0..maxtimer;
(12)    end;
(13)                    // addresses of all 8 Moore-neighbours
(14) group   neighbours={[-1,0],[ 1,0],[0, 1],[ 0,-1],
(15)                     [ 1,1],[-1,1],[1,-1],[-1,-1]};
```

```
(16) colour                    // description of visualisation
(17)    [0  , 255, 0]      ~      *cell.active and      *cell.alarm;
(18)    [255,   0, 0]      ~      *cell.active and not *cell.alarm;
(19)    [*cell.timer * 255 div maxtimer,0,0]   ~ not *cell.active;
(20)
(21) var   neighbour : celladdress;    // local loop variable
(22)
(23) rule begin
(24)    *cell.active := *cell.timer=0;     // is actual timer==0?
(25)    *cell.alarm  :=       // count neighbours in active state
(26)          num(neighbour in neighbours : *neighbour.active)
(27)                                    in {2,4..8};
(28)    if *cell.active and *cell.alarm and (*cell.timer=0)
(29)       then *cell.timer:=maxtimer
(30)       else if *cell.timer!=0 then *cell.timer:=*cell.timer-1;
(31) end;
```

The type celladdress, as used in line (21), is implicitly defined by the compiler from the two constants **dimension** and **distance**. They define how many dimensions the model uses and how far the access to other cells reaches. Both constants must be supplied by the programmer. The type celladdress is a record with as many components as the model has dimensions. Each component can have a value between **-distance** and **+distance**. Lines (14) and (15) show the celladdresses of all eight Moore neighbours. The name of this enumeration does not have any meaning for the compiler. The elements are used in the iterative num–loop in line (26).

4   Transformation into a Hardware Description

Even simulators that are based on specialised hardware are supported by CDL. The CEPRA–1X simulator has been chosen as an example during the design phase of CDL.

The most important restrictions of a hardware simulator are the limited number of cell states and the limitations in the rule complexity. Although floating point numbers are desired and should be included in a cellular language, they are usually not implemented in a specialised hardware simulator because of hardware costs.

Celltype. In the case of CEPRA–1X the states of the cell must be coded with 16 bits. If the **celltype** is a record (as in lines (08)–(12)) it would me more easy to reserve bit groups for the subtypes of this record (one bit for each boolean in lines (09)–(10) and three bits for the integer subrange in line (11)). Usually, this will simplify the logic for the rules, because often the rules access only components of the cell record (e.g. line (28)). On the other hand, this may lead to a state coding, where not all 2^{16} states can be used (e.g. if the integer subrange does not have power of two elements). Enumerating all possible cell states (the power set of the components) will not waste any of the states, but will

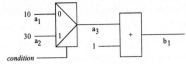

Fig. 3. The implementation of local variables and assignments

increase implementation cost. The CDL compiler decides itself which method to use.

Variables. The classical synthesis approach uses registers to represent variables. The data paths between these registers are controlled by a finite state machine. For the CEPRA-1X machine this is not desired, because it would imply the usage of a clock signal. The number of clocks required to complete the calculation would then depend on the data. The varying time could stall the pipeline and slow down the calculation speed.

To simulate CDL variables, they are represented by local signals. Because a new value can be assigned to signals only once, for each assignment new signals must be created.

The following CDL fragment

```
(1)  a:=10;
(2)  if condition then a:=30;
(3)  b:=a+1;
```

produces the local variables a_1, a_2, and a_3, a multiplexer driven by `condition`, and a following adder which calculates the value of signal b_1 (Fig. 3).

Optimisations. The hardware resources inside a FPGA are limited. Therefore optimisation is necessary. The optimisation supported by the VERILOG compiler is good but not sufficient. The CDL compiler already should keep an eye on the complexity of the description. It should not use too many local signals and avoid generating unused code. To reduce implementation cost, early expression and condition evaluation is necessary and was implemented. The compiler evaluates constant expressions during compilation, taking special properties of the operation into consideration. The **or** operation, for example, with one operand being constant **true** is evaluated during compilation and translated into the constant **true**.

Usually a data type is represented by a fixed number of bytes on common computers. To reduce implementation cost, the compiler should use single bits instead of bytes as the smallest unit. In addition, the size of a data type may vary. For example a variable of an integer subrange type, which is divided by two will need one bit less after the division. Therefore it is useful to know the exact range of possible values for each variable and expression.

Loops. To simulate the behaviour of a loop, hardware must be generated for each iteration. Conditional loops are not available because the number of iterations can not be determined during compilation. (This is equivalent to the demand that calculation must always terminate.)

The **num** expression in line **(30)** can be interpreted as a loop. The constants of the group **neighbours** are assigned to the variable **neighbour** one after the other. After each assignment the expression *neighbour.active is evaluated and the result is assigned to a new local signal. After the eight iterations, the eight signals are connected to a logic, which sums up the conditions that are true. The sum is the result of this expression.

Conditional Statements. The only statement which has a permanent effect is the assignment of a value to a variable or the cell state (e.g. line (28)). For this reason the assignment statement is affected by the corresponding condition. Have a look at line (35). Only if the condition is true, the assignment shall have an effect. Therefore each assignment is implemented as a two–to–one multiplexer, where one input is the old value and the other is the new value. The select signal of this multiplexer is connected to the condition of the surrounding conditional statement. For nested conditional statements their conditions are combined using the logical **and** operation. An **else** part can be realized using the inverted condition and a **case** statement using different cascaded conditions.

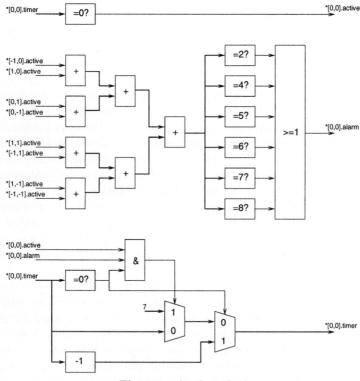

Fig. 4. result of synthesis

Complete Example. The CDL program describing the Belousov–Zhabotinsky reaction from the previous section results in a hardware structure shown in Fig. 4. Obviously program line (24) corresponds with the upper part of the logic. The middle part corresponds to lines (25)–(27). And the lower part of the logic has been synthesised form lines (28)–(30).

Recognise the power of the num() statement. Only three lines of code result in the large amount of the P middle part of the logic.

Colour. The colour definition must be loaded into the CRT controller as a look–up–table. To create this look–up–table during compilation, each possible cell state is associated with the contents of the cell (*[0,0]) and the expressions in the colour definition (lines (17)–(19)+) are evaluated.

5 Conclusion

The CEPRA–1X is a configurable coprocessor which speeds up cellular processing significantly. As it processes data streams it can also be used for other applications. The resulting pixel stream can be coloured and visualised in real-time. Complex rules and time dependent rules can be computed by reloading the FPGA between the generations.

CDL is an implemented language for the concise, readable and portable description of cellular algorithms. One version of the compiler generates C/Java–code for the software simulator. Another version generates logic equations for the field programmable gate arrays of the CEPRA–1X machine. The logic equations are partly minimised by the compiler and partly by a commercial available design system.

Main features of the language are records, unions, groups and the loop construct for testing complex conditions. The language can be used to describe complex cellular algorithms of practical relevance. Based on the experience the language was extended to CDL++[7] for the description of moving objects.

References

[1] Christian Hochberger, Rolf Hoffmann, Klaus-Peter Völkmann, and Stefan Waldschmidt. Cellular processing environment. In Bogusław Butryło, editor, *International Conference on Parallel Computing in Electrical Engineering (PARELEC 98)*, number 1, pages 171–174, Białystok, Poland, 1998. Technical University of Białystok.

[2] Christian Hochberger, Rolf Hoffmann, Klaus-Peter Völkmann, and Jens Steuerwald. The CEPRA–1X cellular processor. In Rainer W. Hartenstein and Viktor K. Prasanna, editors, *Reconfigurable Architectures, High Performance by Configware*. IT Press, Bruchsal, 1997.

[3] Rolf Hoffmann, Klaus-Peter Völkmann, and Marek Sobolewski. The cellular processing machine CEPRA–8L. *Mathematical Research*, 81:179–188, 1994.

[4] R. Hoffmann and K.-P Voelkmann. Hardware support for 3D cellular processing. *Lecture Notes in Computer Science*, 1277:322–??, 1997.

[5] Norman H. Margolus. CAM–8: a computer architecture based on cellular automata. Technical Report 01239, MIT Lab. for Computer Science, December 1993.

[6] A. Zaikin and A. Zhabotinsky. *Nature*, (225):535–, 1970.

[7] Christian Hochberger. *CDL — Eine Sprache für die Zellularverarbeitung auf verschiedenen Zielplattformen*. PhD thesis, Darmstadt University of Technology, 1999.

Loop Pipelining and Optimization for Run Time Reconfiguration[*]

Kiran Bondalapati and Viktor K. Prasanna

Department of Electrical Engineering
University of Southern California
Los Angeles, CA 90089-2562, USA.
{kiran,prasanna}@ceng.usc.edu
http://maarcII.usc.edu

Abstract. *Lack of automatic mapping techniques is a significant hurdle in obtaining high performance for general purpose computing on reconfigurable hardware. In this paper, we develop techniques for mapping loop computations from applications onto high performance pipelined configurations. Loop statements with generalized directed acyclic graph dependencies are mapped onto multiple pipeline segments. Each pipeline segment is executed for a fixed number of iterations before the hardware is reconfigured at runtime to execute the next segment. The reconfiguration cost is amortized over the multiple iterations of the execution of the loop statements. This alleviates the bottleneck of high reconfiguration overheads in current architectures. The paper describes heuristic techniques to construct pipeline configurations which have reduced total execution time including the runtime reconfiguration overheads. The performance benefits which can be achieved using our approach are illustrated by mapping example application loop onto Virtex series FPGA from Xilinx.*

1 Introduction

Reconfigurable computing has demonstrated significant performance gains for several classes of applications[5]. Application mapping onto configurable hardware still necessitates expertise in low-level hardware details. Automatic mapping of applications onto configurable hardware is necessary to deliver high performance for general purpose computing. In this paper we address the issues in mapping application loops onto reconfigurable hardware to optimize the total execution time. Total execution time includes the time spent in actual execution on the hardware and the time spent in reconfiguring the hardware.

Configurable hardware can be utilized to execute designs which are larger than the available physical resources. Run Time Reconfiguration(RTR) between computations facilitates dynamic adaptation of the hardware to suit the design area and computational requirements. But, in current devices, reconfiguration time is still significant compared to the execution time. We focus on developing

---

[*] This work was supported by DARPA Adaptive Computing Systems program under contract DABT63-99-1-0004 monitored by Fort Huachuca.

J. Rolim et al. (Eds.): IPDPS 2000 Workshops, LNCS 1800, pp. 906-915, 2000.

mapping techniques which exploit RTR but attempt to reduce the reconfiguration overhead. This is accomplished by amortizing the the reconfiguration overheads over the execution of large number of iterations of the loop.

Loop statements contribute to a significantly large component of the execution time of an application. Pipelined designs are well structured and map well onto configurable devices. Most reconfigurable architectures, including FPGA devices, provide excellent support for pipelining with their regular logic block layout and large number of registers [17]. Pipelined designs have reduced and predictable delays because they use mostly local interconnections. Hence, mapping loop computations onto pipelined configurations proves to be very effective on configurable hardware.

In this paper, we develop techniques to map computations in a loop onto reconfigurable hardware. The data dependencies in the loop statements constitute a directed acyclic graph (DAG). These loop statements are mapped onto pipelined configurations executing in the reconfigurable hardware. Our mapping techniques attempt to minimize the total execution cost for the computations including the reconfiguration cost. The statements are split into multiple pipeline segments which are executed sequentially for a fixed number of iterations each. Reconfiguration is performed after execution of a pipeline segment to execute the next segment.

Generating optimal schedule from a given task graph is an NP-complete problem. In this paper, heuristic algorithms are utilized to reduce the reconfiguration cost between different pipeline segments. We compare the effectiveness of our heuristics against a greedy heuristic based list scheduling. Our mapping techniques promise potential performance improvement on several classes of FPGAs. We evaluate the performance of our mapping techniques on the Virtex series FPGA from Xilinx [17].

In Section 2, we describe some related research work which addresses similar issues. Our heuristic based algorithms are described in detail in Section 3 and illustrated by using an example. In Section 4, we evaluate the performance benefits achieved using our approach. We draw conclusions based on our approach in Section 5.

2   Related Work

Pipelining of designs has been studied by several researchers in the configurable computing domain. Cadambi et. al. address the issues in mapping virtual pipelines onto a physical pipeline by using incremental reconfiguration in the context of PipeRench [6]. Luk et. al. describe pipeline morphing and virtual pipelines as an idea to reduce the reconfiguration costs [11]. A pipeline configuration is morphed into another configuration by incrementally reconfiguring stage by stage while computations are being performed in the remaining stages. Weinhardt describes the generation of pipelined circuits from parallel-FOR loops in high level programming language [15]. Weinhardt et. al. also developed pipeline vectorization techniques [16].

Other research has addressed related issues in mapping circuits onto reconfigurable hardware [2, 7, 10, 12, 14]. Our prior research has also developed other techniques for mapping application loops [1, 3, 4]. In this paper, the focus is on Run Time Reconfiguration at a different granularity. Our approach is to exploit Run Time Reconfiguration to achieve high performance but schedule it infrequently to minimize the overheads. Algorithmic pipeline construction and partial reconfiguration at runtime are exploited to achieve this goal.

3   Pipeline Construction

The speed-up that can be obtained by using configurable logic increases as the computations in a loop increase. But, the configurable resources that are available can be lower than the required resources to pipeline all the computations in the loop. In this case, the pipeline has to be segmented to run some of the pipeline stages and reconfigured to execute the remaining computations.

In this paper, we consider loops which do not have loop carried dependencies. Such loops do not have any dependencies between different iterations of the loop. Loop transformations can be applied to remove some existing loop carried dependencies. We also assume that the number of iterations to be executed is significantly larger than the number of pipeline stages. Hence, the cycles involved in filling and emptying the pipeline are insignificant compared to the actual execution cycles of the pipeline stages.

The execution of the complete loop can be decomposed into multiple segments, where a fixed number of iterations of each segment are executed in sequence starting from the first segment. Each segment consists of multiple pipeline stages. The logic is reconfigured after each segment to execute the next segment. The intermediate results from each segment execution are stored in memory. The execution of the sequence of segments is repeated until the required number of iterations of the loop are completed. We assume that the reconfiguration of the different segments is controlled by an external controller(e.g. a host processor).

3.1   Definitions

Reconfigurable Architecture A configurable logic array of size $L \times W$ and intermediate memory of size M. One of the basic goals of our approach is to exploit the on chip memory or fast access local SRAM provided in several reconfigurable architectures. M represents the size of this memory.

Input Task Specification A dependency graph $G(V, E)$ of the application tasks of the loop to be executed for N iterations. Each task node v_i denotes the operation to be performed on the inputs specified by the incoming edges to the node. The directed edge e_{ij} from v_i to v_j denotes the data dependency between the two nodes. The weight w_{ij} on each edge denotes the number of bits of data communicated between the nodes.

Output Pipeline Configuration A sequence σ of pipeline segments $\sigma_1, \sigma_2, \ldots, \sigma_p$ where each segment $\sigma_i(1 \leq i \leq p)$ consists of q number of stages $s_{i1}, s_{i2}, \ldots, s_{iq}$.

The pipeline stages are the mapping of the computational task nodes V to configurations of the device. Each of the stages s_{ij} is the configuration which executes a specific task in the input task graph. The size of a pipeline stage is given by the length l_{ij} and the width and w_{ij}. Some of the stages in each segment might be *null* stages which are not actual tasks but are just *place-holders* as explained later in Section 3.6.

Segment Clock Speed Each pipeline segment σ_i can be executed at a different clock speed f_{σ_i} depending on the maximum clock speed at which the stages in that segment can operate.

Segment Data Output A pipeline stage s_{ij} has global outputs if any of the outgoing edges from a task node are to a node that is not mapped to the same pipeline segment. The size of the segment data output $DO_i (1 \leq i \leq p)$ of all the pipeline stages in a segment $\sigma_i (1 \leq i \leq p)$ is given by the sum of all the global outputs of the stages in the segment.

Segment Iteration Count The number of iterations N_σ for which each pipeline segment is executed before reconfiguring to execute the next segment. N_σ depends on the size of the available memory to store the intermediate results. We assume that the initial and final results are communicated from/to external memory.

$$N_\sigma = min_i \left\{ \frac{M}{DO_i + DO_{i+1}} \right\} \quad 1 \leq i \leq p - 1$$

Reconfiguration Cost The reconfiguration cost R_{loop} is the total cost involved in reconfiguring between all the segments of the pipeline configuration. This includes the cost of configuring between the last segment and the first segment if $N > N_\sigma$. The reconfiguration cost between any two segments is given by the difference in the two pipeline configurations. Partial reconfiguration of the device in columns is assumed in our computation. We use the number of logic columns in which the configurations are different as the measure of the reconfiguration cost. When the corresponding stages in different segments are dissimilar, the reconfiguration cost accounts for the multiple adjacent stages that need to be reconfigured.

Total Execution Time The total execution time E is given by the sum of the execution times for each segment and the total reconfiguration time.

$$E = N * \left(\sum_{i=1}^{p} \frac{1}{f_{\sigma_i}} \right) + \frac{N}{N_\sigma} * R_{loop}$$

3.2 Phase 0: Pre-processing and Mapping

In this phase the computation tasks in the input DAG are mapped onto components of the given logic device. The components are chosen from the set of library components available for executing the given application tasks in the task graph. Different components can have different logic-area/execution-time tradeoffs and could potentially have different degrees of pipelining and footprint on the device after layout. The library component of the highest degree of pipelining which satisfies other constraints specified by the task graph(such as precision of inputs) is chosen for a task.

Our proposed approach is illustrated using the mapping and scheduling of the N-body simulation application and the FFT butterfly computation. The resulting task graphs after Phase 0 with the dependency edges are shown in Figure 1. In the graph the operations are represented as A - Addition, M - Multiplication, S - Subtraction and Sh - Shift right by 4 bits(Divide by 16). The operations in the graph are all 16 bits so the weights on the edges are not indicated.

3.3   Phase 1: Partitioning

The partitioning phase generates multiple partitions where size of each partition is smaller than the size of the device. This phase attempts to optimize two criterion - (1) maximize the size of the partition (2) minimize the weight of the edges between partitions. The first criterion improves the logic utilization and the second criterion reduces the memory required to buffer intermediate results generated by each partition (pipeline segment). A sketch of the partitioning algorithm is given below without the intricate details.

A heuristic based multi-way partitioning is used to incrementally generate each of the partitions. The largest size node is chosen from among the list of *Ready* nodes (whose inputs have been computed) to be added to the current partition. When no more nodes can be added to the current partition, a new partition is initiated. For adding a *Ready* node v_i to a partition π_j, the heuristic uses the following sums of weights of edges:

- ω_1: weight of *in* edges to v_i from nodes in π_j
- ω_2: weight of *in* edges to v_i from nodes not in π_j
- ω_3: weight of *out* edges from v_i to nodes in π_j
- ω_4: weight of *out* edges from v_i to neighbors of π_j
  v_k is a neighbor of π_j if there is an edge from a node in π_j to v_k and $v_k \notin \pi_j$
- ω_5: weight of *out* edges from v_i to nodes not in π_j and not neighbors of π_j

The node chosen is the node with maximum value of $\omega_1 + \omega_3 + \omega_4 - \omega_2 - \omega_5$. The primary inputs and outputs are not considered in computing the weights. The largest node which fits in the current partition satisfying the above condition is added to the current partition. Ties are broken by using the height of the node and the different weights of edges listed above. The resulting partitions are illustrated by the partition number on the nodes of the graph in Figure 1.

3.4   Routing Considerations

The algorithm for the partitioning of the task graph assumes that there are enough routing resources to communicate between the different pipeline stages and from pipeline stages to the memory. Some of the pipeline stages might have

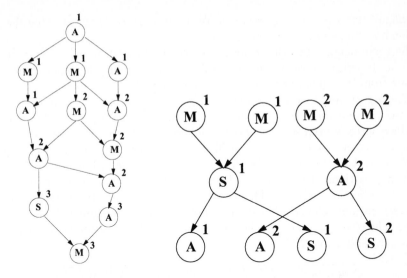

Fig. 1. (a) N-body simulation task DAG and (b) FFT task DAG with partition numbers

global inputs and outputs. These are data inputs and outputs which are not to adjacent pipeline stages, but from/to either non-adjacent stages or from/to memory. Some of the data outputs from the pipeline stages might have to be buffered (using registers) before they are consumed in the later stages.

Routing resources are an important consideration when mapping communication between non-adjacent pipeline stages. In our experiments we have discovered that FPGAs such as Virtex [17] are routing and register rich and can support most pipeline-able designs. The number of bits of data computed in each stage is typically less than or equal to the number of logic cells utilized. Hence, the stage to stage communication has enough routing resources by using nearest neighbor interconnect. Extra routing and logic resources (for buffering and multiplexing) have to be utilized for data values communicated across non-adjacent pipeline stages. In the partitioning algorithm, the remaining area in a partition is reduced to reflect the buffering requirements.

A limitation of our approach is that partitions might have bad memory performance when the computation is highly irregular or there are a large number of data dependencies in the DAG. The approximation of routing resources results in infeasible designs in some cases. But, for most applications, the circuits were finally mapped within the available logic and routing resources.

3.5 Phase 2: Pipeline Segmentation

The configuration of the pipeline is generated from the partitions that are computed in Phase 1 by the algorithm in Figure 2. Each partition is utilized to generate one segment of the pipeline. The goal in the segmentation phase is to generate permutations of the pipeline stages in each segment to reduce the reconfiguration costs across segments. We use the heuristic of matching the corresponding stages of the different pipeline segments. In each partition, the nodes of the same height have the flexibility of being mapped in any order onto the pipeline. In addition, once a node has been mapped onto the pipeline, its successors from the same partition can also be mapped.

The algorithm proceeds by first identifying the list of tasks from each partition that are *Ready* to be scheduled. A task node is *Ready* if all of its predecessors have already been scheduled onto the segment. At the next step, a maximal matching set of task nodes are identified from the set of all *Ready* lists from all *Partitions*. A maximal matching set corresponds to the task node which occurs in most partitions. This step schedules similar nodes from different partitions onto the different segments. This enables the reduction in the reconfigurations costs at runtime. The *Ready* lists are updated before scheduling the next set of nodes. The resulting pipeline schedules with the different segments are shown in Table 1(b) and Table 2(b).

Segment 1	A	M	M	A	A
Segment 2	M	A	M	A	A
Segment 3	S	A	M	*	*

Segment 1	A	M	M	A	A
Segment 2	A	M	M	A	A
Segment 3	A	S	M	*	*

Table 1. Schedules for N-body simulation (a) S_0: Greedy Scheduling (b) S_I: Schedule after Phase 2

Segment 1	M	M	M	*	*
Segment 2	M	S	A	A	A
Segment 3	S	S	*	*	*

Segment 1	M	M	S	A	S
Segment 2	M	M	A	A	S

Table 2. Schedules for FFT (a) S_0: Greedy Scheduling (b) S_I: Schedule after Phase 2

3.6   Reconfiguration of *null* stages

Reconfiguring from a *null* stage to a computation stage can be accomplished by small modifications to the pipeline design. The data values from the previous computation stage are also communicated directly to the output register in addition to flowing through the computational units. 2-input multiplexers are utilized at the output registers to latch one of the two values. Run Time Reconfiguration using partial reconfiguration only needs to modify the SRAM bits controlling the configuration of the multiplexers. This reconfiguration cost is significantly lower than reconfiguring the whole datapath.

4   Results

We evaluate the performance of our techniques by comparing them with a greedy heuristic based on list scheduling. The greedy schedule chooses the largest available *Ready* node as the next stage of the pipeline. A new pipeline segment is

```
1: Function Segmentation(G, Partition)
2: ∀v_i : Mapped(v_i) ← FALSE
3: Num_Partitions ← |Partition|
4: repeat
5:     for i = 1 to i = Num_Partitions do
6:       Ready[i] ← {v_j|v_j ∈ Partition[i] and
7:                       ∀v_k : v_k = Predecessor(v_j) and Mapped(v_k)}
8:     endfor
9:     for i = 1 to i = Num_Partitions do
10:      for all v_j ∈ Ready[i] do
11:          Count(v_j) ← ∑_{l=1}^{Num_Partitions} |{v_k|Type(v_k) = Type(v_j) and
12:                                  v_k ∈ Ready[l]}|
13:      end for
14:    end for
15:    V_curr ← null
15:    for i = 1 to i = Num_Partitions do
16:      v_sel = v_j | v_j ∈ Ready[i] and ∀v_j :  max{Count(v_j)} and v_j ∈ V_curr
17:      if v_sel = null then
18:          v_sel = v_j | v_j ∈ Ready[i] and ∀v_j :  max{Count(v_j)}
19:      end if
20:      Segment[i] ← Segment[i] ⊕ v_sel
21:      if v_sel != null then
22:          V_curr ← V_curr ∪ v_sel
23:          Mapped(v_sel) ← TRUE
24:      end if
25:    end for
26: until (∀i : empty(Partition[i]))
```

Fig. 2. Algorithm to generate the pipeline segments

initiated when no more nodes can be added to the current segment. The resulting schedule is shown in Table 1(a) and Table 2(a). We utilize the modules and the parameters from the Virtex component libraries [17]. Some of the modules utilized are tabulated below in Table 3. The number of pipelined stages, precision of the inputs and the size of the module when mapped onto the device are listed in the table.

For the N-body simulation and FFT examples, the number of slices to be reconfigured for each schedule is shown in Table 4. This is the reconfiguration cost R_{loop} as defined in Section 3.1. The heuristic based algorithms have a significant saving in the reconfiguration cost. This translates to a direct reduction in the total execution time of the configuration. In the worst case, our heuristic algorithms generate a schedule which is at least as good as the greedy heuristic.

The total execution cost was computed for both the applications for a data set size of 4096 data points with the an on-chip memory size of 2KB (M). For the two example applications, reconfiguration cost is the dominant cost in

Module	Stages	Input	Slices	Speed
Add	1	16x16	10	173 MHz
Add	1	32x32	20	157 MHz
Subtract	1	16x16	11	141 MHz
Shift	1	16x16	10	180 MHz
Multiply	1	8x8	39	65 MHz
Multiply	4	8x8	48	131 MHz
Multiply	5	12x12	107	117 MHz
Multiply	5	16x16	168	115 MHz

Table 3. Virtex module characteristics

	Greedy	Our Approach	Speedup
N-body	624	228	2.74
FFT	702	110	6.38

Table 4. Reconfiguration costs in number of Virtex slices

the execution of the application and constitutes more than 95% of the total execution time. The application speedups are of the same order as the speedups in the reconfiguration costs illustrated in Table 4. This shows that our heuristic based approach performs significantly better than the greedy heuristic.

5   Conclusions

Automatic mapping and scheduling of applications is necessary for achieving performance improvement for general purpose computing applications on reconfigurable hardware. These techniques have to address the overheads involved in reconfiguring the hardware. In current architectures the reconfiguration overheads are still significant compared to the execution cost. In this paper, we have proposed algorithmic techniques for mapping and scheduling loops in applications onto reconfigurable hardware. The heuristics we have developed attempt to minimize the reconfiguration overheads by exploiting pipelined designs with partial and runtime reconfiguration. The mapping of example loops from applications illustrates that the proposed algorithms can generate high performance pipelined configurations with reduced reconfiguration cost.

In future work, we will explore the interaction of the proposed techniques with other techniques such as parallelization and vectorization. Reconfigurable hardware specific optimizations such as clock disabling for some pipeline stages and runtime modification of the interconnection to reduce the reconfiguration cost are also being examined.

The work reported here is part of the USC MAARCII project [9]. This project is developing novel mapping techniques to exploit dynamic and self reconfiguration to facilitate run-time mapping using configurable computing devices and architectures. Moreover, a domain-specific mapping approach is being developed to support instance-dependent mapping. Finally, the concept of "active" libraries is exploited to realize a framework for automatic dynamic reconfiguration [8, 13].

References

1. K. Bondalapati. *Modeling and Mapping for Dynamically Reconfigurable Architectures.* PhD thesis, University of Southern California. Under Preparation.
2. K. Bondalapati, P. Diniz, P. Duncan, J. Granacki, M. Hall, R. Jain, and H. Ziegler. DEFACTO: A Design Environment for Adaptive Computing Technology. In *Reconfigurable Architectures Workshop, RAW'99*, April 1999.
3. K. Bondalapati and V.K. Prasanna. Mapping Loops onto Reconfigurable Architectures. In *8th International Workshop on Field-Programmable Logic and Applications*, September 1998.
4. K. Bondalapati and V.K. Prasanna. Dynamic Precision Management for Loop Computations on Reconfigurable Architectures. In *IEEE Symposium on FPGAs for Custom Computing Machines*, April 1999.
5. D. A. Buell, J. M. Arnold, and W. J. Kleinfelder. *Splash 2: FPGAs in a Custom Computing Machine.* IEEE Computer Society Press, 1996.
6. S. Cadambi, J. Weener, S. C. Goldstein, H. Schmit, and D. E. Thomas. Managing Pipeline-Reconfigurable FPGAs. In *Proceedings ACM/SIGDA Sixth International Symposium on Field Programmable Gate Arrays*, February 1998.
7. D. Chang and M. Marek-Sadowska. Partitioning sequential circuits on dynamically reconfigurable fpgas. In *IEEE Transactions on Computers*, June 1999.
8. A. Dandalis, A. Mei, and V. K. Prasanna. Domain specific mapping for solving graph problems on reconfigurable devices. In *Reconfigurable Architectures Workshop*, April 1999.
9. MAARCII Homepage. http://maarcII.usc.edu.
10. R. Kress, R.W. Hartenstein, and U. Nageldinger. An Operating System for Custom Computing Machines based on the Xputer Paradigm. In *7th International Workshop on Field-Programmable Logic and Applications*, pages 304–313, Sept 1997.
11. W. Luk, N. Shirazi, S.R. Guo, and P.Y.K. Cheung. Pipeline Morphing and Virtual Pipelines. In *7th International Workshop on Field-Programmable Logic and Applications*, Sept 1997.
12. K. M. G. Purna and D. Bhatia. Temporal partitioning and scheduling data flow graphs for reconfigurable computers. In *IEEE Transactions on Computers*, June 1999.
13. R. P. Sidhu, A. Mei, and V. K. Prasanna. Genetic programming using self-reconfigurable fpgas. In *International Workshop on Field Programmable Logic and Applications*, September 1999.
14. R. Subramanian, N. Ramasubramanian, and S. Pande. Automatic analysis of loops to exploit operator parallelism on reconfigurable systems. In *Languages and Compilers for Parallel Computing*, August 1998.
15. M. Weinhardt. Compilation and pipeline synthesis for reconfigurable architectures. In *Reconfigurable Architectures Workshop(RAW' 97)*. ITpress Verlag, April 1997.
16. M. Weinhardt and W. Luk. Pipeline vectorization for reconfigurable systems. In *IEEE Symposium on Field-Programmable Custom Computing Machines(FCCM '99)*, April 1999.
17. Xilinx Inc.(www.xilinx.com). *Virtex Series FPGAs.*

Compiling Process Algebraic Descriptions into Reconfigurable Logic

Oliver Diessel and George Milne

Advanced Computing Research Centre
School of Computer and Information Science
University of South Australia
Adelaide SA 5095
{Oliver.Diessel, George.Milne}@unisa.edu.au

Abstract. Reconfigurable computers based on field programmable gate array technology allow applications to be realized directly in digital logic. The inherent concurrency of hardware distinguishes such computers from microprocessor–based machines in which the concurrency of the underlying hardware is fixed and abstracted from the programmer by the software model. However, reconfigurable logic allows the potential to exploit "real" concurrency. We are therefore interested in knowing how to exploit this concurrency, how to model concurrent computations, and which languages allow us to control the hardware most effectively. The purpose of this paper is to demonstrate that behavioural descriptions expressed in a process algebraic language can be readily and intuitively compiled to reconfigurable logic and that this contributes to the goal of discovering appropriate high–level languages for run–time reconfiguration.

1 Introduction

The term *reconfigurable computer* is currently used to denote a machine based on field programmable gate array (FPGA) technology. This chip technology is programmable at the gate level thereby allowing any discrete digital logic system to be instantiated. It differs from the classical von Neumann computing paradigm in that a program does not reside in memory but rather an application is realized directly in digital logic.

For some computing and electronic control applications we are able to exploit the inherent concurrency of digital logic to directly realize algorithms as custom hardware to gain a performance advantage over software executing on conventional microprocessors. Given this observation, we may ask a wide range of questions, such as: how do we exploit this concurrency? How do we harness it to perform computations? How do we model such computation? And what programming languages should we use to help programmers/designers?

This paper demonstrates that we can intuitively and rapidly compile a high–level language that is oriented to describing concurrency and communication into reconfigurable logic. We show how the core features of process algebras

J. Rolim et al. (Eds.): IPDPS 2000 Workshops, LNCS 1800, pp. 916-923, 2000.

[5, 7, 4] and the Circal process algebra in particular [5, 6] can be mapped into reconfigurable logic.

The rationale for focusing on using a process algebra as the basis of a language for specifying reconfigurable logic are that it expresses the behaviour of a design in an abstract, technology–independent fashion and it emphasizes computation in terms of a hierarchical, modular, and interconnected structure. Process algebra have an extensive track record in the expression and representation of highly concurrent systems including digital hardware [1, 6] and are thus a good basis for a high–level language.

A high–level language based on process algebra is quite different from classical hardware description languages, such as VHDL and Verilog, that are oriented towards register–transfer and gate–level descriptions. Instead, this approach provides designers with a design paradigm focussed on behavioural process modules and their interconnection. Because of its modular focus, our approach aids the rapid compilation and partial reconfiguration of designs at run–time. Our approach also presents us with the potential for formally verifying the compilation algorithm. Related research on verifiable compilation to FPGAs was performed by Shaw and Milne [9] while Page and Luk [8] also constructed an Occam to FPGA compiler.

Circal models emphasize the control of and communication between processes. The rapid compilation of Circal models allows assemblies of interacting finite state machines to be implemented quickly. Apart from logic controllers, we may thus be able to build and quickly modify test pattern generators that function at near hardware speed. This project also aims to support dynamic structures that may facilitate the control of dynamically reconfigurable logic.

In the following section we provide an overview of the Circal process algebra and the source language for our compiler. Section 3 introduces our contribution with an overview of the compiler. We describe a technology–independent circuit model of Circal processes in Section 4. The mapping of these circuits to FPGAs, and Xilinx XC6200 chips in particular, is discussed in Section 5. The derivation of the mapping from behavioural Circal descriptions is outlined in Section 6. A summary of the paper and directions for further work are presented in Section 7.

2   The Circal process algebra

Circal is an event–based language; processes interact by participating in events, and sets of simultaneous events are termed actions. For an event to occur, all processes that include the event in their specification must be in a state that allows them to participate in the event. The Circal language primitives are:

State Definition $P \leftarrow Q$ defines process P to have the behaviour of term Q.
Termination $/\backslash$ is a deadlock state from which a process cannot evolve.
Guarding $a\,P$ is a process that synchronizes to perform event a and then behaves as P. $(a\,b)\,P$ synchronizes with events a and b simultaneously and then behaves as P.

Choice $P + Q$ is a term that chooses between the actions in process P and those in Q, the choice depending upon the environment in which the process is executed.

Non–determinism $P \& Q$ defines an internal choice that is determined by the process without influence from its environment.

Composition $P * Q$ runs P and Q in parallel, with synchronization occuring over similarly named events.

Abstraction $P - a$ hides event set a from P, the actions in a becoming unobservable.

Relabelling $P[a/b]$ replaces references to event b in P with the event named a.

3   Overview of compiler operation

This paper describes our efforts to implement a subset of Circal suited to the instantiation of Circal process models as reconfigurable logic circuits. The implementation of the hardware compiler is referred to as HCircal.

An HCircal source file consists of a declaration part, a process definition part, and an implementation part. Events and processes must be declared before use. The definition part consists of a sequence of process definitions adhering to the Circal BNF. The implementation part is introduced with the `Implement` declarative and is followed by a comma–delimited list of process compositions that is to be implemented in hardware. Processes must be defined before they are referred to in an `Implement` statement. HCircal does not currently allow the user to model non–determinism, abstraction, or relabelling. However, implementations of abstraction and relabelling are straightforward extensions to the current system.

In outline, the HCircal compiler operates as follows:

1. The user inputs an HCircal specification of the system to be implemented.
2. A compiler analyses the specification to produce a hardware implementation and a driver program for interacting with the hardware model
   - The current hardware model is in the form of a Xilinx XC6200 FPGA configuration bitstream [10] suitable for loading onto XC6200–based reconfigurable coprocessors such as the SPACE.2 board [3].
   - The driver program is a C program that executes on the host. The program loads the configuration onto the coprocessor and allows the user to interact with the implemented system.
3. The user runs the driver program and interacts with the hardware model by entering event traces and observing the system response.

The following sections describe the mapping from behavioural descriptions to technology–independent circuits, the decomposition of the circuits into modules for which FPGA configurations are readily generated, and the derivation of the module parameters from the Circal specification. The generation of the host program is a straightforward specialization of a general program that obtains appropriate event inputs, loads the input registers, and reads the process state registers. It is not further discussed.

4   A circuit model of Circal

The aim of the model is to represent, as faithfully as possible, Circal semantics in hardware. The design concentrates on the representation of the Circal composition operator, which is of central importance because it is through the composition of processes that interesting behaviour is established. When processes are composed in hardware they are executed concurrently.

The hardware implementation of the Circal system follows design principles that aim to generate fast circuits quickly. The first of these is that, for the sake of speed and scalability, the hardware representation of Circal aims to minimize its dependence upon global computation at the compilation and execution phases. The second principle is that we choose to design for ease of run–time instantiation and computational speed over area minimality. The motivation for these choices is the desire to leverage the speedup afforded by concurrently executing the Circal system in hardware; they are supported by the ability to reconfigure the gate array at run–time in order to provide a limitless circuit area. Finally, we desire a reusable design because we believe that will facilitate design synthesis, circuit reconfiguration, and future investigations into dynamically structured Circal.

4.1   Design outline

A block diagram of a digital circuit that implements a composition of Circal processes in hardware. is shown in Figure 1(a).

The circuit consists of a set of interconnected processes that respond to inputs from the environment by undergoing state transitions. Processes are implemented as blocks of logic with state. In a given state, each process responds to events according to the Circal process definitions. Individual processes examine the event offered by the environment and produce a "request to synchronize" signal if the event is found to be acceptable. The request signals for all processes are then reduced to a single synchronization signal that each process responds to independently.

Implementing Circal in synchronous FPGA circuits leads us to assume that: an event occurs at most once during a clock period; the next state is determined by the events that occurred during the previous clock period; and, if no event occurs between consecutive positive clock edges, then the idling transition $P \rightarrow P$ occurs upon the second clock edge by default.

4.2   Process logic design

Process logic blocks are derived from the process definition syntax and represented as compact localized blocks of logic to simplify the placement and routing of the system. A high–level view of a process logic block is given in Figure 1(b).

A process is designed to respond to events in the environment that are acceptable to all processes in the composed system. In order to perform this function, the process logic first checks whether the event is acceptable to itself. If

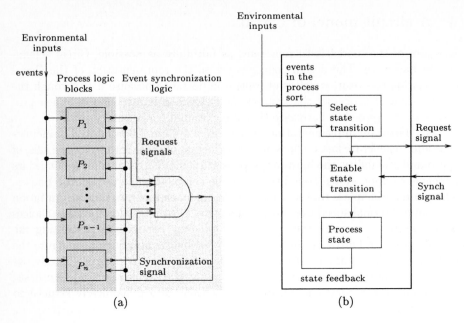

Fig. 1. (a) Circuit block diagram, and (b) Circal process logic block.

all processes find the event acceptable, the event synchronization logic returns a synchronization signal that is used by individual process logic blocks to enable the state transition guarded by the event. The following subsections describe the process logic design in more detail.

Determining the validity of event combinations We construct a combinational circuit that checks whether the events in the sort of the process form a valid guard for the current state. The process also accepts a null event (an event not in its sort) in order to allow other processes to respond to events it does not care about. The current state of the process is recycled if an unacceptable or null event is offered by the environment.

Let us assume at most k possibly recursive definitions $P_0, P_1, \ldots, P_{k-1}$ are necessary to describe the evolution of process P with sort $S = \{e_0, e_1, \ldots, e_{n-1}\}$, and that P_i, with $0 \leq i \leq k-1$, is defined as $P_i \leftarrow g_{i,0} P_{i,0} + \ldots + g_{i,j} P_{i,j} + \ldots + g_{i,j_i} P_{i,j_i}$, where index i refers to the current state, $P_{i,j}$ is the next state, $P_0, \ldots, P_{k-1}$, the state P_i evolves to under guard $g_{i,j} \subseteq S$, and $g_{i,j}$ is interpreted as the simultaneous occurrence of the events in $g_{i,j}$. The definition for P_i consists of $j_i + 1$ guarded terms where the $g_{i,j}$ are all distinct. Note that there may be at most k distinct next states but $2^n - 1$ distinct guards.

If we think of the events and states as *boolean* variables, then in state P_i the process responds to event combinations in the set $\{\gamma_{i,j}\} \cup \{\nu_S\}$, where $\gamma_{i,j} = \varepsilon_0 \varepsilon_1 \ldots \varepsilon_{n-1}$ and $\varepsilon_l = e_l$ or $\varepsilon_l = \overline{e_l}$, for $0 \leq l \leq n-1$, depending upon whether or not $e_l \in g_{i,j}$, and where $\nu_S = \overline{e_0}\,\overline{e_1} \ldots \overline{e_{n-1}}$ is the null event for sort S. Process P in state P_i therefore accepts the boolean expression of events $\nu_S + \sum_{0 \leq j \leq j_i} \gamma_{i,j}$.

The request for synchronization signal, r_P, is thus formed from the expressions for all states: $r_P = \sum_{0 \leq i \leq k-1} (\nu_S + \sum_{0 \leq j \leq j_i} \gamma_{i,j}).P_i$.

Checking the acceptability of an event The request signals for all processes are ANDed together in an AND gate tree that is implemented external to the individual process logic blocks. The output of the tree is fed back to each process as the synchronization signal, s.

Enabling state transitions The state of the process is stored in flip–flops — one for each state. Let $D_{P_l}, 0 \leq l \leq k-1$, denote the boolean input function of the D–type flip–flop for state P_l. Then we can derive the following boolean equations from the process definitions: $D_{P_l} = s.(\nu_S.P_l + \sum_{0 \leq i \leq k-1} \sum_{P_{i,j}=P_l} \gamma_{i,j}.P_{i,j}) + \overline{s}.P_l$, for $0 \leq l \leq k-1$.

In the above equations, the terms in parentheses are enabled when the synchronization signal, s, is high. These terms correspond to the guards on state transitions and to state recycling if a null event was offered to this process. The last term in the equations forces the current state to be renewed if the processes could not accept the event combination offered by the environment. By observing the synchronization signal, the environment can determine whether or not an event was accepted and can thus be constrained by the process composition.

4.3   The complete process logic block

The disjunction of the parenthesized terms in the flip–flop input functions implements the same boolean function as that to obtain the request signal. We therefore use the state selection circuits to form the request signal and use the synchronization signal to enable the selection.

5   Mapping circuits to reconfigurable logic

In this section we consider the placement and routing of the circuits derived in Section 4. The derivation of circuit requirements from the specification is discussed in the next section.

Our primary compilation goal is to generate FPGA configurations rapidly. We also want to be able to replace circuitry at run–time to explore changing process behaviours and to overcome resource limitations. For this reason we're interested in mapping to Xilinx XC6200 technology because its open architecture allows us to produce our own tools and because the chip is partially reconfigurable.

Difficulties with placing and routing the Circal models satisfactorily with XACTStep, the Xilinx APR tool for XC6200, led us to consider decomposing the circuits into modules that can be placed and routed under program control. These modules serve as an attractive intermediate form since they are easily derived from the specification, they completely describe the circuits to be implemented in a hardware–independent manner, and the FPGA configuration can be generated without further analysis.

The circuits described in Section 4 are specified in terms of parameterised modules that communicate via adjoining ports when they are abutted on the array surface. To simplify the layout of the circuits, all modules are rectangular in shape. The internal layout of modules is also simplified by using local interconnects only. The module representation of the circuits is readily mapped to a particular hardware technology by suitable module generators. The compiler can thus be ported to a new FPGA type by implementing a new set of module generators.

We distinguish between 9 module types. Each module type implements a specific combinational logic function using a particular spatial arrangement. Modules are specified in terms of their location on the array, input and/or output wire bit vectors, and the specific function they are to implement, e.g., minterm number. The interested reader is referred to our technical report for a complete description of the module functions, parameters, and circuit generators [2].

6   Deriving modules from process descriptions

For each unique process that is to be implemented, a process template that consists of the modules comprising the process logic is constructed. The module parameters for a process template are independently calculated using relative offsets. Once the size of the logic for each template is known, a copy with absolute offsets (final placement of modules) is made for each process to be implemented. When all the parameters are known, the FPGA configuration is generated.

Currently the compilation is performed off–line and the configurations generated are static. In future implementations we plan to experiment with replacing modules at run–time to overcome resource limitations and implement dynamically changing process behaviours. Minor behavioural changes may simply involve replacing minterms or guard modules which could be done very quickly. The regular shapes and small sizes of modules may allow us to distribute them and finalize the module positioning at run–time in order to maximize array utilization.

For a more detailed description of the steps in the derivation of the module representation please refer to [2].

7   Conclusions

We have shown how to model Circal processes as circuits that can be mapped to blocks of logic on a reconfigurable chip. Modelling system components as independent blocks of logic allows them to be generated independently, to be implemented in a distributed fashion, to operate concurrently, and to be swapped to overcome resource limitations. The model thus exploits the hierarchy and modularity inherent in behavioural descriptions to support virtualization of hardware.

We have shown how to instantiate a circuit by decomposing it into parametric modules that perform functions above the gate level. To simplify the layout, modules are mapped to rectangular regions that are wired together by abutting them on a chip. Since the modules completely describe the circuits to be implemented in a hardware–independent yet readily mapped manner, they could serve as a mobile description of Circal processes that can be transmitted and instantiated remotely.

Future work will investigate developing an interpreter that adapts to resource availability and supports dynamic process behaviour. We also intend assessing the usability of process algebraic specifications for a number of applications. A further direction is to enhance the HCircal language to support stream–oriented and data–parallel computations.

Acknowledgements

We gratefully acknowledge the helpful comments and suggestions made by Alex Cowie, Martyn George, and Bernard Gunther.

References

1. A. Bailey, G. A. McCaskill, and G. Milne. An exercise in the automatic verification of asynchronous designs. *Formal Methods in System Design*, 4(3):213–242, 1994.
2. O. Diessel and G. Milne. Compiling HCircal. Draft manuscript, Advanced Computing Research Centre, University of South Australia, Adelaide, Australia, Spetember 24, 1999.
3. B. K. Gunther. SPACE 2 as a reconfigurable stream processor. In N. Sharda and A. Tam, editors, *Proceedings of PART'97 the 4th Australasian Conference on Parallel and Real–Time Systems*, pages 286 – 297, Singapore, Sept. 1997. Springer–Verlag.
4. C. A. R. Hoare. *Communicating Sequential Processes*. Prentice-Hall International series in computer science. Prentice–Hall, Inc., Englewood Cliffs, NJ, 1985.
5. G. Milne. CIRCAL and the representation of communication, concurrency and time. *ACM Transactions on Programming Languages and Systems*, 7(2):270–298, 1985.
6. G. Milne. *Formal Specification and Verification of Digital Systems*. McGraw–Hill, London, UK, 1994.
7. R. Milner. *Communication and Concurrency*. Prentice–Hall, Inc., New York, NY, 1989.
8. I. Page and W. Luk. Compiling Occam into FPGAs. In W. R. Moore and W. Luk, editors, *FPGAs, Edited from the Oxford 1991 International Workshop on Field Programmable Logic and Applications*, pages 271 – 283, Abingdon, England, 1991. Abingdon EE&CS Books.
9. P. Shaw and G. Milne. A highly parallel FPGA–based machine and its formal verification. In H. Grünbacher and R. W. Hartenstein, editors, *Second International Workshop on Field–Programmable Logic and Applications*, volume 705 of *Lecture Notes in Computer Science*, pages 162–173, Berlin, Germany, Sept. 1992. Springer–Verlag.
10. Xilinx. *XC6200 Field Programmable Gate Arrays*. Xilinx, Inc., Apr. 1997.

Behavioral Partitioning with Synthesis
for Multi-FPGA Architectures
under Interconnect, Area, and Latency Constraints *

Preetham Lakshmikanthan **, Sriram Govindarajan,
Vinoo Srinivasan ***, and Ranga Vemuri
{plakshmi, sriram, vsriniva, ranga}@ececs.uc.edu
Department of ECECS, University of Cincinnati, Cincinnati, OH 45221

Abstract

This paper presents a technique to perform partitioning and synthesis of behavioral specifications. Partitioning of the design is done under multiple constraints – interconnections and device areas of the reconfigurable architecture, and the latency of the design. The proposed Multi-FPGA partitioning technique (FMPAR) is based on the Fiduccia-Mattheyses (FM) partitioning algorithm. In order to contemplate multiple implementations of the behavioral design, the partitioner is tightly integrated with an area estimator and design space exploration engine.

A partitioning and synthesis framework was developed, with the FMPAR behavioral partitioner at the front-end and various synthesis phases (High-Level, Logic and Layout) at the back end. Results are provided to demonstrate the advantage of tightly integrating exploration with partitioning. It is also shown that, in relatively short runtimes, FMPAR generates designs of similar quality compared to a Simulated Annealing partitioner. Designs have been successfully implemented on a commercial multi-FPGA board, proving the effectiveness of the partitioner and the entire design framework.

1 Introduction

Partitioning is essential when designs are too large to be placed on a single device or because of I/O pin limitations. Partitioning of a design can be done at various levels - behavioral, register-transfer level (RTL) or gate-level. Behavioral partitioning is a pre-synthesis partitioning while RTL partitioning is done after high-level synthesis. Various studies [1] show the superiority of behavioral over structural partitioning.

A behavioral partitioner has no a priori knowledge about design parameters such as area and latency. The partitioner must be guided by a high-level estimator that provides the required information. Efficient estimation techniques [2, 3] have been developed for this purpose. The approach presented in [2], presents an efficient design space exploration technique that can be performed dynamically with partitioning. A partitioner can effectively control the trade-off between the execution time and the design space

---

* This work is supported in part by the US Air Force, Wright Laboratory, WPAFB, under contract number F33615-96-C-1912, and under contract number F33615-97-C-1043

** Currently at *Cadence Design Systems* Inc., MA. Work done at University of Cincinnati.

*** Currently at *Intel* Corporation, CA. Work done at University of Cincinnati.

J. Rolim et al. (Eds.): IPDPS 2000 Workshops, LNCS 1800, pp. 924-931, 2000.

explored. We show the effectiveness of integrating the partitioner with a design-space exploration engine in generating constraint satisfying solutions.

There has been a lot of research in multi-FPGA partitioning, as presented in the survey by Alpert and Kahng [4]. In particular, Sanchis [5] extended the FM for multiway partitioning by repeatedly applying standard bi-partitioning. This work attempts to minimize the sum of all the *cutsets* across all partition segments. For a multi-FPGA RC, it is imperative that the pin constraints of the devices are individually satisfied. Therefore, this method of minimizing a summation of cutsets may not produce a constraint satisfying solution. Our goal is to minimize each cutset individually for pin-constraint satisfaction. We present a technique called FMPAR which is an extension of the Fiduccia-Mattheyses algorithm [6]. The results of partitioning are compared against a Simulated Annealing (SA) partitioner that forms part of the SPARCS [7] framework.

The rest of the paper is organized as follows. Section 2 describes the partitioning and synthesis framework. Section 3 presents the FMPAR algorithm in detail and the interaction of FMPAR with an exploration engine. Finally, Section 4 presents results demonstrating the effectiveness of this work.

2 Partitioning and Synthesis Framework

The framework for partitioning and synthesis is shown in Figure 1.

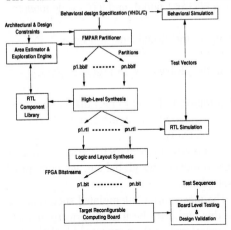

Fig. 1. Partitioning and Synthesis Framework

It consists of the FMPAR partitioner at the front-end and various synthesis phases (High-level, Logic, and Layout) at the back-end. The input behavioral designs are specified in subsets of either VHDL or C. The design descriptions are translated into an equivalent Control-Data Flow Block Graph (CDFG), where the *blocks* contain a simple data-flow graph that captures computation, and the edges between blocks represent both data and control flow.

The FMPAR partitioner views a block in CDFG as an atomic element that cannot be partitioned onto multiple FPGAs. The edges between various blocks are the set of *cutset* constraints for the partitioner. The user can specify any number of logical memory segments modeled as *dummy* blocks in the CDFG. The FMPAR partitioner automatically maps the logical memory blocks onto the physical memory banks. The core of the entire flow is the iterative FMPAR partitioner coupled with an area estimator and exploration engine. The exploration engine performs effective resource sharing across blocks and provides the partitioner with accurate area estimates.

The partitioned behavior segments generated by FMPAR are automatically synthesized by an in-house high-level synthesis tool to generate equivalent RTL designs. Further, the RTL designs are taken through commercial logic (*Synopsys Design Compiler*) and layout (*Xilinx M1*) synthesis tools to generate FPGA bitstreams for the target board. Note that, the communication signals routed across devices are always registered in the RTL designs to ensure that the board interconnect delay does not affect the clock period of the partitioned design.

3   The FMPAR Partitioner with the Exploration Engine

Like the FM, the FMPAR also allows only one block to be moved at a time and the locking option of cells in the standard FM is incorporated here. A block can be moved across the FPGAs, a user-specified number of times, after which it is locked and cannot be moved. We now present the terminology and details of the FMPAR algorithm.

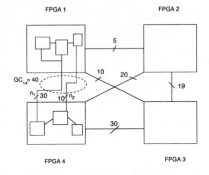

Fig. 2. Cutsets between FPGAs

Global Cut (*GC*) : This is defined as the *cut-set* between the partitions assigned to two FP-GAs. Consider the example shown in Figure 2. The RC board contains 4 FPGAs and it is a fully connected board. There are six global cuts, for example GC_{14} denotes the global cut between FPGAs 1 and 4, and $|GC_{14}|$ denotes the *size* of the global cut.

Current_Max : is the greatest value among all the global cuts. In Figure 2, $|GC_{14}|$ = 40 is the Current_Max.

Current_Min : is the least value among all the global cuts. In Figure 2, $|GC_{12}| = 5$ is the Current_Min.

Net Cut (n_i) : Each global cut is in turn composed of a set of nets that are cut $\{n_1, n_2, ..., n_k\}$. Consider the Current_Max value ($|GC_{14}| = 40$) in Figure 2. It is contributed to by 2 net-cuts n_1 and n_2 of *size* 30 and 10 respectively.

Priority : The net-cuts are prioritized in decreasing order of their sizes. The size of a net-cut is the bit-width of the net. In trying to reduce any global cut, we attempt to eliminate the net-cuts, one at a time, in the sorted order.

Net-Cut Elimination & Move Types : The moves are contemplated such that the worst GC (*Current_Max*) is reduced. For this purpose, the highest priority net-cut of the worst GC is considered. Moves are contemplated on the connected blocks to this net-cut. For example in Figure 2, n_1 is the highest priority net-cut in the worst global cut GC_{14}. Three possible moves can be contemplated to eliminate this net-cut: (1) Move the connected blocks in FPGA 1 into FPGA 4 or vice-versa, (2) Move the connected blocks in FPGA 1 or 4 into FPGA 2, and (3) Move the connected blocks in FPGA 1 or 4 into FPGA 3.

We call Option 1 as *1-degree move* of a netcut, Option 2 as *2-degree move* of a netcut and Option 3 as *3-degree move* of a netcut. In general, for 'n' FPGAs, a 1-degree move is between the pair of FPGAs (say F_i and F_j) associated with the highest priority net-cut. The remaining 'n-2' FPGAs (other than F_i and F_j) are sorted in decreasing order of available *free space*, $F_2, F_3, ..., F_{n-1}$. A *k-degree move* ($2 \leq k \leq$ n-1) is defined as one where blocks on either F_i or F_j are moved to the corresponding k'th FPGA. We define *free space* as the difference between the device area and the estimated area of the partition segment.

3.1 The FMPAR Algorithm

Algorithm 3.1 presents the outline of FMPAR, the proposed multiway FM partitioning technique. The inputs are the design described as a CDFG block graph (BG), the number of FPGAs (N_{fpga}) on the board, the size ($gc\_size[][]$) of interconnections between each pair of FPGAs, the area of each FPGA ($dev\_area[]$), the block locking factor ($lock\_fact$), and the design $latency$. Unlike a standard FM, our algorithm performs a user-specified number of runs (N_{runs}) from different initial partition solutions. During each run, the FM-loop (outer repeat-until loop) is executed until no improvement in cutset is observed for K successive iterations.

During each run of the FM algorithm a *legal* initial partition is generated. A partition is said to be *legal* if and only if all partition segments satisfy the area constraints posed by the individual devices. During each iteration of the FM-loop, all GCs are computed and ordered. If a *constraint satisfying* solution is obtained the entire *FMPAR* algorithm terminates. A *constraint satisfying* solution is a legal partition that satisfies the interconnection constraints as well. *Current_Max* is the worst cutset between all FPGA pairs, and is calculated every time a move is made. *Max* represents the least value of *Current_Max*, over all moves that have been made. *Prev_Max* is the least value of *Max* over all iterations of the FM-loop.

During each iteration of the FM-loop, several *legal moves* are made until no further moves are possible. A move is *legal*

Algorithm 31 (FMPAR Algorithm) FMPAR*(BG, N_fpga · gc_size[][], dev_area[], lock_fact, latency, K, N_runs)*
Begin
  Max ← *Prev_Max* ← ∞; *Current_Max* ← 0;
  For *FM_runs* = 1 to *N_runs*
    New Partition ← *Generate a legal initial partition;*
    Repeat   /* Run FM-loop until no improvement */
      Calculate GCs for all pairs of FPGAs;
      If $(1 \le i, j \le N_{fpga} \cdot |GC_{ij}| \le gc\_size(i,j))$ **Then**
        Output (Constraint Satisfying Solution) and Exit;
      EndIf;
      Repeat   /* Until no moves are possible */
        Calculate Current_Max, Current_Min and order all the GCs;
        If *(Current_Max < Max)* **Then**
          Max ← *Current_Max;*
        EndIf;
        Move ← *Choose_A_Move(N_fpga · Current_Max,*
          Current_Min, dev_area[], lock_fact);
        If *(Move = Φ)* **Then**
          If *(Max < Prev_Max)* **Then**
            Prev_Max ← *Max;*
            Save current partition as **best partition;**
          EndIf;
          Break; /* Out of inner repeat-until loop */
        Else
          Make the move and Increment that block's move_count;
        EndIf;
      Until*(False);*
      If *(Prev_Max hasn't changed over the last K runs)* **Then**
        Output **(best partition solution obtained);**
        Break; /* Out of outer repeat-until loop */
      EndIf;
      Reset move_count of all the blocks to zero;
      New Partition ← **best partition;**
    Until*(False);*
  EndFor; /* Restart FM with a new initial partition */
End.

move only if it leads to a legal partition and does not exceed the locking factor. The *locking factor* is a user-defined upper limit on the number of times a block can be moved. For selecting a legal move, the algorithm contemplates several possible moves in the procedure *Choose_A_Move()*. The contemplated moves are called $k - degree$ moves as explained earlier. The goal is to minimize the worst cutset ($Current\_Max$). If none of the moves decrease the worst cutset, then the least cutset *violating* move is accepted. If no legal move is possible the procedure returns Φ. This terminates the move-making process for one iteration of FM. At this point, each block is *unlocked* (move_count is set to zero) and the *best partition* obtained so far is used as the new partition for the next iteration of the FM.

3.2 Interaction between FMPAR and Exploration Engine

The partitioner is tightly integrated with a high-level exploration engine. The partitioner always communicates any change in the partitioned configuration to the exploration engine and both the tools maintain an identical view of the partitioned configuration. The exploration engine effectively uses the partitioning information by *dynamically* generating implementations that maximize sharing of resources within each partition segment. Further details of the exploration engine can be found in [2].

The partitioner dynamically controls the trade-off between the execution time and the design space explored. The exploration technique provides an Exploration Control Interface (ECI) that facilitates tight integration with the partitioning algorithm. This interface consists of a collection of *exploration methods* that generate new implementations, and *estimation methods* that simply re-compute the design estimates for a modified partition configuration. Algorithm 32 presents the template for the FMPAR algorithm with calls to the exploration engine enclosed in boxes. The FMPAR partitioner calls the area estimator and exploration engine at two places : (1) When moves are being evaluated (line-6), and, (2) when the configuration is reset to the best partition (line-10). A detailed study was conducted to make appropriate usage of the ECI functions at crucial points of the partitioning process.

Algorithm 32 (FMPAR with dynamic exploration)

```
FMPAR()
Begin
 1      Generate random initial partition of blocks;
 2      Repeat
 3          Unlock all blocks;
 4          While (∃ movable blocks) Do
 5              Select a block;
 6              ┌─────────────────┐
                │  Estimate Move;  │
                └─────────────────┘
 7              Make a move and lock;
 8          EndWhile;
 9          Reset to the best partition;
10          ┌──────────────────────────────────┐
            │  Explore Design for best partition; │
            └──────────────────────────────────┘
11      Until (No Cutset Improvement);
End.
```

The *Estimate Move* method evaluates the effect of moving a block from a source partition to a destination partition without performing exploration and hence is not expensive in time. Whereas, the *Explore Design* method attempts to generate area and latency satisfying implementations at the expense of compute time. This way the calls to the exploration engine effectively utilize the trade-off between the exploration time and the amount of design space explored.

Essentially, the partitioner takes care of the interconnection constraints, while the area and the latency constraints are handled by the area estimator and exploration engine. Thus, each time the solution is acceptable in terms of interconnection constraint, the exploration engine ensures the best area and latency satisfying solution.

4   Experimental Results

We first present results to show the effectiveness of the FMPAR algorithm integrated with the exploration engine. Then, the FMPAR is compared with a simulated annealing partitioner. Finally, we report results obtained for designs that were successfully implemented on the *Wildforce* [8], a commercial multi-FPGA board.

4.1   Effectiveness of Dynamic Exploration with FMPAR

We developed two versions of FMPAR, one performing dynamic exploration and another that does not. In the latter case, the exploration engine is used only to obtain area estimates without exploring multiple implementations. For experimentation, we considered the two large DSP benchmarks - the Discrete Cosine Transform (DCT) and the Fast Fourier Transform (FFT). The FFT benchmark has 18 blocks with 2 loops, 152 operations, 1418 nets (data bits) across the blocks, DCT has 66 blocks with 8 loops, 264 operations and 2401 nets and both examples have an extremely large number possible implementations.

We have gathered results by fixing two of the three constraints (design *latency* (L) and RC interconnection *cutset* (C)) and varying the third (device *area* (A)). The results are presented as plots where the x-axis represents the constraint varied (device area) and the y-axis represents the *fitness* value. Fitness is defined as,

$$F = \frac{1}{(1+CutsetPenalty)}, \quad where, \quad CutsetPenalty = \frac{\sum UnroutedNets}{TotalDesignNets}$$ The unrouted nets is the summation of all the nets contributing to GCs that exceed the board cutsize. Fitness is a measure of the solution quality, ranging between 0 and 1. A fitness value of 1 denotes a constraint satisfying solution, while a lower value denotes a poor quality solution because of a violation of cutset constraints.

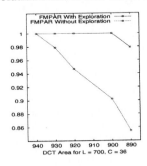

Fig. 3. Plot for DCT

We chose a representation of the *Wildforce* architecture with four FPGA devices and a cutset constraint of 36 interconnections between each pair of FPGAs. Figure 3 plots the fitness of generated solutions for the DCT benchmark. Both versions of the partitioner (with and without the dynamic exploration) generate constraint-satisfying solutions for all area constraints at and greater than 940 CLBs. As we gradually decrease the design area we see that the FMPAR version with dynamic exploration continues to generate constraint-satisfying solutions ($F = 1$), while its counterpart fails ($F < 1$), even after running on a large number random initial configurations.

We have made similar observations for the FFT benchmark, presented in [2]. This clearly demonstrates the effectiveness of interfacing the partitioner with the area estimator and exploration engine.

4.2 Comparison of FMPAR against a Simulated Annealing Partitioner

In this section, we compare the results of the FMPAR algorithm to that of a Simulated Annealing (SA) partitioner that a part of the SPARCS [7] design environment. The SA was also interfaced with the area estimator and design space exploration engine. Both algorithms were implemented and run on the same workstation – a twin processor *UltraSparc* with 384 MB RAM and clocking at 296 Mhz.

The table in Fig.4 provides a comparison of designs partitioned by the FMPAR and SA partitioners. For each design example, both partitioners were run on the same set of device area and design latency constraints. The comparison metrics are: (i) the number of *unrouted nets* (# UN) in the resulting solution and, (ii) the *run time* for each partitioner. The N_{devs} in the first column represents the number of devices on the RC, provided as a constraint to the partitioner. The interconnection constraint ($CutSet$) between each FPGA pair was fixed at 36. The last column in the table presents the *speedup* factor of the FMPAR partitioner over the SA partitioner.

Both the FMPAR and SA partitioners found constraint-satisfying solutions in five cases (Rows 1,2,4,5 and 7). The designs satisfied the cutset constraints as evidenced by '0' unrouted nets. At the same time, we see that the FMPAR algorithm *always* has much lesser run times than that of the SA.

Both partitioners did not find a constraint-satisfying solution for three designs – ELLIP (Row 3), FFT (Row 6) and DCT4x4 (Row 8). This is because the partitioners failed on the a tight cutset constraint. For the DCT4x4 example which is the largest, the SA was run with a slow cooling schedule for 2 hrs and 24 mins and a solution with 21 unrouted nets was obtained. It is observed that for this example, the FMPAR partitioner produced a higher quality solution (only 14 unrouted nets) in a much lesser time (33.4x speedup). In case of the FFT design and the ELLIP examples, the resulting solutions of both partitioner are comparable, yet the *FMPAR* finishes quicker.

From the results, we conclude that FM-PAR produces partitioned solutions whose quality is similar to that of the SA, but, in much lesser run-times. This is because the SA is a stochastic, hill-climbing approach as opposed to the FMPAR which is a move-based algorithm that quickly converges to a constraint satisfying solution. Although FMPAR is highly dependent on the initial solution and could stop at a local optimum, the results are as good as the SA for the constraint satisfaction problem.

Design Name, (N_{devs})	FPGA Area (clbs)	Dsgn Lat. (clks)	Simulated Annealing Partn. Areas (clbs)	# UN	Run Time (h:m:s)	FMPAR Partn. Areas (clbs)	# UN	Run Time (m:s)	Spd Up
ALU (4)	150	18	60 , 123 146, 0	0	0:00 :04	22 , 147 146, 43	0	0: 01	4x
STATS (2)	324	44	287, 60	0	0:00 :10	49, 318	0	0: 02	5x
ELLIP (2)	450	61	337, 362	7	0:00 :55	441, 252	10	0: 12	5x
ELLIP (2)	600	61	536, 92	0	0:00 :12	596, 26	0	0: 02	6x
FIR (3)	290	93	242, 178, 0	0	0:00 :15	23, 86, 288	0	0: 03	5x
FFT (4)	540	104	446, 317 530, 0	4	0:00 :08	387, 494 500, 484	4	0: 16	15x
FFT (4)	580	104	0 , 580 550, 0	0	0:00 :31	480, 564 353, 423	0	0: 09	3x
DCT4x4 (4)	3600	415	3188, 3303 3468, 3266	21	2:24 :09	3338, 3534 3241, 3531	14	4: 19	33x

Fig. 4. Comparison of SA and FMPAR generated designs

4.3   On-Board Implementations

Two designs were executed on the board after logic and layout synthesis. The designs ALU and STATS were successfully implemented and tested on the Wildforce [8], a commercial multi-FPGA board. The ALU is a simple arithmetic unit that has four 16-bit operating modes: addition, subtraction, multiplication and sum of squares of two input operands. The STATS is a statistical analyzer that computes the mean and variance of eight 16-bit numbers.

Information about the synthesized designs are shown in Figure 5. We compare the estimated area and performance measures against the actual values after layout synthesis. Columns 3 and 4 show the estimated and actual area of each partition. In general, we observed in our experiments that the estimated areas are within 10-20% of accuracy. Columns 5 and 6 compare the latency constraint to

Design Name	Partition Number	Area (CLBs) Estimated	Actual	Latency (Clks) Constraint	Actual
ALU	P_1	22	30	18	19
	P_2	147	139		
	P_3	146	179		
	P_4	43	54		
STATS	P_1	49	66	44	46
	P_2	318	335		

Fig. 5. Designs down-loaded onto RC boards

the actual latency of the partitioned design obtained from board-level simulation. We observe that our framework satisfies the latency constraint within a deviation of 5%.

In order to check for functional correctness, the results generated on board were verified against the simulation results. The partitioned designs executed successfully and the results matched that of the simulation.

5 Summary

This paper presents a framework for multi-FPGA partitioning of behavioral designs and their synthesis onto reconfigurable boards. An FM based multiway partitioner was presented, which is integrated with an area estimator and design space exploration engine. By efficiently performing dynamic exploration with partitioning, the partitioner produces good quality solutions in a reasonable amount of time. A limitation of the partitioner is that it can currently handle only fixed interconnection architectures. In the future, we plan to integrate the partitioner with interconnect estimation techniques [9] that can handle programmable interconnection architectures.

Results are provided to demonstrate the advantage of tightly integrating exploration with partitioning. Also, it is shown that the FMPAR produces constraint-satisfying solutions of similar quality to that of the SA, in much lesser run-times. Designs taken down to the Wildforce board proves that the FMPAR algorithm maintains the functionality of the design after partitioning and also shows the effectiveness of the partitioning and synthesis framework.

References

1. F. Vahid. "Functional Partitioning improvements over Structural Partitioning for Packaging Constraints and Synthesis : Tool Performance". In *ACM Transactions on Design Automation of Electronic Systems*, volume 3, pages 181–208, April 1998.
2. S. Govindarajan, V. Srinivasan, P. Lakshmikanthan, and R. Vemuri. "A Technique for Dynamic High-Level Exploration During Behavioral-Partitioning for Multi-Device Architectures". In *Proc, of the 13th IEEE Intl. Conf. on VLSI Design*, January 2000.
3. F. Vahid and D. Gajski. "Incremental Hardware Estimation During Hardware/Software Functional Partitioning". In *IEEE Transactions on VLSI Systems*, volume 3, September 1995.
4. Charles J. Alpert and Andrew B. Kahng. "Recent Directions in Netlist Partitioning". In *Integration, the VLSI Journal*, 1995.
5. L. A. Sanchis. "Multiple-way network partitioning". In *IEEE Transactions on Computers*, pages 62–81. 38(1), January 1989.
6. C. M. Fiduccia and R. M. Mattheyses. "A Linear Time Heuristic for Improving Network partitions". In *Proceedings of the 19th ACM/IEEE DAC*, pages 175–181, 1982.
7. I. Ouaiss, S. Govindarajan, V. Srinivasan, M. Kaul, and R. Vemuri. "An Integrated Partitioning and Synthesis System for Dynamically Reconfigurable Multi-FPGA Architectures". In *Proc. of Reconfig. Arch. Workshop (RAW98)*, pages 31–36., March 1998.
8. Annapolis micro systems, inc. http://www.annapmicro.com/amshhomep.html.
9. V. Srinivasan, S. Radhakrishnan, R. Vemuri and J. Walrath. "Interconnect Synthesis for Reconfigurable Multi-FPGA Architectures". In *Proc. of RAW99*, pages 597–605., April 1999.

Module Allocation for Dynamically Reconfigurable Systems

Xue-jie Zhang and Kam-wing Ng

Department of Computer Science and Engineering
The Chinese University of Hong Kong
Shatin, N. T., Hong Kong
{xjzhang, kwng}@cse.cuhk.edu.hk

Abstract. The synthesis of dynamically reconfigurable systems poses some new challenges for high-level synthesis tools. In this paper, we deal with the task of module allocation as this step has a direct influence on the performance of the dynamically reconfigurable design. We propose a *configuration bundling* driven module allocation technique that can be used for component clustering. The basic idea is to group configurable logic together properly so that a given configuration can do as much work as possible, allowing a greater portion of the task to be completed between reconfigurations. Our synthesis methodology addresses the issues of minimizing reconfiguration overhead by maintaining a global view of the resource requirements at all times during the high-level synthesis process.

1 Introduction

A dynamically reconfigurable system allows hardware reconfiguration while part of the reconfigurable hardware is busy computing, and allows a large system to be squeezed into a relatively small amount of physical hardware[1]. Though very promising, the development of dynamically reconfigurable systems faces many problems.

Since the configuration changes over time, one major problem is that there needs to be some way to ensure that the system behaves properly for all possible execution sequences. For this time-multiplexed reconfiguration to be realized, a new temporal partitioning step needs to be added to the traditional design flow. Some researchers have addressed temporal partitioning heuristically, by extending existing scheduling and clustering techniques of high-level synthesis[2][3][4]. In an earlier work[5], we presented a design model for abstracting, analyzing and synthesizing reconfiguration at the operations level.

In addition to making sure that a temporal partitioning be done correctly and producing a functionally correct implementation of the desired behavior, another important problem is how to produce the best implementation of functionality. With normal FPGA-based systems, one wants to map the configurable logic spatially so that it occupies the smallest area, and produces results as quickly as possible. In a dynamically reconfigurable system one must also consider the time

J. Rolim et al. (Eds.): IPDPS 2000 Workshops, LNCS 1800, pp. 932–940, 2000.

to reconfigure the system, and how this affects the performance of the system. Configuration can take a significant amount of time, and thus reconfiguration should be kept to a minimum. This is in general a challenging problem to address, with almost no current solution[6].

In this paper, we present an efficient high-level synthesis technique which can be used to synthesize and optimize dynamically reconfigurable designs. In particular, we concentrate our investigation on the task of module allocation. Dynamic reconfiguration extends the module allocation space by an additional dimension. The optimizing criteria in dynamic resource allocation also shift from a single static netlist to several configurations of the design. We must account not only for temporal partitioning and scheduling effects but global considerations as well, such as the resource requirements of all configurations, reconfiguration overhead, and the combination of all of the above. We have addressed these issues by using a *configuration bundling* technique that balances the advantages of dynamic reconfiguration against the added cost of configuration time by maintaining a global view of the resource requirements of all temporal partitions at all times during high-level synthesis.

2 Problem Formulation

The contribution of this paper can be seen in the context of our previous work on a design model[5]. Our approach uses an extended control/data flow graph (ECDFG) as the intermediate representation of a design. The CDFG is extended by abstracting the temporal nature of a system in terms of the sensitization of paths in the dataflow. An ECDFG is a behavioral-level model. An ECDFG representation of system behavior consists of three major parts: (1) possible execution paths which are described by the product of the corresponding guard variables, (2)temporal templates which lock several configuration compatible operations into temporal segments of relative schedules, (3) a control and data flow graph (CDFG) describing data-dependency or control-dependency between the operations. Interested readers are referred to the original references for the details about ECDFG.

In high-level synthesis, module allocation is an important task which determines the number and types of RTL components to be used in the design. Since we have encoded the temporal nature of synthesizing such systems by *temporal templates*[5], the module allocation process may be translated into a two-dimensional placement problem of temporal templates. Instead of considering individual CDFG nodes, we restate the dynamic module allocation problem in terms of temporal templates, a given spatial and temporal placement of configurable logic resources used by some temporal templates for a range of time constraints represents a possible configuration. The module allocation problem for dynamically reconfigurable logic involves not only generating the configuration for each of the temporal templates, but also reducing the reconfiguration overhead incurred. Our problem can be formally defined as follows:

Problem 1. Let $F = \{F_1, F_2, ..., F_m\}$ be a set of function units which can be implemented on reconfigurable logic, and $C = \{C_1, C_2, ..., C_n\}$ be a set of possible configurations of the configurable logic units. Given an extended CDFG (ECDFG) $G = (V, E, \xi, \zeta)$ with a set of temporal templates in a given order $TT = (TT_1, TT_2, ..., TT_p)$, where $TT_i \in F$, find an optimal sequence of configurations $CS = (CS_1, CS_2, .., CS_q)$ for temporal template TT, where $CS_i \in C$ which minimizes the reconfiguration cost R. R is defined as

$$R = \sum_{i=2}^{q} \delta_i \qquad (1)$$

Where δ_i is the reconfiguration cost in changing configuration from CS_{i-1} to CS_i.

In the remaining sections, we use a new *configuration bundling driven technique* to address the module allocation problem.

3   Configuration Bundling

The basic idea is to group logic together properly so that a given configuration can do as much work as possible, allowing a greater portion of the task to be completed between reconfigurations.

We illustrate our concept with the help of a motivating example. Consider three temporal templates of an extended CDFG shown in Figure 1. Furthermore, assume that all operations finish in a single cycle and that all temporal templates have to be implemented in three clock cycles. If each temporal template is allocated as a single configuration, the first temporal template (shown in Figure 1(a)) requires a module allocation of *five functional units* namely $\{3\ adders, 1\ multiplier, 1\ subtractor\}$. Similarly, the second and third temporal templates (shown in Figure 1(b)-(c)) can be implemented with module allocations of $\{\ 2\ adders,\ 2\ multipliers,\ 2\ subtractors\}$ and $\{1\ adder, 1\ multiplier, 3\ subtractors\}$ respectively.

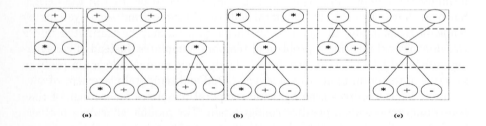

(a)   (b)   (c)

Fig. 1. A Motivating Example

A straightforward approach to optimize the module allocation of the three temporal templates as a dynamically reconfigurable design involves considering the granularity of the reconfiguration. Resource requirements of the temporal templates can be reduced significantly by maintaining a global view of the resource requirements of all temporal templates at all times during the synthesis process. In fact, the three temporal templates can be implemented using a configuration granularity of *two adders, two multipliers and two subtractors*. In this research, we have developed a configuration bundling technique to reduce the reconfiguration overhead. The concept of *configuration bundling* can be defined as follows:

Definition 1. *Given an extended CDFG (ECDFG) $G = (V, E, \xi, \zeta)$ with a set of temporal templates $TT = \{TT_1, TT_2, ..., TT_n\}$, a configuration bundle is a subset of TT such that the hardware resource requirements of individual temporal template in this subset $\{TT_{i_1}, TT_{i_2}, ..., TT_{i_m}\}$ can be implemented by an overall resource allocation schema.*

Configuration bundling is a synthesis technique where n temporal templates are bundled into at most m groups so that each temporal template belongs to at least one bundle and the objective function is optimized. Following configuration bundling, each bundle is synthesized into a separate configuration. The basic idea behind our configuration bundling technique is to attempt to identify and bundle temporal templates with similar computation topology and hardware types into compatible groups, such that these groups may be used to determine the choice of granularity for configurations that optimize the reconfiguration overhead. In particular, the following compatibility issue should be considered during the configuration bundling process.

3.1 Bundling Compatibility of Temporal Templates

If two temporal templates with disparate topologies are implemented in temporally consecutive configurations the attendant configuration overhead will be significant. In the worst case, each functional unit has to be reconfigured and this increases the time of reconfiguration. Therefore, topological similarity between temporal templates should be considered for bundling into the same group. For example, in Figure 2 *Temporal Template 2* can be bundled into a configuration implementing temporal template 4 with almost no reconfiguration overhead.

In addition, resource compatibility is an important issue during *configuration bundling*. For example, in Figure 2, while *Temporal Templates* 2, 3 and 4 use subtractors and multipliers, *Temporal Template* 1 uses adders. Therefore, bundling *Temporal Template* 1 with either *Temporal Template* 2 or 3 or 4 does not yield justifiable benefit for reducing the reconfiguration overhead. On the other hand, based on the compatibility of the functional unit types, *Temporal Templates* 2, 3, 4 are good candidates to be bundled into the same group.

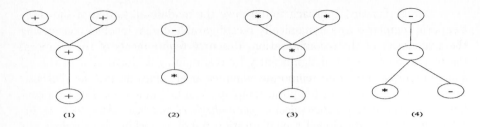

Fig. 2. Compatibility of Temporal Template

3.2   Measure of Configuration Bundling

Configuration bundling should take into account trade-offs between maximizing static resource requirements and minimizing reconfiguration overhead in space. Therefore, a configuration bundle will have the smallest area and the scope for maximum resource usage if the temporal templates in a bundle are compatible with one another. Based on the above observations we have developed a measure to identify bundling compatibility between temporal templates. We first outline the parameters of the function for bundling below.

- B: Set of bundles $B_1, B_2, ..., B_k$ for a given TT that describes a set of possible configuration bundling.
- $N_{F_j}(TT_i)$: the number of configuration of functional unit F_j for temporal template TT_i.
- $Area_{F_i}$: the area of a configuration of a functional unit F_i

Given a temporal template TT_i, the following is an estimate of the area of the temporal templates:

$$Area_{tt_i} = \sum_{f \in F} N_F(TT_i) \cdot Area_f \tag{2}$$

If a bundle B_i has n_i temporal templates, then the area of the bundle is estimated as below.

$$Area_{B_i} = \sum_{f \in F} max_{tt \in B_i} N_f(tt) \cdot Area_f \tag{3}$$

The larger the difference between these areas of temporal templates, the more incompatible the temporal template will be with the remaining temporal templates in the bundle. Given a temporal template TT_j for consideration for bundling in B_i, the incompatibility can be obtained as the following and is used to weigh the candidate solutions.

$$\delta_{B_i, TT_j} = \sum_{f \in F} |max_{TT \in B_i} N_f(TT) - N_F(TT_j)| \cdot Area_f \tag{4}$$

4   Configuration Bundling Driven Module Allocation Algorithm

Since there are several temporal templates in a range of time and module allocation, simultaneously considering all the temporal templates and their respective
constraints is difficult. We propose to allocate the hardware resources from a
range of time by considering one temporal template at a time. In particular, the
following three issues must be taken into account:

- the allocated hardware resource due to the previously considered temporal
  templates
- the estimated hardware resource of the remaining temporal templates
- the hardware resource required by the candidate temporal templates

Here, temporal templates are first bundled randomly. Then, a source configuration bundle is randomly chosen. From such a configuration bundle, an
incompatible temporal template is selected and moved to another configuration
bundle where the temporal templates are compatible with the selected temporal
template. The hardware area of all configurations is then computed, and the current bundling configuration is saved if it is the best so far. The process continues
until no more improvement is obtained for a given number of iterations.

For each configuration bundle $B_i \in B$, the module allocation algorithm is
outlined below. An initial module allocation A_{B_i} for each configuration bundle
B_i is first derived. Starting with a temporal template with the most resource
requirements, a feasible module allocation for the entire bundle is obtained.
From the total resource allocated to the configuration bundle A_{B_i}, the module
allocation R_{TT_i} for each candidate temporal template $TT_i \in B_i$ is obtained.
Then, allocation and scheduling of the design are carried out using this module
allocation technique.

4.1   Initial Module Allocation

Let N_{ij} be the maximum bound on the necessary amount of resource of a certain configuration type C_j of functional unit for the temporal template TT_i of
a configuration bundle. For each resource type C_j and for each *temporal template TT_i* of a configuration bundle B_i, relaxation based scheduling techniques
are used to derive an estimate of N_{ij}. For a configuration bundle B_i, a global
minimum bound of resource requirements N_j is used as the initial allocation for
the configurable logic C_j.

$$N_j = max_{TT_i \in B_i}(N_{ij}) \tag{5}$$

This is based on the fact that there will be at least one *temporal template* in
the *configuration bundle* that requires at least these many hardware configurations of type C_j.

4.2 Ordering and Allocating Temporal Templates

Within our methodology, the ordering of temporal templates in the same config-uration bundle has an impact on resource usage and reconfiguration overhead of the resulting resource allocation. A good order for module allocation of temporal templates is important because this order has a pronounced impact on the final resources allocation and the overall performance of the system. The proposed algorithm for ordering temporal templates include two stages called clustering and scheduling. The objective of the algorithm is to group temporal templates such that they may subsequently be allocated and scheduled.

When considering functional locality in the module allocation process, it is better to schedule and allocate together temporal templates contributing to the same *join node* in the ECDFG, because this could help in the scheduling and allocation of relative temporal templates at higher levels. Therefore, clustering temporal templates is the first step in the temporal templates ordering process. The *cones partitioning algorithm* provides the basis for our clustering stage[8][9].

Once temporal templates are partitioned into clusters, the cluster-based list scheduling and allocation algorithm orders the temporal templates in the same configuration bundle. Our algorithm combines scheduling with module allocation into subsequent configurations for temporal templates in the same configuration bundle, while considering functional locality of the configuration bundle. There are two main steps in our list scheduling algorithms: the formation of clusters and list scheduling temporal templates.

5 Experimental Results

In this section we present results to illustrate the effectiveness of the configu-ration binding technique. In order to experimentally verify the concept of con-figuration bundling driven module allocation, we used three popular high-level benchmarks - elliptical wave filter (EWF), finite impulse response filter (FIR) and bandpass filter (BF) - for optimizing the overall resource allocation as well as the reconfiguration overhead. We assume the following configurations for addi-tion and multiplication operations: look-ahead adder (Area = 1, latency = 1) and a two-stage multiplier (Area = 4, latency = 2). Figure 3 shows the component requirement for the static and configuration bundling driven module allocation.

Bundles	Static module allocation (Area)	Bundling driven module allocation (Area)	Reduction
{EWF,FIR,BF}	24	11	54.2%
{EWF,FIR}, {BF}	24	17	29.2%
{EWF},{FIR,BF}	24	13	45.8%

Fig. 3. Bundling to minimize reconfiguration cost

We have also combined our front-end algorithms with the existing DRL scheduling algorithms[2] back end for demonstrating our results. DRL scheduling algorithms do not consider the module allocation problem. We compare results of the combined algorithms with a single DRL approach[2] as shown in Figure 4, where t_e, n_p, n_f and λ represent the total data-path execution time, the number of partial and the number of full reconfigurations and the graph latency respectively.

Benchmarks	Total area	Combined approach				DRL			
		t_e	n_p	n_f	λ	t_e	n_p	n_f	λ
	15	15	25	0	15	17	1	2	17
	10	15	25	0	15	17	4	2	17
	6	15	24	1	15	17	2	8	17
Elliptic	15	16	12	0	16	17	5	0	17
wave_filter	10	18	13	1	17	18	9	0	18
	6	20	17	0	20	24	19	0	24
	15	19	3	0	19	18	8	0	18
	10	26	9	0	26	21	16	0	21
	6	28	12	0	28	37	1	9	19

Fig. 4. Synthesis result and comparison

The results in Fig.4 have shown that the use of the combined algorithm will lead to a faster execution time compared with a single DRL scheduling implementation, and with considerably smaller area. When the DRL scheduling is used alone, more control steps result but when scheduling together with our module allocation is performed the partial reconfigurations will frequently occur instead of the full reconfigurations. This is expected as our algorithm aims at producing a short reconfiguration time by maintaining a global view of the resource requirements of all temporal templates at all times during the synthesis process.

6   Conclusions and Acknowledgments

We have presented a new module allocation technique in this paper. It is based on a configuration bundling heuristic that tries to allocate configurable logic resources by maintaining a global view of the resource requirements of all temporal templates. The most important value of the configuration bundling driven module allocation technique is that enable trade-offs between the granularity of the configuration and reconfiguration overhead during high-level synthesis process.

The work described in this paper was partially supported by two grants: the Research Grant Council of the Hong Kong Special Administrative Region (RGC Research Grant Direct Allocation - Project ID: 2050196), and Yunnan Province Young Scholar Grant.

References

1. Lysaght and J. Dunlop: Dynamic reconfiguration of FPGAs, More FPGAs, UK:Abingdon EE and CS Books (1994), pp82-94, 1994.
2. M.Vasilko and D.Ait-Boudaoud: Architectural Synthesis Techniques for Dynamically Reconfigurable Logic, Field-Programmable Logic, Lecture Notes in Computer Science 1142, pp290-296
3. J. Spillane and H.Owen: Temporal Partitioning for Partially Reconfigurable Field Programmable Gate, Proceedings of Reconfigurable Architectures Workshop(RAW'98), 1998.
4. M. Kaul and R. Vemuri: Optimal Temporal Partitioning and Synthesis for Reconfigurable Architectures, Proceedings of Design and Test in Europe(DATE'98), 1998.
5. Kam-wing Ng, Xue-jie Zhang, and Gilbert H. Young: Design Representation for Dynamically Reconfigurable Systems, Proceedings of the 5th Annual Australasian Conference on Parallel And Real-Time Systems(PART'98), pp14-23, Adelaide, Australia, September 1998.
6. Scott Hauck and Anant Agarwal: Software Technologies for Reconfigurable Systems, Northwestern University, Dept. of ECE Technical Report, 1996.
7. Ivan Radivojevic and Forrest Brewer: A New Symbolic Technique for Control-Dependent Scheduling, IEEE Trans. on Computer-Aided Design of Integerated Circuit and Systems, vol.15, no.1, pp45-56, Jan. 1996 .
8. D. Brasen, J.P. Hiol and G. Saucier: Finding Best Cones From Random Clusters for FPGA Package Partitioning, IFIP International Conference on VLSI, pp 799-804, Aug. 1995.
9. Sriam Govindarajan and Ranga Vemuri: Cone-Based Clustering Heuristic for List-Scheduling Algorithms. Proceedings of the European Design and Test Conference, Paris, France, March 1997.

Augmenting Modern Superscalar Architectures with Configurable Extended Instructions

Xianfeng Zhou and Margaret Martonosi

Dept. of Electrical Engineering
Princeton University
{xzhou, martonosi}@ee.princeton.edu

Abstract. The instruction sets of general-purpose microprocessors are designed to offer good performance across a wide range of programs. The size and complexity of the instruction sets, however, are limited by a need for generality and for streamlined implementation. The particular needs of one application are balanced against the needs of the full range of applications considered. For this reason, one can "design" a better instruction set when considering only a single application than when considering a general collection of applications. Configurable hardware gives us the opportunity to explore this option. This paper examines the potential for automatically identifying application-specific extended instructions and implementing them in programmable functional units based on configurable hardware. Adding fine-grained reconfigurable hardware to the datapath of an out-of-order issue superscalar processor allows 4-44% speedups on the MediaBench benchmarks [1]. As a key contribution of our work, we present a selective algorithm for choosing extended instructions to minimize reconfiguration costs within loops. Our selective algorithm constrains instruction choices so that significant speedups are achieved with as few as 4 moderately sized programmable functional units, typically containing less than 150 look-up tables each.

1 Introduction

General-purpose instruction sets are intended to implement basic processing functions while balancing the needs of many applications. Complex instructions that might accelerate one application are often unused by several other applications. Worse, their implementation difficulties may impact all programs by degrading clock rates or using up vital chip area.

Configurable hardware allows one to implement complex operations on an as-needed basis, one application at a time. In recent years, configurable computing based on Field-Programmable Gate Arrays (FPGAs) has been the focus of increasing research attention. The circuit being implemented can be changed simply by loading in a new set of configuration bits. Various architectures for FPGA-based computing have been proposed, ranging from co-processor boards accessed via the I/O bus, to relatively fine-grained structures accessed as an integral part of the CPU's data path. The approach we explore here is closest to the latter architecture. We envision programmable functional units (PFUs) with 150 CLBs or less which are built into the datapath of a modern superscalar processor, and which can access the register file and result bus just like other functional units in the machine.

Customized complex or extended instructions have several advantages over traditional instruction sets. First, customization allows one to match the flow of

J. Rolim et al. (Eds.): IPDPS 2000 Workshops, LNCS 1800, pp. 941-950, 2000.

values within an extended instruction to the needs of the particular operation being performed. Second, one can customize the bitwidth of calculations to tightly match the needs of the particular application. Third, one can improve instruction-level parallelism (ILP) by amortizing the per-instruction cost of fetching, issuing, and committing over more work.

While these advantages are compelling, customized extended instructions cannot be applied universally. First, since the PFU is part of the datapath, increasing the number of inputs to a PFU also increases the number of register file ports needed by the processor. This increases machine complexity and may impact the cycle time. Second, reconfiguring a PFU for a particular extended instruction requires fetching configuration bits and sending them to the PFU. This reconfiguration latency warrants care in choosing to implement operations as extended PFU instructions.

With this in mind, we devised and modeled T1000, an out-of-order issue, superscalar processor with programmable functional units. Initial performance studies with a simple instruction selection algorithm shows 4-44% speedups for the MediaBench suite [1] when ignoring the reconfiguration penalties. To improve speedups under more realistic assumptions, we developed a selective approach for determining which extended instructions to implement and when to use them. The key difference of our work from previous work is to check many possibilities of converting an instruction sequence to valid extended instructions. The extended instructions chosen by our selective algorithm can typically fit in a PFU composed of <150 Xilinx look-up tables. Thus, PFUs can help improve performance without adding significantly to the area of the processor.

In the remainder of the paper, Section 2 shows our proposed architecture and its extended instruction-encoding format. Section 3 describes the evaluation methodology. Section 4 discusses a greedy instruction selection algorithm and gives initial performance data. When the number of PFUs is limited, however, greedy instruction selection results in too much reconfiguration overhead. For this reason, Section 5 introduces our selective algorithm which reduces the overhead caused by reconfiguring the PFUs too frequently. We discuss the PFU hardware cost in Section 6. Section 7 describes prior work and Section 8 offers conclusions.

2 T1000 Architecture[1]

The main contribution of our work is developing selective algorithms for harnessing fine-grained configurable hardware units. To evaluate this work, we place it in the following architectural context. The T1000 architecture we model is a reconfigurable computing architecture embedded into a superscalar, out-of-order issue CPU, as illustrated in Figure 1.

2.1 Background

PRISC, the first architecture with fine-grained configurable hardware units, was proposed by Razdan and Smith [2,4]. This architecture includes programmable functional units, or PFUs, attached directly to the datapath of a simple pipelined,

---

[1] T1000 is the name of the "liquid metal" morphing robot in the movie "Terminator 2" starring Arnold Schwartzenegger. We chose the name because, like configurable hardware, the T1000 can change its shape into arbitrary forms.

single-issue in-order processor. The PFU was dynamically configured to implement different extended instructions specialized to the applications.

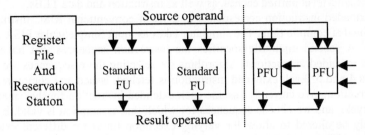

Fig. 1. T1000 architecture. To the left of the dotted line is a fixed RISC processor with out-of-order issue of four instructions per cycle

An extended instruction is created at compile time by converting an appropriate instruction sequence in the compiled code into a single PFU opcode. For example, a sequence of 3 data-dependent logic operations could easily be implemented more efficiently by a PFU based on lookup-tables than by a sequence of RISC instructions. If a PFU's specialized hardware evaluates the equivalent extended instruction in a single cycle, as compared to 3 cycles on traditional hardware, then there is a potential saving of 2 cycles each time the sequence is encountered.

2.2 T1000 Details

Like PRISC, T1000 relies on programmable functional units integrated into the datapath. A key difference from PRISC is that T1000 is assumed to be a 4-issue out-of-order machine. This is a more stringent test of the benefits of PFUs, since out-of-order issue helps tolerate the latencies of some data-dependent instruction sequences.

A T1000 extended instruction is encoded as a register-register operation with a specific opcode. We use a MIPS-like encoding format with an additional *Conf* field to control the loading of configuration bits. It defines the functionality of the PFU corresponding to that extended instruction. In the decode stage, this configuration signal is compared with the ID tag saved in each PFU. If the check shows that the correct extended instruction is currently configured in one of the PFUs, the extended instruction is dispatched normally. (This is akin to a cache hit.) Otherwise, the configuration bits must be loaded into one of the PFUs based on a LRU replacement policy before the extended instruction can be issued. A second key difference from prior work is our selective algorithm for choosing when to use extended instructions. This is described and evaluated in Section 5.

3 Methodology

3.1 Performance evaluation

To evaluate our selection algorithms and architecture, we developed a simulator based on the SimpleScalar tool set [5]. The simulator models an out-of-order issue, superscalar architecture coupled with PFUs. Out-of-order issue is managed using a Register Update Unit (RUU) [6]. The RUU scheme uses a reorder buffer to automatically rename registers and hold the results of pending instructions. The simulated processor can fetch, decode, issue, and commit four instructions per cycle and is simulated with perfect branch prediction. We simulate realistic instruction, data, and second-level unified caches, as well as instruction and data TLBs.

Each extended instruction corresponds to a set of conventional RISC instructions and is defined at compile time. The simulator takes as input SimpleScalar PISA object code files. A second input file specifies the instruction sequences that have been selected as extended instructions. Algorithms for selecting these sequences are given in Sections 4 and 5. Each extended instruction is fetched, issued and committed as a single instruction. We currently assume each extended instruction finishes execution in single cycle, and we choose sequences for which this assumption is valid, but this could easily be altered to allow for varying execution times for different extended instructions. With out-of-order issue hardware, varying execution times do not present a serious implementation obstacle.

In this study, we focus on the MediaBench benchmark suite [1]. The prevalence of narrow-width operations in multimedia processing applications makes them particularly suitable for a reconfigurable computing architecture. To collect data, we run each of the applications to completion.

3.2 Hardware Cost

The extended instructions we choose must fit within the resources of a PFU. We use the Configurable Logic Blocks (CLBs) of Xilinx XC4000-series devices as our target style of programmable hardware. To evaluate the performance and area needs of potential extended instructions, we implemented them in VHDL. The Xilinx Foundation tool then synthesizes and implements the design in an FPGA [7].

4 Potential Performance Payoff of Aggressive Instruction Selection

We evaluate the performance benefits of our selection algorithms in two steps. First, Section 4 explores the performance gain possible when we: (1) aggressively map as many extended instructions as possible, (2) use an unlimited number of PFUs, and (3) ignore any PFU reconfiguration overheads. While such assumptions are clearly not realistic, they help one see the best-case performance. This, in turn, guides the choices we describe in Section 5 to develop a more selective mechanism for choosing extended instructions that works well under more realistic assumptions.

Our greedy selection algorithm chooses all extended instructions that satisfy the following three criteria. First, the sequences are composed of fixed instructions marked as "candidates" by the profiling tool. The profiling tool is based on SimpleScalar's sim_profile, and generates detailed profiles on operand bit-width and instruction execution time. For these experiments candidates are arithmetic and logic instructions with bitwidths of 18 bits or less, but this is a parameter that can be varied. Second, because of limitations on the register file read and write ports, instruction sequences must use at most two input registers and produce at most one output. Third, we look for "maximal" instruction sequences that take as long as possible to execute on the base RISC machine. This characteristic means that the sequences will have as many dependent instructions as possible.

When selecting candidate sequences, the greedy algorithm does not consider the number of PFUs available or the time spent reconfiguring PFUs. Since we evaluate it using an unlimited number of PFUs with zero reconfiguration costs, it represents a best-case performance estimate for this style of compiler and architecture.

4.1 Performance Results Using the Greedy Selection Algorithm

For MediaBench, the greedy algorithm identifies between 6 and 43 distinct extended instructions, and sequence lengths range from 2 to 8 instructions. The first and second bars per program in Figure 2 shows performance results (for unlimited PFUs and no reconfiguration cost) compared to the case of a superscalar processor with no PFUs with the characteristics given in Section 3. Speedups range from 4.5% for *g721-decode* to 44% for *gsm-decode*.

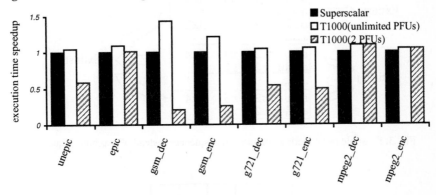

Fig. 2. Speedups using PFUs assuming a greedy selection algorithm. The first bar shows baseline speedup of a superscalar processor with no PFU, normalized to 1. The second bar for each application is the T1000 speedup with unlimited PFUs and zero reconfiguration cost. The third bar shows T1000 speedup with 2 PFUs, each with a 10-cycle reconfiguration penalty

Although the previous paragraph assumed that PFU configurations could be swapped in zero time, real chips have considerable configuration overhead. Thus, the third bar per application in Figure 3 illustrates the performance of the greedy algorithm on more realistic hardware. We consider a superscalar processor with only 2 PFUs, each with a 10-cycle reconfiguration latency. This reconfiguration time is fairly optimistic, but it is sufficient to show the negative impact reconfiguration penalties can have on performance. We use the same set of the extended instructions extracted by the greedy algorithm in the optimal experiment. The results here are quite discouraging. The reconfiguration penalty causes the performance with PFUs to be substantially worse than that of the original processor! Essentially, the PFU is thrashing. Each time an extended instruction is to execute, it causes the PFU to be reconfigured, and the greedy algorithm is too aggressive in selecting extended instructions. These disappointing results motivate the following section in which we present the main contribution of this work: a selective algorithm for choosing extended PFU instructions.

5 A Selective Algorithm for Choosing Extended Instructions

Section 4's results demonstrate that PFUs can improve performance greatly, but only if extended instructions are selected appropriately. In choosing extended instructions, we must not only consider their potential gains, but also minimize the incurred reconfiguration penalties.

5.1 Selective Algorithm Overview

Our selective algorithm works to maximize the benefits from extended instructions subject to the constraint that there are a finite number of PFUs and reconfiguring between different extended instructions takes time. The selective algorithm consists of the steps illustrated in Figure 5.

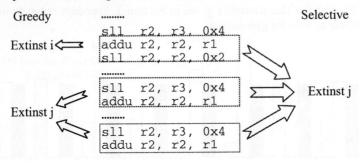

Fig. 3. Example of extracting common subsequences instead of maximal sequences

	i		j	
i	1		0	
j	1		2	

Fig. 4. Matrix for evaluating common subsequences in the selective algorithm

The algorithm has two key characteristics. First, we profile the program and determine which candidate sequences are responsible for more than 0.5% of the total application time. This step allows us to focus on the sequences that offer the highest payoff. It also limits the number of distinct instruction sequences considered, which reduces the likelihood of "thrashing" by too frequently reconfiguring PFUs. Second, the number of extended instructions selected within each loop never exceeds the number of PFUs. This step avoids the frequent reconfigurations that lead to thrashing when the greedy algorithm is used with limited PFUs.

A key difference between the selective algorithm here and the greedy algorithm in Section 4 concerns the treatment of instruction subsequences. Consider the case shown in Figure 3 with three maximal sequences in the same loop. The greedy algorithm would implement all three instruction sequences as extended instructions. Here, we determine when to choose a smaller common subsequence as opposed to instantiating all the possibilities.

Our approach begins by extracting all valid subsequences and adding them to the candidate extended instruction list. The list is organized as a k×k matrix, where k is the total number of distinct candidate instruction sequences in the loop. The [I,J] entry of the matrix corresponds to the number of appearances of candidate sequence I within all candidate sequences J throughout the loop. The sum of entries along the Ith row equals the total number of appearances of sequence I throughout this loop. The [I,I] entry corresponds to how often maximal sequences of I appear: cases where the Ith sequence appears alone and is not a subsequence of a larger instruction.

The matrix for the example in Figure 3 is shown in Figure 4. There is a 1 in the [I,I] entry of the matrix to denote the fact that the first candidate sequence from Figure 3 has one maximal appearance in the loop. Since the latter two sequences in Figure 3 perform the same operation, they share an identical PFU configuration. (This is true in both the greedy and selective algorithms.) They occupy row J in the matrix in Figure 4. A 2 in the [J,J] entry of the matrix indicates that there were two maximal appearances of this extended instruction in the loop. The key leverage of our algorithm is that the [J,I] entry of this matrix reflects the fact that sequence J also appears as a subsequence of I and thus is set to 1.

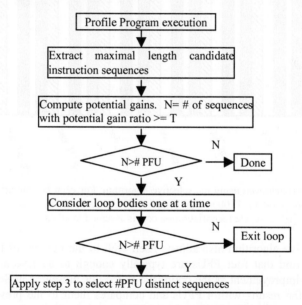

Fig. 5. Flow chart of selective algorithm

The final appropriate extended instructions are then selected from the list by comparing their potential gains. For example, sequence J appears a total of 3 times in the loop, each with a potential gain of 1 cycle. By contrast, sequence I appears only once, but with a potential savings of 2 cycles. If we are working with an architecture with only one PFU, then selecting the sequence with the highest total gain across the loop would lead us to choose sequence J.

5.2 Performance Improvements Using the Selective Algorithm

Figure 6 shows that the selective algorithm successfully chooses extended instructions that offer speedup by avoiding reconfiguration penalties as much as possible. Speedups for these benchmarks now range from 2-27%. Since our approach reduces dramatically the number of PFU reconfigurations, the reconfiguration penalties only account for a small fraction of total potential gains. In fact, our experiments show that we retain our excellent speedups even with reconfiguration times as high as 500 cycles.

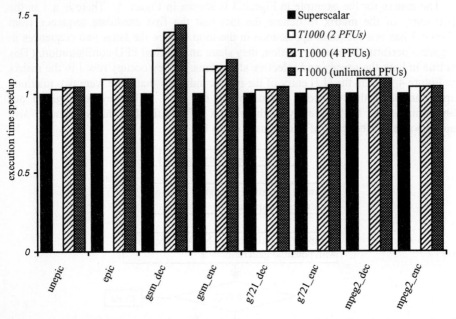

Fig. 6. Speedups achieved using the selective algorithm. For each benchmark, the second and third bars correspond to T1000 with 2 and 4 PFUs, respectively. The fourth bar models unlimited PFUs. A 10-cycle reconfiguration cost is assumed in all cases

Our selective algorithm also adjusts itself well to the number of PFUs available. Overall, we find that four PFUs are typically enough to achieve almost the same performance improvement as the optimistic speed-ups presented in Section 4. Figure 6 illustrates the results with 4 PFUs and compares them to the previous optimistic results achieved with unlimited number of PFUs.

6 Configurable Hardware Cost

The basic component of the PFU is a configurable logic block consisting of look-up tables (LUTs) and flip-flops. An N-input look-up table can implement any Boolean function of N inputs. The LUT propagation delay is independent of the function implemented. In this paper, we use standard CAD tools to map extended instructions to Xilinx devices in order to estimate the PFU hardware cost.

Figure 7 presents the area distribution of instructions chosen by our selective algorithm for the 8 benchmarks. The configurable hardware resources required by an extended instruction depend both on the type of operation and also on the operand widths. Quite a few of the extended instructions need very little hardware, largely due to profiling that indicates when they can be implemented with narrow-bitwidth inputs. On these examples, the most area-intensive extended instruction needs 105 LUTs.

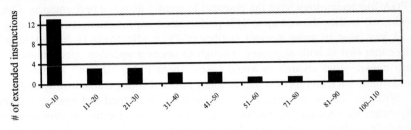

Fig. 7. Distribution of hardware requirements for the extended instructions extracted from 8 MediaBench benchmarks by our selective algorithm

7 Prior Work

There has been a large amount of work on the reconfigurable computing architectures with customizable instruction sets, and an exhaustive summary is difficult. Instead, we present some representative work categorized by the degree of coupling between the configurable hardware resources and the base processor.

Coarse-grained architectures include SPLASH1, SPLASH2 [8] and PAM [9]. In these, the configurable hardware resources are connected as a co-processor on the I/O bus of a standard microprocessor. While appropriate for coarse-grained problems, the disadvantage of these board-based systems is that they have high communication latencies and configurable hardware cost.

Medium-grained architectures include NAPA [10]. In NAPA, the Adaptive Logic Processor (ALP) can access the same memory space as the Fixed Instruction Processor (FIP), so the communication overhead between the ALP and the FIP is reduced compared with the coarse-grained architectures, but this approach still does not give the ALP full access to the register file.

Fine-grained architectures include the PRISC work [2,4]. PRISC was proposed to be a simple, pipelined, single-issue processor augmented with a single PFU. Because of the tight coupling between the PFU and the base CPU, PRISC requires only a small amount of configurable hardware resources and minimizes communication costs. Other representatives of this class include CoMPARE [11] and OneChip [12], etc. CoMPARE explores the impact of multiple PFUs and can execute RISC instructions and customized instructions concurrently. OneChip is an embedded system. It requires more functional modules to be implemented on PFU, which in turn introduces larger communication penalties. All of the above fine-grained architectures were evaluated on simple, in-order-issue, single-issue processors. The impact of PFUs on a superscalar processor's performance is different from that on a simple processor, and our work has quantified these differences.

8 Conclusions and Future Work

This work has explored the use of application-specific instructions in the context of modern superscalar architectures. In particular, we have proposed the T1000 architecture which adds programmable functional units (PFUs) into the datapath of a wide, out-of-order issue processor. These small configurable functional units based on FPGA-like technology have the potential to greatly improve performance. Our initial optimistic studies showed up to 44% performance improvements in some cases.

A key issue in using a small number of PFUs effectively is devising a selection algorithm that is both aggressive enough to uncover speedup opportunities, and yet also conservative enough to avoid cases where PFUs "thrash" as they frequently reconfigure back and forth to handle many selected configurable instructions.

With the goal of avoiding PFU thrashing, we developed and evaluated a selective algorithm for choosing instruction sequences for configurable implementation. Our choice is guided by the number of PFUs available and simple execution profiles of the program loops. This allows us to aggressively select configurable instructions that offer the largest performance savings with the smallest hardware needs. With this algorithm, we have shown performance improvements of up to 28% with 2 PFUs compared to simple superscalar processors without PFUs. Furthermore, our selective algorithm is so successful at avoiding PFU thrashing that these speedups are largely independent of the PFU's reconfiguration overhead. We view our work as a proof-of-concept demonstration that PFUs can offer worthwhile performance improvements in modern high-performance superscalar architectures.

References

1. C. Lee, M. Potkonjak, and W. H. Mangione-Smith, MediaBench: A Tool for Evaluating Multimedia and Communications Systems. *Proc. Micro 30,* 1997
2. R. Razdan, and M.D. Smith: A High-Performance Microarchitecture with Hardware-Programmable Functional Units. *Proc. 27th Intl. Symp. On Micro,* pp. 172-180, Nov., 1994.
3. Xilinx Inc. *The Programmable Logic Data Book,* Xilinx 2100 Logic Dr. San Jose, CA 1998
4. R. Razdan, K. Brace, and M. Smith. PRISC Software Acceleration Techniques. *Proc.Int. Conf. on Computer Design.* Oct.1994.
5. D. Burger, T.M. Austin, and S. Bennett. Evaluating future microprocessors: The SimpleScalar tool set. TR-1308, Univ. of Wisconsin-Madison CS Dept., July 1996
6. G. S. Sohi. Instruction Issue Logic for High-Performance, Interruptible, Multiple Functional Unit, Pipelined Computers. *IEEE Trans. on Computers,* 39(3): 349-359, March 1990
7. Xilinx Inc. Foundation Series Quick Start Guide 1.5, Xilinx 2100 Logic Drive. San Jose, CA
8. J. Arnold et al. The Splash 2 Processor and Applications. *Proc. Int. Conf. on Computer Design,* Oct 1993
9. P. Bertin, D. Roncin, and J. Vuillemin. Introduction to Programmable Active Memories. *Systolic Array Processors,* J. McCanny et al. Eds., Prentice Hall, 1989
10. C. R. Rupp and M. Landguth et al. The NAPA Adaptive Processing Architecture. *Proceedings IEEE Symp. on FPGAs for Custom Computing Machines.* Napa Valley, CA, USA 15-17, April 1998
11. S. Sawitzki, A. Gratz and R.G. Spallek: Increasing Microprocessor Performance with Tightly-Coupled Reconfigurable Logic Array, *Proc. of Field-Programmable Logic and Applications,* Tallinn, Estonia, August 1998
12. R. D. Wittig and P. Chow: OneChip: An FPGA Processor With Reconfigurable Logic, *Proc. IEEE Symp. on FPGAs for Custom Computing Machines,* CA, April 1996

Complexity Bounds for Lookup Table Implementation of Factored Forms in FPGA Technology Mapping

Wenyi Feng[1], Fred J. Meyer[2], and Fabrizio Lombardi[2]

[1] FPGA Software Core Group, Lucent Technologies, 1247 S Cedar Crest Blvd,
Allentown PA   18103
[2] Electrical & Computer Engineering, Northeastern University, 360 Huntington
Avenue, Boston MA   02115

Abstract. We consider technology mapping from factored form (binary leaf-DAG) to lookup tables (LUTs), such as those found in field programmable gate arrays. Polynomial time algorithms exist for (in the worst case) optimal mapping of a single-output function. The worst case occurs when the leaf-DAG is a tree. Previous results gave a tight upper bound on the number of LUTs required for LUTs with up to 5 inputs (and a bound with 6 inputs). The bounds are a function of the number of literals and the LUT size. We extend these results to tight bounds for LUTs with an arbitrary number of inputs.

1   Introduction

We view computer-aided synthesis of a logic circuit in two major steps: (1) the optimization of a technology-independent logic representation, using Boolean and/or algebraic techniques, and (2) technology mapping. Logic optimization is used to transform a logic description such that the resultant structure has a lower cost than the original [1].

Technology mapping is the task of transforming an arbitrary multiple-level logic representation into an interconnection of logic elements from a given library of elements. Technology mapping is very crucial in the synthesis of semicustom circuits for different technologies, such as sea-of-gates, gate arrays, or standard cells. The quality of the synthesized circuit, both in terms of area and performance, depends heavily on this step.

We focus on the problem of technology mapping onto Field-Programmable Gate Arrays (FPGAs). FPGAs are prewired circuits that are programmed by the users to perform the desired functions [13]. In particular, we consider FPGAs where the logic functions are implemented with lookup tables (LUTs). In a LUT-based FPGA, the basic block is a K–input, single-output LUT (K–LUT) that can implement any Boolean function of up to K variables. The technology mapping problem for LUT-based FPGAs is to generate a mapping of a set of Boolean functions onto K–LUTs. Traditional library binding algorithms for

J. Rolim et al. (Eds.): IPDPS 2000 Workshops, LNCS 1800, pp. 951–958, 2000.

standard cells and Mask-Programmable Gate Arrays (MPGAs) are not applicable to FPGAs because the virtual library of a LUT is too large to enumerate (a K–LUT can realize 2^{2^K} logic functions). Many papers have proposed algorithms for LUT-based technology mapping. They can be divided into 3 categories: (1) minimization of number of levels of LUTs in the mapped network [5]; (2) minimization of the number of LUTs used in the mapped network [3, 10, 7, 6], (3) routability of the mapping solution [2, 11], or combinations of these topics [4, 3].

Minimizing the number of levels is solvable in polynomial time in Flow-Map [5]. The key feature in Flow-Map is to compute a minimal height K–feasible cut in the input network. Minimization of the number of LUTs is a much harder problem. It was shown to be $\mathcal{NP}$–hard even for restricted cases [6]. So, heuristics are used in all mapping systems.

In this paper, we restrict our attention to mapping a single-output function onto LUT technology. We specify the input function with a graph, where each node represents a function of 2 inputs. We constrain the problem so that the synthesis must be conducted without being aware of (taking advantage of) the underlying function at each 2–input node.

2 Preliminaries

Definition 1. A *leaf-DAG* is a general case of a tree—the leaves of the tree (primary inputs) are allowed to fan out. If node i is one of the inputs to node j in a leaf-DAG, we say that i is a *child* of j and that j is a parent of i.

In mapping, we will not take any special advantage of leaf-DAGs; instead, we will regard the inputs to the various nodes in the DAG as coming from distinct primary inputs—i.e., we will not take advantage of any knowledge of fan-out at the primary inputs. This yields bounds that are applicable in any case and, in particular, in the worst case of a tree.

- $p(v)$. Apart from the leaves, each node, v, in a leaf-DAG has a unique parent, $p(v)$.
- $l(S)$. The number of literals of the input function, S. This is the sum of the number of inputs to all nodes of the input graph. We simply use l, instead of $l(S)$, whenever S is understood.

Definition 2. The size or complexity, $C(S)$, of a circuit, S, is the number of gates (number of nodes in its DAG). The circuit complexity of a function, f, with respect to a basis, Ω, is $C_\Omega(f)$, which is the minimal number of gates from the set Ω in order to compute f.

- K. The LUTs in the technology to be mapped onto have K inputs. We call them K–LUTs. A K–LUT implements the basis B_K.
- $L_K(f)$. The number of LUTs needed to map function f to K–LUTs. We use $L(f)$, L_K, and L whenever f and/or K are understood.
- $C_K(l)$. This is the circuit complexity for leaf-DAGs mapped onto K–LUTs. It is the worst case, over all functions represented by leaf-DAGs with l literals, of the minimal number of K–LUTs required to implement the function.

Definition 3. A *factored form* of a one-output function is a generalized sum-of-products form allowing nested parentheses and arbitrary binary operations.

A factored form is represented by a binary leaf-DAG (all gates are in B_2).

For example, the function $ab'c' + a'bc' + d$ can be represented in a factored form with 7 literals as $(((ab')c') + ((a'b)c')) + d$, and it can be written more compactly in factored form as $((a \oplus b)c') + d$ with 4 literals.

When all l literals of a factored form are different, its corresponding binary leaf-DAG is a binary tree. The binary tree has l inputs and $l - 1$ internal nodes. Figure 1 shows a binary tree with $l = 7$ inputs and $l - 1 = 6$ internal nodes.

If all inputs of a binary leaf-DAG, D, are different, we have a binary tree, B. So, a realization of B would also serve as a realization of D. Perhaps some other realization of D requires fewer LUTs, using some structural information of D.

Lemma 4. *Suppose a binary tree, B, is obtained from a binary leaf-DAG, D, by viewing all D's inputs as different. $L_K(D) \leq L_K(B)$*

This lemma tells us that, in order to analyze the worst case complexity of binary leaf-DAG mapping, it is enough to analyze binary trees.

In [9], some results are provided on the complexity bound of a function, f, given in a factored form. The results are summarized in the following theorem.

Theorem 5. *For the class of functions with l literals in factored form,*

$$C_2(l) = \quad l - 1 \quad (l \geq 2)$$
$$C_3(l) = \lfloor (2l - 1)/3 \rfloor \; (l \geq 2)$$
$$C_4(l) = \lfloor (l - 1)/2 \rfloor \; (l \geq 3)$$
$$C_5(l) = \lfloor (2l - 1)/5 \rfloor \; (l \geq 4)$$
$$C_6(l) \leq \lfloor (l - 1)/3 \rfloor \; (l \geq 6)$$

Reference [6] presented an optimal algorithm, Tree-Map, for technology mapping where the input is a tree. Tree-Map uses a greedy dynamic programming approach, which happens to guarantee an optimal mapping.

Our approach to determining a tight bound for $C_K(l)$ for all l is to analyze a technology mapping algorithm that is optimal on trees. We use the Tree-Map algorithm [6], because it is easiest to analyze.

Definition 6. For a tree, $T(V, E)$, its *height* is the number of nodes on the longest path from an input to the root. The *level* of the root is the height of the tree. The level of a node (excluding the root) is the level of its parent minus 1.

Definition 7. Consider a tree, $T(V, E)$, with vertex (node) set, V, and edge set, E. Let $V1$ be a subset of V such that a LUT is assigned to precisely those vertices in $V1$. Two quantities are defined for each vertex $v \in V$. These quantities are its *dependency*, $d(v)$, and its *contribution*, $Z(v)$, defined according to:

– Contribution, $Z(v)$:

- For each primary input (or literal), v, $Z(v) = 1$.
- For each $v \in V1$, $Z(v) = 1$.
- For all other vertices $v \in V$, $Z(v) = Z(u_1) + \ldots + Z(u_{c(v)})$—where v has $c(v)$ children: $u_1, \ldots, u_{c(v)}$.

- Dependency, $d(v)$:
  - For each primary input (or literal), v, $d(v) = 1$.
  - For all other vertices $v \in V$, $d(v) = Z(u_1) + \ldots + Z(u_{c(v)})$—where v has $c(v)$ children: $u_1, \ldots, u_{c(v)}$.

Definition 8. In a mapping, if a node is assigned a LUT, we say it is a *LUT node*. Otherwise, we say it is a *free node*.

From Def. 7, we know that, for a free node, its contribution is equal to its dependency, but for a LUT node its contribution is set to 1.

Note that $d(v)$ is the summation of the number of inputs or LUTs that directly or indirectly supply input to vertex v, and it represents the number of signals that would need to be placed at v if the signal at v were implemented with a LUT. The quantity $Z(v)$, on the other hand, represents the contribution of vertex v to the dependency of its parent vertex. Figure 1 shows an example of a tree and the assignment of LUTs to its vertices. The shaded vertices in the figure represent the LUT nodes. The dependency and contribution values for each node, v, in the tree are shown with an ordered pair, $(d(v), Z(v))$.

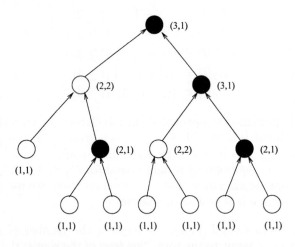

Notation (d,Z) represents the dependency and contribution values for a vertex in the tree.

Fig. 1. The dependency and contribution values for a tree

The Tree-Map algorithm scans from leaves to the root, assigning LUTs as necessary. Whenever it encounters a node with dependency exceeding K, it must assign LUTs. It assigns LUTs to that node's children, starting with the child with largest contribution, until the node's dependency has been sufficiently reduced. This greedy mapping is optimal with respect to the number of LUTs [6].

Our objective is to derive a tight bound for general K–LUT technology mapping. We use Tree-Map [6] as a unified optimal mapping algorithm. Although the dynamic programming algorithm in [7] is also optimal for tree mapping, it is hard to work from it to derive bounds on the circuit complexity. Tree-Map takes a DAG input. In this paper, we constrain it to be a leaf-DAG—i.e., each internal node has fanout 1, the primary inputs may fan out arbitrarily. This is a generalization of trees [6] through allowing the primary inputs to fan out. Generally, the leaf-DAGs will be in factored form, because that is the worst case in terms of LUT complexity.

We do not assume that we know the individual functions used in the formula. For example, if an output of an AND gate goes to another AND gate, we do not allow any inputs to be rearranged between the two gates. In short, the output of the technology mapping must be valid, even if arbitrary functions are substituted for each of the input leaf-DAG's gates.

3   Worst Case Mapping to K–LUTs

Tree-Map proceeds from level to level in the tree. When we deal with vertices at level j, all vertices below level j have dependencies less than or equal to K. Tree-Map had processed all nodes at lower levels. Whenever any of them had dependency more than K, it assigned sufficient LUTs to reduce the dependency to at most K.

Lemma 9. *If K is even,*

$$d(v_i) \geq (K/2 + 1), \qquad 1 \leq i \leq L - 1 \qquad (1)$$

Proof: i ranges up to $L - 1$, so it includes all the LUTs, except the one assigned to the root. According to the Map-Tree algorithm, a node i (except the root) is assigned a LUT only because its parent has dependency larger than K before the assignment. Furthermore, it is selected to be assigned a LUT because its dependency is at least as large as the (only) other child of its parent. So, its dependency must be $\geq K/2 + 1$.   □

Lemma 10. *If K is odd,*

$$d(v_i) \geq (K + 1)/2, \qquad 1 \leq i \leq L - 1 \qquad (2)$$

Proof: Similar to the proof for Lemma 9.   □

Lemma 11. *Suppose K is odd, and v_i ($i \neq L$) is a node with $d(v_i) = (K+1)/2$. Suppose v_j is the first LUT node on the path from v_i to the root. Then:*

$$d(v_j) \geq (K + 3)/2 \qquad (3)$$

and we say v_j is the pair node of v_i.

Proof. Omitted for brevity.   □

Lemma 12. *Suppose K is odd, and v_i and v_j $(i, j \neq L)$ both have dependency $(K + 1)/2$. Then their pair nodes are two different nodes.*

Proof. Contrariwise, suppose v_1 were the pair node of both v_i and v_j. According to the proof of Lemma 11, v_1 must resolve $\geq 2 \cdot (K + 3)/2$ dependency, which cannot be true.   □

Lemma 13. *Suppose $L > 1$ and $v_L = r$. Suppose v_i is a LUT node nearest to r (if there are multiple such nodes, select any one). So, on the path from v_i to r, no other LUT node exists. Then:*

$$d(r) + d(v_i) \geq K + 2 \qquad (4)$$

and we say that v_i is the pair node of r.

Proof. Omitted for brevity.   □

Now we are able to prove our key theorem.

Theorem 14.

$$C_K(l) = \begin{cases} \lfloor (2l - 2)/K \rfloor, & \text{if } K \text{ is even} \\ \lfloor (2l - 1)/K \rfloor, & \text{if } K \text{ is odd} \end{cases}$$

Proof. For brevity, we omit the half of the proof that the bound is always achievable. We only give the half of the proof that the bound is tight.

To show tightness, we need to show some trees that meet the upper bound. We consider two cases.

(1) K is even. Figure 2 shows an example. Node v_a is the root of a binary tree with $K/2 + 1$ inputs; each of the nodes v_b, v_d, v_f, ... is the root of a binary tree with $K/2$ inputs. The shaded nodes show the nodes to which a K–LUT should be assigned according to the Tree-Map algorithm. For example, when node v_c is visited, the dependency $d(v_c)$ is $K + 1$, and we put a K–LUT at node v_a, and so on. The value $K/2 + 1$ beside node v_a represents the amount of dependency resolved at node v_a—i.e., $d(v_a)$. The value $K/2$ beside node v_b represents $d(v_b)$.

Suppose the number of LUTs in the figure is L. The total tree inputs is

$$l = L(K/2) + 1 \qquad (5)$$

So, the tree needs the upper bound number of K–LUTs. Therefore, the bound is tight in this case.

(2) K is odd. We show two subcases in Fig. 3. In the first (second) case, according to the Tree-Map algorithm, an odd (even) number of LUTs is needed.

Suppose the number of K–LUTs needed is L.

In the first subcase (L is odd), the number of inputs is:

$$l = (LK + 1)/2 \qquad (6)$$

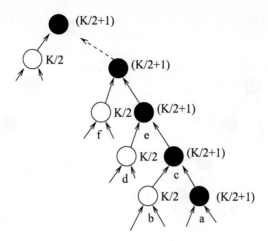

Fig. 2. Proof of tightness when K is even

It meets the upperbound. (In this case, $\lfloor (2l-1)/K \rfloor$ is 1 more than $\lfloor (2l-2)/K \rfloor$, and we need $\lfloor (2l-1)/K \rfloor$ K–LUTs.) Therefore, this also shows that, for each l that makes $(2l-1)/K$ an (odd) integer, there exists a binary tree that needs $(2l-1)/K$ K–LUTs.

In the second subcase (L is even), the number of inputs is:

$$l = L(K/2) + 1 \tag{7}$$

It also meets the upperbound. (In this case, $\lfloor (2l-1)/K \rfloor$ is equal to $\lfloor (2l-2)/K \rfloor$.)
$\square$

4   Conclusion

Arbitrary functions can be mapped onto FPGAs that use lookup tables (LUTs). If the input function is in the form of a tree or leaf-DAG [9], a greedy algorithm can process the input in polynomial time. In the case of a tree, the greedy algorithm minimizes the number of LUTs, subject to the constraint that the algorithm is not allowed to exploit any knowledge of the particular functions represented by the nodes in the input graph. In the case of a leaf-DAG, the number of LUTs needed is bounded by that required for an equivalent tree representation using unique literals.

We differentiate between LUTs by the number of inputs they handle, K. We considered leaf-DAGs where all nodes are 2–input functions. This is the worst case in terms of how many K–LUTs are required. Previous work [9] had obtained bounds on the worst case number of K–LUTs for K up to 6 (tight bounds up to 5). We extended this to tight bounds for all K.

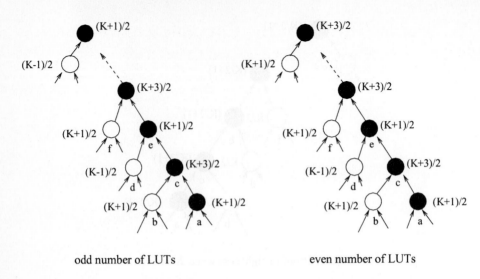

odd number of LUTs   even number of LUTs

Fig. 3. Proof of tightness when K is odd

References

1. Brayton, R. K., Rudell, R., Sangiovanni-Vincentelli, A.: Mis: A multiple-level logic optimization system. IEEE Trans. CAD of Int. Circ. and Sys. **6** (1987) 1062–1081
2. Chan, P. K., Zien, J. Y., Schlag, M.: On routability prediction of FPGAs. IEEE/ACM Des. Auto. Conf. (1993) 326–330
3. Chaudhary, K., Pedram, M.: A near optimal technology mapping minimizing area under delay constraints. IEEE/ACM Des. Auto. Conf. (1992) 492–498
4. Cong, J., Ding, Y.: On area/depth trade-off in LUT-based FPGA technology mapping. IEEE/ACM Des. Auto. Conf. (1993) 213–218
5. Cong, J., Ding, Y.: Flowmap: An optimal technology mapping algorithm for delay optimization in look-up table based FPGA designs. IEEE Trans. CAD of Int. Circ. and Sys. **13** (1994) 1–12
6. Farrahi, A. H., Sarrafzadeh, M.: Complexity of the look-up table minimization problem for FPGA technology mapping. IEEE Trans. CAD of Int. Circ. and Sys. **13** (1994) 1319–1332
7. Francis, R. J., Rose, J., Chung, K.: Chortle: A technology mapping algorithm for lookup table based FPGAs. IEEE/ACM Des. Auto. Conf. (1990) 613–619
8. Francis, R. J., Rose, J., Vranesic, Z.: Chortle-crf: Fast technology mapping for lookup table based FPGAs. IEEE/ACM Des. Auto. Conf. (1991) 227–233
9. Murgai, R., Brayton, R. K., Sangiovanni-Vincentelli, A.: Logic Synthesis for FPGAs. Kluwer Academic Publishers (1995)
10. Murgai, R., Nishizaki, Y., Shenoy, N., Brayton, R. K., Sangiovanni-Vincentelli, A.: Logic synthesis algorithms for programmable gate arrays. IEEE/ACM Des. Auto. Conf. (1990) 620–625
11. Schlag, M., Kong, J., Chan, P. K.: Routability driven technology mapping for look-up table FPGAs. IEEE Int. Conf. Comp. Des. (1992) 89–90
12. Wegener, I.: The Complexity of Boolean Functions. Wiley-Teubner (1987)
13. Xilinx Corporation: Xilinx FPGA Data Book. (1996)

Optimization of Motion Estimator for Run-Time-Reconfiguration Implementation

Camel. Tanougast, Yves. Berviller, Serge. Weber.

Laboratoire d'Instrumentation Electronique de Nancy - Université Henri Poincaré Nancy I
Faculté des Sciences, BP 239
F-54506 Vandoeuvre-lès-Nancy cedex, France
{tanougast, yves.berviller, serge.weber}@lien.u-nancy.fr

Abstract. In this paper, we present a method to estimate the number of reconfiguration steps that a time-constrained algorithm can accommodate. This analysis demonstrates how one would attack the problem of partitioning a particular algorithm into pieces to for run time reconfiguration execution on a Atmel 40K FPGA. Our method consist in evaluating algorithm operators execution time from data flow graph. So, we deduce the reconfiguration number and the algorithm partitioning for RTR implementation. The algorithm used in this work, is a qualitative motion estimator in the Log-Polar plane.

1. Introduction.

The availability of FPGAs which supply fast and partial reconfiguration possibilities, provides a way to dynamically reconfigurable architectures [1]. This new approach enables the successive execution of an algorithms sequence on the same device [2].

This article propose an evaluation method for the determination of the number of successive reconfigurations which can be made for a given algorithm. This evaluation is obtained from the data flow graph in order to optimize its implementation on a run time reconfigurable architecture. This architecture uses Atmel's AT40k FPGAs, which have short configuration times. The evaluation of this number gives us the partitioning of the data flow graph. The aim of this paper is the optimization of hardware resources while satisfying the real time processing constraint. The performances like processing time and resources usage rate of the FPGA are described.

The algorithm is an apparent motion estimator in a Log-Polar images sequence, which estimates the normal optical flow.

Firstly we describe the algorithm. Secondly, we present the method for the determination of the step number for a Run-Time-Reconfiguration (RTR) implementation. Thirdly we give the results compared with a static implementation. Finally we conclude on the contribution of this approach.

J. Rolim et al. (Eds.): IPDPS 2000 Workshops, LNCS 1800, pp. 959-965, 2000.

2. Qualitative motion estimation in the Log-Polar space.

The Log-Polar images are obtained by remapping the Cartesian coordinate images in a Complex Logarithm Mapping [3]. The advantage of this transformation is that the radial and axial motion in the original space becomes mainly horizontal in the new space. Our solution estimates the horizontal displacements of moving objects edges. The method uses OFC (1) (optical flow constraint) of moving points in image sequence.

$$\vec{V} \cdot \overrightarrow{grad} \, I = -\frac{\partial I}{\partial t}. \tag{1}$$

$\vec{V}$ is the apparent velocity vector of an image point and I the intensity of this point. From this Optical Flow Constraint we estimate the normal optical flow by dividing the temporal derivative by the spatial gradient:

$$V_n = -\frac{\dfrac{\partial I}{\partial t}}{\dfrac{\partial I}{\partial x}}. \tag{2}$$

V_n is an estimate of normal optical flow in Log-Polar images.

Before this *computation,* two pre-processing are necessary. The first processing is a gaussian filtering in order to guarantee the existence of the spatial derivative of image intensity $I(x, y)$. The second is a time averaging filter to reduce the noise.

Our apparent motion estimator algorithm in Log-Polar plane, is composed of gaussian and averaging filters, followed by temporal and spatial derivatives and arithmetic divider. The datapath of this algorithm is given on figure 1.

3. Determination of the possible number of steps for RTR implementation.

3.1. Evaluation of the possible number of steps.

The images are acquired at a rate of 25 images per second, this leaves us 40 ms to process the entire image. To satisfy the real time constraint we need to process at a faster rate than that of pixels acquisition. The algorithms are partitioned in N steps corresponding to N execution-reconfiguration pairs. The working frequency of each step needs to verify the following inequality :

$$n^2 \times \sum_{j=1}^{N} te_j \leq Ti - \sum_{j=1}^{N} Trec_j. \tag{3}$$

Where n^2 is the number of pixels in the image, N the number of reconfiguration, Ti is the duration of an image (40 ms), te_j is the elementary processing time of a pixel in the jth steps and $Trec_j$ is the reconfiguration time of the jth steps.

The objective is to make an implementation which requires the minimal logical resources and satisfies the real time constraint.

From equation (3) we obtain the minimal number of steps that we can surely implement :

$$N \geq N_{min} = \frac{T_i}{n^2 \times K \times to_{max} + k_{rec} \times C_{max}}. \tag{4}$$

to_{max} is the maximum execution time of an operator of the data flow graph (without routing), K is a coefficient which take into account the routing delay between operators, k_{rec} is a proportionality constant between the configuration time and the number of used logic cells and C_{max} is the total available logic cells. This evaluation is obtained with the maximal configuration time and the execution time of the slowest operator of each step.

Our method is based on the analysis of the data flow graph of the algorithm in order to deduce the value of these parameters. The determination of N_{min} gives us the number of partitions of the data flow graph which corresponds to the number of reconfiguration steps.

3.2. Modelling and parameters determination.

AT40K's technology enables partial reconfiguration. Each configuration time depends on the quantity of logic cells used for each step [4]. We evaluate the configuration time of the jth step by :

$$Trec_j = k_{rec} \times C_j. \tag{5}$$

Where C_j is the number of Cells of the jth step. In our case, AT40K20's capacity of 819 Cells leads to a total reconfiguration time lower than 0.6 ms at 33 MHz with 8 bits of configuration data [5]. We obtain for k_{rec} a value of 733 ns/ cell.

The maximum execution time of an operator depends on the speed grade of the device and the data size to process (number of bits). The following equation gives this time for a cascaded operator :

$$to_{max} = Dj_{max} \times (Tc + Tr) + Tsetup. \tag{6}$$

Where Dj_{max} is the maximum data size to process, Tc is the logical function path delay, Tr is propagation delay between logical function and $Tsetup$ is setup time. We evaluate these values to $Tc = 1.7$ ns; $Tr = 0.17$ ns and $Tsetup = 1.5$ ns [5].

The maximum working frequency depends on the slowest operator and the routing delays between operators. We determined experimentally that K is constant for a given occupation rate. This coefficient has a value of 1.5 in our application.

The study of the Cell's structure enables the evaluation of the cell usage for each operator. An n bits adder or substractor, latched or not, require n cells. The same cells number applies for n bits multiplexer or register. This allows the evaluation of logical resources needed for each step of the application from its data flow graph.

4. Results.

From the data flow graph (see figure 1), we obtain the size and type of the different operators used (adder, multiplier, multiplexer...). So, in accordance with the technology used, we deduce the slowest operator execution time. With AT40K, adders are the slowest operators of our datapath if we consider identical size operators (number of bits). In our application, the slowest operator is an 15 bits latched adder. Then, the equation (6), give us a value of to_{max} of 29.55 ns.

From the equation (4), and the parameters determination, we estimate the minimal number of reconfiguration-execution (steps) $N_{min} = 3.27$ for our implementation. This result is obtained with a image size of 512 by 512 pixels.

We deduce the data from the following table for a RTR optimized implementation with constant resources usage rate.

Total estimated number of Cells	Mean Cells / step number	Reconfiguration time / step (ms)	te_{max} (ns)
690	212	0.16	44.3

The value N_{min} is calculated by considering that each step require a full device configuration and is executed with a slowest working frequency. In fact, after implementation we obtain reconfiguration and execution time lesser than or equal to evaluated time. That is why four reconfiguration-execution are possible instead of a theoretical value of 3.27.

The partitioning of the data flow graph in four step is made in the following way :

    _ first step   : gaussian filter
    _ second step   : averaging filter and temporal and spatial derivative
    _ third step   : first half of divider
    _ fourth step   : second half divider.

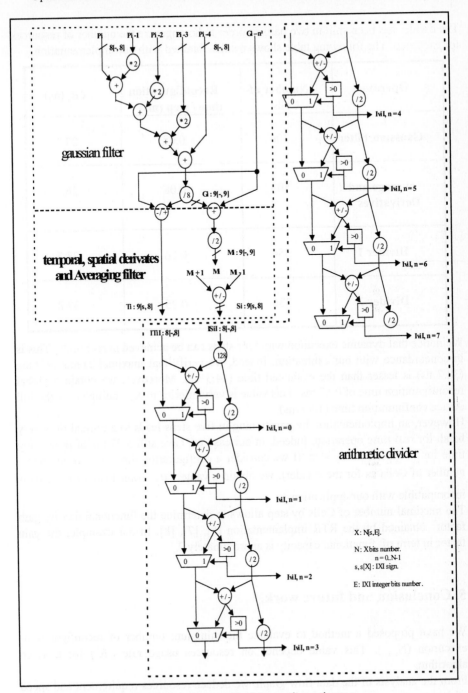

Fig. 1. Data flow graph of the motion estimator.

The divider has been split in two parts in order to homogenize the number of resources for each step. The following table shows results obtained with our implementation.

Operators	Number of Cells	Reconfiguration time / step (ms)	te_j (ns)
Gaussian Filter	106	0.08	27.1
average and Derivatives	103	0.08	26.5
Divider 1	354	0.26	38.7
Divider 2	336	0.25	37.8

We notice that dynamic execution with four steps can be achieved in real time. This is in concordance with our estimation. Indeed, we verify that maximal execution time (38.7 ns) is lesser than the evaluated time (44.3 ns). Moreover, we obtain a global reconfiguration time of 0.67 ms. This value is very inferior to N_{min} multipled by the full device configuration time (1.96 ms).

However, an implementation by partitioning in five steps leads to a critical time very harsh for real time operation. Indeed, in our case we have still 5.22 ms of processing time for a supplementary step. If we consider a configuration time of 0.26 ms (Same number of Cells as for the divider), we obtain a value te_j lower than 19 ns. This is incompatible with our application.

The maximal number of Cells by step allows to determine the functional density gain factor obtained by the RTR implementation [6], [7], [8]. In our example, the gain factor in term of functional capacity is approximately 2.

5. Conclusion and future work.

We have proposed a method to evaluate the minimum number of reconfiguration-execution (N_{min}). This value depends on resources usage rate (K) for a given algorithm.

From the analysis of the data flow graph, we deduce resources requirement and speed of the various operators. This leads to the determination of total processing time, from which we deduce the optimized partitioning of the data flow graph for RTR implementation.

We illustrate our method with an apparent motion estimation algorithm on log-polar images. The results obtained are in accordance with our estimation. The differences between our estimation and experimental results are mainly due to the variations of K (which depends on routing and actual resource occupation rate). The performances obtained are compatible with the requirements of real time processing.

A partitioning which does not rely on the algorithm's functions, enables an implementation very homogeneous in terms of resource used by each step. This would allow to enhance the functional capacity.

References.

1. D. Demigny, M. Paindavoine, S. Weber : Architecture Reconfigurable Dynamiquement pour le Traitement Temps Réel des Images. Revue technique et Sciences de l'information, Numéro Spécial programmation des Architectures Reconfigurables. (1998).
2. H. Guermoud, Y. Berviller, E. Tisserand, S. Weber : Architecture à base de FPGA reconfigurable dynamiquement dédiée au traitement d'image sur flot de données. 16° colloque GRETSI. (1997).
3. M. Tistarelli, G. Sandini : On the advantage of polar and log-polar mapping for direct estimation of time to impact from optical flow. IEEE Transactions on PAMI, vol 15. (1993). 401-410.
4. ATMEL IDS AT40K User' guide.
5. Atmel. *AT40K FPGA*. Data Sheet.
6. M. J. Wirthlin, B.L. Hutchings : Improving functional density through run-time constant propagation. FCCM97 (1997).
7. H. Guermoud : Architecture reconfigurable dynamiquement dédiées aux traitements en temps réel des signaux vidéo. Thèse de l'Université Henri Poincaré. Nancy 1. (1997).
8. J.G. Eldrerge, B.L. Hutchings : Density enhancement of neural network using FPGAs and run-time reconfiguration . FCCM94 (1994).

Constant-Time Hough Transform On A 3D Reconfigurable Mesh Using Fewer Processors

Yi Pan

Department of Computer Science
University of Dayton, Dayton, OH 45469-2160

Abstract. The Hough transform has many applications in image processing and computer vision, including line detection, shape recognition and range alignment for moving imaging objects. Many constant-time algorithms for computing the Hough transform have been proposed on reconfigurable meshes [1, 5, 6, 7, 9, 10]. Among them, the ones described in [1, 10] are the most efficient. For a problem with an $N \times N$ image and an $n \times n$ parameter space, the algorithm in [1] runs in a constant time on a 3D $nN \times N \times N$ reconfigurable mesh, and the algorithm in [10] runs in a constant time on a 3D $n^2 \times N \times N$ reconfigurable mesh. In this paper, a more efficient Hough transform algorithm on a 3D reconfigurable mesh is proposed. For the same problem, our algorithm runs in constant time on a 3D $n \log^2 N \times N \times N$ reconfigurable mesh.

1 Introduction

The Hough transform of binary images is an important problem in image processing and computer vision and has many applications such as line detection, shape recognition and range alignment for moving imaging objects. It is a special case of the Radon transform which deals with gray-level images. The Radon transform of a gray-level image is a set of projections of the image taken from different angles. Specifically, the image is integrated along line contours defined by the equation:

$$\{(x, y) : x \cos(\theta) + y \sin(\theta) = \rho\}, \tag{1}$$

where θ is the angle of the line with respect to positive x-axis and ρ is the (signed) distance of the line from the origin.

The computation of the Radon and Hough transforms on a sequential computer can be described as follows. We use an $n \times n$ array to store the counts which are initialized to zero. For each of the black pixels in an $N \times N$ image and for each of the n values of θ, the value of ρ is computed based on (1) and the sum corresponding to the particular (θ, ρ) is accumulated as given in the following algorithm. In the algorithm, ρ_{res} is the resolution along the ρ direction; and gray-value(x, y) is the intensity of the pixel at location (x, y).

for each black pixel at location (x, y) in an image do
for $\theta = \theta_0, \theta_1, ..., \theta_{n-1}$ do

J. Rolim et al. (Eds.): IPDPS 2000 Workshops, LNCS 1800, pp. 966–973, 2000.

begin
(* parameter computation *)
$\rho := (x \cos \theta + y \sin \theta)/\rho_{res}$
(* accumulation *)
sum[θ, ρ] := sum[θ, ρ] + gray-value(x, y)
end;

Obviously, for an $N \times N$ image, and n values of θ, a sequential computer calculates the Radon (Hough) transform in $O(nN^2)$ time since the number of black pixels is $O(N^2)$. The computation time is too long for many applications, especially for real-time applications, as N and n can be very large.

Recently, several constant-time algorithms for computing the Hough transform have been proposed for the reconfigurable mesh model [1, 5, 6, 7, 9, 10]. Among them, the ones described in [1, 10] are the most efficient. For a problem with an $N \times N$ image and an $n \times n$ parameter space, the algorithm in [1] runs in a constant time on a 3D $nN \times N \times N$ reconfigurable mesh, and the algorithm in [10] runs in a constant time on a 3D $n^2 \times N \times N$ reconfigurable mesh. Besides computing Hough transform, the algorithm in [10] can also compute the Radon transform in a constant time using the same number of processors. In this paper, a more efficient Hough transform algorithm for binary images on a 3D reconfigurable mesh is proposed. For the same problem, our algorithm runs in constant time on a 3D $n \log^2 N \times N \times N$ reconfigurable mesh. We also show that the algorithm can be adapted to computing the Radon transform of gray-level images in constant time on a 3D $n \log^3 N \times N \times N$ reconfigurable mesh. Clearly, our algorithm uses the fewest number of processors to achieve the same objectives and is the most efficient one compared to existing results in the literature [1, 5, 6, 7, 9, 10].

2   The Computational Model

A reconfigurable mesh consists of a bus in the shape of a mesh which connects a set of processors, but which can be split dynamically by local switches at each processor. By setting these switches, the processors partition the bus into a number of subbuses through which the processors can then communicate. Thus the communication pattern between processors is flexible, and moreover, it can be adjusted during the execution of an algorithm. The reconfigurable mesh has begun to receive a great deal of attention as both a practical machine to build, and a good theoretical model of parallel computation.

A 2D reconfigurable mesh consists of an $N_1 \times N_2$ array of processors which are connected to a grid-shaped reconfigurable bus system. Each processor can perform arithmetic and logical operations and is identified by a unique index (i, j), $0 \le i < N_1$, $0 \le j < N_2$. The processor with index (i, j) is denoted by $PE(i, j)$. Each processor can communicate with other processors by broadcasting values on the bus system. We assume that the bus width is $O(\log N)$ and each broadcast takes $O(1)$ time. The arithmetic operations in the processors are performed on $O(\log N)$ bit words. Hence, each processor can perform one logical and arithmetic operation on $O(1)$ words in unit time.

A high dimensional reconfigurable mesh can be defined similarly. For example, a processor in a 3D $N_1 \times N_2 \times N_3$ reconfigurable mesh is identified by a unique index (i, j, k), $0 \leq i < N_1$, $0 \leq j < N_2$, $0 \leq k < N_3$. The processor with index (i, j, k) is denoted by $PE(i, j, k)$. Within each processor, 6 ports are built with every two ports for each of the three directions: i-direction, j-direction, and k-direction. In each direction, a single bus or several subbuses can be established.

A subarray is denoted by replacing certain indices by $*$'s. For example, the ith row of processors in a 2D reconfigurable mesh is represented by $ARR(i, *)$. Similarly, $ARR(*, j, k)$, $0 \leq j < N_2$, $0 \leq k < N_3$, is a 1-dimensional subarray in a 3D reconfigurable mesh, and these $j \times k$ subarrays can execute algorithms independently and concurrently. Finally, a memory location L in $PE(i, j, k)$ is denoted as $L(i, j, k)$.

3   The Constant-Time Algorithm

In this section, we propose a constant time algorithm for computing the Hough transform of an $N \times N$ image on a 3D $n \log^2 N \times N \times N$ reconfigurable mesh. In the following discussion, we partition the image into parallel bands and these bands run at an angle of θ with respect to the horizontal axis, and then sum the pixel values contained in each band. If a pixel is contained in two or more bands, then it will be counted in only the band that contains its center. If the center of a pixel lies on the boundary between two bands, then it is counted only in the uppermost of the two bands. For example, we have computed a $\pi/4$ angle Hough transform for an 8×8 pixel array in Figure 1, where the bands are one pixel-width wide. Clearly, there are 10 different ρ's in the figure. In the figure, the number of 1-pixels contained in each band is displayed at the upperright end of the band. For a particular angle θ and a particular distance ρ, only the values of the pixels lying in the band specified by θ and ρ need to be added together.

In our algorithm, since all pixels in an image are used as the input, we can exploit easily the geometric features and relations of pixels in an image. Clearly, for a given pair of ρ and θ, we do not need to consider all the pixels in an image. Instead, only those pixels that are centered in the band will contribute to the count value of that band. In this way, we can improve the efficiency of the algorithm during computation. Before we describe the algorithms, several observations will be made. In order to speedup the computation, we need to connect together all processors which have computed and stored the same ρ values. In order to do so, we rely on several results obtained in [8]. Although the results are made for θ_i such that $0 \leq \theta_i \leq \pi/4$, they can easily be generalized to other θ values. In the following discussion, we assume that $0 \leq \theta_i \leq \pi/4$.

Lemma 1. For any j, $0 \leq j \leq N-1$, the ρ-distances satisfy $\rho_{i,j} \leq \rho_{i+1,j}$ for $0 \leq i \leq N-2$. It can also be shown that no more than two consecutive values of ρ in row j can be equal.

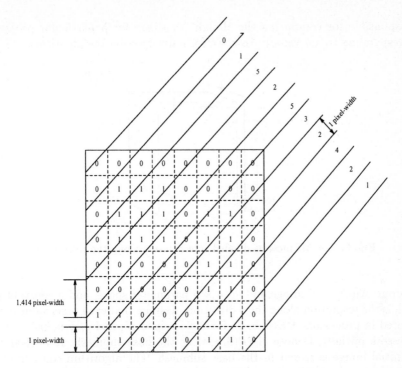

Fig. 1. Parallel Bands for $\theta = \pi/4$ in an 8×8 Image.

Lemma 2. The values of ρ computed using equation (1) by two consecutive processors in a row j, differ by at most 1. More formally, for all $i, j, 0 \le j \le N-1$ and $0 \le i \le N - 2$, $0 \le \rho_{i+1,j} - \rho_{i,j} \le 1$.

Lemma 3. For all values of $i, j, 0 \le i, j \le N - 2$, $\rho_{i,j} \ne \rho_{i+1,j+1}$.

Lemma 4. If $\rho_{i,j} = \rho_{i,j+2}$ for $0 \le i \le N - 1$ and $0 \le j \le N - 3$, then $\rho_{i,j} = \rho_{i,j+1} = \rho_{i,j+2}$. If two ρ-values in a column i are equal and they are placed two rows apart, then the ρ-value in between should have the same value.

The above lemmas will be used in our algorithm to connect related processors together to calculate the number of black pixels in the bands. The following result is also used in our algorithm to compute binary sums efficiently and is due to [?].

Lemma 5. Let a binary sequence of length S stored in the first row of a 2D $S \times \log^2 S$ reconfigurable mesh, the sum of the binary sequence can be computed in a constant time on the array.

For the Radon transform, we need the following result to add integer values. The detailed proof of the lemma is described in [11].

Lemma 6. Given S ($\log S$)-bit integers, these numbers can be added in $O(1)$ time on a 2D $S \times \log^3 S$ reconfigurable mesh.

Assume that the reconfigurable mesh used here is configured as a 3D $n * \log^2 N \times N \times N$ array. The 3D mesh is divided into n layers along the i direction with each layer having a 3D $\log^2 N \times N \times N$ array as shown in Figure 2. Each layer is responsible for computing the Hough transform for a particular projection (corresponding to a θ value). Now, we formally describe the algorithm.

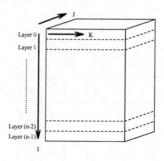

Fig. 2. The 3D mesh is divided into n layers along the i direction.

Input: An $N \times N$ image and an $n \times n$ parameter space, and a constant ρ_{res} which is the resolution along the ρ direction. Assume that each pixel value $a(x, y)$ is stored in processors $PE(0, x, y)$, for $0 \leq x, y < N$, and ρ_{res} is known to all processors initially. Denote $ARR(0, *, *)$ as the base submesh. It is clear that the initial image is stored in the base submesh. The algorithm consists of the following steps.

Step 1. In this step, we copy the whole image from the base submesh to all the other submeshes $ARR(i, *, *)$. All processors $PE(0, j, k)$, $0 \leq j < N$, $0 \leq k < N$, broadcast the image pixels $a(j, k)$ concurrently through its subbuses in direction i such that processors $PE(i, j, k)$, $0 \leq i < n * \log^2 N$, each receives a pixel from $PE(0, j, k)$. At the end of step 1, all processors in subarray $ARR(*, j, k)$, where $0 \leq j < N$, $0 \leq k < N$, contain the pixel value $a(j, k)$ at location (j, k) in the original image. Since only local switch settings and broadcast operations are involved in this step, the time used is $O(1)$.

Step 2. As mentioned before, the whole 3D mesh is divided into n layers with each layer having a 3D $\log^2 N \times N \times N$ submesh. Each layer is responsible for computing the Hough transform for a particular projection. Thus, the top $\log^2 N$ 2D submeshes $ARR(i, *, *)$, $0 \leq i < \log^2 N$, are assigned to computing the Hough transform for θ_0. Similarly, the next $\log^2 N$ 2D submeshes $ARR(i, *, *)$, $\log^2 N \leq i < 2\log^2 N$, are in charge of computing for θ_1, and so on. Thus, each processor can calculates its local θ value easily based on its local index i since it initially knows the resolutions of θ and ρ. This requires $O(1)$ time.

Step 3. In this step, all processors computes its local ρ value independently and in parallel. Here, layer t uses θ_t for $0 \leq t \leq n$ as shown in Figure 2; i.e., submesh $ARR(i, *, *)$ uses $\theta_{\lfloor i/\log^2 N \rfloor}$, for $0 \leq i \leq n\log^2 N$, to calculate their ρ values. In other words, $PE(i, j, k)$ computes $\rho_{j,k} = j\cos\theta_{\lfloor i/\log^2 N \rfloor} +$

$j \sin \theta_{\lfloor i / \log^2 N \rfloor}$. This step involves only local computations, and hence takes $O(1)$ time.

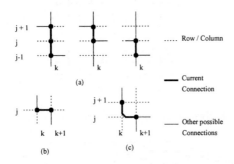

Fig. 3. Possible connections of a processor to neighboring processors for $0 \leq \theta_k \leq \pi/4$.

Step 4. All processors which have computed the same value of the normal distance ρ in the same layer (for a particular angle) can be connected in a 2D submesh. The idea is to count the number of black pixels in the same band (same ρ value for a particular θ value. Since all layers perform a similar job, in the following discussion we concentrate on layer 0. This operation requires only local communications and some setup of local switches. More specifically, the possible cases are depicted in Figure 3. The following connection schemes are based on lemmas 1-4. If $PE(i, j, k)$ computes some value ρ for the normal distance associate to pixel (j, k), and the same value is obtained for pixel $(j, k-1)$ and/or $(j, k+1)$, then $PE(i, j, k)$ should be connected to $PE(i, j, k-1)$ and/or $PE(i, j, k+1)$, as depicted in Figure 3(a). When two adjacent processors in a row have the same ρ value, the connection can be made as shown in Figure 3(b). In case that processors $PE(i, j, k+1)$ and $PE(i, j+1, k)$ have to be connected, a third intermediate processor $PE(i, j, k)$ is used as depicted in Figure 3(c). Using the above rule, a processor in submesh $ARR(i, j, k)$ is connected to at most two buses at a time and no two distinct buses are connected to the same port of a processor in the same submesh. Figure 4 shows the switch and bus configuration for a 11×11 mesh for $\theta_k = \pi/6$. Since all processors in the same layer have the same θ value, the mesh configuration is the same for all 2D submeshes $ARR(i, *, *)$ in the same layer. Thus, $\log^2 N$ 2D submeshes in layer k will have the same configuration as the one depicted in Figure 5. In effect, many 2D vertical submeshes are established. In Figure 5, we show a vertical submesh formed in a layer after the above configurartions and vertical buses are configured along the i direction. In fact, many vertical submeshes exist in the same layer (not shown in the figure). Of course, submeshes in different layers have different shapes. In this step, processors only exchange information with neighboring processors, and then decide on their switch settings. It is obvious that this step also takes constant time.

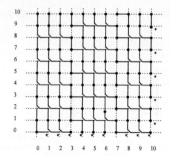

Fig. 4. Switch and bus configuration for $\theta_k = \pi/6$.

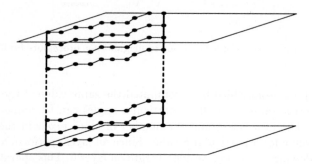

Fig. 5. A 2D vertical submesh is established in a layer after bus reconfiguration in step 4.

Step 5. Accumulate all the pixel values in a band using the corresponding submesh established in the last step in parallel. Notice that each submesh has a size of $\log^2 N \times S$, where S is not fixed and depends its position. As shown in Figure 4, many subbuses of different lengthes are formed and hence their S values are different. However, S is less than $\sqrt{2}N$ and is equal to the number of pixels contained in the band. For S binary values, we can use Lemma 5 to add these binary values in $O(1)$ time on a 2D $\log^2 N \times S$ mesh. Since all submeshes satisfy the above condition, and they can perform the accumulation concurrently, this step uses $O(1)$ time.

Step 6. Each submesh elects a leader and the leader stores the local count from the last step. Notice that this step is necessary since not all boundary processors are the last processors in the reconfigured submesh as indicated in Figure 4. Only those processors with a "*" are leaders. The leaders can be elected easily by simply checking its neighbors and deciding if it should become a leader or not. Clearly, it also takes $O(1)$ time.

The final results are stored in the leaders distributed among different submeshes. Since each step uses $O(1)$ time, the total time used in the algorithm is $O(1)$. To summarize the above discussion, we have:

Theorem 1. For an $N \times N$ binary image and an $n \times n$ parameter space, the

Hough transform can be computed in constant time on a 3D $n \log^2 N \times N \times N$ reconfigurable mesh.

Our result clearly improves the Hough transform algorithms in [1, 10] where a 3D $nN \times N \times N$ reconfigurable mesh and a 3D $n^2 \times N \times N$ reconfigurable mesh are used, respectively, to achieve constant time.

References

1. K.-L. Chung and H.-Y. Lin, "Hough transform on reconfigurable meshes," Computer Vision and Image Understanding, vol. 61, no. 2, 1995, pp. 278-284.
2. P. V. C. Hough, "Methods and means to recognize complex patterns," U.S. Patent 3069654, 1962.
3. H.A.H. Ibrahim, J.R. Kender, and D.E. Shaw, "The analysis and performance of two middle-level vision tasks on a fine-grained SIMD tree machine," Proc. IEEE Computer Society Conf. on Computer Vision and Pattern Recognition, pp. 387-393, June 1985.
4. J. F. Jeng and S. Sahni, "Reconfigurable mesh algorithms for the Hough transform," International Conference on Parallel Processing, vol. III, pp. 34-41, Aug. 12-16, 1991.
5. T.-W. Kao, S.-J. Horng, Y.-L. Wang, "An O(1) time algorithm for computing histogram and Hough transform on a cross-bridge reconfigurable array of processors," IEEE Transactions on Systems, Man and Cybernetics, Vol. 25, No. 4, April 1995, pp. 681-687
6. S.S. Lin, "Constant-time Hough transform on the processor arrays with reconfigurable bus systems," Computing, vol. 52, pp. 1-15, 1994.
7. M. Merry and J. W. Baker, "Constant time algorithm for computing Hough transform on a reconfigurable mesh," Image and Vision Computing, Vol. 14, pp. 35-37, 1996.
8. S. Olariu, J. L. Schwing, and J. Zhang, "Computing the Hough transform on reconfigurable meshes," Image and Vision Computing, vol. 11, no.10, pp.623-628, Dec. 1993.
9. Y. Pan, "A More Efficient Constant Time Algorithm for Computing the Hough Transform," Parallel Processing Letters, vol. 4, no. 1/2, pp. 45-52, 1994.
10. Y. Pan, K. Li, and M. Hamdi, "An improved constant time algorithm for computing the Radon and Hough transforms on a reconfigurable mesh," IEEE Transactions on Systems, Man, and Cybernetics: (Part A), Vol. 29, No. 04, July 1999, pp. 417-421. (A preliminary version also appeared in Proceedings of the 8th International Conference on Parallel and Distributed Computing and Systems, 1996, pp. 82-86.)
11. H. Park, H. J. Kim, and V. K. Prasanna, "An O(1) time optimal algorithm for multiplying matrices on reconfigurable mesh," Information Processing Letters Vol. 47, August 1993, pp. 109-113.

Fifth International Workshop on Formal Methods for Parallel Programming: Theory and Applications FMPPTA 2000

Program and Organizing Chair's Message

It is our pleasure to welcome you to the Fifth International Workshop on Formal Methods for Parallel Programming: Theory and Applications, FMPPTA'2000. This message pays tribute to the many people who have contributed their time and effort in organizing this meeting and reviewing papers.

We are thankful to the IPDPS'2000 committee for accepting the organization of the workshop in cooperation with IPDPS'2000, and especially Viktor K. Prasanna, Mani Chandy and Jose Rolim.

We also would like to thank the authors of all submitted papers, the presenters of accepted papers, the session chairs, the invited speakers and the program committee members.

We hope that every participant will enjoy the workshop.

Beverly Sanders, University of Florida,
and
Dominique Méry, Université Henri Poincaré Nancy I
January 2000

Foreword

The program of FMPPTA 2000 remains focused on the applications of formal methods, particularly for problems involving parallelism and distribution. Seven papers, four contributed and three invited, will be presented, most illustrating the use of techniques that are based on formal concepts and supported by tools. In addition, the workshop will include two tutorials to show how formal techniques can be useful and effective for developing realistic parallel and distributed solutions, for example in telecommunications applications where guaranteeing safety properties, in particular, seems to require the use of formal techniques.

In the first contributed paper, Turner, Argul-Marin, and Laing present the ANISEED method for specifying and analyzing timing characteristics of hardware designs using SDL. Digital hardware is treated as a collection of interacting parallel components. SDL provides a way to validate and to verify digital hardware components. Timing constraints can be studied through SDL specifications.

J. Rolim et al. (Eds.): IPDPS 2000 Workshops, LNCS 1800, pp. 974–976, 2000.

Non-functional requirements are very important aspects of practical systems. The paper by Rosa, Justo and Cunha presents an approach in which transactional and other non-functional requirements are formally incorporated into a special class of software architectures, namely dynamic software architectures. The ZCL framework based on the Z notation is a formal framework which formally incorporates elements of the CL model, a configuration model.

Refinement is a process for developing solutions that satisfy the initial formal specification. In the paper by Filali et. al. refinement is used to develop and validate a termination detection algorithm. The use of UNITY as the development formalism is made easier by the use of PVS, a proof assistant. This work presents a non-trivial case study illustrating the use of a formal method together with mechanized support.

Branco et al. describe their tool Draco-PUC, which automatically generates an implementation in Java for a distributed system described using their formal description technique MONDEL. This approach allows systems to be designed and analyzed at a higher level of abstraction than the implementation language.

The invited presentations will be given by Ganesh Gopalkrishnan, Jean Goubault-Larrecq and Michael Mislove. They will address foundations and applications of formal methods. Ganesh Gopalkrishnan will present verification methods for weak shared memory consistency models; Jean Goubault-Larrecq will address the automatic verification of cryptographic protocols and Michael Mislove will describe the problems encountered in building a semantic model that supports both nondeterministic choice and probabilistic choice.

Two tutorials are summarized by the two abstracts included in the proceedings of the workshop. These are *The Design of Distributed Programs Using the B-Method* by Dominique Cansell, Dominique Méry and Christophe Tabacznyj, and *A Foundation for Composing Concurrent Objects* by Jean-Paul Bahsoun.

We hope that you will enjoy talks and papers.

Beverly Sanders, University of Florida,
and
Dominique Méry, Université Henri Poincaré Nancy I
January 2000

Programme Committee

Flemming Andersen, Tele Danmark R&D, Denmark
Mani Chandy, Caltech, USA
Michel Charpentier, University of New Hampshire, USA
Radhia Cousot, LIX-CNRS, Ecole Polytechnique, France
Mamoun Filali, IRIT, CNRS, Toulouse, France
Pascal Gribomont, Institut MONTEFIORE, Université de LIEGE, Belgium
Dominique Méry, Université Henri Poincaré & IUF, LORIA, France (CoChair)
Lawrence Paulson, Computer Laboratory, Cambridge University, UK

Xu Qiwen, International Institute for Software Technology,
United Nations University, Macau
Joy Reed, Oxford Brookes University, UK
Catalin Roman, Department of Computer Science, Washington University, USA
Beverly Sanders, Department of Computer & Information Science & Engineering, University of Florida (CoChair), USA
Ambuj Singh, Department of Computer Science, University of California at Santa Barbara, USA
David Skillicorn, Department of Computing and Information Science, Queen's University Kingston Canada

A Method for Automatic Cryptographic Protocol Verification (Extended Abstract)

Jean Goubault-Larrecq

G.I.E. Dyade & Projet Coq, Inria, France (Jean.Goubault@dyade.fr)

Abstract. We present an automatic, terminating method for verifying confidentiality properties, and to a lesser extent freshness properties of cryptographic protocols. It is based on a safe abstract interpretation of cryptographic protocols using a specific extension of tree automata, ∨-parameterized tree automata, which mix automata-theoretic techniques with deductive features. Contrary to most model-checking approaches, this method offers actual security guarantees. It owes much to D. Bolignano's ways of modeling cryptographic protocols and to D. Monniaux' seminal idea of using tree automata to verify cryptographic protocols by abstract interpretation. It extends the latter by adding new deductive abilities, and by offering the possibility of analyzing protocols in the presence of parallel multi-session principals, following some ideas by M. Debbabi, M. Mejri, N. Tawbi, and I. Yahmadi.

1 Introduction

It is now well-known that secure cryptographic algorithms (see e.g., [17]) do not suffice in providing system-wide security guarantees, and that one has to be careful in designing cryptographic *protocols*, namely sequences of exchanges of messages purporting to achieve the communication of some piece of data, keeping it confidential or ensuring some level of authentication, to name a few properties of interest [6].

Successful attacks against cryptographic protocols are usually silly, in the sense that they are purely *logical* and do not exploit any weakness in the underlying cryptographic algorithms (e.g., encryption); they are nonetheless difficult to spot. To avoid logical faults, several methods have been designed, based on modal logics of beliefs ([6] and successors), on complexity theory [3] (for specific protocols), on process-algebraic techniques [2], on type disciplines [1], on model-checking [12,13], or on deductive techniques [14,4,16]. While model-checking techniques are fully automated and have been used to find attacks, they cannot directly give actual security guarantees—although reductions to finite-state cases manage to do so in well-behaved cases [18]. On the other hand, the deductive techniques have been designed to give security guarantees, but mechanization is in general partial, as fully automated proof search in general does not terminate. In any case, abstract interpretation (see [8]) can help prepare the grounds for each style of verification. In fact, abstract interpretation alone suffices to verify protocols, as D. Monniaux shows [15], using tree automata to model the set of messages that intruders may build. F. Klay and T. Genet [10] also propose to use tree automata, this time to model the whole protocol itself. Each of the latter two approaches has advantages and disadvantages, but they are automatic, terminate and aim indeed at giving security guarantees, contrarily to standard model-checking tools.

J. Rolim et al. (Eds.): IPDPS 2000 Workshops, LNCS 1800, pp. 977–984, 2000.

Our goal is to present yet another automated technique for guaranteeing the absence of logical faults in cryptographic protocols, which uses tree automata as well. Our contribution is twofold. First, instead of using standard tree automata, we use a refinement (∨PTAs) allowing us to mix enumerative techniques (automata) with deductive techniques (BDDs [5]). The latter will notably help us in modeling freshness and initial states of intruder knowledge. Our ∨PTAs will also be much smaller than standard tree automata, improving the efficiency of verification markedly. Second, we extend the simulation of protocol runs to the case of *parallel multi-session principals*, e.g., key servers, an important case of unbounded parallelism, using ideas from [9].

For space reasons, this paper is only an overview. Moreover, we concentrate on secrecy because it is so fundamental; authentication can be dealt with by simple extensions of the framework presented here, following [10] for example. We describe ∨PTAs in Section 2, and use them to represent and compute states of knowledge in Section 3. We report on practical experience with these techniques in Section 4, showing its practical value, and shedding light on its strengths and weaknesses. We conclude in Section 5.

2   Terms, Formulae, ∨-Parameterized Tree Automata

Let $\mathcal{T}$ be a set of so-called *types* τ. Let $\mathcal{F}$ a set of so-called *function symbols*. A first-order *signature* Σ over $\mathcal{F}$ is a map from $\mathcal{F}$ to the set of expressions of the form $\tau_1 \times \ldots \times \tau_n \to \tau$, where $n \in \mathbb{N}$ and $\tau_1, \ldots, \tau_n, \tau$ are types.

Let $\mathcal{X}_\tau$, for each type τ, be pairwise disjoint non-empty sets, disjoint from $\mathcal{F}$, and $\mathcal{X}$ be $(\mathcal{X}_\tau)_{\tau \in \mathcal{T}}$. The set $T_\tau(\Sigma, \mathcal{X})$ of *terms of type* τ is the smallest set containing $\mathcal{X}_\tau$ and such that for each $f \in \mathcal{F} = \operatorname{dom} \Sigma$, if $\Sigma(f) = \tau_1 \times \ldots \times \tau_n \to \tau$ and $t_1 \in T_{\tau_1}(\Sigma, \mathcal{X})$, $\ldots, t_n \in T_{\tau_n}(\Sigma, \mathcal{X})$, then $f(t_1, \ldots, t_n)$ is in $T_\tau(\Sigma, \mathcal{X})$. We write f instead of $f()$.

We use propositional formulae, up to logical equivalence, to represent (some) sets of terms. Let $\mathcal{A}_\tau$, for each type τ, be a set of so-called *logical variables* of type τ. The intent is that each logical variable α of type τ denotes a set of terms of type τ. *Propositional formulae* F of type τ are defined by the grammar :

$$F ::= A \mid F \wedge F \mid F \vee F \mid \neg F \mid \mathbf{0} \mid \mathbf{1}$$

where A ranges over $\mathcal{A}_\tau$. Formulae F are interpreted as sets $[\![F]\!]\rho_\tau$ in *environments* ρ, where ρ is any family $(\rho_\tau)_{\tau \in \mathcal{T}}$ of maps ρ_τ from $\mathcal{A}_\tau$ to $T_\tau(\Sigma, \mathcal{X})$, by interpreting $\mathbf{0}$ as $\emptyset$, $\mathbf{1}$ as $T_\tau(\Sigma, \mathcal{X})$, $\wedge$ as intersection, $\vee$ as union, $\neg$ as complement.

To deal with term structure, we define the following variant of tree automata. Compared to ordinary tree automata [7], ours integrate propositional formulae at states, and the states are typed (the latter helps in practice limit the size of automata, and does not restrict the generality of the approach). To simplify the following definition, extend Σ to $\mathcal{F} \cup \bigcup_{\tau \in \mathcal{T}} \mathcal{X}_\tau$ by letting $\Sigma(x) \hat{=} \to \tau$ for every $x \in \mathcal{X}_\tau$.

Let Q be a set of so-called *states* q. We assume that each state q has a *type* τ_q, and that Q contains infinitely many states of each type.

An ∨-*parameterized tree automaton*, or ∨*PTA*, of type τ_0, $\mathfrak{A}$, is a 4-tuple $(\mathfrak{Q}, \mathfrak{F}, \mathfrak{R}, \mathfrak{B})$, where $\mathfrak{Q}$ is a finite subset of Q, whose elements are the *states of* $\mathfrak{A}$, $\mathfrak{F} \subseteq \mathfrak{Q}$ is the set of *final states*, $\mathfrak{B}$ maps each state $q \in \mathfrak{Q}$ to a formula of type τ_q, and $\mathfrak{R}$ is a set of rewrites rules $f(q_1, \ldots, q_n) \to q$, the *transitions*, where $f \in \mathcal{F} \cup \bigcup_{\tau \in \mathcal{T}} \mathcal{X}_\tau$

is such that $\Sigma(f)\hat{=}\tau_{q_1} \times \ldots \times \tau_{q_n} \to \tau_q$ ("transitions respect types")—in case f is a variable of type τ, this means $n = 0$, $\Sigma(x) =\to \tau$.

Ordinary tree automata are just $\vee$PTAs without the $\mathfrak{B}$ component (or equivalently, where $\mathfrak{B}$ maps each state to the class of $\mathbf{0}$.) The semantics of $\vee$PTAs is given by defining when a $\vee$PTA $\mathfrak{A}\hat{=}(\mathfrak{Q}, \mathfrak{F}, \mathfrak{R}, \mathfrak{B})$ *recognizes a term t in an environment* ρ *at a state q*; this is so if and only if $t \in [\![\mathfrak{B}(q)]\!]\rho_{\tau_q}$, or t is of the form $f(t_1, \ldots, t_n)$, and there is a transition $f(q_1, \ldots, q_n) \to q$ in $\mathfrak{R}$ such that t_j is recognized by $\mathfrak{A}$ in ρ at q_j for each j, $1 \le j \le n$. A term t is *recognized by* $\mathfrak{A}$ *in* ρ if it so at some final state of $\mathfrak{A}$.

We can compute unions of $\vee$PTAs exactly, and give upper approximants of their intersections by a standard automaton product construction. (This construction gives an exact result in the case of *normal* $\vee$PTAs to be described later.) We can always test whether an $\vee$PTA is *definitely empty*, i.e. whether it cannot recognize any term under any environment ρ: create a Boolean variable ne_q for each state of the $\vee$PTA, produce the clause ne_q if $\mathfrak{B}(q)$ is not equivalent to $\mathbf{0}$ (for each q), the clause $ne_{q_1} \wedge \ldots \wedge ne_{q_n} \Rightarrow ne_q$ for each transition $f(q_1, \ldots, q_n) \to q$ with $\mathfrak{B}(q)$ equivalent to $\mathbf{0}$, and $\neg ne_q$ for each final state q; if the resulting set of clauses is satisfiable, then the given $\vee$PTA is definitely empty; to check it, we use BDDs [5] to represent sets $\mathfrak{B}(q)$ and unit resolution to solve the resulting set of Horn clauses.

We define *assumptions* to be maps $\mathfrak{H}$ from types τ to formulae of type τ. The environment $\rho\hat{=}(\rho_\tau)_{\tau \in \mathcal{T}}$ *satisfies* $\mathfrak{H}$, written $\rho \models \mathfrak{H}$, if and only if $[\![\mathfrak{H}(\tau)]\!]\rho_\tau$ is the set of all terms of type τ, for every type τ. For any two formulae F and G of type τ, we write $F \cap G = \emptyset$ the assumption mapping τ to $\neg(F \wedge G)$ and every other type to $\mathbf{1}$. Given a finite family of assumptions $\mathfrak{H}_i$, $i \in I$, their *conjunction* maps every type τ to $\bigwedge_{i \in I} \mathfrak{H}_i(\tau)$. We reason on $\vee$PTAs $\mathfrak{A}$ modulo assumptions $\mathfrak{H}$ by *reducing* $\mathfrak{A}$, replacing $\mathfrak{B}(q)$ by $\mathfrak{B}(q) \wedge \mathfrak{H}(\tau_q)$ for each state q to get a new $\vee$PTA $\mathfrak{A}_{|\mathfrak{H}}$: under any environment ρ satisfying $\mathfrak{H}$, $\mathfrak{A}$ and $\mathfrak{A}_{|\mathfrak{H}}$ recognize the same terms, and if $\mathfrak{A}_{|\mathfrak{H}}$ is definitely empty, then for no environment ρ satisfying $\mathfrak{H}$, $\mathfrak{A}$ recognizes any term.

3   Messages, What Intruders Know, and Simulating Protocol Runs

To be more specific, our set $\mathcal{T}$ of types contains the type msg of *messages*; the type msglist of tuples of messages, which we shall use to build argument lists to the t tupling operator below; the type K of *raw keys*, e.g. integers of some fixed sizes used to build actual keys, of type key, which we assume to be in $\mathcal{T}$ as well; the type D of *raw data*, e.g. integers, reals, strings, etc. $\mathcal{T}$ may contain other types, which we do not care about. The *basic signature* Σ_0 is:

$$\begin{array}{ll}
\text{symk}, \text{asymk1}, \text{asymk2} : \text{K} \to \text{key} & \quad \text{k} : \text{key} \to \text{msg} \\
\text{d} : \text{D} \to \text{msg} & \quad \text{c} : \text{msg} \times \text{key} \to \text{msg} \\
\text{sk} : \text{msg} \times \text{msg} \to \text{key} & \quad \text{t} : \text{msglist} \to \text{msg} \\
\text{pubk}, \text{privk} : \text{msg} \to \text{key} & \quad \text{nil} : \to \text{msglist} \\
* : \to \text{key} & \quad \text{cons} : \text{msg} \times \text{msglist} \to \text{msglist}
\end{array}$$

The symk constructor builds symmetric keys from raw keys, asymk1 and asymk2 build the two parts of asymmetric keys; sk returns a long-term session key shared between the two principals in argument, pubk and privk return their argument's public and private keys respectively. Any actual key is a message, as represented by the explicit conversion symbol k. Similarly, we use d to convert raw data to messages. The symbol c is used to build *ciphertexts*: $c(M, K)$ is the result of encrypting the *plaintext* M with key K. The special key $*$ is used to model the hash code of M as $c(M, *)$. Finally, any list of messages can be made into a message, using the tupling constructor t that takes a list of messages, of type msglist, in argument: the latter are built using the standard Lisp constructors nil and cons. For legibility we shall abbreviate $cons(M_1, \ldots, cons(M_n, nil) \ldots)$ as $[M_1, \ldots, M_n]$.

We consider as our actual signature Σ any one of the form $\Sigma_0 \uplus \Sigma_1$, where Σ_1 is an unspecified collection of function symbols of signatures $\tau_1 \times \ldots \times \tau_n \to \tau$ where $\tau \notin \{key, msg, msglist\}$. Leaving $\mathcal{T}$ and Σ partly unspecified allows us to deal with extensible types for raw keys and raw data.

We say that, for any keys K and K' (of type key), K' is an *inverse* of K if and only if $K = symk(k)$ or $K = sk(M_1, M_2)$ and $K' = K$; or $K = asymk1(k)$ and $K' = asymk2(k)$; or $K = asymk2(k)$ and $K' = asymk1(k)$; or $K = pubk(M)$ and $K' = privk(M)$; or $K = privk(M)$ and $K' = pubk(M)$. Note that $*$ has no inverse.

Intruders can read on any communication line, and collect what they read. Let E be a set of messages that the intruders have collected (this set might be infinite). These intruders can then forge new messages from E and send them to other principals. Following [4], we model intruders as a deductive system. Write $E \mapsto M$ the predicate "from the set E of messages, the intruders may deduce the message M", defined as follows (E, M denotes the union of E with $\{M\}$):

$$\frac{}{E, M \mapsto M} \, (Ax)$$

$$\frac{E \mapsto M \quad E \mapsto k(K)}{E \mapsto c(M, K)} \, (CryptI) \qquad \frac{E \mapsto c(M, K) \quad E \mapsto k(K') \quad (K' \text{ inverse of } K)}{E \mapsto M} \, (CryptE)$$

$$\frac{E \mapsto M_1 \quad \ldots \quad E \mapsto M_n}{E \mapsto t([M_1, \ldots, M_n])} \, (TupleI) \qquad \frac{E \mapsto t([M_1, \ldots, M_n])}{E \mapsto M_i} \, (TupleE_i), 1 \le i \le n$$

So intruders may replay messages (Ax), *construct* messages by encryption and tupling $((CryptI), (TupleI))$, and *extract* messages by decryption and field selection $((CryptE), (TupleE_i))$—but they cannot crack ciphertexts. Then we may always assume without loss of generality that intruders do all extractions before any construction [4]. That is, let $Ded(E)$ be the set of messages *deducible* from E, i.e., those such that $E \mapsto M$ is derivable; let $Con(E)$ be the constructible ones (derivable using only (Ax), $(CryptI)$, $(TupleI)$), and $Ext(E)$ the extractible ones (derivable using only (Ax), $(CryptE)$, $(TupleE_i)$). Then $Ded(E) = Con(Ext(E))$.

We represent sets of messages E by ∨PTAs, more precisely by *normal* ∨*PTAs*, whose states q of type msg, msglist or key are such that $\mathfrak{B}(q)$ is equivalent to $\mathbf{0}$, and whose transitions $f(q_1, \ldots, q_n) \to q$ are such that f is in the basic signature Σ_0. In particular, computing intersections can be done exactly on normal ∨PTAs.

A central result is that for every normal ∨PTA $\mathfrak{A}$ of type msg, there is a normal ∨PTA that we call $Ded(\mathfrak{A})$ such that, if E is the set of terms recognized by $\mathfrak{A}$ in ρ, then $Ded(\mathfrak{A})$ recognizes at least the terms of $Ded(E)$ in ρ. The idea is by constructing $Ded(\mathfrak{A})$ as $Con(Ext(\mathfrak{A}))$, where the semantics of Con and Ext are as expected.

Building $Ext(\mathfrak{A})$ works by saturating the set $\mathfrak{F}$ of final states by the following two rules: for every transition $\mathtt{t}(ql) \to q$ where q is in $\mathfrak{F}$, add to $\mathfrak{F}$ all states q' of type $\mathtt{msg}$ reachable from ql by following $\mathtt{cons}$-transitions backwards (rule $(TupleE_i)$); for every transition $\mathtt{c}(q', qk) \to q$ with $q \in \mathfrak{F}$, add q' to $\mathfrak{F}$ if for some transition $\mathtt{k}(qk') \to qf$ with $qf \in \mathfrak{F}$, qk' contains possible inverses of qk (rule $(CryptE)$): qk' *contains possible inverses* of qk when there are transitions $f_1(q_{11}, \ldots, q_{1n}) \to qk$ and $f_2(q_{21}, \ldots, q_{2n}) \to qk'$ such that q_{1j} and q_{2j} intersect possibly for every $1 \leq j \leq n$, where $f_1 = \mathtt{symk}$ and $f_2 = \mathtt{symk}$, or $f_1 = \mathtt{asymk1}$ and $f_2 = \mathtt{asymk2}$, etc. (see definition of inverse keys); two states q_1 and q_2 of the same type *intersect possibly* if and only if the intersection of $(\mathfrak{Q}, \{q_1\}, \mathfrak{R}, \mathfrak{B})$ and $(\mathfrak{Q}, \{q_2\}, \mathfrak{R}, \mathfrak{B})$ is not definitely empty.

To build $Con(\mathfrak{A})$, add two fresh states qm of type $\mathtt{msg}$ and ql of type $\mathtt{msglist}$ to $\mathfrak{A}$, mapped to 0 by $\mathfrak{B}$. Then for each transition $f(q_1, \ldots, q_n) \to q'$, where q' is in the set $\mathfrak{F}$ of final states of $\mathfrak{A}$, add a transition $f(q_1, \ldots, q_n) \to qm$, and add transitions $\mathtt{nil}() \to ql$, $\mathtt{cons}(qm, ql) \to ql$, and $\mathtt{t}(ql) \to qm$ (rule $(TupleI)$), and transitions $\mathtt{c}(qm, q) \to qm$ for every transition $\mathtt{k}(q) \to q'$ with q' final in $\mathfrak{A}$ (rule $(CryptI)$).

We simulate protocol runs by describing each principal as a small program. Programs are sequences of instructions, which may either create raw keys, create raw data (nonces), write expressions onto output channels, or read expressions from input channels while pattern-matching them (à la ML). We verify protocols by simulating all possible interleavings (modulo some partial order reductions). The Ded operator handles writes: writing a message M adds M to the set E of messages, and is abstracted by the computation of $Ded(\mathfrak{A})$, where $\mathfrak{A}$ is the normal ∨PTA abstracting E. Reads returns any message M such that $E \mapsto M$ is derivable: we abstract this by having the read instruction return the ∨PTA $\mathfrak{A}$ abstracting E itself as abstract value. Note that abstract values associated with each program variable denote sets of concrete messages, and are represented as normal ∨PTAs again. Pattern-matching is done in the abstract semantics just as in the concrete semantics, replacing equality tests between concrete messages M_1 and M_2 by tests that the ∨PTAs that abstract M_1 and M_2 have an intersection that is not definitely empty after reduction by the current set of assumptions $\mathfrak{H}$. Creating fresh raw data is done as follows. With each instruction creating raw data we associate a *freshness variable* $X \in A_D$; then we insist that $\mathfrak{H}$ be the conjunction of all assumptions $X \cap Y = \emptyset$ for every two distinct freshness variables X and Y, and possibly of other assumptions. ($\mathfrak{H}$ is fixed at the beginning of the simulation and never changes.) Then the abstract value of the variable containing the newly created data is the automaton $(\{q\}, \{q\}, \emptyset, \{q \mapsto X\})$ recognizing exactly those data in (the semantics of) X. Creating fresh keys is done similarly. Note that propositional variables are really needed here to deal with freshness of nonces and keys.

Before we start the simulation, we need to describe the initial set of messages that the intruders know. So let K_0 and D_0 be propositional variables denoting the sets of raw keys that exist (i.e., have been created already), respectively raw data that exist at the start of the run. Let $SSK_0, SAK1_0, SAK2_0$ be variables denoting the sets of raw keys k such that $\mathtt{symk}(k)$, resp. $\mathtt{asymk1}(k)$, resp. $\mathtt{asymk2}(k)$ are initially unknown to the intruders. Let SD_0 be a variable denoting the set of raw data d such that $\mathtt{d}(d)$ is initially unknown to intruders. Assuming for simplicity that every key $\text{sk}(\ldots)$ or $\text{privk}(\ldots)$ is initially unknown to intruders, and that all keys $\text{pubk}(\ldots)$ and $*$ are known, we build an $\vee$PTA $\mathfrak{A}_0$ recognizing the greatest set of terms M known to the intruders validating the secrecy assumptions above. Informally, this is done as follows. Create a state qd of all raw data assumed to exist and initially known; a state qk of all keys assumed to exist and initially known; a state qk^{-1} of all keys assumed to exist but that have no initially known inverse. Then the set E of terms M we look after is given by: M is either $\text{d}(d)$ with d recognized at qd, or $\text{k}(k)$ with k recognized at qk, or a tuple $\text{t}([M_1, \ldots, M_n])$ where each M_i is in E, or $\text{c}(M, K)$, where either M is in E and K is any existing key, or M is any existing message and K is recognized at qk^{-1}. This description can be turned easily into an actual $\vee$PTA $\mathfrak{A}_0$.

We also extend the simulation to handle an unbounded number of copies of any given group of principals. This handles the case of so-called *parallel multi-session* principals S, such as key servers, which actually spawn a new thread after each connection request. (They behave as processes $!S$ in the π-calculus, i.e. they run an unbounded number of copies of S in parallel.) To deal with this case, we use an idea from [9]: such principals S are viewed as accomplices to intruders, and we model them by extending the $Ded(\mathfrak{A})$ automaton by new states and transitions to account for the added computing power that all the copies of S contribute to intruders. This is technical, but let us give a rough idea. First, we assume that each creation (of raw data, of raw keys) done by each copy of S actually returns some unspecified data in the denotation of the freshness variable associated with the creation instruction; so we confuse every copy of S, as far as freshness is concerned. Then, we assume that each instruction of any copy of S executes in any order. Next, we assume that each read succeeds, and pattern-matching is approximated in a crude way: for example, in a $\text{read}\,\text{t}([\text{c}(x, K), y])$ which attempts to read a pair, put the second component in y, decrypt the first component with K and put the resulting plaintext in x, we simply estimate that the value of y will be anything known to intruders, and the resulting value of x will be anything that exists (possibly not known to intruders, because of the enclosing c). We model this by enriching the automaton $Ded(\mathfrak{A})$ with two states, qkn recognizing all known messages, and qx recognizing all existing messages. Writes are then coded by merging these states with other states; e.g., writing $\text{t}([x, \text{c}(y, K)])$ with the same x and y as above implies that $\text{t}([x, \text{c}(y, K)])$ must be recognized at qkn, so that x and $\text{c}(y, K)$ are recognized at qkn, because of $(TupleE_i)$. As far as x is concerned, this means losing any information on existing but unknown messages (merge the qx and qkn states). For $\text{c}(y, K)$, everything depends on whether we assume K to have a known inverse or not: in the first case, then y must exist, otherwise it must become known to the intruders; in any case, since y was already assumed to be known, we do nothing here. In general, the problem of knowing whether K has a known inverse or not matters, and is solved by a fixpoint iteration, which converges because we only deal with finitely many key expressions.

4   Experimental Results

We have implemented these techniques using a bytecode compiler for HimML, a variant of Standard ML incorporating facilities for handling finite sets and maps elegantly and efficiently [11]. We have then tested this implementation on standard cryptographic protocols [6], on a 166MHz Intel Pentium machine running Linux 2.0.30. Each of these protocols are three-party protocols, involving two principals A and B that wish to get a secret key K_{ab} by interacting with a key server S. All of these protocols were tested under an empty assumption $\mathfrak{H}$. Results and running times are as follows:

Protocol	S in mono-session			S in parallel multi-session		
	Result	Time (s.)	#Branches	Result	Time (s.)	#Branches
Needham-Schroeder shared key	p.f.	1.94	4	p.f.	1.56	3
Otway-Rees	OK	1.56	3	OK	1.56	3
Wide-Mouthed Frog	p.f.	0.34	2			
Yahalom	p.f.	1.17	4	p.f.	1.2	3
SimplerYahalom	OK	1.16	3	OK	1.52	3
Otway-Rees2	OK	3.54	4	OK	14.57	15

In the result column, "OK" means the protocol passed, "p.f." means that it contains a possible flaw. The "#Branches" column indicates how much non-determinism is involved in checking all relevant interleavings of the protocol. Times are in seconds, and total the whole exploration of all relevant interleavings; in other words, our tool does not just stop after the first possible flaw.

Note that the Needham-Schroeder protocol was found to be flawed, and indeed our tool finds the standard attack where the intruder plays the second part of the session alone against B, without A or S participating at all. The Yahalom protocol was found to be flawed, too: whether or not our tool has found an attack remains to be examined; indeed, reading attacks off VPTAs is not an easy task! But, as noticed in [6], the Yahalom protocol is a very subtle one, and requires strong assumptions. (By the way, our tool only detects flaws in B's behaviour, so we are guaranteed that A at least cannot be fooled.) On the other hand, the SimplerYahalom protocol (an improved version of the Yahalom protocol given in [6]) is found to be correct by our tool, confirming the opinion of op.cit. that this second version is easier to show correct than the original one.

The last line of the table shows a simulation of two sessions of the Otway-Rees protocol in sequence: OtwayRees2 simulates a principal A_2 playing the role of A twice in a row (with A's identity, and trying to communicate with the same B twice), a principal B_2 that plays the role of B twice in a row (with B's identity, but without checking that its peer is the same A in both sessions), and a server S. The time taken by our tool is still very reasonable, although there should be many more interleavings than for OtwayRees. We are saved by the fact that several interleavings are impossible: our tool discovers that some reads must block (abstract pattern-matching fails).

The worst-case complexity of our algorithms is daunting: abstract pattern-matching in particular takes exponential time and produces VPTAs of exponential size. Nonetheless, the nice news is that verification of actual protocols is quite fast on average, while still maintaining a high level of accuracy.

5 Conclusion

We hope to have convinced the reader that automatic verification of cryptographic protocols was now possible, including some limited form of deduction, and allowing us to prove properties like "M is definitely secret at program point p, whatever the initial messages known to the intruder, provided that assumption $\mathfrak{H}$ is verified". Our technique is natural, provides actual secrecy guarantees—and to a lesser extent freshness guarantees—, and works fast in practice.

Acknowledgments

Many thanks to Dominique Bolignano, David Monniaux, and Mourad Debbabi.

References

1. M. Abadi. Secrecy by typing in cryptographic protocols. *Journal of the Association for Computing Machinery*, 1998. Submitted.
2. M. Abadi and A. D. Gordon. A calculus for cryptographic protocols: The spi calculus. In *Fourth ACM Conference on Computer and Communications Security*. ACM Press, 1997.
3. M. Bellare and P. Rogaway. Provably secure session key distribution–the three party case. In *27th ACM Symposium on Theory of Computing (STOC'95)*, pages 57–66, 1995.
4. D. Bolignano. An approach to the formal verification of cryptographic protocols. In *3rd ACM Conference on Computer and Communication Security*, 1996.
5. R. E. Bryant. Graph-based algorithms for boolean functions manipulation. *IEEE Transactions on Computers*, C35(8):677–692, 1986.
6. M. Burrows, M. Abadi, and R. Needham. A logic of authentication. *Proceedings of the Royal Society*, 426(1871):233–271, 1989.
7. H. Comon, M. Dauchet, R. Gilleron, F. Jacquemard, D. Lugiez, S. Tison, and M. Tommasi. Tree automata techniques and applications. Available on http://www.grappa.univ-lille3.fr/tata/, 1997.
8. P. Cousot and R. Cousot. Abstract interpretation and application to logic programs. *Journal of Logic Programming*, 13(2–3):103–179, 1992. Correct version at http://www.dmi.ens.fr/~cousot/COUSOTpapers/JLP92.shtml.
9. M. Debbabi, M. Mejri, N. Tawbi, and I. Yahmadi. Formal automatic verification of authentication cryptographic protocols. In *1st IEEE International Conference on Formal Engineering Methods (ICFEM'97)*. IEEE, 1997.
10. T. Genet and F. Klay. Rewriting for cryptographic protocol verification (extended version). Technical report, CNET-France Telecom, 1999. Available at http://www.loria.fr/~genet/Publications/GenetKlay-RR99.ps.
11. J. Goubault. HimML: Standard ML with fast sets and maps. In *5th ACM SIGPLAN Workshop on ML and its Applications*, 1994.
12. G. Lowe. Breaking and fixing the Needham-Schroeder public-key protocol using FDR. In *TACAS'96*, pages 147–166. Springer Verlag LNCS 1055, 1996.
13. W. Marrero, E. M. Clarke, and S. Jha. Model checking for security protocols. Technical Report CMU-SCS-97-139, Carnegie Mellon University, 1997.
14. C. A. Meadows. The NRL Protocol Analyzer: An Overview. *Journal of Logic Programming*, 1995.
15. D. Monniaux. Abstracting cryptographic protocols with tree automata. In *6th International Static Analysis Symposium (SAS'99)*. Springer-Verlag LNCS 1694, 1999.
16. L. C. Paulson. The inductive approach to verifying cryptographic protocols. *Journal of Computer Security*, 6:85–128, 1998.
17. B. Schneier. *Applied Cryptography*. John Wiley and Sons, 1996.
18. S. D. Stoller. A bound on attacks on authentication protocols. Technical Report 526, Indiana University, 1999. Available from http://www.cs.indiana.edu/hyplan/stoller.html.

Verification Methods for
Weaker Shared Memory Consistency Models

Rajnish P. Ghughal[1,2] * and Ganesh C. Gopalakrishnan[2]**

[1] Formal Verification Engineer, Intel, Oregon. rajnish.ghughal@intel.com
[2] Department of Computer Science, University of Utah, Salt Lake City, UT
84112-9205. ganesh@cs.utah.edu

Abstract. The problem of verifying finite-state models of shared memory multiprocessor coherence protocols for conformance to weaker memory consistency models is examined. We start with W.W. Collier's architectural testing methods and extend it in several non-trivial ways in
order to be able to handle weaker memory models. This, our first contribution, presents the construction of architectural testing programs similar to those constructed by Collier (e.g. the ARCHTEST suite) suited for
weaker memory models. Our own primary emphasis has, however, been
to adapt these methods to the realm of model-checking. In an earlier
effort (joint work with Nalumasu and Mokkedem), we had demonstrated
how to adapt Collier's architectural testing methods to model-checking.
Our verification approach consisted of abstracting executions that violate memory orderings into a fixed collection of automata (called Test
Automata) that depend only on the memory model. The main advantage
of this approach, called Test Model-checking, is that the test automata
remain fixed during the iterative design cycle when different coherence
protocols that (presumably) implement a given memory model are being
compared for performance. This facilitates 'push-button' re-verification
when each new protocol is being considered. Our second contribution is
to extend the methods of constructing test automata to be able to handle architectural tests for weaker memory models. After reviewing prior
work, in this paper we mainly focus on architectural tests for weaker
memory models and the new abstraction methods thereof to construct
test automata for weaker memory models.
An extended version of this paper is available through
www.cs.utah.edu/formal_verification/ *under 'Publications'*

1 Introduction

Virtually all high-end CPUs are designed for multiprocessor operation in systems
such as symmetric multiprocessor servers and distributed shared memory systems. As processors are getting faster faster than memories are, modern CPUs

---

* The author is currently at Intel, Oregon and was at University of Utah during the
course of the research work presented here.
** Supported in part by NSF Grant No. CCR-9800928.

J. Rolim et al. (Eds.): IPDPS 2000 Workshops, LNCS 1800, pp. 985-992, 2000.

employ *shared memory consistency models* that permit more optimizations at the hardware and compiler levels. As weaker memory models (weaker relative to sequential consistency [7]) permit more hardware/compiler optimizations, virtually all modern processors employ a weak memory model such as *total store ordering* (TSO, [13]), *partial store ordering* (PSO, [13]), or the Alpha Shared Memory Model [11]. Most past work in verifying processors for conformance to memory models has, however, focussed on sequential consistency verification. The upshot of these facts is that there is very limited understanding in the formal verification community on verifying conformance to weaker memory models, and to do it in a way that fits in a modern design cycle in which design changes, and hence verification regressions, are very important.

Contribution 1: Architectural tests for Weaker Memory Models

Our first contribution is in formally characterizing several weaker memory models and presenting new architectural tests for them. In our approach, a formal memory model is viewed as a conjunction of elementary ordering "rules" (relations) such as *read ordering* and *write ordering*, as defined by Collier [1] in conjunction with architectural testing methods for multiprocessor machines developed by him. For example, sequential consistency can be viewed as a conjunction of *computational ordering* (CMP), *program ordering* (PO), and *write atomicity* (WA). This is written "SC=(CMP,PO,WA)" where the right-hand side of the equation is called a *compound rule*, with CMP, PO, WA, etc., then called *elementary rules*. Collier's work was largely geared towards strong memory models, as well as certain atypical weaker memory models. For these memory models, it turns out that it is sufficient to verify for *conjunctions* of 'classical' memory ordering rules such as PO, WA, etc. However, weaker memory models relax these classical ordering rules (often PO and WA) in subtle ways. For example, as we show later, TSO relaxes the write-to-read ordering (WR) aspect of PO. TSO also relaxes WA slightly. Therefore, in a memory system that is supposed to implement TSO, a violation of the classical PO rule does not mean that the memory system is erroneous. The memory system is erroneous with respect to PO only if it violates an aspect of PO *other than* WR orderings. Specifically, given that PO is made up of four sub-rules, namely RO (read ordering), WO (write ordering), WR (write-read ordering), and RW (read-write ordering), it means we must be prepared to look for violations of RO, WO, or RW. Generalizing this idea, to extend Collier's method to cover practical weaker memory models, *pure tests* that test for violations of a single elementary architectural rule or limited combinations of elementary rules would be good to have. In this paper, we outline an example pure test. This example presents a test that checks whether (CMP,RO) (the conjunction of CMP and RO) is violated. We have developed several other such pure tests for other rules to facilate testing for different weak memory models - some of considerably more complexity than the example presented. We will not be presenting all the tests but provide a brief summary of our results at the end of this paper.

In this paper, we explain the technique by which we arrive at pure tests, and examine various aspects of this process, including many non-obvious special cases as well as a few limitations. As one example, we show that sometimes we need to limit the degree to which we leave out rules from a compound rule. For example, we show that the combination (CMP,WO) (WO is "write ordering") is irrelevant in practice; instead, the minimal pure rule worthy of study is (CMP, UPO, WO) where UPO denotes *uniprocessor ordering*. As another example, we show that WO is indistinguishable from WOS if CMP, and a relaxed write atomicity condition WA-S are provided.

The practical implications of these results is that they allow us to explore various tests for a combination of elemental ordering rules and reason about whether an elemental rule is obeyed in presence of other rules. This also enables us to examine a weaker memory model for all aspects of its behavior, come up with different tests to stress these aspects separately, and to correlate the test results. In our work, we have obtained such characterizations for PSO, the Alpha Shared Memory Model, and the IBM 370 memory model. We investigate various pure tests to facilitate verification of conformance to these weaker memory models. In a nutshell, our contribution allows the ARCHTEST methodology to apply to several practical weaker memory models.

Contribution 2: New Abstraction Methods for Architectural Tests

Our second contribution pertains to new abstraction methods in *test model-checking* as explained below. In our earlier work [4, 9], we reported our *test model-checking* approach to verify finite-state models of shared memory systems for conformance to sequential consistency. Test model-checking is basically a reachability analysis technique in which the model of the memory system being verified is closed with test automata playing the roles of the CPUs. The test automata administer a predetermined sequence of write operations involving only a limited range of addresses as well as data values. These writes are interspersed with reads over the same addresses. The test automata were constructed in such a way that when the reads return "unexpected" values, they move to error states, flagging ordering rule violations.

Test model-checking can be carried out in the framework of temporal logic (say, LTL) model-checking by converting each test automaton into a temporal logic formula and checking for the safety property $\Box(\neg inErrorState)$. In a practical setting, however, specialized reachability analysis algorithms may perform better. The fact that the test automata remain the same despite changes in the shared memory system implementation is a significant advantage, as the test model-checking algorithm can be automatically reapplied after each design iteration. In contrast, previous methods required the characterization of the reference specification, namely the desired formal memory model, in terms of very complex temporal logic specifications involving internal details of the memory system under design. This requires the error-prone step of rewriting the temporal logic specification following each design iteration. Many previous efforts also involved manual proofs which are not needed in our approach. For these reasons, test model-checking is eminently suited for use in actual design cycles.

Our earlier reported work on *test model-checking* [4, 9] serves as the background for the work reported here. Our contributions in these works were the following. We demonstrated that test automata can be derived through sound abstractions of architectural tests similar to ARCHTEST. The abstractions were based on *data independence* and *address semi-dependence*. These notions are defined with respect to *executions*, where executions are shared memory programs with reads annotated with the read data values. Under data independence, executions are closed under function applications to the involved data values; in other words, changing the data values does not affect the behavior of the memory system. Under address semi-dependence [5], no other operations may be performed on addresses other than comparison for equality. In our earlier work, we showed that test automata give the effect of running architectural tests for all possible addresses, data values, architectural test-program lengths, and interleavings.

The specific contribution we make with regard to test model-checking is in developing additional abstraction methods that help apply test model-checking for more general varieties of architectural tests. To give a few motivating details, the new *pure tests* we have developed for handling weaker memory models involve architectural tests that examine a finite unbounded history of read values. To handle these situations, we employ data abstraction in conjunction with properties of Boolean operators to derive a finite *summary* of these histories. Details of these abstraction methods and soundness proofs appear in [3].

Another related contribution we make is in handling memory barriers. Given that the test-automata administer a non-deterministic sequence of memory operations, a question that arises in connection with 'membar' instructions is how many membar instructions to consider. We show that under reasonable assumptions – specifically that the memory system does not decode the number of membar instructions it has seen – we need to consider only a limited number of membar instructions. Details appear in [3].

2   Summary of Results

We now summarize our key results in the form of tables and provide an overview (details are in [3]). In Table 1, we summarize the results of test model-checking an operational model of TSO implemented in Berkeley VIS Verilog [12]. This operational model is similar to that used in [2], and usually corresponds to the *reference* specification of TSO. The two 'fail' entries in the table correspond to program ordering, (CMP,PO), and write-to-read orderings, (CMP,WR). Since these orderings are not obeyed in TSO, we obtain 'fail' correctly. The other architectural tests in the tables indicate 'pass' which means that TSO obeys them. These pass/fail results provide added assurance (a 'sanity check') that our characterization of weaker memory models is consistent with the popular understanding of weaker memory models.

Table 2 shows various architecture rules and their *transition templates*. The idea of transition templates introduced in [1] specified a summary of the ordering rule. Many of the entries in this table were specified in [1]. We have defined new architectural rules ($MB - RR$ through $WA - S_{intra}$), defined tests and test automata for them, as well as provided more complete tests for many of the previously existing rules.

Table 3 shows the architecture rules in our discussion and the subrules each of them consists of. In particular, note $WA - S$, which is a relaxed write-atomicity rule that is one of the central sub-rules of TSO. Briefly, write events become visible to the processor issuing the write first, and *then* the events become atomically visible to all other processors. In contrast, in sequential consistency, each write becomes atomically visible (at the same time) to all the processors.

Table 4 shows the memory models in our discussion and their specification in the ARCHTEST framework. These results provide, to the best of our knowledge, the first formal characterization, in one consistent framework, of several practical weaker memory ordering rules. For example, by contrasting TSO and the Alpha Shared Memory Model, it becomes clear that the later is much weaker than the former in terms of read/write orderings, but provides more safety-net operations to recover these orderings.

The Alpha architecture manual [11] describes a number of executions called *Litmus tests* to illustrate which shared memory behavior is allowed and not allowed by the Alpha Shared Memory Model. In [3], we show that all these litmus tests are (often trivially) covered by our characterization of the Alpha Shared Memory architectural compound rule. In addition to sanity-checking our results, these results indicate that a developer of a modern memory system can use our architectural rules to debug the memory system focusing on each *facet* (sub-rule) at a time.

3   Conclusions and Future Work

We formally characterize the problem of verifying finite-state models of shared memory multiprocessor coherence protocols for conformance to weaker memory consistency models in terms of Collier's architectural testing methods. We extend Collier's framework in several non-trivial ways in order to be able to handle

Table 1. Verification results on an operational model of TSO using VIS

test automata	#states	#bdd nodes	runtime (mn:sec)	status
CMP, RO, WO	3819	4872	< 1s	pass
CMP, PO	6.50875e+06	50051	2:38	fail
CMP, WR	6.50875e+06	50051	1:25	fail
CMP,RW	6.50875e+06	50051	3:02	pass
CMP, RO	10187	2463	0:37	pass

Table 2. Architecture rules and their transition templates

Architecture rule	Transition template
SRW	$(P, L, R, V, O, S) <_{SRW} (=, =, W, -, -, -)$
CRW	$(P, L, R, V, O, S) <_{CRW} (-, -, W, -, =, =)$
CWR	$(P, L, W, V, O, S) <_{CWR} (-, -, R, =, =, =)$
CWW	$(P, L, W, V, O, S) <_{CWW} (-, -, W, -, =, =)$
URW	$(P, L, R, V, O, S) <_{URW} (=, -, W, -, =, =)$
UWR	$(P, L, W, V, O, S) <_{UWR} (=, -, R, -, =, =)$
UWW	$(P, L, W, V, O, S) <_{UWW} (=, -, W, -, =, =)$
RW	$(P, L, R, V, O, S) <_{RW} (=, -, W, -, -, -)$
WR	$(P, L, W, V, O, S) <_{WR} (=, -, R, -, -, -)$
RO	$(P, L, R, V, O, S) <_{RO} (=, -, R, -, -, =)$
WO	$(P, L, W, V, O, S) <_{WO} (=, -, W, -, -, -)$
$MB-RR$	$(P, L, R, V, O, S) <_{MB-RR} (=, -, MB - RR, -, -, -)$ $(P, L, MB - RR, -, -, -) <_{MB-RR} (=, -, R, -, -, -)$
$MB-RW$	$(P, L, R, V, O, S) <_{MB-RW} (=, -, MB - RW, -, -, -)$ $(P, L, MB - RW, -, -, -) <_{MB-RW} (=, -, W, -, -, -)$
$MB-WR$	$(P, L, W, V, O, S) <_{MB-WR} (=, -, MB - WR, -, -, -)$ $(P, L, MB - WR, -, -, -) <_{MB-WR} (=, -, R, -, -, -)$
$MB-WW$	$(P, L, W, V, O, S) <_{MB-WW} (=, -, MB - WW, -, -, -)$ $(P, L, MB - WW, -, -, -) <_{MB-WW} (=, -, W, -, -, -)$
$=_{WA-S}$	$(P, L, W, V, O, \neq P) =_{WA-S} (=, =, W, =, =, \neq P)$
$WA-S_{intra}$	$(P, L, W, V, O, = P) <_{WA-S_{intra}} (=, =, W, =, =, \neq P)$
CON	$(P, L, W, V, O, S) <_{CON} (-, -, W, -, =, =)$
WA	$(P, L, A, V, O, S) <_{WA} (-, -, -, -, -, -)$

Table 3. Architecture rules

Architecture rule	Subrules
CMP	SRW, CRW, CWW, CWR
PO	WR, WO, RO, RW
$WA-S$	$=_{WA-S}, WA-S_{intra}, CON$
MB	$MB-WR, MB-WO, MB-RO, MB-RW$

weaker memory models. This permits the construction of architectural testing programs similar to those constructed by Collier (e.g. the ARCHTEST suite). We then adapt these tests to the realm of model-checking, to permit early life-cycle formal verification of design descriptions. Our approach consists of abstracting executions that violate memory orderings into a fixed collection of automata (called Test Automata) that depend only on the memory model. The main advantage of our test model-checking approach is that the test automata remain fixed during the iterative design cycle when different coherence protocols that (presumably) implement a given memory model are being compared for performance. This facilitates 'push-button' re-verification when each new protocol is being considered. Here, we report our new results that extend the methods of constructing test automata to be able to handle architectural tests for weaker memory models. We achieve this as follows: (i) we define new abstraction techniques for summarizing execution histories; (ii) we prove the soundness of these abstractions; and (iii) we provide practical means to reduce the number of memory barrier instructions used in test automata.

We provide a formal characterization, in one consistent framework, of several practical weaker memory ordering rules, including the TSO and PSO models [13], the IBM 370 memory model, and the Alpha Shared Memory Model [11]. We show that 'Litmus tests' that practitioners employ for the Alpha memory model are covered by our formalism. We define a suite of architectural tests and corresponding test model-checking automata to facilitate verification of weaker memory models. We report on VIS based model-checking results that clearly show how developers of modern memory systems can use our architectural rules to debug the memory system focusing on each *facet* (sub-rule) at a time. This helps cut down verification complexity and helps pinpoint errors.

We believe that test automata can provide leverage in attacking the shared memory system verification problem, as they avoid many redundant test cases that would otherwise have been used in a simulation framework. We are working on overcoming a limitation of the present work, namely that we do not yet have complete test automata for the weaker memory models examined here.

Table 4. Memory models specification in the ARCHTEST framework

Memory Model	ARCHTEST specification
Sequential consistency	$A(CMP, PO, WA)$
IBM 370	$A(CMP, UPO, RO, WO,$ $RW, WA, MB-WR)$
Total Sorted Orer (TSO)	$A(CMP, UPO, RO, WO,$ $RW, WA-S, MB-WR)$
Partial Sorted Order (PSO)	$A(CMP, UPO, RO,$ $RW, WA-S, MB-WR, MB-WW)$
Alpha Shared Memory Model	$A(CMP, UPO, ROO,$ $WA-S, MB, MB-WW)$

Promising completeness results obtained in another work [8] may provide the necessary directions to pursue.

We are also investigating ways to mitigate state explosion. Since test model-checking can be cast as *finite-state reachability analysis*, efficient techniques under development by many groups for exact- as well as approximate reachability analysis are believed to be one promising direction to pursue. Once we get a handle on state explosion through reachability analysis based test model-checking, we plan to work on overcoming many other sources of inefficiency, including symmetries that, as yet, stand unexploited. Works on symmetry exploitation (e.g., [6]) and the use of symbolic state descriptors (e.g., [10]) will be examined for possible answers. The integration of these ideas back into the realm of reachability analysis is expected to be a long-term direction.

References

[1] COLLIER, W. W. *Reasoning About Parallel Architectures*. Prentice-Hall, Englewood Cliffs, NJ, 1992.

[2] DILL, D. L., PARK, S., AND NOWATZYK, A. Formal specification of abstract memory models. In *Research on Integrated Systems* (1993), G. Borriello and C. Ebeling, Eds., MIT Press, pp. 38–52.

[3] GHUGHAL, R. Test model-checking approach to verification of formal memory models, 1999. Also available from http://www.cs.utah.edu/formal_verification.

[4] GHUGHAL, R., NALUMASU, R., MOKKEDEM, A., AND GOPALAKRISHNAN, G. Using "test model-checking" to verify the Runway-PA8000 memory model. In *Tenth Annual ACM Symposium on Parallel Algorithms and Architectures* (Puerto Vallarta, Mexico, June 1998).

[5] HOJATI, R., AND BRAYTON, R. Automatic datapath abstraction of hardware systems. In *Conference on Computer-Aided Verification* (1995).

[6] IP, C. N., AND DILL, D. L. Better verification through symmetry. In *Int'l Conference on Computer Hardware Description Language* (1993).

[7] LAMPORT, L. How to make a multiprocessor computer that correctly executes multiprocess programs. *IEEE Transactions on Computers 9*, 29 (1979), 690–691.

[8] NALUMASU, R. *Formal design and verification methods for shared memory systems*. PhD thesis, University of Utah, Salt Lake City, UT, USA, Dec. 1998.

[9] NALUMASU, R., GHUGHAL, R., MOKKEDEM, A., AND GOPALAKRISHNAN, G. The 'test model-checking' approach to the verification of formal memory models of multiprocessors. In *Computer Aided Verification* (Vancouver, BC, Canada, June 1998), A. J. Hu and M. Y. Vardi, Eds., vol. 1427 of *Lecture Notes in Computer Science*, Springer-Verlag, pp. 464–476.

[10] PONG, F., AND DUBOIS, M. New approach for the verification of cache coherence protocols. *IEEE Transactions on Parallel and Distributed Systems 6*, 8 (Aug. 1995), 773–787.

[11] SITES, R. L. *Alpha Architecture Reference Manual*. Digital Press and Prentice-Hall, 1992.

[12] Vis-1.2 release. http://www-cad.eecs.berkeley.edu/Respep/Research/vis/.

[13] WEAVER, D. L., AND GERMOND, T. *The SPARC Architecture Manual – Version 9*. P T R Prentice-Hall, Englewood Cliffs, NJ 07632, USA, 1994.

Models Supporting Nondeterminism and Probabilistic Choice

Michael Mislove*

Department of Mathematics
Tulane University
New Orleans, LA 70118
mwm@math.tulane.edu

Abstract. In this paper we describe the problems encountered in building a semantic model that supports both nondeterministic choice and probabilistic choice. Several models exist that support both of these constructs, but none that we know of satisfies all the laws one would like. Using domain-theoretic techniques, we show how a model can be devised "on top of" certain models for probabilistic choice, so that the expected laws for nondeterministic choice and probabilistic choice remain valid.

1 Introduction

Nondeterminism is a standard component of concurrent programming languages. The most widely employed method for modeling concurrent computation takes sequential composition as a primitive operator, from which it is only natural to use nondeterministic choice to generate an interleaving semantics for concurrent computation. This approach also is well-supported by the models of computation that are available. And, there there are some standard assumptions that one expects to hold in any denotational model supporting nondeterministic choice, $+$. For example, a basic assumption one expects to hold is that $+$ is idempotent, commutative and associative, so, in mathematical parlance, we expect $+$ to be a semilattice operation on whatever denotational model we have devised for our language.

If we view internal nondeterminism from a specification point of view, then $p + q$ represents a "don't care" process – we are just as happy to have our program act like p or like q, since either one will (presumably) have the requisite behavior we are seeking. On the other hand, if we are concerned with such things as network flow, then fairness issues arise, and we now might be more concerned with the possibility that one branch dominates, preventing the other from ever having a chance to execute. This is quite different from the view of nondeterminism as under specification, and so some other operation is needed. A notion of probabilistic choice is natural to consider here, and this leads to

---

* Partial support provided by the National Science Foundation and the US Office of Naval Research

J. Rolim et al. (Eds.): IPDPS 2000 Workshops, LNCS 1800, pp. 993-1000, 2000.

the introduction of probabilistic choice operators. One might think of $p_{.5} + q$ [1] as a probabilistic choice, in which the process acts like p half of the time and like q the other half. Of course, different laws are expected of the operators like $_{.5}+$. For example, we do not expect an associative operation – indeed, we expect $(p_{.5} + q)_{.5} + r = p_{.25} + (q_{.25} + r)$.

Interesting aspects emerge when one combines all these operators within a single model. The existing models for process algebras that support both non-deterministic choice and probabilistic choice do not satisfy all the laws that one might expect to hold. Typically, the nondeterministic choice operator is no longer idempotent for all processes. While arguments have been put forth to justify the fact that $+$ is not idempotent on processes which involve probabilistic choice, our view is that idempotence is a property which would be useful to retain.

And this brings us to the issue we are interested in confronting: how to build denotational models for process algebras which have both nondeterministic choice and probabilistic choice, so that the laws for nondeterministic choice and for probabilistic choice that one expects to hold actually are valid.

2   Domains

In this section, we review some of the basics we need to describe our results. A good reference for most of this can be found in [1]. To begin, a *partial order* is a non-empty set endowed with a reflexive, antisymmetric and transitive relation. If P is a partial order, then a subset $D \subseteq P$ is *directed* if every finite subset of D has an upper bound in D. We say P is *directed complete* if every directed subset of D has a least upper bound, $\sqcup D$, in P. Such partial orders we call *dcpos*, and we use the term *cpo* for a dcpo that also has a least element, usually denoted $\bot$.

The functions of interest between dcpos are those that are *Scott continous*. While this can be stated in topological terms, it's just as easy to describe them as those maps that preserve the order and sups of directed sets. The following result is the basis for the "least fixed point semantics" that often is used in domains to model recursion.

Theorem 1 (Tarksi-Knaster-Scott). *If $f : P \to P$ is a continuous selfmap of a cpo P, then* $\mathrm{FIX}(f) = \sqcup_{n \geq 0} f^n(\bot)$ *is the least fixed point of f.*   □

The class of (d)cpos and Scott continuous maps is a cartesian closed category.

Definition 1. *Let P and Q be cpos. An embedding-projection pair $(e, p) : P \to Q$ is a pair of continuous maps $e : P \to Q, p : Q \to P$ satisfying $x \leq p(y) \Leftrightarrow e(x) \leq y$ for all $x \in P$ and $y \in Q$ and $p \circ e = 1_P$.*

---

[1] We will use the notation $p_{\lambda} + q$ to denote a probabilistic choice in which the process has probability λ of acting like p, and probability $1 - \lambda$ of acting like q, where $0 \leq \lambda \leq 1$.

3 CSP

One of the most widely studied process algebras for concurrent computation is CSP, a process algebra which represents processes by the actions in which they can participate. A representative portion of its BNF is given by:

$$p ::= STOP \mid SKIP \mid a \to p \mid p; p \mid p \setminus A \mid p \Box p \mid p \sqcap p \mid p_A \|_B p \mid p \| \| p \mid x \mid \mu x.p$$

Briefly, $STOP$ is the deadlocked process incapable of any actions, $SKIP$ denotes the process whose only action is normal termination, $a \to p$ is a process which first is willing to participate in the action $a \in \Sigma$, the set of possible actions, and then will act like process p, $p; q$ denotes sequential composition, $p \setminus A$ is the process p with the actions $A \subseteq \Sigma$ hidden from the environment, $p \Box q$ is the process which offers the environment the choice of acting like p or like q, the choice being resolved on the first step, $p \sqcap q$ is the process which internally decides to act like p or like q with no influence from the environment, $p_A \|_B q$ is the parallel composition of p and q synchronizing on all actions in $A \cap B$, and with each branch executing any other actions that it is capable of independently, $p \| \| q$ is the interleaved parallel composition of p and q, x is a process variable, the set of all such being V, and $\mu x.p$ denotes recursion.

A *finite trace* of a process p is a sequence of actions $a_1 a_2 \cdots a_n$ that p can participate in. An *infinite trace* is an infinite sequence of such actions. It is well known that traces are insufficient to distinguish external and internal choice. This shortcoming was overcome for CSP by the introduction of failures [3].

A *failure* of a process p is a pair $\langle t, X \rangle$, where t is a finite trace of p, and X is a set of actions that p might refuse to participate in, once t is completed.

CSP supports the notion of an internal action, τ, that is used to denote a process p executing an action internally and becoming another process. Since CSP also supports hiding of actions - taking visible actions and making them invisible to the environment, the possibility of *divergence* arises: for example, the process $p = (\mu x.a \to x) \setminus \{a\}$ which becomes eternally engaged in internal actions and never responds to the environment. A *divergence* of a process p is a trace t after which p could become divergent.

The *failures-divergences model* $\mathbb{FD}$ for CSP models each process p as a pair (F, D), where F is the set of failures of p, and D the set of divergences. In order to have a well-behaved denotational model for the set of CSP processes, several healthiness conditions are made on these pairs. The failures set F satisfies the conditions that it is non-empty, if (t, X) is a failure of p, then so is $(s, \emptyset)$ for any prefix s of t, and "impossible events can be added to refusals": if (t, X) is a failure of p and $t^\frown a$ is not a trace of p, then $(t, X \cup \{a\})$ is a failure.

The main criterion for divergences is that once a process has diverged on a trace s, then it can never recover. So, if $s \in D$, then $st \in D$ for all traces t, and (st, X) is a failure of p for every trace t and every subset X of actions. The failures-divergences model supports interpretations of the all the standard operators of CSP.

In $\mathbb{FD}$, the *order of nondeterminism* is used, in which the pair $(F, D) \sqsubseteq (F', D')$ if and only if $F \supseteq F'$ and $D \supseteq D'$. Thus, the higher a pair is in this order, the more deterministic is its behavior. In this model, the deterministic processes form the set of maximal elements (cf. [3]).

4   The probabilistic power domain

The construction that allows probabilistic choice operators to be added to a domain is the probabilistic power domain, which was first investigated by Saheb-Djarhomi [12], who showed that the family he defined yields a cpo. It was refined and expanded by Jones [6, 7] who also was showed that the probabilistic power domain of a continuous domain is again continuous.

Definition 2. *If P is a dcpo, then a* continuous valuation *on P is a mapping $\mu\colon \Sigma D \to [0,1]$ defined on the Scott open subsets of P that satisfies:*

1. *$\mu(\emptyset) = 0$.*
2. *$\mu(U \cup V) = \mu(U) + \mu(V) + \mu(U \cap V)$,*
3. *μ is monotone, and*
4. *$\mu(\cup_i U_i) = \sup_i \mu(U_i)$, if $\{U_i \mid i \in I\}$ is an increasing family of Scott open sets.*

We order this family pointwise: $\mu \leq \nu \Leftrightarrow \mu(U) \leq \nu(U)$ $(\forall U \in \Sigma P)$. We denote the family of continuous valuations on P by $\mathcal{P}_{Pr}(P)$.

Among the continuous valuations on a dcpo, the *simple valuations* are particularly easy to describe. They are of the form $\mu = \Sigma_{x \in F}\, r_x \cdot \delta_x$, where $F \subseteq P$ is a finite subset, δ_x represents point mass at x (the mapping sending an open set to 1 precisely if it contains x, and to 0 otherwise), and $r_x \in [0,1]$ satisfy $\Sigma_{x \in F}\, r_x \leq 1$. In this case, the *support* of μ is just the family F.

A basic result of [6] is that the probabilistic power domain of a continuous domain is continuous. The probabilistic power domain extends to an endofunctor on continuous domains, so each continuous map $f\colon P \to Q$ between (continuous) domains can be lifted to a continuous maps $\mathcal{P}_{Pr}(f)\colon \mathcal{P}_{Pr}(P) \to \mathcal{P}_{Pr}(Q)$ by $\mathcal{P}_{Pr}(f)(\mu)(U) = \mu(f^{-1}(U))$. In fact, [6] shows that the resulting functor is a left adjoint, which means that $\mathcal{P}_{Pr}(P)$ is a free object over P in an appropriate category. The category in question can be described in terms of probabilistic choice operators satisfying certain laws (cf. [6]):

Definition 3. *A* probabilistic algebra *is a dcpo A endowed with a family of Scott continuous operators $_\lambda+\colon A \times A \to A$, $0 \leq \lambda \leq 1$ such that $(\lambda, a, b) \mapsto a\,_\lambda+\,b\colon [0,1] \times A \times A \to A$ is continuous and so that the following laws hold for all $a, b, c \in A$:*

- $a\,_\lambda+\,b = b\,_{1-\lambda}+\,a$,
- $(a\,_\lambda+\,b)\,_\rho+\,c = a\,_{\lambda\rho}+\,(b\,_{\frac{\rho(1-\lambda)}{1-\lambda\rho}}+\,c)$ (if $\lambda\rho < 1$).
- $a\,_\lambda+\,a = a$, *and*
- $a\,_1+\,b = a$.

The operations $_\lambda+$ are defined on $\mathcal{P}_{Pr}(P)$ in a pointwise fashion, so for instance, $\mu\,_\lambda+\,\nu = \lambda\mu + (1-\lambda)\nu$. It then is routine to verify that $\mathcal{P}_{Pr}(P)$ is a probabilistic algebra over P for each dcpo P.

4.1   Probabilistic CSP

The syntax of probabilistic CSP investigated in [9] is not much different from that of CSP – PCSP simply adds the family of operators $_\lambda+$ for $0 \leq \lambda \leq 1$. So, we now can reason about processes such as $(a \to Stop)\,_{.5}+ (b \to STOP \ \square \ c \to STOP)$, which will act like $a \to STOP$ half of the time, and offer the external choice of doing a b or a c the other half.

The model for PCSP that was devised in [9] is built simply by applying the probabilistic power domain operator to the failures-divergences model for CSP. The interpretation of the operators of CSP in $\mathcal{P}_{Pr}(\mathbb{FD})$ is accomplished by analyzing the construction itself. Namely, $\mathcal{P}_{Pr}(\mathbb{FD})$ is a set of continuous mappings from the set of Scott open sets of $\mathbb{FD}$ to the unit interval. So, for example, given a unary operator $f \colon \mathbb{FD} \to \mathbb{FD}$, we can extend this to $\mathcal{P}_{Pr}(\mathbb{FD})$ by $\mathcal{P}_{Pr}(f) \colon \mathcal{P}_{Pr}(\mathbb{FD}) \to \mathcal{P}_{Pr}(\mathbb{FD})$ by $\mathcal{P}_{Pr}(f)(\mu)(U) = \mu(f^{-1}(U))$. Similar reasoning shows how to extend operators of higher arity. Two facts emerge from this method:

- If we embed $\mathbb{FD}$ into $\mathcal{P}_{Pr}(\mathbb{FD})$ via the mapping $p \mapsto \delta_p$, then the interpretation of each CSP operator on $\mathbb{FD}$ *extends* to a continuous operator on $\mathcal{P}_{Pr}(\mathbb{FD})$: this means that the mapping from $\mathbb{FD}$ into $\mathcal{P}_{Pr}(\mathbb{FD})$ is compositional for all the operators of CSP. This has the consequence that any laws that the interpretation of CSP operators satisfy on $\mathbb{FD}$ still hold *on the image of* $\mathbb{FD}$ *in* $\mathcal{P}_{Pr}(\mathbb{FD})$.
- The way in which the operators of CSP are extended to the model of PCSP forces all the CSP operators to distribute through the probabilistic choice operators. For example, we have

$$a \to (p\,_\lambda+ q) = (a \to p)\,_\lambda+ (a \to q),$$

for any event a and any processes p and q. This has the result that some of the laws of CSP fail to hold on $\mathcal{P}_{Pr}(\mathbb{FD})$ *as a whole*.

Example 1. Consider the process $(p\,_{.5}+ q) \sqcap (p\,_{.5}+ q)$. The internal choice operator $\sqcap$ is supposed to be idempotent, but using the fact that, when lifted to PCSP, the CSP operators distribute through the probabilistic choice operators, we find that $(p\,_{.5}+ q) \sqcap (p\,_{.5}+ q) = p\,_{.25}+ ((p \sqcap q)\,_{2/3}+ q)$, which means that the probability that the process acts like p is somewhere between .25 and .75, depending on how the choice $p \sqcap q$ is resolved. This unexpected behavior can be traced to the fact that $\sqcap$ distributes through $_{.5}+$. One way to explain this is that the resolution of the probabilistic choice in $p\,_{.5}+ q$ is an internal event, and using the CSP paradigm of *maximal progress* under which internal events happen as soon as possible, the probabilistic choices then are resolved at the same time as the internal nondeterministic one. The processes on either side of $\sqcap$ represent distinct instances of the same process, but because they are distinct, the probabilistic choice is resolved independently in each branch.

Since it is the fact that $\sqcap$ distributes through $_\lambda+$ that causes $\sqcap$ not to be idempotent, one way to avoid this issue would be to craft a model which forces us to resolve $\sqcap$ first, *before* the probabilistic choices are resolved.

5   Constructing a new model

It follows from the method of construction that the lifting of the operations from
$\mathbb{FD}$ to $\mathcal{P}_{Pr}(\mathbb{FD})$ all distribute through the probabilistic choice operators. This
is why certain laws from CSP fail in the extension, such as the failure of the
extension $\sqcap$ to PCSP to be idempotent. We view the fact that nondeterministic
choice is not idempotent on $\mathcal{P}_{Pr}(\mathbb{FD})$ to be problematical. In the example of
the last section, for instance, this leads to the unexpected result that there is
no precise probability that the process $(p_{.5}+q) \sqcap (p_{.5}+q)$ acts like p. We now
show how to construct a domain Q which supports nondeterministic choice and
probabilistic choice, so that the choice operator is idempotent. Moreover, if P is
bounded complete, we can construct Q then there is an e-p pair from P into Q.

One approach to defining an idemptotent nondeterministic choice operator
might be to search for an alternative method for extending $\sqcap$ to $\mathcal{P}_{Pr}(\mathbb{FD})$. The
search is in vain if we also require that the extension be *affine*, since the categorical extension already satisfies this property, and there cannot be two such
extensions (because the image of $\mathbb{FD}$ generates $\mathcal{P}_{Pr}(\mathbb{FD})$). So, we seek to extend
the construction so as to accommodate another internal choice operator.

Another approach would be to apply an appropriate power domain oper-
ator to $\mathcal{P}_{Pr}(\mathbb{FD})$: Indeed, since $\mathcal{P}_{Pr}(\mathbb{FD})$ is coherent, then the classical lower
and upper power domains satisfy $\mathcal{P}_L(\mathcal{P}_{Pr}(\mathbb{FD})), \mathcal{P}_U(\mathcal{P}_{Pr}(\mathbb{FD})) \in$ BCD. However,
this is not exactly what we want, because, if we use the standard approach
to extending the operations from P to $\mathcal{P}_L(P)$ or $\mathcal{P}_U(P)$ in the case P is a
probabilistic algebra, we find that the laws we want no longer are valid. For
example, for $X, Y \in \mathcal{P}_U(P)$, then $X_{\lambda}+Y = \{x_{\lambda}+y \mid x \in X, y \in Y\}$, so
$X_{\lambda}+X = \{x_{\lambda}+y \mid x, y \in X\} \neq X$. In general, $X_{\lambda}+X$ will be larger than X.
To remedy this, we proceed as follows.

Definition 4. *Let P be a probabilistic algebra, and let $X \subseteq P$. We define*

$$\langle X \rangle = \{x_{\lambda}+y \mid x, y \in X \wedge 0 \leq \lambda \leq 1\}.$$

We say that X is affine closed *if $X = \langle X \rangle$. We let $\mathcal{P}_{UA}(P) = \{X \in \mathcal{P}_U(P) \mid X = \langle X \rangle\}$.*

Theorem 2. *Let P be a probabilistic algebra which is also a coherent domain.
Then there is a continuous kernel operator*

$$\kappa \colon (\mathcal{P}_U(P), \supseteq) \to (\mathcal{P}_{UA}(P), \supseteq) \text{ given by } \kappa(X) = \bigcap\{Y \in \mathcal{P}_U(P) \mid X \subseteq Y\}.$$

*Furthermore, $\mathcal{P}_{UA}(P)$ is a continuous, coherent domain which is also a proba-
bilistic algebra. Finally, $\mathcal{P}_{UA}$ extends to a functor $\mathcal{P}_{UA} \colon$ Coh $\to$ BDC which is
continuous.*   □

This leads us to consider the domain $\mathcal{P}_{UA}(\mathcal{P}_{Pr}(\mathbb{FD}))$. We have chosen to focus on
the probabilistic power domain analogous to the upper power domain because
the upper power domain is the power domain of demonic choice, and this is
what underlies the (internal) nondeterministic choice in CSP. We build a model
"over" $\mathbb{FD}$ using the following:

Theorem 3. *Let P be a bounded complete, continuous domain. Then:*

1. *There is an e-p pair from P to $\mathcal{P}_{Pr}(P)$.*
2. *There is an e-p pair from $\mathcal{P}_{Pr}(P)$ to $\mathcal{P}_{UA}(\mathcal{P}_{Pr}(P))$.*

Moreover, the embedding $e\colon \mathcal{P}_{Pr}(P) \to \mathcal{P}_{UA}(P)$ is a morphism of probabilistic algebras.   □

Using these results, we can begin with $\mathbb{FD}$ and generate a bounded complete, continuous domain $\mathcal{P}_{UA}(\mathcal{P}_{Pr}(\mathbb{FD}))$ that also is a probabilistic algebra. This is the model we have been seeking:

Example 2. We show that the domain $\mathcal{P}_{UA}(\mathcal{P}_{Pr}(\mathbb{FD}))$ is a model for PCSP in which internal choice does not distribute over probabilistic choice. We reason as follows. First, using the standard categorical approach, we can extend the interpretation of each CSP operator on $\mathcal{P}_{Pr}(\mathbb{FD})$ to $\mathcal{P}_{UA}(\mathcal{P}_{Pr}(\mathbb{FD}))$, and these extensions all are continuous. Noting that $\mathcal{P}_{UA}(\mathcal{P}_{Pr}(\mathbb{FD}))$ already has an internally defined inf-operation – $(X, Y) \mapsto \kappa(\langle X \cup Y \rangle)$, we then can conclude $\mathcal{P}_{UA}(\mathcal{P}_{Pr}(\mathbb{FD}))$ is a continuous algebra of the same signature as defines the syntax of CSP. Since we can regard CSP as the initial algebra with this signature, it follows that there is a (unique!) algebra homomorphism $[\![\ ]\!] \colon \mathrm{CSP} \to \mathcal{P}_{UA}(\mathcal{P}_{Pr}(\mathbb{FD}))$, and we take this as our semantic map. Actually, this extends to a semantic mapping from PCSP to $\mathcal{P}_{UA}(\mathcal{P}_{Pr}(\mathbb{FD}))$ since the latter is a probabilistic algebra over $\mathcal{P}_{Pr}(\mathbb{FD})$.

   To show that internal choice does not distribute over the probabilistic choices in $\mathcal{P}_{UA}(\mathcal{P}_{Pr}(\mathbb{FD}))$, we simply note that we have chosen the internally defined inf-operation on $\mathcal{P}_{UA}(\mathcal{P}_{Pr}(\mathbb{FD}))$ as our interpretation of $\sqcap$, and since this operator is idempotent on all of $\mathcal{P}_{UA}(\mathcal{P}_{Pr}(\mathbb{FD}))$, we conclude that

$$(p_{.5}+ q) \sqcap (p_{.5}+ q) = p_{.5}+ q$$

for any elements of $\mathcal{P}_{UA}(\mathcal{P}_{Pr}(\mathbb{FD}))$ – in particular, this holds for p and q the denotations of processes from PCSP. But then $\sqcap$ cannot distribute through $_{.5}+$, because we would then have the equality

$$p_{.5}+ q = (p_{.5}+ q) \sqcap (p_{.5}+ q) = p_{.25}+ ((p \sqcap q)_{2/3}+ q),$$

which would imply that $p \sqcap q = p_{.5}+ q$, which certainly does not hold, as easy examples show.   □

6   Summary

We have focused on an anomaly with the model for probabilistic CSP devised in [9], namely that the interpretation of internal choice on this model is not idempotent. The anomaly is a result of the way in which the operators of CSP are defined on this model. This motivates the construction of an alternative model for probabilistic CSP. We have given a construction of a model using a new power domain – the power domain of affine, compact upper subsets of a coherent continuous domain, and we showed that this family also is a probabilistic algebra. We then showed that, when applied to the model for PCSP from [9], this model satisfies the property that the interpretation of nondeterministic choice is idempotent, and' the model also satisfies all the laws of a probabilistic algebra. This is the only model of this type we know of. In particular, the models defined in several of the papers listed in the references (except, of course, that of [6, 7]) seem not to address this issue.

A question we have left unaddressed is what laws our new model satisfies. This is a very important issue, especially given the tradition of algebraic semantics for CSP and its related languages. An obvious example is the external choice operator □ of CSP, since this operator becomes internal choice if both branches offer the same event.

Acknowledgements: The author wishes to thank JOËL OUAKNINE and STEVEN SHALIT for stimulating conversations about this work, and also DOMINIQUE MERY for his incredible patience waiting for this long overdue manuscipt.

References

1. S. Abramsky and A. Jung, *Domain Theory*, In: S. Abramsky, D. M. Gabbay and T. S. E. Maibaum, editors, *Handbook of Logic and Computer Science*, **3**, Clarendon Press (1994), pp. 1–168.
2. C. Baier and M. Kwiatkowska, *Domain equations for probabilistic processes*, *Electronic Notes in Theoretical Computer Science* **7** (1997), URL: http://www.elsevier.nl/locate/entcs/volume7
3. S. D. Brookes and A. W. Roscoe, *An improved failures model for communicating processes*, *Lecture Notes in Computer Science* **197** (1985), pp. 281 - 305.
4. J. I. den Hartog and E. P. de Vink, *Mixing up nondeterminism and probability: A preliminary report*, *Electronic Notes in Theoretical Computer Science* **22** (1998), URL: http://www.elsevier.nl/locate/entcs/volume22.html.
5. R. Heckmann, *Probabilistic domains*, *Proceedings* of CAAP '94, *Lecture Notes in Computer Science* **787** (1994), pp. 142–156.
6. C. Jones, "Probabilistic Non-determinism," PhD Thesis, University of Edinburgh, 1990. Also published as Technical Report No. CST-63-90.
7. C. Jones and G. Plotkin, *A probabilistic powerdomain of evaluations*, *Proceedings* of 1989 Symposium on Logic in Computer Science, IEEE Computer Society Press, 1989, pp. 186–195.
8. A. Jung and R. Tix, *The troublesome probabilistic powerdomain*, *Electronic Notes in Theoretical Computer Science* **13** (1997), URL: http://www.elsevier.nl/locate/entcs/volume7.html
9. C. Morgan, A. McIver, K. Seidel and J. Sanders, *Refinement-oriented probability for CSP*, University of Oxford Technical Report, 1994.
10. C. Morgan, A. McIver, K. Seidel and J. Sanders, *Argument duplication in probabilistic CSP*, University of Oxford Technical Report, 1995.
11. J. J. M. M. Rutten and D. Turi, *On the foundations of final semantics: Nonstandard sets, metric spaces and partial orders*, *Proceedings* of the REX'92 Workshop, *Lecture Notes in Computer Science* **666** (1993), pp. 477–530.
12. N. Saheb-Djahromi, *CPOs of measures for nondeterminism*, *Theoretical Computer Science* **12** (1980), pp. 19–37.

Concurrent Specification And Timing Analysis
of Digital Hardware using SDL

Kenneth J. Turner, F. Javier Argul-Marin and Stephen D. Laing

Computing Science and Mathematics, University of Stirling
kjt@cs.stir.ac.uk, javargul@bbvnet.com, stephenl@reading.sgi.com

Abstract. Digital hardware is treated as a collection of interacting parallel components. The ANISEED method (Analysis In SDL Enhancing Electronic Design) uses SDL (Specification and Description Language) to specify and analyse timing characteristics of hardware designs. A library contains specifications of typical components in single/multi-bit and untimed/timed forms. Timing may be specified at an abstract, behavioural or structural level. Consistency of temporal and functional aspects may be assessed between designs at different levels of detail. Timing characteristics of a design may also be inferred from validator traces.

1 Introduction

Digital hardware can be viewed as a concurrent system whose components operate in parallel but synchronised via exchange of electrical signals. Although SDL (Specification and Description Language [9]) was developed for specifying communications systems, it is a general-purpose language of wide applicability. It is the contention of this paper that SDL is appropriate and useful for specifying and analysing digital hardware as collections of interacting parallel components. The approach particularly focuses on timing aspects, which are often tricky in hardware design. SDL is of interest to hardware specifiers because it offers rigorous specification, good system structuring features, high-level communication, and the possibility of hardware-software co-design.

Most uses of SDL for hardware description have aimed at synthesis using standard engineering tools (e.g. [2, 5, 6]). The authors are engaged in the project ANISEED (Analysis In SDL Enhancing Electronic Design [1, 4]). Its goals are complementary to those of others who have used SDL for hardware description. Specifically, translation to VHDL and/or C is assumed to be dealt with by other tools. Instead, the authors have concentrated on timing aspects of hardware specification and analysis. Timing constraints on circuits and components can be specified and analysed at various levels: *abstract* (overall sequencing constraints), *behavioural* (black-box viewpoint), and *structural* (internal design).

2 Approach

The behaviour of hardware components and their timing characteristics can be modelled naturally using SDL processes, since these run in parallel and communicate via signals. For timing analysis, ANISEED can use a standard SDL simulator or validator. However, the authors have also implemented a discrete event simulation by automatically modifying the scheduling strategy of a standard SDL simulator.

A hardware signal is modelled as an SDL signal with parameters:

J. Rolim et al. (Eds.): IPDPS 2000 Workshops, LNCS 1800, pp. 1001-1008, 2000.

Time-stamp optionally records the time at which a signal is considered to have been generated. This is necessary partly because an output signal may not be consumed immediately; it still, however, carries the time of its generation.

Value is mandatory and may simply be a bit. However, in general it may be multi-bit (i.e. a list of bits). This is appropriate at a high level of specification where a bus or group of wires is to be specified as a whole. In a very abstract specification it might even be desirable to carry arbitrary values such as data structures.

ANISEED allows single-bit/multi-bit and untimed/timed specifications. Library components are available in all four bit/time combinations. It is useful to write an untimed specification first in order to check the functional correctness of a design. Timing constraints can then be added (by a change of library component names), allowing timing issues such as race conditions and hazards to be studied.

ANISEED is supported by a library of common components and circuits (i.e. designs consisting of a number of components.) These are stored in SDL packages, forming a modular and easily extended library. Some of the packages depend on others (for example, all packages use bits). A type of component is modelled as a block type in SDL. The motivation for choosing block types rather than process types is mainly that the internal construction of a component should be invisible. A further consideration is that library block types are instantiated statically. A block type is parameterised by its signal names and its gates (in the SDL sense). Timed components are also parameterised by characteristics such as their propagation delay or setup time.

Components are interconnected by no-delay channels. Like real wires these are considered to convey signals instantaneously. If it is necessary to model the propagation delay of a wire, a delay component can be used. Where multi-bit components are interconnected with single-bit components, a split 'component' is used to separate the bits. Correspondingly a merge 'component' is used to combine single-bit signals into a multi-bit signal.

It would have been possible to specify all the library components individually. However this would have been very tedious. As a more pragmatic solution, all variants are generated automatically from an SDL template that is parameterised by the logic function, the number of inputs, whether timed and whether multi-bit. The template is an outline PR (SDL Phrasal Representation) specification that is pre-processed to yield the required variants. The approach using templates makes the library much smaller (10%–15% in size compared to the generated PR) and more maintainable.

The current ANISEED library contains over 400 standard components such as might be found in a typical logic family. The library includes packages for arithmetic units, bit values, decoders and encoders, demultiplexers and multiplexers, flip-flops and latches, logic gates, sequencing constraints, and tri-state devices. Explanatory comments are automatically generated by the templates. Since GR (SDL Graphical Representation) is often preferred by SDL specifiers, the library templates can also automatically generate comments in the style of CIF (SDL Common Interchange Format).

The authors use version 3.5 of the TAU/SDT toolset. This has some restrictions that affect its suitability for ANISEED. The library modules generate two PR variants: one has axioms, and the other has C code for simulation/validation. A more severe problem is that commercial SDL tools (SDT, ObjectGeode) do not currently support context parameters fully. These are essential for block types since the actual timing parameters and signal names are not known until a block type is instantiated in a particular context. To overcome tool limitations, ANISEED automatically instantiates context parameters in the PR generated from a graphical description.

3   Validation and Verification

SDL tends to be used in pragmatic ways, so validation is usually the method of choice. What would normally be termed verification (proof, model-checking) is comparatively rare for SDL [3, 7]. The SDL community uses the term validation to mean automated checking of an SDL specification. A specification may be validated in isolation; such a check is useful but does not confirm functional correctness. Alternatively a specification may be validated against an MSC (Message Sequence Chart [8]). This may be written by the specifier, derived from a validation run, or generated automatically from a higher-level specification. An MSC can be used to confirm correct refinement of a specification.

Most SDL validation is oriented towards checking functional correctness. However, SDL allows timing aspects to be specified and validated. In ANISEED, interactive simulation can be used to check for the expected timing behaviour. However, automated validation is preferred using exhaustive analysis or random-walk validation.

For hierarchical timing specifications, the validator is useful in checking consistency between different design levels. For example the abstract and behavioural specifications may be compared, or the behavioural and structural specifications. The MSC traces produced by the validator are useful in deriving timing characteristics. When many components are interconnected, a range of high-level timing properties for a circuit can be derived from the validator output.

During SDL simulation, the signal with the earliest time-stamp must be consumed first, even if other signals have been placed before it in the input queue of a process. The SDT Master Library was modified to achieve this effect, with ANISEED supplying its own scheduling functions for discrete event simulation.

4   Abstract Sequencing Constraints

Constraints at the highest level may be given without regard to functionality. This defines gross sequencing relationships among the inputs and outputs of a component. The constraints may be given in untimed form, but are most useful when used to express timing restrictions. It is valuable to check for timing inconsistencies before any more detailed functional design is undertaken. Once high-level sequencing properties have been validated, their specifications are replaced by those of the real behaviour.

The following examples of sequencing constraints are drawn from the ANISEED library; the constraints exist in untimed and timed forms. An *N-Of* constraint requires an input event to occur N times before output occurs. As an example without a timing constraint, a divide-by-4 counter produces one output pulse for every four input pulses. A period during which counting occurs may optionally be given. A *One-Of* constraint accepts just one input before producing output. For example, a bus arbiter must service just one client request during a bus cycle of some period. A second input within this period is retained until the next cycle. A variant of this constraint discards additional inputs before the period has elapsed. An *All-Of* constraint requires all inputs of a component to be received before output is produced. For example, the inputs to an adder must be received before its output can be calculated. The order in which inputs arrive is unimportant, but all inputs may be required in some period. Unless all the inputs arrive in time, the whole constraint is re-enforced.

5   A Typical Component: A Delay Flip-Flop

As an example of hierarchical specification, a DFF (Delay or D Flip-Flop) is the basic storage element in many hardware designs. The conventional symbol for a delay flip-flop is shown in figure 1 (a). The variant to be described here is positive edge-triggered, which means that the data input *D* is stored when the clock *C* goes from 0 to 1. After the data has been clocked in, it appears at the output *Q* after some propagation delay that depends on the hardware family. Apart from the obvious propagation delay *TProp*, a D flip-flop also imposes two other timing constraints. It is required that the data input be steady for a period *TSetup* before it can be clocked in. Immediately after a clock trigger, the data input must remain steady for a period *THold*. The timing constraints are shown graphically in figure 1 (b).

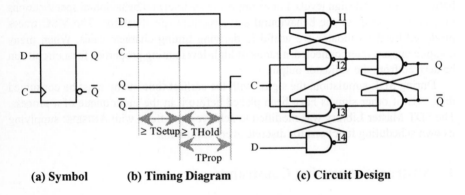

| (a) Symbol | (b) Timing Diagram | (c) Circuit Design |

Fig. 1. D Flip-Flop

The timing rules of figure 1 (b) can be readily represented in an abstract SDL specification. Once the flip-flop has committed to output (i.e. after setup and hold periods), a separate process instance is created to produce output after the propagation delay. During this period it is necessary to allow for further inputs in parallel with the output delay. The flip-flop therefore consists of a single input process instance plus output process instances as required.

In hardware description terminology, a behavioural specification treats a circuit or component as a black box. Since the D flip-flop has very little functionality, timing considerations dominate its specification. The behavioural specification of the flip-flop thus differs little from the abstract one. The main addition for the flip-flop behavioural specification is how to calculate the new output value.

A structural specification concerns the internal design of a component. Structural specifications may form a hierarchy of designs at progressive levels of detail. At each level of the design hierarchy, ANISEED may be used to specify both timing and functionality.

As an example, a typical design for a D flip-flop is shown in figure 1 (c). The internal signals *I1* to *I4* are shown. This circuit uses a number of *nand* gates (the D-shaped symbols); other flip-flop designs are possible. The SDL equivalent of the flip-flop design is shown in figure 2. This is closely related to the circuit diagram in figure 1 (c). Since SDL allows channel details to be inferred from block inputs and outputs, these do not strictly need to be drawn and hence are shown as gray in figure 2. For convenience the instantiations of each block type are listed separately in the figure. The context parameters of a *nand* gate are *delays, inputs, output*. For a junction the parameters are: *input, outputs*. Primed signals refer to the outputs of a junction (e.g. output *I`* would correspond to input *I*).

N1 : Nand2T < CMOSDel1, CMOSDel0, I4`I2`, I1>
N2 : Nand2T < CMOSDel1, CMOSDel0, I1, C`, I2>
N3 : Nand3T < CMOSDel1, CMOSDel0, I2`, C`, I4`, I3>
N4 : Nand2T < CMOSDel1, CMOSDel0, I3`, D, I4>
N5 : Nand2T < CMOSDel1, CMOSDel0, I2`, I6`, I5>
N6 : Nand2T < CMOSDel1, CMOSDel0, I5`, I3`, I6>

J1 : Junction2T <C, C`, C`>
J2 : Junction3T <I2, I2`, I2`, I2`>
J3 : Junction2T <I3, I3`,I3`>
J4: Junction2T <I4, I4`, I4`>
J5 : Junction2T <I5, I5`, Q>
J6 : Junction2T <I6, I6`, QBar>

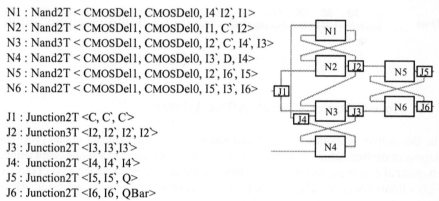

Fig. 2. SDL Specification of D Flip-Flop Design

Like each of the ANISEED library components, the D flip-flop was simulated and validated using the SDT toolset. Using the validator it was shown that the different levels of abstraction for the D flip-flop are equivalent in the sense that they respect the same traces. Exhaustive validation of the gate-level specification takes about one minute and 412 states. Random-walk validation of the gate-level specification takes about two minutes to explore over 211,000 states.

6   A Simple Circuit: The Single Pulser

The Single Pulser is a standard hardware verification benchmark circuit [10]. It is a clocked device with a one-bit input and a one-bit output. The purpose of the circuit is to debounce a push-button. The circuit must sense the depression of the button and assert an output signal for one clock pulse. The circuit does not allow further output until the operator has released the button.

The circuit specification was interactively simulated and mechanically checked. An example of the circuit behaviour appears in figure 3, generated automatically from a validator trace (MSC). The time base corresponds to 100 ns per clock cycle. Although the functionality is correct, it was found that there is a flaw in the supposedly proven benchmark circuit! The first output pulse after power-up is longer than expected (110 ns, from 67 ns to 177 ns) instead of lasting one clock cycle. The reason for this is that on the first pulse, one of the flip-flops in the design does not need to complete its setup time. This causes the first output pulse to appear 10 ns early. On subsequent pulses this flip-flop allow for the setup delay so the output pulse length is correct at 100 ns.

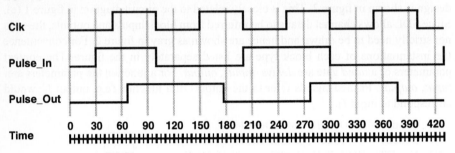

Fig. 3. Timing Behaviour of Single Pulser

7   A More Complex Circuit: A Bus Arbiter

The Bus Arbiter is another standard hardware verification benchmark circuit [10]. The purpose of the Bus Arbiter is to grant access on each clock cycle to a single client among a number of clients requesting use of a bus. The inputs to the arbiter are a set of request signals from each client. The outputs are a set of acknowledge signals, indicating which client is granted access during a clock cycle.

As shown in the structural specification of figure 4, each cell of the arbiter is moderately complex. The whole circuit consists of a number of such cells connected cyclically, e.g. three as shown in figure 5.

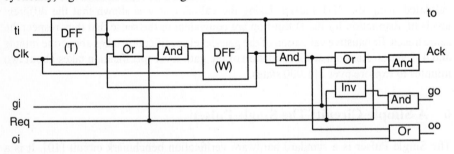

Fig. 4. Design of A Bus Arbiter Cell

The design of the circuit will be explained only briefly here. The *ti* (token in) and *to* (token out) signals are for circulation of the token. The *to* output of the last cell is connected to the *ti* input of the first cell to form a token ring. The *gi* (grant in) and *go* (grant out) signals are related to priority. The grant of cell *i* is passed to cell *i+1*,

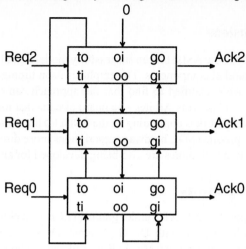

Fig. 5. Interconnection of Multiple Bus Arbiter Cells

indicating that no client of index less than or equal to i is requesting. Hence a cell may assert its acknowledge output if its grant input is asserted. The *oi* (override in) and *oo* (override out) signals are used to override the priority. When the token is in a persistently requesting cell, its corresponding client will get access to the bus; the *oo* signal of the cell is set to 1. This signal propagates down to the first cell (numbered 0) and resets its grant signal through an inverter. As a consequence the *gi* signal of every cell is reset, in other words the priority has no effect during this clock cycle. Within each cell, flip-flop *T* stores 1 when the token is present, and flip-flop *W* (waiting) is set to 1 when there is a persistent request. Initially the token is assumed to be in the first cell.

This circuit is relatively challenging. In the detailed design, the SDL specification contains 56 components (over 60 concurrent processes) and 93 signals. Nonetheless, the structure of the SDL specification closely resembles the circuit diagram and was easily produced. A behavioural specification of the intended behaviour was also written, so that it might be compared to the structural specification. The behavioural specification reflects the arbitration algorithm of the circuit.

Validation of the supposedly proven benchmark circuit uncovered a problem. Suppose that client 0 requests the bus in the first three clock cycles. In the fourth cycle, client 0 cancels its request but client 1 begins to request access. At this point the structural specification does not grant access to client 1 whereas the behavioural specification does. After interactive simulation of this case, it was discovered that the circuit provided in the benchmark (figure 4) may not properly reset the *oo* (override out) signal. The required correction is to connect the *Req* signal to the *And* gate that follows flip-flop *W*. The output of the *And* gate guarantees that the *oo* signal is always correctly set or reset according to the request signal in the current clock cycle.

A further problem was then discovered during automated validation using the random-walk approach. Analysis with the interactive timing simulator showed that this is due to the arbiter misbehaving when three clients simultaneously request access. In such a case the given arbiter design grants requests to two of the clients concurrently!

8   Conclusions

It has been seen how ANISEED can successfully model digital hardware as a collection of interacting parallel components. The emphasis is on timing specification and analysis. The authors were gratified to find that the approach can genuine problems with the Single Pulser and the Bus Arbiter – standard circuits that might have been supposed to be well verified. Work is continuing on the ANISEED library and tools. A GUI editor will be written to produce SDL hardware descriptions more directly from circuit diagrams. SDL verification techniques are also being developed for analysing hardware timing.

Acknowledgements

Financial support from NATO under grant HTECH.CRG974581 is gratefully acknowledged for collaboration with Dr. G. Adamis, Dr. Gy. Csopaki (who contributed to section 4 of this paper) and Mr. T. Kasza of the Technical University of Budapest. F. J. Argul-Marin thanks the Faculty of Management, University of Stirling, for supporting his work. Mrs. Ji He, University of Stirling, discovered and analysed the first arbiter design problem mentioned in section 7.

References

1. F. J. Argul Marin and K. J. Turner. Extending hardware description in SDL. Technical Report CSM-155, Department of Computing Science and Mathematics, University of Stirling, UK, Feb. 2000.
2. I. S. Bonatti and R. J. O. Figueiredo. An algorithm for the translation of SDL into synthesizable VHDL. *Current Issues In Electronic Modeling*, 3, Aug. 1995.
3. E. Bounimova, V. Levin, O. Başbuğoğlu, and K. İnan. A verification engine for SDL specification of communication protocols. In S. Bilgen, M. U. Çağlayan, and C. Ersoy, editors, *Proc. 1st. Symposium on Computer Networks*, pages 16–25, Istanbul, Turkey, 1996.
4. G. Csopaki and K. J. Turner. Modelling digital logic in SDL. In T. Mizuno, N. Shiratori, T. Higashino, and A. Togashi, editors, *Proc. Formal Description Techniques X/Protocol Specification, Testing and Verification XVII*, pages 367–382. Chapman-Hall, London, UK, Nov. 1997.
5. J.-M. Daveau, G. F. Marchioro, C. A. Valderrama, and A. A. Jerraya. VHDL generation from SDL specifications. In C. Delgado-Kloos and E. Cerny, editors, *Proc. Computer Hardware Description Languages and their Applications XIII*, pages 20–25. Chapman-Hall, London, UK, Apr. 1997.
6. T. Hadlich and T. Szczepanski. The ODE system – An SDL based approach to hardware-software co-design. In C. Müller-Schlör, F. Geerinckx, B. Stanford-Smith, and R. van Riet, editors, *Embedded Microprocessor Systems*, pages 269–281. IOS Press, Amsterdam, Netherlands, 1996.
7. G. J. Holzmann. Practical methods for the formal validation of SDL. *Computer Communications*, 15(2):129–134, 1992.
8. ITU. *Message Sequence Chart (MSC)*. ITU-T Z.120. International Telecommunications Union, Geneva, Switzerland, 1996.
9. ITU. *Specification and Description Language*. ITU-T Z.100. International Telecommunications Union, Geneva, Switzerland, 1996.
10. J. Staunstrup and T. Kropf. IFIP WG10.5 benchmark circuits. http: //goethe. ira. uka.de/hvg/ benchmarks.html, July 1996.

Incorporating Non-functional Requirements into Software Architectures

Nelson S. Rosa[12*], George R. R. Justo[1], and Paulo R. F. Cunha[2]

[1] University of Westminster, Centre for Parallel Computing,
115 New Cavendish Street, London W1M 8JS, UK
[2] Universidade Federal de Pernambuco, Centro de Informática,
Av. Prof. Luiz Freire, s/n - Cidade Universitária
CEP 50732-970 Recife, Pernambuco, Brasil

Abstract. The concept of software architecture has created a new scenario for incorporating non-functional and transactional requirements into the software design. Transactional and non-functional requirements can be included in an architecture-based software development through formal approaches in which first-order and temporal logic are utilised to deal with them. In this paper, we present an approach in which transactional and non-functional requirements are formally incorporated into a special class of software architectures, known as dynamic software architectures. In order to demonstrate how this proposal can be utilised in a real application, an appointment system is presented.

1 Introduction

Functional requirements define what a software is expected to do. Non-functional requirements (NFRs) specify global constraints that must be satisfied by the software. These constraints, also known as software global attributes, typically include performance, fault-tolerance, availability, security and so on. Closely related to NFRs[1], transactional requirements state the demand for a consistent, transparent and individual execution of transactions by the system. The well known ACID properties, Atomicity, Consistency, Isolation and Durability, summarise these transactional requirements.

During the software development process, functional requirements are usually incorporated into the software artifacts step by step. At the end of the process, all functional requirements must have been implemented in such way that the software satisfies the requirements defined at the early stages. NFRs, however, are not implemented in the same way as functional ones. To be more realistic, NFRs are hardly considered when a software is built. There are some reasons that can help to understand why it is too difficult to consider these requirements into

---

* This research was supported by the Brazilian Government Agency – CAPES Foundation – under process BEX1779/98-2.
[1] Throughout the paper the term NFRs is used to refer to both non-functional and transactional requirements.

J. Rolim et al. (Eds.): IPDPS 2000 Workshops, LNCS 1800, pp. 1009-1018, 2000.

the software development: firstly, NFRs are more complex to deal with; secondly, they are usually very abstract and stated only informally, rarely supported by tools, methodologies or languages; thirdly, it is not trivial to verify whether a specific NFR is satisfied by the final product or not; fourthly, very often NFRs conflict with each other, e.g. availability and performance; and finally, NFRs commonly concern environment builders instead of application programmers.

In spite of the difficulties mentioned above, the concept of software architecture and new developments in component-based technologies offers a new perspective as NFRs can be incorporated in software development. The software architecture principles facilitate the incorporation and analysis of transactional requirements [9] and certain NFRs such as security [5], fault-tolerance [6] and multiply NFRs [1] at early stage of the development. Component-based technologies like EJB (Enterprise JavaBeans [8]) support NFRs and transactional requirements adopting the strategy of separation of concerns, in which functional and NFRs are clearly separated in the implementation.

This paper presents a formal model for specifying NFRs and show how this model has been incorporated into an existing formal framework [3] for specifying dynamic software architectures. Dynamic software architectures are usually applied to dynamic reconfigurable distributed systems, the main feature of which is the possibility of changing the system structure (configuration) during its execution. Models for describing this kind of system contain abstractions and mechanisms that allow dynamic configuration to be explicitly stated in the model. These models also form a basis for describing dynamic software architectures [3].

This paper is organised as following: Section 2 describes the configuration model CL and its formal framework ZCL. Section 3 presents the formal model for describing non-functional and transactional requirements and describes its incorporation into the ZCL framework. In order to illustrate the use of the formal model, a case study is shown in Section 4. Finally, the last section presents the conclusions and some directions for future work.

2   ZCL Framework

This section presents the CL model and the ZCL framework, which are the basis for describing dynamic software architectures, that we will adopt.

2.1   CL Model

CL is a configuration model for describing dynamic configuration systems, which are systems that require its structure (or configuration) to change during execution without having to interrupt the entire system [2]. In order to model this kind of systems, a set of abstractions is defined in the CL model: modules, ports, instances, connections and configurations. Modules are software components and may be classified into primitive (task) and composite (group). Each component has an interface, which is a set of ports (entry and exit ports), that allows the component to communicate with its external environment. Components can be interconnected each other through its ports, defining the (system) configuration.

Using the CL model, a dynamic software architecture (or configuration) is built in five main steps: storing the components in the component library; selecting components from a component library; creating instances of these components; linking the ports of the instances; and finally activating the instances.

2.2 ZCL Framework

The ZCL framework is a formal framework based on the CL model, which formally incorporates the elements of the model. The framework consists of the CL abstractions, configuration commands and the execution model specified in the Z language [7]. Observe that in this paper we only consider structural aspects of the ZCL framework. It means that the execution model is not covered in our formal model.

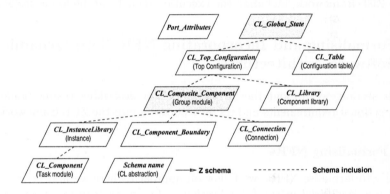

Fig. 1. Overview of the ZCL framework schemas

Figure 1 shows an overview of the Z schemas defined for modelling the main concepts of the ZCL model. The *CL_Component* schema is included in the definition of the *CL_InstanceLibrary* schema. The *CL_InstanceLibrary* schema together with *CL_Connection* schema and *CL_Composite_Boundary* schema make up the *CL_Composite_Component* (group module) schema.

$$
\begin{array}{l}
\rule{0pt}{0pt}\ \underline{\ CL\_Component\ }\rule{4cm}{0.4pt} \\
component\_attr : Indices \nrightarrow Attributes \\
interfaces : \mathbb{F}\,PortNames \\
port\_attr : PortNames \nrightarrow Port\_Attributes \\
\rule{7cm}{0.4pt} \\
\mathrm{dom}\ port\_attr \subseteq interfaces
\end{array}
$$

The *CL_Composite_Component* schema is defined as a state schema. It keeps information about its task components and group components, links between them (connections) and a set of virtual ports[2] that belongs to its interface. The *Composite_Component* and the *CL_Library* (component library) schemas, in which predefined components are stored, form the top configuration (see Fig. 1). Finally, the *CL_Top_Configuration* schema and the *CL_Table* schema are used for storing information about the context, instances and their status and form the global state of a configuration defined by the *CL_Global_State* schema.

$$
\begin{array}{l}
\underline{\text{— } CL\_Global\_State \text{ —————————————————}} \\
CL\_Top\_Configuration \\
CL\_Table \\
\hline
(InContext \subseteq \text{ dom } tasks \lor InContext \subseteq \text{ dom } groups) \\
InstNodes \subseteq \text{ dom } node\_parent
\end{array}
$$

In addition to the state schemas described above, the ZCL framework also defines operations to be applied to these schemas. A more detailed description of the ZCL framework, including the execution model, can be found in [3].

3   Formalising and Incorporating NFRs into Dynamic Software Architectures

In this section, we present the formal model for describing transactional and non-functional requirements and its incorporation into the ZCL framework.

3.1   Formalising NFRs

In order to formalise NFRs, we follow an approach where we first select which NFRs are considered in the formalisation, define a strategy for specifying the NFRs considering that the formal notation to be utilised is Z, and finally specify the NFRs.

Selecting NFRs. The first step to be carried out in the formalisation of NFRs is to decide which ones will be considered. It is a consequence of a very distinctive nature of NFRs, in which a wide variety of aspects such as modifiability and fault-tolerance are categorised as non-functional properties. The IEEE/ANSI 830-1993, IEEE Recommended Practice for Software Requirements Specifications, defines thirteen non-functional requirements that must be included in the software requirements document: performance, interface, operational, resource, verification, acceptance, documentation, security, portability, quality, reliability, maintainability and safety. Kotonya [4] classifies these requirements into three main categories: Product requirements, Process requirements and External requirements. Product requirements specify the desired characteristics that a system or subsystem must possess. Process requirements put constraints on the development process of the system. External requirements are constraints

---

[2] Virtual ports define the interface of a composite module.

applied to both the product and the process and which are derived from the environment where the system is developed.

Adopting this classification, we define the following criteria for selecting NFRs:

- The first criteria adopted is to consider only the first category of NFRs, namely Product requirements. This decision is based on the fact that we are interested in NFRs related to runtime properties, because the nature of dynamic systems demands special considerations during the system execution. For instance, if a component must be replaced dynamically, it is necessary to know if the new component satisfies the same requirements of performance, availability, fault-tolerance and so on of the original one.
- The second important decision relates to Product requirements. Our decision is based on the principle of how precisely these requirements could be described. Essentially, requirements were chosen according to the possibility of formulating them precisely and thus quantify them.
- The third criteria is directly related to the increasing support provided by implementation environments, e.g. Enterprise JavaBeans [8], for implementing NFRs. NFRs potentially provided by implementation environments are strong candidates because their implementation is more concrete.

Following above three criteria, three Product requirements were chosen, namely performance, reliability, security, in addition to the transactional requirements. For each selected requirement, we also identified its key attributes. Two attributes related to performance were defined: processed transaction per second; and the response time that a system should respond a user request. In terms of reliability, we define the rate of occurrence of failure and mean time to failure as the key attributes. For safety, user identification has been identified as the key attribute to protect the system against unauthorised access. Finally, the transactional requirements are represented by the ACID properties: Atomicity, Consistence, Isolation and Durability.

Specification Strategy. After the definition of which NFRs are considered in the context of dynamic systems, it is necessary to define how these requirements must be specified using the Z language. It raises two main questions: firstly, the Z notation is based on set theory, i.e. the NFRs must be specified with the strong notion of types and sets of Z. Secondly, the specification of the NFRs must be as generic as possible to allow its assignment to different elements of the CL model.

The first point focuses on the fact that the specification of performance, availability and safety must be expressed in terms of sets, operations and constraints on these sets. Essentially, first-order logic expressions combined with the definition of sets are sufficient for modelling NFRs, instead of temporal logic [9] or first-order predicates.

Secondly, the proposed specification of NFRs is defined as Z type schemas. In this way, NFRs may be assigned to different elements of the CL model, i.e. it is possible to assign NFRs to different abstractions of the CL model, whatever its granularity. For instance, a single NFRs may be assigned to a port, a simple component or a composite component. Additionally, it is possible to refine the NFRs specification defining additional constraints on the original type.

NFR's Specification. Following the criteria defined previously, a Z specification has been defined for three NFRs, namely performance, reliability, security and the transactional requirements represented by the ACID properties. The general overview of these Z schemas is shown in Fig. 2.

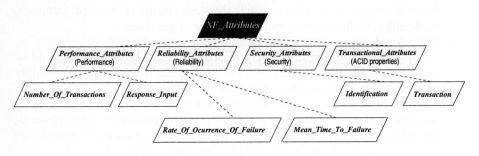

Fig. 2. Overview of Z schemas of NFRs

For each NFRs and transactional requirements a type schema was defined. For lack of space, we illustrate only one of NFRs, namely the performance. Performance is characterised by two attributes specified by the schemas *Number_Of_Transactions* and *Response_Input*, respectively. The number of transactions is specified in the *Number_Of_Transactions* schema using four variables. The variable *number_unt* defines the type of metric unit and the variable *number_value* specifies the number of processed transactions. Additionally, *number_min* and *number_max* variables define a range for the *number_value*. The predicate specifies that *number_value* should be in the range defined by *num_min* and *num_max*.

$$
\begin{array}{|l}
\hline
\_\ Number\_Of\_Transactions _____ \\
number\_unt : Transaction\_Unit \\
number\_value : \mathbb{N}_1 \\
number\_min : \mathbb{N}_1 \\
number\_max : \mathbb{N}_1 \\
\hline
number\_min \leq number\_value \leq number\_max \\
\hline
\end{array}
$$

In a similar way, the attribute response time is defined by the *Response_Input* schema.

$$
\begin{array}{|l}
\underline{\ Response\_Input\ }\rule{4cm}{0.4pt}\\
response\_unt : Time\_Unit\\
response\_value : \mathbb{N}_1\\
response\_min : \mathbb{N}_1\\
response\_max : \mathbb{N}_1\\
\hline
response\_min \leq response\_value \leq response\_max\\
\end{array}
$$

The NFR performance is fully specified by combining its attributes in a single schema type *Performance_Attributes*.

$$
\begin{array}{|l}
\underline{\ Performance\_Attributes\ }\rule{3cm}{0.4pt}\\
attr1 : Number\_Of\_Transactions\\
attr2 : Response\_Input\\
\end{array}
$$

The schemas *Reliability_Attributes* and *Security_Attributes* are specified in a similar manner, while transactional attributes are characterised by two values defining whether the attribute is required (*ON*) or not (*OFF*). The schemas specifying each NFR are combined in the *NF_Attributes* schema type.

$$
\begin{array}{|l}
\underline{\ NF\_Attributes\ }\rule{4cm}{0.4pt}\\
performance : Performance\_Attributes\\
reliability : Reliability\_Attributes\\
security : Security\_Attributes\\
transaction : Transactional\_Attributes\\
\end{array}
$$

3.2   Integrating NFRs into the ZCL Framework

The description of dynamic software architectures demands models that address specific questions related to dynamic configuration. The CL model has been widely utilised for this purpose. Additionally, the formalisation of this model as presented in the ZCL framework enables the formal specification and verification of dynamic reconfigurable systems. Using the ZCL framework as a tool to formally describe dynamic software architectures, we present a solution for integrating the formal specification of the NFRs into the ZCL framework: the NfZCL framework. The NfZCL framework is presented in two parts: firstly, the NFRs are integrated to the general structure of the ZCL framework; secondly, the configuration operations are extended to deal with NFRs integrated to the architecture.

Non-functional Architecture. The definition of NfZCL consists of defining how the NFRs specified in Sect. 3.1 can be assigned to the elements of the CL model. Basically, it allows the definition of dynamic software architectures with non-functional attributes. We call this kind of architecture a non-functional architecture. The non-functional architecture is specified extending the ZCL framework by assigning NFRs to each port. Observe that the decision of assigning non-functional attributes to port, rather than component, offers more flexibility to component development. The flexibility comes from the fact that the global non-functional attributes of a component may be defined as the conjunction of the specification of its ports. If non-functional attributes were assigned to the entire component, it would not be possible to make any assumption about the performance of individual ports.

The *NF_Architecture* state schema defines a non-functional architecture as an extension to the description of dynamic architecture defined by the ZCL framework, namely *CL_Global_State* (see Fig. 1). The extension basically defines a function relating a component port with its non-functional attributes.

__ *NF_Architecture* _____

CL_Global_State
$nf\_port : (ID\_Component \times PortNames) \nrightarrow NF\_Attributes$

ran (dom $nf\_port) \subseteq interfaces$
dom (dom $nf\_port) \subseteq$ dom $tasks$

The partial function *nf_port* is the key element that assigns each port belonging to each component to the NFRs. In practical terms, this function creates a library containing ports and NFRs assigned to them. The schema asserts that each port belonging to *interface* of the component has assigned *NF_Attributes*. It is also asserted that the component must belong to the set of *tasks* of the dynamic architecture.

Non-functional Operations. As NFRs were assigned to ports, the configuration operations must be modified. For example, the operation used to link two ports needs to be extended to deal with NFRs. This is necessary because ports with NFRs incompatible can not be linked. To carried out this check, we defined one checking function for each NFR (*check_performance*, *check_reliability*, *check_security* and *check_transactional*). Each function checks whether non-functional values of *port1?* and *port2?* of distinct components *node1?* and *node2?* are compatible or not.

__ *NF_Check* _____

$\Xi NF\_Architecture$
$node1? : Nodes$
$node2? : Nodes$
$port1? : PortNames$
$port2? : PortNames$

$(node\_parent(node1?), port1?) \in$ dom nf_port
$(node\_parent(node2?), port2?) \in$ dom nf_port
(**let** $elem1 == nf\_port(node\_parent(node1?), port1?)$
•**let** $elem2 == nf\_port(node\_parent(node2?), port2?)$
•$((check\_performance(elem1.performance\_attr, elem2.performance\_attr) = ok)$
$\wedge(check\_reliability(elem1.reliability\_attr, elem2.reliability\_attr) = ok)$
$\wedge(check\_security(elem1.security\_attr, elem2.security\_attr) = ok)$
$\wedge(check\_transactional(elem1.transactional\_attr, elem2.transactional\_attr) = ok)))$

---

4 Case Study: an Appointment System

In order to illustrate our approach, we specify in this section an appointment system. The appointment system follows the client-server style and realises a simple distributed appointment scheduling system. The server component provides operations that can be used to add and remove an appointment, and also an operation that returns the current time schedule. The non-functional properties of the appointment system require the set of server operations to be atomic and the isolated execution of clients that manipulate shared time schedules. In addition, it is important to ensure that changes made to the time schedules will survive subsequent system failures.

Software Architecture. The system architecture is composed by the components *AppointmentServer*, *AppointmentClient* and *DataBase*. The *AppointmentClient* component makes requests to the *AppointmenServer*, the *AppointmentServer* models the server and the *DataBase* represents the data base in which the information about appointments are stored.

Z specification. As mentioned in Sect. 2.1, five steps are usually necessary to build a configuration. The first step consists of initialising the state schemas and composing, using the Z schema composition operator $\substack{9 \\ 9}$, with the creation of instances. The ZCL operation *CL_Create_Component* schema creates the *AppointmentServer*, *AppointmentClient* and *DataBase* components and also stores them in the library for future instantiations. After the components have been stored in the library, it is necessary to assign the non-functional properties for their ports. For example, the NfZCL operation *Associate_NF_Attributes* schema allows us to assign performance attributes (number of transactions) to the *provide* port of the *AppointmentServer*.

$assigning \widehat{=} Associate\_NF\_Attributes[c? := AppointmentServer, port? := provide,$
$number\_unt? := transactions\_per\_milliseconds, number\_value? := 8,$
$number\_min? := 1, number\_max? := 10]$

The second step consists of defining the context (defining the component types that can be used by the architecture) of the dynamic software architecture using the ZCL operation *CL_Define_Context* schema. After the definition of the context, in the third step, it is possible to create instances of components that will form the architecture. The ZCL operation *CL_Create_Instance* schema instantiates a component and places it to execute on a particular machine. Instances can be linked to define the structure of the architecture (fourth step). In this case, the ZCL operation *CL_NF_Link* schema is defined as a composition ($\substack{9 \\ 9}$) of the NfZCL operation *NF_Check* schema and the ZCL operation *CL_Link* schema. Finally, the instances can be activated using the ZCL operation *CL_Activate* schema.

5   Conclusion and Future Works

This paper has illustrated how NFRs can be formally specified and incorporated into dynamic software architectures. Using the ZCL framework as a basis, two main contributions have been proposed: the formal specification of some NFRs and their incorporation into software architectures. From the definition of some criteria, three NFRs and transactional properties have been chosen, namely safety, availability and performance. These NFRs were formally defined in Z and incorporated into ZCL framework according to a strategy that assigns non-functional attributes to ports. Additionally, configuration operations have been extended to enable the ZCL framework to deal with NFRs.

The introduction of NFRs into the software architecture is an important step in the software design. In addition, the use of a formal model enables us to verify properties of the software in the early stages of the software design. In terms of dynamic systems, it assumes a very important role because at configuration time it is possible to check when two components have compatible non-functional properties.

As we said in the beginning of the paper, NFRs are usually considered only during the implementation. This paper has show how NFRs can be successfully incorporated at early stages of the design. It is very important, however, to relate the NFRs defined during the design to those used by the implementation environment. In this direction, our future work will focus on a refinement calculus (set of formally defined rules) for the transformation of NFRs-based software architectures.

References

[1] Issarny, V., Bidan, C., Saridakis, T.: Achieving Middleware Customization in a Configuration-based Development Environment: Experience with the Aster Prototype. Fourth International Conference on Configurable Distributed Systems, Annapolis Maryland USA (1998) 207-214

[2] Justo, G. R. R., Cunha, P. R. F.: An Architectural Application Framework for Evolving Distributed Systems. Journal of Systems Architecture 45 (1999) 1375-1384

[3] Justo, G. R. R., Paula, V. C. C. de, Cunha, P. R. F.: Formal Specification of Evolving Distributed Software Architectures. International Workshop on Coordination Technologies for Information Systems(CTIS'98), Viena Austria (1998) 548-553

[4] Kotonya, G. and Sommerville, I. Requirements Engineering: Process and Techniques. John Wiley & Sons, Inc 8 (1998) 190-213

[5] Moriconi, M., Qian, X., Riemenschneider, R. A., Gong, L.: Secure Software Architectures. IEEE Symposium on Security and Privacy. Oakland CA (1997)

[6] Saridakis, T., Issarny, V.: Fault Tolerant Software Architectures. INRIA 3350 (1998)

[7] Spivey, J.M.: Understanding Z: a Specification Language and Its Formal Semantics. Cambridge University Press (1988)

[8] Matena, V., Hapner, M.: Enterprise JavaBeans^{TM}. Sun Microsystems (1997)

[9] Zarras, A., Issarny, V.: A Framework for Systematic Synthesis of Transactional Middleware. Middleware'98. The Lake District England (1998) 257-272

Automatic Implementation of Distributed Systems Formal Specifications

Luiz Henrique Castelo Branco[1], Antonio Francisco do Prado[1], Wanderley Lopes de Souza[1], and Marcelo Sant'Anna[2]

[1] Departamento de Computação, Universidade Federal de São Carlos, Av. Washington Luiz - 235, 04499-610 São Paulo, Brazil
{branco, prado, desouza}@dc.ufscar.br
[2] Departamento de Informática, Pontifícia Universidade Católica do Rio de Janeiro, R. Marquês de S. Vicente - 225, 22453-900 Rio de Janeiro, Brazil
santanna@inf.puc-rio.br

Abstract. The increasing demand for Distributed Systems(DS's) raised the need of a quality-assured development process, which could not only address the issue of requirement compliance, but also could help the construction of tools able to derive implementations automatically. In order to attend such a need, some Formal Description Techniques (FDT's) have been proposed. This paper defends the transformational approach as a good strategy to carry out the automatic implementation of DS's expressed in FDTs, focusing Mondel as FDT, and the DRACO-PUC environment as transformational system.

Introduction

Distributed Systems (DSs) are becoming increasingly popular. They have been used to meet the need of natural distribution of people and information, to allow for improved and more cost effective performance, to facilitate maintenance and to make computer systems highly reliable. To achieve all these objectives, DSs have become very complex and there is no consensus regarding an exact definition of DSs. According to Bochmann [1] a DS can be classified according to the four types of distribution. These types of distribution can be combined, increasing the system complexity, and must be carefully taken into through the different phases of a DS development.

In the specification phase the system is designed based on the user's requirements. Frequently many steps are necessary until a final specification of the system can be reached. In the implementation phase an instance of the system specification is produced in a high level programming language (e.g., Pascal, C, C++, Java) using software engineering techniques.

This paper concentrates on the implementation of DS's formal specifications. Section 2 deals with the subject of Formal Description Techniques and presents the specification language employed in this work. Section 3 introduces the approach and the tool used to derive automatic implementations from DS's formal specifications and the implementation issues, like a communication infra-structure. Finally, some concluding remarks are made in section 4.

J. Rolim et al. (Eds.): IPDPS 2000 Workshops, LNCS 1800, pp. 1019-1026, 2000.
© Springer-Verlag Berlin Heidelberg 2000

Formal Description Techniques

Distributed Systems may be structured as a set of hierarchical layers [4] where each layer uses the services provided by the layers below it in order to provide its service to the next layer up. A service layer can be modeled as a black box. The service specification describes the external behavior for this black box when it interacts with the environment (users) through its interfaces (interaction points).

The main goal of the formal specifications is to produce unambiguous system descriptions. A formal specification can also be useful in other phases of a software production process. Some desired properties can be verified, (semi-)automatic implementations can be generated and the formal specification can be used as a reference for system implementation conformance testing.

A Formal Description Technique (FDT) is needed to produce formal specifications for distributed systems. FDTs are self-contained specification languages, which means that the system specification given in an FDT need not to refer to any informal knowledge about the system. A FDT must also have a mathematical basis that can be used to demonstrate a specification correctness.

Mondel Language

Mondel is the result of a joint research project involving the Centre de Recherche Informatique de Montréal (CRIM), the Université de Montréal (UdeM) and Bell Northern Recherche (BNR). This project focuses on modeling of operational and management aspects of communication networks, in particular the detection and recovery of failures. Despite its initial focus, however, its basic principles appear to be sufficiently generic to apply it to other types of distributed systems [3,2] such as real time control systems and open distributed processing. Because it is an object-oriented specification language, it uses terminology that is very similar to terminology used in the object-oriented paradigm.

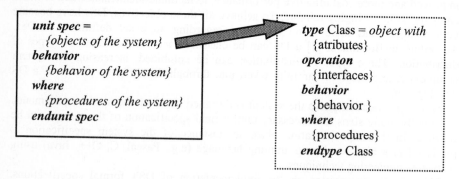

Fig. 1. The Mondel structure. An unit spec, as shown in figure, represents the specification properly said. In Mondel, the entities that communicate are represented as objects whose interaction points are represented by object interfaces.

A Mondel object is defined by the reserved word type, followed by its attributes. The signature and behavior of the methods offered by the object are indicated by the interface and behavior clauses, respectively.

In Mondel, states are represented by procedures. The current state is represented by the name of the procedure, and the next state is represented by the name of another procedure, or the same in case it continues in the same state.

The following section justifies the transformational approach as a language independent means to generate DS's implementations from their formal specifications. At the same time, the transformational environment DRACO-PUC is introduced, showing how it was successfully used in the implementation of DS's specified in the Mondel FDT.

Implementation Approaches

Along the last decade, several FDT's have been proposed, yelding a plethora of language dependent tools. At the same time, research has been done in the software engineering area to overcome the burden of having to develop a new compiler as a new FDT or implementation language arises. In this sense, the transformational approach comes as a natural choice, as it represents the state-of-the-art in compiler technology.

The DRACO-PUC environment, presented in the following sub-section, is a meta-compiler comprised of a domain network containing the domains Mondel, Estelle, C++ and Java, among others. In this environment, the language descriptions are kept apart from the semantic and transformation rules. This modularity saves effort when defining a new edge in the domain network.

This work produced Java implementations of DS's written in Mondel. Reusing the Mondel and Java domains, it is now possible to describe transformations between these high-level domains with a great economy in effort and time.

The DRACO-PUC Environment

The DRACO-PUC environment uses the ideas of the DRACO paradigm [6], which states that it is possible to develop software based on the reuse of high level abstractions. It is a domain-oriented transformational system where system or software descriptions, in high level specification or programming languages, may be automatically transformed into executable code. The following parts define a domain for DRACO:

- a **language**: a syntax must be well defined in order to make it possible to write programs and to build a parser.;
- a **prettyprinter**: this is an unparser responsible for mapping the internal Draco representation into concrete syntax representations of the domain language, that is, exhibits the application DAST in the domain language; and
- the **transformation libraries**: The handling of the DAST is carried out by formally defined steps called transformations. Each transformation rule has a Left

hand side (Lhs) describing the recognition pattern, and a Right hand side (Rhs) describing the rewriting pattern. The normal execution order of the transformation rules can be changed by tasks executed in association with the check-points available in the Draco machine. The main transformation control points are: Pre-Match (executed every time a Lhs rule is tested on the input description), Post-Match (executed just after the Lhs rule matches some piece of the input description), Pre-Apply (executed immediately before the Rhs rule is applied, replacing the selected piece of the input description) and Post-Apply (executed just after the input space selected on the Lhs is replaced by the pattern declared in the Rhs).

| Transform **GetTypeDef**
Lhs:
{{dast mondel.type_definition
  type [[type_id **ID**]] =
    [[type_conjunction **conj**]] with
    [[opt_attribute_decl **attrib_decl**]]
    [[opt_private_attributes **priv_attrib**]]
    [[opt_operation_signatures **sig**]]
    [[opt_behavior_definition **behavior**]]
  endtype [[type_id **ID**]]   }}
Post-Match: {{ dast txt.decls .... }} | TEMPLATE **T_Class_Constructor**
Rhs: {{dast java.compilation_unit
  import java.lang.*; import java.util.*;
  import java.rmi.*;
  public class [[Identifier **ID**]] extends
    UnicastRemoteObject
    implements
      [[interface_name **conj**]]   {
  [[field_declaration* **states**]]
  [[constructor_declaration **constru**]]
  [[field_declaration* **behavior**]]
  }   }} |

Fig. 2. Internal aspects of the DRACO Transforms. Figure shows one of the transformations that is used to transform an Mondel type declaration into its corresponding Java class. The left side is related to the Mondel domain, while the right one is related to the Java domain.

The **Lhs** of the **GetTypeDef** *Transform* contains the pattern to be recognized according to the **type_definition** Mondel grammar rule. The word **ID** is a meta variable of **type_id** type, **conj** is a pattern variable of type **type_conjunction**. This *Transform* does not have an **Rhs**. Instead, it does some data manipulation in its *Post-Match* control point and calls the **T_Class_Constructor** *Template*.

The **Rhs** in **T_Class_Constructor** contains the pattern to be written in the Java domain. The word **ID** is a meta variable of **Identifier** type and **conj** is a meta variable of **interface_name** type.

It should be noted, by Figure 3, that the transformations are generic. The meta variables are placeholders that hold grammatical patterns independently of the specification. Therefore, the Mondel operational semantics implemented in the transformations are reused across implementations.

The principal steps taken to use the Draco Domains in the automatic implementation of distributed systems specifications (Mondel) into executable languages (Java), using the Draco machine are presented in the next section.

Setting up the Implementation Environment

Using the strategy proposed by Prado [9], it is possible to rebuild software by direct porting of the source code to languages of other domains. According to the Draco machine, a domain consists of three parts, i.e. a *parser*, a *prettyprinter*, and one or more *transformers*.

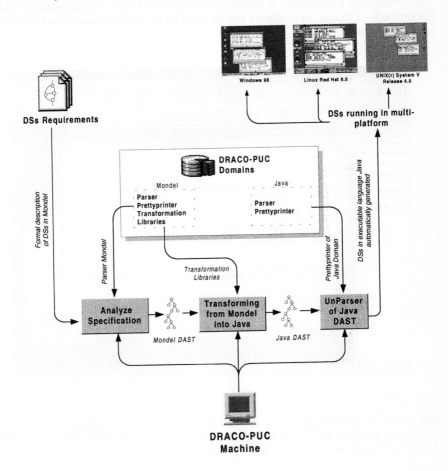

Fig. 3. Model of Automatic Transformation from Mondel to Java. The figure presents the principal steps taken to use the Mondel and Java Domains in the automatic implementation of the Mondel specifications into Java, using the Draco machine.

In **Analyze Specification**, the Draco Machine using the Mondel parser analyzes the Mondel specification. The output of this activity is the code in the internal representation; the DAST Mondel guided by the concrete Mondel syntax thus obtaining the syntactically correct Mondel specification.

In **Transforming from Mondel into Java**, the syntactically correct Mondel specification is automatically transformed by the Draco machine, which uses the *mondelTojava.tfm* inter-domain transformer. This activity generate a Java DAST guided by the concrete Java syntax what is the entry to the **Unparser of Java DAST** which makes textual again - this activity is guided by the Java *PrettyPrinter*.

The Java programs generated automatically are based on Java/RMI structure. The RMI enables the programmer to create distributed Java-to-Java applications, in which the methods of remote Java objects can be invoked from other Java virtual machines, possibly on different hosts.

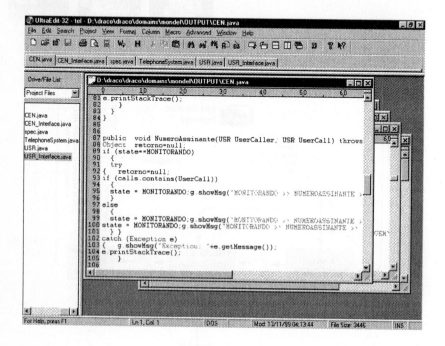

Fig. 4. Part of source-code in Java obtained automatically for the Telephone System. The source-code obtained include the library java.rmi.* to support the objects distributition from Distributed System and it's running in multiplatform.

The major solutions for distributed applications (e.g. CORBA [7] and PVM [10]) rely on a run-time mechanism to supply services such as message passing between remote objects, process creation and remote procedure calls (or remote method invocations). The solutions adopted in this work also consider some kind of run-time support through of the Remote Method Invocation (RMI) Java library .

The methods of an RMI object can send messages to other objects in a remote Java Virtual Machine (JVM), usually using the net, as if it was calling a local object. In that sense, RMI is very similar to CORBA. And like CORBA, RMI allows the clients interact with remote objects by public interfaces. The clients do not interact directly with the classes that implement the interfaces.

Conclusions

FDTs are becoming increasingly important in the development of distributed systems. They are not restricted to the specification phase since they are also used as a reference for the production process phases of these systems. The use of supporting tools is very important to reap the full benefit from an FDT when developing different kinds of distributed systems.

One of the results of this work was the definition of a strategy for the automatic implementation of Mondel specifications using the DRACO-PUC tool. The importance of this strategy lies in the use of a transformation system that allows for the implementation of Mondel to other different Java languages. It is possible, for instance, to define a library of transformations that transform Mondel specifications into Pascal or any language whose domain is defined in the DRACO-PUC tool.

The use of the DRACO-PUC tool in software development is important because it allows for automatic generation of the code based on high level specifications. Maintenance of the DRACO-PUC generated systems is easier, even in Distributed Systems with different protocols, architectures, operational systems and databases, because the designer can perform the task of maintenance at the level of Mondel language specifications.

Although some Mondel development environments already exist that permit semi-automatic generation of the code based on specifications, our contribution consists of the use of an up-to-date technology [8] that can be broadened and improved upon to make the entire automation process feasible for different hardware and software platforms.

Some significant results have already been obtained using the strategy presented herein. The Draco environment has proved its effectiveness in enabling automatic transformation of object-oriented models of Mondel specifications into Java, as shown in the *Telephone System* case (300 source code lines in Mondel specification and 600 source code lines in Java generated automatically). Other case study, the specification of the *ODP Trader Function* [5] (1500 source code lines in Mondel specification and 3000 source code lines in Java generated automatically), was submitted to the library of transformations of the *mondelTojava* transformer. A large library of transformations has already been built that currently allows for the automatic transformation of a large part of the Mondel specifications into Java. Further research work is underway to enlarge this library with the aim of implementing all Mondel specifications.

Future research work includes the following: enlargement of the transformation library to allow for implementation of more complex systems, creation of other libraries for implementation in new languages and operational systems, and the use of other architectures, such as CORBA[7] and DCE [8] to support distributed computation. Another research work, very important, to validate all process of transformations using the DRACO-PUC environment is related to the studies that guaranteed the semantic maintained of the Formal Specification, in the language of the implementation, when applied the transformations.

1026   L.H. Castelo Branco et al.

Others programs transformations systems were compared with DRACO-PUC. The PetDingo tool that transform Estelle specifications to C++ language produce 90% of the process transformation automatization. In DRACO-PUC the rate of the automatization to C++ language and other languages was 100% without anyone manual interference.

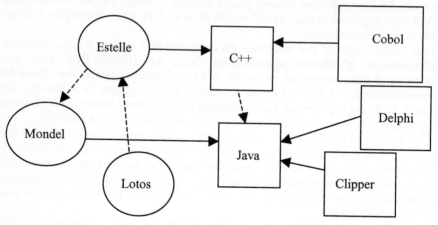

Fig. 5. Part of the Draco Domain Network. In figure some possible transformations are shown. The full lines represent transformation libraries under development or already built. The dashed lines represent transformation libraries to be built.

References

1. BOCHMANN, G. V.: Concepts for Distributed Systems Design. Springer-Verlag, 1983.
2. BOCHMANN, G. v., Poirier, S., Mondain-Monval, P.: Object-Oriented Design for Distributed Systems: The OSI Directory Example, Université de Montréal, 1991.
3. BOCHMANN, G. v., Barbeau, M., Erradi, M., Lecomte, L, Mondain-Monval, P., Willians: An Object-Oriented Specification Language, Université de Montréal, 1990.
4. ISO IS 7498: Information processing systems - Open Systems Interconnection - Basic Reference Model, 1984.
5. ITU-T Draft Rec X.904, ISO/IEC DIS2 13235: ODP Trading Function - Part 1 - Specification, Document approved for standardization, Jan 1997.
6. NEIGHBORS, J.: Software Construction Using Components, Tese de Doutorado, University of California at Irvine (EUA), Set/1980.
7. Object Management Group and X/Open: The Common Object Request Broker: Architecture and Specification, Document Number 91.12.1, 1992.
8. Open Software Foundation: DCE Application Development Guide, Cambridge, MA (EUA), 1992.
9. PRADO, A.F.: Estratégia de Re-Engenharia de Software Orientada a Domínios, Tese de Doutorado, Pontifícia Universidade Católica, Rio de Janeiro, Ago/1992.
10. SOUZA, P.S.Lopes de: O Impacto do Protocolo TCP/IP na Computação Paralela Distribuída no Ambiente Windows95, Proceedings of the 15th Brazilian Symposium on Computer Networks, São Carlos(SP), 19-22/05/1997, pp.118-134.

Refinement Based Validation of an Algorithm for Detecting Distributed Termination

Mamoun Filali, Philippe Mauran, Gérard Padiou,
Philippe Quéinnec, and Xavier Thirioux

Institut de Recherche en Informatique de Toulouse
118 route de Narbonne, 31062 Toulouse cedex 4, France

Abstract. We present the development and the validation of an algorithm for detecting the termination of diffusing computations. To the best of our knowledge, this is the first one which is based on the maximal paths generated by a diffusing computation. After an informal presentation of the algorithm, we proceed to its rigorous development within the framework of the UNITY formalism and the assistance of the PVS proof system. The correctness of the algorithm is established through a refinement of an abstract model.

1 Introduction

Termination detection of diffusing computations has been extensively studied [13]. Many algorithms [5, 9, 10] have been published to deal with this problem. We propose an algorithm which is based on the observation of final paths of a tree where the nodes represent the local computations, and the edges represent message communication.

The main features of the algorithm are: no assumption about message ordering is made; algorithm messages are piggybacked over those of the computation, and the algorithm greatly reduces the number of control messages. Termination messages are generated once a maximal path of the dynamic tree has been reached. With respect to memory requirements, each node only needs a local counter, the piggybacked data has a fixed size, as well as the collector's data structures. Termination is detected as soon as the collector has received all the maximal path messages.

In order to validate the proposed algorithm, we adopt the UNITY formalism and a refinement based development methodology. The correctness of the algorithm is established through a refinement between a concrete model (associated with the algorithm) expressed as a UNITY program and an abstract model also expressed as a UNITY program. Algorithm properties are derived from those of the abstract model. The development has been mechanized and expressed as theories of the PVS assistant theorem prover [12].

2 Description of the Algorithm

The algorithm observes a diffusing computation according to a specific viewpoint. Instead of reasoning on messages, we interpret communication as a transfer of control. This leads us to introduce path vectors as a new counting mechanism of maximal paths in a tree. After defining a diffusing computation, we describe the actual algorithm.

J. Rolim et al. (Eds.): IPDPS 2000 Workshops, LNCS 1800, pp. 1027-1036, 2000.

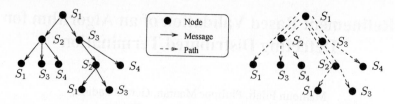

Fig. 1. A message versus path oriented interpretation of a diffusing computation

2.1   Diffusing Computation

We consider a set of processes or nodes $\{P_i \ : 0 \le i < N\}$ exchanging messages. An initial process P_0 starts the computation by sending a message to a subset of the processes. Then, every process forever repeats the following steps :

`receive(m); handle m; skip, or send up to N messages;`   Our model is based on the following assumptions:

- the communication network is connected, asynchronous and reliable;
- processes continuously wait for an input message in passive state, and perform an active step for each available input message;
- each active step is considered to be an atomic action. Such an atomic step ends either "silently" terminating without any applicative message, or by sending one or several messages to other computation processes.

2.2   Termination Detection

We say that a diffusing computation is terminated when all the processes are waiting to receive a message, and no messages are in transit. Due to the atomicity hypothesis, the diffusing computation is terminated when no messages are in transit.

We interpret a diffusing computation as a tree structured spreading of paths. This more abstract viewpoint allows to detect the termination through an algorithm that counts path creations and endings (see Figure 1). More precisely, the algorithm is based the counting of maximal paths starting from the root of the computation tree. A maximal path is defined as a sequence of adjacent edges leading to a leaf. Henceforth, maximal paths starting from the root are called paths.

According to a well-known pattern, a collector process receives the information required to detect the termination.

The algorithm computes the difference $\Delta(i)$ between the number of edges and the number of vertices created by the computation on site i. A process collects these differences, and records the current number T of actual final leaves. By Euler's theorem on planar graphs, the following relation holds at termination (only): $T = \sum_{i=0}^{N} \Delta(i) + 1$

The algorithm evaluates the $\Delta(i)$ terms thanks to the handling of path vectors.

2.3   Path Vectors

Path vectors provide a mechanism to count the number of path creations. Path vectors are vectors of non decreasing integer counters, and their handling is based on the following rules:

- a local counter C_i in each process records the number of locally created paths. These counters are initialized to 0;
- the size of the vector is equal to the number of processes;
- each underlying computation message piggybacks a path vector. A vector of a given path is updated along the spreading of the path. This path vector records the number of new paths issued by the current path, and carries the causal history of these paths.

An initiator process P_{i_0} performs a **Begin** action to assign the first path vector V_0 and the local counter C_{i_0}: $\langle \forall k \neq i_0 : V_0[k] = 0 \rangle \wedge V_0[i_0] = 1$

Let V be a path vector received by a P_i process. This vector is updated according to the following actions:

- **Split(p)**: the computation step ends by sending p messages with $(p > 1)$: new paths are created, and the incoming one follows its way. In this case, the local counter increases by the number of new paths. If p messages are sent, $p - 1$ paths are new. Therefore, the following increment is performed: $C_i := C_i + (p - 1)$. Then, the i'th component of the vector V is assigned the resulting value: $V[i] := C_i$. This updated vector is piggybacked in all sent messages. Every vector leaving a process exactly knows the number of paths created by this process.
- **End**: the computation step ends by sending no message: the path ends. The V vector is sent to the collector. It reports the path ending, and states that a set of paths has been created along its way.

Algorithm of the Collector Thanks to the gathered path vectors, the collector progressively evaluates the number of created and ended paths. It handles a path vector MT and a counter T. In the initial state, they are equal to zero. The collector gathers the path vectors until it can decide the termination.

```
process Collector
    integer MT[N] = [0, ... , 0];    /* maximal vector */
    integer T = 0;                   /* counter of terminated paths */
repeat
    integer V[N];
    receive(V);
    MT, T := max(MT, V), T + 1;
    Term := (∑ MT = T);
until (Term) /* Detect */
```

The timing diagram of Fig. 2 illustrates the computation behavior.

3   The UNITY Formalism

The UNITY formalism consists of two parts: a programming language, based on transition systems, and a specification language, based on a linear temporal logic [8], to express behavioral properties of these transition systems. A comprehensive presentation of the UNITY formalism can be found in [3] or [11].

As UNITY is a modeling language, it accepts any notation, or data type, with a clearly defined semantics. In this paper, we will mostly use:

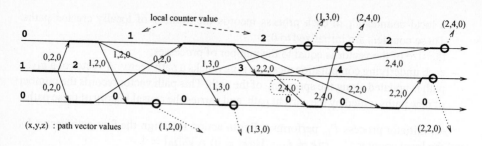

Fig. 2. A computation example

- the bag type, i.e. multisets. The operators on bags are defined by overloading the usual set operators ($\cup$, $\emptyset$, ϵ^1, ...), $k * e$, where k is a natural, is the multiset consisting of k occurrences of e.
- the Vector type, which corresponds to integer vectors, indexed by process identifiers. V with [(i):=e] denotes a vector that is identical to V, except for the ith component, where the value is e.

3.1 UNITY Logic Predicates

Reasoning about UNITY programs generally involves two kinds of predicates:

- properties whose definition refers to the program text. UNITY's original (inductive) temporal operators are defined in this way [3].
- properties which are defined from the set of execution sequences that result from the execution of a UNITY [14, 15, 11] program. An execution sequence σ resulting from a run of a program F verifies the following properties:
  - valid initial state: $F.init(\sigma(0))$
  - transition by statement s: $\langle \forall i :: \langle \exists s : s \in F.assign :: \sigma(i+1) = s(\sigma(i)) \rangle \rangle$
  - equity: $\langle \forall s : s \in F.assign :: \langle \exists_i^\infty :: \sigma(i+1) = s(\sigma(i)) \rangle \rangle$

  Let F be a program, and $\mathcal{S}(F)$ the set of such sequences, we define:

  > **stable** p **in** $F \equiv \langle \forall \sigma, i : \sigma \in \mathcal{S}(F) :: p(\sigma(i)) \Rightarrow p(\sigma(i+1)) \rangle$
  > > If p is true, it remains true.
  >
  > **alwt** p **in** $F \equiv \langle \forall \sigma, i : \sigma \in \mathcal{S}(F) :: p(\sigma(i)) \rangle$
  > > p is always true.
  >
  > $p \mapsto q$ **in** $F \equiv \langle \forall \sigma, i : \sigma \in \mathcal{S}(F) :: p(\sigma(i)) \Rightarrow \langle \exists j : j \geq i :: q(\sigma(j)) \rangle \rangle$
  > (leadsto) If p is true, q will become true.

3.2 Refinements

Several definitions of refinement have been proposed in the field of reactive programs [1, 7, 2]. In this paper, our approach is similar to Abadi and Lamport's [1], as their formal framework is very close to the UNITY formalism; We mostly used the results about

---

[1] ϵ denotes the choice of a *given* element in a non-empty set.

the preserving of safety properties when an "abstract" model is refined by a "concrete model". Informally, [1] defines a refinement as a mapping φ from states of the concrete model, to states of the abstract model such that:

- φ takes initial states of the concrete model into initial states of the abstract model.
- φ maps state transitions in the concrete model into state transitions in the abstract model, i.e.: if (c, c') is a pair of states of the concrete model that corresponds to a transition, then $(\varphi(c), \varphi(c'))$ corresponds to a transition of the abstract model.

4   Validation

In this section, we study a model for the detection termination algorithm. After presenting the basic properties to be satisfied by a detection algorithm, we present the validation structure within the UNITY framework.

4.1   Specification of the Termination

In the following, we take the *bag* data type as the basic data structure for modeling the communication medium. To each process p, we assign a bag B[p]. Inserting an element into B[p] corresponds to sending a message to p, extracting an element from B[p] corresponds to the reception of message by p. The elements of B[p] represent the messages that have been sent to p but not yet received by p. Thanks to the atomicity hypothesis (§2.2), we can specify termination as follows:

- safety: there is no false detection: **alwt** Term $\Rightarrow \langle \forall p : B[p] = \emptyset \rangle$ (term-0)
- liveness: a termination is eventually detected: $\langle \forall p : B[p] = \emptyset \rangle \mapsto$ Term   (term-1)

In the following development, we are mainly concerned by safety. A sketch of the liveness proof is given in Sect. 4.7.

4.2   Structure of the Validation

During the validation process, we have been concerned by two aspects:

- first, defining a sound model of the underlying communication network as well as of the underlying diffusing computation. We have chosen the bag data structure and the UNITY superposition mechanism.
- secondly, stating the basic idea of the algorithm in a simple way. We have elaborated an abstract model, whose invariant allows to conclude when termination has occurred.

Figure 3 illustrates the structure of the validation, where: dc is a model for a diffusing computation, t is an abstraction of the basic idea of the algorithm, and Term_Max is a model of the termination algorithm. The program aux_Term_Max introduces an auxiliary variable and corresponding statements updating it. This program is used for the proof.

The validation consists in showing that the concrete model, namely the program Term_Max is an instance of a diffusing computation, and then, satisfies the termination properties. The termination properties will be established by "inheriting" those of the abstract and auxiliary programs.

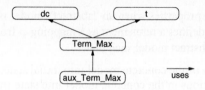

Fig. 3. Structure of the validation

4.3 Diffusing Computation Pattern

In the following, we give the pattern of a diffusing computation as an incomplete UNITY program. It belongs to the user to fill in the pattern to obtain a program[2]. In this pattern, message type (<M>) must be instantiated, new variables can be declared (<v>) and initialized (<iv>). It is also possible to superpose new statements either synchronously (<is>) or asynchronously (<ia>).

```
Program  dc
declare        B : array Process of bag of <M>
               <v>
initially      B = [<d>,∅, ... ]
               <iv> --  superposed variables initialization
assign -- Split
   ⟨▯ p,D : D ≠ ∅ ∧ B[p] ≠ ∅ :
       B[p]:= B[p] − ε(B[p]) if p ∉ D
              ~   B[p] − ε(B[p]) + <m> if p ∈ D
       ‖ ⟨‖ f : f ∈ D − p : B[f] := B[f] + <m> ⟩
       ‖ <is>            -- synchronous superposition
   ⟩
▯ ⟨▯ p : B[p] ≠ ∅ : B[p]:= B[p] − ε(B[p]) ‖ <is> ⟩ --  End
▯ <ia>       -- asynchronous superposition
end  dc
```

This pattern is a model for the program in Sect. 2.1:

- the first statement (commented *Split*) is a model for the atomic sequence:
  `receive(m); handle(m); send up to N Messages`
  where p is the sending process, and D is the set of destination processes[3]: receiving a message consists in extracting a given message ($\epsilon(B[p])$) from the input box of p; sending a new message to a process f consists in adding it to its input box.
- the second statement (commented *End*) is a model for the atomic sequence:
  `receive(m); handle(m); skip`

4.4 The Concrete Model

The concrete model is obtained by instantiating the diffusing computation pattern. Since the properties of the concrete model are established from those of the abstract model and the auxiliary model, we just give here the UNITY expression of the concrete model.

---

[2] Such a pattern could be implemented by a syntax editor.
[3] Two cases must be distinguished, since a process can send a message to itself.

Program Term_Max
declare   B : **array** Process **of** bag **of** Vector
-- *variables superposition*
  C,MT : Vector;   BT : bag **of** Vector
  T: nat   Term: bool
always   nc(p,D) = C[p]+| D| −1   -- *new counter*
        nv(v,p,D) = v with [(p):= nc(p,D)] -- *new vector*
initially   ⟨∃ i, V :: (∀ k: k ≠ i ⇒ V[k] = 0 ∧ C[k] = 0) ∧
                V[i] = 1 ∧ C[i] = 1 ∧ B = {v} ⟩
-- *superposed variables initialization*
  MT,BT,T,Term= [0, ...],∅,0,false
assign   -- *Split*
  ⟨[] p,D : D ≠ ∅ ∧ B[p] ≠ ∅ :
      B[p]:= B[p] − ε(B[p]) **if** p ∉ D
        ~ B[p] − ε(B[p]) + nv(ε(B[p]),p,D) **if** p ∈ D
    || ⟨||f: f ∈ D − p: B[f]:=B[f] + nv(ε(B[p]),p,D) ⟩
    || C[p]:= nc(p,D) ⟩-- *synchronous superposition*
[] ⟨[] p: B[p]≠∅:B[p]:= B[p]−ε(B[p]) || BT:=BT+ε(B[p])⟩ -- *End*
-- *asynchronous superposition*
[] BT,T,MT:= BT−ε(BT),T+1,max(MT,ε(BT)) **if** BT ≠ ∅   -- *Term*
[] Term := Σ MT = T **if** T > 0   -- *Detect*
end Term_Max

In this UNITY program:

- The **always** section should be understood as a declaration of macros.
- The statements commented by *Split* and *End* are the actions of a computation process. The statements commented by *Term* and *Detect* are the actions of the collector process.
- The bags B[_] and BT represent the communication medium. MT, T, Term are the variables of the Collector process, each element of the vector C represents the local counter which is updated by computation processes (Sect. 2.3).
- nv is the new vector sent, and nc is the new value of the local counter.

In order to have a program structure similar to the model described in terms of processes of the first part of the paper, we could have expressed this program as follows:

$$\text{Term\_Max} = \langle [] p :: \text{Computation}[p] \rangle [] \text{Collector}$$

We have chosen the monolithic structure to stay within a classic framework for refinements and to reason in a simple way with respect to liveness properties.

4.5   The Abstract Model

The abstract model expresses the basic mechanism of the algorithm which consists in counting the elements of the bag g through a distributed counter : the vector C.

Program t
declare   c: Vector   g: bag **of** Vector
always   nc(p,d) = c[p] + d − 1
           nv(v,p,d) = v with [(p):= nc(p,d)]
invariant
  $\langle \forall A::(A \neq \emptyset \land A \subset g \land \Sigma \max(A) \leq |A|) \Rightarrow A = g\rangle$-- *maximal*
$\land$ g $\neq \emptyset \land$ c = max(g) $\land$ |g| $= \Sigma$ max(g)   -- *card_max*
initially |g| = 1 $\land$ Σ max(g) = 1
assign $\langle [\![v,p,d: v \in g \land 0<d: c[p],g:= nc(c,p),g-v \cup d*nv(v,p,d)\rangle$
end t

In the abstract model[4], we have represented the leaves of the tree by the bag g. In the full report [6], we give the correctness proof for the invariant.

4.6   The Auxiliary Model

The auxiliary model enriches the concrete model with an auxiliary variable a : bag of Vector which represents the bag of the messages which have been received and handled by the collector. This variable is initialized to the empty bag, and is updated each time a message is delivered to the collector. The following property holds in the auxiliary program (its Unity program text is in the full report [6]):

invariant
  $\langle \forall$ A::(A $\neq \emptyset \land$ A $\subset$ g $\land \sum\max(A) \leq |A|) \Rightarrow$ A = g$\rangle$
$\land$ g $\neq \emptyset \land$ C = max(g) $\land$ |g| $= \sum\max(g)$
$\land$ MT = max(a) $\land$ T = |a|
$\land$ a = $\emptyset \Rightarrow \neg$Term
$\land$ Term $\Rightarrow \sum$MT = T

The properties of the auxiliary model are inherited from those of the abstract model. For this purpose, we have shown that the auxiliary model is a refinement of the abstract model through the functional relation:

$$(g = \cup_p B[p] \cup BT \cup a) \land \text{aux\_term\_max}.C = t.c \tag{1}$$

The abstract bag g is the union of B[p] the processes bags, BT the bag containing the messages sent to the collector but not yet accepted, and a the bag of the messages sent to the collector and accepted.

In order to show the refinement, bag insertions and extractions of the auxiliary model have to be mapped to operations on the abstract variable g such that (1) is preserved. Thus, we map the statement Split to the unique explicit statement of the abstract model, and we map the statements End, Term and Detect to the implicit statement Skip. The main property established on the auxiliary program is:

alwt Term $\Rightarrow \langle \forall p : B[p] = \emptyset \rangle$ in aux_term_max

In a similar way, the concrete model can inherit the properties of the auxiliary model since the concrete model refines the auxiliary model. It follows, that the safety of the detection (term-0) is satisfied by the concrete model.

---

[4] Following the path initiated by [4], we put together a Unity program and its properties.

4.7   Liveness

Informally, liveness relies on the following arguments:

- once the bags of the computation processes are empty, they remain empty;
- when the bags of the computation processes are empty, the bag of the collector eventually decreases;
- once the bags of the computation processes are empty and the bag of the collector is empty, termination is eventually detected.

It follows that the liveness proof relies mainly on the transitivity of $\mapsto$ and on the fact that finite bag inclusion is well-founded. The following lemmas, combined with the **alwt** properties of the algorithm, establish the liveness of the termination (term-1).

stable $\langle \forall p : B[p] = \emptyset \rangle$

$(\langle \forall p : B[p] = \emptyset \rangle \wedge BT = \mathcal{B} \wedge BT \neq \emptyset) \mapsto (\langle \forall p : B[p] = \emptyset \rangle \wedge BT \subsetneq \mathcal{B})$

$(\langle \forall p : B[p] = \emptyset \rangle \wedge BT = \emptyset) \mapsto \text{Term}$

4.8   Mechanizing the Development

We have encoded the semantics of the preceding models in the typed logic of PVS [12]. The typed logic allowed us to encode the generic aspects of patterns as well as the parameterized aspects of the problem. The main theories of this development and their relationships are given in Fig. 4. `programs` and `refinements` are generic theories and contain results on UNITY logic. The PVS development can be found in [6].

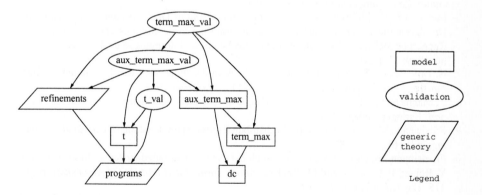

Fig. 4. Theories of the algorithm's validation

The expression of UNITY programs in the PVS system is straightforward: the syntax of the PVS specification language is close to the imperative description of models.

5   Conclusion

This paper presents an original case study, in that it shows the use and the integration of formal models in the design process of a new algorithm. The other specificity of this case study lies in its area of application : distributed algorithms and systems.

As regards the first point, the UNITY formalism provided a sound grounding for the specification and design of the algorithm, prior to its validation. The design followed a refinement based approach: a first (abstract) model describes the main idea of the algorithm, and a second (concrete) model refines and maps the algorithm to a distributed environment. The formal validation was carried out with the PVS assistant theorem prover [12]. PVS turned out to be an appropriate tool, w.r.t. the ability to easily express UNITY programs, properties and their development. The validation provided key elements for a deeper understanding of the algorithm, and for improving the model and the informal proof.

As regards the second point, the modeling reflects the hypotheses on the communication network by using multisets, instead of sequences, to represent pending messages.

References

1. M. Abadi and L. Lamport. The existence of refinement mappings. Technical Report 29, DEC, August 1988.
2. J.R. Abrial. *The B-Book Assigning programs to meanings*. Cambridge University Press, 1996.
3. K. Mani Chandy and Jayadev Misra. *Parallel Program Design: A Foundation*. Addison-Wesley, 1988.
4. Michel Charpentier. *Assistance à la Répartition de Systèmes Réactifs*. Thèse de doctorat, Institut National Polytechnique de Toulouse, France, November 1997.
5. E.W.D. Dijkstra and C.S. Scholten. Termination detection for diffusing computations. *Information Processing Letters*, 11(4):1–4, 1980.
6. Mamoun Filali, Philippe Mauran, Gérard Padiou, Philippe Quéinnec, and Xavier Thirioux. Un algorithme de terminaison de calcul diffusant par évaluation des chemins maximaux. Rapport de recherche 99-22-R, Institut de Recherche en Informatique de Toulouse, November 1999. 35 pages.
7. N. Lynch and F. Vaandrager. Forward and backward simulations – part I: Untimed systems. *Information and Computation*, 121(2):214–233, September 1995.
8. Z. Manna and A. Pnueli. *The temporal logic of reactive and concurrent systems: specification*. Springer-Verlag, 1992.
9. Friedemann Mattern. Algorithms for distributed termination detection. *Distributed Computing*, 2(3):161–175, 1987.
10. Friedemann Mattern. Virtual time and global state in distributed systems. In M. Cosnard, Y. Robert, P. Quinton, and M. Raynal, editors, *Parallel and Distributed Algorithms*, pages 215–226. North-Holland, 1989.
11. Jayadev Misra. A logic for concurrent programming. Technical report, The University of Texas at Austin, April 1994.
12. Sam Owre, John M. Rushby, and Natarajan Shankar. PVS: A prototype verification system. In *11th Int'l Conf. on Automated Deduction*, volume 607 of *Lecture Notes in Computer Science*, pages 748–752. Springer-Verlag, June 1992.
13. Michel Raynal. *Synchronisation et État Global dans les Systèmes Répartis*. Collection Direction Etudes-Recherches EDF. Edition Eyrolles, 1992.
14. Beverly A. Sanders. Eliminating the substitution axiom from UNITY logic. *Formal Aspects of Computing*, 3(2):189–205, April–June 1991.
15. Rob T. Udink and Joost N. Kok. On the relation between Unity properties and sequence of states. Technical Report RUU-CS-93-07, Utrecht University, Dept. of Computer Science, February 1993.

Tutorial 1 : Abstraction and Refinement of Concurrent Programs and Formal Specification A Practical View

Dominique Cansell[2,3]
cansell@loria.fr, Dominique Méry[2]
mery@loria.fr, and Christophe Tabacznyj[1,2]
tabaczny@loria.fr

[1] CEA
CENTRE D'ETUDES DE SACLAY
DEIN / SLA - Bt 528
F 91191 GIF-SUR-YVETTE CEDEX
FRANCE
[2] LORIA Université Henri Poincaré Nancy 1
BP 239
54506 Vandoeuvre-ls-Nancy Cedex
[3] Universit de Metz
Ile du Saulcy
57045 Metz
France

Formal methods allow to model systems and systems properties by providing accurate mathematical notations (type theory, set theory, ...). Implementations can be derived from a formal specification using methods based on refinement. Therefore, from a pragmatic industrial point of vue, the dual work based on abstraction is very important too for verifying safety critical systems, but also for addressing questions like maintenance, reverse ingineering of codes, modifications of programming language, code evolution, inspection of open codes to ensure their correctness with respect to the specification, program comprehension... The tutorial will sketch practical issues related to abstraction and refinement techniques for concurrent programming.

Following P. Cousot's Abstract Interpretation[6, 7, 5], abstraction is a theory of approximation of semantics. It formalizes the idea that a projection of the semantics on a simpler domain can help to compute properties (rather than to prove them). Properties given by the effective computation of the approximate semantics are closely related to the projection

J. Rolim et al. (Eds.): IPDPS 2000 Workshops, LNCS 1800, pp. 1037-1038, 2000.
© Springer-Verlag Berlin Heidelberg 2000

domain. All questions can not be solved, but all answers are always correct. Static analysis uses abstract interpretation to derive a computable semantics from the standard semantics. We will present such a framework specialized to the comprehension of sequential programs and study both the geometry of computation and how abstraction is related to it. Then, we will give some insights on how to extend the framework to concurrent programs.

Refinement of programs or refinement of specifications are techniques used in real case studies and tools[1, 8]; the B method developed by Abrial, is a suitable framework for developing distributed algorithms[4, 3, 2]. We will show how the B method may be used to develop distributed solutions to classical problems of distributed computations and we will give relationship to others approaches like actions systems, UNITY, TLA. The tutorial will be illustrated by demonstrations using the Atelier B environment.

References

1. J.-R. Abrial. *The B book - Assigning Programs to Meanings.* Cambridge University Press, 1996.
2. J.-R. Abrial. Extending b without changing it (for developing distributed systems). In H. Habrias, editor, *1st Conference on the B method*, pages 169–190, November 1996.
3. J.-R. Abrial and L. Mussat. Introducing dynamic constraints in B. In D. Bert, editor, *B'98 :Recent Advances in the Development and Use of the B Method*, volume 1393 of *Lecture Notes in Computer Science*. Springer-Verlag, 1998.
4. J.R. Abrial. Development of the abr protocol. ps file, february 1999.
5. P. Cousot. *Calculational System Design*, chapter The Calculational Design of a Generic Abstract Interpreter. NATO ASI Series F. Amsterdam: IOS Press, 1999.
6. P. Cousot and R. Cousot. Abstract interpretation frameworks. *Journal of Logic and Computation*, 2(4):511–547, 1992.
7. P. Cousot and R. Cousot. Refining model checking by abstract interpretation. 6(1):69–96, January 1999.
8. STERIA - Technologies de l'Information, Aix-en-Provence (F). *Atelier B, Manuel Utilisateur*, 1998. Version 3.5.

Tutorial 2: A Foundation for Composing Concurrent Objects

Jean-Paul Bahsoun

IRIT - Universit Paul Sabatier
118 route de Narbonne
F-31062 Toulouse
email: bahsoun@irit.fr

In recent years, a lot of new languages and new concepts have been conceived in order to promote parallelism in the object-oriented framework. These proposals were investigated using different concepts related to parallelism and object orientation. Among these concepts, we can find shared variables/message passing, inheritance/delegation, reflection... The degrees of a good cohabitation may be appreciated by combining the above concepts. In order to have significant criteria we have to determine how languages fit some requirements. These requirements should cover the different phases of program development i.e. specification, design, implementation and verification.

The underlying metaphor of object orientation is that of largely autonomous objects with encapsulated states interacting by message passing. This has led many researchers to design concurrent object-oriented programming models. It is well known that concurrent programs require a much more careful analysis to prove them to be correct. We propose to adapt formal methods used for the specification and verification of conventional concurrent programs to an object-oriented framework. The usual definition of an object consists of an identity, a state, and methods which modify the state. A software system is composed of several objects which interact by means of messages. Methods are usually partial, i.e. they may be safely executed only if certain preconditions hold. In a distributed environment, the caller usually cannot guarantee that the precondition of the called method is met. Hence, preconditions are made explicit as enabling conditions, and messages may be blocked if the corresponding method is disabled. Our aim is to define a formalism for proving properties of agents and agent systems, reflecting their structural definition by inheritance and parallel composition. It therefore has

J. Rolim et al. (Eds.): IPDPS 2000 Workshops, LNCS 1800, pp. 1039-1041, 2000.

to be compositional w.r.t. both these means of software construction. To make the model intuitive and avoid unnecessary complexity, parallelism occurs only between different agents, whereas each agent executes his methods strictly sequentially. All communication among the agents as well as between the environment and the agents occurs by explicit message passing. Therefore, we propose to distinguish between three levels of reasoning about systems built from concurrently executing agents, each focusing on one particular aspect or view of agent systems. The action level is concerned with the local effects of single methods or actions offered by an agent. The execution of a method or action transforms the agent's local state. Therefore, a simple formalism based on pre- and post-conditions of methods and actions is sufficient at this level. At the agent level, we reason about behaviors of individual agents of a given class, dealing with both safety and liveness properties. Proof rules take advantage of the encapsulation of an agent's private state which may only change by executing methods or actions defined for the agent. In contrast to the action level, initialization and fairness conditions are taken into account. Aspects of communication and interaction with other agents are not dealt with except in the form of environment assumptions. Finally, the system level models the top-level view of the entire system by an external observer. In particular, it cannot refer to the internal state of any object. Only the execution of actions by agents and the transmission of messages are visible. Typical properties of interest include synchronization and liveness involving several agents. We will define three different formal languages and proof rules, one for each aspect of reasoning. These languages will be related by "interface rules". Care will be taken to obtain simple and intuitive transitions from one level to the next. The conflict between inheritance and synchronization constraints is nothing else than a particular case of the more general problem raised when method constraints and inheritance are put together. Indeed, a synchronization constraint is a method constraint with a different semantics. Whereas a non-satisfied precondition raises an exception in the sequential case, it requires waiting in the parallel case. In our approach [BMS95], the problems of synchronization constraints caused by parallelism between methods disappears, since the agent itself executes the methods. Inheritance is treated at the sequential level, and all the aspects related to parallelism are treated at the composition level. The three levels of logic proposed correspond well to our intuition. The properties expressed with the logic of actions are automatically inherited (if, of course, the concerned method is not redefined or extended in the subclass). Therefore, only the proofs of agent properties have to be checked. But, under some conditions, we can reuse the existing proof of a property in a subclass. An environment to develop applications using this model has been implemented [BFS00]. On the one hand, this environment allows us to produce programs coded in Java from the model. On the other hand we have elaborated two decisions procedures to verify properties on the agent level and on the system level [BEBY99].

References

[BEBY99] Jean-Paul Bahsoun, Rami El Baida, and Hugues-Olivier Yar. Decision procedure for temporal logic of concurrent objects. In P. Amestoy and alt, editors, *Europar'99*, volume 1685 of *LNCS*, pages 1344–1352. Springer, 1999.

[BFS00] Jean-Paul Bahsoun, Pascal Fars, and Corinne Servires. *Object-Oriented Parallel and Distributed Programming*, chapter Multilevel Proof System for Concurrent Object-Oriented Systems, pages 31–52. Herms Science publications. Herms, 2000.

[BMS95] Jean-Paul Bahsoun, Stephan Mertz, and Corinne Servires. *Protocoles et programmation dans les rseaux*, volume 4 of *Paralllisme, rseaux et rpartition*, chapter A Framework for Programming and Proving Concurrent Objects, pages 65–98. Herms, 1995.

Workshop on Optics and Computer Science (WOCS 2000)

Organizers:

Fouad Kiamilev, University of Delaware, USA
Jeremy Ekman, University of Delaware, USA
Afonso Ferreira, CNRS, LIP–ENS, France
Sadik Esener, University of California, San Diego, USA
Yi Pan, University of Dayton, USA
Keqin Li, State University of New York, USA

Preface

The research projects described in this section were presented at the 5th Workshop on Optics and Computer Science (WOCS) at the International Parallel and Distributed Processing Symposium (IPDPS 2000) in Cancun, Mexico.

Optical interconnects are already a viable and proven technology at the serial and parallel data–link levels. The papers presented here focus on chip–to–chip optical interconnections. This approach enables high–density and high–speed optical interconnection between multiple integrated circuits (ICs) in a system.

The unique capabilities offered by chip–level optical interconnects (vs. electrical wiring) enable efficient implementation of new computer architectures and algorithms. Several papers in this section focus on such developments.

At present time the device technology is mature enough to allow demonstrator/prototype systems to be implemented. The purpose of these systems is to demonstrate the technology potential and to serve as a testbed for computer architectures and algorithms (specifically developed for chip–level optical interconnects). A number of papers in this sections describe such demonstrator systems and their underlying technology platform.

On the technology front, advancement in chip–level optical interconnection has been rapid. This is evidenced by the increasing optoelectronic device complexity and rise in the number of system demonstrators. However; to fully exploit the capabilities of chip–level optical interconnects, computer scientists must collaborate with device researchers and system integrators. The papers presented at the WOCS workshop represent an effort to bring these groups together and facilitate their collaboration.

Primary focus areas for WOCS 2000 include, but are not limited to:
    High–Speed Interconnections
    Optical Interconnects
    Algorithms using Optical Interconnects
    Parallel Optical Architectures
    Reconfigurable Optical Interconnects and Architectures

J. Rolim et al. (Eds.): IPDPS 2000 Workshops, LNCS 1800, pp. 1042-1043, 2000.
© Springer-Verlag Berlin Heidelberg 2000

Applications of Optical interconnects
Modeling of Optical Systems and Applications
Performance Analysis and Comparisons
Packaging of optical interconnects
System Demonstrations
Routing in Optical Networks

Program Chair

Fouad Kiamilev, University of Delaware

Steering Committee

Afonso Ferreira, CNRS, LIP–ENS, France [Chair]
Sadik Esener, University of California, San Diego
Yi Pan, University of Dayton, USA
Keqin Li, State University of New York, USA

Program Committee

Selim G. Akl, Queen's University, Canada
Hyeong–Ah Choi, George Washington University, USA
Allen Cox, Honeywell, USA
Jeremy Ekman, University of Delaware, USA
Joseph W. Goodman, Stanford University, USA
Mounir Hamdi, Hong Kong University of Science and Technology
Michael Haney, George Mason University
Yao Li, NEC Research Institute, Princeton, USA
Ahmed Louri, University of Arizona, USA
Rick Lytel, SUN Microsystems, USA
Philippe Marchand, University of California, San Diego, USA
Rami Melhem, University of Pittsburgh, USA
John Neff, University of Colorado, USA
Haldun Ozaktas, Bilkent University, Turkey
Tim Pinkston, University of Southern California, USA
Dennis Prather, University of Delaware, USA
Sanguthevar Rajasekaran , University of Florida, USA
Richard Rozier, Bell Labs Lucent Technologies, USA
Sartaj Sahni, University of Florida, USA
Hong Shen, Griffith University, Australia
Paul Spirakis, University of Patras, Greece
Ted Szymanski, McGill University, Canada
Jun Tanida, Osaka University, Japan
Jerry L. Trahan, Louisiana State University, USA
Ramachandran Vaidyanathan, Louisiana State University, USA

Fault Tolerant Algorithms for a Linear Array with a Reconfigurable Pipelined Bus System

Anu G. Bourgeois and Jerry L. Trahan

Department of Electrical and Computer Engineering
Louisiana State University
Baton Rouge, Louisiana 70803
{anu, trahan}@ee.lsu.edu

Abstract. Recently, many models using reconfigurable optically pipelined buses have been proposed in the literature. All algorithms developed for these models assume that a healthy system is available. We present some fundamental algorithms that are able to tolerate up to $N/2$ faults on an N-processor LARPBS (one particular optical model). We then extend these results to apply to other algorithms in the areas of image processing and matrix operations.

1 Introduction

Currently, optical fiber is the preferred medium for telecommunication networks of long distances, due in part to its high bandwidth, reliability, low distortion, and low attenuation [5]. This wide use of optical interconnects and advances in optical and optoelectronic technologies indicate successful use of these technologies for shorter distances, such as interconnecting processors in parallel computers. The advantages of using an optically pipelined bus rather than an electrical bus to transfer information among processors are that the optical waveguides have unidirectional propagation and predictable delays. These two properties enable synchronized concurrent access to an optical bus in a pipelined fashion [15]. Combined with the abilities of the bus structure to broadcast and multicast, this architecture suits many communication-intensive applications.

As a result, several processor arrays based on pipelined optical buses have been proposed as practical parallel computing platforms [4, 13–15, 20]. Many parallel algorithms exist for arrays with pipelined buses [5, 8, 12, 16, 19], indicating that such systems are very efficient for parallel computation due to the high bandwidth available by pipelining messages. This work will focus on the *Linear Array with a Reconfigurable Pipelined Bus System* (LARPBS) [9, 19], one such model.

The number of processors involved in the systems considered raises the probability of a fault occurring. Researchers have proposed fault tolerant algorithms for many parallel architectures, such as the hypercube, mesh, and torus [2, 3, 10, 11]. They have not, however, addressed the issue of fault tolerance for reconfigurable models, and more specifically, for any of the optically pipelined models.

J. Rolim et al. (Eds.): IPDPS 2000 Workshops, LNCS 1800, pp. 1044-1052, 2000.

In this paper we present several basic fault tolerant algorithms for the LARPBS. Specifically, we have developed algorithms to calculate binary prefix sums, perform compression, sort, and perform a general permutation routing step on an N-processor array that can have up to $N/2$ static faults. We then extend these results to other fault tolerant algorithms in the areas of image processing and matrix operations.

In the next section, we describe the LARPBS and the fault model used. Section 3 explains the preprocessing phase for fault tolerant algorithms. Section 4 details the basic fault tolerant algorithms and extends the results to other more complex algorithms. Finally, in Section 5, we make some concluding remarks.

2   Model Descriptions

As mentioned in the introduction, many models exist that use an optically pipelined bus system. This work centers on the *Linear Array with a Reconfigurable Pipelined Bus System* (LARPBS), as described by Pan and Li [9]. In this section, we briefly describe this model and also explain the assumptions made for the fault model used in the following sections.

2.1   LARPBS Model

An LARPBS comprises a linear array of N processors, connected by an optically pipelined bus. Let an optically pipelined bus have the same length of fiber between consecutive processors, so propagation delays between consecutive processors are the same; we refer to this delay as one *petit cycle*. Let a *bus cycle* be the end-to-end propagation delay on the bus. We specify time complexity in terms of a step comprising one bus cycle and one local computation. For more details on the time complexity issue, see Guo *et al.* [4] and Pan and Li [9].

The optical bus of the LARPBS is composed of three waveguides, one for carrying data (the *data waveguide*) and the other two (the *reference* and *select waveguides*) for carrying address information (see Figure 1). (For simplicity, the figure omits the data waveguide, as it resembles the reference waveguide.) Each processor connects to the bus through two directional couplers, one for transmitting and the other for receiving [4, 15]. The receiving segments of the reference and data waveguides contain an extra segment of fiber of one unit pulse-length, Δ, between each pair of consecutive processors (shown as a delay loop in Figure 1). The transmitting segment of the select waveguide also has a switch-controlled conditional delay loop of length Δ between processors R_i and R_{i+1}, for each $0 \leq i \leq N - 2$ (Figure 1).

To allow segmenting, the LARPBS has optical switches on the transmitting and receiving segments of each bus for each processor. With all switches set to *straight*, the bus system operates as a regular pipelined bus system. Setting the switches at R_i to *cross* segments the whole bus system into two separate pipelined bus systems, one consisting of processors $R_0, R_1, \cdots, R_i$ and the other consisting of $R_{i+1}, R_{i+2}, \cdots, R_{N-1}$. We will refer to a processor that set its

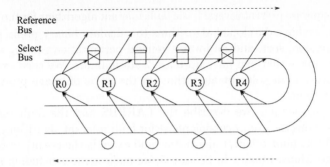

Fig. 1. Structure of a Linear Array with a Reconfigurable Pipelined Bus System (LARPBS).

segment switch to cross (the one closest to the U-turn) as the *head* of its segment. The processor furthest from the U-turn within a segment, is called the *tail* of its segment.

The LARPBS uses the *coincident pulse technique* [15] to route messages by manipulating the relative time delay of *select* and *reference* pulses on separate buses so that they will coincide only at the desired receiver. Each processor has a *select frame* of N bits (*slots*), of which it can inject a pulse into a subset of the N slots. The coincident pulse technique admits broadcasting and multicasting of a single message by appropriately introducing multiple select pulses within a select frame. When multiple messages arrive at the same processor in the same bus cycle, it receives only the first message and disregards subsequent messages that have coinciding pulses at the processor. (This corresponds to priority concurrent write.)

2.2   Fault Model

Let a *processing element* consist of a single processor, its conditional delay switches, and its directional couplers. We consider a processing element to be faulty if any one of its components is faulty, and refer to it as a faulty processor for short. Faults on any of the three optical waveguides are not considered. A fault-free processor is able to detect if either of its neighbors is faulty.

All faults that are considered are static and occur prior to the execution of any algorithm. Therefore, faults occurring during execution of a algorithm are not tolerated. The algorithms presented in Section 4 can tolerate up to $N/2$ faults on an N-processor LARPBS. These assumptions are consistent with those described by Parhami and Yeh [11].

If a conditional delay switch is faulty, that is, if it is stuck in either the cross or straight position, it remains that way for the remainder of the algorithm. Faulty segment switches are not considered, since this would result in a shorter available working array, and thus, would be a scaling problem rather than a fault tolerance problem. (For work on scalable algorithms for the LARPBS, refer to Trahan *et al.* [17, 19].)

3   Preprocessing Phase

Prior to running any algorithm on a faulty LARPBS, we perform some initial processing to ensure proper execution. Each working processor, p_i, determines the number of faulty processors p_j, where $i < j < N$, that have their conditional delay switches stuck at cross. Call the value of this suffix sum f_i. This is because any stuck delays that a select pulse travels through will alter the destination of the message sent by a working processor. By determining the total number of stuck delays ahead of it on the bus, each working processor can adjust its reference pulse to avoid miscommunication. Processor p_i shifts its reference pulse to the left by f_i slots. Then, provided each working processor has its conditional delay switch set to straight, the message sent by p_i reaches the intended destination.

First, each working processor segments the bus if it detects a faulty processor to its left. This results in two cases: 1) a segment has one or more good processors at the lower end, or 2) a segment ends with one or more faults and the head is the only good processor in the segment.

Consider the first case and one segment. The head of the segment, p_h, broadcasts its index to the segment. Any other fault-free processor, p_l, with a fault to its right, broadcasts its index to the head. The head of the segment can now determine the number of faults in the segment, call this value k. Processor p_l injects a select pulse into its highest $k/2$ slots and sends its index. Processor p_h then broadcasts a message indicating whether or not it received the message. (The segment head would receive the message if there were at most $k/2$ stuck delay switches.) If it did, p_l repeats this by injecting a select pulse into its highest $k/4$ slots. If not, then p_l injects a select pulse into slot $(N-1) - 3k/4$ through slot $(N-1) - k/2$. Repeat this process a total of $\log k$ times to determine the number of conditional delay switches that have faulted in the cross position. Worst case time complexity is when $k = N/2$, resulting in $O(\log N)$ steps.

Now consider a segment that fits the second case. The head of the segment needs the index of the head of the previous segment to determine the number of faults within its own segment. There could, however, be a string of such segments. We proceed in $\log N$ phases to relay information between these heads of segments, with each phase corresponding to one bit position of the processor indices. During phase i, where $1 \leq i \leq \log N$, each segment head with a '0' in bit position $i - 1$ of its index segments the bus and and listens for those with a '1' to broadcast their index within the segment. This step is then repeated with the writers now reading, and the readers now writing. Once the preceding index is known, the segments determine the number of stuck delays within the segment, as in the first case. With $\log N$ phases, and each phase taking $O(\log N)$ steps, the total time to determine the number of stuck delay switches in each segment takes $O(\log^2 N)$ steps. This is done in $\log N$ phases rather than a simple odd/even phase, because the two processors communicating could possibly both have odd or even indices. Eventually, each segment head will receive the proper index since the two must differ in at least one bit position. In addition, the first index the segment head receives is the proper index, since the previous segment head would be segmenting the bus for each of the phases until the two communicate.

With the number of stuck delays computed for each segment, we can now determine the number of stuck delays ahead of each working processor on the array. We perform a prefix sums operation as on a tree-like structure. The head of each segment holds the data for its segment, and each other working processor holds a value of '0'. Using the indices, the processors can determine when to write and read on the bus. For each pair of processors communicating, the higher indexed processor segments the bus, in order for the two to broadcast. A working processor can determine if it is paired with a faulty processor. If the working processor is the higher indexed, then it will not receive a message from the faulty processor. If the faulty processor is the higher indexed, it will not be able to segment the bus, and the working processor will receive the index from a processor in another segment. Once a working processor determines that it is paired with a faulty processor, then the working processor continues on to the next phase. After $\log N$ phases, the head of the array broadcasts the total, so that each processor can then locally determine the number of stuck delay switches ahead of it on the bus. The prefix sum can be computed in $O(\log N)$ steps.

The next item to consider is the mapping, since each good processor will need to simulate up to two processors. We use a compaction mapping, such that the working processor with rank i simulates processors with index $2i$ and $2i + 1$, for $i \leq f$, where f is the total number of faults. The remaining working processors simulate the processor with index $i + f$. In order to do this, we determine the ranks of all fault-free processors. Set the data value to '1' for each good processor and compute prefix sums in $O(\log N)$ steps. With this ranking, each working processor can determine which processor(s) it simulates. It can also properly send a message to any specific processor by adjusting its reference frame as described earlier in this section.

Combining the time to determine information on the number of stuck delay switches and to determine the mapping results provides us with the following result.

Theorem 1. *An N-processor LARPBS with up to $N/2$ faults is able to compute the number of stuck delay switches succeeding each working processor and determine the mapping of all processors to working processors in a total of $O(\log^2 N)$ preprocessing steps.*

4   Fault Tolerant Algorithms

In this section we describe some basic algorithms for an N-processor LARPBS that can tolerate up to $N/2$ faults. Using these fundamental algorithms, we can then extend the results to develop other more complex fault tolerant algorithms for the LARPBS.

Theorem 2. *Binary prefix sums of N elements can be computed on an N-processor LARPBS with up to $N/2$ faults in $O(\log N)$ steps.*

Proof. Proceed in the same manner as ranking the good processors in the previous section. In this case, however, each processor can have up to two elements to handle. With the ranking of the working processors known as well as the number of stuck delays ahead of each processor, it is possible perform the operation in $O(\log N)$ steps.

First, each good processor locally determines the total sum for the one or two elements it is simulating. Next, using the rankings of the good processors, perform prefix sums as in the tree method. (At each step, the appropriate working processors segment the bus so the corresponding processors can communicate.) Once the prefix sums is complete, each working processor can locally determine the prefix sum for each of the elements it is simulating.   □

Assume that each processor of an N-processor LARPBS is either marked or unmarked based upon a local variable. Let there be x marked processors. The compression algorithm shifts all marked processors to the lower end of the array, namely processor p_0 through processor p_{x-1}.

Theorem 3. *Compression of x elements, where $x \leq N$, can be performed on an N-processor LARPBS with up to $N/2$ faults in $O(\log N)$ steps.*

Proof. First the working processors rank the marked processors, using the prefix sums algorithm of the previous theorem, in $O(\log N)$ steps. Call this the marked rank. The processor with marked rank i determines the index of the processor simulating p_i as follows and routes its data to that processor.

Let $x = 4i$. The processor, p_k, with marked rank $2i$ broadcasts its index to all processors. Next, the processor simulating processor p_{2i} broadcasts its index, j, to all processors. As a result, all processors receive the index of the processor simulating the processor with the middle rank. Next, the processor with marked rank i ($3i$) multicasts its index to $p_0, p_1, \ldots, p_{j-1}$ ($p_{j+1}, p_{j+2}, \ldots, p_{N-1}$). Similar to the previous phase, the processor simulating p_i (p_{3i}) multicasts its index to the segment of processors below (above) p_k. Repeat this phase $\log x$ times, until all ranked processors can determine the corresponding indices.

Repeat these steps for unmarked processors. These processors will determine the indices of processors starting after the last ranked processor in the previous phase, however. Once all indices of the simulating processors have been determined, send messages in two steps. First, send messages destined for an even numbered processor, then those for odd numbered processors. Routing messages this way will prevent messages from colliding at any processor.   □

Theorem 4. *Sorting N k-bit integers can be performed on an N-processor LARPBS with up to $N/2$ faults in $O(k \log N)$ steps.*

Proof. We use the radix sort method and the compression algorithm to sort the N integers. The algorithm proceeds in k phases, one for each bit position of the integers. During execution of phase j, where $j \leq k$, perform compression based upon the jth bit position (Theorem 3). Each phase takes $O(\log N)$ steps, for a total of $O(k \log N)$ steps.   □

Table 1. Fault Tolerant LARPBS Algorithms

Algorithm	Time on Faulty	Time on Fault-Free	No. of Processors
median row	$O(\log N)$	$O(1)$	$O(N)$
image area	$O(\log^2 N)$	$O(1)$	$O(N)$
image perimeter	$O(\log^2 N)$	$O(1)$	$O(N)$
histogram	$O(\log h \log N)$	$O(\log h)$	$O(N)$
matrix transposition	$O(\log^2 N)$	$O(1)$	$O(N^2)$
matrix multiplication	$O(N \log^2 N)$	$O(N)$	$O(N^2)$

A generalized permutation routing step is one in which each processor sends at most one message and is the intended destination for at most one message.

Theorem 5. *Any generalized permutation routing step can be performed on an N-processor LARPBS with up to N/2 faults in $O(\log^2 N)$ steps.*

Proof. We proceed by first sorting the messages by their destinations. Since some processors may not be receiving messages, the messages are in order after the sort, but not necessarily at their final destination. So we then shift the messages to the intended processors. Perform the algorithm in two phases, one for messages destined to even numbered processors, and one for messages destined to odd numbered processors.

To perform the shifting, the processors holding the messages before the shifting determine the indices of the destination processors. Since all messages are in proper order, we can proceed in $O(\log N)$ phases broadcasting the indices of midpoints of segments, as we did for the compression algorithm (Theorem 3). The algorithm runs in $O(\log^2 N)$ steps.   □

We have extended these results to apply to other algorithms in the areas of image processing and matrix operations. Table 1 lists the algorithms considered, the time complexity on a faulty and a fault-free LARPBS, and the number of processors required. The algorithms listed tolerate at most $N/2$ faults for an N-processor LARPBS. Our fault tolerant algorithms combine the techniques of the previous fundamental algorithms presented and build upon existing algorithms for the LARPBS. The image processing algorithms follow the approach of Pan and Li [9]. The matrix operation algorithms follow the approach of Li *et al.* [6, 7].

Lemma 1. *Any algorithm executed on an N-processor LARPBS with $O(1)$ faults will result in a constant factor slowdown.*

Proof. Broadcast the indices of the faulty processors in $O(1)$ steps. Each working processor locally keeps a table of the faulty processors and the working processors that are simulating them. The algorithms then run as required, with a constant number of straightforward steps to accommodate the faulty processors.   □

It is important to note that the preprocessing stage presented in Section 3 is not necessary before execution of each algorithm. If there is a sequence of algorithms to be executed, the preprocessing need only be done once. Once the mapping and information on the number of stuck delays has been established, it will apply to all algorithms run thereafter on the LARPBS.

5   Conclusions

The work presented in this paper is the first to address the issue of fault tolerance for reconfigurable models, and more specifically, those with optically pipelined buses. We have shown that it is possible to design fault tolerant algorithms with low overhead.

One possible extension is to develop other complex fault-tolerant algorithms for the LARPBS, as well as minimize the time complexity for the preprocessing stage. Another possibility is to extend the results to the *Pipelined Reconfigurable Mesh* (PR-Mesh) [1, 18], a multi-dimensional extension of the LARPBS, and other optically pipelined models.

References

1. A. G. Bourgeois and J. L. Trahan, "Relating Two-Dimensional Reconfigurable Meshes with Optically Pipelined Buses," to appear in *Int'l. Jour. Found. of Comp. Science.*
2. H.-L. Chen and S.-H. Hu, "Distributed Submesh Determination in Faulty Tori and Meshes," *Proc. Int'l. Par. Processing Symp.*, (1997).
3. G.-M. Chiu and S.-P. Wu, "A Fault-Tolerant Routing Strategy in Hypercube Multicomputers," *IEEE Trans. Comput.*, vol. 45, (1996), pp. 143–154.
4. Z. Guo, R. Melhem, R. Hall, D. Chiarulli, and S. Levitan, "Array Processors with Pipelined Optical Busses," *J. Parallel Distrib. Comput.*, vol. 12, (1991), pp. 269–282.
5. K. Li, Y. Pan, and S. Q. Zheng, *Parallel Computing Using Optical Interconnections*, Kluwer Academic Publishers, Boston, MA, 1998.
6. K. Li, Y. Pan, and S. Q. Zheng, "Fast and Efficient Parallel Matrix Operations Using a Linear Array with a Reconfigurable Pipelined Bus System," in *High Performance Computing Systems and Applications*, J. Schaeffer and R. Unrau, eds., Kluwer Academic Publishers, Boston, MA, 1998.
7. K. Li, Y. Pan, and S. Q. Zheng, "Fast and Efficient Parallel Matrix Multiplication Algorithms on a Linear Array with a Reconfigurable Pipelined Bus System," *IEEE Trans. Parallel Distrib. Systems*, vol. 9, (1998), pp. 705–720.
8. M. Middendorf and H. ElGindy, "Matrix Multiplication on Processor Arrays with Optical Buses," to appear in *Informatica.*
9. Y. Pan and K. Li, "Linear Array with a Reconfigurable Pipelined Bus System: Concepts and Applications," *Information Sciences – An International Journal*, vol. 106, (1998), pp. 237–258.
10. B. Parhami, "Fault Tolerance Properties of Mesh-Connected Parallel Computers with Separable Row/Column Buses," *Proc. Midwest Symp. on Cir. and Syst.*, (1993).

11. B. Parhami and C.-H. Yeh, "The Robust-Algorithm Approach to Fault Tolerance on Processor Arrays: Fault Models, Fault Diameter, and Basic Algorithms," *Proc. Int'l. Par. Processing Symp.*, (1998), pp. 742–746.

12. S. Pavel and S. G. Akl, "Matrix Operations Using Arrays with Reconfigurable Optical Buses," *Par. Algs. and Appl.*, vol. 8, (1996), pp. 223–242.

13. S. Pavel and S. G. Akl, "On the Power of Arrays with Optical Pipelined Buses," *Proc. Int'l. Conf. Par. Distr. Proc. Techniques and Appl.*, (1996), pp. 1443–1454.

14. C. Qiao, "On Designing Communication-Intensive Algorithms for a Spanning Optical Bus Based Array," *Parallel Proc. Letters*, vol. 5, (1995), pp. 499–511.

15. C. Qiao and R. Melhem, "Time-Division Optical Communications in Multiprocessor Arrays," *IEEE Trans. Comput.*, vol. 42, (1993), pp. 577–590.

16. S. Rajasekaran and S. Sahni, "Sorting, Selection and Routing on the Arrays with Reconfigurable Optical Buses," *IEEE Trans. Parallel Distrib. Systems*, vol. 8, (1997), pp. 1123–1132.

17. J. L. Trahan, A. G. Bourgeois, Y. Pan, and R. Vaidyanathan, "Optimally Scaling Permutation Routing on Reconfigurable Arrays with Optically Pipelined Buses," *Proc. 13th Int'l. Par. Process. Symp. & 10th Symp. Par. Distr. Process.*, (1999), pp. 233–237.

18. J. L. Trahan, A. G. Bourgeois, and R. Vaidyanathan, "Tighter and Broader Complexity Results for Reconfigurable Models," *Parallel Proc. Letters*, vol. 8, (1998), pp. 271–282.

19. J. L. Trahan, Y. Pan, R. Vaidyanathan, and A. G. Bourgeois, "Scalable Basic Algorithms on a Linear Array with a Reconfigurable Pipelined Bus System," *Proc. Int'l. Conf. on Parallel and Distributed Computing Systems*, (1997), pp. 564–569.

20. S. Q. Zheng and Y. Li, "Pipelined Asynchronous Time-Division Multiplexing Optical Bus," *Optical Engineering*, vol. 36, (1997), pp. 3392–3400.

Fast and Scalable Parallel Matrix Computations with Optical Buses (Extended Abstract*)

Keqin Li

Department of Mathematics and Computer Science
State University of New York
New Paltz, New York 12561-2499
li@mcs.newpaltz.edu

Abstract. We present fast and highly scalable parallel computations for a number of important and fundamental matrix problems on linear arrays with reconfigurable pipelined optical bus systems. These problems include computing the Nth power, the inverse, the characteristic polynomial, the determinant, the rank, and an LU- and a QR-factorization of a matrix, and solving linear systems of equations. These computations are based on efficient implementation of the fastest sequential matrix multiplication algorithm, and are highly scalable over a wide range of system size. Such fast and scalable parallel matrix computations were not seen before on distributed memory parallel computing systems.

1 Introduction

Parallel matrix computations using optical buses have recently been addressed in [9], where a number of matrix manipulation problems are considered, including computations of the Nth power, the inverse, the characteristic polynomial, the determinant, the rank, and an LU- and a QR-factorization of an $N \times N$ matrix, and solving linear systems of equations. It is shown in [9] that compared with the best known results on distributed memory systems, the parallel complexities of these problems can be reduced by a factor of $O(\log N)$ on linear arrays with reconfigurable pipelined bus systems (LARPBS).

While speed is an important motivation of parallel computing, there is another issue in realistic parallel computing, namely, scalability, which measures the ability to maintain speedup linearly proportional to the number of processors [5]. In this paper, we present fast and highly scalable parallel computations for the above problems. These computations are based on efficient implementation of the fastest $O(N^\alpha)$ sequential matrix multiplication algorithm [2, 11], and are highly scalable (i.e., constant efficiency, cost-optimality, and linear speedup can be achieved) over a wide range of system size p. Such fast and scalable parallel matrix computations were not seen before on distributed memory parallel computing systems.

2 Scalable Parallelization

The time complexity of a parallel computation can be represented as $O(T(N)/p + T_{\text{comm}}(N, p))$, where N is the problem size, p is the number of processors available, $T(N)$ is the time complexity of the sequential algorithm being parallelized,

---

* A complete version of the paper is available as Technical Report #00-100, Dept. of Mathematics and Computer Science, SUNY at New Paltz, January 2000. See http://www.mcs.newpaltz.edu/tr.

J. Rolim et al. (Eds.): IPDPS 2000 Workshops, LNCS 1800, pp. 1053-1062, 2000.

and $T_{\mathrm{comm}}(N, p))$ is the overall communication overhead of a parallel implementation. We say that a parallel implementation is *scalable* in the range $[1..p^*]$ if linear speedup and cost-optimality can be achieved for all $1 \leq p \leq p^*$. Clearly, p^* should be as large as possible, since this implies that the parallel implementation has the ability to be scaled over a large range of system size p. A parallel implementation is *highly scalable* if p^* is as large as $\Theta(T(N)/(T^*(N)(\log N)^k))$ for some constant $k \geq 0$, where $T^*(N)$ is the best possible parallel time. It is clear that not every sequential algorithm can be parallelized (by using sufficient processors) so that constant parallel execution time is achieved. If $T^*(N)$ is the best possible parallel time, the largest possible value for p^* is $\Theta(T(N)/T^*(N))$. High scalability means that p^* is very close to the largest possible except for a polylog factor of N. A highly scalable parallel implementation is *fully scalable* if $k = 0$, which means that the sequential algorithm can be fully parallelized, and communication overhead $T_{\mathrm{comm}}(N, p)$ in parallelization is negligible.

3   Optical Buses

An LARPBS is a distributed memory system which consists of p processors P_1, P_2, ..., P_p linearly connected by a reconfigurable pipelined optical bus system. Each processor can perform ordinary arithmetic and logic computations, and interprocessor communication. All computations and communications are synchronized by bus cycles, so that an LARPBS is similar to an SIMD machine. In addition to the tremendous communication capabilities supported by a pipelined optical bus, an LARPBS can also be partitioned into several independent subarrays by reconfiguring an optical bus, and all subarrays can be used independently for different computations without interference (see [12] for detailed exposition). Due to reconfigurability, an LARPBS can also be viewed as an MIMD machine.

A computation on LARPBS is a sequence of alternate global communication and local computation steps. The time complexity of an algorithm is measured in terms of the total number of bus cycles in all the communication steps, as long as the time of the local computation steps between successive communication steps is bounded by a constant.

A rich set of basic communication, data movement, and aggregation operations on the LARPBS model implemented using the coincident pulse processor addressing technique [1] have been developed [8, 12]. The following primitive operations on LARPBS are used in this paper, and our algorithms are described using these operations as building blocks.

One-to-One Communication. Processor P_{i_k} sends a value to P_{j_k}, for all $1 \leq k \leq q$ simultaneously.

Multiple Multicasting. Assume that we have g disjoint groups of destination processors G_k and g senders P_{i_k}, where $1 \leq k \leq g$. Processor P_{i_k} broadcasts a value to all the processors in G_k, for all $1 \leq k \leq g$ simultaneously.

Global Aggregation. Suppose processor P_j holds a value v_j, $1 \leq j \leq p$. We need to calculate the summation $v_1 + v_2 + \cdots + v_p$, which is saved in P_1. It is assumed that the v_j's are integers or floating-point values with finite magnitude and precision, or boolean values (where $+$ is replaced by logical operations). [1]

---

[1] It has been a common practice in algorithm analysis to assume that a single manipulation takes constant time. This essentially implies that numerical values have finite magnitude and precision; otherwise, a manipulation either takes non-constant time or requires extra hardware support. Therefore, our assumption in the global aggregation operation is quite reasonable.

All these communication, data movement, and global aggregation primitives can be performed on an LARPBS in constant number of bus cycles [8, 12]. We would like to point out that if each value is replaced by a block of s data, the execution time of the above primitive operations is simply increased to $O(s)$ by executing an operation for s times.

The above primitive operations can be directly used for some basic matrix manipulations. For instance, one-to-one communication can be used for transposing a matrix. We divide an $N \times N$ matrix $A = (a_{ij})_{N \times N}$ into submatrices $A = (A_{ij})_{\sqrt{p} \times \sqrt{p}}$, where each submatrix A_{ij} is of size $N/\sqrt{p} \times N/\sqrt{p}$. Assume that A is stored in a p-processor LARPBS in the row-major order, that is, A_{ij} is hold by $P_{(i-1)\sqrt{p}+j}$, for all $1 \leq i, j \leq \sqrt{p}$. By using one-to-one communication with blocks of data of size $s = N^2/p$, processor $P_{(i-1)\sqrt{p}+j}$ sends A_{ij} to processor $P_{(j-1)\sqrt{p}+i}$ simultaneously for all $1 \leq i, j \leq \sqrt{p}$.

Theorem 1. *Transposing an $N \times N$ matrix can be done on a p-processor LARPBS in $O(N^2/p)$ time. Our implementation for matrix transposition is fully scalable.*

The aggregation operation can be used to calculate the summation of N vectors. Given N N-dimensional vectors $\mathbf{v}_i = (v_{i1}, v_{i2}, ..., v_{iN})$, where $1 \leq i \leq N$, the vector chain addition problem is to calculate $\mathbf{v} = (v_1, v_2, ..., v_N)$, where $v_j = v_{1j} + v_{2j} + \cdots + v_{Nj}$, for all $1 \leq j \leq N$. By using one-to-one communication and the global aggregation operation, it is known [9] that the summation of q N-dimensional vectors can be calculated in $O(1)$ time on a qN-processor LARPBS. If $p < N$, each processor holds at most $\lceil N/p \rceil$ vectors. It takes $O(N^2/p)$ time for each processor to compute locally the summation of the vectors it holds. Then, the summation of the p partial sums can be obtained in $O(N)$ time by the p processors. If $p \geq N$, we will assign p/N processors to each vector $\mathbf{v}_i = (\mathbf{v}_{i,1}, \mathbf{v}_{i,2}, ..., \mathbf{v}_{i,p/N})$, such that each processor holds a vector segment $\mathbf{v}_{i,j}$ of size N^2/p. Then, the method in [9] is applied to these vector segments, and the execution time is increased by a factor of $O(N^2/p)$.

Theorem 2. *The summation of N N-dimensional vectors can be calculated on a p-processor LARPBS in $O(N^2/p)$ time. Our implementation for vector chain addition is fully scalable.*

Given N matrices $A_1, A_2, ..., A_N$, where $A_k = (a_{ij}^{(k)})_{N \times N}$, the matrix chain addition problem is to calculate $A = (a_{ij})_{N \times N}$, where $a_{ij} = a_{ij}^{(1)} + a_{ij}^{(2)} + \cdots + a_{ij}^{(N)}$, for all $1 \leq i, j \leq N$. By using one-to-one communication and the global aggregation operation, it is known [9] that the summation of N $s \times s$ matrices can be computed in $O(1)$ time on an s^2N-processor LARPBS. If $p < N$, each processor holds at most $\lceil N/p \rceil$ matrices. It takes $O(N^3/p)$ time for each processor to compute locally the summation of the matrices it holds. Then, the summation of the p partial sums can be obtained in $O(N^2)$ time by the p processors. If $p \geq N$, we will assign p/N processors to each matrix $A_k = (A_{ij}^{(k)})_{\sqrt{p/N} \times \sqrt{p/N}}$ such that each processor holds a submatrix $A_{i,j}$ of size $(N^{1.5}/\sqrt{p}) \times (N^{1.5}/\sqrt{p})$. Then, the method in [9] is applied to these submatrices, and the execution time is increased by a factor of $O(N^3/p)$.

Theorem 3. *The summation of N $N \times N$ matrices can be computed on a p-processor LARPBS in $O(N^3/p)$ time. Our implementation for matrix chain addition is fully scalable.*

In [9], it is shown that by using one-to-one communication, multiple multicasting, and the global aggregation operation, the product of an $N \times N$ matrix $A = (a_{ij})_{N \times N}$, and an N-dimensional vector $\mathbf{v} = (v_1, v_2, ..., v_N)$ can be obtained on an N^2-processor LARPBS in $O(1)$ time. It is clear that we can divide $A = (A_{ij})_{\sqrt{p} \times \sqrt{p}}$ into submatrices of size $(N/\sqrt{p}) \times (N/\sqrt{p})$ and $\mathbf{v} = (\mathbf{v}_1, \mathbf{v}_2, ..., \mathbf{v}_{\sqrt{p}})$ into vector segments of size $N/\sqrt{p}$. Then, the method in [9] is applied to these submatrices and vector segments, and the execution time is increased by a factor of $O(N^2/p)$.

Theorem 4. *The product of an $N \times N$ matrix and an N-dimensional vector can be obtained on a p-processor LARPBS in $O(N^2/p)$ time. Our implementation for matrix-vector multiplication is fully scalable.*

4   Matrix Multiplication, Chain Product, and Powers

Matrix multiplication is definitely the most important subproblem in many other matrix manipulations. A number of parallel matrix multiplication algorithms have been developed on LARPBS [8, 10]. The following most noteworthy result, which leads to fast and scalable algorithms for many matrix operations, was shown in [7].

Theorem 5. *For all $1 \le p \le N^\alpha$, multiplying two $N \times N$ matrices can be performed on a p-processor LARPBS in*

$$T_{\mathrm{mm}}(N, p) = O\left(\frac{N^\alpha}{p} + \frac{N^2}{p^{2/\alpha}} \log p\right)$$

time. In particular, multiplying two $N \times N$ matrices can be performed in $O(\log N)$ time on an LARPBS with N^α processors. Our implementation for matrix multiplication is scalable for $p = O(N^\alpha/(\log N)^{\alpha/(\alpha-2)})$.

The last statement in the above theorem needs explanation. Assume that $p = N^\alpha/(\log N)^\beta$. Then, $T_{\mathrm{mm}}(N, p) = O((\log N)^\beta + (\log N)^{1+2\beta/\alpha})$. It is clear that when $\beta \ge 1+2\beta/\alpha$, i.e., $\beta \ge \alpha/(\alpha-2)$, the first term dominates $T_{\mathrm{mm}}(N, p)$. That is, $T_{\mathrm{mm}}(N, p) = O(N^\alpha/p)$, which results in linear speedup and cost-optimality.

Given N matrices $A_1, A_2, ..., A_N$ of size $N \times N$, the matrix chain product problem is to compute $A_1 \times A_2 \times \cdots \times A_N$. Given an $N \times N$ matrix A, the matrix powers problem is to calculate the first N powers of A, i.e., $A, A^2, A^3, ..., A^N$. For both problems, the sequential time complexity is $O(N^{\alpha+1})$.

Based on Theorem 5, the following result has been established in [6].

Theorem 6. *For all $1 \le p \le N^{\alpha+1}/2$, the product of N matrices of size $N \times N$ can be computed on a p-processor LARPBS in*

$$T_{\mathrm{chain}}(N, p) = O\left(\frac{N^{\alpha+1}}{p} + \frac{N^{2(1+1/\alpha)}}{p^{2/\alpha}} \log \frac{p}{N} + (\log N)^2\right)$$

time. In particular, the product of N matrices of size $N \times N$ can be computed in $O((\log N)^2)$ time on an LARPBS with $N^{\alpha+1}/(\log N)^{\alpha/2}$ processors. Our implementation for matrix chain product is scalable for $p = O(N^{\alpha+1}/(\log N)^{\alpha/(\alpha-2)})$.

For matrix powers, the following result has been established in [6].

Theorem 7. *For all $1 \leq p \leq N^{\alpha+1}$, the first N powers of an $N \times N$ matrix can be computed on a p-processor LARPBS in*

$$T_{\text{power}}(N, p) = O\left(\frac{N^{\alpha+1}}{p} + \frac{N^{2(1+1/\alpha)}}{p^{2/\alpha}} \log p + \log N \log p\right)$$

time. Consequently, the first N powers of an $N \times N$ matrix can be computed in $O((\log N)^2)$ time on an LARPBS with $N^{\alpha+1}/(\log N)^{\alpha/2}$ processors. Our implementation for matrix powers is scalable for $p = O(N^{\alpha+1}/(\log N)^{\alpha/(\alpha-2)})$.

5   Inversion of Lower and Upper Triangular Matrices

Let A be an $N \times N$ lower triangular matrix that is invertible, i.e., all the elements on A's main diagonal are nonzeros. We partition A into four submatrices of equal size $N/2 \times N/2$, $A = \begin{bmatrix} A_1 & 0 \\ A_3 & A_2 \end{bmatrix}$. Since A_1 and A_2 are also invertible lower triangular matrices, we have $A^{-1} = \begin{bmatrix} A_1^{-1} & 0 \\ -A_2^{-1}A_3A_1^{-1} & A_2^{-1} \end{bmatrix}$. Therefore, A^{-1} can be obtained by inverting A_1 and A_2 recursively, and then multiplying A_1^{-1} and A_2^{-1} with A_3. Similarly, if A is an $N \times N$ upper triangular matrix that is invertible, we partition A into four submatrices of equal size $N/2 \times N/2$, $A = \begin{bmatrix} A_1 & A_3 \\ 0 & A_2 \end{bmatrix}$. Then, A_1 and A_2 are also invertible upper triangular matrices, and $A^{-1} = \begin{bmatrix} A_1^{-1} & -A_1^{-1}A_3A_2^{-1} \\ 0 & A_2^{-1} \end{bmatrix}$. The above discussion yields the following method for inverting a lower (upper) triangular matrix.

A Method for Lower (Upper) Triangular Matrix Inversion.

(1) Recursively calculate A_1^{-1} and A_2^{-1};
(2) Compute $-A_2^{-1}A_3A_1^{-1}$ $(-A_1^{-1}A_3A_2^{-1})$ by using the matrix multiplication algorithm MM.

This method reduces the lower/upper triangular matrix inversion problem to matrix multiplication. Without loss of generality, we assume that $N = 2^n$ is a power of two. The recursion can be unwound into $n + 1$ iterations, such that a sequence of matrices $A_0, A_1, A_2, ..., A_n$ are calculated. Initially, A_0 is obtained from A by inverting the elements on the main diagonal. This is the base of the recursion. In general, A_k is obtained from A_{k-1} by further calculating 2^{n-k} submatrices of size 2^{k-1}, where $1 \leq k \leq n$. Finally, $A_n = A^{-1}$. The above method implies that the sequential time complexity is $\sum_{k=1}^{n} 2^{n-k}O((2^{k-1})^\alpha) = O(2^{n-\alpha} \sum_{k=1}^{n} 2^{(\alpha-1)k}) = O(2^{n-\alpha} \cdot 2^{(\alpha-1)(n+1)}) = O(2^{n\alpha-1}) = O(N^\alpha)$, which is the same as matrix multiplication. We need to develop scalable parallelization of the above method for the number of processors p in the range $[1..N^\alpha]$.

It is clear that in calculating A_k, $1 \leq k \leq n$, there are 2^{n-k} submatrices of size 2^{k-1} that can be computed in parallel, and each requires only two matrix multiplications. To calculate A_k, an LARPBS with p processors is reconfigured into 2^{n-k} subarrays for 2^{n-k} simultaneous invocation of submatrix multiplications. To this end, certain data movements are required such that the entries of a submatrix are packed together. Fortunately, the communication patterns for this purpose are quite regular. If the p processors are evenly distributed to the 2^{n-k} subarrays, the time complexity for computing A_k is $T_{\mathrm{mm}}(2^{k-1}, p/2^{n-k})$, this gives the overall time complexity $T_{\mathrm{tri}}(N, p) = \sum_{k=1}^{n} T_{\mathrm{mm}}(2^{k-1}, p/2^{n-k})$. The following result is obtained by algebraic manipulations (see the complete version of the paper). Compared to matrix multiplication, our implementation only introduces a small overhead.

Theorem 8. *For all $1 \le p \le N^\alpha$, the inverse of a lower/upper triangular matrix can be calculated by a p-processor LARPBS in*

$$T_{\mathrm{tri}}(N, p) = O\left(\frac{N^\alpha}{p} + \frac{N^2}{p^{2/\alpha}} \log p + \log N \log \frac{p}{N}\right)$$

time. In particular, the inverse of a lower/upper triangular matrix can be calculated in $O((\log N)^2)$ time on an LARPBS with $N^\alpha/(\log N)^{\alpha/2}$ processors. Our implementation for lower/upper triangular matrix inversion is scalable for $p = O(N^\alpha/(\log N)^{\alpha/(\alpha-2)})$.

6   Determinants, Characteristic Polynomials, and Ranks

Let the determinant of a matrix A be denoted by $\det(A)$. The characteristic polynomial of a matrix A is defined as $\phi_A(\lambda) = \det(\lambda I_N - A) = \lambda^N + c_1 \lambda^{N-1} + c_2 \lambda^{N-2} + \cdots + c_{N-1}\lambda + c_N$, where I_N is the $N \times N$ identity matrix. The trace $\mathrm{tr}(A)$ of a matrix $A = (a_{ij})_{N \times N}$ is the sum of the entries on A's main diagonal, i.e., $\mathrm{tr}(A) = a_{11} + a_{22} + \cdots + a_{NN}$.

The following classical result is the basis of a parallel algorithm for obtaining $\phi_A(\lambda)$ and $\det(A)$.

Leverrier's Lemma. *The coefficients c_1, c_2, ..., c_N of the characteristic polynomial of a matrix A satisfy*

$$S \begin{bmatrix} c_1 \\ c_2 \\ c_3 \\ \vdots \\ c_{N-1} \\ c_N \end{bmatrix} = \begin{bmatrix} s_1 \\ s_2 \\ s_3 \\ \vdots \\ s_{N-1} \\ s_N \end{bmatrix}, \quad \text{where } S = \begin{bmatrix} 1 & 0 & 0 & \cdots & 0 & 0 \\ s_1 & 2 & 0 & \cdots & 0 & 0 \\ s_2 & s_1 & 3 & \cdots & 0 & 0 \\ \vdots & \vdots & \vdots & \ddots & \vdots & \vdots \\ s_{N-2} & s_{N-3} & s_{N-4} & \cdots & N-1 & 0 \\ s_{N-1} & s_{N-2} & s_{N-3} & \cdots & s_1 & N \end{bmatrix},$$

and $s_k = \mathrm{tr}(A^k)$, for all $1 \le k \le N$.

Based on Leverrier's Lemma, Csanky devised the following method for calculating the characteristic polynomial of a matrix A [3]. Since $\phi_A(0) = \det(-A) = c_N$, i.e., $\det(A) = (-1)^N c_N$, this algorithm can also be used to calculate $\det(A)$.

Csanky's Strategy for Characteristic Polynomial.

(1) Calculate A^2, A^3, A^4, ..., A^N;

(2) Calculate $s_k = \text{tr}(A^k)$, for all $1 \le k \le N$;

(3) Find S^{-1}, where S is the lower triangular matrix in Leverrier's Lemma;

(4) Calculate the matrix-vector product $S^{-1}[s_1, s_2, ..., s_N]^T$ to obtain $c_1, c_2, ...,$ c_N.

Step (1) involves the calculation of the first N powers of A (Theorem 7). Step (2) can be implemented using the aggregation operation plus certain data movements. Step (3) invokes the lower triangular matrix inversion algorithm (Theorem 8). Finally, Step (4) is a matrix-vector multiplication (Theorem 4). It is clear that the time complexity of Csanky's method is dominated by Step (1).

Theorem 9. *For all $1 \le p \le N^{\alpha+1}$, the characteristic polynomial of an $N \times N$ matrix can be obtained on a p-processor LARPBS in*

$$T_{\text{poly}}(N, p) = O\left(\frac{N^{\alpha+1}}{p} + \frac{N^{2(1+1/\alpha)}}{p^{2/\alpha}} \log p + \log N \log p\right)$$

time. In particular, the characteristic polynomial of an $N \times N$ matrix can be obtained in $O((\log N)^2)$ time on an LARPBS with $N^{\alpha+1}/(\log N)^{\alpha/2}$ processors. Our implementation for characteristic polynomial is scalable for $p = O(N^{\alpha+1}/ (\log N)^{\alpha/(\alpha-2)})$.

Theorem 10. *For all $1 \le p \le N^{\alpha+1}$, the determinant of an $N \times N$ matrix can be obtained on a p-processor LARPBS in*

$$T_{\text{det}}(N, p) = O\left(\frac{N^{\alpha+1}}{p} + \frac{N^{2(1+1/\alpha)}}{p^{2/\alpha}} \log p + \log N \log p\right)$$

time. Consequently, the determinant of an $N \times N$ matrix can be obtained in $O((\log N)^2)$ time on an LARPBS with $N^{\alpha+1}/(\log N)^{\alpha/2}$ processors. Our implementation for determinant is scalable for $p = O(N^{\alpha+1}/(\log N)^{\alpha/(\alpha-2)})$.

The rank of a matrix A, $\text{rank}(A)$, is the number of nonzero rows (or columns) in the row-reduced (or column-reduced) echelon form of A. It is well known that $\text{rank}(A) = \text{rank}(A^T A)$, where A^T is the transpose of A (or conjugate transpose of A for complex matrices), and $A^T A$ is similar to a diagonal matrix whose elements are the roots of the characteristic polynomial $\phi_{A^T A}(\lambda)$. Therefore, $\text{rank}(A)$ is the number of nonzero roots of $\phi_{A^T A}(\lambda)$. This leads to the following algorithm for finding $\text{rank}(A)$ [4].

An Algorithm for Calculating Matrix Rank.

(1) Get the matrix $A^T A$;

(2) Calculate $\phi_{A^T A}(\lambda) = c_0 \lambda^N + c_1 \lambda^{N-1} + c_2 \lambda^{N-2} + \cdots + c_{N-1}\lambda + c_N$;

(3) Find $\text{rank}(A) = N - i$, where i, $0 \le i \le N$, is the largest integer such that $c_{N-i} \ne 0$, and $c_{N-i+1} = c_{N-i+2} = \cdots = c_N = 0$.

In the above algorithm, Step (1) performs a matrix transposition (Theorem 1) and a matrix multiplication (Theorem 5). Step (2) invokes Csanky's method for computing characteristic polynomial (Theorem 9). Step (3) can be implemented in $O(N/p)$ time by simple data testing, comparison, and movement.

Theorem 11. *For all $1 \leq p \leq N^{\alpha+1}$, the rank of an $N \times N$ matrix can be obtained on a p-processor $\overline{LARPBS}$ in*

$$T_{\text{rank}}(N, p) = O\left(\frac{N^{\alpha+1}}{p} + \frac{N^{2(1+1/\alpha)}}{p^{2/\alpha}} \log p + \log N \log p\right)$$

time. Consequently, the rank of an $N \times N$ matrix can be obtained in $O((\log N)^2)$ time on an LARPBS with $N^{\alpha+1}/(\log N)^{\alpha/2}$ processors. Our implementation for matrix rank is scalable for $p = O(N^{\alpha+1}/(\log N)^{\alpha/(\alpha-2)})$.

7 Inversion of Arbitrary Matrices

Inverting an arbitrary matrix A is closely related to the calculation of the characteristic polynomial $\phi_A(\lambda)$, as revealed by the following well known theorem from linear algebra.

Cayley-Hamilton Theorem. *Let $\phi_A(\lambda) = \lambda^N + c_1 \lambda^{N-1} + c_2 \lambda^{N-2} + \cdots + c_{N-1}\lambda + c_N$ be the characteristic polynomial of matrix A. Then $\phi_A(A) = A^N + c_1 A^{N-1} + c_2 A^{N-2} + \cdots + c_{N-1}A + c_N A^0$ is the $N \times N$ zero matrix.*

Cayley-Hamilton Theorem implies that $A(A^{N-1} + c_1 A^{N-2} + c_2 A^{N-3} + \cdots + c_{N-1}I_N) = -c_N I_N$. Hence, the inverse of a matrix A can be calculated using the following identity, $A^{-1} = -(1/c_N)(A^{N-1} + c_1 A^{N-2} + c_2 A^{N-3} + \cdots + c_{N-2}A + c_{N-1}I_N)$. Csanky's method for calculating matrix inversion can be described as follows [3].

Csanky's Strategy for Matrix Inversion.

(1) Calculate the characteristic polynomial of A, that is, $c_1, c_2, ..., c_N$;
(2) Compute A^{-1} by using the above identity.

The time complexity of Step (1) is given in Theorem 9. Step (2) involves the computation of the first N powers of A (Theorem 7) and matrix chain addition (Theorem 3). Hence, we have the following result.

Theorem 12. *For all $1 \leq p \leq N^{\alpha+1}$, the inverse of an $N \times N$ matrix can be obtained on a p-processor $\overline{LARPBS}$ in*

$$T_{\text{inverse}}(N, p) = O\left(\frac{N^{\alpha+1}}{p} + \frac{N^{2(1+1/\alpha)}}{p^{2/\alpha}} \log p + \log N \log p\right)$$

time. Consequently, the inverse of an $N \times N$ matrix can be obtained in $O((\log N)^2)$ time on an LARPBS with $N^{\alpha+1}/(\log N)^{\alpha/2}$ processors. Our implementation for matrix inversion is scalable for $p = O(N^{\alpha+1}/(\log N)^{\alpha/(\alpha-2)})$.

8   Linear Systems of Equations

Let A be a nonsingular $N \times N$ matrix $(a_{ij})_{N \times N}$, and $\mathbf{b} = (b_1, b_2, ..., b_N)^T$ be an N-dimensional vector. The problem of solving a linear system of equations is to find a vector $\mathbf{x} = (x_1, x_2, ..., x_N)^T$ such that $A\mathbf{x} = \mathbf{b}$. It is clear that $\mathbf{x} = A^{-1}\mathbf{b}$, i.e., solving the above linear system of equations can be accomplished by a matrix inversion (Theorem 12) and a matrix-vector multiplication (Theorem 4).

Theorem 13. *For all $1 \leq p \leq N^{\alpha+1}$, a linear system of equations $A\mathbf{x} = \mathbf{b}$ can be solved on a p-processor LARPBS in*

$$T_{\text{equation}}(N, p) = O\left(\frac{N^{\alpha+1}}{p} + \frac{N^{2(1+1/\alpha)}}{p^{2/\alpha}} \log p + \log N \log p \right)$$

time. Consequently, a linear system of equations $A\mathbf{x} = \mathbf{b}$ can be solved in $O((\log N)^2)$ time on an LARPBS with $N^{\alpha+1}/(\log N)^{\alpha/2}$ processors. Our implementation for solving a linear system of equations is scalable for $p = O(N^{\alpha+1}/(\log N)^{\alpha/(\alpha-2)})$.

Let $\mathbf{k}_j = A^j \mathbf{b}$, where $0 \leq j \leq N - 1$. Then, the matrix $K(A, \mathbf{b}, N) = [\mathbf{b}, A\mathbf{b}, A^2\mathbf{b}, ..., A^{N-1}\mathbf{b}]$ is called the Krylov matrix defined by the matrix A, the vector $\mathbf{b}$, and the integer N. By Cayley-Hamilton Theorem, we have $A^N\mathbf{b} + c_1 A^{N-1}\mathbf{b} + c_2 A^{N-2}\mathbf{b} + \cdots + c_{N-1}A\mathbf{b} + c_N\mathbf{b} = 0$, that is, $A(A^{N-1}\mathbf{b} + c_1 A^{N-2}\mathbf{b} + c_2 A^{N-3}\mathbf{b} + \cdots + c_{N-1}\mathbf{b}) = -c_N\mathbf{b}$, which implies that $\mathbf{x} = -(1/c_N)\mathbf{k}_{N-1} - (c_1/c_N)\mathbf{k}_{N-2} - (c_2/c_N)\mathbf{k}_{N-3} - \cdots - (c_{N-1}/c_N)\mathbf{k}_0$. In other words, $\mathbf{x}$ is a linear combination of the column vectors of the Krylov matrix $K(A, \mathbf{b}, N)$. Thus, we can first calculate the first N powers of A (Theorem 7), and then compute the Krylov matrix $K(A, \mathbf{b}, N)$ using matrix-vector multiplication (Theorem 4). Once $K(A, \mathbf{b}, N)$ is available, $\mathbf{x}$ can be obtained via vector chain addition (Theorem 2). This proves the following result.

Theorem 14. *For all $1 \leq p \leq N^{\alpha+1}$, the Krylov matrix defined by the matrix A, the vector $\mathbf{b}$, and the integer N, can be calculated on a p-processor LARPBS in*

$$T_{\text{Krylov}}(N, p) = O\left(\frac{N^{\alpha+1}}{p} + \frac{N^{2(1+1/\alpha)}}{p^{2/\alpha}} \log p + \log N \log p \right)$$

time. Consequently, the Krylov matrix can be calculated in $O((\log N)^2)$ time on an LARPBS with $N^{\alpha+1}/(\log N)^{\alpha/2}$ processors. Our implementation for Krylov matrix is scalable for $p = O(N^{\alpha+1}/(\log N)^{\alpha/(\alpha-2)})$.

9   LU- and QR-Factorizations

The LU-factors of matrix A contain a nonsingular lower triangular matrix L and a nonsingular upper triangular matrix U such that $A = LU$. Suppose that the LU-factors of A exist. We divide A, L, and U into $N/2 \times N/2$ blocks as follows: $\begin{bmatrix} A_{11} & A_{12} \\ A_{21} & A_{22} \end{bmatrix} = \begin{bmatrix} L_{11} & 0 \\ L_{21} & L_{22} \end{bmatrix} \times \begin{bmatrix} U_{11} & U_{12} \\ 0 & U_{22} \end{bmatrix}$. The above identity implies that $A_{11} = L_{11}U_{11}$, $A_{12} = L_{11}U_{12}$, $A_{21} = L_{21}U_{11}$, $A_{22} = L_{21}U_{12} + L_{22}U_{22}$.

Pan's Method for LU-Factorization [11].

(1) Compute A_{11}^{-1};

(2) Calculate $X_1 = A_{11}^{-1} A_{12}$, $X_2 = A_{21} A_{11}^{-1}$, and $X_3 = A_{21} A_{11}^{-1} A_{12}$;

(3) Set $X_4 = A_{22} - X_3$; (It can be verified that $X_4 = L_{22} U_{22}$.)

(4) Recursively LU-factorize A_{11} to get L_{11} and U_{11}, and recursively LU-factorize X_4 to get L_{22} and U_{22};

(5) Calculate $L_{21} = X_2 L_{11}$, and $U_{12} = U_{11} X_1$.

If $T(N)$ is the sequential time complexity of the above method, then we have $T(N) = 2T(N/2) + O((N/2)^{\alpha+1})$. It can be verified that $T(N) = O(N^{\alpha+1})$. Let $T_{\text{LU}}(N,p)$ be the parallel time complexity of the above algorithm on an LARPBS. Then $T_{\text{LU}}(N,p) = T_{\text{LU}}(N/2, p/2) + T_{\text{inverse}}(N/2, p) + 5T_{\text{mm}}(N/2, p)$. The following result is obtained by algebraic manipulations (see the complete version of the paper).

Theorem 15. *For all $1 \le p \le N^{\alpha+1}$, LU-factorization of an $N \times N$ matrix can be performed on a p-processor LARPBS in*

$$T_{\text{LU}}(N,p) = O\left(\frac{N^{\alpha+1}}{p} + \frac{N^{2(1+1/\alpha)}}{p^{2/\alpha}} \log p + (\log N)^3 \right).$$

time. Consequently, LU-factorization of an $N \times N$ matrix can be performed in $O((\log N)^3)$ time on an LARPBS with $N^{\alpha+1}/(\log N)^\alpha$ processors. Our implementation for LU-factorization is scalable for $p = O(N^{\alpha+1}/(\log N)^{\alpha/(\alpha-2)})$.

(The part for QR-factorization is omitted due to space limitation.)

References

1. D. Chiarulli, R. Melhem, and S. Levitan, "Using coincident optical pulses for parallel memory addressing," *IEEE Computer*, vol. 30, pp. 48-57, 1987.
2. D. Coppersmith and S. Winograd, "Matrix multiplication via arithmetic progressions," *Journal of Symbolic Computation*, vol. 9, pp. 251-280, 1990.
3. L. Csanky, "Fast parallel matrix inversion algorithms," *SIAM Journal on Computing*, vol. 5, pp. 618-623, 1976.
4. O.H. Ibarra, S. Moran, and L.E. Rosier, "A note on the parallel complexity of computing the rank of order n matrices," *Information Processing Letters*, vol. 11, no. 4,5, p. 162, 1980.
5. V. Kumar, *et al.*, *Introduction to Parallel Computing*, Benjaming/Cummings, 1994.
6. K. Li, "Fast and scalable parallel algorithms for matrix chain product and matrix powers on optical buses," in *High Performance Computing Systems and Applications*, Kluwer Academic Publishers, Boston, Massachusetts, 1999.
7. K. Li and V.Y. Pan, "Parallel matrix multiplication on a linear array with a reconfigurable pipelined bus system," *Proc. IPPS/SPDP '99*, pp. 31-35, April 1999.
8. K. Li, Y. Pan, and S.-Q. Zheng, "Fast and processor efficient parallel matrix multiplication algorithms on a linear array with a reconfigurable pipelined bus system," *IEEE Trans. on Parallel and Distributed Systems*, vol. 9, no. 8, pp. 705-720, 1998.
9. K. Li, Y. Pan, S.-Q. Zheng, "Parallel matrix computations using a reconfigurable pipelined optical bus," *Journal of Parallel and Distributed Computing*, vol. 59, no. 1, pp. 13-30, October 1999.
10. K. Li, Y. Pan, and S.-Q. Zheng, "Scalable parallel matrix multiplication using reconfigurable pipelined optical bus systems," *Proc. of 10th Int'l Conf. on Parallel and Distributed Computing and Systems*, pp. 238-243, October 1998.
11. V. Pan, "Complexity of parallel matrix computations," *Theoretical Computer Science*, vol. 54, pp. 65-85, 1987.
12. Y. Pan and K. Li, "Linear array with a reconfigurable pipelined bus system – concepts and applications," *Information Sciences*, vol. 106, no. 3-4, pp. 237-258, 1998.

Pulse-Modulated Vision Chips with Versatile-Interconnected Pixels

Jun Ohta, Akihiro Uehara, Takashi Tokuda, and Masahiro Nunoshita

Graduate School of Materials Science,
Nara Institute of Science and Technology
8916-5 Takayama, Ikoma, Nara 630-0101 JAPAN
phone: +81-743-72-6051, fax: +81-743-72-6059
e-mail: ohta@ms.aist-nara.ac.jp

Abstract. This paper proposes and demonstrates novel types of vision chips that utilize pulse trains for image processing. Two types of chips were designed using 1.2 μm double-metal double-poly CMOS process; one is based on a pulse width modulation (PWM) and the other is based on a pulse frequency modulation (PFM). In both chips the interaction between the pixels were introduced to realize the image pre-processing. The basic experimental and simulation results are shown for the PWM and PFM chips, respectively. Also the comparison between two types is discussed.

1 Introduction

A vision chip is a kind of image sensors in which some functional circuits are integrated with a photodetector in each pixel. Several types of vision chips have been reported so far [1]. Because of employing pre-processing in the pixel stage, a vision chip has advantages over a charge-coupled device (CCD) in a viewpoint of fast and versatile image pre-processing.

Viewed in a point of signal processing, they are classified into an analog type and a digital type. The former is more sensitive to noise and has less precision, while the latter has more space-consumed. Pulse modulation is another type [2], where an output signal from a photodiode is converted into a pulse train such as pulse amplitude modulation (PAM), pulse width modulation (PWM), pulse frequency modulation (PFM), pulse phase modulation (PPM), and so on. Except in PAM, all the others are used digital values for output representation as shown in Fig. 1. Thus, in nature the pulse modulation is robust in the signal transmission and well compatible with digital logic circuits. These features are very effective in vision chips where an image pre-processing function is integrated in each pixel.

J. Rolim et al. (Eds.): IPDPS 2000 Workshops, LNCS 1800, pp. 1063-1071, 2000.
© Springer-Verlag Berlin Heidelberg 2000

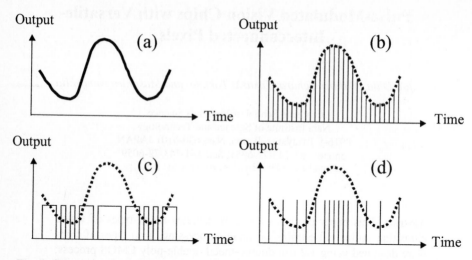

Fig. 1. Schematic of pulse modulation scheme. (a) original analog signal, (b) pulse amplitude modulation, (c) pulse width modulation, (d) pulse frequency modulation.

It is thus crucial issue for vision chips in the next generation to realize the interaction between pixels using pulse modulation scheme. To explore this issue, we propose and demonstrate two types of vision chips employing PWM and PFM. Focusing on implementing inter-pixel connections using pulse trains, we discuss the comparison between two types.

2   Vision Chip Based on PWM

Figure 2 shows circuit diagram of the fabricated PWM vision chip. The circuits consist of a photodiode (PD), a reset transistor M1, a comparator (COMP), and resistive interconnections. The voltage V_{PD} of the floated photodiode PD, which is charged to near V_{dd}, through M1, decreases because the input light discharges the stored charge in the photodiode capacitor C_{PD}. At $t = t_o$ when $V_{PD} < V_{ref}$, the comparator COMP turns on. Here V_{ref} is the reference voltage in COMP. If the output keeps high state or "HI" during the interval, the pulse width T corresponds to the input light intensity. If the change of C_{PD} by the bias voltage is neglected, T is given as follows,

$$T = C_{PD}(V_{reset} - V_{ref})/I_{ph}. \tag{1}$$

Here I_{ph} is the photocurrent through PD. While in most vision chips as well as conventional image sensors the photodiode voltage V_{PD} is sensed and outputted in a given time, in the PWM the time for reaching a given voltage is sensed and outputted [3], [4].

An inverter with canceling the threshold is used as the comparator COMP as shown in Fig. 3. The reference voltage V_{ref} corresponds to the threshold voltage of the inverter, which is equal to about a half of the supply voltage to the inverter V_{dd}.

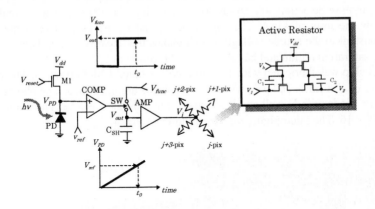

Fig. 2. Circuit diagram of one pixel in a PWM chip.

We employed arbitrary input-output transformation by controlling the switch SW with the output voltage of COMP. The transformation results from the time-dependent voltage of $V_{func}=V_{func}(t)$. When COMP is "HI", SW is ON and C_{SH} is charged through V_{func}. At $t = t_0$ when $V_{PD} < V_{ref}$, the output of COMP is low state or "LO", then SW is OFF and V_{out} is equals to the value of V_{func} at $t = t_0$. In the case of the inset in Fig. 2, we choose the transformation function as binarization by choosing the step function as V_{func}. We can realize any transformation in the relationship between the output voltage and input light intensity by choosing the time-dependence of $V_{func}(t)$.

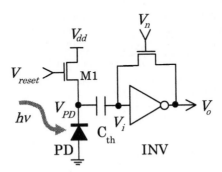

Fig. 3. Circuit diagram of the inverter with canceling the threshold. This is used as the comparator.

As described in the introduction, processing between the nearest neighboring pixels is essential to vision chips. It is difficult to use the output of PWM as it is for the nearest neighboring processing. Thus each pulse width is converted into an analog value by using the sample-and-hold circuits and then transferred into resistive network through the buffered amp AMP. The capacitance C_{SH} in the sample-and-hold is 220 fF. Resistive network is effective in image processing such as blurring, edge extraction, motion detection, and so on [5]. In our chip, the active resistor is realized as shown in the inset of Fig.4. The capacitors C_1 and C_2 are used to keep the gate voltage constant to the source voltage, which makes the linearity better. This resistor acts in the saturation region and has around 10 kΩ from 0 to 1 V.

The test chip was fabricated with 1.2 μm 2-poly 2-metal CMOS process. The pixel has the area of 158 μm × 163 μm, where 49 transistors are integrated. The photodiode is a parasitic diode with N-well and P-substrate. The fill-factor is 5 %. The number of pixels is 16 × 16. The microphotographs are shown in Fig. 4.

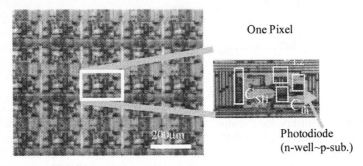

Fig. 4. The microphotographs of the PWM chip and one pixel.

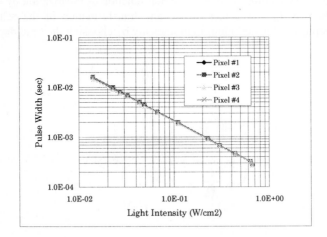

Fig. 5. The pulse width as a function of the input light intensity in different pixels.

The pulse width was measured as a function of the light intensity in the fabricated chip. The pulse width decreases in the inverse proportion to the light intensity, which agrees with Eq. (1). A red LED with the central wavelength of 630 nm was used as the light source. V_{reset} and V_{ref} are set 3.16V and 1.8 V, respectively. Figure 5 shows the experimental results within 4 pixels, which demonstrates the relative variation of the pulse width was less than 5 %.

Figure 6 shows the similar experimental results as in Fig. 5. In this case the parameter is the power supply voltage to the comparator. As mentioned above, the inverter was used as the comparator, thus the reference voltage was about a half of the supply voltage. This results shows the variation of the pulse width changes only less than 10 % when V_{dd} changes from 2 to 5 V, which shows the effectiveness of this circuit. We are now testing the transformation circuits and the neighboring processing circuits.

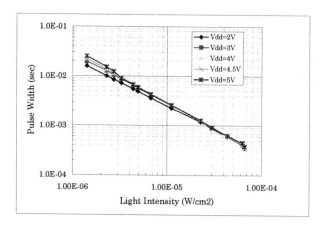

Fig. 6. The pulse width as a function of the input light intensity in different power supply voltages.

3 Vision Chip Based on PFM

A PFM is an output representation that an analog output is converted into pulse frequency, and is used in the output from the nerve cells as spikes. To employ a PFM in a vision chip, relaxation oscillation circuits [6, 7] are constructed as shown in Fig.7. The photodiode PD acts as a variable current source controlled by the input light intensity and is charged through the transistor M1. The gate of M1 is switched by the feedback from the Schmidt trigger ST and the inverter INV. In such a configuration the stronger the light intensity is, the higher the pulse frequency is. The analog value of the light intensity is consequently naturally converted into a pulse train as digital signal.

The circuits in Fig. 7 are designed using the same 1.2 μm CMOS process as in the PWM chip. The layout of one pixel is shown in Fig. 8. The size of one pixel is 50 μm × 50 μm. The number of the pixel are 32 × 32. The photodiode consists of N-diffusion and P-substrate, which makes the area reduced compared to the PWM chip. The capacitance of the photodiode is 19 fF without the bias.

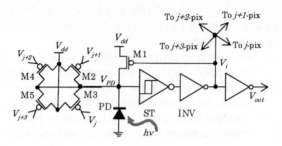

Fig. 7. The circuits diagram of the PFM chip.

Figure 9 shows HSPICE simulation results of output pulse trains from the pixel as shown in Fig. 7. The parasitic capacitances are extracted from the layout shown in Fig. 8. The frequency of the pulse trains is modulated according to the photocurrent, which changes in a ramped manner. The feedback loop to M1 consists of the Schmidt trigger ST that binarizes the output from PD, and the inverter INV for the delay. In this case, the oscillation frequency f is given as

$$f = \frac{I_{ph}}{C_{PD}(V_o - V_{th})}, \tag{2}$$

where V_o is the charge voltage of PD, V_{th} is the threshold voltage of ST. As can be seen from Eq. (2), the frequency f or the firing rate increases if the input light intensity becomes large and the size of PD becomes small.

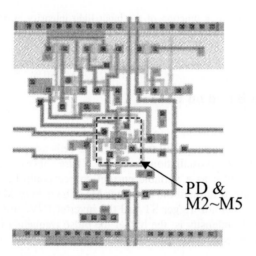

Fig. 8. The layout of one pixel in a PFM chip.

An inhibitory interconnection, which is one of the important functions between neighboring pixels and is much effective in image preprocessing, is realized in a PFM scheme with very simple circuits as shown in Fig. 7. The additional reset transistors M2 ~ M5 are connected to the PD as well as M1. M2 is switched by the output pulse train from the neighboring pixels and causes the PD charged according to the firing frequency of the neighboring pixels. Thus, the pulse frequency of a pixel is inhibited through the firing frequency of the neighboring pixels. An excitatory connection could be easily realized if the gate signals of M2~M5 are inverted.

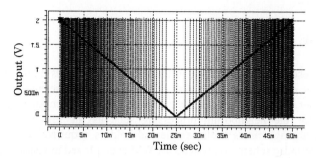

Fig. 9. HSPICE simulation of the pulse stream in a PFM chip. The solid line is the photocurrent, and the dashed line is the output pulses.

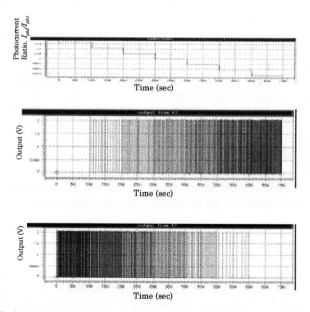

Fig. 10. HSPICE simulation of the mutual inhibitory interconnection in a PFM chip. Upper: photocurrent, middle: output from #1 pixel, bottom: output from #2.

The results of HSPICE simulation for two inhibitory connected pixels are shown in Fig. 10. The power supply voltage is 2 V. In the figure I_{ph1}, the photocurrent in pixel

#1 is kept constant at 1.8 nA while I_{ph2}, one in pixel #2 varying from 2.1 nA to 1.4 nA in a step-like manner. The output pulse trains from pixel #1 and pixel #2 are shown in the middle and the bottom of Fig. 10, respectively. The maximum firing frequency is around 10 kHz. The firing rate in pixel #1 is completely suppressed when $I_{ph2}/I_{ph1}=1.2$, and gradually increases as I_{ph2}/I_{ph1} decreases. This figure clearly demonstrates that the both firing rates are suppressed each other.

4 Discussion

The PWM represents the information in the rise and fall points of the pulse, which leads to make it easy to use PWM with other logic circuits because of the similar operation manner. Although this is a great advantage over conventional image sensing methods [4], for the neighboring processing such as the resistive network, the conversion to analog values with sample-and-hold is required. This implies the advantage of the pulse scheme is reduced if PWM is employed in vision chips.

As for PFM, we have proposed and demonstrated effective architecture of a vision chip employing PFM. In the architecture, the pulse trains directly affect the state of the neighboring pixels without converting into the analog value. This holds the advantages in the digital manner such as robustness against noise.

Another important feature of the proposed architecture is that decreasing the area of the photodiode little affects the dynamic range. This is confirmed from Eq. (2). It shows that the pulse frequency, which determines the dynamic range, increases if the photodiode capacitance decreases. Consequently it is possible to integrate more functional circuits in each pixel with smaller fill factor. In our chip, the photodiode is designed with the minimum area.

Finally, we discuss the next step of the PFM chip. We have a plan to fabricate the next chip using a deep sub-micron process such as a feature size of 0.25 or 0.18 μm. In such a process, two inevitable issues arise [8]; one is the decrease of the power supply to reduce an electric field applying on a thin oxide layer. The other is that the junction depth for the photodiode becomes shallower. These are detrimental to the dynamic range and the sensitivity for conventional image sensing method. The PFM is one of the candidates for vision chips applicable to a deep sub-micron technology. In that case, parasitic capacitance would be considered more carefully, because it could cause the deviation of the oscillation frequency between pixels.

5 Conclusion

In conclusion, vision chips employing two types of pulse modulation are designed with 1.2 μm CMOS process. In the PWM chip, the interaction between the neighboring pixels is incorporated using the resistive network with the sample-and-hold circuits. The fabricated chip is tested and found to work well in fundamental characteristics. In the PFM chip, mutual inhibitory interconnections are designed and demonstrated in the simulation. The architecture is very simple and suitable to versatile image processing. All digital circuits except for PD makes it possible to use a conventional logic VLSI fabrication process in the proposed pulsed vision chip. Moreover, the deep sub-micron technology would be applicable to the PFM vision chip.

Acknowledgments

This work was partially supported by the Ministry of Education, Science, Sports, and Culture of Japan under Grant-in-Aid for Scientific Research on Priority Area #11167256. The VLSI chip in this study has been fabricated in the chip fabrication program of VLSI Design and Education Center (VDEC), the University of Tokyo with the collaboration by Nippon Motorola LTD., Dai Nippon Printing Corporation, and KYOCERA Corporation.

References

1. For example, Koch, C., Li, H.: VISION CHIPS, Implementing Vision Algorithms with Analog VLSI Circuits. IEEE Computer Society Press, CA (1995).
2. For example, Mass, W., Bishop, C.M. (eds.): Pulsed Neural Networks. The MIT Press, Mass. (1998).
3. Ni, Y., Devos, F., Boujrad, M., Guan, H.,: Histogram-Equalization-Based Adaptive Image Sensor for Real-Time Vision. IEEE J. Solid-State Circuits, 32 (1997) 1027-1036.
4. Nagata, M., Homma, M., Takeda, N., Norie, T., Iwata, A.: A Smart CMOS Imager with Pixel Level PWM Signal Processing. IEEE 1999 Symposium on VLSI Circuits, (1999) 141-144.
5. Mead, C.: Analog VLSI and Neural Systems. Addison-Wesley Pub. Inc., Mass. (1989).
6. Yang, W.: A Wide-Dynamic-Range, Low-Power Photosensor Array. 1994 IEEE Int'l Solid-State Circuits Conference (1994) 230-231.
7. Andoh, F. Nakayama, M., Shimamoto, H., Fujita, Y.: A Digital Pixel Image Sensor with 1-bit ADC and 8-bit Pulse Counter in each Pixel. 1999 IEEE Workshop on Charge-Coupled Devices and Advanced Image Sensors, (1999) 44-47.
8. Wong, H.-S.: IEEE Trans. Electron Devices: Technology and Device Scaling Considerations for CMOS Imagers. 43 (1996) 2131-2142.

Connectivity Models for Optoelectronic Computing Systems

Haldun M. Ozaktas

Bilkent University
Department of Electrical Engineering
TR-06533 Bilkent, Ankara, Turkey

Abstract. Rent's rule and related concepts of connectivity such as dimensionality, line-length distributions, and separators have found great
use in fundamental studies of different interconnection media, including
superconductors and optics, as well as the study of optoelectronic computing systems. In this paper generalizations for systems for which the
Rent exponent is not constant throughout the interconnection hierarchy are provided. The origin of Rent's rule is stressed as resulting from
the embedding of a high-dimensional information flow graph to two- or
three-dimensional physical space. The applicability of these traditionally
solid-wire-based concepts to free-space optically interconnected systems
is discussed.

1   Connectivity, Dimensionality, and Rent's Rule

The importance of wiring models has long been recognized and they have been
used not only for design purposes but also for the fundamental study of interconnections and communication in computing. A central and ubiquitous concept
appearing in such contexts is the connectivity of a circuit graph or computer network. Connectedness has always been a central concept in mathematical graph
theory [1–3], whose extensions play a central role in graph layout [4]. The purpose
of these concepts is to quantify the communication requirements in computer circuits. This paper aims to discuss several concepts related to the connectivity of
circuits (such as Rent's rule, dimensionality, line-length distributions, and separators), provide certain extensions, and briefly discuss some of their applications.
We first discuss generalizations for systems for which the Rent exponent is not
constant throughout the interconnection hierarchy, and present a number of related results. Special emphasis is given to the role of discontinuities and the
origin of Rent's rule. The applicability of these concepts to free-space optically
interconnected systems and the role of Rent's rule in fundamental studies of different interconnection media, including superconductors and optics, are briefly
reviewed.

Graph layout deals with the problem of how to situate the nodes and edges
of an abstract graph in physical space. Optimal graph layout [5] is in general an
NP-complete problem [6]. However, if a hierarchical decomposition of a graph
is provided, this graph can be laid out following relatively simple algorithms. A

J. Rolim et al. (Eds.): IPDPS 2000 Workshops, LNCS 1800, pp. 1072-1088, 2000.

hierarchical decomposition of a graph consisting of N nodes and the associated decomposition function $k(\bar{N})$ are obtained as follows: First, we remove $k(N)$ edges in order to disconnect the graph into 2 subgraphs, each of approximately $N/2$ nodes. Roughly speaking, we try to do this by removing as few edges as possible. We repeat this procedure for the subgraphs thus created. The subgraphs will in general require differing numbers of edges to be removed from them to be disconnected into subsubgraphs of $\simeq N/2^2$ nodes each. We denote the largest of these numbers as $k(N/2)$. Continuing in this manner until the subgraphs consist of a single node each, we obtain the function $k(\bar{N})$, the (worst case) number of edges removed during decomposition of subgraphs of $\bar{N}$ nodes. Once such a decomposition is found, it is possible to lay out the graph in the intuitively obvious manner by working upwards [7, 4, 6]. Whereas one can always find some such decomposition, finding the decomposition that leads to a layout with some optimal property (such as minimal area) is not a trivial problem. We will assume that we agree on a particular decomposition obtained by some heuristic method.

Now, let us define the *connectivity* $p(\bar{N})$ and *dimensionality* $n(\bar{N})$ associated with the hierarchical level of a decomposition involving subgraphs of $\bar{N}$ nodes by [8]

$$p(\bar{N}) = \frac{n(\bar{N}) - 1}{n(\bar{N})} = \log_2 \frac{k(\bar{N})}{k(\bar{N}/2)}. \tag{1}$$

In general the defined quantities satisfy $0 \le p(\bar{N}) \le 1$ and $1 \le n(\bar{N}) \le \infty$. It is possible to find many examples of graphs for which the values of $k(\bar{N})$ and $p(\bar{N})$ for different values of $\bar{N}$ are totally erratic and have no correlation whatsoever. However, both computer circuits and natural systems are observed to exhibit varying degrees of continuity of the functions $k(\bar{N})$ and $p(\bar{N})$.

Let the geometric derivative $\tilde{f}$ of a function f at the point x be defined analogous to the usual arithmetic derivative:

$$\tilde{f}(x) = \lim_{\nu \to 1^+} \log_\nu \frac{f(\nu x)}{f(x)}. \tag{2}$$

If $k(\bar{N})$ is slowly varying, we may pretend that it is a continuous function and write

$$p(\bar{N}) = \frac{n(\bar{N}) - 1}{n(\bar{N})} = \tilde{k}(\bar{N}), \tag{3}$$

which may be inverted as

$$k(\bar{N}) = k(1) \exp\left(\int_1^{\bar{N}} \frac{p(\bar{N}')}{\bar{N}'} \, d\bar{N}' \right), \tag{4}$$

where $\bar{N}'$ is a dummy variable.

Of course, since $k(\bar{N})$ is actually a function of a discrete variable, we cannot actually let $\nu \to 1$. The smallest meaningful value of ν in our context is 2. Hence the geometric derivative should be interpreted in the same sense as we interpret the common derivative in the form of a finite difference for discrete functions.

For our definition to make sense, $k(\bar{N})$ must be a slowly varying function. As already noted, this is indeed observed over large variations of $\bar{N}$ in both computer circuits and natural systems. In fact, in many cases it is found that $p(\bar{N})$ and $n(\bar{N})$ are approximately constant over a large range of $\bar{N}$. Such systems are said to exhibit self similarity [9, 10], or scale invariance.

Assuming that $p(\bar{N}) = p = $ constant, we find from equation 4 that $k(\bar{N}) = k(1)\bar{N}^p$. Apart from a constant coefficient, this is nothing but Rent's rule [11, 12] which gives the number of graph edges $k(\bar{N})$ emanating from partitions of computer circuits containing $\bar{N}$ nodes (such as the number of pinouts of an integrated circuit package containing $\bar{N}$ gates). $k(1)$ is interpreted as the average number of edges per node and p is referred to as the Rent exponent.

Conversely, by taking Rent's rule as a starting point, however allowing the exponent to be a function $p(\bar{N})$ of $\bar{N}$, we can derive equation 4 by working up the hierarchy. (Readers wishing to skip this reverse derivation given below may move directly to the paragraph following equation 9.) We consider a system with a total of N primitive elements and express N in the form

$$N = N_m = \prod_{i=1}^{m} \nu_i, \tag{5}$$

i.e., we have ν_m groups of ν_{m-1} groups of ... of ν_1 primitive elements. We are assuming all subgroups of any group to be identical (it is of course possible to go one step further and remove this restriction as well). The ν_i are sufficiently small so that the connectivity requirements within each level of the hierarchy can be assumed constant. The subtotals at any level are similarly expressed as

$$N_j = \prod_{i=1}^{j} \nu_i, \qquad 1 \le j \le m, \tag{6}$$

with $N_0 = 1$. If we let i and j approach continuous variables we can write this as

$$N(j) = \exp\left(\int_1^j \ln \nu(i) \, di\right). \tag{7}$$

Let it be the case that the ν_j subgroups forming one of ν_{j+1} groups have connectivity requirements characterized by a Rent exponent of p_j. The number of edges k_j emanating from each of the ν_{j+1} groups (each containing ν_j subgroups) is

$$k_j = k_{j-1}\nu_j^{p_j} = k_0 \prod_{i=1}^{j} \nu_i^{p_i}, \tag{8}$$

where k_0 is the number of edges of the primitive elements. In continuous form we have

$$k(j) = k(0) \exp\left(\int_1^j p(i) \ln \nu(i) \, di\right). \tag{9}$$

Now, it is possible to combine this with equation 7 to eliminate $\nu(i)$ to obtain equation 4, completing the derivation.

The dimensionality and Rent exponent will depend not only on the graph, but also on the layout of the graph. However, since there must be some layout of the graph which results in the smallest exponent, this smallest exponent may be considered intrinsic to the graph and representative of the intrinsic information flow requirements of the computational problem. The minimum information flow requirements can be quantified for several relatively structured and simple problems, such as sorting, fast Fourier transforms, etc. [4]. However, it should not be forgotten that there may be no efficient method of finding the layout resulting in the smallest exponent, so that it may not be possible to determine this intrinsic minimum Rent exponent.

We now briefly discuss the concept of *separators*. A graph of N nodes is said to have an $S(\bar{N})$ separator (or to be $S(\bar{N})$ separable) if the graph can be disconnected into two (roughly equal) subgraphs by removal of $S(N)$ edges and if the subgraphs thus created are also $S(\bar{N})$ separable [4]. Although we do not go into the details, we note that a graph with connectivity function $p(\bar{N})$ has a separator of the form $S(\bar{N}) \propto \bar{N}^{\max[p(\bar{N})]}$ where the maximum is taken over the whole domain of $\bar{N}$. Separators play a central role in combinatoric approaches to graph layout [7, 4], sometimes referred to as area-volume complexity theory. (An alternative way of describing the communication requirements of graphs is based on what are called *bifurcators* [6].)

Thus, we see that both Rent's rule and separators of the form $S(\bar{N}) \propto \bar{N}^p$ are special cases of the more general formalism we have introduced. Apart from minor technicalities involved in their definition, all are essentially equivalent when $p(\bar{N}) = $ constant. In general, $p(\bar{N})$ and $n(\bar{N})$ will be functions of $\bar{N}$.

The dimensionality $n(\bar{N})$ defined in equation 3 is a fractal dimension [10, 13–15]. Fractal dimensions of natural systems, just as those of computer circuits, may also vary as we ascend or descend the hierarchical structure of a system. In any event, Rent's rule, fractal geometry and separators are tied together by the notions of self similarity, scale invariance, and continuity in the relationships between the volume-like (number of nodes) and surface-like (number of edges) quantities.

To clarify this point, we offer the following explanation of why the quantity defined as $n = 1/(1 - p)$ is referred to as a "dimension." The perimeter of a square region is proportional to the 1/2 power of its area. The surface area of a cube is proportional to the 2/3 power of its volume. In general, the hyperarea enclosing a hyperregion of e dimensions is proportional to the $(e-1)/e$ power of its hypervolume. Let us now make an analogy between "hyperarea" $\leftrightarrow$ "number of graph edges emanating from a region," and "hypervolume" $\leftrightarrow$ "number of nodes in the region." According to Rent's rule, the number of edges emanating from the region is proportional to the pth power of the number of nodes in the region. Thus, it makes sense to speak of the quantity n defined by the relation $p = (n-1)/n$ as a "dimension." Note that in general n need not be an integer.

Now, let us assume that a graph with $n(\bar{N}) = n = $ constant is laid out in e-dimensional Euclidean space according to the divide-and-conquer layout algorithm (i.e., as intuitively suggested by its decomposition) [7, 4]. (e is often

= 2 but always $\leq$ 3.) Such a layout will internally satisfy Rent's rule. Donath [16] and Feuer [17] had shown that such a layout has a distribution $g(r)$ of line lengths of the form

$$g(r) \propto r^{-\frac{e}{n}-1}, \qquad r \leq \sqrt{e}\,N^{\frac{1}{e}}, \tag{10}$$

where r denotes line lengths in units of node-to-node grid spacing of the layout. (We assume the nodes are situated on a regular Cartesian grid.) The relationship between such inverse-power-law distributions and fractal concepts was discussed by Mandelbrot [18–20], closing the circle. Using the above distribution, or by combinatoric methods, we can show that when $n > e$ the average connection length $\bar{r}$ of such a layout of N elements is given by [9]

$$\bar{r} = \kappa(n, e)N^{\frac{1}{e}-\frac{1}{n}}, \tag{11}$$

where $\kappa(n, e)$ is a coefficient of the order of unity. The accuracy of this expression requires that $N^{1/e-1/n} \gg 1$. This result has a simple interpretation. The average connection length is simply the ratio of the linear extent $N^{1/e}$ of the system in e-dimensional space to the linear extent $N^{1/n}$ in n-dimensional space. The node-to-node grid spacing necessary to lay out a graph of dimensionality n is given by $\propto \bar{r}^{1/(e-1)}\lambda \approx N^{(n-e)/ne(e-1)}\lambda$, where λ is the line-to-line spacing of whatever interconnection technology is being used [21–23]. Thus, when $n > e$, the area (or volume) per node grows with N. This has been referred to as *space dilation* [24]. Examples of graphs with well-defined structures which exhibit large values of n are hypercubes, butterflies, and shuffle-exchange graphs. It is also easily verified that the given definition of dimension is consistent with that for multidimensional meshes [25].

Before closing this section, we discuss two further results regarding the calculation of the average and total connection lengths of a layout. Readers wishing to skip these may directly go to the next section.

First, we discuss the invariance of the total connection length under different grain-size viewpoints. One might view a computing system as a collection of its most primitive elements, for instance gates or transistors. Alternately, one might prefer to view it as a collection of higher-order elements, such as chips or processors which are simply taken as black boxes with a certain number of pinouts. Both viewpoints are perfectly valid, however one must interpret the average connection length with care so as to maintain consistency. The average connection length will be higher when calculated with reference to the higher-order picture, as compared to the lower-order picture. This is because the shorter interconnections inside the black boxes are not being taken into account while computing the average. Let us quantify this situation by considering N/N_1 black boxes (blocks) with N_1 primitive elements in each, representing a simple partitioning of the total N elements. Let the grid spacing of the black boxes be d_1 and that of the primitive elements be d_0. In an e-dimensional space we have $d_1 = N_1^{1/e}d_0$. Let us consider $n = \text{constant} > e$. We let $\bar{\ell}$ denote the average interconnection length in real units, as opposed to $\bar{r}$ which is the average connection length in grid units. ℓ_{total} will denote the total interconnection length. Furthermore, assume $N_1^{1/e-1/n} \gg 1$ and $(N/N_1)^{1/e-1/n} \gg 1$. The fine-grain picture yields

$$\bar{\ell} = \bar{r}d_0 = \kappa N^{\frac{1}{e}-\frac{1}{n}} d_0, \qquad (12)$$

whereas the large-grain picture yields

$$\bar{\ell} = \kappa(N/N_1)^{\frac{1}{e}-\frac{1}{n}} d_1 = \kappa N^{\frac{1}{e}-\frac{1}{n}} N_1^{\frac{1}{n}} d_0, \qquad (13)$$

from which we see that indeed the large-grain picture yields a larger value of $\bar{\ell}$ (unless $n \to \infty$). However, in calculating the total system size, it is not $\bar{\ell}$ but rather ℓ_{total} that is the significant quantity. For the fine grain picture we have

$$\ell_{\text{total}} = k_0 N \bar{\ell} = k_0 \kappa N^{p+\frac{1}{e}} d_0, \qquad (14)$$

since $p = 1 - 1/n$. For the large-grain picture we use the number of pinouts as given by $k_1 = k_0 N_1^p = k_0 N_1^{1-1/n}$, so that

$$\ell_{\text{total}} = k_1 (N/N_1)\bar{\ell} = k_0 \kappa N^{p+\frac{1}{e}} d_0, \qquad (15)$$

identical to what we found with the fine grain picture. So whatever way we choose to look at it we always will end up calculating the total system size consistently.

This result means that we can ignore the contributions of the local interconnections in calculating the total area (or volume) required for wiring. The longer interconnections, although much fewer in number, constitute most of the wiring volume. The total interblock wiring length is, once again

$$k_0 N_1^{1-\frac{1}{n}} (N/N_1)\kappa(N/N_1)^{\frac{1}{e}-\frac{1}{n}} d_1, \qquad (16)$$

whereas the total local wiring length is

$$(N/N_1)k_0 N_1 \kappa N_1^{\frac{1}{e}-\frac{1}{n}} d_0. \qquad (17)$$

The ratio of the former to the latter is

$$\frac{\text{Total interblock wire length}}{\text{Total local wire length}} = (N/N_1)^{\frac{1}{e}-\frac{1}{n}}, \qquad (18)$$

which we had assumed to be $\gg 1$ from the beginning. When n is bounded away from e and when N_1 and N/N_1 are large, it is the higher level of the interconnection hierarchy that limits how dense the elements can be packed. This conclusion can also be traced to the fact that the integrand $rg(r)$ in the first moment integral of $g(r)$ decays slower than $1/r$ when $n > e$.

Once ℓ_{total} is obtained, calculation of the system linear extent $N^{1/e}d_0$ is easy. We simply equate the total available area (volume) to the total wire area (volume) [26]

$$N d_0^e = \ell_{\text{total}} \lambda^{e-1}, \qquad (19)$$

where λ is the line spacing. Thus, within a wiring inefficiency factor we obtain

$$N^{\frac{1}{e}} d_0 = (k_0 \kappa N^p)^{\frac{1}{(e-1)}} \lambda. \qquad (20)$$

Next, we derive an expression for the layout area for a system with arbitrary, possibly discontinuous $k(\bar{N})$ in two dimensions. So as to simplify the representation of the results, we will restrict ourselves to $p(\bar{N}) > 1/2$. First, consider a group of ν_1 primitive elements. This group can be laid out with linear extent $d_1 = k_0\kappa_1\nu_1^{p_1}\lambda = k_1\kappa_1\lambda$ where κ_1 is the coefficient corresponding to p_1. Thus, the total system linear extent must be at least $(N/\nu_1)^{1/2} = (N/N_1)^{1/2}$ times the extent of this group, where in general N_j is given by equation 6. Now consider a supergroup of ν_2 such groups. Taking $\lambda = 1$, the linear extent of this supergroup satisfies

$$\max(k_2\kappa_2, \nu_2^{\frac{1}{2}}d_1) \leq d_2 \leq k_2\kappa_2 + \nu_2^{\frac{1}{2}}d_1. \tag{21}$$

In general,

$$\max(k_j\kappa_j, \nu_j^{\frac{1}{2}}d_{j-1}) \leq d_j \leq k_j\kappa_j + \nu_j^{\frac{1}{2}}d_{j-1}. \tag{22}$$

$k_j\kappa_j$ is the wiring requirement obtained at the jth level. $\nu_j^{1/2}d_{j-1}$ is the requirement inherited from lower levels. The maximum and summation represent best case and worst case assumptions on how these requirements interact. Taking $d_0 = 1$, and expanding the recursion on both sides leads to the following bounds for the linear extent d_m of the complete system:

$$\max\left[k_i\kappa_i\left(\frac{N}{N_i}\right)^{\frac{1}{2}}\right]_{i=1}^m \leq d_m \leq \sum_{i=1}^m k_i\kappa_i\left(\frac{N}{N_i}\right)^{\frac{1}{2}}. \tag{23}$$

The right hand side of the above can be at most m times greater than the left hand side. Since $m \leq \log_2 N$, the system linear extent is given by the left hand side within this logarithmic factor. In continuous form, the linear extent is given by

$$N^{\frac{1}{2}}\max\left[\frac{k(\bar{N})\kappa(\bar{N})}{\bar{N}^{\frac{1}{2}}}\right]_{\bar{N}}, \tag{24}$$

where $\kappa(\bar{N}) \sim 1$ is only weakly dependent on $\bar{N}$.

If $k(\bar{N})$ is slowly varying, it may be expressed by equation 4 over the whole range of $\bar{N}$. Now if $p(\bar{N}) > 1/2$ as we have assumed, it is possible to show that $k(\bar{N})$ grows faster than $\bar{N}^{1/2}$ so that the expression in the square brackets above is maximized for $\bar{N} = N$. Thus, the system linear extent is given by

$$k(N)\kappa(N) \sim k(N). \tag{25}$$

That is, the system size is set by the highest level of interconnections if $k(\bar{N})$ is a slowly varying function and $p(\bar{N}) > 1/2$. This implies that the choice of interconnection technology for the highest level is the most critical.

2   Discontinuities and the Origin of Rent's Rule

Whereas it is observed that the function $k(N)$ exhibits considerable continuity over large variation of N, it is also observed that it occasionally exhibits sharp discontinuities. In other words, it no longer becomes possible to predict the value of the function $k(N)$ for certain N by knowing its values at nearby N. For instance, in the context of Rent's rule, it may not be possible to predict the number of pinouts of a VLSI chip by observing its internal structure, or vice versa [13]. However, this does not imply that Rent's rule (in its generalized form, as given by equation 4) is useless. Consider a multiprocessor computer. Rent's rule may be used to predict the wiring requirements internal to each of the processors. It may also be used for similar purposes for the interconnection network among the processors. In fact, the Rent exponent may even be similar in both cases. However, the function $k(N)$ may exhibit a steep discontinuity (often downward), as illustrated in figure 1 [8]. As is usually the case, a finite number of discontinuities in an otherwise smooth function need not inhibit us from piecewise application of our analytical expressions. Such discontinuities are often associated with the *self-completeness* of a functional unit [12, 13]. Similar examples may be found in nature. For instance, mammalian brains seem to satisfy $n > 3$ (i.e. $p > 2/3$), since the volume per neuron has been found to be greater in species with larger numbers of neurons [27]. The human brain has 10^{11} neurons each making about 1000 connections [28]. Thus, we would expect at least $1000(10^{11})^{2/3} \sim 10^{10}$ "pinouts." However, we have only about 10^6 fibers in the optic nerve and 10^8 fibers in the *corpus callosum*.

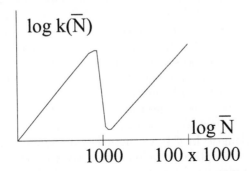

Fig. 1. $k(\bar{N})$ for a system of $N = 100 \times 1000$ primitive elements consisting of 100 processors of 1000 elements each. The number of "pinouts" of the processors bears no relationship to their internal structure. Equation 4 may be used directly for the range $1 < \bar{N} < 1000$, and with a shift of origin for the range $1000 < \bar{N} < 100 \times 1000$.

In the context of microelectronic packaging, a quote from C. A. Neugebauer offers some insight as to why such discontinuities are observed: "Since the I/O capacity (of the chip carrier) is exceeded, a significant number of chips can be interconnected only if the pin/gate ratio can be drastically reduced, normally well below that predicted by Rent's rule. Rent's rule can be broken at any level of integration. The microprocessor chip is an example of the breaking of Rent's rule in its original form for gate arrays on the chip level. Being able to delay the breaking of Rent's rule until a much higher level is always an advantage because it preserves many parallel data paths even at very high levels of integration, and thus offers higher systems performance and greater architectural flexibility." [29] The breaking of Rent's rule seems to be a technological necessity, and undesirable from a systems viewpoint. We will later discuss studies which indicate that superconducting or optical interconnections may allow the maintainment of a large dimensionality and Rent exponent throughout higher levels of the hierarchy.

The origin of Rent's rule has intrigued many researchers. Donath had shown that Rent's rule is a consequence of the hierarchical nature of the logic design process [30,31]. Some have viewed it merely as an empirical observation obtained from an examination of existing circuits. Others have suggested that it is as natural as the branching of trees or the human lung (a consequence of their growth process), or that it represents the adaptation of computer circuits to serve the needs of percolation of information. Fractal concepts have been quite successful in describing natural phenomena. However, it is often more challenging to explain why fractal forms come up so often. Why do computer circuits lend to such a description? One suspects that fractal forms may exhibit certain optimal properties. For instance, bitonic (divide-and-conquer) algorithms can be viewed as elementary fractal forms. Is it possible to postulate general principles (such as the principle of least action in mechanics) regarding optimal information flow or computation that would lead to an inverse-power-law distribution of line lengths (a constant fractal dimension)? Mandelbrot has postulated maximum entropy principles to predict the observed inverse-power-law distribution of word frequencies (linguistics) [19] and monetary income (economics) [20]. Christie has pursued the idea that the wires in a computing system should obey Fermi-Dirac statistics, based on the observation that the wires are indistinguishable (any two wires of same length can be exchanged) and that they obey an exclusion principle (only one wire need connect two points) [32,33]. Keyes [27] has shown how the number of distinct ways one can wire up an array of elements increases with average wire length. In [34] we showed that the number of distinct ways one can "wire up" an optical interconnection system increases similarly with a fundamental quantity known as the space-bandwidth product of the optical system, and thus the average interconnection length.

The author finds the following viewpoint especially illuminating. At the microscopic level, all information processing involves the distributed manipulation and back-and-forth transfer of pieces of information. There is a certain requirement on the amount of information that must flow or percolate depending on the particular problem we are trying to solve. This requirement can be embodied in an information flow graph. The dimensionality of this graph can then be taken as a measure of the information flow requirements of the problem. For some problems which require little transfer of information, this dimension may be small. For others, it may be large. When the dimensionality associated with the problem exceeds the dimensions of the physical space in which we construct our circuits (often 2 but at most 3), we are faced with the problem of embedding a higher-dimensional graph into a lower-dimensional space. This is what leads to Rent's rule: the fact that we try to solve problems with inherently higher dimensionality of information flow than the two- or three-dimensional physical spaces we build our computers in.

Several structured problems, such as sorting and discrete Fourier transforming, are known to have global information flow requirements leading to separators which are $\propto \bar{N}$, corresponding to large dimensions and nearly unity Rent exponents. The dimensionality associated with general purpose computing may also be presumed to be large. In any event, it certainly seems that quite a fraction of interesting problems have dimensions higher than two or three, so that the space dilation effect associated with Rent's rule is expected.

Despite these considerations, Rent's rule may not apply to a particular circuit we examine. The challenges involved in dealing with greater numbers of interconnections may lead designers to reduce the number of physical ports and channels, and to shift the "communication burden" to other levels of the computational hierarchy [35]. Careful examination often reveals that the price of reducing the number of wires is often paid in terms of computation time, intermediated by techniques such as multiplexing or breaking the transfer of information into multiple steps. Clever schemes can reduce the number of wires that are apparently needed, but these often essentially amount to reorganizing the processing of information in such a way that the same information is indirectly sent in several pieces or different times. Ultimately, a certain flow and redistribution of information must take place before the problem is solved.

Several levels of graphs can come between the $n \gg 1$ dimensional graph characterizing the information flow requirements of the problem to be solved, and the $e \leq 3$ dimensional physical space. These graphs correspond to different levels of the computational hierarchy, ranging from the abstract description of the problem to the concrete physical circuits. The dimensionality of these graphs provide a stepwise transition from n dimensions to e dimensions (figure 2). Level transitions involving large steps (steep slopes) are where the greatest implementation burden is felt. For line a in figure 2, this burden is felt at the relatively concrete level, and for line c at the relatively abstract level. The burden is more uniformly spread for line b. Shifting the burden from one level to the others may be beneficial because of the different physical and technological limitations associated with each level. Techniques such as algorithm redesign, multiplexing, parallelism, use of different kinds of local or global interconnection networks, use of alternative interconnection technologies such as optics, can be used to this end. Better understanding and deliberate exploitation of these concepts and techniques may be expected to translate into practical improvements.

A particular question that may be posed in this context is whether the burden should lean primarily towards the software domain or primarily towards the hardware domain. An embodiment of the first option may be a nearest-neighbor connected mesh-type computer in which the physical interconnect problem is minimized. Global flows of information are realized indirectly as pieces of information propagate from one neighbor to the next. The second option, in contrast, might rely on direct transfer of information through dedicated global lines which

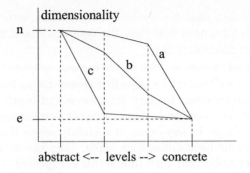

Fig. 2. The dimensionality of graphs corresponding to different levels for a hypothetical system with four levels.

result in heavy physical interconnect congestion. Although determination of the proper balance between these two extremes is in general a very complex issue, it has been addressed in a specific context in [36]. The conclusion is that use of direct global lines is more beneficial than simulating the same information flow on a locally connected system. This conclusion assumes the use of optical lines to overcome the severe limitations associated with resistive interconnections.

Contexts in which the nature of the problem to be solved does require global information flows, but only at a relatively low rate, may result in poor utilization of dedicated global lines, which nevertheless contribute significantly to system area or volume. This situation can be especially common with optical interconnections which can exhibit very high bandwidths which are difficult to saturate. For this reason, techniques have been developed for organizing information flow such that distinct pairs of transmitters and receivers can share common high-bandwidth channels to make the most of the area or volume invested in them [37].

3   Free-Space Optical Interconnections

The concepts discussed in this paper are immediately applicable to three-dimensional layouts [38–40], including those based on optical waveguides or fibers. However, the extension of results originally developed for "solid wires" to free-space optics, which can offer much higher density than waveguides and fibers, is not immediate.

Since optical beams can readily pass through each other, it has been suggested that optical interconnections may not be subject to area-volume estimation techniques developed for solid wires. However, proper accounting for the effects of diffraction leads to the conclusion that from a global perspective, optical interconnections can also be treated as if they were solid lines for the purpose of area and volume estimation, so that most of the concepts discussed in this paper are applicable to free-space optical systems as well.

This conclusion is based on the following result [41]: *The minimum total communication volume required for an optical system whose total interconnection length is ℓ_{total} is given by $\ell_{\text{total}}\lambda^2$*. This result is stated globally; it does not imply that each optical channel individually has cross-sectional area λ^2, but only that the total volume must satisfy this minimum. Indeed some channels may have larger cross-sectional areas but share the same extent of space with other channels which pass through them. The bottom line is that even with the greatest possible amount of overlap and space sharing, the global result is as if each channel required a cross-sectional area of λ^2, as if they were solid wires. If the average connection length in grid units is given by $\bar{r} = \kappa N^{p-2/3}$ as before, then the minimum grid spacing d must satisfy $Nd^3 = Nk\bar{r}d\lambda^2$, leading to a minimum system linear extent of $N^{1/3}d = (k\kappa N^p)^{1/2}\lambda$, just as would be predicted for solid wires of width λ (equation 20 with $e = 3$ and the subscript 0 suppressed) [42].

In many optical systems, the devices are restricted to lie on a plane, rather than being able to occupy a three-dimensional grid. Although in general these systems are subject to the same results, certain special considerations apply [43–46].

The above does not imply that there is no difference between optical and electrical interconnections. Optical interconnections allow the realization of three-dimensional layouts. Optical beams can pass through each other, making routing easier. Furthermore, the linewidth and energy dissipation for optical interconnections is comparatively smaller for longer lines. (This latter advantage is also shared by superconducting lines.)

4   Fundamental Studies of Interconnections

Rent's rule and associated line-length distributions have been of great value in fundamental studies of integrated systems [47–50]. Two considerations are fundamental in determining the minimum layout size and thus the signal delay: interconnection density and heat removal [51–54]. Both considerations are interrelated since, for instance, the energy dissipation on a line also depends on its length, which in turn depends on the grid spacing, which in turn depends on both the total interconnection length and the total power dissipated. The complex interplay between the microscopic and macroscopic parameters of the system must be simultaneously analyzed. Rent's rule and line-length distributions are indispensable to this end. However, it is necessary to complement these tools with physically accurate models of interconnection media. Such analytical models for normally conducting, repeatered, superconducting, and optical interconnections which take into account the skin effect, both unterminated and terminated lines, optimization of repeater configurations, superconducting penetration depth and critical current densities, optical diffraction, and similar effects have been developed in [43, 44] and subsequently applied to determine the limitations of these interconnection media and their relative strengths and weaknesses [43, 44, 40, 55–57, 36, 58]. Treating inverse signal delay S and bandwidth B as performance parameters, these studies characterize systems with N elements by surfaces of physical possibility in S-B-N space, which are to be compared with surfaces of algorithmic necessity in the same space.

This approach has allowed comparative studies of different interconnection media to move beyond comparisons of isolated electrical and optical lines, to evaluation of the effects of their different characteristics at the system level. These studies clearly show the benefit of optical and superconducting interconnections for larger systems. One of the most striking results obtained is that there is an absolute bound on the total rate of information that can be swapped from one side of an electrically connected system to the other, and that this bound is independent of scaling. Such a bound does not exist for optics and superconductors [43, 59].

An interesting extension is to allow the longer lines in a system to be of greater width to keep their RC delays within bounds. Use of the calculus of variations has shown that the widths of lines should be chosen proportional to the cube root of their length for two-dimensional layouts and to the fourth root of their length for three-dimensional layouts [60]. Staircase approximations to these analytical expressions can serve as practical design guidelines.

These studies have also been extended to determine how electrical and optical interconnections can be used together. It is generally accepted that optics is favorable for the longer lines in a system whereas the shorter lines should be electrical. Results based on comparisons of isolated lines may not be of direct relevance in a system context. The proper question to ask is not "Beyond what length must optical interconnections be used?", but "Beyond how many logic elements must optical interconnections be used?". Studies have determined that optical interconnections should take over around the level of 10^4-10^6 elements [61–63].

This body of work has demonstrated that inverse-power-law type line-length distributions are very suitable for such studies. This is because distributions which decay faster, such as an exponential distribution, effectively behave like fully local distributions in which connections do not reach out beyond a bounded number of neighbors. Such layouts are essentially similar to nearest-neighbor connected layouts, and are already covered by Rent's rule when we choose $n = e$. On the other hand, for any layout in which the number of connections per element is bounded, the behavior is at worst similar to that described by a Rent exponent of unity. Thus, although all systems may not exhibit a precise inverse-power-law distribution of line lengths, Rent's rule is nevertheless sufficient to represent the range of general interest.

5   Conclusion

We believe that many criticisms of Rent's rule are a result of not allowing the Rent exponent and dimensionality to vary as we ascend the hierarchy and a failure to recognize discontinuities. It seems that in most cases of practical interest, the decomposition function $k(\bar{N})$ is piecewise smooth with a finite number of discontinuities. The role of discontinuities in an otherwise smooth decomposition function, and whether it is beneficial to construct systems in the form of a hierarchy of functionally complete entities, are less understood issues. Is it functionally desirable to construct systems that way, or do physical and technical limitations force us to?

Parts of this work appeared in or were adapted from [8].

References

1. B. Bollobas. *Graph Theory: An Introductory Course*. Springer, Berlin, 1979.
2. G. Strang. *Introduction to Applied Mathematics*. Wellesley-Cambridge Press, Wellesley, Massachusetts, 1986.
3. H. N. V. Temperley. *Graph Theory and Applications*. Ellis Horwood Ltd., Chichester, 1981.
4. J. D. Ullman. *Computational Aspects of VLSI*. Computer Science Press, Rockville, Maryland, 1984.
5. T. C. Hu and E. S. Kuh. *VLSI Circuit Layout: Theory and Design*. IEEE Press, New York, 1985.
6. S. N. Bhatt and F. T. Leighton. A framework for solving VLSI layout problems. *J Computer System Sciences*, 28:300–343, 1984.
7. C. E. Leiserson. *Area-Efficient VLSI Computation*. The MIT Press, Cambridge, Massachusetts, 1983.
8. H. M. Ozaktas. Paradigms of connectivity for computer circuits and networks. *Optical Engineering*, 31:1563–1567, 1992.
9. W. E. Donath. Placement and average interconnection lengths of computer logic. *IEEE Trans Circuits Systems*, 26:272–277, 1979.
10. L. Pietronero. Fractals in physics: Introductory concepts. In S. Lundqvist, N. H. March, and M. P. Tosi, eds., *Order and Chaos in Nonlinear Physical Systems*. Plenum, New York, 1988.
11. B. S. Landman and R. L. Russo. On a pin versus block relationship for partitions of logic graphs. *IEEE Trans Computers*, 20:1469–1479, 1971.
12. R. L. Russo. On the tradeoff between logic performance and circuit-to-pin ratio for LSI. *IEEE Trans Computers*, 21:147–153, 1972.
13. D. K. Ferry. Interconnection lengths and VLSI. *IEEE Circuits Devices Mag*, pages 39–42, July 1985.
14. B. B. Mandelbrot. *Fractals: Form, Chance and Dimension*. W. H. Freeman, San Francisco, 1977.
15. P. Christie, J. E. Cotter, and A. M. Barrett. Design and simulation of optically interconnected computer systems. In *Interconnection of High Speed and High Frequency Devices and Systems, Proc SPIE*, 947:19–24, 1989.
16. W. E. Donath. Wire length distribution for placements of computer logic. *IBM J Research Development*, 25:152–155, 1981.
17. M. Feuer. Connectivity of random logic. *IEEE Trans Computers*, 31:29–33, 1982.
18. B. B. Mandelbrot. *The Fractal Geometry of Nature*. W. H. Freeman, New York, 1983.
19. B. B. Mandelbrot. Information theory and psycholinguistics: A theory of word frequencies. In P. F. Lazarsfeld and N. W. Henry, eds., *Readings in Mathematical Social Science*. The MIT press, Cambridge, Massachusetts, 1968.

20. B. B. Mandelbrot. The Pareto-Levy law and the distribution of income. *Int Economic Review*, 1:79–106, 1960.
21. I. E. Sutherland and D. Oestreicher. How big should a printed circuit board be? *IEEE Trans Computers*, 22:537–542, 1973.
22. W. R. Heller, W. F. Mikhail, and W. E. Donath. Prediction of wiring space requirements for LSI. *J Design Automation Fault Tolerant Computing*, 2:117–144, 1978.
23. A. El Gamal. Two-dimensional stochastic model for interconnections in master slice integrated circuits. *IEEE Trans Circuits Systems*, 28:127–134, 1981.
24. A. C. Hartmann and J. D. Ullman. Model categories for theories of parallel systems. In G. J. Lipovski and M. Malek, eds,, *Parallel Computing: Theory and Experience*. Wiley, New York, 1986.
25. W. J. Dally. *A VLSI Architecture for concurrent data structures*. Kluwer, Norwell, Massachusetts, 1987.
26. R. W. Keyes. The wire-limited logic chip. *IEEE J Solid State Circuits*, 17:1232–1233, 1982.
27. R. W. Keyes. Communication in computation. *Int J Theoretical Physics*, 21:263–273, 1982.
28. R. F. Thompson. *The Brain*. W. H. Freeman and Company, New York, 1985.
29. C. A. Neugebauer. Unpublished manuscript.
30. W. E. Donath. Stochastic model of the computer logic design process. Tech Rep RC 3136, IBM T. J. Watson Research Center, Yorktown Heights, New York, 1970.
31. W. E. Donath. Equivalence of memory to 'random logic'. *IBM J Research Development*, 18:401–407, 1974.
32. P. Christie and S. B. Styer. Fractal description of computer interconnection distributions. In *Microelectronic Interconnects and Packaging: System and Process Integration, Proc SPIE*, 1390, 1990.
33. P. Christie. Clouds, computers and complexity. In S. K. Tewksbury, ed., *Frontiers of Computing Systems Research, Volume 2*, pages 197–238. Plenum, New York, 1991.
34. H. M. Ozaktas, K.-H. Brenner, and A. W. Lohmann. Interpretation of the space-bandwidth product as the entropy of distinct connection patterns in multifacet optical interconnection architectures. *J Optical Society America A*, 10:418–422, 1993.
35. H. M. Ozaktas. Levels of abstraction in computing systems and optical interconnection technology. In P. Berthomé and A. Ferreira, eds., *Optical Interconnections and Parallel Processing: Trends at the Interface*, chapter 1. Kluwer, Dordrecht, The Netherlands, 1998.
36. H. M. Ozaktas and J. W. Goodman. Comparison of local and global computation and its implications for the role of optical interconnections in future nanoelectronic systems. *Optics Communications*, 100:247–258, 1993.
37. H. M. Ozaktas and J. W. Goodman. Organization of information flow in computation for efficient utilization of high information flux communication media. *Optics Communications*, 89:178–182, 1992.
38. A. L. Rosenberg. Three-dimensional VLSI: a case study. *J Assoc Computing Machinery*. 30:397–416, 1983.
39. F. T. Leighton and A. L. Rosenberg. Three-dimensional circuit layouts. *J Computer System Sciences*, 15:793–813, 1986.
40. H. M. Ozaktas and M. F. Erden. Comparison of fully three-dimensional optical, normally conducting, and superconducting interconnections. In *2nd Workshop on Optics and Computer Science*, April 1, 1997, Geneva. Submitted to *Applied Optics*.

41. H. M. Ozaktas and J. W. Goodman. Lower bound for the communication volume required for an optically interconnected array of points. *J Optical Society America A*, 7:2100–2106, 1990.

42. H. M. Ozaktas, Y. Amitai, and J. W. Goodman. A three dimensional optical interconnection architecture with minimal growth rate of system size. *Optics Communications*, 85:1–4, 1991.

43. H. M. Ozaktas and J. W. Goodman. The limitations of interconnections in providing communication between an array of points. In S. K. Tewksbury, ed., *Frontiers of Computing Systems Research, Volume 2*, pages 61–124. Plenum, New York, 1991.

44. H. M. Ozaktas. *A Physical Approach to Communication Limits in Computation*. PhD thesis, Stanford University, California, 1991.

45. H. M. Ozaktas, Y. Amitai, and J. W. Goodman. Comparison of system size for some optical interconnection architectures and the folded multi-facet architecture. *Optics Communications*, 82:225–228, 1991.

46. H. M. Ozaktas and D. Mendlovic. Multi-stage optical interconnection architectures with least possible growth of system size. *Optics Letters*, 18:296–298, 1993.

47. R. W. Keyes. *The Physics of VLSI Systems*. Addison-Wesley, Reading, Massachusetts, 1987.

48. R. W. Keyes. Fundamental limits in digital information processing. *Proc IEEE*, 69:267–278, 1981.

49. R. W. Keyes. The evolution of digital electronics towards VLSI. *IEEE Trans Electron Devices*, 26:271–279, 1979.

50. H. B. Bakoglu. *Circuits, Interconnections and Packaging for VLSI*. Addison-Wesley, Reading, Massachusetts, 1990.

51. H. M. Ozaktas, H. Oksuzoglu, R. F. W. Pease, and J. W. Goodman. Effect on scaling of heat removal requirements in three-dimensional systems. *Int J Electronics*, 73:1227–1232, 1992.

52. W. Nakayama. On the accomodation of coolant flow paths in high density packaging. *IEEE Trans Components, Hybrids, Manufacturing Technology*, 13:1040–1049, 1990.

53. W. Nakayama. Heat-transfer engineering in systems integration—outlook for closer coupling of thermal and electrical designs of computers. *IEEE Trans Components, Packaging, Manufacturing Technology, Part A*, 18:818–826, 1995.

54. A. Masaki. Electrical resistance as a limiting factor for high performance computer packaging. *IEEE Circuits Devices Mag*, pages 22–26, May 1989.

55. H. M. Ozaktas. Fundamentals of optical interconnections—a review. In *Proc Fourth Int Conf Massively Parallel Processing Using Optical Interconnections*, pages 184–189, IEEE Computer Society, Los Alamitos, California, 1997. (Invited paper, June 22–24, 1997, Montreal.)

56. H. M. Ozaktas. Toward an optimal foundation architecture for optoelectronic computing. Part I. Regularly interconnected device planes. *Applied Optics*, 36:5682–5696, 1997.

57. H. M. Ozaktas. Toward an optimal foundation architecture for optoelectronic computing. Part II. Physical construction and application platforms. *Applied Optics*, 36:5697–5705, 1997.

58. H. M. Ozaktas and J. W. Goodman. The optimal electromagnetic carrier frequency balancing structural and metrical information densities with respect to heat removal requirements. *Optics Communications*, 94:13–18, 1992.

59. D. A. B. Miller and H. M. Ozaktas. Limit to the bit-rate capacity of electrical interconnects from the aspect ratio of the system architecture. *J Parallel Distributed Computing*, 41:42–52, 1997.

60. H. M. Ozaktas and J. W. Goodman. Optimal linewidth distribution minimizing average signal delay for RC limited circuits. *Int J Electronics*, 74:407–410, 1993.
61. H. M. Ozaktas and J. W. Goodman. Elements of a hybrid interconnection theory. *Applied Optics*, 33:2968–2987, 1994.
62. H. M. Ozaktas and J. W. Goodman. Implications of interconnection theory for optical digital computing. *Applied Optics*, 31:5559–5567, 1992.
63. A. V. Krishnamoorthy, P. J. Marchand, F. E. Kiamilev, and S. C. Esener. Grain-size considerations for optoelectronic multistage interconnection networks. *Applied Optics*, 31:5480–5507, 1992.

Optoelectronic-VLSI Technology: Terabit/s I/O to a VLSI Chip

Ashok V. Krishnamoorthy

Bell Labs, Lucent Technologies, Holmdel, NJ 07733

The concept of a manufacturable technology that can provide parallel optical interconnects directly to a VLSI circuit, proposed over 15 years ago in [1], now appears to be a reality. One such optoelectronic-VLSI (OE-VLSI) technology is based on the hybrid flip-chip area-bonding of GaAs/AlGaAs Multiple-Quantum Well (MQW) electro-absorption modulator devices directly onto active silicon CMOS circuits. The technology has reached the point where batch-fabricated foundry shuttle incorporating multiple OE-VLSI chip designs are now being run [2]. These foundry shuttles represent the first delivery of custom-designed CMOS VLSI chips with surface-normal optical I/O technology. From a systems point of view, this represents an important step towards the entry of optical interconnects in that: the silicon integrated circuit is state-of-the-art; the circuit is unaffected by the integration process; and the architecture, design, and optimization of the chip can proceed independently of the placement and bonding to the optical I/O.

To date, over 5760 MQW modulator devices have been integrated onto a single CMOS IC with a device yield exceeding 99.95%. Each bonded device has a load capacitance of approximately 50fF (65fF including a 15μmx15μm bond pad) and can be driven by a CMOS inverter to accomplish the electrical-to-optical interface. Compact CMOS transimpedance receiver circuits have been developed to execute the photocurrent-to-logic-level voltage conversion. Operation of single-ended receivers [3] (one diode per optical input) fabricated in a 0.35μm linewidth CMOS technology, has been demonstrated over 1Gigabit/s with a measured bit-error-rate less than 10^{-10}. Differential two-beam receiver, have similarly been operated to over 1Gbit/s. The

J. Rolim et al. (Eds.): IPDPS 2000 Workshops, LNCS 1800, pp. 1089-1091, 2000.

receiver circuits mentioned above have static power dissipation in the range of 3.5-8mW per receiver. More recently, arrays of up to 256 active light sources known as Vertical-Cavity Surface-Emitting Lasers (VCSELs) have also been bonded directly to CMOS VLSI chips [4], with each VCSEL capable of over 1Gigabit/s modulation by the CMOS circuits.

Before such a technology can be deployed on a large scale, several issues related to the scalability of the optoelectronic technology and its compatibility with deep sub-micron CMOS technologies must be addressed. In terms of the modulator technology, the challenges are in reducing the drive voltages of the modulators to stay compatible with sub-micron CMOS technologies, and to continue to improve the yield in the manufacturing and hybridizing of the MQW diodes. In terms of the VCSELs, the challenge will be in producing arrays of power-efficient VCSELs that can attached to CMOS circuits with high-yield, and be simultaneously operated at high speeds [5]. In terms of the circuits, the challenges will be to continue to improve receiver sensitivity while reducing power dissipation and cross-talk. A final consideration is that of the systems integration, where the challenge will be to package systems that can efficiently transport large arrays of light-beams to and from such chips.

Based on relatively conservative assumptions on how these components will evolve, a general conclusion is that it appears this hybrid optical I/O technology has substantial room for continued scaling to large numbers of higher-speed interconnects [6]. Indeed, future OE-VLSI technologies (whether modulator-based or VCSEL-based) can be expected to provide an I/O bandwidth to a chip that is commensurate with the processing power of the chip, even in the finest linewidth silicon: a task that cannot be expected from conventional electrical interconnect technologies. Initial work on space-division crossbar OE-VLSI switches have suggested that terabit capacities are achievable. The availability of optical access to high-speed RAM [7] will also permit the development of shared-memory (SRAM)-based switches: a goal that cannot be achieved with conventional space-division photonic switching technologies. It is anticipated that the availability of such an OE-VLSI technology will enable terabit-per-second throughput switches with power dissipations on the order of 20-50mW per Gigabit/s of switch throughput.

References:

1. J. W. Goodman, F. J. Leonberger, S.-Y, Kung, and R. A. Athale, "Optical interconnections for VLSI systems," *Proceedings of the IEEE,* vol. 72, no. 7, pp. 850-866, July 1984.
2. A. V. Krishnamoorthy and K. W. Goossen, "Optoelectronic-VLSI: photonics integrated with VLSI circuits," *IEEE Jour. Sel. Topics in Quantum Elec.,* Vol. 4, pp. 899-912, December 1998.
3. A. L. Lentine et al., "Optoelectronic VLSI switching chip with over 1Tbit/s potential optical I/O bandwidth," *Electronics Letters,* Vol. 33, No. 10, pp. 894-95, May 1997.
4. A. V. Krishnamoorthy et al., "Vertical cavity surface emitting lasers flip-chip bonded to gigabit/s CMOS circuits," *Photonics Technology Letters,* Vol. 11, pp. 128-130, January 1999.
5. A. V. Krishnamoorthy et al., "16x16 VCSEL array flip-chip bonded to CMOS," *OSA Top. Meet. Optics in Computing,* (Snowmass) Postdeadline PD3, April 1999.
6. A. V. Krishnamoorthy and D. A. B. Miller, "Scaling Optoelectronic-VLSI circuits into the 21st century: a technology roadmap," *IEEE J. Special Topics in Quant. Electr.,* Vol. 2, pp. 55-76, April 1996.
7. A. V. Krishnamoorthy et al., "CMOS Static RAM chip with high-speed optical read-write," *IEEE Photonics Technology Letters,* Vol. 9, pp. 1517-19, November 1997.

Three Dimensional VLSI-Scale Interconnects

Dennis W. Prather

University of Delaware
Department of Electrical and Computer Engineering
Newark, DE 19716
email: dprather@ee.udel.edu

Abstract. As processor speeds rapidly approach the Giga-Hertz regime, the disparity between process time and memory access time plays an increasing role in the overall limitation of processor performance. In addition, limitations in interconnect density and bandwidth serve to exacerbate current bottlenecks, particularly as computer architectures continue to reduce in size. To address these issues, we propose a 3D architecture based on through-wafer vertical optical interconnects. To facilitate integration into the current manufacturing infrastructure, our system is monolithically fabricated in the Silicon substrate and preserves scale of integration by using meso-scopic diffractive optical elements (DOEs) for beam routing and fan-out. We believe that this architecture can alleviate the disparity between processor speeds and memory access times while increasing interconnect density by at least an order of magnitude. We are currently working to demonstrate a prototype system that consists of vertical cavity surface emitting lasers (VCSELs), diffractive optical elements, photodetectors, and processor-in-memory (PIM) units integrated on a single silicon substrate. To this end, we are currently refining our fabrication and design methods for the realization of meso-scopic DOEs and their integration with active devices. In this paper, we present our progress to date and demonstrate vertical data transmission using DOEs and discuss the application for our architecture, which is a multi-PIM (MPM) system.

Introduction

As modern day technologies continue to develop an increasing number of applications are resorting to computational based simulations as a tool for research and development. However, as simulation tools strive to incorporate more realistic properties their computational requirements quickly increase and in many cases surpass that which is currently available. As a result, a seemingly perpetual demand to process more information in shorter time frames has resulted. Moreover, while current computer architectures are steadily improving they are not keeping pace with the requirements of more sophisticated applications and in fact for some applications they are falling behind. To this end, new paradigm computer architectures need to be developed.

J. Rolim et al. (Eds.): IPDPS 2000 Workshops, LNCS 1800, pp. 1092-1103, 2000.

The current paradigm for addressing this shortcoming is to simply incorporate smaller devices into larger die. However, while this does enable the design and realization of more sophisticated circuits it also exacerbates an already serious problem, namely the interconnection and packaging of the devices and components within the system. For example, according to the National Technology Roadmap for Semiconductors, processors based on 1µm fabrication have a ratio of transistor -to-interconnect delay of 10:1 (assuming a 1mm long interconnect), whereas that for the same processor based on 0.1µm fabrication is 1:100. This represents a shift in emphasis of more than three orders of magnitude. As a result alternative interconnect and packaging technologies need to be developed. Therefore, in this paper we report on our work in addressing these technological barriers by designing an embedded processor-in-memory (PIM) architecture realized using an optically interconnected three-dimensional (3D) package.

While conventional 3D packaging increases circuit density, decreases interconnect delay, and reduces critical interconnect path lengths, their full potential has yet to be realized. This is due mainly to the associated capacitive and inductive loading affects of vertical vias, which reduce bandwidth and allow for only a 1-to-1 interconnect. To overcome these limitations we propose an alternate approach that is based on recent advances in micro-optical technology.

Our approach uses vertical cavity surface emitting lasers (VCSELs) that are flip-chip bonded onto CMOS drivers. The VCSELs have a 1.3µm wavelength which is transparent to the silicon wafer. The VCSELs are oriented such that the output beam is directed vertically through the silicon wafer. However, before the beam enters the wafer it is incident on a VLSI-scale diffractive optical element (DOE) that not only focuses the beam to a subsequent wafer, but also performs a 1-to-N fanout (N can range from 1 to 50 depending on the area used for the DOE). This allows for nearly real time data routing and distribution, which is essential to overcome conventional computational bottlenecks. However, before presenting further details in our approach we first motivate our PIM-based architecture.

PIM Motivation

A current trend in computer system design is to develop architectures based on the integration of a large number of smaller and more-simple processing cores that work together in unison. The idea here is that such processors can be integrated directly into random access memory (RAM) to simplify the memory hierarchy, i.e., level-1 and level-2 CACHE, and thereby streamline processor to memory communication. Such systems have been named Intelligent RAM (IRAM), Flexible RAM (Flex-RAM) and PIM, as we refer to it.

Currently, several high profile research initiatives (sponsored by federal agencies, e.g., the HTMT-PIM project [1,2], the DIVA project [3], the FlexRAM project [4] are investigating many of the architectural and system design issues related with the implementation of PIM-based systems. In fact, IBM recently announced the introduction of the Blue-Gene project [5], which anticipates an industry investment on the order of a $100M dollars to produce a petaFLOPS scale machine based on thousands of PIM components. Therefore, even though PIM-based architectures are not currently being used in commercial machines, they promise to overcome the limitations of conventional computer architectures.

However, in general the amount of memory and the processing capability of individual PIMs is limited, therefore the construction of PIM-based high performance systems will require the integration of upwards of tens of thousands of PIMs. Thus the integration of multiple PIMs into a single package will be absolutely essential to reduce latencies, increase communication bandwidth between PIMs, reduce power consumption, and reduce the integration cost of the entire system.

Therefore the problem addressed by this research proposal is the implementation of a Multiple PIM Module (MPM) to harness the processing capability and the memory storage capability of multiple PIMs into a single computational module. A MPM can be used as the building block to implement mobile computers as proposed by the MIT RAW project. It can be used as the basic building block for computer systems specialized in data intensive computation, as proposed in the DIVA project. And it can be a building block for the DPIM region of a large scale, high performance computer such as the one proposed in the HTMT project.

Some of the open research problems in the implementation an MPM and in its use in a system architecture are: (1) How the multiple PIMs, that form the MPM, communicate and synchronize with each other. (2) Is it possible to design and implement a fast and versatile interconnection between the multiple PIMs in the MPM. (3) How MPMs can be programmed and how the interconnection can be adapted for new communication pathways. And (4) how does the runtime system control MPMs to ensure the communications/synchronizations are performed in the most efficient way according to the needs of the application program.

To address these issues, we are developing a technology based on the interconnection of multiple PIMs within a single MPM via an array of vertical cavity surface emitting laser (VCSEL) and SiGe detector arrays that are vertically interconnected through the silicon wafer using a DOE. This technology allows for fast, abundant, and distributed interconnections amongst the PIMs in a given module. Also, because this approach allows for data distribution at the 2-5GHz rate it reduces the latency in communication between PIMs to unprecedented levels and because optical beams can essentially pass right through each other without exchanging information it all but eliminates the place and route problem. Also, each interconnect link in our design would consume approximately 50mW of power, which when applied to a full 16×16 interconnection would consume on the order of 10 Watts of power. This is nearly an order of magnitude less than current architectures that are limited to only 4×4 interconnections.

To realize this architecture three critical technologies must be used: long wavelength VCSELs (1.3μm), high speed (2-5 Gbits) CMOS drivers for the VCSELs, and VLSI-scale DOEs. To this end we have been working with Gore Photonics for the 1.3μm VCSELs and developing our own high-speed CMOS drivers, VLSI-scale DOEs, and system integration techniques at UD. Thus, in the remainder of the paper we report on our progress in this effort. We begin by motivating the 3D architecture and the describe the component optical technologies needed to realize it.

Optoelectronic Technologies

Whereas the use of optical interconnects in long haul and local area networks has proven extremely successful, its use on the VLSI-scale has been limited. This is due in large part to the continual increase in speed and performance of conventional electronic devices. However, the issues associated with next generation PIM architectures cannot be adequately addressed with speed alone. Instead, such systems will require not only the ability to share or distribute information among PIM modules (signal fan-out) but also a significant increase in interconnect density. While the issues of increased bandwidth, interconnect densities, and signal fan-out are individually compelling reasons for considering optical interconnects, when combined together they become persuasive. For example, one possible electronic solution to increasing interconnect density is to use flip-chip, or bump, bonds, which can require approximately $20\mu m^2$ of chip area while offering only a 1-to-1 interconnect. In comparison, we have designed VLSI-scale diffractive optical elements (DOEs) that within the same area provide a 1-to-16 interconnect. Currently we have experimentally demonstrated a 1-to-4 and are in the process of fabricating the 1-to-16. For this reason we propose the use of an optoelectronic 3D architecture that uses monolithically integrated VLSI-scale DOEs for application to PIM architecture, as shown in Fig. 1.

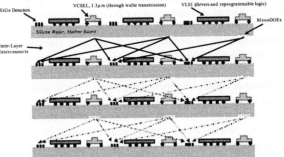

Fig. 1. Monolithic interconnect architecutre that uses 3D diffractive optical interconnects on the VLSI-scale for through wafer fan-out interconnects for data or clock distrbution. Various modules of this architecture can also be stacked together to realize more complex systems.

While the notion of 3D architectures is appealing, due to the efficient use of power and increased processing and interconnect densities, few of the systems proposed in the literature have received wide spread use. Reasons for this depend on the technology being used. For instance, all electronic architectures suffer from either reduced communication bandwidth, due to routing the inter-layer interconnects through the periphery of the 3D stack, or reduced interconnection density, due to the inability to distribute data between layers using 1-to-1 bump bonds. Along the same lines, optical architectures suffer from input/output coupling efficiencies, for wave guide based approaches, interconnect density and distribution, for 1-to-1 emitter-receiver-based approaches, and scale of integration for bulk optical systems. Thus, we believe that in order for an optical interconnect system to be viable it *must* satisfy the following conditions: (1) It must have a scale of integration comparable to VLSI, to preserve scales of integration. (2) The optical system must be monolithic in the Silicon substrate, in order to alleviate alignment issues and improve system reliability. And (3) the fabrication methods and materials used must be compatible with the current manufacturing infrastructure, in order to reduce cost of implementation. In the design of our architecture we will are strictly adhering to these conditions.

Our approach is based on our recent progress in the development of suitable design tools, which enable the design of VLSI-scale DOEs for monolithic integration with active devices. As a result, we have been able to significantly increase the interconnection density as compared to all electronic vertical interconnections as illustrated in Fig. 2, which illustrates a DOE that occupies $10\mu m^2$ and provides a 1-to-4 fan-out. If this DOE is tiled over a $20\mu m^2$ area, equivalent to that of a bump bond, it would provide a 1-to-16 fan-out in comparison to a 1-to-1, which represents more than an order of magnitude increase in interconnect density. In addition to increasing density this approach significantly simplifies the place and route problem because optical beams do not exchange information and can therefore accommodate overlap in the routing process.

In order to realize optical interconnections within a Silicon wafer and on a scale comparable with VLSI circuits, one must be able to heterogeneously integrate active and passive optical devices together on a scale comparable to microelectronic devices. This must also be done in such a way that the ability to control and redirect light in a general fashion is preserved, e.g., off-axis focusing, mode shaping, and beam fan-out. Whereas active optical devices, such as emitters, detectors, and modulators are readily designed and fabricated with dimensions on the micron scale, until recently passive optical elements capable of such general behavior were not. However, recent advances in both the design and fabrication of diffractive structures [6] now enables the integration of active and passive optical devices on the VLSI-scale and the ability to efficiently control and redirect light in a general fashion, see Fig. 3. Thus, the integration of VCSELs with wavelength scale fan-out DOEs on the VLSI-scale offer not only an order of magnitude improvement (in terms of density, bandwidth, and power consumption) but also the ability to design architectures that heretofore have not been possible. As a result new optical interconnect architectures can now be developed.

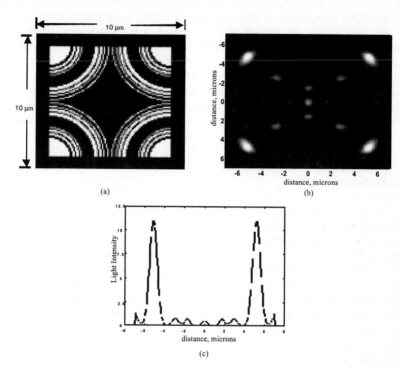

(a)

(b)

(c)

Fig. 2. Illustration of a three dimensional subwavelength off-axis lenslet array used for 1-to-4 fanout on the VLSI-scale (a) DOE, (b) intensity image in the focal plane, and (c) line scan thorough the focal plane. Results were generated using a 3D FDTD diffraction model.

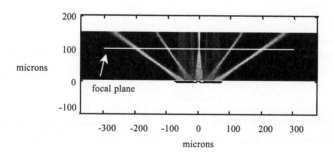

Fig. 3. Illustration of a VLSI-scale 1-to-5 fan-out DOE, computed using the boundary element method. The width of the DOE is 120 microns and the focal length is 100 microns.

Recently we have fabricated and experimentally validated these elements and are currently preparing them for system level integration [7]. However, critical to the successful completion of this effort is the ability to fabricate DOEs that have features sizes on the nanometer scale. Although many fabrication techniques for DOEs exist, by far the most general and widely used is that of the microelectronics photolithographic process. In this technique the profile of a DOE is realized by etching micro-relief patterns into the surface of either conducting or dielectric substrates. A curved surface profile is realized by using a multi-step process which produces a stair-step approximation. Using this fabrication process DOEs that have diffraction efficiencies on the order of 95% have been fabricated. Unfortunately, as the scale of a DOE is reduced the alignment process, needed for multi-step profiles, becomes exceedingly difficult. As a result, alternate fabrication methods based on single step gray-scale lithography and direct electron beam (e-beam) exposure have been developed.

In the gray-scale process one wishes to realize continuous profiles, or structures. However, for devices on the VLSI-scale current fabrication technology limits us to a discrete number of levels, typically 4-8 levels. Thus, we can currently fabricate our DOEs using a gray-scale technique which results in multilevel structures from a single processing step, as shown in Fig. 4. To this end, we designed our multi-level masks in the lab and used an outside vendor [8] to provide the gray-scale mask. Once we have the mask we deposit an initial height of the photoresist on the substrate, i.e., silicon wafer, which can be precisely controlled by adjusting the spin rate at the time of deposition. Through experimentation, we have characterized the response of photoresist to various degrees of UV exposure. This allows us to precisely designate the correct transmission levels of the mask to create our multi-level DOE profiles in the photoresist.

After the grayscale photolithography, the pattern is transferred into the surface of the Silicon substrate using a Plasmatherm 790 series reactive ion etching (RIE) system. Careful calibration of the RIE process is required to achieve structures with smooth surfaces and submicron feature resolution while preserving the height of the initial profile.

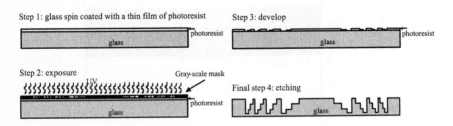

Fig. 4. Graphical illustration of the gray-scale photolithographic fabrication process.

An alternative fabrication method based on direcet e-beam write can also be used to fabricate VLSI-scale DOEs. In this appraoch a high energy electron beam is used to expose a photoresist coated substrate. As the substrate is exposed the energy level of the e-beam is varied in accordance with the desired DOE profile. Once developed the substrate is etched, using techniques such as reactive ion etching, to transfer the continuous photoresist profile into the substrate, see Fig. 5. This process is capable of fabricating binary DOE profiles that have feature sizes on the order of 60nm, which is several times smaller than the wavelength of illumination. As a result efficiencies exceeding those predicted by scalar diffraction theory can be achieved [9]. Through collaboration with Axel Scherer of CalTech we have recently had several DOEs fabricated, as shown in Fig. 6.

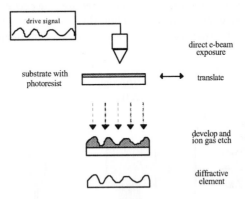

Fig. 5. Fabrication process for continuous profile DOEs based on direct electron beam write.

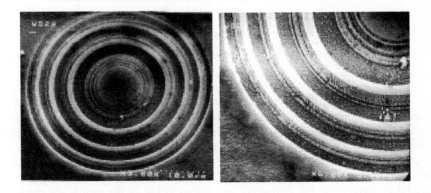

Fig. 6. Illustration of a mesoscopic diffractive lens having a diameter of 36µm, a focal length of 65µm and a minimum feature size of 60nm. The element was fabricated by Dr. Axel Scherer, of the California Institute of Technology.

In addition to developing the theoretical and experimental framework necessary to design and realize DOEs we have developed a novel system for characterizing their performance.

Our system consists of a microscope objective (20X) and a 1inch diameter lens. The system has an overall magnification of 4.2 (based on the ratio of the two focal lengths, f_2 / f_1), and is able to resolve 1 micron minimum features. The entire imaging system is mounted on an x,z translation stage, as shown in Fig. 7. Because the object and image planes, in this system, are fixed and well defined they can be used to determine the axial location relative to the DOE, i.e., the reference plane for $z=0$. This is achieved by translating the imaging system toward the DOE until the surface is imaged on to the CCD. Subsequently, the translation stage, with the entire imaging system on it, is translated back to the plane of interest, i.e., $z=z_o$. Because the microscope objectives have large numerical apertures the performance of the imaging system, i.e., its modulation transfer function (MTF), reproduces the intensity profile in the object plane, i.e, the observation plane, with excellent fidelity.

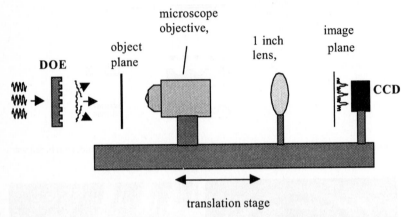

Fig. 7. Micro 4f imaging system for characterizing mesoscopic diffractive optical elements.

To validate our electromagnetic design models we used the system to measure the diffracted light from a precision pin-hole of 71μm in diameter, from a collimated incident wave of 0.633nm. We then calculated the diffracted light using both scalar diffraction theory and using our electromagnetic model, results for $z=350$μm are shown in Fig. 8. Additional measurements were made along the z-axis and showed the same level of agreement. To illustrate the utility of this system we used it to characterize the diffractive lens shown in Fig. 6, the results are shown in Fig. 9. Once confident that our design and fabrication methods were working we then applied them to the realization of through silicon wafer DOEs [7].

Integration

In order to achieve optical interconnects on a single Silicon die, we must be able to integrate emitters, detectors, drivers, and DOEs on the VLSI scale. Our approach toward integration will be to construct a hybrid system using flip-chip bonding. For this part of the project we will use a SEC Omnibonder 860 flip-chip bonding machine to construct a multichip module for the integration of the active and passive optical devices with their electronic counterparts. Figure illustrates the integration of an 8 × 8 CMOS driver array with an 8 × 8 980nm VCSEL array.

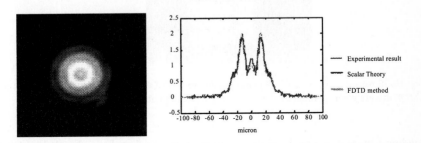

Fig. 8. Comparison between experimental results and theoretical predictions for the diffraction from a precision pin-hole that had a diameter of 71 microns at an axial location of 350 microns.

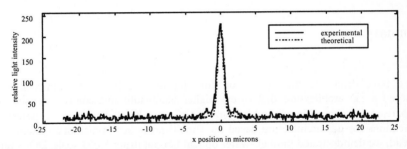

Fig. 9. Overlay of the experimental characterization of a mesoscopic diffractive lens and the results predicted from our electromagnetic models. Data was taken from our system using a 40X magnification objective at the loacation of z=65μm, the design focal length.

Ultimately, we plan to use 1.3 micron VCSELs as emitters and a Silicon substrate as the medium of propagation. However, such long wavelength VCSELs are not currently available in die form at present, so we have begun the construction of a pre-prototype system using an 850nm and 980nm VCSELs on a glass substrate. In this preliminary system, the VCSEL is bonded to a CMOS driver circuit and directed through the DOE as shown in Fig. 10.

Our main concern associated with bonding the VCSEL over a DOE, is the air gap spacing between the VCSEL and the backside of the glass substrate. Since the VCSEL will be flip-chip bonded to the glass surface, the solder bump size, bond pressure and bond temperature profile will affect the resultant air gap. Additionally, the proximity of the CMOS driver and the VCSEL will be a guiding parameter of the

Fig. 10. Illustration of a VCSEL flipchip bonded to a CMOS driver circuit. The VCSELs and CMOS drivers were supplied by the U.S. Army Research Laboratory.

bonding temperature profile, since we do not want the first device bonded to detach during the second bond. Most likely, we will choose to bond the CMOS driver first in order to maximize control over the air gap spacing. That way heating during the bonding of the driver will not affect the final VCSEL position.

Summary

We have discussed the motivation for chip-level optical interconnects, and proposed a 3D architecture that offers higher bandwidth interconnect density in comparison to conventional architectures. Also, we have discussed a potential applications for our architecture based on a multi-processor-in-memory system. To this end, we demonstrated through-wafer optical fan-out using VLSI-scale DOEs and long wavelength VCSELs (courtesy of Gore Photonics). Flip-chip bonding gives us the ability to integrate active and passive devices on a single die, and we are currently building a prototype system to demonstrate this integration. The significance of our approach lies in the ability to design optical elements that efficiently control, or redirect, light on a VLSI-scale and can be directly integrated into the current VLSI-based manufacturing infrastructure. As such this technology lends itself nicely to 3D interconnect schemes and facilitates the trend toward higher levels of parallelism in computer architectures.

References

[1] T. Sterling, "Achieving petaflops-scale performance through a synthesis of advanced device technologies and adaptive latency tolerant architectures," in *Supercomputing 99*, (Portland, OR), Novermber 1999.

[2] P.M. Kogge, J.B. Brockman, T. Sterling, and G. Gao, "Processing-in-memory: chips to petaflops," in *International Symposium on Computer Architecture*, (Denver, CO), June 1997.

[3] M. Hall, P. Kogge, J. Koller, P. Diniz, J. Chame, J. Draper, J. LaCoss, J. Granacki, A. Srivastava, W. Athas, J. Brockman, V. Freeh, J. Park, and J. Shin, "Mapping irregular applications to DIVA, a PIM-based data-intensive architecture," in *Supercomputing 99*, Portland OR, Novermber 1999.

[4]   Y. Kang, M. Huang, S.M. Yoo, Z.Ge, D. Keen, V. Lam, P. Pattnaik, and J. Torrellas, "Flexram: toward an advanced intelligent memory system," in International Conference on Computer Design, October 1999.

[5]   IBM, "IBM unveils $100 million research initiative to build world's fastest Semiseek, December 1999.

[6]   D.W. Prather, M.S. Mirotznik, and S. Shi, *Mathematical Modeling in Optical Science*, Ch. Electromagnetic models for finite aperiodic diffractive optical elements, *in print*, SIAM Frontier Book Series, Society for Industrial and Applied Mathematics, 2000.

[7]   M. LeCompte, X. Gao, H. Bates, J. Meckle, S. Shi, and D.W. Prather, Three-dimensional through-wafer fan-out interconnects," in Optoelectronics Interconnects VII, SPIE 3952, The International Society Optical Engineering, Bellingham WA, January 2000.

[8]   Canyon Materials, Inc., San Diego, CA.

[9]   J.N.Mait, D.W. Prather, and M.S. Mirotznik, "Binary subwavelength diffractive-lens design," *Opt. Lett.*, **23,** pp. 1343-1345, September 1998.

Present and Future Needs of Free-Space Optical Interconnects

Sadik Esener and Philippe Marchand

Electrical and Computer Engineering Department
University of California, San Diego, La Jolla, CA 92093, USA

Abstract. Over the last decade significant progress in optoelectronic devices and their integration techniques have made Free-Space Optical Interconnects (FSOI) one of the few physical approaches that can potentially address the increasingly complex communication requirements at the board-to-board and chip-to-chip levels. In this paper, we review the recent advances made and discuss future research directions needed to bring FSOI to the realm of practice.
Keywords: Optical Interconnects, Optical Packaging, Micro-optics, OptoElectronics, Free-Space Optical Interconnects

1 Introduction

Exchanging data at high speed over sufficiently long distances is becoming a bottleneck in high performance electronic processing systems [1,2,3]. New physical approaches to dense and high-speed interconnections are needed at various levels of a system interconnection hierarchy starting from the longest interconnections: board to board, MCM to MCM on a board, chip-to-chip on a multi-chip module (MCM), and on-chip. For the next decade, FSOI when combined with electronics offer a potential solution [4,5,6,7,8,9] at the inter and intra-MCM level interconnects promising large interconnection density, high distance-bandwidth product, low power dissipation, and superior crosstalk performance at high-speeds [10,11,12,13].

2 Present Status of FSOI

Opto-Electronic (OE) devices including Vertical Cavity Surface Emitting Lasers (VCSELs), light modulators, and detectors have now been developed to a point that they can enable high speed and high-density FSOI [14,15,16]. Flip-chip bonding offers a convenient approach to their integration with silicon. For example, members of the 3-D OESP consortium (Honeywell Technology Center and University of California, Santa Barbara) have demonstrated FSOI links operating up to 2.5Gb/s between VCSEL arrays and suitable detector arrays. These developments occurred at an opportune time when high performance workstation manufacturers struggle to resolve communication bottlenecks at the board-to-board level. As a result, high efficiency FSOI links between VCSEL and detector arrays has sparkled the interest of

J. Rolim et al. (Eds.): IPDPS 2000 Workshops, LNCS 1800, pp. 1104–1109, 2000.

high performance workstation manufacturers such as Sun Microsystems. While board-to-board interconnect solutions using FSOI are now being evaluated by the computer industry, chip-to-chip interconnects are being investigated at a more fundamental level at several universities including UCSD. One of the key issues that needs to be addressed at this level is packaging. Indeed a packaging architecture and associated technologies need to be developed to integrate OE devices and optical components in a way that is fully compatible with conventional electronic multi-chip packages.

Recently at UCSD, we developed and demonstrated the operation of a fully packaged FSOI system for multi-chip interconnections capable of sustaining channel data rates as high as 800Mb/s. A picture of this system is shown in Figure 1. A conventional PCB/ceramic board is populated with silicon and OE chips and mated to a FSOI layer that is assembled separately. Design considerations, packaging approaches as well as testing results indicate that it is now possible to build FSOI electronic systems that are compatible in packaging techniques, physical dimensions and used materials with conventional electronics.

Figure 1. Fully packaged FSOI system

The overall packaging approach consists of the assembly of two different packaging modules: the opto-electronic module (multi-chip carrier and the OE chips (VCSEL, MSM and silicon chips), and the optics (FSOI) module. In our approach both modules are assembled separately then snaped on together. A mechanical pin-pinhole technique combined with alignment marks makes the alignment of the two modules a rather straightforward task. The optics module is built out of plastic except for the glass optical lenses that were commercially available. In the current demonstration system, four one-dimensional (1D) proton implanted VCSEL arrays (1'12 elements each) and four 1D Metal-Semiconductor-Metal (MSM) detector arrays (1'12) are used as light sources and photodetectors, respectively. The lasers and detectors are on a 250µm pitch. The VCSELs operate at 850nm with 15o-divergence angle (full angle at $1/e2$), and the detector aperture is 80'80µm. Laser drivers, receiver (amplifiers), and router circuits are integrated on three silicon chips and included into the system. VCSEL arrays are optically connected to their corresponding detector arrays. Data can be fed electrically to any one of the silicon chips and routed to the VCSELs through driver circuits. The silicon chips also contain receiver circuits directly connected to the detectors; thus, data can also be readout electrically from each silicon chip independently.

In this FSOI demo system, 48 optical channels each operating up to 800Mb/s with optical efficiencies exceeding 90% and inter-channel crosstalk less than -20dB were implemented in a package that occupied less than 5x5x7 cm3. All channels were operational. This packaging technique is now being applied to demonstrate an FSOI connected board that is populated by three 3-D stacks of silicon chips. Each stack contains 16 silicon chips each hosting a 16x16 crossbar switch. In addition each stack is flip-chip bonded to a 16x16 array of VCSELs and detectors and communicates with other stacks via these devices. Thus with this package of very small footprint, 48 silicon chips will be interconnected via FSOI with each other.

OE Array

Chip stack

Figure 2. Application of UCSD's chip-to-chip FSOI packaging technique to 3-D stack-to-stack communication

3   Present limitations in FSOI and future directions

Although the demonstrations described above are important milestones in the quest for using optics within the board, it also underlines some of the present limitations of FSOI. These shortcomings include the:
• height of the optical package
• signal integrity and synchronization issues
• thermal stability of the assembly
• effective CAD tools
• ultra low voltage light modulation
• costs associated with FSOI.
To reduce the height of the package micro-optical elements compatible with oxide confined VCSELs need to be developed and become commercially available. Presently commercially available micro-optical components do not provide simultaneously the necessary high efficiency, low F# and spatial uniformity. In addition, communication within the box requires very low bit error rates. It is therefore critical to use extensive encoding techniques to minimize the error rates in FSOI. To this end there is a need for more silicon real estate and power consumption.

As the power in the package is increased passive alignment techniques may not be sufficient. Active alignment techniques based for example on MEMs components or special alignment facilitating OE Array Chip stack OE Array.

Chip stack.optical components must be examined. Also, in order to build more complex optoelectronic systems and packages, it is now clear that powerful CAD systems capturing both electronic circuits and sub systems as well as optoelectronic and optical components and sub-systems must be made available. Such a CAD system is not only essential for the optoelectronics sub-system designer but also for the electronics system designer. Furthermore, with the scaling of CMOS circuits, in order to conserve drive voltage compatibility, optoelectronic devices that require very low drive voltages are required. Finally, the cost associated with FSOI is of prime concern. The main cost factors include the optoelectronic devices and their integration as well as the overall packaging. The device costs can only be reduced with manufacturing volume. Therefore it is critical to direct the use of optoelectronic arrays to markets with large volumes including optical data storage and bio-photonics. Further in the future, flip-chip bonding with its associated parasitics and high cost should be replaced with heterogeneous integration technologies at the device and material levels rather than at the chip level. Such technologies have the potential to relieve present layout constraints and ultimately reduce cost.

4   Conclusions

Significant progress both at the device and sub-system levels has been made in FSOI to the point where FSOI can now be considered to push the envelope in computing hardware at the board to board interconnect level. However, at the chip to chip level considerable amount of research and development effort still needs to be conducted. Some of the promising new directions that are being investigated at UCSD include the use of 3-D silicon stacks in conjunction with MEMs devices, Conical tapered lens arrays for increased alignment tolerance [17] ,Chatoyant as a versatile CAD system for optoelectronics [18], Ultra low drive surface normal light modulators based on the VCSEL structure [19] and Electric-field assisted micro-assembly and pick and place for advanced integration [20].

References

1. Krishnamoorthy, A.V., Miller, D.A.B. "Firehose architectures for free-space optically interconnected VLSI circuits". *Journal of Parallel and Distributed Computing*, vol.41, (no.1), Academic Press,. pp.109-14. 25 Feb. 1997
2. P. J. Marchand, A. V. Krishnamoorthy, G. I. Yayla, S. C. Esener and U. Efron, "Optically augmented 3-D computer: system technology and architecture." *J.*

Parallel Distrib.Comput. Special Issue on Optical Interconnects, vol.41, no.1, pp.20-35, February 1997

3. Betzos, G.A.; Mitkas, P.A. "Performance evaluation of massively parallel processing architectures with three-dimensional optical interconnections," *Applied Optics*, vol.37, (no.2), pp.315-25, 10 Jan. 1998.

4. J. W. Goodman, F. J. Leonberger, S. C. Kung, and R. A. Athale, "Optical Interconnections for VLSI Systems, " *Proc. IEEE*, vol. 72, no. 7, pp. 850-66, Jul. 1984

5. L. A Bergman, W. H. Wu, A. R. Johnston, R. Nixon, S. C. Esener, C.C Guest, P. Yu, T.J. Drabik, M. Feldman, S. H. Lee, "Holographic Optical Interconnects in VLSI," *Opt. Eng.*, vol. 25, no. 10, pp. 1109-18, Oct. 1986

6. W. H. Wu, L. A Bergman, A. R. Johnston, C. C. Guest, S.C Esener, P.K.L Yu,. M. R. Feldman, S. H. Lee, "Implementation of optical Interconnections for VLSI," *IEEE Trans. Electron Devices*, vol. ED-34, no. 3, pp. 706-14, Mar. 1987

7. R. K. Kostuk, J. W. Goodman, and L. Hesselink, "Optical Imaging Applied to Microelectric Chip-to-Chip Interconnections," *Appl. Opt.*, vol. 24, no. 17, pp. 2851-8, Sep. 1985.

8. D. A. B. Miller, "Physical reasons for optical interconnection," *Intl. J. of Opto-electronics*, vol. 11, no.3, pp. 155-68, 1997.

9. A. Krishnamoorthy and D. A. B. Miller, " Scaling opto-electronic-VLSI circuits into 21st century: a technology roadmap," *IEEE JST in Quantum Opto-electronics*, Vol.2, No.1 , pp.55-76, Apr. 1996.

10. M. R. Feldman, S. C. Esener, C. C. Guest, and S. H. Lee, "Comparison between optical and electrical interconnects based on power and speed considerations," *Appl. Opt.*, 27, no.9, pp. 1742-51, May 1988.

11. F. Kiamilev, P. Marchand, A. Krishnamoorthy, S. Esener, and S. H. Lee, "Performance comparison between opto-electronic and VLSI multistage interconnection networks," *IEEE J. Lightwave Technol.*, vol. 9, no. 12, pp.1674-92, Dec. 1991.

12. A. V. Krishnamoorthy, P. Marchand, F. Kiamilev, K. S. Urquhart, S. Esener, "Grain-size consideration for opto-electronic multistage interconnection network," *Appl. Opt.*, 31 (26), pp. 5480-5507, 1992.

13. G. Yayla, P. Marchand, and S. Esener, "Speed and Energy Analysis of Digital Interconnections: Comparison of On-chip, Off-chip and Free-Space Technologies," *Appl. Opt.*, 37, pp. 205-227, January 1998.

14. Morgan, R.A.; Bristow, J.; Hibbs-Brenner, M.; Nohava, J.; Bounnak, S.; Marta, T.; Lehman, J.; Yue Liu "Vertical cavity surface emitting lasers for spaceborne photonic interconnects," *Proceedings of the SPIE* – The International Society for Optical Engineering, vol.2811, (Photonics for Space Environments IV, Denver, CO, USA, 6-7 Aug. 1996.) SPIE-Int. Soc. Opt. Eng,. pp.232-42.1996.

15. A. Krishnamoorthy, "Applications of opto-electronic VLSI technologies," Optical Computing 1998, Bruges, Belgium , June 1998.

16. A. V. Krishnamoorthy, L. M. F. Chirovsky, W. S. Hobson, R. E. Leibenguth, S. P. Hui, G. J. Zydzik, K. W. Goosen, J. D. Wynn, B. J. Tseng, J. A. Walker, J. E. Cunningham, and L. A. D'Asaro, "Vertical-Cavity Surface-Emitting Lasers Flip-Chip Bonded to Gigabit-per-Second CMOS Circuits", IEEE Phot. Tech. Lett., Vol.11, No.1, pp.128-130, 1999.

17. Cornelius Diamond, Ilkan Cokgor, Aaron Birkbeck and Sadik Esener, " Optically Written Conical Lenses for Resonant Structures and Detector Arrays" *Optical Society of America, Spatial Light Modulators and Integrated Optoelectronic Arrays, Technical Digest, Salt Lake City, Snowmas*s, April 1999.

18. S.P. Levitan, T.P. Kurzweg, P. Marchand, M.A. Rempel, D.M. Chiarulli, J.A. Martinex, C. Fan, and F.B. McCormick, "Chatoyant, a Computer-Aided Design Tool for Free-Space Optoelectronic Systems," *Appl. Opt.*, January 1998.

19. O. Kibar and S. Esener "Sub-threshold operation of a VCSEL structure for ultra-low voltage, high speed, high contrast ratio spatial light modulation" *Optical Society of America, Spatial Light Modulators and Integrated Optoelectronic Arrays, Technical Digest, Salt Lake City, Snowmas*s, April 1999.

20. S. C. Esener, D. Hartmann, M. J. Heller and J. M. Cable, " DNA Assisted Micro-Assembly: A Heterogeneous Integration Technology For Optoelectronics, " Proc. SPIE Critical Reviews of Optical Science and Technology, Heterogeneous Integration, Ed. A. Hussain, CR70-7, Photonics West 98, San Jose, January-98.

Fast Sorting on a Linear Array with a Reconfigurable Pipelined Bus System*

Amitava Datta, Robyn Owens, and Subbiah Soundaralakshmi

Department of Computer Science
The University of Western Australia
Perth, WA 6907
Australia
email:{datta,robyn,laxmi}@cs.uwa.edu.au

Abstract. We present a fast algorithm for sorting on a linear array with a reconfigurable pipelined bus system (LARPBS), one of the recently proposed parallel architectures based on optical buses. Our algorithm sorts N numbers in $O(\log N \log \log N)$ worst-case time using N processors. To our knowledge, the previous best sorting algorithm on this architecture has a running time of $O(\log^2 N)$.

1 Introduction

Recent advances in optical and opto-electronic technologies indicate that optical interconnects can be used effectively in massively parallel computing systems involving electronic processors [1]. The delays in message propagation can be precisely controlled in an optical waveguide and this can be used to support high bandwidth pipelined communication. Several different opto-electronic parallel computing models have been proposed in the literature in recent years. These models have opened up new challenges in algorithm design. We refer the reader to the paper by Sahni [8] for an excellent overview of the different models and algorithm design techniques on these models.

Dynamically reconfigurable electronic buses have been studied extensively in recent years since they were introduced by Miller *et al.* [3]. There are two related opto-electronic models based on the idea of dynamically reconfigurable optical buses, namely, the *Array with Reconfigurable Optical Buses* (AROB) and the *Linear Array with Reconfigurable Pipelined Bus Systems* (LARPBS). The LARPBS model has been investigated in [2,4–6] for designing fast algorithms from different domains. There are some similarities between these two models. For example, the buses can be dynamically reconfigured to suit computational and communication needs and the time complexities of the algorithms are analyzed in terms of the number of bus cycles needed to perform a computation, where a bus cycle τ is the time needed for a signal to travel from end to end along a bus. However, there is one crucial difference between these two models. In the AROB model, the processors connected to a bus are able to count optical pulses within a bus cycle, whereas in the LARPBS model counting is not allowed during a bus cycle. In the LARPBS model, processors can set switches at the start of a bus cycle and take no further part during a bus cycle. In other words, the basic assumption of the

---

* This research is partially supported by an Australian Research Council (ARC) grant.

J. Rolim et al. (Eds.): IPDPS 2000 Workshops, LNCS 1800, pp. 1110-1117, 2000.
© Springer-Verlag Berlin Heidelberg 2000

AROB model is that the CPU cycle time is equal to the optical pulse time since the processors connected to a bus need to count the pulses. This is an unrealistic assumption in some sense since the pulse time is usually much faster than the CPU time of an electronic processor. On the other hand, the LARPBS model is more realistic since the basic assumption in this model is that the bus cycle time is equal to the CPU cycle time.

Sorting is undoubtedly one of the most fundamental problems in computer science and a fast sorting algorithm is often used as a preprocessing step in many other algorithms. The first sorting algorithm on the LARPBS model was designed by Pan *et al.* [7]. Their algorithm is based on the sequential quicksort algorithm and runs in $O(\log N)$ time on an average and in $O(N)$ time in the worst case on an N processor LARPBS. To our knowledge, the best sorting algorithm for this model is due to Pan [4]. His algorithm sorts N numbers in $O(\log^2 N)$ worst-case time. We present an algorithm for sorting N numbers in $O(\log N \log \log N)$ time on an LARPBS with N processors. Our algorithm is based on a novel deterministic sampling scheme for merging two sorted arrays of length N each in $O(\log \log N)$ time.

2   Fast sorting on the LARPBS

We refer the reader to [2, 5, 6] for further details of the LARPBS model. The measure of computational complexity on an LARPBS is the number of bus cycles used for the computation and the amount of time spent by the processors for local computations. A bus cycle is the time needed for end to end message transmission over a bus and assumed to take only $O(1)$ time. In most algorithms on the LARPBS model, a processor performs only a constant number of local computation steps between two consecutive bus cycles and hence the time complexity of an algorithm is proportional to the number of bus cycles used for communication.

We use some basic operations on the LARPBS in our algorithm. In a *one-to-one communication*, a source processor sends a message to a destination processor. In a *broadcasting operation*, a source processor sends a message to all the other $N - 1$ processors in an LARPBS consisting of N processors. In a *multicasting operation*, a source processor sends a message to a group of destination processors. In a *multiple multicasting* operation, a group of source processors perform multicasting operations. A destination processor can only receive a single message during a bus cycle in a multiple multicasting operation. In the *binary prefix sum* computation, each processor in an LARPBS with N processors stores a binary value, with processor P_i, $1 \le i \le N$ storing the binary value b_i. The aim is to compute the N prefix sums S_i, $1 \le i \le N$, where $S_i = \sum_{j=1}^{i} b_j$.

Suppose each processor in an N processor LARPBS is marked either as *active* or as *inactive* depending on whether the processor holds a 1 or a 0 in one of its registers R_i. Also, each processor holds a data element in another of its registers R_j. In the *ordered compression* problem, the data elements of all the active processors are brought to consecutive processors at the right end of the array, keeping their order in the original array intact.

The following lemma has been proved by Li *et al.* [2] and Pan and Li [5].

Lemma 1. *One-to-one communication, broadcasting, multicasting, multiple multicasting, binary prefix sum computation and ordered compression all can be done in $O(1)$ bus cycles on the LARPBS model.*

Given a sequence of N numbers $k_1, k_2, \ldots, k_N$, the sorting problem is to arrange these numbers in nondecreasing order. Our sorting algorithm on the LARPBS is based on the well known sequential *merge sort* algorithm. We use an algorithm for merging two sorted arrays of length N each in $O(\log \log N)$ time on an LARPBS with N processors. We now give some definitions and properties which are necessary for designing our merging algorithm.

2.1   Definitions and properties

Suppose we have two arrays $L = \{l_1, l_2, \ldots, l_N\}$ and $R = \{r_1, r_2, \ldots, r_N\}$ each having N elements and each sorted according to ascending order. We assume for simplicity that all the elements in $L \cup R$ are distinct. It is easy to modify our algorithm for the case when an element may occur multiple times. For an element $l_i \in L$, we denote its predecessor and successor in L by $pred(l_i)$ and $succ(l_i)$. Successors and predecessors are denoted similarly for an element in R.

The *rank* of l_i in L is its index i in the array L and denoted by $rank_L(l_i)$. Similarly, the rank of r_i in R is its index i in the array R and denoted by $rank_R(r_i)$. The rank of l_i in R, denoted by $rank_R(l_i)$, is $rank_R(r_j)$ of an element $r_j \in R$ such that $r_j < l_i$ and there is no other element $r_k \in R$ such that $r_j < r_k < l_i$. Sometime we will write $rank_R(l_i) = r_j$ by abusing the notation.

Similarly, the rank of r_i in L, denoted by $rank_L(r_i)$, is $rank_L(l_j)$ of an element l_j in L such that $l_j < r_i$ and there is no other element $l_k \in L$ such that $l_j < l_k < r_i$. For an element $l_m \in L$, the rank of l_m in $L \cup R$ is denoted by $rank(l_m)$. The following lemma is a direct consequence of definitions of these three kinds of ranks.

Lemma 2. *For an element $l_m \in L$, $1 \leq m \leq N$, rank$(l_m) =$rank$_L(l_m)+$ rank$_R(l_m)$. Similarly, for an element $r_n \in R$, $1 \leq n \leq N$, rank$(r_n) =$ rank$_R(r_n) +$ rank$_L(r_n)$.*

It is clear from Lemma 2 that if we compute $rank_R(l_i)$ for each element $l_i \in L$, we can compute $rank(l_i)$. Note that, we already know $rank_L(l_i)$ since L is already sorted and $rank_L(l_i)$ is simply the index i. Similarly, if we compute $rank_L(r_j)$ for each element $r_j \in R$, we can compute $rank(r_j)$. We refer to these two problems as *ranking of L in R* and *ranking of R in L*.

We do the ranking of L in R recursively in several stages. When every element in L is ranked in R, we say that L is *saturated*. Consider a stage when L is still unsaturated. In other words, some elements in L are already ranked in R and some are yet to be ranked.

Definition 3 *Consider two consecutive ranked elements l_m and l_n, $m < n$. All the elements between l_m and l_n, i.e., $succ(l_m), \ldots, pred(l_n)$ are unranked and these elements are called the gap between l_m and l_n and denoted by $Gap(l_m, l_n)$.*

Definition 4 *Consider two consecutive ranked elements l_m and l_n in L. Suppose, $rank_R$ $(l_m) = r_p$ and $rank_R(l_n) = r_q$. The elements $succ(r_p), \ldots, r_q$ are collectively called the cover of $Gap(l_m,\ l_n)$ and denoted as $Cover(l_m,\ l_n)$. See Figure 1 for an illustration.*

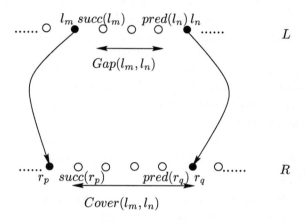

Figure 1. Illustration for *gap* and *cover*.

Lemma 5. *For an element $l_i \in Gap(l_m, l_n)$, **either** $rank_R(l_i) = rank_R(l_m)$ **or** $rank_R(l_i)$ $= r_m$ such that, $r_m \in Cover(l_m, l_n)$.*

Definition 6 *For two ranked elements $l_m,\ l_n \in L$, if $l_n \neq succ(l_m)$, we say that $Gap(l_m,\ l_n)$ is non-empty.*

Definition 7 *Consider a non-empty gap $Gap(l_m,\ l_n)$ and its $Cover(l_m,\ l_n)$. We say that $Gap(l_m,\ l_n)$ has an **empty cover** if $rank_R(l_m) = rank_R(l_n)$, i.e., if l_m and l_n are ranked at the same element in R.*

The following two lemmas are crucial for our algorithm.

Lemma 8. *If $Cover(l_m, l_n)$ is the non-empty cover for $Gap(l_m, l_n)$, an element $r_j \in Cover(l_m, l_n)$ must be ranked in $Gap(l_m, l_n)$.*

Lemma 9. *If $Gap(l_m,\ l_n)$ and $Gap(l_o,\ l_p)$ are two arbitrary and distinct non-empty gaps in L, then $Gap(l_m,\ l_n) \cap Gap(l_o,\ l_p) = \emptyset$. Similarly, if $Cover(l_m,\ l_n)$ and $Cover(l_o,\ l_p)$ are two arbitrary and distinct non-empty covers in R, then $Cover(l_m,\ l_n) \cap Cover(l_o,\ l_p) = \emptyset$.*

We assume that the sorted sequences L and R have N and M elements respectively. First, we choose every $\sqrt{N}$-th element, i.e, the elements $l_{\sqrt{N}}, l_{2\sqrt{N}}, \ldots, l_{\sqrt{N}\sqrt{N}}$ from L. We denote the set $\{l_{\sqrt{N}}, l_{2\sqrt{N}}, \ldots, l_{\sqrt{N}\sqrt{N}}\}$ as $Sample_L$. Similarly, we choose the elements $r_{\sqrt{M}}, r_{2\sqrt{M}}, \ldots, r_{\sqrt{M}\sqrt{M}}$ from R and denote this set of elements as $Sample_R$. Note that there are $\sqrt{N}$ elements in $Sample_L$ and $\sqrt{M}$ elements in $Sample_R$.

The elements $l_{i\sqrt{N}}$ (resp. $r_{i\sqrt{N}}$), $1 \le i \le \sqrt{N}$ in $Sample_L$(resp. $Sample_R$) impose a block structure on the sequence L(resp. R). Consider two consecutive elements $l_{i\sqrt{N}}$ and $l_{(i+1)\sqrt{N}}$ in $Sample_L$. The elements $\{succ(l_{i\sqrt{N}}), \ldots, l_{(i+1)\sqrt{N}}\}$ are called the i-th block in L imposed by $Sample_L$ and denoted by $Block_i^L$. The superscript L indicates that it is a block in the sorted sequence L. The elements $l_{i\sqrt{N}}$ and $l_{(i+1)\sqrt{N}}$ are called the *sentinels* of $Block_i^L$. Similarly, we define the jth block $Block_j^R$ imposed by two consecutive elements $r_{j\sqrt{M}}$ and $r_{(j+1)\sqrt{M}}$ of $Sample_R$.

Consider the ranking of $Sample_L$ in $Sample_R$. When an element $l_{i\sqrt{N}} \in Sample_L$ is ranked in $Sample_R$, we denote this rank by a superscript S, i.e., $rank_R^S(l_{i\sqrt{N}})$. Note that, $rank_R^S(l_{i\sqrt{N}})$ is only an approximation of the true rank $rank_R(l_{i\sqrt{N}})$ of $l_{i\sqrt{N}}$ in R.

Assume that for two consecutive elements $l_{k\sqrt{N}}$ and $l_{(k+1)\sqrt{N}}$ in $Sample_L$, $rank_R^S$ $(l_{k\sqrt{N}}) = r_{m\sqrt{M}}$ and $rank_R^S(l_{(k+1)\sqrt{N}}) = r_{n\sqrt{M}}$, where $r_{m\sqrt{M}}$ and $r_{n\sqrt{M}}$ are two elements in $Sample_R$. In the following lemma, we estimate the true ranks of the elements in $Block_k^L$ in R.

Lemma 10. *If an element $l_r \in L$ is in $Block_k^L$, i.e., in between the two elements $l_{k\sqrt{N}}$ and $l_{(k+1)\sqrt{N}}$, l_r must be ranked in $Block_m^R \cup Block_{m+1}^R \cup \ldots \cup Block_n^R$, i.e., in $Cover(l_{k\sqrt{N}}, l_{(k+1)\sqrt{N}})$.*

2.2   An $O(\log \log N)$ time merging algorithm on the LARPBS

A variant of the following lemma has been proved by Pan *et al.* [7].

Lemma 11. *Given two sorted sequences A and B of length $\sqrt{N}$ each, all the elements of A can be ranked in B in $O(1)$ bus cycles on an LARPBS with N processors.*

Our algorithm is recursive and at every level of recursion, our generic task is to set up appropriate subproblems for the next level of recursion. In the following description, we explain how all the subproblems associated with $Gap(l_m, l_n)$ and $Cover(l_m, l_n)$ are set up for the next level of recursion. We assume that $Gap(l_m, l_n)$ has N' elements and $Cover(l_m, l_n)$ has M' elements.
Step 1.

We take a sample from $Gap(l_m, l_n)$ by choosing every $\sqrt{N'}$-th element from Gap (l_m, l_n). We denote this sample by $Sample_L(Gap(l_m, l_n))$ Similarly, we take a sample from $Cover(l_m, l_n)$ by choosing every $\sqrt{M'}$-th element from $Cover(l_m, l_n)$ and denote it by $Sample_R(Cover(l_m, l_n))$. We explain how to take the sample from $Gap(l_m, l_n)$. The sample from $Cover(l_m, l_n)$ is taken in a similar way.

First, each processor holding an element in $Gap(l_m, l_n)$ writes a 1 in one of its registers. Next, a parallel prefix computation is done in one bus cycle to get N', the total number of elements in $Gap(l_m, l_n)$ in the processor holding l_n. This processor computes $\sqrt{N'}$ and broadcasts $\sqrt{N'}$ to all the processors in $Gap(l_m, l_n)$. We assume for simplicity that $\sqrt{N'}$ is an integer. Each processor in $Gap(l_m, l_n)$ determines whether its prefix sum is an integer multiple of $\sqrt{N'}$ and marks itself as a member of

$Sample_L(Gap(l_m, l_n))$ accordingly. Note that, $Sample_L(Gap(l_m, l_n))$ consists of the sentinels of the blocks in L.

Step 2.

In this step, we assume that $\sqrt{N'} < \sqrt{M'}$ and we rank $Sample_L(Gap(l_m, l_n))$ in $Sample_R(Cover(l_m, l_n))$. This ranking is done by the method in Lemma 11 in $O(1)$ bus cycles.

Step 3.

After the ranking in Step 2 is over, for every sentinel $l_{k\sqrt{N'}} \in Sample_L(Gap(l_m, l_n))$, we know $Block_m^R$, the block of $\sqrt{M'}$ elements in R in which $l_{k\sqrt{N'}}$ should be ranked.

Next, we determine all the sentinels in $Sample_L(Gap(l_m, l_n))$ ranked in $Block_m^R$ in the following way. After the ranking in Step 2 is over, each processor holding a sentinel $l_{i\sqrt{N'}}$ gets $rank_R^S(l_{(i+1)\sqrt{N'}})$ from its neighbor in the sample through a one-to-one communication. After this, a group of consecutive sentinels in $Sample_L(Gap(l_m, l_n))$ which are ranked at the same block of $Sample_R(Cover(l_m, l_n))$ can be determined.

We consider two cases depending on whether a single sentinel or multiple sentinels from $Sample_L(Gap(l_m, l_n))$ are ranked in the same block of $Sample_R(Cover(l_m, l_n))$.

Case i. In this case, only one sentinel $l_{k\sqrt{N'}}$ in $Sample_L(Gap(l_m, l_n))$ is ranked in $Block_m^R$. The processor holding $l_{k\sqrt{N'}}$ broadcasts $l_{k\sqrt{N'}}$ to all the processors in $Block_m^R$ and the processors in $Block_m^R$ determine $rank_R(l_{k\sqrt{N'}})$. This takes $O(1)$ bus cycles. We determine $rank_R(l_{(k+1)\sqrt{N'}})$ in $Block_n^R$ in a similar way. Note that, the elements $succ(rank_R(l_{k\sqrt{N'}})), \ldots, rank_R(l_{(k+1)\sqrt{N'}}))$ are the elements in $Cover(l_{k\sqrt{N'}}, l_{(k+1)\sqrt{N'}})$.

It follows from Lemma 5 that all the elements in $Gap(l_{k\sqrt{N'}}, l_{(k+1)\sqrt{N'}})$ must be ranked either at $rank_R(l_{k\sqrt{N'}})$ or among the elements in $Cover(l_{k\sqrt{N'}}, l_{(k+1)\sqrt{N'}})$. Similarly, it follows from Lemma 8 that all the elements in $Cover(l_{k\sqrt{N'}}, l_{(k+1)\sqrt{N'}})$ must be ranked at the elements in $Gap(l_{k\sqrt{N'}}, l_{(k+1)\sqrt{N'}})$. Hence we recursively call our algorithm with elements in $Gap(l_{k\sqrt{N'}}, l_{(k+1)\sqrt{N'}})$ and elements in $Cover(l_{k\sqrt{N'}}, l_{(k+1)\sqrt{N'}})$. In this recursive call, all the elements from L are within a block of size $\sqrt{N'}$. The processors holding the elements in $Gap(l_{k\sqrt{N'}}, l_{(k+1)\sqrt{N'}})$ and the elements in $Cover(l_{k\sqrt{N'}}, l_{(k+1)\sqrt{N'}})$ participate in this recursive call.

Case ii. In this case, multiple sentinels $l_{j\sqrt{N'}}, \ldots, l_{k\sqrt{N'}}$ are ranked in $Block_m^R$. In two bus cycles, first $rank_R(l_{j\sqrt{N'}})$ and then $rank_R(l_{k\sqrt{N'}})$ are determined by broadcasting first $l_{j\sqrt{N'}}$ and then $l_{k\sqrt{N'}}$ to all the processors in $Block_m^R$. We then recursively call our algorithm with the elements in $Gap(l_{j\sqrt{N'}}, l_{k\sqrt{N'}})$ and the elements in $Cover(l_{j\sqrt{N'}}, l_{k\sqrt{N'}})$. Note that, all the elements from R are within a block of size $\sqrt{M'}$ in this recursive call.

These two types of recursive calls are illustrated in Figure 2.

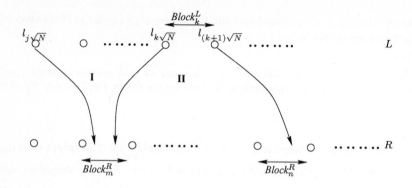

Figure 2. The two types of recursive calls are indicated by **I** and **II**. In the first type, the elements from R are within the same block of size $\sqrt{M'}$. In the second type, the elements from L are within the same block of size $\sqrt{N'}$.

Note that, the inputs to each level of recursion are disjoint subsets of processors holding elements of L and R and hence all the one-to-one communication, broadcasting and multiple multicasting operations at each level of recursion for each of the subproblems can be done simultaneously in parallel.

Once the recursive calls return, an element $l_i \in L$ knows $rank_R(l_i)$ and it knows $rank_L(l_i)$ since L is already sorted. Hence the processor holding l_i can compute $rank(l_i)$ and sends l_i to the processor with index $rank(l_i)$ through a one-to-one communication. This can be done in one bus cycle. Similarly the overall rank of each element in R can be computed and the elements can be sent to the appropriate processors. Hence each processor P_i will hold the ith element in $L \cup R$ after the merging algorithm terminates.

This concludes the description of our merging algorithm.

Lemma 12. *The merging algorithm terminates in $O(\log \log N)$ bus cycles with all the elements of L ranked in R and all the elements of R ranked in L.*

Proof. (sketch) Suppose in the ith level of recursion, each block in L and R is of size $\sqrt{N}$ and $\sqrt{M}$ respectively. Suppose, the input to one of the recursive calls at the $(i+1)^{th}$ level of recursion are the elements in two groups of processors G^L from L and G^R from R. From the description of the algorithm, it is clear that either G^L is within a block of size $\sqrt{N}$ or G^R is within a block of size $\sqrt{M}$. Hence, due to this recursive call, at the $(i+1)^{th}$ level of recursion, either we get new blocks of size $N^{1/4}$ in L or we get new blocks of size $M^{1/4}$ in R. This gives a recurrence of : $T(N) = T(\sqrt{N}) + O(1)$ or a recurrence of : $T(M) = T(\sqrt{M}) + O(1)$, since each level of recursion takes $O(1)$ bus cycles. Hence, the recursion stops after $2 \log \log N$ levels and all the elements in L and R are ranked at that stage.

2.3   The sorting algorithm

Phase 1. Initially, each processor in an N processor LARPBS holds one element from the input. The complete LARPBS with N processors is recursively divided in this phase.

Consider a subarray with processors P_i, $P_{i+1}, \ldots$, P_j to be divided into two equal parts. Each processor writes a 1 in one of its registers and a prefix computation is done to renumber the processors from 1 to $j - i$. Now, the last prefix sum is broadcast to all the processors and the processor with index $\lfloor (j + i)/2 \rfloor$ splits the bus to divide the original subarray into two subarrays of equal size. This process is repeated for all the subarrays recursively until each subarray contains only one processor and one element which is trivially sorted. This phase can be completed in $O(\log N)$ bus cycles.

Phase 2. The merging is done in this phase using the algorithm in Section 2.2. In the generic merging step, a pair of adjacent subarrays of equal size merge their elements to form a larger subarray of double the size. Each subarray participating in this pairwise merging first renumber its processors starting from 1 and then the merging algorithm is applied. At the end, processor P_i, $1 \leq i \leq N$ in the original array holds the element with rank i from the input set.

Since there are $O(\log N)$ levels in the recursion and the merging at each level can be performed in $O(\log \log N)$ bus cycles, the overall algorithm takes $O(\log N \log \log N)$ bus cycles and hence $O(\log N \log \log N)$ time since each bus cycle takes $O(1)$ time.

Theorem 1. *N elements can be sorted in $O(\log N \log \log N)$ deterministic time on an LARPBS with N processors.*

References

1. Z. Guo, R. Melhem, R. Hall, D. Chiarulli, S. Levitan, "Pipelined communication in optically interconnected arrays", *Journal of Parallel and Distributed Computing*, **12**, (3), (1991), pp. 269-282.
2. K. Li, Y. Pan and S. Q. Zheng, "Fast and processor efficient parallel matrix multiplication algorithms on a linear array with a reconfigurable pipelined bus system", *IEEE Trans. Parallel and Distributed Systems*, **9**, (8), (1998), pp. 705-720.
3. R. Miller, V. K. Prasanna Kumar, D. Reisis and Q. F. Stout, Parallel computations on reconfigurable meshes. *IEEE Trans. Computers*, **42**, (1993), 678-692.
4. Y. Pan, "Basic data movement operations on the LARPBS model", in *Parallel Computing Using Optical Interconnections*, K. Li, Y. Pan and S. Q. Zheng, eds, Kluwer Academic Publishers, Boston, USA, 1998.
5. Y. Pan and K. Li, "Linear array with a reconfigurable pipelined bus system - concepts and applications", *Journal of Information Sciences*, **106**, (1998), pp. 237-258.
6. Y. Pan, M. Hamdi and K. Li, "Efficient and scalable quicksort on a linear array with a reconfigurable pipelined bus system", *Future Generation Computer Systems*, **13**, (1997/98), pp. 501-513.
7. Y. Pan, K. Li and S. Q. Zheng, "Fast nearest neighbor algorithms on a linear array with a reconfigurable pipelined bus system", *Journal of Parallel Algorithms and Applications*, **13**, (1998), pp. 1-25.
8. S. Sahni, "Models and algorithms for optical and optoelectronic parallel computers", *Proc. 1999 International Symposium on Parallel Architectures, Algorithms and Networks*, IEEE Computer Society, pp. 2-7.

Architecture Description and Prototype Demonstration of Optoelectronic Parallel-Matching Architecture

Keiichiro Kagawa, Kouichi Nitta, Yusuke Ogura, Jun Tanida,
and Yoshiki Ichioka

**Department of Material and Life Science, Graduate School of Engineering,
Osaka University

Abstract. We propose an optoelectronic parallel-matching architecture
(PMA) that provides powerful processing capability for distributed algorithms comparing with traditional parallel computing architectures.
The PMA is composed of a parallel-matching (PM) module and multiple processing elements (PE's). The PM module is implemented by
a large-fan-out free-space optical interconnection and parallel-matching
smart-pixel array (PM-SPA). In the proposed architecture, each PE can
monitor the other PE's by utilizing several kinds of global processing by
the PM module. The PE's can execute concurrent data matching among
the others as well as inter-processor communication. Based on the state-
of-the-art optoelectronic devices and a diffractive optical element, a prototype of the PM module is constructed. The prototype is assumed to
be used in a multiple processor system composed of 4×4 processing elements, which are completely connected via 1-bit optical communication
channels. On the prototype demonstrator, the fundamental operations of
the PM module such as parallel-matching operations and inter-processor
communication were virified at 15MHz.

1   Introduction

Parallel distributed processing is an effective method to accelerate the performance of computing system. In the parallel distributed processing, a task is
divided into a number of processes executable concurrently. The processes are
distributed and executed over multiple processing elements (PE's), so that the
total processing time can be reduced.

A heuristic optimization described by a distributed algorithm is a good application of a parallel computing system. In the algorithm, the solution space is
divided into multiple pieces of segments, in which the candidates of the solution
are sought concurrently by multiple PE's. In the framework of the traditional
parallel computing architecture, global processing to calculate multiple data from
all the PE's can be a processing bottleneck. Because communication between the
PE's and processing are implemented separately, the heavy traffic occurs on the

---

** kagawa@mls.eng.osaka-u.ac.jp

J. Rolim et al. (Eds.): IPDPS 2000 Workshops, LNCS 1800, pp. 1118-1125, 2000.

network path to or from the PE that executes the global processing. The bottleneck causes throughput reduction of the whole parallel computing system. This bottleneck can not be eliminated by simply increasing the communication capacity of network. Therefore, the traditional parallel computing architectures are not always suitable for the distributed algorithms.

In this paper, we propose an optoelectronic parallel-matching architecture (PMA) which is an effective parallel computing architecture suitable for the distributed algorithms. The PMA is based on an optoelectronic heterogeneous architecture formerly presented by Tanida *et al.*,[1] which is composed of electronic parallel processors for local processing and an optical network processor for interconnection and global processing between the electronic processors. The optical network processor is assumed to be embodied by the optical interconnection and the smart-pixes[2] for wide communication bandwidth and dense connectivity between the PE's. In the architecture, both electronic and optical processors work in complementary manner. An electronic processor shows high performance in the local processing, whereas an optical processor is good at the global processing. The system based on the PMA also has ability to execute the global processing without degrading the throughput of network. Detection of the PE's satisfying a given condition and summation of absolute differences over the multiple PE's are typical examples of the global processing. The optical network processor of the PMA is called a parallel-matching (PM) module, which consists of a large-fan-out free-space optical interconnection and a parallel-matching smart-pixel array (PM-SPA). The proposed architecture can reduce the execution time for the fundamental global data processing: global data matching, detection of the maximum (minimum) data, and ranking of the data, compared with the other traditional architectures with photonic networks.

2   Parallel Matching Architecture

We assume a multiple-instruction multiple-data stream (MIMD) parallel computing system consisting of N PE's embodied by the smart-pixel technology. The PE's are connected each other via a photonic network. A heuristic optimization algorithm based on the distributed algorithm is a good application of the parallel computing systems, which can be applied to the problems that do not always have a rigorous solving method. A general procedure of the distributed optimization algorithm is composed of distribution of the data, parallel processing, and integration of the calculated data. First, the candidates of solutions are distributed to the PE's. Second, each PE locally calculates the fitness function of candidate. Finally, good candidates are selected among the candidates based on the values of the fitness function. Note that this operation is achieved by global processing over the multiple PE's.

Figure 1 shows the system compositions for the distributed algorithms by the traditional MIMD parallel computing system and the parallel matching architecture. The traditional architecture has a hierarchy composed of a master PE and multiple slave PE's as shown in Fig. 1(a). The rolls of the master PE are

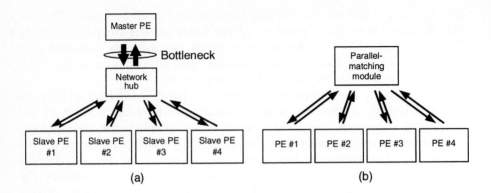

Fig. 1. Configurations of parallel computing architectures: (a) a traditional master-slave architecture and (b) the parallel-matching architecture (PMA)

data distribution, integration, and global processing. The master PE distributes the data to the slave PE's and integrates the resulting data from them through the network. After data integration, the master PE executes global processing locally. Because the amount of the network traffic in the data distribution and integration is very large, these procedures can be processing bottlenecks. This bottleneck can not be eliminated by simply increasing the communication capacity of network, for the total amount of the fanned-in data from N slave PE's to the master PE is N times as large as the bandwidth of the communication path between the network communication module and the PE's.

On the other hand, the PMA has a different composition as shown in Fig. 1(b). The PMA is composed of the PM module and the multiple PE's. In the PMA, the fitness of each candidate is compared with the candidates on the other PE's by using the global processing mechanism of the parallel-matching (PM) module. The PM module offers both networking and global processing, so that the master PE for data distribution and integration is not required. The PE's in the system have the same priority because the global processing is executed inside the PM module; that is the system has no hierarchy. As a result, there is no bottleneck in the proposed architecture in global processing.

The PM module consists of large-fan-out free-space optical interconnection and a parallel-matching smart-pixel-array (PM-SPA). The PM module is regarded as a kind of the network hub in which a specific mechanism for the global processing is built-in. The global processing in the PMA is data comparison among the data sent from the PE's. The PM module monitors the output data from all PE's, and concurrently compares the datum from each PE with the data from the other PE's. When a PE requires the compared result, it is sent back to the PE through the network communication channel. As mentioned above, data distribution and integration increase the network traffics and the processing overheads at a PE. However, because the global processing is ex-

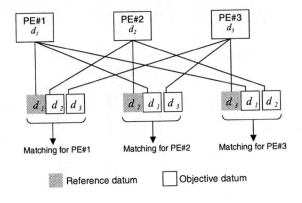

Fig. 2. Reference datum and objective data in the parallel matching. d_1, d_2, and d_3 denote the output data of PE#1-#3, respectively

ecuted inside the PM module without occupying the network bandwidth, the throughput of the total system does not become decreased.

We define the datum from each PE as the reference datum and the one from the other PE's as the objective data as shown in Fig. 2. The reference datum and the objective datum to be compared are called a matching pair. The PM module tests the reference datum and each of the objective data for the following conditions: 1) the reference datum is equivalent to the objective datum, 2) the reference datum is less than the objective datum, and 3) the reference datum is more than the objective datum. The result of the global comparison is expressed by a set of logical values. When the condition is satisfied, the returned value is 1 (true), otherwise 0 (false). These operations are called parallel-matching operations, which are denoted by **pEQU**, **pMORETHAN**, and **pLESSTHAN**, respectively. (The prefix **p** means 'parallel.') We also define the fourth parallel-matching operation: summation of the absolute differences denoted by **pDIFF**. This operation provides the summation of the absolute difference between the reference datum and the objective datum. Utilizing the **pDIFF** operation, each PE can obtain the quantitative value of the difference.

Figure 3 shows a schematic diagram of the parallel matching with 5 PE's. In the figure, PE's A, B, and C obtain 4-bit binary values representing the results of parallel matching: **pEQU**, **pMORETHAN**, and **pLESSTHAN**, respectively. PE-D obtains the result of the **pDIFF** operation. The numbers in the boxes of PE's are the output data from the PE's. After the output data are fanned out and exchanged, they are concurrently compared by the parallel-matching operations in the PM module. Then, one of the parallel-matching results or the objective datum is selected by the multiplexer on the request from the PE's. In general, for m-bit data format, up to $(m + 1)$ PE's can be compared at the same time. Finally, the selected result is sent back to each PE. In Fig. 3, example values of the parallel-matching results are shown. The operation mode of PE-E

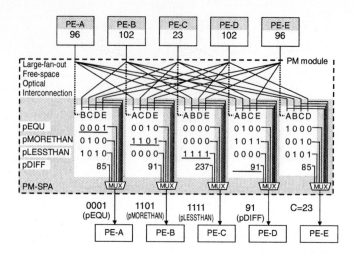

Fig. 3. Fundamental operations of the parallel-matching architecture. MUX means a multiplexer

is different from the others. That is the communication mode in which the data from PE-C is sent to PE-E transparently.

3 Experimental prototype system

We construct a prototype system of the PM module to demonstrate its fundamental operations. In designing the prototype, we assume the parallel computing system shown in Fig. 4. The parallel computing system consists of 4×4 PE', which are completely connected via the PM module. The PE's are located on a two-dimensional grid, and each of them is connected to the PM module with bit-serial optical fiber channels. Each PE is embodied by smart-pixels coupled with an optical fiber. The data from the PE's are sent to the PM module by the optical fibers. As mentioned below, a complete-connection network is implemented by optical data fanning. With the optically fanned-out signals, the parallel-matching operations and the processing for inter-PE communication are executed by the PM-SPA. The resulting data are emitted from the PM-SPA, and returned to the PE's through the optical fibers.

Figure 5 shows the schematic diagram of the optoelectronic complete-connection. As shown in Fig. 5(a), the optical signals from 4×4 PE's in the bit-serial format are assumed to be aligned on a two-dimensional grid as an input image toward the PM module. Because the whole image of the light signals is required for one PE, 4×4 replica images shown in Fig. 5(b) are prepared for 4×4 PE's. In the prototype, an 8×8-VCSEL array (Gigalase^{TM}; Micro Optical Devices; emitting wavelength, 850nm; pixel pitch, 250μm) is used as a light emitter array. In the prototype, the function of the PM-SPA is emulated by a CPLD

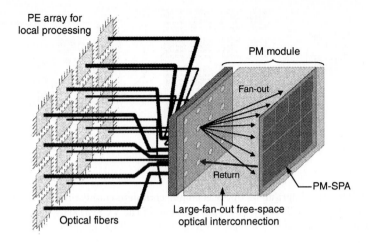

Fig. 4. Target prototype system of the PMA

(Model FLASH374i, Cypress) coupled with a 4 × 4-complementary-metal-oxide-semiconductor photodetector (CMOS-PD) array (Model N73CGD) supplied by United States-Japan Optoelectronic Project (JOP). As shown in Fig. 5(c), one of the replicas is detected by a CMOS-PD array, then transferred to the CPLD, and the fundamental operations of the PMA are executed.

For the large-fan-out optical interconnection, a conventional $4f$ optical correlator was adopted. We constructed a Fourier transform lens system whose focal length is 160.0mm for wavelength 850nm. In designing the lens system, CodeV^{TM} of Optical Research Associates was used.

As an optical fan-out element that generates complete-connection pattern shown in Fig. 5(b), we designed a phase-only computer-generated hologram (CGH) filter with two-level phase modulation based on the Gerchberg-Saxton algorithm.[3] Figure 6(a) shows the ideal mapping on the output plane of the interconnection optics. The output pattern contains 16 replicas of the VCSEL image arranged on a grid, in which each quadrant contains 2 × 2 replicas of the VCSEL image. Each replica corresponds to the optical signals for a single PE. Because the equipments used in fabrication of the CGH filter do not have enough fabrication accuracy to eliminate the 0th light spot, the copied images are located not to be overlapped with the 0th image in the design. The pitch and the margin of adjacent replicas of the VCSEL image are 2.5mm and 1.5mm, respectively. Figure 6(c) shows the filter pattern with two-level phase modulation. The CGH filter was fabricated by the electron beam (EB) lithography. Figure 6(b) shows the reconstructed interconnection pattern of the fabricated CGH filter for 4 × 4 VCSEL's when the filter was incorporated in the $4f$ optical correlator.

Finally, we operated the prototype system without the CGH filter to verify the fundamental parallel-matching operations and inter-PE communication. The

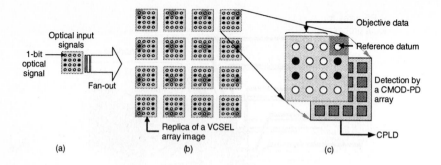

Fig. 5. Schematic diagram of optoelectronic complete-connection: (a) Output data displayed on the VCSEL array, (b) replica images of the VCSEL array for the complete-connection network, and (c) a replica image of the VCSEL image for one PE

data transfer was in the bit serial format, and the word length of the data was set to 4. From the experimental results, we have verified that the fundamental operations of the prototype were executed exactly at 15MHz. The operational speed was limited by the one of the CMOS-PD array. The bit rate of communication per PE and the total bit rate of the prototype were 15Mbps (bit per second) and 240Mbps, respectively. The frequencies of the parallel-matching operation for each PE and the whole system were 0.68M operations/sec and 11M operations/sec, respectively.

4   Conclusions

We have proposed an optoelectronic parallel-matching architecture (PMA) as an effective parallel computing architecture. The fundamental operations of the PMA, **pEQU**, **pMORETHAN**, **pLESSTHAN**, and **pDIFF**, have been defined. This architecture is specialized for the global data processing and has capability to accelerate execution of distributed algorithms, because the PMA has a specific mechanism for parallel-matching operations over multiple processing elements. The prototype system of the PMA was constructed to demonstrate the fundamental global operations of the PMA based on the state-of-the-art optoelectronic devices and a phase-only CGH filter. In the prototype, the PM-SPA, which was the core module of the PM module, was emulated by the CPLD and the CMOS-PD array. The prototype was assumed to be used with 4 × 4 PE's that are completely connected via the PM module with 1-bit optical channels. For optical interconnection of the prototype, a Fourier transform lens system was designed. As a fan-out element, the phase-only CGH filter with two-level phase modulation was designed based on the Gerchberg-Saxton algorithm, and was fabricated by the EB lithography. We confirmed that the prototype performed the fundamental parallel-matching operations and the inter-PE communication at 15MHz. For the whole system, the bit rate of inter-PE communication and the

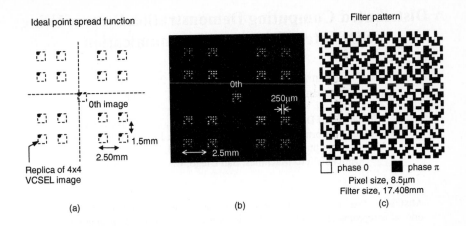

Fig. 6. (a) Designed optical interconnection pattern for complete-connect network composed of 4×4 PE's, (b) a part of the obtained CGH filter with two-level phase modulation, and (c) experimental result of the optical interconnection by the CGH filter

frequency of the parallel-matching operation were 240 Mbps and 11M operations per second, respectively. The operational speed of the prototype was limited by the CMOS-PD array. The performance can be improved by using high-speed photodetectors with high sensitivity such as MSM photodetectors coupled with transimpedance photo-amplifiers.

Acknowledgment

This research was supported by the JOP user funding under the Real World Computing Partnerchip (RWCP). The authors would like to appreciate the activities of the JOP. This work was also supported by Development of Basic *Tera* Optical Information Technologies, Osaka Prefecture Joint-Research Project for Regional Intensive, Japan Science and Technolgy Corporation.

References

1. P. Berthomé and A. Ferreira, *Optical interconnections and parallel processing: trends at the interface* (Kluwer Academic Publishers, London, 1998).
2. T. Kurokawa, S. Matso, T. Nakahara, K. Tateno, Y. Ohiso, A. Wakatsuki, and H. Tsuda, "Design approaches for VCSEL's and VCSEL-based smart pixels toward parallel optoelectronic processing systems," Appl. Opt. **37**, 194–204 (1996).
3. R. W. Gerchberg and W. O. Saxton, "A Practical Algorithm for the Determinaion of Phase from Image and Diffraction Plane Pictures," OPTIK **35**, 237 – 246 (1972).

A Distributed Computing Demonstration System Using FSOI Inter–Processor Communication

J. Ekman[1], C. Berger[2], F. Kiamilev[1], X. Wang[1],
H. Spaanenburg[3], P. Marchand[4], S. Esener[2]

[1]University of Delaware, ECE Dept., Newark, DE 19716, USA
[2]University of California San Diego, ECE Dept., La Jolla, CA 92093, USA
[3]Mercury Computer Systems Inc., Chelmsford, MA 01824, USA
[4]Optical Micro Machines, San Diego, CA 92121, USA

Abstract. Presented here is a computational system which uses free–space optical interconnect (FSOI) communication between processing elements to perform distributed calculations. Technologies utilized in the development of this system are integrated two–dimensional Vertical Cavity Surface Emitting Lasers (VCSELs) and MSM–photodetector arrays, custom CMOS ASICs, custom optics, wire–bonded chip–on–board assembly, and FPGA–based control. Emphasis will be placed on the system architecture, processing element features which facilitate the system integration, and the overall goals of this system.

1 Introduction

The area of optical interconnects is continually growing with many advances in optoelectronic devices, integration of CMOS ICs with these devices, and integration of hybrid electrical/optical devices into functional systems. It is clear that the flexibility in terms of scalability, and optical bandwidth which can be achieved by using optical interconnects will lead to changes in system architectures as designers move to to take advantage of this flexibility. As a part of the 3–D OptoElectronic Stacked Processor program[1], a demonstration system is being developed which illustrates the ability to construct distributed computational systems which use optical communication for passing data between processing elements. In this system, the distribution takes the form of linear chains of processors with nearest neighbor communication.

Communication between processors in a multiprocessor system quickly becomes the bottleneck and is therefore an ideal target for the integration of optical communication. One of the goals in developing this system was that of illustrating the use of optical communication in a low cost distributed system as a step toward validation of such architectures.

2 System Topology

This demonstration system consists of two linear chains of five processors each. Three processors in each chain are configured to perform computation and the two remaining (one on each end of the chain) are configured to bring data into and out of each chain. This is accomplished by converting between electrical–domain (digital)

J. Rolim et al. (Eds.): IPDPS 2000 Workshops, LNCS 1800, pp. 1126-1131, 2000.

and optical–domain (analog) signals at the ends of each chain (see figure 1). The two chains operate independently, but based on available optoelectronic device arrays, share OptoElectronic (OE) chips for communication. In addition to the ability to lengthen each chain, there is flexibility to scale the number of chains to yield a larger system. The optical chip–to–chip communication is achieved through the use of two dimensional VCSEL and MSM–photodetector arrays provided by Honeywell Technology Center[2] and custom optics designed at UCSD.

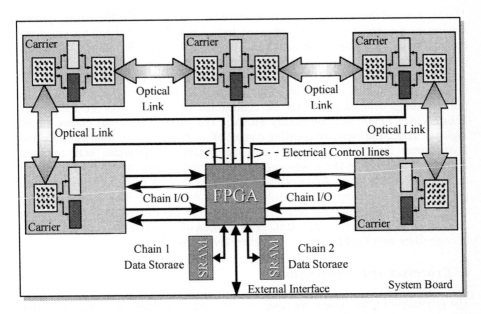

Figure 1. System diagram showig five carrier boards placed on system board. OptoElectronic arrays are shown on left and right sides of carrier boards and processing elements in the center of the carrier boards. The upper *(light)* PEs indicate one chain and the lower *(dark)* PEs indicate the second chain

2.1 Carrier Boards

Each unit in the chain is assembled onto a small "carrier board" where each of these carrier boards contains two processing elements (PEs) and two OE arrays. The OE arrays consist of sixteen VCSELs and sixteen photodetectors in an inter–digitated 4 x 4 array. These parts originally fabricated as a part of the GMU Co–Op program[3]. Each of the chips on the carrier boards are bare die, wire–bonded to contacts on the carrier board. One PE belongs to each of the two chains and the OE arrays are shared among the two chains with dedicated array elements for each chain. These carrier boards are then mounted onto a "system board" which also supports the optics, additional chips to provide control and system interface, and power connectors, etc. For the system described here, there are five carrier boards mounted onto one system board. This is illustrated in figure 1. Another goal of this demo system is to experiment with different opto–mechanics in an effort to demonstrate the ability to scale–down what has traditionally been a (physically) large part of such systems through the use of "plug–on–top" optical assemblies[4]. The construction of the carrier board modules facilitates this by allowing independent units to be rotated or moved according to a particular optical arrangement.

2.2 System Board

The purpose of the system board is to serve as a substrate for the entire system, supporting the carrier boards and opto–mechanics as well as providing the necessary control to the processing elements, and interfacing with the "outside world" to provide power, data, and system diagnostics. The board itself is a multi–layer printed circuit board (PCB) fabricated commercially. There is electrical and precision mechanical connection of the carrier boards to the system board. The primary components that perform the control and interfacing tasks are a high–end Xilinx Virtex FPGA and commodity SRAM. The Virtex FPGA was chosen for its high pin–count and capacity allowing control of the entire system from one chip and giving great flexibility to re–configure the system. It provides both the data necessary to configure the processing elements initially and control their operation throughout calculations. Additionally, it will provide data to the processor chains, gather results and monitor the results checking for errors. This approach helps reduce risk by allowing for reprogramming of the FPGA and also helps during assembly of such a prototype system. An extension of this system would have built–in controllers with the PEs and allow higher–level programming.

3 Processor Interconnection

The processors in this system are connected in two linear chains with each processor communicating with the one to its left and the one to its right. On the ends of the chain, there is only optical communication in one direction. Data is brought in and taken out of the ends of the chain electrically. The interconnection scheme chosen is meant to facilitate construction of this prototype system and serve as a starting point which can lead to more complex connection schemes which may provide additional benefit to specific applications. The logical connection of the processors in this system and the connection to the FPGA control unit is shown in figure 2. With this connection scheme, all data is brought into the processor chain from the two ends. All data communication within the processor chain is through the FSOI links. This both helps illustrate the viability of optical communication in a multiprocessor system as well as ensure that the links will be heavily utilized. The impact on the system architecture is of course that data must be passed to processors in the center of the chains before they can begin calculations. This is not seen as a serious drawback in this system as it adds only some latency to the beginning of calculations. It should be mentioned here that the application chosen for demonstration on this system is that of a radix–2 butterfly engine as a part of an FFT calculation. With this application, data points are brought into the chain and bounced back and forth between the processors in the chain during calculation and finally output from the ends of the chain. The two chains of this system are utilized to compute real and imaginary points simultaneously.

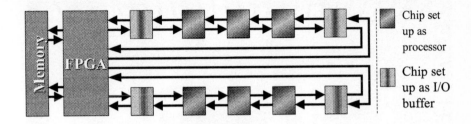

Figure 2. Diagram of logical connection between the multiple processors and the FPGA controller and memory. The upper and lower chains here illustrate the two independant chians in this system

4 Processing Element

The processing element itself is a custom ASIC designed and fabricated for this demo system. It is a 0.5 micron CMOS chip of roughly 10,000 transistors comprising both digital and analog circuitry (shown in figure 3). Some of the design goals for this chip were the ability to interface with the optoelectronic devices to be used in the system, that it provide digital signal processing capability, facilitate system construction and debugging, allow for possible changes to the optical system, and provide the capability to use the chip as an electrical/optical interface at the ends of each chain.

The design of the processing element is divided into the following functional units: Input/Output switching, arithmetic and logic units, optical I/O buffering, and a control interface. Input and output switching provides for the re–organization of data as it is received from, or transmitted over the optical chip–to–chip data links. The arithmetic and logic units provide the calculation capability based on a small instruction set. Translation between the optoelectronic analog domain and the digital domain is accomplished through on–chip receiver and VCSEL driver circuits. The control interface provides for configuration of the input/output switches and the selection of the function performed by the chip.

Input and output switches provided give much of the flexibility achieved in this design. The primary function of the input switches is the correction to input data words that may be necessary due to changes in the optical communication between chips or system I/O. The chip–to–chip communication links are all eight bits wide and the internal datapaths of the PE are six bits wide. The two remaining links out of every eight are devoted to fault tolerance. In the event that a data link is non–operational for any reason, the data being sent over that link can be diverted to one of these two redundant links. In such a case, the input switch would re–assemble the data word before calculation begins. In this manner the calculation is not corrupted or impeded by a loss of a link between chips. This fault tolerance is important in a demo system to ensure that a faulty link does not deteriorate the demonstration, but will also be important in future systems to provide reliability.

The output switches compliment the fault tolerance achieved with the input switches by providing the capability to re–route outgoing data onto a redundant link in the event that a link is known bad. Additionally, the output switches are used to select between outputs of the arithmetic and logic units, the receiver outputs in order

Figure 3. Microphotograph of the CMOS ASIC used as the processing element in this multiprocessor system. Eighty–six wire bonded pads are shown at the chip perimeter. Other unbonded pads are for probe–testing

to completely by–pass the processing functionality, and an auxiliary set of inputs which allow the chip to be used simply as a parallel VCSEL driver. Complete by–pass functionality is included in the PE chip to add flexibility and aid in system construction and debugging as it will allow chips to logically be removed from the chain without changes to the optics and also isolate the optical path from the digital functionality. The dataflow through the PE is shown in the diagram of figure 4.

In addition to the possible loss of an optical data link, changes to the optical system may result in a flipping of the data word during transmission. In order to allow different optical systems to be explored with this system, the ability to account for such flipping is included in the input switches. A final feature of the input switches is the ability to interchange the two inputs before sending them to the Arithmetic and Logic Units.

The arithmetic and logic unit (ALU) is a custom developed component which provides the capability to perform addition, subtraction, and multiplication of signed or unsigned numbers as well as a variety of common logic functions and comparisons for maximum/minimum determination. The unit is a three stage pipeline to increase achievable clock rates which gives the PE its characteristic three–cycle latency on all instructions except complete by–pass. Scan chain registers are used in the ALU and include the capability to generate pseudo–random data to provide testability.

The on–chip analog receiver and VCSEL driver cells included on the CMOS ASIC are previously verified designs from UCSD and UNCC/UDel respectively and were designed to operate with the specific OE elements used in this system. As an additional testability feature, stand–alone copies of these cells have also been placed on the ASIC connected to probe pads.

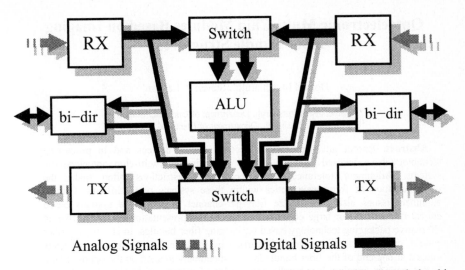

Figure 4. Architecture diagram of the processor element showing dataflow through the chip. *(Thinner)* lines indicate dataflow pattern when by–passing the computational portion of the chip

5 Conclusion

The current state of integration of optical communication with digital CMOS logic affords the ability to build functional systems from which new processing architectures can evolve. We have taken advantage of this to build a prototype multiprocessor demonstration system which utilizes FSOI data communication. This system is currently in the final stages of development and additional results will be presented at the conference.

References

1. 3D–OESP Consortium website: http://soliton.ucsd.edu/3doesp/
2. George Mason University Consortium for Optical and Optoelectronic Technologies in Computing website: http://co–op.gmu.edu/
3. Honeywell Technology Center: http://www.htc.honeywell.com/photonics/
4. C. Berger, J. T. Ekman, P. J. Marchand, F. E. Kiamilev, H. Spaanenburg: Parallel distributed free–space optoelectronic compute engine using flat "plug–on–top" optics package, accepted for presentation at the International Topical Meeting on Optics in Computing in Quebec, Canada, June 2000

Effort sponsored by the Defense Advanced Research Projects Agency (DARPA) and Air Force Research Laboratory under agreement number F30602–97–2–0122. The US government is authorized to reproduce and distribute reprints for governmental purposes notwithstanding any copyright annotation thereon.

Optoelectronic Multi-Chip Modules Based on Imaging Fiber Bundle Structures

Donald M. Chiarulli[1], Steven P. Levitan[2]

[1]University of Pittsburgh, Department of Computer Science
[2]University of Pittsburgh, Department of Electrical Engineering

Abstract. Recent advances in optoelectronic (OE) devices and in processing technology have focused attention on the packaging of multi-chip optoelectronic systems. Alignment tolerances and geometrical restrictions often make the implementation of free space optics within these systems quite difficult. Critical alignment issues also characterize fiber-per-channel guided wave systems based optical ribbon cable or large core fiber arrays. In this presentation I will describe an alternative packaging technology based on imaging fiber bundles. In an imaging fiber bundle, each optical data channel is carried by multiple fibers. An array of spots imaged at one end of the fiber bundle is correspondingly imaged on the opposite end. In this manner, imaging fiber bundles are capable of supporting the spatial parallelism of free space interconnects with relaxed alignment and geometry constraints. We have demonstrated a 16-channel point-to-point link between two VCSEL arrays that were directly butt coupled to an imaging fiber bundle. No other optical elements were used in the setup. We have also investigated a number of multi-chip interconnection module designs using both rigid and flexible imaging fiber bundles. Our basic approach to multipoint interconnect is to fabricate structures in which individual regions of the image at the input surface of a fiber bundle (or a fiber bundle array) are passively routed to different output surfaces. Opto-electronic devices, such as flip-chip bonded GaAs on silicon can be mounted on metal traces plated on to each surface of the module. The resulting network provides for spatially resolved bidirectional channels between each of the OE chips.

J. Rolim et al. (Eds.): IPDPS 2000 Workshops, LNCS 1800, pp. 1132-1132, 2000.
© Springer-Verlag Berlin Heidelberg 2000

VCSEL-Based Smart Pixel Array Technology Enables Chip-to-Chip Optical Interconnect

Yue Liu

Honeywell International
3660 Technology Drive, Minneapolis, MN 55418

Abstract. This paper describes most recent development and demonstration of a VCSEL-based smart pixel array (SPA) technology for chip-to-chip interconnect. This technology is based on Honeywell's commercial successful 850nm VCSEL components, incorporates both monolithic and hybrid integration techniques, and aims to address anticipated interconnect bottleneck in networking interconnect fabric and between processors and memories. Following features of this technology makes it not only technically feasible but also practically viable for system insertion in very near future. First, new generating of oxide VCSEL technology provides key characters that high density 2D optical interconnect systems desire, such as high speed, high efficiency, low power dissipation and good array uniformity. Secondly, monolithically integration VCSEL and photodetector provides system with flexible bi-directional optical I/O solutions, and advantages in adopting new system architectures. Third, the 2D-optoelectronic array can be seamlessly merged with state-of-the-art Si-based VLSI electronics, and micro-optics using hybrid integration techniques such as solder bump bonding and wafer scale integration. Last, and perhaps most importantly, all of our technology implementations follow the guideline of being compatible with mainstream and low cost manufacturing practices. Device performance characteristics, integration approach, and results of up to 34x34 SPA prototype demonstration will be presented.

Run-Time Systems for Parallel Programming

4th RTSPP Workshop Proceedings

Cancun, Mexico, May 1, 2000

Organizing Committee

General Chair – Laxmikant V. Kale
Program Chair – Ron Olsson

Program Committee

Pete Beckman	Los Alamos National Laboratory, USA
Greg Benson	University of San Francisco, USA
Luc Bougé	École Normale Supérieure of Lyon (ENS Lyon),France
Matthew Haines	Inktomi, USA
Laxmikant V. Kale	University of Illinois at Urbana Champaign, USA
Thilo Kielmann	Vrije Universiteit, The Netherlands
Koen Langendoen	Delft University of Technology, The Netherlands
David Lowenthal	University of Georgia, USA
Frank Müller	Humboldt-Universitaet zu Berlin, Germany
Ron Olsson	University of California, Davis, USA
Raju Pandey	University of California, Davis, USA
Alan Sussman	University of Maryland, USA

J. Rolim et al. (Eds.): IPDPS 2000 Workshops, LNCS 1800, pp. 1134-1135, 2000.
© Springer-Verlag Berlin Heidelberg 2000

Preface

Runtime systems are critical to the implementation of parallel programming languages and libraries. They support the core functionality of programming models and the glue between such models and the underlying hardware and operating system. As such, runtime systems have a large impact on the performance and portability of parallel programming systems.

Despite the importance of runtime systems, there are few forums in which practitioners can exchange their ideas, and these are typically forums showcasing peripheral areas, such as languages, operating systems, and parallel computing. RTSPP provides a forum for bringing together runtime system designers from various backgrounds to discuss the state-of-the-art in designing and implementing runtime systems for parallel programming.

The RTSPP workshop will take place on May 1, 2000 in Cancun, Mexico, in conjunction with IPDPS 2000. This one-day workshop includes technical sessions of refereed papers and panel discussions. The 8 paper presentations were selected out of 11 submissions after a careful review process; each paper was reviewed by at least four members of the program committee. Based on the reviewers' comments, the authors revised their papers for inclusion in these workshop proceedings.

We thank the RTSPP Program Committee (see previous page) and the following additional people for taking part in the review process: Gabriel Antoniu (LIP, ENS Lyon, France) Yves Denneulin (IMAG, Grenoble, France) Emmanuel Jeannot (LaBRI, University of Bordeaux, France) and Loïc Prylli (LIP, ENS Lyon, France).

We also thank the previous Organizing Committees for initiating this workshop and the participants in the previous workshops for making this forum successful and lively. We hope that this year's Workshop will be equally interesting and exciting.

Ron Olsson
Laxmikant V. Kale

A Portable and Adaptive Multi-Protocol Communication Library for Multithreaded Runtime Systems

Olivier Aumage, Luc Bougé, and Raymond Namyst

LIP, ENS Lyon, France*

Abstract. This paper introduces *Madeleine II*, an adaptive multi-protocol extension of the *Madeleine* portable communication interface. *Madeleine II* provides facilities to use multiple network protocols (VIA, SCI, TCP, MPI) and multiple network adapters (Ethernet, Myrinet, SCI) within the same application. Moreover, it can dynamically select the most appropriate transfer method for a given network protocol according to various parameters such as data size or responsiveness user requirements. We report performance results obtained using Fast-Ethernet and SCI.

1 Efficient Communication in Multithreaded Environments

Due to their ever-growing success in the development of distributed applications on clusters of SMP machines, today's multithreaded environments have to be highly portable and efficient on a large variety of architectures. For portability reasons, most of these environments are built on top of widespread message-passing communication interfaces such as PVM or MPI. However, the implementation of multithreaded environments mainly involves *RPC-like* interactions. This is obviously true for environments providing a RPC-based programming model such as Nexus [2] or PM2 [4], but also for others which often provide functionalities that can be efficiently implemented by RPC operations.

We have shown in [1] that message passing interfaces such as MPI, do not meet the needs of RPC-based multithreaded environments with respect to efficiency. Therefore, we have proposed a portable and efficient communication interface, called *Madeleine*, which was specifically designed to provide RPC-based multithreaded environments with *both* transparent and highly efficient communication. However, the internals of this first implementation were strongly message-passing oriented. Consequently, the support of non message-passing network protocols such as SCI or even VIA was cumbersome and introduced some unnecessary overhead. In addition, no provision was made to use multiple network protocols within the same application. For these reasons, we decided to design *Madeleine II*, a full multi-protocol version of *Madeleine*, efficiently portable on a wider range of network protocols, including non message-passing ones.

---

* LIP, ENS Lyon, 46, Allée d'Italie, F-69364 Lyon Cedex 07, France. Contact: Raymond.Namyst@ens-lyon.fr.

J. Rolim et al. (Eds.): IPDPS 2000 Workshops, LNCS 1800, pp. 1136-1143, 2000.

`mad_begin_packing`	Initiates a new message
`mad_begin_unpacking`	Initiates a message reception
`mad_end_packing`	Finalize an emission
`mad_end_unpacking`	Finalize a reception
`mad_pack`	Packs a data block
`mad_unpack`	Unpacks a data block

Table 1. Functional interface of *Madeleine II*.

2   The *Madeleine II* Multi-Protocol Communication Interface

The *Madeleine II* programming interface provides a small set of primitives to build RPC-like communication schemes. Theses primitives actually look like classical message-passing-oriented primitives. Basically, this interface provides primitives to send and receive *messages*, and several *packing* and *unpacking* primitives that allow the user to specify how data should be inserted into/extracted from messages (Table 1).

A message consists of several pieces of data, located anywhere in user-space. They are constructed (resp. de-constructed) incrementally using *packing* (resp. *unpacking*) primitives, possibly at multiple software levels, without losing efficiency. The following example illustrates this need. Let us consider a remote procedure call which takes an array of unpredictable size as a parameter. When the request reaches the destination node, the header is examined both by the multithreaded runtime (to allocate the appropriate thread stack and then to spawn the server thread) and by the user application (to allocate the memory where the array should be stored).

The critical point of a send operation is obviously the series of *packing* calls. Such packing operations simply *virtually* append the piece of data to a message under construction. In addition to the address of data and its size, the packing primitive features a pair of *flag* parameters which specifies the semantics of the operation. The available emission flags are the following:

send_SAFER This flag indicates that *Madeleine II* should pack the data in a way that further modifications to the corresponding memory area should not corrupt the message. This is particularly mandatory if the data location is reused before the message is actually sent.

send_LATER This flag indicates that *Madeleine II* should not consider accessing the value of the corresponding data until the `mad_end_packing` primitive is called. This means that any modification of these data between their packing and their sending shall actually update the message contents.

send_CHEAPER This is the default flag. It allows *Madeleine II* to do its best to handle the data as efficiently as possible. The counterpart is that no assumption should be made about the way *Madeleine II* will access the data. Thus, the corresponding data should be left unchanged until the send operation has completed. Note that most data transmissions involved in parallel applications can accommodate the send_CHEAPER semantics.

The following flags control the reception of user data packets:

receive_EXPRESS This flag forces *Madeleine II* to guarantee that the corresponding data are immediately available after the *unpacking* operation. Typically, this flag is mandatory if the data is needed to issue the following *unpacking* calls. On some network protocols, this functionality may be available for free. On some others, it may put a high penalty on latency and bandwidth. The user should therefore extract data this way only when necessary.

receive_CHEAPER This flag allows *Madeleine II* to possibly defer the extraction of the corresponding data until the execution of **mad_end_unpacking**. Thus, no assumption can be made about the exact moment at which the data will be extracted. Depending on the underlying network protocol, *Madeleine II* will do its best to minimize the overall message transmission time. If combined with **send_CHEAPER**, this flag guarantees that the corresponding data is transmitted as efficiently as possible.

Figure 1 illustrates the power of the *Madeleine* interface. Consider sending a message made of an array of bytes whose size is unpredictable on the receiving side. Thus, on the receiving side, one has first to extract the size of the array (an integer) before extracting the array itself, because the destination memory has to be dynamically allocated. In this example, the constraint is that the integer must be extracted EXPRESS *before* the corresponding array data is extracted. In contrast, the array data may safely be extracted CHEAPER, striving to avoid any copies.

Sending side   Receiving side

```
conn = mad_begin_packing(...);        conn = mad_begin_unpacking(...);
mad_pack(conn,&size,sizeof(int),      mad_unpack(conn,&size,sizeof(int),
  send_CHEAPER,receive_EXPRESS);        send_CHEAPER,receive_EXPRESS);
                                      array = malloc(size);
mad_pack(conn, array, size,           mad_unpack(conn, array, size,
  send_CHEAPER,receive_CHEAPER);        send_CHEAPER,receive_CHEAPER);
mad_end_packing(conn);                mad_end_unpacking(conn);
```

Fig. 1. Sending and receiving messages with *Madeleine II*.

Madeleine II aims at enabling an efficient and exhaustive use of underlying communication software and hardware functionalities. It is able to deal with several network protocols within the same session and to manage multiple network adapters (NIC) for each of these protocols. The user application can dynamically and explicitly switch from one protocol to another, according to its communication needs. The multi-protocol support of *Madeleine II* relies on the concept of *channel*.

Channels in *Madeleine II* are pretty much like radio channels. They are allocated at run-time. The communication on a given channel does not interfere with the communication on another one. As a counterpart, in-order delivery is not guaranteed among distinct channels. In-order delivery is only enforced for

```
text_chan  = mad_open_channel(TCP_ETH0);
video_chan = mad_open_channel(SISCI_SCIO);
text_conn  = mad_begin_packing(text_chan, video_client);
video_conn = mad_begin_packing(video_chan, video_client);
mad_pack(text_conn, text_dataptr, text_len, ...);
mad_pack(video_conn, video_dataptr, video_len, ...);
...
```

Fig. 2. Example of a video server simultaneously sending video information using a SISCI channel and translation text data using TCP channel.

point-to-point connections within the same channel. In this respect, they look like MPI communicators, but different *Madeleine II* channels can be bound to different protocols as well as adapters (Fig. 2). Of course, several channels may share the same protocol, and even the same adapter.

3   Inside *Madeleine II*: from the Application to the Network

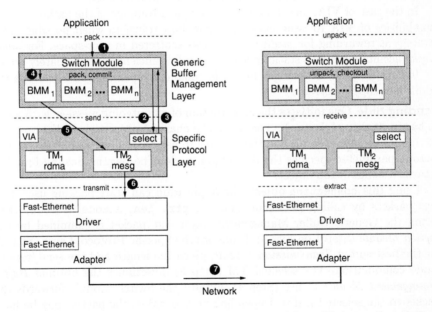

Fig. 3. Conceptual view of the data path through *Madeleine II*'s internal modules.

The transmission of data blocks using *Madeleine II* involves several internal modules. We illustrate its internals in the case of an implementation on top of VIA (Fig. 3).

Protocols such as VIA provide several methods to transfer data, namely regular message passing and remote DMA write (and optionally RDMA-read).

Moreover, there are several ways to use these transfer methods, as VIA requires registering the memory blocks before transmission. It is for instance possible to dynamically register user data blocks, or to copy them into a pool of pre-registered internal buffers. Their relative efficiency crucially depends on the size of the blocks. The current implementation of *Madeleine II* on top of VIA supports the three following combinations:

Small blocks: message-passing + static buffer pool.
Medium-sized blocks: message-passing + dynamically registered buffers.
Large blocks: RDMA-write + dynamically registered buffers.

Each transfer method is encapsulated in a protocol-specific *Transmission Module* (TM, see Fig. 3). Each TM is associated with a *Buffer Management Module* (BMM). A BMM implements a generic, protocol-independent management policy: either the user-allocated data block is directly referenced as a buffer, or it is copied into a buffer provided by the TM. Moreover, each BMM implements a specific scheme to aggregate successive buffers into a single piece of message. Each TM is associated with its optimal BMM. However, observe that several TM (even from different protocols) may share the same BMM, which results in a significant improvement in development time and reliability.

In the case of VIA, one can for instance take advantage of the gather/scatter capabilities of VIA to issue one-step burst data transfers when possible. This strategy is rewarding for *medium-size blocks* scattered in user-space. For *small blocks* accumulated into static buffers, it is most efficient to immediately transfer buffers as soon as they get full: this enhances pipelining and overlaps the additional copy involved.

Sending Side One initiates the construction of an outgoing message with a call to `begin_packing(channel, remote)`. The `channel` object selects the protocol module (VIA in our case), and the adapter to use for sending the message. The `remote` parameter specifies the destination node. The `begin_packing` function returns a `connection` object.

Using this `connection` object, the application can start packing user data into packets by calling `pack(connection, ptr, len, s_mode, r_mode)`. Entering the Generic Buffer Management Layer, the packet is examined by the *Switch Module* (Step 1 on Fig. 3). It queries the Specific Protocol Layer (Step 2) for the best suited *Transmission Module*, given the length and the send/receive mode combination. The selected TM (Step 3) determines the optimal *Buffer Management Module* to use (Step 4). Finally, the Switch Module forwards the packet to the selected BMM. Depending on the BMM, the packet may be handled as is (and considered as a buffer), or copied into a new buffer, possibly provided by the TM. Depending on its aggregation scheme, the BMM either immediately sends the buffer to the TM or delays this operation for a later time. The buffer is eventually sent to the TM (Step 5). The TM immediately processes it and transmits it to the Driver (Steps 6). The buffer is then eventually shipped to the Adapter (Step 7).

Special attention must be paid to guarantee the delivery order in presence of multiple TM. Each time the Switch Step selects a TM differing from the previous one, the corresponding previous BMM is flushed (*commit* on Fig. 3) to ensure that any delayed packet has been sent to the network. A general *commit* operation is also performed by the end_packing(connection) call to ensure that no delayed packet remains waiting in the BMM.

Receiving Side Processing an incoming message on the destination side is just symmetric. A message reception is initiated by a call to begin_unpacking(channel) which starts the extraction of the first incoming message for the specified channel. This function returns the connection object corresponding to the established point-to-point connection, which contains the remote node identification among other things.

Using this connection object, the application issues a sequence of unpack(connection, ptr, len, s_mode, r_mode) calls, symmetrically to the series of pack calls that generated the message. The Switch Step is performed on each unpack and must select the same sequence of TM as on the sending side. For instance, a packet sent by the DMA Transmission Module of VIA must be received by the same module on the receiving side. The *checkout* function (dual to the *commit* one on the sending side) is used to actually extract data from the network to the user application space: indeed, just like packet sending could be delayed on the sending side for aggregation, the actual packet extraction from the network may also be delayed to allow for burst data reception. Of course, the final call to end_unpacking(connection) ensures that all expected packets are made available to the user application.

Discussion This modular architecture combined to packet-based message construction allows *Madeleine II* to be efficient on top of message-passing protocols as well as put/get protocols. Whatever the underlying protocol used, *Madeleine II*'s generic flexible buffer management layer is able to tightly adapt itself to its particularities, and hence deliver most of the available networking potential to the user application. Moreover, the task of implementing a new protocol into *Madeleine II* is considerably alleviated by re-using existing BMM.

4   Implementation and Performances

We now evaluate *Madeleine II* on top of several network protocols. All features mentioned above have been implemented. Drivers are currently available for TCP, MPI, VIA, SISCI [3] and SBP [6] network interfaces.

Testing Environment The following performance results are obtained using a cluster of dual Intel Pentium II 450 MHz PC nodes with 128 MB of RAM running LINUX (Kernel 2.1.130 for VIA, and Kernel 2.2.10 for TCP and SISCI). The cluster interconnection networks are 100 Mbit/s Fast Ethernet for TCP and VIA, and Dolphin SCI for SISCI. The tests run on the TCP/IP protocol use the standard UNIX sockets. The tests run on the VIA protocol use the M-VIA 0.9.2 implementation from the NERSC (National Energy Research Scientific Computing Center, Lawrence Berkeley Natl Labs).

	Latency		Bandwidth	
Protocol	TCP	SISCI	TCP	SISCI
Raw performance	59.8 μs	2.3 μs	11.1 MB/s	76.5 MB/s
Madeleine	77.4 μs	5.9 μs	10.5 MB/s	70.0 MB/s
Madeleine II	67.2 μs	7.9 μs	11.0 MB/s	57.0 MB/s

Table 2. Latency (left) and bandwidth (right) on top of TCP and SISCI.

TCP Surprisingly enough, *Madeleine II* outperforms *Madeleine* (Table 2). *Madeleine* used to require attaching a short header to each transfered message, whereas *Madeleine II* gives the user finer control on the message structure. The difference in performance between raw TCP and *Madeleine II* on top of TCP is the result of the current software overhead of *Madeleine II*. The bandwidth of *Madeleine II* on top of TCP is very close to the raw bandwidth of TCP.

SISCI The new SISCI Specific Protocol Layer of *Madeleine II* is not yet as optimized as the one used by *Madeleine*. This is why the bandwidth measured with *Madeleine II* on top of SISCI is not as good as the one obtained with *Madeleine* (Table 2). The difference in latency between *Madeleine II* and *Madeleine* is due to some additional processing in the internals of *Madeleine II*. Future optimizations will hopefully solve this problem.

Dynamic Transfer Method Selection We mentioned above the capability of *Madeleine II* to dynamically choose the most appropriate transfer paradigm within a given protocol. Figure 4 shows the dramatic influence of dynamic transfer paradigm selection on performance using VIA. VIA requires the memory areas involved in transfer to be registered. Such dynamic registration operations are expensive. This cost is especially prohibitive for short messages, and using a pool of

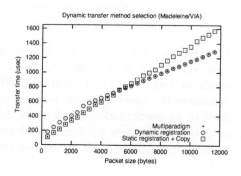

Fig. 4. Multi-Paradigm support.

pre-registered buffers help circumventing the problem. Instead of registering the memory area where the messages are stored, one can copy the messages into these buffers. This amounts to exchanging registration time for copying time. This is obviously inefficient for long messages. The two curves are plotted on Figure 4. The *Multi-Paradigm* curve is obtained by activating the dynamic paradigm selection of *Madeleine II*. It is optimal *both* with short messages and long messages!

5   Related work

Many communication libraries have recently been designed to provide portable interfaces and/or efficient implementations to build distributed applications.

However, very few of them provide an efficient support for RPC-like communication schemes, support for multi-protocol communications and support for multithreading.

Illinois Fast Messages (FM) [5] provides a very simple mechanism to send data to a receiving node that is notified upon arrival by the activation of a handler. Releases 2.x of this interface provide interesting gather/scatter features which allow an efficient implementation of zero-copy data transmissions. However, it is not possible to issue a transmission with the semantics of the `receive_CHEAPER` *Madeleine II* flag: only `receive_EXPRESS`-like receptions are supported, and it is not possible to enforce aggregated transmissions.

The Nexus multithreaded runtime [2] features a multi-protocol communication subsystem very close to the one of *Madeleine II*. The messages are constructed using similar *packing* operations except that no "high level" semantics can be associated to data: there is no notion of `CHEAPER` specifications, which allows *Madeleine II* to choose the best suited strategy. Also, as for FM, *unpacking* operations behave like `receive_EXPRESS` *Madeleine II* transmissions.

6   Conclusion

In this paper, we have described the new *Madeleine II* communication interface. This new version features full multi-protocol, multi-adapter support as well as an integrated new dynamic *most-efficient transfer-method* selection mechanism. We showed that this mechanism gives excellent results with protocols such as VIA. We are now actively working on having *Madeleine II* running across clusters connected by heterogeneous networks.

References

1. Luc Bougé, Jean-François Méhaut, and Raymond Namyst. Efficient communications in multithreaded runtime systems. In *Proc. 3rd Workshop on Runtime Systems for Parallel Programming (RTSPP '99)*, volume 1586 of *Lect. Notes Comp. Science*, pages 468–182, San Juan, Puerto Rico, April 1999. Springer-Verlag.
2. I. Foster, C. Kesselman, and S. Tuecke. The Nexus approach to integrating multithreading and communication. *Journal on Parallel and Distributed Computing*, 37(1):70–82, 1996.
3. IEEE. *Standard for Scalable Coherent Interface (SCI)*, August 1993. Standard no. 1596.
4. Raymond Namyst and Jean-François Méhaut. PM2: Parallel Multithreaded Machine. a computing environment for distributed architectures. In *Parallel Computing (ParCo'95)*, pages 279–285. Elsevier, September 1995.
5. S. Pakin, V. Karamcheti, and A. Chien. Fast Messages: Efficient, portable communication for workstation clusters and MPPs. *IEEE Concurrency*, 5(2):60–73, April 1997.
6. R.D. Russell and P.J. Hatcher. Efficient kernel support for reliable communication. In *13th ACM Symposium on Applied Computing*, pages 541–550, Atlanta, GA, February 1998.

CORBA Based Runtime Support for Load Distribution and Fault Tolerance

Thomas Barth, Gerd Flender, Bernd Freisleben, Manfred Grauer, and
Frank Thilo

University of Siegen, Hölderlinstr.3, D–57068 Siegen, Germany
{barth, grauer, thilo}@fb5.uni-siegen.de,
{freisleb, plgerd}@informatik.uni-siegen.de

Abstract. Parallel scientific computing in a distributed computing environment based on CORBA requires additional services not (yet) included
in the CORBA specification: load distribution and fault tolerance. Both
of them are essential for long running applications with high computational demands as in the case of computational engineering applications. The proposed approach for providing these services is based on
integrating load distribution into the CORBA naming service which in
turn relies on information provided by the underlying WINNER resource
management system developed for typical networked Unix workstation
environments. The support of fault tolerance is based on error detection and backward recovery by introducing proxy objects which manage
checkpointing and restart of services in case of failures. A prototypical
implementation of the complete system is presented, and performance
results obtained for the parallel optimization of a mathematical benchmark function are discussed.

1   Introduction

Object–oriented software architectures for distributed computing environments
based on the *Common Object Request Broker Architecture* (CORBA) have started
to offer real-life production solutions to interoperability problems in various business applications, most notably in the banking and financial areas. In contrast,
most of todays applications for distributed scientific computing traditionally use
message passing as the means for communication between processes residing on
the nodes of a dedicated parallel multiprocessor architecture. Message passing
is strongly related to the way communication is realized in parallel hardware
and is particularly adequate for applications where data is frequently exchanged
between nodes. Examples are data–parallel algorithms for complex numerical
computations, such as in computational fluid dynamics where essentially algebraic operations on large matrices are performed.

The advent of networks of workstations (NOW) as cost effective means for
parallel computing and the advances of object-oriented software engineering
methods have fostered efforts to develop distributed object-oriented software
infrastructures for performing scientific computing applications on NOWs and

J. Rolim et al. (Eds.): IPDPS 2000 Workshops, LNCS 1800, pp. 1144-1151, 2000.

also over the WWW [7]. Other computationally intensive engineering applications with different communication requirements, such as simulations and/or multidisciplinary optimization (MDO) problems [3] [5] typically arising in the automotive or aerospace industry, have even strengthened the need for a suitable infrastructure for distributed/parallel computing.

Two essential features of such an infrastructure are load distribution and a certain level of fault tolerance. Load distribution improves the effectiveness of the given resources, resulting in reduced computation times. Fault tolerance is important especially for long–running engineering applications like MDO software systems. It is obviously crucial to provide mechanisms to prevent the whole computation from failing due to a single error on the server side.

In this paper, CORBA based runtime support for parallel applications is presented. This support encompasses load distribution as well as fault tolerance for parallel applications using CORBA as communication middleware.

2   Integrating Load Distribution into CORBA

In general, CORBA applications consist of a set of clients (applications objects) requesting a set of services. These services can either be other application objects within a distributed application, or commonly available services (object services) providing e.g. name resolution (naming service) or object persistence (persistence service). There are different approaches to integrate load distribution functionality into a CORBA environment:

- Implementation of an explicit service (e.g. a "trader", [12]) which returns an object reference for the requested service on an available host (centralized load distribution strategy) or references for all available service objects. In the latter case, the client has to evaluate the load information for all of the returned references and has to make a selection by itself (decentralized load distribution strategy).
- Integrating the load distribution mechanism into the ORB itself, e.g. by replacing the default locator by a locator with an integrated load distribution strategy [6] or using an IDL–level approach [13]

The drawbacks of these approaches are either that the source code of clients has to be changed (as in the first approach) or that load distribution depends on a specific ORB implementation or IDL compiler and can thus not be utilized when other ORBs are used (as in the second approach). To integrate load distribution transparently into a CORBA environment, our proposal is based on integrating it into the naming service. This ensures transparency for the client side and allows the reuse of the load distribution naming service in any other CORBA–compliant ORB implementation. In almost every CORBA–based implementation the naming service is utilized. In the case of applications which do not make use of the naming service, it would be useful to implement load distribution as an explicit service.

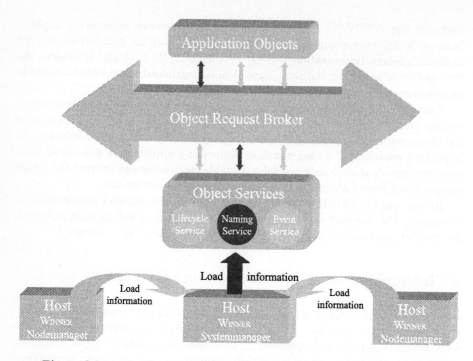

Fig. 1. *Schema for the integration of load distribution in a naming service.*

Our concept is illustrated in Fig. 1; it relies on the WINNER resource management system [1] [2]. Basically, WINNER provides load distribution services for a network of Unix workstations. Its components of interest here are the central system manager and the node managers. There is one node manager on each participating workstation, periodically measuring the node's performance and system load, i.e. data like CPU utilization which is collected by the host operating system. This data is sent to the system manager, which has functionality to determine the machine with the currently best performance. Requests from application objects to the naming service are resolved using this load information for the selection of an appropriate server.

The naming service is not an integral part of a CORBA ORB but is always implemented as a CORBA service. The OMG specifies the interface of a naming service without making assumptions about implementation details of the service. Therefore, every ORB can interoperate with a new naming service as long as it complies to the OMG specification.

3   Runtime Support for Fault Tolerance in CORBA Based Systems

The CORBA specification as well as the Common Object Services Specification offer no adequate level of fault tolerance yet. Due to the need for fault tolerance in more complex distributed systems, various approaches were developed. The Piranha system [8] for example, is based on an ORB supporting object groups, failure detection etc. Using these facilities provided by the ORB, Piranha is implemented as an CORBA Object Service for monitoring distributed applications and managing fault tolerance via active or passive replication. The major drawback of Piranha is its dependency on non–standard ORB features like object groups. Another approach avoids this drawback by complying completely to the CORBA standard: IGOR (Interactive–Group Object–Replication) [9] realizes fault tolerance also by managing groups of objects providing redundant services. In contrast to the Piranha system, IGOR is portable and interoperable with today's ORB implementations. Lately, there is also a proposal for the integration of redundancy, fault detection and recovery into the CORBA standard [10].

Unlike the previously mentioned approaches, our concept is not based on replicated services in object groups but on the integration of checkpointing and restarting functionality only. Especially for applications with a maximum degree of parallelism (e.g. scalable optimization algorithms) it is not desirable to use a large amount of the computational resources (i.e. hosts in the network) exclusively for availability purposes as in the case of active replication. Thus, in the case of parallel, long running applications it is a good compromise to restrict fault tolerance to checkpointing and restarting. Similar to the concept of passive replication, frequently (i.e. after each method call on the server side) generated checkpoints are used to restart a failed service.

Currently, the only way to detect an error on the client side of a CORBA application is the exception `CORBA::COMM_FAILURE` thrown when a CORBA client tries to call a service which is not available anymore (e.g. due to a network failure, a crashed server process or machine). Using the concepts for the naming service already described, it is possible to request a new reference to a service if a call to a server object fails. This approach is sufficient for services without an internal state. In the more general case of services depending on an internal state of the server object, it is inevitable to (a) save the state (checkpoint) of the server object e.g. after each successful call to a server's method and (b) have the opportunity to restore this state in a newly created server object.

We evaluated the following alternatives to integrate checkpointing and restarting functionality on the client side assuming that the service object provides a method to create a checkpoint for restarting the service if an error occurs: (a) modification of the client–side code to handle the `CORBA::COMM_FAILURE` exception and to restart a service, (b) extending the client–side stub code generated by the IDL–compiler with exception handling etc., and (c) introduction of proxy-classes derived from the stub classes on the client–side.

The major drawback of alternative (a) is the amount of code to be inserted on the client side: every single call from a client to a method of the server must at first get a checkpoint from the server, then handle the exception, and start a new server (using the checkpoint) in case of a failure. It would be useful if the automatically generated stub code comprises this code as in alternative (b). But this means changing the IDL–compiler itself, and thus this solution would be specific for a certain CORBA implementation providing its own IDL–compiler. Alternative (c) is a compromise between the amount of modifications to be made on the client side and the targeted platform independence of the concept: the modifications on the client side are limited to the use of a proxy class instead of the stub class. This proxy class is derived from the stub class and therefore provides all of the methods of the stub class. The additional methods handle the creation of a checkpoint and the restoring of an object's state according to a checkpoint. If a class offers this functionality for checkpointing and restoring a certain internal state it is in principle possible to migrate a service from host to another one not only when an error occured but also due to a changing load situation on a host.

With the current implementation, the proxy class for each service class has to be implemented manually. This could be easily automated by parsing the class definition. For each method, code to call the parent class (the stub) method along with exception handling code and a call to the server object's checkpoint and restore functions would have to be generated.

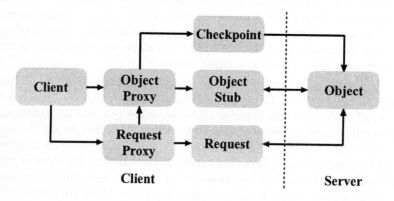

Fig. 2. *Scheme of client, server, proxy objects and their call relationship.*

As a proof of concept, a simple service for storing checkpointing data has been implemented. It simply provides functions to store/retrieve arbitrary values to the server object. No real persistency like storing checkpoints on disk media has been implemented, yet. Furthermore, the current implementation is rather inefficient. In addition to transparent synchronous method calls, CORBA provides asynchronous method invocations via DII (Dynamic Invocation Interface). When a client wants to utilize DII, it does not call the server object's methods directly, but uses so-called *request* objects instead. These request objects offer methods to asynchronously initiate methods of the server object and fetch the corresponding results at a later time. To enable fault tolerance in this case, *request proxies* are used just like the object proxies. The relationship between the described objects is shown in Fig. 2.

4   Experimental Results

To investigate the benefits of an integrated load distribution mechanism in CORBA, a test case from mathematical optimization was taken. The well known Rosenbrock test function [14] is widely used for benchmarking optimization algorithms because of its special mathematical properties. In our experiments, the function is only used to demonstrate the benefits of an adequate placement of computationally expensive processes on nodes of a NOW. It is not intended to present a new approach to the solution of the benchmark problem. To compute the function in parallel, a decomposed formulation of the Rosenbrock function has been taken. In the decomposed formulation, several (sub-)problems with a smaller dimension than the original n–dimensional problem are solved by workers, and the subproblems are then combined for the solution of the original problem in a manager.

In Fig. 3, the results of the different test scenarios are compared. All test cases were computed using multiple instances of a sequential implementation of the Complex Box algorithm [4] on a network of 10 workstations. The ORB used was omniORB 2.7.1 [11]. For the comparison of the different implementations of the naming service, a background load was generated on 0, 2, 4, 6 or 8 hosts. The

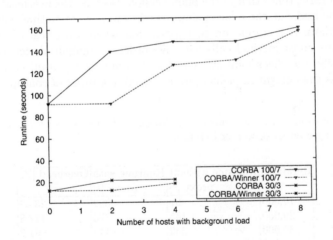

Fig. 3. *Different test cases of a decomposed 30– and 100–dimensional Rosenbrock function with 3 and 7 worker problems under different load situations.*

two lower curves show the computation times for a 30–dimensional Rosenbrock function with 3 worker problems (problem dimension 10, 9 and 9) and a 2–dimensional manager problem. In this scenario, 6 workstations were available for the 4 processes. The effect of load distribution is obvious when 2 hosts had background load. The selection of hosts with the new naming service avoided these hosts and hence the computation time was the same as in the case without background load. The two upper curves compare the computation times for a 100–dimensional Rosenbrock function with 7 worker problems. With increasing background, load the advantage diminishes because both implementations of the naming services are forced to select services on hosts with background load. To summarize, the benefit of load distribution for the test cases mentioned above can be estimated by ca. 40% runtime reduction in the best case. Even in the worst case it yields at least the same results as the unmodified naming service. The mathematical properties of the test cases as mentioned above result in an average reduction of computation time of about 15%.

Providing fault tolerance by proxy classes introduces an additional level of indirection. Additionally, storing the state of the server objects upon each method invocation causes some overhead. To quantify to what extent this overhead affects application runtimes, the above experiment has been repeated, this time using fault tolerant proxy classes. In Table 1 computation times for a 100–dimensional Rosenbrock problem are shown for the proxy and non–proxy case, respectively. The measurements have been conducted for different numbers of iterations of the worker's algorithm. The increasing number of iterations results in longer runtimes of the worker problems because it is a stopping criterion of the algorithm. Table 1 demonstrates that fault tolerance comes at quite a cost in this scenario. In the worst case, the application runtime using proxy objects is more than three times that of the plain version. Because the overhead is constant for each method call, the relative slowdown is lower the more time is spent in the called method. It is important to remark that when using real life engineering applications, most method calls will take orders of magnitude longer to finish. Additionally, the checkpoint storage class has not been optimized for speed in any way as the current implementation is merely a proof of concept.

Table 1. Runtimes for a 100–dimensional Rosenbrock function with 7 worker problems and a varying number of worker iterations.

Iterations	Runtime without proxy [s]	Runtime with proxy [s]	Overhead [%]
10,000	92	309	235.9
20,000	165	376	127.8
30,000	232	445	91.8
40,000	299	505	68.9
50,000	383	594	55.1

5   Conclusions

The design and implementation of a CORBA naming service providing load distribution and basic fault tolerance services based on proxy objects was presented. These services are essential for long–running computational engineering applications in distributed computing environments. Experiments demonstrated the feasibility of both concepts. Areas of future work are: (a) improving, optimizing, and stabilizing the prototype implementation of the proposed CORBA load distribution and fault tolerance services, (b) evaluating its benefits in real-life engineering MDO applications, and (c) extending the WINNER load measurement and process placement features for wide-area networks to enable CORBA based distributed/parallel meta-computing over the WWW. Additionally, the proposed extensions to the CORBA specification concerning redundancy, fault detection and recovery must be evaluated.

References

1. Arndt, O., Freisleben, B., Kielmann, T., Thilo, F., Scheduling Parallel Applications in Networks of Mixed Uniprocessor/Multiprocessor Workstations, Proc. Parallel and Distributed Computing Systems (PDCS98), p.190–197, ISCA, Chicago, 1998
2. Barth, T., Flender, G., Freisleben, B., Thilo, F. Load Distribution in a CORBA Environment, in: Proc. of Int'l Symposium on Distributed Object and Application 99, p. 158–166, IEEE Press, Edinburgh 1999
3. Barth, T., Grauer, M., Freisleben, B., Thilo, F. Distributed Solution of Simulation-Based Optimization Problems on Workstation Networks. Proc. 2nd Int. Conf. on Parallel Computing Systems, pp. 152–159, Ensenada, Mexico, 1999
4. Boden, H., Gehne, R., Grauer, M., Parallel Nonlinear Optimization on a Multiprocessor System with Distributed Memory, in: Grauer, M., Pressmar, D. (eds.), Parallel Computing and Mathematical Optimization, Springer, 1991, p.65–78.
5. Grauer, M., Barth, T., Cluster Computing for treating MDO–Problems by OpTiX, to appear in: Mistree, F., Belegundu, A. (eds.), Proc. Conference on Optimization in Industry II, Banff, Canada, June 1999
6. Gebauer, C., Load Balancer $\mathcal{LB}$ – a CORBA Component for Load Balancing, Diploma Thesis, University of Frankfurt, 1997
7. Livny, M., Raman, R., High-Throughput Resource Management, in: The GRID: Blueprint for a New Computing Infrastructure, Foster, I., Kesselman, C. (eds.), pp. 311–337, Morgan Kaufmann, 1998
8. Maffeis, S., Piranha: A CORBA Tool for High Availability, IEEE Computer, Vol. 30, No.4, p. 59–66, April 1997
9. Modzelewski, B., Cyganski, D., Underwood, M., Interactive–Group Object–Replication Fault Tolerance for CORBA, 3rd Conf. on Object–Oriented Techniques and Systems, Portland, Oregon, June 1997, pp. 241–244
10. Fault tolerant CORBA, Object Management Group TC Document Orbos/99-12-08, December 1999
11. omniORB – a Free Lightweight High–Performance CORBA 2 Compliant ORB, (http://www.uk.research.att.com/omniORB/omniORB.html), AT&T Laboratories Cambridge, 1998
12. Rackl, G., Load Distribution for CORBA Environments, Diploma Thesis, (http://wwwbode.informatik.tu-muenchen.de/~rackl/DA/da.html), University of Munich, 1997
13. Schiemann, B., Borrmann, L., A new Approach for Load Balancing in High–Performance Decision Support System, Future Generation Computer Systems, Vol: 12, Issue: 5, April 1997, pp. 345-355
14. Schittkowski, K., Nonlinear Programming Codes, Springer, 1980

Run-time Support for Adaptive Load Balancing

Milind A. Bhandarkar, Robert K. Brunner, and Laxmikant V. Kalé

Parallel Programming Laboratory,
Department of Computer Science,
University of Illinois at Urbana-Champaign, USA
{milind,rbrunner,kale}@cs.uiuc.edu,
WWW home page: http://charm.cs.uiuc.edu/

Abstract. Many parallel scientific applications have dynamic and irregular computational structure. However, most such applications exhibit persistence of computational load and communication structure. This allows us to embed measurement-based automatic load balancing framework in run-time systems of parallel languages that are used to build such applications. In this paper, we describe such a framework built for the Converse [4] interoperable runtime system. This framework is composed of mechanisms for recording application performance data, a mechanism for object migration, and interfaces for plug-in load balancing strategy objects. Interfaces for strategy objects allow easy implementation of novel load balancing strategies that could use application characteristics on the entire machine, or only a local neighborhood. We present the performance of a few strategies on a synthetic benchmark and also the impact of automatic load balancing on an actual application.

1 Motivation and Related Work

An increasing number of emerging parallel applications exhibit dynamic and irregular computational structure. Irregularities may arise from modeling of complex geometries, and use of unstructured meshes, for example, while the dynamic behavior may result from adaptive refinements, and evolution of a physical simulation. Such behavior presents serious performance challenges. Load may be imbalanced to begin with due to irregularities, and imbalances may grow substantially with dynamic changes. We are participating in physical simulation projects at the Computational Science and Engineering centers of University of Illinois (Rocket simulation, and Simulation of Metal Solidification), where such behaviors are commonly encountered.

Load balancing is a fundamental problem in parallel computing, and a great deal of research has been done in this subject. However, a lot of this research is focussed on improving load balance of particular algorithms or applications. General purpose load balancing research deals mainly with process migration in operating systems and more recently in application frameworks. C++ libraries such as DOME [1] implement the data-parallel programming paradigm as distributed objects and allow migration of work in response to varying load conditions. Systems such as CARMI [10] simply notify the user program of the load

J. Rolim et al. (Eds.): IPDPS 2000 Workshops, LNCS 1800, pp. 1152-1159, 2000.

imbalance, and leave it to the application process to explicitly move its state to a new processor. Multithreaded systems such as PM^2 [9] require every thread to store its state in the specially allocated memory, so that the system can migrate the thread automatically. An object migration system called ELMO [3], built on top of Charm [6, 7], implements object migration mainly for fault-tolerance. Applications in areas such as VLSI, and Computational Fluid Dynamics (CFD) use graph partitioning programs such as METIS [8] to provide initial load balance. However, every such application has to specifically provide code for monitoring load imbalance and to invoke the load balancer periodically to deal with dynamic behavior.

We have developed an automatic measurement-based load balancing framework to facilitate high-performance implementations of such applications. The framework requires that a computation be partitioned into more pieces (typically implemented as objects) than there are processors, and letting the framework handle the placement of pieces. The framework relies on a "principle of persistence" that holds for most physical simulations: computational load and communication structure of (even dynamic) applications tends to persist over time. For example, even though the load of some object instance changes at adaptive refinement drastically, such events are infrequent, and the load remains relatively stable between such events.

The framework can be used to handle application-induced imbalances as well as external imbalances (such as those generated on a timeshared cluster). It cleanly separates runtime data-collection and object migration mechanisms into a distributed database, which allows optional *strategies* to plug in modularly to decide which objects to migrate where. This paper presents results obtained using our load balancing framework. We briefly describe the framework, then the strategies currently implemented and how they compare on a synthetic benchmark, and finally results on a crack-propagation application implemented using it.

2   Load Balancing Framework

Our framework [2] views a parallel application as a collection of computing objects which communicate with each other. Furthermore, these objects are assumed to exhibit temporal correlation in their computation and communication patterns, allowing effective measurement-based load balancing without application-specific knowledge.

The central component of the framework 1 is the load balancer distributed database, which coordinates load balancing activities. Whenever a method of a particular object runs, the time consumed by that object is recorded. Furthermore, whenever objects communicate, the database records information about the communication. This allows the database to form an object-communication graph, in which each node represents an object, with the computation time of that object as a weight, and each arc is a communication pathway representing communication from one object to another object, recording number of messages and total volume of communication for each arc.

The design of Charm++ [5] offers several advantages for this kind of load balancing. First, parallel programs are composed of many coarse-grained objects, which represent convenient units of work for migration. Also, messages are directed to particular objects, not processors, so an object may be moved to a new location without informing other objects about the change; the run-time system handles the message delivery with forwarding. Furthermore, the message-driven design of Charm++ means that work is triggered by messages, which are dispatched by the run-time system. Therefore, the run-time knows which object is running at any particular time, so the CPU time and message traffic for each object can be deposited with the framework. Finally, the encapsulation of data within objects simplifies object migration.

However, the load balancing framework is not limited to Charm++ only. Any language implemented on top of Converse can utilize this framework. For this purpose, the framework does not interact with object instances directly. Instead, interaction between objects and the load balancing framework occurs through object managers. Object managers are parallel objects (with one instance on each processor) that are supplied by the language runtime system. Object managers are responsible for creation, destruction, and migration of language-specific objects. They also supply the load database coordinator with computational loads and communication information of the objects they manage. Object managers register the managed objects with the framework, and are responsible for mapping the framework-assigned system-wide unique object identifier to the language-specific identifier (such as thread-id in multithreaded systems, chare-id in Charm++, processor number in MPI etc.)

We have ported a CFD application written using Fortran 90 and MPI with minimal changes to use our framework using MPI library called ArrayMPI on top of the Converse runtime system. The ArrayMPI library allows an MPI program to create a number of virtual processors, implemented as Converse threads, which are mapped by the runtime system to available physical processors. The application program built using this MPI library then executes as if there are as many physical processors in the system as these virtual processors. The LB framework keeps track of computational load and communication graph of these virtual processors. Periodically, the MPI application transfers control to the load balancer using a special call MPI_Migrate, which allows the framework to invoke a load balancing strategy and to re-map these virtual processors to physical processors thus maintaining load balance.

3   Load Balancing Strategies

Load balancing strategies are a separate component of the framework. By separating the data collection code common to all strategies, we have simplified the development of novel strategies. For efficiency, each processor collects only a portion of the object-communication graph, that is, only the parts concerning

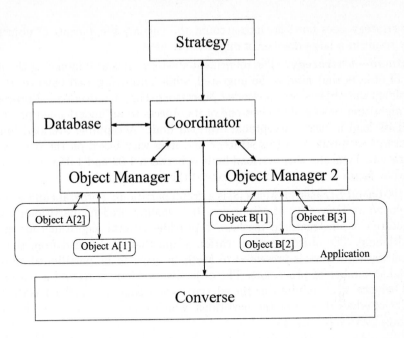

Fig. 1. Components of the load balancing framework on a processor.

local objects. This gives the strategy the freedom to ignore or locally analyze part of the graph (to minimize load-balancing overhead), or to collect the graph all in one place for a more thorough, centralized analysis. The strategy chooses a number of objects to migrate to improve program efficiency, and those decisions are handed back to the framework, which packs and migrates the objects to their new locations.

Once the run-time instrumentation has captured running times and communication graph, it is necessary to have a re-mapping strategy in place, which will attempt to produce an improved mapping. This is a multi-dimensional optimization problem, as it involves minimizing both the communication times and load-imbalances. Producing an optimal solution is not feasible, as it is an NP-hard problem. We have developed and experimented with several preliminary heuristic strategies, which we describe next.

Greedy Strategy: The simplest strategy is a greedy strategy. It organizes all objects in decreasing order of their computation times. All the processors are organized in a min-heap based on their assigned loads. The algorithm repeatedly selects the heaviest un-assigned object, and assigns it to the least loaded processor, updating the loads, and re-adjusting the heap. Although this strategy is capable of taking the communication costs into account while computing processor loads, it does not explicitly aim at minimizing communication. For N objects, this strategy has the re-mapping complexity of $O(N \log N)$. Also, since this strategy does not take into account the current assignments of objects, it may result in a large number of migration requests.

Refinement Strategy: The refinement strategy aims at minimizing the number of objects that need to be migrated, while improving load balance. It only considers the objects on overloaded processors. For each overloaded processor, the algorithm repeatedly moves one of its objects to an underloaded processor, until its load is below acceptable overload limit. Acceptable overload limit is a parameter specified to this strategy and may vary based on the overhead of migration. Typically this overload limit is between 1.02 and 1.05 which governs by what factor any processor may exceed the average load.

Metis-based Strategy: Metis [8] is a graph partitioning program and a library developed at University of Minnesota. It is mainly used for partitioning large structured or unstructured meshes. It provides several algorithms for graph-partitioning. The object communication graph that is obtained from the load balancing framework is presented to Metis in order to be partitioned onto the available number of processors. The objective of Metis is to find a reasonable load balance, while minimizing the edgecut, where edgecut is defined as the total weight of edges that cross the partitions, which in our case denotes number of messages across processors.

Figure 2 shows time taken per iteration of a synthetic benchmark when run with load balancing strategies described above. This benchmark consists of 32 objects with different loads and relatively low communication, initially mapped in a round-robin fashion to 8 processors. Load balancing is performed after every 500 iterations. All strategies improve performance, with Metis-based strategy leading to the best performance.

A load balancing strategy may improve performance of a parallel application, but if the load balancing step consumes more time than is gained by load redistribution, it may not be worthwhile. Today's parallel scientific applications run for hours. Thus it may be possible for the load balancers to spend more time in finding a better load distribution. All the three load balancing strategies described above take less than 0.5 seconds for load balancing 1024 objects on 8 processors. Thus a moderate decrease in time per iteration justifies use of any of these strategies. Also, owing to the principle of persistence, load balance deteriorates very slowly with drastic changes occurring very infrequently. Thus it may be possible to employ multiple strategies in such situations: One thorough load re-distribution in case of drastic changes, and a refinement strategy for slower load variations. We are currently experimenting with such combined strategies.

Also, note that all the strategies presented above take into consideration the application performance characteristics across all the processors. For ease of implementation, we used a global synchronizing barrier. Thus, all objects are made to temporarily stop computation while the load balancer re-maps them. However, this is usually not necessary. One can use a local barrier (barrier synchronization among objects on a single processor) for load database update, and another local barrier for performing load re-distribution, thus reducing the overheads associated with global synchronization. We are also implementing load balancing strategies that take only a partial object communication graph (based on a few neighboring processors) into account.

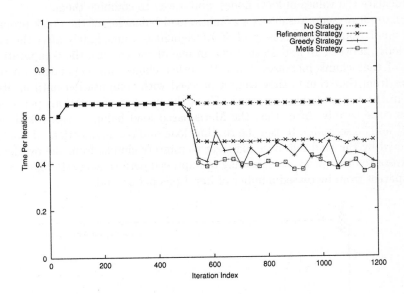

Fig. 2. Comparison of various load balancing strategies

4   Application Performance

In order to evaluate the framework, we implemented a Finite Element application that simulates pressure-driven crack propagation in structures. The physical domain is discretized into a finite set of triangular *elements*. Corners of these elements are called *nodes*. In each iteration, displacements are calculated at the nodes from forces contributed by surrounding elements. Typically, the number of elements is very large, and they are split into a number of *chunks* distributed across processors. In each iteration of simulation, forces on boundary nodes are communicated across chunks, where they are combined in, and new displacements are calculated. To detect a crack in the domain, more elements are inserted between some elements depending upon the forces exerted on the nodes. These added elements, which have zero volume, are called *cohesive* elements. At each iteration of the simulation, pressure exerted upon the solid structure may propagate cracks, and therefore more cohesive elements may have to be inserted. Thus, the amount of computation for some chunks may increase during the simulation. This results in severe load imbalance.

This application, originally written in sequential Fortran90, was converted to a C++-based FEM framework being developed by authors. This framework presents a template library, which takes care of all the aspects of parallelization including communication and load balancing. The application developer simply provides the data members of the individual nodes and elements, and a function to calculate the values of local nodes, and a way to combine them.

Figure 3 presents results of automatic load balancing of the crack propagation simulation on 8 processors of SGI Origin2000. Immediately after the crack develops (between 10 and 15 seconds) in one of the chunks, the computational load of that chunk increases. Since the other chunks are dependent on node values from that chunk, they cannot proceed with computation until an iteration of the *heavy* chunk is finished. Thus, the number of iterations per second drops considerably. After this, the Metis-based load balancer is invoked twice (at 28 and 38 seconds). It uses the runtime load and communication information collected by the load database manager to migrate chunks from the overloaded processor to other processors, leading to improved performance. (In figure 3, this is apparent from increased number of iterations per second.)

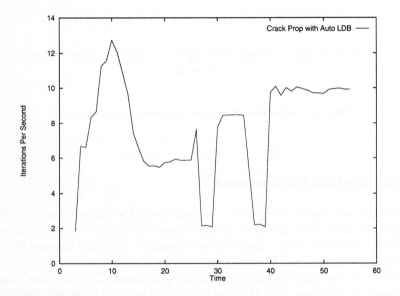

Fig. 3. Crack Propagation with Automatic Load Balancing. Finite Element Mesh consists of 183K nodes.

5   Conclusion

In this paper, we described a measurement-based automatic load balancing framework implemented in the Converse interoperable runtime system. This framework allows for easy implementation of novel load balancing strategies, while automating the tasks of recording application performance characteristics as well as load redistribution. A few strategies have been implemented and their performance on a synthetic benchmark have been compared. A real finite element method application was ported to use our load balancing framework, and its performance improvement has been demonstrated. Based on the encouraging results with such real applications, we are currently engaged in developing a more comprehensive suite of load balancing strategies, and in determining suitability of different strategies for different kinds of applications.

References

1. Jose Nagib Cotrim Arabe, Adam Beguelin, Bruce Lowekamp, Erik Seligman, Mike Starkey, and Peter Stephan. Dome: Parallel programming in a heterogeneous multiuser environment. Technical Report CS-95-137, Carnegie Mellon University, School of Computer Science, April 1995.
2. Robert K. Brunner and Laxmikant V. Kalé. Adapting to load on workstation clusters. In *The Seventh Symposium on the Frontiers of Massively Parallel Computation*, pages 106–112. IEEE Computer Society Press, February 1999.
3. N. Doulas and B. Ramkumar. Efficient Task Migration for Message-Driven Parallel Execution on Nonshared Memory Architectures. In *Proceedings of the International Conference on Parallel Processing*, August 1994.
4. L. V. Kale, Milind Bhandarkar, Narain Jagathesan, Sanjeev Krishnan, and Joshua Yelon. Converse: An Interoperable Framework for Parallel Programming. In *Proceedings of the 10th International Parallel Processing Symposium*, pages 212–217, April 1996.
5. L. V. Kale and Sanjeev Krishnan. Charm++: Parallel Programming with Message-Driven Objects. In Gregory V. Wilson and Paul Lu, editors, *Parallel Programming using C++*, pages 175–213. MIT Press, 1996.
6. L. V. Kalé, B. Ramkumar, A. B. Sinha, and A. Gursoy. The CHARM Parallel Programming Language and System: Part I – Description of Language Features. *IEEE Transactions on Parallel and Distributed Systems*, 1994.
7. L. V. Kalé, B. Ramkumar, A. B. Sinha, and V. A. Saletore. The CHARM Parallel Programming Language and System: Part II – The Runtime system. *IEEE Transactions on Parallel and Distributed Systems*, 1994.
8. George Karypis and Vipin Kumar. A fast and high quality multilevel scheme for partitioning irregular graphs. TR 95-035, Computer Science Department, University of Minnesota, Minneapolis, MN 55414, May 1995.
9. R. Namyst and J.-F. Méhaut. PM^2: Parallel multithreaded machine. A computing environment for distributed architectures. In E. H. D'Hollander, G. R. Joubert, F. J. Peters, and D. Trystram, editors, *Parallel Computing: State-of-the-Art and Perspectives, Proceedings of the Conference ParCo'95, 19-22 September 1995, Ghent, Belgium*, volume 11 of *Advances in Parallel Computing*, pages 279–285, Amsterdam, February 1996. Elsevier, North-Holland.
10. J. Pruyne and M. Livny. Parallel processing on dynamic resources with CARMI. *Lecture Notes in Computer Science*, 949:259–??, 1995.

Integrating Kernel Activations in a Multithreaded Runtime System on top of LINUX

Vincent Danjean[1], Raymond Namyst[1], and Robert D. Russell[2]

[1] Laboratoire de l'Informatique du Parallélisme
École normale supérieure de Lyon
46, Allée d'Italie
F-69364 Lyon Cedex 07, France
{Vincent.Danjean, Raymond.Namyst}@ens-lyon.fr

[2] Computer Science Department
Kingsbury Hall
University of New Hampshire
Durham, NH 03824, USA
rdr@unh.edu

Abstract. Clusters of SMP machines are frequently used to perform heavy parallel computations, and the concepts of multithreading have proved suitable for exploiting SMP architectures. Generally, the programmer uses a thread library to write this kind of program. Such a library schedules the threads or asks the OS to do it, but both of these approaches have problems. Anderson et al. have introduced another approach which relies on cooperation between the OS scheduler and the user application using *activations* and *upcalls*. We have modified the LINUX kernel and adapted the MARCEL thread library (from the programming environment PM^2) to use activations. Improved performance was observed and problems caused by blocking system calls were removed.

1 Kernel Support for User Level Thread Schedulers

The increasing popularity of clusters of SMP machines creates a need for multithreaded programming environments able to fully exploit such architectures. Indeed, the thread model naturally helps to make efficient use of all available processors and to overlap I/O operations with computations. Furthermore, threads are often considered as "virtual processors" and are targeted as such by compilers or runtime support systems for portability purposes. However, these runtime systems are built on top of thread libraries that do not all have the same properties, and thus do not provide the same functionalities. Moreover, these properties directly depend on how much control the thread scheduler has over the architecture's resources. There are two principle kinds of threads: user-level and kernel-level, each with its own advantages and inconveniences.

Efficiency is the main advantage of user-level thread libraries, whose scheduler is completely implemented in user space. Most operations on threads (creations,

J. Rolim et al. (Eds.): IPDPS 2000 Workshops, LNCS 1800, pp. 1160-1167, 2000.

context switches, etc.) can be done without any call to the operating system. As a result, some computations utilizing these threads may perform one or two orders of magnitude better than kernel-level threads. Furthermore, user threads are much more efficient in terms of kernel resource consumption, which means there can often be many more of them per application. Finally, since user-level threads are implemented in user space, they can be tailored to each user's application. The disadvantage is that user-level threads are "ignored" by the OS and thus cannot be scheduled correctly in many cases. For instance, since user threads within the same process cannot be scheduled concurrently on multiple processors, no real parallelism can be achieved. Similarly, when a thread makes a blocking system call (for example, a `read()` on an empty socket), all the threads in that process are blocked.

Obviously, kernel-level threads do not suffer from these drawbacks, since their scheduling is realized within the OS kernel, which handles them the same way it handles processes, except that multiple threads may share the same address space. It is therefore possible on an SMP machine for the kernel to simultaneously assign processors to multiple threads in the same application, thus achieving true parallelism. Furthermore, when one thread makes a blocking system call, the kernel can give control to another thread in the same application. However, even if operations such as thread context switching are more efficient than those related to processes, they still require system calls to be performed.

1.1   The MARCEL Mixed Thread Scheduler

To try to obtain the best properties of the two kinds of threads, some libraries mix them together: there are a fixed number of kernel threads each running a number of user threads. This approach retains the efficient scheduling of user threads, but is able to take advantage of parallelism between threads on SMP machines. One such library is MARCEL[5], which was developed for use by PM^2[4] (*Parallel Multithreaded Machine*), a distributed multithreaded programming environment. MARCEL delivers good performance by eliminating some features from the POSIX pthreads specification that are not useful for scientific applications (*e.g.*, per-thread signal handling). In addition, it supports multiple optimizations as well as dynamic thread migration across a homogeneous cluster. MARCEL has been ported to a number of different platforms. It utilizes a fixed number of kernel threads, each managing a pool of user-level threads.

1.2   Better Support: Kernel Activations

Although the two-level version of MARCEL achieves better performance than the earlier user-level version, it still suffers from some of the problems discussed earlier. The first problem is that when a user thread makes a blocking system call, the underlying kernel thread is stopped too. It is possible with a few blocking user threads to block all the kernel threads, thereby blocking the whole application, even if some other user threads are ready to run. Another problem is that even if MARCEL can control the scheduling of user-level threads in each pool, it cannot do anything between the different pools. So, if thread A in pool 1 holds a lock and is preempted by the system, then when thread B in another pool wants the lock, it has to wait for the OS to give control back to pool 1 so that thread A can release the lock.

These problems could be avoided if the OS scheduler reported its scheduling decisions to the application. One mechanism to achieve this cooperation is based on the concept of *activations*, which was first proposed in an article by Anderson et al.[1] Its authors implemented this mechanism with the FASTTHREAD library on the TOPAZ system. However, this system is no longer running, and the sources were never released. All the terms (*activation, upcall*, etc.) used in this paper come from this article.

This mechanism enables the kernel to notify a user-level process whenever it makes a scheduling decision affecting one of the process's threads. This mechanism is implemented as a set of *upcalls* and *downcalls*. A traditional system call is a *downcall*, from the user-level down into a kernel-level function. The new idea is a corresponding *upcall*, from the kernel up into a user-level function. An upcall can pass parameters, just as system calls do. An *activation* is an execution context (*i.e.*, a task control block in the kernel, similar to a kernel-level thread belonging to the process) that the kernel utilizes to make the upcall. The key point is that each time the kernel takes a scheduling action affecting any of an application's threads, the application receives a report of this fact and can take action to (re)schedule the user-level threads under its control.

We have modified the LINUX kernel by adding activations and changing the existing kernel scheduler to use upcalls to report some scheduling events to the MARCEL scheduler running in user space. Upcalls are mainly used to report that a new activation has been created, that an activation has blocked in a system call, that a previously blocked activation has just been unblocked, or that an activation has been preempted. We have also modified MARCEL to utilize this mechanism efficiently, as discussed in the next section.

2   MARCEL on Top of LINUX Activations

The user-level MARCEL thread scheduler utilizes the new mechanism as follows: MARCEL begins by making an `act_new()` system call to notify the kernel that it wants to utilize activations. The scheduler provides parameters that include a vector of entry points for a fixed set of user-level management functions to which the kernel will make upcalls. Whenever the kernel makes a scheduling decision affecting any of this process's activations, such as creating, blocking or unblocking it, the kernel informs the process by choosing one of its activations and using it to make the appropriate upcall, such as `upcall_new`, `upcall_block`, or `upcall_unblock`. In order to guarantee exclusive access to management information while executing one of these functions, the kernel maintains an internal mutual exclusion lock that allows only one upcall at a time to be outstanding per process. Therefore, the management function must make an `act_resume()` system call to release that lock after making its management decision but before

Table 1. Upcalls made by the LINUX kernel to the user-level thread scheduler

Upcall	Description
upcall_new	a new activation is starting
upcall_block	an activation blocked
upcall_unblock	an activation unblocked. The scheduler has its state, so it can restart the activation's thread when it wants.
upcall_preempt	an activation was preempted. The scheduler has its state, so it can restart the activation's thread when it wants.
upcall_restart	Used by the kernel to make an upcall (*e.g.*, in response to an act_send() system call) when it has no scheduling event to report.

executing application specific code. If the kernel scheduler decides that an activation holding this lock should be preempted, the kernel will preempt another activation instead (via upcall_preempt) and will simply reschedule the original activation without an upcall.

Our implementation of the activations within the LINUX kernel is close to the one proposed by Anderson et al. It is described more fully in [2]. The next section presents some general characteristics that are referred to in the following sections.

The programming interface provides a few new system calls, and the targeted thread library must be prepared to handle several kinds of upcalls. Table 1 describes the upcall interface used by the kernel to notify the user thread scheduler about certain scheduling events.

2.1 How it works

Figure 1 illustrates how MARCEL uses activations to keep both processors on a dual-processor SMP platform actively executing application threads, even when some threads are blocked in the kernel.

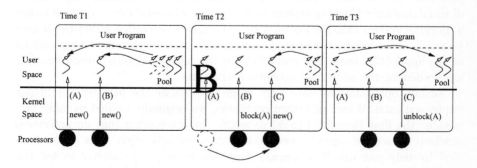

Fig. 1. A blocking system call with activations

At time T1, the kernel creates two activations "A" and "B" and makes an `upcall_new` to each. In each activation, the MARCEL scheduler will choose a ready application thread and give it control.

At time T2, the application thread running in activation "A" makes a blocking I/O system call. The kernel determines the process to which this activation belongs and creates a third activation, "C", into which it makes an `upcall_new`. In this activation, the MARCEL scheduler will choose a third application thread and then call `act_resume()` to release the mutual exclusion lock. The kernel next chooses one of the activations, say "B", and makes an `upcall_block` to it, providing "A" as the parameter to indicate which activation was just blocked. The MARCEL scheduler uses this information to keep track of the status of the corresponding application thread.

At time T3, the I/O request completes. The kernel then chooses one of the activations, say "C", and makes an `upcall_unblock` to it, providing "A" as the parameter to indicate which activation was just unblocked. The MARCEL scheduler now chooses whether to return the application thread previously assigned to "A" to the pool and continue running the application thread already assigned to "C", or vice versa. In either case, activation "A" remains idle until needed by the kernel to make another upcall.

2.2 Extensions to the original proposal

Although this work is mainly based on the *Scheduler Activation* model proposed by Anderson et al, we have developed a number of improvements which extend the set of supported system calls and increase efficiency in some situations.

One important point with activations is that the number of running activations at the application level is constant. In Anderson's implementation, this also meant that the number of activation structures for that user in the kernel was constant. This has the advantage of using a constant amount of kernel resources. However, it does not allow the kernel to handle blocking system calls properly, since a kernel activation structure is tied up during the time its thread is blocked, thereby preventing the kernel from running another user-level thread in that activation. Our implementation does not keep constant the number of activation structures for one user within the kernel. This allows us to handle any number of simultaneously blocking system calls, because whenever one activation issues a blocking system call, the kernel will create a new activation structure, if necessary, in order to keep constant the number of concurrently running activations at the application level. The cost of this is the additional kernel resources that are needed for the additional activation structures.

Several optimizations have been made to increase the performance of our implementation. When an activation blocks, we originally needed to make two upcalls: the first (`upcall_block`) to notify the application that an activation blocked, the second (`upcall_new`) to launch a new activation. This is now handled by only one upcall to `upcall_new`, which uses a parameter to tell the application whether another activation has blocked. An additional optimization has been made as far as preemption is concerned. In the original model, when an activation is preempted, an `upcall_preempt` upcall occurs, and an `upcall_new` upcall is made when the kernel is ready to restart an activation. Now, the application can tell the kernel at the end of the `upcall_preempt` upcall (with a parameter to the system call `act_resume()`) that instead of calling the upcall `upcall_new`, it can continue this activation directly.

2.3 Modifications to MARCEL

Surprisingly, integration of LINUX Activations within the MARCEL library required almost no rewriting of existing code. We needed only a few localized extensions.

The major issue that we had to address was related to the "ready-threads" queue. The problem was to opt either for a global pool (as in a user-level version of MARCEL) or for a collection of activation-specific local pools (as in the mixed version). We have opted for the global pool implementation because maintaining separate pools introduces a number of synchronization problems. In particular, when an activation gets blocked within the kernel, the other activations must retrieve the running threads that were kept in its ready-threads pool. Such a step requires a costly synchronization scheme and the associated overhead may become important in the presence of frequent I/O operations. The drawback of our strategy is that the global pool may become a bottleneck on a large number of processors.

MARCEL uses a special lock to prevent concurrent access to its internal data structures. Our implementation of activations ensures that if the kernel preempts the MARCEL thread which is holding this lock, then it is relaunched immediately (instead of the one running on the activation that receives the `upcall_preempt` upcall). This allows us to avoid contention situations in the presence of busy waiting threads. Note that a related problem can occur with the upcall `upcall_new`. Indeed, when a new activation is created, it may not succeed in acquiring the aforementioned lock. Since it is mandatory to run a regular MARCEL thread when calling `act_resume()`, the activation must schedule a "dummy" thread. To this end, we have added a pool of preallocated "dummy" threads (together with their stacks) into MARCEL.

3 Performance and Evaluation

The new version of MARCEL on top of LINUX Activations is completely operational, although we did not yet implement all the optimizations we discussed in the previous sections. To investigate the gain or the overhead generated by activations and upcalls, we have compared the new version of MARCEL to the two existing versions (one purely user-level, one mixed two-level) as well as to native LINUX kernel-level threads [3]. The tests were run on an Intel Pentium II 450 MHz platform running LINUX v2.2.13. On this platform, we ran a microbenchmark program to measure the time taken by an *upcall* from the kernel up to user-space. This test reported an average time of $5\mu s$ per upcall.

Table 2. Performance of various thread libraries

Library	Single processor		Dual processor
	Basic	With I/O	With computation
MARCEL user-level	0.308ms	119.959ms	6932ms
MARCEL mixed two-level	0.435ms	23.241ms	3807ms
MARCEL with activations	0.417ms	10.118ms	3551ms
LINUXTHREAD (kernel-level)	13.319ms	14.916ms	3566ms

The test programs used to compare these libraries are all based on a common synthetic program. The basic program implements a *divide and conquer* algorithm to compute the sum of the first N integers. At each iteration step, two threads are spawned to compute the two resulting sub-intervals concurrently, unless the interval to compute contains one element. The "parent" of the two threads waits for their completion, gets their results, computes the sum and, in turn, returns it to its own parent. This program generates a tree of threads and involves almost no real computation but a lot of basic thread operations such as creation, destruction and synchronization.

In order to evaluate the different thread libraries in the presence of blocking calls, we have extended the previous program so as to make extensive use of UNIX I/O operations. In this case, we have simply replaced all the thread creation calls by a write into a UNIX *pipe*. At the other end of the pipe, a dedicated server thread simply transforms the corresponding requests into thread creations.

Finally, we also extended the basic version of the program by adding some artificial computation into each thread so that some speedup can be obtained on a multiprocessor platform.

3.1 Performance

Table 2 reports the performance obtained with the three aforementioned program versions for each thread library. The first two programs were run on a uniprocessor machine whereas the last one was run on a dual-processor.

The basic version of the divide and conquer program makes heavy use of thread creations and synchronizations. As one may expect on a uniprocessor, the user-level MARCEL library is obviously the most efficient, while the LINUX-THREAD library exhibits poor performance, because kernel thread operations are much more inefficient than those related to user threads. It is interesting to note that the version using activations achieves good performance. The difference with the user-level version is due to the MARCEL lock acquire/release primitives that are a little more complex in the presence of activations.

With the version involving many I/O operations, things change significantly. The most noticeable result is the *huge* amount of time taken by the program with the user-level version. It is, however, not surprising: each time a user thread makes a blocking call, it blocks the entire UNIX process until a timer signal forces a preemption and schedules another thread (in this case, every 20*ms*).

The activation version has the best execution time. The mixed MARCEL library does not behave as well because two underlying kernel threads are needed to handle the blocking calls properly. Thus, it introduces overhead due to additional synchronization and preemption costs.

When the program containing substantial computation is executed on a dual-processor machine, we observe that the activation version has approximately the same execution time as the MARCEL "mixed" and LinuxThread versions. It reveals that the activation version is perfectly able to exploit the underlying architecture by using two activations simultaneously within the application. The user-level version obviously performs poorly, because only one processor is used in this case.

4 Conclusion

This work augmented the design of activations, a new technique to handle thread support in an OS, then implemented and tested their use under Linux. We wrote a new version of the MARCEL thread library that utilizes activations while preserving the existing user interface, so that existing MARCEL programs still work with this new model. We have demonstrated that for applications using threads that make blocking system calls, performance of the new version of MARCEL on both single and dual processor platforms is superior to the best previous version of MARCEL and to kernel-level threads. Furthermore, since our new library is implemented in user space, we do not need to change the kernel to add new thread features, such as thread migration.

A two-level thread library based on activations seems to be a very attractive way to manage application threads. This work shows that this model is a valid one, in particular for application threads that utilize blocking system calls, which often happens within a communication library, for example.

References

1. T. Anderson, B. Bershad, E. Lazowska, and H. Levy. Scheduler Activations: Effective Kernel Support for the User-Level Management of Parallelism. *ACM Transactions on Computer Systems*, 10(1):53–79, February 1992.
2. Vincent Danjean, Raymond Namyst, and Robert Russell. Linux kernel activations to support multithreading. In *Proc. 18th IASTED International Conference on Applied Informatics (AI 2000)*, Innsbruck, Austria, February 2000. IASTED. To appear.
3. Xavier Leroy. The LinuxThreads library. http://pauillac.inria.fr/~xleroy/linuxthreads.
4. R. Namyst and J.F. Mehaut. PM2: Parallel Multithreaded Machine. a computing environment for distributed architectures. In *ParCo'95 (PARallel COmputing)*, pages 279–285. Elsevier Science Publishers, Sep 1995.
5. R. Namyst and J.-F. Méhaut. MARCEL : *Une bibliothèque de processus légers*. Laboratoire d'Informatique Fondamentale de Lille, Lille, 1995.

DyRecT: Software Support for Adaptive Parallelism on NOWs

Etienne Godard   Sanjeev Setia   Elizabeth White

Department of Computer Science, George Mason University

Abstract. In this paper, we describe DyRecT (Dynamic Reconfiguration Toolkit) a software library that allows programmers to develop adaptively parallel message-passing MPI programs for clusters of workstations. DyRecT provides a high-level API that can be used for writing adaptive parallel HPF-like programs while hiding most of the details of the dynamic reconfiguration from the programmer. In addition, DyRecT provides support for making a wider variety of applications adaptive by exposing to the programmer a low-level library that implements many of the typical tasks performed during reconfiguration. We present experimental results for the overhead of dynamic reconfiguration of several benchmark applications using DyRecT.

1   Introduction

Parallel applications executing on clusters of workstations have to be able to "withdraw" from a workstation if its owner returns. This is because workstation owners are typically unwilling to share their workstation with parallel applications while they are using it for doing interactive tasks. Thus, it is necessary to ensure that parallel applications execute only on idle workstations.

To address this issue, several run-time libraries and environments provide mechanisms for process migration [1]. When owner activity is detected on a workstation being used by a parallel application, the process executing on that workstation is migrated to an idle workstation. If no idle workstation is available, the parallel application is either suspended until more resources are available or multiple processes that compose the parallel application are scheduled on the same processor.

Several studies [5, 7] have shown that a more desirable approach from the performance viewpoint would be to dynamically reconfigure the parallel application so that its parallelism matched the number of processors available for execution. Such dynamically reconfigurable applications have been referred to as adaptive parallel or malleable parallel applications. Unlike conventional parallel applications, adaptive parallel applications can adapt to changes in the availability of underlying resources by dynamically shrinking or expanding their degree of parallelism.

While the performance benefits of supporting adaptively parallel applications seem clear, most parallel programming environments do not provide mechanisms for dynamically changing the degree of parallelism of executing applications. In

J. Rolim et al. (Eds.): IPDPS 2000 Workshops, LNCS 1800, pp. 1168-1175, 2000.

this paper, we describe DyRecT (Dynamic Reconfiguration Toolkit), a software library that allows programmers to develop adaptively parallel message-passing MPI programs for clusters of workstations.

Ideally, writing adaptive parallel applications should be no more difficult than developing conventional parallel applications. To this end, several run-time systems [1, 6] have been designed that support adaptive parallel applications in a user-transparent fashion. Some of these systems, however, require all applications to be written using a master-slave programming paradigm. This can lead to poor performance for several classes of applications [3]. Other systems support adaptive parallelism for specific classes of applications, e.g., Adaptive Multiblock Parti [2] supports adaptive parallel structured and block-structured parallel applications.

Recently two systems have been developed that have a wider applicability than the systems discussed above. DRMS [3] supports adaptive parallelism for grid-based message-passing programs on the IBM SP2, while in [4], Scherer et al describe a system for adaptively parallel shared memory programs that use the OpenMP programming model. The wider applicability of these systems arises from the fact that they support the OpenMP and HPF programming models that are used for several classes of applications.

DyRecT resembles DRMS in that one of its goals is to support grid-based message-passing programs. To this end we provide a high level API that can be used by the programmer for writing adaptive parallel HPF-like programs. It differs from DRMS in two important ways. First, we provide support for making a wider variety of applications adaptive by exposing to the programmer a low-level library that implements many of the typical tasks performed during reconfiguration. Second, we support adaptive parallelism on NOWs consisting of potentially heterogeneous workstations by providing support for saving and restoring the stack of an executing process in an architecture-independent fashion.

Our approach is motivated by the observation that while the details of the actions that need to be taken during reconfiguration depend upon the application, there are common tasks that typically need to performed, e.g., spawning processes, synchronizing the application, capturing and restoring the stack, exchanging data, etc. For example, to move from the first configuration in Figure 1 to the second, the four starting processes must synchronize at some point in the computation where a consistent grid exists across the processes. At that point, data must be moved so that it is distributed across three of the processes. The process leaving the computation must be terminated. Finally, any required changes to the communication bindings must be made. At this point, the grid computation can continue.

In the case of regular grid-based iterative applications, most of these reconfiguration related tasks are performed by our high-level library and are hidden from the programmer. However, the high-level API provided with DyRecT is only suitable for certain classes of grid-based applications. Using the low-level library, discussed in Section 3, a programmer can develop reconfiguration code

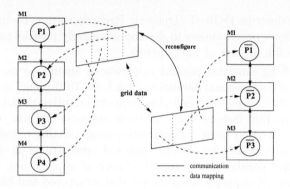

Fig. 1. Changing the level of parallelism by moving between configurations in a grid-based parallel application.

for other classes of applications with considerably less effort than if they had to develop the code from scratch.

2   High-Level Primitives

There are several different types of high-level primitives provided by the toolkit: initialization and finalization, synchronization, data distribution registration, runtime data support and reconfiguration data distribution. These primitives hide many of the details that user would typically have to deal with when making iterative grid-based applications adaptive; describing how the partitioning is related to the number of processes, moving data between processes at reconfiguration time, performing some data initialization, spawning and terminating processes and synchronizing to ensure that a consistent grid is repartitioned.

As an example, consider a typical iterative grid-based application as shown in Figure 1. For each process, every iteration consists of doing a local computation, exchanging information with neighboring processes, and synchronizing to decide convergence. When the global grid is uniformly distributed across the participating processes, this application can be made adaptive by instrumenting the source code with our high-level primitives. These calls provide to the runtime system basic information about how the grids are partitioned across any number of processes. The code for this, described below, is shown in Figure 2.

The data partitioning high-level primitives allow users to define uniform partition schemes over multi-dimensional data. In the example, the user specifies a *block* partition along the first (row) dimension (DYR_Block()) combined with a *collapsed* (non-partitioned) partition (DYR_Collapsed()) for the second (column) dimension. Two grids, one for the current iteration and one for the previous, that are partitioned using this scheme, are registered with the library using the DYR_Register_data() calls.

After providing information about the data to be repartitioned, the user decides where in the component source code it is legal for repartitioning to

```
int main (int argc, char *argv[]) {
  int local_dims[2],dims[2];
  double **mydata, **mydata_next;
  DYR_Disttype dist_types[2];
  DYR_Disthandle strips;
  MPI_Comm Compute_context;
  MPI_Init(&argc, &argv);
  DYR_Init(&Compute_context);                             /* initialize DYRECT */
  DYR_Save((void *) &iter, 1, MPI_INT);
     /* save variable(s) needed across all nodes */
  DYR_Block(&dist_types[0]);                              /* globally distributed data */
  DYR_Collapsed(&dist_types[1]);
  DYR_Borders_uniform (1, 1, &dist_borders[0]);
  dist_borders[1] = dist_borders[0];                      /* define borders */
  DYR_Create_distribution (2, dist_types, dist_borders, &strips);
  DYR_Register_data (&mydata,2, dims,MPI_DOUBLE,0,0,strips);
  DYR_Register_data (&mydata_next,2,dims,MPI_DOUBLE,0,0,strips);
  if (DYR_Init_node()) {
     DYR_Local_shape (&mydata, local_dimens);   /* new local size */
        /* Put standard initialization calls, etc. from original program */
     iter = 0; init_data (local_dimens, mydata_next);    /* initialize data area */
  }

  do { /* iterate using Jacobi relaxation until block has converged */
        /* check for reconfiguration */
     if (DYR_Check_reconf(0)) {
        DYR_Reconfigure(1, &Compute_context); /* reconfigure */
        DYR_Local_shape (&mydata, local_dimens);
     } /* if */
     DYR_Update_borders ( &mydata_next, 0, 0);
     copy_data (local_dimens, mydata_next, mydata);
     calc_area (local_dimens, mydata, mydata_next);
     iter++;
  } while (cont_iter (local_dimens, mydata, mydata_next);
  MPI_Finalize(); DYR_Final();
}

cont_iter(...) {
  /* compute local norm */
  DYR_Sync_MS(...,result,comp_norm,set_flag);
  return result;
}
```

Fig. 2. Abbreviated source code for the Jacobi Application. Code added for dynamic reconfiguration is shown in boldface.

occur. The start of each iteration is used for the Jacobi application. At that point, the user adds an invocation to DYR_Reconfigure() guarded by a call to DYR_Check_reconfig(). The DYR_Reconfigure() function uses the data registration information to take care of all of the repartitioning calculations, data exchange, and process creation and termination required for the new set of processes.

The toolkit provides two different synchronization mechanisms, both of which assume that the application is iteration based and that reconfiguration must occur when all processes are at the same iteration. Both synchronization functions are responsible for setting a flag that is used by the DYR_Check_reconfig() function. The synchronization mechanism used in the example extends the existing global synchronization at the end of an iteration. In the master process, this function takes over the details of receiving the data and computing convergence using a user-provided function. It determines if a reconfiguration is needed and and informs the other processes about both convergence and reconfiguration in the return message.

If a parallel application has a variable that needs to hold the same value across all participating processes, (such as iter in Figure 2), it is registered with the toolkit using DYR_Save(). If a process joins the application at reconfiguration time, the toolkit ensures that it is initialized appropriately. It is sometimes necessary to transform the control flow of the components depending upon whether or not the process was one of the initial processes. Function DYR_Init_node() only returns true for processes that were part of the application at start time. In Figure 2, this function is used so that the initial processes can initialize their local data and variables. When new processes enter later in the application execution, they skip this code and immediately enter the loop, perform their reconfiguration and get information using DYR_Local_shape() about their data set. Then they execute normally. This primitive can also be used to guard code that only new processes should execute.

3   Low-Level Primitives

In addition to primitives tailored toward one class of parallel applications, our toolkit also provides to the user a set of low-level primitives. The primary reason for providing these primitives is to allow programmers to more easily handle situations where the standard high-level functionality is not sufficient. There are several different types of primitives we provide for specialized partitioning, physical resource control, tailoring of work done at reconfiguration points, and dealing with data on the runtime stack. We have found these types of low-level primitives useful for several different types of applications.

As an example, consider the case where there is variation in the relative processor speeds in the workstation cluster. In this situation, it makes sense to give processes on faster processors larger local grids than processes on slower processors. While high-level functions may provide solutions for some aspects of the problem (synchronization for example), support for non-uniform partitioning schemes, e.g., recursive bisection, are not supported by the high-level primitives. However, using the low-level primitives provided by DyRecT, the user can tailor the actions taken during reconfiguration such that non-uniform partitioning schemes can be handled.

The default assumption is that the given reconfiguration points are placed in the main program. While this is not atypical of this class of applications, for some members of this class, more efficient reconfiguration can be achieved by placing reconfiguration points in other locations in the source code where they are encountered more frequently. For example, a multigrid V-cycle can be implemented recursively and one logical place for reconfiguration is inside the recursive function. However, this placement of reconfiguration points raises the question of how to create the correct runtime stack for new processes and how to update data (typically variables tied to grid size and pointers to intermediate grids) that may be on the stack in existing processes. Our low-level primitives include functions to deal with these problems and some rudimentary source-to-source transformation tools that deal with some of the difficult issues of the placement of these functions.

4   Performance Results

In this section, we describe the results of experiments in which we measured the cost of dynamically reconfiguring several parallel applications. The main goals of these experiments were to demonstrate the feasabilty of using DyRecT for supporting adaptive parallelism on NOWs and to identify the various components that contribute to the overhead of dynamic reconfiguration.

Our experimental environment consists of 16 PCs connected by a switched 100 Mbps Ethernet. Each machine has one or two 200 MHz Intel Pentium Pro processors and between 128 and 256 MB RAM. The computers run Linux 2.2.10. Our reconfiguration software was built on top of the LAM (version 6.2b) implementation of MPI.

We measured the cost of reconfiguration for five benchmark applications. The first two applications (referred to as Jacobi and RB) use the Jacobi relaxation method to solve Poisson's equation on a square grid. In Jacobi, a strip partitioning scheme is used to distribute the grid among the processors, while RB uses recursive bisection to partition the grid. The third benchmark (BC) employs a block cyclic data decomposition technique to allocate grid data to processors. The next two applications, Multigrid and Integer Sort are taken from the NAS parallel benchmarks.

We reconfigured each application several times and measured the adaptation time under different scenarios. These scenarios are representative of fluctuations in resource availability that can occur in non-dedicated clusters of workstations such as new nodes joining the computation, nodes leaving the computation, and migration of a process from one node to another.

In our experiments, an executing parallel application reconfigures itself when it receives a signal sent via the LAM "doom" command. The delay before the application resumes execution after reconfiguration consists of two components. The main component is the actual cost of reconfiguration itself (as discussed below). In addition, before the reconfiguration can be initiated each process in the computation needs to reach the next "safe" point in its execution. This synchronization delay is application-specific since it depends on the location and frequency of occurrence of reconfiguration points. For example, in the case of the Jacobi, RB, and BC benchmarks, the reconfiguration point occurs at the end of each iteration whereas in the case of multigrid, reconfiguration points occur at each level of the multigrid V-cycle. For our benchmark applications, the synchronization delay varied from 0.07 to 3.77 seconds depending on the number of processors and the data set size of the application.

The reconfiguration cost can be broken down into several components corresponding to the different steps involved in the dynamic reconfiguration of parallel applications. These steps are: (i) spawning any new processes (ii) re-establishing the logical configuration of the application (iii) figuring out the new logical data partitioning, e.g. by invoking the recursive bisection algorithm (iv) allocating memory for any newly assigned data (v) figuring out the overlap of the current data assignment with the future data assignments, and (vi) exchanging data between nodes to account for the new configuration. Figure 3 shows the costs for each benchmark for two reconfiguration scenarios: changing the parallelism from 8 to 16 nodes, and vice versa. The time for steps (i) through (vi) is labeled spawn, init, part, alloc, overlap, and redist respectively.

Our experiments showed that the main component of the total reconfiguration time was the data redistribution time, which is proportional to the amount of data that needs to be redistributed between the processors. The reconfiguration time for our benchmarks ranged from hundreds of milliseconds to around 15 seconds depending mainly on the data set size of the application. For a more thorough discussion of our performance results, the reader is referred to [8].

5   Conclusion

Efficient and non–intrusive use of NOWs for parallel applications requires easy to use mechanisms for providing adaptive behavior. This paper describes research into providing both high– and low–level functionality for achieving this. The high–level primitives, tailored to iterative grid–based applications, provide simple to use mechanisms for many of the common tasks in this domain. When this functionality does not capture some required feature of the application, the user can use the provided low–level functions to provide additional flexibility.

This work is ongoing in that we are still refining both the API and the functionality provided by the API. One natural next step is to look at how high–level APIs for other classes of applications can be constructed on top of our low–level primitives. Research into efficient algorithms for data exchange within this framework is also of interest.

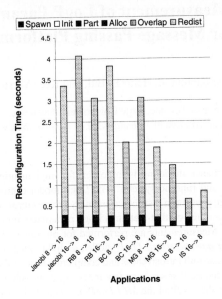

Fig. 3. The components of the reconfiguration overhead for five benchmark applications. The data set sizes for the benchmarks are as follows: Jacobi, RB, and BC – 144 MB, MG – 55 MB, IS – 24 MB.

References

1. J. Pruyne and M. Livny. Interfacing Condor and PVM to harness the cycles of Workstation Cluster s. In *Journal of Future Generation Computer Systems*, Vol. 12, 1996.
2. G. Edjlali et al. Data Parallel Programming in an Adaptive Environment. Technical Report CS-TR-3350, University of Maryland, 1994.
3. J. Moreira, V. Naik and M. Konuru. Designing Reconfigurable Data-Parallel Applications for Scalable Parallel Computing Enviromnents. Technical Report RC 20455, IBM Research Division, May 1996.
4. A. Scherer, H. Lui, T. Gross, W. Zwaenepoel. Transparent Adaptive Parallelism on NOWs using OpenMP. In Proc. of PPoPP'99, May 1999.
5. A. Acharya, G. Edjlali, J. Saltz. The Utility of Exploiting Idle Workstations for Parallel Computation. In *Proc. of ACM Sigmetrics '97*, 1997.
6. N. Carriero, E. Freeman, D. Gelernter. Adaptive Parallelism and Piranha. *IEEE Computer*, pp. 40-49, Jan 1995.
7. A. Chowdhury, L. Nicklas, S. Setia, E. White Supporting Dynamic Space-sharing on Non-dedicated clusters of Workstati ons. In *Proc. of ICDCS '97*, 1997.
8. E. Godard, S. Setia, E. White. DyRecT: Software Support for Adaptive Parallelism on NOWs. Technical Report GMU-TR00-01, Department of Computer Science, George Mason University, January 2000.

Fast Measurement of LogP Parameters
for Message Passing Platforms

Thilo Kielmann, Henri E. Bal, and Kees Verstoep

Department of Computer Science, Vrije Universiteit, Amsterdam, The Netherlands
kielmann@cs.vu.nl   bal@cs.vu.nl   versto@cs.vu.nl

Abstract. Performance modeling is important for implementing efficient parallel applications and runtime systems. The LogP model captures the relevant aspects of message passing in distributed-memory architectures. In this paper we describe an efficient method that measures LogP parameters for a given message passing platform. Measurements are performed for messages of different sizes, as covered by the *parameterized LogP* model, a slight extension of LogP and LogGP. To minimize both intrusiveness and completion time of the measurement, we propose a procedure that sends as few messages as possible. An implementation of this procedure, called the *MPI LogP benchmark*, is available from our WWW site.

1   Introduction

Performance modeling is important for implementing efficient parallel applications and runtime systems. For example, *application-level schedulers* (AppLeS) [2] aim to minimize application runtime based on application-specific performance models (e.g., for completion times of given subtasks) which are parameterized by dynamic resource performance characteristics of CPUs and networks. An AppLeS may, for example, determine suitable data distributions and task assignments based on the knowledge of message transfer times and computation completion times.

Another example for the use of performance models is our MagPIe library [8, 9] which optimizes MPI's collective communication. Based on a model for the completion times of message sending and receiving, it optimizes communication graphs (e.g., for broadcast and scatter) and finds suitable segment sizes for splitting large messages in order to minimize collective completion time.

The LogP model [4] captures the relevant aspects of message passing in distributed-memory systems. It defines the number of *processors* P, the network *latency* L, and the time (*overhead*) o a processor spends sending or receiving a message. In addition, it defines the *gap* g as the minimum time interval between consecutive message transmissions or receptions at a processor, which is the reciprocal value of achievable end-to-end bandwidth. Because LogP is intended for short messages, o and g are constant. The LogGP model extends LogP to also cover long messages [1]. It adds a parameter G for modeling the *gap per byte* for long messages, which are typically handled more efficiently. Other variants of LogP have also been proposed where the overhead at the sender and the receiver side is treated separately as o_s and o_r, and where some parameters depend on the message size [5, 7, 8].

J. Rolim et al. (Eds.): IPDPS 2000 Workshops, LNCS 1800, pp. 1176-1183, 2000.

For practical use of LogP, the actual parameters of a parallel computing platform have to be measured. Inside a supercomputer or workstation cluster, the network performance characteristics remain constant, except for possible changes in system software. In this case, the respective LogP parameters may be measured offline, and measurement efficiency hardly matters. Our MagPIe library, however, targets multiple clusters connected via wide-area networks. In this context, off-line measurements are not feasible due to two reasons, so measurement efficiency is very important. First, intrusiveness on other ongoing communication has to be kept as small as possible. Second, the performance of wide-area networks may change during application runtime [11], causing measurements also to be performed regularly.

The main problem with measurement efficiency is how to accurately measure the *gap* parameter. The measurement methods described in [5, 7] measure the *gap* by sending large sequences of messages in order to saturate the communication links in which case the link capacity (as expressed via the gap) can be observed. This measurement procedure has two drawbacks. It is highly intrusive and may disturb other ongoing communication. Also, it is time consuming when measuring long messages, especially when the network has high latency and/or low bandwidth, as is the case with wide-area connections as targeted by MagPIe.

In this paper, we present a procedure that measures LogP parameters without saturating the network with long messages. Only for empty messages (with zero bytes of data), the gap has to be determined by saturating the network. This can be achieved in reasonable time even across wide area links. For all other message sizes, simple message roundtrips (and the gap for empty messages) are sufficient to determine the corresponding LogP parameters. In the remainder of the paper, we briefly clarify the LogP variant we use (*parameterized LogP* [8]), then we describe our measurement procedure and compare our measurements with results obtained by saturation-based measurements.

2   Parameterized LogP

The *parameterized LogP* model defines five parameters, in analogy to LogP. P is the number of processors. L is the end-to-end latency from process to process, combining all contributing factors such as copying data to and from network interfaces and the transfer over the physical network. $o_s(m)$, $o_r(m)$, and $g(m)$ are send overhead, receive overhead, and gap. They are defined as functions of the message size m. $o_s(m)$ and $o_r(m)$ are the times the CPUs on both sides are busy sending and receiving a message of size m. For sufficiently long messages, receiving may already start while the sender is still busy, so o_s and o_r may overlap. The gap $g(m)$ is the minimum time interval between consecutive message transmissions or receptions. It is the reciprocal value of the end-to-end bandwidth from process to process for messages of a given size m. Like L, $g(m)$ covers all contributing factors. From $g(m)$ covering $o_s(m)$ and $o_r(m)$, follows $g(m) \geq o_s(m)$ and $g(m) \geq o_r(m)$. A network N is characterized as $N = (L, o_s, o_r, g, P)$.

To illustrate how the parameters are used, we introduce $s(m)$ and $r(m)$, the times for sending and receiving a message of size m when both sender and receiver simultaneously start their operations. $s(m) = g(m)$ is the time at which the sender is ready to send the next message. Whenever the network itself is the transmission bottleneck, $o_s(m) < g(m)$, and the sender may continue computing after $o_s(m)$ time. But because $g(m)$ models the time a message "occupies" the network, the next message cannot be sent before $g(m)$. $r(m) = L + g(m)$ is the time at which the receiver has received the message. The latency L can be seen as the time it takes for the first bit of a message to travel from sender to receiver. The message gap adds the time after the first bit has been received until the last bit of the message has been received. Figure 1 (left) illustrates this modeling. When a sender transmits several messages in a row, the latency will contribute only once to the receiver completion time but the gap values of all messages sum up. This can be expressed as $r(m_1, m_2, \ldots, m_n) = L + g(m_1) + g(m_2) + \ldots + g(m_n)$.

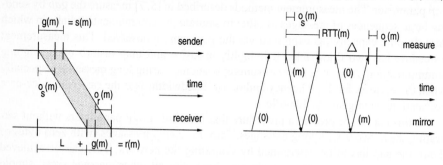

Fig. 1. Message transmission as modeled by parameterized LogP (left); fast measurement procedure (right)

For completeness, we show that *parameterized LogP* subsumes the original models LogP and LogGP. In Table 1, LogGP's parameters are expressed in terms of *parameterized LogP*. We use 1 byte as the size for short messages; any other reasonable "short" size may as well be used instead. Note that neither LogP nor LogGP distinguishes between o_s and o_r. For short messages, they use $r = o + L + o$ to relate the L parameter to receiver completion time which gives L a slightly different meaning compared to *parameterized LogP*. We use this equation to derive LogP's L from our own parameters.

3   Fast parameter measurement

Previous LogP micro benchmarks [5, 7] measure the gap values by saturating the link for each message size. Our method has to use saturation only for obtaining $g(0)$. As we use $g(0)$ for deriving other values, we measure it first. We measure the time RTT_n for a roundtrip consisting of n messages sent in a row by *measure*, and a single, empty reply message sent back by *mirror*. The procedure starts with $n = 10$. The number of messages n is doubled until the gap per message changes only by $\epsilon = 1\%$. At this point, saturation is assumed to be reached. We take the time measured for sending the so-far largest number of messages (without reply) as $n \cdot g(0)$. We start with a small number of messages in a row in order to speed up the measurement. So we have to ensure that the messages are sufficiently many such that the roundtrip time is dominated by bandwidth rather than latency. Therefore, we also keep doubling n until the inequality $RTT_1 < \epsilon \cdot RTT_n$ holds. By waiting for a reply we enforce that the messages are really sent to *mirror* instead of just being buffered locally.

Table 1. LogGP's parameters expressed in terms of *parameterized LogP*

LogP/LogGP		parameterized LogP
L	$=$	$L + g(1) - o_s(1) - o_r(1)$
o	$=$	$(o_s(1) + o_r(1))/2$
g	$=$	$g(1)$
G	$=$	$g(m)/m$, for a sufficiently large m
P	$=$	P

All other parameters can be determined by the procedure shown in Fig. 1 (right). It starts with a synchronization message by which the so-called *mirror* process indicates being ready. For each size m, two message roundtrips are necessary from *measure* to *mirror* and back. (We use $RTT(m) = RTT_1(m)$.) In the first roundtrip, *measure* sends an m-bytes message and in turn receives a zero-bytes message. We measure the time for just sending and for the complete roundtrip. The send time directly yields $o_s(m)$. $g(m)$ and L can be determined by solving the equations for $RTT(0)$ and $RTT(m)$, according to the timing breakdown in Fig. 1 (left):

$$RTT(0) = 2(L + g(0)) \qquad RTT(m) = L + g(m) + L + g(0)$$
$$g(m) = RTT(m) - RTT(0) + g(0) \qquad L = (RTT(0) - 2g(0))/2$$

In the second roundtrip, *measure* sends a zero-bytes message, waits for $\Delta > RTT(m)$ time, and then receives a m-bytes message. Measuring the receive operation now yields $o_r(m)$, because after waiting $\Delta > RTT(m)$ time, the message from *mirror* is available at *measure* immediately, without further waiting.

For each message size, the roundtrip tests are initially run a small number of times. As long as the variance of measurements is too high, we successively increase the amount of roundtrips. We keep adding roundtrips until the average error is less than ϵ, or until an upper bound on the total number of iterations is reached (60 for small messages, 15 for large messages).

Initially, measurements are performed for all sizes $m = 2^k$ with $k \in [0, k_m]$. The value of k_m has to be chosen big enough to cover any non-linearity caused by the tested software layer. In our experiments, we used $k_m = 18$ to cover all changes in send modes of the assessed MPI implementation (MPICH).

After measuring the initial set of message sizes, we check whether the gap per byte $(g(m)/m)$ has stabilized for large m. If this is not the case, sending larger messages may achieve lower gaps (and hence higher throughput). So k_m is incremented and the next message size is tested. This process is performed until $g(2^{k_m})$ is close (within ϵ) to the value linearly extrapolated from $g(2^{k_m-2})$ and $g(2^{k_m-1})$.

So far, the "interesting" range of message sizes has been determined. Finally, possible non-linear behavior remains to be detected. For any size m_k, we check whether the measured values for $o_s(m_k)$, $o_r(m_k)$, and $g(m_k)$ are consistent with the corresponding, predicted values for size m_k, extrapolated from the measurements of the previous two (smaller) message sizes, m_{k-1} and m_{k-2}. If the difference is larger than ϵ, we do new measurements for $m = (m_{k-1} + m_k)/2$, and repeat halving the intervals until either the extrapolation matches the measurements, or until $m_k - m_{k-1} \leq \max(32 \text{ bytes}, \epsilon \cdot m_k)$.

3.1   Limitations of the method

Except for measuring $g(0)$, all parameters are derived from pairs of single messages sent between the *measure* and *mirror* processes. The correctness of timing these messages relies on the independence of the message pairs from each other: the time it takes to send a message from *measure* to *mirror* and back must always be the same, whether or not other messages have been exchanged before. Whenever *measure* issues several messages in a row, sending is slowed down to the rate at which the message pipeline is drained. This exactly is the effect used to measure $g(0)$. For all other measurements, we avoid this effect by always sending messages in pairs from *measure* to *mirror* and back. Before *measure* may send the next message, it first has to receive from *mirror*. This procedure enforces that pipelines will always be drained between individual message pairs, assuming that message headers carry "piggybacked" flow control information that resets senders to their initial state after each message roundtrip. This assumption may fail for communication protocols which update their flow control information in a more lazy fashion. So far, we found our assumption to be reasonable, as it works both with TCP and with our user-level Myrinet control software LFC [3].

In some cases, our measurements reveal values for the receive overhead such that $o_r(m) > g(m)$ which seems to contradict *parameterized LogP*. This phenomenon is caused by different behavior of the receive operation depending on whether the incoming message is *expected* to arrive. Messages are expected to arrive whenever the application called a matching receive operation before the message actually arrives at the receiving host. The treatment of expected messages may be more efficient because unexpected messages, for example, may have to be copied to a separate receive buffer, before they can later be delivered to the application. In our measurement procedure, $o_r(m)$ is measured with unexpected messages whereas $g(m)$ is measured while receiving expected messages. Whenever $o_r(m) > g(m)$, $g(m)$ gives an upper bound for processing expected messages. With synchronous receive operations, this measurement setup is unavoidable, because otherwise the measured receive overhead cannot be separated from the time waiting for the message to arrive. (With our MPI-based implementation, we can also measure the receive overhead of expected messages for the asynchronous receive operation, *MPI_Irecv*, in combination with *MPI_Wait*.)

The measurement procedure described above assumes that network links are symmetrical, such that sending from *measure* to *mirror* has the same parameters as for the reverse direction. However, this assumption may not always be true. On wide area networks, for example, the achievable bandwidth (the gap) and/or the network latency may be different in both directions, due to possibly asymmetric routing behavior or link speed. Furthermore, if the machines running the *measure* and *mirror* processes are different (like a fast and a slow workstation), then also the overhead for sending and receiving may depend on the direction in which the message is sent. In such cases, the parameters o_s, o_r, and g may be measured by performing our procedure twice, while switching the roles of *measure* and *mirror* in between. Asymmetric latency can only be measured by sending a message with a timestamp t_s, and letting the receiver derive the latency from $t_r - t_s$, where t_r is the receive time. This requires clock synchronization between sender and receiver. Without external clock synchronization (like using GPS receivers or specialized software like the *network time protocol*, NTP), clocks can only be synchronized up to a granularity of the roundtrip time between two hosts [10], which is useless for measuring network latency. Unfortunately, as we can not generally assume the clocks of (possibly widely) distributed hosts to be tightly synchronized, we can not measure asymmetric network latencies within our measurement framework.

4   Result evaluation

We implemented the measurement procedure on our experimentation platform called the DAS system, which consists of four cluster computers. Each cluster contains Pentium Pros that are connected by Myrinet. The clusters are located at four Dutch universities and are connected by dedicated 6 Mbit/s ATM networks. (The system is more fully described on *http://www.cs.vu.nl/das/*.)

For the measurements presented in Fig. 2, we have used our MPI message passing system (described in [8, 9]) which can send messages inside clusters over Myrinet and between clusters over the ATM links, using TCP. We implemented the procedure as an MPI application, called the *MPI LogP benchmark*. We measured the LogP parameters for MPI_Send and MPI_Recv as described above, except for $g(m)$ which was measured with our fast method, and by the link saturation method [5, 7]. The graphs in Fig. 2 show o_s (for comparison) and g, as measured by both methods. In general, on both networks, the curves for g are rather close to each other, confirming the efficacy of our method. There is a general trend that the new, fast method measures slightly larger gaps. This can partially be explained by the systematic error of the saturation method which has to be stopped heuristically based on the increase rate ϵ of the measured gap values, causing part of the gap being missed. However, there is a region (64 byte—1 Kbyte over TCP, and 128 byte—4 Kbyte over Myrinet) where the saturation method measures significantly less (up to 50%) than the fast method. We could attribute the majority of this effect to a cache sensitivity of the *mirror* process which has better data locality with the saturation-based method as it does not send messages while draining the link. So, cache misses occur with somewhat larger messages, compared to the fast, roundtrip-based measurement.

Table 2 provides a breakdown of the measurement completion times shown in Fig. 2 for measuring $g(0)$, o_s/o_r (with implicit $g(m > 0)$), and $g(m > 0)$ (with saturation) over both networks. With our fast measurement procedure, only the first two measurements are necessary, yielding a performance gain of a factor of 10 over Myrinet, and a factor of 17 over the TCP link.

5   Conclusions

We presented a new, fast micro benchmark for measuring LogP parameters for messages of various sizes. We used the *parameterized LogP* [8] performance model. The

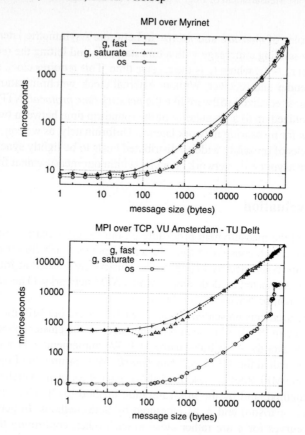

Fig. 2. Measured send overhead and gap; over Myrinet (top) and over TCP (bottom)

major improvement of our measurement procedure is that the minimal *gap* between two messages can be observed without saturating the network for each message size. Furthermore, our procedure adapts itself to the network characteristics in order to measure parameters for all relevant message sizes.

We implemented the new measurement procedure, called the *MPI LogP benchmark*, for our MPI platform and verified on two different networks that it gets the same results as a saturation-based measurement. The improvements in measurement time are significant. However, the time needed for a full measurement with various message sizes still takes too long to be performed during application runtime. As our ultimate goal is to enable applications to react to changing WAN conditions, we will need to restrict the

Table 2. Breakdown of measurement completion times (seconds)

	Myrinet	TCP
$g(0)$	0.05	12.3
o_s/o_r (with implicit $g(m > 0)$)	0.16	102.7
$g(m > 0)$ (with saturation)	1.96	2018.7

measurements to only a few message sizes and extrapolate the others by a technique like the one in [6]. The *MPI LogP benchmark* is available from

http://www.cs.vu.nl/albatross/

Acknowledgements

This work is supported in part by a USF grant from the Vrije Universiteit. The wide-area DAS system is an initiative of the Advanced School for Computing and Imaging (ASCI). We thank Rutger Hofman for his contributions to this research. We thank John Romein for keeping the DAS in good shape, and Cees de Laat (University of Utrecht) for getting the wide area links of the DAS up and running.

References

1. A. Alexandrov, M. F. Ionescu, K. E. Schauser, and C. Scheiman. LogGP: Incorporating Long Messages into the LogP Model — One Step Closer Towards a Realistic Model for Parallel Computation. In *Proc. Symposium on Parallel Algorithms and Architectures (SPAA)*, pages 95–105, Santa Barbara, CA, July 1995.
2. F. Berman, R. Wolski, S. Figueira, J. Schopf, and G. Shao. Application-Level Scheduling on Distributed Heterogeneous Networks. In *Proc. Supercomputing'96*, Nov. 1996. Online at http://www.supercomp.org/sc96/proceedings/.
3. R. Bhoedjang, T. Rühl, and H. Bal. User-Level Network Interface Protocols. *IEEE Computer*, 31(11):53–60, 1998.
4. D. Culler, R. Karp, D. Patterson, A. Sahay, K. E. Schauser, E. Santos, R. Subramonian, and T. von Eicken. LogP: Towards a Realistic Model of Parallel Computation. In *Proc. Symposium on Principles and Practice of Parallel Programming (PPoPP)*, pages 1–12, San Diego, CA, May 1993.
5. D. E. Culler, L. T. Liu, R. P. Martin, and C. O. Yoshikawa. Assessing Fast Network Interfaces. *IEEE Micro*, 16(1):35–43, Feb. 1996.
6. M. Faerman, A. Su, R. Wolski, and F. Berman. Adaptive Performance Prediction for Distributed Data-Intensive Applications. In *Supercomputing'99*, Nov. 1999. Online at http://www.supercomp.org/sc99/proceedings/.
7. G. Iannello, M. Lauria, and S. Mercolino. Cross–Platform Analysis of Fast Messages for Myrinet. In *Proc. Workshop CANPC'98*, number 1362 in Lecture Notes in Computer Science, pages 217–231, Las Vegas, Nevada, January 1998. Springer.
8. T. Kielmann, H. E. Bal, and S. Gorlatch. Bandwidth-efficient Collective Communication for Clustered Wide Area Systems. In *Proc. International Parallel and Distributed Processing Symposium (IPDPS 2000)*, Cancun, Mexico, May 2000.
9. T. Kielmann, R. F. H. Hofman, H. E. Bal, A. Plaat, and R. A. F. Bhoedjang. MAGPIE: MPI's Collective Communication Operations for Clustered Wide Area Systems. In *Proc. Symposium on Principles and Practice of Parallel Programming (PPoPP)*, pages 131–140, Atlanta, GA, May 1999.
10. V. Paxson. On Calibrating Measurements of Packet Transit Times. In *Proc. SIGMETRICS'98/PERFORMANCE'98*, pages 11–21, Madison, Wisconsin, June 1998.
11. R. Wolski. Forecasting Network Performance to Support Dynamic Scheduling Using the Network Weather Service. In *Proc. High-Performance Distributed Computing (HPDC-6)*, pages 316–325, Portland, OR, Aug. 1997. The network weather service is at http://nws.npaci.edu/.

Supporting Flexible Safety and Sharing in Multi-threaded Environments*

Steven H. Samorodin[1] and Raju Pandey[2]

[1] Marimba, Inc. Mountain View, Ca.
shs@marimba.com
[2] Computer Science Department, University of California at Davis
pandey@cs.ucdavis.edu

Abstract. There is increasing interest in extensible systems (such as extensible operating systems, mobile code runtime systems, Internet browsers and servers) that allow external programs to be downloaded and executed directly within the system. While appealing from system design and extensibility points of view, extensible systems are vulnerable to aberrant behaviors of external programs. External programs can interfere with executions of other programs by reading and writing into their memory locations. In this paper, we present an approach for providing safe execution of external programs through a safe threads mechanism. The approach also provides a novel technique for safe sharing among external programs. The paper also describes the design and implementation of the safe threads.

1 Introduction

There is increasing interest in extensible systems that allow external programs to be downloaded and executed directly within a local system. Examples of such systems include extensible operating systems [3, 7], the Java runtime system [1], mobile code runtime systems [6], Internet browsers and web servers. While appealing from both system design and extensibility points of view, extensible systems are vulnerable to aberrant behaviors of external programs. External programs can interfere with executions of other programs by accessing their memory. They can corrupt system-dependent data, force a program into an inconsistent state, and crash the system. They can write into another program's memory, thereby corrupting system-dependent data, force a program into an

---

* This work is supported by the Defense Advanced Research Project Agency (DARPA) and Rome Laboratory, Air Force Materiel Command, USAF, under agreement number F30602-97-1-0221. The U.S. Government is authorized to reproduce and distribute reprints for Governmental purposes notwithstanding any copyright annotation thereon. The views and conclusions contained herein are those of the authors and should not be interpreted as necessarily representing the official policies or endorsements, either expressed or implied, of the Defense Advanced Research Project Agency (DARPA), Rome Laboratory, or the U.S. Government.

inconsistent state, and overwrite other programs. Clearly, system software must provide *safety* against malicious or buggy external programs.

The notion of safety has been studied quite extensively in the operating system research and, recently, in type-safety based approaches [8, 9]. Most operating systems implement the notion of safety through address containment as in UNIX [12]. Address containment schemes provide safety by ensuring that a program cannot address the memory used by another program. The problem with the address containment-based approaches is that, in general, they enforce a rigid notion of safety and do not adequately support flexible sharing of data between processes. Sharing mechanisms, such as inter-process communication (IPC) or shared memory, are either inefficient (due to data copying) or require coordination of addresses among processes. Work in single address operating systems (SASOS) [5, 10] have proposed the notion of address spaces that support safety among threads of execution, while providing sharing through address pointers. SASOS provide a nice solution but requires a specialized operating system.

While the above approaches do provide mechanisms for safety and sharing, the mechanisms are either too inflexible or difficult to use for the kind of application we are building. We are interested in developing a mobile code runtime system that creates a thread of execution for every mobile code. Our focus is on developing an execution environment that protects the runtime system and the mobile programs from each other. Further, since data sharing among mobile programs may be dynamic and flexible, the system software must support sharing mechanisms that can be customized dynamically to reflect these sharing patterns.

We, thus, need a protection mechanism that provides protection as well as flexible and dynamic sharing among threads of execution at the user level. This paper presents such a notion of protection and sharing for threads. We present a threads package, called Safe Threads, that supports the notion of threads whose stacks and data elements are completely protected. The thread package contains a novel mechanism for specifying flexible and dynamic sharing and protection among threads. In this approach, the notion of protection is represented by an abstract entity, called *protected domain*. Sharing is defined by *permission* relationships among protected domains. Applications can bind threads and data elements to different protected domains in order to implement different sharing relationships dynamically. We have implemented the thread package through mprotect system calls which make threads context switches quite expensive. Performance analysis of the thread package shows that protected thread creation is approximately 1.5 times more expensive. Context switch times are more expensive as well, but vary depending upon the number of protected domains involved.

The rest of this paper is organized as follows: In Section 2, we describe the notion of safe threads and sharing among them. We also present an implementation of the threads package in this section. In Section 3, we present the performance characteristics of our system. Section 4 discusses related work and we conclude in Section 5.

2   Safe Threads package

In this section, we present the notion of safety and sharing within a thread package that we have developed. The thread package supports creation of multiple threads, provide fundamental safety guarantees, and supports mechanism for safe sharing among threads. We first briefly describe the notion of threads and then discuss how safety and sharing is defined in the thread package

2.1   Support for Threads

User-level threads packages provide creation, deletion, and management of multiple threads of execution. Threads are execution contexts and share an address space and other per-process resources. Unlike processes which may require a large amount of state information, threads generally need only a program counter, a set of registers, and a stack of activation records. Context switching costs for threads are, therefore, much lower.

Typical user-level threads packages, such as Pthreads [4],are implemented by constructing a separate stack for each thread, while sharing the code and heap data segments. In these thread packages, any thread can access any memory location, including code, scheduler thread stack, other thread stacks, and heap segments.

We have developed a thread package that provides for safe execution of external programs. The thread package provides two levels of safety guarantees. The first is an absolute safety guarantee for data that must always be protected. A thread's per-thread data (including stack and code) are completely protected from other threads. The second guarantee concerns data whose safety and sharing properties can be defined dynamically by the threads themselves. The thread package supports this through the notion of *protected domains* and *permission relationships*.

2.2   Protected domains and Permission relationships

A protected domain aggregates regions of memory that have similar sharing properties. A thread cannot access a protected domain and therefore any of the data contained in that the protected domain unless the thread has been bound to the protected domain. A thread can define a binding relationship with a protected domain explicitly or implicitly. An explicit binding between a thread, T_1, and a protected domain, P_1, denoted $T_1 \rightarrow P_1$, can occur in two ways: (i) When T_1 creates P_1, T_1 is said to be the owner of P_1 and can access all entities bound to P_1. (ii) When T_1, the owner thread of P_1, explicitly binds a thread T_2 with P_1, denoted $T_1(T_2 \rightarrow P_1)$. This explicit binding allows T_2 to access any data entities associated with P_1. Note that such bindings allow T_1 to share any data contained within P_1 with other threads. Only the owner can change bindings to allow other threads access or permit other protected domains access.

Implicit binding, occurs as a result of thread bindings and permission relationships among protected domains. A *permission* relationship $\mapsto$ between two protected domains captures an asymmetric sharing relationship between threads bound to the protected domains. For instance, the relation $P_1 \mapsto P_2$ (read P_1 is permitted by P_2) specifies that threads bound to P_1 can access data entities bound to P_2, but not vice versa.

We represent threads, protected domains and permission relationships in terms of a directed graph called a *sharing relationship graph* in which a node denotes a thread or a protected domain and an edge denotes a permission relationship. Each permission relationship indicates a chaining of access to the contents of the protected domain for threads bound to the permitting protected domain. The access associated with each permission relationship is labeled read, write, or read/write, indicating the kind of permission that is allowed. Each protected domain has an access list of (thread ID, access type) pairs associated with it.

The notions of protected domain and permission relationship allow one to define complex and dynamic sharing relationships between threads and data. An example of such a relationship is the hierarchical notion of trust and safety implemented in many systems. In these systems, a multi-level information sharing specification is created where entities (for instance, workers) at level L can access any information that exists at levels $\leq L$. However, they cannot access any information that exists above level L. Such a sharing relationship can be easily represented through protected domain and permission relationships. Thus, protected domain and permission relationships allow one to capture patterns of accesses and restriction among cooperating threads.

2.3 Implementation

We have implemented the Safe Threads package on top of the QuickThreads [11] library on the FreeBSD 2.2.6 operating system. The QuickThreads library supports non-preemptive user level threads. Safe Threads implements basic threading functionality on top of protection mechanisms. Our current implementation runs inside a single UNIX process virtual address space. Protection is enforced through use of the mprotect(2) system call. mprotect changes the access restrictions for the calling process on specified regions of memory within that processes' virtual address space. Utilizing this mechanism allows for the flexibility to protect any page-sized region of memory.

One important design decision involved whether a thread's stack should be protected from other threads. In order that a thread truly be safe from other threads, stacks must be inaccessible to other threads. There are two important implications, however, on impact the performance of the threads package. Firstly, the context switching code cannot be executed on either thread's stack since at some point in the algorithm each stack is not accessible. This makes context switches more expensive than if the switching code could be executed directly on the stack of the thread that was previously executing. Secondly, and perhaps more importantly, because the stacks are not visible to all threads, parameters passed between threads must be copied.

Optimizations: Since systems calls require crossing the user/kernel protection boundary, system calls are more expensive than normal procedure calls. The method for implementing the thread context switch described above may potentially require many system calls per thread context switch. Therefore we have developed methods to speed up a protected context switch. There are two kinds of optimization possible: the first reduces the number of memory regions that must be protected and the second reduces the number of times the user/kernel boundary is crossed. The first can be achieved by combining protection domains of a thread if they do not export data to different threads, and by placing protected regions in contiguous regions so that such regions can be protected through one system call. The current version of the package does not include these optimizations yet as we are still formulating a general algorithm for using the sharing relationship graph to generate optimal[1] protected memory region layouts. Further, since the context switching code is usually very small, it is not clear if there are large benefits to be derived from implementing complex memory layout algorithms.

The second optimization involves reducing the number of system calls. During a context switch, the threads package determines which protected domains need to be protected and unprotected. In our initial implementation, the package makes one mprotect call for each protected region that needed to be protected or unprotected. This results in $O(n)$ system calls per context switch, where n is the number of protected regions. To reduce this number, we extended the FreeBSD kernel to include a new system call, multiMprotect(), which takes a vector of (address, length, protection type) triples. We, therefore, make one system call per context switch by packing all of the data into an array of triples. multiMprotect is a simple wrapper that takes each argument from the parameter vector and calls mprotect.

3   Performance Analysis

In this section, we focus on analyzing the costs associated with providing the safety and sharing model. Two benchmarks were performed: a thread creation benchmark and a context switching benchmark.

3.1   Thread Creation

The thread creation benchmark compares the cost of creating protected and unprotected threads. Beyond what is required to create an unprotected thread, protected thread creation requires creating a protected domain, adding its stack as a data item, and protecting the stack. Table 1 shows results which indicate that for large numbers of threads creating protected threads is about 1.5 times as expensive.

---

[1] It is our belief that the general algorithm is at least NP-hard, but we have not proven it yet.

Thread creation times on Pentium 120 w/32mb			
# Threads	Time (No Protection)	Time (With Protection)	% Difference
100	6.85	11.82	173
500	8.89	13.40	151
750	9.45	13.75	146
1000	9.66	14.01	145
Thread creation times on Pentium II 300 w/128mb			
# Threads	Time (No Protection)	Time (With Protection)	% Difference
100	2.27	4.18	184
500	2.72	4.72	174
750	4.16	6.65	160
1000	3.94	6.37	162

Table 1. Data for thread creation times is given for two different machines. All times are in micro seconds and are the average of 20 runs of creating the number of threads specified. All machines run FreeBSD 2.2.6-STABLE. g++ v.2.7.2.1 with -O2 and -m486 optimizations was used to compile all test programs.

3.2   Context Switch

Context switch times for Safe Threads are highly dependent upon the number of protected domains and the number of data elements contained within those protected domains. Figures 1(a) and 1(b) show the cost for context switches with different numbers of protected domains. The cost of an unprotected context switch is, as expected, a constant value. This number was determined by using the Safe Threads package with protection turned off.

As mentioned in Section 2.3, our optimization goal with multiMprotect was to reduce the number of system calls from $O(n)$ to 1 per context switch. In this we were successful, but we found that additional overhead introduced minimizes the performance advantage gained by reducing the number of system calls. For all but the smallest numbers of protected domains, our new system call multiMprotect outperforms mprotect. However, the performance benefit from using multimprotect is not as great as we expected. We feel that this is largely due to inefficient implementation. With different data structures and other optimizations these numbers could be significantly reduced.

While the times for individual context switches can be very high for large numbers of protected domains, the tests were constructed to show worst case behavior where no protected domains are shared between threads. We believe that many applications will share protected domains and thereby incur lower context switch costs, even for large number of context switches.

4   Existing Safety Solutions

Existing solutions to the safety problem function at three levels of abstraction: hardware/OS, software and language. Hardware-based solutions address the problem at the lowest level. These solutions rely upon hardware to enforce

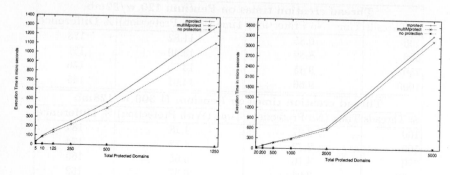

(a) Context switch times for various Safe Threads protection options for 5 threads.

(b) Context switch times for various Safe Threads protection options for 20 threads.

Fig. 1. Overhead Cost of context switching for safe threads

safety [12]. Hardware protection has the advantage that it physically guarantees protection.

The problem of safely executing untrusted code can also be addressed at the software level. Software safety solutions work at the user-level modifying the compiler, runtime system, and sometimes the untrusted code itself to ensure that software modules do not misbehave. Software Fault Isolation (SFI) [13] and Protected Shared Libraries (PSL) [2] are examples of software safety solutions.

Finally, type-safe languages, such as Java, use language semantics to provide safety. Name space encapsulation ensures that private variables and methods cannot be accessed by other classes. Language-based protection schemes have the advantage that often a cross protection domain call can be as inexpensive as a procedure call. Several systems have been built using these languages including the SPIN extensible operating system [3] and the J-Kernel system [8]. The J-Kernel [8] protection system provides a general framework for supporting multiple protection domains within a single process address space. This work is similar to Safe Threads in that both develop a mechanism for allowing multiple protection domains to exist within a single address space. However, since J-Kernel relies upon Java to enforce its protection, it is limited to creating safety solutions for Java programs.

Opal [5], Mungi [10], and other single address space operating systems (SASOS) address many of the same problems as Safe Threads on an operating system level. Specifically, providing protection and sharing within a single address space.

5   Conclusion

We have presented the design and implementation of a threads package that supports provides safety among threads. The package supports creation of threads, provides isolation among them, and includes mechanisms for protected sharing among threads. We have implemented the thread package and initial performance analysis suggests that thread creation is approximately 1.5 times more expensive for creating protected threads. Context switching times depend upon the number of protected domains involved. We are currently looking at different techniques for optimizing the cost of thread creation and context switching.

References

1. K. Arnold and J. Gosling. *The Java Programming Language*. Addison Wesley, 1996.
2. A. Banerji, J. M. Tracey, and D. L. Cohn. Protected shared libraries-a new approach to modularity and sharing. In *Proceedings of the USENIX 1997 Annual Technical Conference*, pages 59–75, Anaheim, CA, January 1997.
3. B. Bershad et al. Extensibility, safety and performance in the SPIN operating system. *15th Symposium on Operating Systems Principles*, pages 267–283, December 1995.
4. D. R. Butenhof. *Programming with POSIX Threads*. Addison Wesley Longman, Inc., 1997.
5. J. Chase, H. Levy, M. Feeley, and E. Lazowska. Sharing and protection in a single address space operating system. *ACM Transactions On Computer Systems*, 12(4):271–307, May 1994.
6. D. Chess, C. Harrison, and A. Kershenbaum. Mobile Agents: Are they a good idea? In *Mobile Object Systems: Towards the Programmable Internet*, pages 46–48. Springer-Verlag, April 1997.
7. D. R. Engler, M. F. Kaashoek, and J. O'Toole Jr. Exokernel: An operating system architecture for application-level resource management. In *15th Symposium on Operating Systems Principles*, pages 251–266, December 1995.
8. C. Hawblitzel, C. Chang, G. Gzajkowski, D. Hu, and T. von Eicken. Implementing multiple protection domains in Java. In *Proceedings of the USENIX 1998 Annual Technical Conference*, pages 259–272, New Orleans, La., June 1998.
9. C. Hawblitzel and T. von Eicken. A case for language-based protection. Technical Report 98-1670, Cornell University, Ithaca, NY, 1998.
10. G. Heiser, K. Elphinstone, J. Vochteloo, and S. Russell. Implementation and performance of the Mungi single-address-space operating system. Technical Report UNSW-CSE-TR-9704, The University of New South Wales, Sydney, Australia, June 1997.
11. D. Keppel. Tools and techniques for building fast portable threads packages. Technical Report UWCSE 93-05-06, University of Washington, 1993.
12. U. Vahalia. *UNIX Internals: The New Frontiers*. Prentice Hall, Upper Saddle River, New Jersey 07458, 1996.
13. R. Wahbe, S. Lucco, T. E. Anderson, and S. L. Graham. Efficient software-based fault isolation. *14th Symposium on Operating Systems Principles*, pages 203–216, 1993.

A Runtime System for Dynamic DAG Programming

Min-You Wu[1], Wei Shu[1], and Yong Chen[2]

[1] Department of ECE, University of New Mexico
{wu,shu}@eece.unm.edu
[2] Department of ECE, University of Central Florida

Abstract. A runtime system is described here for dynamic DAG execution. A large DAG which represents an application program can be executed on a parallel system without consuming large amount of memory space. A DAG scheduling algorithm has been parallelized to scale to large systems. Inaccurate estimation of task execution time and communication time can be tolerated. Implementation of this parallel incremental system demonstrates the feasibility of this approach. Preliminary results show that it is superior to other approaches.

1 Introduction

Task parallelism is essential for applications with irregular structures. With computation partitioned into tasks, load balance can be achieved by scheduling the tasks, either dynamically or statically. Most dynamic algorithms schedule independent tasks, that is, a set of tasks that do not depend on each other. On the other hand, static task scheduling algorithms consider the dependences among tasks. The Directed Acyclic Graph (DAG) is a task graph that models task parallelism as well as dependences among tasks. As the DAG scheduling problem is NP-complete in its general form [4], many heuristic algorithms have been proposed to produce satisfactory performance [6, 3, 9].

Current DAG scheduling algorithms have drawbacks which may limit their usage. Some important issues to be addressed are:

- They are slow since they run on a single processor machine.
- They require a large memory space to store the graph and are not scalable thereafter.
- The quality of the obtained schedules relies heavily on the estimation of execution time. Accurate estimation of execution time is required. Without this information, sophisticated scheduling algorithms cannot deliver satisfactory performance.
- The application program must be recompiled for different problem sizes since the number of tasks and the estimated execution time of each task varies with the problem size.
- They are static as the number of tasks and dependences among tasks in a DAG must be known at compile-time. Therefore, they cannot be applied to dynamic problems.

J. Rolim et al. (Eds.): IPDPS 2000 Workshops, LNCS 1800, pp. 1192-1199, 2000.

These problems limit applicability of current DAG scheduling techniques and have not yet received substantial attention. Thus, many researchers consider the static DAG scheduling unrealistic.

The memory space limitation and the recompiling problem can be eliminated by generating and executing tasks at runtime, as described in PTGDE [2], where a scheduling algorithm runs on a supervisor processor, which schedules the DAG to a number of executor processors. When a task is generated, it is sent to an executor processor to execute. This method solves the memory limitation problem because only a small portion of the DAG is in the memory at a time. However, the scheduling algorithm is still sequential and not scalable. Because there is no feedback from the executor processors, the load imbalance caused by inaccurate estimation of execution time cannot be adjusted. It cannot be applied to dynamic problems either. Moreover, a processor resource is solely dedicated to scheduling. If scheduling runs faster than execution, the supervisor processor will be idle; otherwise, the executor processors will be idle.

We have proposed a *parallel incremental* scheduling scheme to solve these problems [5]. A scheduling algorithm can run faster and is more scalable when it is parallelized. By incrementally scheduling and executing DAGs, the memory limitation can be alleviated and inaccurate weight estimation can be tolerated. It can also be used to solve dynamic problems. This parallel incremental DAG scheduling scheme is based on general static scheduling and is extended from our previous project, Hypertool [6]. The new system is named Hypertool/2. Different from runtime incremental parallel scheduling for independent tasks, Hypertool/2 takes care of dependences among tasks and uses DAG as its computation model.

2   DAG and Compact DAG

A DAG, or a macro dataflow graph, consists of a set of nodes $\{n_1, n_2, ..., n_n\}$ connected by a set of edges, each of which is denoted by $e_{i,j}$. Each node represents a task, and the weight of node n_i, $w(n_i)$, is the execution time of the task. Each edge represents a message transferred from node n_i to node n_j and the weight of edge $e_{i,j}$, $w(e_{i,j})$, is equal to the transmission time of the message. Figure 1 shows a DAG generated from a parallel Gaussian elimination algorithm with partial pivoting, which partitions a given matrix by columns. Node n_0 is the INPUT procedure and n_{19} the OUTPUT procedure. The size of the DAG is proportional to N^2, where N is the matrix size.

In a static system, a DAG is generated from the user program and scheduled at compile time. Then this *scheduled DAG* is loaded to PEs for execution. In a runtime scheduling system, the DAG is generated incrementally and each time only a part of the DAG is generated. For this purpose, a compact form of the DAG *(Compact DAG, or CDAG)* is generated at compile time. It is then expanded to the DAG incrementally at runtime. The CDAG is similar to the parameterized task graph in [2]. The size of a CDAG is proportional to the program size while the size of a DAG is proportional to the problem size or the matrix size.

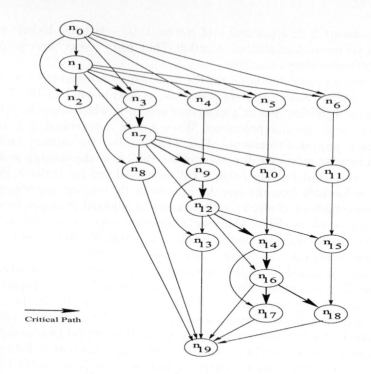

Fig. 1. A DAG (Gaussian elimination).

A CDAG is defined by its communication rules. A communication rule is in the format of

$$source\ node \rightarrow destination\ node:\ message\ name\ |\ guard.$$

The communication rules in Figure 2 is generated from an annotated C program of Gaussian elimination. For details, refer to [8]. The corresponding CDAG is shown in Figure 3. The runtime system takes the CDAG as its input.

$INPUT \rightarrow FindMax(i) : vector[0], matrix[0,0]|i = 0$
$INPUT \rightarrow UpdateMtx(0,j) : matrix[0,j]|0 \leq j \leq N$
$FindMax(i) \rightarrow FindMax(i+1) : vector[i+1]|0 \leq i \leq N-2$
$FindMax(i) \rightarrow OUTPUT : vector[N]|i = N-1$
$FindMax(i) \rightarrow UpdateMtx(i,j) : vector[i+1]|0 \leq i \leq N-1, i \leq j \leq N$
$UpdateMtx(i,j) \rightarrow UpdateMtx(i+1,j) : matrix[i+1,j]|0 \leq i \leq N-2, i+1 \leq j \leq N$
$UpdateMtx(i,j) \rightarrow FindMax(i+1) : matrix[i+1,j]|0 \leq i \leq N-2, j = i+1$
$UpdateMtx(i,j) \rightarrow OUTPUT : matrix[i+1,j]|0 \leq i \leq N-1, j = i$
$UpdateMtx(i,j) \rightarrow OUTPUT : matrix[N,N]|i = N-1, j = N$

Fig. 2. Communication rules for the Gaussian elimination code.

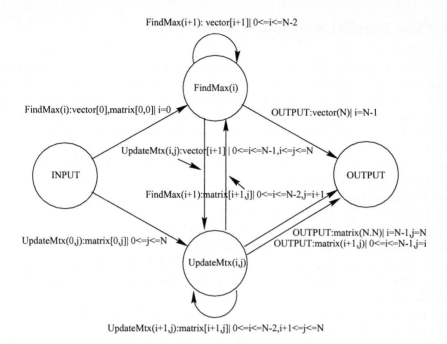

FindMax(i+1): vector[i+1]| 0<=i<=N-2

FindMax(i)

FindMax(i):vector[0],matrix[0,0]| i=0

OUTPUT:vector(N)| i=N-1

UpdateMtx(i,j):vector[i+1] | 0<=i<=N-1,i<=j<=N

INPUT

OUTPUT

FindMax(i+1):matrix[i+1,j]| 0<=i<=N-2,j=i+1

UpdateMtx(0,j):matrix[0,j]| 0<=j<=N

OUTPUT:matrix(N.N)| i=N-1,j=N
OUTPUT:matrix(i+1,j)| 0<=i<=N-1,j=i

UpdateMtx(i,j)

UpdateMtx(i+1,j):matrix[i+1,j]| 0<=i<=N-2,i+1<=j<=N

Fig. 3. A CDAG (Gaussian elimination).

3   The Incremental Execution Model

Different models can be used to execute a CDAG at runtime. We proposed an incremental execution model. In this model, only a subgraph is scheduled in each system phase. The size of the subgraph to be generated each time is normally limited by the available memory space. The system scheduling activity alternates with the underlying computation work. It starts with a system phase where only a part of the DAG is generated and scheduled. It is followed by a user computation phase to execute the scheduled tasks. The PEs will execute until most tasks have been completed and transfer to the next system phase to generate and schedule the next part of the DAG. For details, refer to [8]. The incremental execution model has many advantages:

- This approach can relax the memory space requirement because the task graph is generated incrementally.
- The demand for an accurate estimation of execution time becomes less critical. While an inaccurate estimate may result in load imbalance, the load imbalance can be adjusted when scheduling the next set of tasks.
- The application programs need not be recompiled for each problem size since this scheme adapts to different problem sizes.
- It can be applied to dynamic problems. In general, the structure of dynamic problems cannot be known at compile-time. An incremental scheduler can adapt to this property by scheduling subgraphs for execution when partial information becomes available.

4   The Parallel Scheduling Algorithm

Since the scheduling time is a part of the runtime system time, minimizing scheduling time becomes extremely important in incremental scheduling. Here we describe a simple DAG scheduling algorithm, the MCP algorithm [6]. This algorithm was designed to schedule a DAG on a bounded number of PEs. First, the *as-late-as-possible (ALAP)* time of a node is defined as $T_L(n_i) = T_{critical} - level(n_i)$, where $T_{critical}$ is the length of the critical path counting both node and edge weights, and $level(n_i)$ is the length of the longest path from node n_i to the end point, including node n_i.

MCP Algorithm

1. Calculate the ALAP time of each node.

2. Sort the node list in an increasing ALAP order. Ties are broken by using the smallest ALAP time of the successor nodes.

3. Schedule the first node in the list to the PE that allows the earliest *start_time* with *insertion*. Delete the node from the list and repeat Step 3 until the list is empty.

A comparison study has shown that MCP performed better than many other scheduling algorithms [1]. The parallel scheduling algorithm, used in our system is a parallel version of the MCP algorithm. The PEs that execute a parallel scheduling algorithm are called the *physical PEs (PPEs)* in order to distinguish them from the *target PEs (TPEs)* to which the DAG is to be scheduled. The quality and speed of a parallel scheduler depend on data partitioning. There are two major data domains in a scheduling algorithm, the source domain and the target domain. The source domain is the DAG and the target domain is the schedule for TPEs. A *horizontal* scheme [7] is illustrated in Figure 4, where three PPEs schedule the graph to six TPEs and each PPE holds a portion of schedules of six TPEs. In this scheme, each PPE is assigned a set of graph nodes using time domain partitioning. The resultant schedule is also partitioned so that each PPE maintains a portion of the schedule of every TPE. Each PPE schedules its

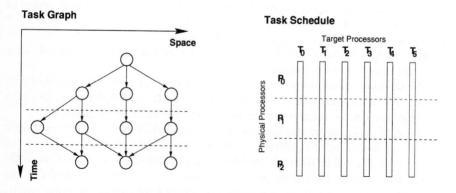

Fig. 4. The Horizontal Scheme.

own portion of the graph before all PPEs exchange information with each other to determine the final schedule.

The horizontal parallel MCP is named the HPMCP algorithm [7], which is shown in Figure 5. After the graph is partitioned, each PPE uses MCP to schedule its partition to produce its sub-schedule. When applying MCP, we ignore the dependences between partitions so that each partition can be scheduled independently. Then adjacent sub-schedules are concatenated to form the final schedule.

---

1. Partition the nodes and each partition is assigned to a PPE.
2. Each PPE applies the MCP algorithm to its partition to produce a sub-schedule, ignoring the edges between a node and its remote parent nodes. Schedule the first node in the list to the TPE that allows the earliest start time. Delete the node from the list and repeat this scheduling step until the list is empty.
3. Concatenate each pair of adjacent sub-schedules.

Fig. 5. The HPMCP Algorithm.

5   Runtime System Organization

The runtime system of Hypertool/2 is organized as modules including graph generation, scheduling, node execution, communication handling, and incremental execution handling modules.

Graph generation
The CDAG is expanded incrementally. In the first system phase, a subset of nodes is generated from the CDAG. Assume that the subset has B nodes and there are P PPEs. Each PPE generates B/P nodes independently. Then, this subset is scheduled and executed. In the next system phase, the newly expanded nodes together with the residual nodes from the last phase form a new subset. Any outgoing edge from a node in the subset becomes a *future* message if its destination node is not in the subset. This partial DAG is then scheduled to PEs and executed in a new user phase.

Scheduling
The scheduling module will establish a mapping for each node from its logic ID to its physical ID, which consists of a target PE number and a local ID. Once a PPE generates its subset of nodes, it independently schedules these nodes to form a subschedule using the MCP algorithm. Obviously, an accurate ALAP time cannot be obtained if there exists any future edge in the subset. Instead, we use quasi-ALAP binding based on the subgraph only and ignoring all future edges. We expect that strategy to achieve sub-optimal results. These subschedules are then concatenated to form a schedule for the current phase.

Execution
In the scheduled DAG, the nodes in the list are to be executed in order. The dispatch routine picks the next node in the list and check its incoming

messages. When all of its incoming messages have arrived, the dispatch routine allocates memory and prepares parameters for the node's execution. The node procedure is then invoked. On completion of node execution, output parameters are processed by communication handling module. All memory space allocated for the node is also deallocated.

Communication handling

Message receiving is handled at the idle time or between completion of one node execution and start of the next ready node execution. A message is sent for each outgoing edge after node execution. If a message has a single destination, it is classified as *unicast*. If a message has multiple receipts, it is classified as a *multicast*.

Incremental execution handling

Every PE that has received a *pause* massage will finish its current node execution and pauses its current user phase. Before entering the next system phase, the residual nodes as well as corresponding messages need to be processed to incorporated into the new phase. Once a system phase is completed, a new mapping from logic IDs to physical IDs is made available to all target PEs.

6   Experimental Study

In this experimental study, we use the Gaussian elimination code as an example. It has been partitioned by defining a grain-size parameter SN. that is, SN columns are merged into a single task. This parameter controls the grain size so that the overhead to handle a task is relatively smaller than the computation time of the task. The problem size (matrix size) is $2K \times 2K$. SN is set to be 8 and the total number of tasks is 32966.

The Hypertool/2 has been implemented on an Intel Paragon machine. First, we study the system performance, especially the system overhead, which includes task generation, scheduling, and other overhead. Table 1 shows an analysis of execution time on 16 TPEs. The time unit is second. The computation time is the time of executing tasks. By parallelizing graph generation and scheduling, the system overhead has been reduced from 126.1 seconds on a single PE to 64.7 seconds on eight PEs.

Table 1. Timing on Various Numbers of PPEs (Number of TPE is 16).

Number of PPEs	1	2	4	8	16
Task generation	16.8	8.6	4.6	3.2	2.8
Scheduling	31.2	15.1	9.9	6.0	5.8
Other overhead	78.1	61.0	58.2	55.5	57.9
Computation	347	351	354	355	356
Total	473	436	427	420	423

PTGDE is a work similar to ours [2]. PTGDE runs on a simulator, which does not reflect the real situation. For comparison purpose, we have re-implemented it on Intel Paragon based on description in [2]. Table 2 compares the execution time and speedup of the Gaussian elimination code. For two PEs, only one of them executes the user program and there is no speedup. PTGDE uses a dynamic scheduling algorithm, PTGDS, to schedule tasks at runtime, which is in a depth-first-search style. This scheduling algorithm does not balance the load well, especially for a large number of PEs. Its performance is not as good as the MCP algorithm.

Table 2. Comparison of Gaussian Elimination on Hypertool/2 and PTGDE.

Number of PEs		1	2	4	8	16	32
Hypertool/2	Time (S)	5542	2788	1431	766	423	268
	Speedup	1.00	1.99	3.87	7.23	13.1	20.7
PTGDE	Time(S)	—	5574	1998	1042	808	590
	Speedup	—	1.00	2.79	5.35	6.90	9.45

7 Conclusion

This paper described a new approach, which schedules DAGs in parallel and incrementally executes DAGs at runtime. Our experimental results showed that DAGs can be used for large problems, parallel scheduling can scale well, and inaccurate estimation can be tolerated.

Acknowledgments

This research was partially supported by NSF grants CCR-9505300 and CCR-9625784.

References

1. I. Ahmad, Y.K. Kwok, and M. Y. Wu. Performance comparison of algorithms for static scheduling of DAGs to multiprocessors. In *Second Australasian Conference on Parallel and Real-time Systems*, pages 185–192, September 1995.
2. M. Cosnard, E. Jeannnot, and L. Rougeot. Low memory cost dynamic scheduling of large coarse grain task graphs. In *International Parallel Processing Symposium*, April 1998.
3. H. El-Rewini and T. G. Lewis. Scheduling parallel program tasks onto arbitrary target machines. *Journal of Parallel and Distributed Computing*, June 1990.
4. M.R. Gary and D.S. Johnson. *Computers and Intractability: A Guide to the Theory of NP-Completeness*. W.H. Freeman and Company, 1979.
5. M. Y. Wu. Parallel incremental scheduling. *Parallel Processing Letters*, 5(4):659–670, December 1995.
6. M. Y. Wu and D. D. Gajski. Hypertool: A programming aid for message-passing systems. *IEEE Trans. Parallel and Distributed Systems*, 1(3):330–343, July 1990.
7. M. Y. Wu and W. Shu. On parallelization of static scheduling algorithms. *IEEE Transactions on Software Engineering*, 23(8):517–528, August 1997.
8. M. Y. Wu, W. Shu, and Y. Chen. Incremental scheduling and execution of dags. In *IASTED International Conference on Parallel and Distributed Computing Systems*, November 1999.
9. T. Yang and A. Gerasoulis. DSC: Scheduling parallel tasks on an unbounded number of processors. *IEEE Trans. Parallel and Distributed System*, 5(9):951–967, September 1994.

Workshop on Fault-Tolerant Parallel and Distributed Systems (FTPDS '00)

Workshop Chair

Dimiter R. Avresky, Network Computing Lab, Boston University, USA

Invited speakers

Jean-Claude Laprie
I. Levendel

Papers

Computing in the RAIN: A Reliable Array of Independent Nodes
V. Bohossian, C. Fan, P. LeMahieu, M. Riedel, L. Xu, and J. Bruck

Fault-Tolerant Wide-Area Parallel Computing
J. Weissman

Transient Analysis of Dependability/Performability Models by Regenerative Randomization with Laplace Transform Inversion
J. Carrasco

FANTOMAS: Fault Tolerance for Mobile Agents in Clusters
H. Pals, S. Petri, and C. Grewe

Metrics, Methodologies, and Tools for Analyzing Network Fault Recovery in Real-Time Distributed Systems
P. Irey, B. Chappell, R. Hott, D. Marlow, K. O'Donoghue, and T. Plunkett

Consensus Based on Strong Failure Detectors: Time and Message-Efficient Protocols
F. Greve, M. Hurfin, R. Macêdo, and M. Raynal

Implementation of Finite Lattices in VLSI for Fault-State Encoding in High-Speed Networks
A. Döring and G. Lustig

J. Rolim et al. (Eds.): IPDPS 2000 Workshops, LNCS 1800, pp. 1200-1201, 2000.

Certification of System Architecture Dependability

I. Levendel
Director High Availability Technology Program
Motorola
Schaumburg – Illinois
E-mail: i.levendel@motorola.com

Wireless systems are particularly vulnerable to failures and malfunctions for several reasons. First, wireless architectures are naturally distributed over a wide geographic area. Secondly, equipment pricing pressures prohibit massive fault-tolerance similar to that in the slowly deregulating wireline business. Thirdly, the responsibility for insuring dependable functioning of the systems is generally distributed among several independent collaborating entities (wireless equipment owners and leased lines providers). Fourthly, the access medium of wireless (RF) is much more vulnerable than the access medium of wireline (wire to the home).

As a result of this situation, service quality has been much below the standards, which were previously set by the wireline during the decades of regulation and are still in effect in spite of the current deregulation process. Both system availability and call completion are lower than in the wireline by several orders of magnitude. This situation has two down sides. First, it causes significant dissatisfaction for end customers and loss of revenue for service providers. In addition, it limits wireless communication to lower quality voice communication and is a significant barrier for the wireless industry to expand its scope to other lucrative forms of communication. In order for the wireless industry to broaden its span, system dependability is a necessary ingredient that would allow the deployment of extensive high capacity dependable infrastructures and services.

In order to alleviate these problems, it is necessary to introduce in the industry a better discipline for dependable system design in all phases of the design process. The speaker will focus on the introduction of a quantitative approach to the certification of system architecture. The quantitative side of the approach strongly contrasts with the common practice of architecture peer review. Our approach is based on modeling and analysis of the system architecture and includes the following steps:

1) Construct a statistical model for the architecture. The model is driven by outages due to the following components:
a) Hardware
b) Software
c) Geographically distributed links
d) Hardware and software upgrades

J. Rolim et al. (Eds.): IPDPS 2000 Workshops, LNCS 1800, pp. 1202-1203, 2000.
© Springer-Verlag Berlin Heidelberg 2000

2) Evaluate availability, and, if appropriate, its tradeoffs with performance and capacity.
3) Identify potential architectural deficiencies in relation with customer requirements
4) Propose architectural remedies
5) Define the architectural improvement roadmap

This approach can also be applied to legacy systems as well as to new designs and gives a rational character to the architecture certification process.

Our experience has spanned several voice-based systems, and we are engaged in the definition of packet network architecture certification. Examples and templates will be discussed during the talk, and open issues will be brought up. Of course, other phases of the development process deserve the same degree of rationalism.

Computing in the RAIN:
A Reliable Array of Independent Nodes*

Vasken Bohossian, Charles C. Fan, Paul S. LeMahieu, Marc D. Riedel, Lihao
Xu, and Jehoshua Bruck

California Institute of Technology, Pasadena, CA 91125, USA
{vincent, fan, lemahieu, riedel, lihao, bruck}@paradise.caltech.edu
http://www.paradise.caltech.edu

Abstract. The RAIN project is a research collaboration between Caltech and NASA-JPL on distributed computing and data storage systems
for future spaceborne missions. The goal of the project is to identify
and develop key building blocks for reliable distributed systems built
with inexpensive off-the-shelf components. The RAIN platform consists
of a heterogeneous cluster of computing and/or storage nodes connected
via multiple interfaces to networks configured in fault-tolerant topologies. The RAIN software components run in conjunction with operating system services and standard network protocols. Through software-
implemented fault tolerance, the system tolerates multiple node, link,
and switch failures, with no single point of failure. The RAIN technology has been transfered to RAINfinity, a start-up company focusing on
creating clustered solutions for improving the performance and availability of Internet data centers. In this paper we describe the following
contributions: 1) fault-tolerant interconnect topologies and communication protocols providing consistent error reporting of link failures; 2)
fault management techniques based on group membership; and 3) data
storage schemes based on computationally efficient error-control codes.
We present several proof-of-concept applications: highly available video
and web servers, and a distributed checkpointing system.

1 Introduction

The Reliable Array of Independent Nodes (RAIN) project is a research collaboration between Caltech's Parallel and Distributed Computing Group and the Jet
Propulsion Laboratory's Center for Integrated Space Microsystems, in the area
of distributed computing and data storage systems for future spaceborne missions. The goal of the project is to identify and develop key building blocks for
reliable distributed systems built with inexpensive off-the-shelf components. The
RAIN platform consists of a heterogeneous cluster of computing and/or storage
nodes connected via multiple interfaces to networks configured in fault-tolerant

* Supported in part by an NSF Young Investigator Award (CCR-9457811), by a Sloan
Research Fellowship, by an IBM Partnership Award and by DARPA through an
agreement with NASA/OSAT.

J. Rolim et al. (Eds.): IPDPS 2000 Workshops, LNCS 1800, pp. 1204-1213, 2000.

topologies. The RAIN software components run in conjunction with operating system services and standard network protocols.

Features of the RAIN system include scalability, dynamic reconfigurability, and high availability. Through software-implemented fault tolerance, the system tolerates multiple node, link, and switch failures, with no single point of failure. In addition to reliability, the RAIN architecture permits efficient use of network resources, such as multiple data paths and redundant storage, with graceful degradation in the presence of faults. The RAIN technology has been transfered to RAINfinity, a start-up company focusing on creating clustered solutions for improving the performance and availability of Internet data centers [18].

We have identified the following key building blocks for distributed computing systems.

- Communication: fault-tolerant interconnect topologies and reliable communication protocols. We describe network topologies that are resistant to partitioning, and a protocol guaranteeing a consistent history of link failures. We also describe an implementation of the MPI standard [21] on the RAIN communication layer.
- Fault Management: techniques based on group membership. We describe an efficient token-based protocol that tolerates node and link failures.
- Storage: distributed data storage schemes based on error-control codes. We describe schemes that are optimal in terms of storage as well as encoding/decoding complexity.

We present three proof-of-concept applications based on the RAIN building blocks:

- A video server based on the RAIN communication and data storage components.
- A Web server based on the RAIN fault management component.
- A distributed checkpointing system based on the RAIN storage component, as well as a leader election protocol.

This paper is intended as an overview of our work on the RAIN system. Further details of our work on fault-tolerant interconnect topologies may be found in [14]; on the consistent-history protocol in [15]; on the leader election protocol in [11]; and on data storage schemes in [25], [26] and [27].

1.1 Related Work

Cluster computing systems such as the NOW project at the University of California, Berkeley [1] and the Beowulf project [2] have shown that networks of workstations can rival supercomputers in computational power. Packages such as PVM [23] and MPI [21] are widely used for parallel programming applications. There have been numerous projects focusing on various aspects of fault management and reliability in cluster computing systems. Well-known examples are the Isis [4] and Horus [24] systems at Cornell University, the Totem system at the University of California, Santa Barbara [17], and the Transis system at the Hebrew University of Jerusalem [9].

1.2   Novel Features of RAIN

The RAIN project incorporates many novel features in an attempt to deal with faults in nodes, networks, and data storage.

- **Communication**: Since the network is frequently a single point of failure, RAIN provides fault tolerance in the network via the following mechanisms.
  - *Bundled interfaces*: Nodes are permitted to have multiple interface cards. This not only adds fault tolerance to the network but also gives improved bandwidth.
  - *Link monitoring*: To correctly use multiple paths between nodes in the presence of faults, we have developed a link-state monitoring protocol that provides a consistent history of the link state at each endpoint.
  - *Fault-tolerant interconnect topologies*: Network partitioning is always a problem when a cluster of computers must act as a whole. We have designed network topologies that are resistant to partitioning as network elements fail.
- **Group membership**: A fundamental part of fault management is identifying which nodes are healthy and participating in the cluster. We give a new protocol for establishing group membership.
- **Data Storage**: Fault tolerance in data storage over multiple disks is achieved through redundant storage schemes. Novel error-correcting codes have been developed for this purpose. These are *array codes* that encode and decode using simple XOR operations. Traditional RAID codes generally only allow mirroring or parity (i.e., one degree of fault tolerance) as options. Array codes can be thought of as data partitioning schemes that allow one to trade off storage requirements for fault tolerance. These codes exhibit optimality in the storage requirements as well as in the number of update operations needed. Although some of the original motivation for these codes came from traditional RAID systems, these schemes apply equally well to partitioning data over disks on distinct nodes (as in our project) or even partitioning data over disks at remote geographic locations.

2   Communication

The RAIN project addresses fault tolerance in the network with fault-tolerant interconnect topologies and with bundled network interfaces.

2.1   Fault-Tolerant Interconnect Topologies

We were faced with the question of how to connect compute nodes to switching networks to maximize the network's resistance to partitioning. Many distributed computing algorithms face trouble when presented with a large set of nodes that have become partitioned from the others. A network that is resistant to partitioning should lose only some constant number of nodes (with respect to the total number of nodes) given that we do not exceed some number of failures.

After additional failures we may see partitioning of the set of compute nodes, i.e., some fraction of the total number of compute nodes may be lost. By carefully choosing how we connect our compute nodes to the switches, we can maximize a system's ability to resist partitioning in the presence of faults.

Our main contributions are: (i) a construction for degree-2 compute nodes connected by a ring network of switches of degree 4 that can tolerate any 3 switch failures without partitioning the nodes into disjoint sets, (ii) a proof that this construction is optimal in the sense that no construction can tolerate more switch failures while avoiding partitioning, and (iii) generalizations of this construction to arbitrary switch and node degrees and to other switch networks, in particular, to a fully-connected network of switches. See [14] for further details.

2.2 Consistent-History Protocol for Link Failures

When we bundle interfaces together on a machine and allow links and network adapters to fail, we must monitor available paths in the network for proper functioning. In [15] we give a modified *ping* protocol that guarantees that each side of the communication channel sees the same history. Each side is limited in how much it may lead or lag the other side of the channel, giving the protocol *bounded slack*. This notion of identical history can be useful in the development of applications using this connectivity information. For example, if an application takes error recovery action in the event of lost connectivity, it knows that both sides of the channel will see the exact same behavior on the channel over time, and will thus take the same error recovery action. Such a guarantee may simplify the writing of applications using this connectivity information.

Our main contributions are: (i) a simple, *stable* protocol for monitoring connectivity that maintains a *consistent history* with *bounded slack*, and (ii) proofs that this protocol exhibits *correctness*, *bounded slack*, and *stability*. See [15] for further details.

2.3 A Port of MPI

A port of MPI [21] (using the MPICH implementation from Argonne Labs [12]) was done on the RAIN communication layer. This port involved creating a new communications device in the MPICH framework, essentially adapting the standard communication device calls of MPICH to those presented by the RAIN communication layer, called RUDP (Reliable UDP). RUDP is a datagram delivery protocol that monitors connectivity to remote machines using the consistent history link protocol presented in detail in [15]. The port to MPI was done to facilitate our own analysis and use of the RAIN communication layer.

MPI is *not* a fault-tolerant API, and as such the best we can do is mask network errors to the extent redundant hardware has been put in place. For example, if all machines have two network adaptors and one link fails, the MPI program will proceed as if nothing had happened. If a second link fails, the MPI application may hang until the link is restored. There is no possibility to return errors related to link connectivity in the MPI communications API. Thus, although the RUDP communication layer knows of the loss of connectivity, it can do nothing about it and must wait for the problem to be resolved.

The implementation itself has a few notable features:

- It allows individual networking components to fail up to the limit of the redundancy put into the network.
- It provides increased network bandwidth by utilizing the redundant hardware.
- It runs entirely in user space. This has the important impact that all program state exists entirely in the running process, its memory stack, and its open file descriptors. The result is that if a system running RUDP has a checkpointing library, the program state (including the state of all communications) can be transparently saved without having to first synchronize all messaging. The communications layer only uses the kernel for unreliable packet delivery and does not rely on any kernel state for reliable messaging.
- It illustrates an experimental communication library can make the step to a practical piece of software easily in the presence of standards such as MPI.

The MPI port to RUDP has helped us use our own communication layer for real applications, has helped us argue the importance of keeping program state out of the kernel for the purposes of transparent checkpointing, and has highlighted the importance of programming standards such as MPI.

3   Group Membership

Tolerating faults in an asynchronous distributed system is a challenging task. A reliable group membership service ensures that the processes in a group maintain a consistent view of the global membership.

In order for a distributed application to work correctly in the presence of faults, a certain level of agreement among the non-faulty processes must be achieved. There are a number of well-defined problems in an asynchronous distributed system, such as consensus, group membership, commit, and atomic broadcast that have been extensively studied by researchers. In the RAIN system, the group membership protocol is a critical building block. It is a difficult task, especially when a change in the membership occurs, either due to failures or to voluntary joins and withdrawals.

In fact, under the classical asynchronous environment, the group membership problem has been proven impossible to solve in the presence of any failures [7], [10]. The underlying reason for the impossibility is that according to the classical definition of an asynchronous environment, processes in the system share no common clock and there is no bound on the message delay. Under this definition, it is impossible to implement a reliable fault detector, for no fault detector can distinguish between a *crashed* node and a *very slow* node. Since the establishment of this theoretic result, researchers have been striving to circumvent this impossibility. Theorists have modified the specifications [3], [8], [19], while practitioners have built a number of real systems that achieve a level of reliability in their particular environment [4], [17].

3.1 Novel Features

The group membership protocol in the RAIN system differs from that of other systems, such as the Totem [17] and Isis [4] projects, in several respects. Firstly, it is based exclusively on unicast messages, a practical model given the nature of the Internet. With this model, the total ordering of packets is not relevant. Compared to broadcast messages, unicast messages are more efficient in terms of CPU overhead. Secondly, the protocol does not require the system to freeze during reconfiguration. We do make the assumption that the mean time to failure of the system is greater than the convergence time of the protocol. With this assumption, the RAIN system tolerates node and link failures, both permanent and transient. In general, it is not possible to distinguish a slow node from a dead node in an asynchronous environment. It is inevitable for a group membership protocol to exclude a live node, if it is slow, from the membership. Our protocol allows such a node to rejoin the cluster automatically.

The key to this fault management service is a token-based group membership protocol. Using this protocol, it is possible to build the fault management service. It is also possible to attach to the token application-dependent synchronization information. For example, in the SNOW project described in Section 5.2, the HTTP request queue is attached to the token to ensure mutual exclusion of service.

4 Data Storage

Much research has been done on improving reliability by introducing data redundancy (also called information dispersity) [13], [22]. The RAIN system provides a distributed storage system based on a class of error-control codes called array codes. In Section 4.2, we describe the implementation of distributed store and retrieve operations based upon this storage scheme.

4.1 Array Codes

Array codes are a class of error-control codes that are particularly well-suited to be used as erasure-correcting codes. Erasure-correcting codes are a mathematical means of representing data so that lost information can be recovered. With an (n, k) erasure-correcting code, we represent k symbols of the original data with n symbols of encoded data ($n - k$ is called the amount of redundancy or parity). With an m-erasure-correcting code, the original data can be recovered even if m symbols of the encoded data are lost [16]. A code is said to be Maximum Distance Separable (MDS) if $m = n - k$. An MDS code is optimal in terms of the amount of redundancy versus the erasure recovery capability. The Reed-Solomon code [16] is an example of an MDS code.

The complexity of the computations needed to construct the encoded data (a process called encoding) and to recover the original data (a process called decoding) is an important consideration for practical systems. Array codes are ideal in this respect [6]. The only operations needed for encoding and decoding are simple binary exclusive-or (XOR) operations, which can be implemented efficiently in hardware and/or software. Several MDS array codes are known. For example, the EVENODD code [5] is a general (n, k) array code. Recently, we described two classes of $(n, n-2)$ and $(n, 2)$ MDS array codes with an optimal number of encoding and decoding operations [26], [27].

4.2   Distributed Store/Retrieve Operations

Our distributed store and retrieve operations are a straight-forward application of MDS array codes to distributed storage. Suppose that we have n nodes. For a store operation, we encode a block of data of size d into n symbols, each of size $\frac{d}{k}$, using an (n, k) MDS array code. We store one symbol per node. For a retrieve operation, we collect the symbols from any k nodes, and decode them to obtain the original data.

This data storage scheme has several attractive features. Firstly, it provides reliability. The original data can be recovered with up to $n - k$ node failures. Secondly, it permits dynamic reconfigurability and hot-swapping of components. We can dynamically remove and replace up to $n - k$ nodes. In addition, the flexibility to choose any k out of n nodes permits load balancing. We can select the k nodes with the smallest load, or in the case of a wide-area network, the k nodes that are geographically closest.

5   Proof-of-Concept Applications

We present several applications implemented on the RAIN platform based on the fault management, communication, and data storage building blocks described in the preceding sections: a video server (RAINVideo), a web server (SNOW), and a distributed checkpointing system (RAINCheck).

5.1   High-Availability Video Server

For our RAINVideo application, a collection of videos are encoded and written to all n nodes in the system with distributed store operations. Each node runs a client application that attempts to display a video, as well as a server application that supplies encoded video data. For each block of video data, a client performs a distributed retrieve operation to obtain encoded symbols from k of the servers. It then decodes the block of video data and displays it. If we break network connections or take down nodes, some of the servers may no longer be accessible. However, the videos continue to run without interruption, provided that each client can access at least k servers.

5.2   High-Availability Web Server

SNOW stands for Strong Network Of Web servers. It is a proof-of-concept project that demonstrates the features of the RAIN system. The goal is to develop a highly available fault-tolerant distributed web server cluster that minimizes the risk of down-time for mission-critical Internet and Intranet applications.

The SNOW project uses several key building blocks of the RAIN technology. Firstly, the reliable communication layer is used to handle all of the message passing between the servers in the SNOW system. Secondly, the token-based fault management module is used to establish the set of servers participating in the cluster. In addition, the token protocol is used to guarantee that when a request is received by SNOW, one and only one server will reply to the client. The latest information about the HTTP queue is attached to the token. Thirdly, the distributed storage module can be used to store the actual data for the web server.

SNOW also uses the distributed state sharing mechanism enabled by the RAIN system. The state information of the web servers, namely, the queue of HTTP requests, is shared reliably and consistently among the SNOW nodes. High availability and performance are achieved without external load balancing devices, such as the commercially available Cisco LocalDirector. The SNOW system is also readily scalable. In contrast, the commercially available Microsoft Wolfpack is only available for up to two nodes per cluster.

5.3   Distributed Checkpointing Mechanism

The idea of using error-control codes for distributed checkpointing was proposed by Plank [20]. We have implemented a checkpoint and rollback/recovery mechanism on the RAIN platform based on the distributed store and retrieve operations. The scheme runs in conjunction with a leader election protocol, described in [11]. This protocol ensures that there is a unique node designated as leader in every connected set of nodes. The leader node assigns jobs to the other nodes. As each job executes, a checkpoint of its state is taken periodically. The state is encoded and written to all accessible nodes with a distributed store operation. If a node fails or becomes inaccessible, the leader assigns the node's jobs to other nodes. The encoded symbols for the state of each job are read from k nodes with a distributed read operation. The state of each job is then decoded and execution is resumed from the last checkpoint. As long as a connected component of k nodes survives, all jobs execute to completion.

6   Conclusions

The goal of the RAIN project has been to build a testbed for various building blocks that address fault-management, communication, and storage in a distributed environment. The creation of such building blocks is important for the development of a fully functional distributed computing system. One of the fundamental driving ideas behind this work has been to consolidate the assumptions required to get around the "difficult" parts of distributed computing into several basic building blocks. We feel the ability to provide basic, provably correct services is essential to building a real fault-tolerant system. In other words, the difficult proofs should be confined to a few basic components of the system. Components of the system built on top of those reliable components should then be easier to develop and easier to establish as correct in their own right. Building blocks that we consider important and that are discussed in this paper are those providing reliable communication, group membership information, and reliable storage. Among the future and current directions of this work are:

- Development of API's for using the various building blocks. We should standardize the packaging of the various components to make them more practical for use by outside groups.
- The implementation of a real distributed file system using the data partitioning schemes developed here. In addition to making this building block more accessible to others, it would help in assessing the performance benefits and penalties from partitioning data in such a manner.

We are currently benchmarking the system for general assessment of the performance of the algorithms and protocols developed in the project.

References

1. T.E. Anderson, D.E. Culler and D.A. Patterson, "A Case for NOW (Networks of Workstations)," *IEEE Micro*, Vol. 15, No. 1, pp. 54–64, 1995.
2. D. J. Becker, T. Sterling, D. Savarese, E. Dorband, U.A. Ranawake , and C.V. Packer, "BEOWULF: A Parallel Workstation for Scientific Computation," *Proceedings of the 1995 International Conference on Parallel Processing*, pp. 11–14, 1995.
3. M. Ben-Or, "Another Advantage of Free Choice: Completely Asynchronous Agreement Protocols," *Proceedings of the Second ACM Symposium on Principles of Distributed Computing*, ACM Press, pp. 27–30, August 1983.
4. K.P. Birman and R. Van Renesse, "Reliable Distributed Computing with the Isis Toolkit," *IEEE Computer Society Press*, 1994.
5. M. Blaum, J. Brady, J. Bruck and J. Menon, "EVENODD: An Efficient Scheme for Tolerating Double Disk Failures in RAID Architectures," *IEEE Trans. on Computers*, Vol. 44, No. 2, pp. 192–202, 1995.
6. M. Blaum, P.G. Farrell and H.C.A. van Tilborg, "Chapter on Array Codes," *Handbook of Coding Theory*, V.S. Pless and W.C. Huffman eds., to appear.
7. T.D. Chandra, V. Hadzillacos, S. Toueg and B. Charron-Bost, "On the Impossibility of Group Membership," *Proceedings of the Fifteenth ACM Symposium on Principles of Distributed Computing*, ACM Press, pp. 322–330, 1996.
8. T. D. Chandra and S. Toueg, "Unreliable Failure Detectors for Reliable Distributed Systems," *Journal of the ACM*, Vol. 43, No. 2, pp. 225–267, 1996.
9. D. Dolev and D. Malki "The Transis Approach to High Availability Cluster Communication," *Communications of the ACM*, Vol. 39, No. 4, pp. 64–70, 1996.
10. M.J. Fischer, N.A. Lynch and M.S. Paterson, "Impossibility of Distributed Consensus with One Faulty Process," *Journal of the ACM*, Vol. 32, No. 2, pp. 374i–382, 1985.
11. M. Franceschetti and J. Bruck, "A Leader Election Protocol for Fault Recovery in Asynchronous Fully-Connected Networks," *Paradise Electronic Technical Report #024*, http://paradise.caltech.edu/ETR.html, 1998.

12. W. Gropp, E. Lusk, N. Doss and A. Skjellum, "A High-Performance, Portable Implementation of the MPI Message Passing Interface Standard," *Parallel Computing*, Vol. 22, No. 6, pp. 789–828, 1996.

13. T. Krol, "(N,K) Concept Fault Tolerance," *IEEE Trans. on Computers*, Vol. C-35, No. 4, 1986.

14. P.S. LeMahieu, V.Z. Bohossian and J. Bruck, "Fault-Tolerant Switched Local Area Networks," *Proceedings of the International Parallel Processing Symposium*, pp. 747–751, 1998.

15. P.S. LeMahieu and J. Bruck, "Consistent History Link Connectivity Protocol," *Proceedings of the International Parallel Processing Symposium*, pp. 138–142, 1999.

16. F.J. MacWilliams and N.J.A. Sloane, *The Theory of Error Correcting Codes*, North-Holland, 1977.

17. L.E. Moser, P.M. Melliar-Smith, D.A. Agarwal, R.K. Budhia and C.A. Lingley-Papadopoulos, "Totem: A Fault-Tolerant Multicast Group Communication System," *Communications of the ACM*, Vol. 39, No. 4, pp. 54–63, 1996.

18. "NASA-Funded Software Aids Reliability," *Network World*, Vol. 16, No. 51, Dec. 20, 1999, http://www.nwfusion.com/news/1999/1220infra.html.

19. G. Neiger, "A New Look at Membership Services," *Proceedings of the Fifteenth ACM Symposium on Principles of Distributed Computing*, ACM Press, pp. 331–340, 1996.

20. J.S. Plank and K. Li, "Faster Checkpointing with N+1 Parity," *IEEE 24th International Symposium on Fault-Tolerant Computing*, pp. 288–297, 1994.

21. M. Snir, S. Otto, S. Huss-Lederman, D. Walker, and J. Dongarra, "MPI: The Complete Reference," MIT Press, http://www.netlib.org/utk/papers/mpi-book/mpi-book.ps, 1995.

22. H.-M. Sun and S.-P. Shieh, "Optimal Information Dispersal for Increasing the Reliability of a Distributed Service," *IEEE Trans. on Reliability*, Vol. 46, No. 4, pp. 462–472, 1997.

23. V. Sunderam, "PVM: A Framework for Parallel Distributed Computing," *Concurrency: Practice and Experience*, Vol. 2, No. 4, pp. 315–339, http://www.netlib.org/ncwn/pvmsystem.ps, 1990.

24. R. van Renesse, K.P. Birman and S. Maffeis, "Horus: a Flexible Group Communication System," *Communications of the ACM*, Vol. 39, No. 4, pp. 76–83, 1996.

25. L. Xu and J. Bruck, "Improving the Performance of Data Servers Using Array Codes," *Paradise Electronic Technical Report #027*, http://paradise.caltech.edu/ETR.html, 1998.

26. L. Xu and J. Bruck "X-Code: MDS Array Codes with Optimal Encoding," *IEEE Trans. on Information Theory*, Vol. 45, No. 1, pp. 272–276, January 1999.

27. L. Xu, V. Bohossian, J. Bruck and D. G. Wagner, "Low Density MDS Codes and Factors of Complete Graphs," *IEEE Trans. on Information Theory*, Vol. 45, No. 6, pp. 1817–1826, September 1999.

Fault Tolerant Wide-Area Parallel Computing

Jon B. Weissman

Department of Computer Science and Engineering

University of Minnesota

Minneapolis, MN 55455

(jon@cs.umn.edu)

Abstract. Executing parallel applications across distributed networks introduces the problem of fault tolerance. A viable solution for fault tolerance must keep overhead manageable and not compromise the high performance objective of parallel processing. In this paper, we explore two options for achieving fault tolerance for a common class of parallel applications, single-program-multiple-data (SPMD). We quantitatively compare checkpoint-recovery and wide-area replication as a means of achieving fault tolerance. The experimental results obtained for a canonical SPMD application suggest that checkpoint-recovery may be preferable for small problems if local parallel disks are available, but wide-area replication outperforms checkpoint-recovery for larger-grain problems, precisely the problems most suited for the wide-area network environment. The results also show that it possible to accurately model and predict the overheads of the two methods[1]

1.0 Introduction

High-performance distributed computing across wide-area networks has become an active topic of research [1][3][4][11]. Metasystem and grid software infrastructure projects, most notably, Legion [4] and Globus [3], have emerged to support this new computational paradigm. Achieving large-scale distributed computing in a seamless manner introduces a number of difficult problems. This paper examines one of the most critical problems, fault tolerance. A large wide-area system that contains hundreds to thousands of machines and multiple networks has a small mean time to failure. The most common failure modes include machine faults in which hosts go down and get rebooted, and network faults where links go down. A single monolithic solution for fault tolerance that is acceptable to all user applications is unlikely. For example, some applications may require continuous availability, or may require protection from byzantine failures, or require light-weight, low overhead fault tolerance. The most appropriate method for fault tolerance clearly may be application-specific. This follows the current trend in distributed systems and operating systems in which generic functions once performed within the "system" are now being moved to user-space for increased flexibility and performance. Because general purpose systems often impose a high cost on applications that do not fit their assumptions, the maxim "pay for what you need" has been proposed as a guiding principle for application-centric policy decisions in metacomputing systems such as Legion.

---

1. This work was partially funded by grants NSF ACIR-9996418 and CDA-9633299, AFOSR-F49620-96-1-0472.

J. Rolim et al. (Eds.): IPDPS 2000 Workshops, LNCS 1800, pp. 1214-1225, 2000.

We believe that the relative performance of fault tolerance methods is a key piece of information needed to enable users to select the most appropriate method for their application. This is particularly true for high-performance applications. To this end, we have examined two fault tolerance two options, *checkpoint-recovery* and *wide-area replication*, for a common class of high-performance parallel applications, single-program-multiple-data (SPMD). To compare these approaches, performance models that characterize the overheads have been developed. These models enable quantitative comparisons of the two methods as applied to SPMD applications. We have not implemented complete fault tolerance solutions, but a sufficient subset to capture the associated overheads. In particular, we do not consider fault detection or fault recovery.

We compare wide-area replication across the Internet to application-level checkpointing for a canonical SPMD application. The wide-area replication is provided by the Gallop system, a wide-area scheduler, and has the advantage that no source code changes are required. The checkpointing version of the application required the insertion of checkpointing code. It is possible to perform application checkpoints with operating system support or binary code rewriting avoiding source changes, but we speculate that the checkpoint overheads would not differ dramatically. Two modes of checkpointing were compared with wide-area replication: a single NFS-mounted checkpoint disk and a parallel array of locally attached disks. The results indicate that the performance models are accurate and predictive, and that both methods appear to be appropriate under different conditions: checkpoint-recovery performs best for small problems providing fast local disks are available, while wide-area replication is generally more efficient for bigger problems.

2.0 Related Work

Numerous research groups are examining the problem of fault tolerance for parallel applications in distributed networks. Globus provides a heartbeat service to monitor running processes to detect faults [4]. The application is notified of the failure and expected to take appropriate recovery action. Legion provides mechanisms to support fault tolerance such as checkpointing, but the policy must be provided by the application. A component-based reflexive architecture allows fault tolerance methods to be encapsulated in reusable components that user applications may choose from [7]. Condor [6][13] provides system-level checkpoints of distributed applications but it not geared to high-performance parallel programs. A number of PVM-based checkpointing systems have also appeared over the past few years [2]. Our work is different from these and related projects as we are focused on the problem of deciding which fault tolerance method can be expected to perform best, not in a specific implementation of fault tolerance methods.

3.0 Fault Tolerance Options for SPMD Applications

In a canonical data parallel SPMD application, a set of identical tasks are created one per processor with each assigned a portion of a data domain such as a grid or a matrix. The tasks alternate between computing on their portion of the data domain and communications, typically in an iterative style. The state of each task is the updated data domain that changes with each iteration. A transient network or single processor fault are the most common events that will cause such an application to fail. Executing SPMD applications in a fault tolerant manner can be achieved by checkpointing or replication (Figure 1). For the purposes of a direct quantitative comparison, a simple checkpoint model is adopted in which each SPMD task saves its portion of the data domain on disk at a set of pre-determined iterations. Recomputing lost iterations starting from the last checkpoint would be straightforward given the last iteration index. We studied two configurations for checkpointing: a single NFS-mounted disk from a file server (Figure 1a) and a parallel array of locally attached disks (Figure 1b). The dominant overhead for checkpoint-recovery is the cost of stopping the application and writing the checkpoints to disk. The local checkpoint model is useful only if the failed processor is expected to recover and pick up where it left off since it is the only processor that can access the checkpoint. The network checkpoint model would allow another processor to pick up the checkpoint since it is stored on a common server.

In option (c), each application replica runs in a different site to avoid overloading the available computation and communication resources of a particular site. The use of wide-area computing offers a solution to the resource demand of replication because some sites in a wide-area system will likely be underutilized. We have developed a wide-area scheduling system called Gallop [11] that remotely executes SPMD applications for improved performance. Gallop picks the best site to run an application based

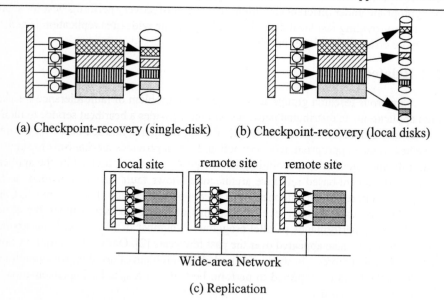

(a) Checkpoint-recovery (single-disk)   (b) Checkpoint-recovery (local disks)

local site   remote site   remote site

Wide-area Network

(c) Replication

Fig. 1: Fault tolerance options for SPMD applications. Four tasks (circles) on four processors (squares) operate on rectangular piece of data. In (a-b), the pieces are written to disk periodically. In (c), the local site is where the application request originates and is responsible for coordinating the remote replicas.

on an estimate of how well the site can run the job. Gallop performs this task by utilizing a local data parallel scheduling system called Prophet [12]. Prophet generates a performance prediction based on application and site resource information for each site in which it is running (refer to [12] for additional details).

We have modified Gallop to support the scheduling of application replicas in multiple sites to compare with checkpointing. Gallop will create a fixed number of application replicas determined by a user-specified runtime parameter. A complete independent copy of each application will run in the chosen number of sites. Important issues relating to replica consistency, dynamic replica creation, etc., are not germane to the performance study, but are discussed elsewhere [10]. The dominant overhead for wide-area replication is the scheduling protocol used by Gallop which includes Prophet overhead, multiple messages exchanged between the local and remote sites, and (potentially) application file transmission to enable execution in remote sites.

4.0 Performance Models

We have developed performance models for checkpoint-recovery (CR) and wide-area replication (WR) to enable quantitative comparisons. We consider only checkpoint costs for CR as recovery costs are more difficult to model accurately due to the large variety of potential failure modes each with different characteristics. For example, the recovery time for a network failure vs. a machine failure may in fact be very different. However, some preliminary work with single machine failures and immediate restart indicated that the inclusion of recovery overhead did not change the relative performance of the two methods for the vast majority of problem instances.

In our CR model, checkpoints occur at a single place in the applications execution and checkpoints from all SPMD tasks are written atomically to disk. For SPMD applications each task performs a checkpoint of its data domain and the current iteration after it has computed on its data, and before any messages are sent during that iteration. The checkpoints occur every k iterations. The WR model assumes that the set of replicas are *static* and unchanging. When a replica fails, it is not restarted nor is a new replica scheduled elsewhere. A model that includes *dynamic* replica creation is the subject of future work. The performance of CR and WR may be defined in terms of the following parameters:

N: problem size

P: number of processors

I: number of iterations

k: checkpoint frequency

m: number of wide-area replicas (or sites)

λ: average rate of site failure (failures/minute)

β: desired reliability (desired probability of application success)

We make the assumption that the failure rate is exponentially distributed as is commonly done and that it is identical for all sites. Different failure rates for different sites could easily be accommodated, but are omitted for simplicity (disk-config is NFS or local).

The following cost functions can be defined (disk-config is NFS or local):

T_{CT} (N, P): completion time for NxN problem using P processors

T_{CP} (N, P, disk-config): checkpoint time for single checkpoint on disk-config

T_{CR} (N, P, disk-config): total time to perform the optimal number of checkpoints

T_{WR}(N, m): time to schedule m replicas remotely

P_f (N, P, λ): probability of application failure (single site failure)

The cost functions can be expressed in terms of these parameters and several system-dependent constants:

T_{CT} (N, P) = I * T_c(N, P)

- T_c is the average time per iteration and includes computation and communication

$$T_{CP} \text{ (N, P, NFS)} = T_{NFS\_latency} + \frac{N^2}{P} T_{NFS\_bw}$$

- The latency and bandwidth terms include disk and network overhead; $\frac{N^2}{P}$ is the per processor data size

$$T_{CP} \text{ (N, P, local)} = T_{disk\_latency} + \frac{N^2}{P} T_{disk\_bw}$$

- Local checkpointing only involves disk overhead; $\frac{N^2}{P}$ is the per processor data size

T_{CR} (N, P, disk-config) = k_{opt} * T_{CP} (N, P, disk-config)

- k_{opt} is the optimal number of checkpoints

T_{WR}(N, m) = $T_{WR\_scheduling}$ (N, m) + T_{WR_upload} (m, bsize)

- WR consists of scheduling costs and file upload costs (which depends on binary and input file sizes)

$T_{WR\_scheduling}$ (N, m) = $T_{WR\_sched\_latency}$ + m ($T_{WR\_sched\_overhead}$)

- Scheduling overhead consists of a base cost and a per site cost

T_{WR_upload} (m, bsize) = $T_{WR\_upload\_latency}$ + m ($T_{WR\_upload\_overhead}$)

- Upload consists of a base cost and a per site file transfer cost

$$P_f \text{ (N, P, } \lambda) = 1 - e^{-\lambda T_{CT}(N, P)}$$

- Exponential failure distribution - this gives the probability the application will NOT fail (in a site)

We have experimentally derived the constants for these cost functions in a local-and wide-area network testbed environment. In the next section, we show that these functions accurately predict the real overhead costs. We model two modes each for CR and WR. For CR, we experimented with local and NFS-mounted file systems. For WR, we experimented with remote sites in which application files (binaries and input files) were either pre-staged or required uploading at runtime[2]. For pre-staged files,

---

2. Another option is to transfer the much smaller source files and compile them on the remote site. This option may be examined in the future.

T_{WR_upload} is 0. WR overhead consists of scheduling overhead and file transmission overhead. Scheduling overhead includes Prophet and protocol overhead, the latter is dominated by wide-area message passing cost. Both components of WR, $T_{WR\_scheduling}$ and T_{WR_upload} have a constant latency term and a per site term. Although the scheduling protocol and file transmission are largely parallel activities, we have empirically observed a dependence on the number of sites that is accurately modelled by a small linear constant. For $T_{WR\_scheduling}$: as the number of sites increases, the number of protocol messages handled by the local site daemon increases in a linear fashion, and the probability that a message is delayed from a remote site back to the local site, also increases. For T_{WR_upload}: as the number of sites increases, the local ftp server will have to serve a proportionally larger number of sites. Note that WR depends only on the number of sites and not directly on the problem size N because the application creates the initial data domain internally, while CR depends strongly on N.

The function P_f gives the probability that a site (or application running in the site) will not remain "up" through a time t. A site is up if all constituent machines and connecting networks applied to the application are up. This function has the property that a longer running application (hence a larger T_{CT}) incurs a larger probability of failure. This probability failure model is needed to enable a meaningful and fair performance comparison between the two methods. Without a model for failure, it is unclear how many checkpoints or replicas are needed to obtain a desired level of reliability. We use the parameter P_f to construct the cost functions for both CR and WR.

First, we derive the cost equation for CR. For a given P_f, we determine the optimal number of checkpoints to perform. The optimal number of checkpoints balances the cost of checkpointing with the cost of re-executing old iterations in the event that the application fails and needs to be restarted from the last stored checkpoint. Finding this value requires minimization of the following expression for total overhead experienced using CR (we omit some of the function parameters for brevity):

$$(1 - P_f) \left[\frac{kT_{CT}}{2I} \right] + T_{CP} \cdot \frac{I}{k} \qquad (1)$$

The first term in brackets is the average cost to re-execute iterations from the past checkpoint and the second term is the checkpoint cost. The factor of $\frac{k}{2}$ reflects a failure which occurs midway between checkpoints on average. Differentiating with respect to k, and solving for the minimum k yields:

$$k_{opt} = \left\lceil \sqrt{\frac{2I^2 \cdot T_{CP}}{T_{CT} \cdot (1 - P_f)}} \right\rceil \qquad (2)$$

This agrees with similar results in the literature [5][9] and provides a mechanism to determine the minimum overhead, T_{CR}, for a given P_f.

For WR, P_f plays a different role. Given P_f, we can determine the number of replicas m that achieve a desired level of reliability β, where β is the probability that at least one replica finishes. The probability that at least one replica finishes is $1 - P_f^m$

assuming independent site failures each with probability P_f. If sites have different failure probabilities then the equation becomes only slightly more complex. Solving for m yields the following:

$$m = \left\lceil \frac{\log(1-\beta)}{\log P_f} \right\rceil \tag{3}$$

Given a desired value of β, we can easily compute m and T_{WR}, and compare it with T_{CR} for a given problem instance. The dependent parameter β can be used to select between different fault tolerance methods (CR or WR) and provides a way to adjust the level of fault tolerance for WR. For example, if the number of available sites is less than the number of sites required to achieve the desired β, then CR becomes a more desirable option. Similarly, if $\beta = 1$, then CR is the logical choice since it is not possible to guarantee any replicas will finish if $P_f > 0$ (for static WR). However, if the user is willing to accept $\beta < 1$, then WR may offer some performance advantages as we will show. These models give us a way to predict the performance of CR and WR, and ultimately to enable the user to select the most suitable method given their preferences and constraints in the network environment.

5.0 Results

We performed an evaluation of CR and WR by simulating a wide range of failure rates (λ) to answer the following questions. Given a problem instance (N, P, I), network characteristics (local/NFS, m, P_f), and a desired degree of reliability (β) will CR or WR perform best? Which method might be expected to perform best as the problem size grows? What is the impact of uploading on the suitability of WR? We answer these questions for two specific experimental environments using a canonical SPMD application (STEN) that solves Poisson's equation on a NxN grid using an iterative method (Figure 1). STEN creates the initial data domain internally and does not use any input files, hence only application binary files need to be transferred under WR. We have two versions of STEN: (1) a CR version with user-level checkpoint code inserted and (2) a WR version that uses the Gallop wide-area scheduler as described earlier. Gallop was run using an experimental testbed containing Internet sites at the University of Texas at San Antonio (UTSA), Southwest Research Institute in San Antonio, University of Virginia, University of Kentucky, Argonne National Laboratories, and the University of California, San Diego. The local site is UTSA and contains 15 ethernet-connected Sparc 5s both configured with local disks (local), and with a NFS-mounted file system (NFS). The other sites contain Sparcs of similar capability. The local site is where the CR data was gathered. The experimental testbeds were used to gather data to construct the cost functions of Section 4.0. We first show that the experimentally derived cost functions are accurate. We then compare the performance of CR and WR using these cost functions for a wide range of parameter values.

5.1 Validating the Models

We performed a set of experiments with STEN using our local- and wide-area testbed to determine the cost functions for CR and WR. We ran STEN using a large number of values for N and P and used linear regression to derive the constants from this experimental data:

T_{CP} (N, P, NFS) = 30 + (N^2/P)0.005 msec

T_{CP} (N, P, local) = 0.2 + (N^2/P)0.0005 msec

T_{WR_upload} (m, bsize) = 4821 + m (0.023) msec

$T_{WR\_scheduling}$ (N, m) = 135 + m (28.4) msec

We then performed another separate set of runs for CR and WR using specific values of N (128, 256, 512, 1024) and P (1, 2, 4), and plotted these observed values (ten scatter points are shown for each x value plotted) against the derived cost functions (straight line). The results for WR (Figure 2) and for CR (Figure 3) are shown. For CR we show results for NFS (top row) and local file systems (bottom row), for P=1, 2, and 4 respectively.

In the majority of cases, the error between the overhead predicted by the cost function and the experimental overhead falls within 10% for CR with the exception of N=1024, P=1,2, local file system configuration. We speculate that disk buffer cache effects were exposed by the large write requests. The CR data indicates that the checkpoint cost functions for both local and NFS file systems provide a good fit as N and P vary. We observed that the cost of performing a single checkpoint depends on the total checkpoint data size and appears to be invariant to the number of processors. This was surprising for NFS as we expected more processors would create additional server load. It is more difficult to accurately model WR costs due to the high variance in wide-area communications. However, the cost functions give a reasonably good fit (within 10-30%) that is sufficient for quantitative comparisons. For a different network environment, the cost equations will have different constants which can be easily obtained by our test programs. The results indicate that it is possible to predict the overhead costs for a given network environment with sufficient fidelity to enable quantitative comparison of the two methods.

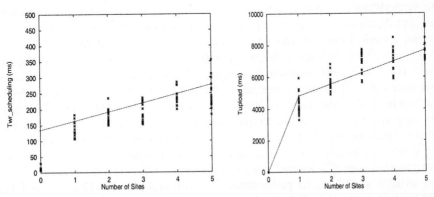

Fig. 2: Results for WR: scheduling and uploading. Upload results correspond to a 300 KB bina for STEN with ftp used to transfer the file. 0 sites corresponds to the use of the local site only.

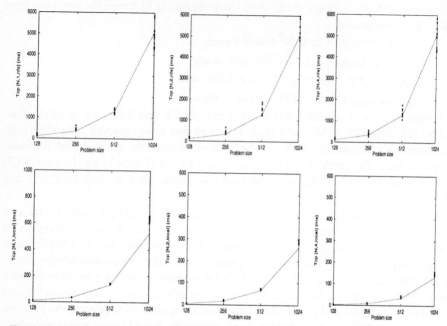

Fig. 3: Results for CR shown for NFS (top row) and local file system (bottom row), as a function of problem size. Each graph corresponds to a different P selected (1, 2, and 4). The scatter plots show 10 points per (N, P) pair: in some cases nearly identical points are plotted on the same space and appear darker.

5.2 Head-to-head Comparison

We now compare the *cost functions* T_{CR} and T_{WR} head-to-head to see how well the methods can be expected to perform under different failure and reliability parameters. For P_f, we vary λ to be 1 site failure per the following time intervals: 6 months, 1 month, 1 week, 1 day, half day, and 1 hour (corresponding to failure rates = 11, 12, ..., 16 respectively on the graphs). We vary β to be .999, .999999, .99999999 (corresponding to B = 1, 2, 3, respectively on the graphs). We also vary the problem size as before and since CR performance is invariant to P (for a fixed failure rate), we pick P=4 for the data plots (the other values of P yield similar graphs). We obtained the T_{CT} values for each (N, P) pair from the Prophet scheduler. Unless otherwise stated, we set I=1000 iterations. We show the total overhead under either method (T_{WR} vs. T_{CR}) on the y-axis as a function of varying failure rate for all graphs. The first set of results compares CR with WR without uploading (Figure 4). CR with a local disk configuration exhibits slightly better performance for small problems (N=128, 256, and 512; 128 is not shown), while WR is a clear winner for large problems (N=1024). WR performance does not vary much as β increases which suggests that a very high rate of reliability appears to be affordable. The results also indicate that for small problems, WR and CR offer similar performance. When a NFS-mounted disk configuration is used, WR is clearly superior for virtually all problem sizes (N=256, 512, and 1024;

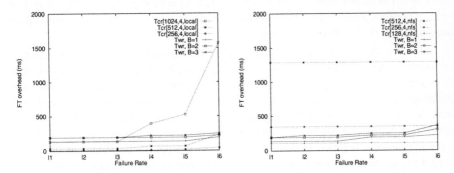

Fig. 4: Comparing WR w/o uploading to CR for local and NFS respectively.

1024 is off the graph). The flat-line indicates that the checkpoints are so expensive, that only a single checkpoint is affordable. When WR incorporates uploading, the results change fairly significantly (Figure 5). For both configurations, CR is clearly superior to WR due to high cost of uploading the STEN binaries. However as compared to the NFS configuration, WR is competitive for the largest problem (N=1024) for β = .999. Since our largest problem (N=1024 with I=1000) is small by some standards (T_c = 350 ms, which gives a total time of 350 sec), we wanted to know what would happen if the problem was scaled to the sizes one might expect in a metacomputing environment (and uploading was required). In particular, would WR become more attractive for larger problems that ran longer? To model a larger longer-running problem, we fixed T_c at 350 ms and simply increased I, keeping all other values such as T_c and T_{CP} constant. In reality, a larger problem also means that N and P both increase. If we assume that P grows in proportion to the increased computation then T_c could reasonably remain unchanged. Similarly, T_{CP} for the local configuration depends on the amount of data each processor writes to disk. If this is unchanged, then T_{CP} could also reasonably remain unchanged. However for the NFS configuration an increase in N will surely increase T_{CP} independent of P. Consequently, the T_{CR} results for NFS should be viewed as a lower-bound. When I is increased, WR begins to

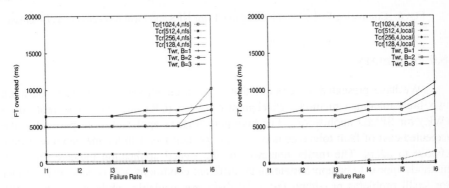

Fig. 5: Comparing WR with uploading to CR for local and NFS respectively.

exhibit better behavior than CR under the NFS and local configurations (Figure 6). For the local configuration (Figure 6b), the problem must be scaled significantly for WR to outperform CR (I=50000 or ~ 12 hours!).

We have shown that a meaningful quantitative comparison between WR and CR is possible for a typical SPMD application. For this application, our results suggest that CR is generally a less expensive method for small problems provided local disks are available, WR is cheaper for larger problems provided binaries are pre-staged, but when the problems become very long-running, CR may again be better. To apply our approach to a different application and network environment, some benchmarking to determine the cost function constants is a necessary precondition. This approach can be used to give the user some guidance in the selection of fault tolerance methods and determine whether affordable fault tolerance is possible given their application and network environment (provided estimates of P_f are known or can be approximated). But the precise benefit depends on the application and network environment at hand. Another interesting possibility is to provide both fault tolerance implementations for an application and allow the system to pick the best one automatically at run-time. Similarly, the cost models could be used by a metacomputing scheduler in deciding where to run an application. For example, if CR is the desired method for an application, then the scheduler should consider the predicted cost of CR in choosing the application's location.

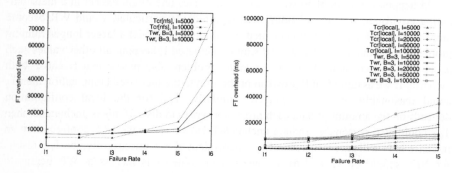

Fig. 6: Comparing WR with uploading to CR for NFS and local respectively as I increases

6.0 Summary

We have presented a technique that enabled quantitative comparisons between two fault tolerance methods: checkpoint-recovery (CR) and wide-area replication (WR) for SPMD applications. For high-performance applications in particular, the expected cost of fault tolerance may be an important factor in the method a user may choose to adopt. The results obtained for the stencil application indicate that both methods appear to be appropriate under different conditions: CR is generally cheaper for small problems providing fast local disks are available, while WR is generally cheaper for bigger problems provided binaries are pre-staged. We believe our technique can be used to support the "pay for what you need" policy currently advocated in metacomputing systems. However, experimentation with additional SPMD applications is needed to confirm our assertion. Future work also includes an investigation into site reliability (the parameter P_f) and the development of tools to gather this information. Finally, we plan on incorporating the cost of recovery and dynamic replicas into our performance models.

7.0 References

[1]   Bal, H. et al, "Optimizing Parallel Applications for Wide-Area Clusters," *Twelfth International Parallel Processing Symposium,*" March 1998.

[2]   Casas, J. et al, "Adaptive Load Migration systems for PVM," *Supercomputing* 1994.

[3]   Foster, I. and Kesselman, C., "Globus: A Metacomputing Infrastructure Toolkit," *International Journal of Supercomputing Applications*, 11(2), 1997.

[4]   Grimshaw, A.S. and Wulf, W. A., "The Legion Vision of a Worldwide Virtual Computer," Communications of the ACM, Vol. 40(1), 1997.

[5]   Jalote., P. , "Fault Tolerance in Distributed Systems," Prentice-Hall Publishers, Englewood Cliffs, New Jersey, 1994.

[6]   Litzkow, M.J. et al., "Condor - a hunter of idle workstations," In *Proceedings of the 8th International Conference on Distributed Computing Systems*, June 1988.

[7]   Nguyen-Tuong, A. and Grimshaw, A.S., "Using Reflection to Incorporate Fault-Tolerance Techniques in Distributed Applications," Computer Science Technical Report, University of Virginia, CS 98-34, 1998.

[8]   Stelling, P. et al., "A Fault Detection Service for Wide Area Distributed Computations," *Proceedings of the Seventh IEEE International Symposium on High Performance Distributed Computing*, August 1998.

[9]   Vaidya, N.H., "Impact of Checkpoint Latency on Overhead Ratio of a Checkpointing Scheme," *IEEE Transactions on Computers*, Vol. 46(8), August 1997.s

[10]   Weissman, J.B. and Womack, D. "Fault Tolerant Scheduling in Distributed Networks," UTSA Technical Report, CS-96-10, October 1996.

[11]   Weissman, J.B., "Gallop: The Benefits of Wide-Area Computing for Parallel Processing," *Journal of Parallel and Distributed Computing*, Vol. 54(2), November 1998.

[12]   Weissman, J.B., "Prophet: Automated Scheduling of SPMD Programs in Workstation Networks," *Concurrency: Practice and Experience,* Vol. 11(6), May 1999.

[13]   Zandy, V., Miller, B. and Livny, M., "Process Hijacking," *Proceedings of the Eighth IEEE International Symposium on High Performance Distributed Computing*, August 1999.

Transient Analysis of Dependability/Performability Models by Regenerative Randomization with Laplace Transform Inversion [*]

Juan A. Carrasco

Departament d'Enginyeria Electrònica
Universitat Politècnica de Catalunya
Diagonal 647, plta. 9, 08028 Barcelona, Spain
carrasco@eel.upc.es

Abstract. In this paper we develop a variant of a previously proposed method (the regenerative randomization method) for the transient analysis of dependability/performability models. The variant is obtained by developing a closed-form expression for the solution of the truncated transformed model obtained in regenerative randomization and using a Laplace transform inversion algorithm. Using models of moderate size of a 5-level RAID architecture we compare the new variant with the original randomization method, with randomization with steady-state detection for irreducible models, and with the standard randomization method for transient models (models with absorbing states). The new variant seems to be competitive for models of moderate size.

1 Introduction

Homogeneous continuous time Markov chains (CTMCs) are frequently used for performance, dependability and performability modeling. Commonly used methods for the transient analysis of CTMCs are ODE (ordinary differential equation) solvers and randomization. Good recent reviews of these methods can be found in [6] and [12]. The randomization method (also called uniformization) is attractive because it has guaranteed numerical stability, since it involves additions of positive numbers, and the computation error can be specified in advance. Let Λ be the maximum output rate of the CTMC in consideration. Then, the number of steps required by the method is roughly equal to Λt when Λt is large. For models of repairable fault-tolerant systems the t of interest makes typically Λt very large and, then, randomization is very inefficient.

Several variants of the (standard) randomization method have been proposed to improve its efficiency. Miller has used selective randomization to solve reliability models with detailed representation of error handling activities [7]. Reibman

---

[*] This work was supported by the "Comisión Interministerial de Ciencia y Tecnología" (CICYT) under the research grant TAP99-0443-C05.

J. Rolim et al. (Eds.): IPDPS 2000 Workshops, LNCS 1800, pp. 1226-1235, 2000.

and Trivedi [12] have proposed a more general approach based on the multistep concept. However, that method introduces fill-in in the transition probability matrix of the randomized discrete-time Markov chain (DTMC). In adaptive uniformization (randomization) [8] the randomization rate is adapted depending on the states in which the randomized DTMC can be at a given step. Adaptive randomization seems to be faster than the standard randomization method for small and medium mission times. In addition, it can be used to solve models with infinite state spaces and not uniformly bounded transition rates. Recently, it has been proposed the combination of adaptive and standard randomization [9]. Another recent proposal to speed up the randomization method when X is irreducible is steady-state detection [6]. Recently, a method based on steady-state detection which gives error bounds has been developed [14]. Regenerative randomization [1, 2] is another recent proposal.

In this paper we develop a variant of the regenerative randomization method described in [1, 2]. The state space of the CTMC X is assumed to be $\Omega = S \cup \{f_1, f_2, \ldots, f_A\}$, where f_i are absorbing states and all states in S are strongly connected and have paths to f_i (for $A = 0$ X is irreducible). We will assume $P[X(0) = f_i] = 0$, $1 \leq i \leq A$. In addition, we assume a reward rate structure $r_i \geq 0$, $i \in \Omega$, with different reward rates assigned to the A absorbing states. We consider two measures, the transient reward rate at time t, $TRR(t)$, and the mean reward rate during the interval $[0, t]$, $MRR(t)$. The rest of the paper is organized as follows. Section 2 describes the variant. Section 3 compares the variant with the original randomization method, randomization with steady-state detection, and standard randomization. Section 4 concludes the paper.

2   The New Variant

The regenerative randomization method is described in detail in [1, 2]. In that method, a transformed truncated CTMC $V_{K,L}$ (V_K) is obtained in terms of which can be expressed with a preespecified accuracy the measures $TRR(t)$ and $MRR(t)$. Regenerative randomization requires the selection of a regenerative state r and its performance will be good when r is visited often in the DTMC $\widehat{X}$ obtained by randomizing X with rate Λ. Let $\alpha_i = P[X(0) = i]$. Figure 1 illustrates the state transition diagram of $V_{K,L}$, truncated transformed model for the case $\alpha_r < 1$; the initial probability distribution of $V_{K,L}$ is $P[V_{K,L}(0) = s_0] = \alpha_r$, $P[V_{K,L}(0) = s_0'] = 1 - \alpha_r$, $P[V_{K,L}(0) = i] = 0$, $i \neq s_0, s_0'$. The reward rate structure of $V_{K,L}$ is $r_{s_k} = b(k)$, $r_{s_k'} = b'(k)$, $r_a = 0$. Then, the $TRR(t)$ and $MRR(t)$ measures for the original CTMC X can be computed with given error bound $\epsilon/2$ (the remaining $\epsilon/2$ is reserved for the solution of the truncated transformed models) as the $TRR(t)$ and $MRR(t)$ measures of $V_{K,L}$, called $TRR_{K,L}^a(t)$ and $MRR_{K,L}^a(t)$. In the particular case $\alpha_r = 1$, the truncated transformed model, V_K, has identical structure as $V_{K,L}$, except that states s_k' disappear. The approximated values for the desired measures are denoted in this case by $TRR_K^a(t)$ and $MRR_K^a(t)$.

The parameters q_i, w_i, v_i^j, q_i', w_i', $v_i'^j$, $b(k)$ and $b'(k)$ on which the truncated transformed models depend can be obtained by stepping DTMCs of about the

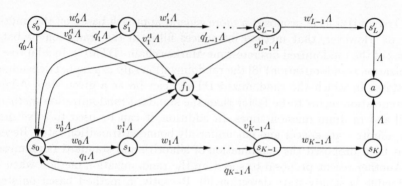

Fig. 1. State transition diagram of the CTMC $V_{K,L}$ for the case $A = 1$.

same size as $\widehat{X}$ $K + L$ steps when $\alpha_r < 1$ and K steps when $\alpha_r = 1$. The parameters w_k obey the relation $w_k = a(k+1)/a(k)$, where $a(0) = 1$. Similarly, the parameters w'_k satisfy $w'_k = a'(k+1)/a'(k)$, where $a'(0) = 1 - \alpha_r$. The parameters $a(k)$ and $a'(k)$ are easily computable while stepping the same DTMCs as above.

In the regenerative randomization method, $V_{K,L}$ or V_K is solved using standard randomization, which can be relatively expensive when Λt is large and the transformed truncated model is not significantly smaller than X. The new variant proposed here is called regenerative randomization with Laplace transform inversion and combines a closed-form solution of the $V_{K,L}$ (V_K) model in the Laplace transform domain with the usage of a numerical Laplace inversion algorithm.

2.1   Closed form solution in the Laplace transform domain

Given the structure of $V_{K,L}$, it is possible to obtain closed-form expressions for the Laplace transforms of the transient probabilities of the states of the CTMC. Details can be found in [3]. From them, it is possible to find a closed-form expression for the Laplace transform of $TRR_{K,L}^a(t)$. Let $c(k) = a(k)b(k)$ and $c'(k) = a'(k)b'(k)$, we have

$$
\widetilde{TRR}_{K,L}^a(s) = \sum_{k=0}^{K} b(k)\widetilde{p}_k(s) + \sum_{k=0}^{L} b'(k)\widetilde{p}'_k(s) + \sum_{i=1}^{A} r_{f_i}\widetilde{p}_{f_i}(s)
$$

$$
= \left[\sum_{k=0}^{K} c(k) \left(\frac{\Lambda}{s+\Lambda} \right)^k + \frac{\Lambda}{s} \sum_{k=0}^{K-1} \left(\sum_{i=1}^{A} r_{f_i} v_k^i \right) a(k) \left(\frac{\Lambda}{s+\Lambda} \right)^k \right] \widetilde{p}_0(s)
$$

$$
+ \sum_{k=0}^{L} c'(k) \frac{\Lambda^k}{(s+\Lambda)^{k+1}} + \sum_{k=0}^{L-1} \left(\sum_{i=1}^{A} r_{f_i} v_k'^i \right) a'(k) \frac{\Lambda^{k+1}}{s(s+\Lambda)^{k+1}},
$$

where

$$
\widetilde{p}_0(s) = \frac{A(s)}{B(s)},
$$

$$A(s) = 1 - \frac{s}{s+\Lambda} \sum_{k=0}^{L} a'(k) \left(\frac{\Lambda}{s+\Lambda}\right)^k - \frac{\Lambda}{s+\Lambda} \sum_{k=0}^{L-1} \left(\sum_{i=1}^{A} v_k'^i\right) a'(k) \left(\frac{\Lambda}{s+\Lambda}\right)^k$$
$$-a'(L) \left(\frac{\Lambda}{s+\Lambda}\right)^{L+1},$$

and

$$B(s) = s \sum_{k=0}^{K} a(k) \left(\frac{\Lambda}{s+\Lambda}\right)^k + \Lambda \sum_{k=0}^{K-1} \left(\sum_{i=1}^{A} v_k^i\right) a(k) \left(\frac{\Lambda}{s+\Lambda}\right)^k$$
$$+a(K)\Lambda \left(\frac{\Lambda}{s+\Lambda}\right)^K.$$

The Laplace transform of $MRR_{K,L}^a(t)$ can be easily obtained noting that $MRR_{K,L}^a(t) = \int_0^t TRR_{K,L}^a(\tau)\,d\tau/t$. Then defining $C_{K,L}(t) = tMRR_{K,L}^a(t)$ we have $\widetilde{C}_{K,L}(s) = \widetilde{TRR}_{K,L}^a(s)/s$.

In the particular case $\alpha_r = 1$ we have $\widetilde{p}_0(s) = 1/B(s)$,

$$\widetilde{TRR}_K^a(s) = \left[\sum_{k=0}^{K} c(k) \left(\frac{\Lambda}{s+\Lambda}\right)^k + \frac{\Lambda}{s} \sum_{k=0}^{K-1} \left(\sum_{i=1}^{A} r_{f_i} v_k^i\right) a(k) \left(\frac{\Lambda}{s+\Lambda}\right)^k\right] \widetilde{p}_0(s),$$

and defining $C_K(t) = tMRR_K^a(t)$, $\widetilde{C}_K(s) = \widetilde{TRR}_K^a(s)/s$.

2.2   Numerical Laplace inversion

There are several numerical Laplace inversion algorithms. We have experimented with the methods proposed in [4] and [11]. Both are based on Durbin's approximation for $f(t)$ [5]:

$$f_a(t) = \frac{1}{T} e^{at} \left[\frac{\widetilde{f}(a)}{2} + \sum_{k=1}^{\infty} \Re\left\{\widetilde{f}\left(a + \frac{ik\pi}{T}\right) e^{\frac{ik\pi t}{T}}\right\}\right], \tag{1}$$

where $i = \sqrt{-1}$. The approximation error is:

$$f_\epsilon(t) = \sum_{k=1}^{\infty} f(2kT + t) e^{-2akT}.$$

The method described in [4] takes $T = t$ and accelerates the convergence of the series of (1) using the epsilon algorithm. We have found that the method is fast, but it is sometimes unstable. On the other hand, the method described in [11], which only differs from the method described in [4] in that it takes $T = 16t$ is very stable but significantly slower. Thus, we decided to experiment with several choices for T, increasing from $T = t$ to $T = 16t$. We found that $T = 8t$ gave enough stability, and we use that selection of T.

To control the error of the Laplace inversion algorithm we proceed as follows. The total error on $TRR^a_{K,L}(t)$ $(MRR^a_{K,L}(t))$ must be $\leq \epsilon/2$. There are two kinds of errors: the approximation error and the truncation error (resulting from the truncation of the convergent series) and we allocate $\epsilon/4$ to each of them. For the measure $TRR(t)$ we have $TRR^a_{K,L}(t) \leq r_{\max}$, where $r_{\max} = \max_{i \in \Omega} r_i$, and

$$(TRR^a_{K,L})_\epsilon(t) \leq \sum_{k=1}^{\infty} r_{\max} e^{-2akT} = r_{\max} \frac{e^{-2aT}}{1 - e^{-2aT}}.$$

To bound the approximation error we take the a satisfying

$$r_{\max} \frac{e^{-2aT}}{1 - e^{-2aT}} = \frac{\epsilon}{4},$$

i.e.

$$a = -\frac{1}{2T} \log \left(\frac{1}{1 + \frac{4r_{\max}}{\epsilon}} \right).$$

Regarding the truncation error we only have control over the tolerance between consecutive approximations of the (accelerated) convergent series and we decide achieved the convergence when that difference is $\leq \epsilon/100$, i.e. we leave a factor 25 to account for the difference between the tolerance between consecutive values and the actual truncation error.

For the measure $MRR(t)$ we in fact invert $\widetilde{C}_{K,L}(s)$, where $C_{K,L}(t) = tMRR^a_{K,L}(t)$. Then, to have an error $\epsilon/2$ in $MRR^a_{K,L}(t)$ we must require an error $t\epsilon/2$ in the inversion of $\widetilde{C}_{K,L}(s)$. We allocate $t\epsilon/4$ for the approximation error and $t\epsilon/4$ for the truncation error, with a factor 25 as before. We have $C_{K,L}(t) \leq r_{\max}t$ and

$$(C_{K,L})_\epsilon(t) \leq \sum_{k=1}^{\infty} r_{\max}(2kT + t)e^{-2akT} = r_{\max}t \sum_{k=1}^{\infty} e^{-2akT} + 2r_{\max}T \sum_{k=1}^{\infty} ke^{-2akT}$$

$$= r_{\max}t \frac{e^{-eaT}}{1 - e^{-2aT}} + 2r_{\max}T \frac{e^{-2aT}}{(1 - e^{-2aT})^2} = r_{\max} \frac{(t + 2T)e^{-2aT} - te^{-4aT}}{(1 - e^{-2aT})^2}.$$

To bound the approximation error we take the a with $e^{-2aT} < 1$ satisfying

$$r_{\max} \frac{(t + 2T)e^{-2aT} - te^{-4aT}}{(1 - e^{-2aT})^2} = \frac{\epsilon}{4},$$

i.e.

$$a = \frac{1}{2T} \log \left(\frac{1}{x} \right),$$

$$x = \frac{\frac{\epsilon}{2} + (t + 2T)r_{\max} - \sqrt{(\frac{\epsilon}{2} + (t + 2T)r_{\max})^2 - (\frac{\epsilon}{4} + tr_{\max})\epsilon}}{\frac{\epsilon}{2} + 2tr_{\max}}. \tag{2}$$

Regarding the truncation error we only have control over the tolerance between consecutive approximations of the (accelerated) convergent series and we decide achieved the convergence when that difference is $\leq \epsilon t/100$. Expression (2) has severe cancellation errors when $y = \sqrt{(\epsilon/4 + tr_{\max})\epsilon}/(\epsilon/2 + (t + 2T)r_{\max}) \ll 1$. The problem can be solved by taking the Taylor series on y and use it when y is small, say $y < 10^{-3}$. In that case, the Taylor series gives

$$x \approx \frac{(\frac{\epsilon}{4} + tr_{\max})\epsilon}{(\frac{\epsilon}{2} + (t + 2T)r_{\max})(\frac{\epsilon}{2} + 2tr_{\max})}.$$

Up to now we have implicitly consider the case $\alpha_r < 1$. The case $\alpha_r = 1$ is treated identically. Efficient algorithms to compute the Laplace transforms required by the inversion algorithm can be found in [3].

3   Analysis and Comparison

In this section we analyze the performance of the proposed variant of the regenerative randomization method (RRL). For irreducible models ($A = 0$) we will compare RRL with the original regenerative randomization method (RR) and randomization with steady-state detection (RSD) [14]. For models with absorbing states ($A \geq 1$), we will compare RRL with RR and standard randomization (SR). The analysis and comparison will be made using dependability models of a level 5 RAID architecture [10]. The models we will consider are similar to a model described in [13]. Our models consider hot spares for controllers, which were not considered in [13] and encompass availability measures.

Figure 2 shows the architecture of the considered level 5 RAID system. The system includes $G \times N$ disks and N controllers. The disks are organized in G parity groups, each with N disks. Each controller controls a string of G disks. The system also includes C_{H} hot spare controllers and D_{H} hot spare disks. The system is operational if there is access to at least $N - 1$ available disks of each parity group. When there is a failed controller all disks of the associated string become unavailable. When a failed disk is replaced by a good one and if all disks of the parity group are available, the parity group starts the reconstruction of data in the replaced disk. The reconstruction process also starts when a disk of a parity group which was not available due to failure of one controller becomes available due to the replacement of the failed controller. All disks of the parity group involved in a reconstruction are "overloaded" and have a higher failure rate. Non-overloaded disks fail with rate λ_{D}. Overloaded disks fail with rate λ_{S}. Controllers fail with rate λ_{C}. The reconstruction process has an exponential duration with rate μ_{DRC}. Failed disks and controllers are replaced, if respective hot spares are available, by a repairman with rates μ_{DRP} and μ_{CRP}, respectively, with priority given to controllers. Lacking spares and failed disks and controllers for which there are not spares are replaced with rate μ_{SR} by an unlimited number of repairmen. A reconstruction process is successful with probability P_{R}. Failure in a reconstruction process causes the failure of the system. Finally, when the system is failed, it is returned to its original state, with all disks and hot spares available, by a global repair action which has rate μ_G. The exact model gives very large CTMCs for moderate values of G and N. Instead, we will use a pessimistic approximated model giving CTMCs with much smaller size. Unavailable disks are said to be aligned if they belong to the same string. The approximation consists in assuming that if unavailable disks are not aligned, when one of them becomes available the remaining disks would still be unaligned whenever their number is ≥ 2. Using that approximation it is possible to describe the state of the CTMC using the following state variables: NFD (number of failed disks), NDR (number of disks under reconstruction), NWD (number of disks waiting for reconstruction), NSD (number of hot spare disks), AL (a boolean variable which is YES when unavailable disks are aligned and NO otherwise), NFC (number of failed controllers), NSC (number of hot spare controllers), and F (a boolean value which is YES when the system is failed and NO otherwise). We will vary the model parameters G, C_H, and D_H and will fix the other parameters of the model to the values: $N = 5$, $\lambda_D = 10^{-5}$, $\lambda_S = 2 \times 10^{-5}$, $\lambda_C = 5 \times 10^{-5}$, $\mu_{DRC} = 1$, $\mu_{DRP} = 4$, $\mu_{CRP} = 4$, $\mu_{SR} = 0.25$ and $\mu_G = 0.25$, with all rates in h^{-1}. We will consider two measures. The first them is a particular case of $TRR(t)$ when the model is irreducible $(A = 0)$, is point unavailability $UA(t)$ and is obtained by assigning a reward rate 0 to the operational states and a reward rate 1 to the failed state. The second of them is a particular case of $TRR(t)$ when the model has absorbing states, is the unreliability $UR(t)$, and is obtained by making the system failed state absorbing (and thus $A = 1$) and assigning a reward rate 1 to the absorbing state and a reward rate 0 to the transient states. For all measures we will assume that the initial state is the state without failed components and all hot spares available, which will be taken as regenerative state for the methods RR and RRL. For all methods we will take $\epsilon = 10^{-12}$, which gives enough accuracy for all measures and values of t.

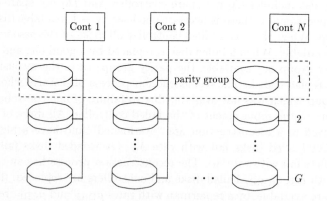

Fig. 2. Architecture of the considered level 5 RAID system.

To analyze the performance of the proposed RRL method and compare it with that of RR, RSD and SR we will use two instances of the parametric model defined by $C_H = 1$, $D_H = 3$ and $G = 20$ for one instance and $G = 40$ for the second instance. The irreducible models have 3,841 states and 24,785 transitions for $G = 20$ and 14,081 states and 94,405 transitions for $G = 40$. The models with absorbing state have the same number of states and one transition less. We start comparing RRL, RR, and RSD for the measure $UA(t)$. Table 1 gives the number of steps required by RRL and RR (both require the same number of steps) and RSD for several values of t. Figure 3 plots the corresponding CPU times. We can note that RRL and RR require fewer steps than RSD up to a certain value of t. Regarding CPU times, there are crosspoints between RR and RRL in one hand and RSD on the other hand. RRL is about as fast as RSD and significantly faster than RR for large t. The numerical Laplace transform inversion is fast and consumes a very small percentage of the time of the RRL method (about 2 % for the example with $G = 20$ and 1 % for the example with $G = 40$). The number of required abscissae varied from 105 to 329.

Table 1. Number of steps required by RR, RRL and RSD for the measure $UA(t)$ for several values of t.

t (h)	$G = 20$		$G = 40$	
	RR/RRL	RSD	RR/RRL	RSD
1	56	66	86	99
10	323	355	554	594
100	2,234	2,612	4,187	4,823
1,000	2,708	2,612	5,123	4,823
10,000	2,938	2,612	5,549	4,823
100,000	3,157	2,612	5,957	4,823

We next compare RRL, RR and SR using the example with the measure $UR(t)$. Table 2 and Figure 4 give the results. For small t, SR is slightly faster than both RR and RRL. Similarly, for models with $A = 0$ such as the one considered previously, SR should be slightly faster than RRL, RR and RSD. We can note that SR is extremely expensive for large t. For those t, RR performs better than SR and the fastest method is the proposed RRL, which outperforms RR significantly. We note that for the largest t considered ($t = 100,000$ h), $UR(t)$ is 0.50480 for the model with $G = 20$ and 0.74750 for the model with $G = 40$. Thus, the selection $\epsilon = 10^{-12}$ is a very stringent one and translates to require about 14 digits of accuracy to the numerical Laplace inversion algorithm. Thus, that algorithm seems to be very stable.

4   Conclusions

We have proposed a new variant of the regenerative randomization method for the transient analysis of dependability/performability models. For irreducible

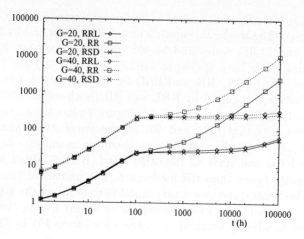

Fig. 3. CPU times in seconds required by RRL, RR and RSD for the measure $UA(t)$ as a function of t.

Table 2. Number of steps required by RR, RRL and SR for the measure $UR(t)$ for several values of t.

t (h)	$G = 20$		$G = 40$	
	RR/RRL	SR	RR/RRL	SR
1	56	65	86	98
10	323	354	554	593
100	2,233	2,726	4,186	4,849
1,000	2,708	24,844	5,122	45,234
10,000	2,937	240,958	5,547	442,203
100,000	3,157	2,386,068	5,955	4,390,141

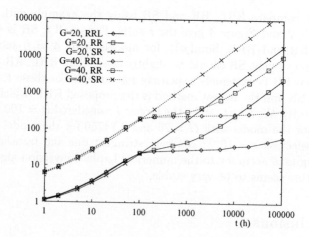

Fig. 4. CPU times in seconds required by RRL, RR and SR for the measure $UR(t)$ as a function of t.

models, the new variant seems to be about as fast as randomization with steady-state detection and, for large t and models of moderate size, significantly faster than the original regenerative randomization. For models with absorbing states and large t, the new variant is much faster than standard randomization and significantly faster than the original regenerative randomization for models of moderate size.

References

1. J. A. Carrasco, "Solving Large Reliability Models of Repairable Fault Tolerant Systems using Regenerative Randomization," Technical Report DMSD_99_2, Universitat Politècnica de Catalunya, February 1999, available at `ftp://ftp-eel.upc.es/techreports`.

2. J. A. Carrasco, "Efficient Computation of Bounds for Transient Performability Measures using Regenerative Randomization," Technical Report DMSD_99_4, Universitat Politècnica de Catalunya, May 1999, available at `ftp://ftp-eel.upc.es/techreports`.

3. J, A. Carrasco, "Transient Analysis of Dependability/Performability Models by Regenerative Randomization with Laplace Transform Inversion," Technical Report DMSD_2000_1, Universitat Politècnica de Catalunya, May 1999, available at `ftp://ftp-eel.upc.es/techreports`.

4. K. S. Crump, "Numerical Inversion of Laplace Transforms Using a Fourier Series Approximation," *Journal of the ACM*, vol. 23, pp. 89–96, 1976.

5. F. Durbin, "Numerical inversion of Laplace transforms: an efficient improvement to Dubner and Abate's method," *Computer Journal*, vol. 17, no. 4, 1974, pp. 371–376.

6. M. Malhotra, J. K. Muppala and K. S. Trivedi, "Stiffness-Tolerant Methods for Transient Analysis of Stiff Markov Chains," *Microelectron. Reliab.*, vol. 34, no. 11, 1994, pp. 1825–1841.

7. D. R. Miller, "Reliability Calculation using Randomization for Markovian Fault-Tolerant Computing Systems," *in Proc. 13th IEEE Int. Symp. on Fault-Tolerant Computing, FTCS-13*, June 1983, pp 284–289.

8. A. P. Moorsel and W. H. Sanders, "Adaptive Uniformization", *Communications in Statistics - Stochastic Models*, vol. 10, no. 3, 1994, pp. 619–647.

9. A. P. Moorsel and W. H. Sanders, "Transient Solution of Markov Models by Combining Adaptive and Standard Uniformization," *IEEE Trans. on Reliability*, vol. 46, no. 3, 1997, pp. 430–440.

10. D. A. Patterson, P. M. Chen, G. A. Gibson, and R. H. Katz, "Introduction to redundant arrays of inexpensive disks (RAID," in *Proc. IEEE Computer Society International Conference (COMPCON)*, 1989, pp. 112–117.

11. R. Piessens and R. Huysmans, "Algorithm 619: Automatic Numerical Inversion of the Laplace Transform [D5]," *ACM Transactions on Mathematical Software*, vol. 10, no. 3, pp. 348–353, September 1984.

12. A. Reibman and K. S. Trivedi, "Numerical Transient Analysis of Markov Models," *Comput. Operations Res.*, vol. 15, 1988, pp. 19–36.

13. V. Santoja, M. Alonso, J. Molero, J. J. Serrano, P. Gil, and R. Ors, "Dependability Models of RAID using Stochastic Activity Networks," *Dependable Computing-EDCC-2*, Lecture Notes in Computer Science, 1996.

14. B. Sericola, "Availability Analysis of Repairable Computer Systems and Stationarity Detection," *IEEE Trans. on Computers*, vol. 48, no. 11, November 1999, pp. 1166–1172.

FANTOMAS
Fault Tolerance for Mobile Agents in Clusters

Holger Pals, Stefan Petri, and Claus Grewe

Medical University of Lübeck
Institute of Computer Engineering
Ratzeburger Allee 160, D-23538 Lübeck
{pals, petri, grewe}@iti.mu-luebeck.de

Abstract. To achieve an efficient utilization of cluster systems, a proper programming and operating environment is required. In this context, mobile agents are of growing interest as base for distributed and parallel applications. As mobile and autonomous software units, mobile agents can execute tasks given to the system and allocate independently all the needed resources. However, with growing cluster sizes, the probability of a failure of one or more system components and therewith the loss of mobile agents rises. While fault tolerance issues for applications based on "traditional" processes have been extensively studied, current agent environments provide only insufficient, if at all, extensions for a capable reaction on such kinds of failures.

We examine fault tolerance with regard to properties and requirements of mobile agents, and find that independent checkpointing with receiver based message logging is appropriate in this context. We derive the FANTOMAS (**Fa**ult-Tolerant approach for **Mo**bile **A**gent**s**) design which offers a user transparent fault tolerance that can be activated on request, according to the needs of the task. A theoretical analysis examines the advantages and drawbacks of FANTOMAS.

1 Introduction and Motivation

Clusters of standard components, PCs or workstations, can (almost) reach the performance of "real" parallel computers, but at much better price-performance ratio. They have already entered the Top 500 list of supercomputers.

The agent concept, originally developed in the domain of artificial intelligence, attracts growing interest in many application areas. According to a simple definition [20], *mobile agents* are active and autonomous software entities, which are able to migrate through the network to get from one node to the other. During migration, a mobile agent keeps its program code, data and possibly execution state. Despite their autonomy, mobile agents are no loners but they try potentially to reach common goals by interacting. Furthermore, mobile agents are proactive, reactive and cognitive [26]. So, mobile agents can take subtasks of the parallel application and travel independently within the cluster to allocate all the resources that are needed to perform their tasks. Furthermore, they can respond to dynamically changing load situations in the cluster by traveling from highly loaded nodes to lightly loaded ones. Therefore, not the user but the mobile agents are responsible for managing the cluster utilization. Also, migration has been shown to be more light weighted than with "traditional" programs [13,5].

J. Rolim et al. (Eds.): IPDPS 2000 Workshops, LNCS 1800, pp. 1236-1247, 2000.

All these capabilities predestine mobile agents for many applications in distributed environments, e. g. data collecting, searching and filtering, network management, entertainment, e-commerce or parallel processing [2, 24]. One example application we are working on is mapping of data flow graphs for mobile robot control onto agents [8]. In contrast to alternative implementations, where in general only data can be transported to processing units, mobile agents bring operators to the sources of data. This can improve the efficiency and simultaneously reduce the network traffic in a drastic manner [4].

Cluster, and therefore parallel applications running on them, are very susceptible for failures of components of the cluster. However, programmers and users of distributed applications are interested in their algorithms and solutions. They expect fault tolerance as a service from the underlying run time system. These considerations show the necessity for a design, which enables user-transparent fault tolerance in agent environments. Current agent systems, and also the underlying operating systems, provide this feature only insufficiently, if at all.

In this paper we introduce an approach for such a design. It can be applied to different agent systems, if they fulfill certain requirements as discussed below. The approach (FANTOMAS, Fault-Tolerant approach for mobile agents) offers a user-transparent fault tolerance which can be selected by the user for every single application given to the environment. Thereby, the user can decide for every application wether it has to be treated fault-tolerant or not.

In the next sections, we briefly introduce agent systems, and give overview of related work. Section 3 examines fault tolerance with regard to properties and requirements of agent environments.

2   Related Work: Fault Tolerance for Mobile Agents

Current operating systems do not have support for mobile agents. Thence, special Agent systems are needed to provide suitable infrastructure as middleware beween mobile agents and operating system. The particular functions agent systems have to provide can be found in [9]. [17] presents an overview of existing agent systems.

In the area of mobile agents, only few work can be found relating to fault tolerance. Most of them refer to special agent systems or cover only some special aspects relating to mobile agents, e. g. the communication subsystem. Nevertheless, most people working with mobile agents consider fault tolerance to be an important issue [5, 25].

Johansen et al. [7] detect and recover from faulty migrations inside the TACOMA [6] agent system. When an agent migrates, a *rear-guard agent* is created that stays on the origin node. It monitors the migrated agent on the destination node. This very simple concept does not tolerate network partitioning.

In the scope of the TACOMA project, Minsky et al. [10] propose an approach based on an itinerary concept, i. e. a plan, which nodes the mobile agent has to visit and what it has to do there. They assume that the itinerary is known at start time and the order of visited nodes is fixed. Fault tolerance is obtained by performing every itinerary step on multiple nodes concurrently and sending the results (resp. the Mobile Agent) to all nodes of the next step. The majority of the received inputs from the last step becomes the input of the new task for this step and so on. Thus, a certain number of faults can be tolerated in each step. Disadvantages of this fault tolerance concept are the very inflexible description of the itinerary and the simple model of Mobile Agents. For example, communication between different mobile agents is not included in this concept, so it is not suited for distributed or parallel applications.

Strasser and Rothermel present a more flexible itinerary approach for fault tolerance [22] within the Mole system [1]. Independent items in this enhanced itinerary can be reordered. Each itinerary stage comprises the action that has be to done on one node. When the mobile agent enters a new stage by migrating to the next node of the itinerary, called a *worker* node, it is also replicated onto a number of additional nodes, called *observers*. If the *worker* becomes unavailable (due to a node or network fault), the *observer* with the highest priority is selected as the new worker by a special *selection protocol*. A *voting protocol* ensures the abortion of wrong multiple *workers* in case of a network fault. The *voting protocol* is integrated into a 2-phase commit protocol (2PC) that encloses every stage execution. These protocols cause a significant communication overhead. Just as in the work of Minsky et al., nothing is said about the interaction between different mobile agents that are executed concurrently.

Vogler et al. [23] introduce a concept for reliable migrations of mobile agents based on distributed transactions. The migration protocol is derived from known transaction protocols like the already mentioned 2PC. Besides the migrations, no other fault tolerance aspects of mobile agents are treated.

Murphy and Picco [12] also treat only a special aspect of agent systems, the problem of messages that never reach their destination because of the high agent mobility, not because of failures in the agent system or the network.

These problems can arise in almost every message delivery scheme. Their approach is based on the Chandy-Lamport algorithm for consistent global snapshots [3]. This algorithm is used to force all messages to be delivered at least once. The network graph can grow dynamically, but nodes that fail at runtime cannot be removed. Due to the underlying snapshot algorithm, messages can be initiated at any time, but delivery at each node is restricted to one message per source.

Concordia [11], a commercial agent system from the Mitsubishi Electric Information Technology Center America (ITA), advertises fault-tolerant execution of mobile agents, but no further information about the tolerated failures and their treatment is available.

3   Concepts for a Fault Tolerance Approach for Mobile Agents

As shown in the previous sections, despite the generally agreed-upon necessity, existing agent systems contain only limited provisions for fault tolerance, if at all. Especially treatment of node or agent failures with support for communicating agents is covered only insufficiently. However, such support is essential for distributed and/or parallel applications. This section discusses possibilities and approaches to augment an agent system to achieve fault tolerance, with focus on these fault and application types.

First, the goals are described, then the fault model for an agent system is explained, and the faults examined with respect to their occurrence probability and treatment in existing systems. From this, a set of faults is determined, for which further treatment is still needed. After that, an overview over fault tolerance approaches in known environments is given, and examined, if and how they are suited for mobile agents. From these investigations, the FANTOMAS concept is developed.

3.1   Goals and Requirements

The goals for a fault tolerance approach for mobile agents result from the the properties of agents and agent systems, and from the intended application area, parallel computation.

- **Support for communicating agents:** this is essential for parallel distributed applications, but not covered by some of the related work.
- **Autonomy:** If the actions to achieve fault tolerance was dictated by a supervisor instance, the autonomy were limited. All decisions that are made by an agent according to its normal autonomous behavior would need to be coordinated with such a supervisor. Besides fault tolerance, also other aspects, e. g. load management, would be involved, because otherwise they could contradict the decisions of the fault tolerance supervisor.
- **Fault tolerance as optional feature:** Not every application in an agent environment is that important that it requires fault-tolerant execution. The user, the agent environment, or the application itself, should be able to decide individually, if and when fault tolerance is to be activated. Also, the activation of fault tolerance for one application should have no influence on other already running applications, so that the concurrent execution of fault-tolerant and non-fault-tolerant applications is possible.
- **User transparency:** Programmers and users of applications are usually interested in coding their algorithms and getting their problems solved. They do not want to care for fault tolerance in the applications. Thus, to enable fault-tolerant execution, it should not be necessary to change the application code.
- **Efficiency:** One major goal of using mobile agents for execution of parallel applications in cluster systems is efficient resource usage. Thus, fault tolerance measures should have a very low overhead during the fault free operation. This is also one more reason to require fault tolerance as optional feature. Also, of course, recovery after a fault should be fast.
- **No modifications to hardware, operating system, or run time environment:** The necessity for such modifications would limit the usability of fault tolerance measures to dedicated system environments, and portability would be severed.
- **Portability and reusability:** Changes to an existing agent environment should be an augmentation, not a redesign or reimplementation. Already existing function units should remain available at least for the non-fault-tolerant execution of applications. Reusing existing function units also guarantees that existing applications stay usable.

From these goals, we can already derive some requirements for the selection of an agent system, which can be used as base for a realization of the FANTOMAS concept:

- **Modular, exchangeable architecture:** With a monolithic agent kernel, it would not be possible to adapt specific functionalities to the requirements of fault tolerance measures. To enable activation of fault tolerance properties during run time, the complete agent would have to be replaced with one that already carries those functionalities with it, all the time. This would increase memory and computing time demands even in the case of non-fault-tolerant execution. To avoid this disadvantage, a modular composition of mobile agents is required, in which the functionalities are contained in specific modules which can be exchanged at run time.

- **Separation between application and agent kernel:** The partial tasks of a parallel application, which are submitted to the agent environment, should be integrated seamlessly into the modular composition of the environment. The separation between application and agent kernel, on one hand, facilitates user transparency, and, on the other hand, makes it possible that one agent can execute different applications modules in its life cycle. Nevertheless, the application must have the possibility to influence the agents behavior.

- **Function modules working in parallel:** The required modularity and separation between application and the rest of the agent suggests that the functional units should be contained in parallel working modules. This enables independent and concurrent execution of different strategies, e. g. concerning migration or resource usage. This increases the agent's degrees of reactivity and proactivity.

- **Adaptivity:** It must be possible to influence the behavior of a mobile agent during run time. Without this possibility, adding fault tolerance (e. g. on demand from the user) would not be possible.

- **Automatic detection of dependencies:** Some of the functional units in a mobile agent expect certain functionality and services from other units, which are mutually usable. For example, the fault tolerance unit could initiate a migration by using a service of the migration unit. As explained above, each agent should contain only those modules, that are actually needed. However, when a module is exchanged, it might happen that the new module expects more services than are already present in the agent. To maintain transparency, such dependencies must be detected and satisfied automatically.

3.2   Fault Model

Several types of faults can occur in agent environments. Here, we first describe a general fault model, and focus on those types, which are for one important in agent environments due to high occurrence probability, and for one have been addressed in related work only insufficiently.

- **Node failures:** The complete failure of a compute node implies the failure of all agent places and agents located on it. Node failures can be temporary or permanent.
- **Failures of components of the agent system:** Failures of agent places, or components of agent places become faulty, e. g. faulty communication units or incomplete agent directory. These faults can result in agent failures, or in reduced or wrong functionality of agents.

- **Failures of mobile agents:** Mobile agents can become faulty due to faulty computation, or other faults (e. g. node or network failures).
- **Network failures:** Failures of the entire communication network or of single links can lead to isolation of single nodes, or to network partitions.
- **Falsification or loss of messages:** These are usually caused by failures in the network or in the communication units of the agent systems, or the underlying operating systems. Also, faulty transmission of agents during migration belongs to this type.

Especially in the intended scenario of parallel applications, node failures and their consequences are important. Such consequences are loss of agents, and loss of node-specific resources. In general, each agent has to fulfill a specific task to contribute to the parallel application, and thus, agent failures must be treated. In contrast, in applications where a large number of agents are sent out to search and process information in a network, the loss of one or several mobile agents might be acceptable.

In the following, we assume that only one agent place is located on each network node, which is typical for parallel execution environments. Generalization to multiple agent places per node is easily derived. Also, currently only crash faults are regarded. The reason is the observation that other node faults are significantly less frequent, but require much higher effort to detect.

Failures of the entire network occur relatively rarely, and standard network protocols provide tolerance against loss or falsification of messages. Thus, those fault types are not regarded in the following. Treatment of network partitioning will be addressed in future work.

An 1-fault assumption is used, i. e., during the recovery after a node failure, no second node failure does occur. After recovery, further node failures can be tolerated, until only one node remains intact. Tolerating more than one simultaneous node failure would require using additional resources, which conflicts with the efficiency requirements. Also, the intended short recovery times yield a very low probability for the occurrence of further failures in that phase.

It is further assumed that failed nodes stay unavailable, since repair and re-use is not generally possible. Many agent systems or environments do not support dynamic reconfiguration during run time.

Also, intermittent faults are not regarded. They are difficult to diagnose, and occur with only very low probability.

Failures of single agents, independent from node failures, are regarded and should be tolerated. Here, more specific, the failures of agents which are currently being executed on a node are regarded. In contrast, the problem of agent failures during migration was already investigated in other work [23] and is not regarded here.

Further, correct communication is assumed. Messages are not falsified or intercepted during their transmission. These faults are already investigated in the literature, e. g. [12].

3.3 Discussion of Fault Tolerance Methods

Fault tolerance methods for distributed applications have been widely studied in the context of "traditional" processes. As shown above, agent environments and mobile agents have some special properties and requirements, which must be regarded, but also can be exploited, when fault tolerance is introduced.

Recovery can be divided into *backward recovery* and *forward recovery*. Forward recovery avoids the rollback and re-execution phase and thus can provide fast recovery with (almost) seamless computation progress. However, for our primarily intended application area, this is not so important. To avoid the overhead that is incurred by forward recovery during fault-free execution, we choose backward recovery.

The state of an agent-based distributed application consists of the states of the agents, and of the communication links between them. When saving the state of the communication links, the *Domino Effect* [18] has to avoided. Because of the high dynamicity of agent systems, *independent checkpointing* techniques are beneficial against *coordinated checkpointing* algorithms.

To avoid the Domino Effect with these techniques, the exchanged messages must be logged, so that they can be regenerated during the re-execution phase. Though receiver based logging has more time overhead during fault-free operation, its advantage against sender based logging is faster recovery (see also e. g. [19]). Also, important for agents, recovery and pruning of the message log can be done autonomously, without interaction with other user agents.

3.4   The FANTOMAS Concept

From these considerations, we choose independent checkpointing with receiver based logging as base for our fault tolerance approach for mobile agents. Adhering to the agent paradigm, and exploiting the already available facilities of the mobile agent resp. the agent environment, an agent is used as the stable storage for the checkpointed state and the message log. For each mobile agent (called *user agent* in the following), for that fault tolerance is enabled, a *logger agent* is created. A user agent and its logger agent form an agent pair[1] (figure 1). The logger agent does not participate actively in the application's computation, and thus needs only a small fraction of the available CPU capacity. It follows the user agent at a certain, non-zero, distance on its migration path through the system. They must never reside on the same node, so that not a single fault destroys both of them. User and logger agent monitor each other, and if a fault is detected by one of them, it can rebuild the other one from its local information.

The creation of the agent pair is readily derived from the already existing migration facilities. To create a logger agent, the user agent serializes its state in the same way as for a migration, and sends it to a remote agent place. There, a new agent is created from this data. Different from migration, the new agent does not start the application module that was sent with the state information, and the user agent continues normal execution. Further, the communication unit of the agent is exchanged against a version that first forwards each incoming message to the logger agent before delivering it locally.

The checkpointing strategy in the fault tolerance unit is responsible for chosing apropriate times to save checkpoints. Besides doing it in periodic intervals, migration is a convenient occasion, since it already involves serialization of the agent state.

---

[1] Note that generalization to agent groups can also be derived.

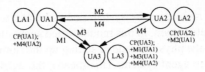

Fig. 1. Example for an application with three user and logger agents with checkpoints (CP) and messages (M) [16]

This approach supports tolerance against node and agent failures, even for communicating agents, and thus for distributed applications, and it does not rely on a predetermined itinerary for the agents.

3.5 Diagnosis

Following the agent concept, diagnosis should be done autonomously and co-operatively. A user transparent, generally applicable approach implies that the fault tolerance unit of an agent does not have further knowledge about the application, especially about the meaning of its outputs, and thus cannot check them for correctness. So far, the user and logger agents of an agent pair monitor each other through I-am-alive messages and timeouts to detect crash faults.

To exploit application knowledge and enable detection of erroneous computation results, the modularity enables the optional addition of acceptance tests to the application, if the programmer cares to provide them. Via an interface between agent kernel, resp. the fault tolerance unit in the agent kernel, and the application module, the acceptance tests are invoked, and their results given to the fault tolerance unit.

Modification of the standby logger agent to a hot spare, and comparison of messages or checkpoints to detect erroneous computation via differences in the output is possible. However, when this is done in a user transparent way, this requires that no location dependent information (network addresses, time stamps, ...) must be contained in the output [15]. Since the serialization is left to the agent system, this requirement cannot be generally assumed as fulfilled. Thus, this approach is currently not pursued further.

4 Analytic Evaluation

This section presents first analytic evaluation results of the FANTOMAS approach based on a Markov model [21]. For the model, the system behavior is abstracted as follows:

The cluster system consists of n nodes. On each node, one agent place exists. Failures of these places caused by failures of the physical node or failures of the runtime environment (e. g. the agent system) will be tolerated. After such a failure, all affected user and logger agents will be recovered. During this recovery time the system is in an insecure state and a second failure generally cannot be tolerated. This assumption is very pessimistic, because a second failure during the recovery could be tolerated if from every agent pair at least one agent is not effected from these failures. The detection of this would need an additional protocol step and is therefor not regarded here.

After completion of all recoveries the system comes back into a secure state and the next failure will be tolerated again. In this model, failed agent places are not assumed to be repaired by repairing the physical node or setting up the runtime environment again. Repairs only apply to the agent system. This means, the agent system is repaired in such way that all affected mobile agents are recovered and all running applications are executed on the remaining agent places.

Two different sets of system states can be distinguished:

1. The states s_j, $j \in [0, 2, \ldots, 2n - 4]$ are the secure system states after $\frac{j}{2}$ failures of agent places.
2. The states s_j, $j \in [1, 3, \ldots, 2n - 3]$ characterize the insecure system states during the recovery after $\frac{(j+1)}{2}$ failures of agent places.

In principle, the state s_{2n-2} can be counted as a secure system state. However, in this state only one agent place remains and therefore no other failure can be tolerated. The last state (s_{2n-1}) is the final (absorbing) state. This state can be reached by two ways:

1. A failure of an agent place occurs in an insecure system state.
2. A failure of the last remaining agent place occurs.

The rate of failures of agent places is denominated by λ, the rate of repair by μ. Therewith, the transition rate from a secure state s_j, $j \in [0, 2, \ldots, 2n - 4]$ to the insecure state s_{j+1} is $\left(n - \frac{j}{2}\right) \cdot \lambda$, because $\frac{j}{2}$ agent places have already failed and $\left(n - \frac{j}{2}\right)$ agent places that can fail are left. The transition rate from an insecure state s_j, $j \in [1, 3, \ldots, 2n - 3]$ to the following secure state occurs with transition rate μ. The transition rate from an insecure state s_j, $j \in [1, 3, \ldots, 2n - 3]$ to the final state is $\left(n - \frac{(j+1)}{2}\right) \cdot \lambda$, because $\frac{(j+1)}{2}$ agent places already failed and $\left(n - \frac{(j+1)}{2}\right)$ agent places can fail during this recovery phase. Figure 2 shows this Markov model.

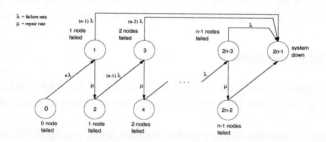

Fig. 2. Markov model of the FANTOMAS approach

Let $P(t)$ be the vector of probabilities $P_i(t)$, $i = 0, \ldots, 2n - 1$, for the system being in state i at time t. $R(t)$ is the reliability of the system, i. e. the probability of not being in state s_{2n-1} at time t, $R(t) = P_{2n-1}(t)$. To compute it, the Laplace transform is used, with $\widetilde{P}(s)$ denoting the transform of vector $P(t)$, and I the identity matrix. In general, $\widetilde{P}(s) = (s \cdot I - Q)^{-1} \cdot P(0)$. At start time $t = 0$, the system is in state s_0. The transition matrix Q is obtained from the given Markov model. With this knowledge, we do not need to invert $(s \cdot I - Q)$ entirely, but calculate only the transform of being in the final state, $\widetilde{P}_{2n-1}(s)$. It can be calculated recursively and results to:

$$\widetilde{P}_{2n-1}(s) = 1 + \left[\frac{1}{s} \sum_{i=1}^{n-1} \left(-(n-1)\,\lambda\,\frac{\frac{n!}{(n-i)!}\,\lambda^i\,\mu^{(i-1)}}{A(i)\,B(i)} \right) - \frac{n!\,\lambda^n\,\mu^{(n-1)}}{(s^2 + s\lambda)\,A(n-1)\,B(n-1)} \right]$$

with:

$$A(l) := \prod_{j=1}^{l} (s + (n+1-j)\,\lambda), \quad B(l) := \prod_{j=1}^{l} (s + (n-j)\,\lambda + \mu).$$

The reverse transformed $P_{2n-1}(t)$ cannot be written in a closed form. Here, we show curves for different parameter sets.

Figure 3 compares the reliability of a simplex system with a system using the FANTOMAS concept, with varied number of nodes. The assumed MTTF of 1000 hours for one agent place can be regarded as a realistic estimation. The MTTR of 2 minutes was chosen after first measurements with the exemplary implementation of the FANTOMAS concept within the FLASH environment. A significantly better reliability for the system using the FANTOMAS concept can be seen, increasing with rising system sizes.

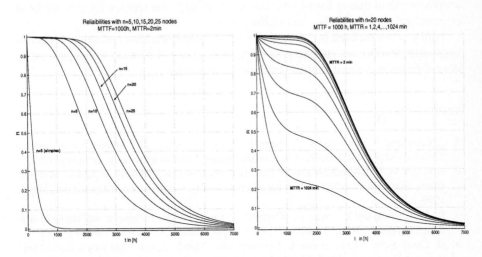

Fig. 3. Comparison of systems with varied size (MTTF=1000h)   **Fig. 4.** Comparison of system behavior with varied MTTR

Additional system components like the agent system or other used middleware can raise the probability of a failure of an agent place, leading to a worse MTTF. Remembering the possible reasons for a total system shutdown mentioned above, the system's reliability is determined by both kinds of failures. However, decreasing MTTF/MTTR ratios heighten the danger of two or more failures of agent places within a temporal distance less than the MTTR, leading to a worse reliability in the initial phase. After a certain time, the reliability is more influenced by failures of the second kind and the reliability graphs look like those in figure 3. Figure 4 shows this effect of the relation between the two possibilities of system failures. It shows the reliabilities of systems with varied MTTR, because rising MTTR values lead to the same effect.

5   Conclusions and Future Work

We have motivated the use of mobile agents as base for parallel and distributed applications in clusters, and shown the need for user transparent fault tolerance measures for such application environments. From a discussion of the possible fault types, their coverage in related work, and their relevance for agent environments, this work's focus on node and agent failures was derived. General possibilities for achieving fault tolerance in such cases were regarded, and their respective advantages and disadvantages for mobile agent environments, and the intended parallel and distributed application scenarios shown. This leads to an approach based on warm standby and receiver side message logging. It can exploit and augment the facilities, which are already present in the agent to provide the mobility. The analytical evaluation of the approach with a Markov model shows that our approach is feasible and beneficial. First measurements on a prototype implementation within the FLASH environment have been performed and showed a notable overhead during fault-free execution. Currently, the reasons for this overhead are examined closer with the aim to reduce it.

Further work includes the investigation of user transparent hot replication for better diagnosing possibilities and the possibility to employ roll-forward techniques. Also, forward recovery without explicit replication is examined.

References

1. J. Baumann, F. Hohl, and K. Rothermel. Mole – Concepts of a Mobile Agent System. Technical Report TR-1997-15, Universität Stuttgart, Fakultät Informatik, Germany, 1997.
2. L. F. Bic, M. B. Dillencourt, and M. Fukuda. Mobile Network Objects. In John G. Webster, editor, *Encyclopedia of Electrical and Electronics Engineering*. John Wiley & Sons, March 1999.
3. K. M. Chandy and L. Lamport. Distributed snapshots: Determining global states of distributed systems. *ACM Transactions on Computer Systems*, 3(1):63–75, February 1985.
4. D. Chess and C. Harrison and A. Kershenbaum. Mobile Agents: Are they a Good Idea? Research Report RC 19887 (88465), IBM T.J. Watson Center, March 1995.
5. M. Izatt, P. Chan, and T. Brecht. Ajents: Towards an Environment for Parallel, Distributed and Mobile Java Applications. In *Proc. ACM 1999 Conference on Java Grande*, pages 15–24, June 1999.
6. D. Johansen, R. van Renesse, and F. B. Schneider. An Introduction to the TACOMA Distributed System, Version 1.0. Report, Institute of Mathematical and Physical Science, Department of Computer Science, University of Tromsø, Norway, 1995.
7. D. Johansen, R. van Renesse, and F. B. Schneider. Operating System Support for Mobile Agents. In *Proceedings of the 5th. IEEE Workshop on Hot Topics in Operating Systems, Orcas Island, USA, May 1994*, pages 42–45, May 1995.

8. R. Kluthe, W. Obelöer, and C. Grewe. Agent-based Load Balancing for Mobile Robot Application. In *Distributed and Parallel Embedded Systems – Proc. of the Int. Workshop IFIP WG 10.3/WG 10.5, October 1998 (DIPES'98)*, pages 117–126. Kluwer Academic Press, Boston, 1999.

9. D. S. Milojičić, S. Guday, and R. Wheeler. Old Wine in New Bottles, Applying OS Process Migration Technology to Mobile Agents. In *Proceedings of the 3rd Workshop on Mobile Object Systems, 11th European Conference on Object-Oriented Programming*, June 1997.

10. Y. Minsky, R. van Renesse, F. B. Schneider, and S. D. Stoller. Cryptographic Support for Fault-Tolerant Distributed Computing. In *Proc. 7th ACM SIGOPS European Workshop*, pages 109–114. ACM Press, September 1996.

11. Mitsubishi Electric ITA, Horizon Systems Laboratory. Mobile Agent Computing. White Paper, 1998.

12. A. L. Murphy and G. P. Picco. Reliable communication for highly mobile agents. Report (WUCS-99-15), Washington University in St. Louis, 1999.

13. W. Obelöer, C. Grewe, and H. Pals. Load Management with Mobile Agents. In *Proceedings of the 24th EUROMICRO Conference*, volume II, pages 1005–1012. IEEE Computer Society, August 1998.

14. H. Pals. Parallel, Fault-Tolerant Applications with Mobile Agents. Diploma thesis, Institut für Technische Informatik der MU Lübeck, December 1999. (in German).

15. S. Petri. A Common Framework for Transparent Checkpointing, Replication and Migration in Clusters. In *ARCS'99: Architektur von Rechensystemen, Vorträge der Workshops im Rahmen der 15. GI/ITG-Fachtagung*, pages 79–88, Jena, October 1999.

16. S. Petri and C. Grewe. A Fault-Tolerant Approach for Mobile Agents. In *Dependable Computing – EDCC-3, Third European Dependable Computing Conference, Fast Abstracts*. Czech Technical University in Prague, September 1999.

17. V. A. Pham and A. Karmouch. Mobile software agents: An overview. *IEEE Communications Magazine*, 36(7):26–37, July 1998.

18. B. Randall. System Structure for Software Fault Tolerance. *IEEE Transactions on Software Engineering*, SE-1(2):220–232, June 1975.

19. S. Rao, L. Alvisi, and H.M. Vin. Hybrid Message Logging Protocols for Fast Recovery. In *Digest of FastAbstracts: FTCS-28 – The Twenty-Eighth Annual International Symposium on Fault-Tolerant Computing*, pages 41–42. IEEE Computer Society, June 1998.

20. K. Rothermel and R. Popescu-Zeletin, editors. *Mobile Agents*, volume 1219 of *LNCS*. Springer, 1997.

21. R. A. Sahner, K. S. Trivedi, and A. Puliafito. *Performance and Reliablity Analysis of Computer Systems*. Kluwer Academic Publishers, Boston, 1996.

22. M. Strasser and K. Rothermel. Reliability concepts for mobile agents. *International Journal of Cooperative Information Systems (IJCIS)*, 7(4):355–382, 1998.

23. H. Vogler, T. Kunkelmann, and M.-L. Moschgath. An Approach for Mobile Agent Security and Fault Tolerance using Distributed Transactions. In *Proc. 1997 International Conference on Parallel and Distributed Systems (ICPADS '97)*. IEEE Computer Society, December 1997.

24. J. White. Mobile Agents. White Paper, General Magic, Inc., Mountain View, 1996.

25. D. Wong, N. Paciorek, and D. Moore. Java-based mobile agents. *Communications of the ACM*, 42(3):92–102, March 1999.

26. M. J. Wooldridge and N. R. Jennings. Agent theories, architectures and languages: A survey. In *ECAI-94 Workshop on Agent Theories, Architectures and Languages*, number 890 in LNAI, pages 1–39. Springer, August 1994.

Metrics, Methodologies, and Tools for Analyzing Network Fault Recovery Performance in Real-Time Distributed Systems

P. M. Irey IV, B. L. Chappell, R. W. Hott, D. T. Marlow,
K. F. O'Donoghue, and T. R. Plunkett

System Research and Technology Department
Combat Systems Branch
Naval Surface Warfare Center, Dahlgren Division
Dahlgren, Virginia 22448-5100, U.S.A

Abstract. The highly distributed computing plants planned for deployment aboard modern naval ships will serve real-time, mission-critical applications. The high-availability requirements of the distributed applications targeted for these platforms require mechanisms to rapidly recover from system faults (e.g., battle damage). Providing network fault recovery mechanisms that support rapid recovery in the network infrastructure used to compose these systems is critical. This paper examines how metrics used for general network performance analysis can be used to analyze fault recovery performance. A methodology for applying these metrics is presented. A testing toolset that implements the metrics and complies with the testing methodology is presented. Finally, test data collected using the toolset is presented to show the utility of the metrics and testing methodology for evaluating network fault recovery performance.

1   Introduction

Distributed computing systems composed of hundreds of processors interconnected by a network infrastructure are planned for deployment aboard modern naval ships. Applications required to enable the ship to carry out its missions and to keep the ship at sea (e.g., navigation, steering, sonar processing, command and control, etc.) are assigned to processors within the distributed computing system. A wide range of performance requirements exist for these applications. The real-time mission critical nature of the processing performed in the distributed computing system requires a fault-tolerant architecture as a key design element. An important aspect of this architecture is that the performance requirements of the system must be maintained even during system faults.

This paper focuses on a key performance requirement of the fault-tolerant distributed computing architecture: network fault recovery performance. Fault tolerance is an important concern in the design of the network infrastructure since it interconnects all of the processors in the distributed system. We use the term network fault recovery performance to refer to the time required to detect and recover from a network fault (e.g., a broken cable or connector, a failed interface adapter, a failed network switch, etc). If the maximum time required to perform a network fault recovery can be bounded and the system performance requirements are still met at

J. Rolim et al. (Eds.): IPDPS 2000 Workshops, LNCS 1800, pp. 1248-1257, 2000.

this bound, then that network fault recovery mechanism can be used as the foundation in the hierarchical fault-tolerant architecture of the distributed computing system. Naval anti-air warfare systems have network fault recovery requirements on the order of hundreds of milliseconds. This requirement is several orders of magnitude more stressing on the network design than requirements typically found in today's commercial systems.

In this paper, we use metrics defined previously [IREY97, IREY98] for measuring network fault recovery performance in a generalized networking environment. Next, the paper defines a testing methodology appropriate for the real-time distributed computing environment based on these metrics. The paper concludes by providing the results of measurements made by a toolset developed by the Naval Surface Warfare Center, Dahlgren Division, for evaluating network fault recovery performance using the metrics and testing methodology defined.

2   Network Fault Recovery Technologies

In the initial efforts to use Commercial off the Shelf (COTS) networking products aboard Navy ships, the standard COTS Fiber Distributed Data Interface (FDDI) [FDDI] technology was used. Complex FDDI-based architectures were developed which supported the loss of multiple interconnects. To evaluate the fault recovery time provided by a highly survivable FDDI architecture, [HILES95] used the FDDI Station Management (SMT) [SMT] standard to define a set of network fault recovery time metrics to compute the theoretical FDDI minimum fault recovery time. The fault recovery times measured varied widely from just under 100 milliseconds to over 500 milliseconds, with typical values in the 200 millisecond range. The paper alerted the FDDI community to the fact that a large variance in fault recovery performance was to be expected in COTS implementations and that the performance varied by product.

As the Navy moves from FDDI towards follow-on COTS solutions, new network fault recovery approaches will be used. Each approach considered must be rigorously tested to evaluate its fault recovery performance. FDDI is the only COTS networking technology that incorporated dual-attachment functionality within its defining standards. Vendors are supplying proprietary versions of fault recovery capabilities in other technologies (e.g., Ethernet); however additional testing is needed to ensure that the fault recovery behavior of these mechanisms is understood. The metrics used in this paper for evaluating network fault recovery performance are defined in general terms and thus may be applied to any type of networking technology.

3   Network Fault Recovery Performance

Network fault recovery performance is used to evaluate the ability of a given networking technology to detect and recover from network faults such as a broken cable or a failed network switch. Network fault recovery performance is a critical design parameter for systems such as the mission critical distributed computing systems of interest to the United States Navy.

Previous characterizations of FDDI fault recovery performance were obtained through the use of either theoretical analysis or hardware specific measurement techniques [HILES95] [RALPH]. While these approaches are valid for a technology such as FDDI, they are not general enough to evaluate technologies such as Fast

Ethernet where no standardized fault recovery mechanisms exist. Although work is ongoing to standardize some of these fault recovery mechanisms [802.1D], all implementations today are proprietary and may or may not interoperate. [HUANG] defines fault recovery performance metrics for a fault-tolerant Ethernet architecture based on specific fault-tolerant middleware.

We propose that the network fault recovery performance of a component, system, or architecture should be characterized by a set of metrics rather than as a single number. The metrics we propose are independent of specific hardware capabilities or network architectures. They allow network fault recovery performance to be evaluated from an application point of view. While these metrics have been used before in other contexts [IREY97] [IREY98], their application to the evaluation of network fault recovery performance and the measurement methodology required to obtain accurate measurements for this testing domain is unique. Little has been published on the evaluation of network fault recovery performance which is independent of any particular networking technology or concept.

3.1   Testing Model

The network fault recovery performance metrics used in this paper are based on a constant stream of test data to a receiving host through a network infrastructure. In a fault-tolerant environment, it is likely that the transmitting and receiving hosts are interconnected using redundant network connections and paths. These details, however, are transparent to the metrics we use to evaluate network fault recovery performance.

To measure network fault recovery performance, the messages in the test data stream are transmitted at a constant rate. The offset in time between when a message is received and when it was transmitted reflects the latency incurred in delivering the message. At a controlled point during the test, one or more faults are injected into network infrastructure connecting the transmitter and receiver. The faults cause perturbations in the normally regular data stream as observed at the receiving host (e.g., lost messages, increased message latency, etc.).

3.2   Fault Recovery Performance Metrics

There are several metrics used to describe the performance of a fault recovery mechanism in a network. These metrics are inter-send time, inter-arrival time, one-way latency, percent data received, and number of duplicates. Inter-send time and inter-arrival time are measures of the time between the sending of two consecutive packets at the transmitting host or the receiving of two consecutive packets at the receiving host respectively. One-way latency is the measure of the time period between sending a packet at the transmitting host and receiving that packet at the receiving host. Percent data received is the number of received packets divided by the number of transmitted packets. Finally, the number of duplicate packets is the number of packets received that are in excess of the number sent. All of these metrics can be combined to give a comprehensive picture of the fault-recovery performance.

4   Testing Methodology

A testing methodology must be defined specific to this problem domain. The testing methodology includes a calibration test procedure for tuning test runs for

---

1)   Select a test message size, S.

2)   Minimize Is_i so that $PDR = 100\%$.

3)   Maximize Th so that $M(Ev) - E(Ev)$ approaches or equals zero for all metrics of interest.

4)   Select N so that $(Is_i * N) >> F$.

5)   Run test measurement with computed Is_i, Th, N, and S.

Figure 1: Calibration Test Procedure

maximum accuracy and a high-fidelity testing procedure. At first glance, it may appear that a test tool as simple as Ping [KESSLER] would suffice to measure the network fault-recovery performance. Tools such as this are not adequate, however, as they lack the fidelity to make accurate measurements for the real-time distributed computing environment of interest. Also, many of these tools use transaction based measurements rather than one-way measurements which again reduces the fidelity of the measurements as shown in [IREY98].

Figure 1 defines a calibration test procedure for tuning the measurement process to maximize the accuracy of network fault-recovery performance measurements. Several new terms are used in the calibration test procedure: S specifies the size in bytes of each message sent on the test data stream during a test; Th specifies a measurement threshold used to filter out non-event related measurements; N specifies the number of messages transmitted on the data stream during a test; and F specifies the time elapsed between the injection of the first failure into the network and when the network has recovered completely from all injected failures. It is expected that multiple test runs will be conducted to select values for the parameters in each of the steps of the calibration test proceedure. When a value is selected for a given step, additional test runs should be conducted to ensure that the new value selected didn't unexpectedly impact values previously selected for other steps.

Tests of network performance by the authors (e.g., [IREY98]) in the real-time distributed environment of interest to the Navy has shown that the testing methodology used must support high-fidelity measurements. To satisfy this requirement, the network fault recovery performance tests must: 1) gather a large number of test samples to determine the range of performance; 2) examine the maximum and minimum (e.g., worst case) values rather than only mean values with standard deviations; and 3) provide visualization tools which allow the large data sets to be analyzed and iteratively reduced.

5   Network Fault Recovery Performance Measurement Toolset

A toolset was developed to measure network fault recovery performance using the metrics and testing methodology defined here. The tools in the toolset fall into three classes: test orchestration tools, data collection tools, and analysis/visualization tools. The relationships among these tools are shown in Figure 2.

5.1   Test Orchestration Tools

The main test orchestration tool in the toolset is called *nettest*. The main functions of *nettest* are: 1) to activate the data collection tools (e.g., transmitters and receivers) on specified nodes; 2) to initiate the test data stream at a specified time; 3) to inject faults in the network at specified times; 4) to gather results from the experiment; and 5) to repeat the process from step 1 until a specified number of iterations have been completed. The *nettest* program is generally run on a control host which is usually a system other than the sending and receiving hosts involved in an experiment. The complete set of results gathered by *nettest* is passed to the analysis/visualization tools.

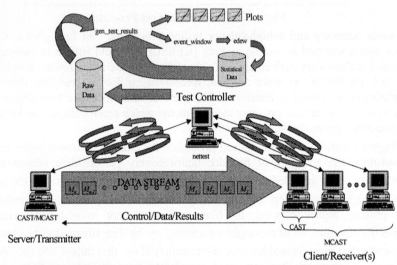

Figure 2: Network Fault Recovery Performance Toolset

One of the unique features of *nettest* specifically related to the measurement of network fault recovery performance is its fault injection capability. A scripting language was developed for *nettest* which allows the user to invoke actions on network components and to specify faults to be injected to specific network components (e.g., cables, switches, etc.) at specified times. When the power is removed from a media converter, it appears to the network components or hosts connected by the media converter that a cable has been broken. This provides one mechanism that enables automated testing needed for performing numerous tests which enables the high-fidelity testing previously described. A large library of reusable test scripts has been developed for testing failure scenarios for a variety for fault-tolerant network architectures.

For each test iteration, nettest first runs a user provided reset script to force the components under test into a known state. Next, the test data stream is started in parallel with a fault injection script which injects faults into specified network components at specified times. We have found that very complex reset scripts may be needed for particular architectures to be tested. It is best to assume that the network infrastructure is in an unknown state and to reset the state to a known state before a test run is conducted. Events external to a particular test procedure (e.g., a previous test) may have left the network infrastructure in an undesirable state.

5.2   Data Collection Tools

The main data collection tool used in the toolset is called *CAST* (Communications Analysis and Simulation Tool). *CAST* was developed by the authors to support end-to-end network performance analysis for unicast communications. It gathers a large number of performance metrics of which the network fault recovery performance metrics presented in Section 3 are a subset. *CAST* performs preliminary filtering on the data using *Th* defined in Section 3.

5.3   Analysis/Visualization Tools

To decrease the time required for a user to analyze various *CAST* test runs, a set of analysis and visualization tools were developed. The *nettest* tool generates large data files containing the results of iterative *CAST* test runs. In most cases, the data is best interpreted through visual plots. A tool was developed called *gen_test_results* that extracts the data of interest from the raw data files and then plots the extracted data.

To aid in reducing the data gathered during a network fault recovery test (i.e., analysis of events), two other tools were developed, *edew* and *eventwindow*. *Edew* computes statistical functions on the set of recorded events, such as the mean and standard deviation. *Eventwindow* filters the data from a network fault recovery test to obtain the event data. During the analysis of a network fault recovery test, these two tools are used to process and format the data. The output of these tools is provided as input to the *gen_test_results* tool.

6   Applying the Metrics, Tools, and Testing Methodology

The previous sections described the metrics, the tools that are used to measure those metrics, and some methods of how those tools can be used to perform an experiment. This section demonstrates the utility of the metrics in evaluating the performance of a survivable network by examining experimental data collected.

6.1   General Test Setup

This testing methodology is applied to both an FDDI based network and a survivable Fast Ethernet based network. The FDDI network consists of two host machines and a FDDI concentrator. One of the hosts is dual-homed to the FDDI concentrator. This is achieved by connecting both ports of a dual-attached FDDI NIC, which is installed on the host, to two ports on the FDDI concentrator. The second host uses only one of the ports on its installed FDDI NIC. The dual-homed machine serves as the transmitting host during the test, while the single-homed machine performs the function of the receiving host.

In the Fast Ethernet test, two host machines are used in conjunction with two Fast Ethernet switches. Each network host is dual-homed to each of the two Ethernet switches. The two dual-homed interfaces on the hosts may be on the same Network Interface Card (NIC) or may be on multiple NICs depending on the type of network fault recovery scheme being used. In either case, both network ports appear as a single network interface to the applications running on the network host. The two Ethernet switches are directly interconnected by one or more Fast Ethernet links.

6.2   Example FDDI Test Results

The results of performing a network fault recovery performance test on FDDI are located in the plots in Figure 3. The plot on the left shows the distribution of inter-arrival times measured throughout the length of the test. Notice that a majority of the inter-arrival times occur around the expected value of 200 milliseconds, which agrees with the results found by other measurement techniques [HILES95]. All other values of inter-arrival times appear to be well below this range. It is difficult to determine the significance of these values until the plot located on the right in Figure 3 is examined. In this plot, the inter-arrival times are examined with respect to the sequence numbers of the packets being received. This shows at what point in the test that the events occurred. Notice that this plot shows there are two distinct sets of events occurring during the test. This result agrees with the expectation that two events occur during each iteration of the test. These expected events are 1) the primary port on the transmitting host machine is brought down, and 2) the primary port is subsequently brought back up again.

The second event is expected since FDDI has one port that is the default when both ports are active. Other fault tolerant solutions including some of the fault tolerant Fast Ethernet solutions do not exhibit this behavior because they do not have a default port. In this case a failover only occurs if the active port goes down..

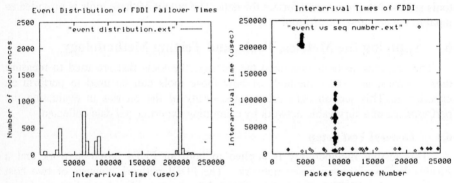

Figure 3: Measured FDDI Interarrival Times

6.3   Example Fast Ethernet Test Results

The results of performing network fault recovery performance tests on Fast Ethernet are located in figures 4-6. The first set of data represented by the left hand plot of Figure 4 shows measured inter-send times in a test where the inter-send interval has been set to 30 milliseconds. Notice that most of the values are very near the 30-millisecond setting. The outlying values are useful in determining the threshold for the inter-arrival metrics in the absence of network failures. For instance, if an inter-arrival time above 35 milliseconds is considered a network fault, then it is likely that one network failure would be erroneously detected due to the one inter-send value above 37 milliseconds. Normally the variance in the inter-send values will be taken into account if the calibration test procedure described in Section 4 has been performed properly. However, it is wise to check the inter-send data from a test to

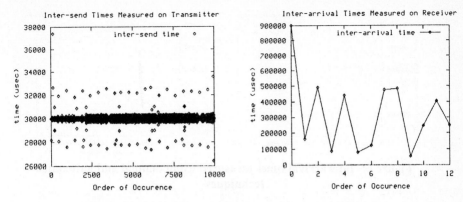

Figure 4: Measured inter-send and inter-arrival times

insure that the test results are due to the performance of the network and not a result of an anomaly on the transmitter.

The next few sets of data shown in the right hand plot of Figure 4 and both plots of Figure 5 are from a test in which one of the two Ethernet switches is powered off and the clients have to fail over to the second Ethernet switch. The right hand plot of Figure 4 shows the inter-arrival times measured during this particular test. The data set indicates that the network fault recovery performance of this survivable Ethernet solution is similar to the fault recovery time of FDDI. All the network fault recovery times are less than one second, and all but one of the measurements are below 500 milliseconds. If this were the only type of measurement made during the test, as was the case with [HILES95], then some very undesirable network behavior would have been undetected.

The left hand plot of Figure 5 shows the percent of the transmitted data which was received by the receiving host. Since there is a period of time in which the network is inaccessible due to the powering down of the Ethernet switch, one would expect that only a fraction of the total messages would make it to the destination. Instead more messages were received than were transmitted (approximately 1.5 times as many). This indicates that messages are being duplicated by the network fault recovery process. The data set shown in right hand plot of Figure 5 confirms this. This data set shows the actual number of duplicate messages that have been received. In this case, conclusions drawn from the data in the left hand plot of Figure 5 lead to a

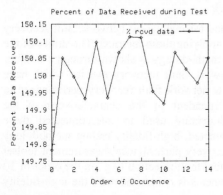

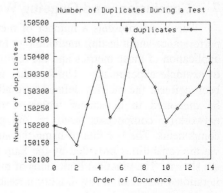

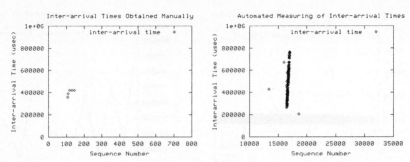

Figure 6: Interarrival times measured using manual and automated techniques

closer examination of the data in the right hand plot of Figure 5. The order of magnitude of the number of duplicate messages is important to the results of this test. Even if each individual receiving host knows to ignore duplicate packets, the health of the network as a whole may be in danger. If the switches used in this particular survivable Ethernet configuration had been fully populated with clients, then the duplicate packets could have saturated the bandwidth available on the switches, thereby eliminating any chances of recovering from the network failure in a timely manner. Further analysis would be needed to pinpoint the source of the duplicate packets (e.g., transmitting host NIC, switch, etc.)

The plots in Figure 6 show the importance of a high-fidelity testing methodology for measuring network fault recovery performance. The plot on the left shows measurement of inter-arrival times obtained by manually inserting network faults into the test. After performing the measurement five times, it appears that the network fault recovery time is consistently around 400 milliseconds. The right hand plot of Figure 6 shows the results an automated, high-fidelity measurement made using *nettest*. The network fault recovery performance measured using this testing methodology shows that the network fault recovery times can vary within a broader range of 300 to 800 milliseconds. This shows why an automated, high-fidelity testing methodology is preferable since many more iterations are possible. This leads to a better view of the performance being obtained. Typical *nettest* runs last for hours or days with hundreds or thousands of test iterations.

7   Conclusions and Ongoing Work

This paper presents a number of metrics for evaluating network fault recovery performance and a testing methodology for applying these metrics. The utility of the application of these metrics and the testing methodology is shown through a number of example experiments. Unlike other network fault recovery metrics which have been defined, the metrics defined here allow network fault recovery performance to be measured in a manner which is independent of the characteristics of the networking components used or the architecture used to interconnected these components. The benefits of using the automated, high-fidelity testing methodology via the capabilities provided by the fault recovery performance measurement toolset are shown. The practical limitations of manual fault injection (e.g. on the order of 10 iterations) that lead to low-fidelity measurements is contrasted with the high-fidelity measurements possible using the automated fault injection capabilities of the *nettest* test orchestration tool to run hundreds or thousands of test iterations.

Experiments are ongoing to evaluate components to provide a fault tolerant networking alternative that uses readily obtainable COTS components rather than FDDI components that will no longer be produced. The experiments are looking at the network fault recovery performance of Network Interface Cards (NICs) and switching components as well as architectures based on these components. Seven network architectures have been identified and testing on them is in-progress. A number of failure scenarios have been developed which are applicable to one or more of these architectures. The network fault recovery performance toolset presented in Section 5 will be used to perform these experiments based on the scenarios developed. In addition to helping the Navy transition to a new approach for configuring shipboard networks, the lessons learned through this use of the toolset will be applied to improving these tools and testing methodologies.

8   References

[802.1D] IEEE Draft P802.1w/D2, Supplement to ISO/IEC 15802-3 (IEEE Std 802.1D) Information technology – Telecommunications and information exchange between systems – Local and metropolitan area networks – Common specifications – Part 3: Media Access Control (MAC) Bridges: Rapid Reconfiguration.

[FDDI]   ISO-9314-1, Information Processing Systems – Fibre Distributed Data Interface (FDDI) – Part 1: Token Ring Physical Layer Protocol (PHY).

[SMT]   *ISO-9314-6, Information Processing Systems – Fibre Distributed Data Interface (FDDI) – Part 6: Station Management.*

[HILES95] Hiles, William S.,   Marlow, David T.,   *Approximation of FDDI Minimum Reconfiguration Time*, Proceedings of the IEEE Computer Society 20th Conference on Local Computer Networks, September 1995[†].

[HUANG] Huang, J., Song, S., Li, L., Kappler, P., Freimark, R., Gustin, J., Kozlik, T., *An Open Solution to Fault-Tolerant Ethernet: Design, Prototyping, and Evaluation*, 18th IEEE International Performance, Computing, and Communications Conference, February 1999.

[IREY97] Irey, Philip M., Marlow, David T., Harrison, Robert D., *Distributing Time Sensitive Data in a COTS Shared Media Envoronment*, 5th International Workshop on Parallel and Distrbuted Real-Time Systems, pp. 53-62, April 1997[†]

[IREY98] Irey IV, Philip M., Harrison, Robert D., Marlow, David T., *Techniques for LAN Performance Analysis in a Real-Time Environment*, Real-Time Systems - International Journal of Time Critical Computing Systems, Volume 14, Number 1, pp. 21-44, January 1998.[†]

[KESSLER] Kessler, G., Shepard, S., *RFC 1739 - A Primer On Internet and TCP/IP Tools*, December 1994.

[MILLS] Mills, David L., *RFC 1305 - Network Time Protocol (Version 3) Specification, Implementation and Analysis*, March 1992.

[RALPH] Ralph, Stanley F., Ukrainsky, Orest J., Schellak, Robert H., Weinberg, Leonard, *Alternate Path FDDI Topology*,   Proceedings of the IEEE Computer Society   17th Conference on Local Computer Networks, September 1992.

[†] Documents available at http://www.nswc.navy.mil/ITT.

Consensus Based on Strong Failure Detectors: A Time and Message-Efficient Protocol

Fabíola GREVE[†]   Michel HURFIN[†]   Raimundo MACÊDO[‡]   Michel RAYNAL[†]

[†] IRISA - Campus de Beaulieu, 35042 Rennes Cedex, France,
[‡] LaSiD-CPD-UFBA, Campus de Ondina, CEP 40170-110 Bahia, Brazil
{fgreve|hurfin|raynal}@irisa.fr   macedo@ufba.br

Abstract. The class of *strong* failure detectors (denoted S) includes all failure detectors that suspect all crashed processes and that do not suspect some (a priori unknown) process that never crashes. So, a failure detector that belongs to S is intrinsically unreliable as it can arbitrarily suspect correct processes. Several S-based consensus protocols have been designed. Some of them systematically require n computation rounds (n being the number of processes), each round involving n^2 or n messages. Others allow early decision (i.e., the number of rounds depends on the maximal number of crashes when there are no erroneous suspicions) but require each round to involve n^2 messages.

This paper presents an early deciding S-based consensus protocol each round of which involves $3(n - 1)$ messages. So, the proposed protocol is particularly time and message-efficient.

Keywords: Asynchronous Distributed System, Consensus, Crash Failure, Perpetual Accuracy Property, Unreliable Failure Detector.

1   Introduction

Several crucial practical problems (such as *atomic broadcast* and *atomic commit*) encountered in the design of reliable applications built on top of unreliable asynchronous distributed systems, actually belong to a same family: the family of *agreement problems*. This family can be characterized by a single problem, namely the *Consensus* problem, that is their *"greatest common subproblem"*. That is why the consensus problem is considered as a fundamental problem. This is practically and theoretically very important. From a practical point of view, this means that any solution to consensus can be used as a building block on top of which solutions to particular agreement problems can be designed. From a theoretical point of view, this means that an agreement problem cannot be solved in systems where consensus cannot be solved.

Informally, the consensus problem can be defined in the following way. Each process proposes a value and all correct processes have to decide the same value, which has to be one of the proposed values. Solving the consensus problem in asynchronous distributed systems where processes may crash is far from being a trivial task. It has been shown by Fischer, Lynch and Paterson [3] that

J. Rolim et al. (Eds.): IPDPS 2000 Workshops, LNCS 1800, pp. 1258-1265, 2000.

there is no deterministic solution to the consensus problem in those systems as soon as processes (even only one) may crash. This impossibility result comes from the fact that, due to the uncertainty created by asynchrony and failures, it is impossible to precisely know the system state. So, to be able to solve a-greement problems in asynchronous distributed systems, those systems have to be "augmented" with additional assumptions that make consensus solvable in such improved systems. A major and determining advance in this direction has been done by Chandra and Toueg who have proposed [1] (and investigated with Hadzilacos [2]) the *Unreliable Failure Detector* concept.

A failure detector can informally be seen as a set of *oracles*, one per process. The failure detector module (oracle) associated with a process provides it with a list of processes it guesses to have crashed. A failure detector can make mistakes by not suspecting a crashed process, or by erroneously suspecting a correct process. In their seminal paper [1], Chandra and Toueg have defined two types of property to characterize classes of failure detectors. A class is defined by a *Completeness* property and an *Accuracy* property. A completeness property is on the actual detection of crashes. The completeness property we are interested in basically states that "every crashed process is eventually suspected by every correct process". An accuracy property limits the mistakes a failure detector can make.

In this paper, we are interested in solving the consensus problem in asyn-chronous distributed systems equipped with a failure detector of the class S. A failure detector of this class suspects all crashed processes (completeness) and guarantees that there is a correct process that is never suspected, but this pro-cess is not a priori known (perpetual weak accuracy). Several S-based consensus protocols have been proposed. They all assume $f \leq n - 1$ (where f is the max-imal number of processes that may crash), and consequently are optimal with respect to the number of crash failures they can tolerate. They all proceed in asynchronous "rounds".

The S-based consensus protocol proposed in [1] requires exactly n rounds, each round involving n^2 messages (each message being composed of n values). The S-based protocols presented in [7,8] also require n rounds, but each round involves only n messages carrying a single value. It is important to emphasize that these three protocols require n rounds whatever the value of f, the number of actual crashes and the occurrences of erroneous suspicions are.

To our knowledge, very few *early deciding* S-based consensus protocols have been proposed, more precisely, we are only aware of the generic protocol present-ed in [6][1]. When instantiated with a failure detector $\in S$, this generic protocol

---

[1] The generic dimension of the protocol introduced in [6] lies in the class of the failure detector it relies on. This generic protocol can be instantiated with any failure de-tector of S (provided $f \leq n-1$) or $\Diamond S$ (provided $f < n/2$). A failure detector that belongs to $\Diamond S$: (1) eventually suspects permanently all crashes processes, and (2) guarantees that there is a time after which there is a correct process that is never suspected.

provides a S-based consensus protocol that terminates in at most $(f + 1)$ rounds when there are no erroneous suspicions. So, when the failure detector is tuned to very seldom make mistakes, this protocol provides early decision. Each round of this protocol involves n^2 messages (each message being made of a proposed value plus a round number) and one or two communication steps.

This paper presents an *early deciding* S-based consensus protocol. When there are no erroneous suspicions, the proposed protocol requires $(f + 1)$ rounds, in the worst case. When there are neither crashes nor erroneous suspicions, it requires a single round, a round being made up of two communication steps. Each round involves $3(n - 1)$ messages, and each message carries at most three values: a round number, a proposed value and a timestamp (i.e., another round number). So, the protocol is both time and message-efficient. Moreover, a generalization of the protocol exhibits an interesting tradeoff between the number of rounds and the number of messages per round.

The paper is made up of five sections. Section 2 introduces the asynchronous system model, the class S of failure detectors, and the consensus problem. Then, Section 3 presents the S-based consensus protocol. Section 4 discusses its cost. Finally, Section 5 concludes the paper.

2   Asynchronous Distributed Systems, Failure Detectors and the Consensus Problem

The system model is patterned after the one described in [1, 3]. A formal introduction to failure detectors is provided in [1].

2.1   Asynchronous Distributed System with Process Crash Failures

We consider a system consisting of a finite set Π of $n > 1$ processes, namely, $\Pi = \{p_1, p_2, \ldots, p_n\}$. A process can fail by *crashing*, (i.e., by prematurely halting). It behaves correctly (i.e., according to its specification) until it (possibly) crashes. By definition, a *correct* process is a process that does not crash. Let f denote the maximum number of processes that can crash ($f \leq n - 1$).

Processes communicate and synchronize by sending and receiving messages through channels. Every pair of processes is connected by a channel. Channels are not required to be FIFO, they may also duplicate messages. They are only assumed to be reliable in the following sense: they do not create, alter or lose messages. This means that a message sent by a process p_i to a process p_j is assumed to be eventually received by p_j, if p_j is correct[2].

The multiplicity of processes and the message-passing communication make the system *distributed*. There is no assumption about the relative speed of pro-

---

[2] The "no message loss" assumption is required to ensure the Termination property of the protocol. The "no creation and no alteration" assumptions are required to ensure its Validity and Agreement properties.

cesses or the message transfer delays. This absence of timing assumptions makes the distributed system *asynchronous*.

2.2   The Class S of Unreliable Failure Detectors

Informally, a failure detector consists of a set of modules, each attached to a process: the module attached to p_i maintains a set (named *suspected$_i$*) of processes it currently suspects to have crashed. Any failure detector module is inherently unreliable: it can make mistakes by not suspecting a crashed process or by erroneously suspecting a correct one. Moreover, suspicions are not necessarily stable: a process p_j can be added to and removed from a set *suspected$_i$* according to whether p_i's failure detector module currently suspects p_j or not. As in [1], we say "process p_i suspects process p_j" at some time t, if at time t we have $j \in suspected_i$.

As indicated in the introduction, a failure detector class is defined by two abstract properties, namely a *Completeness* property and an *Accuracy* property. In this paper we are interested in the following properties [1]:

- **Strong Completeness:** Eventually, every crashed process is permanently suspected by every correct process.
- **Perpetual Weak Accuracy:** Some correct process is never suspected.

The failure detectors that satisfy these properties define the class S (*Strong* failure detectors). It is important to note that a failure detector $\in S$ can make an arbitrary number of mistakes: at any time all (but one) correct processes can be erroneously suspected. Moreover, a process can alternatively suspect and not suspect some correct processes.

2.3   The Consensus Problem

In the consensus problem, every correct process p_i *proposes* a value v_i and all correct processes have to *decide* on some value v, in relation with the set of proposed values. More precisely, the *Consensus* problem is defined by the three following properties [1, 3]:

- **Termination:** Every correct process eventually decides on some value.
- **Validity:** If a process decides v, then v was proposed by some process.
- **Agreement:** No two correct processes decide differently.

The agreement property applies only to correct processes. So, it is possible that a process decides on a distinct value just before crashing. *Uniform Consensus* prevents such a possibility. It has the same Termination and Validity properties plus the following agreement property:

- **Uniform Agreement:** No two processes (correct or not) decide differently.

In the following we are interested in the *Uniform Consensus* problem.

3   The $\mathcal{S}$-Based Consensus Protocol

3.1   The Protocol

3.2   Underlying Principles

As other failure detector-based consensus protocols, the proposed protocol uses the *rotating coordinator* paradigm and processes proceed in asynchronous rounds [1]. There are at most n rounds. Each round r $(1 \le r \le n)$ is managed by a predetermined coordinator, namely, p_r. Moreover, during r, the coordinator of the next round (namely, p_{r+1}) acts also a particular role. Each process p_i manages three local variables: the current round number (r_i), its current estimate of the decision value (est_i), and a timestamp (ts_i) that indicates the round number during which it adopted its current estimate est_i.

As in [6-8], during a round, the current coordinator tries to impose its current estimate as the decision value. To attain this goal, each round r is made of two steps (see Figure 1).

- During the first step (lines 4-6) the current coordinator p_r broadcasts a message carrying its current estimate, namely, the message PHASE1(r, est_r). When a process p_i receives such a PHASE1(r, v) message, it adopts v as its new estimate and consequently updates ts_i to the current round number (line 6). If p_i suspects p_r, its "state variables" est_i and ts_i keep their previous values.
- During the second phase (lines 7-13), each process sends its "current state" to the current round coordinator (p_r) and the next round coordinator (p_{r+1}). This "state" is carried by a PHASE2 message (line 8). The triple (r, est_i, ts_i) indicates that during r, (1) the estimate of the decision value considered by p_i is est_i and (2) this value has been adopted during the round ts_i.
  Then, p_r and p_{r+1} follow the same behavior: each waits until it has received PHASE2 messages from all the processes it does not suspect (let us note that due to the completeness property of the underlying failure detector, all crashed processes are eventually suspected).
  If all the PHASE2 messages the process p_r (resp. p_{r+1}) has received have a timestamp equal to the current round number (r), p_r (resp. p_{r+1}) decides on its current estimate (lines 11-12). This means that the current estimate of p_r has been imposed as decision value.
  Whether the process p_{r+1} decides during r or proceeds to $r+1$, it must have a correct estimate (in order not to violate the consensus agreement property). This is ensured by requiring it to update its local estimate (est_{r+1}) to the estimate it has received with the highest timestamp in a PHASE2(r, est, ts) message (line 10).

3.3   Structure

The protocol is fully described in Figure 1. A process p_i starts a consensus execution by invoking Consensus(v_i), where v_i is the value it proposes. The protocol terminates for p_i when it executes the statement *return* which provides it with the decided value (at line 12 or 15).

It is possible that distinct processes do not decide during the same round. To prevent a process from blocking forever (i.e., waiting for a value from a process that has already decided), a process that decides, uses a reliable broadcast [5] to disseminate its decision value. To this end, the Consensus function is made of two tasks, namely, $T1$ and $T2$. $T1$ implements the previous discussion. Line 12 and $T2$ implement the reliable broadcast.

Function Consensus(v_i)

Task $T1$:

(1) $r_i \leftarrow 0$; $est_i \leftarrow v_i$; $ts_i \leftarrow 0$;
(2)   **while** $r_i < n$ **do**
(3) $r_i \leftarrow r_i + 1$;
          % p_{r_i} is the coordinator of the current round %
          % p_{r_i+1} is the coordinator of the next round %

——————— Phase 1 of round r: p_{r_i} proposes to all ———————
(4)   **if** $(i = r_i)$ **then** $\forall j$: send PHASE1(r_i, est_i) to p_j **endif**;
(5)   **wait until** (PHASE1(r_i, v) has been received from $p_{r_i} \vee r_i \in suspected_i$);
(6)   **if** (PHASE1(r_i, v) received from p_{r_i}) **then** $est_i \leftarrow v$; $ts_i \leftarrow r_i$ **endif**;

——————— Phase 2 of round r: each process replies to p_{r_i} and p_{r_i+1} ———————
(7)   **let** $X = \{r_i, r_i + 1\}$ **if** $(r_i < n)$, $X = \{r_i\}$ **otherwise**;
(8) $\forall j \in X$: send PHASE2(r_i, est_i, ts_i) to p_j;
(9)   **if** $(i \in X)$ **then wait until** (PHASE2(r_i, est, ts) messages have been
                                received from all non suspected processes);
(10)   **if** $i = r_i + 1$ **then** $est_i \leftarrow est$ rec. with highest ts **endif**;
(11)   **if** (all PHASE2 messages are such that $ts = r_i$) **then**
(12) $\forall j \neq i$: send DECISION(est_i) to p_j; return(est_i) **endif**
(13)   **endif**
(14)   **endwhile**

Task $T2$:
(15)   **upon** *the reception of* DECISION(est) **from** p_k:
          $\forall j \neq i, k$: send DECISION(est) to p_j; return(est)

Fig. 1. The $\mathcal{S}$-Based Consensus Protocol

3.4 Proof

The protocol satisfies the Termination, Validity, and Uniform Agreement properties defining the Consensus problem (these properties have been stated in Section 2.3). The reader interested in the proof will consult [4].

4   Cost of the Protocol

Time complexity The number of rounds of the protocol is $\leq n$. Differently from the S-based consensus protocols described in [1, 7, 8] that always require n rounds, the actual number of rounds of the proposed protocol depends on failures occurrences and erroneous suspicions occurrences. So, to analyze the time complexity of the protocol, we consider the length of the sequence of messages (number of communication steps) exchanged during a round. Moreover, as we do not master the quality of service offered by the underlying failure detector, but as in practice failure detectors can be tuned to very seldom make mistakes, we do this analysis considering the underlying failure detector behaves reliably. In such a context, the time complexity of a round is characterized by a pair of integers [6]. Considering the most favorable scenario that allows to decide during the current round, the first integer measures its number of communication steps (without counting the cost of the reliable broadcast implemented by the task $T2$). The second integer considers the case where a decision cannot be obtained during the current round and measures the minimal number of communication steps required to progress to the next round. Let us consider these scenarios.

- The first scenario is when the current round coordinator is correct and is not suspected. In that case, 2 communication steps are required to decide. During the first step, the current coordinator broadcasts a PHASE1 message (line 4). During the second step, each process sends a PHASE2 message (line 8). So, in the most favorable scenario that allows to decide during the current round, the round is made up of two communication steps.
- The second scenario is when the current round coordinator has crashed and is suspected by all processes. In that case, as processes correctly suspect the coordinator (line 5), the first communication step is actually skipped. Processes only send PHASE2 message (line 8) and proceed to the next round. So, in the most favorable scenario to proceed to the next round, the round is made up of a single communication step.

So, when the underlying failure detector behaves reliably, according to the previous discussion, the time complexity of a round is characterized by the pair $(2, 1)$ of communication steps.

Message complexity of a round During each round, the round coordinator broadcasts a PHASE1 message and each process sends two PHASE2 messages. Hence, the message complexity of a round is bounded by $3(n - 1)$.

Message type and size There are three types of message: PHASE1, PHASE2 and DECISION. A DECISION message carries only a proposed value. A PHASE1 message carries a proposed value plus a round number. A PHASE2 message carries a proposed value plus two round numbers. As the number of rounds is bounded by n, the size of the round number is bounded by $log_2(n)$.

Let $|v|$ be the bit size of a proposed value. According to the previous discussion, $3n(n - 1)(|v| + log_2 n)$ is an upper bound of the bit complexity of the protocol.

5   Conclusion

The paper has studied the consensus problem in the setting of asynchronous distributed system equipped with a failure detector of the class S. This class includes all the failure detectors that suspect all crashed processes and that do not suspect some (a priori unknown) correct process. The proposed protocol proceeds in asynchronous rounds and allows early decision. If there are neither failures nor false suspicions the decision is obtained in a single round. A round is made up of two communication steps and involves $3(n-1)$ messages.

The proposed protocol compares very favorably to the previous S-based consensus protocols, as those require n rounds or n^2 messages per round.

References

1. Chandra T. and Toueg S., Unreliable Failure Detectors for Reliable Distributed Systems. *Journal of the ACM*, 43(2):225-267, March 1996.
2. Chandra T., Hadzilacos V. and Toueg S., The Weakest Failure Detector for Solving Consensus. *Journal of the ACM*, 43(4):685-722, July 1996.
3. Fischer M.J., Lynch N. and Paterson M.S., Impossibility of Distributed Consensus with One Faulty Process. *Journal of the ACM*, 32(2):374-382, April 1985.
4. Greve F., Hurfin M., Macêdo R. and Raynal M., Consensus Based on Strong Failure Detectors: Time and Message-Efficient Protocols. *Tech Report #1290*, IRISA, Université de Rennes, France, January 2000. http://www.irisa.fr/EXTERNE/bibli/pi/1290/1290.html.
5. Hadzilacos V. and Toueg S., Reliable Broadcast and Related Problems. In *Distributed Systems*, ACM Press (S. Mullender Ed.), New-York, pp. 97-145, 1993.
6. Mostéfaoui A. and Raynal M., Solving Consensus Using Chandra-Toueg's Unreliable Failure Detectors: a General Quorum-Based Approach. *Proc. 13th Int. Symposium on Distributed Computing (DISC'99) (formerly, WDAG)*, Springer-Verlag LNCS 1693, pp. 49-63, (P. Jayanti Ed.), Bratislava (Slovaquia), September 1999.
7. Mostéfaoui A. and Raynal M., Consensus Based on Failure Detectors with a Perpetual Weak Accuracy Property. *Proc. Int. Parallel and Distributed Processing Symposium (IPDPS'2k), (14th IPPS/11th SPDP)*, Cancun (Mexico), May 2000.
8. Yang J., Neiger G. and Gafni E., Structured Derivations of Consensus Algorithms for Failure Detectors. *Proc. 17th ACM Symposium on Principles of Distributed Computing*, Puerto Vallarta (Mexico), pp.297-308, 1998.

Implementation of Finite Lattices in VLSI for Fault-State Encoding in High-Speed Networks

Andreas C. Döring, Gunther Lustig

Medizinische Universität zu Lübeck
Institut für Technische Informatik
Ratzeburger Allee 160
23538 Lübeck, Germany
{doering,lustig}@iti.mu-luebeck.de

Abstract. In this paper the propagation of information about fault states and its implementation in high-speed networks is discussed. The algebraic concept of a lattice (partial ordered set with supremum and infimum) is used to describe the necessary operation. It turns out that popular algorithms can be handled this way. Using the properties of lattices efficient implementation options can be found.

1 Introduction

The constantly decreasing prices of computer hardware have made the building of ever larger computer systems more attractive. The interconnection between the components like storage media, memory, processing elements and graphics system has consequently migrated from busses to networks. Even inside a single computer the I/O-subsystem will consist of a high-speed network in the near future. The high bandwidth and low latency of modern network switches require an operating mode without or only with small software protocols. To achieve this, the network has to act very reliably. Hence, some of the actions to maintain operation in the presence of faults of some network components (routers or 'links', i.e. wires) have to be performed in the network itself. Toward this aim intensive research has been done. The task of allowing a network to circumnavigate faults consists mainly of two independent problems:

1. Detect the faults and propagate the knowledge about them through the network, and
2. use this information to select fault free routes for the messages.

In this paper the first problem is considered from a general point of view. The high variety of applications of networks makes a universal router desirable that can be configured for a given problem. In particular, both tasks mentioned have to be implemented in a configurable way. The investigation of hardware structures fulfilling this, leads to a description and implementation method for a wide range of routing algorithms [DOLM98].

Though failure of a network component is expected to be rare, the reaction of the network has to be very fast. This is because a new fault leaves the network for some time in an inconsistent state. This problem has been excluded in most algorithms by assuming a separation of the diagnosis phase (item 1 above) and the message transport. Many applications may not allow this procedure requiring additional methods like transporting affected messages to the nearest node and retransmitting them. The faster the fault state propagation in the network is, the fewer messages will have to be dealt with this way. Hence in this paper the necessary operation is considered with the aim of a hardware implementation.

J. Rolim et al. (Eds.): IPDPS 2000 Workshops, LNCS 1800, pp. 1266-1275, 2000.

With respect to the knowledge base needed to provide connectivity in a faulty network, fault-tolerant routing algorithms can be divided into local- and global-information-based. Global-information-based algorithms rest their routing decision upon fault-information that contains the location of any faulty component in the network. Hence optimal routing decisions can be made at intermediate nodes in the presence of faults. The problem of those algorithms is to ensure a consistent information base among all nodes of the network. This leads mostly to an off-line re-configuration of the whole system after failure detection.

The routing decision of local-information-based algorithms uses only knowledge about the router's own and the neighbors' fault states. This could result in a backtracking for messages trapped in a dead-end [TW98, CS90]. In order to avoid backtracking despite of the topological irregularities resulting from faulty components, some algorithms relay on a convex fault model. To achieve this with only local knowledge a router is allowed to pretend a fault although it is fault-free. Each connected set of faults is completed in this way to a convex shape (e.g. rectangular in 2D meshes/tori or cubic in 3D meshes/tori). This transforms the remaining topology to a more regular one by adding fault states over a certain distance and enables messages to bypass the faulty regions on detour routes.

Alternative approaches e.g. [Wu98, CW96, CA95] use limited global information in their routing decision. Beside the local node state, the routing decision uses information from neighbor nodes that allows conclusion about the reachability of even non-adjacent regions in the network. To generate this knowledge, each router calculate its own state dependent on local faults and its neighbors' states. Clearly, it takes some time until all nodes have adjusted their own state to a new fault and several state exchanges and updates are needed.

The larger the set of possible states a router can take is, the better the knowledge about the distribution of faults in the whole network can be. Better knowledge allows tolerating more faults, keeping the performance in presence of faults higher or reducing the number of nodes that cannot be reached by the routing algorithm (like those in the convex fault model that have to be marked faulty). A larger state set requires more hardware (memory) to store it, bandwidth to exchange it, and processing power to handle it. The processing of state information is the topic of this paper.

It appears that no general method or framework dedicated for the fault state propagation has been developed, because routers implemented so far are either done entirely in software, or use no or only one fault tolerance method[AGSY94]. The flexible implementation of the fault state propagation requires special methods due to the strong speed requirements.

A large number of routing algorithms working with a limited global knowledge reveals the observation that the set of states can be described by the mathematical concept called *Lattice*. Especially the update of a node's state is essentially the computation of the supremum operation in the lattice. This concept is described in section 2. To illustrate the approach, the fault state propagation of two algorithms is used in section 3. Some proposals for a hardware implementation are given in 4 followed by the conclusion (section 5).

2   Lattices and Fault-Tolerance

The notion of a lattice is well known in algebra for a long time. For different purposes advanced theories have been created. There are two meanings for the term lattice, namely a discrete subset of a vector space (a grid) and a set with a partial ordering. In this paper the second meaning is understood.

Fault-tolerant systems are usually modeled as construction of components which are all or partially vulnerable to defects (faults). Hence, the state with respect to the faults is described by a combination of fault states from the individual components. The individual fault states have different implications to the functionality of the whole system. With respect to this influence and the methods applied in which the system reacts to the state a partial order can be defined. If some situation is strictly "worse" than another, it has a higher order of "defectness". More precisely, if all consequences for mal-function or poorer performance from the first situation apply to the second and if at least the same actions (repair etc.) have to be taken, than both situations can be compared. Of course there are situations where two different properties of the whole system are affected and where the necessary actions also vary. This notion of "worse" justifies the application of the concept of lattices to the handling of fault states.

Since a certain fault state of a certain router is less critical for its neighbors, the transmitted value has to be converted. To incorporate geometric information all neighbors get different information, see for instance figure 3 in section 3. The state of total fault of a node is transformed into North-,South-, East- or West-Failure respectively. These mappings are lattice-homomorphisms. Since they are less complex than the generation of the new fault state they are not further considered in this paper.

Definition 1. *A lattice is a tuple $(M, \vee, \wedge)$ of a set M and tow functions $\vee, \wedge :$ $M \times M \mapsto M$ which fulfill*

$$\forall m, n \in M : \wedge(m, n) = \wedge(n, m), \vee(m, n) = \vee(n, m)$$

$$\forall l, m, n \in M : \wedge(l, \wedge(m, n)) = \wedge(\wedge(l, m), n), \vee(l, \vee(m, n)) = \vee(\vee(l, m), n)$$

$$\forall m, n \in M : \wedge(m, \vee(m, n)) = m, \vee(m, \wedge(m, n)) = m$$

The two functions are also called supremum sup *and infimum* inf *.*

The intuitive meaning is that the elements of a set M are partially ordered, where all pairs of elements (m, n) have a unique smallest upper $\vee(m, n)$ and lower $\wedge(m, n)$ bound. In the following only lattices with a finite set M are considered. This is motivated by a system model with a finite number of points of failure. In every finite lattice a unique smallest $\bot$ and a unique largest element $\top$ exists. Furthermore the supremum $\vee_S$ and infimum $\wedge_S$ is well defined for arbitrary subsets S of M. A chain in $\mathcal{L}$ is a sequence of distinct elements $m_1, \ldots, m_k$ where $\vee(m_i, m_{i+1}) = m_{i+1}$. It is usual to illustrate partially ordered sets as a Hasse-diagram which is a directed Graph (M, E). The edges E in this graph are given by directly comparable relations: $(m, n) \in V \Leftrightarrow \wedge(m, n) = n$ and $\forall l \in M : (\wedge(m, l) = l$ and $\wedge(l, n) = n) \Rightarrow l = n$

Edges of the Hasse-diagram can be viewed as generators of the lattice. The nodes can be labeled ($\alpha : M \to \mathbb{N}$) with the distance to to the top ($\top$) which results in a layered representation. From $\wedge(m, n) = m$ follows that m is in a higher level than n. More generally, $\alpha(\wedge(m, n) \geq \max(\alpha(m), \alpha(n))$.

For the application to fault-tolerance a special class of lattices is of interest, lowest-level generated lattices:

Definition 2. *A lattice $(M, \vee, \wedge)$ with Hasse-diagram (M, E) is called lowest-level generated (llg) iff $M = \{\bot\} \cup G \cup \{\vee_S | S \subset G\}$ where $G = \{g \in M | (\bot, g) \in E\}$*

From an algebraic point of view this means that the set of points which are immediately larger than the bottom element generate the whole lattice already by the $\wedge$, of course except the bottom element. The importance of this class is that it exactly reflects the situation of fault-states sketched before. The single faults of components in the system represent the generating set G. The only better situation than two different single faults is a system in order. Furthermore if the set of faults are the only parameter that induces the fault state of the system all elements of the lattice have to be generated by the set G.

There is a simply criterion whether a lattice is llg or not.

Lemma 1. *A lattice* $\mathfrak{L} = (M, \vee, \wedge)$ *with Hasse-diagram* (M, E) *is llg iff the in-degree (number of edges to a node) of all elements from* M *except* G *and* $\perp$ *is greater than one.*

$$\forall x \in (M \backslash G) \backslash \{\perp\} : |\{(y, x) \in E\} > 1$$

Some important examples are given now, where all but the first one are generally llg:

1. $\mathfrak{B}(n) := (\{\text{True}, \text{False}\}^n, |, \&)$ Here $\&$ denotes the conjunction and $|$ the disjunction, i.e. digit-wise AND respectively OR. The top element is the vector with all True and the bottom element with all False. Clearly, this lattice is llg, since the lowest level consists of all vectors with just one place True. It has 2^n elements.

2. $\mathfrak{N}(n) := (\{0, \ldots, n - 1\}, \max, \min)$ Every finite linearly ordered set with n elements is isomorphic to this lattice. If $n > 2$ it is not llg, since the generating set consists of only a single element.

3. $\mathfrak{M}_k(n) = (\{T \subseteq \{1, \ldots, n\}, |T| \leq k\} \cup \{\{1, \ldots, n\}, \cap, \cup_k\})$
   where $S_1 \cup_k S_2 := S_1 \cup S_2$ if $|S_1 \cup S_2| \leq, \{1, \ldots, n\}$ otherwise
   This the the frequently-found max-k-faults failure model. More than k faults make the whole system faulty.

Though there has been intensive research on lattices the only information about the occurrence frequency seems to be found in [Kyu79]. In this paper the number of different lattices with up to 9 elements is elaborated by algorithmic enumeration and isomorphy checking. We re-implemented this algorithm in order to get further information. Surprisingly some more graphs have been found. The results are given in the following table. It appears that the number of llg lattices is only a small fraction of the lattices of a given size.

$|M|$	1	2	3	4	5	6	7	8	9	10	11	$\cdots$
#lattices	1	1	1	2	5	15	53	223	1100	6330	42155	$\cdots$
# llg	1	1	0	1	1	2	4	9	22	60	193	$\cdots$

Up to $l = 4$ elements in the lattice there are only the Boolean lattices which are llg. Hence, there is no llg lattice with three elements. Though the table does not suggest it the number of llg lattices may also grow strongly. Any lattice can be made llg with adding some elements. The resulting lattices have too many elements to be found in the table. How many of them are non-isomorphic remains open.

3   Application to Selected Fault-Tolerant Routing Algorithms

By exchanging status information with its neighbors, a router node determine whether it is part of a faulty region. Figure 3 illustrates the lattice for routing algorithms which allow the overlapping of detour routes on the border of fault regions. The router is a neighbor node of a fault region if it detects only one of his neighbors to be faulty or in the state $\top$. A combination of two faults in the same dimension leads to a state that indicates the router to be adjacent to distinct fault regions. The case that two or more failures are detected in different dimensions results in the supremum. This means that the node itself is part of a fault region and reaches the state $\top$.

The approach of convex fault regions offers a simple solution for low dimensional topologies. In higher dimensional topologies the link redundancy enables

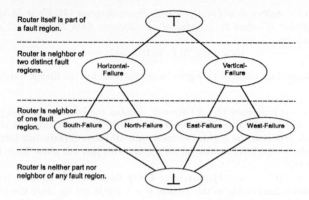

Fig. 1. Fault-lattice for convex regions in a 2D-mesh.

fault-tolerance concepts that allow the avoidance of detours in advance. As an example the formation of a fault lattice in hypercubes to the concept of [Wu98] is used. Within this approach the fault information is captured in a safety vector of n bits. A node where the k-th bit of the safety vector is set, guarantees message transportation on a minimal path for all destinations with Hamming distance k. To calculate the vector within each node, an information exchange of $n - 1$ rounds is necessary. Based on a topological property of the binary hypercube, the k-th bit of the safety vector (signed with S in the following) is determined from the $(k - 1)$th bit of the neighbors' safety vectors. The first bit of each safety vector is initialized with respect to the own node state.

Figurge 3 shows the resulting lattice for determining the second bit of a safety vector in a fault free node. In order to concentrate on the concept, a hypercube with only three dimensions has been chosen.

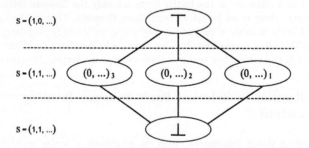

Fig. 2. Fault-lattice for determination of safety vectors 2nd Bit.

For calculating the second bit of the safety vector only the first bit of the neighbors' vectors $(0, ...)$ or $(1, ...)$ are essential. The safety vector of the i-th neighbor is indicated with the index i. As neighbors' safety vectors with the first bit is set to one do not change the status of the own safety vector, these transitions are dispensable for the fault-lattice. After receiving one safety vector containing a zero at its first bit, the resulting vector does not differ from its initial state. Only the receipt of two or three safety vectors where the first bit was set to zero results in an unsetting of the second bit in the own safety vector.

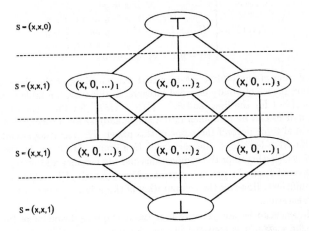

Fig. 3. Fault-lattice for determination of a safety vector's 3rd Bit.

Calculation of the third safety bit in Figure 3 is similar to the second one, except that the result has only a zero if all adjacent nodes have set their second bit to zero. Many other routing algorithms can be treated this way but as the resulting lattices are larger an abstract approach to their description is needed, allowing an automated processing. The rule-based approach presented in [DOLM98] serves this purpose.

4   Implementation

In this section the configurable implementation of finite lattices $\mathfrak{L} = (L, \vee, \wedge)$ in hardware is considered. The description of the implementation is scalable with respect to the size of the lattice $l = |L|$. Furthermore only the implementation of one function, say $\vee L \times L \to L$ is discussed, because this is sufficient for fault-tolerant routing algorithms. All proposed implementations can be easily extended to also include $\wedge$. Another assumption is that the encoding of the elements of L into a bit vector can be freely chosen.

Only few results seem to be known about the complexity of implementing the supremum in lattices. For the two trivial cases – the Boolean lattice and sets of integers – the implementation is straight-forward: n-bit OR and MAX respectively. Both need $O(\log l)$ gates and $O(1)$ respectively $O(\log \log l)$ depth. Layout questions are widely discussed since both are frequently used functions. However both circuits are quite different and there is no obvious configurable super-circuit which could implement both problems.

4.1   Simple Table-Based Method

A reference design is given by a look up table with l^2 entries. It has an area of $O(l^2 \log l)$ and a depth of $O(\log l)$. Reflexivity ($\vee(x, x) = x$) and commutativity

($\vee(x,y) = \vee(y,x)$) can be used to reduce the size of the table t. For the latter, any order $<$ can be used to "reflect on the matrix diagonal":

$$\vee(x,y) = \begin{cases} x & \text{if } x = y \\ t(x,y) & \text{if } x > y \text{ and } l/2 > y \\ t(l-x, l-y) & \text{if } x > y \text{ and } l/2 > y \\ t(y,x) & \text{if } y > x \text{ and } l/2 > x \\ t(l-y, l-x) & \text{otherwise} \end{cases}$$

The indexing of the table t assumes that the encoding of the elements uses a continuous sequence of binary encoded integers. In this way, the table t has dimensions $0, \ldots, l-1$ for the first, and $0, \ldots, \lceil ((l-1)/2) \rceil$ for the second index. How these two indices can be combined into a single address for a memory in a chip is a more general topic beyond the scope of this paper. The resulting circuit is dominated by the $l(l-1)/2 \lceil \log l \rceil$ memory cells. In a limited fan-in model the depth of this circuit is $O(\log \log l)$ for the comparisons and the memory addressing. Since the order relation of $\mathcal{L}$ can be presented as an adjacency matrix, $l(l-1)/2$ memory bits would be sufficient. However the computation of the $\vee$ function is rather difficult in such a presentation.

Since it seems to be not yet sure that the asymptotic behavior of the number of lattices for a given l is considerably smaller than $O(c^{l^2})$ this might be already an asymptotically optimal implementation up to constant factors. Hence, in the following only implementations which do not necessarily cover all lattices for a given l are considered.

4.2   Implementation with Boolean Lattice

The Boolean lattice is universal, which means that every lattice can be presented as a sub-lattice of the boolean lattice. However this can be quite inefficient: the lattice $\mathfrak{N}(n)$ with n elements would have to be implemented by the Boolean lattice with 2^{n-1} elements. Hence, a reduction is desirable for an implementation on base of $\mathfrak{B}$ (l-1). For most lattices there already exists an inclusion into a smaller Boolean lattice $\mathfrak{B}(m)$. The circuit consists of three parts:

1. Two encoders which generate an injective mapping $\iota : L \to \mathfrak{B}(m)$ from L into the Boolean lattice of m bits(configurable).
2. The circuit for $\vee$ in the Boolean lattice (m OR-gates).
3. A decoder $\mu : \mathfrak{B}(m) \to L$ which generates again elements in L (also configurable).

Since the correspondence of the top- or bottom element is given, it is clear that $m = l - 1$ if all possible lattices with l elements shall be implementable. For doing this, the digits of the vectors in $\mathfrak{B}(l-1)$ are indexed with the elements of $\mathcal{L}$ except $\top$. The image $\iota(x)$ of an element x of $\mathcal{L}$ is the vector where the digit d_y is True iff $x \vee y = y$, i.e. for all elements greater or equal than x in $\mathcal{L}$. Hence conjunction results in an element of $\iota(m)$.

Encoding of the elements of M can be chosen in a way that minimizes the size of the encoder circuit. Since the codes for the top and bottom elements are fixed there are $l - 2$ codes to be chosen. One further element can be fixed: there has to be at least one element s in M which is immediately larger than bottom, i.e. no other element but bottom is smaller than s. Hence, this element can be mapped to the vector $(1, 0, \ldots, 0)$ in $\mathfrak{B}(n)$. Looking at the both lattices $\mathfrak{N}(n)$ and $\mathfrak{M}_2(n-2)$ it is obvious that there is no further similar simplification, since in the first case all other elements are comparable to each other while in the second example no two of the remaining elements are comparable.

The implementation of μ uses a standard circuit namely a priority encoder. This is possible because every element of $M\backslash\{\top\}$ refers to a digit in $\mathfrak{B}(l-1)$. The order of M induces an order on the digits. Finding the right element in $\iota(M)$ is identical with finding the lowest digit in the resulting vector which is True. In consequence the decoder (consisting of an priority encoder) does not need to be configurable. Since every partial order can be embedded in a total order this is fairly possible.

Because the bottom element has no corresponding digit, the highest input of the priority encoder (of l bits) is fed with constant True, reflecting that $\top$ is larger than any element of M. A similar simplification can be done for $\bot$ if $l > 2$. In this case the test for the result $\vee(\bot,\bot)$ saves another bit per vector. It remains to discuss options for the implementation of ι. If two tables are used (with diagonal reflection), the resulting circuit needs $(l-2)(l-3)/2$ memory bits. The memory has to be dual ported, that is the multiplexers for output selection are needed twice.

However, for many applications including fault-tolerant routing algorithms a much smaller m can be sufficient because typical applications have shorter chains. Especially llg lattices can be included in $\mathfrak{B}$ (g), where g is the number of generating elements.

The interesting question is the implementation of the encoder and decoder circuit. For instance the top and bottom elements of the lattice can be assigned with fixed codes removing any need for configuration. Though the encoder is found twice in the design, the problematic part is the decoder because it has to map from the much larger $\mathfrak{B}(n)$ onto L. The mapping has to be consistent with the order of L, i.e. it has to be an partial-order endomorphism. In most cases $C = \vee(\iota(a), \iota(b))$ will be no element of $\iota(M)$ for arbitrary $a, b \in M$. More precisely, if for all pairs a, b C is in M then ι is an lattice endomorphism which is an unnecessary strong restrictions for computing $\vee$ alone.

Hence, in the more general case the implementation of $\mu(C)$ has to find the smallest element of $\iota(M)$ which is larger than C with respect to the order in $\mathfrak{B}(m)$. For an efficient implementation ι has to be chosen in a way which makes this step easy. Fortunately there are only the two restrictions of monotony and injectivity on ι.

4.3   Hybrid Implementations

Since lattices are a category of universal algebras, constructs like direct products are available. These constructions carry over to the implementation: direct products of lattices can be computed by the independent calculation of its factors. A direct product represents the combined state of two independent systems. A more interesting option is the identification of sub-lattices, because this occurs much more frequently. In this context two different ways of embedding a sub-lattice are distinguished. A sub-lattice $\mathfrak{A} = (A, \vee, \wedge)$ is said to be included point-like in a lattice $\mathfrak{L}$ if the only edges in the Hasse-diagram to nodes outside the sub-lattice are adjacent to $\mathfrak{A}$'s top and bottom elements. Formally

$$\forall x \in M\backslash A, \forall a \in A : \vee(a, x) = \vee(\top_\mathfrak{A}, x) \text{ and } \wedge(a, x) = \wedge(\bot_\mathfrak{A}, x)$$

In Figure 4 two sub-lattices are shaded where only one (denoted "Lattice B") is embedded point-like.

At least small point-like sub-lattices can be found in every lattice. Especially with respect to fault-tolerance it can be expected that a lot of typical point-like included sub-lattices can be found. This is interesting from an implementation point of view because optimized subcircuits for the special sub-lattice can be used. To allow more lattices to exploit these fixed implementations by adding only a small amount of hardware some selected points with edges out off the sub-lattice can be permitted.

For a configurable implementation for large lattices with a suffient fraction of typical sub-lattices an implementation is suggesting which consists of two parts:

1. A pool of circuits for the implementation of typical sub-lattices, e.g. $\mathfrak{M}_k(n)$, $\mathfrak{N}(n)$, etc. These components can be configured only in a very restricted way, e.g. the constants n,k and the selection of some special elements.
2. An implementation of the "super-lattice" by a universal methods like those described before.

A configurated implementation of the example lattice can be seen at the right in figure 4. It is interesting to observe that the resulting architecture resembles strongly the implementation approach for the whole routing algorithm given in [DOL98].

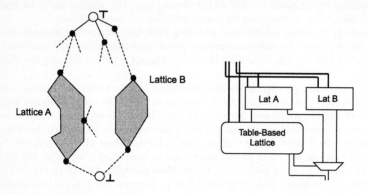

Fig. 4. A composed lattice and its implementation

5   Conclusion

In this paper it has been shown that the fault state propagation of many routing algorithms can be described by a lattice. A special class of lattices has been identified which is especially important for this application. In a discussion of implementation methods some VLSI architectures have been proposed. Of course, this method can also be used in a software implementation. In the latter context it is interesting to look for special machine instructions for a network processor to support these kind of problems.

Since most routing algorithms do not provide complexity measures for the state propagation, the overhead implied by the interpretation as a lattice can not be answered directly. Conversely, properties of the lattice (largest point-like sub-lattice, number of chains, level sizes etc.) can be used to judge various approaches.

The main disadvantage of the approach in comparison to an ad-hoc solution is that the sequence of arriving fault information can only hardly be exploited. On the other hand this gives robustness when some of the exchange messages get lost. Furthermore different routing algorithms may profit from the placement of the 're-ducing homomorphism' which transforms the state of one node into an appropriate value for the neighbor's state calculation. Further questions remaining open are:

- asymptotic frequency of lattices and lowest-level generated lattices
- For networks with some hundreds of nodes routing algorithms with global know-ledge can be still practical. Does the better behavior justify the higher demand on bandwidth, memory and processing effort in contrast to limited-global rout-ing algorithms. Furthermore routing algorithms with global knowledge tend to do a lot of redundant calculations.

– For many topologies only a few schemes of reduced fault state representation are known. More detailed methods like the one in NAFTA ([CA95]) could not be found in the literature for many important topologies, like the star graph.

The common framework of lattices can be integrated into tools for the evaluation simulation and optimization of routing algorithms and routers. This is simplified by using results, algorithms and tools from algebra.

In the project RuBIN ("Rule Based Intelligent Networks") the authors have implemented a translation tool for a high-level description method onto dedicated hardware structures. These hardware structures are implemented in a FPGA-based prototype using 1 million gates of programmable logic [Xil98] and myrinet link technology [BCF+95].

References

[AGSY94]   James D. Allen, Patrick T. Gaughan, David E. Schimmel, and Sudhakar Yala-manchili. Ariadne — an adaptive router for fault-tolerant multicomputers. In *Proceedings of the 21st Annual International Symposium on Computer Architecture*, pages 278–288, Chicago, Illinois, April 18–21, 1994. IEEE Computer Society TCCA and ACM SIGARCH.

[BCF+95]   Nanette J. Boden, Danny Cohen, Robert E. Felderman, Alan E. Kulawik, Charles L. Seitz, Jakov N. Seizovic, and Wen-King Su. Myrinet: A gigabit-per-second local-area network. *IEEE Micro*, 15(1), Februrary 1995.

[CA95]   Chris M. Cunningham and Dimiter R. Avresky. Fault-Tolerant Adaptive Routing for Two-Dimensional Meshes. In *Proceedings of the First International Symposium on High-Performance Computer Architecture*, pages 122–131, Raleigh, North Carolina, January 1995. IEEE Computer Society.

[CS90]   Ming-Syan Chen and Kang G. Shin. Depth-First Search Approach for Fault-Tolerant Routing in Hypercube Multicomputers. *IEEE Transactions on Parallel and Distributed Systems*, 1(2):152–159, April 1990.

[CW96]   Ge-Ming Chiu and Shui-Pan Wu. A Fault-Tolerant Routing Strategy in Hypercube Multicomputers. *IEEE Transactions on Computers*, 45(2):143–155, February 1996.

[DOL98]   A. C. Döring, W. Obelöer, and G. Lustig. Programming and Implementation of Reconfigurable Routers. In *Proc. Field Programmable Logic and Applications. 8th International Workshop, FPL '98*, volume 1482 of *Lecture Notes in Computer Science*, pages 500–504, 1998.

[DOLM98]   Andreas C. Döring, Wolfgang Obelöer, Gunther Lustig, and Erik Maehle. A Flexible Approach for a Fault-Tolerant Router. In *Proc. Parallel and Distributed Processing - Proceedings of 10 IPPS/SPDP '98 Workshops*, volume 1388 of *Lecture Notes in Computer Science*, pages 693–713, 1998.

[Kyu79]   Shoji Kyuno. An Inductive Algorithm to Construct Finite Lattices. *Mathematics of Computation*, 33(145):409–421, January 1979.

[TW98]   Ming-Jer Tsai and Sheng-De Wang. A fully adaptive routing algorithm. *IEEE Transactions on Parallel and Distributed Systems*, 9(2):163–174, February 1998.

[Wu98]   Jie Wu. Adaptive Fault-Tolerant Routing in Cube-Based Multicomputers Using Safety Vectors. *IEEE Transactions on Parallel and Distributed Systems*, 9(4):321–334, April 1998.

[Xil98]   Xilinx, Inc. XC4000XV Family Field Programmable Gate Arrays. Data Sheet, May 1998.

Building a Reliable Message Delivery System Using the CORBA Event Service

Srinivasan Ramani[1], Balakrishnan Dasarathy[2], and Kishor S. Trivedi[1]

[1]Center for Advanced Computing and Communication
Department of Electrical and Computer Engineering
Duke University, Durham, NC 27708-0291, USA
{sramani, kst}@ee.duke.edu
[2]Telcordia Technologies
445 South Street
Morristown, NJ 07960, USA
bdasarat@telcordia.com

Abstract. In this paper we study the suitability of the CORBA Event Service as a reliable message delivery mechanism. We first show that products built to the CORBA Event Service specification will not guarantee against loss of messages or guarantee order. This is not surprising, as the CORBA Event Service specification does not deal with Quality of Service (QoS) and monitoring issues. The CORBA Notification Service, although it provides much of the QoS features, may not be an option. Therefore, we examine application-level reliability schemes to build a reliable communication means over the existing CORBA Event Service. Our end-to-end reliability schemes are applicable to management applications where state resynchronization is possible and sufficient. The reliability schemes proposed provide resilience in the face of failures of the supplier, consumer, and the Event Service processes.

1 Introduction

CORBA [1], spearheaded by the Object Management Group (OMG), provides a basis for portable distributed object-oriented computing applications [2] and is a marriage of the object-oriented paradigm with a client/server architecture [3]. Before the advent of CORBA, clients had to resort to sockets or remote procedure calls such as those provided by DCE or Sun RPC to communicate with objects on remote servers [6, 7]. CORBA abstracts away the complexities of distributed object communication between clients and servers.

The Event Service, one of the Common Object Services (COS) [1] provided by CORBA, is intended to support de-coupled, asynchronous communication among objects. For many applications, a synchronous communications model will not scale well. An elegant solution for these applications is to have the server push messages into a "channel" and let any interested client connect to the channel and pick up the messages. In Event Service parlance, the server is called a *supplier*, the clients are called *consumers*, and the channel is called the *event channel*.

J. Rolim et al. (Eds.): IPDPS 2000 Workshops, LNCS 1800, pp. 1276-1280, 2000.

The Event Service however is not reliable. This is because the CORBA Event Service specification does not deal with QoS (Quality of Service) or run-time management issues. There exist applications such as Telcordia's mission critical network management system for which reliable, in order message delivery is important. In addition, these applications should also be resilient to failures of the supplier, consumer and the Event Service itself. In an attempt to remedy this situation, OMG has come out with the Notification Service specification [5] that explicitly deals with QoS issues. But the Notification Service implementation is not widely available for all ORBs (Object Request Brokers) and also there is no commitment from vendors to make an implementation available for several platforms. Moreover, the Notification Service does not guarantee against loss of messages, as there is no pro-active monitoring.

Fault tolerance via replicated objects, such as in [4] provides some resilience in the face of event channel failures. In this approach, suppliers retain a copy of messages pushed in to the event channel in a backup queue. But this scheme could still drop messages, as it does not prevent queue overflows. Also the use of ORB-specific API's prevents the adaptability of the scheme to other ORBs.

2 Log files and retry policies – Are they adequate?

Vendors have sought to provide resilience by providing additional features in their Event Service implementations. These include, a retry mechanism that uses a log file to deal with failed message deliveries or messages displaced from overflowing queues. The ordering of messages is no longer guaranteed. This scheme is also not guaranteed to work as the log files might reach their maximum size limit. There is no programming or administrative interface to monitor the queues. Some vendors have mechanisms to deal with object failures (for example, the Visibroker [8] ORB has an Object Activation Daemon to provide some resiliency). But this is not a CORBA standard and hence implementation-dependent.

3 Application-level reliability mechanism to provide resilience

In an experiment that we setup, we found that the maximum supplier throughput is 49.9 messages per second, if no message loss is to occur. Having identified the lack of adequate guarantees in the CORBA Event Service specification, we investigated application-level reliability mechanisms at the supplier and the consumer.

3.1 Model for reliability: Resynchronization

Many applications for which the Event Service is applicable need to propagate messages that resemble status updates. If and when messages are detected to be missing, what is required is not the history of changes but the current status. The same is true after failures of the supplier, consumer or the Event Service lead to message loss. Our reliability scheme illustrated in Figure 3 exploits this property and is based on resynchronization of states.

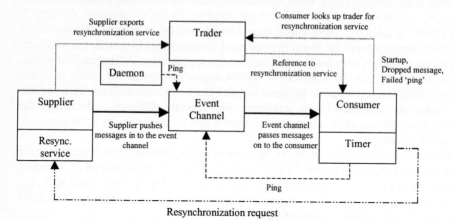

Fig. 1. Scheme to provide resilience to the CORBA Event Service

With this scheme, no log/retry mechanism of the Event Service is used, nor is the rebinding of the restarted objects dependent on any implementation-dependent facility. The highlights of our application-level reliability scheme are:

- When the supplier connects to the event channel for the first time or is restarted after a failure, it does a resynchronization to bring all the consumers up to date. The supplier puts some sequencing information in each message.
- The supplier provides a "resynchronization request service" that it announces through the Trader Service[1]. This service is to be used by consumers to send a resynchronization request to the supplier whenever required.
- When a consumer connects to the event channel – either for the first time or when it comes back up after a failure, it requests a resynchronization.
- When the consumer notices that a message has been lost (using the sequence information), it requests a resynchronization.
- A daemon to "ping" periodically and restart the Event Service if necessary.

We now explain how our scheme deals with failure scenarios.
Supplier failure:
When the supplier reboots, the recovery is simple – it simply resynchronizes.
Consumer failure:
When a consumer goes down, because of the de-coupled nature of the link between the supplier and the consumer, the supplier remains unaware of the failure. When the consumer comes up again, it reconnects to the event channel and requests a resynchronization. A consumer may, of course, choose not to act on the status updates, if it already has the latest status.

---

[1] The Trader Service (one of the COS) acts as the "yellow pages" for object services. The provider of a service exports the location and description of the service to a "trader". A client requiring a service can look up the trader by supplying the description.

Event Service failure:

To detect the failure of the Event Service, a daemon is used to "ping" the Event Service periodically. If the daemon notices that the Event Service has died, it restarts the Event Service. In a "push" model, a publisher notices that the event channel it is connected to has died when it tries to push a message, and tries to rebind with the new Event Service using the trader. The consumer can be unaware of the failure forever since in the push model, the Event Service initiates a message transfer. In our scheme, we make the consumer "ping" the event channel whenever a timer times out. The timer restarts whenever a message reaches the consumer. If the ping operation fails, the consumer tries to rebind to the event channel. The resynchronization strategy also obviates the need for mechanisms to deal with messages lost in the event channel.

Queue overflow:

Finally, if there is a queue overflow, the consumer will detect the dropped message(s) and request a resynchronization.

4 Effectiveness of the reliability mechanism - Experiments

In our earlier experiments, we determined that a supplier throughput of 49.9 messages per second or lower was needed to avoid message loss. We use this rough estimate while doing a resynchronization. Of course, if a resynchronization message were dropped, then the consumer would request a new resynchronization. The results in Table 1 report the mean and standard deviation for each measure, obtained by repeating each experiment 25 times.

Table 1. Effect of Supplier Speed on Resynchronization Overhead

Supplier throughput (#messages per sec)		% of normal messages received by consumer		#resynchronization messages pushed per 10000 normal messages	
Effective throughput	Normal messages				
Mean	S.Dev	Mean	Mean	S.Dev	Mean	S.Dev
265.6	29.8	247.8	26.3	10.5	580.9	224.7
98.1	1.3	96.9	96.1	3.9	122.4	127.0
49.9	0.0	49.9	99.9	0.1	2.0	5.0

Table 1 shows that when the supplier attempts a throughput higher than 49.90 messages per second, there are several resynchronizations before the supplier finishes pushing 10000 messages. Because of the additional resynchronization messages that are pushed and the overheads associated with the reliability mechanism, it now took longer to push 10000 normal messages. In Table 1, the column titled "normal messages" lists the supplier throughput for normal messages alone (that is, without considering the resynchronization messages). This is calculated as "*1 / (Total time to push 10000 normal messages (in sec)/10000)*". If the supplier throughput is substantially higher than 49.90 messages per second, then it can be seen from the table that a low percentage of these normal messages eventually make it to the consumer. This means that the (slower) resynchronization process is doing most of the status updates. As the supplier throughput approaches the recommended value, the effective throughput of the supplier matches the throughput of normal messages from the supplier. This is because almost all the normal messages are successfully delivered to the consumers and the resynchronization messages are rarely required.

The general guideline for effectively putting into practice our reliability scheme is as follows:

- Determine the supplier throughput that almost always delivers messages without loss in the event channel. Use a much lower value as the throughput for the messages.

For the particular application and hardware platform on hand, optimization studies can be done to establish safe and most efficient operating ranges.

5   Summary and Concluding Remarks

In this paper we have studied the suitability of the CORBA Event Service as a reliable message delivery mechanism. To deal with its lack of reliability, we examined application-level schemes to build a reliable communication means over the existing CORBA Event Service. The reliability schemes, which are general enough to be applicable to any Event Service implementation, deal with message losses and also provide resilience in the face of failures of the supplier, consumer, and the Event Service processes. Our end-to-end reliability scheme is suitable for applications for which status resynchronization is possible and sufficient.

References

1. Object Management Group. "The Common Object Request Broker: Architecture and Specification". Revision 2.3, 1998. ftp://ftp.omg.org/pub/docs/formal/98-12-01.pdf.
2. M. Henning and S. Vinoski. Advanced CORBA Programming with C++. Addison Wesley, Reading, Massachusetts, 1999.
3. D. Pedrick, J. Weedon, J. Goldberg and E. Bleifield. Programming with Visibroker – A Developer's Guide to Visibroker for Java. John Wiley & Sons, 1998.
4. T. Luo et al. "A Reliable CORBA-Based Network Management System". Int. Conf. on Communications, ICC99, Vancouver, BC, Canada, June 1999.
5. Object Management Group. "Notification Service: Joint Revised Submission". January 25, 1999. http://www.omg.org/cgi-bin/doc?telecom/98-11-01.pdf.
6. J. Shirley, W.Hu and D. Magid. Guide to Writing DCE Applications. O'Reilly & associates, Inc., May 1994.
7. W.R. Stevens. UNIX Network Programming, Vol. 1, Second Edition, Prentice-Hall, Upper Saddle River, New Jersey, 1998.
8. Visibroker. http://www.inprise.com/visibroker

Acknowledgements
Our thanks to Mark Segal, Brian Coan, Michael Skurkay, Neal Bickford and Sarah Tisdale of Telcordia Technologies for their support and review of this paper.

Network Survivability Simulation of a Commercially Deployed Dynamic Routing System Protocol

Abdur Chowdhury[1,2], Ophir Frieder[1], Paul Luse[2], Peng-Jun Wan[1]

`{abdur, wan, ophir}@cs.iit.edu, pluse@iitri.org`
Department of Computer Science
Illinois Institute of Technology[1] and
IIT Research Institute[2]

Abstract. With the ever-increasing demands on server applications, many new server services are distributed in nature. We evaluated one hundred deployed systems and found that over a one-year period, thirteen percent of the hardware failures were network related. To provide end-user services, the server clusters must guarantee server-to-server communication in the presence of network failures. In prior work, we described a protocol to provide proactive dynamic routing for server clusters architectures. We now present a network survivability simulation of the Dynamic Routing System (DRS) protocol. We show that with the DRS the probability of success for server-to-server communication converges to 1 as N grows for a fixed number of failures. The DRS's proactive routing policy performs better than traditional routing systems by fixing network problems before they effect application communication.

INTRODUCTION

Traditional supercomputers are becoming scarce and distributed server clusters are becoming the solution of choice. These smaller computers are coupled by networks to achieve the same objective at a substantially lower cost. The Berkley NOW (Network Of Workstations) project was one of the first projects pushing this solution [2]. PVM (Parallel Virtual Machine) [3] and MPI (Message Passing Interface) [4] libraries provide messaging and synchronization constructs that are needed for distributed parallel computing with NOW solutions. Projects like Beowulf [5] for Linux are continuing the distributed computing approach. All of these approaches have one common resource, the network. While the network is very important, no strong push has been made to provide fault tolerance for network failures in a server cluster solution.

We developed a network routing algorithm to provide fault-tolerance for server-to-server communication by proactively monitoring network communication links between servers. This is different from reactive routing techniques [6] that wait for a failure to occur and then react by finding an alternative route. Our proactive algorithm constantly looks for errors via continuous ICMP echo requests. When a failure is identified, a new route is selected around the failed portion of the network. This new

J. Rolim et al. (Eds.): IPDPS 2000 Workshops, LNCS 1800, pp. 1281-1285, 2000.

route is often found in the time of a TCP retransmit, so server applications are unaware that a network failure has occurred.

Our algorithm, the Dynamic Routing System (DRS) [1], improves reliability by providing a second network interface card for each server thus providing an alternate method of physical communications in the case of hardware failure. The DRS works by frequent link checks between all pairs of nodes to determine if the link between pairs of computers is valid. This algorithm uses the redundant network link between two servers to provide multiple communication channels. When one link fails, the second direct link is checked and used. However, if no link exists, a broadcast is made to identify whether or not some other server is able to act as a router to create a new path between the sender and the proposed recipient. Our algorithm discovers the failure before server-to-server communication is affected. The essential goal of our algorithm is to hide network failures from distributed applications.

The DRS was deployed in 27 local voice mail server clusters by MCI WorldCom, each cluster contains between 8 and 12 servers. Thus understanding the reliability supported is not only of theoretical interest but of practical interest as well. In prior work [1] we showed that, over a one year period, 13% of hardware failures for 100 compute servers were network related, i.e., network interface cards, hubs, etc. This likelihood of failure provides motivation to improve the resilience of server clusters where services need to be guaranteed. We show that, with the DRS, the probability of success of server-to-server communication converges to 1 as N grows for a fixed number of failures. The proactive routing policy of the DRS performs better than traditional routing systems by fixing network problems before they effect the server-to-server communication.

DRS ALGORITHM

RIP [7], OSPF [8], EGP and BGP [9] are routing solutions to many different routing problems, however, they do not address the needs of a high availability server cluster environment [11]. Their primary goal is to provide routing updates to other routers on the network to find alternative routes to the same network. The general design goal is based on reactively rerouting when a specified timeout period has been reached. So if a destination network does not respond to a route query, after some time quantum, it is considered down and a new route is sought after.

The DRS works with IP networks unlike some telecommunication approaches using specialized hardware [10] and improves fault tolerance via proactive failure recognition and the use of a redundant network. Thus, each computer has two network interface cards connected to two separate networks. It is the task of the DRS routing demons to monitor the connections between two servers. If a failure occurs, the demons set up new point-to-point routes around the problem before network applications are aware that a problem occurred.

The DRS runs on every node in the server array. Each DRS demon is configured to monitor hosts on the networks and executes a two stage run process. In the first phase, the communications links between the local host and all other hosts that is it has been configured to monitor are checked. These checks are accomplished using the ICMP (Internet Control Message Protocol) [13] echo request. Host "A" sends an ICMP echo request to host "B" via the first network. If the echo is returned, the DRS can assume that the hub, wiring, network interface card, device driver, network protocol stack and host kernel, are operational. The DRS continues to test all known hosts on all known networks in the same manner.

Each demon keeps track of which hosts to monitor and the state that they are in (i.e., "up", "down"). If a failure occurs, the DRS demon must determine a new route of communication between host "A" and "B". The DRS demon loops through a cycle of monitoring communication links, answering requests, and fixing problems as they occur, for the life of the server cluster. The DRS algorithm avoids routing loops and other issues involved in distributed routing. For a detailed presentation and proof of correctness see [1].

DRS PROACTIVE COST

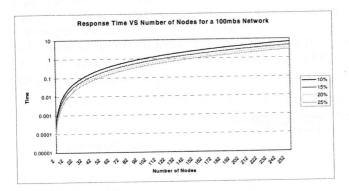

Figure 1: 100Mb Network Performance

The DRS's proactive monitoring of network links comes at a cost of network bandwidth. To find errors before they effect network communication, the links must be checked frequently. If the links were not checked frequently, the DRS would become equivalent to a reactive routing protocol. As the number of nodes increase, the bandwidth required to support the frequent checks likewise increases. In Figure 1, we present the maximum number of servers in the cluster that the DRS supports given a requirement for error resolution in X time units and the percentage of network bandwidth useable by the DRS. As show in Figure 1, ninety hosts are supported in less than 1 second with only 10% of the bandwidth usage.

NETWORK SURVIVABILITY ANALYSIS

We now present a conditional probability model like [12] to quantitatively evaluate networking systems with a given number of network failures occurring at any given instance. This model yields the probability of success, independent of time, of a system with N nodes and f failures.

We assume that in a system with N nodes, there are exactly 2N interface connections and two non-meshed back planes, each with equal probability of failure, say q, for $0 \leq q \leq 1$. Therefore, the probability of 2 failures in any system will be q^2, the probability of 3 failures will be q^3, and the probability of f failures will be q^f. It follows that $\lim_{f \to \infty} q^f = 0$. Therefore, the probability of multiple failures in a system decreases exponentially.

Now we develop the equation for the probability of success by counting the number of possible failure combinations for a system with N nodes and f failures. We represent this number by the combinatorial function $F(N,f)$, with

$$F(N,f) = \binom{2N}{f-2} + 2 \cdot \left[\binom{2N}{f-1} - \binom{2N-2}{f-1} \right] + \binom{2N-2}{f-2} + 2 \cdot \left[\binom{2N-4}{f-3} + \binom{N-2}{f-N} \cdot 2^{2N-f-2} \right] + \binom{2N-4}{f-2}$$

Because the total number of combinations in a networking system is $\binom{2N+2}{f}$, the probability of success can be written, as shown in Equation 1.

$$P[\text{Success}] = \frac{\binom{2N+2}{f} - F(N,f)}{\binom{2N+2}{f}}$$

Equation 1: Probability of Success

By graphing Equation 1 for fixed values of f, it is evident that as the number of nodes in a system increases, the probability of that system maintaining a successful connection between any two nodes at any given time will approach 1 using the DRS. More specifically, for f=2 the P[S] surpasses 0.99 at 18 nodes. For f=3 the P[S] surpasses 0.99 at 32 nodes, and for f=4 the P[S] surpasses 0.99 at 45 nodes. Given that $\lim_{f \to \infty} q^f = 0$ and that $\lim_{N \to \infty} P[S] = 1$, a system implementing the DRS has a high probability of resilience to network failure, as show in Figure 2.

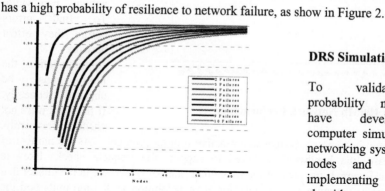

Figure 2: Convergence of P[Success] to 1

DRS Simulation

To validate our probability model, we have developed a computer simulation of a networking system with N nodes and f failures implementing the DRS algorithm. Given a specified number of iterations and a fixed f, the simulation output consists of randomly generated success probability values for f<N<64. The graph in Figure 3 displays the convergence of the simulation outputs to the actual equation values for two through ten network failures as we increase the number of iterations. The y-axis represents the mean absolute difference between the simulation output and the equation value for f<N<64. The x-axis represents the number of iterations in log10 scale. With 1,000 iterations, the mean absolute difference is less than 0.009 for each of the fixed f values, and as the number of iterations increases the mean absolute difference converges to zero. Therefore, the simulation results support the probability model of Equation 1 given in the prior section.

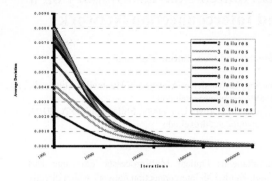

**Figure 3: Convergence of Simulation Results
to Equation Results**

CONCLUSIONS

The DRS algorithm provides a reactive routing protocol for tightly coupled server clusters of the given topology. These server clusters do not have elaborate network topologies since server-to-server communication of the cluster is of concern. We provided a brief review of the DRS algorithm as described in detail in [1]. We provided a probability model to quantitatively evaluate the DRS algorithm resilience to network failures. The model gives a conditional failure probability of the entire system. Using Equation 1, we showed that the probability of success converges to 1 as N gets large for fixed values of f. More specifically, for f=2 the P[S] surpasses 0.99 at 18 nodes. For f=3 the P[S] surpasses 0.99 at 32 nodes, and for f=4 the P[S] surpasses 0.99 at 45 nodes. Given that $\lim_{f \to \infty} q^f = 0$ and that $\lim_{N \to \infty} P[S] = 1$, a system implementing the DRS has a high probability of resilience to network failure. To validate this model we present a validation simulation of the DRS algorithm.

References

1 A. Chowdhury, et. al., "Dynamic Routing System (DRS): Fault tolerance in network routing", Computer Networks And ISDN Systems (31) 1-2 (1999).
2 T.Anderson, et.al, "A Case for NOW (Networks of Workstations)", IEEE Micro 15 (1995).
3 J. Casas, et.al. "Adaptive Load Migration Systems for PVM". Supercomputing 94, November 1994.
4 M. Snir, et.al, "MPI: The Complete Reference", The MIT Press, 1996
5 C. Reschke, et. al, "A Design Study of Alternative Network Topologies for the Beowulf Parallel Workstation,", IEEE High Performance Distributed Computing, 1996.
6 G. R. Ash, Dynamic Routing in Telecommunications Networks, McGraw Hill, 1998.
7 C. Hedrick, Request For Comment 1058, "Routing Information Protocol", 06/01/1988, http://ds.internic.net/ds/dspg2intdoc.html
8 J.Moy, Request For Comment 1583, "OSPF Version 2", 03/23/1994, http://ds.internic.net/ds/dspg2intdoc.html
9 K. Varadhan, Request For comment 1503, "BGP OSPF Interaction", 01/14/1993, http://ds.internic.net/ds/dspg2intdoc.html
10 B. R. Hurley, C. J. R. Seidl, and W. F. Sewell. "A Survey of Dynamic Routing Methods for Circuit-Switched Traffic". IEEE Communications Magazine, 25(9), September 1991.
11 S. Low, P. Varaiya, "Stability of a class of dynamic routing protocols (IGRP)". In IEEE Proceedings of the INFOCOM, volume 2, pages 610--616, March 1993.
18 R. Talbott, "Network Survivability Analysis", Fiber & Integrated Optics, Vol. 8, 1988.
19 J. Postel, "Internet Control Message Protocol (ICMP)", RFC 792, 1981

Fault-tolerant Distributed-Shared-Memory on a Broadcast-based Interconnection Network

Diana Hecht[1] and Constantine Katsinis[2]

[1]Electrical and Computer Engineering, University of Alabama in Huntsville, Huntsville, AL 35899, hecht@ece.uah.edu
[2]Electrical and Computer Engineering, Drexel University, Philadelphia, PA 19104, ckatsini@ece.drexel.edu

Abstract. The Simultaneous Optical Multiprocessor Exchange Bus (SOME-Bus) is a low-latency, high-bandwidth interconnection network which directly links arbitrary pairs of processor nodes without contention, and can efficiently interconnect over one hundred nodes. Each node has a dedicated output channel and an array of receivers, with one receiver dedicated to every other node's output channel. The SOME-Bus eliminates the need for global arbitration and provides bandwidth that scales directly with the number of nodes in the system. Under the distributed shared memory (DSM) paradigm, the SOME-bus allows strong integration of the transmitter, receiver and cache controller hardware to produce a highly integrated system-wide cache coherence mechanism. Backward Error Recovery fault-tolerance techniques can rely on DSM data replication and SOME-Bus broadcasts with little additional network traffic and corresponding performance degradation. This paper uses extensive simulation to examine the performance of the SOME-Bus architecture under DSM and Backward Error Recovery.

1 Introduction

In distributed shared memory (DSM) systems, an important objective of current research is the development of approaches that minimize the access time to shared data, while maintaining data consistency. The success of DSM depends on its ability to free the programmer from any operations that are necessary for the only purpose of supporting the memory model, and therefore it is critical that interconnection networks be developed, connecting hundreds of nodes with high-bisection bandwidth and low latency, that result in the least possible adverse impact on DSM performance.

As the number of nodes in the system implementing DSM increases, so does the likelihood that the system will experience node failures. For this reason, tolerating node failures becomes essential for parallel applications with large execution times. Coherence protocols for fault-tolerant DSM systems are described in [3]. A popular approach to fault tolerance, Backward Error Recovery (BER) enables an application which encounters an error to restart its execution from an earlier, error-free state. Much research is currently being conducted in order to exploit the data replication mechanism already existing in a DSM to reduce or hide the overhead of implementing BER. An example is the ICARE system [2], a software-based recoverable DSM for a network of workstations.

J. Rolim et al. (Eds.): IPDPS 2000 Workshops, LNCS 1800, pp. 1286-1290, 2000.

Due to advances in fiber-optics and VLSI technology it is possible to design an architecture that relies on broadcasts to support hardware-based DSM, allowing the implementation of coherence protocols at the cache block level through interactions of cache, memory and network interface controllers. Fault-tolerance protocols on this system rely on data replication and broadcasts to implement Backward Error Recovery resulting in little additional cost in terms of network traffic.

The Simultaneous Optical Multiprocessor Exchange Bus (SOME-Bus) [1] is such a network. One of its key features is that each node has a dedicated broadcast channel, realized by a specific group of wavelengths in a specific fiber, and an input channel interface based on an array of receivers shown in Fig. 1, which simultaneously monitors all channels. This design results in an effectively fully-connected network.

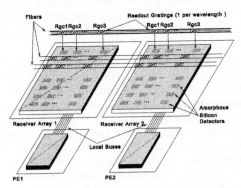

Fig. 1. SOME-bus Parallel Receiver Array and Output Coupler

Although the SOME-bus can utilize software techniques for implementing cache coherence, it allows strong integration of the transmitter, receiver and cache controller hardware to produce a highly integrated system-wide cache coherence mechanism.

2   Fault Tolerant DSM on the SOME-bus

A multicomputer system based on the SOME-Bus can be composed of hundreds of nodes. As the number of nodes in the system increases, however, so does the likelihood that the system will experience node failures or reboots. For this reason, the ability to tolerate node failures becomes essential for parallel applications with large execution times.

Backward Error Recovery is a mechanism for providing fault tolerance through the periodic saving of system state, which is known as checkpointing. Typically, in order to ensure that the checkpoint is consistent over the entire system, the processes are synchronized before the checkpoint is established and then resume normal execution after the checkpoint has been saved. Care must be taken that the checkpoint is saved to a storage medium that can be accessed after a failure occurs. In [2], checkpoints are saved in the memory of another node rather than to disk thereby reducing the amount of time that is required to save the checkpoint. We have adopted this approach for our simulations.

This paper explores the performance of a fault tolerant DSM system based on the SOME-bus. An event simulator was developed to evaluate the performance of the proposed system. The simulated multicomputer consists of a set of 64 nodes organized as a CC-NUMA system interconnected by the SOME bus architecture. The shared virtual address space is distributed across local memories which can be accessed both by the local processor and by processors from remote nodes, with different access latencies.

A multithreaded execution model is assumed and each processor executes a program which contains a set of parallel threads. A sequential consistency model is adopted and statically distributed directories are used to enforce a write-invalidate protocol. Each node contains a processor with cache, memory, an output channel and a receiver which can receive messages simultaneously on all N channels. Each node contains a processor controller, directory controller, cache controller and channel controller. There is a separate input queue associated with cache and directory controllers and the channel controller is associated with the output queue for the node. In addition, the directory controller contains an internal waiting-message queue used to hold the requests that are waiting for invalidation or downgrade acknowledge messages to be received.

We examine two mechanisms for providing fault tolerance in our DSM system. In both approaches, each node keeps a full copy of its local memory which will be referred to as its as recovery data. In addition, every node also contains a copy of the recovery data of another node. It is not necessary to recreate an entire copy of the memory at each checkpoint interval. Instead the checkpoints can be incrementally updated since it is only necessary to save the data that has been modified since the last checkpoint[2]. After the processors are synchronized in preparation for taking a checkpoint, all of the cache controllers write back any exclusively owned cache blocks. Once the directory controllers have received all of the cache writeback data, information that will be used to update the recovery memory is compiled. A copy of the compiled updates must be sent to the node with the backup copy of the recovery data. When all copies of the recovery data have been updated successfully, the processors synchronize again and then resume normal operation.

As the checkpoint interval increases, the total number of checkpoints taken over the course of the application execution decreases. Depending upon the data access pattern, however, the time required to perform the cache writeback, assemble the recovery data updates, and send them to the backup node will increase. This approach might lead to a burst of traffic appearing on the bus during the establishment of the checkpoint.

Another approach would be for a node to contain a copy of another node's local memory as well as the recovery data. If node X contains the recovery data and copy of node Y's data, we can refer to node Y as the home location for its local memory and node X as the home2 location for node Y's memory. If the cache coherence protocol was modified so that both copies of Node Y's memory remain consistent, updating the copy of Node Y's recovery data could be handled locally on Node X. Instead of sending all the data that was modified since the previous checkpoint, the cache controller will multicast the writeback blocks to both the home and home2 nodes.

This would reduce the burst of network traffic that occurs in the first approach during the establishment of a checkpoint. In order to keep both copies of the memory consistent, additional coherence messages will be required. The number of invalidation, downgrade and writeback messages will have to be increased to include the home2 node. Since these messages can be multicast, the additional overhead for the SOME-bus will not be as high as it would for other types of interconnection architectures. Furthermore, the home2 data can be used to fill read requests. This increases the probability of having a local miss rather than a remote one. For example, it is possible to service a cache read miss on node X with either its own local memory or the copy of node Y's local memory.

Two algorithms, FT1 and FT2, were implemented based on the ideas described in the preceding paragraphs. The FT1 algorithm maintains two consistent copies of the local memory and the corresponding recovery data for each node. Updating both copies of the recovery data can be accomplished without additional traffic during the creation of the checkpoint. The FT2 algorithm keeps copies of the recovery data only, and must transmit updates to the node containing the remote copy during the creation of a new checkpoint.

In each of the experiments reported below, the simulation was executed for a period of 100,000 simulation time cycles. Data was collected for five checkpoint intervals (10000, 20000, 30000, 40000 and 50000), which resulted in a total of 9, 4, 3, 2, and 1, checkpoints for the FT1 algorithm and 8, 4, 3, 2, and 1 for the FT2 algorithm. Fig. 2 provides the average time requred to create a new checkpoint.

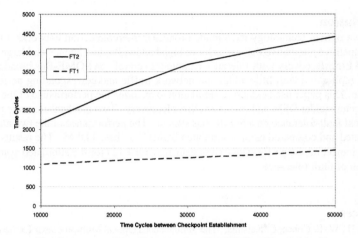

Fig. 2. Average Time Required to Establish a Checkpoint

As the checkpoint interval is increased from 10000 to 50000, the time required to create the checkpoint for the FT2 algorithm more than doubles while the time for the FT1 algorithm increases by less than 500 time cycles. The recovery data update messages are responsible for the large difference. The recovery data update messages will vary in size based on the number of distinct data elements modified since the previous checkpoint was established. As the checkpoint interval is increased, more data will be placed in the update messages. The message size affects the message transfer time and the time it takes for the directory controller to write the data to memory.

The FT1 algorithm involves some overhead due to the fault tolerance during normal operation. This is due to the necessity of keeping the home2 copy of a node's memory consistent with the home copy. For this reason, the processor and channel utilization for the FT1 algorithm does not change significantly when the checkpoint interval is varied. The overhead for the FT2 algorithm occurs only when establishing a checkpoint. The processor utilization for the FT2 algorithm decreased 12% for a 20% increase in checkpoint frequency. The channel utilization results indicate that the message transfer time is not the major cause of the decrease in performance for the FT2 algorithm. Although the processor utilization declines when the checkpoint interval is reduced, the channel utilization remains almost the same.

The service time for the output queues and directory waiting-message queues does not change dramatically for either of the algorithms when the checkpoint interval is changed. The service time for the FT1 algorithms, however, is double that of the FT2 algorithm for both queues. The increase in the output queue is due to the increase in the number of messages required to keep the second copy of memory consistent. When a message is removed from the directory input queue or waiting-message queue, the directory begins to process the message. The Directory service time represents the time necessary to make the required accesses to the local memory. The directory service time is the major source of overhead for the FT1 algorithm. The increase in service time is due to the additional time each node must spend reading or writing to the home2 copy of another node's memory. The additional memory access time required for the coherence messages reduces the ability of the directory controller to process the normal data and ownership requests in a timely manner.

3 Conclusion

Advances in optical technology have made broadcast interconnection networks a realistic, highly competitive alternative that can achieve high bandwidth, low latency and large fan-out. Backward Error Recovery fault-tolerance techniques can rely on SOME-Bus broadcasts and DSM data replication with little additional network traffic and corresponding performance degradation. This paper used extensive simulation to examine the performance of the SOME-Bus architecture under DSM and Backward Error Recovery. Two mechanisms were proposed to implement fault-tolerance on a SOME-bus system. The performance of the two algorithms was evaluated and compared using a simulated SOME-bus-based DSM. The results showed a minimal impact on the performance (less than 16% for FT1) for the simulated system with the addition of fault tolerance.

References

1. Kulick J. H., W. E. Cohen, C. Katsinis, "The Simultaneous Optical Multiprocessor Exchange Bus," IEEE Conference on Massively Parallel Processor Optical Interconnects, 1995.
2. Kermarrec, A-M, Morin, C., Banatre, M., "Design, implementation and evaluation of ICARE: an efficient recoverable DSM", Software - Practice and Experience, vol. 28, no. 9, pp. 981-1010, 25 Jul 1998.
3. Theel O., Fleisch B., "A dynamic coherence protocol for distributed shared memory enforcing high data availability at low costs", IEEE Transactions on Parallel and Distributed Systems", vol.7, no.9, p. 915-30, Sept. 1996.

An Efficient Backup-Overloading for Fault-Tolerant Scheduling of Real-Time Tasks

R. Al-Omari, G. Manimaran, and Arun K. Somani

Dept. of Electrical and Computer Engineering
Iowa State University, USA
{romari,gmani,arun}@iastate.edu

Abstract. Many time-critical applications require dynamic scheduling with predictable performance. Tasks corresponding to these applications have deadlines to be met despite the presence of faults. In this paper, we propose a technique called *dynamic grouping*, to be used with backup overloading in a primary-backup based fault-tolerant dynamic scheduling algorithm in multiprocessor real-time systems. In dynamic grouping, the processors are dynamically grouped into logical groups in order to achieve efficient overloading of backups, thereby improving the schedulability. We compare the performance of dynamic grouping with that of static grouping and no-grouping schemes through extensive simulation studies and show the effectiveness of dynamic grouping.

1 Introduction

Due to the critical nature of tasks in a hard real-time system, it is essential that every task admitted in the system completes its execution even in the presence of faults. Therefore, fault tolerance is an important requirement in such systems. Scheduling multiple versions of tasks on different processors can provide fault tolerance. One of the models that is used for fault-tolerant scheduling of real-time tasks is the Primary-Backup (PB) model, in which two versions of a task are scheduled on two different processors and an acceptance test is used to check the correctness of the execution result [1, 2]. The backup version is executed only if the output of the primary version fails the acceptance test, otherwise it is deallocated from the schedule. A concept, called *backup overloading*, was introduced in [1] and has been extended in [2] to capture the trade-off between the number of faults in the system and system utilization.

In this paper, we address the problem of dynamically scheduling real-time tasks with PB fault-tolerance onto a set of processors in such a way that the versions of the tasks are feasible in the schedule. In order to improve the performance of backup overloading, we propose a technique called dynamic grouping, which dynamically divides the processors of the system into logical groups as tasks arrive into the system and finish executing.

Task Model: (i) Tasks are aperiodic. Every task Ti has the attributes arrival time (a_i), ready time (r_i), worst case computation time (c_i) and a deadline (d_i). **(ii)** Each task T_i has two versions, namely primary copy (Pr_i) and backup copy

J. Rolim et al. (Eds.): IPDPS 2000 Workshops, LNCS 1800, pp. 1291-1295, 2000.
© Springer-Verlag Berlin Heidelberg 2000

(Bk_i). They have identical attributes. **(iii)** Tasks are non-preemptable. **(iv)** All tasks arrive to a centralized scheduler.

Fault Model: *Assumption 1:* Processor faults can be transient or permanent and are independent. *Assumption 2:* The maximum number of faulty processor at any instant of time in a "group" is limited to one (group is defined later). *Assumption 3:* The minimum interval between two successive faults within a group of processors $>$ Max $\{d_i - r_i\}$ $\forall T_i \in T$, where T is the set of tasks. *Assumption 4:* There exists a fault-detection mechanism to detect processor faults. The scheduler does not assign tasks to a faulty processor.

Related Work and Motivation for Our Work

Backup overloading concept was proposed in [1] to allow backups of different tasks to overlap in time on the same processor. This is valid only if the following conditions are satisfied: *Condition 1:* The primaries should be scheduled on different processors. *Condition 2:* At most one of the primaries can encounter a fault. *Condition 3:* At most one version of a task can encounter a fault.

Condition 1 is needed to handle permanent faults. Condition 2 is needed to ensure that at most one backup can be executed among the overloaded backups. Condition 3 is needed to ensure that at least one version of each task will execute without any fault.

The algorithm proposed in [1] assumes at most one fault at any instant of time in the entire system in order to satisfy condition 2, which is optimistic. Another PB based algorithm for tasks with resource requirements was proposed in [2]. This algorithm can tolerate more than one fault at a time by employing a technique called flexible backup overloading. Though this algorithm can tolerate more than one fault at a time, it statically divides the processors into groups of three or more. This static division restricts the flexibility of backup overloading, thus reducing the guarantee ratio (% of tasks accepted).

2   Dynamic Logical Groups

In this section, we propose a technique called dynamic grouping of processors which overcomes the limitations of static grouping [2] and no-grouping [1]. Dynamic logical grouping is defined as the process of dynamically dividing the processors of the system into logical groups as tasks arrive into the system and finish executing. The logical groups are determined when the scheduler decides where to schedule the two versions of a task. The number of groups and the size of the groups will vary with time, in contrast to static grouping where the number and size of the groups are fixed and are known a priori. Moreover, in static grouping a processor can be a member of only one group. Whereas, in dynamic grouping a processor can be a member of more than one group which allow efficient use of backup overloading.

We will show that dynamic grouping offers better schedulability than static grouping due to its flexible nature of overloading backups. We will also show that dynamic grouping offer higher fault-tolerance degree than the static grouping due to its dynamic nature of forming the groups. However, dynamic grouping involves more scheduling overhead compared to static grouping. The dynamic grouping concept works under the following propositions:

Proposition 1: At most one version of a task will encounter a fault.

Proof: Since the two versions (Pr_i and Bk_i) of a task T_i are scheduled on two different processors in the same group in the interval $[r_i, d_i]$. Assumptions 1 and 2 of the fault model will ensure that at most one version of the task will encounter a fault, which satisfy condition 3.

Proposition 2:. Two primaries (Pr_1 and Pr_2) of tasks T_1 and T_2 that are scheduled on two different processors ($Proc(Pr_1)$ and $Proc(Pr_2)$) in the same group can overload their backups (Bk_1 and Bk_2) on a third processor ($Proc(Bk_1, Bk_2)$) in the same group if $|st(Pr_1) - ft(Pr_2)| < \text{Max}\ \{d_i - r_i\}\ \forall T_i \in T$, where T is the set of tasks and $|X|$ is the absolute value of X.

Proof: Since these three processors ($Proc(Pr_1)$, $Proc(Pr_2)$), and $Proc(Bk_1, Bk_2)$) are within the same group, this group can have only one fault at a time (Assumption 1), and since $|st(Pr_1) - ft(Pr_2)| <$ minimum interval between successive faults (assumption 2), at most the fault can be in one of these primaries which satisfy condition 2.

Group Dynamics: The creation, deletion, expansion, and shrinking of logical groups is controlled by the following rules:

Rule 1: A logical group G_k is dynamically formed from two processors ($\text{Proc}(Pr_i)$, $\text{Proc}(Bk_i)$) when the two versions of a task T_i are scheduled to these two processors. This logic group stays either until Pr_i is successfully executed and Bk_i is de-allocated or the Pr_i fails and the Bk_i is executed.

Rule 2: If another primary (Pr_j) scheduled on a third processor $\text{Proc}(Pr_j)$ whose backup (Bk_j) overloads with Bk_i, then the logical group (G_k) will expand to have three processors ($\text{Proc}(Pr_i)$, $\text{Proc}(Pr_j)$, $\text{Proc}(Bk_i)=\text{Proc}(Bk_j)$). This logical group will stay until the two tasks (T_i and T_j) are successfully executed.

Rule 3: If primary (Pr_i) is successfully executed then its backup (Bk_i) will be de-allocated from its processor $\text{Proc}(Bk_i)$, and the logical group (G_k) will shrink to have two processors ($\text{Proc}(Pr_j)$, $\text{Proc}(Bk_j)$). The same is true for Pr_j.

Rule 4: If primary (Pr_i) is failed then the group will stay until primary (Pr_j) and (Bk_i) are successfully executed. The same is true for Pr_j.

Figure 1 shows an example that illustrates how the groups are formed and removed dynamically as tasks arrive into the system and finish executing. Figure 1a shows that primary and backup copies of task T_1 are scheduled on processors 1 and 2, respectively. Then, these two processors will form a logical group (group 1) that will stay until one of these copies executes successfully. Figure 1b shows the same situation as in Figure 1a, but this time the primary of T_2 is scheduled on processor 3, and its backup is overloaded with Bk_1 on processor 2. This results in expanding the group to have three processors. Figure 1c shows the same situation as in Figure 1b, but now the scheduler has decided to schedule the primary of T_3 on processor 4, and its backup on processor 3, then processors 3 and 4 will form a logical group (group 2). Figure 1d shows the situation when Pr_1 has executed successfully. Therefore Bk_1 is de-allocated, which results in shrinking group 1 to have two processors 1 and 2.

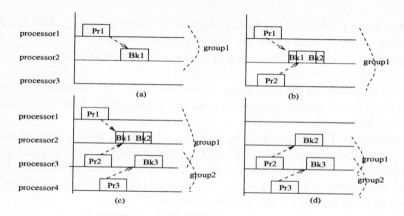

Fig. 1. Dynamic grouping

On the other hand, if static grouping is employed for the same example, processors 1, 2, and 3 will form group 1 and processor 4 will be in group 2. In that case, the situations shown in Figures 1c and 1d cannot occur because the groups are disjoint and their sizes are fixed. It can be seen that dynamic grouping results in higher utilization for processor 3 (in Figure 1c), because in static grouping Bk_3 can not be scheduled to processor 3 since Pr_3 is in group 2. Also, dynamic grouping increases the number of faults tolerable in these processors as tasks T_2 and T_3 can be executed successfully even in the presence of faults in processors 2 and 4 (in Figure 1d). Similarly, if no-grouping was employed for the same example, all the processors would be in one group. In this case the scheduling will be the same except that only one fault can be tolerated in the entire system.

3   Performance Study

We compare the performance of the proposed dynamic grouping based fault-tolerant scheduling algorithm with that of the static grouping based algorithm [2] and also with the no-grouping algorithm [1] using guarantee ratio as the performance metric. In simulation, the number of processors is taken to be 12.

Effect of Task Load: The task load (L) has been varied in Figure 2. The figure shows that increasing L decreases the guarantee ratio for all the algorithms. From the figure, the difference in the performance for the dynamic and static grouping algorithms is maximum for medium task load. This is explained as follows: For higher L, the guarantee ratio is lower for all the algorithms because the task load is much more than the capacity of the system. On the other hand, for lower L, the guarantee ratio is higher for all the algorithms because the task load is less than the system capacity, which means that most of the tasks are schedulable by all the algorithms. Also, note that the difference in performance between no-grouping and dynamic grouping is small which means that the dynamic grouping algorithm increases the utilization of the system to a point equal to the no-grouping algorithm and the difference in performance is partly due to the overhead cost associated with dynamic grouping.

Effect of Number of Faults: Figure 3 shows the effect of varying number of fault occurrences in the system for task loads of 0.25, 0.5 and 1. The figure shows that the difference in the guarantee ratio between the algorithms is significant at low load ($L = 0.25$) and medium load ($L = 0.5$) compared to at high load ($L = 1$). This is explained as follows. For light load (L=0.25), the dynamic grouping can compensate the degradation in guarantee ratio due to faults by rearranging the groups which is not possible in the static grouping. For full load (L=1), the dynamic grouping tends to behave similar to static grouping as all the processors are heavily loaded. The figure also shows that the guarantee ratio offered by the static grouping algorithm decreases more rapidly than the dynamic grouping as the number of faults increases due to the same reason.

4   Conclusions

In this paper, we have proposed a concept called dynamic logical grouping of processors for overloading of backups in a PB-based fault-tolerant dynamic scheduling algorithm. Our simulation studies show that the dynamic grouping offers significantly better guarantee ratio than the static grouping under all the interesting conditions that we have simulated in the system.

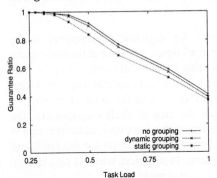

Fig. 2. Effect of task load

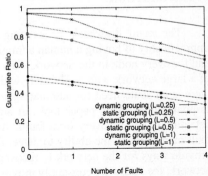

Fig. 3. Effect of number of faults

References

1. S. Ghosh, R. Melhem, and D. Mosse, "Fault-tolerance through scheduling of aperiodic tasks in hard real-time multiprocessor systems," *IEEE Transaction on Parallel and Distributed Systems,* vol. 8, no. 3, pp. 272-284, Mar. 97.
2. G. Manimaran and C. Siva Ram Murthy, " A fault-tolerant dynamic scheduling algorithm for multiprocessor real-time systems and its analysis," *IEEE Transaction on Parallel and Distributed Systems,* vol. 9, no. 11, pp. 1137-1152, Nov. 98.
3. R. Al-Omari, G. Manimaran, and A. K. Somani, " A Fault- tolerant dynamic scheduling algorithm for multiprocessor real- time systems," *Fault-tolerant Computing Symp. (FTCS), FAST ABSTRACTS,* pp. 63-64, Jun. 99.

Mobile Agents to Automate Fault Management in Wireless and Mobile Networks[1]

Niki Pissinou, Bhagyavati, Kia Makki

The University of Louisiana at Lafayette
{pissinou/bxb8329/makki@cacs.usl.edu}

Abstract. This paper studies the automation of fault management of wireless and mobile networks. A network management protocol similar to SNMP with an integrated architecture using mobile agents will prove effective in fault management of such networks. In view of the above, we design a wireless network management protocol to support fault management. In particular, we use mobile agents to detect, diagnose and recover from faults in wireless and mobile networks.

1. Introduction

The main drawback of existing protocols for network management is the impracticality of having a human administrator at all times, working at the same pace, and for every node in the network. This shortcoming is especially arduous in the case of a large network with many users with different needs and expectations. Therefore, there is an acute need for automating some or, ideally, all of the network manager's responsibilities. This paper looks at automating the area of fault management, the detection of and recovery from faults. An important task of network administrators is fault management: the ability to detect faults, diagnose the cause(s) for the fault, and provide ways for the network to recover from them. In a typical wireless and mobile network, the nodes are constantly moving and are called mobile units.

This work looks at networks that have wireless capabilities as well as mobility. Such networks are expected to provide the same type of access and quality of services to customers on the move that customers using wireline networks now possess. The main problem addressed in this paper is the extension of the architecture of a wireless and mobile network by adding mobile agents to automate recovery from faults. Since SNMP is considered the standard network management protocol for the wireline networks of today [8], extending it to wireless and mobile networks and integrating mobile agents into this new architecture is bound to work as well, if not better, than wireline networks. Since there is no currently existing network management protocol

[1] Partially funded by NSF grant number DUE-9751414, NSF EPSCOR grant, NASA grant number NAG5-7127, BoRSF RD-A-39, ENH-TR-95, and LEQSF Industrial ties grant.[1]

J. Rolim et al. (Eds.): IPDPS 2000 Workshops, LNCS 1800, pp. 1296-1300, 2000.

for such networks, an SNMP-like protocol would best serve the needs of customers because of inherent simplicity and widespread usage.

2. The Fault-Tolerant Wireless Network Management Architecture

This section looks at the architecture of a wireless and mobile network extended by using mobile agents in an effort to automate fault management. Dependability is the primary weakness of these networks [1]. The main liability of existing protocols for network management is the impracticality of having a human manager at all times. Therefore, there is an enormous need for automating network management. As networks become more and more complex, a human administrator is in danger of getting overwhelmed and not being able to meet the service needs. In such a scenario, automation of fault management in networks is a pressing necessity.

A brief description of the components of the architecture is presented. In this paper, we consider systems where the base stations are connected through wires, and the channel between the mobile units and base stations is wireless. If a mobile unit wants to communicate with another unit, it can do so via the underlying base stations. A mobile unit can have three states [2]: active (on), sleep, disconnected (off). Even while the mobile unit is off, control signals are still communicated to and from the base station. If the mobile unit is in the off state, then there is a waiting period until it goes to the on state. If it is in sleep state, then the super agent decides that there is no fault, and tries again after some time. We consider in this paper a network wherein only the communication between mobile unit and base unit is wireless.

3. Overall Description of Methodology

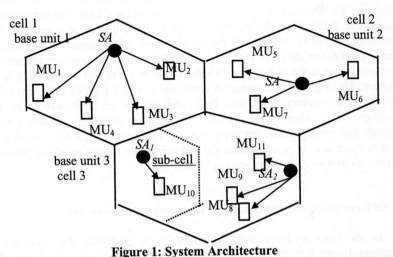

Figure 1: System Architecture
MU = Mobile Unit, SA = Super Agent, BU = Base Unit, MA = Mobile Agent

The super agent consists of three modules: history, diagnostic and recovery modules. The mobile agent reports to the super agent after copying the information from the mobile unit. If there was no fault found, then the super agent records that fact, along with the time, in the log, updates its VLR entry, and transmits the information to the home base unit, which updates its HLR entry. When a mobile unit, shown in the architecture, enters a cell, it notifies the base unit of its arrival. This is done by periodically transmitting short messages and judging the closest base unit by the fastest response time of its acknowledgment [2]. The base unit broadcasts this information to all super agents in the cell. The super agent of that sub-cell acknowledges and sends a mobile agent to the mobile unit. The mobile agent copies the parameter values of the mobile unit's profile and reports back to the super agent, which then performs a comparison with the history. If there is no mismatch or discrepancy, the super agent concludes that there is no fault and performs the necessary updates, including an update of this record in the history module.

If there is a mismatch, however, the possibility of a fault cannot be ruled out. The super agent records this fact in the history module and performs other updates (for example, in the VLR). The history module passes on the mismatched entry to the diagnostic module, and the super agent does a table lookup to find the type of fault that most likely caused the mismatch. Then the super agent passes these results to the recovery module as parameters, looks up the table and determines which procedures to initiate for recovery. The super agent then launches a repair agent, which migrates to the mobile unit and repairs the fault using the recovery procedure embedded in it.

4. Fault Management

A typical network management protocol provides information on the number of nodes in the network, which base stations are homes to which mobile units, and so on. This information aids in fault management. Typical procedures in fault management include the maintenance and periodic examination of error logs. The periodicity of the examination depends on the network traffic and the degree of robustness that the system is required to maintain. If the mobile agent is not "smart" enough to recover from or even detect the fault, then a human will be called in, who can recover from it. The human then updates the logs, which the super agent can use to learn from past behavior. The logs have information about the possible type of fault that led to the problem, and recovery from that fault, and how much time was taken for this procedure. Thus, after being operational for some time, the system can detect, diagnose and recover from any type of fault. The amount of time it takes a mobile agent to learn how to recover from the different types of commonly occurring faults depends on its speed, capacity and sophistication.

4.1 An Example of the Steps in Fault Correction and Recovery

In the fault recovery task, a mobile agent performs the necessary repair procedure. Hence it is called a repair agent. An example is considered next. Assume that the user informs the network manager of a problem in running an application.

The manager will retrieve the information about that node without even physically being present at the site. We can access diagnostic information such as utilization rate and system uptime at the network manager's workstation. For instance, the repair agent, disk and buffer usage can be checked to see if they are exceeded. Furthermore, this check admits of two kinds of failures: allocation failures and near-the-limits failures [7]. If this check unearths such a failure, then the repair agent, or disk or buffer should be erased, and thus recovery from this problem is possible.

If the repair agent, disk or buffer usage is not exceeded, a check is performed to see if there are too many users or too many processes. If either case were true, then clearly the CPU (Central Processing Unit) of that remote site/node is overloaded. Then recovery may be accomplished by killing some processes -- either the longest running or the ones started first, or the ones started last, or the ones with the least priority. This decision depends on the algorithm used for killing the process. The fault can be assumed to be a software problem if both the above checks turn nothing out of the ordinary. The ordinary behavior for that node is again determined by its profile and history, obtained from its host base station. A software problem might be one of incompatibility, as, for example, the incompatibilities that exist between the operating system, the windowing wrapper used, the network file system, and the application software [8]. Recovery from such software problems as incompatibilities is possible by notifying the user of the problem, and giving a set of options, such as conversion from one format to another, or continuing in spite of the incompatibility. In the case of SNMP, the MIB (Management Information Base) called HostResources allows the manager to remotely determine what versions of software have been installed.

4.2 A High-Level View of the System

In the following section, we present a high-level view of our system, including the algorithm of its operation.

```
A mobile unit arrives inside a cell and alerts the base unit.
The base unit updates its registers and logs, including time of entry.
The base unit broadcasts to all super agents in that cell.
Super agent in the particular cell acknowledges transmission from the base unit.
The super agent sends a mobile agent to the mobile unit.
The mobile agent copies the parameter values stored in the mobile unit.
The mobile agent reports back to the super agent with the values.
Super agent performs updates to logs and compares values with unit profile.
    if mismatch
    fault detected
    query diagnostic module to determine cause(s) for fault
    lookup table in recovery module for loading procedures onto repair agent
    super agent launches repair agent, which performs recovery at mobile unit
    the repair agent reports back to the super agent
    else
    no fault detected
The super agent makes updates to history module, VLR and HLR.
The super agent kills the mobile agent (and repair agent, if any).
```

5. Conclusion

We have designed an architecture for network management of wireless and mobile networks. It is SNMP-like and is extended by using mobile agents (aglets) to automate the fault management functionality in network management. The system was simulated successfully on UNIX and Windows NT platforms using ASDK from IBM. The system offers attractive features such as low memory usage, fast and efficient information gathering and automated recovery from faults. In addition, the super agent learns from past experiences. After implementation of our system is completed, we plan to test exhaustively for all kinds of typical faults and to evaluate the performance of our system against existing standards.

References

1. Bennington, B.J., Bartel, C.R.: Wireless Andrew: experience building a high speed, campus-wide wireless data network. Proceedings of the Third Annual ACM/IEEE International Conference on Mobile Computing and Networking, IEEE MOBICOM (1997) 55-65
2. Garg, V.K., Smolik, K., Wilkes, J.E.: Applications of Mobile Agent in Wireless/Personal Communications. Prentice Hall PTR, Upper Saddle River NJ (1997)
3. Geier, J.: Wireless Networking Handbook. New Riders Publishing, Indianapolis Indiana (1996)
4. Goodman, D.J.: Wireless Personal Communications Systems. Addison-Wesley Publishing Co., Reading Massachusetts (1997)
5. Magedanz, T., Rothermel, K., Krause, S.: Intelligent Agents: An Emerging Technology for Next Generation Telecommunications? Proceedings of the Fifteenth Annual IEEE INFOCOM (1996) 464-472
6. Rappaport, T.S.: Wireless Communications: Principles and Practice. Prentice Hall PTR, Upper Saddle River NJ (1996)
7. Stallings, W.: SNMP, SNMPv2, and CMIP: The Practical Guide to Network Management Standards. Addison-Wesley Publishing Co., Reading Massachusetts (1993)
8. Zeltserman, D.: A Practical Guide to SNMPv3 and Network Management. Prentice Hall PTR, Upper Saddle River NJ (1999)

9th Heterogeneous Computing Workshop (HCW 2000)

The 9th Heterogeneous Computing Workshop (HCW 2000) is a forum to discuss latest findings in heterogeneous computing and promising work in progress. Heterogeneous computing systems range from diverse elements within a single computer to coordinated, geographically distributed machines with different architectures. A heterogeneous computing system provides a variety of capabilities that can be orchestrated to execute multiple tasks with varied computational requirements. Applications in these environments achieve performance by exploiting the affinity of different tasks to different computational platforms or paradigms, while considering the overhead of inter-task communication and the coordination of distinct data sources and administrative domains. Such computing systems support information infrastructure and other terms including Cluster Computing and Grid Computing are also used to describe heterogeneous computing. HCW 2000 is co-sponsored by the IEEE Computer Society, through the technical committee of Parallel Processing, and the U. S. Office of Naval Research.

42 submissions were received of which 32 papers were selected for presentation. The papers are grouped into 10 sessions along two parallel tracks. The following is the technical program.

Session 1-A
Grid Environment

Master/Slave Computing on the Grid
Gary Shao and Fran Berman, University of California San Diego, Rich Wolski, University of Tennessee, USA

Heterogeneity as Key Feature of High Performance Computing: the PQE1 Prototype
P.Palazzari, L. Arcipiani, M.Celino, A. Mathis, P. Novelli, and V. Rosato, ENEA - HPCN project, R. Guadagni, ENEA-Funzione Centrale Informatica, Italy

The NRW-Metacomputer - Building Blocks for A Worldwide Computational Grid
Claus Bitten, University of Cologne, Joern Gehring, Paderborn Center for Parallel Computing, U. Schwiegelshohn and R. Yahyapour, University Dortmund, Germany

J. Rolim et al. (Eds.): IPDPS 2000 Workshops, LNCS 1800, pp. 1301-1305, 2000.
© Springer-Verlag Berlin Heidelberg 2000

Session 1-B
Resource Discovery and Management

Agent-Based Resource Discovery
K. Jun, L. Boloni, K. Palacz, and Dan Marinescu, Purdue University, USA

Evaluation of PAM's Adaptive Management Services
Yoonhee Kim, Syracuse University, Salim Hariri, and Muhamad Djunaedi, University of Arizona, USA

Load Balancing Across Near-Homogeneous Multi-Resource Servers
William Leinberger, George Karypis, and Vipin Kumar, University of Minnesota, USA

Session 2-A
Communication and Data Management

Evaluation of Expanded Heuristics in a Heterogeneous Distributed Data Staging Network
Noah B. Beck, H.J. Siegel, Purdue University, Mitchell D. Theys, University of Illinois at Chicago, and Michael Jurczyk, University of Missouri - Columbia, USA

Fast Heterogeneous Binary Data Interchange
Greg Eisenhauer and Lynn K. Daley, Georgia Institute of Technology, USA

A Heuristic Algorithm for Mapping Communicating Tasks on Heterogeneous Resources
Kenjiro Taura and Andrew Chien, University of California San Diego, USA

Design of an Infrastructure for Data-Intensive Wide-Area Applications
Michael D. Beynon, T. Kurc, M. Uysal, A. Sussman, and J. Saltz, University of Maryland, USA

Session 2-B
Modeling and Metrics

Quality of Security Service in a Resource Management System Benefit Function
Tim Levin, Anteon Corporation, Cynthia Irvine, Naval Postgraduate School, USA

Optimizing Heterogeneous Task Migration in the Gardens Virtual Cluster Computer
Ashley Beitz, S. Kent, and Paul Roe, Queensland University of Technology, Australia

Linear Algebra Algorithms in Heterogeneous Cluster of Personal Computers
Jorge Barbosa, J. Tavares, and A. Padilha, FEUP-INEB, Portugal

New Cost Metrics for Iterative Task Assignment Algorithms in Heterogeneous Computing Systems
Raju Venkataramana and N. Ranganathan, University of South Florida, USA

Session 3-A
Theory and Modeling

Task Execution Time Modeling for Heterogeneous Computing Systems
Shoukat Ali, H.J. Siegel, Purdue University, USA, Muthucumaru Maheswaran, University of Manitoba, Canada, and Debra Hengsen, Naval Postgraduate School, USA

Distributed Quasi Monte-Carlo Methods in a Heterogeneous Environment
E. deDoncker, R. Zanny, M. Ciobanu, and Y. Guan, Western Michigan University, USA

Session 3-B
Scheduling I

Scheduling Multi-Component Applications in Heterogeneous Wide-area Networks
Jon B. Weissman, University of Minnesota, USA

Application-Aware Scheduling of a Magnetohydrodynamics Application in the Legion Metasystem
Holly Dail, G. Obertelli, Francine Berman, University of California San Diego, R. Wolski, University of Tennessee, and A. Grimshaw, University of Virginia, USA

Fast and Effective Task Scheduling in Heterogeneous Systems
Andrei Radulescu and A. Gemund, Delft University of Technology, Netherlands

Session 4-A
Grid Applications

Combining Workstations and Supercomputers to Support Grid Applications: The Parallel Tomography Experience
Shava Smallen, W. Cirne, and F. Berman, University of California San Diego, J. Frey, S. Young, and M. Ellisman, National Center for Microscopy and Imaging Research, R. Wolski, University of Tennessee, M. Su and C. Kesselman, Information Science Institute/University of Southern California, USA

Cluster Performance and the Implications for Distributed, Heterogeneous Grid Performance
Craig A. Lee, C. Dematteis, J. Stepanek, and J. Wang, The Aerospace Corporation, USA

A Debugger for Computational Grid Applications
Robert Hood and G. Jost, NASA Ames Research Center, USA

Session 4-B
Resource Management

A Framework for Resource Co-Allocation in Heterogeneous Computing Systems
Ammar Alhusaini, Viktor K. Prasanna, and C.S. Raghavendra, University of Southern California, USA

Heterogeneous Resource Management for Dynamic Real-Time Systems
Eui-Nam Huh and L. Welch, Ohio University, Behrooz Shirazi, C. Cavanaugh, and S. Anwar, University of Texas at Arlington, USA

Dynamic Resource Sharing Mechanisms for High-performance Heterogeneous Clusters
Dimitrios Katramatos, Deepak Saxena, Nehal Mehta, and Steve Chapin, Syracuse University, USA

Session 5-A
Design tools

The Harness PVM-Proxy: Gluing PVM Applications to Distributed Objects Environments and Applications
Mauro Migliardi and Vaidy Sunderam, Emory University, USA

MoBiDiCK: A Tool for Distributed Computing on the Internet
Moez A. Dharsee and Christopher W.V. Hogue, Samuel Lunenfeld Research Institute, Canada

RsdEditor: A Graphical User Interface for Specifying Metacomputer Components
R. Baraglia and D. Laforenza, Instituto del Consiglio Nazionale delle Ricerche, Italy, A. Keller, Paderborn Center for Parallel Computing, Germany, A. Reinefeld, Konrad-Zuse-Zentrum Berlin, Germany

Session 5-B
Scheduling II

Heuristics for Scheduling Parameter Sweep Applications in Grid Environments
Henri Casanova, D. Zagorodnov, and F. Berman, University of California San Diego, A. Legrand, Ecole Normale Superieure de Lyon, France

Parallel Program Execution on a Heterogeneous PC Cluster Using Task Duplication
Yu-Kwong Kwok, University of Hong Kong

Segmented Min-Min: A Static Mapping Algorithm for Meta-tasks on Heterogeneous Computing Systems
Min-You Wu, Wei Shu, and Hong Zhang, University of Central Florida, USA

The HCW 2000 proceedings are published by the IEEE Computer Society Press. The ISBN is 0-7695-0556-2. The IEEE Computer Society Order Number is PR00556. The proceedings may be ordered online, at http://www.computer.org/cspress/

Cauligi S. Raghavendra
Program Chair

Viktor K. Prasanna
General Chair

Author Index

Lecture Notes in Computer Science

For information about Vols. 1–1719
please contact your bookseller or Springer-Verlag